Living in a Dynamic Tropical Forest Landscape

Living in a Dynamic
Tropical Forest
Landscape

Edited by

Nigel E. Stork and Stephen M. Turton

Blackwell
Publishing

First published 2008 by Blackwell Publishing Ltd

1   2008

Library of Congress Cataloging-in-Publication Data
Living in a dynamic tropical forest landscape / edited by Nigel E. Stork and Stephen M. Turton.
p. cm.
Includes bibliographical references and index.
ISBN 978-1-4051-5643-1 (pbk. : alk. paper)
1. Rain forest ecology--Australia--Queensland. 2. Queensland--Social life and customs. I. Stork, Nigel. II. Turton, Steve.
QH197.2.Q8L58 2008
577.3409943--dc22

2007038972

ISBN: 978-1-4051-5643-1 (hardback)

A catalogue record for this title is available from the British Library.

Set in 10/13 pt Palatino Linotype
by Newgen Imaging Systems (P) Ltd, Chennai, India
Printed and bound in Singapore
by Markono Print Media Pte Ltd

The publisher's policy is to use permanent paper from mills that operate a sustainable forestry policy, and which has been manufactured from pulp processed using acid-free and elementary chlorine-free practices. Furthermore, the publisher ensures that the text paper and cover board used have met acceptable environmental accreditation standards.

For further information on
Blackwell Publishing, visit our website at
www.blackwellpublishing.com

CONTENTS

Colour plates appear between pages 320 and 321

CONTRIBUTORS

Many of the contributors were participants in the Cooperative Research Centre for Tropical Ecology and Management and these are identified with an asterisk.

EDITORS

Nigel E. Stork* School of Resource Management and Geography, Faculty of Land and Food Resources, University of Melborne, Burnley Campus, Richmond, Victoria, Australia.

Stephen M. Turton* Australian Tropical Forest Institute, School of Earth and Environmental Sciences, James Cook University, Cairns, Queensland, Australia.

CONTRIBUTORS

Angela Arthington* Griffith University, Nathan, Queensland, Australia.

Jacqueline Balston Department of Primary Industries and Fisheries, Cairns, Queensland, Australia.

Tom Barrett* CSIRO Sustainable Ecosystems, Tropical Forest Research Centre, Atherton, Queensland, Australia.

Joan Bentrupperbäumer* School of Earth and Environmental Sciences, James Cook University, Cairns, Queensland, Australia.

K. Rosalind Blanche* CSIRO Entomology, Tropical Forest Research Centre, Atherton, Queensland, Australia.

Mike Bonell Formerly UNESCO, Section on Hydrological Processes and Climate, Division of Water Sciences, 1 rue Miollis, Paris, France.

Sarah L. Boulter* Griffith School of Environment , Griffith University, Nathan, Queensland, Australia.

Luz Boyero School of Tropical Biology, James Cook University, Cairns, Queensland, Australia.

Matt G. Bradford* CSIRO Sustainable Ecosystems, Atherton, Queensland, Australia.

Jeff Callaghan Severe Weather Section, Bureau of Meteorology, Brisbane, Queensland, Australia.

Carla P. Catterall* Griffith School of Environment, Griffith University, Nathan, Queensland, Australia.

Faye Christidis School of Marine and Tropical Biology, James Cook University, Townsville, Queensland, Australia.

Campbell Clarke* Wet Tropics Management Authority, Cairns, Queensland, Australia.

Bradley C. Congdon* School of Marine and Tropical Biology, James Cook University, Cairns, Queensland, Australia.

Niall M. Connolly* School of Marine and Tropical Biology, James Cook University, Cairns, Queensland, Australia.

Saul A. Cunningham* CSIRO Entomology, Canberra, ACT, Australia.

Ian Curtis*   Nature Conservation Trust of New South Wales; SMEC Australia Pty Limited, Townsville, Queensland, Australia; and School of Earth and Environmental Sciences, James Cook University, Queensland, Australia.

Allan Dale   Terrain Natural Resource Management, Innisfail, Queensland, Australia.

Andrew J. Dennis*   CSIRO Sustainable Ecosystems, Herberton, Queensland 4887, Australia.

Mark Disher*   CSIRO Land and Water, Davies Laboratory, Townsville, Queensland, Australia.

Peter Erskine*   School of Integrative Biology, University of Queensland, St Lucia, Queensland, Australia.

Peter Fitch*   CSIRO Land and Water, Christian Laboratory, Black Mountain, ACT, Australia.

Andrew J. Ford*   CSIRO Sustainable Ecosystems, Tropical Forest Research Centre, Atherton, Queensland, Australia.

David Gillieson*   School of Earth and Environmental Sciences, James Cook University, Cairns, Queensland, Australia.

Kylie L. Goodall*   Griffith University, Nathan, Queensland, Australia.

Miriam Goosem*   School of Earth and Environmental Sciences, James Cook University, Cairns, Queensland, Australia.

Stephen Goosem*   Wet Tropics Management Authority, Cairns, Queensland, Australia.

Catherine Graham   Department of Ecology and Evolution, Stony Brook State University, New York, USA.

Caroline L. Gross   Ecosystem Management, University of New England, Armidale, New South Wales, Australia.

Graham N. Harrington*   CSIRO Sustainable Ecosystems, Atherton, Queensland, Australia.

Debra A. Harrison*   Griffith School of Environment, Griffith University, Nathan, Queensland, Australia.

Steve Harrison*   School of Natural and Rural Systems Management, University of Queensland, Gatton, Queensland, Australia.

Alex Held*   Division of Land and Water, CSIRO, Canberra, ACT, Australia.

John Herbohn*   School of Natural and Rural Systems Management, University of Queensland, Gatton, Queensland, Australia.

David W. Hilbert*   CSIRO Sustainable Ecosystems, Herberton, Queensland, Australia.

Rosemary Hill*   School of Earth and Environmental Sciences, James Cook University, Cairns, Queensland, Australia; and CSIRO Sustainable Ecosystems, Cairns, Queensland, Australia.

Bradley G. Howlett*   Griffith University, Nathan, Queensland, Australia; and New Zealand Institute for Crop and Food Research Ltd, Christchurch, New Zealand.

Joanne L. Isaac   Centre for Tropical Biodiversity and Climate Change, School of Marine and Tropical Biology, James Cook University, Townsville, Queensland, Australia.

Kasper Johansen   Biophysical Remote Sensing Group, School of Geography, Planning and Architecture, University of Queensland, Australia.

John Kanowski*   Centre for Innovative Conservation Strategies, School of Environment, Griffith University, Nathan, Queensland, Australia.

Mark Kennard*   Australian Rivers Institute, Griffith University, Nathan, Queensland, Australia.

Jiro Kikkawa*   School of Integrative Biology, University of Queensland, Australia.

Roger L. Kitching*   Griffith School of Environment, Griffith University, Nathan, Queensland, Australia.

David Lamb*   School of Integrative Biology, University of Queensland, St Lucia, Queensland, Australia.

William F. Laurance   Smithsonian Tropical Research Institute, Balboa, Ancon, Panama, Republic of Panama; and Biological Dynamics of Forest Fragments Project, National Institute for Amazonian Research (INPA), Manaus, Amazonas, Brazil.

Tina Lawson*   CSIRO Sustainable Ecosystems, Tropical Forest Research Centre, Atherton, Queensland, Australia.

Geoff McDonald*   CSIRO Sustainable Ecosystems, St Lucia, Queensland, Australia.

David McJannet*   CSIRO Land and Water, 120 Meiers Rd, Indooroopilly, Queensland, Australia.

Adam McKeown*   CSIRO Sustainable Ecosystems, Atherton, Queensland, Australia.

Brendan McKie   Department of Ecology and Environmental Science, Umeá University, Sweden.

Jeffrey A. McNeely   International Union for Conservation of Nature and Natural Resources (IUCN), Gland, Switzerland.

Chris R. Margules*   CSIRO Sustainable Ecosystems; and Tropical Forest Research Centre, Atherton, Queensland, Australia.

Daniel J. Metcalfe*   CSIRO Sustainable Ecosystems, Tropical Forest Research Centre, Atherton, Queensland, Australia.

Geoff B. Monteith*   Queensland Museum, South Brisbane, Queensland, Australia.

Craig Moritz*   Museum of Vertebrate Zoology, University of California, Berkeley, California, USA.

Joanne Nightingale*   Biophysical Remote Sensing Group, School of Geography, Planning and Architecture, University of Queensland, Australia.

Sandra Pannell*   Discipline of Anthropology and Archaeology, School of Arts and Social Sciences, James Cook University, Townsville, Queensland, Australia.

Philip L. Pearce*   Tourism Program, School of Business, James Cook University, Australia.

Richard Pearson*   School of Marine and Tropical Biology, James Cook University, Townsville, Queensland, Australia.

Petina L. Pert*   CSIRO Sustainable Ecosystems, Tropical Forest Research Centre, Atherton, Queensland, Australia.

Stuart Phinn*   Biophysical Remote Sensing Group, School of Geography, Planning and Architecture, University of Queensland, Australia.

Mandy Price   Natural and Rural Systems Management, University of Queensland, Gatton, Queensland, Australia.

Brad Pusey*   Australian Rivers Institute, Griffith University, Nathan, Queensland, Australia.

Paul Reddell*   CSIRO Land and Water, Tropical Forest Research Centre, Atherton, Queensland, Australia.

Joseph Reser*   School of Psychology, Griffith University, Gold Coast, Queensland, Australia.

Jeffrey Sayer   Forest Conservation Programme, International Union for Conservation of Nature and Natural Resources (IUCN), Gland, Switzerland.

Peter Scarth*   Biophysical Remote Sensing Group, School of Geography, Planning and Architecture, University of Queensland, Australia.

Les Searle   School of Earth and Environmental Sciences, James Cook University, Cairns, Queensland, Australia.

Luke P. Shoo*   Centre for Tropical Biodiversity and Climate Change, School of Marine and Tropical Biology, James Cook University, Townsville, Queensland, Australia.

Catherine Ticehurst*   Division of Land and Water, CSIRO, Canberra, ACT, Australia.

Nigel Tucker Biotropica Australia Pty Ltd, Malanda, Queensland, Australia.

David J. Turton Department of Humanities, School of Arts and Social Sciences, James Cook University, Cairns, Queensland, Australia.

Peter S. Valentine* School of Earth and Environmental Sciences, James Cook University, Townsville, Queensland, Australia.

Jim Wallace* CSIRO Land and Water, Davies Laboratory, Townsville, Queensland, Australia.

Grant W. Wardell-Johnson* Griffith School of Environment, Griffith University, Nathan, Queensland, Australia; and Natural and Rural Systems Management, University of Queensland, Gatton, Queensland, Australia.

David A. Westcott* CSIRO Sustainable Ecosystems, Atherton, Queensland, Australia.

Nigel Weston* Griffith University, Nathan, Queensland, Australia.

Stephen E. Williams* Centre for Tropical Biodiversity and Climate Change, School of Marine and Tropical Biology, James Cook University, Townsville, Queensland, Australia.

Kristen J. Williams* CSIRO Sustainable Ecosystems; and Tropical Forest Research Centre, Atherton, Queensland, Australia.

S. Joseph Wright Smithsonian Tropical Research Institute, Balboa, Panama, Republic of Panama.

David Yeates* The Australian National Insect Collection, CSIRO Entomology, Canberra, ACT, Australia.

FOREWORD

The world's tropical rainforests, which occupy no more than 7% of the Earth's land mass, sequester within them about 40% of all carbon that is not held in the oceans. Importantly, they are home to a large part of global biodiversity, with perhaps as many as half of the world's total species found nowhere else. In addition, they play a key role in the Earth's atmospheric circulation and in the determination of climate, including precipitation, at a local and regional scale. Located almost entirely within developing countries, these forests are heavily impacted by legal and illegal logging, destructive mining, clearing for agriculture and plantations and shifting cultivation. A majority of Indigenous people living in rainforest areas have been removed from their traditional lands, and the megafauna in these forests, essential to their regular functioning, is being devastated by hunting.

Despite strong efforts for more than three decades, it has proved extraordinarily difficult to develop sustainable land-use systems in the moist tropics. Their resources have proved attractive for exploitation by corporations and individuals within their own countries, and the speed of their destruction has been increased by the demands of an emerging global economy. Industrialized countries have, as a whole, exhibited insufficient will to secure the protection of resources outside of their boundaries, despite continued lamentation about the situation.

Tropical rainforests are found on the mainland of only one industrialized nation, Australia, and it is in the so-called 'Wet Tropics' of that nation that major progress has been made in achieving sustainable systems for these forest ecosystems. The local scientific community has played a major, long-term role, particularly in driving the creation of the Wet Tropics of Queensland World Heritage Area in the 1980s. Impressive advances have been made in the past 10–15 years through the creation of a multidisciplinary science-based partnership – the Cooperative Research Centre for Tropical Rainforest Ecology and Management – that unites universities, the Commonwealth Scientific Industrial Research Organisation (CSIRO), other research organizations, local communities and local people, the Indigenous community, governments at all levels, industry, particularly the tourism industry, and non-governmental organizations in an effort to manage these ecosystems sustainably. This book provides an in-depth analysis of how this progress has been achieved.

It is fitting that we should pay respect to the research pioneers of the Wet Tropics and in particular to Len Webb, whose botanical studies in the 1960s and 1970s and later, often with Geoff Tracey, laid out the path for others to follow. Len was passionate about Indigenous people, and would be pleased to see the recent strength of engagement with Rainforest Aboriginal peoples, evidenced by numerous chapters in this book. It is also good to see a few of those pioneers as authors in this book – Jiro Kikkawa, Mike Bonell, and many more. Also included as authors are some of those who made the conservation and protection of rainforests in North Queensland happen, including Aila Keto, Rosemary Hill and Mike Berwick. The battles to preserve Australia's rainforests up and down the east coast and in south-west Tasmania have been fierce and have received much international attention.

In the final chapter, editors Nigel Stork and Steve Turton ask whether there are lessons from the Australian Wet Tropics that can be applied elsewhere. There certainly are! It is essential in pursuing sustainability anywhere to engage all the stakeholders in debates about the way rainforests can be managed, to make science-based decisions and to work across disciplines and ecosystems. The ways in which our landscapes are managed directly affect the health of waterways, estuaries, wetlands, coral reefs and oceans. This book takes a uniquely comprehensive and therefore exemplary holistic approach to landscape science and sustainable management, and is a valuable contribution that will certainly attract interest throughout the world.

Peter H. Raven
Missouri Botanical Garden
St Louis, Missouri, USA

ACRONYMS AND ABBREVIATIONS

AATSE	Australian Academy of Technological Sciences and Engineering
ABA	additive basal area
ACF	Australian Conservation Foundation
ACIUCN	Australian Committee for the World Conservation Union
AGB	above-ground biomass
AHC	Australian Heritage Commission
AIMS	Australian Institute of Marine Science
ALP	Australian Labour Party
ANN	artificial neural network
ATSIC	Aboriginal and Torres Strait Islander Commission
AWS	automatic weather station
BA	basal area
BK	Bellenden Ker
BMB	Black Mountain Barrier
BMC	Black Mountain Corridor
BP	before present
BRDF	bidirectional reflectance distribution function
CAFNEC	Cairns and Far North Environment Centre
cal. yr BP	calculated year before present
CAPE	convective available potential energy
CCA	Community Conserved Areas
CDM	clean development mechanism
CEO	chief executive officer
CMA	Catchment Management Authorities
CNVF	complex notophyll vine forest
CRC	Cooperative Research Centre
CRRP	Community Rainforest Revegetation Program
CSIRO	Commonwealth Scientific and Industrial Research Organisation
CTCC	Cape Tribulation Community Council
CVM	contingent valuation method
Cwlth	Commonwealth
CYCC	Cape York Conservation Council
D	Recharge
DASETT	Department of Arts, Sports, the Environment, Tourism and Territories
DBH	diameter at breast height
DEC	Department of Environment and Conservation
DEM	digital elevation model
DFG	disperser functional groups
DN	digital numbers
DNRM	Department of Natural Resources and Mines
DNRMW	Department of Natural Resources, Mines and Water
DOGIT	deed of grant in trust
DPI	Queensland Department of Primary Industries
DPIF	Department of Primary Industries and Fisheries
EIA	environmental impact assessment
EMS	environmental management systems
ENSO	El Niño Southern Oscillation
EOS	experience opportunity spectrum
EPA	Environmental Protection Agency
EPBC Act	Environment Protection and Biodiversity Conservation Act 1999
ERS	European Remote Sensing satellite
E_s	forest floor evaporation
ET	evapotranspiration
EVI	enhanced vegetation index
FANN	forest artificial neural network
FFG	fruit functional groups
FIS	forest inventory survey
FLR	forest landscape restoration
FNQ NRM	Far North Queensland Natural Resource Management Ltd.
FNQEB	Far North Queensland Electricity Board
FPQ	Forestry Plantations Queensland
FWPRDC	Forest and Wood Products Research and Development Corporation
GAM	generalized additive models
GBR	Great Barrier Reef
GBRMPA	Great Barrier Reef Marine Park Authority
GCM	global climate models
GCP	ground control points
GDR	Great Dividing Range

GIS	geographical information systems
GLM	generalized linear models
GPS	global positioning systems
HCO	Holocene climatic optimum
HoA	heads of agreement
I	canopy interception
IBRA	Interim Biogeographic Regionalisation for Australia
IFOV	instantaneous field of view
ILUA	Indigenous land use agreement
IPA	Indigenous Protected Areas
IPCC	Intergovernmental Panel on Climate Change
IPCC TAR	International Panel for Climate Change Third Assessment Report
IPM	integrated pest management
ITSG	Indigenous technical support group
IUCN	International Union for Conservation of Nature and Natural Resources (World Conservation Union)
IWG	Indigenous working group
JCU	James Cook University
JERS	Japanese Earth Resource Satellite
JI	joint implementation
K^*	satiated (saturated) hydraulic conductivity
LAI	leaf area index
LGM	Last Glacial Maximum
MDI	mean daily intensity
MEA	millennium ecosystem assessment
MHR	Member of the House of Representatives
MIS	managed investment schemes
MJO	Madden–Julian Oscillation
ML1	Mount Lewis
MP	Member of Parliament
MSL	mean sea level
MVF	mesophyll vine forest
NAP	National Action Plan
NAPSWQ	National Action Plan for Salinity and Water Quality
NCAR	National Centre for Atmospheric Research
NCEP	National Centre for Environmental Prediction
NDVI	normalized difference vegetation index
NGO	non-government organization
NHT	Natural Heritage Trust

NIR	near infra-red
NORMA	Northern Rainforest Management Agency
NPP	net primary production
NQAA	North Queensland Afforestation Association
NQTC	North Queensland Timber Cooperative
NRM	natural resource management
NRM & E	natural resources, mines and energy
NSW	New South Wales
NT	Northern Territory
OC	Oliver Creek
OECD	Organisation for Economic Co-operation and Development
P	total precipitation
PAR	photosynthetically active radiation
P_c	cloud interception
PFANN	palaeo-forest artificial neural network rainfall
P_g	rainfall
P_{ga}	rainfall corrected for slope effects and wind losses
PHT	Pleistocene/Holocene transition
PJVS	plantation joint venture scheme
PSG	programme support groups
PSIA	psychosocial impact assessment
PV	potential vorticity
QBVR	quantifying the biodiversity values of reforestation
QCC	Queensland Conservation Council
QDMR	Queensland Department of Main Roads
QF	quickflow
QFD	Queensland Forestry Department
QPWS	Queensland Parks and Wildlife Service
QRR	quickflow response ratios
R	runoff
RAAF	Royal Australian Air Force
RAIN	Rainforest Information and Action Network
RCSQ	Rainforest Conservation Society of Queensland
RE	regional ecosystem
RF	return flow
RFID	Rainfall intensity–frequency–duration
RIS	regional investment strategy
ROS	recreation opportunity spectrum
RPAC	Regional Planning Advisory Committee
SAP	structural adjustment package
SAR	Synthetic Aperture Radar

SCP	Smithfield Conservation Park
S_f	stemflow
SIA	social impact assessment
SLATS	Statewide Landcover and Trees Study
SNSM	simple notophyll and simple microphyll forests and thickets
SoE	state of the environment
SOF	saturation overland flow
SoWT	State of the Wet Tropics
SPOT	Systeme Pour l'Observation de la Terre
spp.	species (plural)
SSF	subsurface stormflow
SVI	spectral vegetation indices
T	transpiration
TEK	traditional ecological knowledge
T_f	throughfall
TIN	triangulated irregular network
TOAC	Traditional Owner Advisory Committee
TOFTW	tall open forests and tall woodlands
TREAT	Trees for the Evelyn and Atherton Tablelands
TRS	Tropical Rainforest Society
TWS	The Wilderness Society
UB	Upper Barron

UNESCO	United Nations Educational, Scientific and Cultural Organisation
VIM	visitor impact management
VMS	visitor monitoring system
VP	vertical percolation
VPD	vapour pressure deficit
WA	Western Australia
WAG	Douglas Shire Wilderness Action Group
WHA	World Heritage Area
WHC	World Heritage Committee
WMC	Western Mining Corporation
WTAPPT	Wet Tropics Aboriginal Plan Project Team
WTMA	Wet Tropics Management Authority
WTP	willingness to pay
WTQWHA	Wet Tropics of Queensland World Heritage Area
WTTPS	Wet Tropics Tree Planting Scheme
WTVPRAS	Wet Tropics Vertebrate Pest Risk Assessment Scheme
WTWHA	Wet Tropics World Heritage Area
WTWHPM Act	Wet Tropics World Heritage Protection and Management Act 1993
$\delta\theta$	soil water storage

EDITORS

Nigel Stork holds the Chair of Resource Management and is Head of School of Resource Management and Geography, Head of the Burnley Campus and Associate Dean for Knowledge Transfer at the University of Melbourne. Formerly the CEO of the Cooperative Research Centre for Tropical Rainforest Ecology and Management, he has studied tropical forest ecology with particular interest in insect diversity in many tropical regions of the world. He has edited or co-edited ten books and written more than 150 scientific papers. Nigel is a Director of Earthwatch Australia, Member of Council for Association for Tropical Biology and Conservation and was the former Chair of the Wet Tropics Management Authority Community Consultative Committee.

Stephen Turton is the Executive Director for the James Cook University/Commonwealth Scientific and Industrial Research Organisation Tropical Landscapes Joint Venture at James Cook University in Cairns, Australia. Previously, he was Associate Professor in Geography and Director of Research for the Rainforest Cooperative Research Centre. His research interests include tropical climatology, rainforest ecology, urban ecology, recreation ecology and natural resource management. Steve has published over 100 scientific papers in these fields of study, comprising refereed journal articles, book chapters and research monographs. Steve is a former Councillor of the Institute of Australian Geographers and a member of the Wet Tropics Management Authority's Scientific Advisory Committee. He is also the honorary treasurer and council member of the Association for Tropical Biology and Conservation, Asia-Pacific Chapter.

INTRODUCTION

Nigel E. Stork[1]* and Stephen M. Turton[2]*

[1]School of Resource Management and Geography, Faculty of Land and Food Resources, University of Melbourne, Burnley Campus, Richmond, Victoria, Australia
[2]Australian Tropical Forest Institute, School of Earth and Environmental Sciences, James Cook University, Cairns, Queensland, Australia
*The authors were participants of Cooperative Research Centre for Tropical Ecology and Management

This book is a compendium of what we have learnt about the so-called 'Wet Tropics' landscapes of north-east Australia and brings together a wealth of scientific findings and traditional ecological knowledge. These forested landscapes, although only a very small part of Australia in geographical terms, are home to a high proportion of the continent's species and ecosystems, and have a special significance both nationally and internationally. These tropical forest landscapes have also been the home for Indigenous Australians for thousands of years. In recognition of the global significance of the natural history of the region the Wet Tropics was World Heritage listed by UNESCO in 1988.

Like other regions of eastern Australia (and the humid tropics in general), the Wet Tropics has experienced widespread clearing for agriculture, notably along the coastal plain between Mossman and Ingham and on the Atherton Tablelands inland from Cairns (Figures I.1 and I.2). Despite these major land use impacts, the region still contains large tracts of intact forest and wetlands that, elsewhere in eastern Australia, have been severely fragmented. In recent decades there has been increasing pressure for further agricultural, urban, peri-urban and tourism development in the Wet Tropics and these and other uses compete with nature conservation in what is a highly contested landscape.

This has provided regional planners with both challenges and opportunities for sustainable use of Australia's most biologically complex landscape. Many of these impacting forces are discussed in this volume.

Although a few scientists had worked for many years on various aspects of the natural history of the Wet Tropics, until quite recently our understanding of the region was patchy. This changed with the significant funding of the Cooperative Research Centre for Tropical Rainforest Ecology and Management (the Rainforest CRC) from 1993 to 2006. The Rainforest CRC, driven by the wide-ranging needs of its stakeholders, encouraged long-term foundational research and supported multidisciplinary projects often emphasizing the importance of linking social and ecological systems. It is doubtful that such an integrated, concerted and broad-scale research effort has ever been achieved before for a tropical forest landscape anywhere in the world. All those involved in the Rainforest CRC were keen to acknowledge that the important lessons gained from this living research laboratory should be used to guide future research efforts in tropical and sub-tropical Australia and elsewhere in the world. We therefore felt compelled to bring together this knowledge and the lessons learnt in a single comprehensive volume of work. In doing this we were well aware of the paucity

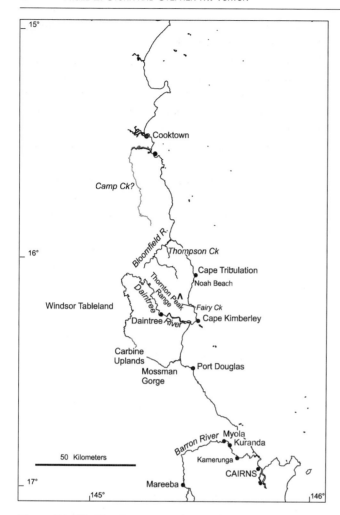

Figure I.1 Wet Tropics region of Australia – northern costal section.

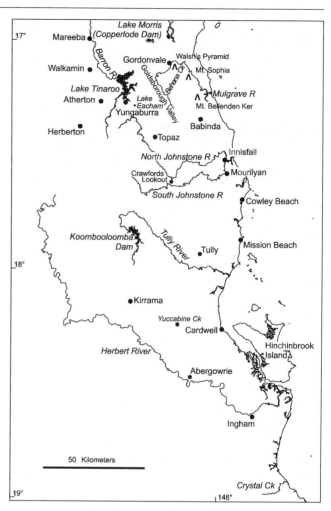

Figure I.2 Wet Tropics region of Australia – southern costal section and Atherton Tablelands.

of information for other tropical forest landscapes around the world. Internationally there are no texts that provide such a holistic view of any tropical forest landscape, including the social, cultural and economic dimensions. Furthermore, no other texts provide such a breadth of understanding and linkages among different fields of study. Other texts focusing on individual tropical forests really only take a biological view and usually lack an Indigenous and management focus (e.g. Gentry 1990; McDade *et al.* 1994; Leigh *et al.* 1996; Laurance & Bierregaard 1997). Recent books by Bermingham *et al.* (2005) examining the history and ecology of tropical forests and by Laurance and Peres (2006) on the threats to tropical forests draw heavily on examples from the Wet Tropics.

Australia has a reputation for its environmental science and its application to improve natural resource management, conservation and sustainability at the landscape scale. It is also the only developed country with tropical rainforest on its mainland. The rainforest science carried out in many fields of study over the past 10–15 years has been world class and there have been many internationally significant scientific breakthroughs, such as those in conservation genetics, vegetation modelling, agroforestry and revegetation techniques, biodiversity assessment and modelling of the impacts of climate change on tropical biodiversity, and the integration of science with natural resource management, to name but a few. In this book authors have been encouraged to place their chapters in an international context.

Since the Australian Wet Tropics rainforests and their adjacent agricultural and urban landscapes are now as well understood as or better understood than any other tropical forest landscapes in the world, we hope that this book also will be of interest to a wide range of readers, including students, scientists, policy-makers and natural resource managers, especially in the humid tropics. The book is presented in six parts, with part summaries being written by international luminaries who have tried to place the chapters in a global context. Part I looks at the history and biodiversity of the Wet Tropics region and includes chapters on Indigenous cultures and European settlement as well as the establishment of the World Heritage Area. Part II examines ecological processes and other ecosystem services and includes chapters on seed dispersal, pollination and economic valuation of the region. Part III looks at the threats to the environmental values of the region, including biological and human-induced threats, such as climate and land-use change. Part IV examines the social and cultural dimensions of living in a World Heritage Area, including reference to the Indigenous People and their ancient links with this landscape. Part V tackles various approaches to restoring tropical forest landscapes, including production versus biodiversity trade-offs. Part VI is concerned with how science can inform policy, conservation and management of tropical forest landscapes. Most authors have included a summary at the end of their chapters and many have also included text boxes highlighting significant issues or case studies.

In writing and editing this book we have been influenced and assisted by a large number of people. We are grateful for the inspirational leadership and encouragement provided by Ralph Slatyer and Sydney Schubert, who chaired the Rainforest CRC from 1993 to 2002 and 2002 to 2006, respectively, the inaugural CEO of the CRC, Jiro Kikkawa, and many Directors of the CRC, including Mike Berwick, David Butcher, Guy Chester, John Courtenay, Josh Gibson, Daniel Gschwind, Brian Keating, John Mullins, Norman Palmer, Julia Playford, David Siddle, Vicki Pattemore and Russell Watkinson. Working in government-funded research programmes means that your research is often subjected to endless reviews! However, we found these to be very useful in guiding our research, with an increased likelihood of useful outcomes for our stakeholders. Here we would like to acknowledge the wise advice provided by some of those reviewers, which often led to significant changes in direction and scientific advances. In particular, we thank Keith Boardman, Henry Nix, Andrew Beattie and Graham Kelleher, all of whom particularly influenced our thinking.

Our editorial assistant, Annette Bryan, performed miracles transforming draft chapters into ready to go text and working with the authors. Adella Edwards similarly transformed the figures provided by authors into a uniform and polished style. We also acknowledge and thank Shannon Hogan, David Knobel and Trish O'Reilly of the Rainforest CRC for their support in the production of the book. Our thanks are also extended to Ward Cooper, Delia Sandford and Rosie Hayden from Blackwell Publishing for their assistance and guidance.

Finally, we wish to acknowledge the remarkable contribution that the late Geoff McDonald made to our own understanding of tropical landscapes and the involvement of indigenous and non-indigenous communities in sustainable management. He was a true visionary.

References

Bermingham, E., Dick, C. W. & Moritz, C. (eds) (2005). *Tropical Rainforests: Past, Present, and Future.* Chicago University Press, Chicago. 745 pp.

Gentry, A. H. (1990). *Four Neotropical Rainforests.* Yale University Press, New Haven, CT. 627 pp.

Laurance, W. F. & Bierregaard, R. O. (eds) (1997). *Tropical Forest Remnants: Ecology, Management, and Conservation of Fragmented Communities.* University of Chicago Press, Chicago. 616 pp.

Laurance, W. F. & Peres, C. A. (eds) (2006). *Emerging Threats to Tropical Forests.* Chicago University Press, Chicago. 520 pp.

Leigh, E. G., Rand, A. S. & Windsor, D. M. (eds) (1996). *The Ecology of a Tropical Rainforest: Seasonal Rhythms and Long-term Changes,* 2nd edn. Smithsonian Institution, Washington, DC. 503 pp.

McDade, L. A., Bawa, K. S., Hespenheide, H. A. & Hartshorn, G. S. (eds) (1994). *La Selva: Ecology and Natural History of a Neotropical Rainforest.* University of Chicago Press, Chicago. 486 pp.

1 AUSTRALIAN RAINFORESTS IN A GLOBAL CONTEXT

Nigel E. Stork[1]*, Stephen Goosem[2]* and Stephen M. Turton[3]*

[1]School of Resource Management and Geography, Faculty of Land and Food Resources, University of Melbourne, Burnley Campus, Richmond, Victoria, Australia
[2]Wet Tropics Management Authority, Cairns, Queensland, Australia
[3]Australian Tropical Forest Institute, School of Earth and Environmental Sciences, James Cook University, Cairns, Queensland, Australia
*The authors were participants of Cooperative Research Centre for Tropical Ecology and Management

Introduction

Moist tropical rainforests cover approximately 6–7% of the surface of the globe and occur in a band about 15–20° either side of the equator. Typically they receive more than 2000 mm precipitation a year and although they may frequently experience a dry season, this is often punctuated by periods of heavy rainfall. These forests are typified by their evergreen nature, although some species of trees can be deciduous. Longer and drier dry seasons inevitably produce tropical dry forests, with most tree species being deciduous. Throughout this book when authors refer to rain-forests they are referring to moist tropical rainforests (Figure 1.1).

Rainforests are renowned for their immense biodiversity. It is often said that tropical rainforests house more than half of the world's biodiversity. At least 44% of the world's vascular plants and 35% of the world's vertebrates (Sechrest *et al.* 2002) are endemic to 25 'global biodiversity hotspots' (Myers *et al.* 2000) more than half of which are tropical rainforest sites. Much less is known about the diversity of non-vertebrate animals in tropical rainforests, although some would consider that there are possibly tens of millions of species in these ecosystems.

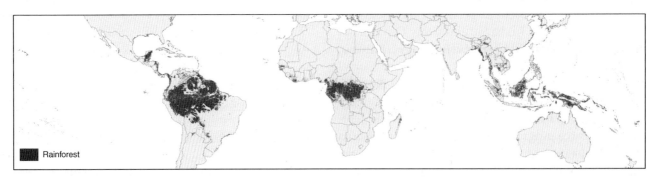

Figure 1.1 The global distribution of tropical rainforest. *Source*: from Primack and Corlett (2005), original figure provided by UNEP-WCMC.

Australia's rainforests comprise only a minuscule proportion of the global total but are vitally important for their unique biodiversity, for the unique ecosystem they represent in what is a very dry continent and because they are the last vestiges of what was an ancient and extensive ecosystem once covering perhaps as much as a third of Australia (Bowman 2000).

Rainforests as contested landscapes

Rainforests throughout the world are highly contested landscapes as governments and the commercial sector seek to increase economic benefits from what are seen as largely unproductive forests. Major threats include logging, both legal and illegal, fire and general encroachment through increased road and rail access (Laurance & Peres 2006). Laurance *et al.* (2001) highlighted the fate of tropical rainforests when they showed how the Amazonian rainforest might be reduced to 40% of its current coverage if proposed infrastructure projects were to come to fruition in Brazil. Earlier Myers (1993) brought to the attention of the world the decline in rainforests (and tropical dry forests) due to the actions of growing numbers of shifting cultivators who were taking advantage of new access roads into previously inaccessible areas. Population growth is seen by many as a major threat to tropical forests, leading to the extinction of tropical forest species (but see Wright & Muller-Landau 2006a, b). In many parts of the world, and in particular in African rainforests, the bush-meat trade is devastating the mega-fauna of rainforests (Bennett & Robinson 2000). The loss of these large vertebrates will result in many changes to the structure and composition of tropical rainforests. These changes may well be exacerbated by the impact of climate change.

Australian rainforests and their significance

With the exception of Antarctica, Australia is the driest continent on Earth. However, northern Australia receives monsoonal rains in north Queensland, the Northern Territory and the Kimberley region of Western Australia, with patches of rainforest occurring there (McKenzie *et al.* 1991; Bowman 2000). Most rainfall occurs along the east coast in places where the Great Dividing Range meets the coast. Although much of the east coastline is or was forested, rainforest now only occurs where the rainfall is high and where there is sufficient rain during the dry season to maintain this forest type. As a result, rainforests are scattered throughout tropical, subtropical, warm temperate and cool temperate areas of Queensland, New South Wales, Victoria and Tasmania, with small patches also found in north coastal Northern Territory and Western Australia. Rainforests occur from sea level to high altitudes, usually within 100 km of the coast in areas receiving more than 1200 mm of rainfall that are climatic and fireproof refuges. Drier, semi-deciduous vine thickets are also found in the Brigalow Belt and monsoonal vine thickets are scattered over parts of the seasonal tropics of northern Australia. Figure 1.2 (after Bowman 2000) shows the distribution of rainforests in Australia and demonstrates how fragmented these forests are. Not surprisingly, these forests have been the focus of much research on forest fragmentation (Laurance & Bierregaard 1997; and see Laurance & Goosem, Chapter 23, this volume). Only about 20% or 156 million hectares of Australia has a native forest cover of which just over 3.0 million hectares is rainforest (Table 1.1). Rainforests are located in 31 of Australia's 80 Interim Biogeographic Regionalisation for Australia (IBRA) biogeographical regions (Thackway & Cresswell 1995). The largest area of remaining rainforest in Australia is located in the so-called Wet Tropics region (27.6%), where most of the larger blocks are contained within the boundaries of the Wet Tropics of Queensland World Heritage Area (WHA) (Table 1.2).

It is estimated that about 30% (~13 000 km^2) of the pre-European extent of rainforests has been cleared (National Land and Water Resources Audit 2001). Most accessible lowland and tableland rainforests have been cleared and/or have become highly fragmented, while most remaining larger blocks of rainforest inhabit steep or rugged terrains. Historically, rainforests were among the earliest Australian native vegetation communities to be exploited for timber and agriculture. Examples of extensive past rainforest clearing include the decimation of the 'Big Scrub' rainforests in northern New South Wales (Floyd 1987), the Illawarra rainforests, the hoop pine scrubs of south-east Queensland (Young & McDonald 1987), the rainforests of the Atherton and Eungella Tablelands, the coastal floodplain rainforests of the Daintree, Barron, Johnstone, Tully–Murray, Herbert, Proserpine and Pioneer rivers in north-east

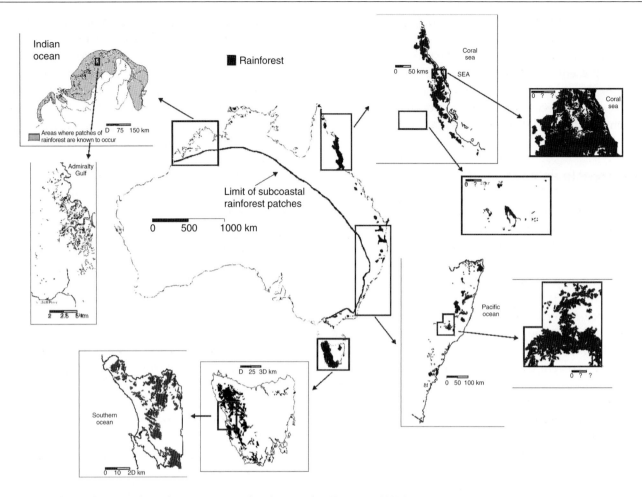

Figure 1.2 Map of extent of rainforests in Australia. *Source*: after Bowman (2000).

Table 1.1 Area of pre-European and present-day rainforest in Australia (km^2)

	State						Continent	Region
	Vic	*WA*	*NSW*	*NT*	*Tas*	*Qld*	*Australia*	*Wet Tropics*
Present area	407	16	2 218	977	7 055	19 558	30 231	8 340
Pre-European estimate	445	18	4 836	978	7 161	30 055	43 493	10 974
Per cent remaining	91.5	88.9	45.9	99.9	98.5	65.1	69.5	76

Source: National Land and Water Resources Audit (2001), WTMA (2002).

Table 1.2 Percentage contribution to Australia's present-day extent of rainforest

	State							Continent	Region	Subregion
ACT	*SA*	*Vic*	*WA*	*NSW*	*NT*	*Tas*	*Qld*	*Australia*	*Wet Tropics*	*WHA*
0	0	13	0.1	7.3	3.2	23.3	64.7	100	27.6	22.1

Source: National Land and Water Resources Audit (2001), WTMA (2002).

Queensland and extensive areas of Brigalow Belt vine thickets in Queensland and New South Wales (Sattler & Williams 1999). In recent years there has been strong opposition to such clearing from the conservation sector and the battles have featured prominently in the media. Recent Regional Forest Agreements in various key locations in Australia have resulted in the protection of large areas of rainforest. Valentine and Hill (Chapter 6, this volume) discuss further the conflict that resulted in the World Heritage listing of much of the Wet Tropics.

The broad range of ecological community types classified under the umbrella term 'rainforest' masks the level of regional depletion of some rainforest and vine thicket types. In the Wet Tropics, for example, the escarpment and highland rainforest communities remain largely intact, whereas the coastal lowland and tableland rainforest communities have been severely depleted. Of 24 endangered Wet Tropics regional ecosystems 18 occur on the coastal lowlands as fragmented remnants, while a further five are from basalt landscapes on the Atherton Tableland. The status of endangered, in general, refers to those regional ecosystems that have been reduced to less than 10% of their pre-European extent (Sattler & Williams 1999).

Studies have shown that rainforests are remnants of the oldest types of vegetation in Australia. Many species have ancestors dating back to the Cretaceous or early Tertiary Period, over 65 million years ago (Keto & Scott 1986; DASETT 1987; BRS 2003). For this reason, Australian rainforests have major historical and scientific significance. Australian rainforests are more important for the maintenance of biodiversity than their small area would imply (e.g. Table 1.1). Five of the 13 centres of plant diversity identified in Australia are dominated by rainforest, while a further three have rainforest components (BRS 2003).

The total Australian rainforest area might be small in global terms but the forests are unique in a number of important ways: their variety is unusual; the range of climates in which they have survived is exceptionally wide; and the number of plants and animals that are endemic to them and are still identifiable as being of very primitive stocks is scientifically exciting.

The Wet Tropics rainforests

The largest fragment of rainforest in Australia occurs as a narrow strip along the east coast from 15° 30' S to almost 19° 25' S and covers approximately 2 million hectares. Such is the biological significance of the region that a large part of this was inscribed on the World Heritage List on 9 December 1988 as the Wet Tropics of Queensland World Heritage Area. The listing was the direct consequence of the accumulated scientific research and understanding of the region's rainforests up to that time (see McDonald & Lane 2000). The tenure of the land within the World Heritage Area is complex and changing (Table 1.3). Although the rainforests of Queensland's Wet Tropics are small in size when compared to the rainforests of other parts of the world, the World Heritage Area covers such a

Table 1.3 Size comparisons of World Heritage tropical rainforest properties

Major rainforest type	Country	Size (ha)
Lowland rainforest		
Salonga National Park	Zaire	3 600 000
Lorentz	Indonesia	2 500 000
Okapi Faunal Reserve	Zaire	1 372 625
Wet Tropics of Queensland	Australia	894 420
Virunga National Park	Zaire	790 000
Thungyai-Huai Kha Khaeng	Thailand	622 200
Kahuzi-Biega National Park	Zaire	600 000
Sian Ka'an	Mexico	528 000
Dja Faunal Reserve	Cameroon	526 000
Rio Platano Biosphere Reserve	Honduras	500 000
Tai National Park	Côte d'Ivoire	330 000
Ujung Kulon National Park	Indonesia	78 359
Los Katios National Park	Colombia	72 000
Tikai National Park	Guatemala	57 600
Sinharaja Forest Reserve	Sri Lanka	8 864
Montane rainforest		
Canaima National Park	Venezuela	3 000 000
Talamanca/Amistad	Costa Rica/Panama	791 592
Sangay National Park	Equador	271 925
Machu Picchu	Peru	32 592
Mount Nimba Reserves	Côte d'Ivoire	18 000
Mome Trois Pitons	Dominica	6 857
Sub-montane rainforest		
Manu National Park	Peru	1 532 806

Table 1.4 The changing nature of land tenure in the Wet Tropics of Queensland World Heritage Area (WHA)

Tenure	Percentage of WHA							
	1992	1995	1997	1998	1999	2000	2002	2003
National parks	28	28	30	30	30	32	32	32
Forest reserves	–	–	–	–	–	–	29	38.9
State forests	36	38	38	38	38	39	10	0.1
Timber reserves	9	8	8	8	8	8	7	7
Various reserves and dams	1	1	1	1	1	1	1	1
Unallocated state land	7	7	8	8	8	7	7	7
Leasehold	16	15	12	12	12	10	11	11
Freehold and similar	2	2	2	2	2	2	2	2
Rivers, roads, railways and esplanades	1	1	1	1	1	1	1	1

high proportion of the total rainforested area that it ranks highly in size among other rainforest World Heritage Areas (Table 1.4).

The Wet Tropics contains the richest variety of animals and plants in the country, including two-thirds of the butterfly species, half of the birds and a third of the mammals. A very high proportion of the fauna and flora is endemic to the Wet Tropics (Commonwealth of Australia 1986), including 70 vertebrate species (Williams & Hilbert 2006). More than 400 plant and 76 animal species are officially listed as rare, vulnerable or endangered (WTMA 1999). The Wet Tropics also provides an unparalleled living record of the ecological and evolutionary processes that shaped the flora and fauna of Australia over the past 400 million years when it was first part of the Pangaean landmass and then of the ancient Gondwana continent. For example, the rainforests of the Wet Tropics have more plant taxa with primitive characteristics than any other area on Earth. Of the 19 angiosperm families described as the most primitive (Walker 1976), 12 occur in the Wet Tropics, giving it the highest concentration of such families on Earth. These families are: Annonaceae, Austrobaileyaceae, Eupomatiaceae, Himantandraceae, Myristicaceae and Winteraceae of the order Magnoliales; Atherospermataceae, Gyrocarpaceae, Hernandiaceae, Idiospermaceae, Lauraceae and Monimiaceae of the order Laurales (DASETT 1987).

The cool temperate rainforests of Tasmania contain several primitive conifers and flowering plants (Adam 1994; BRS 2003).

Until the 1970s, it was thought that rainforests were 'alien' to the Australian landscape, while sclerophyll types of vegetation, such as eucalypts and acacias, were considered the quintessential Australian vegetation. Rainforests were considered to be recent invaders across the land bridge that, in fairly recent geological times, connected Australia with New Guinea. Ecological and taxonomic research, however, gradually provided evidence that radically changed this view (Webb 1959; Webb *et al.* 1976, 1984; Webb & Tracey 1981). It was not just that rainforests had evidently adapted themselves to various climatic conditions (tropical, monsoonal, subtropical, temperate) that bore witness to longer local habitation than was commonly believed. It was other irrefutable evidence, such as the discovery of many families of primitive ancient angiosperms in the Wet Tropics, that confirmed these rainforest ecosystems as among the oldest rainforests on Earth. Although many of these elements also occur in New Caledonia and to a smaller extent in New Guinea, the Wet Tropics also displays a co-evolution with related sclerophyll floras and faunas. The setting of extensive tropical rainforests adjacent to the world's largest fringing reef, the Great Barrier Reef, is another unusual feature, with rainforest meeting the reef found elsewhere only in a few Pacific islands, Indonesia and Belize.

The varied topography of the region and its effect on rainfall is the reason behind this enormous biodiversity. A combination of high elevated mountains and plateaus that run roughly perpendicular to the prevailing south-east trade winds results in the Wet Tropics being the wettest region in Australia (Turton *et al.* 1999). About one-third of the Wet Tropics bioregion is more than 600 m above sea level, giving rise to cooler meso-thermal climates, where annual average temperatures are below 22°C. Many plant and animal species are adapted to these cooler climates that occur in largely continuous areas of rainforest in mountainous regions. Paradoxically, it is these very species that are severely threatened by climate change (Williams *et al.* 2003; Williams *et al.*, Chapter 22, this volume), as less than 5% of the total protected area is more than 1000 m above sea level. One of the remarkable features of the Wet Tropics is the sharp rainfall gradient from east to west: across the lowlands rainfall may be 2500–4000 mm per annum, while in places across the montane region rainfall may be as high as 6000–8000 mm per annum, with the savanna region experiencing annual rainfall as low as 1500 mm per annum. All these changes can occur in less than 40 km from the coast. The hydrology and climate of the region is discussed further by Bonnell and Callaghan, Chapter 2, this volume.

Historical aspects of the Wet Tropics and changes in human perceptions

Australia has a long history of settlement and use by Aboriginal people. Precisely when Aboriginal people arrived from Southeast Asia is hotly debated, with Flannery (1995) suggesting this might be 40 000–60 000 years before present. All parts of Australia were occupied by these Aborigines, often referred to as Traditional Owners, and most of these people were displaced or killed by European settlement. In the Wet Tropics region there are at least 20 tribal groups and many of these people were displaced from their traditional lands, known to them as country (see Parnell, Chapters 4 and 30, this volume). To rainforest Aboriginal people or 'Bama', the Wet Tropics is a series of complex living cultural landscapes. This means that natural features are interwoven with rainforest Aboriginal people's religion, spirituality and economic use (including food, medicines and tools), as well as their social and moral organization. The landscape identifies rainforest Aboriginal peoples' place within their country and reinforces their ongoing customary laws and connection to country. The country is therefore embedded with enormous meaning and significance to its Traditional Owners.

The history of the Wet Tropics from European settlement is discussed in more detail by Turton (Chapter 5, this volume) but here we discuss how human perceptions of rainforest values in the Wet Tropics have changed over time and how these differing value systems influence decisions concerning management of the environment. Although environmental management decisions are influenced by available scientific information, such decisions are ultimately based on community values held at the time, which also largely influence the prevailing political climate at the local, national and often international level. For example, rainforests in the Wet Tropics, in the late 1880s and early 1900s, were viewed as fertile soil deemed more valuable if converted to pasture or crop lands. Even up to the late 1950s landholders could not receive government incentive funding until the land had been cleared of trees. Between the 1930s and the 1960s the perceived value of rainforests shifted towards timber resources and rainforests were retained as Crown land in state forests for the purpose of timber production (see Valentine & Hill, Chapter 6, this volume). In the late 1970s and 1980s the perceived values shifted again from a strictly utilitarian view as leisure time increased for Australians and as international tourism started growing (see Pearce, Chapter 7, this volume). Since World Heritage listing, the perception has progressively changed to emphasize the non-market values of rainforests – scientific, cultural and aesthetic.

The late 1880s was also when scientific interest in the region's rainforests first emerged, with several expeditions mounted to collect and record plants and animals, such as those led by, among others, Walter Hill, Carl Lumholtz, Archibald Meston, Cairn and Grant. The first comprehensive exploration of the wet tropical coast was the 1873 Dalrymple expedition, whose primary purpose was discovery of agricultural lands, especially those suitable for sugarcane and other tropical crops. North of Mission Beach, Dalrymple (1874) reported on a 'great coast basin' – densely forested and with half a million acres of soil 'unsurpassed by any in the world – all fitted for tropical agriculture'. Referring to the 'Northern Eldorado', Dalrymple contributed to an image used by promoters of land settlement in the area.

There were also clear expressions of antipathy towards the rainforest, such as in the article from a local 'gentleman' printed in the *Herberton Advertiser* (2 August 1889) after a visit to Lake Eacham on the Atherton Tableland: 'Most of your readers know Atherton, and I look on this small settlement as marking the first skirmish in the coming war between the pioneers of civilization and the vast wilderness that stretches N S and E over hundreds of square miles. This war between man and the scrub has begun – and will never cease till the axe has laid the enemy low and smiling pastures have taken the place of the heavy scrub.'

Although the dominant view of the rainforests of the Wet Tropics was utilitarian and directed towards agriculture and associated timber getting, there was also some degree of scientific interest from amateur natural historians and a limited number of government botanists. The Bellenden Ker Range between Innisfail and Cairns attracted special interest and in 1889 a government-funded expedition led by Archibald Meston explored the area, ascending all the peaks and returning with an extensive plant and animal collection (Meston 1889). Early scientific work was essentially taxonomic, with the first ecological studies appearing from the 1920s, especially the work of W. D. Francis, culminating in the 1929 publication of the illustrated field guide *Australian Rain Forest Trees* (Francis 1929).

For most of the twentieth century, the narrow utilitarian image of rainforest remained dominant. In Queensland, where there was most of the accessible, potentially useful rainforest land, politicians, public servants and local promoters proposed huge schemes to develop most of the north Queensland 'scrub lands' into small family farms (Frawley 1983, 1987). The Queensland Forestry Department argued determinedly against this proposed land alienation, and for the reservation of forest lands for timber production, as well as for some national parks. The foresters were arguing for professional management of the forests for produc-

tion forestry purposes, consistent with the utilitarian conservation philosophy of the 'wise use' of resources. During this same period major land clearing was being undertaken on the coastal plain and on the gentle terrain of the fertile Atherton Tableland, which generated conflict between those who valued rainforests as agricultural and pastoral land to be cleared as extensively and as quickly as possible and those who valued rainforests for their longer-term timber resource. State economic prosperity was closely identified with rural development and closer settlement became the accepted political objective. Growth of the dairying industry after 1890 was a major driver of landscape change in the region, expanding rapidly into rainforested areas with their supposedly fertile soils (especially basalt landscapes on the tablelands and well drained, alluvial landscapes on the coastal lowlands). In pursuit of this policy of the government of that time, rainforest clearing proceeded without any assessment of its suitability. This policy of closer settlement and freeholding of Crown land continued until the late 1950s with the post-Second World War soldier settlement schemes.

Legislation passed in 1959 established the Department of Forestry as a separate entity to the Lands Department. The Department of Forestry included in its responsibilities the reservation and management of national parks until the creation of a separate National Parks and Wildlife Service in 1975. The post-1960 period was very significant in the history of rainforest management, planning and utilization for two reasons: first, the expansion in effort and expenditure by the government into long-term management planning; second, the evolution of the conservation movement, which successfully challenged the pre-1960 management models in favour of rainforest preservation and strict rainforest conservation models (Valentine & Hill, Chapter 6, this volume). This radical change in the way society values the region's rainforests was due in large measure to the changes in our knowledge and appreciation of the international scientific significance of the rainforests resulting from research (culminating in World Heritage inscription in 1988). This conflict in social and political values continued until as recently as the 1960s, when the last large-scale clear-felling of forests (42 900 ha) for pastoral purposes occurred in the Tully River lowlands (King Ranch).

Table 1.5 summarizes the main socio-economic characteristics of the Wet Tropics Natural Resource Management (NRM) region, which represents about 80% of the bioregion (Dale, Chapter 32, this volume). While the Wet Tropics NRM Region only covers 1% of

Table 1.5 Wet Tropics NRM Region statistics

Land area	2.2 million ha
Current population	220 000 people
Projected population growth by 2025	300 000 people
Land area under cropping	130 000 ha
Land under horticulture	47 000 ha
Land under improved pasture	65 000 ha
Land under grazing	600 000 ha
Total value of tourism industry	$2 billion
Number of visitors to region per year	3 million
Tourism sector employment	20% of regional total

Source: McDonald & Weston (2004).

Queensland, it contributes 10% and 23% of its agricultural and tourism activity, respectively (C. Margules pers. comm.). Over the past decade, population growth rates have been among the highest in the state outside the south-east corner, with this adding to increasing pressure for resource use that has resulted in environmental degradation, particularly along the coastal plain.

Recent change in perception of rainforests

The Wet Tropics occupied a central position in national environmental politics throughout the 1980s. The events surrounding the World Heritage listing of the Wet Tropics were beset in controversy, characterized by protest campaigns for and against rainforest logging, including a political battle between the Queensland and Australian governments (Box 1.1). There was conflict between the then Queensland government, which supported logging of the rainforests, and the Australian federal government, which proposed to nominate the Wet Tropics for the World

---

Box 1.1 Synopsis of recent historical events leading to the protection of rainforests in the Wet Tropics (a detailed description is presented in McDonald & Lane 2000)

1980

- Second World Wilderness Congress held in Cairns (June 1980).
- Australian Heritage Commission listed a number of rainforest areas (Greater Daintree region – 350 000 ha) on the Register of the National Estate (October 1980).

1981

- Mt Windsor logging operations blockaded by conservationists (13 people arrested).
- The Australian Conservation Foundation launched the Rescue the Rainforest campaign in Cairns, and the Cairns and Far North Environment Centre was formed. Rainforests were the major priority of the Australian Conservation Foundation in 1981.
- The original proposal by the conservation movement was for a *Greater Daintree National Park* including only the

Cape Tribulation National Park, Roaring Meg/Alexandra Creek catchments, Daintree River catchment, Mt Windsor Tableland, Mt Spurgeon, Mossman Gorge, Mt Lewis, Cedar Bay area and Walker Bay area.

1982

- The impetus for World Heritage listing of the Wet Tropics came with the 1982 publication *The World's Great Natural Areas* and its inclusion on IUCN's 1982 list of places deserving world heritage protection.

1983

- Clearing commenced for a new Cape Tribulation–Bloomfield Road: construction started in December 1983 to be met by a blockade of protesters, which elevated the campaign to national and international levels and drew the federal government into the debate.

(Continued)

Box 1.1 *(Continued)*

1984

- The Australian Heritage Commission engaged the Rainforest Conservation Society of Queensland to evaluate and report on the international conservation significance of the Wet Tropics between Cooktown and Townsville. The report concluded that the area met all four natural heritage criteria and this finding was supported by several international referees.
- Conservationists from around Australia met in Brisbane to form a national coalition to seek listing of the Wet Tropics as a World Heritage site.
- The General Assembly for IUCN passed a resolution recognizing the value of the Wet Tropics.

1985

- Downey Creek logging blockade.

1986

- The Commonwealth established a $22.5 million National Rainforest Conservation Program but the Queensland government refused to participate.

1987

- The Commonwealth announced that it would proceed immediately and unilaterally towards nomination of the Wet Tropics to the World Heritage list.
- A social impact assessment (SIA) was commissioned by the Commonwealth.
- The nomination was presented to the Bureau of the World Heritage Committee on 23 December 1987.

1988

- The Commonwealth made a regulation under the World Heritage Properties Conservation Act 1983 to prevent activities associated with commercial forestry operations in the area covered by the nomination.
- The Commonwealth implemented a structural adjustment package (SAP) to address the potential negative social impacts identified in the SIA.
- The shire councils made a submission against listing.
- The Commonwealth established the Wet Tropics of Queensland SAP to offset the impacts of the cessation of logging ($75.3 million).
- At a meeting in Brasilia in December 1988, the World Heritage Committee formally accepted the Commonwealth nomination and the Area was officially inscribed on the World Heritage list (9 December 1988): Twelfth Session of the World Heritage Committee meeting in Brasilia, Brazil, 5–9 December 1988.

1989

- The state government's legal challenge to the constitutional validity of the listing was rejected by the High Court (30 June 1989).
- A Labour government elected in Queensland (2 December 1989), which withdrew the challenge in the Federal Court that selection logging did not detract from World Heritage values.

1990

- A Ministerial Council was established comprising two federal ministers and two state ministers.

1992

- Establishment of the Wet Tropics Management Authority, which became a statutory authority following gazettal of the Wet Tropics World Heritage Protection and Management Act 1993.

Heritage List. All local governments (shire councils) and the major representatives of Rainforest Aboriginal people in the region also opposed the listing.

New scientific research and understanding regarding the origin and evolution of Australia's rainforests (Box 1.1) and events such as the Second World Wilderness Congress held in Cairns in 1980 drew national and international attention to the significance of the rainforests of the Wet Tropics of Queensland and the threats to their internationally significant values. In the early 1980s strong pressure was being mounted by conservation groups to protect the rainforests from logging operations. The primary focus of early campaigns was confined to the northern, 'Greater Daintree' section of the region. In 1982, the Wet Tropics was included on the World Conservation Union's (IUCN) list of places deserving World Heritage protection. This provided the impetus for World Heritage listing of the Wet Tropics.

A significant event in the campaign for rainforest protection came in November 1983 when a developer, supported by the then Queensland government, constructed a road through the lower Daintree rainforests (the Cape Tribulation–Bloomfield Road). This resulted in a blockade by protestors, which although unsuccessful in stopping the construction of the road, focused significant national and international attention to the area. In 1984 the Australian Heritage Commission engaged the Rainforest Conservation Society of Queensland to evaluate the international conservation significance of the area between Townsville and Cooktown. Their report concluded that 'From the information compiled in this study, we conclude that the Wet Tropics region of North-East Queensland is one of the most significant regional ecosystems in the world. It is of outstanding scientific importance and natural beauty and adequately fulfils all four of the criteria defined by the World Heritage Convention for inclusion in the "World Heritage List"' (Rainforest Conservation Society of Queensland 1984; report published by the Australian Heritage Commission in 1986).

During 1985 the federal government developed the National Rainforest Conservation Program for the long-term protection of North Queensland rainforests. The programme ear-marked $22.24 million of Commonwealth funds for a review of the rainforest timber industry, acquisition of rainforest on private lands, preservation of virgin rainforest and the establishment of a national rainforest research institute. The Queensland government rejected the programme and refused to participate. It was not until the 1987 federal election that the Commonwealth government announced its commitment to the World Heritage listing of the rainforests of the Wet Tropics of Queensland despite the objections of the Queensland state government, all shire councils and several Aboriginal representative groups within the region.

In April 1988, the Commonwealth government announced the $75.3 million Wet Tropics of Queensland Structural Adjustment Package (SAP) for job creation, labour adjustment and assistance and business compensation to offset the impacts of the cessation of logging in the area. The package included:
• compensation of business directly related to and dependent on the logging industry;
• payment of a dislocation allowance to retrenched timber workers;

• payment of early retirement assistance to eligible retrenched workers;
• employment and training assistance;
• local council projects to develop infrastructure in the area;
• tree planting projects (see this volume);
• grants and subsidies to businesses to employ retrenched timber workers;
• community initiatives.

At the Twelfth Session of the World Heritage Committee meeting in Brasilia, Queensland's Wet Tropics nomination was endorsed and the Area was officially inscribed on the World Heritage list on 9 December 1988.

Tropical rainforest studies in Australia

Consistent with the prevailing social and political values held at various times, the earliest studies of the region's vegetation were undertaken for the purposes of assessing the potential of the land for development. From 1882, when naturalist explorer Carl Lumholtz explored the Herbert River district, until the 1960s, sporadic expeditions laid the foundation of a general knowledge of the region's fauna and its distribution. The Archbold expedition of 1948 was the largest of these. With the publication in 1961 of a series of papers by Darlington describing the remarkable flightless beetle fauna from the summits of Mount Bartle Frere and Mount Bellenden Ker, a new era of intensive exploration of the fauna of the region commenced. For many faunal groups, most or a very large proportion of the species have been described since that time.

While a succession of early botanists made general observations and limited collections, the classification and ecological relationships of the rainforest vegetation were largely neglected, except for forestry and timber extraction purposes. The foundations for the study of the Wet Tropics rainforests were laid by Len Webb (1959, 1968), who published the first systematic classification of Australian rainforest vegetation from Tasmania to the monsoon tropics. He also considered the environmental forces that limited the distribution of rainforest. A plethora of papers by Len Webb and Geoff Tracey followed. The first overall classification and mapping of the region's rainforests by Tracey and Webb was in 1975, and this was followed by detailed descriptions by Tracey in 1982.

The real stimulus for increased research effort to understand the Wet Tropics rainforests came from those concerned with the impact of logging, clearing of forest and other impacts on the rainforest. The marker for this was the Daintree protests in 1983. This stimulated researchers through the Rainforest Conservation Society of Queensland to prepare a landmark publication on the conservation significance of the Wet Tropics forests for the Australian Heritage Commission (Commonwealth of Australia 1986). The report persuaded the Commonwealth government to apply for World Heritage Listing in 1988. The report by the Australian Rainforest Society not only highlighted the biological significance of the region but highlighted the need for improved information in order to allow governments to make informed policy and management decisions with respect to the rainforest and its biodiversity. In this context two edited volumes summarize the state of knowledge in the late 1980s (Kitching *et al.* 1991; Goudberg *et al.* 1991). Two other single-authored books also provide an insight into tropical forests. The first, by Adam (1992), provides an overview of Australia's rainforests, and the second, by Bowman (2000), is a highly insightful view of the role of fire and other abiotic and biotic factors in determining the distribution of tropical rainforests.

History of monitoring

Although the Sub-Department of Forestry commenced strip-line assessments of commercial forest stands in the 1930s it was not until the 1960s that monitoring plots were established to estimate the extent and yield of the timber resource. Their initial Forest Inventory Survey (FIS) involved the establishment and periodic remeasurement of permanent plots on a systematic grid. Most of the major rainforest logging areas were inventoried between 1961 and 1969 and were maintained and remeasured up to 1987. These data are still available today but are little used. Between 1971 and 1980, twenty 0.5 ha reference plots were established by the CSIRO across the Wet Tropics and in forests further south (Eungella approximately 21° S) and north (McIlwraith Range and Iron Range approximately 12–13° S). Although these have been resurveyed

several times the data have yet to be analysed and published apart from the initial recent description of the sites and methods. This is a very important resource that has yet to be maximized and made available to the scientific community and the public. Small plots have an important role to play in understanding plant communities but internationally 25–50 ha plots, made famous by the first such plot in Panama (Hubbell & Foster 1983), have become a standard for examining plant dynamics. There are now 17 such large-scale plots around the world that are recensured every five years. The lack of similar plots in New Guinea and Australia is a notable gap in this global network.

Evolution of regional rainforest research capability and infrastructure capacity

The early 1940s represents the watershed when scientific research into the region's rainforests became firmly established. There was an increase in professional foresters stationed in the region and the CSIRO (then the CSIR, Council for Scientific and Industrial Research) established a phytochemical survey of rainforest plants (e.g. by Len Webb). This research provided the catalyst for cooperative rainforest studies and led to the establishment of a small rainforest ecology unit within CSIRO in Brisbane (led by Webb and Tracey). The studies on the community ecology of rainforests initiated by this unit still continue and have had a profound influence on our scientific understanding and appreciation of Australian rainforests. The CSIRO research presence in the Wet Tropics expanded with the establishment of a Tropical Forest Research Station in Atherton in 1971. Also in the early 1970s, the first zoologist attached to Queensland National Parks was stationed in the Wet Tropics to undertake faunal inventory studies.

A further stimulus was the opening of a new university, James Cook University, in Townsville in 1970. This university has become world renowned for its work on coral reefs and marine science because of the proximity to the Great Barrier Reef. Since that time we have seen a rapid growth in another city, Cairns, which has become the main centre for ecotourism in the region. This change has been so rapid that Cairns was not even mentioned in the *Australian Science and Technology* review of Australian tropical science in

1993, yet it is now recognized internationally as one of the leading centres for tropical forest and landscape science. The origins of this were formed through the Institute for Tropical Rainforest Studies, a Cairns-based collaborative partnership between James Cook University and CSIRO, which lasted from 1990 to 1992. This entity was then replaced by the broader-scale Cooperative Research Centre for Tropical Rainforest Ecology and Management (Rainforest CRC), which was funded by the Commonwealth government from 1993 to 2006 and grew from 5 to 12 partners during that time. By 1998 the Rainforest CRC was ranked by an international review panel as being among the top three institutions of its kind in the world. The legacy of the Rainforest CRC is the novel science that it created and the way that it has been successfully applied in the region by organizations such as the Wet Tropics Management Authority (WTMA), State agencies, Far North Queensland Natural Resource Management Ltd (FNQ NRM Ltd), the tourism industry and the public. The Rainforest CRC was created at a time when the new campus of James Cook University was being established in Cairns. The location of the headquarters of the Rainforest CRC, which opened in 1995 on the new campus, and is surrounded by rainforest, has resulted in this campus having a strong rainforest focus.

Existing and emerging threats

Here we return to a theme that we discussed earlier in this chapter – how the Wet Tropics rainforests are changing through existing and emerging threats. Here we have much to learn from studies of this problem in rainforests across the world (Laurance & Peres 2006). This is also clearly a major focus for organizations such as WTMA, FNQ NRM Ltd and various state and Commonwealth departments and is addressed in a variety of reports (e.g. WTMA 2004). These reports identify a number of direct and underlying threats to natural values of the region, including internal fragmentation and community infrastructure (Laurance & Goosem, Chapter 23, this volume), climate change (see Balston, Chapter 21, this volume), the introduction and spread of weeds (Goosem, Chapter 24, this volume), feral animals (Congdon & Harrison, Chapter 25, this volume) and pathogens, altered fire regimes, water

quality, flow regimes and drainage patterns (Bonell & Callaghan, Chapter 2, this volume). Figure 1.3 (from Stork 2005) shows examples of human activities associated with pressures at a range of scales with regard to the Wet Tropics WHA. We recognize that some processes, such as climate change, are likely to result in long-term and pervasive transformation of the Wet Tropics landscape, while others such as walking tracks will cause mostly local but cumulative minor impacts. Many of the ongoing threats are intermediate to the above, and are undoubtedly interfering with natural processes with widespread and/or long-term consequences (Figure 1.4, from Stork 2005). We also recognize that declining water quality entering the Great Barrier Reef Lagoon, associated with agricultural and urban run-off, is a major threat to the ecological and

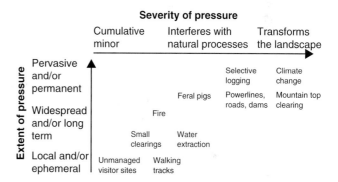

Figure 1.3 Examples of human pressures on the Wet Tropics WHA at a range of spatial scales. *Source*: after Stork (2005).

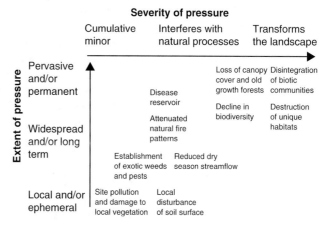

Figure 1.4 Examples of consequences of human pressures on the Wet Tropics WHA at a range of spatial scales. *Source*: after Stork (2005).

economic sustainability of the Great Barrier Reef WHA (Pearson & Stork, Chapter 45, this volume).

The main environmental impacts of sustained population growth in the Wet Tropics include those associated with:

- infrastructure development, such as the creation of new resorts and tourist attractions;
- linear infrastructure developments, such as roads and walking tracks needed to transport tourists and residents in the region or power lines and water lines to provide power and water (Goosem & Turton 2000; Goosem 2004; Turton 2005);
- urban and peri-urban growth, particularly population increases in the areas adjacent to the Wet Tropics WHA;
- water supply and increasing demand for more water storage;
- waste disposal in environmentally sensitive areas.

Many of the regional-scale processes are interrelated, with the growth in tourism and associated service industries being the principal driver for all of them (Turton & Stork 2006).

Tourism and recreation activities and their associated environmental impacts in the Wet Tropics WHA have been largely associated with visitor use of walking tracks and trails, old forestry roads and tracks, day use areas, camping areas and water holes and rivers (Turton 2005). While most visitor activities result in highly localized impacts at a small number of sites, threats such as the spread of weeds, feral animals and soil pathogens as a consequence of tourism and recreation activities potentially affect much larger areas, largely because of the extensive network of old forestry (logging) roads in the WHA (Goosem & Turton 1998, 2000; Goosem 2004). While demand for recreational use of forestry roads and long-distance walking tracks is much lower than demands for use of other visitor settings in the region, the threats to World Heritage values are considerably greater owing to the more dispersed nature of these activities (WTMA 2003).

Wet Tropics rainforests are threatened by a large and increasing number of invasive species and these create a huge management problem (Grice & Setter 2003). Many environmental weeds affect large areas of rainforest or former rainforest lands that have been converted for agriculture. There are many feral vertebrates in the Wet Tropics, including pigs (Johnson 2001), cats,

dogs, cattle, deer, some bird species and numerous introduced fish. In an analysis of the risk posed by these feral animals, one species that has not yet been seen in the area, the fox, was identified as posing the greatest threat (Harrison & Congdon 2002; Chapter 25, this volume). Recently, another introduced species, the rabbit, has moved north into agricultural areas and it is expected that its main predator, the fox, will soon follow and will present a major threat to many of the native vertebrates. Other introduced mammals, such as feral cats, as yet have had little reported impact on rainforest species. Cane toads introduced from Hawaii are now very widespread in lowland rainforest and little is known of their impacts on native fauna.

The problem of pest species is likely to increase for the WHA with the globalization of trade and the demand for access to more areas of rainforest sought by the tourism industry. The spread of soil pathogens by hikers along walking trails and by vehicle tyres on forestry roads is a potentially serious threat. Of particular concern is the spread or activation of the root rotting fungus *Phytophthora cinnamomi*, which is known to cause forest dieback (Worboys & Gadek 2004). Research within the Tully Falls–Koombooloomba section of the WHA has shown a strong association between distributions of *P. cinnamomi* in mapped dieback polygons and the location of roads and old forestry tracks in the area (Gadek & Worboys 2002). Recently, Worboys and Gadek (2004) found similar associations between dieback polygons and distance from roads and tracks for the Mt Lewis and Kirrama/Cardwell range sections of the WHA. Recreational use of long-distance walking tracks by hikers and old logging tracks by off-road enthusiasts has the potential to spread the fungus spores to susceptible areas currently unaffected by dieback, or to activate them there. But one unknown factor is how this disease might be spread by native and introduced fauna such as cassowaries and feral pigs.

Old forestry roads and off-road tracks produce a multitude of biophysical impacts on rainforest ecosystems, including linear barrier effects on arboreal and ground-dwelling fauna, road kill and biotic and abiotic edge effects, which may extend more than 200 m into the adjacent forest (Goosem 2004; Chapter 36, this volume). Other environmental effects include provision of alien habitat along road verges that are often colonized by non-rainforest fauna and flora, feral animals and weed species (Goosem 2004). Roads and old logging tracks also act as conduits for feral animals, weeds and soil pathogens, facilitating their movement and colonization of core rainforest habitat (Goosem & Turton 2000). Rainforest roads disrupt canopy cover, leading to increased rates of erosion and sedimentation along road sections with open forest canopies compared with closed canopies (Goosem & Turton 1998). Linear clearings created by roads also create significant barriers to the movement of many rainforest animals, leading to sub-division of animal populations and associated demographic and genetic problems for many rare and threatened species (Goosem 2004).

There is now widespread agreement among climate scientists and ecologists that climate change poses the greatest single threat to the ongoing integrity of natural values in the Wet Tropics WHA and surrounding landscapes (Krockenberger et al. 2004; Williams & Hilbert 2006). Table 21.1 in Balston, this volume, summarizes the current climate of Cairns and what we might expect by 2030 and 2070 as a consequence of climate change. The Wet Tropics region may expect a 1°C warming by 2030 and as much as 2°C by 2070, together with more hot days with maxima over 35°C. Rainfall is predicted to decline by as much as 10%, with more pronounced dry seasons and higher inter-annual variability. Evaporation is expected to increase in response to higher average temperatures and slightly lower rainfall.

Rainforest CRC researchers have been global leaders in the field of climate change impacts on tropical rainforests and their biota (Krockenberger et al. 2004; Williams & Hilbert 2006). Greatest impacts of climate change are likely to be in the upland forest (above 600 m elevation), where cool-adapted species are found. Modelling studies by Hilbert et al. (2001) have shown that these upland forests are particularly sensitive to changes in temperature and rainfall. Changes as small as a 1°C increase in temperature and a 10% decline in rainfall result in a significant contraction of rainforest above 600 m in elevation, resulting in less suitable and more fragmented habitat. Problems are exacerbated by the fact that many regionally restricted (endemic) species occur in the Wet Tropics with restricted climate ranges. Particularly alarming is modelling work by Williams et al. (2003) that shows that high elevation species may become progressively more restricted as their already confined habitat declines or even disappears under global warming. Much of this modelling work formed the basis of a global review (Thomas et al. 2004) that showed that about one-third of all the world's species are threatened with extinction within 50 years due to global warming, with many endemic species from far north Queensland likely to face an even higher loss.

Ecosystem function is also likely to be severely affected by climate change in far north Queensland (cf. Crimp et al. 2004; Krockenberger et al. 2004). In particular, ecosystem services provided by our rainforests, such as carbon storage and water supply, are predicted to change over the next 50–100 years. At present our cool upland forests contain a large storage 'pool' of carbon compared with the warmer lowlands where there is a much higher turnover of biomass (D. Hilbert, unpublished). Under global warming scenarios there is a real risk that much of this carbon will be released into the atmosphere, thereby enhancing anthropogenic emissions of greenhouse gases. Our upland rainforests also act as 'cloud strippers', with field studies showing that up to 30% extra water is delivered to Wet Tropics catchments from this process (McJannet et al., Chapter 15, this volume). Under global warming, the cloud-base will increase about 100 m for every 1°C rise in temperature, thus reducing water input to our catchments, particularly in the dry season.

Climate change may also create new or modified environments that will favour the expansion of weeds, feral animals and diseases adapted to warmer climes (Krockenberger et al. 2004). With warmer temperatures, slightly lower rainfall and higher evaporation there is a real threat of increased fire activity in the Wet Tropics. Altered fire regimes that lead to more open forest environments may also favour the spread of certain weed types, with long-term consequences for our regional ecosystems.

Summary

• Moist tropical rainforests are renowned for their immense biodiversity and cover approximately 6–7% of the surface of the globe, occuring in a band about 15–20° on either side of the equator. Rainforests throughout the world are highly contested landscapes as governments and the commercial sector seek to increase economic benefits from what are seen as largely unproductive forests. Major threats include logging, both legal and illegal, fire and general encroachment through increased access.

- Australia's rainforests comprise only a minuscule proportion of this total but are vitally important for their unique biodiversity. Only about 20% or 156 million hectares of Australia has a native forest cover, of which just over 3.0 million hectares is rainforest.

- The largest fragment of rainforest in Australia occurs as a narrow strip along the east coast from 15° 30' S to almost 19° 25' S and covers approximately 2 million hectares. Such is the biological significance of the region that a large part of this was inscribed on the World Heritage List on 9 December 1988 as the Wet Tropics of Queensland World Heritage Area.

- To the Bama Aboriginal people, the Traditional Owners of the region, the Wet Tropics is a series of complex living cultural landscapes. Their country is therefore embedded with enormous meaning and significance.

- European settlement has brought about radical change to the Wet Tropics in line with the perceptions and needs of the time. The scientific understanding of the region has increased dramatically in the past 50 years and with it an improved base for sustainable management of the region.

References

Adam, P. (1992). *Australian Rainforests*. Clarendon Press, Oxford.

Adam, P. (1994). *Australian Rainforests*. Oxford Biogeography Series No. 6. Oxford University Press, Oxford.

Australian Heritage Commission (1987). *The Rainforest Legacy, Vol. 1. The Nature, Distribution and Status of Rainforest Types*. Australian Government Publishing Service, Canberra.

Bennett, E. L. & Robinson, J. G. (2000). Hunting for sustainability: the start of a synthesis. In *Hunting for Sustainability in Tropical Forests*, Bennett, E. L. & Robinson, J. G. (eds). Columbia University Press, New York, pp. 499–520.

Bowman, D. M. J. S. (2000). *Australian Rainforests: Islands of Green in a Land of Fire*. Cambridge University Press, Cambridge. 345 pp.

BRS (2003). *Australia's State of the Forests Report*. National Forest Inventory, Bureau of Rural Sciences (www.affa.gov.au/content/output.cfm?ObjectID=1F434DF7-3882-42C6-9BD9F1ED1336D03E&contType=outputs).

Commonwealth of Australia (1986). *Tropical Rainforests of North Queensland. Their Conservation Significance*. Australian Heritage Commission Special Australian Heritage Publication Series No. 3. Australian Government Publishing Service, Canberra. 195 pp.

Crimp, S., Balston, J. & Ash. A. (2004). *Climate Change in the Cairns and Great Barrier Reef Region: Scope and Focus for an Integrated Assessment*. Australian Greenhouse Office, Canberra.

DASETT (1987). *Nomination of the Wet Tropical Rainforests of North-east Australia by the Government of Australia for Inclusion in the World Heritage List*. Department of Arts, Sport, the Environment, Tourism and Territories, Canberra, December.

Dalrymple, G. E. (1874). *Narrative and Report of the Queensland Northeast Coast Expedition 1873*. Government Printers, Brisbane.

Flannery, T. (1995). *The Future Eaters: An Ecological History of Australasian Lands and People*. Reed Books, Port Melbourne. 432 pp.

Floyd, A. G. (1987). Status of rainforests in northern New South Wales. In *The Rainforest Legacy, Vol. 1*. Werren, G. L. & Kershaw, A. P. (eds). Australian Government Publishing Service, Canberra, pp. 95–118.

Francis W. D. (1929). *Australian Rain Forest Trees*. Government Printer, Brisbane.

Frawley, K. (1983). Forest and land management in north-east Queensland: 1859–1960. PhD Thesis, Australian National University.

Frawley, K. J. (1987). *The Maalan Group Settlement, North Queensland, 1954: An Historical Geography*. Monograph Series No. 2, Dept of Geography and Oceanography, University College, Australian Defence Force Academy, Canberra.

Gadek, P. A. & Worboys, S. J. (2002). *Rainforest Dieback Mapping and Assessment:* Phytophthora *Species Diversity and Impacts of Dieback on Rainforest Canopies*. School of Tropical Biology, James Cook University and Rainforest CRC, Cairns.

Goosem, M. W. (2004). Linear infrastructure in tropical rainforests: mitigating impacts on fauna of roads and powerline clearings. In *Conservation of Australia's Forest Fauna*, 2nd edn, Lunney, D. (ed.). Royal Zoological Society of NSW, Mosman, pp. 418–34.

Goosem, M. W. & Turton, S .M. (1998). *Impacts of Roads and Powerline Corridors on the Wet Tropics of Queensland World Heritage Area, Stage 1*. Wet Tropics Management Authority and Rainforest CRC, Cairns.

Goosem, M. W. & Turton, S. M. (2000). *Impacts of Roads and Powerline Corridors on the Wet Tropics of Queensland World Heritage Area, Stage 2*. Wet Tropics Management Authority and Rainforest CRC, Cairns.

Goosem, S. (2002). *Update of Original Wet Tropics of Queensland Nomination Dossier*. Wet Tropics Management Authority. Cairns (www.wettropics.gov.au/res/downloads/periodic_report/WT_Periodic_Report_Attachments.pdf).

Goudberg, N., Bonell, M. & Benzaken, D. (eds) (1991). *Tropical Rainforest Research in Australia. Present Status and Future Directions for the Institute for Tropical Rainforest Studies*. Institute for Rainforest Studies, Townsville. 210 pp.

Grice, W. A. C. & Setter, M. J. (eds) (2003). *Rainforests and Associated Ecosystems*. Rainforest CRC and Weeds CRC, Cairns.

Harrison, D. A. & Congdon, B. C. (2002). *Wet Tropics Vertebrate Pest Risk Assessment Scheme*. Cooperative Research Centre for Tropical Rainforest Ecology and Management, Cairns.

Hilbert, D., Ostendorf, B. & Hopkins, M. (2001). *Austral Ecology* **26**: 590–603.

Hubbell, S. P. & Foster, R. B. (1983). Diversity of canopy trees in a Neotropical forest and implications for conservation. In *Tropical Rain Forest: Ecology and Management*, Sutton, S. L., Whitmore, T. C. & Chadwick, A. C. (eds). Blackwell Scientific Publications, Oxford, pp. 25–41.

Johnson, C. N. (ed.). (2001). *Feral Pigs: Pest Status and Prospects for Control*. Proceedings of a feral pig workshop, James Cook University, Cairns, March 1999. Cooperative Research Centre for Tropical Rainforest Ecology and Management, Cairns.

Keto, A. & Scott, K. (1986). *Tropical Rainforests of North Queensland: Their Conservation Significance*. A report to the Australian Heritage Commission by the Rainforest Conservation Society of Queensland. Australian Government Publishing Service, Canberra.

Kitching, R. L. (ed.) (1988). The ecology of Australia's Wet Tropics. In *Proceedings of the Ecological Society of Australia, Vol. 15*. Surrey Beatty & Sons, Sydney. 326 pp.

Krockenberger, A .K., Kitching, R. L. & Turton, S. M. (2004). *Environmental Crisis: Climate Change and Terrestrial Biodiversity in Queensland*. Rainforest CRC, Cairns.

Laurance, W. F. & Bierregaard, R. O. (eds). (1997). *Tropical Forest Remnants: Ecology, Management and Conservation of Fragmented Communities*. University of Chicago Press, Chicago. 616 pp.

Laurance, W. F., Cochrane, M. A., Bergen, S. *et al.* (2001). The future of the Brazilian Amazon. *Science* **291**: 438–9.

Laurance, W. F. & Peres, C. A. (eds). (2006). *Emerging Threats to Tropical Forests*. Chicago University Press, Chicago. 520 pp.

McDonald, G. & Lane. M. (2000). *Securing the Wet Tropics?* The Federation Press, Sydney.

McDonald, G. & Weston, N. (2004). *Sustaining the Wet Tropics: A Regional Plan for Natural Resource Management, Vol. 1. Background to the Plan*. Rainforest CRC and FNQ NRM Ltd, Cairns.

McKenzie, N. L., Johnston, R .B. & Kendrick, P. G. (eds) (1991). *Kimberley Rainforests of Australia*. Surrey Beatty & Sons, Sydney. 490 pp.

Meston, A. (1889). *Report by A. Meston on the Government Scientific Expedition to the Bellenden-Ker Range (Wooroonooran), North Queensland*. Votes and Proceedings of the Legislative Assembly, 3, Brisbane.

Myers, N. (1993). Tropical forests: the main deforestation fronts. *Environmental Conservation* **20**: 9–16.

Myers, N. Mittermeier, R. A., Mittermeier, S. G., da Fonseca, G. A. & Kent, J. (2000). Biodiversity hotspots for conservation priorities. *Nature* **403**: 853–5.

National Land and Water Resources Audit (2001). CSIRO Land and Water Technical Report Number 15/01.

Primack, R. & Corlett, R. (2005). *Tropical rain forests: An ecological and biogeographical perspective*. Malden, MA. Blackwell.

Sattler, P. S. & Williams, R. (1999). *The Conservation Status of Queensland's Bioregional Ecosystems*. Environmental Protection Agency, Queensland Government, Brisbane.

Sechrest, W., Brooks, W. M., da Fonseca, G. A. B. *et al.* (2002). Hotspots and the conservation of evolutionary history. *Proceedings of the National Academy of Sciences* **99**: 2067–71.

Stork, N. E. (2005). The theory and practice of planning for long-term conservation of biodiversity in the Wet Tropics of Australia: integrating ecological and economic sustainability. In *Tropical Rainforests: Past, Present, and Future*, Bermingham, E., Dick, C. W. & Moritz, C. (eds). Chicago University Press, Chicago, pp. 507–26.

Thackway, R. & Cresswell, I. D. (1995). *An Interim Biogeographic Regionalisation for Australia: A Framework for Setting Priorities in the National Reserves System Cooperative Program*. Australian Nature Conservation Agency, Canberra.

Thomas, C. D., Williams, S. E., Cameron, A. *et al.* (2004). Biodiversity conservation: uncertainty in predictions of extinction risk/Effects of changes in climate and land use/Climate change and extinction risk (reply). *Nature* **430**: 145–8.

Turton, S .M. (2005). Managing environmental impacts of recreation and tourism in rainforests of the Wet Tropics of Queensland World Heritage Area. *Geographical Research* **43**: 140–51.

Turton, S. M., Hutchinson, M. F., Accad, A., Hancock, P. E. & Webb, T. (1999). Producing fine-scale rainfall climatology surfaces for Queensland's wet tropics region. In *Geodiversity: Readings in Australian Geography at the Close of the 20th Century*, pp. 415–28. Kesby, J. A., Stanley, J. M.,

McLean, R. F. & Olive L. J. (eds). Special Publication Series No. 6, ACT, School of Geography and Oceanography, University College, ADFA, Canberra, 630 pp.

Turton, S. M. & Stork, N. E. (2006). Tourism and tropical rainforests: opportunity or threat? In *Emerging Threats to Tropical Forests*, Laurance, W. F. & Peres, C. A. (eds). University of Chicago Press, Chicago, pp. 337–91.

Tracey, J.G., (1982). The vegetation of the humid tropical region of North Queensland. CSIRO, Melbourne, 124 pp.

Tracey, J. G. & Webb, L. J. (1975). *Vegetation of the Humid Tropical Region of North Queensland.* (15 maps at 1:100 000 scale + key). CSIRO Long Pocket Laboratory, Indooroopilly.

Walker, J. W. (1976). Comparative pollen morphology and phylogeny of the Ranalean Complex. In. *Origin and Early Evolution of Angiosperms*, Beck, C. B. (ed.). Colombia University Press, New York, pp. 241–99.

Webb, L. J. (1959). A physiognomic classification of Australian rain forests. *Journal of Ecology* **47**: 551–70.

Webb, L. J. & Tracey, J. G. (1981). The rainforests of northern Australia. In *Australian Vegetation*, Groves, R. H. (ed.). Cambridge University Press, Cambridge, pp. 67–101.

Webb, L. J., Tracey, J. G. & Williams, W. T. (1976). The value of structural features in tropical forest typology. *Australian Journal of Ecology* **1**: 3–28.

Webb, L. J., Tracey, J. G. & Williams, W. T. (1984). A floristic framework of Australian rainforests. *Austral Ecology* **9**: 169–98.

Williams, S. E., Bolitho, E. E. & Fox, S. (2003). Climate change in Australian tropical rainforests: an impending environmental catastrophe. *Proceedings of the Royal Society Lond on B* **270**: 1887–92.

Williams, S. E. & Hilbert, D. W. (2006). Climate change as a threat to the biodiversity of tropical rainforests in Australia. In *Emerging Threats to Tropical Forests*, Laurance, W. F. & Peres, C. (eds). Chicago University Press, Chicago, pp. 33–52.

Worboys, S. J. & Gadek, P. A. (2004). *Rainforest Dieback: Risks Associated with Roads and Walking Track Access in the Wet Tropics World Heritage Area.* School of Tropical Biology, James Cook University and Rainforest CRC, Cairns.

Wright, S. J. & Muller-Landau, H. C. (2006a). The future of tropical forests. *Biotropica* **38**: 287–301.

Wright, S. J. & Muller-Landau, H. C. (2006b). The uncertain future of tropical forests. *Biotropica* **38**: 443–5.

WTMA (1999). *Wet Tropics Management Authority Annual Report 1998–99.* WTMA, Cairns, 42 pp.

WTMA (2002). State of the Wet Tropics. In: *Wet Tropics Management Authority Annual Report* 2001–02. Wet Tropics Management Authority, Cairns, Australia.

WTMA (2003). *Wet Tropics Management Authority Annual Report 2002–03.* Wet Tropics Management Authority, Cairns.

WTMA (2004). *Wet Tropics Conservation Strategy.* Wet Tropics Management Authority, Cairns.

Young P. A. R. & McDonald W. J. (1987). The distribution, composition and status of rainforests in southern Queensland. In *The Rainforest Legacy, Australian National Rainforest Study, Vol. 1.* Australian Heritage Commission, AGPS, Canberra.

Part 1

History and Biodiversity of the Wet Tropics

INTRODUCTION

Nigel E. Stork[1]* and Stephen M. Turton[2]*

[1]School of Resource Management and Geography, Faculty of Land and Food Resources, University of Melbourne, Burnley Campus, Richmond, Victoria, Australia
[2]Australian Tropical Forest Institute, School of Earth and Environmental Sciences, James Cook University, Cairns, Queensland, Australia
*The authors were participants of Cooperative Research Centre for Tropical Ecology and Management

The first half of this section looks at the climatic, social and historical influences on the Wet Tropics. Tropical rainforests are restricted to only a very small part of the East coast of Australia and are largely present as a result of region's climate and hydrology (Bonnell and Callaghan, Chapter 2). A further important influence on the region is the impact of cyclones and as a result these forests are rather different in structure to some of the much taller tropical rainforests further north in the south-east Asian archipelago (Turton and Stork, Chapter 3). The Wet Tropics has a diverse and rich cultural history that is addressed in several chapters in this book but is first discussed here by Pannell (Chapter 4). European settlement in the region is much more recent but has undergone many transformations from the early pioneers seeking farm land, through a highly developed forestry phase to what is now a very diverse use of the landscape. This is summarized by D. Turton (Chapter 5). One of the more recently developed and most successful industries is tourism and, in particular, rainforest tourism. Pearce (Chapter 6) discusses how if such an industry to be sustainable there is a need to identify and understand a range of the forces impacting on it and address these.

The second part of this section discusses the biodiversity of the Wet Tropics and provides an understanding of how history has shaped this biodiversity. In particular, Hilbert (Chapter 8) shows the dynamic nature of the floristic rainforest landscape through the past, present and future and Metcalfe and Small (Chapter 9) described the unique flora of the region. Williams *et al.* (Chapter 10) examine the genetic and species diversity of the vertebrates across the landscape. It is often easy to focus solely on the terrestrial fauna and flora and to forget the rich biodiversity of streams and rivers. This gap is addressed for the intensively studied fish fauna (Pusey *et al.*, Chapter 11) and invertebrates (Connolly *et al.*, Chapter 12). The biodiversity of this area of tropical rainforest is easily the best known of any comparable area in the world. One of the outstanding examples is our knowledge of the upland invertebrates (Yeates & Monteith, Chapter 13). It is fitting that a personal perspective and summary of this section should be one of Australia's outstanding biologists, Jiro Kikkawa (Chapter 14).

2 THE SYNOPTIC METEOROLOGY OF HIGH RAINFALLS AND THE STORM RUN-OFF RESPONSE IN THE WET TROPICS

Mike Bonell[1]* and Jeff Callaghan[2]

[1]Formerly UNESCO, Section on Hydrological Processes and Climate, Division of Water Sciences, 1 rue Miollis, Paris, France
[2]Severe Weather Section, Bureau of Meteorology, Brisbane, Queensland, Australia
*Now at the UNESCO HELP Water Centre, University of Dundee, Dundee DD1 4HN, Scotland, UK and the Department of Environmental Science, Lancaster University, Lancaster LA1 4YW, UK

Introduction

The nature of the synoptic meteorology and climatology of the Wet Tropics makes rainfall a key driver in differentiating the hydrological response to storm events from those that occur elsewhere in many of the forests of the humid tropics as well as the humid temperate areas. This chapter selectively highlights the dynamic features of the synoptic climatology and meteorology that have a bearing on the more extreme rainfalls that occur along the wet tropical coast; and subsequently the work provides a succinct overview of the impacts of such rainfalls on the storm run-off generation process within the closed tropical forests. When concerning the hydrology, most attention is given to the principal research outputs from the long-term experimental research campaign that took place in the paired, headwater catchments (North Creek, South Creek) near Babinda (1969–94) located at 17° 20′ S, 145° 58′ E. Detailed descriptions of these catchments can be found in several sources (e.g. Bonell 1991, 2004; Bonell *et al.* 1991, 1998).

The recent monograph by Bonell and Bruijnzeel (2004) included comprehensive details of both the synoptic and rainfall climatology as well as the storm run-off generation process of the north-east Queensland area, as part of a global review of tropical forest hydrology. The hydrology component elaborated and updated earlier reviews lodged mostly within the Australian literature (e.g. Bonell 1991; Bonell *et al.* 1991), except Bonell *et al.* (1998). Consequently the reader is directed to these sources for essential background information of which space limitations here preclude such mention. The current work, however, offers an opportunity to take a more retrospective position on recent research achievements biased to the authors' own work. Definitions of meteorological and hydrological terms used in this account were presented in Bonell and Bruijnzeel (2004) under Callaghan and Bonell (2004), Bonell *et al.* (2004), Bonell (2004) and Chappell *et al.* (2004).

Climatology and meteorology

One of the outstanding features of the wet tropical coast (including the adjacent mountain ranges) between Cooktown and Cardwell is the high annual rainfall (commonly 2000–8000 mm) and the marked concentration of rain in a few months of the year. Typically in excess of 60% of mean annual rainfall occurs in the months December to March.

On the basis of short-term intensity rainfalls, Bonell and Gilmour (1980) divided up the year into four rain season types: summer monsoon (mid-December to March), post-monsoon (April to mid-June), winter 'dry' season (mid-June to September or October) and pre-monsoon (October to early December). The beginning and end times of these seasons vary from year to year depending on inherent climatic variability (e.g. the phase of the El Niño Southern Oscillation (ENSO)). Moreover, the winter 'dry' season can continue up to December with no significant pre-monsoon storm events. In some years the summer monsoon season can commence abruptly with the southward advance of a tropical cyclone, which affects the Wet Tropics either by taking a track across the south-east Gulf of Carpentaria (see example of the rainfall impacts from tropical cyclone Ted in Gilmour and Bonell 1979) or the western Coral Sea (e.g. tropical cyclone Joy, late December 1990).

Howard (1993) (summarized in Bonell 2004, pp. 347–8) later extended the work of Bonell and Gilmour (1980) by undertaking a statistical analysis of both the rainfall and stream hydrograph characteristics of the Babinda research basins over the period 1974 to 1981 across the above mentioned four rain seasons. The magnitude of the quickflow component (as defined by Chorley 1978) of the stream hydrograph was used as a more realistic criterion for separating each rain season by year that also took into account climatic variability. As part of this analysis, the quickflow component by storm event was divided into the categories 2–5 mm, 5–10 mm, 10–50 mm, 50–100 mm and greater than 100 mm using the hydrograph separation technique of Lyne and Hollick (1979) (Table 2.1). During the monsoon season, the average maximum rain intensities (I_6) by event were in the range 14–82 mm h^{-1} across the preceding categories of quickflow. For the post-monsoon season, the corresponding I_6 reduced to 15–26 mm h^{-1} (by which time the quickflow had contracted to the categories, 1–5 mm, 5–10 mm, >10 mm only). There was a further marginal relaxation of I_6 to 17–24 mm h^{-1} (such relaxation also apply to higher maximum rain depths up to I_{60}) in the winter 'dry' season leading to a further reduction of the number of quickflow categories; that is, 1–5 mm and >5 mm. For the pre-monsoon season I_6 can attain monsoon levels up to 110 mm h^{-1} or occasionally higher.

During the Austral summer, tropical northern Australia becomes a very active meteorological area and merges with the maritime continent of Ramage (1968) further north, to become the most important of the three global hot spots (the others being the Amazon and Congo basins in respective order) in terms of the global redistribution of latent energy (e.g. Gunn et al. 1989). As part of this activity, the meteorological drivers at the synoptic scale that trigger major storm events over north-east Queensland are:

• the southern monsoon shearline (locally known as the monsoon trough), which is the confluence of cross-equatorial flow (originally Northern hemisphere NE trades now deflected by the Coriolis force near the equator as the NW monsoon) with the Southern hemisphere south-east trades;

• tropical depressions and occasionally tropical cyclones, most of which (but not all, e.g. tropical cyclone Larry as mentioned below) have their origins connected with the more active phases of the monsoon trough;

• the northward penetration within the upper levels of the atmosphere of upper troughs embedded within the westerly air flow that overlay the lower trade wind easterlies. Some of these upper troughs are extensions of disturbances originating from the higher latitude temperate westerly winds (see Box 2.1);

• perturbations embedded in the E to SE trades, of which some have links with the preceding upper trough phenomena.

Other drivers that impact on the summer monsoon meteorology are external to tropical Australia. These include the eastward progression of the 30–60-day Madden-Julian Oscillation (MJO) which coincides with some of the more active phases of the monsoon trough; the varying strength of cross-equatorial flow from the NW Pacific and east Asia (which is the origin of the NW monsoon); the phases of the ENSO; and inter-decadal climatic variability (see reviews in Manton & Bonell 1993; Callaghan & Bonell 2004). It is pertinent to highlight that the genesis of the recent category 4/5, tropical cyclone Larry,[1] which had severe impacts on the tropical forest centred on the Innisfail area at landfall (00:00 UTC, 20 March 2006), is an example that does not conform with the usual combination of drivers mentioned above. Larry developed from an initial vortex within the middle levels of the atmosphere from a Southern hemisphere influence associated with a trough system extending into the tropics from the south (see Box 2.1). The genesis of this tropical cyclone also occurred during a suppressed phase of the MJO (Callaghan 2006).

Table 2.1 Descriptive statistics of the rainfall and runoff variables for the undisturbed South Creek (25.7 ha) for the monsoon, post-monsoon and dry seasons, 1 January 1974 to 31 May 1981

Number of storms	Storm class Q (mm)		D	I	P6	P12	P18	P24	P30	P60	Q_p	Q_F	Q	P
A Monsoon															
32	2–5	Mean	8.3	20.4	1.4	2.3	2.9	3.2	3.4	4.0	0.5	0.6	2.9	6.3
		Max.	39	214.4	7.9	14.6	18.6	19.6	20.5	20.8	3.4	1.6	5.0	32.0
		Min.	1	2.2	0.2	0.3	0.4	0.5	0.5	0.5	0.2	0.2	2.0	0.5
30	5–10	Mean	15.6	22.0	3.0	5.0	6.4	7.2	8.0	10.4	0.9	2.1	7.5	17.5
		Max.	56	242.5	7.4	12.0	16.3	19.6	21.4	24.8	3.9	4.8	9.9	81.5
		Min.	2	1.4	0.4	0.6	0.8	0.8	0.9	1.0	0.3	0.6	5.1	1.0
62	10–50	Mean	33.0	26.2	3.5	5.8	7.4	8.6	9.4	12.1	2.1	8.8	23.5	33.7
		Max.	163	107.9	11.5	16.9	19.3	22.4	26.6	36.7	14.5	36.1	49.7	194.8
		Min.	6	1.1	0.7	1.2	1.5	1.8	2.0	2.8	0.4	1.6	10.1	6.5
27	50–100	Mean	56.7	24.9	6.3	11.4	15.2	18.2	20.9	29.2	8.9	36.9	71.0	96.6
		Max.	156	264.4	12.1	21.3	30.0	38.4	45.9	72.5	30.7	70.6	99.5	265.8
		Min.	10	1.7	2.7	5.5	6.7	7.8	8.7	10.9	1.8	11.2	53.1	20.7
34	>100	Mean	73.9	39.7	7.2	13.2	18.1	22.1	25.4	39.0	18.3	117.4	197.3	245.8
		Max.	220	291.2	16.7	31.6	45.3	55.9	64.1	104.1	53.9	415.1	547.8	628.0
		Min.	22	0.8	2.9	5.1	7.4	9.1	9.9	11.9	4.4	44.6	105.4	86.5
21	>150	Mean	73.1	43.6	8.1	14.7	20.6	25.4	29.7	46.1	23.8	163.2	257.0	329.3
		Max.	220	291.2	16.7	31.6	45.3	55.9	64.1	104.1	53.9	415.1	547.8	628.8
		Min.	22	0.8	2.9	5.1	7.4	9.3	11.7	21.3	7.1	85.0	163.9	134.1
16	>200	Mean	73.6	49.5	8.2	14.8	20.7	25.7	30.2	47.6	24.7	181.4	288.2	362.1
		Max.	220	291.2	16.7	31.6	45.3	55.9	64.1	104.1	53.9	415.1	547.8	628.8
		Min.	22	0.8	2.9	5.1	7.4	9.3	11.7	21.3	7.1	85.0	208.5	134.1
11	>250	Mean	75.2	88.6	7.4	13.3	18.7	23.3	27.7	45.4	23.9	193.4	323.1	345.9
		Max.	220	291.2	13.3	24.8	33.8	42.6	51.4	81.0	53.9	415.1	547.8	628.8
		Min.	22	14.8	2.9	5.1	7.4	9.3	11.7	21.3	7.1	85.0	250.2	134.1
B Post-monsoon															
26	2–5	Mean	9.3	18.3	1.4	2.3	2.8	3.1	3.3	4.1	0.4	0.6	2.8	7.6
		Max.	36	38.5	3.5	6.8	7.5	8.0	8.1	11.6	0.7	1.9	5.0	31.8
		Min.	4	8.0	0.2	0.3	0.3	0.3	0.3	0.3	0.2	0.2	2.0	0.3
15	5–10	Mean	25.6	13.7	2.3	3.7	4.4	5.1	5.6	6.4	0.4	1.8	6.8	19.3
		Max.	43	27.1	6.3	10.3	10.3	10.3	11.9	12.0	1.2	4.2	9.8	59.8
		Min.	15	7.8	1.0	1.7	1.8	1.8	2.0	2.0	0.3	1.0	5.0	5.0
36	>10	Mean	45.8	20.8	2.4	4.2	5.3	6.0	6.7	9.4	1.7	8.8	24.9	33.4
		Max.	138	48.0	11.9	18.1	23.3	26.9	31.8	58.1	11.4	55.8	121.2	130.6
		Min.	11	5.9	0.8	1.3	1.5	1.5	1.9	2.2	0.4	1.7	10.1	2.4
C Dry season															
48	1–5	Mean	27.5	3.4	1.7	2.8	3.4	3.8	4.1	5.2	0.2	0.7	2.1	14.2
		Max.	77	13.5	8.8	17.4	25.1	32.0	37.3	44.5	3.1	3.8	5.0	98.3
		Min.	8	0.5	0.1	0.3	0.3	0.3	0.3	0.3	0.01	0.2	1.0	0.5
24	>5	Mean	50.1	5.3	2.4	4.1	5.2	6.2	6.8	9.6	0.7	4.4	10.5	39.5
		Max.	158	20.3	12.4	23.1	29.7	37.3	43.1	50.5	5.8	21.5	32.4	212.6
		Min.	16	0.1	0.4	0.8	0.9	1.0	1.0	1.7	0.1	0.7	5.1	1.9

Q (total run-off (mm)); D (storm duration (hours)); I (antecedent streamflow prior to storm event (l s^{-1})); P6 to P60 (maximum rainfall depths by event (mm) for the respective periods, 6–60 minutes); Q_p (peak hourly runoff (mm)); Q_F (quickflow or stormflow (mm) using the hydrograph separation procedure of Lyne & Hollick, 1979); P (total precipitation by event (mm)). The means are the antilogs of logarithmic means.
Source: Howard (1993) using the database of the Babinda catchments, courtesy of D. Gilmour, D. Cassells and M. Bonell.

Box 2.1 The north-west penetration of an upper trough leads to a record severe thunderstorm in north Queensland, 19 January 2001

A band of storms passed through the Ayr/Home Hill area just after 6 p.m. on 19 January 2001, damaging houses, downing powerlines and causing widespread damage. More than 26 200 lightning strikes were recorded within 1 hour, associated with a series of storms in the region, resulting in widespread damage to the electricity grid. Rainfall from the storms was: Alva Beach 56 mm, Guru 55 mm, Home Hill 61 mm, Clare 78 mm and Townsville 24mm.

Mean sea level and other analyses

The mean sea level analyses (Figure 2.1) show a trough, separating cooler drier air to the south from very warm and moist air to the north, moving up towards north Queensland on 19 January 2001. The trough was linked to a low pressure system developing off the south Queensland coast east of a tropopause undulation (see Figure 2.2 and Plate 2.2 at the 200 hPa level). The undulation was associated with an upper trough that brought diffluent flow across north Queensland. Finally the 700 hPa charts (Figure 2.2 and Plate 2.2) indicate that as the upper trough extended into North Queensland, so did cold middle level air.

Instability

Cold middle level air overlying hot humid air allows for very unstable conditions. Convective available potential energy (CAPE) represents the amount of buoyant energy available to accelerate a parcel vertically. The higher the CAPE value, the more energy there is available to foster thunderstorm development. The CAPE in this case was 6515 J kg^{-1}, which is extremely large. Usually a CAPE of 2000 J kg^{-1} would be considered large enough to help trigger severe thunderstorms. As the cold 700 hPa thermal trough reached Townsville that afternoon, the pre-storm environment would have become even more unstable due to cooling at middle levels (see Shin *et al.* 2006 for more details on this event).

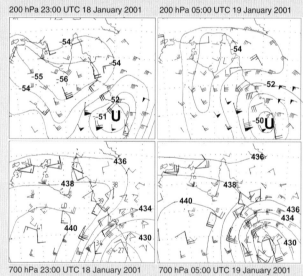

200 hPa 23:00 UTC 18 January 2001   200 hPa 05:00 UTC 19 January 2001

700 hPa 23:00 UTC 18 January 2001   700 hPa 05:00 UTC 19 January 2001

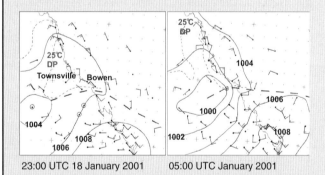

23:00 UTC 18 January 2001   05:00 UTC January 2001

Figure 2.1 Mean sea level pressure distribution and surface wind, wind plots for 23:00 UTC 18 January 2001 (left frame) and 05:00 UTC 19 January 2001 (right frame). Note that the areas with dewpoints 25°C or greater lie to the north of the dashed line. UTC is Coordinated Universal Time (equal to Greenwich Mean Time).

Figure 2.2 Corresponding to the same times as for Figure 2.1, the 200 hPa wind observations (large plots) and the NCEP/NCAR reanalysis winds (smaller plots). The blue lines are the 200 hPa isotherms with the tropopause undulation marked by the 'U'. The lower frames are the 700 hPa winds and the 850–500 hPa shears, with the blue contours the 850–500 hPa thickness in decametres (*see also* Plate 2.2).

Extreme rainfalls associated with an active monsoon trough, tropical cyclones and a winter cold-cored cyclone

During the course of describing the synoptic-scale and meso-scale features of a persistent, active monsoon trough over the Cairns region in January 1981, which produced rainfalls well in excess of 2000 mm over 13 days (see Bonell et al. 1991 for a discussion of the hydrological response); Callaghan (1985, also reviewed in Manton & Bonell 1993: 21–2) noted that the heaviest rainfall occurred just to the south of the monsoon trough and analyses elsewhere showed that such characteristics were beginning to emerge as a pattern common to other similar meteorological situations in north-east Queensland. Callaghan (1985) attributed this high rainfall zone to two processes: (i) a frictionally induced convergence zone along the coast, just south of the monsoon trough; (ii) subsequent upward motion of moist air backing (anti-clockwise) with height that then transported this moist air into the upper north-westerlies, causing the zone of heaviest rainfall. The later work of Bonell et al. (2004) elaborated on these phenomena during the course of outlining tropical cyclone formation, intensification and rain structure using case studies in both north-east Queensland and Fiji. They showed that extreme rainfall occurred under two common circumstances: (i) when upper troughs extend into the north-east Australian region and impact on poleward moving tropical lows; (ii) the warm air advection–vertical ascent mechanism (noted above) backing with height (for the southern hemisphere) that was identified by Holton (1972) and Hoskins et al. (1978).

Subsequently, observational studies on tropical cyclone intensification for the south-west Pacific (Callaghan 2004) have shown that tropical cyclones all intensified while the vortex was located over a thermal gradient at the 700 hPa (hectoPascal) level at about 3000 m elevation associated with warm air advection. Generally the vertical wind structure across the storms was backing (anti-clockwise) with height on one side of the storm centre (that is, within the eastern half of the vortex) and veering (clockwise) with height on the other (western) side, which for the southern hemisphere is related to warm (upward) and cold (downward) air advection respectively (Holton 1972). Such observations are also in line with the work of Raymond and Jiang (1990). Other dynamic mechanisms concur-

rently occur. From Hirschberg and Fritsch (1993a, b) this wind structure implies additional forcing from vorticity (defined in Callaghan & Bonell 2004) advection. Moreover, such a wind structure is also associated with upslope flow along isentropic[2] surfaces as long as the isentropic surfaces are not moving faster than the flow (Saucier 1955).

Other studies have noted that this process of shearing causing isentropic ascent leading to the release of convection (e.g. Fritsch et al. 1994; see figure 6 in Raymond & Jiang 1990) was typically illustrated by a 700 hPa surface where upward (downward) motion was correlated with warm (cold) air advection. In reality, however, within the warm air advection regions isentropic ascent and cooling tends to cancel any actual warming and the thermal effects are more complex. Nonetheless, within the meteorology literature, the terminology of warm air advection is widely used and understood based on the previous discussion. Boxes 2.2, 2.3 and 2.4 provide examples of this mechanism.

Operationally the preceding thermal patterns are detected by simply differencing the flow patterns at 850 hPa (elevation 1500 m) and 500 hPa (elevation 5600 m), which is intrinsically linked with potential vorticity (PV) (defined in Callaghan & Bonell 2004) in the upper atmosphere. Using quasi-geostrophic and semi-geostrophic theory, Hoskins et al. (2003) identified a vertical velocity term that may link the upper PV with the development of the thermal pattern identified at 700 hPa in the vicinity of intensifying tropical cyclones.

It is significant that subsequent analysis of extreme rainfall that occurred in Mumbai, India (944.2 mm in 24 hours ending 0330UTC, 27 July 2005), showed that the same warm air advection, vertical ascent mechanism could be easily detected at the 700 hPa level, with the low level W/SW monsoon air flow veering (opposite in the northern hemisphere) to northerlies with height towards the middle levels of the atmosphere (Callaghan, pers. comm., 2006, for the Mumbai soundings).

Elsewhere the recording of the first ever tropical cyclone in the south Atlantic (the storm locally known as 'Catarina' because it hit the southern region of Brazil, State of Santa Catarina, and there is no naming system in that region – other reports refer to it as 'Caterina') was notable for having developed where a low level east to south-east trade wind is prevalent and thus outside a monsoon region (where no low

Box 2.2 A tropopause undulation event, an intense winter cyclone developing off North Queensland

Very heavy winter rain was recorded in the north-east tropics as an intense low pressure system developed west of Willis Island. Record rainfall was observed in the 24 hours to 9 a.m. on 6 June 1990. The following registrations were all time June records, Townsville 93 mm, Ingham 141 mm, Cardwell 115 mm. Other notable totals were Tully 118 mm, Crawfords Lookout 87 mm, Topaz 83 mm, Cairns 52 mm and Kuranda Forestry 89 mm.

Synoptic pattern June 1990

At mean sea level (Figure 2.3) this low developed at the same time as a very large high moved into the Tasman Sea. As the storm began to intensify (top right panel in Figure 2.3) an offshore automatic weather station recorded 10-minute average wind speed of 50 knots (26 m s^{-1}). The 700 and 200 hPa charts (Figure 2.4 and Plate 2.4) show 700 hPa warm air advection (associated with the heavy rain) over the north-east tropics at 23:00 UTC 4 June 1990 and clearing from the north-east tropics (with the rain) at 23:00 UTC 5 June 1990. Overlying this warm air advection was warm air advection at 200 hPa from a tropopause undulation over southern Australia. The mean sea level low intensified as it moved under this area of overlapping warm air advection in the model of Hirschberg and Fritsch (1993a, b, 1994). Sea surface temperatures where the Low passed over up to 23:00 UTC 6 June 1990 were warmer than 26°C. Note that in this event the coastal rainfall totals were comparable with the elevated stations like Topaz and Crawfords Lookout, the latter of which usually receive much heavier rainfall totals.

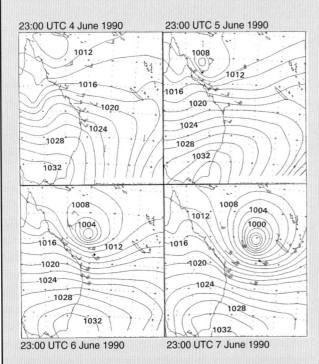

Figure 2.3 The mean sea level pressure (MSL) distribution and some surface wind plots for 23:00 UTC 4 June 1990 to 23:00 UTC 7 June 1990.

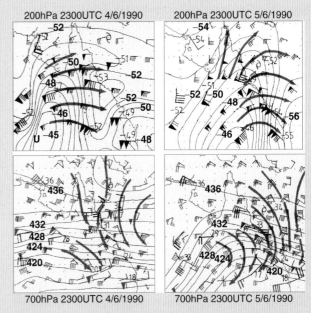

Figure 2.4 At the corresponding times shown in Figure 2.3 (top frames), the 200 hPa charts showing wind observations (large plots) and NCEP/NCAR reanalyses winds (smaller plots); the lines are the 200 hPa isotherms with the tropopause undulation marked by the 'U'. The lower frames are the 700 hPa winds and 850–500 hPa shears with the contours the 850–500 hPa thickness in decametres. Large streamlines highlight zones of warm (cold) air advection (*see also* Plate 2.4).

Box 2.3 Tropical cyclones: the impacts of contrasting thermal patterns on rainfall

In this study we contrast the different thermal patterns and consequent rainfall distributions associated with four tropical cyclones. The Australian category scale for tropical cyclones is:

- 77 knots 1 minute mean or 70 knots 10 minute mean (Australian category 3);
- 100 knots 1 minute mean or 90 knots 10 minute mean (Australian category 4);
- 110 knots 1 minute mean or 100 knots 10 minute mean (Australian category 4);
- 127 knots 1 minute mean or 115 knots 10 minute mean (Australian category 5);
- 138 knots 1 minute mean or 125 knots 10 minute mean (Australian category 5).

Tropical cyclones Rona and Steve

Rona was a very rapidly developing system that was named at 23:00 UTC on 10 February 1999 and reached category 3 intensity 12 hours later. It crossed the coast north of Port Douglas around 13:00 UTC on 11 February, while moving towards the west at around 11 knots.

Steve was also a rapidly developing system that reached tropical cyclone intensity at 21:00 UTC on 26 February 2000 and crossed the coast at Cairns just below category 3 intensity at 09:00 UTC on 27 February, while moving towards the west at 10 knots.

Both systems were moving at more or less the same speed until 18:00 UTC (4 a.m. local), after landfall, when in both cases the heaviest rain eased in coastal areas of north Queensland. Therefore these are good examples to compare the thermal structures and resultant rainfall distributions, as the rate movement did not play a crucial role in producing vastly different rainfall distributions.

Rainfall and flooding with Rona

The rainfall associated with Rona was much heavier and more widespread than with Steve. Associated with Rona was major flooding in the Tully, Barron, Johnstone and Herbert Rivers and in the Barron River delta on the northern side of Cairns, where the highest floods were recorded since January 1979. Figure 2.5 shows the widespread distribution of heavy rain associated with Rona. Intense 1-hour rainfalls in the Barron River Catchment include 66 mm in the hour to 13:10 UTC on 11 February 1999 at Copperlode Dam just to

the south-west of Cairns and 57 mm in the hour to 15:15 UTC on 11 February at Bolton Road (18 km north-east of Mareeba).

Barron River levels began to rise throughout the catchment from about 02:00 UTC on 11 February. At Myola, the river rose quickly from 0.64 m at 02:00 UTC to a peak of 11.40 m at 21:00 UTC. This level is the highest since 1977. The peak of 8.65 m occurred at Kamerunga at 22:00 UTC, only one hour later. This was the highest flood since 1977 and 1979, when the flood peaks were 9.5 and 9.4 m respectively.

In the Johnstone River at Innisfail a peak of 6.37 m was reached at 19:30 UTC on 11 February, and was the second highest flood there in the twentieth century.

Thermal pattern with Rona

As Figure 2.6 and Plate 2.6 shows, Rona had large areas of warm air advection at 700 hPa on the western and southern sides of the cyclone and the heavy rain began around 23:00

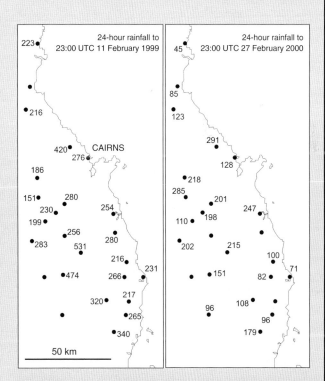

Figure 2.5 The rainfall distribution over 24 hours (mm) for tropical cyclone Rona (left) and tropical cyclone Steve (right).

(*Continued*)

Box 2.3 (*Continued*)

UTC on 10 February 1999 or about 14 hours before landfall, and continued up to 17:00 UTC (3 a.m. local), after landfall. The Cairns radar still showed heavy rain in the Cairns area at 17:00 UTC on 11 February (3 a.m., 12 February local time), when the winds still showed a strong warm air advection signal of 090 (degrees from north) and 28 knots at 943 hPa, backing to 075/34 knots at 850 hPa, backing to 065/42 knots at 700 hPa and to 050/26 knots at 500 hPa.

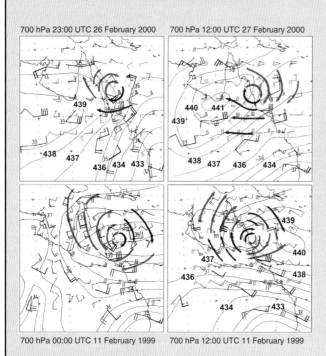

Figure 2.6 The NCEP/NCAR reanalyses 700 hPa winds and 850–500 hPa vertical wind shears for tropical cyclone Steve (top two frames) and tropical cyclone Rona (lower two frames). The contours are the 850–500 hPa thickness in decametres. The larger wind and shear plots are actual measurements taken from radiosonde (upper air sounding) stations. Large streamlines highlight zones of warm (cold) air advection (*see also* Plate 2.6).

Rainfall and flooding with Steve

As Figure 2.5 shows, in most locations the rainfall associated with Steve was much less than that associated with Rona, the only exception being to the west–south-west of Cairns, where Walkamin (285 mm) and Mareeba (218 mm) exceeded the Rona totals. This contributed to a record local flood at Mareeba that experienced its highest flood since records began in 1921. The highest one hourly fall was 72 mm in the hour to 18:50 UTC on 27 February 2000 at Bolton Road (18 km north-east of Mareeba).

With Steve there was no warm air advection on the southern side of the cyclone leading up to landfall. Around Cairns the heaviest rain in the vicinity of tropical cyclones is found on the southern side. At Cairns the heavy rainfall began at and following landfall, when winds with a warm air advection profile reached Cairns.

Radar

Radar images of Rona and Steve as they both approached the coast (Figure 2.7) illustrate the rainfall distributions discussed above. With Rona, extensive areas of rain echoes are evident west and south of the centre. With Steve, apart from heavy rain in the western eyewall, most of the rain was located over the ocean.

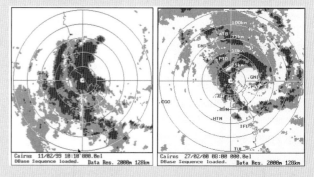

Figure 2.7 Radar rain echoes for tropical cyclone Rona at 10:10 UTC 11 February 1999 (left frame) and tropical cyclone Steve at 08:00 UTC 27 February 2000 (right).

level monsoon shearline ever exists). Further the prevailing upper, equatorial westerlies normally are able to shear any such latent storms. This cyclone crossed the coast of Brazil on 28 March 2004 (McTaggart-Cowan *et al.*, 2006; Pizza & Simmonds, 2005; see Bonell *et al.* 2004; Callaghan 2004, 2005), and

cited websites: www.bom.gov.au/bmrc/clfor/clfor/cfstaff/jmb/synoptic.html for UK model analysis, discussion and associated charts; www.ucar.edu/communications/quarterly/summer05/catarina.html and http://severe.worldweather.org/iwtc/document/Topic_2a_Pedro_Silva_Dias.pdf.).

Box 2.4 Heavy rain in the Cairns region from tropical cyclones that formed in the south-east Gulf of Carpentaria

We compare two Gulf cyclones. One, Sadie, produced major flooding in the Johnstone, Tully and Herbert Rivers and the other, Abigail, produced only light rain in the Cairns region.

Heavy rainfall with Sadie

The 700 hPa thermal pattern associated with Sadie (Figure 2.8 and Plate 2.8) covers the 24-hour period up to 9 a.m. on 31 January 1994 (23:00 UTC 30 January 1994). The heaviest known rainfall was recorded at Crawfords Lookout, where the most intense rain fell in the 9 hours up to 23:00 UTC 30 January 1994 (Table 2.2).

Light rainfall with Abigail

The 700 hPa thermal pattern associated with Abigail (Figure 2.8 and Plate 2.8) covers the 24-hour period up to 9 a.m. on 26 February 2001 (23:00 UTC 25 February 2001). Rainfall registrations for the 24-hours to 9 a.m. on both 25 and 26 February are shown in Table 2.3.

Warm air advection and heavy rain at Cairns with Sadie

The structure of the atmospheric environmental around Sadie included a strong low-level ridge to the south, with an upper trough at 500 hPa trough also to the south and a strong 500 hPa ridge to the east. Such circumstances produced a 700 hPa thermal pattern with warm air to the north and cooler air to the south; which resulted at the 700 hP level in a warm advection flow east of the centre and over the

Table 2.2 The 24-hour totals to 9 a.m. 31 January 1994

Location	Rainfall (mm)
Crawfords Lookout	472
Cardwell	400
Mt Sophie	381
Tully	351
Cairns Airport	346
South Johnstone	345
Babinda	336
Ingham	309
Topaz	297
Abergowrie Bridge	271
Innisfail	264
Mourilyan	189

Table 2.3 Rainfall registrations for cyclone Abigail

Location	25 Feb. (mm)	26 Feb. (mm)
Cairns	41	18
Gordonvale	50	2
Mt Sophia	35	1
Topaz	57	3
Innisfail	26	1
Crawfords Lookout	22	11
Mourilyan	76	21
Tully	70	1
Cardwell	58	25
Abergowrie Bridge	48	12
Ingham	41	13

700 hPa 23:00 UTC 29 January 1994   700 hPa 23:00 UTC 30 January 1994

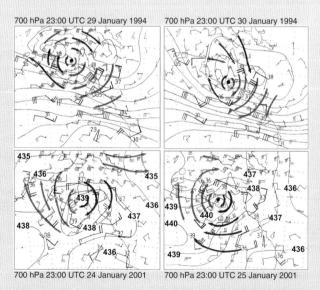

700 hPa 23:00 UTC 24 January 2001   700 hPa 23:00 UTC 25 January 2001

Figure 2.8 As in Figure 2.6, except for tropical cyclone Sadie (top two frames) and tropical cyclone Abigail (lower two frames) (*see also* Plate 2.8).

(*Continued*)

Box 2.4 (*Continued*)

Cairns region and the drier subsiding cold air advection flow east and north of the cyclone.

Cold air advection pattern and light rain at Cairns with Abigail

In contrast, the structure of the atmospheric environment around Abigail had a strengthening ridge with height between 850 and 500 hPa and southerly 500 hPa flow east of Queensland, which produced at the 700 hPa level a thermal pattern with warm air to the west and cooler air to the north and south-east of the cyclone centre. These circumstances produced the 700 hPa warm advection flow west of the centre and drier subsiding cold air advection flow south and east of the cyclone, including over the Cairns region.

The above event has caused a re-evaluation of the mechanisms for storm development in synoptic meteorological situations outside a monsoon region, as well as within a monsoon region such as the northern Coral Sea, at times when a low-level monsoon shearline is absent. Bonell *et al.* (2004) had provided case studies of the development of a tropical cyclone and a tropical depression, respectively, over the Coral Sea, both of which developed and were wholly embedded within low-level easterlies and in the absence of a monsoon trough.

The mechanisms associated with the development of extra-tropical and tropical cyclones were re-analysed over eastern Australia coastal waters of the Coral and Tasman Seas, taking into view some of the dynamic mechanisms identified with the description for Caterina (Callaghan 2004). This re-analysis suggested that there is a continuum between some of the underlying mechanisms associated with the development of intense, extra-tropical cyclones (Tasman Sea) and tropical cyclones (Coral Sea) in terms of a commonly occurring synoptic scale feature in the upper tropopause around 200 hPa (elevation 11 800 m). In addition, the mentioned warm air advection–vertical ascent mechanism at lower levels was also commonly identified to be concurrently in place. For both high and low latitude storms, the low-level cyclone develops to the east of an area of strong 200 hPa warm temperature advection (associated with a tropopause undulation), which subsequently moves over the heavy rainfall area near the centre of the cyclone. This mechanism is in line with the approach described by Hirschberg and Fritsch (1993a, b, 1994) and is also consistent with the potential vorticity arguments of Hoskins *et al.* (1985). The heavy rain area is also concurrently characterized by the existence of the warm air advection–vertical ascent mechanism close to the 700 hPa level.

Callaghan (2004, 2005) noted that the above pattern was associated with every major east coast Australian low over the past 40 years that has occurred equatorward of the sub-tropical high pressure ridge. Consequently the role of the tropopause undulation and warm air advection–vertical ascent mechanisms in tropical cyclone intensification and associated occurrence of heavy rainfall is even more critically important than was appreciated at the time of the review of Bonell *et al.* (2004). Box 2.2 provides a case study that demonstrates the existence of both mechanisms linked with an intense, winter cold-cored[3] cyclone. This system developed over the northern Coral Sea and impacted on the north-east Queensland coast rainfall, and illustrates the influence of a tropopause undulation overlapping with the warm air advection – vertical ascent mechanism. These type of storms commonly occur off the coasts of south-east Queensland and northern New South Wales as well.

Elsewhere Bonell *et al.* (2004) provided contrasting examples of tropical cyclones for the Fijian area where the warm air advection–vertical mechanism was either present or not, and the corresponding impacts on rainfall over the islands. Boxes 2.3 and 2.4 illustrate similar outcomes on the rainfall distribution for the Wet Tropics area linked with two separate tropical cyclones that made landfall respectively in the Cairns and south-east Gulf of Carpentaria regions.

Causal factors affecting rain producing perturbations in the SE trades

Topographic (orographic) uplift over the mountain ranges of moist east to south-east trade winds is the attribute most commonly cited to explain the high rainfalls along the wet tropical coast and adjacent hinterland.

Work in the past decade (Connor & Bonell 1998; Connor & Woodcock 2000) has shown that trade wind rainfall is dependent on a more complex combination of parameters and that it is unwise to use a single forecast parameter. To explain rainfall variability in the trade regime, these writers noted that atmospheric stability alone displayed a poor correlation with tradewind rainfall. In contrast, much stronger links with precipitation occurrence emerged for different mean pressure levels of the atmosphere such as onshore wind components, wind magnitude and bearing, and air-stream confluence components. Most interesting, the warm air advection–vertical ascent mechanism, particularly in the layer between 950 hPa (elevation 600 m) and 800 hPa (elevation 2000 m) was also detected, which contributed significantly to tradewind precipitation amounts along the north-east Queensland coast (Connor & Bonell 1998). Nonetheless, the stability of the lower atmosphere still remains an important consideration when studying the air flow characteristics and the dynamic blocking effect that the coastline mountain ranges present, due to their alignment more perpendicular to the trade wind flow. In that regard, the vertical position of the tradewind temperature

inversion controls the depth of instability and moisture in the lower atmosphere. For example, it is well established that there is a general association of a low inversion with low rainfall occurrence such that the drier, warm air aloft immediately above the inversion will inhibit or decouple convection taking place in the moist, low levels over the Coral Sea. This is a situation common in winter and spring when the cooler sea surface temperatures make it more difficult for a middle- level trough systems to force a corresponding trough system to develop at mean sea level (MSL) (Hoskins et al. 1985); thus Cairns, for example, has an all time record 24-hour rainfall record for July of only 31 mm.

Any weakening of the inversion or its displacement to higher levels can produce heavier rainfalls where a lifting mechanism is in place (e.g. upper trough overlaying the trades, see Bonell & Gilmour 1980; Boxes 2.5 and 2.6). Such lifting is sufficient to allow low-level convection to break through the inversion and release large amounts of convective available potential energy (CAPE) (as defined in Callaghan & Bonell 2004). Specific examples of lifting mechanisms are identified in the case studies shown in Boxes 2.5 and 2.6. It is pertinent to observe that the rain areas identified with Box 2.5 initially developed offshore over the sea

---

Box 2.5 Upper trough overlying trade winds

Heavy rain case

Warm air advection was indicated from the vertical profile of winds at Cairns during February 2000 as follows (angle in degrees from north/wind speed in knots/hPa):

- 11:00 UTC 7 February, 105/22/850 hPa; 090/16/700 hPa; 080/08/600 hPa
- 17:00 UTC 7 February, 100/36/850 hPa, 100/28/700 hPa, 085/22/600 hPa, 045/26/500 hPa

The warm air advection was associated with heavy rain at Cairns: at 17:00 UTC 7 February , 31 mm was recorded from rain in 6 hours and at 20:00 UTC, 103 mm of rain was recorded in 3 hours.

The mean sea level (MSL) sequence (Figure 2.9) showed a trough developing just offshore from Cairns over the period, with the offshore automatic weather station (AWS) winds

indicating easterly to east–north-easterly winds east of the trough. From the National Centre for Environmental Prediction (NCEP) and National Centre for Atmospheric Research (NCAR) website the sea surface temperature in the Cairns region was 29.0°C and the temperature at Cairns was near this before the heavy rain commenced.

The heavy rain event appears mostly to begin as convection in small areas, which then spread out into rain areas. This can be seen in the radar sequence in Figure 2.10 and Plate 2.10, which shows the development period of the heavy rain as viewed from the Cairns radar. Therefore, to start the rain beneath the warm air advection it appears that there must be low-level convergence, with enough forcing to initiate convection through the trade inversion.

(Continued)

Box 2.5 (*Continued*)

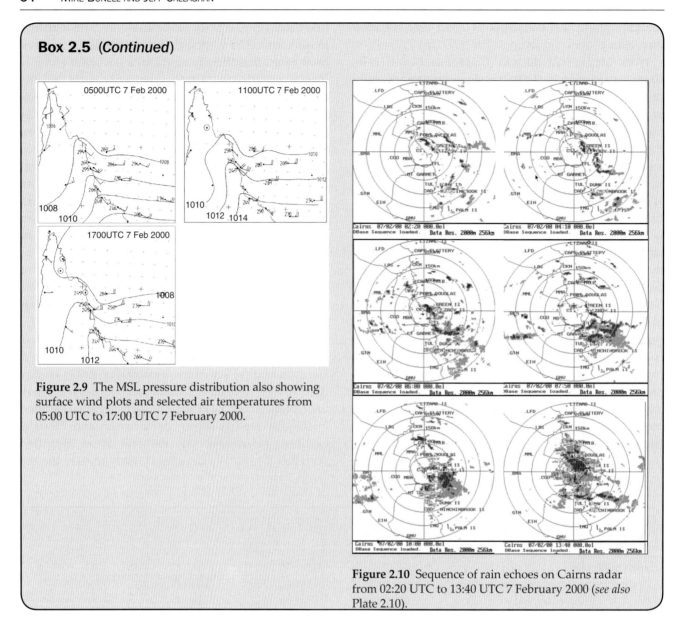

Figure 2.9 The MSL pressure distribution also showing surface wind plots and selected air temperatures from 05:00 UTC to 17:00 UTC 7 February 2000.

Figure 2.10 Sequence of rain echoes on Cairns radar from 02:20 UTC to 13:40 UTC 7 February 2000 (*see also* Plate 2.10).

(and so were not orographically induced). These rain-fields subsequently expanded and moved on to the coast where orographic uplift could then accentuate precipitation.

Aspects of rainfall intensity–duration–frequency (RIDF) of the Wet Tropics

Bonell *et al.* (2004) highlighted the lack of RIDF information for the global humid tropics, especially for short-term rainfalls of less than one hour. Such knowledge is critical to the later understanding of the storm run-off generation process. These writers gathered selected RIDF for previous durations and return periods over different periods of record for rain stations in both tropical cyclone-prone and non-tropical cyclone affected areas. When comparing 24-hour rainfall for different return periods, the tropical cyclone-prone stations in La Reunion (SW Indian Ocean) and the Babinda experimental basins (south of Cairns) conform with Jackson's (1989) observations of much higher rain amounts for longer durations in cyclone-prone areas due to the spatial and temporal organization of tropical vortices. Less conclusive from that review (Bonell *et al.* 2004) was support for much higher equivalent hourly rain intensities for durations less than one hour (which impact on the storm run-off generation process) in non-tropical cyclone affected areas, which Jackson (1989) attributed to the severity of convection associated with more persistent thunderstorms, especially in near-Equatorial stations. Bonell and Gilmour (1980) noted that the monsoon NW (a favoured air mass for thunderstorms) contributed only a small percentage of rain to the annual total recorded in the Babinda experimental basins in contrast to locations further west extending into the semi-arid areas of tropical-central Queensland (see, e.g., Bonell & Williams 1986; Williams & Bonell 1988). Nonetheless, for the Babinda research basins, the

6-minute equivalent hourly rain intensities are high and range from 127 to 247 mm h^{-1} for return periods of 2–14 years. Table 2.5 shows that, with the exception of Mareeba (due to its inland location), all coastal rain stations show high rainfall amounts across different durations and return periods, irrespective of significant differences in annual rainfall totals. Such characteristics are due to the dominance on the summer precipitation regime of the more extreme rainfall-producing systems outlined above in this account (i.e. tropical lows and tropical cyclones). For individual years, it is quite common for either one or two episodes of long duration high rainfall to account for up to 50% of the annual rainfall (Gilmour & Bonell 1979; Bonell & Gilmour 1980; Bonell *et al.* 1991).

Two other aspects need mention. The above rain characteristics for the Wet Tropics are in line with the

---

Box 2.6 Occurrence of upper warm air–advection in absence of a surface trough and only light rainfall

Warm air advection was indicated from the vertical profile of winds at Cairns during August 1995 as follows (wind speed in knots) (Table 2.4).

Over the period 08:00 UTC 17 August to 05:00 UTC 18 August (when strong warm air advection was indicated over Cairns) only 10.8 mm of rain was recorded, mostly in the form of drizzle with some showers. However there was heavy rain at Topaz, south-west of Cairns (elevation 686 m), where 178 mm was recorded in the 24 hours to 23:00 UTC 17 August and 130 mm in the 24 hours to 23:00 UTC 18 August.

A MSL sequence leading up to and including the period of warm air advection is shown in Figure 2.11. A sharp trough

Table 2.4 Wind direction (in degrees from north)/wind speed (in knots) recorded at 850, 700, and 500 hPa levels during 17–18 August 1995

Date	850 hPa	700 hPa	500 hPa
11:00 UTC 17 August	115/30	075/26	055/20
17:00 UTC 17 August	095/30	070/26	050/20
23:00 UTC 17 August	115/30	090/22	015/32
05:00 UTC 18 August	125/32	110/30	355/24

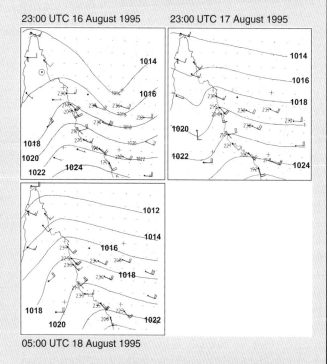

05:00 UTC 18 August 1995

Figure 2.11 The MSL pressure distribution also showing surface wind plots and selected air temperatures from 23:00 UTC 16 August 1995 to 23:00 UTC 18 August 1995.

(Continued)

Box 2.6 *(Continued)*

east of Cairns at 23:00 UTC 16 August weakened, and following this, a firm ridge extends along the coast near Cairns, with fresh to strong trade winds flowing on to the coast. In Figure 2.12 and Plate 2.12, the warm air advection and large-scale ascent is highlighted by the red arrows at 500 hPa. The system was distorted in the vertical, with a 700 hPa low further west over the Gulf of Carpentaria.

Therefore, although there was strong warm air advection pattern over Cairns at 23:00 UTC 17 August 1995, associated with the 500 hPa cyclonic circulation, there was no large-scale trough reflection of this circulation at low levels. Ridging dominated over Cairns at low levels. From the MSL charts the temperatures in the Cairns region were relatively cool and in the low twenties. From the NCEP/NCAR website the sea surface temperature in the Cairns region was 23.5°C. The 850 hPa air stream directed into the coast near Cairns was relatively warm, with Willis Island reporting 19°C and then 15°C over the two days at 850 hPa. Therefore, quite stable conditions were probably prevailing at Cairns, with cool trade winds overlain by warm 850 hPa winds. The combination of the surge and the elevation at Topaz helped the moist onshore flow to break through the inversion there and into the up motion region, producing heavy rain there. The satellite imagery indicated cloud tops of −7°C or near the 500 hPa level in the general Topaz area.

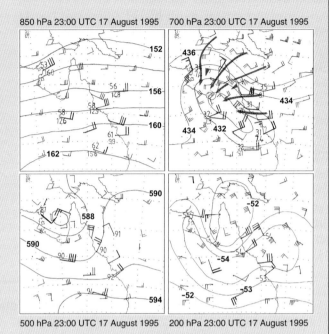

850 hPa 23:00 UTC 17 August 1995   700 hPa 23:00 UTC 17 August 1995

500 hPa 23:00 UTC 17 August 1995   200 hPa 23:00 UTC 17 August 1995

Figure 2.12 The NCEP/NCAR reanalyses at 850, 700, 500 and 200 hPa for 23:00 UTC 17 August 1995. Winds are plotted to 850–500 hPa vertical wind shears. Contours are the 850 and 500 hPa heights (decametres), 850–500 hPa thickness in decametres and 200 hPa isotherms. Larger wind and shear plots are actual from radiosonde stations. The large streamlines highlight zones of warm (cold) air advection (*see also* Plate 2.12).

conclusions from statistical analyses undertaken by Jackson (1986, 1988). Using both a tropical global network and an analysis of northern Australian coverage of rain stations, this work used a mean daily intensity parameter (MDI = monthly averages/average number of rain days). In comparison with most other global rain stations (but not all), the northern Australian stations showed consistently positive deviations (i.e. higher mean daily falls) in residuals for all months; that is, rainfall is more concentrated, with fewer rain days and higher mean daily intensities than predicted from monthly totals and worldwide regressions. Jackson (1988) attributed such results to the persistence of the southern monsoon shearline, and associated tropical lows and occasional cyclones, as a principal causal explanation.

The above analyses (and the later work, Jackson & Weinand 1994, 1995), however, cautioned against the simple division between tropical cyclone-prone and non-tropical cyclone areas in terms of rainfall intensity characteristics. Bonell *et al.* (2004) review summarized examples of where there was no clear distinction between cyclonic and non-cyclonic regions. For example, parts of the Western Ghats of India record similar 24-hour rain totals to those of the Wet Tropics but short-term intensities are much weaker in the absence of persistent tropical vortices (which are further north) and the rain emanates primarily from the orographic uplift of the SW monsoon.

Nonetheless, the Wet Tropics has the distinction of being one of the most globally energetic areas in terms of intense rainfalls. Bonell *et al.* (2004) concluded that

Table 2.5 Rainfall (mm) for various durations and return periods for selected stations along the wet tropical coast

Return period (years)	Duration (hours)		
	1	6	24
Cairns			
2	60	140	250
5	75	180	335
10	85	210	385
20	100	245	455
50	115	295	550
100	130	330	620
Mareeba			
2	45	85	135
5	55	105	175
10	65	125	195
20	70	135	230
50	85	155	270
100	90	170	300
Innisfail			
2	60	135	265
5	75	180	360
10	80	205	420
20	90	240	495
50	105	290	600
100	115	330	680
Ingham			
2	60	130	240
5	75	170	320
10	85	195	370
20	100	230	435
50	115	275	525
100	130	310	595
Townsville			
2	55	105	185
5	70	140	250
10	80	160	290
20	95	190	345
50	110	225	415
100	125	255	470

the synoptic meteorology and associated rainfall climatology of the south-west Indian Ocean, incorporating Madagascar, La Reunion and Mauritius, was one of the most similar to that of north-east Queensland. Moreover, the islands of the south-west Pacific are not too dissimilar from the descriptions here, except there is the additional influence of the mid-South Pacific, South Pacific Convergence Zone (see Callaghan & Bonell 2004). Certainly the Wet Tropics meteorology and its rainfall characteristics are very different from

the more publicized hurricane-prone, tropical North Atlantic basin, as well as the Amazon basin further south (see discussion in Bonell *et al.* 2004).

The storm run-off hydrology response within the closed tropical forests to extreme rainfalls: selected results from the Babinda headwater research basins

Tropical forest hydrological research in the Wet Tropics is one of the few, if not the only one, to recognize the strong linkage between the seasonal change in the synoptic climatology-prevailing rainfall intensities with a corresponding change in the dominant stormflow pathways within tropical forest hillslopes (Gilmour *et al.* 1980).

During the summer monsoon season, the high prevailing short-term rainfall intensities (as shown in Table 2.1 for the storm classes exceeding 50 mm), notably experienced in the active phases of the southern monsoon shearline, are capable of producing extensive saturation overland flow (SOF) (defined by Chorley 1978) over the tropical forest floor. The first descriptions of this process appeared in Bonell and Gilmour (1978) and at the time were contrary to the widely held view that the high surface infiltration properties of forest soils (due to continued incorporation of organic matter) inhibited any overland flow. Until then most hillslope hydrology work had been undertaken in the humid temperate forests, notably in the United States of America and Western Europe where rainfall intensities are much lower, to support the preceding notion (see the benchmark review by Kirkby 1978). The fundamental difference is that the Wet Tropics receive short-term rain intensities (say from 0.1 to 1 hour) of up to two orders of magnitude higher than humid temperate forests. While the surface permeabilities (technically known as field, satiated (saturated) hydraulic conductivity, K^*) (see Bonell 2004) of the tropical forests are high (>1000 mm h^{-1}), which is consistent with measurements in other tropical and temperate forests. A field survey of several Great Soil Groups in collaboration with the then CSIRO Division of Soils, incorporating soils from granites, metamorphic (basic and acidic), basalts and colluvium across the Wet Tropics, established that there is a rapid decline in K^* with depth, such that below 0.2 m, the K^* can be up to three orders of magnitude lower (<1 mm h^{-1}) (Bonell *et al.* 1983a). Measurements later taken by Herwitz (1986) on the slopes of Mt Bellenden Ker also confirmed the same vertical trends in K^* in soils developed from granites. These findings conformed with the data collected in different projects within the Babinda research basins (Bonell & Gilmour 1978; Bonell *et al.* 1981, 1983b, 1984, 1987, 1998; Bonell 1991), and so implied that some of the research outputs from within the concentrated research basin studies near Babinda could be extrapolated to elsewhere in the Wet Tropics. Consequently soil profiles below 0.2 m depth are incapable of accepting the prevailing high rainfalls so ponding occurs within the top 0.2 m of the soil profile leading to subsurface stormflow (SSF) and extensive SOF as well as return flow (RF) (as defined in Chorley 1978). RF is a combination of SOF and SSF that has passed into and then re-emerged out of the soil profile. Such descriptions are most appropriate to the high rainfall, long duration events associated with tropical lows and tropical cyclones.

A current modelling project for the Babinda research basins (Barnes & Bonell 2004) suggests that SOF occurrence is threshold or rain intensity dependent, whereby rainfall is routed over the forest floor as SOF when the equivalent hourly rain intensity exceeds a certain level (estimated to be around 6 mm h^{-1}). Once prevailing rainfalls decline below this threshold SOF does not occur and gives way to SSF as the dominant lateral pathway as well as vertical percolation (VP) (defined in Chorley 1978) towards the lower groundwater table (note that SSF and VP are always present during SOF).

Such findings were already empirically appreciated from collected field data (Gilmour *et al.* 1980), although no such modelling had been undertaken to quantify a threshold rain intensity. This was shown by examining the hillslope run-off trough collectors and stream hydrograph responses to storm events across the summer monsoon, post-monsoon and winter seasons (reproduced in Bonell 2004). SOF occurrence progressively decreased in line with the seasonal relaxation of maximum rain intensities by event, and corresponding storm run-off amounts (see Table 2.1), such that by the winter season, SOF only intermittently occurs and the dominant pathway is VP, supplemented by minimal contributions of SSF. Thus, at this time, the forests are reacting to rainfall more like those in the humid temperate areas as well in some other tropical forests

(Bonell 2004) and the subsoils are no longer acting as impeding layers to precipitation.

The consequences of the above interactions between prevailing rainfalls and soil hydraulic properties is that the stream hydrograph characteristics of heavy rainfall events are unusually responsive for a tropical forest environment. For individual events the quickflow (Q_F, defined in Chorley 1978) response ratios (QRR) (total quickflow/total storm precipitation, Q_F/P) during the monsoon season can exceed 50%, irrespective of whether a drainage basin has been disturbed by partial forest clearing or not. For example, during the period 1974–81, and for quickflow volumes in excess of 250 mm per event in the Babinda experimental basins, the median QRR was 57% for the undisturbed forest (South Creek) and 54.5% for the disturbed forest basin (North Creek) (Table 2.6). Overall, the QRR was 50.5% and 54.5% (disturbed) for storms with quickflow exceeding 100 mm for the larger monsoon season events. Thus at such times the storm run-off generation process remains unchanged, irrespective of land cover, due to the dominance of long duration high rainfall (Howard 1993; see Bonell 2004: 348). Such findings are in contrast to many other studies in the humid tropics where land disturbance clearly increases Q_F (Bonell & Bruijnzeel 2004). On moving into the post-monsoon and winter dry seasons, the QRRs progressively reduce in line with the corresponding weakening of prevailing rain intensities (Bonell 2004), and trend towards values commonly noted for other tropical forest and temperate environments, especially by the time the winter dry season is attained (Gilmour *et al.* 1980). Parallel time series analysis by event of rainfall-SOF and rainfall-stream hydrograph within the closed forest (Bonell *et al.* 1979, 1981) also demonstrated high responsiveness, with respective time lags being 0.1 and 0.4 hour using 6-minute time units. Consequently, a combination of high rainfalls, steep topography and optimal antecedent basin wetness (where soil moisture storage in the top 3 m can attain between 1.2 and 1.5 m equivalent depth of rain (Gilmour 1975; see also Bonell *et al.* 1998) all favour rapid routing of SOF to organized drainage and produce flashy storm hydrographs.

Of particular relevance to biogeochemical cycling, a later campaign for detecting storm run-off sources and related processes using concentrations of environmental tracers (isotopic, i.e. deuterium, and non-isotopic

Table 2.6 The variations in mean and median quickflow response ratios (QRR) for each total storm run-off (Q, mm) class by event for the monsoon, post-monsoon and 'dry' seasons at North Creek (18.3 ha, disturbed) and South Creek (25.7 ha, undisturbed) during the period January 1974 to May 1981

Storm class (total Q) (mm)	North Creek			South Creek		
	No. of obs.	Mean (%)	Median (%)	No. of obs.	Mean (%)	Median (%)
Monsoon						
2–5	43	11.3	10.0	32	9.5	9.0
5–10	21	10.0	8.0	30	12.1	12.0
10–50	48	27.0	29.0	62	26.1	25.0
50–100	29	39.0	41.0	27	38.1	34.0
>100	42	52.2	53.5	34	47.7	50.5
>150	30	55.7	55.0	21	49.5	52.0
>200	22	57.0	55.0	16	50.1	52.0
>250	17	57.3	55.0	11	56.0	57.0
Post-monsoon						
2–5	27	7.5	5.0	26	7.9	6.0
5–10	16	10.1	8.0	15	9.0	8.0
>10	33	24.9	28.0	36	26.4	22.0
Dry						
1–5	65	5.0	4.0	48	4.7	5.0
>5	16	11.6	12.0	24	11.3	10.0

Source: Howard (1993) using the database of the Babinda catchments, courtesy of D. Gilmour, D. Cassells and M. Bonell.

chemical species) within rainfall, soil water, groundwater and streamwater in the Babinda research basins (Elsenbeer *et al*. 1994, 1995; Bonell *et al*. 1998) showed that the storm hydrographs during heavy rain events in the monsoon season were dominated by new water (i.e. event water) based on a principal of first-in/first-out, in contrast to other studies elsewhere, where first-in/last-out is more typical, such as in humid temperate environments where old water (pre-event water) is more dominant (see various contributions in Kendall & McDonnell 1998). Modelling by Barnes and Bonell (2004) and comparison of two events under similar conditions of antecedent basin wetness (Elsenbeer *et al*. 1995) showed that contributions of new water were highly sensitive to rain intensity. As the latter declines temporally during the monsoon season (or more persistently during the post-monsoon season), the volumes of new water decline in line with a corresponding decrease in SOF occurrence. These findings confirmed

the earlier hydrometric research campaigns that established the extensive occurrence of SOF, as well as the potential for considerable nutrient transfer from the forested hillslopes into organized drainage.

The same environmental tracer studies also established that the lower groundwater aquifer was much more coupled to the storm run-off generation process than had previously been suggested (Barnes & Bonell 1996, 2004; Bonell *et al*. 1998). Although the subsoil below 0.2 m acts as an impeding layer to high rainfalls, the random detection of new water immediately after storm events in selected soil water extractors and wells indicated that the subsoil is leaky. A bi-phasic nature of soil water drainage and recharge to the lower water table is indicated, whereby rapid response occurs through preferential flows within a macropore system, with a much slower response through a micropore system within the soil matrix (Bonell *et al*. 1983b, 1984; Bonell 2004;). An analogy here is preferential flow through random holes in a slice of cheese. By way of example, the monsoon case study in February 1991, which was associated with a developing tropical low east of the Babinda research basins (shown in Bonell *et al.* 1998; and included in Bonell 1998) and produced 259.8 mm of rain, showed two important features: (i) the domination of new water at the stream hydrograph peaks from both the undisturbed and disturbed basins; (ii) a protracted recovery to pre-storm background concentrations of deuterium in both streams, which was interpreted as contributions of different groundwater residence times as well as the draining of shallow SSF from the hillslopes. Contiguous measurements of the water table profiles also showed that the maximum hydraulic gradients occurred near or soon after the stream hydrograph peaks that could provide significant groundwater effluent contributions to streamflow at such times (Bonell *et al.* 1998).

More important to the general hydrological functioning of these tropical forests is that a current modelling campaign of the environmental tracer data is suggesting that beneath the Babinda research basins, there exists a large groundwater reservoir of at least 2500–3000 mm equivalent of rain in storage capacity (Barnes & Bonell 1996, 2004). A surprise in the collected data was that despite more than 3000 mm of rain (incorporating a wide range of deuterium concentrations) over the 1990–1 wet season, the background concentration of deuterium of streamflow between rain events (i.e. at low flows) had scarcely changed. This remarkable observation could only be explained by the above modelling work, which also indicated that these large groundwater reservoirs are well buffered from new water inputs.

Moreover, when examining Table 2.1, it is interesting that the mean antecedent flows (I) are commonly less than $30 \, l \, s^{-1}$ between storms, even during the monsoon season when peak storm flows can be up to $5000 \, l \, s^{-1}$ during the major events of that season. The sustained low discharges, as represented by (I), are linked with the unvarying background concentrations of deuterium highlighted above and the corresponding (effluent) discharge from the deep, groundwater reservoir of large capacity. Thus despite the high responsiveness of the stream hydrographs to the heavy rainfalls of the monsoon season, the underlying fractured bedrock is still capable of being significantly recharged and of storing large volumes of groundwater with residence times >1 year (Barnes & Bonell 2004). Such recharge is indicated in the water balance study of Gilmour (1975), and the hydrograph separations presented for a prolonged monsoon event (2564.5 mm) in January 1981 (Bonell *et al.* 1991); both showed respectively that for the undisturbed forest, the delayed flow component of the stream hydrograph, ($Q_{delayed}$)/total precipitation (P) was 33% annually and 29% for the monsoon event. It is these groundwater aquifers in the fractured bedrock that sustain low flows in the network of higher stream orders of the closed forest as well as their transpiration requirements in the dry season. Moreover, the same groundwater reservoirs must be a major store of nutrients within the nutrient cycling of these closed forests, in addition to sustaining the nutrient requirements of in-stream aquatic ecology.

Some hydrological implications for forest-land management

On the basis of the above outcomes from the research basin studies, it is not surprising that extensive flooding during major monsoon rain events occurs within the major rivers of the Wet Tropics. Even more significant, a principal source of such flooding in the narrow coastal corridor is from the forest-covered steep slopes of the adjacent mountain ranges, immediately to the west. Such circumstances are contrary to the commonly accepted belief (i.e. myth) in many developing countries of the humid tropics, which mistakenly credit forests as acting if they are infinite sponges by absorbing incoming precipitation and later releasing this moisture as run-off over intervening drier periods. The occurrence of devastating floods is then simplistically blamed on forest clearance *per se* (several contributions in Bonell & Bruijnzeel 2004 challenge the preceding myth). It is clear that under optimal conditions of drainage basin wetness, even the Wet Tropics forests have an upper limit for vertical percolation, beyond which the high prevailing rainfalls far exceed the soil hydraulic properties (e.g. K^*) and so lead to extensive SOF. Moreover, annual precipitation over the adjacent mountain ranges (about 8000 mm) that is at least double the annual rainfalls measured along the coast (e.g. the Babinda research basins, about 4000 mm) implies that there is even greater potential for widespread SOF to occur over the high topography during long duration, high intensity storm events. Bonell (1991) highlighted almost sustained short-term rain intensities (over 1-minute intervals) of between 30 and 60 mm h^{-1} with peaks up to 150 mm h^{-1} at the top of Mt Bellenden Ker (1561 m elevation) during landfall of tropical cyclone Ivor north of Cooktown, 19–20 March 1990 (24-hour rain totals were 482.5 mm (19 March) and 182.0 mm (20 March)). Further, Herwitz (1986) showed that the closed montane forests of Mt Bellenden Ker are capable of funnelling very high volumes of stem flow down the tree boles, which produce localized infiltration-excess overland flow and add to the prevailing SOF. During the proximity of tropical cyclone Peter to Mt Bellenden Ker in January 1979, when record rainfall occurred (see Bonell *et al.* 2004), there were visual reports from Australian Telecom staff at the lower station of the cableway of widespread overland flow emerging from the forest, up to 0.15 m deep.

Because of the difficulties of gaining good spatial and temporal estimates of precipitation over these mountain ranges (the Mt Bellenden communication tower cableway offers a unique exception) attempts to provide mean annual percentages for total run-off (*Q*)/total precipitation (*P*) for the larger rivers will be suspect and probably will underestimate the true figures. Nonetheless, Bonell *et al.* (1991), drawing on data from the Queensland Water Resources Commission, indicated that for selected rivers (22–930 km^2), the mean annual *Q*/mean annual *P* was in the range 58–90%.

It is evident that because of the nature of the storm run-off hydrology, any disturbance to these Wet Tropics landscapes by forest logging operations or sugarcane monoculture results in considerable erosion. It is beyond the scope of this account to mention all the intricate research details connected with erosion research; these have been summarized in a variety of sources (e.g. Gilmour *et al.* 1982; Capelin & Prove 1983; Cassells *et al.* 1984, 1985; Prove *et al.* 1986; Prove 1991; and reviewed in Bonell 1991; Bonell *et al.* 1991). Nonetheless, the harsh climatic environment of the Wet Tropics enabled research outputs from the Babinda research basins to stimulate new standards for tropical forest management guidelines (described by Cassells *et al.* 1984). These guidelines were adopted by the International Tropical Timber Organisation for application in other countries in the humid tropics. Paradoxically, the same research outputs from the

Babinda research basins significantly contributed a large amount of evidence towards the case made by the federal government of Australia for the World Heritage Listing of the Wet Tropics in 1988.

Some retrospective remarks: the importance of field observations

The early work undertaken in the Babinda research basins was recently recognized in the *International Association of Hydrological Sciences (IAHS) Benchmark Papers in Hydrology Series* (McDonnell 2006, ed.), where Beven (2006) included a commentary on the article by Bonell and Gilmour (1978), among 30 other papers written on hillslope hydrology between 1933 and 1984. His basis for the inclusion was that 'the papers are very much those I consider have shaped perceptions of hillslope processes, rather than those that have provided for example, modelling techniques' (Beven 2006: 1). Subsequently, Beven (2006: 2) remarked that in the current era of modelling, for work undertaken before 1945, 'we have simply forgotten that the scientists of that time were also serious observers of nature, equally perceptive of their responses to their observations'. Such comments continued to apply in the Babinda work highlighted here (as well as in other papers presented by Beven 2006, for the post-1945 period), which was undertaken in an era when there was an unfortunate separation between the field and modelling communities in hydrology, and a lack of the user-friendly modelling tools that are now routine practice. Moreover, along with other early field-based, research basins work, the early Wet Tropics work was carried out in a period of discovery when collected data showed that the real world of water transfer through drainage basins was much more complicated than the assumptions implicit in the hydrological models of that era, for example, preferential flow pathways and spatial heterogeneity of soil and rock properties. With the decline of the spirit of the UNESCO International Hydrological Decade, 1965–74 (Batisse 2005), several writers more recently have expressed concern at the corresponding decline in field-based observations to improve hydrological modelling (Grayson *et al.* 1992; Bonell 1993; Beven 2002, 2006), which recently caused Sidle (2006: 1441) to pose the question 'are we spawning a new generation of computer hydrologists?'

In this context it is pertinent to emphasize that the material presented in Bonell and Gilmour (1978) was much influenced by our observations of what happened during monsoon storms, as the following extracts from a field diary describe for 25 February 1976:

We had just taken manual readings of all the instruments on run-off hillslope site 2 as well as having checked the recording equipment; when in the distance we could hear the advancing roar over the forest canopy of a torrential downpour. On its arrival it was like a domestic shower suddenly being turned on full. The entire hillslope was transformed into a saturated state, with throughfall splash attaining up to near-knee height. We went under the tarpaulin covering the run-off troughs and tipping buckets and witnessed prolific volumes of SOF being collected in the surface trough, with much less volumes in the troughs for 0.25 m and 0.5 m. During the downpour it was impossible to differentiate the throughfall splash from SOF, water was everywhere as two excited scientists and a technician moved over the site re-checking instruments during the deluge. The most exciting phenomenon to witness was thereafter when the downpour stopped as suddenly as it had started. In the following thirty seconds to one minute, there was the remarkable scene of widespread SOF advancing downslope. This SOF then quickly contracted into rivulets controlled by the micro topography on the slope, before subsequently disappearing beneath the forest floor. At the same time, the tipping-bucket connected to the surface run-off trough ceased to function. The one connected to the 0.25 m trough continued tipping for several minutes longer, reflecting the drainage of SSF and what we later appreciated as the dissipation of the temporal (within storm event) perched water table in the upper soil profile.

Gilmour *et al.* (1980) provided technical details of that small (in total rain) but intense storm. This early field observation was highly influential in the development of our hypotheses, but at that time we did not have the data to provide a full explanation of what we had witnessed. A better understanding only later emerged after a soil K^* survey had been undertaken adjacent to site 2, the results of which were presented in Bonell and Gilmour (1978). Certainly we must have been some of the first scientists, if not the first, in a hillslope hydrology project within a research basin to observe widespread overland flow in a forest. Significantly, subsequent work in other land cover types of the Wet Tropics, for example, sugarcane lands (Prove 1991) and riparian buffers (McKergow *et al.* 2004), outlined run-off generation mechanisms similar to those established within the closed forest.

Summary

The Wet Tropics is located within a very highly energetic meteorological area in which tropical lows and the occasional tropical cyclones dominate the rainfall amount–frequency–duration relationships. Moreover, very recent work has detected the recurrence of common meteorological, dynamic mechanisms associated with the highest rainfall events. For this area, there is a remarkably close linkage between the synoptic meteorology, rainfall climatology and storm run-off hydrology, which has been insufficiently acknowledged elsewhere in the global, tropical hydrology literature. Elsewhere the work has suggested that the south-west Indian Ocean region is probably the closest in meteorological characteristics to those of the Wet Tropics.

For a typical summer monsoon that incorporates episodes of long duration, high intensity rainfalls, the hillslope hydrology (and associated drainage basin QRRs) moves to the extreme end of the wet, most responsive part of the run-off spectrum. These characteristics make the Wet Tropics quite distinct from in humid temperate forests and in many tropical forest environments. A review by Elsenbeer (2001) (and later further elaborated by Chappell *et al.*, 2007) however, indicated that since the Babinda research basin first reported overland flow occurrence in tropical forest, there has been an emerging pattern of such flow being associated with other tropical forests within the Acrisol-type End Member group (using the FAO-UNESCO soil classification 1974); the Babinda basins also fit within this group. Bonell (2004) further included more recent results that further supported Elsenbeer's concept. A fundamental difference between Babinda and many of the other Acrisol-type studies, however, is that the weaker, prevailing rain intensities (e.g. in Peru, Elsenbeer *et al.*, 1994) occurring in the latter rainforest studies are compensated by even lower subsoil K^* to be able to produce overland flow, for example, La Cuenca, Peru (see Bonell 2004). Thus while overland flow occurrence is not globally exclusive to the forests of the Wet Tropics, the volumes of run-off certainly are exceptional due to the nature of the meteorology, aided by the steep topography of the area.

Two final points need mention. Chiew and McMahon (2003) showed a strong link between rainfall and run-off for Queensland with the ENSO phenomenon. Thus, during negative ENSO phases, there is a corresponding temporal shift in responsiveness away from the extreme end of the run-off generation spectrum, for example in 1982–3, 1991–2 (see example for January–March 1983 in Bonell 1988). Nonetheless, under the commonly occurring optimal conditions of catchment wetness, in the midst of the summer monsoon season, the modelling of rainfall and run-off for long duration, high intensity events allows the application of much older and simpler (less parameter demanding), spatially lumped models. The high rainfalls, combined with near-saturated soils, reduce the traditional concerns of soil heterogeneity impacts on modelling (and the need for their parameterization) and thus allow the testing of quasi-linear models (Barnes & Bonell 1996, 2004) with reasonable success.

Notes

1   Post-storm analysis of Larry is still continuing because the actual intensification of the system at landfall is still not fully understood due to the complex interaction between the tropical cyclone and the extreme topography (high relief) adjacent to the coast. Thus there remains debate whether the storm was a category 4 or 5 at landfall. The storm surge along the coast, south of Innisfail, is indicative of a category 5, but structural damage reports from engineers suggest category 4 winds. The described warm air advection (outlined later in this chapter) mechanism was evident within this tropical cyclone and intensification of the system was also noted from the 85 GHz H-polarization image at the time of landfall (Callaghan 2006).

2   Isentropic surfaces are surfaces of constant potential temperature, and air parcels follow these surfaces during adiabatic processes. The isobars and temperature isotherms on these surfaces are equivalent and parallel to the thermal wind so that a parcel moving upwards (i.e. from higher to lower pressure) is equivalent to warm air advection. As soon as the parcel reaches the cloud level, the parcels follow moist isentropic surfaces.

3   The term 'cold-cored' contrasts with a typical tropical cyclone where little temperature differences across the storm at the surface occur and their winds are derived from the release of energy due to cloud/rain formation

from the warm moist air of the tropics. The warmest air associated with tropical cyclones is near the centre (i.e. warm-cored) throughout the troposphere. Thus the thermal properties of the eyewall are much warmer than the surrounding atmosphere, due to the excess release of latent heat of condensation (a key driver of tropical cyclones), during the vertical uplift of saturated air. In cold-cored (extra-tropical) cyclones, these storm systems primarily obtain energy from the horizontal temperature contrasts (baroclinic) that exist in the atmosphere which include associated fronts (warm, cold, occluded).

Strictly, all low pressure systems must be warm cored in an integrated sense as the low surface pressure is a consequence of less dense air aloft. For an extra-tropical cyclone, the warm air is found in the lower stratosphere at the bottom of a tropopause undulation. Thus at the warm seclusion phase of an extra tropical cyclone, this represents the intense, mature stage of the extra-tropical cyclone where there are some parallels with tropical cyclones. Characteristics include eye-like features near the centre, dramatic pressure falls, hurricane force winds along the bent-back warm front periphery and vigorous convective precipitation. Examples include the December 1998 Sydney to Hobart yacht race storm in the Tasman Sea and the 1979 Fastnet yacht race storm, in the southwest approaches of the British Isles.

Tropical cyclones, in contrast, typically have little temperature difference across the storm at the surface and their winds are derived from the release of energy due to cloud/rain formation from the warm moist air of the tropics. The warmest air associated with tropical cyclones is near the centre throughout the troposphere.

References

Barnes, C. & Bonell, M. (1996). Application of unit hydrograph techniques to solute transport in catchments. *Hydrol Process* **10**: 793–802.

Barnes, C. & Bonell, M. (2004). How to choose an appropriate catchment model. In *Forests, Water and People in the Humid Tropics: Past, Present and Future Hydrological Research for Integrated Land and Water Management*, Bonell, M. and Bruijnzeel, L. A. (eds). International Hydrological Series, Cambridge University Press, Cambridge, pp. 717–41.

Batisse, M. (2005). *The UNESCO Water Adventure from Desert to Water 1948–1974. From the 'Arid Zone Programme' to the 'International Hydrological Decade'*. AFUS History Papers 4, IHP Essays on Water History 1, Joint International Hydrological Programme and History Club, Association of Former UNESCO Staff, Paris. 191 pp.

Beven, K. J. (2002). Towards an alternative blueprint for a physically-based digitally simulated hydrologic response modelling system. *Hydrol Process* **16**: 189–206.

Beven, K. J. (2006). Streamflow generation processes: selection, introduction and commentary. In *Benchmark Papers in Hydrology Series no. 1*, McDonnell, J. J. (series ed.). IAHS Press, Centre for Ecology and Hydrology, Wallingford. 431 pp.

Bonell, M. (1988). Hydrological processes and implications for land management in forests and agricultural areas of the wet tropical coast of north-east Queensland. In *Essays in Australian Fluvial Geomorphology*, Warner, R. F. (ed.). Academic Press, New York, pp. 41–68.

Bonell, M. (1991). Progress and future research needs in water catchment conservation within the wet tropical coast of NE Queensland. In *Tropical Rainforest Research in Australia: Present Status and Future Directions for the Institute for Tropical Rainforest Studies*, Proceedings of a Workshop held in Townsville, Australia, 4–6 May, 1990. Institute for Tropical Rainforest Studies, James Cook University, Townsville, pp. 59–86.

Bonell, M. (1993). Progress in the understanding of runoff generation dynamics in forests. *J Hydrol* **150**: 217–75 (special issue).

Bonell, M. (1998). Selected challenges in runoff generation research in forests from the hillslope to headwater drainage basin scale. *J Amer Water Res Assoc* **34**: 765–85.

Bonell, M. (2004). Run-off generation in tropical forests. In *Forests, Water and People in the Humid Tropics: Past, Present and Future Hydrological Research for Integrated Land and Water Management*, Bonell, M. and Bruijnzeel, L. A. (eds). International Hydrological Series – Cambridge University Press, Cambridge UK, pp. 314–406.

Bonell, M., Barnes, C. J., Grant, C. R., Howard, A. & Burns, J. (1998). High rainfall response-dominated catchments: a comparative study of experiments in tropical north-east Queensland with temperate New Zealand. In *Isotope Tracers in Catchment Hydrology*, Kendall, C. and McDonnell, J. J. (eds). Elsevier, Amsterdam, pp. 347–90.

Bonell, M. & Bruijnzeel, L. A. (eds) (2004). *Forests, Water and People in the Humid Tropics: Past, Present and Future Hydrological Research for Integrated Land and Water Management*. International Hydrology Series, Cambridge University Press, Cambridge. 925 pp.

Bonell, M., Callaghan, J. & Connor, G. (2004). Synoptic and mesoscale rain producing systems in the humid tropics. In *Forests, Water and People in the Humid Tropics: Past, Present and Future Hydrological Research for Integrated Land and Water Management*, Bonell, M. & Bruijnzeel, L. A. (eds). International Hydrological Series, Cambridge University Press, Cambridge, pp. 194–266.

Bonell, M., Cassells, D. S. & Gilmour, D. A. (1983a). Vertical soil water movement in tropical rainforest of north-east Queensland. *Earth Surf Proc Landforms* **8**(3): 253–72.

Bonell, M., Cassells, D. S. & Gilmour, D. A. (1987). Spatial variations in soil hydraulic properties under tropical rainforest in north-eastern Australia. In *Proceedings of the International Conference on Infiltration Development and Application*, Fok Yu-Si (ed.). Water Resources Research Center, University of Hawaii at Manoa, January, pp. 153–65.

Bonell, M. & Gilmour, D. A. (1978). The development of overland flow in a tropical rainforest catchment. *J Hydrol* **39**: 365–82.

Bonell, M. & Gilmour, D. A. (1980). Variations in short-term rainfall intensity in relation to synoptic climatological aspects of the humid tropical northeast Queensland coast. *Sing J Trop Geog* **1**(2): 16–30.

Bonell, M., Gilmour, D. A. & Cassells, D. S. (1991). The links between synoptic climatology and the run-off response of rainforest catchments on the wet tropical coast of north-eastern Queensland. In *Australian National Rain Forests Study Report 2*, Kershaw, P. A. & Werran, G. (eds). Australian Heritage Commission, Canberra, pp. 27–62.

Bonell, M., Gilmour, D. A. & Cassells, D. S. (1983b). A preliminary survey of the hydraulic properties of rainforest soils in tropical north-east Queensland and their implications for the runoff process. *Catena Suppl* **4**: 57–78.

Bonell, M., Gilmour, D. A. & Cassells, D. S. (1984). Tritiated water movement in clay soils of small catchment under tropical rainforest in north-east Queensland. In *Proceedings of the ISSS Symposium on Water and Solute Movement in Heavy Clay Soils*, Bouma, J. and Raats, P. A. C. (eds). ILRI, Wageningen, The Netherlands, pp. 197–204.

Bonell, M., Gilmour, D. A. & Sinclair, D. (1979). A statistical method for modelling the fate of rainfall in a tropical rainforest catchment. *J Hydrol* **42**: 241–57.

Bonell, M., Gilmour, D. A. & Sinclair, D. F. (1981). Soil hydraulic properties and their effect on surface and subsurface water transfer in a tropical rainforest catchment. *Hydrol Sci Bull* **26**: 1–18.

Bonell, M. & Williams, J. (1986). The generation and redistribution of overland flow on a massive oxic soil in a eucalypt woodland within the semi-arid tropics of Australia. *Hydrolo Proc* **1**: 31–46.

Callaghan, J. (1985). *The North Queensland Monsoon of January 1981*. Meteorological Note 163, Australian Bureau of Meteorology, Dept. Science, July. 30 pp.

Callaghan, J. (2004). *Tropical Cyclone Intensification*, Preprints from The International Conference on Storms, Brisbane, Queensland, July. Available from the author.

Callaghan, J. (2005). Brief overview of tropical cyclones and related systems and a discussion on warm air advection and heavy rainfall in the tropics. Internal Notes, Bureau of Meteorology, Brisbane, 8 pp. Available from the author. Selected material from these notes was also included in Bonell *et al.* (2004).

Callaghan, J. (2006). Paper submitted for Topic 2.3, Operational forecasting of tropical cyclone formation. *Sixth International Workshop on Tropical Cyclones*, 20 November to 1 December, Costa Rica. 16 pp (available from the author).

Callaghan J. & Bonell M. (2004). An overview of the meteorology and climatology of the humid tropics. In *Forests, Water and People in the Humid Tropics: Past, Present anf future Hydrological Research for Integrated Land and Water Management*, Bonell, M. & Bruijnzeel, L. A. (eds). International Hydrological Series, Cambridge University Press, Cambridge, pp. 158–93.

Capelin, M. A. & Prove, B. G. (1983). Soil conservation problems of the humid coastal tropics of north Queensland, *Proceedings of the Australian Society of Sugar Cane Technology Conference*, pp. 87–93.

Cassells, D. S., Gilmour, D. A. & Bonell, M. (1984). Watershed forest management practices in the tropical rainforests of north-eastern Australia. In *Proceedings of the IUFRO Symposium on Effects of Forest Land Use on Erosion and Slope Stability*, O'Loughlin, C. L. and Pearce, A. J. (eds). East-West Center, Honolulu, NZ Forest Research Institute Publication, pp. 289–98.

Cassells, D. S., Gilmour, D. A. & Bonell, M. (1985). Watershed management in tropical forests – some experiences from north-eastern Australia. *Forestry Ecol Manag* **10**: 155–75.

Chappell, N. A., Bidin, K., Sherlock, M. D. & Lancaster, J. W. (2004). Parsimonous spatial representation of tropical soils within dynamic rainfall-runoff models. In *Forests, Water and People in the Humid Tropics: Past, Present and Future Hydrological Research for Integrated Land and Water Management*, Bonell, M. and Bruijnell, L. A. (eds). International Hydrology Series, Cambridge University Press, Cambridge, pp. 756–69.

Chappell, N. A., Sherlock, M., Bidin, K., Macdonald, R., Najman, Y. & Davies, G. (2007). Runoff processes in Southeast Asia: role of soil, regolith, and rock type. In *Forest Environments in the Mekong River Basin*, Swada, H., Araki, M., Chappell, N. A., LaFrankie, J. V. and Shimizu, A. (eds). Springer-Verlag, in press. (Available from http://www.es.lancs.ac.uk/people/nickc/ChappellSherlocketal107.pdf)

Chiew, F. H. S & McMahon, T. A. (2003). Australian rainfall and streamflow and El Niño/Southern Oscillation. *Aust J Water Resour* **6**: 115–29.

Chorley, R. J. (1978). Glossary of terms. In *Hillslope Hydrology*, Kirkby, M. J. (ed.). Wiley, Chichester, pp. 365–75.

Connor, G. J. and Bonell, M. (1998). Air mass and dynamic parameters affecting trade wind precipitation on the northeast Queensland tropical coast. *Int J Climatol* **18**: 1357–72.

Connor, G. J. & Woodcock, F. (2000). The application of synoptic stratification to precipitation forecasting in the trade wind regime. *Weather Forecasting*, **15**: 276–97.

Elsenbeer, H. (2001). Hydrologic flowpaths in tropical rainforest soilscapes – a review. *Hydrol Process* **15**: 1751–9.

Elsenbeer, H., Cassel, D. K. & Zuniga, L. (1994). Throughfall in the Terra Firme forest of western Amazonia. *J Hydrol (New Zealand)* **32**(2): 30–44.

Elsenbeer, H., Lorieri, D. & Bonell, M. (1995). Mixing model approaches to estimate storm flow sources in an overland flow-dominated tropical rainforest catchment. *Water Resour Res* **31**: 2267–78.

Elsenbeer, H., West, A. & Bonell, M. (1994). Hydrologic pathways and storm-flow hydrochemistry at South Creek, northeast Queensland. *J Hydrol* **162**: 1–21.

FAO-UNESCO (1974). *FAO-UNESCO Soil Map of the World Vol.1 Legend*. UNESCO Press, Paris.

FAO-UNESCO (1988). *Soil Map of the World, Revised Legend*. World Soil Resources Report 60, FAO, Rome.

Fritsch, J. M., Murphy, J. D. & Kain, J. S. (1994). Warm core vortex amplification over land. *J Atmos Sci* **51**: 1780–807.

Gilmour, D. A. (1975). Catchment water balance studies on the wet tropical coast of North Queensland. Unpublished PhD Thesis, Dept of Geography, James Cook University of North Queensland, Townsville.

Gilmour, D. A. & Bonell, M, (1979). Run-off processes in tropical rainforests with special reference to a study in north-east Australia. In *Geographical Approaches to Fluvial Processes*, Pitty, A. F. (ed.). Geo Books, Norwich, pp. 73–93.

Gilmour, D. A., Bonell, M. & Sinclair, D. F. (1980). An investigation of storm drainage processes in a tropical rainforest catchment. Australia Water Processes Council Technical, Paper 56, Canberra, Australian Government Publishing Service. 93 pp.

Gilmour, D. A., Cassells, D. S. & Bonell, M. (1982). Hydrological research in the tropical rainforests of north Queensland: some implications for land use management. In *First National Symposium on Forest Hydrology*, O'Loughlin, E. M. and Bren, L. J. (eds). Instit Engrs, Australia, Nat Conf Public, No. 82/6, Melbourne, May, pp. 145–52.

Grayson, R. B., Moore, I. D. & McMahon, T. A. (1992). Physically based hydrologic modelling, 2. Is the concept realistic? *Water Resour Res* **26**: 2659–66.

Gunn, B. W., McBride, J. L., Holland, G. J., Keenan, T. D., Davidson, N. E. & Hendon, H. H. (1989). The Australian Summer Monsoon Circulation during AMEX Phase II. *Mon Weath Rev* **117**: 2554–74.

Herwitz, S. R. (1986). Infiltration excess caused by stem-flow in a cyclone-prone tropical rainforest. *Earth Surf Proc Landforms* **11**: 401–12.

Hirschberg, P. A. & Fritsch, J. M. (1993a). On understanding height tendency. *Mon Weath Rev* **121**(9): 2646–61.

Hirschberg, P. A. & Fritsch, J. M. (1993b). A study of the development of extratropical cyclones with an analytic model. Part I: The effects of stratospheric structure. *J Atmos Sci* **50**: 311–27.

Hirschberg, P. A. & Fritsch, J. M. (1994): A Study of the development of extratropical cyclones with an analytic model. Part II: Sensitivity to tropospheric structure and analysis of height tendency dynamics. *Mon Weath Rev* **122**: 2312–30.

Holton, J. R. (1972). *Introduction to Dynamic Meteorology*. Academic Press, New York. 319 pp.

Hoskins, B. J., Draghici, I. & Davies, H. C. (1978). A new look at the omega equation. *Q J R Met Soc* **104**: 31–8.

Hoskins, B. J., McIntyre, M. E. & Robertson, A. W. (1985). On the use and significance of isentropic potential vorticity maps. *Q J R Met Soc* **111**: 877–946.

Hoskins, B. J., Pedder, M. & Wyn Jones, D. (2003). The omega equation and potential vorticity. *Q J R Met Soc* **129**: 3277–303.

Howard, A. J. (1993). The effect of rainfall intensity on storm-flow peak flow within the humid tropics of Northeastern Queensland. Unpublished honours thesis, Dept of Geography, James Cook University of North Queensland. 141 pp.

Jackson, I. J. (1986). Relationships between rain-days, mean daily intensity and monthly rainfall in the tropics. *J Climatol* **6**: 117–34.

Jackson, I. J. (1988). Daily rainfall over northern Australia: deviations from the world pattern. *J Climatol* **8**: 463–76.

Jackson, I. J. (1989). *Climate, Water and Agriculture in the Tropics*, 2nd edn. Longman, London.

Jackson, I. J. & Weinand, H. (1994). Towards a classification of tropical rainfall stations. *Int J Climatol* **14**: 263–86.

Jackson, I. J. & Weinand, H. (1995). Classification of tropical rainfall stations: a comparison of clustering techniques. *Int J Climatol* **15**: 985–94.

Kendall, C. & McDonnell, J. J. (1998). *Isotope Tracers in Catchment Hydrology*. Elsevier, Amsterdam. 839 pp.

Kirkby, M. J. (ed.) (1978). *Hillslope Hydrology*. Chichester, Wiley. 389 pp.

Lyne, V. D. & Hollick, M. (1979). Stochastic time-varying rainfall-run-off modelling. In *Proceedings of the Hydrology and Water Resources Symposium, Perth*. Inst. Engrs, Canberra, pp. 89–92.

Madden, R. A. & Julian, P. R. (1972). Description of global-scale circulation cells in the tropics with a 40–50 day period. *J Atmos Sci* **29**: 1109–23.

McDonnell, J. J. (ed.) (2006). *Benchmark Papers in Hydrology Series No. 1*. IAHS Press, Centre for Ecology and Hydrology, Wallingford, UK. 431 pp.

McKergow, L. A., Prosser, I. P., Grayson, R. B. & Heiner, D. (2004). Performance of grass and rainforest riparian buffers in the Wet Tropics, Far North Queensland. 1. Riparian hydrology. *Aust J Soil Res* **42**: 473–84.

McTaggart-Cowan, R., Bosart, L. F., Davis, C. A., Atallah, E. H., Gyakum, J. R. & Emanual, K. A. (2006). Analysis of Hurricane Catarina (2004). *Monthly Weather Review*, **134**(11): 3029–53.

Manton, M. J. & Bonell, M. (1993). Climate and Rainfall variability in the humid tropics. In *Hydrology and Water Management in the Humid Tropics*, Bonell, M., Hufschmidt, M. M. & Gladwell, J. S. (eds). International Hydrology Series - Cambridge University Press, Cambridge, UK pp. 13–33.

Pizza, A. & Simmonds, I. (2005). The first south Atlantic hurricane: Unprecedented blocking, low shear, and climate change. *Geophysical Research Letters* **32** (L15712): 12 August 2005.

Prove, B. G. (1991). A study of the hydrological and erosional processes under sugar cane culture on the wet tropical coast of north eastern Australia. Unpublished PhD thesis, Department of Geography, James Cook University of North Queensland. 273 pp.

Prove, B. G., Truong, P. N. & Evans, D. S. (1986). Strategies for controlling caneland erosion in the wet tropical coast of Queensland. In *Proceedings of the Australian Society of Sugar Cane Technology Conference*, pp. 77–84.

Ramage, C. S. (1968). Role of a 'maritime continent' in the atmospheric circulation. *Mon Wealth Rev* **96**: 365–70.

Raymond, D. J. & Jiang, H. (1990). A theory for long lived mesoscale convective systems. *J Atmos Sci* **47**: 3067–77.

Saucier, W. J. (1955). *Principles of Meteorological Analysis*. University of Chicago Press, Chicago. 438 pp.

Sidle, R. C (2006). Field observations and process understanding in hydrology: essential components in scaling. *Hydrol Proc* **20**(6): 1439–45.

Shin, S. E., Smith, R. K. & Callaghan, J. (2006). Severe thunderstorms over north eastern Queensland on 19 January 2001: the influence of an upper level trough on convective destabilisation of the atmosphere. *Aust Meteorolog Mag* **54**: 333–46.

Williams, J. & Bonell, M. (1988). The influence of scale of measurement on the spatial and temporal variability of the Philip infiltration parameters: an experimental study in an Australian savannah woodland. *J Hydrolog* **104**: 33–51.

3 IMPACTS OF TROPICAL CYCLONES ON FORESTS IN THE WET TROPICS OF AUSTRALIA

Stephen M. Turton[1]* *and Nigel E. Stork*[2]*

[1]Australian Tropical Forest Institute, School of Earth and Environmental Sciences, James Cook University, Cairns, Queensland, Australia
[2]School of Resource Management and Geography, Faculty of Land and Food Resources, University of Melbourne, Burnley Campus, Richmond, Victoria, Australia
*The authors were participants of Cooperative Research Centre for Tropical Ecology and Management

Introduction

Disturbances, both anthropogenic and natural, shape forest ecosystems by controlling their species composition, structure and functional processes (Dale *et al.* 2001). The tropical forests of the Wet Tropics of north-east Australia have been moulded by their land-use and disturbance history over many millennia but particularly over the past 100 years (see Pannell, Chapter 4; Turton, Chapter 5; Laurance & Goosem, Chapter 23, this volume). Forests of the Wet Tropics region are subjected to a plethora of natural disturbances, including fire, drought, native pathogen outbreaks, floods, occasional landslides and tropical cyclones. All these natural disturbances interact in complex ways with anthropogenic disturbances across the Wet Tropics landscape, such as land-use change resulting from forest conversion to agricultural systems, earlier logging practices, urban and peri-urban development and expansion of the tourism industry (see Part III, this volume).

As in many parts of the non-equatorial tropics (Whitmore 1974; Boose *et al.* 1994), tropical cyclones (a.k.a. hurricanes and typhoons) are significant disturbance phenomena for forest ecosystems in the Wet Tropics of north-east Australia, especially those near the coast. For this reason, even continuous forests in the region have been described as hyper-disturbed ecosystems with patches of damaged forest constantly recovering from previous cyclonic events, often in concert with floods, droughts and fires (Webb 1958). Cyclones are part of the ecosystem dynamics of these forested landscapes and recovery of canopy cover following such events is often remarkably rapid, although forest structure and composition may take many decades to recover. The same cannot be said for fragmented forests in the Wet Tropics, located within either an agricultural or grassland matrix. These forest fragments are particularly vulnerable to impacts of tropical cyclones and their associated strong winds, largely due to their high forest edge to area ratios (Laurance 1991, 1997). These fragments are also more prone to post-disturbance weed invasion and bush fires than nearby areas of continuous forest.

We begin this chapter by providing an overview of impacts of tropical cyclones on forest ecosystems, with an emphasis on landscape- and local-scale patterns and processes. We then provide examples of these scale effects on forests for documented cyclones of differing intensities that have impacted on the Wet Tropics region of north-east Australia in recent decades. We conclude by providing a brief overview of the predicted impacts of climate change on cyclone frequency and intensity in the region and likely ecological effects on our rainforest ecosystems.

Landscape and local scale impacts of cyclones on tropical forest ecosystems

Tropical cyclones range in intensity from comparatively weak systems, where maximum wind speeds do not generally exceed 160 km h^{-1}, to extremely destructive, where maximum wind gusts have been recorded in excess of 350 km h^{-1} (McGregor and Nieuwolt 1998). Severe cyclones cause widespread defoliation of rainforest canopy trees, removal of vines and epiphytes, along with the breakage of crown stems and associated tree falls (Lugo *et al.* 1983; Brokaw & Walker 1991; Tanner *et al.* 1991; Boose *et al.* 1994; Bellingham *et al.* 1995; Everham & Brokaw 1996; Lugo & Scatena 1996). These catastrophic impacts typically result in significant changes in forest microclimates in the understorey and canopy (Fernandez & Fetcher 1991; Turton 1992; Bellingham *et al.* 1996; Lugo & Scatena 1996; Turton & Siegenthaler 2004), and complex vegetation responses to newly created light, temperature and humidity regimes (Bellingham *et al.* 1994; Vandermeer *et al.* 1995; Harrington *et al.* 1997). Cyclonic disturbance has also been shown to accelerate invasion by exotic tree species, leading to a decline in biodiversity of native trees (Bellingham *et al.* 2005). Tropical cyclones also play a significant role in the synoptic meteorology and catchment hydrology of the Wet Tropics region (see Bonell & Callaghan, Chapter 2, this volume).

Given the relatively high frequency of tropical cyclones in many tropical and sub-tropical regions (McGregor & Nieuwolt 1998), there is a general consensus that they contribute to the structure and function of tropical rainforests in cyclone-prone areas, with ecosystem impacts and recovery processes occurring at several spatial and temporal scales (Webb 1958; Whitmore 1989, 1991; Bellingham 1991; Yih *et al.* 1991; Bellingham *et al.* 1992, 1994, 1995; Lugo & Scatena 1996; Harrington *et al.* 1997; Grove *et al.* 2000; Tanner & Bellingham 2006). Boose *et al.* (1994) examined the impacts of severe hurricanes on forested landscapes in the north-eastern United States and Puerto Rico and concluded that wind damage served as an important source of landscape-scale (~10 km) patterning in

---

Box 3.1 The role of scale in determining forest impacts from tropical cyclones (after Boose *et al.* 1994)

Three main factors control forest damage at the landscape scale (>10 km):

1 wind velocity gradients resulting from cyclone size, speed of forward movement, cyclone intensity and proximity to the storm track, complicated by local convective-scale effects;

2 variations in site exposure and other effects of local topography (e.g. severe lee waves or leeward acceleration, windward exposure, topographic shading);

3 differential response of individual ecosystems to wind disturbance as a function of species composition and forest structure.

Two main factors control forest damage at the local scale (<1 km):

1 impacts are generally associated with individual wind gusts created by turbulent eddies and with wind gradients associated with the intense convective cells and sometimes tornadoes;

2 changes in the wind flow may result from small-scale topographic features and from the structure of the forest itself (e.g. lowland versus montane forests).

---

forests and was a major factor in initiating vegetation dynamics at that scale. On the other hand, at the local scale (approx. 1 km), extensive toppling of trees and canopy thinning was considered to be an important factor regulating hydrological, energy and nutrient regimes (Boose *et al.* 1994). Box 3.1 summarizes tropical cyclone impacts on forests at the landscape and local scales.

Impacts of tropical cyclones on Australian tropical forests

Cyclone frequency and impacts in Queensland's Wet Tropics region

Table 3.1 summarizes the frequency and intensity of tropical cyclones crossing the wet tropical coast

Table 3.1 Tropical cyclones that have made east to west landfall along the wet tropical coast of Queensland (Cooktown–Ingham) over the period 1858–2006

Australian category	Peak gusts (km h⁻¹)	Central pressure (hPa)	Frequency (number recorded)	Return interval (years)*
1	90–124	986–995	18	5
2	125–169	971–985	11	9
3	170–224	956–970	10	15
4	225–279	930–955	0	35
5	≥280	≤929	2	75

Source: Turton (in press).
*Derived using a quadratic function.

Table 3.2 Notable east to west moving tropical cyclones by code name that have impacted along the wet tropical coast (Cooktown–Ingham) over the period 1858–2006

Code name	Date	Area most impacted	Estimated lowest barometric pressure (hPa)
Unnamed	8 March 1878	Cairns	Not available
Unnamed	16 March 1911	Port Douglas	Not available
Unnamed	10 March 1918	Innisfail region	910
Unnamed	3 February 1920	Port Douglas	962
Unnamed	9 February 1927	Cairns-Port Douglas	971
Unnamed	12 March 1934	Cape Tribulation	978
Unnamed	18 February 1940	Cardwell	965
Agnes	6 March 1956	Cairns-Townsville	961
Winifred	1 February 1986	Cairns-Tully	958
Rona	11 February 1999	Daintree-Cape Tribulation	970
Larry	20 March 2006	Innisfail region	915

Source: Turton (in press).

(Cooktown–Ingham) over the period 1858–2006 according to the Australian Cyclone Severity Scale (Turton submitted). These data suggest that a weak cyclone (Australian System: categories 1–2) is likely to cross the wet tropical coast with a frequency interval of about 5 years, compared with a frequency of about 15 years for a moderate to severe cyclone (category 3) and about 75 years for a very severe cyclone (categories 4–5). Table 3.2 provides a summary of notable

severe cyclones (categories 3–5) that have impacted on the Wet Tropics region from 1858 to 2006.

Over the past 20 years three significant tropical cyclones have impacted on the forest ecosystems of north-east Australia, cyclones Winifred (1986), Rona (1999) and Larry (2006). Characteristics of these three severe cyclones are summarized in Table 3.2 and Box 3.2. These three events have generated a number of ecological impact and recovery studies at landscape and local scales that form the basis of the following discussion.

Landscape scale impacts of cyclones in the Wet Tropics

Impacts of tropical cyclones on forests at the landscape scale are the result of the complex interaction of anthropogenic, meteorological, topographical and biotic factors. Given that no cyclone is identical to another in terms of intensity, radius of maximum winds, forward speed and direction of movement, it is impossible to generalize beyond the main factors summarized in Box 3.1. In the aftermath of severe cyclone Larry (Box 3.2) a helicopter aerial survey was conducted across the Wet Tropics to examine landscape-scale impacts from a compact category 4 system (Turton, in press).

Adopting a ground-based rapid assessment methodology, developed by Unwin et al. (1988) following cyclone Winifred (see Box 3.2), four classes of initial forest damage were identified from the aerial surveys (Table 3.3).

Severe forest damage (classes 1 and 2) from cyclone Larry appeared to be more extensive than that caused by cyclone Winifred that impacted on same area in February 1986 (for details see Box 3.2). In the case of cyclone Larry, moderate and severe forest damage (classes 2 and 3) extended much further north, south and west than that reported for cyclone Winifred despite their similar paths across the landscape (Unwin et al. 1988). However, slight forest damage (Class 4) for cyclone Larry appeared to be less extensive than that reported for cyclone Winifred. The latter may be explained by the compact nature of cyclone Larry's area of destructive winds, compared with the less intense but much larger diameter of cyclone Winifred (Unwin et al. 1988). Compared with previous cyclones, Larry moved inland unusually quickly while maintaining its intensity. This caused extensive damage to upland

Box 3.2 Cyclone characteristics

Characteristics of cyclone Winifred (February 1986)

Severe tropical cyclone Winifred (Australian category 3), with a central pressure of 958 hPa, crossed the coast at Cowley Beach, to the south of Innisfail, on 1 February 1986. The cyclone was tracking slowly west at about 10 km h^{-1} when it crossed the coast with maximum wind gusts of 220 km h^{-1} near the centre. Winifred was a fairly large system with destructive winds extending north to Cairns, south to Cardwell and as far west as Atherton. The westerly (offshore) winds, to the north of the eye, were unusually strong and were implicated in extensive forest damage along west-facing slopes in areas north of Innisfail. The onshore easterlies were comparatively weak for such an intense system. Although preceded by heavy rain, resulting in significant tree falls, Winifred was followed by dry conditions and reports of fires near forest edges (Unwin et al. 1988).

Characteristics of cyclone Rona (February 1999)

Severe tropical cyclone Rona (Australian category 3), with a central pressure of 970 hPa, crossed the north-eastern Australian coast in the Daintree region between Cape Tribulation and Cape Kimberley on 11 February 1999. The cyclone was tracking north-west at about 15 km h^{-1} when it crossed the coast, with maximum wind gusts of 180 km h^{-1} near the centre (Bureau of Meteorology 1999). Sustained (1-min average) winds were above 120 km h^{-1}, but varied significantly due to effects of steep terrain in the Daintree–Cape Tribulation area.

The zone of very destructive (hurricane force) winds extended only 15 km from the centre of the cyclone, while destructive winds (storm force) extended some 80 km from the centre (Bureau of Meteorology 1999). As a consequence, all the rainforest areas of the Greater Daintree region were impacted by the cyclone to varying degrees (Grove et al. 2000).

Characteristics of cyclone Larry (March 2006)

Severe tropical cyclone Larry (estimated central pressure of 915 hPa) crossed the coast near Innisfail as a category 4 storm on 20 March 2006, causing extensive damage to human communities, primary industries, infrastructure and ecosystems across a 100 km strip of coastal lowlands and uplands. Maximum wind gusts were near 280 km h^{-1}, making Larry the most severe cyclone to threaten a populated area of Queensland since the unnamed Innisfail cyclone of 1918 (Table 3.2). The system was moving at 25 km h^{-1} when it crossed the coast, and was still category 3 status when it crossed over the Atherton Tablelands some 60 km from the coast. Cyclone Larry produced numerous tornado-type features within the system's eyewall that have been linked to patches of catastrophic forest damage. Tornadoes are frequently associated with severe cyclones (categories 4 and 5) and are often implicated in severe local structural damage to human infrastructure. Larry was a very 'compact' system, with its radius of maximum winds extending only 20–30 km from the centre.

forests as far inland as 100 km that normally escape the worst impacts of cyclones (Turton, in press).

In the Wet Tropics region, topography plays a significant role in accounting for forest impact patterns and recovery processes after cyclones at the landscape scale (Grove et al. 2000). Cyclonic winds exhibit complex interactions when they encounter steep topography. Depending on the location of the cyclone centre and the direction and speed of movement of the cyclone, forests in mountainous areas may experience severe windward exposure, topographic shading (leeward protection) and, under certain circumstances, severe lee turbulence. The latter explains observed areas of severe forest damage in parts of the landscape that would normally be expected to be protected from cyclonic winds due to topographic shading.

Figures 3.1–3.5 provide examples of the four classes of forest damage ranging from severe to slight (Table 3.3), obtained during the aerial surveys after cyclone Larry (Turton, in press). Figures 3.2 and 3.3 provide examples of windward and leeward forest damage (both class 2), respectively. These images demonstrate the patchy nature of forest damage at the landscape scale, with many areas of forest remaining intact despite being close to the centre of the cyclone. Intact forest areas will become important refuges for many non-volant fauna, while adjacent areas of damaged forest recover. In the Wet Tropics this includes the endangered frugivorous cassowary and several rare folivorous marsupials.

Following cyclone Rona (see Box 3.2), Grove et al. (2000) also noted the strong effects of topography on

Table 3.3 Initial effects of severe cyclone Larry on forests in the Wet Tropics of north-east Australia

Forest damage classes (areal extent of damage)	Definition (by essential characteristics)	Occurrence in rainforests	Occurrence in open forests
1 Severe and extensive (mostly within 30 km from centre of cyclone)	Boles or crowns of most trees broken, smashed or wind-thrown. Impact multi-directional	Extensive among coastal forests, riparian forests, forest remnants and foothills in area along track of the cyclone's core, including foothills as far west as Millaa Millaa	Rarely evidenced
2 Severe and localized (mostly 30–50 km from centre of cyclone)	Severe canopy disruption (as above) restricted to windward aspects. Direction of destructive winds clearly identifiable	Defoliation, tree falls and stem breakage common on windward aspects only. Leeward forest areas more or less intact except for patches of lee wave damage	As above, (transitional understoreys, defoliated as for rainforests)
3 Moderate canopy disturbance (mostly 50–75 km from centre of cyclone)	Structural injury mostly branch and foliage loss. Some treefalls; most stems erect	Extensive areas on exposed uplands and tablelands, beyond most destructive winds. Widespread canopy thinning, exposure of epiphytes and understorey. Heavy litter and branch fall	Most common injury in open forests of *Eucalyptus* and *Melaleuca*, on lowlands, foothills and uplands near Ravenshoe. Widespread defoliation and loss of twigs
4 Slight canopy disturbance (mostly 75–120 km from centre of cyclone)	Partial foliage loss on forest edges only. Subsequent leaf desiccation and heavy litter fall. Occasional stem or branch breakage	Common on exposed margins of cyclone affected area. Canopy interior intact	On main ridges and spurs peripheral to damage Class 3 areas above. Foliage desiccation short-lived and inconspicuous

Source: derived from aerial surveys Turton (in press); forest damage classes adopted from Unwin *et al.* (1988).

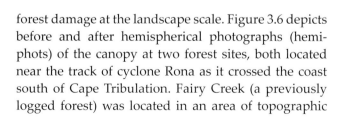

Figure 3.1 Severe and extensive (Class 1) forest damage along the lower North Johnstone River. Photo: S. M. Turton.

Figure 3.2 Severe and localized (Class 2) forest damage near Kurrimine Beach. Note severe windward forest damage and intact forest in leeward (sheltered) areas. Photo: S. M. Turton.

forest damage at the landscape scale. Figure 3.6 depicts before and after hemispherical photographs (hemiphots) of the canopy at two forest sites, both located near the track of cyclone Rona as it crossed the coast south of Cape Tribulation. Fairy Creek (a previously logged forest) was located in an area of topographic shading with no discernible change in canopy openness after the cyclone, compared with Thompson Creek (old growth forest) where the forest was severely impacted by strong lee waves originating off Mt Sorrow to the immediate west (Grove *et al.* 2000). Other forest sites in the region impacted by cyclone Rona exhibited

Figure 3.3 Severe and localized (Class 2) forest damage on the east face of the Murray-Prior Range. This is an example of forest damage owing to severe leewaves propagating over the mountain range from the west, leaving an adjacent area of intact forest. Photo: S. M. Turton.

Figure 3.4 Moderate (Class 3) forest damage along the upper South Johnstone River. Photo: S. M. Turton.

Figure 3.5 Slight (Class 4) forest damage near Lake Tinaroo. Photo: S. M. Turton.

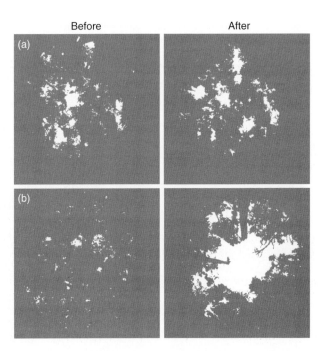

Figure 3.6 Hemispherical photographs (hemiphots) of two forest sites before and after cyclone Rona: (a) insignificant change in canopy openness (Fairy Creek); (b) highly significant increase in canopy openness (Thompson Creek). From Grove *et al.* (2000). Copyright Cambridge University Press, Reproduced with permission.

degrees of canopy damage intermediate to Fairy and Thompson Creeks (Table 3.4). Some of these patterns may be attributed to previous management history but results were inconclusive due to compounding effects of topography and site location with respect to the cyclone's track (Grove *et al.* 2000). This would suggest that cyclones provide a spectrum of disturbance regimes across forested landscapes, a process that undoubtedly promotes biodiversity at that scale. However, cyclonic disturbance also promotes the risk of forest fires and the spread of invasive species, including soil pathogens, weeds and feral animals (see Goosem, Chapter 24, and Congdon & Harrison, Chapter 25, this volume).

Table 3.4 A site level comparison of canopy openness before and after cyclone Rona (February 1999) derived from hemispherical photographs (n = 9 per site). Refer to Figure 3.6 for canopy hemiphot examples from Fairy Creek and Thompson Creek. Site exposure is defined according to local topography and the track of the cyclone

Site name	Site exposure with respect to cyclone track	Mean canopy openness (%) (before cyclone)	Mean canopy openness (%) (after cyclone)	Mean change in canopy openness (%)
Cooper Creek (OG)	Leeward sheltered	3.2	5.6	2.4
Thompson Creek (OG)	Leeward exposed	3.6	15.6	12.0
Pimms Block (L)	Windward sheltered	3.5	5.5	2.0
Fairy Creek (L)	Windward sheltered	3.8	5.6	1.8
Mangrove Road (RG)	Windward exposed	5.8	8.4	2.6
Noah Beach (RG)	Leeward exposed	3.8	8.0	4.2

Source: after Grove *et al.* (2000).
OG = old growth, L = logged, RG = regrowth.

Figure 3.7 Stand level forest damage caused by cyclone Rona (February 1999) at the Australian Canopy Crane Research Facility near Thompson Creek. Refer to Figure 3.6(b) for canopy changes. *Source*: Rainforest CRC Digital Library.

Local scale impacts of cyclones in the Wet Tropics

Impacts of cyclonic winds at the local scale are largely controlled by forest composition and structure as well as smaller scale topographic features (Box 3.1). Typically, higher elevation (montane) tropical forests are more resistant to strong winds compared with lowland tropical forests, mainly due to their lower stature and aerodynamically smooth canopies. Lowland forests are known for their higher stature and generally uneven canopy architecture. Their rougher canopy surface results in more turbulent wind flow under cyclonic conditions, with resultant damage in the form of branch and vine removal, occasional tree falls and widespread defoliation. Despite their lower resistance to cyclonic winds, lowland tropical forests are more resilient than their less productive montane counterparts due to their high rates of primary productivity.

Figure 3.7 shows local-scale damage to complex lowland rainforest at the Australian Canopy Crane Research Facility, near Thompson Creek (Figure 3.6b), following cyclone Rona in 1999 (see Box 3.2). The forest at this location was severely damaged by leeward acceleration of gravity waves originating from a high plateau (about 1000 m) some 5 km to the west. Although cyclone Rona was only rated as a category 3 system (Box 3.2, Table 3.2), the amount of forest damage at this site suggests wind gusts more usually associated with a category 4 or 5 cyclone, such as Larry (Box 3.2). The importance of leeward wind acceleration during cyclones as a mechanism for catastrophic forest damage in the Wet Tropics region is poorly understood.

Turton (1992) was the first to document the effects of a tropical cyclone on the understorey light environments in a tropical forest. The study was undertaken at the Curtain Fig Forest, near Yungaburra on the Atherton Tablelands. Cyclone Winifred (see Box 3.2) caused moderate damage to the canopy of this forest in February 1986 (refer to Figure 3.4 for an example of moderate canopy damage). Immediately after the cyclone there was a two- to threefold increase in light availability in the understorey of the Curtain Fig forest, mainly due to defoliation but also as consequence of some branch and stem breakages. Light conditions were more spatially homogeneous after the cyclone than before and there was a greater seasonal difference in light availability in the understorey after the cyclone. The forest canopy recovered rapidly at this site, with understorey light levels approaching their pre-cyclone state about 11 months later (Turton 1992).

While we have some knowledge of microclimate patterns across rainforest understoreys following canopy disturbance we know very little about microclimate changes within the main canopy following cyclonic disturbance. Given that the canopy is the powerhouse of the rainforest in terms of productivity (Parker 1995) as well as containing significant biodiversity, understanding the magnitude of changes associated with a catastrophic natural disturbance within the canopy will contribute to our understanding of long-term dynamics of such complex ecosystems. Turton and Siegenthaler (2004) were the first to document the immediate effects of severe defoliation, with stem and branch breakages, on the microclimate of a tropical rainforest canopy at the Australian Canopy Crane Research Facility following cyclone Rona (Box 3.2).

Forest surrounding the Canopy Crane site suffered moderate to severe damage as a result of the hurricane-force winds (Figure 3.7). Observations after the storm

found about 10% of the trees in the 1ha plot covered by the Canopy Crane were completely toppled, about 30% of the trees lost their crowns but remained standing and most remaining canopy trees were severely defoliated. Many vines and epiphytes were removed. Figure 3.8 shows mean hourly photosynthetically

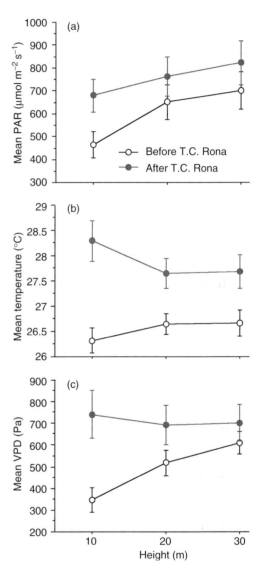

Figure 3.8 (a) Mean hourly photosynthetically active radiation (PAR), (b) air temperature and (c) vapour pressure deficit (VPD) at three heights on the canopy crane tower 6 days before Cyclone Rona (5–10 February 1999) and 6 days afterwards (12–17 February 1999). Values shown are means (± 1 SE) for daylight hours only. Data for the actual days of the cyclone (11, February 1999) are not included. From Turton & Siegenthaler (2004). Copyright Cambridge University press, Reproduced with permission.

active radiation (PAR), air temperature and vapour pressure deficit (VPD) (daylight hours only) for six days immediately before and after the cyclone. Turton and Siegenthaler (2004) demonstrated that the most profound changes in microclimate conditions affected the sub-canopy of the forest at the Canopy Crane site. Increases in PAR, air temperature and VPD in the lower canopy following the cyclone undoubtedly impacted on plant taxa normally tolerant of dim, cool and relatively humid conditions. In particular, increases in PAR and the amounts of red light in the lower canopy and understorey during the first few weeks after the cyclone favoured germination of pioneer species in the soil seed bank, and the rapid growth of normally suppressed understorey seedlings and saplings. Moderate increases in PAR between 20 and 30 m above the ground encouraged rapid growth rates of mid-storey trees in the first few months after the cyclone, mainly due to changes in lateral shading among neighbouring canopy trees. Over longer time periods (12–36 months after the cyclone), changes in net radiation, PAR and red/far red light ratios are likely to have affected nutrient cycling, plant reproduction and survivorship, plant architecture and growth and leaf turnover in the forest at the Canopy Crane site (Turton & Siegenthaler 2004).

Likely impacts of climate change on cyclone frequency and intensity in the Wet Tropics

Global warming is predicted to increase the intensities of tropical cyclones by 10–20% by the end of the twenty-first century, while cyclone frequency is predicted to remain unchanged (Henderson-Sellers et al. 1998). Two factors that contribute to more intense tropical cyclones, ocean heat content and water vapour, have both increased over the past several decades. This is primarily due to human activities, such as the burning of fossil fuels and the clearing of forests, which have significantly elevated carbon dioxide levels in the atmosphere. Carbon dioxide and other heat-trapping gases, such as methane, act like an insulating blanket that warms the land and ocean and increases evaporation from the ocean surface. Emanuel (2005) has produced some evidence for a substantial increase in the power of tropical cyclones (denoted by the inte-

gral of the cube of the maximum winds over time) during the past 50 years. This result is supported by the findings of Webster et al. (2005) that there has been a substantial global increase (nearly 100%) in the proportion of the most severe tropical cyclones (categories 4 and 5 on the Saffir–Simpson scale) in the period from 1970 to 1995, which has been accompanied by a similar decrease in weaker systems (categories 1 to 3, same scale). The findings from both studies correlate with the rise in sea surface temperatures in regions where tropical cyclones typically originate, including the oceans around Australia.

What are the implications of more intense cyclones for our tropical rainforests? Will biodiversity decline? Will forest structure change over time? Given that cyclones are a significant disturbance agent for the forests of the Wet Tropics region (Webb 1958), there is a plausible link between periodic cyclonic disturbance and forest composition and structure at the landscape scale. The Intermediate Disturbance Hypothesis proposes that biodiversity is highest when ecosystem disturbance is neither too rare nor too frequent (Connell 1978). With low disturbance, competitive exclusion by dominant species arises, while with high disturbance, only species tolerant of the stress agent can persist. A shift in the severity spectrum of tropical cyclones will tend to favour species more resistant to strong winds, such as palms, and those more tolerant of post-disturbance environmental stresses, such as pioneers, vines and weeds (Lugo and Scatena 1996). Increases in cyclone intensity, even with cyclone frequency remaining stable, may lead to shifts in forest succession direction, higher rates of species turnover and hence opportunities for species change in tropical forests (Dale et al. 2001). Such changes will be in concert with other ecological and physiological effects of global warming on forest biota and ecosystem function (see Williams et al., Chapter 22, this volume).

Summary

- Tropical cyclones are significant disturbance phenomena for forest ecosystems in the Wet Tropics of Australia, especially those near the coast. Continuous forests in the region have been described as hyper-disturbed ecosystems with patches of damaged forest constantly recovering from previous cyclonic events, often in concert with floods, droughts and fires. Cyclones are part of the ecosystem dynamics of these forested landscapes and recovery following such events is often remarkably rapid. However, compared with continuous forest areas, forest fragments are particularly vulnerable to impacts of tropical cyclones and their associated strong winds, largely due to their high forest edge to area ratios.

• Given the relatively high frequency of tropical cyclones in many tropical and sub-tropical regions there is a general consensus that they contribute to the structure and function of tropical rainforests in cyclone-prone areas, with ecosystem impacts and recovery processes occurring at several spatial and temporal scales. Wind damage is an important source of landscape-scale (about 10 km) patterning in forests and is a major factor initiating vegetation dynamics at that scale. At the local scale (about 1 km), extensive toppling of trees and canopy thinning is considered to be an important factor regulating hydrological, energy and nutrient regimes at that scale.

• The frequency and intensity of tropical cyclones crossing the wet tropical coast (Cooktown–Ingham) over the period 1858–2006, according to the Australian Cyclone Severity Scale, shows that a weak cyclone (Australian System: categories 1, 2) is likely to cross the wet tropical coast with a frequency interval of about 5 years, compared with a frequency of about 15 years for a moderate to severe cyclone (category 3) and about 75 years for a very severe cyclone (categories 4, 5). Over the past 20 years three significant tropical cyclones have impacted on the forest ecosystems of the Wet Tropics of Australia.

• Impacts of tropical cyclones on forests at the landscape-scale are the result of the complex interaction of anthropogenic, meteorological, topographical and biotic factors. In the Wet Tropics region, topography plays a significant role in accounting for forest impact patterns and recovery processes after cyclones at the landscape scale, where winds exhibit complex interactions when they encounter steep topography. Depending on the location of the cyclone centre and the direction and speed of movement of the cyclone, forests in mountainous areas may experience severe windward exposure, topographic shading (leeward protection) and, under certain circumstances, severe lee wind turbulence associated with gravity waves.

• Cyclones provide a spectrum of disturbance regimes across forested landscapes, a process that undoubtedly promotes biodiversity at that scale. However, cyclonic disturbance also promotes the risk of forest fires and the spread of invasive species, including soil pathogens, weeds and feral animals, especially in fragmented forests.

• Impacts of cyclonic winds at the local scale are largely controlled by forest composition and structure as well as smaller scale topographic features. Higher elevation forests are generally more resistant to cyclonic winds than lowland forests, mainly due to their lower stature and aerodynamically smooth canopies, but have low rates of recovery. Despite their lower resistance to cyclonic winds, lowland tropical forests are more resilient than their less productive montane counterparts due to their high rates of primary productivity.

• Global warming is predicted to increase the intensities of tropical cyclones by 10–20% by the end of the century, while cyclone frequency is predicted to remain unchanged. There is strong evidence that cyclone intensity has increased significantly over the past 50 years, while frequency has remained stable.

• Increases in cyclone intensity in the future, even with cyclone frequency remaining stable, may lead to shifts in forest succession direction, higher rates of species turnover and hence opportunities for species change in tropical forests. Such changes will be in concert with other ecological and physiological effects of global warming on forest biota and ecosystem function.

References

Bellingham, P. J. (1991). Landforms influence patterns of hurricane damage: evidence from Jamaican montane forests. *Biotropica* **23**: 427–33.

Bellingham, P. J., Kapos, V., Varty, N. *et al.* (1992). Hurricanes need not cause high mortality: the effects of hurricane Gilbert on forests in Jamaica. *Journal of Tropical Ecology* **8**: 217–23.

Bellingham, P. J., Tanner, E. V. J. & Healey, J. R. (1994). Sprouting of trees in Jamaican montane forests after a hurricane. *Journal of Ecology* **82**: 747–58.

Bellingham, P. J., Tanner, E. V. J. & Healey, J. R. (1995). Damage and responsiveness of Jamaican tree species after disturbance by a hurricane. *Ecology* **76**: 2562–580.

Bellingham, P. J., Tanner, E. V. J. & Healey, J. R. (2005). Hurricane disturbance accelerates invasion by the alien tree *Pittosporum undulatum* in Jamaican monatane rain forests. *Journal of Vegetation Science* **16**: 675–84.

Bellingham, P. J. Tanner, E. V. J., Rich, P. M. & Goodland, T. C. R. (1996). Changes in light below the canopy of a Jamaican montane rainforest after a hurricane. *Journal of Tropical Ecology* **12**: 699–722.

Boose, E. R., Foster, D. R. & Fluet, M. (1994). Hurricane impacts to tropical and temperate forest landscapes. *Ecological Monographs* **64**: 369–432.

Brokaw, N. V. L. & Walker, L. R. (1991). Summary of the effects of Caribbean hurricanes on vegetation. *Biotropica* **23**: 442–7.

Bureau of Meteorology (1999). *Annual Report 1998–99.* Commonwealth of Australia, Canberra. http://www.bom.gov.au/inside/services_policy/public/sigwxsum/sigw0299.shtml

Connell, J. H. (1978). Diversity in tropical rain forests and coral reefs. *Science* **199**: 1302–10.

Dale, V. H., Joyce, L. A., McNulty, S. *et al.* (2001). Climate change and forest disturbances. *Bioscience* **51**: 723–34.

Emanuel, K. (2005). Increasing destructiveness of tropical cyclones over the past 30 years. *Nature* **436**: 686–8.

Everham, E. M. & Brokaw, N. V. L. (1996). Forest damage and recovery from catastrophic wind. *Botanical Review* **62**: 113–73.

Fernandez, D. S. & Fetcher, N. (1991). Changes in light availability following Hurricane Hugo in a subtropical montane forest in Puerto Rico. *Biotropica* **23**: 393–9.

Grove, S. J., Turton, S. M. & Siegenthaler, D. T. (2000). Mosaics of canopy openness induced by tropical cyclones in lowland rain forests with contrasting management histories in northeastern Australia. *Journal of Tropical Ecology* **15**: 883–94.

Harrington, R. A., Fownes, J. H., Scowcroft, P. G. & Vann, C. S. (1997). Impact of hurricane Iniki on native Hawaiian *Acacia koa* forests: damage and two-year recovery. *Journal of Tropical Ecology* **13**: 539–58.

Henderson-Sellers, A., Zhang, H., Berz, G. *et al.* (1998). Tropical cyclones and global climate change: a post-IPCC assessment. *Bull Amer Meteor Soc* **79**: 19–38.

Laurance, W. F. (1991). Edge effects in tropical forest fragments: application of a model for the design on nature reserves. *Biological Conservation* **57**: 205–19.

Laurance, W. F. (1997). Hyper-disturbed parks: edge effects and ecology of isolated rainforest reserves in tropical Australia. In *Tropical Forest Remnants: Ecology, Management and Conservation of Fragmented Communities*, Laurance, W. F. & Bierregaarrd, R. O. (eds). University of Chicago Press, Chicago, pp. 71–83.

Lugo, A. E., Applefield, M., Pool, D. J. & Mcdonald, R. B. (1983). The impact of hurricane David on the forests of Dominica. *Canadian Journal of Forest Research* **13**: 201–11.

Lugo, A. E. & Scatena, F. N. (1996). Background and catastrophic tree mortality in tropical moist, wet, rain forests. *Biotropica* **28**: 585–99.

McGregor, G. R. & Nieuwolt, S. (1998). *Tropical Climatology: An Introduction to the Climates of the Low Latitudes,* 2nd edn. John Wiley & Sons, Chichester. 339 pp.

Parker, G. G. (1995). Structure and microclimate of forest canopies. In *Forest Canopies*, Lowman, M. D. & Nadkarni, M. N. (eds). Academic Press, San Diego, CA, pp. 73–106.

Tanner, E. V. J., Kapos, V. & Healey, J. R. (1991). Hurricane effects on forest ecosystems in the Caribbean. *Biotropica* **23**: 513–21.

Tanner, E. V. J. & Bellingham, P. J. (2006). Less diverse forest is more resistant to hurricane disturbance: evidence from montane rain forests in Jamaica. *Journal of Ecology* **94**: 1003–10.

Turton, S. M. (1992). Understorey light environments in a north-east Australian rain forest before and after a tropical cyclone. *Journal of Tropical Ecology* **8**: 241–52.

Turton, S .M. (in press) How severe was cyclone Larry? A comparative evaluation with previous cyclones impacting the north tropical coast of Australia: 1858–2006. *Austral Ecology.*

Turton, S. M. & Siegenthaler, D. T. (2004). Immediate impacts of a severe tropical cyclone on the microclimate of a rain-forest canopy in north-east Australia. *Journal of Tropical Ecology* **20**: 583–6.

Unwin, G. L., Applegate, G. B., Stocker, G. C. & Nicholson, D. I. (1988). Initial effects of tropical cyclone 'Winifred' on forests in north Queensland. *Proceedings of Ecological Society of Australia* **15**: 283–96.

Vandermeer, J., Mallona, M. A., Boucher, D., Yih, K. & Perfecto, I. (1995). Three years of ingrowth following catastrophic hurricane damage on the Caribbean coast of Nicaragua: evidence in support of the direct regeneration hypothesis. *Journal of Tropical Ecology* **11**: 465–71.

Webb, L. J. (1958). Cyclones as an ecological factor in tropical lowland rainforest, north Queensland. *Australian Journal of Botany* **6**: 220–8.

Webster, P. J., Holland, G. J., Curry, J. A. & Chang, H.-R. (2005). Changes in tropical cyclone number, duration, and intensity in a warming environment. *Science* **309**: 1844–6.

Whitmore, T. C. (1974). Change with time and the role of cyclones in tropical rain forest on Kolombangara, Solomon Islands. Commonwealth Forestry Institute paper 46.

Whitmore, T. C. (1989). Changes over twenty-one years in the Kolombangara rain forests (Solomon Islands). *Journal of Ecology* **77**: 469–515.

Whitmore, T. C. (1991). Tropical rain forest dynamics and its implications for management. In *Rain Forest Regeneration and Management*, Gomez-Pompa, A., Whitmore, T. C. & Hadley, M. (eds). UNESCO, Paris, pp. 67–89.

Yih, K., Boucher, D. H., Vandermeer, J. H. & Zamora, N. (1991). Recovery of the rain forest of southeastern Nicaragua after destruction by Hurricane Joan. *Biotropica* **23**: 106–13.

4 ABORIGINAL CULTURES IN THE WET TROPICS

*Sandra Pannell**

Discipline of Authropology and Archaeology, School of Arts and Social Sciences, James Cook University, Townsville, Queensland, Australia
*The author was participant of Cooperative Research Centre for Tropical Ecology and Management

Introduction

In late April 2005, a ceremony was held in Far North Queensland to mark the signing of a regional agreement between Rainforest Aboriginal people and state and commonwealth environmental agencies regarding Indigenous involvement in the future management of the Wet Tropics. Promoted as a historic moment in the nearly 20-year struggle for the recognition of Aboriginal culture and rights in the World Heritage-listed property, the agreement signing-ceremony culminated in a performance of dance and music featuring members of the 18 Rainforest Aboriginal tribal groups. These tribal groups are Banjin, Djabugay, Djiru, Girramay, Gugu Badhun, Gulngay, Gunggandji, Jirrbal, Koko Muluridji, Kuku Yalanji, Ma:Mu, Ngadjon-Jii, Nywaigi, Warrgamay, Warungu, Wulgurukaba, Yidinji and Yirrganydji (WTMA 2005).[1]

Rather than drawing attention to their shared history of dispossession and marginalization, or their common experience of racism and social and economic disadvantage, in a very public coming together of politics and identity, the Aboriginal signatories to the Agreement chose to put culture centre-stage in the form of a corroboree. For the performers and the Aboriginal audience alike, this dramatic visual and aural performance reflected the coming together of Rainforest Traditional Owners as one voice.

This self-consciousness of culture is part of what could be considered an Indigenous counterculture. As the anthropologist Marshall Sahlins points out, the current emphasis upon culture goes hand-in-hand with the 'demand of [Indigenous] peoples for their space within the world cultural order' (Sahlins 2000: 513). An integral part of this movement is the reclamation of the symbols of Aboriginal culture that all too often have been appropriated as markers of a more generalized Australian national identity. For example, a version of the shield that stands as a powerful image of Rainforest Aboriginal culture on the cover of the Agreement document also features on the 1953/4-issued Australian five pound note. Set amidst a bustling scene of agricultural and pastoral industries, the presence of two rainforest shields and a single boomerang represented the first time that Aboriginal culture was symbolized on Australian currency notes. While it could be argued that the use of these artefacts reduces the complexity of Aboriginal society and culture to simple and somewhat clichéd symbols, their very presence in this bucolic landscape of Anglo-Australian settlement says something more about Indigenous cultural survival at the height of the era of assimilation. The persistence of Aboriginal culture, its performance by Indigenous people and its representation by Anglo-Australians over the course of the past 100 years comprises the subject of this and a number of other chapters in this volume.

The invention of culture

Roy Wagner (1975) once suggested that culture is something invented by anthropologists as a foil to assist them in arranging their fieldwork experiences. Certainly, anthropologists have spent more than a hundred years debating the nature and meaning of culture in the course of trying to study it (Appadurai 1996). For example, in 1871, E. B. Tylor, one of the founding fathers of anthropology, defined culture as: 'That complex whole which includes knowledge, belief, art, morals, law, custom, and any other capabilities and habits acquired by man as a member of society' (cited in Kuper 2000: 56).

Tylor's original idea of culture as a list of disparate traits has been refined, developed and at times rejected by subsequent generations of social scientists. Indeed, by the early 1950s, 164 anthropological definitions of culture existed in the literature (see Kroeber and Kluckhohn 1952). Notwithstanding this plethora of definitions, Wagner's idea of culture as a product of a particular discursive tradition, rather than as a description of an objective thing that exists in the world, is useful. It throws up questions about the universality of the concept and whether non-Western people even have a notion of culture. It also makes us think about what are some of the political and moral effects of invoking the concept of culture. Certainly, the idea of culture as an invention allows us to follow its rocky trajectory in North Queensland, from its very denial by settlers and colonial authorities in the late nineteenth century to the contemporary 'deployment of a rhetoric of culture' (Cowan *et al.* 2001: 2), where 'culture is on everyone's lips'. While Wagner (1975: 1) once remarked that 'we could define an anthropologist as someone who uses the word culture habitually', it is clear that culture is no longer an academic term of art. Following the implications of Wagner's remarks, we can see how ideas about culture are created and transformed in the course of colonial and post-colonial entanglements.

In Far North Queensland, these ideas are inextricably linked to colonial perceptions of the forested spaces they initially encountered.

The invasion of culture

The first European explorers and adventurers, men such as Edmund Kennedy and Christie Palmerston, characterized the rainforests of this region as scrub, jungle, bush and myall woods. From the mid-nineteenth to the start of the twentieth century, these forested spaces were frequently described in journals, government reports and newspaper articles of the day as inhospitable, dank and dark. After the red cedar stands had been extensively logged in the 1880s, the rainforests were largely regarded as an impenetrable tangle of economically useless timber. As Terry Birtles (1995: 3) observes, scrub 'constituted an obstacle to agricultural settlement … a nuisance to be removed to expose the soil for cultivation'. For the European colonizers, the rainforests of the region, or jungle as they were more commonly known, stood as uncultivated spaces in contrast to the desired and gradually emerging patchwork of settled places.

As Raymond Williams (1976) reminds us, the term culture derives in a roundabout way from the Latin verb *colere*, to cultivate, and is thus associated with the tilling of the soil. In this sense then, the forests of Far North Queensland were construed as being without culture; brutish enclaves lacking refinement and the other human-created characteristics attributed to agricultural vistas. In the late 1800s, a certain cultural framework emerged for perceiving this environment. Based upon a series of binary oppositions, scrub was seen as wild and natural, while the developing farms and towns came to epitomize a European sense of domesticated, cultural spaces. Given this ecological model, it is not so surprising that the Aborigines who inhabited the rainforests of the region were spoken of in the same terms reserved for these dark and repugnant jungles (Savage 1992). The writings of Christie Palmerston and Archibald Meston provide numerous examples of this usage.

Christie Palmerston is widely credited in European accounts as being the first white person to breach the heavily forested ranges that separated the coast from the interior in North Queensland. As well as opening the country for settlement, some also saw Palmerston's role as assisting the authorities in their endeavours to 'civilise the blacks' (James Mulligan, cited in Savage 1992: 94). While Palmerston speaks of his Aboriginal boy Pompo with paternalistic concern (Savage 1992: 32), his diaries also reveal that he was particularly prone to giving local Aborigines 'a taste of the rifle' (Savage 1992: 34) or the 'contents of his revolver' (Savage 1992: 87). Palmerston shot and killed Aboriginal people with the same absence of hesitation and remorse he showed the local fauna, and at times, it seems with the same frequency. Indeed, it appears from the way in which Palmerston describes the Aborigines he encounters as foul brutes, savages, hostile and fierce tribes, bucks and cannibals, that they are regarded as being as wild and as uncultured as the forests they occupy. There is no suggestion in Palmerston's writings that anything akin to an Indigenous culture or society exists in these forested, but soon to be cleared, spaces of the 1880s. Notwithstanding the amount of time that Palmerston spent with Aboriginal people on his rainforest explorations, he does not identify by name any one Aboriginal group. In keeping with the practice of the day, Aborigines are identified with reference to prominent landforms, i.e. the Russell River blacks, the Johnstone [River] blacks etc.

Three years after Palmerston's final expedition in the region in 1886, Archibald Meston claimed the distinction of being the first European to reach the summit of Bellenden Ker in June 1889. Meston's report of this trip is arguably the first scientific account of the Aborigines of the area and he devotes a section to a description of the Bellenden Ker blacks. Meston's attempt to classify scientifically the Aborigines of the Russell, Mulgrave and Barron Rivers into a broader taxonomic scheme is informed by evolutionary ideas of the time. Accordingly, the Australian native is characterized as a savage, separated from Europeans by 'the many thousand years of cumulative civilization' (Meston 1895: 24). For the likes of Meston and others of his era, it is apparent that the history of the Aboriginal occupiers of the region only really begins with the arrival of the Europeans. Prior to this moment of enlightenment, the story of the Aborigine is concealed within the dark recesses of time.

From his 1889 report, it is clear that Meston subscribes to the Darwinian notion of survival of the fittest as an explanation for why the Australian Aborigine, like all 'savage and inferior races', is 'destined to disappear' (Meston 1889: 9). However, in his 1895 report to the Colonial Secretary of Queensland on the Queensland Aboriginals, Meston (1895) dismisses the doomed race theory to its deserved oblivion. Instead, he advocates a System for their Improvement and Preservation, based upon the establishment of isolated reserves and Aboriginal settlements, overseen by government-appointed protectors. Critical to Meston's (1895) proposition is the view that stone-age Aborigines could, in time, be assisted forward through the various stages of civilization until they attained the agricultural stage, when they would be of use to 'themselves and mankind' (Meston 1895: 25). As part of this process, Meston envisaged that on the proposed reserves

The village would be laid out in regular order, with streets and squares, special sanitary arrangement and water supply. Each married couple would have their own allotment of land, and cultivate their own vegetables and fruit. The single people would cultivate and share the proceeds of their lands in common. Habits of cleanliness and industry would be taught regularly and enforced when necessary. (Meston 1895: 26)

For Meston, and others so disposed, civility and progress are fundamentally linked to the cultivation of the land.[2] In the eyes of the Europeans, the attainment of culture by Aboriginal people was linked to the transition from a 'wild free hunter's life' to that of an 'industrious farmer' (Meston 1895: 25). It is clear from Meston's report that the activity of tilling the land was seen by Europeans as distilling, constituting and communicating the cultural values of European society at that time. However, as history shows, farming constitutes a hard cultural form (Appadurai 1996) that resists 'reinterpretation as it crosses social boundaries'. Notwithstanding the government's varied and concerted efforts to this effect, the pursuit of agriculture proved to be so hard that few, if any, Aboriginal people in Far North Queensland became or remained cultivated by this means.

In 1897, the Aboriginal Protection and Restriction of the Sale of Opium Act was passed in Queensland to bring into effect Meston's vision and W. E. Parry-Okeden, Commissioner of Police, duly undertook the first census of Aborigines in North Queensland. Based on the distribution of blankets, this census also represented the first systematic attempt by the government to map tribal names (Parry-Okeden 1897). Nearly 60 named Aboriginal groups, consisting of about 3000 members, were identified in the region east of the Great Dividing Range between Rockingham Bay in the south and Cooktown in the north. Some of the names recorded on Parry-Okeden's map refer to clan names (Dixon 1972), while other names are barely recognizable in terms of the identity of contemporary tribal groups.[3] Prior to 1897, E. M. Curr's four-volume work on the Australian Race (1886–9) represented the only other systematic attempt to piece together the nature of Aboriginal groups and their distribution in this region.[4]

The collection of culture

In 1898 Meston and Walter E. Roth were appointed as the respective Aboriginal Protectors for the Southern and Northern Divisions (Lees 1902).[5] In his role as Protector, Meston returned to Bellenden Ker in 1904, and in his report of this expedition he proceeds to provide names for the tribes and languages he had previously described in 1889 by reference to rivers and mountains (i.e. Russell River Tribe). Some of these groups – for example, Yeedinjee and Matcham (Meston 1904) – are also described in Roth's publications.

While Meston might have had a deep administrative interest in his Indigenous charges, Walter E. Roth exhibited a strong ethnographic interest in Aboriginal customs and he subsequently published a number of bulletins outlining details about the tribes in his division. His appointment ushered in a short-lived era of anthropological enquiry, which gradually introduced to the general public details of the names and distribution of tribes in the rainforest region. For example, in his scientific report of 1900, Roth identifies the tribes that occupied the area from the lower Tully River to Clump Point as the Mallan-para, Walmal, Kirinja and Chiru (Djiru). Based upon his research in 1898, Roth provides a sketch map of Cairns and the surrounding district showing the location of the three

tribes Yirkanji (Yirrganydji), Kungganji (Gunggandji) and Yidinji (Roth 1984). On the Atherton Tableland, he associates the Ngachan-ji tribe (Ngadjon-Jii) with the area from Atherton to the headwaters of the Russell River, while the Chirbal-ji (Jirbal) are said to have their 'main camp in the vicinity of Carrington at Scrubby Creek, traveling to Atherton … and the Herberton Ranges' (Roth 1984, Vol. II: 91–2). Further north, the Koko Yellanji (Kuku Yalanji) are associated with the Bloomfield River.

In his bulletins and reports to the government of the day, Roth describes a range of customs and beliefs that, when considered together, were thought to comprise the entire repertoire of a culture. Roth's accounts constitute a form of cultural butterfly collecting in which practices are described merely for the sake of having a record of 'the lowest type of savage' (Roth 1900: private note). There is little attempt to relate the customs and beliefs described to each other in a systematic way or even to suggest that together they comprise a regional system. This latter point is evident in Roth's bulletins, where specific traits are presented according to known areas of ethnographic investigation at that time. For example, in the rainforest region of North Queensland, the same social and cultural traits are listed under different area headings, such as the lower Tully River, Russell River, Atherton, Cairns. There is a strong emphasis in Roth's description of Aboriginal customs upon items of material culture, e.g. fighting weapons, huts and shelters, bodily decorations and clothing. As these comments suggest, Roth's ideas about culture as a series of acts, institutions and material forms closely conform to Tylor's (1871) definition in the opening lines of his volume, *Primitive Culture* (1871).

By the turn of the twentieth century, Meston and Roth, together with a number of amateur observers, had recorded and published the names and some details about more than half of the 18 Rainforest Aboriginal tribal groups (WTMA 2005) today acknowledged as the Traditional Owners of the Wet Tropics. It was also around this time, and in the decade following, that many of the Aboriginal people occupying this rainforested region were physically removed from their customary lands and transported to one of the reserves and missions established under the 1897 Act.

In the period from 1900 to 1930, as the Indigenous occupiers of the region were removed from their traditional lands or commenced work under the Act for Europeans on the plantations and farms that had rapidly transformed the rainforested landscapes of the region, little appears in the written record about local Aboriginal society and culture.[6] Ironically, while the 1897 Act generated the possibilities for, and limits of, European knowledge about Aboriginal society and culture, to many Europeans of the day it also signalled the start of the detribalization and acculturation of the Indigenous occupiers of the region. While Meston (1895: 26) had originally proposed that under the reserve system Aborigines would be encouraged to 'preserve their language, and maintain all their own laws and customs which do not interfere with the harmonious management of the [Reserve] community', in reality the European superintendents and missionaries who controlled these segregated communities systematically prohibited the public expression of the practices and beliefs they regarded as manifestations of native culture. For Aboriginal people of the Wet Tropics, no sooner had their culture come into view as part of their surveillance under the Act than it was deemed to have disappeared under the very same instrument of law. As Sahlins (2000: 502) observes, this is a familiar story: 'in too many narratives of Western domination, the indigenous victims appear as neo-historyless peoples: their own agency disappears, more or less with their culture, the moment Europeans irrupt on the scene'.

In the mid- to late 1930s, believing that a traditional way of life and an ancient culture were about to disappear as a result of continued government efforts to protect and assimilate Aboriginal people, a number of anthropologists visited the Wet Tropics region. Ursula McConnel, Lauriston Sharp, Joseph Birdsell and Norman Tindale were among the new phalanx of trained professionals who briefly visited regional towns, missions and settlements recording information about Aboriginal social life and customary practices. For Norman Tindale (1974) his 1938–9 expedition to North Queensland was part of a life-long endeavour to record and map the Aboriginal tribes of Australia. The wealth of sociological (as well as physiological) data amassed by Tindale and Birdsell on the 1938–9 expedition indicates that while incarcerated on missions or indentured on farms in the regions, Aboriginal people in the Wet Tropics had somehow managed to maintain enough of what anthropologists

of the era acknowledged as culture for it to be worth recording.

A collection of cultures

As a result of their investigations in North Queensland, Tindale and Birdsell identified not one culture, but the existence of what they regarded as a distinctive bloc of cultures. Discussing the results of their 1938–9 expedition, Tindale and Birdsell (1941) speak of the existence of a Tasmanoid group in the rainforests of North Queensland. The dozen small or pygmoid tribes belonging to this group were said to exhibit physical characteristics similar to those of Tasmanian Aborigines, hence their use of the label. Tindale and Birdsell also referred to this aggregation of tribes as the Barrineans, in reference to their territorial proximity to Lake Barrine. Both authors believed the rainforest-dwelling Barrineans to be 'a separate small-framed type of modern man forming one of the earliest stocks in southern Asia' (Tindale 1974: 89), and thus the Barrineans supposedly represented the 'first wave of the Aboriginal occupation of Australia' (DASETT 1987).

According to Tindale and Birdsell (1941: 3), the 12 groups comprising the Tasmanoid or Barrinean tribes are: Ngatjan, Mamu, Wanjuru, Tjapukai, Barbaram, Idindji, Kongkandji, Buluwai, Djiru, Djirubal, Gulngai and Keramai.

While Tindale and Birdsell focus on the physical characteristics of these groups (e.g. diminutive stature, crisp curly hair), they also identify a number of linguistic, social and cultural features shared by the rainforest-dwelling Tasmanoids. These include

- a patrilineal moiety system or a four-class system;
- partial mummification of the dead;
- carrying the skulls and jaw bones of the deceased for a lengthy period before disposal by burning;
- food cannibalism;
- the use of large decorated fighting shields;
- the production and wearing of beaten bark blankets;
- the manufacture of lawyer cane-sewn fig tree baskets;
- specialized techniques of food gathering and preparation, including the leaching of alkaloid-loaded nuts (Tindale and Birdsell 1941: 6–8).

This idea of a distinctive cultural assemblage also emerges in the writings of Lauriston Sharp, an American anthropologist who conducted research in the region between 1933 and 1935.[7] Unlike Tindale and Birdsell, who largely focus upon material culture, Sharp's classification of more than one hundred tribes into nine groups is based upon a combination of totemic traits. Sharp's interest in totemic organization is not so surprising given the widespread anthropological interest in Aboriginal totemism around this time. Indeed, Sharp is joined in his efforts to discern totemic cult patterns in north-eastern Queensland by fellow researchers Ursula McConnel and Donald Thomson. Sharp's attempt to classify the tribes of north-eastern Queensland into supra-totemic categories is reminiscent of the work of earlier anthropologists. For example, in 1898, R. H. Mathews reports upon the 'organization of a community of tribes in North Queensland' based upon similar class divisions, namely sections and moieties, and initiation ceremonies. Mathews identifies these tribal aggregations as nations.

The anthropological classification of local Aboriginal groups into larger cultural assemblages is also reflected in the writings of a number of linguists. For example, of Tindale and Birdsell's 12 Tasmanoid tribes, R. M. W. Dixon (1976) identifies six (Ngadjan, Mamu, Djirbal, Djiru, Gulngay and Giramay) as speaking dialects belonging to the Dyirbal language group. Dixon's observations about the existence of a rainforest-based, multi-dialectal language group reflects a growing linguistic interest in recording and classifying endangered Aboriginal languages in this part of Queensland. Prior to the development of this professional interest in the early 1960s, much of the material on Aboriginal languages in this area had been incidental and amateurish (Dixon 1972). As a result of the research undertaken by trained linguists, European knowledge of Indigenous languages was transformed from a haphazard collection of unrelated word lists to a more complex picture of linguistic homogeneity and diversity, presented in terms of more inclusive language categories.

In his 1976 article on tribes, languages and other boundaries in north-east Queensland, Dixon is at pains to point out that linguistic classifications do not necessarily correspond to the way Aboriginal people themselves identify and categorize languages or, for that matter, groups. This observation is particularly evident in the comments made by Ernie Raymont, a Ngadjon-Jii elder, at the 1993 Julayinbul conference:

He [R. M. W. Dixon] found that we're all one people but our language was just slightly different and … all that time we were thinking we were all strangers and we were all enemies and that's the attitude I was brought up when I was a kid in the camp in Malanda from the old people. … So it's only in the last ten years as Professor Dixon went amongst our people and wrote books about it, that we have come together and start talking to one another and all those years we thought we all enemies talking different tribal dialects. (Rainforest Aboriginal Network 1993: 23)

While it is apparent that the collectivities identified by a number of anthropologists in the 1930s and by linguists in the 1960s do not readily overlap, or are not necessarily recognized by Indigenous people themselves, as Ernie Raymont's comments suggest, by the late 1970s the idea of a single rainforest Aboriginal cultural bloc was already part of a wider discourse about Aboriginality in North Queensland. For example, the geographer David Harris (1978) identifies certain material cultural elements as unique to the Aboriginal groups that inhabit the rainforest area of North Queensland. In the 1990s, the historian Timothy Bottoms describes the Aboriginal groups that inhabit the rainforest environment between Ingham and Cooktown as the rainforest Bama or the Bama nation – *bama* being the indigenous term used throughout some of this region to denote an Aboriginal person (Bottoms 1992, 1993). As a number of linguists note, *bama* is the term for 'man' in the Yidiny, Gunggay, Madjay and Wangur group of dialects (Dixon 1977: 547). It is also found among the Djabugay group of dialects (Patz 1991: 332), and among the 12 dialects of the Kuku Yalanji language (Patz 2002: 20).[8]

The growing emphasis upon and presentation of Aboriginal identity and heritage in terms of a unique cultural bloc has to be seen against a broader social and political backdrop. The 1970s heralded a new era of multiculturalism in Australia and also saw the introduction of federal Aboriginal land rights legislation. These two developments not only placed the concept of culture and the issue of rights on the national agenda, they also highlighted the emergence of a global discourse that linked rights and culture. For Aboriginal people, in order to avoid being characterized as just another ethnicity in the sea of multiculturalism now washing over the suburbs of Australia, they needed to position themselves within this discourse as a distinctive group with exclusive cultural rights.

Rights to culture and culture as rights

As Cowan and her colleagues observe (2001: 2), the 1970s also signalled a shift in identity politics from 'claims of social equality to claims of group difference'. Previously, the acknowledgement of so-called human rights often entailed a 'denial, rejection or overriding of culture' (Cowen *et al.* 2001: 4). Cultural difference was seen as an impediment to the recognition of universal individual rights. Indeed, in the human rights discourse that emerged after the Second World War with the establishment of the United Nations, the Declaration of Human Rights and the process of decolonization, rights and culture were formulated as opposing concepts.

In Australia and elsewhere, the legal acknowledgement of Indigenous people as citizens with fundamental human rights did little to address the social and economic inequalities arising from historical dispossession and racial vilification. Nor did it seem that decades of assimilation and the bureaucratic denial of difference had delivered on the promised lifestyles and livelihoods of acceptability and affluence.

In the early 1970s, the policy of self-determination was promoted as the answer to these political fables and social failings. In their different ways, the policies of self-determination and multiculturalism signalled a change in the public psyche. Diversity was no longer a problem in search of a final solution. Instead, difference was conceived as a 'richness to be celebrated' (Cowan *et al.* 2001: 3). Fundamental to both self-determination and multiculturalism was the idea of a universal right to have and enjoy culture and that certain groups, notably non-Europeans, are defined by a distinctive culture. The conceptualization of culture promoted in both discourses is one where culture is seen as 'discrete, clearly bounded and internally

homogenous, with relatively fixed meanings and values' (Cowan *et al.* 2001: 3). Ironically, this essentialist view of culture closely resembles the nineteenth-century anthropological definition espoused by E. B. Tylor and operationalized by colonial agents, such as Meston and Roth.

In north Queensland, the 1970s saw increasing demands for Aboriginal cultural autonomy and land rights, resulting in the establishment of the North Queensland Aboriginal Land Council in 1976 (Bama Rainforest Aboriginal Corporation 1996). As in other parts of Aboriginal Australia, country and culture became inextricably linked in contemporary rights processes in the region. For example, in the vision statement of the 1998 review of Aboriginal involvement in the management of the Wet Tropics of Queensland World Heritage Area (WHA) the assertion of customary-law rights in land is clearly linked to 'the protection and preservation of cultural survival' (Review Steering Committee 1998: iv).

In the 1980s, the invocation of Indigenous rights to culture is also increasingly linked to the language of resistance, as Aboriginal groups, conservationists, developers, local councils and the state and federal governments battled over the future of the wet tropical forests of the region. However, in the campaign to stop construction of the Daintree to Bloomfield road and save the area from loggers and developers, and in the allied proposal to nominate the threatened rainforests for World Heritage listing, Aboriginal culture plays a minor role in relation to the perceived natural values of the region. For example, the 1988 listing of the Wet Tropics of North Queensland skates over Indigenous opposition to the inclusion of Aboriginal lands within the boundaries of the World Heritage property without consultation and consent from the affected Aboriginal people. While listed for its natural values, the original World Heritage nomination document presented by the Government of Australia makes passing reference to an 'extant Aboriginal rainforest culture' (DASETT 1987: 19). Interestingly, this reference to a single rainforest culture is based upon Tindale and Birdsell's long-discredited concept of the Barrinean tribes, and in keeping with their description, largely depicts culture in evolutionary (i.e. the tri-hybrid theory) and material terms (i.e. unique weapons). However, unlike Tindale and Birdsell's depiction of the Barrineans, the nomination document fails to mention the different Aboriginal groups said to share this culture, nor does it speak of regional diversity, or draw attention to some of the social and cultural similarities with non-rainforest groups. Moreover, in speaking of a single Aboriginal rainforest culture, the nomination document assumes the a priori and decontextualized existence of culture, rather than viewing the Barrineans as an anthropological construct, the product of a historically specific argument about human evolutionary origins and linkages.

In many ways, the assumptions about culture embedded in the Australian government's World Heritage nomination served Indigenous political action in the 1980s and 1990s. Invocation of a distinctive Rainforest Aboriginal people and culture, represented by similarly named political groups, such as Bama Wabu or the Rainforest Aboriginal Network, focused Indigenous opposition to the listing of the wet tropical forests. Reference to the existence of a unique rainforest people and culture also served to consolidate Aboriginal objections to the 1995 draft plan of management for the listed World Heritage property (Bama Rainforest Aboriginal Corporation 1996; Review Steering Committee 1998). As environmental consciousness developed in Australia in the 1970s and 1980s, Aboriginal people, like Anglo-Australian conservationists, capitalized on the symbolic and political value of staking a claim to rainforest. Previously demonized by Europeans as useless scrub and jungle, in the intense political battles throughout the 1980s to secure the Wet Tropics (McDonald & Lane 2000), the forests of the region took on mythic and somewhat romanticized characteristics. In the closing decades of the twentieth century, both the concept and the physicality of rainforest demarcated a new political space for the assertion and contestation of culture.

As the struggle for the recognition of Aboriginal cultural values in the management of the WHA proceeded into and throughout the 1990s, we see that claims to country and culture are increasingly articulated in terms of intellectual property rights (Janke 1998). In North Queensland, these claims came together in the 1993 *Julayinbul Statement on Indigenous Intellectual Property Rights* (Rainforest Aboriginal Network 1993). This declaration reaffirmed the right of Indigenous Nations and Peoples to self-determination and the

right to control their intellectual property. Drawing heavily upon the terminology of a number of international conventions, the Julayinbul Statement speaks of nations instead of cultures, highlighting the idea of Indigenous sovereignty and harking back to the short-lived colonial notion of autonomous Aboriginal political entities akin to states (see Hiatt 1996).[9]

Considered in conjunction with the legal recognition of native title in Australia, developments in the 1990s point to a significant change in Indigenous rights and culturalist discourses. The Native Title Era, as the period since the Mabo High Court determination in 1992 has often been called, certainly effected a shift from claims to a right to culture to a scenario where increasingly rights, or more correctly, a bundle of rights, are regarded as constituting culture. In this new era, the 'essentializing proclivities of the law' (Cowan *et al*. 2001: 10–11) produces a situation in which Indigenous identities, knowledges and practices are reduced to and codified in terms of a rights-based, normative discourse. Increasingly, culture in this legal scenario is 'based on a quasi-biological analogy by which a group of people has a culture in the way that animals have fur' (Merry 2001: 42).

While there was an initial proposal from some Aboriginal people to lodge 'one native title claim on the whole of the Wet Tropics' (Rainforest Aboriginal Network 1993: 20), the enactment of native title legislation has resulted in a more fragmented view of the Aboriginal landscape of North Queensland. As the content and meaning of native title evolves with each court determination, the idea of a single rainforest Aboriginal culture and people is gradually being subsumed by a mosaic of claims (and counter claims) made by different language, tribal, clan and family groups. In many ways, the emerging native title landscape reflects Indigenous ideas about their society and culture, and the spatial limits of these local understandings. However, as the most recent version of the map of Aboriginal Australia takes shape as a series of bounded spaces and discrete societies, trapped in a pre-sovereignty snapshot of laws and customs, native title also appears to articulate a legal nostalgia for familiar nineteenth-century ideas about primitive culture. Indeed, it seems that the twenty-first-century native title judgements about Aboriginal society and culture owe much to E. B. Tylor's original definition of this 'most complex whole'.

Conclusions, implications and management considerations

While Tylor and the many others who subscribed to the unilear evolutionism of the nineteenth century predicted an end to primitive culture or cultural diversity, it is apparent at the beginning of the third millennium that culture 'is not disappearing' (Sahlins 2000: 524). If the closing scenes of the recent Regional Agreement signing ceremony in North Queensland are anything to go by, Indigenous ideas about what constitutes culture and how it is to be presented are indeed complex and encompassing, though perhaps not in the way that Tylor envisaged. As the historic day's proceedings drew to a close, the Kuranda-based Traditional Owner band, Willie and the Poor Boys, came on stage and started singing about Aboriginal Dreamings and story places in a mixture of Djabugay and Murri English, played to a reggae beat. Welcome to country and culture in the twenty-first century!

As this performance illustrates, Aboriginal cultural self-representations both challenge and embrace earlier views of culture as fixed, uniform and unchanging. As Cowan and her colleagues (2001: 42) point out, Indigenous people's appropriation of essentialist views of culture often joins together 'an interest in cultural renaissance with the political constraints of a society willing to recognize claims on the basis of cultural authenticity and tradition'. While presenting culture in these terms may be politically effective in certain contexts, in other situations it may deny the interactive and inventive nature of Aboriginal social constructs and actions. As Lisa Meekison (2000: 11) points out, the boomerang-throwing image of Aboriginal culture 'tends to reinforce what non-Indigenous people think about Indigenous people rather than what the latter think about themselves'. While there is no denying the fact that Aboriginal cultural identities are often turned into marketable commodities for tourist consumption or reduced to stereotypical images in the media, contemporary Indigenous representations and forms of communication point to more dynamic and innovative

processes at work. Appropriating new technologies and drawing upon the inter-connectedness offered by an increasingly globalized world, these new cultural forms are often used to revivify 'local languages, traditions and histories, and [articulate] community identity and concerns' (Ginsburg 2000: 30). In this sense we see how Aboriginal traditions can serve as both 'a means and measure of innovation' (Sahlins 2000: 512).

In the light of these comments, we can conclude, along with Arjun Appadurai (1996: 12–13), that it is perhaps more useful to think of culture as a 'heuristic device' to talk about difference than to focus upon the noun form of culture, so often regarded as an object, thing or substance. Rather than trying to define what culture is or is not, a more interesting exercise involves looking at the 'relationship between the word and the world' (Appadurai 1996: 51). Put another way, the invocation of culture is a far more interesting area of investigation than its definition.

Given the problems of culture as a technical term, and the futility of trying to define it once and for all time, the question to ask here is not what are some of the considerations for management, but perhaps more to the point, what are the implications of management upon a world rediversified by Indigenous adaptations to modernity and globalization? In other words, what are some of the effects and consequences for Aboriginal people, and their varied use of culture, of the economic rationalist discourse that increasingly frames contemporary environmental management of protected areas?

While the recent signing of the Regional Agreement is seen by some as a step towards addressing the many problems with whitefella management of the WHA identified in *Which Way Our Cultural Survival* (Review Steering Committee 1998), resolving these problems will not simply be achieved by increasing Aboriginal involvement or adding Indigenous cultural values to the equation. Rather than trying to fit Indigenous people and claims within conventional management scenarios, we need to stretch or rethink the cultural frameworks that harbour these conventional understandings. As Tsing and other researchers (see Zerner 2003) point out, bureaucratic management regimes, which view forests as 'impersonal, passive and context-free' (Tsing 2003: 31), are at odds with the charismatic and personalized claims of Indigenous people to forests and forest resources. In order to recognize different ways of understanding and making forest landscapes, we need to use other frameworks than those based on science or founded upon commodity-property systems. Presented in the guise of neutral knowledge, the cool logic of economic efficiency or the technical magic of science, current management frameworks work to conceal their own cultural basis and assumptions.

The implication of these comments for future sustainable management of the region is clear. If scientists and environmental management agencies are serious about new partnerships with Aboriginal Traditional Owners and want truly to understand the cultures of the Wet Tropics, then perhaps it is time to reflect upon our own cultural assumptions about humans, rights and nature.

Notes

1   Throughout this chapter I retain the orthography used by the authors cited.
2   The concept of *terra nullius*, which underscored more than two centuries of Aboriginal dispossession, is based upon similar assumptions about what constitutes landed productivity in the eyes of the colonists.
3   For example, Norman Tindale identifies 'Eaton' and 'Hucheon' as alternative appellations for the 'Ngatjan' tribe (Tindale 1974: 183), while Tuffelcey is interpreted by Tindale as referring to 'Tjapukai'.
4   It appears that some of the 'group' names on Parry-Okeden's 1897 map for area 'A' are based upon some of the 'tribal' names given by Curr's correspondent, Thomas Hughes.
5   Roth resigned from his position as the Northern Protector of Aborigines on 10 June 1906.
6   Interestingly, in their review of the development of the anthropological idea of culture, Kroeber and Kluckhohn found that after Tylor's 1871 definition, no new definitions appeared for 32 years. In the period between 1900 and 1918, they were only able to identify six new definitions of culture (Kroeber & Kluckholm 1952).
7   It was also around this time that A. P. Elkin (1931a, 1934) identified the 'Aluridja community' of western South Australia and the 'Ungarinyin community' of tribes in the north-west Kimberley (1931b).

8   Dixon records *bama* as the Wargamay term for a male (animal and human), but also lists *ma:l* and *yamarra* as the Wargamay terms for man (Dixon 1981: 124). *Djinga* is the Nyawaygi word for man (Dixon 1983: 512), while in the Dyirbal language group, *yarra* is the term for man (Dixon 1972: 42).
9   In his criticism of 'nations', Norman Tindale states that 'in the early days of white contact there was a compulsion to try and find major units in Australia of the kinds familiar to the people of Europe' (1974: 156).

References

Appadurai, A. (1996). *Modernity at Large: Cultural Dimensions of Globalization*. University of Minnesota Press, Minneapolis.

Bama Rainforest Aboriginal Corporation (1996). *Reasonable Expectations or Grand Delusions? Submission to the Draft Wet Tropics Plan*. Bama Rainforest Aboriginal Corporation, Cairns.

Birtles, T. (1995). Colonial European culture conflict with the rainforest environment and Aborigines of north Queensland. Paper presented at the Ninth International Conference of Historical Geographers, Perth, Western Australia, 3–7 July.

Bottoms, T. (1992). *The Bama: People of the Rainforest*. Gadja Enterprises, Cairns.

Bottoms, T. (1993). The world of the Bama. *Journal of the Royal Historical Society of Queensland* **15**(1): 1–15.

Cowan, J. K., Dembour, M.-B. & Wilson, R. A. (2001). *Culture and Rights: Anthropological Perspectives*. Cambridge University Press, Cambridge.

Curr, E. M. (1886–9). *The Australian Race: Its Origin, Languages, Customs, Place of Landing in Australia, and the Routes by which It Spread Itself over that Continent*. Four volumes. Government Printer, Melbourne.

Department of Arts, Sports, the Environment, Tourism and Territories (DASETT) (1987). *Nomination of Wet Tropical Rainforests of North-east Australia by the Government of Australia for Inclusion in the World Heritage List*. Department of Arts, Sports, the Environment, Tourism and Territories, Canberra.

Dixon, R. M. W. (1972). *The Dyirbal Language of North Queensland*. Cambridge University Press, Cambridge.

Dixon, R. M. W. (1976). Tribes, languages and other boundaries in northeast Queensland. In *Tribes and Boundaries in Australia*, Peterson, N. (ed.). Australian Institute of Aboriginal Studies, Canberra, pp. 207–39.

Dixon, R. M. W. (1977). *A Grammar of Yidiny*. Cambridge University Press, Cambridge

Dixon, R. M. W. (1981). Wargamay. In *Handbook of Australian Languages: Volume 2*, Dixon, R.M.W. & Blake, B. J. (eds). The Australian National University Press, Canberra, pp. 1–144.

Dixon, R. M. W. (1983). Nyawaygi. In *Handbook of Australian Languages: Volume 3*, Dixon, R.M.W. & Blake, B. J. (eds). The Australian National University Press, Canberra, pp. 431–525.

Elkin, A. P. (1931a). The social organization of South Australian tribes. *Oceania* 2(1): 44–73.

Elkin, A. P. (1931b). Social organization in the Kimberley Division, North-Western Australia. *Oceania* 2(3): 296–333.

Elkin, A. P. (1934). Cult totemism and mythology in northern South Australia. *Oceania* 5(2): 171–92.

Elkin, A. P (1943). *The Australian Aborigines: How to Understand Them*. Angus and Robertson, Sydney.

Ginsburg, F. (2000). Resources of hope: learning from the local in a transnational era. In *Indigenous Cultures in an Interconnected World*, Smith, C. & Ward, G. K. (eds). Allen and Unwin, Sydney, pp. 27–49.

Harris, D. (1978). Adaptation to a tropical rain-forest environment: Aboriginal subsistence in northeastern Queensland. In *Human Behaviour and Adaptation*, Blurton-Jones, N. G. & Reynolds, V. (eds). Taylor & Francis, London, p. 113.

Hiatt, L. R. (1996). *Arguments about Aborigines: Australia and the Evolution of Social Anthropology*. Cambridge University Press, Cambridge.

Janke, T. (1998). *Our Culture: Our Future. Report on Australian Indigenous Cultural and Intellectual Property Rights*. Michael Frankel & Company Solicitors, AIATSIS and ATSIC, Surry Hills, NSW.

Kroeber, A. L. & Kluckhohn. C. (1952). Culture: a critical review of concepts and definitions. *Papers of the Peabody Museum*, Vol. 47, No. 1. Harvard University Press, Cambridge, MA.

Kuper, A. (2000). *Culture: The Anthropologist's Account*. Harvard University Press, Cambridge, MA.

Lees, W. M. (1902). *The Aboriginal Problem in Queensland, How It Is Being Dealt with: A Story of Life and Work under the New Acts*. The City Printing Works, Brisbane.

McDonald, G. & Lane, M. (2000). *Securing the Wet Tropics?* The Federation Press, Sydney.

Mathews, R. H. (1898). The group divisions and initiation ceremonies of the Barkunjee tribes. *Journal and Proceedings of the Royal Society of New South Wales* 32: 241–55.

Meekison, L. (2000). Indigenous presence in the Sydney Games. In *Indigenous Cultures in an Interconnected World*, Smith, C. and Ward, G. K. (eds). Allen and Unwin, Sydney, pp. 109–29.

Merry, S. E. (2001). Changing rights, changing culture. In *Culture and Rights: Anthropological Perspectives*, Cowan, J. K., Dembour, M.-B. & Wilson, R. A. (eds). Cambridge University Press, Cambridge, pp. 31–56.

Meston, A. (1889). Report by A. Meston on the government scientific expedition to the Bellenden-Ker Range (Wooroonooran), North Queensland. *Votes and Proceedings of the Legislative Assembly*, 3, Brisbane.

Meston, A. (1895). *Queensland Aboriginals: Proposed System for Their Improvement and Preservation*. Edmund Gregory, Government Printer, Brisbane.

Meston, A. (1904). Report by Mr A. Meston on expedition to the Bellenden-Ker Range. *Votes and Proceedings of the Legislative Assembly*, Brisbane.

Parry-Okeden, W. E. (1897). Report on the North Queensland Aborigines and the Native Police. *Votes and Proceedings of the Legislative Assembly*, 2, Brisbane.

Patz, E. (1991). Djabugay. In *Handbook of Australian Languages: Volume 4*, Dixon, R.M.W. & Blake, B. J. (eds). Oxford University Press, Oxford, pp. 245–347.

Patz, E. (2002). *A Grammar of the Kuku Yalanji Language of North Queensland*. Pacific Linguistics, Research School of Pacific and Asian Studies, The Australian National University, Canberra.

Rainforest Aboriginal Network (1993). *Julayinbul Conference Proceedings. Aboriginal Intellectual and Cultural Property: Definitions, Ownership and Strategies for Protection in the Wet Tropics World Heritage Area*. Rainforest Aboriginal Network, Cairns.

Review Steering Committee (1998). *Which Way Our Cultural Survival? The Review of Aboriginal Involvement in the Management of the Wet Tropics World Heritage Area*. Review Steering Committee, Cairns.

Roth, W. E. (1900). *Scientific Report to the Under Secretary with an Index on the Natives of the (Lower) Tully River*. Unpublished report. Queensland Government, Brisbane.

Roth, W. E. (1984). *The Queensland Aborigines*. Facsimile edition. Hesperian Press, Carlisle.

Sahlins, M. (2000). *Culture in Practice: Selected Essays*. Zone Books, New York.

Savage, P. (1992). *Christie Palmerston, Explorer*. James Cook University Press, Townsville.

Tindale, N. B. (1938–9). Harvard and Adelaide Universities Anthropological Expedition, Australia. Field Notes. Journal 1938, Volume 1. Unpublished manuscript. South Australian Museum, Adelaide.

Tindale, N. B. (1974). *Aboriginal Tribes of Australia: Their Terrain, Environmental Controls, Distribution, Limits, and Proper Names*. University of California Press, Berkeley.

Tindale, N. B. & Birdsell, J. (1941). Results of the Harvard–Adelaide Universities Anthropological Expedition,

1938–39. Tasmanoid tribes in North Queensland. *Records of the South Australian Museum* **7**(1): 1–9.

Tsing, A. L. (2003). Cultivating the wild: honey-hunting and forest management in southeast Kalimantan. In *Culture and the Question of Rights: Forests, Coasts, and Seas in Southeast Asia*, Zerner, C. (ed.). Duke University Press, Durham, NC, pp. 24–56.

Tylor, E. B. (1871). *Primitive Culture*. John Murray, London.

Wagner, R. (1975). *The Invention of Culture*. University of Chicago Press, Chicago.

Wet Tropics Management Authority (WTMA) (2005). *The Wet Tropics of Queensland World Heritage Area Regional Agreement*. Wet Tropics Management Authority, Cairns.

Williams, R. (1976). *Keywords: A Vocabulary of Culture and Society*. Fontana Croom Helm, London.

Zerner, C. (ed.) (2003). *Culture and the Question of Rights: Forests, Coasts, and Seas in Southeast Asia*. Duke University Press, Durham, NC.

5 EUROPEAN SETTLEMENT AND ITS IMPACT ON THE WET TROPICS REGION

David J. Turton

Department of Humanities, School of Arts and Social Sciences, James Cook University, Cairns, Queensland, Australia

Introduction

The Wet Tropics has been the site of much environmental change following European settlement from the 1870s onwards. Many industries were established in the area, all with various consequences for the environment. These included the mining and dairy industries, sugarcane farming, as well as economic forays into coffee, coconuts, pineapples, mangoes, tea and maize ventures, *inter alia*. Identifying appropriate crops for this region led to much experimentation, not necessarily commercial successes. However, it was the discovery of gold west of the area to be known as the Wet Tropics that provided the initial impetus for European settlement. Gold was discovered on the Palmer River in 1873 (Kirkman 1980), then in the Hodgkinson River area in 1876 (Kirkman 1982). Both mining fields were inland from the coastal towns of Cooktown and Cairns respectively. Minor fields such as Mt Spec near Ingham were subsequently exploited for tin. As mining opportunities declined and Europeans sought more permanent development in Far North Queensland, infrastructure was built to ensure this, particularly roads and railways. In addition, early tourism brought about further environmental impacts to improve access to areas of interest to visitors and projected an image of the environment for marketing purposes.

Maize

Maize was introduced to Queensland in the 1820s (Gilmore 2002) and was the first crop to be planted in the Wet Tropics for several reasons. Maize was regarded as a poor man's crop and was appealing because it was easy to sow and required little input from farmers once the land was cleared (Fitzgerald 1982). In addition, it had the advantages of providing two crops a year (Wegner 1984) and could be sown among the burnt ashes of previously existing patches of rainforest (Birtles 1982). Preparing the land for and then cultivating maize was deemed unsuitable for Europeans and consequently the industry became monopolized by the Chinese in coastal areas, with a majority stake in the industry; for example, in the land around Atherton (May 1984). Maize was, however, susceptible to a number of diseases, especially Cob Rob (*Diplodia*). On the Atherton Tableland weed infestation caused further problems, as this necessitated intensive cultivation to create efficient growing conditions for the crop (Gilmore 2002). The humid conditions on the Atherton Tableland also made the maize grown there vulnerable after harvesting, exposing it to mould, weevil infestation and rotting (Gilmore 2002). Maize was also the first crop grown in the southern half of the present Wet Tropics in the Herbert River district in the 1870s

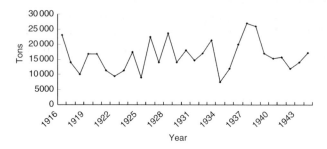

Figure 5.1 Production of maize on the Atherton Tablelands (XE 'maize') (tons), 1916–45. *Source*: Gilmore (2005: 120).

(Wegner 1984). In both these areas, however, the humid climate interfered with the storage of the corn, so expensive drying arrangements were necessitated (Wegner 1984). However, by 1895, maize had become established and by 1899 the area under maize cultivation had increased from 1590 to 2756 acres (Gilmore 2002) (see Figure 5.1 for tons produced). It would seem that maize, at least until the turn of the twentieth century, was virtually the only large-scale agricultural venture on the Atherton Tableland (May 1984). Ultimately, maize proved to be a highly successful crop in the initial years of European settlement in the region.

Timber

The potential of the timber resources of the region was first discussed in 1874 within a report by George Dalrymple to the Queensland government, following the completion of a government-sponsored expedition between Cardwell and the Endeavour River (Bolton 1972). Initially, clearing was centred on the Johnstone and Tully Rivers. The Palmer River gold rush brought attention to the land around Cooktown and interest in the Cairns hinterland, Atherton Tableland, Mossman and Daintree Rivers soon grew (Frawley 1983). In 1881, a timber rush took place after the founding of Herberton on the Atherton Tableland, red cedar being exhausted from the coastal rivers by this time (Frawley 1983). By the end of that decade, several land openings took place as farming came to assume importance in the face of declining gold and tin fields (Frawley 1983). The 1890s were a time of depression for the Queensland economy and this, combined with transport difficulties, ensured that clearing was limited to some extent (Frawley 1983). Weed infestation followed the clearing of the land for agricultural purposes and was a grave

concern to farmers at this time. The industry recovered, however, with the building of railways and by 1906 the most extensive clearing was concentrated around Atherton (Frawley 1983). Following the separation of Queensland from New South Wales in 1859, land development in Queensland followed a policy of closer settlement to fill the lands of the newly founded state (Frawley 1983).

In the Depression that characterized the late 1920s and early 1930s, there were calls from various quarters for the Forestry Department to permit relief work in the form of timber cutting, a suggestion that was rejected (Frawley 1983). The retention of remaining forests became important to the Queensland government from the 1920s onwards. This preservation was not without a practical purpose for future use, however (Frawley 1983). The Second World War impacted on both the ability of the Forestry Department to supervise logging in the future Wet Tropics area and their ability to control the numbers of skilled timber workers. During this time it was likely that the theft of timber from National Parks did take place (Frawley 1983). From the 1930s it became obvious that native softwoods would not be sustainable indefinitely, and in response the Forestry Department initiated quotas upon these species from 1948 (Frawley 1983). In the 1960s, clearing forest for agricultural purposes was stopped, with remaining rainforest largely confined to the residual lands between the clearings on the plateaux and the lowlands of the Great Dividing Range (Frawley 1983). At this time, the motivation behind the conversion of the rainforest was almost wholly utilitarian in nature (Frawley 1983). It was during the 1970s that the importance of managing and maintaining rainforest came to national and international prominence (Frawley 1983). Ultimately this increased level of concern and interest in the rainforest of the region became the catalyst for the declaration of the present Wet Tropics of Queensland World Heritage Area (WHA) in 1988 (see Valentine & Hill, Chapter 6, this volume).

Sugar

Following the initial push for maize and timber industries, sugar emerged as Far North Queensland's principal crop along the coast within a few years of settlement, the region's winds, heavy rainfall and warm climate being highly suitable for cultivating cane. Sugar was first planted within the Wet Tropics area in 1866, at the Bellendeen Plains plantation (north of present-day Cardwell at the Murray River) by John Ewen Davidson (Griggs 2000). During the 1860s, farmers were initially forced to diversify into alternative crops to expand their market base. In the Herbert River district alone by 1875, crops as varied as passion fruit, granadillas, custard apples, cherimoyas, pawpaws and mulberries were being trialled; few, however, advanced to the status of a commercial crop (Wegner 1984). Plantations were generally situated along rivers, which secured access for the export of their produce in an area that was often devoid of transport services. The rivers also supplied the alluvial soils with a regular water supply and provided the steam necessary to operate the machinery within the sugar mills (Griggs 2000). As soil fertility was gauged by growers at this time to be proportionate to the thickness of the vegetation, it is not surprising that river frontages were seen as the most fertile sites upon which to grow cane (Griggs 2004) and that many of these were prone to flooding. Although prudent cane growers often sought the rewards that proximity to a river could provide to a plantation, this did not mean that rivers were the sole problem they had to surmount. Other difficulties, such as high rainfall, scrub-clearing difficulties and long grass, were common problems for all planters entering the Wet Tropics for the first time.

Plantations established in the Hinchinbrook Shire area during the 1870s lacked technological sophistication and despite the fact that vast areas were taken up for planting, only a small percentage of this land was cultivated. However, this did not mean the land was useless, and on the Hambledon Plantation (near Innisfail) pineapples, oranges and lemons were grown on the marginal land (Griggs 2000) (see Figure 5.2). In the first year, sugar was often unproductive. Following the usual clearing of the landscape, large stumps were left in the ground to rot, with cane being planted by hoe around these. Encroachment upon the surrounding environment was also limited by the fact that the chipping of weeds, the trashing of cane and all harvesting was completed by hand, making the overall acreage set aside for sugar commensurate with the available labour source (Wegner 1984). By this time, there was also a shift in the structure of sugar plantations away from large, single sugarcane estates in the hands of wealthy individuals (Griggs 2000), to a

system of small farmers near centralized mills (Wegner 1984). All of this was accompanied by large-scale investment (Griggs 2000). It was this investment and capital that allowed acreages of cane to be extended, thereby impinging on the environment even further. The use of fertilizer was infrequent for many years and initially cane growers ignored the health of the soil to the point where its productivity declined before any counter-measures to restore fertility were made (Griggs 2004) (see Table 5.1). Ultimately it was the advent of the Central Mills system that ensured the use of fertilizers from the 1890s, the Mulgrave mill in Cairns being one example of supplying lime and mill products to encourage the use of fertilizer by the canegrowers (Griggs 2004). Mechanization of the sugar industry in the Wet Tropics also brought about great changes. Tractors were being adopted in the region from the 1920s onwards, further increasing the acreage that could be planted with sugarcane (see Table 5.2), as well as reducing the need for a labour force and also the time needed for cultivation (Wegner 1984). All of this increased the severity and extent of erosion. Some of the other practices adopted by cane farmers were also environmentally damaging, such as the burning of

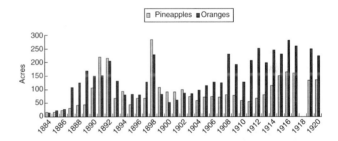

Figure 5.2 Acreage under pineapples and oranges in Cairns by the Chinese, 1884–1920. *Source:* May (1984: 251).

Table 5.1 Estimated gross value of production of selected base industries, far north Queensland, 1957 and 2002/3

	1957		2002/3	
	$m	*% of total*	*$m*	*% of total*
Sugarcane	33.9	42.2	241	5.6
Other agricultural/pastoral	21.6	26.9	639	14.9
Total agriculture	55.5	69.1	880	20.6

Source: Cairns Business Research Manual (2005: 21).

Table 5.2 Acreages of sugar plantations in the Wet Tropics established in 1879–85

Name of estate	District	Original owners	Size (thousand acres in 1888)	Area (thousand acres under cane in 1888)
Vilele	Bloomfield River	Bloomfield River Sugar Co.	6.0	5.8
Brie Brie	Port Douglas	Harriet Parker	1.2	2.0
Hop Wah	Cairns	Hop Wah Co.	0.6	3.0
Hambledon	Cairns	Swallow and Ariell Ltd	6.0	9.0
Pyramid	Cairns	J.B. Loridan and Co.	3.4	6.7
Mourilyan	Johnstone River	Mourilyan Sugar Co.	5.0	8.5
Innishowen	Johnstone River	Queensland Sugar Co.	3.8	6.0
Goondi	Johnstone River	Colonial Sugar Refining Co.	12.5	23.0
Innisfail	Johnstone River	T.H. Fitzgerald	2.9	4.0
Ripple Creek	Herbert River	Wood Bros and R.M. Boyd	1.6	7.2
Victoria	Herbert River	Colonial Sugar Refining Co.	20.0	2.9
Hamleigh	Herbert River	Hamleigh Sugar Co.	4.7	5.0

Source: Griggs (2000: 646).

cane: by 1907, 60% of all cane grown in the Mossman district (north of Cairns) was being burnt (Kerr 1979). Following the expansion of the industry during the 1920s, other environmental problems began to emerge. One of the most harmful agents introduced into the Wet Tropics by the sugar industry was the cane-toad (*Bufo marinus*) in 1935, in response to the unwelcome attentions of the indigenous grey-beetle (*Lepidiota caudata*). From the 1950s to the present, the industry has increased in value (Table 5.1). It should not be considered that the region was concerned solely with sugar, as several other minor crops emerged at this time in quick succession.

Coffee

The establishment of the Kamerunga State Nursery (near Cairns) in 1889 as an experimental station for agriculture enhanced the development of such varied products as pasture grasses and rubber for use in tropical areas (Bolton 1972). Although coffee growing was trialled around Cairns prior to 1890, it wasn't until the Kamerunga State Nursery had successfully planted Arabian and Liberian varieties of coffee that experimentation with the crop began to be widely attempted (Bolton 1972). In addition to Cairns, plantations were established at Russell River, Cooktown, Kuranda and Atherton. Coffee was usually planted in small clearings surrounded by rainforest, and this shielded the plants from wind and frost. Coffee plantations relied on cheap Aboriginal and Kanaka labour to function; thus when the White Australia Policy was introduced in 1901 the importation of Kanakas to Australia for assistance with the harvesting of coffee beans was forbidden. This disaster combined with severe frosts on the Tableland and cyclones on the coast in 1900–1 combined to force most coffee growers to switch to alternative crops (Tranter 1998). There were exceptions to this trend. For example, Alfred Street owned a plantation, known as Fernhill (Jones 1976), near Kuranda after this period and made a commercial success of the product, labelled Barron Falls Coffee. Coffee enjoys a niche market on the Atherton Tablelands in the present era and was valued at $3.78 million in 2003 (BRM Partnership 2005).

Tea

The foundations of the Far North Queensland tea industry were laid on April Fool's Day 1882 with the arrival of the four Cutten brothers at Bingil Bay (near Cardwell). Seeking suitable land upon which to establish themselves, James, Leonard, Sidney and Herbert Cutten settled upon the location of their property (which they called Bicton) and then set to work clearing the surrounding area (Taylor 1982). Once the forest was cleared, bananas and pineapples (generally rapid-growing crops) took its place, aided in no small way through the efforts of local Aborigines. In addition to this, various tropical seeds were planted, such as tea, oranges, jack fruit, coffee, chicory, coconuts and mangoes (Taylor 1982). However, the tea was not planted until 1885, when it was situated on the northern end of the selection. The tea nursery comprised possibly two acres and was protected from the sea by a small hill. Remnants of the plantation can still be found, with some of the surviving tea trees in the forest serving as a testament to natural selection. The Cutten tea plantation declined following the horrendous cyclone of 1918, when it was found that the now elderly siblings could no longer maintain their estate of Bingil Bay (Taylor 1982). In spite of the problems, inspired by their example and the possibilities for profit, modern tea manufacturing takes place in the present era in Far North Queensland.

Cotton

The first attempt to grow cotton in the Wet Tropics was made by the Chinese businessman Andrew Leon at the Hop Wah plantation, south of Cairns. Unfortunately this crop was not a success (Jones 1976). In the aftermath of this failure, Leon and his syndicate decided to move into the sugar market (May 1984), leaving the cotton unattended to the point where it self-seeded and became wild. The environmental consequences were dramatic and harmful as this cotton acquired a resistance to diseases and pests and became acclimatized (Jones 1976).

Dairy farming

The potential for the Atherton Tableland as a dairy-producing area was recognized early on when the unique climate of the region was discovered. In 1888 the idea of establishing a weekly market for the sale of

European dairy produce at Herberton was first contemplated. This was done in part as a protective economic measure against more successful Chinese vegetable growers in the area. However, the logistics of such a business proposition proved to be uneconomical and did not warrant further discussion (Statham 1998). Despite the obstacles to large-scale cooperative marketing, by 1889 several farmers had pursued dairy farming, often as side-line to other crops (Gilmore 2005). Notwithstanding the productivity of the land upon which farmers placed so much faith for their future, the first farms were not managed in a particularly efficient manner. Once the forest was removed and the pastures were stocked, the early dairy farmers entrusted their cattle to the less than constant food source of the indigenous grasses, thus causing milk production to be erratic at times (Gilmore 2005). Other problems emerged with the realization that the soil's fertility was not commensurate with the thick rainforest. With the removal of decomposing forest matter when the landscape was cleared, soil fertility decreased and the landscape was impacted upon by infestations of milkweed, wild tobacco and other weeds. Since these pests had no place in the diet of cattle, their removal had to be implemented before suitable pastures could be established in their place, such as paspalum grass (*Paspalum dilatatum*) and Rhodes grass (*Chloris gayana*) for use in winter (Gilmore 2005). These actions then introduced further exotic species into the ecology of the region. Unfortunately land degradation set in from the 1930s with rapid erosion following closely after the destruction of the rainforest. This issue was complicated further with the damage caused to pastures by white grubs (*Scarabeidae*) and grass grubs (*Hepialidae*) (see Goosem, Chapter 24, this volume). The grub species responsible (*Lepidiota caudata*) became an enormous problem. The solution proposed by the White Grub Committee, brought together in August 1935 to examine the problem, recommended two actions: (i) the introduction of the Giant American Toad (*Bufo marinus*); and (ii) allowing pigs free range over pasturelands during the flight season of the beetle to reduce their numbers, thus adding to existing feral pig populations present since European settlement (Statham 1998). The introduction of these species continues to play a role in the destruction of native flora and fauna in the Wet Tropics area (see Congdon, Chapter 25, this volume).

Tourism

While many people might consider the development of tourism in the Wet Tropics area as relatively recent, this is not so. The region served several interests of the prospective tourist early on, whether in search of health or mountain resorts (Thorp 2005). By the 1890s the Barron Falls was one of the most popular tourist sites in Queensland. More specifically, waterfalls in the Wet Tropics region were seen as tourist attractions prior to 1900. Needless to say, waterfalls alone do not woo tourists to a locality; instead it is the combined scenery that sparks a tourist's interest. One of the most popular tourist-orientated waterfalls was Millaa Millaa Falls, outside the town of the same name. As a result of delayed road and rail connections and links to a quarry in the 1920s, the development of tourism there was slowed and the approaches to Millaa Millaa Falls impacted upon substantially. From 1922 to 1927 rock was removed from a quarry adjacent to the Falls and used on the construction of the Palmerston Highway. The Millaa Millaa Progress Association furiously wrote letters against these activities to prevent the dynamiting of the Falls itself. It would also appear that the Falls was included in a day-trip by car that took in such sites as Lake Eacham and Malanda Falls. However, a disused quarry would not have aided the aesthetic value of the Falls in the eyes of tourists. However, it was still used to some extent, in spite of the fact that access to the Falls was via a road down to the former quarry and the fact that safety fences were not constructed until the late 1950s. This indicates that the Millaa Millaa Falls did not assume the role of a tourist attraction completely until that time (Thorp 2005).

Mining

The mining industry was confined to a series of small fields at the Mulgrave River, Mount Bartle Frere (where gold was found) and Mount Spec (mined for tin). Machinery was acquired for the extraction of the tin and the Mount Spec Tin Mines Company was founded in 1892. Ultimately, however, transport and mining problems arose in the rainforest and, combined with the low price of tin, forced leases to be abandoned. The field revived somewhat in 1901 when the company (relocated to Charters Towers) took up lodes once

again and built a battery for the site. However, this mine venture also collapsed due to its inaccessibility and abundance of water, which caused pumping machines to be overwhelmed (Wegner 1984). Mount Bartle Frere's gold deposits were discovered in 1936–7 and it became the last and least successful discovery of the region (De Havelland 1989). Other fields included Jordan Creek and Russel River (De Havelland 1989). Pollution from mining affected the Herbert River and occurred during the 1940s, in the form of a colloidal clay that was caused through tin dredging in the river's tributaries (specifically the Return, Battle, Smith's and Nettle Creeks) from mining operations at Mount Garnet. The problem was first identified in 1944, with the pollution proving difficult to remove for many years afterwards (Wegner 1984).

Railways

Railways often meant the difference between a ghost town or thriving centre for many embryonic Far North Queensland towns. The Cairns–Kuranda railway line was first considered during the early 1880s to link up with the mining centre of Herberton. The greatest obstacle to this ambition was the coastal range between Mourilyan Harbour and Port Douglas. Finally, a route was established from Cairns to the base of the coastal range, up the escarpment of the coastal range via the Barron gorge. Surveyors on the site had to be slung over the mountainside on ropes to put in their markers (Smith 1996). Explosives usage was mediated somewhat through the unstable nature of the rock and the drop down the side of the Barron gorge. The line opened up as far as Myola in 1891 after five years of construction, eventually reaching Mareeba in 1893 (Kerr 1990). In 1898 a proposal was brought forth by George Phillips to connect Innisfail (then called Geraldton) and Herberton by rail. Construction began in October 1899, with the link being completed towards the end of 1900 (Kerr 1990).

Roads

The Gillies, Cook, Kuranda and Palmerston highways all traverse the present Wet Tropics area and their construction caused strong environmental impacts.

The Gillies Highway was constructed after much lobbying from the Eacham Shire Council and local MP (Member of Parliament) W. Gillies for better transport facilities, with work starting in 1923. The project was given a very small budget and consequently 613 curves in a 19 km stretch of road had to be created to economize on the construction. When it was finished in 1926 it became a one-lane road with specified hours for driving up and alternate hours for driving down, with toll-keepers at either end to ensure effective enforcement (Tranter & Tranter 1999).

Part of the future Palmerston Highway was originally a track carved by the North Queensland explorer Christie Palmerston during his employment with the Queensland government to locate a suitable route for a railway line between Innisfail and Herberton in 1882 (Savage 1989). In 1932, Rollo Gallop and Douglas MacArthur were entrusted with the task of constructing a road between Innisfail and Millaa Millaa as part of a relief project for the unemployed during the Depression. The clearing of the scrub was accomplished by hand and then a pegging team moved in to set out the lines for the cuts and fills for the earthworks following in their path. These earthworks consisted of the use of groups of men with picks and shovels gradually moulding the road into existence, as powered machinery was not applicable to the conditions and climate. Gelignite too was utilized in the clearing stage of construction, often causing great harm to tree-kangaroos (*dendrolagus lumholtzi*) in the vicinity of a blast. By 1934 the road was completed and followed part of Palmerston's original route (Gallop 1979).

An intention to establish a main road between Cairns and Port Douglas was first expressed in 1921 by the Main Roads Board of Queensland, but due to various arguments over cost, surveys for the road did not begin until 1928, the project being named the Cook Highway on 25 October 1930 (Richards 1996). Construction of the road began the following year and was lauded as a potential tourist route within a short period. It was also deemed one of the most important roads in Queensland for the promise it offered for removing the isolation of the Daintree. The Cook Highway was officially opened on 17 December 1933 at Hartley's Creek by the Queensland Minister for Works, H. Bruce (Richards 1996).

The Kuranda Highway started life as an abominable bullock track for the passage of goods between the township of Smithfield (north of Cairns) and the Hodgkinson mining field in the west after much surveying work had been done to locate the best route for supplies between there and Cairns (established in 1876). The population shifted significantly away from Cairns to Smithfield with the announcement that a road would be constructed in 1877 following suggestions from the Engineer for Northern Roads, Archibald Macmillan. The population of Cairns stabilized somewhat during the early 1890s (see Figure 5.3 for details). Consequently, the town of Smithfield sprang up next to this road, providing services to the bullock teams and packers. Unfortunately for Smithfield and Cairns, when yet another route to the goldfields to the west was discovered in late 1877 by the explorer Christie Palmerston near the future site of Port Douglas, both centres declined in size. Smithfield became a ghost town and Cairns survived long enough due to its harbour facilities to win the economic boon of a railway in 1886. However, the future Kuranda Road faded into obscurity until an election promise was made to provide the people of Kuranda with access to Cairns in 1938. Survey work began in 1939 with the construction starting on 18 January 1940. The methods used in the building of this road called for the services of four bulldozers. These machines increased the road's progress and were among the first of their type in North Queensland used for this purpose. Each stage of construction in the rainforest above Cairns had to be bored and blasted with gelignite before the men and

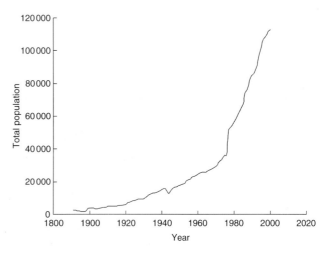

Figure 5.3 Population figures for Cairns 1891–2000. *Source:* Cairns Business Research Manual (2005: 500–1).

bulldozers could proceed. It was also decided between the engineer, O. Anderson, and foreman, Henry Ross, to construct the highway to a two-lane standard, unusual for the budgetary constraints imposed in wartime. The Barron Falls Hydro-electricity Board also assisted with construction, providing its air-compressing plant and the power necessary to operate the 30 jackhammers in use at the job site. In July 1941, problems emerged when the lower and upper sections of the road started to slide on to each other, forcing the hurried erection of a retaining wall to prevent slippage before the wet season set in. The Kuranda highway was not officially opened because it had importance as a defence road with war looming against Japan and was surrounded by secrecy. However, it would appear from the change in language in the local press from future to present tense that it was completed in early December 1941. The road subsequently became the main thoroughfare through which Australian and US troops travelled to the Atherton tablelands for jungle-training exercises prior to operations against the Japanese. Indeed, the road was of such importance to the war effort that mines were placed on the road during the Battle of the Coral Sea, as part of a wider strategy of denial that called for the obliteration of communication networks if an invasion appeared imminent from a Japanese victory. Thus, it could be argued that further ecological damage to this part of the Wet Tropics was prevented through the victory of the Allied military forces. In the post-war era the Kuranda highway was transformed into a popular tourist drive, a future hinted at several years earlier in the annual reports of the Main Roads Commission (Turton 2006).

Hydro-electricity

It was the Queensland Railway Department that first examined the possibility of harnessing the Barron Falls for hydro-electricity for use on an electrified Cairns–Herberton Railway (Blackhurst 1992). The Railway Commissioner ordered a survey that indicated 2400 horsepower would be generated, but no further action was taken by the Department (Pearson 2000). The next suggestion for hydro-electric power came from Chillagoe Proprietary Limited, in their application for land adjacent to the Barron Falls for a reduction works

for the treatment of ores from the Chillagoe Copper Mines. As a result of the community's protests against this development near what was then a popular tourist destination, the plant was moved to near Mareeba (Pearson 2000). In 1906 William Corin was commissioned to investigate the Barron Falls as a site for the utilization of electricity and it was partly in response to his report that the Cairns Hydro-Electric Power Investigation was formed in 1923 (Pearson 2001). This organizational body called upon Mr Corin once more to undertake an examination of the potential and expected costs of developing the Barron Falls for hydro-electrical purposes. Further engineering surveys were made over the following years, but finally, the first meeting of the body overseeing the construction of the Barron Falls Hydro-Electric scheme was convened in May 1930 (Pearson 2001). Construction on the project started on 20 August 1932 (Pearson 2001).

During the construction phase, a channel was cut through the rock on the northern bank of the Barron River to divert water flow. The threat of landslides was an ever-present one as workmen cut away at the toe of embankments, causing slips and erosion. Further difficulties emerged when flooding struck the Barron River in 1933, washing away trestle bridging and some of the piping used in the compression of air for tunnelling purposes (Pearson 2001). Despite all these problems the Barron Falls Hydro-Electric scheme was officially opened on 20 November 1935 in the presence of the Queensland Governor and with much celebration: 'a milestone in the progress of Queensland' (Pearson 2001: 94), with a high environmental impact (Pearson 2001). Proposals for the other major hydro-scheme in the Wet Tropics, at the Tully Falls, emerged during the 1940s to meet increasing demand (and to save the Queensland government up to 225 000 tons in coal per year), but did not become a reality until the Kareeya Power Station at Tully Falls was put into operation with the completed Koombooloomba Dam during 1957–8 (Pound 1995).

Summary

It would seem that European settlement in and around the present Wet Tropics area had enormous environmental consequences. Clearing the land initially for the exploitation of timber and then later for agricultural reasons removed nutrients from the soil and destroyed indigenous habitats. These processes were further expanded by the introduction of pigs and cane toads, among numerous other animals. Dairy farmers brought new species of grass to feed their cattle, and various agricultural experiments that went awry subsequently grew wild (tea and cotton, for example), impacting on the ecology of the Wet Tropics still further. Wildlife prospecting for opportunities afforded by these agricultural crops was swiftly neutralized by the use of pesticides, with the sugar industry being one example of this practice.

Mining operations damaged river systems, while the vast clearing performed for the provision of roads and railway lines increased general erosion processes. The use of hydro-electric power altered water courses and necessitated the further clearance of land. European settlement and associated actions damaged the ecological integrity (and in some cases, existence) of the region in and around the present Wet Tropics area, a situation that only began to be remedied on a large scale in land management strategies adopted for the preservation of specific industries in the post-war era.

References

Birtles, T. (1982). Trees to burn: settlement in the Atherton–Evelyn rainforest, 1880–1900. *North Australia Research Bulletin No. 8*, North Australia Research Unit, Australian National University, Darwin.

Blackhurst, E. (1992). Telegraphic address hydro: a history of hydro-electricity in far north Queensland. MA Thesis, James Cook University.

Bolton, G. (1972). *A Thousand Miles Away: A History of North Queensland to 1920*. Australian National University Press, Canberra.

Business Research Manual (BRM) Partnership (2005). *Cairns 2020–2050*. Business Research Manual, Cairns.

De Havelland, D. (1989). *Gold and Ghosts: A Prospector's Guide to Metal Detecting in Australia. Volume 4: Northern and North-western Districts of Queensland*. Hesperian Press, Carlesle, Victoria.

Fitzgerald, R. (1982). *From the Dreaming to 1915: A History of Queensland*. University of Queensland Press, St Lucia.

Frawley, K. (1983). Forest and land management in north-east Queensland: 1859–1960. PhD Thesis, Australian National University.

Gallop, A. (1979). *The Bush Engineer: Early Road Construction and Development in Northern and Western Queensland*. Self-published, Cairns.

Gilmore, M. (2002). Faith, hope and charity: the history of the Atherton Tableland maize industry, 1895–1945. BA (Hons) thesis, James Cook University.

Gilmore, M. (2005). Kill, cure, or strangle: the history of government intervention in three key agricultural industries on the Atherton Tablelands, 1895–2005. PhD thesis, James Cook University.

Griggs, P. (2000). Sugar plantations in Queensland, 1864–1912: origins, characteristics, distribution and decline. *Agricultural History* **74**(1): 3.

Griggs, P. (2004). Improving agricultural practices: science and the Australian sugarcane grower, 1864–1915. *Agricultural History* **78**(1): 1.

Jones, D. (1976). *Trinity Phoenix: A History of Cairns and District*. Cairns Centenary Committee, Cairns.

Kerr, J. (1979). *Northern Outpost*. Mossman Central Mill Company Ltd, Mossman.

Kerr, J. (1990). *Triumph of the Narrow Gauge: A History of Queensland Railways*. Australian Railway Historical Association, Brisbane.

Kirkman, N. (1980). The Palmer Goldfield. In *Readings in North Queensland Mining History Volume I*, Kennedy, K. (ed.). James Cook University Press, Townsville.

Kirkman, N. (1982). The Hodgkinson Goldfield. In *Readings in North Queensland Mining History Volume II*, Kennedy, K. (ed.). James Cook University Press, Townsville.

May, C. (1984). *Topsawyers: The Chinese in Cairns, 1870–1920*. James Cook University Press, Townsville.

Pearson, L. (2000). *The Cairns Electric Authority. History of the Early Electricity Supply in the Cairns Region. Part 1*. Self-published, Cairns.

Pearson, L. (2001). *The Hydro: The Barron Falls Hydro-Electricity Board. History of the Early Electricity Supply in the Cairns Region. Part 2*. Self-published, Cairns.

Pound, C. (1995). Engineering marvel or ecological disaster? Attitudes towards the damming of wilderness areas for hydro-electric schemes with particular reference to the Tully Millstream Hydro-Electric Scheme. BEc (Hons) thesis, James Cook University.

Richards, J. (1996) A 'Captain Cook' at the Coast Road: histories of the Cook Highway in north Queensland. BA (Hons) thesis, Griffith University.

Savage, P. (1989). *Christie Palmerston, Explorer*. James Cook University Press, Townsville.

Smith, A. (1996). Money will force a railway anywhere: construction of the Cairns–Kuranda railway line, 1886–1891. In *Lectures on North Queensland History*, Dalton, B. (ed.). James Cook University Press, Townsville.

Statham, A. (1998). *Cows in the Vine Scrub: A History of Dairying on the Atherton Tableland.* Malanda Dairy Foods, Malanda.

Taylor, R. (1982). *The Lost Plantation: A History of the Australian Tea Industry.* G. K. Bolton, Cairns.

Thorp, J. (2005). The development of the tourism cultural landscape of the Cairns region, 1890–1970. PhD thesis, James Cook University.

Tranter, H. (1998). The coffee industry on the Tableland. *Eacham Historical Society Bulletin* no. 235.

Tranter, H. & Tranter, M. (1999). *A Bend Too Many: The Story of the Gillies Highway.* Eacham Historical Society, Atherton.

Turton, D. (2006). From dray track to eco-road: a history of the Kuranda Range road 1877–2005. Unpublished manuscript in progress.

Wegner, J. (1984). Hinchinbrook: the Hinchinbrook Shire Council, 1879–1979. MA thesis, James Cook University.

6 THE ESTABLISHMENT OF A WORLD HERITAGE AREA

Peter S. Valentine[1]* and Rosemary Hill[2]*

[1]School of Earth and Environmental Sciences, James Cook University, Townsville, Queensland, Australia
[2]School of Earth and Environmental Sciences, James Cook University, Carins, Queensland, Australia; and CSIRO Sustainable Ecosystems, Cairns, Queensland, Australia
*The author was participant of Cooperative Research Centre for Tropical Ecology and Management

Introduction

Tropical rainforest was often seen as a barrier to progress in northern Queensland and the descriptive terminology used most frequently reflected that (scrub, jungle; see Stork and Turton, Chapter 1, this volume). Clearing the scrub to develop agriculture was a principal task of the early European settlers (see Turton, Chapter 5, this volume), as was the continuing attempt to remove the timber resources. The notion that the rainforest might need protecting from the impacts of human activity was quite weird until very recent times. In a historical review of attitudes towards rainforest in northern Queensland, Valentine (1980) documented numerous antagonistic and utilitarian views from explorers, settlers and government agencies. Indigenous Australians appreciated the qualities of tropical rainforests, which were integral to their culture, economy and society. The wider community has also seen the development of a much more positive view about tropical rainforests over time. One of the first to advocate its protection was John Büsst, a local Bingil Bay resident who is best known for his work to save the Great Barrier Reef. He also formed the Rainforest Protection Society in 1962 and was at that time thought somewhat quirky for these ideas. His perspicacity is now acknowledged.

In this chapter we outline the social and political processes that led to the formal protection of the rainforests, its listing as a World Heritage Area (WHA) and the development of a modern management framework. One of the most interesting elements of the story is that the outstanding universal value of the Wet Tropics was never in doubt. Nevertheless, societal attitudes towards management of the rainforests changed dramatically during the period from 1980 to 1988 when listing occurred (Cassells & Valentine 1988), moving rapidly through the alpha, omega and lambda phases of McDonald and Lane's (2000) model: the development of a crisis, search for alternatives and reorganization of policies (Figure 6.1). The kappa phase, implementation of the new policy to give primacy to the outstanding heritage values, followed and was essentially finalized with the adoption of the Wet Tropics Management Plan in 1998. The Australia-wide community-based environment movement, a loose collection of individuals and organizations with a common theme of environmental protection, played an important role in the alpha, omega and lamba phases, with the focus shifting to state and federal governments for the implementation phase. Community involvement in all aspects remains vital to management of the Wet Tropics of Queensland World Heritage Area.

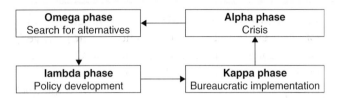

Figure 6.1 Cycles of change. *Source*: after McDonald and Lane (2000).

Crisis phase

As the wider community became more aware of the incredible biodiversity within the tropical rainforests, questions began to be asked about the way forests were being exploited. The main target of concern was the logging of high quality timbers undertaken under the control and management of the Queensland Forestry Department. The usual process was to allocate crown quotas based on quantities of the timber to be extracted. Thus, for most of the 50 years up to 1980 the Wet Topics had been logged at a rate of around 200 000 m³ of timber per annum. By 1980 it was already clear to forestry professionals that the allowable cut was far in excess of sustainable yields (Stocker *et al.* 1977). A primary reason for this situation was that Crown logging was still based on initial cuts (old growth forests). However, it was clear that the future of rainforest logging would be based on 40–50-year cycles of previously cut forests, a situation that was very different. It was considered that at a certain point there would be a significant reduction in cut and a much reduced (and perhaps sustainable) yield would be implemented. In 1981 the Crown quota of timber from northern Queensland rainforests was reduced to 152 000 m³ but even this figure could not be achieved by the logging companies (Department of Forestry Queensland 1981). In 1983 Crown quotas were set at 130 000 m³ but only 112 000 m³ could be found. Although the crown quota remained at 130 000 m³, in 1985 the industry harvested only 88 000 m³). By 1987 the quota was reduced to 60 000 m³ and still could not be met when the Commonwealth government introduced a unilateral decision to ban logging. In the decade leading up to World Heritage listing, the industry based on Crown timber had collapsed from 200 000 m³ per year to less than 60 000 m³. Associated with this demise was the loss of perhaps two-thirds of the workforce

(from about 2000 in 1981 to around 760 in 1987) and a reduction in logging contractors associated with the reduced yield.

The community concerns about lack of sustainability of rainforest logging in New South Wales (NSW) had already led to dramatic confrontations between conservationists and loggers at Terania Creek in 1979. Here, hundreds were arrested in protests that brought the issue of rainforest protection to the national limelight and resulted in a moratorium on logging by NSW Premier Neville Wran. The idea of direct action as an effective means of highlighting the crisis quickly spread to Far North Queensland. In 1980, the Cape York Conservation Council (CYCC), supported by local groups including the Cape Tribulation Community Council (CTCC), had hosted more than 500 people to the Second World Wilderness Congress in Cairns. The Congress was opened by Prime Minister Malcolm Fraser, and (among other resolutions) called on both state and federal governments to protect all remaining Australian rainforest as World Heritage (O'Grady 1980). Despite this, logging was gathering pace, particularly with the opening up of the virgin Mt Windsor Tableland area. Queensland Premier Joh Bjelke-Petersen's announcement of his government's intention to create the Cape Tribulation National Park, a 'living museum of plants and animals', did nothing to slow the logging (Hill & Grespan 1980: 4; Toyne 1994: 65).

The Australian Conservation Foundation (ACF) and the Queensland Conservation Council (QCC) launched the Rescue the Rainforest Campaign in 1981 by hosting a national/local meeting in Cairns in July (ACF 1981; *Cairns Post* 5 May 1981: 10). Papers on the role of direct action were circulated at the meeting and a small working group of people agreed to blockade logging operations on Mt Windsor Tableland later that year (McCabe 1981; Russell 1981). Cairns and Far North Environment Centre (CAFNEC) was incorporated as a regional umbrella organization for both new and established regional groups, including the North Queensland Land Council. An office was established in Cairns as a central coordination point for rainforest conservationists. ACF Vice-President Milo Dunphy, together with CAFNEC, developed a rainforest campaign that included a tour by national media, politicians and parliamentary committees, alternative policy generation and strategies for the proliferation of local media and letters to the editor (Dunphy 1981; Mosley 1982).

The Mt Windsor logging truck blockade took place on 11 November when 13 people were arrested, generating national media, but failing to attract a flood of supporters or any interest from the Queensland Premier, Bjelke-Petersen (Elvery 1981). However, the Qld Forestry Department's (QFD) claim that logging cycles of between 40 and 50 years were sustainable was exposed to intense media scrutiny and public contestation from that time. The QFD never conceded unsustainability despite the accelerating decline in available supply throughout the 1980s outlined above (Dansie 1980; Allworth *el al*. 1981; Department of Forestry Queensland 1981, 1983; Keto & Scott 1986; *Cairns Post* 16 October 1985). Mr James Blair, a *National Geographic* photographer who was nearing the end of two years of global travel for a special edition on rainforest, came to Australia especially to cover the protest; he had encountered no other protest action anywhere in the world despite a great deal of concern about rainforest destruction. Two further direct actions played a significant role in the crisis, the blockades during the construction of a road through the Cape Tribulation (now Daintree) National Park in 1983/4, and protests of the logging of Downey Creek in 1985 (Figure 6.2).

A commitment to blockade any construction of the Daintree road had been made by conservationists who organized the Mt Windsor direct action, coordinated through CAFNEC. ACF Rainforest Project Officer

David Allworth, based in Brisbane, stimulated formation of a new state-based group who became involved, the Rainforest Conservation Society of Queensland (RCSQ), formed in May 1982 simultaneously with related but independent branches in Townsville and Cairns (the Cairns branch became the Tropical Rainforest Society (TRS) in 1984). Local conservationists in Mossman, inspired by a film night presented by CAFNEC in October 1983, formed the Douglas Shire Wilderness Action Group (WAG), and decided to be centrally involved in the blockade, becoming the local mainstay for intense confrontations during December 1983 that brought national political focus (Douglas Shire WAG 1984). Nationwide networks ensured both donations and people flowed in to assist with the substantial costs borne by CAFNEC. Road works were halted in late December 1983 due to the wet season. Members of The Wilderness Society (TWS), the newly badged version of the Tasmanian Wilderness Society, fresh from the success of the Franklin River blockade, travelled north, met the local groups and subsequently decided to adopt the Daintree as their first campaign outside Tasmania (Washington 1984). TWS brought new skills in market research to assist in honing the public campaign messages. A second phase of the blockade in August 1984 attracted many more supporters. Significant management challenges arose in response to heavy-handed police tactics, including dog attacks on protesters. Because of poor media coverage and the loss of focus on the rainforest issues CAFNEC decided to close down the direct action. The International Union for the Conservation of Nature (IUCN) had already included the region on an indicative list of World Heritage sites (IUCN 1982). Despite significant and growing political pressure, Federal Environment Minister Barry Cohen maintained that state rights prohibited any federal intervention (Toyne 1994).

Two legal actions were mounted by the conservationists against the road in this crisis phase. The first, led by the Aboriginal Development Commission, which owned land through which the road passed, achieved a small delay in proceedings to obtain the proper permits (Anderson 1989). Further engagement between Aboriginal people and the environment movement stalled after the Chairman of the Wujal Wujal Community Council subsequently announced his support for the road (Anderson 1989). Dale *et al.* (2000)

Figure 6.2 Protesters at the Daintree Road blockade, December 1983. Mike Berwick, now Mayor of Douglas Shire, is seated in the centre of the photo. *Source*: James Cook University Cairns Campus Special Collection.

describe how the challenges of addressing Indigenous interests were taken up again at the beginning of the policy implementation stage. The second legal action, led by Cape Tribulation land-owner Cliff Truelove, arguing that the road construction was illegal in the absence of impact assessment, was lost on the basis that government action was precluded from judicial review, a technical argument that Fowler (1985) believes would have been overturned on appeal.

The search for alternatives

Conservationists were strongly involved in a search for alternatives to logging and other rainforest destruction, generating and responding to many new policy initiatives throughout the period from 1980 to 1988. Three examples will give a flavour of the work:
- the Cape Tribulation Management Plan;
- the search for economic alternatives;
- the National Rainforest Conservation Program and associated World Heritage assessment.

The concept of a plan for Cape Tribulation arose out of failure to prevent the subdivision of the private land containing most of the remaining lowland rainforest in 1981. Although the Douglas Shire Council had initially rejected a rezoning application, on the grounds that there were no water, sewerage or other services, subsequent intervention by the Queensland Minister for Local Government, Russ Hinze, forced its approval (Douglas Shire Council 1981; *Cairns Post* 7 May 1981: 3). A Cape Tribulation Rainforest Trust had been established to raise funds to buy private land in 1980, but attracted little money. The new subdivision enormously raised the cost of acquisition in the single stroke of a pen and created an environmental policy problem that is still being addressed today. ACF granted the CTCC $5000 to compile a plan that presented the conservation and cultural values of the area, and recommended, with maps, a combination of acquisitions for national parks and tree preservation orders (Hill 1982). Mike Berwick, who was on the steering committee for the Plan, subsequently became the Mayor of the Douglas Shire. He fostered many of these ideas during the 1990s into the Daintree Rescue Package, the Daintree Futures Studies (Rainforest CRC 2001) and the current phase of a local planning instrument preventing development on some of the remaining forested blocks.

ACF and CAFNEC led the search for economic alternatives, jointly funding Griffith University to undertake a study of alternative employment creation and timber supply industries (Morison *et al.* 1983). The report found that the strong growth of the regional economy in tourism, agriculture and horticulture meant that workers put out of a job if logging ceased would have alternative job opportunities, but that government assistance would be needed to facilitate the change (Hundloe 2000). A seminar in June 1982 established a North Queensland Rainforest Working Committee with membership from timber workers, Aboriginal people, the forestry department and conservationists (*Cairns Post* 21 June 1982). After six months of discussions they were unable to find common ground and disbanded, failing to take up the recommendations of the Griffith study (North Queensland Rainforest Working Committee 1982). The Innisfail Branch of the Wildlife Preservation Society strongly promoted a tourism canopy walkway as a more economically viable option than logging for Downey Creek in 1985. Regional studies demonstrated that tourism was eclipsing primary production as the main industry (Toyne 1994; Doyle 2000). The subsequent extraction of timber from Downey Creek was perhaps the worst decision from this era, this magnificent 6000 ha catchment of lowland tropical rainforest was one of the unspoiled wonders of the Wet Tropics in 1985 when the government announced its intention to log some or all of the catchment.

Despite these initiatives, effective policy proposals to support the transition were not in place when the government made the decision to end the logging in 1987, leading to a great deal of anxiety about the possible social impact. Lynch (2000) and Toyne (1994) describe how government support via a structural adjustment package enabled the regions expected to be worst affected to emerge thriving, despite misappropriation of some of the funds and the associated closure of the Ravenshoe mill.

Despite their refusal to intervene in the logging and Daintree road issues, the federal government took several actions to initiate new, more conservation-oriented rainforest policy. A National Summit of conservationists held in Brisbane in January 1984 resolved to pursue World Heritage listing for the entire Wet Tropics with a focus on the Daintree (Washington 1984). Federal Environment Minister Cohen held a workshop in Cairns in February 1984 that established the multisectoral Working Group on Rainforest Conservation (Winter 1991). A consensus report formed the basis of the National Rainforest Conservation Program, which achieved some significant outcomes outside Queensland, including the nomination of the Central Eastern Rainforest of Australia for World Heritage listing (Commonwealth Minister for Arts, Heritage and Environment and NSW Minister for Planning and Environment 1986). A major conference on tropical forests at Griffith University and other research had highlighted the outstanding natural values of the region (e.g. Webb 1966; Webb & Tracey 1981; Tracey 1982; Werren & Allworth 1982). The Australian Heritage Commission contracted the Rainforest Conservation Society to investigate the global significance of the region. Dr Aila Keto led the study that was to prove seminal in establishing the outstanding universal value of the region (RCSQ 1986). Despite the strong case made by Keto, the Federal Minister was unable to obtain Queensland's participation in the suggested conservation measures through the National Rainforest Conservation Program, and the federal government eventually decided to take unilateral action in 1987.

Policy development phase

Political carriers, those conservationists who possessed the skills to make good connections with those who exercise political power, played a critical role in eventually securing policy uptake by the Federal government (Hutton & Connors 1999). Keto from RCSQ had strong relationships with the ALP (Australian Labor Party) at the state level, and endorsed a vote for them at the 1986 state election, but did not gain sufficient environment movement support to make this action effective, as QCC was fostering an issues focus without political endorsement (Henry & Olson 1986). Philip Toyne, ACF Director, and Jonathan West, TWS Director, developed strong working relationships with Senator Graham Richardson, and through him Prime Minister Bob Hawke (Broadbent 1999). Toyne (1994) describes how negotiations leading up to the 1987 election persuaded the Federal Labor Party to commit, if re-elected, to act unilaterally on World Heritage listing and stop the logging. A National Summit held in Townsville in 1986 had established the Rainforest Action and Information Network, with ACF, TWS, RCSQ and QCC endorsed as the core decision-making group, in consultation with a large network of other organizations. Commitment to cessation of logging throughout the region was established as the bottom line required for any political support (Jeffries 1986). The Rainforest Information and Action Network (RAIN) functioned effectively throughout 1986/7 and ensured that when TWS and ACF subsequently endorsed the vote for the ALP at the 1987 federal election, a flood of supporters handed out how-to-vote cards at polling booths. Market research demonstrated very strong support for federal action to end logging, and a favourable electoral impact in 11 marginal seats, with no negative impact in Brisbane or Far North Queensland where Member of the House of Representatives (MHR) John Gayler kept his strong margin (Brown 1985; *Cairns Post* 29 July 1987; Toyne 1994). Two further aspects of the policy uptake initiatives are worthy of note: the role of media/publications; and the support of opinion leaders.

ACF funded a photographic expedition and facilitated publication of two books on the Daintree region that gave a very positive portrayal of the conservation case, as well as numerous articles in *Habitat Australia* (Sokolich 1980; Borschmann 1984; Russell 1985). Borschmann was also employed by RCSQ as a publicist during 1986/7 and succeeded in having long opinion pieces published in the influential *Age* newspaper and facilitating publication of the first photos of the white possums, threatened by logging, in the popular *Woman's Day* magazine (Borschmann 1986; Hill 1987). The *Port Douglas and Mossman Gazette* was part-owned and managed during the campaign period by Jane King, partner of Mike Berwick, who was also employed there as a journalist. This paper's independent status (one of the few in Australia) was critical: despite being dubbed the *Greenie Gazette*, the paper gained a strong reputation for fairness and credibility in reporting. Berwick developed strong networks with the rapidly growing tourism industry at Port Douglas and facilitated three tourism leaders to identify themselves publicly as National Party supporters and announce that the Queensland government was out of step in opposing the listing, which was strongly supported by business (*Courier Mail* 10 July 1987).

Mike Graham, a geologist with excellent mapping skills and the founding CAFNEC President, produced an array of maps that included the first compilation of the rainforests of the region into a single map, a Wet Tropics National Park, a full boundary description for National Estate Listing proposed by TRS, and the proposal for the Greater Daintree National Park, which he promoted to the Cairns Chamber of Commerce as the economy of the future (Hutton & Connors 1999). Keto used her strong scientific networks to facilitate a host of public and private statements in support of the World Heritage listing by the world's top rainforest scientists. The scientists included Dr Norman Myers, whose attraction of a sell-out crowd at the Cairns Civic Centre in March 1987 gave important signals to the national government about the strength of regional conservation support (Myers 1987). Numerous information nights, rallies, special events, concerts and seminars hosted by the environment movement over that decade had developed a keen and supportive audience.

Community response to the federal intervention

Clear government proposals to support the economic transition, and to involve the broader community and Indigenous peoples in management, were not in place at the time the intended World Heritage listing was announced, leading to a great deal of community anxiety. The Concerned Ladies Action Group, the World Heritage Opposition Group and the Association for Real Balance in our Rainforests were among groups formed between 1987 and 1989 in response to real concerns about the social consequences of listing. Queensland Premier Joh Bjelke-Petersen played on their fears, urging groups to 'set yourselves up like an army and fight … we won't do it unless you fight properly' (Woodward 1987: 4). The next day Ravenshoe residents physically assaulted Federal Environment Minister Senator Graham Richardson and greeted him a week later with swastikas and Nazi salutes (Prismall 1987). Horsfall and Fuary (1988) reported that Aboriginal groups were divided between those who supported the listing because cultural maintenance was strongly linked to rainforest protection and those who opposed it as a limitation on their rights, including groups who wanted to undertake logging (Toyne 1994). Aboriginal leaders were among groups

funded by the state government on a trip to Paris to lobby the World Heritage Bureau against the listing (*Cairns Post* 9 December 1988). The League of Rights began to organize in the community, circulating videos and warning that the World Heritage listing was an international conspiracy (*Courier Mail* 14 March 1988). National Party Senator Ron Boswell warned that the League was evil, racist, anti-Semitic and neo-Nazi (*Cairns Post* 28 April 1988). The League's claim to be on the side of Queen and Country was weakened when Prince Phillip announced his support for the listing (Murray 1988).

Partially as a result of this politicization process, there has been an unfortunate and incorrect association that World Heritage listing destroyed the rainforest logging industry. As previously stated, the Wet Tropics logging industry was on its last legs when World Heritage was declared. The major loss of employment and income occurred due to the running out of logs as a result of quotas set at unsustainable levels for so long. The logging ban implemented by the federal government in association with the World Heritage nomination was applied to a significantly reduced timber industry. The question of what level of logging from the Crown forests of northern Queensland may be sustainable remains unclear but it is certainly much lower than any level actually logged over the 50 years leading up to 1987. Even official Queensland Forestry Department estimates of sustainable logging levels were reducing rapidly from the 1970s onwards in the face of increased scientific awareness of timber production levels and of the significance of environmental impacts of logging (Cassells *et al.* 1988). Traditionally areas of the Wet Tropics were not logged where the cost of building access roads was greater than the income the government would receive from the extremely low royalty rates charged. However, towards the end of Crown logging, the Windsor Tableland forests were logged despite the cost of the road possibly exceeding royalties.

The social impact assessment commissioned by the federal government first predicted and later confirmed that impacts at a community level would be negligible (Lynch-Blosse *et al.* 1991). However, many people were caught up in the emotions generated by political manipulation, such as the call to make war by Premier Bjelke-Petersen. Leading conservationists who lived in the region bore the brunt of this feeling through death threats, and public and private harassment (e.g. *Cairns Post* 9 September 1987: 7), while conservationist leaders in the south received global and national recognition and awards.

While the role of these political carriers was absolutely critical, the credit for the achievement of World Heritage listing lies with many conservationists who, through the broad environment movement, were able to generate the key kernels of lasting policy solutions and to foster their adoption through persistent activist campaigning and networking over a much longer period than that associated with political carriage. Leaders like Mosley, Dunphy, McCabe, Henry and Wilcox were ahead of their time in fostering empowerment of, and national and international networking with, local conservationists. They helped to transform the movement into its current phase, conservation as collaboration. Better development of the economic, social and cultural aspects of the new environmental policy by governments prior to the political decision on World Heritage listing for the Wet Tropics would have enabled a much easier transition into the kappa implementation phase of McDonald and Lane's (2000) model for societal change.

Implementation phase: state and Commonwealth government tensions

The implementation phase of establishing the World Heritage listing was led by the state and federal governments, rather than from the community environment movement. Tension and political conflict between the Queensland and federal governments greatly complicated the implementation process. While some of this related to firmly held views about the rights of state governments under the Australian Constitution, there was also an underlying antagonism towards international conventions that was frequently expressed by the Queensland government, especially under Premier Joh Bjelke-Petersen. In a transparent attempt to retain the ability to log the rainforests of northern Queensland, the Queensland government established its own organization in August 1987, the Northern Rainforest Management Agency (NORMA). The task set was to coordinate rainforest management planning of the Wet Tropics but central to the mission was to provide protection through the existing

National Parks while promoting sustainable logging in the State Forests. In what was a distinctly ironic development, the state government proposed the establishment of a Wet Tropics Biosphere Reserve, something that had previously been considered heretical (no Ramsar Site, no Biosphere Reserve and no World Heritage Area had ever been agreed by the then Bjelke-Petersen government or its predecessors). The Biosphere Reserve was seen as the lesser of two evils, for it would not necessarily inhibit logging activities.

One mechanism employed by the Queensland government was to harness the antagonism towards World Heritage by the many local governments in the region. Both at local government association meetings and at special meetings coordinated by NORMA, there was an attempt to get strong local government opposition targeting World Heritage listing. In September 1987 the Queensland government prepared a document that painted a very negative picture about the listing (Northern Rainforests World Heritage Listing, The Consequences for Local Authority, I.G.R. 67) and circulated it throughout the region. Although most local governments in the region supported the state government's opposition to the listing, unanimity was denied by the strong support given to World Heritage listing by the Townsville City Council (Valentine 1987).

The political tensions spilled over into quite bizarre behaviour at times. In June 1988 the Queensland government sent a team of people to Paris to oppose the listing of the Wet Tropics, then being considered by a meeting of the World Heritage Bureau at UNESCO headquarters. The Queensland group included the Minister of the Environment, two Indigenous people and 16 other delegates. In accordance with proper UN procedures, the group could not be heard by the Bureau and the exercise was a largely political one for domestic consumption. At the formal meeting of the World Heritage Committee in Brazil in December 1988, the slightly revised boundaries were accepted and the property was listed under its formal name as the Wet Tropics of Queensland World Heritage Area.

The actual listed area is quite complex, involving 620 parcels of land including some private property, some Commonwealth land, many National Parks and State Forests as well as other state lands (Table 6.1). This complexity has added to the challenges of managing the WHA. On-ground management of the World Heritage site was therefore shared among many agencies and

Table 6.1 National parks and other state lands in the Wet Tropics World Heritage Area

National parks and state forests	Forest reserves
Girringun National Park	Alcock Forest Reserve
Wooroonooran National Park	Paluma Forest Reserve
Daintree National Park	Mount Windsor Forest Reserve
Edmund Kennedy National Park	Koombooloomba Forest Reserve
Paluma Range National Park	Palmerston Forest Reserve
Cedar Bay National Park	Japoon Forest Reserve
Russell River National Park	Kuranda Forest Reserve
Ella Bay National Park	Dinden Forest Reserve
Hull River National Park	Mount Lewis Forest Reserve
Barron Gorge National Park	Meunga Forest Reserve
Tully Gorge National Park	Tully Falls Forest Reserve
Eubenangee Swamp National Park	Mount Fox Forest Reserve
Grey Peaks National Park	Kirrama Forest Reserve
Crater Lakes National Park	Little Mulgrave Forest Reserve
Kurrimine Beach National Park	Gadgarra Forest Reserve
Black Mountain National Park	Mowbray Forest Reserve
Moresby Range National Park	Garrawalt Forest Reserve
Mount Hypipamee National Park	Danbulla Forest Reserve
Clump Mountain National Park	Herberton Range Forest Reserve
Murray Upper State Forest	Rollingstone Forest Reserve
Kuranda State Forest	Cardwell Forest Reserve
Cardwell State Forest	Malbon Thompson Forest Reserve
Paluma State Forest	Clemant Forest Reserve
Herberton Range State Forest	Tam O'Shanter Forest Reserve
Abergowrie State Forest	Lannercost Forest Reserve
Clemant State Forest	Mount Mackay Forest Reserve
Danbulla State Forest 1	Trinity Forest Reserve
Mount Fox State Forest	Mount Fisher Forest Reserve
Koombooloomba State Forest	Mount Spurgeon Forest Reserve
Ravenshoe State Forest 1	Riflemead Forest Reserve 2
Danbulla State Forest 2	Riflemead Forest Reserve 1
Murray Upper State Forest	Gillies Highway Forest Reserve
Goldsborough Valley State Forest	Dirran Forest Reserve
Monkhouse Timber Reserve	Curtain Fig Forest Reserve
Jumrum Creek Conservation Park	Graham Range Forest Reserve
Meingan Creek Conservation Park	

individuals, although the main areas were captured by land managed by state government agencies, especially Queensland Parks and Wildlife Service and the then Queensland Forestry Department.

World Heritage listing: the formal process for the Wet Tropics

The World Heritage Convention, adopted by UNESCO in 1972, was designed to protect those cultural and natural properties that are of such universal value they form part of the heritage of all people. The Convention came into force in 1975 and Australia has been a participant since its beginning. The Convention itself has never been amended but it allows for the World Heritage Committee to establish Operational Guidelines that may be revised as needed by the Committee. Article 2 of the Convention defines the natural heritage recognized for the purposes of the Convention. This is as follows:

• natural features consisting of physical and biological formations or groups of formations, which are of outstanding universal value from the aesthetic or scientific point of view;

• geological and physiographical formations and precisely delineated areas which constitute the habitat of threatened species of animals and plants of outstanding universal value from the point of view of science or conservation;

• natural sites or precisely delineated natural areas of outstanding universal value from the point of view of science, conservation or natural beauty (UNESCO 1972).

The Operational Guidelines in place at the time of the nomination of the Wet Tropics were agreed by the Committee in 1984. Those Guidelines identified four criteria for the determination of outstanding universal value. The Committee requires that a nominated site must meet at least one of these four criteria and fulfils other conditions before it can be considered for inscription on the World Heritage List. The criteria in use at the time were:

(i) be outstanding examples representing major stages of the Earth's evolutionary history; or

(ii) be outstanding examples representing significant on-going geological processes, biological evolution and man's interaction with his natural environment;

as distinct from the periods of the Earth's development, this focuses upon ongoing processes in the development of communities of plants and animals, landforms and marine areas and fresh water bodies; or

(iii) contain superlative natural phenomena, formations or features, for instance, outstanding examples of the most important ecosystems, areas of exceptional natural beauty or exceptional combinations of natural and cultural elements; or

(iv) contain the most important and significant natural habitats where threatened species of animals or plants of outstanding universal value from the point of view of science or conservation still survive. (World Heritage Committee 1984: paragraph 24).

Following the Paris Bureau meeting, the natural heritage advisory body, the IUCN, was asked to review the nomination. The evaluation by IUCN experts included an extremely experienced site inspection team consisting of Dr Jim Thorsell (chief scientist from IUCN's natural heritage programme), Mr Bing Lucas, one of the outstanding members of the World Commission on Protected Areas and a Vice-Chair for World Heritage, and Professor Larry Hamilton from the East–West Center in the USA. The political tension and local conflict made the inspection quite remarkable and included formal protection for the members of the team. While there was no dispute about the outstanding universal value of the proposed site there were concerns about management arrangements and also about boundary selection. The latter concerns were addressed by the Australian government later in the year and the minor revisions made prior to listing. During this period the Queensland government was still arguing for a 40% reduction in area to allow continuation of the logging industry.

The Wet Tropics of Queensland World Heritage Area was nominated and listed on all four criteria. Some of the other requirements (including a management plan) were not immediately met and were the subject of ongoing discussions between the Committee and the Australian government.

The formal processes and calendar involved with World Heritage listing in operation at the time that the Wet Tropics of Queensland was nominated by Australia is shown in Figure 6.3. The World Heritage Committee consists of 21 members (State Party representatives) elected by a meeting of all State Parties. The World Heritage Bureau is an executive group (seven members)

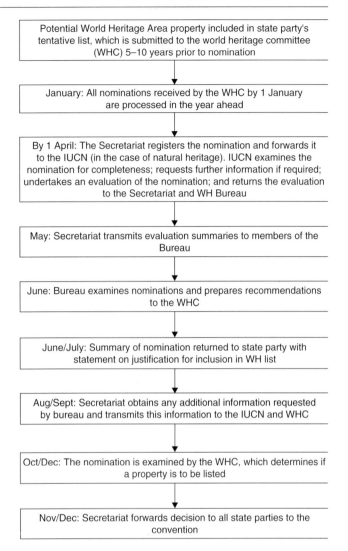

Figure 6.3 Procedure and calendar for the processing of nominations (as per the Operational Guidelines for the Implementation of the World Heritage Convention, 1984).

who traditionally met between the annual Committee meetings. The Secretariat is the World Heritage Centre based in the UNESCO building in Paris.

World Heritage listing: responsibilities and action

Under the World Heritage Convention, State Parties take on a number of core responsibilities. These are in essence identification, protection, conservation, presentation, transmission and rehabilitation. Having identified the Wet Tropics World Heritage Area the

State Party (in this case the Australian government) is required to offer:

- protection (legislative action);
- conservation (management action);
- presentation (sharing the outstanding universal value of the place with the community);
- transmission to future generations (usually met through ensuring no damage, which implies a formal monitoring and reporting programme);
- rehabilitation (repairing any damage to the WHA).

The period since listing of the Wet Tropics has been very much concerned with delivery of these responsibilities.

Initially the tension between state and federal governments inhibited progress in meeting responsibilities, especially regarding management. Legal challenges by the Queensland government further soured relations but these were eventually resolved, at least in part, by a change of government in Queensland. In November 1990 the state and federal governments signed the Wet Tropics World Heritage Area Management Scheme, enabling the appointment of a Director of the Wet Tropics Management Authority (WTMA) and the allocation of up to $10 million of federal funds over the first three years. The Scheme also allowed for the establishment of the Community Consultative Committee and for a Scientific Advisory Committee. At the IUCN General Assembly in 1990, as part of the World Heritage Session, some of the failings were identified, especially the delays in establishing a management authority and threats to integrity in the form of a proposed Tully-Millstream dam (Valentine 1991). Subsequently the World Heritage Bureau meeting in June 1991 expressed concern about these and other issues. In 1992 the WTMA was established in the form of an Executive Director and additional staff. This led to development of the Wet Tropics Plan, initially as the Strategic Directions (August 1992), and then a long process of community deliberation, development of policies (Protection Through Partnerships document: WTMA 1997) and development of the statutory Wet Tropics Management Plan (1998) as a regulation under the Wet Tropics World Heritage Protection and Management Act 1993. This Queensland legislation also established a statutory authority, the Wet Tropics Management Authority.

Since the establishment of the Wet Tropics Plan, the operations of the WTMA (under the direction of its independent board) have been principally concerned with delivering the outcomes envisaged by the plan. Strong community engagement has dominated the processes but tensions over budgets remain. The level of financial support from both Commonwealth and state governments is very low, especially in the context of the extremely large income generated through tourism and other economic benefits of the area. By comparison with other World Heritage Areas that earn large economic benefits for their region (for example, Yellowstone National Park World Heritage Area) the funds invested in management and protection of the Wet Tropics are about 20% of what they should be. Yellowstone has a total budget of around $30 million per annum (Wet Tropics, perhaps $7 million) but the areas are roughly the same size, have about the same number of visitors per annum and contribute about the same amount to their respective regional economies. One effect of this is the contrasting ability to fund staff: Yellowstone has 400 permanent staff and another 400 seasonal staff while the Wet Tropics probably has fewer than 80 staff (only 20 of these are WTMA staff). Yellowstone also raises about US$5 million in user fees versus virtually none from the Wet Tropics. In these circumstances the achievements of the small band of professionals managing the Wet Tropics World Heritage Area are outstanding.

Looking forward

One characteristic of the WHA has been its close engagement with the community. This has been expressed through strong participation. The Community Consultative Committee was established from the beginning and has been joined by advisory groups including those from the conservation and tourism sectors. Close links have been established with the Wet Tropics Traditional Owners, most recently through the recognition of the Aboriginal Rainforest Council as a formal advisory body to the board. As part of the Regional Agreement signed in 2005 a second Aboriginal Board member will be appointed, further increasing the influence of Traditional Owners. In June 2006 an appointment was made by the Queensland government to meet this obligation. Another aspect of this work has been the participation by the WTMA in the development of Indigenous Land Use Agreements

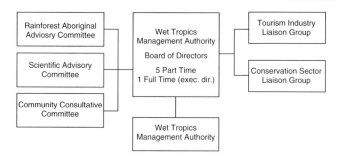

Figure 6.4 WTMA management and community interface.

(ILUAs). These agreements generally relate to the recognition of rights and agreements about Traditional Owner access, use and management of the WHA. Figure 6.4 shows the current structure of the WTMA and its formal interface with the community. Additional information about the community engagement processes will be found in Clarke (Chapter 29, this volume).

One of the key elements of World Heritage today, and a core component of the processes of nomination and management of World Heritage Areas, is engagement with the community. A particular responsibility of State Parties is to give World Heritage 'a function in the life of the community' (Article 5, World Heritage Convention 1972). The Wet Tropics World Heritage Area began amidst community conflicts and has progressed through many phases of values review and changes to exemplify community-based management. It can be seen as a great success.

References

Allworth, D., Keto, A. &. Morisonm, J. (1981). *A Critical Review of the Qld Forestry Department Timber Production from North Queensland Rainforests Sept 1981*. Australian Conservation Foundation, Petrie Terrace.

Anderson J. C. (1989). Aborigines and conservationism: the Daintree–Bloomfield road. *Australian Journal of Social Issues* **24**(3): 214–27.

Australian Conservation Foundation (ACF) (1981). *Forestry and Tropical Rainforest – Conservation or Degradation?* Australian Conservation Foundation Environmental Education Leaflet, Cairns.

Borschmann, G. (1984). *Greater Daintree World Heritage Tropical Rainforest at Risk*. Australian Conservation Foundation, Melbourne.

Borschmann, G. (1986). Cohen's rainforest deal close to collapse. *The Age* 19 December: 1.

Broadbent, B. (1999). *Inside the Greening. 25 Years of the Australian Conservation Foundation*. Insite Press Elwood, Victoria.

Brown, W. (1985). Brisbane people want protection for rainforest. *Courier Mail* 26 January: 3.

Cassells, D. S., Bonell, M. B., Gilmour, D. A. & Valentine, P. S. (1988). Conservation and management of Australia's tropical rainforests: local realities and global responsibilities. *Proceedings of the Ecological Society of Australia* **15**: 313–26.

Cassells, D. S. & Valentine, P. S. (1988). From conflict to consensus: towards a framework for community control of public forests and wildlands. *Australian Forestry*, **51**(1): 47–56.

Commonwealth Minister for Arts, Heritage and Environment, and NSW Minister for Planning and Environment (1986). *National Rainforest Conservation Program*. Commonwealth Minister for Arts, Heritage and Environment, and NSW Minister for Planning and Environment, Canberra and Sydney.

Dale, A., Lane, D., Yarrow, M. & Bigelow, A. (2000). Aboriginal participation in management: reconciling local interests with World Heritage. In *Securing the Wet Tropics?* McDonald, G. & Lane, M. (eds). The Federation Press, Annandale, NSW, pp. 187–99.

Dansie, S. (1980). *Marketing in the Eighties*. Paper presented to Department of Forestry, Queensland Seminar, Gympie.

Department of Forestry Queensland (1981). *Timber Production from North Queensland's Rainforests. A Position Paper*. Department of Forestry Queensland, Brisbane.

Department of Forestry Queensland (1983). *Rainforest Research in North Queensland. A Position Paper*. Department of Forestry Queensland, Brisbane.

Douglas Shire Council (1981). *Minutes of 10 March 1981*. Minutes held in CAFNEC archives box 1, James Cook University Cairns Campus North Queensland Collection.

Douglas Shire WAG. (1984). *The Trials of Tribulation*. The Douglas Shire Wilderness Action Group, Mossman.

Doyle, T. (2000). The campaign to save the Wet Tropics. In *Securing the Wet Tropics?* McDonald, G & Lane, M. The Federation Press, Annandale, NSW, pp. 103–16.

Dunphy, M. (1981). Letter to John Sinclair, Vice-President Australian Conservation Foundation, 27 June. Copy of letter held in R. Hill, personal papers.

Elvery, L. (1981). *Cairns Post* 7 November: 1.

Fowler, R. J. (1985). Rainforests, roads and private clauses: the Daintree case. *Environment and Planning Law Journal* **2**: 345.

Henry, D. & Olson, M. (1986). *Queensland's National Parks, Losing Ground*. Queensland Conservation Council, Brisbane (leaflet).

Hill, R. (ed.) (1982). *Cape Tribulation Management Plan.* Cape Tribulation Community Council, Cairns.

Hill, R. (1987). A rare and beautiful find. *Woman's Day*: pp. 18–19.

Hill, R. &. Grespan. G. (1980). Threat to rare rainforests. *Courier Mail* 10 November: 4.

Horsfall, H. &. Fuary, M. (1988). *The Cultural Heritage Values of Aboriginal Archaeological Sites and Associated Themes in and Adjacent to the Area Nominated for World Heritage Listing in the Wet Tropics Rainforest Region of North East Queensland.* Unpublished report to the State of Queensland, 12 December. Permit No. 31, EIS. 88.

Hundloe, T. (2000). Behind the Wet Tropics: politics, institutions and sustainable development. In *Securing the Wet Tropics?* McDonald, G. and Lane, M. (eds). The Federation Press, Annandale NSW, pp. 150–67.

Hutton, D. & Connors, L. (1999). *A History of the Australian Environment Movement.* Cambridge: Cambridge University Press.

International Union for the Conservation of Nature (IUCN) (1982). *The World's Greatest Natural Areas.* IUCN, Gland, Switzerland.

Jeffries, A. (1986). *Rainforest Strategy Meeting Follow-up No 1.* Unpublished minutes of the National Summit held in Townsville on 12–13 July, circulated to the Rainforest Action and Information Network, held in R. Hill personal papers.

Lynch, M. (2000). The social impacts of World Heritage listing. In In *Securing the Wet Tropics?* McDonald, G. and Lane, M. (eds). The Federation Press, Annandale NSW, pp. 113–17.

Lynch-Blosse, M., Turrell, G. & Western, J. (1991). *Impact Assessment World Heritage Listing. SRCU Report No. 22.8.* Consultancy Report Commissioned by the Resources Assessment Commission, Canberra.

McCabe, J. (1981). *A Proposal for Direct Action to Save Key Rainforest Areas in North Queensland.* Document held in CAFNEC archives box 1, James Cook University Cairns Campus North Queensland Collection.

McDonald, G. &. Lane, M. (eds) (2000). *Securing the Wet Tropics?* The Federation Press, Annandale, NSW.

Morison, J., Driml, S. & Hudson, D. (1983). *Alternative Employment Creation and Timber Supply Industries in North Queensland.* Institute of Applied Social Research, School of Australian Environmental Studies, Griffith University, Nathan, Queensland.

Mosley, G. J. (1982). *Director's Report to the 40th Council Meeting.* Minutes of the Australian Conservation Foundation 40th Council Meeting, Melbourne.

Murray, R. (1988). New right leader says royal forest stand subversive. *Courier Mail* 8 April: 3.

Myers, N. (1987). Letter to Hon. Barry Cohen, Minister for Arts Heritage and Environment, 28 March.

North Queensland Rainforest Working Committee (1982). *Minutes of Meetings Held at CSIRO Division of Forest Research.* Minutes folder held in R. Hill personal papers.

O'Grady, W. (1980). *Report on Resolutions from the Second World Wilderness Congress from the Chairman.* File note held in CAFNEC archives box 1, held at James Cook University Cairns Campus North Queensland Collection.

Prismall, B. (1987). Minister greeted with swastikas and nazi salutes. *Courier Mail* 8 October: 2.

Rainforest CRC (2001). *Daintree Futures Study.* Cooperative Research Centre for Tropical Rainforest Ecology and Management: Cairns.

RCSQ (1986). *Tropical Rainforests of North Queensland: Their Conservation Significance.* Australian Government Publishing Service, Canberra.

Russell, R. (1981). *Paper on Direct Action. Herberton.* Unpublished paper, held in R. Hill personal papers.

Russell, R. (1985). *Daintree.* Kevin Weldon & Australian Conservation Foundation, Sydney.

Sokolich, W. (1980). Daintree–Bloomfield: symbol of Australia's rainforest at risk. *Habitat Australia* **8**(6): 14–18.

Stocker, G. C., Gilmour, D. A. & Cassells, D. S. (1977). The future of our northern rainforests in the face of economic and political reality. *Focus on the Forester, Volume 1, Proceedings of the 8th Institute of Foresters of Australia.* 10 pp.

Toyne, P. (1994). *Reluctant Nation.* ABC Books, Sydney.

Tracey, J. G. (1982). *The Vegetation of the Humid Tropical Region of North Queensland.* CSIRO, Melbourne.

TRS (1984). *Cairns Situation Report 3 October 1984.* Document held in the CAFNEC Library, Cairns.

UNESCO (1972). *Convention Concerning The Protection of the World Cultural and Natural Heritage.* UNESCO, Paris.

Valentine, P. S. (1980). Tropical rainforest and the wilderness experience. Paper presented at the Second World Wilderness Congress, Cairns. Subsequently published in *Wilderness* (1982), Martin, V. (ed.). Findhorn Press, Findhorn, Scotland, pp. 123–32.

Valentine, P. S. (1991). Over the hill: progress in management of the Queensland Wet Tropics. In *Critical Issues for Protected Areas, Part I: World Heritage Session.* IUCN, Gland, Switzerland, pp. 53–8.

Valentine, V. M. (1987). An analysis of the Queensland State Government Document, *Northern Rainforests World Heritage Listing: The Consequences for Local Authority* I.G.R. 67, September. Townsville City Council.

Washington, H. (1984). *The Daintree Campaign Strategy and Tactics, State and National Directions.* ACF and TWS. Document held in CAFNEC archives box 3, James Cook University Cairns Campus North Queensland Collection.

Webb, L. J. (1966). The identification and conservation of habitat-types in the wet tropical lowlands of north

Queensland. *Proceedings of the Royal Society of Queensland* **78**(6): 59–86.

Webb, L. J. & Tracey, J. G. (1981). Australian rainforests: patterns and change. In *Ecological Biogeography of Australia*, Keast, A. E. (ed.). Dr W. Junk, The Hague, pp. 607–94.

Werren, G. L. & Allworth, D. (1982). *Australian Rainforests: A Review.* Monash Publications in Geography Number 28, Melbourne.

Winter, J. W. (1991). The Wet Tropics of north east Queensland. In *Environmental Research in Australia Case Studies*, Australian Science and Technology Council (ed.). Australian Government Publishing Service, Canberra, pp. 151–72.

Woodward, L. (1987). Joh crusades against the listing. *Cairns Post* 29 August: 3.

World Heritage Committee (1984). *Operational Guidelines for the Implementation of the World Heritage Convention.* UNESCO, Paris.

Wet Tropics Management Authority (WTMA) (1997). *Protection through Partnerships.* Wet Tropics Management Authority, Cairns.

7 THE NATURE OF RAINFOREST TOURISM: INSIGHTS FROM A TOURISM SOCIAL SCIENCE RESEARCH PROGRAMME

*Philip L. Pearce**

Tourism Program, School of Business, James Cook University, Australia
*The author was participant of Cooperative Research Centre for Tropical Ecology and Management

Introduction

Tourism in and around Australia's Wet Tropics World Heritage Area can be briefly characterized by describing its structure and by providing a succinct description of images and services. The Wet Tropics Area is moderately accessible with a number of highways, minor roads, rivers and tracks providing multiple pathways and points of entry (Tourism Queensland 2005; Tourism Tropical North Queensland 2005). There is a clear seasonality to the region's rainfall (highest from December to March) but visitors are spread throughout the year, with the cooler drier winter months (April to October) being the busiest (Moscardo & Pearce 2000). Visitor activities in the region include self-drive tours, commercial four wheel drive and coach tours, short walks (and, increasingly, longer walks), white water rafting, shopping in small towns and visits to sites featuring wildlife and aboriginal culture. Much visitor activity is focused in the most northerly part of the region, especially the nationally well-known Daintree area where the commercial tours are concentrated (Young 1998).

Small towns adjacent to the rainforest itself provide some tourism services including cafes, accommodation and shopping but the largest number of visitors use Cairns as a tourism dormitory. The area is ubiquitously green and generally rugged with deep valleys, waterfalls and fast flowing short rivers. There is much diversity in the fauna and flora although wildlife is not so abundant as to be easily seen during daylight tours and travel (Woods 2003). The region is richly featured in Australian painting, poetry and literature. The images of Skyrail, a rainforest cableway, and Cape Tribulation where the tropical rainforest merges with white beaches blending to adjacent reefs, are core icons for local, state and national promotion (Tourism Australia 2005). For international visitors to the region the rainforest is a secondary attraction to that of the Great Barrier Reef, and the rainforest is often a surprise bonus for their regional experience (Pearce *et al.* 1996a).

The rainforest businesses that operate in the area tend to be small or micro-businesses, locally owned and with modest financial resources. Attractions, tour companies, accommodation and shopping facilities characterize the regional tourism business types (Moscardo *et al.* 1996). The opportunities for and style of tourism to the Wet Tropics World Heritage Area have grown and expanded since the declaration of the area in 1988. In the past decade more attractions, local markets, interpretation and walking tracks have been put in place. Total tourist numbers to the rainforest region are of the order of 3.5 million and the numbers have grown modestly in the past decade.

The indigenous communities with longstanding connections to the area were displaced from the rainforest during white settlement and the accompanying violence (see Pannell, Chapter 4, this volume). The consequences of this aggression are that Australia's wet tropical rainforests, unlike many other tropical rainforests, are not used or lived in extensively by traditional groups. There is, though, an important resurgence and struggle by the local indigenous community to share in tourism's economic benefits and to use tourism to sustain cultural traditions. These efforts are manifest in some tourist attractions with indigenous themes and a limited number of tours conducted by Indigenous people as well as themed souvenirs and shopping (Moscardo & Pearce 1999).

Towards rainforest tourism research

There is a schizophrenic public view of rainforest tourism and its role in conservation. On the one hand tourism can play the role of environmental saviour because its economic dimensions can justify the protection of species and the preservation of settings (Butler 1991). In this view tourism can provide communities with substitute, additional or totally new livelihoods, especially when conservation efforts limit other economic activities such as logging or farming (Archer & Cooper 1994; Buhalis & Fletcher 1995). This outlook on tourism with its focus on economic and employment benefits may be contrasted with an alternate perspective built around tourism's negative impacts. In this second view tourism is a form of local environmental terrorism and, in particular, biophysical impacts are feared and stressed (Butcher 2003; Bramwell & Lane 2005; Turton & Stork, Chapter 27, this volume). It has been apparent for some time that these perspectives arise out of the large scale social representations or views of the world held by community groups rather than being keenly fought debates on the scientific evidence of tourism's dimensions and impacts (Pearce et al. 1996; Hall 2005). Public discussions of the value and importance of tourism in Australia's Wet Tropics rainforests oscillates between these schizophrenic appraisals. Research undertaken in the

tourism social science area seeks to close some of the knowledge gaps that give rise to false expectations and unproductive confrontations (Machlis 1996; Ruhanen & Cooper 2004; Bentrupperbäumer & Reser, Chapter 31, this volume).

The schizophrenic opinions about tourism are due in part to a prevailing ignorance of what tourism really is, how it works and what it can achieve for natural settings and regions (Hall 2005). The ignorance is extensive and is shared by travellers themselves, a number of public officials, some sector-specific business personnel and at times scientific researchers. There are two broad views of tourism needed to redress the prevailing knowledge gaps and to foster an appreciation of tourism's role in natural environments in general and rainforest settings in particular. An understanding of sustainability as a holistic force in shaping thinking about development is one requirement. A second necessity is an appreciation of the multiplicity of interacting forces that govern the system of tourism in which local businesses and operations are located. In this chapter these two broad contextual views of tourism will be considered. Once this context is established, a tourism research agenda for Australia's wet tropical rainforests will be reviewed. The applicability of this agenda for other settings such as South East Asia will be noted. A 10-year programme of tourism social science research directed at key elements of the agenda will be outlined. A second related programme of research focusing particularly on tourism's impacts occupies a separate chapter in this volume (Reser & Bentrupperbäumer, Chapter 34, this volume). It will be argued that the task of studying rainforest tourism is certainly not complete but that the pathways explored to date have been promising both in and beyond Australia's Wet Tropics rainforests.

The sustainability context

The emergence of sustainability as a key political topic guiding both communities and research activities arose at almost the same time as the declaration of the Wet Tropics World Heritage Area. The defining approaches and the action agendas were set in the late 1980s although many of the key issues had substantial roots in earlier formulations (see Valentine & Hill, Chapter 6, this volume). The emphasis on sustainability shapes rainforest tourism research in some important ways. First there is the fundamental notion that sustainability involves multiple and potentially compatible outcomes. A guiding expression in this field is that of the triple bottom line, a term deriving from the work of Elkington (1997) and his imaginatively entitled book *Cannibals with Forks*, which is a metaphor for civilizing the behaviour of the corporate and business world. The three components of the triple bottom line are the economic, socio-cultural and environmental dimensions of well-being; the positive contributions to all of these dimensions define the goals for sustainable tourism (World Tourism Organization 2004).

Tourism research conducted within the rainforests of the Wet Tropics World Heritage Area sits comfortably within the larger framework of sustainability. It is not research aimed only at promoting tourism and tourism businesses nor is it seeking to develop an anti-tourism and exclusively pro-environment cause. The research is viewed as contributing to some fundamental and new appraisals of tourism in the Wet Tropics by understanding the interaction of settings, markets, communities, businesses and management, with a particular focus on assessing outcomes. It thus seeks primarily to describe, understand and help to manage sustainable tourism within the larger context of sustainable development.

Rainforest tourism as a system

The earlier claim that many people do not understand the complexity of tourism is an assertive one. What is there to understand? Basic and early appraisals of tourism identified the importance and centrality of tourist attractions, such as the rainforest itself, to the system of tourism. It was suggested further that visits to attractions are fostered by direct marketing and local information sources and strongly influenced by accessibility, governed in particular by transport options (Mill & Morrison 1992; Gunn & Var 2002). Accommodation, restaurants and shopping opportunities can all be seen as part of the supporting infrastructure for attractions-led tourism. In this early view changes in any one part of the system were seen as routinely generating changes in other components, such as the expectation that when road access is developed visitor flows to the newly developed area will be fostered. This generic model of

tourism serves some useful purposes, such as identifying the centrality of attractions, but has been described as inadequate in dealing with the organization, chaos and complexity of contemporary tourism (Farrell & Twining-Ward 2004). Many regional destinations have found to their cost that visitors do not necessarily arrive simply because new access is provided or additional attractions are constructed (Moscardo 2005).

The power of tourism wholesalers, travel agents and inbound tour companies is frequently misunderstood in predicting the likely impacts of emerging international markets where there is much competition from other international destinations. In order to appreciate a more holistic view of rainforest tourism many of the key elements driving site selection and in turn the outcomes of rainforest tourism are depicted in Figure 7.1. In reviewing the elements of this figure some of the subtleties influencing site selection begin to emerge, including an array of influential factors such as the controlling variables of travel experience and motivation, the role of tourism distribution systems and the regulatory environment in which tourism businesses must locate their tourism product.

The notion that changes to parts of this rainforest tourism system will have impacts on site selection and outcomes does remain an important organizing principle but it should be noted that some effects are likely to be weakened as strong filters may shape their influence. For example, rainforest tourism businesses that seek to position the style and appeal of their tourism offering with an accreditation scheme may receive only a small boost in visitor numbers if problems of access, the actions of travel wholesalers and local tour desk directives persuade visitors to make other choices.

A rainforest tourism research agenda

The preceding foray into the multiple determinants of tourist activity in tourism destinations, and rainforest tourism in particular, quickly implicates many topics for researcher attention. The course of study undertaken during a 10-year period of tourism rainforest research was designed to address key parts of the system identified in Figure 7.1.

A brief historical review of the evolution of the research undertaken is worth attention both to demonstrate the reality of how the research evolved and as a

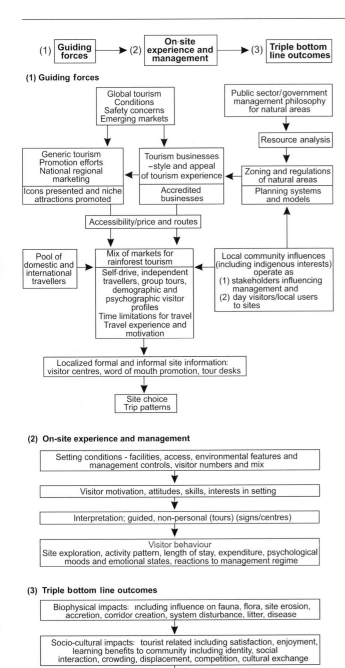

Figure 7.1 A guide to rainforest tourism.

tourism. It also proposed that tourism planning models and systems could be developed and that community views of rainforest tourism, including indigenous views, were worthy of investigation. Some achievements were realized in all of these areas but not all of the topics were exhaustively examined due to funding limitations and shifts in the priorities of the research funding scheme.

After an initial and productive period of academic-led activity on the existing markets, a greater voice was given to the government and tourism industry representatives in determining issues for research. The role of other rainforest scientists in competing for research funds was also an influential factor. The work on tourism markets continued, some initiatives in interpretation, the presentation of rainforest environments for tourism, were added, while the emphasis on planning models and community perspectives was discontinued. Local media and publicity concerning specific biophysical impacts of tourist business activities, especially the supposed spread of the fungus *Phytophthora cinnamoni* on the tyres of tourist vehicles, assumed a particular prominence and attention. As a consequence, funding was reduced for social science tourism study in a bid to explore the specific biophysical impacts (Gadek & Worboys 2002). This fluctuation in what to study over what time scale and with what relevance to what audiences prompted the negotiation of a new social sciences research agenda for rainforest study. The documents produced built on ideas from the tourism literature, as well as international, national and local consultative processes including summary documents of user needs derived from workshops. The agenda that resulted from these multiple inputs is presented in Table 7.1. No priorities were attached to this list of research requirements because the intention of the listing was to set out a broad array of issues for the long-term understanding of rainforest tourism.

Not surprisingly given the scale of the research agenda outlined in Table 7.1, only parts of this programme of studies were completed. The efforts of a kindred social science research group studying the impacts of tourism are reported in separate chapters in this volume (Bentrupperbäumer & Reser, Chapter 31, this volume). This work is less concerned with tourism markets and tourism businesses and systems, and more specifically focused on predominantly environmental social impacts (Bentrupperbäumer *et al.* 1998;

reflexive account of interest to other researchers seeking to establish programmes of study in rainforest and natural environment systems. The initial work was determined largely by academic and researcher initiatives and focused on the existing markets for rainforest

Table 7.1 The tourism research agenda for Australia's Wet Tropics Rainforests

Future research requirements

1 Research into resources	• Audit and identification of natural and cultural resources for tourism • Assessment of the attractiveness of natural and cultural resources to different tourist groups (or market segments) • Identify limits of acceptable change
2 Research into patterns of tourism	• Assessment of existing and potential tourism to regions and sites • Existing and required infrastructure • Factors influencing global and national patterns of tourism includes technological, sociodemographic, political, economic and environmental factors (such as global climate change) • Analysis of competitive destinations
3 Research into communities (includes indigenous communities)	• Audit and identification of natural and cultural resources for tourism • Assessment of the attractiveness of natural and cultural resources to different tourist groups (or market segments) • Identify limits of acceptable change
4 Research into tourists	• Patterns and modes of travel within regions • Activity participation • Use of facilities • Demographic and sociocultural characteristics, images, expectations, and motivations • Evaluations of experiences and satisfaction • Expenditure
5 Research into impacts	• Identify cultural, social, economic and environmental impacts of tourism (and recreation) • Monitor impacts • Identify the factors that contribute to impacts
6 Research into management	• Evaluate management actions, including effectiveness at achieving objectives, cost, and user responses • Evaluate interpretive/education activities • Examine mechanisms for cooperation
7 Research into economic issues	• Determine value of protected areas for regional, state and national economies • Examine economic impacts of fees and privatisation
8 Research into business issues	• Analyse workplace practices, including cost effectiveness assessments, and environmental audits • Identify factors contributing to profitability

Bentrupperbäumer & Reser 2002; Turton 2005; see also Bentrupperbäumer & Reser, Chapter 31, this volume; Turton & Stork, Chapter 27, this volume).

In reviewing the achievements of the major social science tourism group studying rainforest tourism during the Cooperative Research Centre (Rainforest CRC) funding period, two important but hidden issues deserve commentary. First, many of the topics reviewed in Table 7.1 can be assessed at different scales or levels of analysis. Rainforests in general, and the Wet Tropics World Heritage Area in particular, are physically large regions often spread over several hundred kilometres and with varying points of access and specific concentrations of tourist activity. It is therefore possible to frame the tourism analysis at several levels or scales such as individual sites, sub-regional clusters or districts, or as a treatment of the whole area. Much of the

work done on tourist markets in this programme of studies was conceived at the holistic or total area scale while recognizing that individual sites or sub-regions may have subtly different markets. A second issue pertaining to this rainforest research agenda lies in the distinction between the everyday, the scientific and the legislative definitions of the rainforest. For the public management agencies and for scientific researchers, the rainforest area is defined by formal boundaries and specific ecological characteristics. Visitors, by way of contrast, have a much broader and more inclusive conception of the rainforest and the World Heritage Area, which can include small towns and areas of farming as well as commercial services such as attractions and accommodation. All of these settings can be a part of 'visiting the rainforest'. For the visitor there is indivisibility between the physical and socio-cultural settings and unless researchers specifically indicate what is meant by the term rainforest when seeking visitor responses, it is likely that a 'whole of area' response will be given rather than reference to specific vegetation patterns or designated World Heritage boundaries.

The wider applicability of the tourism research agenda outlined here deserves attention for rainforest tourism researchers in other countries and settings. Throughout South East Asia, across the Pacific Islands, in parts of equatorial Africa and in Central America there are rainforest regions that also need tourism research effort (cf. Turton & Stork 2006). Several researchers have noted the importance of designing and implementing tourism social science agendas for natural environment management. Machlis (1996: 1) argues: 'As every park superintendent comes to know, the management of national parks is necessarily the management of people.' The need for comprehensive social science analysis is reinforced by Brake and Williams (1990: 1), who assert: 'Traditionally there has been very little behavioural research on visitors. In the absence of an objective data base, issue analysis degenerates into no more than a debate about personal perceptions and values. In this context it is not surprising that planners are constrained by the biases of the current user groups.'

It is useful to reflect here that one of the user groups is in fact rainforest scientists themselves and their use of the setting involves study and analysis to further knowledge and, not unimportantly, their own careers. As was outlined earlier it may well be the interests of this group, rather than the interests of users, that limit social science research due to the increasing university pressures to fight for and find research funds (Page 2005).

A particular impetus for using the new tourism social science research agenda outlined above in international settings arises from the rapid growth of tourism in countries where there are substantial rainforests. In Thailand, for example, the Tourism Authority of Thailand has a target of 20 million international visitors in 2007, more than four times that of Australia (Tourism Authority of Thailand 2005). While much of this tourism is beach- and resort-based, the value of rainforests for enhancing the style of Thailand tourism has been frequently advocated (Nakjan 2004). Proposed activities and types of tourism

in the rainforests of Thailand include bird and butterfly watching, trekking, river rafting, elephant treks, sea fog and mist viewing and wildlife watching (Kekul 1999; Eggington 2003). Backpackers in particular, who are a strong element in Thailand tourism, have substantial mobility as well as the time to visit rainforest settings and the lack of research-based planning and management for these visitors generates not only physical environmental impacts but substantial sociocultural concerns (Pearce 2005b).

There are two specific applications of the present studies in the Wet Tropics context to these international destinations. First, the research agenda presented above may usefully guide research efforts in other settings as key planners can examine the listings and determine for their own settings where the most significant and immediate gaps exist. That is, by using the research agenda in conjunction with Figure 7.1 the trouble spots existing in a setting can be highlighted and decisions made on where to plan for and how to promote, manage and assess the local tourism system. In many other rainforest environments, for example, assessing community responses and behaviours towards rainforest use and tourism will be critical to tourism's success.

A second contribution of the present work to international settings is more conceptual. Some of the research findings, as well as the underlying ideas and conceptual schemes used to explain tourist behaviour and its management in the Wet Tropics, are also transportable. In particular some of the ways to describe tourist markets are worthy of export, as are the theoretical formulations of mindfulness and its role in interpretation. Additionally, the value of social representations and their importance in understanding community responses to tourism could be used in many contexts. Further, the benchmarking approach used to understand satisfaction in this programme of studies also has wider applicability for many natural environment settings. These and other research contributions will now be the focus of attention.

Research achievements

The principle achievements of the major tourism research effort in the 10-year period from 1993 to 2003 can be summarized under five headings. There are over 30 linked research papers built on findings and conceptual developments of note in the area of:

- understanding rainforest tourism markets;
- the operation and appreciation of interpretation in influencing visitor behaviour;
- community attitudes to tourism;
- issues in managing tourism rainforest businesses;
- advances in understanding tourist satisfaction.

Achievements and findings pertaining to these areas are not uniform since greater attention was devoted to the analysis of markets and interpretation than to the other topics. The present review will concentrate on highlights from the research efforts.

The appraisal of tourism markets to the Wet Tropics rainforests developed a new scale of analysis in Australian tourism studies. By collecting two tailored large-scale data sets in the region and by substantial reanalysis of existing state and national data, a reliable multifaceted base for considering the rainforest visitors to the Wet Tropics World Heritage Area was established.

The total Far North Queensland region received between 3 and 3.5 million visitors a year during the study period (WTMA 2003; Tourism Queensland 2005). This figure includes local regional travel, business travel and visiting friends and relatives travel as well as repeat visit behaviour. It is difficult to be definitive on the actual number of visitors to the Wet Tropics World Heritage area itself but across all sites and including rainforest tourist attractions and towns, a scale of two million visitors is a base figure. For comparative purposes the number of commercial visitors taking paid trips to the Great Barrier Reef from 2000 to 2004 from the Far North Queensland region averaged 750 000. The latter figure was obtained on the basis of the environmental levy information collected by reef operators and is consistent with the view that there were approximately two million visitors to the more readily accessible rainforest environment.

From the total holiday market to Queensland, which was shown to consist of six activity groups, two groups in particular were and remain highly relevant to the Wet Tropics region. One was an older predominantly Australian touring and sightseeing group and a second segment was a younger international budget accommodation group. There are also some younger domestic Australian visitors who use local resorts and hotels (Pearce *et al.* 1996a). These broad groups were explored more fully in specialist papers where characteristics of the senior self-drive market to northern Australia and international differences in young budget (backpacker) groups were intensively studied. Some of the findings from these studies included the perspective that senior self-drive travellers consisted of international visitors, domestic travellers touring without children and some domestic travellers with families. The older domestic unaccompanied self-drive markets are the most satisfied with the region as a whole, visit more places and are often on a second and third visit to the region (Pearce 1999).

Using the distinction between convergence and divergence in market development, one of the achievements of the detailed studies of backpacker behaviour in the region was to highlight the similar motivational and activity preferences of backpackers from different nationalities' dominant source markets (Pearce 2005c). This finding favours the convergence theory of market evolution and thus suggests that business operators need not stress unique experience opportunities in the rainforest for backpacker nationality groups. Recent evidence suggests that other nationality groups but especially Israeli and Asian backpackers may have some different requirements and styles in their travel behaviour (Maoz 2004). Better facilities and catering for ethnic and religious needs may become more important in these emerging markets.

The theme of nationality difference in market segments and visitor behaviour was also explored in some other ways. The very basis of how to segment tourism markets was addressed, with comparisons being drawn between geographical (place of origin) approaches and activity-based analyses (Moscardo *et al.* 2001). It was demonstrated that activity and interest segments offer more subtle understanding but that geographical segments may be more readily understood.

Studies of the nationality of visitors were also developed in the latter phases of the market research analyses. Using digitally altered photographs of visitors at rainforest settings Yagi (2004) revealed that Japanese visitors preferred to be with Western visitors, a finding contrary to some stereotypes about Japanese tourists. Western visitors had no clear preferences for the types of other visitors present, simply preferring that few other visitors were in the setting. Yagi (2004) explained the Japanese preference to be with Western visitors in two ways: there is the concept of *akogare* or respect for cool Western styles that informs the behaviour of some Japanese; and being with Westerners is seen as offering a greater personal freedom in how to behave outside of the strongly restrictive Japanese social norms of conduct. From a managerial perspective this finding of how groups of Japanese tourists respond is an easy to accommodate result because it means that this Asian group can easily be mixed with Westerners who showed no particular fellow visitor preferences. Additionally, the growth in numbers of Asian tourists to Australia emerged as a major dimension of Australia's tourism future during the study period. This growing importance reflects the input of global external forces affecting rainforest tourism, as foreshadowed in Figure 7.1. China, in particular, was identified as a major new market for regional and rainforest tourism, with the implication that much work remains to be done to interpret settings and accommodate the new waves of Mandarin speaking visitors (Moscardo 2004).

The role and influence of interpretation on visitor responses to the rainforest represented an important second line of work in the tourism social science programme. The efforts in researching interpretation were directed towards three areas: interpretation at leading attractions; visitor centres and their role in interpretation; and signs and symbols in interpretation. Based on this work and related studies in reef tourism and other settings, a full model of interpretation was developed that has power for numerous settings. A planned study of tourist guides in the rainforest setting was not undertaken due to funding limitations but information from other sources on this topic was used to help develop the mindfulness model of interpretation.

Some highlights from the effort at researching interpretation at leading rainforest tourism-related attractions included a study of the effectiveness of commercially based interpretation at Skyrail, the area's iconic cableway tourist attraction that traverses the rainforest for 7.5 km. Using regression analysis based on survey results of nearly 400 visitors, it was found that three of the four most important and significant factors affecting satisfaction were related to the interpretive experiences offered. This finding consolidated the perspective that interpretation matters to commercial businesses in natural environments and confers business advantages rather than being a practice restricted to public management agencies seeking to educate the visitors.

Another tourism context using interpretation was also explored; in this case the Tjapukai Cultural Park was the venue for a study of visitor attitudes to the rainforest aboriginal community and presentation of that culture. It was found that four market segments appreciated the cultural park to different degrees. There was an enthusiastic culturally aware international market; an interested but cautious group who avoided direct contact with the aboriginal staff; a commercially important group who collected the symbols and signs of the visit but whose interest was dominated by fun and entertainment; and finally an accompanying persons group for whom the park had modest appeal. This kind of result, it can be argued, highlights the need not to make sweeping generalizations about the appeal of indigenous tourism displays and performances but instead suggests that, in a postmodern sense, tourist groups exhibit much diversity and difference in interests (Ryan & Huyton 1998; Moscardo & Pearce 1999).

One further study of a significant tourist attraction was that by Woods (2000), who investigated tourist interests in wildlife at the Rainforest Habitat, an enclosed richly vegetated micro-zoo and dining facility. Woods's (2000) findings, like other regional studies of tourists' responses to wildlife, indicated that most visitors do have a keen interest in wildlife, that accessing wildlife is seen as a consistent problem and that the needs of many visitors are well met by structured wildlife experiences such as those offered in enclosed, capture settings.

The studies of visitor centres developed the notion that such centres have five functions: promotion, enhancement, control, substitution and civic roles. All are important. Different centres fulfil these roles to different degrees and the planning of centres in a region should be strategic, avoiding repetition of purposes and themes (Pearce 2004). Another aspect of interpretation studied was good practices in sign design, noting issues of clarity, readability and layout (Ballantyne *et al.* in press).

An important contribution of the tourism research programme arising from and, by the end of the programme, helping to define interpretation studies was the evolution of a mindfulness model (Moscardo 1999; Pearce 2005a). Mindfulness is a mentally active state where individuals are processing information and forming new connections between what they know and what they are processing. By way of contrast, mindlessness is a cognitive state where individuals may look as if they are paying attention but in reality are responding with pre-existing scripts, behaving according to fixed categories and not genuinely integrating what they are hearing, reading or viewing (Langer 1989). The mindfulness model of interpretation is presented in Figure 7.2 as a summary of the forces contributing to effective tourist learning and satisfaction.

The value of the mindfulness interpretation model is that it provides a systematic tool for investigating what is likely to be good design and practice in interpretation. The application of this model to other natural environment settings has been an important outcome of the rainforest tourism research programme. In particular the $7.5 million visitor centre and interpretation space built at Flinders Chase National Park on South Australia's Kangaroo Island was partly designed using mindfulness principles and its success with the public vindicates the nexus between research, conceptual development and practice in this area (Pearce & Moscardo 2005).

Tourism businesses and management practices, while identified as an important element of the research agenda, were studied in only a limited way in the programme of research. In addition to the interpretation and satisfaction studies at the three sites already mentioned, specialist accommodation businesses were investigated (Moscardo *et al.* 1996). Evidence was provided that visitors staying in the more specialist boutique rainforest resorts, farm stays and bed and breakfast options in the northern part of the Wet Tropics World Heritage area were more likely to exhibit an interest in sustainability practices. Another business and managerial study of note was directed at an understanding of public attitudes towards user fees (Lee & Pearce 2002). In this doctoral study attitudes of both international tourists and local residents towards fees and user pays philosophies were explored, with a key finding being some acceptance of the practice for World Heritage areas but a strong resentment of the practice for most other natural settings. There remains much scope to explore further the community, domestic tourist and international tourist views on tourism businesses in general. Additionally, attitudes towards the current and possible management practices for the rainforest region could be expanded and be a powerful realization of Machlis's perspective that social science research results are usable knowledge for management.

Two further contributions with broad implications completed the Wet Tropics rainforest tourism social science research effort. A new perspective on community attitudes towards tourism was developed, built in part on case studies from both the northern and southern shires in the Wet Tropics region (Pearce *et al.* 1996b). This approach suggested that attitudes to tourism depend not as much on an equity or personal trade-off approach by community members as has often been advanced (Ap 1990), but arise from people's understandings of tourism as a phenomenon. These understandings constitute organized systems of knowledge

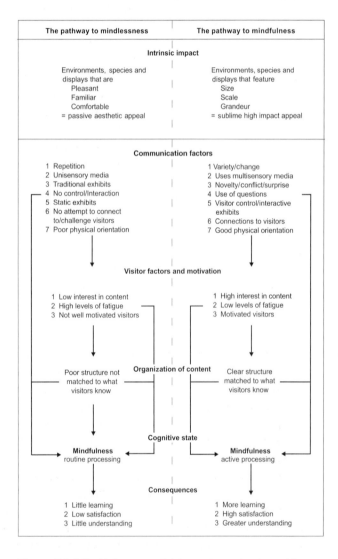

Figure 7.2 Mindfulness model for communicating with visitors. *Source*: after Moscardo (1999), Pearce (2005a).

that individuals share with others. The term to define these knowledge systems is social representation and such clusters of attitudes, knowledge and understanding are linked to media reports and the communication of scientific research. The particular value of this approach lies in reshaping research and public communication efforts about tourism, since it is less important to explore, say, age differences in community attitudes than to seek commonalities across demographic variables for the ways of seeing the world. An important implication here is that attempts to inform and educate communities about tourism should no longer be homogeneous but tailored to the prevailing myths and understandings that dominant sections of the community share. It is an important reflection of the innovative nature of this work on community attitudes to tourism that we now see this style of analysis in the major texts and references on tourism community analysis (Murphy & Murphy 2004).

A final direction of substance established during the rainforest tourism research programme was a conceptual readjustment in assessing one of the key visitor outcome variables pertaining to sustainability, that of visitor satisfaction. Visitor responses to their destination experiences are embedded in most monitoring and sustainability systems for tourism. For example, the Australian derived Tourism Optimisation Management Model uses visitor satisfaction as one of its key social outcome measures (McCarthur 2000). The particular contribution of the research conducted in the rainforest settings and the Wet Tropics region generally was to shift the assessment of satisfaction away from expectancy disconfirmation models and towards post-visit only assessments supported by benchmarking data (Pearce in press). In brief, the argument here is that the expectancy models are unworkable for unpredictable, unfamiliar experiences and produce blunt non-insightful research into what visitors think of tourism settings and services. It is better to seek only post-experience satisfaction scores on expanded ten point rating scales and be able to refer the answers obtained to other data scores obtained in other places, at previous times and in previous seasons. In this view a satisfaction score of 8.5 out of 10 on a Likert scale is no longer judged according to visitor expectations but according to other satisfaction scores from other tourism experiences, businesses and regions. Additionally, the rainforest tourism research was the

beginning of a more rounded, fuller treatment of the satisfaction outcome measure in tourism study as related concepts such as enjoyment and eudaimonia (human flourishing or overall well-being) were identified as adding additional interpretive power to the more functional assessments of satisfaction (Noe 1999; Pearce 2005a).

The progress of tourism research

Tourism research itself has not stood still while these efforts at understanding visitors and tourism to the Wet Tropics rainforest have been under way. In the past five years in particular, some shifts in the paradigms for investigating tourism have flourished.

One particularly powerful and recent trend in the design of tourism studies has been the growth of sophisticated research efforts in systems thinking, mobility and qualitative research (Ritzer 1999; Farrell & Twining-Ward 2004; Phillimore & Goodson 2004; Woehler 2004). This new qualitative emphasis in particular is not small sample, low power, quantitative work but intensively researched analysis of people's experiences in the best sociological traditions (Pine & Gilmour 1999; Franklin 2003; Bryman 2004; Cohen 2004; Maoz 2004; Matos 2004; Hall 2005). It is important to reflect on what thick qualitative work can and cannot do for rainforest tourism studies. It is inadequate in describing the distribution and patterns of tourist and community attitudes, responses and behaviour. That is, it is not insightful or reliable in assessing how many people from a particular group behave or think. Instead its value lies in obtaining new insights about a sub-group, such as emerging Israeli backpackers or local politicians. For broader visitor studies, thick qualitative work needs the buttress of well executed, externally valid and large sample quantitative visitor appraisals.

In conclusion, the opportunities for insightful analysis of many parts of the rainforest tourism system remain. Such analysis is a challenge to researchers, it is required for informed scientific management and it could assist the triple bottom line growth of tourism businesses. The Wet Tropics in Australia have been fortunate to receive a surge of fundamental tourism social science research attention during the first period of their management. In order to build on this foundation, and to promote the sustainability of the region for all parties, the role of people-based research must be extended and expanded in the next decade.

Summary

The tourism social science research reported here is set within two frameworks. It exists within the ambit of sustainable development studies thus stressing that it seeks to provide information on knowledge gaps for long-term regional performance rather than narrowly supporting specific business models or conversely anti-tourism and exclusively pro-environment agendas. The approach also stresses that a focus on natural attributes of the rainforest alone does not ensure an understanding of sustainable tourism and that the operation of contemporary tourism depends on a complex international system of demand, local management practices and community support.

The key achievements of this tourism social science programme for Australia's Wet Tropics do have a broad and international relevance in the following areas:
- the formulation of a full social science research agenda for tourism studies in rainforest destinations;
- the development of a multifaceted guide to understanding the forces impinging on rainforest tourism sustainability;
- the creation and synthesis of a visitor database and resource to analyse existing markets;
- the identification of details about visitor characteristics, needs and patterns for the two dominant markets, the self-drive older Australian visitors and the international backpacker market;
- insights into important preferences of Japanese visitors and the identification of the need for a research understanding of emerging Asian markets and, in particular, Chinese visitors;
- the full development of an explanatory yet pragmatic mindfulness model of interpretation that describes success factors influencing the presentation of rainforest settings;
- a new perspective on interpreting community attitudes to tourism where the social representations or macro-views of tourism are seen as the basis for public communication rather than stressing demographic differences in the community;
- a focused and revised approach to assessing satisfaction in natural environments where the emphasis is on benchmarking studies and post-visit assessments rather than expectation-based assessments;
- a concluding view that social science rainforest research needs to maintain a fully up-to-data awareness of recent advances in tourism studies more generally, including the balance needed between small-scale insightful studies and large-scale survey work that tracks patterns.

References

Ap, J. (1990). Residents' perceptions research on the social impacts of tourism. *Annals of Tourism Research* **17**(4): 610–16.

Archer, B. & Cooper, C. (1994). The positive and negative impacts of tourism. In *Global Tourism*, Theobald, W. F. (ed.). Butterworth-Heinemann, Oxford, pp. 73–91.

Ballantyne, R., Hughes, K. & Moscardo, G. (in press). Interpretive signs: principles and practice (www.talm. uq.edu.au/signage/signage-interpretation.html).

Bentrupperbäumer, J. M. & Reser, J. P. (2002). *Measuring and Monitoring Impacts of Visitation and Use in the Wet Tropics World Heritage Area 2001/2: A Site Based Regional Perspective*. James Cook University and Rainforest CRC, Cairns.

Bentrupperbäumer, J. M, Reser, J. P. & Turton, S. (1998). *Impacts of Visitation and Use in the Wet Tropics of Queensland World Heritage Area Stage 1*. Rainforest CRC, James Cook University, Cairns.

Brake, L. & Williams, M. (1990). Managing visitor impacts: a conflict in perception. In *Managing Conflict in Parks and Recreation*. Royal Australian Institute of Parks and Recreation, Canberra.

Bramwell, B. & Lane, B. (2005). Sustainable tourism research and the importance of societal and social science trends. *Journal of Sustainable Tourism* **13**(1): 103.

Bryman, A. (2004). *The Disneyization of Society*. Sage. London.

Buhalis, D. & Fletcher, J. (1995). Environmental impacts on tourism destinations. In *Sustainable Tourism Development*, Coccossis, H. & Nijkamp, P. (eds). Avebury Ashgate Publishing, Aldershot, pp. 3–25.

Butcher, J. (2003). *The Moralisation of Tourism. Sun, Sand … and Saving the World?* Routledge, London.

Butler, R. W. (1991). Tourism, environment and sustainable development. *Environmental Conservation* **18**(3): 201–9.

Cohen, E. (2004). Backpacking: diversity and change. In *The Global Nomad*, Richards, G. & Wilson, J. (eds). Channel View, Clevedon, pp. 43–59.

Eggington, J. (2003). *Spiral Guide: Thailand*. Duncan Baird, London.

Elkington, J. (1997). *Cannibals with Forks. The Triple Bottom Line of 21st Century Business*. Capstone, Oxford.

Farrell, B. & Twining-Ward, L. (2004). Reconceptualising tourism. *Annals of Tourism Research* **31**(2): 274–95.

Franklin, A. (2003). *Tourism: An Introduction*. Sage, London.

Gadek, P. A. & Worboys, S. J. (2002). *Rainforest Dieback. Mapping and Assessment:* Phytophthora *Species Diversity and Impacts on Rainforest Canopies*. School of Tropical Biology, James Cook University and Rainforest CRC, Cairns.

Gunn, C. A. & Var, T. (2002). *Tourism Planning: Basics, Concepts, Cases*, 4th edn. Taylor and Francis, New York.

Hall, C. M. (2005). *Tourism: Rethinking the Social Science of Mobility*. Pearson Education, London.

Kekul, L. B. (1999). *Wildlife in the Kingdom of Thailand*. WKT Publishing, Bangkok.

Langer, E. J. (1989). *Mindfulness*. Addison-Wesley, Reading, MA.

Lee, D. & Pearce, P. L. (2002). Community attitudes to the acceptability of user fees in natural settings. *Tourism and Hospitality Research* **4**(2): 158–73.

McCarthur, S. (2000). Beyond carrying capacity: introducing a model to monitor and manage visitor activity in forests. In *Forest Tourism and Recreation: Case Studies in Environmental Management*, Font X. & Tribe J. (eds). CABI Publishing, Wallingford, pp. 259–78.

Machlis, G. E. (1996). *Usable Knowledge: A Plan for Furthering Social Science and the National Parks*. United States Department of the Interior, National Park Service, Washington, DC.

Maoz, D. (2004). The conquerors and the settlers: two groups of young Israeli backpackers in India. In *The Global Nomad*, Richards, G. & Wilson, J. (eds). Channel View, Clevedon, pp. 109–22.

Matos, R. (2004). Can slow tourism bring new life to Alpine regions? In *The Tourism and Leisure Industry: Shaping the Future*, Weiermair, K. & Mathies, C. (eds). Haworth, New York, pp. 93–102.

Mill, R. C. & Morrison, A. M. (1992). *The Tourism System: An Introductory Text*, 2nd edn. Prentice Hall, Englewood Cliffs, NJ.

Moscardo, G. (1999). *Making Visitors Mindful: Principles for Creating Sustainable Visitor Experiences through Effective Communication*. Sagamore Publishing, Champaign, IL.

Moscardo, G. (2004). Exploring change in Asia Pacific tourism markets. In *Globalization and Tourism Research: East meets West*, Chon, K., Hsu, C. & Okamoto, N. (eds). Conference Proceedings Asia Pacific Tourism Association Tenth Annual Conference, Nagasaki, Japan, 4–7 July, pp. 369–78.

Moscardo, G. (2005). Peripheral tourism development: challenges, issues and success factors. *Tourism Recreation Research*, **30**(1): 27–43.

Moscardo, G., Morrison, A. M. & Pearce, P. L. (1996). Specialist accommodation and ecologically-sustainable tourism. *Journal of Sustainable Tourism*, **4**(1): 29–52.

Moscardo, G. & Pearce, P. L. (1999). Understanding ethnic tourists. *Annals of Tourism Research* **26**(2): 416–34.

Moscardo, G. & Pearce, P. L. (2000). Seasonality and tropical tourism. In *Tourism: A Strategic Industry in Asia and Pacific. Defining Problems and Creating Solutions*, McKellar, B. (ed.). Proceedings of The Sixth Asia Pacific Tourism Association Annual Conference, 28 June to 1 July , Phuket, Thailand, pp. 91–100.

Moscardo, G., Pearce, P. L. & Morrison, A. (2001). Evaluating different bases for market segmentation: a comparison of geographic origin versus activity participation for generating tourist market segments. *Journal of Travel and Tourism Marketing* **10**(1): 29–49.

Murphy, P. & Murphy, A. (2004). *Strategic Management for Tourism Communities: Bridging the Gaps*. Channel View Publications, Clevedon.

Nakjan, S. (2004). *English for Tourism in Phetchaburi Maung*. Phetchaburi Rajabaht University, Phetchaburi.

Noe, F. P. (1999). *Tourism Service Satisfaction*. Sagamore, Champaign, IL.

Page, S. (2005). Academic ranking exercises – do they achieve anything meaningful? A personal view. *Tourism Management* **26**: 663–6.

Pearce, P. L. (1999). Touring for pleasure: studies of the senior self-drive travel market. *Tourism Recreation Research* **24**(1): 35–42.

Pearce, P. L. (2004). The functions and planning of visitor centres in regional tourism. *Journal of Tourism Studies* **15**(1): 8–17.

Pearce, P. L. (2005a). *Tourist Behaviour: Themes and Conceptual Schemes*. Channel View Publications, Clevedon.

Pearce, P. L. (2005b). Sustainability research and backpacker studies: intersections and mutual insights. Keynote speech, ATLAS Backpacker Conference, Bangkok, Thailand, 1–3 September.

Pearce, P. L. (2005c). Great divides or subtle contours? Contrasting British, North American/Canadian and European backpackers. In *Down the Road: Exploring Backpacker and Independent Travel*, West, B. (ed.). Perth: API Network, Perth, pp. 131–51.

Pearce, P. L. (in press). The value of a benchmarking approach for assessing service quality satisfaction in environmental tourism. In *Tourism and Hospitality Services Management: An International Case Book*, Laws, E. and Prideaux, B. (eds). Continuum, London.

Pearce, P. L., Morrison, A., Scott, N., O'Leary, J., Nadkarni, N. & Moscardo, G. (1996a). The holiday market in Queensland: building an understanding of visitors staying in commercial accommodation. In *Tourism and Hospitality Research: Australian and International Perspectives*, Prosser, G. (ed.). Proceedings from the Australian Tourism and Hospitality Research Conference, Coffs Harbour. Bureau of Tourism Research, Canberra, pp. 427–39.

Pearce, P. L. & Moscardo, G. (2005). *The Flinders Chase National Park Visitor Centre: An Evaluation*. Report to the South Australian National Parks and Wildlife Service, September 2005.

Pearce, P. L., Moscardo, G. M. & Ross, G. F. (1996b). *Tourism Community Relationships*. Pergamon Press, Oxford.

Phillimore, J. & Goodson, L. (eds) (2004). *Qualitative Research in Tourism*. Routledge, London.

Pine, B. J. II & Gilmour, J. (1999). *The Experience Economy*. Harvard Business School Press, Boston.

Ritzer, G. (1999). *Enchanting a Disenchanted World: Revolutionising the Means of Consumption*. Pine Forge Press, Thousand Oaks, CA.

Ruhanen, L. & Cooper, C. (2004). Applying a knowledge management framework to tourism research. *Tourism Recreation Research* 29(1): 83–8.

Ryan, C. & Huyton, J. (1998). Who is interested in Aboriginal tourism in the Northern Territory, Australia? Paper given at the Council of Australian Universities in Tourism and Hospitality Education (CAUTHE) National Tourism and Hospitality Conference, Jupiters Casino, Gold Coast, February.

Tourism Australia (2005). www.australia.com/

Tourism Authority of Thailand (2005). *Sawadee Thailand: Phetchaburi* (www.tourismthailand.org/).

Tourism Queensland (2005). www.tq.com.au/index.cfm

Tourism Tropical North Queensland (2005). www.tropicalaustralia.com.au

Turton, S. M. (2005). Managing environmental impacts of recreation and tourism on rainforests of the Wet Tropics of Queensland World Heritage Area. *Geographical Research* 43(2): 140–51.

Turton, S. & Stork, N. (2006). Tourism and tropical rainforests: opportunity or threat? In *Emerging Threats to Tropical Rainforests*, Laurance, W. F. & Peres, C. A. (eds). University of Chicago Press, Chicago.

Wet Tropics Management Authority (WTMA) (2003). *Wet Tropics Management Authority Annual Report 2002–3*. Wet Tropics Management Authority, Cairns.

Woehler, K. (2004). The rediscovery of slowness, or leisure time as one's own and as self-aggrandizement? In *The Tourism and Leisure Industry: Shaping the Future*, Weiermair, K. & Mathies, C. (eds). Haworth, New York, pp. 83–90.

Woods, B. (2000). Beauty and the beast: preferences for animals in Australia. *Journal of Tourism Studies* 11(2): 25–35.

Woods, B. A. (2003). Examining the characteristics of wildlife tourists and their responses to Australian wildlife tourism experiences. Unpublished doctoral dissertation, James Cook University, Townsville.

World Tourism Organization (2004). *Indicators of Sustainable Development for Tourism Destinations. A guidebook*. World Tourism Organization, Madrid.

Yagi, C. (2004). Tourists' views of other tourists. Unpublished PhD thesis. Tourism Program, James Cook University, Townsville.

Young, M. (1998). The Daintree region. Unpublished doctoral dissertation, James Cook University, Townsville.

8 THE DYNAMIC FOREST LANDSCAPE OF THE AUSTRALIAN WET TROPICS: PRESENT, PAST AND FUTURE

David W. Hilbert*

CSIRO Sustainable Ecosystems, Herberton, Queensland, Australia
*The author was participant of Cooperative Research Centre for Tropical Ecology and Management

Introduction

Under cursory examination for short time-scales, for example, a human generation, the natural landscape may appear to be relatively unchanging. However, ecological research shows that change in landscapes is constantly occurring at many temporal and spatial scales and that this dynamic is driven by geological and evolutionary processes, climate change and human impacts of various kinds. Today, the rate of landscape change is so rapid and widespread that the phrase global change has come into currency. Describing, understanding, predicting and managing this rapid change have become a major scientific pursuit.

This chapter focuses on change in the dynamic landscape of the Wet Tropics bioregion (Sattler & Williams 1999), especially the complex mosaic of forest types that is the context within which the diversity of all biodiversity flourishes. The goal of this research is an understanding of the dynamic distributions of ecological properties and processes in time and space, especially forest types and other land-cover classes. Published and previously unpublished results are presented along with a partial review of other relevant research.

The changing geographical distribution of vegetation is an essential feature of the region that influences its matter and energy exchange with the atmosphere, evolutionary processes, biodiversity and many other ecological parameters. Measuring, mapping and understanding these distributions are a fundamental prerequisite for informed management and sustainable use. The research presented here relies extensively on the high quality vegetation maps produced by Tracey and Webb (1975). These maps, and other spatial information about climate, geology and terrain, provide the essential basis for the analysis of the controls on the landscape mosaic and an understanding of the main drivers of change in the present, past and future.

Using this spatial information and a variety of modelling techniques, it is possible to infer the environmental factors that determine the present land-class distributions, estimate their potential distributions, indicate the pre-clearing pattern, predict the conditions that are favourable for land-clearing, estimate changes in forest distributions in the recent geological past and assess the sensitivity of forests to future climate change.

This chapter begins by describing the landscape as it existed at the time of the Tracey and Webb mapping (based on aerial photography in the mid-1970s) and the results from analyses of what environmental variables best explain these distributions. A modelling methodology is then briefly reviewed, followed by estimates of the impact of clearing since European settlement and the conditions that favour(ed) land-clearing. Factors causing the conversion of tall open forests (wet sclerophyll forests) to rainforest, a major dynamic at the present, are briefly discussed. Estimates of forest biomass and carbon are given, then estimated distributions of forest environments since last glacial maximum (LGM at *c*.18 000 yr BP) are presented. This is followed by analyses of the sensitivity of the Wet Tropics vegetation to future climate change and an assessment of spatial constraints that may limit the ability of vegetation to track environmental change in the future. After landscape change from LGM into the future is summarized, management implications are discussed.

Present vegetation

The combination of high topographic, edaphic and climatic variability in the Wet Tropics bioregion of north-east Queensland (Sattler & Williams 1999) results in a diverse regional mosaic of many forest types that were mapped at 1 : 100 000 scale (Tracey & Webb 1975). The extent of this vegetation map (about 1 700 000 hectares of forest) and subsequent mapping of rainforests in the Paluma range defines the extent of the Wet Tropics region for the purposes of this chapter. A structural typology for rainforests (Webb 1959, 1968, 1978) and non-rainforest vegetation (Specht 1970) was used in the vegetation mapping. Detailed descriptions of the rainforest types are given in Tracey (1982). The vegetation classes are largely based on structural attributes such as canopy height, degree of canopy closure, complexity of the forest profile, the relative abundance of epiphytes and lianas and average leaf size of the canopy trees. Fine-grained vegetation mosaics that could not be mapped in detail at the 1 : 100 000 scale are classified as several classes of coastal complexes or mountain rock pavements.

The structural types can be broadly correlated with climatic zones and soil patterns, with some effect of disturbance, especially fire (Webb & Tracey 1981). While there are floristic correlations with structural types (Graham, 2006), and sometimes diagnostic species (Webb & Tracey 1981), the floristic composition at different locations of the same structural type can be very distinct. For example, two half hectare

plots of simple notophyll vine forest (on the same soil type and in very similar climates) have 70 and 77 tree species respectively (CSIRO permanent plots, unpublished data). Of the total of 110 species (914 stem >10 cm DBH (diameter at breast height)) found in the two plots, only 37 are found in both plots and the most common species in each plot are entirely different.

Occupation of the Wet Tropics by aboriginal people began at least 40 000 years ago (Singh *et al.* 1980), while the first European exploration occurred in 1848. Settlement by Europeans resulted in cattle grazing, timber cutting, clearing for sugarcane and tin mining. Approximately 15% of the study area, as defined here, was cleared of natural vegetation at the time of vegetation mapping. Winter *et al.* (1987) report that by 1985 approximately 25% of the Wet Tropics bioregion was cleared of natural vegetation. The discrepancy in these two estimates is mainly due to the inclusion of a large area of woodland in the study area as defined here.

Table 8.1 describes the extent and spatial characteristics of the region's vegetation in 1975. Medium and low woodlands were the most extensive class in this mapping of the Wet Tropics. Woodlands are the landscape context (largely in the dry west) within which rainforests are embedded as long-term refugia. Clearing of native vegetation after European settlement has had a significant influence on the region's forests and produces the second most common vegetation class. The most extensive rainforests today are complex mesophyll vine forests and simple vine forest with *Agathis microstachya*. The other most extensive forests are largely dominated by types such as vine forests with *Eucalyptus* and *Acacia* that indicate transitions from fire-dominated sclerophyll forests to non-fire rainforests. The cooler climate forests that support the majority of the region's local endemic species represent a small proportion of the overall area.

Environmental controls on forest distributions

The importance of bioclimatic variables, topography and geology for the spatial distribution of all of the vegetation types was assessed through several statistical analyses (Ostendorf & Hilbert, unpublished). The significance of relationships was assessed using generalized linear models (GLM), generalized additive

Table 8.1 The extent and spatial characteristics of the region's vegetation, ordered by area and based on an analysis of the available digital map

Forest class	Class number	Area (ha)	Percentage of landscape	Number of patches	Mean patch size (ha)
Medium and low woodlands	16	8360903	37.18	300	2787
Cleared areas	24	326905	14.54	299	1093
Complex mesophyll vine forests	1	308243	13.70	227	1358
Simple vine forest often with *Agathis microstachya*	8	195308	8.67	58	3367
Vine forests with eucalyptus and acacia	13	94815	4.22	613	155
Mesophyll vine forests	2	73667	3.28	365	202
Tall open forests and tall woodlands	14	61120	2.72	191	320
Vine forests with *Acacia*	12	49011	2.18	337	145
Complex notophyll vine forest with emergent *Agathis robusta*	6	48178	2.14	264	182
Saline littoral zone	22	42783	1.90	115	372
Medium open forests and medium woodlands	15	40203	1.79	160	251
Texture-contrast soils with impeded drainage on coastal plains	20	33067	1.47	78	424
Coastal floodplains and Piedmont slopes	19	25671	1.14	34	755
Complex notophyll vine forests	5	23199	1.03	52	446
Coastal beach ridges and swales	17	19639	0.87	95	207
Mountain rock pavements	21	18521	0.82	83	223
Simple microphyll vine-fern thicket	10	12189	0.54	21	580
Deciduous microphyll vine thicket	11	7578	0.34	72	105
Swampy coastal plains	18	7645	0.34	25	306
Simple microphyll vine-fern forest often with *Agathis montana*	9	7062	0.31	10	706
Semi-deciduous mesophyll vine forests	4	4778	0.21	99	48
Araucarian notophyll vine forest	25	3943	0.18	39	101
Coastal plains and foothills	23	2735	0.12	18	152
Notophyll semi-evergreen vine forest	27	2734	0.12	53	52
Mesophyll vine forests with dominant palms	3	2405	0.11	30	80
Notophyll vine forest rarely without *Acacia* emergents	7	610	0.03	19	32
Low microphyll vine forest and *Araucarian* microphyll vine forest	26	563	0.03	16	35

models (GAM) and the G-model approach (Ostendorf & Reynolds 1998). These analyses used the full set of 36 bioclimatic variables from BIOCLIM and 43 topographic variables. The latter were derived from a regional digital elevation model (DEM) with 100 metre pixel size and derived from 1 : 50 000 topographic maps. These variables include elevation, slope, aspect, flow accumulation, curvature and these variables smoothed at various spatial resolutions and directions.

The analysis indicates that the overall vegetation pattern of the Wet Tropics is most strongly influenced by mean annual precipitation and mean annual temperature. Topographic location and horizontal redistribution of water play a minor role. Individually, most forest types are best predicted by a combination of some temperature and a precipitation variable, with quadratic terms often being significant. Soil parent material and topography play a more significant role for some vegetation types, such as coastal complexes.

Overall, the analyses demonstrate the strong climatic control of the tropical forests in the region.

Modelling approach

While many techniques can be used to quantify vegetation–environment relationships (discriminate analysis, general linear and general additive models, classification trees, genetic algorithms etc.), I used a feedforward artificial neural network trained by backpropogation (Rumelhart & McClelland 1986). Advantages of this method include that it makes no assumptions about the statistical distributions of the variables (Brown *et al*. 1998), it can use categorical independent variables such as soil type, it is non-linear, it can simultaneously model a large number of vegetation classes, it can effectively discount outliers (Haung & Lippman 1988) and it often outperforms other techniques when faced with complex classification problems.

FANN (Forest Artificial Neural Network; Hilbert & van den Muyzenberg 1999) characterizes the relative suitability of environments for 15 forest classes based on the 26 mapping types (see Table 8.2). Inputs to the model include seven climate variables, nine soil parent material classes and seven terrain variables (see Table 8.3). A separate model, called Palaeo-Forest Artificial Neural Network (PFANN), was also developed that excluded distance to the coast as an input but was otherwise the same in all respects as FANN. These models and the methods used in their application are further described in Hilbert and Ostendorf (2001).

Table 8.2 Vegetation classes used in the modelling, abbreviations used, and the corresponding types in the classification of Tracey and Webb (1975). Detailed descriptions of rainforest classes are given in Tracey (1982)

Abbreviation	FANN class	Tracey and Webb (1975) class	Description
MVF	1	1, 2	Mesophyll vine forests – rainforest with a complex structure, high diversity of vascular epiphytes, and a canopy dominated by mesophyllous species
MVFP	2	3	Mesophyll vine forests with palms – rainforest with the canopy dominated by palms, occurring on poorly drained soils near the coast
SDMVF	3	4	Semideciduous mesophyll vine forest – mesophyll vine forest with canopy emergents often being deciduous
CNVF	4	5, 6	Complex notophyll vine forests – rainforests of cooler uplands, structurally complex, canopy dominated by notophyll species
NVF	5	7	Notophyll vine forests – rainforest found in drier coastal zones, simple structure, low canopy
SNSMi	6	8–10	Simple notophyll and simple microphyll forests and thickets – diverse group of montane rainforest in the coolest and wettest parts of the study area, simple structure
DMiVT	7	11	Deciduous microphyll vine thicket – rainforest with a low canopy of mainly drought deciduous species
VFAE	8	12, 13	Vine forest with *Acacia* and/or *Eucalyptus* – rainforest with sclerophyll canopy emergents, successional communities
TOFTW	9	14	Tall open forest and tall woodland – sclerophyll forests with tall canopies, occurring in moist environments
MOFW	10	15	Medium open forest and woodlands – medium height sclerophyll forests, including poorly drained coastal locations
MLW	11	16	Medium and low woodlands – dry, open sclerophyll woodlands
CC	12	17–20, 22, 23	Coastal complexes – large variety of fine grained vegetation mosaics of several rainforest and sclerophyll types, occurring near the coast
MRP	13	24	Mountain rock pavements – fine grained mosaic of dry rainforest and sclerophyll types on steep mountain slopes with thin soils
AVF	14	25	Araucarian vine forests – rainforest and woodland with dominant *Araucaria* spp., in the drier southern part of the study area
NSEVF	15	26, 27	Notophyll semi-evergreen vine forests – notophyll vine forest where many tree crowns become sparse in the dry season, true deciduous species generally absent

Table 8.3 List of the environmental variables used in FANN and their ranges in the region

	Minimum	Maximum
Annual mean temperature (°C)	16.1	25.7
Min. temp. of coldest period (°C)	5.1	18.4
Mean temp. of warmest quarter (°C)	20.9	27.9
Mean temp. of coldest quarter (°C)	10.8	23.1
Annual precipitation (mm)	775	7986
Precip. of wettest quarter (mm)	552	3751
Precip. of driest quarter (mm)	0	913
Slope (%)	0	61.7
Soil water index (logarithmic scale)	−9.7	26.4
Aspect NS	−59.2	54.4
Aspect EW	−50.4	56.2
Distance to nearest drainage line (m)	0	1613
Distance to nearest perennial stream (m)	0	50700
Distance to coastline (m)	0	83922

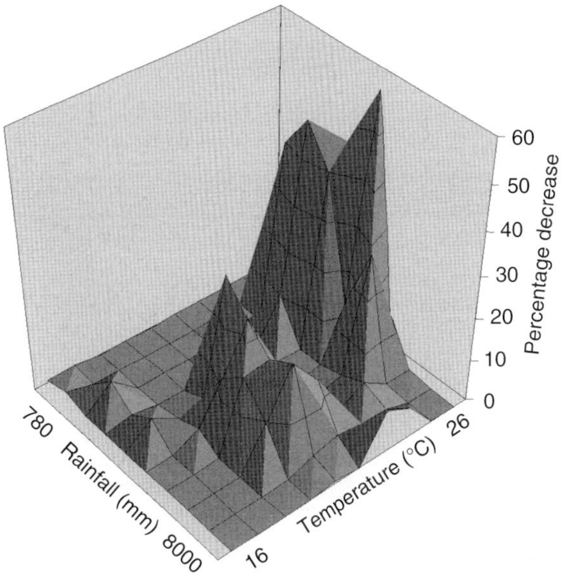

Figure 8.1 The proportion of climatic habitats, defined as having particular combinations of mean annual temperature (°C) and rainfall (mm), which are no longer available to most native species due to land-clearing in the region.

Estimated effects of clearing

One of the most direct and useful applications of predictive vegetation models is the estimation of the extent and distribution of vegetation classes before extensive clearing for agriculture or other purposes. European clearing removed nearly twice as much area in the lowlands (211 291 ha) as in the uplands (107 780 ha). Using FANN, I estimate that in the lowlands (<300 m), coastal complexes were the most cleared (99 401 ha), followed by mesophyll vine forests (69 470 ha), and medium open forests and woodlands (*Melaleuca* swamps) (30 668 ha). In proportion to pre-European areas, mesophyll vine forests with palms (64% cleared), coastal complexes (45% cleared) and *Melaleuca* swamps (45% cleared) were most affected by clearing. Notophyll vine forests (39% cleared) were also extensively cleared. In the uplands and highlands (>300 m), complex notophyll vine forests were the most cleared (38 487 ha), followed by mesophyll vine forests (31 070 ha), and medium and low woodlands (26 868 ha). In proportion to pre-European areas, two rainforest types, complex notophyll vine forests (40% cleared) and mesophyll vine forests (14% cleared) were most affected. Forest clearing has not only

removed forest types but also altered the overall area of climates available to most native species, as can be seen in Figure 8.1.

Where small-scale vegetation fragments have been mapped they provide an ideal, independent test of the vegetation model. While maps of small fragments exist for the Atherton Tablelands, the classification of forest types within these fragments did not follow the Tracey and Webb typology. However, rainforest and open, sclerophyllous forest types can be distinguished and compared with an aggregation of FANNs predictions. On this basis, FANN's overall accuracy is 88%. Prediction of rainforests is 90% correct and sclerophyll forests are predicted with 72% accuracy. Qualitatively, the predicted map predicts the overall proportions of forest types quite well, although the precise location of forest patches is somewhat less accurate. Predictions for the extensively cleared coastal plains appear to be less accurate, although a quantitative test based on fragment data has not yet been possible. The difficulty in these areas is that the spatial resolution of the soil parent material data, in particular the lumping of all alluvium into a single class irrespective of its source, may underestimate the area of rainforest that was cleared.

Properties of cleared land

Fourteen environmental variables were used to predict land-clearing in the Wet Tropics using a general additive model (GAM). These variables were: annual mean temperature, minimum temperature for the coldest period, mean temperature of the warmest quarter, mean temperature of the coldest quarter, annual precipitation, precipitation of the wettest period, precipitation of the driest period, geology, slope, flow accumulation, distance to coast, distance to drainage, distance to perennial streams and land tenure.

These variables were included in a forward/backward GAM stepwise selection process. The data set used in the stepwise procedure was a random sample totalling 50 000 points consisting of cleared and uncleared land in equal proportions. Applying the model to the entire region results in 75% overall accuracy (kappa statistic). The best predictors of land-clearing in the Wet Tropics are: minimum temperature of the coldest period, mean temperature of the coldest quarter, annual mean precipitation, geology, slope, distance to coast, distance to perennial streams and land tenure.

Further analyses using simpler models show that geology, tenure and slope are important predictors of cleared land, with tenure being the most important. Land that has a high probability of being cleared is freehold with alluvial-colluvial or basic volcanic soil parent material on a slope of less than 6.5%. Comparing model results with recent remnant mapping of the Atherton Tablelands, the model can predict future clearing (79% correct) but regrowth (abandonment) areas are predicted poorly, with only 15% accuracy.

Effects of altered fire patterns

The potential rates of change of forest structural types, in response to climatic change, are generally not known for the forests of our region. However, it is clear that rainforest can rapidly establish within more open sclerophyll forests (Harrington & Sanderson 1994). That this process has been happening for some time is evident from the large area of vine forests with *Acacia* or *Eucalyptus* emergents (94 815 hectares) compared with 61 120 hectares of tall open forest and woodland that has not yet been invaded by rainforest (see Table 8.1). Harrington and Sanderson (1994) estimated that a very

large percentage, up to 70%, of tall open forests in the Wet Topics have been invaded by rainforest species in the past 200 years. One hypothesis for this dynamic is the change from Aboriginal to European fire management (Harrington & Sanderson 1994). In the former, fires were lit throughout the dry season, while European graziers more often lit fires only after the first storm at the end of the dry period (Hill *et al.* 2001). Hill *et al.* (2001) propose that the interaction between grazing (horses and/or cattle) and fire is more important than fire *per se*.

Carbon in forests

The region's terrestrial carbon (C) pools are primarily plant biomass (mainly wood in forests) and soil carbon in both natural ecosystems and cultivated land. Analyses of CSIRO's long-term plots in undisturbed rainforests (Hilbert, unpublished) provide some insight into the C pools. The Wet Tropics' cooler, highland rainforests have some of the highest densities of biomass found anywhere in the world. Undisturbed upland forests contain as much as 800 Mg ha^{-1} of above-ground biomass, which translates to approximately 450 Mg C ha^{-1} (assuming 20% of total biomass is below ground and C is 45% of biomass by weight). Undisturbed lowland rainforests may contain approximately 280 Mg C ha^{-1}. Very rough guesses at the C content in the biomass of all Wet Tropics forests give an order of magnitude approximation of 0.3 Gt C. The Intergovernmental Panel on Climate Change (IPCC 2001) estimated that the world's vegetation contained 550 Gt C in 1980. So C in Wet Tropics forest may represent about 0.05% of the Earth's C in vegetation in about 0.001% of the global land surface. Certainly, the density of the vegetative C pool in the Wet Tropics is equal to or much greater than anywhere else in Australia and, excluding peatlands, most of the globe. Soil carbon in forests and agricultural systems increase the total terrestrial C pool. Because of the high amounts of C in these rainforests, forest clearing can add substantially to emissions of CO_2 to the atmosphere. Similarly, afforestation has the potential to decrease net emissions. Some coastal forests (e.g. paperbark swamps) may stock high amounts of carbon in the soil but this has not been quantified.

Analysis of biomass in the rainforest–wet sclerophyll transition (Ostendorf *et al.* 2004) showed that the spatial patterns of precipitation, and to a lesser degree temperature, are the strongest controls on forest biomass. Soil parent material and soil depth play a lesser role.

Past forests and climates

It is now widely recognized that climates changed appreciably in the tropics throughout the glacial cycles of the Pleistocene (Farrera *et al.* 1999). In the Neotropics, climate changes during the late Tertiary and Quaternary Periods indicate low-latitude temperature fluctuations of up to 5°C or 6°C (Colinvaux *et al.* 1996; Burnham & Graham 1999; Heine 2000) and similar cooling was widespread throughout the tropics (Vanderkaars & Dam 1997). In general, glacial cooling and aridity restricted the extent and altered the spatial distribution of tropical rainforests and depressed altitudinal zones, while warmer and wetter conditions during the Holocene allowed marked expansion of rainforests (Walker & Chen 1987; Flenley 1998). African lowland rainforest, e.g., may have contracted to 25% of its present area at the LGM, *c*.20–18 kyr BP, and expanded to three times present during the Holocene climatic optimum (HCO), *c*.5 kyr BP (Hamilton 1976). During glacial periods, these rainforests may have been replaced by tropical seasonal forest, the seasonal or dry forests were replaced by savanna or steppe and mountain forests occurred at lower elevations than today (Elenga *et al.* 2000).

During the late Pleistocene in Central America, lowland rainforest species may have been limited to riparian habitats and expanded with increased temperature and rainfall approximately 12 kyr BP (Aide & Rivera 1998). In contrast, rainforest remained in the Amazonian lowlands throughout the Pleistocene and the main effect of climate changes may have been on the distribution of heat intolerant plants responding to Holocene warming (Colinvaux *et al.* 2000; Colinvaux & De Oliveira 2001; Haberle 1999). In the Sunda shelf of Southeast Asia, drier climates during the peak of the last ice age led to a reduction in the extent of rainforests (Taylor *et al.* 1999). Rainforests in central New Guinea contracted to 75% of their present area at LGM (Walker & Chen 1987) and the tree line in the central highlands was 1500 m lower than today (Walker & Flenley 1979; Walker & Hope 1982).

In Australia, during the last glacial cycle, the most important climatic feature has been variation in precipitation, with the driest conditions occurring during the transition from the peak of the last glacial to the Holocene (Kershaw & Nanson 1993). The major change in the vegetation of the region, occurring within the past 140 000 years, involved the replacement of extensive moist rainforest by open eucalypt woodland, postulated to have been caused by the burning activities of Aboriginal people (Kershaw 1994). Over long-term and continental scales, all the extant rainforests in Australia can be thought of as refugia (Webb & Tracey 1981).

At the spatial scale of the Wet Tropics, it is now well established that several rainforest refugia have expanded and contracted throughout the Pleistocene. Charcoal collected from the Windsor Tableland, now dominated by complex humid rainforest, demonstrates the existence of pyrophytic forests dominated by *Eucalyptus* in the late Pleistocene, 26 860–12 750 yr BP (Hopkins *et al.* 1990). Further evidence from charcoal fragments collected throughout the rainforests of the Wet Tropics suggest that *Eucalyptus* woodlands or forests reached a maximum extent towards the end of the Pleistocene, between 13 and 8 kyr BP (Hopkins *et al.* 1993). Palynological evidence from Lynch's crater indicates sclerophyll forests and woodlands in that location from as early as 38 to *c*.8 kyr BP (Kershaw 1985, 1989). This expansion of sclerophyll forest has been attributed to both increasing aridity in the late Pleistocene and burning by Aborigines (Kershaw 1986). Further analysis of pollen and charcoal from Lynch's Crater suggests an increase in fire frequency at 45 ^{14}C kyr BP, supporting the view that human occupation of Australia occurred by at least 45–55 kyr BP (Turney *et al.* 2001). Analysis of a sediment core from Lake Xere Wapo, New Caledonia, shows a decline in *Araucaria* species at around 45 000 kyr BP, as is seen at Lynch's Crater, that could not have been caused by human fire use because humans had not yet reached this island (Stevenson & Hope 2005). The pollen and charcoal record from Lake Euramoo suggests the presence of dry sclerophyll woodland between 23 and 16.8 kyr BP, wet sclerophyll woodland between 16.8 and 8.6 kyr BP, warm temperate rainforest between 8.6 and 5 kyr BP, dry subtropical rainforest between 5000 and 70 years BP, followed by degraded dry subtropical rainforest (Haberle 2005).

Unlike the early Holocene re-expansion of upland rainforests, the sclerophyll forests in the lowland, Daintree region were present until at least 1400 years BP (Hopkins *et al.* 1996). These authors suggest that rapid sea level rise concentrated aboriginal populations in the area of the present coastline and that their burning activities were sufficient to re-establish sclerophyll forests during the latter part of the Holocene from approximately 4000 years BP, following a warmer and wetter period that would have been conducive to rainforest re-expansion. This expansion may have accelerated since the arrival of Europeans and the concomitant decrease in fire.

Webb and Tracey (1981) subjectively identified a large number of Pleistocene rainforest refugia in the Wet Tropics. These include all the presently cloudy and wet mountains, high tablelands and the very wet lowlands that now receive greater than 2500 millimetres of annual rainfall. Gallery forests were assumed to occur in drier areas of the permanent rivers, while wet coastal gorges on most of these rivers are also likely to be refugia.

Modelled spatial changes from the late Pleistocene to the present

Knowledge of the past distribution of vegetation and climates is most often based on palynology. While this is often the only direct empirical approach possible, biogeographical interpretations from pollen records are limited by the typically sparse spatial samples and the difficulties associated with the variable production and spatial redistribution of pollen from various taxa. Macrofossils, such as the charcoal studied by Hopkins *et al.* (1993), have the advantage that taxa can be placed at specific locations at specific times. My modelling approach is more like the bioclimatic analyses of Nix (1991), who used similar palaeoclimate reconstructions and mapped the spatial distribution of temperature classes (mesothermal and megathermal) throughout the Wet Tropics. He used these bioclimatic patterns to infer qualitatively the extent and distribution of rainforests at various periods in the past. Unlike Nix's model, PFANN is a quantitative model that classifies environments based on the present associations between forest structural classes and environmental variables. Unlike in purely bioclimatic analyses, soil parent material and terrain-related indices are also

included. This is important because past changes in vegetation pattern are best interpreted in relation to the total habitat, including edaphic and topographic factors (Webb & Tracey 1981). The result is conceptually similar to Nix's bioclimatic analyses but is more refined in the sense that the environments of many forest classes are distinguished and the classification is objective and quantitative.

Palaeoclimate scenarios were developed from previously published estimates (Nix & Kalma 1972; Nix 1991) and are presented in Table 8.4. These scenarios represent estimates of rainfall and temperature in three different climates thought to have been present in the Wet Tropics since the late Pleistocene. The earliest period is the LGM, when the region was the driest and coolest during the past *c*.100 000 years. During the Pleistocene/Holocene transition (PHT) the climate had warmed somewhat and is believed to have been wetter than today's climate. Following this, the HCO or altithermal was both warmer and wetter than present conditions.

Results for the Wet Tropics are presented in Figure 8.2 along with the potential distributions for today's climate (from FANN). Comparing the broad classes of woodland, tall open forest and rainforests, roughly 75% of the region had environments most suitable to pyrrhic, sclerophyll vegetation at LGM. Tall open forest environments were approximately three times as extensive at this time as in any of the other climates, including today. The extent of woodland environments responds strongly to reductions in rainfall and occupies 49% of the area at LGM, while they are least extensive in the cool, wet climate following the PHT when effective rainfall is the highest. Not surprisingly,

Table 8.4 Palaeo-climate scenarios for the Wet Tropics. Changes are relative to today's climate and rainfall changes are given as a multiplier

	Time period		
	LGM	PHT	HCO
All temp. variables	−3.5	−2	+2
Mean annual precipitation	0.5	1.2	1.25
Precipitation of the wettest quarter	0.5	1.2	1.25
Precipitation of the driest quarter	0.5	2	1.5

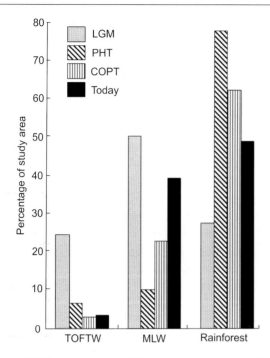

Figure 8.2 The percentage of the study area occupied by tall open forests and tall woodlands, medium and low woodlands, and rainforests (FANN types 1–8, 14 and 15) in each of the three palaeoclimate scenarios and the potential distribution in today's climate.

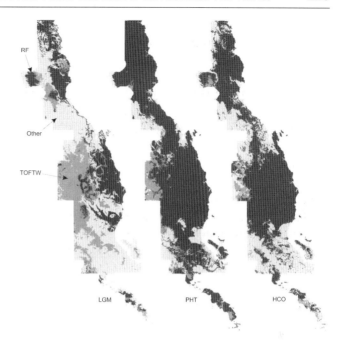

Figure 8.3 Maps of forest environments in three past climate scenarios.

rainforest environments as a whole respond inversely to woodlands, with the smallest extent at LGM and maximum at PHT. Specific rainforest environments respond individually to the climate changes studied here. Mesophyll vine forest environments cover 47% of the area in the warm, wet climate (HCO), while they are the most restricted rainforest environment (3% of the study area) at LGM. Complex notophyll vine forest environments are rare in all climates, with a maximum extent of 7.7% at LGM and a minimum of 2.3% at HCO. Simple notophyll and microphyll environments are less restricted at LGM (6.7%) than at HCO, when they have a minimum extent of 4% of the region. Their maximum occurs in the cool, wet climate of PHT (41%). Thus each of these three rainforest environments has a maximum extent in a different climate scenario. While it is difficult to test these predictions with a high degree of spatial and temporal resolution, they are qualitatively consistent with the past distributions of sclerophyll forests identified through carbon dating and electron microscopy of charcoal fragments taken from soil pits throughout the study area (Hopkins *et al*. 1993, 1996).

The spatial pattern of forest environments is strongly influenced by climate and different responses are predicted for each forest class. The distributions of rainforest, tall open forest and all remaining classes combined (mostly medium and low woodland) for each of the climate scenarios are presented in Figure 8.3. At LGM, rainforest environments occurred in three relatively distinct groups of patches in the northern, central and southern Wet Tropics. In the south, rainforest environments are mostly restricted to the higher elevations of the Paluma Range and are well isolated from the central block. A tenuous connection between the northern and central blocks exists along the so-called Black Mountain corridor. North of Cape Tribulation, much of the predicted rainforest environment is suitable for semi-deciduous mesophyll vine forest. Today, this forest type is common across northern Australia but rare in the Wet Tropics. The mapping suggests that it would have extended further north than the study region at LGM. It is likely that gallery forest also occurred on the extensive coastal plain that then extended many kilometres to the east. Tall open forest environments are predicted to have been very extensive in the west-central part of the Wet Tropics (probably extending west of the study area) and also occur on the Carbine and Windsor Tablelands and the Seaview Range.

In the cool, wet PHT scenario, rainforest environments expand to form a more or less continuous block from the northern limits of the region to the Walter Hill Range, with discontinuous patches extending south through the Seaview and Paluma Ranges. Tall open forest environments are drastically reduced. With further warming to the HCO, rainforest environments become somewhat more fragmented. Tall open forest environments are most restricted at this time.

The most important refugia for rainforests can be identified from their extent and persistence in the various subregions of the Wet Tropics bioregion (see Table 8.5). With the exception of the Bloomfield-Helenvale Lowlands, the subregions that retain greater than 100 km^2 of rainforest in all of the climate scenarios have a minimum rainforest extent at LGM. The Mossman Lowlands retain roughly the same extent of rainforest in all scenarios, which indicates a very high degree of stability in response to large climate change. The Bellenden-Ker Bartle-Frere subregion is also very stable and, as a proportion of its area, retains the most rainforest at LGM. The Thornton Lowlands are also quite stable. However, in terms of total area in rainforest at LGM, the Cairns-Cardwell Lowlands and Atherton Uplands are the two most important refugia, followed by the Mossman and Thornton Lowlands.

Climate change and future forests

Current projections for warming in the Wet Tropics by 2100 are one degree per degree of global warming (i.e. 1.4–5.8°C by 2100). Rainfall scenarios for areas as small as the Wet Tropics are uncertainty but several models predict from +4% to −10% changes in rainfall per degree of global warming in North Queensland. The CSIRO's (1996) climate model predicts more frequent El Niño-like conditions for the Pacific through this century and tropical cyclone intensity may increase. El Niño conditions cause lower rainfall and longer dry seasons in the Wet Tropics. Warming can have a particularly strong impact on mountainous regions like the Wet Tropics. The mountain tops and tablelands are cool islands in a sea of warmer climates. These islands are separated from each other by the warmer valleys and form a scattered archipelago of habitat for organisms that appear to be unable to survive and reproduce in warmer climates. Several analyses demonstrate the great sensitivity of Wet Tropics biota to climate change (Williams & Hilbert 2006).

Rainforests

Using FANN, the distribution and extent of environments suitable for 15 structural forest types were estimated in ten climate scenarios that include warming

Table 8.5 Analysis of the presence of rainforests (RF) of any type in Wet Tropics subregions where the minimum area of rainforest present in the climate scenarios is greater than 100 km^2

| Subregion | % of area in RF | | | Min. area (km^2) | Max. area (km^2) | Total area (km^2) |
	LGM	PHT	HCO			
Cairns–Cardwell Lowlands	50.40	80.96	74.32	2,087.24	3,352.88	4,237.66
Atherton Uplands	15.09	85.09	81.13	575.26	3,243.77	3,909.00
Mossman Lowlands	71.51	71.24	73.49	462.98	477.59	662.31
Thornton Lowlands	62.03	96.60	91.71	329.79	513.56	544.65
Bellenden Ker Bartle–Frere	79.16	100.00	100.00	244.52	308.89	308.89
Mt Finnegan Uplands	33.23	65.90	94.17	168.17	476.64	523.71
Bloomfield–Helenvale Lowlands	60.55	26.05	64.81	142.32	354.11	560.31
Carbine Uplands	17.33	89.71	88.53	130.96	678.00	806.88
Thornton Uplands	34.11	100.00	100.00	102.96	301.89	301.89

up to one degree and altered precipitation from −10% to +20% (Hilbert *et al.* 2001b). Forest type is an important habitat variable, although many trees and other organisms are not restricted to a single forest type. Therefore, the extent and distribution of forest types in the future is one important indicator of potential changes in biodiversity. Overall, the location and extent of rainforests are determined by rainfall and its seasonality, with some influence of soil fertility and water holding capacity. But the type of rainforest and the kinds of organisms found there depend more on temperature.

In the Wet Tropics, one degree of warming can increase the potential area of rainforests as a whole, as long as rainfall does not decrease. However, large changes in the distribution of forest environments are likely with even minor climate change and the relative abundance of some types could decrease significantly. Increased rainfall favours some rainforest types, while decreased precipitation increases the area suitable for woodlands and forests dominated by sclerophyllous genera like *Eucalyptus* and *Allocasuarina*.

Rainforest environments respond differentially to increased temperature (Figure 8.4). The area of lowland, mesophyll vine forest environments increases with warming, while upland, complex notophyll vine forest environments respond either positively or negatively

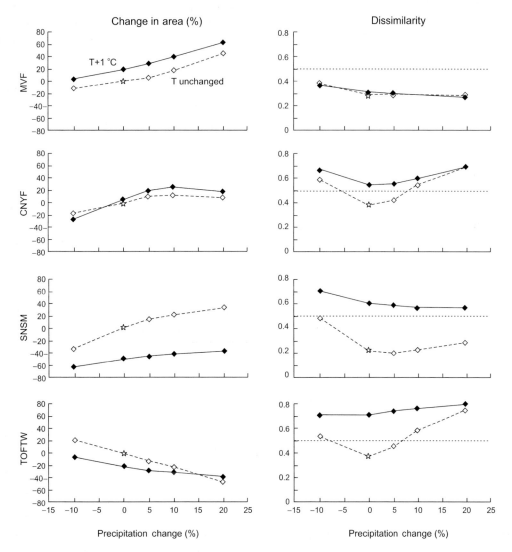

Figure 8.4 Changes in area and stress, measured by environmental dissimilarity, for the major rainforest types (MVF, mesophyll vine forest; CNVF, complex notophyll vine forest; SNSM, simple notophyll and simple microphyll forests and thickets) and for tall open forests and tall woodlands (TOFTW). *Source*: modified from Hilbert *et al.* (2001b).

to temperature, depending on precipitation. Highland rainforest environments (simple notophyll and simple microphyll vine fern forests and thickets), the habitat for many of the region's endemic vertebrates, decrease by 50% with only one degree of warming.

Estimates of the stress to present forests (Figure 8.4), resulting from spatial shifts of forest environments (assuming no change in the present forest distributions), indicate that several forest types would be highly stressed by one degree of warming and most are sensitive to any change in rainfall. Most forests will experience climates in the near future that are more appropriate to some other structural forest type (dissimilarity >0.5). Thus, the propensity for ecological change in the region is high and, in the long term, significant shifts in the extent and spatial distribution of forests are likely. A detailed, spatial analysis of the sensitivity to climate change indicates that the strongest effects of climate change will be experienced at boundaries between forest classes and in ecotonal communities between rainforest and open woodland.

The potential future distributions (1°C warming and −10% rainfall) of upland and highland rainforest types not only decline but also become very fragmented in this climate change scenario. Obviously, if the upper range of predicted warming occurs (5.8°C), no appropriate environments would remain within the Wet Tropics. Whether and where appropriate climates might come to exist further to the south, say in the Border Ranges, is unknown. However, regional rainfall patterns and topographic constraints imply that such new habitat would be very far removed from the Wet Tropics.

The high sensitivity of the region to only one degree of warming can be seen in the large number of transitions among forest classes (Figure 8.5). For the most part, several wetter rainforest types and medium open forests and woodland convert to mesophyll vine forest environments. Actual change in forest structure could occur in these cases without disturbance but the potential rates of change are unknown. The drier rainforest types and tall open forests and woodlands primarily are converted to medium and low woodland environments. In these cases, fire or other catastrophic disturbance, such as severe cyclones, would be required for these transitions.

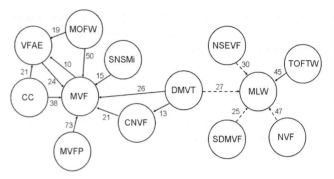

Figure 8.5 Transitions (percentage of current extent given on the arrows) among forest types with 1°C warming and no change in rainfall. Only transitions >10.0% are shown. Dashed lines indicate transitions from closed forest to woodland that would require fire or other severe disturbance. See Table 8.2 for descriptions of the classes.

Spatial and ecological constraints to forest change

Linking FANN with a cellular automaton-like model allowed investigation of how the spatial arrangement of forest types may limit their response to future climate change and how transitions might be constrained by geographical, anthropogenic, biological and environmental factors (Ostendorf et al. 2001). In this model, climate change can result in a transition to a new forest class only if the new, more suitable type exists in its proximity. Neighbourhood sizes from 0.1 to 5 km were tested using distance-weighted and presence/absence rules. In addition to the spatial constraints, the means to include ecological constraints into the model were explored. The cellular automata approach is an iterative process. For every iteration, a probability of change ($p_{i,j}$) from the current vegetation class (*i*) to a new type (*j*) is computed at each location from the environmental suitability for type (*j*) (found directly from the ANN output) and the constraint that the forest type (*j*) must be present in the neighbourhood of the location to provide a sources of propagules. The neighbourhood shape is approximately circular, defined as the set of pixels whose midpoints lie within the circle defined by given radius. The number of neighbours of type (*j*) and their distances from the centre of the neighbourhood are used to weight linearly the probability of transition. The probabilities were further modified by ecological constraints that might limit the range of possible transition. Results indicate that the spatial arrangement of vegetation may impose relatively little constraint on the region's potential change in response to small changes in climate. Depending on the neighbourhood size included into the model, the predicted future vegetation patterns differ by only 1–10% of all locations compared to the pattern expected from the ANN model. In contrast, ecological constraints on transitions can potentially have a marked influence on landscape change. But, for the most part, the nature of these constraints is unknown.

Summary of long-term dynamics

The extent and distribution of sclerophyll forests and rainforests is primarily dependent on rainfall distributions and seasonality, with an additional influence of human-controlled fire frequency and intensity. Within rainforests, temperature variables are the primary control on rainforest types, while soil drainage and fertility have a secondary effect. Thus, changes in global and regional climate, along with human activities, have created a constantly shifting mosaic of forest types within the Wet Tropics landscape.

Perhaps the largest change coincided with the arrival of humans c.45 000 yr BP that resulted in the expansion of phyric, scleropyll forests and the decline of Araucarian rainforests. But, in the 18 000 years covered by this chapter, European settlement and subsequent land-clearing certainly caused the most rapid change to the landscape overall. The largest changes, while slower, were driven by climate change. Anthropogenic climate change is likely to be much more rapid than in the past and is likely to pose a significant threat to the region's biodiversity.

Each of the major forest types responds uniquely to climate change in the past and in the future (see Figure 8.6). At LGM, the region is dominated by woodlands, as is the case today, and the expectation is that they will expand further in the future. Tall open forests were extensive at LGM but have declined since then, accelerated by human-caused changes in fire regime since European settlement but they may re-expand somewhat in the future. Lowland rainforests (MVF) are very limited in small refugia at LGM but expand during the Holocene to their peak during the HCO.

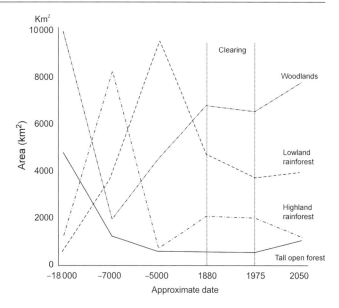

Figure 8.6 A synthesis of landscape change in the Wet Tropics. Dates are approximate and the time axis is not linear. Forest clearing has continued since 1975 but is not represented in this figure.

Clearing of lowland rainforests caused a very rapid decline in their extent but they may expand at the expense of other rainforest types in the future. Highland rainforests (CNVF and SNSMi) are restricted to refugia at LGM but less so than lowland rainforest types. In contrast to other rainforest types their minimum extent occurs during the warm-wet HCO. For these forests, interglacial, rather than glacial, refugia are perhaps the most important. While they were largely unaffected by clearing, only one degree of future warming will reduce these environments considerably and the hot-dry future is likely to eliminate them entirely. Unfortunately, these upland and highland rainforests are the habitat of most of the bioregion's local endemic species.

Implications for management

Research by Mike Hopkins, Peter Kershaw and others into the past distributions of Wet Tropics forests and the ways they were affected by glacial cycles of climate change may appear obscure to management agencies. However, this research provided the knowledge that forests in the region are highly sensitive to climate change and helped to explain current patterns of biodiversity. This research motivated me to study the implications of future climate change for the region at a time when managers, and most scientists, believed that global warming was of no significance in the Wet Tropics or in any tropical rainforests. Several years later and with additional studies, the Wet Tropics Management Authority (WTMA), which is responsible for the management of the Wet Tropics of Queensland World Heritage Area (WHA), Queensland and other commonwealth managers recognize climate change as one of the most significant threats to biodiversity in Australian rainforest. The lesson is that, while it is important that scientists respond to their end users' immediate needs, it is also very important that scientists have the opportunity to take the lead and identify issues of which managers are not yet aware.

Since climate change is a global phenomenon, driven by global patterns of population growth, fossil fuel use and deforestation, it is obviously impossible for the Wet Tropics to reduce the rate or extent of global warming. However, management policies can and must be developed that attempt to minimize its negative impacts on the region's ecosystems and unique biodiversity. A fundamental difficulty is that political boundaries like national parks or the WHA are static, while environments and habitats are dynamic, and especially so with rapid climate change. Thus, conservation of ecosystems and the biodiversity within them is not ensured by a static network of reserves. Consequently, policy and management need to be on a large, biogeographical scale and to consider land outside reserves. It is possible that suitable habitat for many Wet Tropic species will only occur thousands of kilometres to the south in 50–100 years time. Managers need to know where this habitat might occur and to begin considering the implications of such changes. Assuming that research identifies regions within the Wet Tropics that might act as greenhouse refugia (restricted regions where biota can survive despite warming), these must be protected and managed to enhance their stability. Similarly, connectivity among suitable habitat areas could be improved and efforts made to minimize the interacting effects of other, more tractable, global change processes (i.e. land clearing, linear barriers, weeds and feral animals). Finally, proactive management of the species that are most threatened by warming must be considered. The possibility and desirability of translocating species to distant suitable areas must be debated now. However, these management issues and actions cannot be discussed or implemented before research has begun to fill the current information gaps.

It is imperative that we continue to improve our understanding of ecological patterns and processes over large spatial and temporal scales and develop predictive tools to enable realistic conservation planning for the continued preservation of the unique biota and ecosystems of the Wet Tropics and other rainforests. We must begin monitoring for ecological changes and continue to refine our projections of impacts. With the resulting knowledge, management and policy can begin to address this challenge creatively and work to preserve as much of the Wet Tropics' biodiversity as possible. The results presented here emphasize the need for current conservation efforts to consider climate change in both *in situ* conservation and reintroduction efforts.

Acknowledgements

Many people contributed to this research. In particular, I acknowledge Mike Hopkins, Andrew Graham and Bertram Ostendorf. Dr Hopkins was especially significant in pointing me in this research direction and supporting my group's efforts. Matt Bradford, Brett Buckley, J. van den Muyzenberg, Trevor Parker and Warwick Sayers provided valuable technical assistance. CSIRO's collaboration with the Cooperative Research Centre for Rainforest Ecology and Management was the primary source of funding.

References

Aide, T. M. & Rivera, E. (1998). Geographic patterns of genetic diversity in *Poulsenia armata* (Moraceae): implications for the theory of Pleistocene refugia and the importance of riparian forest. *Journal of Biogeography* **25**: 695–705.

Brown, D. G., Lusch, D. P. & Duda, K. A. (1998). Supervised classification of types of glaciated landscapes using digital elevation data. *Geomorphology* **21**: 233–250.

Burnham, R. J. & Graham, A. (1999). The history of Neotropical vegetation: new developments and status. *Annals of the Missouri Botanical Garden* **86**: 546–89.

Colinvaux, P. A. & De Oliveira, P. E. (2001). Amazon plant diversity and climate through the Cenozoic. *Palaeogeography Palaeoclimatology Palaeoecology* **166**: 51–63.

Colinvaux, P. A., De Oliveira, P. E. & Bush, M. B. (2000). Amazonian and Neotropical plant communities on glacial time-scales: the failure of the aridity and refuge hypotheses. *Quaternary Science Reviews* **19**: 141–69.

Colinvaux, P. A., Liu, K. B., De Oliveira, P. E., Bush, M. B., Miller, M. C. & Kannan, M. S. (1996). Temperature depression in the lowland tropics in glacial times. *Climatic Change* **32**: 19–33.

CSIRO (1996). *Climate Change Scenarios for Australia*. CSIRO Division of Atmospheric Sciences, Melbourne.

Elenga, H., Peyron, O., Bonnefille, R. *et al.* (2000). Pollen-based biome reconstruction for Southern Europe and Africa 18,000 yr BP. *Journal of Biogeography* **27**: 621–34.

Farrera, I., Harrison, S. P., Prentice, I. C. *et al.* (1999). Tropical climates at the last glacial maximum: a new synthesis of terrestrial palaeoclimate data. I. Vegetation, lake levels and geochemistry. *Climate Dynamics* **15**: 823–56.

Flenley, J. R. (1998). Tropical forests under the climates of the last 30,000 years. *Climatic Change* **39**: 177–97.

Graham, A.W. (2006). The CSIRO Rainforest Permanent Plots of North Queensland -- Site, Structural, Floristic and Edaphic Descriptions. CSIRO and the Cooperative Research Centre for Tropical Rainforest Ecology and Management. Rainforest CRC, Cairns. 252pp.

Haberle, S. G. (1999). Late Quaternary vegetation and climate change in the Amazon Basin based on a 50,000 year pollen record from the Amazon Fan, Odp Site 932. *Quaternary Research* **51**: 27–38.

Haberle, S. G. (2005). A 23,000-yr pollen record from Lake Euramoo, Wet Tropics of NE Queensland, Australia. *Quaternary Research* **64**: 343–56.

Hamilton, A. C. (1976). The significance of patterns of distribution shown by forest plants and animals in tropical Africa for the reconstruction of Upper Pleistocene palaeoenvironments: a review. *Palaeoecology of Africa and the Surrounding Islands* **9**: 63–97.

Harrington, G. N. & Sanderson, K. D. (1994). Recent contraction of wet sclerophyll forest in the Wet Tropics of Queensland due to invasion by rainforest. *Pacific Conservation Biology* **1**: 319–27.

Haung, W. Y. & Lippman, R. P. (1988). Comparisons between neural net and conventional classifiers. *Proceedings of the International Joint Conference on Neural Networks*, 4: 485–93. San Diego, CA.

Heine, K. (2000). Tropical South America during the last glacial maximum: evidence from glacial, periglacial and fluvial records. *Quaternary International* **72**: 7–21.

Hilbert, D. W., Graham, A. &. Parker, T. (2001a). Tall open forest and woodland habitats in the wet tropics: responses to climate and implications for the northern bettong (*Bettongia tropica*). *Tropical Forest Research Series* No. 1 (www.TFRSeries.jcu.edu.au).

Hilbert, D. W. & Ostendorf, B. (2001). The utility of artificial neural networks for modeling the distribution of vegetation in past, present and future climates. *Ecological Modelling* **146**: 311–27.

Hilbert, D. W., Ostendorf, B. &. Hopkins, M. (2001b). Sensitivity of tropical forests to climate change in the humid tropics of North Queensland. *Austral Ecology* **26**: 590–603.

Hilbert, D. W. & van den Muyzenberg, J. (1999). Using an artificial neural network to characterise the relative suitability of environments for forest types in a complex tropical vegetation mosaic. *Diversity and Distributions* **5**: 263–74.

Hill, R., Smyth, D., Shipton, H. &. Fischer, P. (2001). Cattle, mining or fire? The historical cause of recent contractions of open forest in the wet tropics of Queensland through invasion by rainforest. *Pacific Conservation Biology* **7**: 187–94.

Hopkins, M. S., Ash, J., Graham, A. W., Head, J. & Hewett, R. K. (1993). Charcoal evidence of the spatial extent of the eucalyptus woodland expansion and rainforest contractions in north Queensland during the late Pleistocene. *Journal of Biogeography* **20**: 357–72.

Hopkins, M. S., Graham, A. W., Hewett, R., Ash, J. & Head, J. (1990). Evidence of late Pleistocene fires and eucalypt forest from a North Queensland humid tropical rainforest site. *Australian Journal of Ecology* **15**: 345–7.

Hopkins, M. S., Head, J., Ash, J., Hewett, R. K. & Graham, A. W. (1996). Evidence of a Holocene and continuing recent expansion of lowland rain forest in humid, tropical north Queensland. *Journal of Biogeography* **23**: 737–45.

Intergovernmental Panel on Climate Change (IPCC) (2001). *Climate Change 2001: The Scientific Basis*. Contribution of Working Group I to the Third Assessment Report of the Intergovernmental Panel on Climate Change. Cambridge University Press, New York, 881 pp.

Kershaw, A. P. (1985). An extended late Quaternary vegetation record from north-eastern Queensland and its implications for the seasonal tropics of Australia. *Proceedings of the Ecological Society of Australia* **13**: 179–89.

Kershaw, A. P. (1986). Climate change and Aboriginal burning in north-eastern Australia during the last two glacial/interglacial cycles. *Nature* **322**: 47–9.

Kershaw, A. P. (1989). Was there a 'great Australian arid period?' *Search* **20**: 89–92.

Kershaw, A. P. (1994). Pleistocene vegetation of the humid tropics of northeastern Queensland, Australia. *Palaeogeography Palaeoclimatology Palaeoecology* **109**: 399–412.

Kershaw, A. P. & Nanson, G. C. (1993). The last full glacial cycle in the Australian region. *Global and Planetary Change* **7**: 1–9.

Nix, H. A. (1991). Biogeography: pattern and process. In *Rainforest Animals: Atlas of Vertebrates Endemic to Australia's Wet Tropics*, Nix, H. A. & Switzer, M. A. (eds). Australian National Parks and Wildlife Service, Canberra, 11–40.

Nix, H. A. & Kalma, J. D. (1972). Climate as a dominant control in the biogeography of northern Australia and New Guinea. In *Bridge and Barrier: the Natural and Cultural History of Torres Strait*, Walker, D. (ed.). Australian National University, Canberra, 61–91.

Ostendorf, B., Bradford, M. G. & Hilbert, D. W. (2004). Regional analysis of forest biomass at the rainforest/sclerophyll boundary in Northern Queensland, Australia. *Tropical Ecology* **45**(1): 31–41.

Ostendorf, B., Hilbert, D. W. &. Hopkins, M. S. (2001). The effect of climate change on tropical rainforest vegetation pattern. *Ecological Modelling* **145**: 211–24.

Ostendorf, B. &. Reynolds, J. F. (1998). A model of arctic tundra vegetation derived from topographic gradient. *Landscape Ecology* **13**: 187–200.

Rumelhart, D. E. & McClelland, J. L. (1986). *Parallel Distributed Processing: Explorations in the Microstructures of Cognition*. MIT Press, Cambridge, MA. 1208 pp.

Sattler, P. S. & Williams, R. D. (1999). *The Conservation Status of Queensland's Bioregional Ecosystems*. Environmental Protection Agency, Brisbane.

Singh, G., Kershaw, A. P. & Clark, R. L. (1980). Quaternary vegetation and fire history in Australia. In *Fire and the Australian Biota*, Gill, A. M., Groves, R. A. & Noble, I. R. (eds). Australian Academy of Science, Canberra, 23–54.

Specht, R. L. (1970). Vegetation. In *The Australian Environment*, Leeper, G. W. (ed.). CSIRO and Melbourne University Press, Melbourne, 44–67.

Stevenson, J. & Hope, G. (2005). A comparison of late Quaternary forest changes in New Caledonia and northeastern Australia. *Quaternary Research* **64**: 372–83.

Taylor, D., Saksena, P., Sanderson, P. G. & Kucera, K. (1999). Environmental change and rain forests on the Sunda Shelf of Southeast Asia: drought, fire and the biological cooling of biodiversity hotspots. *Biodiversity and Conservation* **8**: 1159–77.

Tracey, J. G. (1982). *The Vegetation of the Humid Tropical Region of North Queensland*. CSIRO, Melbourne.

Tracey, J. G. & Webb, L. J. (1975). *Vegetation of the Humid Tropical Region of North Queensland* (15 maps at 1 : 100 000 scale + key), CSIRO Australia Long Pocket Labs, Indooroopilly, Qld.

Turney, C. S. M., Kershaw, A. P., Moss, P. *et al.* (2001). Redating the onset of burning at Lynch's Crater (North Queensland): implications for human settlement in Australia. *Journal of Quaternary Science* **16**: 767–71.

Vanderkaars, S. & Dam, R. (1997). Vegetation and climate change in West-Java, Indonesia during the last 135,000 years. *Quaternary International* **37**: 67–71.

Walker, D. &. Chen, Y. (1987). Palynological light on tropical rainforest dynamics. *Quaternary Science Reviews* **6**: 77–92.

Walker, D. & Flenley, J. R. (1979). Late Quaternary vegetational history of the Enga Province of upland Papua New Guinea. *Philosophical Transactions Royal Society of London, B* **286**: 265–344.

Walker, D. & Hope, G. S. (1982). Late Quaternary vegetation history. In *Biogeography and Ecology of New Guinea*, Gressitt, J. L. (ed.). W. Junk, The Hague, pp. 263–85.

Webb, L. J. (1959). A physiognomic classification of Australian rain forests. *Journal of Ecology* **47**: 551–70.

Webb, L. J. (1968). Environmental relationships of the structural types of Australian rain forest vegetation, *Ecology* **49**: 296–311.

Webb, L. J. (1978). A general classification of Australian forests. *Australian Plants* **9**: 349–63.

Webb, L. J. & Tracey, J. G. (1981). Australian rainforest: patterns and change. In *Ecological Biogeography of Australia*, Keast, A. (ed.). Junk, The Hague, pp. 607–94.

Williams, S .E. & Hilbert, D. W. (2006). Climate change threats to the biodiversity of tropical rainforests in Australia. In *Emerging Threats to Tropical Forests*, Laurance, W. F. and Peres, C. A. (eds). University of Chicago Press, Chicago.

Winter, J. W., Bell, F. C., Pahl, L. I. & Atherton, R. G. (1987). Rainforest clearing in North Queensland. *Proceedings of the Royal Society of Queensland* **98**: 41–57.

9 FLORISTICS AND PLANT BIODIVERSITY OF THE RAINFORESTS OF THE WET TROPICS

Daniel J. Metcalfe* and Andrew J. Ford*

CSIRO Sustainable Ecosystems, Tropical Forest Research Centre, Atherton, Queensland, Austalia
*The authors were participants of Cooperative Research Centre for Tropical Ecology and Management

Introduction

The Wet Tropics bioregion (*sensu* Environmental Australia 2005) is a narrow belt of coastal plains and ranges, and the adjacent section of the Great Dividing Range and its associated tablelands, stretching some 450 km from Black Mountain (15° 40′ S), south of Cooktown to Bluewater (19° 15′ S), north of Townsville. To the south of Bluewater, the upland forest on Mount Elliot (19° 30′ S) probably bears more similarity to the northern tropical forests than the southern subtropical forests, placing the boundary between tropical and subtropical rainforest in eastern Australia to the north of the geographical tropic at around 20° S. Scattered areas of subtropical rainforest with some floristic affinities to the Wet Tropics forests occur to a little south of about 35° S in New South Wales. North of Cooktown the vegetation of Cape York is dominated by open forests and woodlands with savannah understoreys (*sensu* Specht & Specht 1999), with small areas of mostly seasonal rainforest, such as on the Iron and McIlwraith Ranges. West of the bioregion the forests become more open and give way to woodlands and savannah as the influence of the coastal rainfall diminishes and fire becomes a more important driver of vegetation pattern (see Bowman 2000).

The Wet Tropics bioregion contains a remarkable diversity of vegetation types reflecting the varied habitats provided by the considerable topographic diversity and high rainfall gradients. Consequently the region consists of a mosaic of vegetation types, many associated with particular landforms. The remaining natural vegetation in the lowlands is dominated by swamp or seasonally inundated communities, including mangroves, beach scrub, *Melaleuca* woodlands and palm-swamp forests. Most of the lowland area has been cleared, primarily for agriculture, and today is largely managed for sugar, bananas, pasture and orchard crops. The foothills of the coastal ranges retain most of their original vegetation cover, a mixture of rainforest and tall open forest communities, as do the foothills of the other mountainous areas, including the Paluma, Seaview, Cardwell, Malbon Thompson, Lamb, Thornton and Great Dividing Ranges. The highest mountains are typically clothed in closed rainforest types up to their summits, though on steep slopes and where skeletal soils occur mountain rock pavements and treeless communities exist. On the tablelands inland of the coastal range much less of the original vegetation cover remains, with agriculture again dominating the landscape of many areas (especially the Atherton and Evelyn Tablelands). Classification of the natural forest types in the Wet Tropics has been attempted a number of times, most notably by the pioneering work of Geoff Tracey and Len Webb in the 1970s (e.g. Tracey & Webb 1975; Tracey 1982), which resulted in a physiognomic classification, and recently by the more comprehensive treatment of Peter and David Stanton (Stanton & Stanton 2005), whose highly detailed classification takes into account geology or substrate, species composition and disturbance history. This later classification has been adapted by the Queensland Environmental Protection Agency (EPA) to arrive at the current legislatively informing Regional Ecosystem (RE) framework (Queensland Government Environmental Protection Agency 2005).

Climatic effects on vegetation

The main climatic factors affecting the distribution of different floristic assemblages are rainfall and temperature, with wind affecting structure rather than composition. Rainfall within the bioregion varies from the very wet and effectively aseasonal (e.g. Cape Tribulation, Babinda and Tully >3500 mm per annum) to the highly seasonal and drier areas (e.g. Kairi and Bluewater <1300 mm per annum). Except in the areas of highest rainfall, the climate of most of the Wet Tropics rainforests is not ever-wet, with a distinct five-to six-month dry season. It is probably of great ecological significance that the season of lowest rainfall is also that of low temperature, so that the vegetation may be less affected by water stress than in typical seasonal tropical climates in which temperatures are usually higher in the dry season than in the wet season (Richards 1996). A consequence of the seasonality of the rainfall is a range of rainforest types from lush, epiphyte-rich communities in areas with appreciable 'dry season' rainfall through to semi-deciduous forests and drier woodland communities as the length and dryness of the period of lowest rainfall increases. Periods of several weeks without significant rain may challenge plants, leading to wilting of shallow-rooted understorey species, and in extreme cases leading to defoliation and tree death. The drought associated with the 1992 El Niño Southern Oscillation had such effects, causing canopy thinning, death of large trees even on some of the wetter mountains (e.g. *Leptospermum wooroonooran* (F. M. Bailey) on Mount Lewis;

Bruce Gray, pers. comm.) and recruitment of weeds into the understorey of closed forest where the canopy had been removed (M. S. Hopkins, A. W. Graham, R. Jensen, J. Maggs, B. Bayly & R. K. Hewett, unpublished data). Seasonality in rainfall at more usual scales drives considerable phenological pattern within the vegetation, with many species flushing new leaves and fruiting in the summer wet season (e.g. Hopkins & Graham 1989; Spencer *et al.* 1996; Boulter *et al.* 2006).

Temperature grades with altitude and seasonality, with the highest temperatures recorded from the lowlands in the summer (November to January) and the lowest from the highlands in winter (June to August) (Commonwealth Bureau of Meteorology 2006). The annual temperature range is greatest in the drier areas, where temperature fluctuations are least buffered by high humidity (calculated using the BIOCLIM module (Rainforest CRC 2003) within ANUCLIM version 5.1 for Windows (Houlder *et al.* 2000), the AUSLIG (2001) and Wet Tropics Management Authority (WTMA) (Accad 1999) digital elevation models). Highest daily temperatures may reach 45°C, but seldom for more than a day or two (see Balston, Chapter 21, this volume); however, hot conditions on dry days in summer may lead to considerable vapour pressure deficits, which challenge plant photosynthetic abilities and may lead to leaf scorching. Freezing temperatures are also rare, of short duration and typically only 1–2°C below freezing; frost-maintained open forest occurs in various hollows and low places on the tablelands, where cold air can pool. Rainforest species do not tolerate frost well, but frost is rare in closed forest except along existing edges (van Steenis 1972; J. G. Tracey, pers. comm.; Bernie Hyland, pers. comm.), and therefore these frost-maintained woodland sites may have been cleared of their rainforest cover at some time in the past (either by natural disturbance or through Aboriginal management practices) or are relics of drier periods.

Prevailing winds are from the south-east, although wind appears to have a limited effect on community structure except when tropical cyclones irregularly cross the coast, when high wind speeds, gusting, very high rainfall events and storm surges may cause various effects on the vegetation (see Turton & Stork, Chapter 3, this volume). These effects are most apparent in their impact on the vegetation of the coastal ranges, where repeated canopy disturbance and subsequent growth of vine towers on remaining trees causes a distinctive 'cyclone scrub' (*sensu* Webb 1958), which is dominated by a few species (e.g. *Merremia peltata* (L.) Merr.), and under which there is very limited regeneration of tree and shrub species. Other impacts of cyclones include canopy shredding, leading to removal of most of the leaves and exposure of the understorey to high light levels. For example, canopy shredding in the Barong Logging Area (now Wooroonooran National Park) after Cyclone Winifred in 1986 led to massive recruitment of light-demanding species such as *Dendrocnide moroides* (Wedd.) Chew (unpublished CSIRO data). Cyclones may also trigger a mass flowering of shrubs and small trees, possibly through increased light levels to the understorey (Hopkins & Graham 1987). Storm surges and flooding also affect plant communities and their distribution; a cyclone in 1918 caused extensive flooding of the Tully area and storm surges deposited the sand dune on which South Mission Beach is largely built at this time. This is the latest of an extensive series of dunes that run inland up to 10 kilometres in places, and are successively colonized by littoral scrub and *Melaleuca* swamp, and then by various types of woodland, mesophyll or notophyll vine forest.

Edaphic and topographic effects of vegetation

Soil fertility alone is implicated in explaining the distribution of remarkably few widespread regional endemic species. While it is true that many localized species in the Wet Tropics occur on nutrient-poor substrates at altitude (especially north of the Black Mountain Corridor), it would be erroneous to suggest they were confined to such substrates as all substrates at altitude (greater than 800 metres) north of the Lamb Range are nutrient poor. Many species are usually found, for example, on nutrient poor substrates but can also occur on the high-fertility basalts and vice versa. *Balanops australiana* F. Muell. and *Crispiloba disperma* (S. Moore) (Steenis) are examples of species that are restricted to nutrient poor substrates, whereas *Syzygium erythrocalyx* (C. T. White) B. Hyland is one of the few species that are known to occur only on the high-fertility basalts.

Edaphic considerations are also important in determining the location and stability of rainforest/tall open

forest boundaries (see Tracey 1969; Webb 1969). In general, Australian rainforest trees have higher nutrient requirements (especially of phosphorus) than eucalypt forest trees; hence in more seasonal areas there is a greater luxuriance of rainforest on basalts, schists and alluvium, whereas open forests and woodlands typically exist on the poorer granitic sandstones and similar substrates and are more influenced by fire regimes (but see Bowman 2000). The boundary may be abrupt, but there is often a transitional community with eucalypt overstorey and rainforest understorey, an ecotonal community also seen in the *Eucalyptus deglupta* Bl. forests of New Guinea (Richards 1996).

Ash (1988) showed that rainforest/tall open forest boundaries frequently coincided with major changes in geological substrate, with basalts, alluvia and scoria frequently supporting rainforest, while more rugged granitic and metamorphic rocks supported tall open forest and woodland communities. This generalization, especially in rugged areas, will also be influenced by rainfall patterns. Conversely, significant geological boundaries are also noted for their rapid species turnover and change in rainforest structural types where rainfall is equal and not limiting. Possibly the most dramatic example of this is seen in Wooroonooran National Park, in the headwaters of the Russell and North Johnstone Rivers (annual rainfall >4000 mm). The basalt areas support species-rich, luxuriant mesophyll forest, whereas the adjacent fine grained metamorphic substrate supports relatively species-poor notophyll forest, with species turnover estimated at 80% (A. J. Ford, unpublished data).

On hill slopes where both rainforest and tall open forest exist, the rainforest is normally downslope of the more fire-prone open forest, presumably utilizing deeper soils and water resources accumulated through seepage to support lush, fire-inhibiting vegetation (Ash 1988). However, rainforest may also exist above open forest, typically on mountains with slightly sloping summits where soil has been able to accumulate, and where occult precipitation and low cloud maintain the water balance at a higher level than downslope in open woodland (Ash 1988; Hutley *et al.* 1997; and see McJannet *et al.*, Chapter 15, this volume). Rainforest also appears on hill slopes in gullies; again these are wetter habitats with deeper soil profiles and often rocky protection from wildfire. Equivalent gullies in closed rainforest still maintain higher humidity and ground water levels than surrounding forest in the dry season, and so are notable for supporting rich communities of herbs, ferns, epiphytes and lithophytes.

Although much of the lowland forest has been cleared, relict patches of swamp forest and riverine forest on seasonally or permanently flooded sites where there is impeded drainage occur around the major river systems, and grade into mangrove communities at the coast. Extensive topographic modification of the lowlands makes it difficult to ascertain what proportion of the original extent of these communities remains, but some of these communities, particularly the *Licuala*-dominated fan-palm swamps, provide almost iconic images of the bioregion, seen extensively in tourist and publicity materials.

Anthropogenic modification of vegetation

The area now occupied by rainforests in Australia is believed to be only about a quarter of what was present at European settlement some 200 years ago (Webb & Tracey 1981). The area of the Wet Tropics still covered by native vegetation is nearer 70% (interpreted from Queensland Government Environmental Protection Agency 2005), but this is unequally distributed, with the rainforests of the lowlands and more fertile uplands having borne the brunt of clearance for agriculture. Most remnant forest is above 500 metres above sea level, with remaining natural vegetation in the lowlands dominated by prograding dune successions on sand, *Melaleuca* and other types of swamp forest in areas too wet to drain easily and various open forest and woodland types on thin, poor soils. However, the fragments that remain provide a tantalizing glimpse of some structurally and floristically interesting communities without recognizable upland analogues (Metcalfe & Ford, unpublished data), and deserve further study. The tableland areas have also seen extensive clearance of distinct community types associated with specific environmental conditions. For example, the mid-elevation basalt plains have been cleared to such an extent that 'Mabi forest' (type 5b *sensu* Tracey 1982, REs 7.3.37 and 7.8.3) is the bioregion's only 'critically endangered ecological community', with an estimated 4% of its former extent remaining (Department of the Environment and Heritage 2005). Poorer soils

derived from granite and metamorphic rocks are better represented, as are the higher elevation forests, although virtually all accessible forest has been logged at some point in the past. The impacts of logging are not always clear, as different logging areas were harvested under different extraction rules, and some forestry districts presided over much more aggressive extraction regimes than others. However, the Queensland Forestry Act (1959) and subsequent amendments imposed rules governing forest management, including permanent reservation of forests, protection of watersheds and management to permit production of timber and associated products in perpetuity. Continued implementation, review and amendment of these rules may have resulted in one of the few truly sustainable extraction forestry systems in the world, had World Heritage listing in 1988 not put an end to the experiment (Crome *et al.* 1992; Whitmore 1998; though see Valentine & Hill, Chapter 6, this volume). Post World Heritage listing, an increase in the pace of change to public attitudes and legislation such as the Vegetation Management Act (1999) have ensured the protection of virtually all of the significant remnant forest areas left in the bioregion, with the main remaining threats being posed by weeds and feral animals, which are addressed in subsequent chapters, and by the mismanagement of tall open forest.

Fire is important in determining the distribution of rainforest in Australia, with much of the tropical region dominated by fire-prone vegetation, but even within these pyrophytic communities, in fire-protected habitats, dry or monsoonal rainforest types exist (Bowman 2000). In the Wet Tropics the western boundary of the rainforest with the open forest and woodlands is largely determined by fire. Soils and rainfall permit rainforest taxa to persist to the west of the existing boundary, while fires of varying intensity running in from the west inhibit seedling establishment and permit eucalypts and other fire-tolerant, or fire-maintaining, species to persist. This ecotonal community forms distinct forest types termed wet sclerophyll forest (see Webb 1968; various REs in 7.8, 7.11 and 7.12) in a discontinuous strip rarely more than 4 km wide but extending some 400 km along the western margin of the Wet Tropics. Dominated by a eucalypt overstorey, characteristically of *Eucalyptus grandis* (W. Hill) or a mixture of *Corymbia intermedia* (R. T. Baker) K. D. Hill

& L. A. S. Johnson, *E. pellita* F. Muell., *E. resinifera* Sm., *E. tereticornis* Sm. and *Syncarpia glomulifera* (Sm.) Nied. (Harrington *et al.* 2005), the wet sclerophyll forests typically have a continuous grass or sedge understorey if regularly burned, and an increasing abundance of regenerating rainforest trees in the understorey if not burned. Current concern over the increasingly fragmented nature of the wet sclerophyll forest type due to rainforest encroachment and canopy closure is focused on threats to the endemic and dependent fauna associated with them (e.g. Bradford & Harrington 1999; Vernes 2000). Natural oscillations of the rainforest boundary were reported over long periods of time by Hopkins *et al.* (1993; and see Hilbert, Chapter 8, this volume) but more widespread invasion of wet sclerophyll forests was documented by Harrington and Sanderson (1994) over a 50-year period in a study addressing northern, central and southern parts of the Wet Tropics' western boundary. Changes in natural and Aboriginal fire regimes induced by European introduction of extreme fire control as a part of pastoral land management have been surmised as a major cause of these changes (Ash 1988; Unwin 1989; Hopkins *et al.* 1996; and see Hill *et al.* 2001).

Fire is likely to be an infrequent phenomenon in rainforest types, presumably occurring naturally only when lightning ignites dry fuel before the onset of the wet season. Large accumulations of leaves and woody litter are most likely to occur after cyclone-stripping of canopy foliage (Webb 1958), or extended dry periods during which trees have shed leaves in a drought response (Unwin *et al.* 1985). When fire has been observed to run through areas of rainforest many trees survive its passage and resprout (27 species from 17 families, Mount Lewis; P. T. Green & M. G. Bradford, pers. comm.; 65 out of 72 species of seedling resprouted following a fire on the coastal range north of Cairns; Marrinan *et al.* 2005), and coppicing is a common response to felling and burning (82 species, Python Logging Area, Danbulla; Stocker 1981). Considerable evidence exists of burning in rainforest areas in pre-European times (Bowman 2000 and references therein), and Aborigines are known to have deliberately burned areas to keep them free of rainforest in the Wet Tropics (e.g. at Behana Creek, Gordonvale; Tony Irvine, pers. comm.), though little is known of fire frequency or intensity. Repeat fires may kill

coppice shoots and seedlings if too closely spaced (Stocker & Mott 1981, cited in Stocker 1981), leading to transition to a community dominated by fire-adapted species. However, there are suggestions that infrequent fire may actually promote forest recovery from cyclonic damage due to suppression of vines (especially *Merremia peltata* and *Calamus* spp.), which otherwise overtop and smother trees (Webb 1958; Stocker 1981), or may open up the dense shrub layer, permitting regeneration of overstorey species in some of the more seasonal rainforest types such as Mabi forest (type 5b *sensu* Tracey 1982; also type 6 complex notophyll vine forest with emergent *Agathis robusta* (C. Moore ex F. Muell.) F. M. Bailey; RE 7.12.7). Consequently, it seems likely that fire should be considered as a very sparingly used management tool in some situations and in some rainforest types. Fire seems likely to have an important role in maintaining diversity in heath communities too (Williams *et al.* 2005), with recently burnt heath supporting greater diversity than long unburnt heath. Williams *et al.* (2005) suggest that a fire return rate of not less than eight years may be necessary to allow the replenishment of soil seed reserves and the development of mature canopies of some woody species.

Floristic biodiversity and origins of the rainforests

The Wet Tropics' rainforests are mostly evergreen, although deciduous and semi-deciduous species are not uncommon, particularly in the forests on fertile but thin basalt soils of the tablelands (Type 5b *sensu* Tracey 1982); conifers occur but are not abundant and erect palms are common. Characteristic features are abundant tree ferns and cycads, and a sparsity of herbs in the understorey except in gaps; epiphytes are abundant but not usually conspicuous, and epiphyllae are generally scarce. Forests are usually composed of a mix of species (51–190 vascular species in 0.1 ha plots; A. J. Ford, unpublished data), but occasionally species such as *Argyrodendron polyandrum* L. S. Sm., *Backhousia bancrofti* F. M. Bailey & F. Muell. ex F. M. Bailey, *Ceratopetalum virchowii* F. Muell. and *Lindsayomyrtus racemoides* (Greves) Craven achieve local dominance. Species diversity is not especially high at the plot level by tropical standards, but stem density and standing biomass, particularly in upland forests, are notably higher than in most other tropical forests (Lewis *et al.* 2004).

The rainforests of the Wet Tropics are for the most part structurally similar to the Indo-Malayan forests, especially those of New Guinea, with abundant climbing palms (Richards 1996). However, floristically the Australian rainforests contain a large number of endemic species, and also a range of species with affiliations from much further afield. There is some overlap of families, genera and, rarely, species with forests in Southeast Asia; the most notable difference is the absence from the Wet Tropics (and Australia) of the important timber-tree family Dipterocarpaceae. Analysis of the affiliations of the flora (Table 9.1) reveals considerable overlap with other regions that may be loosely termed Gondwanan, suggesting that ancestral taxa occurred on the super continent before its break-up and separation (120–50 million years ago). Families with Australian-Gondwanan origins presumably include Cunoniaceae, Elaeocarpaceae,

Epacridaceae (now part of Styphelioideae in Ericaceae), Monimiaceae (including Atherospermataceae), Proteaceae and Winteraceae (Thorne 1986); all of these are represented in the rainforests of the Wet Tropics. Whiffin and Hyland (1986) showed for *Syzygium* and allied genera (Myrtaceae), and for *Cryptocarya* (Lauraceae), that two major centres of diversity and endemism exist, one in the Wet Tropics and a second in the south-east Queensland and northern New South Wales region. Minor centres of diversity were identified on Cape York and in the Northern Territory; together these locations probably represent centres of isolation and long-term refugia during climatic fluctuations. The ancient origins of the rainforest flora (see Box 9.1) and the extended close proximity to other tropical regions through their evolutionary history probably explain the relatively low level of generic endemism (6% compared with early-isolated Madagascar's 32% generic endemism in the tree and shrub flora; Schatz 2001). However, the rainforests contain 38 monotypic genera endemic to the Wet Tropics and a further 24 monotypic genera with a broader distribution (including eight Australian endemics and four that are exotic). Considerably more genera are only represented in the rainforests by a single species (though fewer than the 700 suggested by Gross 2005). The persistence and later isolation of the Wet Tropics' rainforests has resulted in high levels of speciation in some families, including the primitive Lauraceae (82 spp./8 genera) as well as the Myrtaceae (124 spp./33 genera) and Proteaceae (49 spp./26 genera). This accounts, in part, for the high level of species endemism, with nearly a third of the rainforest species endemic to the bioregion, and nearly two-thirds restricted to Australia (Table 9.1). Of concern is that nearly 9% of the current flora is made up of exotic species (see Goosem, Chapter 24, this volume).

Current understanding of the origins of the flora of the Wet Tropics' rainforests has developed over the past few decades to recognize the important local nature of many of the taxa. While many other taxa share common ancestry with floristic elements from other tropical regions, the Wet Tropics may justifiably be considered to contain both a unique mixture of widespread species and a significant number of local endemics. These, and the significant role that the Australian land mass has played in providing a cradle for a number of more widespread taxa (see Thorne 1986),

Table 9.1 Geographical distribution of species, genera and families of vascular plants found in the rainforests of the Wet Tropics bioregion. Classification follows Stevens (2006) for cycads, gymnosperms and angiosperms, and Henderson (2002) for ferns and lycopods

	Species		Genera		Families	
	No.	Σ%	No.	Σ%	No.	Σ%
Wet Tropics endemic	674	31.4	53	6.2	1	0.5
Australian endemic	637	61.2	55	12.6	0	0.5
Australo-Papuan*	189	70.0	36	16.8	1	1.1
Gondwanan[†]	14	70.7	47	22.3	7	4.8
Cosmopolitan[‡]	627	100	664	100	179	100
Total	**2141**	–	**855**	–	**188**	–
Exotic[§]	206	8.8	122	12.5	10	5.1
Total including exotics	2347	–	977	–	198	–

* Includes Australia, the island of New Guinea and associated archipelagos.

[†] Includes Australasia and New Caledonia, India, Madagascar, southern Africa and southern South America, as any combination of countries.

[‡] Includes Asia, Philippines, China, northern South America and combinations not as * or [†] above.

[§] Includes only those exotic species that have naturalized in the rainforests of the Wet Tropics; for a longer list of exotics that have been recorded in the Bioregion see Goosem, Chapter 24, this volume.

Box 9.1 Wet Tropics rainforests support a biodiverse primitive plant flora

World Heritage listing of the Wet Tropics was in part achieved through recognition of the high diversity of primitive plant groups present in the rainforests, some of which are endemic to Australia. We reappraised the Wet Tropics' flora using a modern phylogenetic taxonomy based on molecular evidence (Stevens 2006). Such approaches have substantially revised our understanding of which families are old and which families are much younger, but may retain primitive features. Taking the eudicot clade as representing a significant evolutionary break from the more primitive magnoliid groups and their ancestors, there are 28 primitive dicot families recognized today, of which the bioregion can claim 13, including the Wet Tropics endemic Austrobaileyaceae (Table 9.2). A similar approach shows that Borneo also has 13/28 primitive families, New Caledonia supports 12, the Congo basin 11 and Amazonia 10 families. The 13 families in the Wet Tropics contain 46 genera and 174 species, respectively 5 and 8% of those that comprise the bioregion's flora. The rainforests also contain representatives of five of the 14 extant gymnosperm families and 39 of the 93 monocot families (*sensu* Stevens 2006). Basal angiosperms are thought likely to have evolved in the wet understorey of tropical rainforests (Field *et al.* 2004), conditions that have persisted, albeit in small refugia for some periods, in the Wet Tropics region for millions of years. A relatively high incidence of species pollinated by beetles may also imply long associations, as beetles are implicated in the pollination of some of the first flowering plants (Ervik & Knudsen 2003). Gross (2005) suggested that inefficient pollinators such as beetles may have favoured the maintenance of monoecy (having male and female flowers on the same plant) in the Australian tropics, where it is more common than in the Palaeo- or Neotropics; the families Euphorbiaceae and Sapindaceae contribute about 50% of the monoecious taxa in the Wet Tropics. Beetles are implicated in the pollination of species from both primitive (Eupomatiaceae, Myristicaceae; Armstrong & Irvine 1988, 1989) and more recent families (Rhamnaceae, Rutaceae; Armstrong & Irvine 1990) (see also Boulter *et al.*, Chapter 17, this volume, on pollination landscapes).

Table 9.2 Orders and genera of primitive angiosperms. Families and genera endemic to the rain forests of the Wet Tropics are in bold text

Order	Family	Genera in the Wet Tropics
Amborellales	Amborellaceae	—
Nympheales	Cabombaceae	—
	Nymphaceae	—
Austrobaileyales	**Austrobaileyaceae**	***Austrobaileya***
	Schisandraceae	—
	Trimeniaceae	—
Ceratophyllales	Ceratophyllaceae	—
Chloranthales	Chloranthaceae	—
Magnoliales	Myristicaceae	*Myristica*
	Magnoliaceae	—
	Degeneraceae	—
	Himantandraceae	*Galbulimima*
	Eupomatiaceae	*Eupomatia*
	Annonaceae	*Cananga, Desmos, Fitzalania, Goniothalamus, Haplostichanthus, Meiogyne, Melodorum, Miliusa, Polyalthia, Pseuduvaria, Uvaria, Xylopia*
Laurales	Atherospermataceae	*Daphnandra, Doryphora, Dryadodaphne*
	Calycanthaceae	***Idiospermum***
	Gomortegaceae	—
	Hernandiaceae	*Hernandia*
	Lauraceae	*Beilschmiedia, Cassytha, Cinnamomum, Cryptocarya, Endiandra, Lindera, Litsea, Neolitsea*
	Monimiaceae	***Austromatthaea*, Gen. (Aq. 20546), Gen (Aq. 385424), Gen. (Aq. 63687),** *Hedycarya, Kibara, Levieria, Palmeria, Steganthera, Tetrasynandra, Wilkiea*
	Siparunaceae	—
Canellales	Canellaceae	—
	Winteraceae	*Bubbia, Tasmannia*
Piperales	Aristolochiaceae	*Aristolochia, Pararistolochia*
	Hydnoraceae	—
	Lactoridaceae	—
	Piperaceae	*Peperomia, Piper*
	Saururaceae	—

Source: Stevens (2006).

emphasize the longevity and significance of rainforest as a vegetation type in an Australian context. The dynamics of change over the past dozen or so millennia, and the future of Australian tropical rainforests, are still current research questions, and will be dealt with elsewhere (see Hilbert, Chapter 8, this volume).

Summary

- The Wet Tropics bioregion consists of a mosaic of natural and human-modified landscapes, with the natural plant communities determined by the combined effects of various climatic and environmental factors.
- The rainforest communities range from effectively aseasonal luxuriant wet lowland types to highly seasonal semi-deciduous forests, with substrate and parent geology having significant effects on floristic composition. Understanding of factors affecting community distribution and functioning is poor, with research to date largely focusing at the species level or at the distribution of biomes.
- Natural community types are modified by disturbances involving cyclones and fire, and in the past 150 years by increasing human modification of the landscape. Threats from weeds and feral animals are likely to become more significant in the coming years.
- Although the majority of the bioregion still supports native vegetation, there is unequal representation of the original community types, with some forests on the best agricultural land reduced to <10% of their former extent.
- The bioregion is remarkable for its high generic endemism and diversity, relatively low speciation within genera (with a few notable exceptions) and an abundance of gymnosperms and primitive angiosperms. Together these hint at the long history of rainforest on the Australian continent, together with the impacts of relatively late separation from some sections of the Gondwana super continent. Better understanding of the local and regional biogeography would aid understanding of existing patterns in the vegetation, and enhance our abilities to manage future change.
- The majority of the bioregion's key floristic elements are protected through state or federal legislation, but continued refinement of the Regional Ecosystem framework is needed, particularly of the complex forest communities, to ensure that adequate representation of different communities is achieved, particularly of those with the most limited or threatened distributions.

Acknowledgements

We are grateful to the Wet Tropics Management Authority for pre-release copies of the Stanton and Stanton data, to Caroline Bruce for BIOCLIM mapping and interpretation of the Regional Ecosystems and to Matt Bradford and Pete Green for unpublished data on the Mount Lewis fire of November 2002. Bernie Hyland, Pete Green, Matt Bradford and Nigel Stork made valuable comments on early drafts of this chapter.

References

Accad, A. (1999). *Wet Tropics 80m Digital Elevation Model*. Wet Tropics Management Authority, Cairns.

Armstrong, J. E. & Irvine, A.K (1988). Beetle pollination in Australian tropical rainforests. *Proceedings of the Ecological Society of Australia* **15**: 107–13.

Armstrong, J. E. & Irvine, A. K. (1989). Floral biology of *Myristica insipida* (Myristicaceae), a distinctive beetle pollination syndrome. *American Journal of Botany* **76**: 86–94.

Armstrong, J. E. & Irvine, A. K. (1990). Functions of staminodia in the beetle-pollinated flowers of *Eupomatia laurina*. *Biotropica* **22**: 429–31.

Ash, J. (1988). The location and stability of rainforest boundaries in northeastern Queensland, Australia. *Journal of Biogeography* **15**: 619–30.

Australian Surveying and Land Information Group (AUSLIG) (2001). *GEODATA 9 Second DEM Version 2 User Guide*, 2nd edn. AUSLIG, Canberra.

Boulter, S. L., Kitching, R .L. & Howlett, B. G. (2006). Family, visitors and the weather – patterns of flowering in tropical rain forests of northern Australia. *Journal of Ecology* **94**: 369–82.

Bowman, D. M. J. S. (2000). *Australian Rainforests: Islands of Green in a Land of Fire*. Cambridge University Press, Cambridge.

Bradford, M. G. & Harrington, G. N. (1999). Aerial and ground survey of sap trees of the yellow-bellied glider (*Petaurus australis reginae*) near Atherton, North Queensland. *Wildlife Research* **26**: 723–9.

Commonwealth Bureau of Meteorology (2006). Climate averages for Australian sites (www.bom.gov.au/climate/map/climate_avgs /a31.shtml).

Crome, F. H. J., Moore, L. A. & Richards, G. C. (1992). A study of logging damage in upland rainforest in north Queensland. *Forest Ecology and Management* **49**: 1–29.

Department of Environment and Heritage (2005). Mabi Forest: Advice to the Minister for the Environment and Heritage from the Threatened Species Scientific Committee (TSSC) on Amendments to the List of Ecological Communities under the Environment Protection and Biodiversity Conservation Act 1999 (EPBC Act). Version last updated 14 February 2005 (www.deh.gov.au/biodiversity/threatened/communities/ mabi-forest.html).

Environment Australia (2005). Revision of the Interim Biogeographic Regionalisation for Australia (IBRA) and Development of Version 5.1 – Summary report (2000). Updated, IBRA Version 6.1 (Digital Data, metadata) (www.deh.gov.au/parks/nrs/ibra/version6-1/index.html).

Ervik, F. & Knudsen, J.T. (2003). Water lilies and scarabs: faithful partners for 100 million years? *Biological Journal of the Linnean Society* **80**: 539–43.

Field, T. S., Arens, N. C., Doyle, J. A., Dawson, T. E. & Donoghue, M.J. (2004). Dark and disturbed: a new image of early angiosperm ecology. *Paleobiology* **30**: 82–107.

Gross, C. L. (2005). A comparison of the sexual systems in the trees from the Australian tropics with other tropical biomes – more monoecy but why? *American Journal of Botany* **92**: 907–19.

Harrington, G. N., Bradford, M. G. & Sanderson, K. (2005). *The Wet Sclerophyll and Adjacent Forests of North Queensland: A Directory to Vegetation and Physical Survey Data.* CSIRO and the Cooperative Research Centre for Tropical Rainforest Ecology & Management. Rainforest, Cairns.

Harrington, G. N. & Sanderson, K. D. (1994). Recent contraction of wet sclerophyll forest in the wet tropics of Queensland due to invasion by rainforest. *Pacific Conservation Biology* **1**: 319–27.

Henderson, R. J. F. (ed.) (2002). *Queensland Plants, Algae and Lichens: Names and Distributions.* Queensland Herbarium Environmental Agency, Indooroopilly, Queensland.

Hill, R., Smyth, D., Shipton, H. & Fischer, P. (2001). Cattle, mining or fire? The historical causes of recent contractions of open forest in the wet tropics of Queensland through invasion by rainforest. *Pacific Conservation Biology* **7**: 185–94.

Hopkins, M. S., Ash, J., Graham, A. W., Head, J. & Hewett, R. K. (1993). Charcoal evidence of the spatial extent of the eucalyptus woodland expansions and rainforest contractions in North Queensland during the late Pleistocene. *Journal of Biogeography* **20**: 357–72.

Hopkins, M. S. & Graham, A. W. (1987). Gregarious flowering in a lowland tropical rainforest: a possible response to disturbance by Cyclone Winifred. *Australian Journal of Ecology* **12**: 25–9.

Hopkins, M. S. & Graham, A. W. (1989). Community phenological patterns of a lowland tropical rainforest in north-eastern Australia. *Australian Journal of Ecology* **14**: 399–413.

Hopkins, M. S., Head, J., Ash, J., Hewett, R. K. & Graham, A. W. (1996). Evidence of a Holocene and continuing recent expansion of lowland rain forest in humid, tropical north Queensland. *Journal of Biogeography* **23**: 737–45.

Houlder, D., Hutchinson, M., Nix, H. & McMahon, J. (2000). *ANUCLIM User's Guide.* Australian National University, Canberra.

Hutley, L. B., Doley, D., Yates, D. J. & Boonsaner, A. (1997). Water balance of an Australian subtropical rainforest at altitude: the ecological and physiological significance of intercepted cloud and fog. *Australian Journal of Botany* **45**: 311–29.

Lewis, S. L., Phillips, O. L., Sheil, D., *et al.* (2004). Tropical forest tree mortality, recruitment and turnover rates: calculation, interpretation and comparison when census intervals vary. *Journal of Ecology* **92**: 929–44.

Marrinan, M. J., Edwards, W. & Landsberg, J. (2005). Resprouting of saplings following a tropical rainforest fire in north-east Queensland, Australia. *Austral Ecology* **30**: 817–26.

Queensland Government Environmental Protection Agency (2005). Regional ecosystem framework (www.epa.qld.gov.au/nature_conservation/biodiversity/regional_ecosystems/regional_ecosystem_framework/).

Rainforest CRC (2003). *35 Bioclimatic Surfaces at 80 m Grid Cell Resolution.* Rainforest Cooperative Research Centre, Cairns.

Richards, P. W. (1996). *The Tropical Rain Forest*, 2nd edn. Cambridge University Press, Cambridge.

Schatz, G. E. (2001). *Generic Tree Flora of Madagascar.* Royal Botanic Gardens, Kew and Missouri Botanical Garden.

Specht, R. L. & Specht, A. (1999). *Australian Plant Communities: Dynamics of Structure, Growth and Biodiversity.* Oxford University Press, Oxford.

Spencer, H., Weiblen, G. & Flick, B. (1996). Phenology of *Ficus variegata* in a seasonal wet tropical forest at Cape Tribulation, Australia. *Journal of Biogeography* **23**: 467–75.

Stanton, P. & Stanton, D. (2005). *Vegetation of the Wet Tropics of Queensland Bioregion.* Wet Tropics Management Authority, Cairns.

Stevens, P. F. (2006). Angiosperm Phylogeny Website. Version 6, May 2005 (www.mobot.org/MOBOT/research/APweb/).

Stocker, G. C. (1981). Regeneration of a north Queensland rain forest following felling and burning. *Biotropica* **13**: 86–92.

Thorne, R. F. (1986). Antarctic elements in Australasian rainforests. *Telopea* **2**: 611–17.

Tracey, J. G. (1969). Edaphic differentiation of some forest types in eastern Australia. I. Soil Physics factors. *Journal of Ecology* **57**: 805–16.

Tracey, J. G. (1982). The vegetation of the humid tropical region of north Queensland. CSIRO, Melbourne.

Tracey, J. G. & Webb, L. J. (1975). *Vegetation of the Humid Tropical Region of North Queensland* (15 maps at 1:100 000 scale + key). CSIRO Long Pocket Laboratory, Indooroopilly, Australia.

Unwin, G. L. (1989). Structure and composition of the abrupt rainforest boundary in the Herberton Highland, north Queensland. *Australian Journal of Botany* **37**: 413–28.

Unwin, G. L., Stocker, G. C. & Sanderson, K. D. (1985). Fire and the forest ecotone in the Herberton highland, north Queensland. *Proceedings of the Ecological Society of Australia* **13**: 215–24.

van Steenis, C. G. G. J. (1972). *The Mountain Flora of Java*. Brill, Leiden.

Vernes, K. (2000). Immediate effects of fire on survivorship of the northern bettong (*Bettongia tropica*): an endangered Australian marsupial. *Biological Conservation* **96**: 305–9.

Webb, L. J. (1958). Cyclones as an ecological factor in tropical lowland rainforest, North Queensland. *Australian Journal of Botany* **6**: 220–8.

Webb, L. J. (1968). Environmental relationships of the structural types of Australian rain forest vegetation. *Ecology* **49**: 296–311.

Webb, L. J. (1969). Edaphic differentiation of some forest types in eastern Australia. II. Soil chemical factors. *Journal of Ecology* **57**: 817–30.

Webb, L. J. & Tracey, J. G. (1981). Australian rainforests: patterns and change. In *Ecological Biogeography of Australia*, Keast, A. (ed.). Junk, The Hague, pp. 605–94.

Whiffin, T. & Hyland, B. P. M. (1986). Taxonomic and biogeographics evidence on the relationships of Australian rainforest plants. *Tolepea* **2**: 591–610.

Whitmore, T. C. (1998). *An Introduction to Tropical Rain Forests*. Oxford University Press, Oxford.

Williams, P., Kemp, J., Parsons, M., Devlin, T., Collins, E. & Williams, S. (2005). Post-fire plant regeneration in montane heath of the Wet Tropics, north-eastern Queensland. *Proceedings of the Royal Society of Queensland* **112**: 63–70.

10 TOWARDS AN UNDERSTANDING OF VERTEBRATE BIODIVERSITY IN THE AUSTRALIAN WET TROPICS

Stephen E. Williams[1]*, Joanne L. Isaac[1], Catherine Graham[2] and Craig Moritz[3]*

[1]Centre for Tropical Biodiversity and Climate Change, School of Marine and Tropical Biology, James Cook University, Townsville, Queensland, Australia
[2]Department of Ecology and Evolution, Stony Brook State University, New York, USA
[3]Museum of Vertebrate Zoology, University of California, Berkeley, California, USA
*The authors were participants of Cooperative Research Centre for Tropical Ecology and Management

Introduction

Few other fields within ecology have received as much attention as the study of the generation and maintenance of patterns of biodiversity. Biodiversity, the diversity of life, can be viewed as a concept, a measurable entity or a socio-political phenomenon that embodies concern over the degradation of the natural environment (see Gaston 1996 for review). The broad concept of biodiversity has been defined many times, such as the definition of Williams *et al.* (1994) that biodiversity is 'the irreducible complexity of the totality of life'. Determination of the processes behind patterns of biodiversity is difficult and complex, both in theory and in practice. Many possible processes need to be considered and there are many biases and confounding influences. Nevertheless, understanding diversity is a vital part of ecology and conservation biology. Schluter and Ricklefs (1993) describe seven types of processes that contribute to patterns of diversity, including local ecological interactions, movement of individuals within, and between, habitats, the spread of taxa within regions, speciation within the region, long-term exchange of taxa between regions and unique events. Ricklefs (1987) proposed that the production of an integrated theory of the determinants of

patterns of diversity requires the integration of all spatial scales and fields of study from genetics to biogeography. Biodiversity can be treated as a measurable, albeit complex, entity and can be described using four hierarchical dimensions: the diversity of genes, species, ecological functions and ecosystems.

Tropical rainforests are the most diverse terrestrial biome (Wilson 1988; Joseph *et al.* 1995), covering only 7% of the Earth's surface, but estimated to contain over half of the world's species (Wilson 1988). Global biodiversity is concentrated in the tropics, where there are also high levels of vulnerable species and restricted endemics. Tropical mountain systems, like the Australian Wet Tropics, also represent hotspots of biodiversity and endemism due to the compression of climatic zones over the elevational gradient (Körner 2002). In a national context, although the Wet Tropics covers only 0.1% of Australia (Figure 10.1) (Keto & Scott 1986; Winter *et al.* 1987), the region contains approximately 30% of the Australian terrestrial vertebrate fauna, including 83 species that are regionally endemic. This is despite the fact that a significant proportion of the tropical rainforest has been cleared (estimates vary from 20% by Winter *et al.* (1987) to over 60% by Webb and Tracey (1981)). There are about 783 000 ha of tropical rainforest in the Australian Wet Tropics

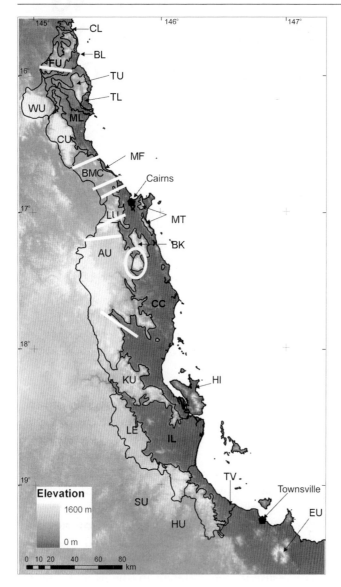

Figure 10.1 Map showing the Wet Tropics bioregion with outlines of the major subregions (CL, Cooktown Lowlands; FU, Finnegan Uplands; BL, Bloomfield-Helenvale Lowlands; TU, Thornton Uplands; TL, Thornton Lowlands; WU, Windsor Uplands; CU, Carbine Uplands; ML, Mossman Lowlands; BMC, Black Mountain Corridor; MF, McAlister Foothills; LU, Lamb Uplands; AU, Atherton Uplands; BK, Bellenden Ker/Bartle-Frere; KU, Kirrama Uplands; HI, Hinchinbrook Island; CC, Cairns-Cardwell Lowlands; MT, Malbon-Thompson Uplands; LE, Lee Uplands; SU, Spec Uplands; HU, Halifax Uplands; IL, Ingham Lowlands; TV, Townsville Lowlands; EU, Elliot Uplands) major phylogeographical breaks (in white) and elevation.

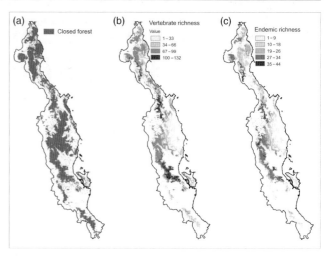

Figure 10.2 Map of the Wet Tropics bioregion showing: (a) the distribution of closed canopy rainforest; (b) the distribution of rainforest vertebrate species diversity; and (c) the distribution of endemic vertebrate species diversity.

bioregion (Figure 10.2a), representing about 0.09% of the total rainforest in the world (Winter *et al.* 1987; Wilson 1988; Goudberg & Bonell 1991). However, while this is a small proportion of the world's rainforests, the Wet Tropics are of great scientific interest owing to their latitudinal spread and high levels of faunal endemicity (Winter *et al.* 1987). These high levels of diversity, coupled with high levels of endemicity, are especially important to the conservation of biodiversity (Gentry 1992) and are vital as evolutionary 'hot spots', as discussed by Myers (1988).

The Wet Tropics is an ideal location in which to study the determinants of biodiversity for a variety of reasons, including

1 The region has a unique fauna of global significance, with high levels of regional diversity and endemism.

2 The levels of taxonomic, genetic and ecological knowledge of the terrestrial vertebrates are better than for many other tropical rainforests around the world.

3 There has been considerable research on the long-term history of the region, especially with respect to the effects of climate and habitat fluctuation during the Pleistocene glaciations.

4 The habitats and environments of the region have been mapped in detail using GIS technology, making the analysis of spatial patterns of environmental variables possible at a regional scale.

5 The rainforests within the region are distributed along a series of disjunct mountain ranges, resulting in a number of relatively discrete subregions with differing characteristics.

This chapter reviews some of the fundamental concepts in understanding patterns of biodiversity, with regard to research on a model system, of the rainforests of the Wet Tropics. In the Wet Tropics bioregion, extensive studies have examined patterns of biodiversity and community structure over a range of spatial scales, considering the interfaces between the traditional local approach (niche theory) and the related fields of evolutionary biology, genetic diversity, phylogeography, biogeography and unique events that have affected the evolution of the assemblage structure and biodiversity of the region. The importance of these studies in terms of understanding patterns of biodiversity in general, and in conservation plans aiming to maintain high levels of biodiversity in tropical rainforests worldwide, are also discussed.

Current knowledge of patterns of biodiversity in the Wet Tropics

The current knowledge of the patterns of biodiversity of vertebrates within the Wet Tropics is now one of the best in the world for a regional fauna in a tropical ecosystem. Although the state of knowledge is variable across taxonomic groups, species distributions, patterns of species richness and assemblage structure are now well documented (Williams 2006). Additionally, the patterns of geographical variation in population genetic structure, abundance and morphological variation have been examined for many species of birds, mammals, reptiles and frogs. These data continue to be refined, updated and analysed within the new Centre for Tropical Biodiversity and Climate Change at James Cook University.

Winter *et al.* (1984) proposed that for mammals, there are two subregions of faunal assemblage occurring in the Wet Tropics: a northern subregion (Thornton Peak Range, Finnegan Range and surrounding lowlands) and a southern subregion (northern end of Atherton Uplands south to Townsville), with an overlap zone including the Carbine and Windsor Uplands. This was supported and refined by multivariate analyses of assemblage structure that suggested five distinct

mammal assemblages (Williams 1997). Finer resolution maps of species richness of rainforest mammals show a clear pattern of highest diversity in the Atherton Uplands and in uplands in general (Williams 2006). Mammalian species richness is also high in the open forest adjacent to rainforest along the western edge of the region where there is a high spatial heterogeneity of vegetation structure (Williams *et al.* 2002; Williams 2006).

Detailed data on the distribution of individual bird species across the Wet Tropics region are now relatively well known and mapped in Williams (2006), although the level of knowledge is best in the upland rainforests and poorer in the complex coastal vegetation mosaic. Generally, rainforest bird assemblages north of Cooktown are similar to lowland rainforest assemblages in New Guinea, suggesting that these are relicts from a previous land connection with New Guinea, or from an older, cool temperate, Australian fauna (Kikkawa *et al.* 1981). South of Cooktown there is a high level of regional endemism (Kikkawa *et al.* 1981). Most species of birds are broadly distributed across the region, with species richness being highest at mid-range elevations of 600–800 metres, and the degree of endemism is highest in the uplands (Williams 2006).

The Wet Tropics contains the most diverse rainforest frog assemblage in Australia, with very high levels of regional endemism, including several species (mostly microhylids) restricted to just one or two mountain-tops. Core areas of frog diversity include the Atherton Uplands, Carbine Uplands and Thornton Uplands, as well as other high altitude areas characterized by high rainfall, granite parent rock and high historical habitat stability (McDonald 1992; Williams & Hero 2001; Graham *et al.* 2006). Frog diversity and species richness decrease to the north and south of the central uplands.

The richness of endemic species of reptiles in the Wet Tropics is the highest of any Australian rainforest (Moritz *et al.* 2005). Recent surveys and taxonomic revisions have revealed several new narrowly endemic species (Hoskin *et al.* 2003; Hoskin & Couper 2004; Couper *et al.* 2005), especially in the southern and northern extremes of the region. Williams *et al.* (1996) determined that the most diverse reptile assemblages are found in the Atherton Uplands, Cooktown lowlands and Cairns-Cardwell lowlands. Generally, species richness of reptiles is somewhat greater in lowland zones of the Wet Tropics and this is thought to be due to the heterogeneous mixture of sclerophyll habitats and thermal microhabitats (Williams *et al*. 1996). More recent analyses suggest that the mid-altitudinal band (400–800 m) is the most diverse area for rainforest reptiles, and as in the other taxa, there is a tendency for the regionally endemic species richness to be higher in the uplands than the lowland rainforests.

A second significant component of diversity in this system is morphologically cryptic genetic diversity among areas. Molecular (mostly mtDNA) analyses of rainforest-restricted species that are widespread within the Wet Tropics have consistently revealed strong phylogeographic structure (Hugall *et al*. 2002), this being more pronounced for species of reptiles and amphibians than for birds (Joseph *et al*. 1995; Schneider *et al*. 1998). Most of these phylogeographical breaks are centred on the Black Mountain Corridor (Figure 10.1), but there is variation among species in the spatial scale of genetic diversity. While this strong genetic structuring of populations does not increase local richness, it does represent a significant component of the overall endemism and beta-diversity in this system (Moritz 2002).

Overall, vertebrate biodiversity, as measured by species richness, in the Wet Tropics bioregion is greatest in upland areas, including the Atherton Uplands, Carbine Uplands and Thornton Uplands (Figure 10.2b), although patterns differ somewhat between vertebrate groups. Endemic species of all vertebrate groups tend to be mostly confined to the island-like cool upland rainforest patches of >300 metres above sea level (Figures 10.1 and 10.2c) and the Atherton Uplands is most species rich in endemics (Figures 10.1 and 10.2c) (Moritz *et al*. 2001). When examined in detail, patterns of abundance and diversity generally peak in mid-altitudes (400–800 metres) (Figure 10.2), where there are both large areas of rainforest and high levels of primary productivity. The potential causes of these patterns in biodiversity and endemism are discussed in more detail in the following sections.

Evolution and assembly of the Wet Tropics vertebrate fauna

Given the growing interest in the effects of biogeographical and speciation history, and of niche conservatism on macroecological patterns of diversity (Hubbell

2001; Webb *et al*. 2002; Wiens & Graham 2005), it is important to understand the evolutionary processes that have generated the high species endemism of the Wet Tropics. The antiquity of these rainforests, together with the late Tertiary isolation of the system from the remainder of the mesotherm archipelago (reviewed in Moritz *et al*. 2000; Morley 2000; Kershaw *et al*. 2005), leads to the prediction that many species now endemic to the system will be phylogenetic relics or have deep connections to other forests in this system (i.e. PNG highlands, southern or middle-east Queensland). Indeed, previous overviews of vertebrate diversity in the system have stressed this aspect (Keto & Scott 1986; Goosem 2002). Another possibility is that the climatic oscillations of the Quaternary have resulted in speciation among rainforest refuges within the Wet Tropics, as originally envisaged for other rainforest systems (Haffer 1969; Mayr & O'Hara 1986; Rull 2006).

Tests of these predictions require molecular phylogenies with thorough sampling of congeners of endemic species, and such information is available for 42 of the 83 vertebrates endemic to the system (Table 10.1). As expected, there are indeed several species that represent phylogenetic relics in the sense of being basal to large clades – the musky rat kangaroo (*Hypsipromnodon moschatus*) is basal to the Macropodidae, the golden bowerbird (*Priodonura newtoniana*) to the Australopapuan bowerbirds, the chameleon gecko (*Carphodactylus laevis*) to the leaftail geckos (*Phyllurus* and *Saltuarius*) and the two Wet Tropics skinks (*Eulamprus frerei* and *Gnypetoscincus queenslandiae*) to the east coast water skinks (*Eulamprus* and allied genera). In addition, among amphibians, it has been shown that *Taudactylus* is basal to other myobatrachine frogs (Read *et al*. 2001), but molecular data on relationships and divergence times among the species endemic to the Wet Tropics and their congeners in forests to the south are lacking and, given recent severe declines, may be unobtainable. As predicted, many species endemic to the Wet Tropics have deep (mostly mid-Miocene to early Pliocene) connections to congeners inhabiting rainforests elsewhere in the mesotherm archipelago (New Guinea and central-southern Queensland) (Table 10.1). In some cases, highly conservative morphology has obscured true evolutionary histories. The chowchilla (*Orthonyx spaldingii*) was supposed to be the sister species to the logrunner (*O. temminickii*), which itself had disjunct populations

Table 10.1 Phylogenetic evidence on origins of selected vertebrates endemic to the Wet Tropics, and species within which there is significant genetic divergence across the Black Mountain Corridor (BMC)

Origin	Mammals	Birds	Reptiles	Amphibians
Relictual	Hypsiprimnodon moschatus[1]	Prionodura newtoniana[6]	Carphodactylus laevis,[11] Eulamprus frerei,[12] Gnypetoscincus queenslandiae[12]	
Old, PNG	Pseudochirops archeri[2]	Orthonyx spaldingii,[7] Scenopoeetes dentirostris[6]	Hypsilurus boydii[13]	Litoria genimaculata[4]
Old, E Aus	Antechinus godmani,[3] Hemibelideus lemuroides[2]		Phyllurus amnicola,[11] Phyllurus gulbaru,[11] Saltuarius cornutus,[11] Eulamprus tigrinus,[12] Carlia rubrigularis,[14] Saproscincus czechuri[15]	
Recent, PNG		Tyto multipunctata[7]		
Recent, E Aus		Sericornis keri,[8] Acanthiza katherina[9]		
Within AWT, recent	Pseudochirulus cinereus & P. herbertensis;[2,4] Dendrolagus lumholtzi & D. bennettianus[5]			Litoria spA[18]
Within AWT, old			Saproscincus basiliscus & S. lewisi,[15] S. tetradactyla[15]	Mixophyes (3 species),[19] Cophixalus (10 spp., see text)[20]
Deep gene splits across BMC			C. laevis,[16] G. queenslandiae,[16] S. cornutus,[16] C. rubrigularis[17]	L. genimaculata,[16] Litoria nannotis,[16] Cophixalus ornatus[20]
Recent splits across BMC	Bettongia tropica[21]	O. spaldingii,[10] Heteromyias albispecularis,[10] Sericornis citreogularis[10]		Litoria rheocola[16]

Source: [1]Burk et al. 1998, [2]Springer et al. 1992, [3]Armstrong et al. 1998, [4]Moritz et al. 1997, [5]Bowyer et al. 2003, [6]Endler et al. 2005, [7]Norman et al. 2002, [8]Joseph & Moritz 2003, [9]Nicholls 2001, [10]Joseph et al. 1995, [11]Hoskin et al. 2003, [12]O'Connor & Moritz 2003, [13]Schulte et al. 2003, [14]Stuart-Fox et al. 2003, [15]Moussalli et al. 2005, [16]Schneider et al. 1998, [17]Dolman & Moritz 2006, [18]Hoskin et al. 2005, [19]Mahoney et al. 2006, [20]Hoskin 2004, [21]Pope et al. 2000.

in the rainforests of New Guinea and southern Queensland; however, molecular analysis demonstrated that the latter is paraphyletic, with the Wet Tropics species being sister to (but still highly divergent from) the New Guinea taxon. The northern rainbow skink *Carlia rubrigularis* has been shown to consist of two divergent lineages of which the southern one is more closely related to the mid-east Queensland congener *C. rhomboidalis* than it is to northern conspecific populations (Dolman & Moritz 2006). As yet, only few of the species endemic to the Wet Tropics have been shown to have recent (i.e. Pleistocene) phylogenetic connections to taxa elsewhere, and these are all birds: the lesser sooty owl (*T. multipunctata*) has essentially identical mtDNA to populations of *T. tenebricosa* from southern Queensland and New Guinea, and the Mountain thornbill (*Acanthiza katherina*) and Atherton scrubwren (*Sericornis keri*) are closely related (mtDNA divergences of c.4–6%) to widespread species inhabiting multiple forest types (*A. pusilla* and *S. frontalis*, respectively). In addition to the species endemic to the Wet Tropics, there are several morphologically defined subspecies of birds and mammal, each with conspecific populations in disjunct rainforests (primarily southern Queensland). Molecular data are available for only two of these and reveal different histories of isolation – Wet Tropics populations of the satin bowerbird (*Ptilonorhynchus violaceus minor*) are highly distinct genetically from southern populations (Nicholls & Austin 2005; Nicholls et al. 2006), whereas the Wet Tropics subspecies of the tiger quoll (*Daryurus maculatus gracilis*) is genetically similar to populations from southern forests (Firestone et al. 1999). Conversely, comparisons of morphologically similar populations of birds (*Sericornis*) and the skink *C. rhomboidalis* between Wet Tropics and southern rainforest populations have revealed relatively strong genetic differentiation (Joseph & Moritz 1994; Dolman & Moritz 2006), attesting to the isolation of the Wet Tropics populations since the late Quaternary (or earlier).

The above examples illustrate one common phenomenon – most vertebrate species that are endemic to the Wet Tropics have their closest relatives in rainforests elsewhere. The corollary is that few vertebrates have speciated within the Wet Tropics, certainly fewer than is the case for insects (Bouchard *et al.* 2005). The major exceptions are microhylid frogs, especially the genus *Cophixalus*, with 11 species endemic to the Wet Tropics, 8 being narrowly endemic to one or a few adjacent mountain-tops. Here, species from the southern Wet Tropics tend to be distantly related with no clear geographical pattern, whereas the narrow endemics from the northern Wet Tropics show more geographical adjacency of sister species. But once again, these are mostly old (mid-Miocene to Pliocene) species. Deep molecular divergence is also evident between two parapatric sister species of *Saproscincus* skinks (*S. basiliscus* and *S. lewisi*) and among three newly described species of *Mixophyes* frogs. Of the few examples of recent (potentially Pleistocene) speciation within the Wet Tropics, two involve allopatric sister taxa separated by the Black Mountain Corridor (*Dendrolagus* tree kangaroos, and *Pseudochirulus* ringtail possums). The third is a newly described species of *Litoria* frog restricted to a hybrid zone (Hoskin in press), discussed in more detail below.

Contrasting with the limited extent of vertebrate speciation within the Wet Tropics is a common pattern of deep phylogeographical divisions within widespread, morphologically conservative species (Table 10.1) (Schneider & Moritz 1999; Hoskin *et al.* 2005). The genetic breaks, mostly identified via mtDNA analysis, but supported by nuclear loci where this has been examined (Moritz *et al.* 1993; Dolman & Moritz 2006), are of a scale suggesting pre- or early-Pleistocene divergence. Most are located between major Quaternary refugia (Hugall *et al.* 2002), in particular between the Atherton Uplands in the south and the Carbine Uplands in the north (Figures 10.1 and 10.3). Conversely, populations within regions where little rainforest persisted during glacial maxima (e.g. Windsor and Spec Uplands) tend to be closely related genetically to those in adjacent refugia (e.g. Thornton, Kirrima/Atherton Uplands, respectively), from which they presumably

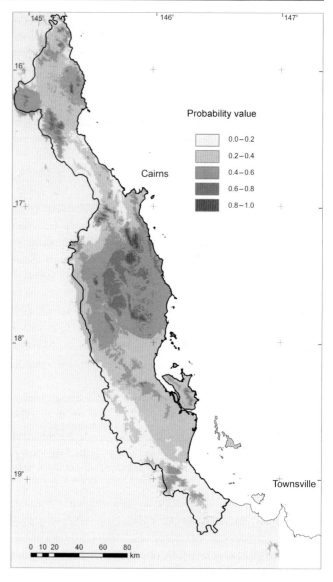

Figure 10.3 Map of relative rainforest stability over the Pleistocene using MaxEnt (Phillips *et al.* 2006) to model rainforest distribution over different climates. The probability scale refers to the probability that rainforest persisted throughout the time period modelled.

recolonized during the cool-wet conditions of the early Holocene (Schneider *et al.* 1998; Hugall *et al.* 2002).

Given that genetic divergences among these phylogeographical units are sometimes on par with those among morphologically distinct congeners, it is reasonable to ask whether the former are in fact distinct species. The answer depends on how you chose to delimit species (Sites & Marshall 2003). According to some criteria (e.g. reciprocal monophyly of mtDNA;

Wiens & Penkrot 2002) they would be, in which case the number of species endemic to the Wet Tropics would increase dramatically. However, we prefer a more holistic view that requires evidence for corresponding phenotypic divergence and/or reproductive isolation among parapatric or sympatric populations. As already noted, the genetically divergent lineages tend to be morphologically indistinguishable, even though there is variation in body size across environmental gradients within lineages (Schneider & Moritz 1999; Hoskin *et al.* 2005). Another approach is to examine parapatric contact zones for reproductive isolation. Several such studies in the centrally located suture zone (Figure 10.3) have produced varying results. In the tropical bettong (*Bettongia tropica*) no genetic disequilibrium was observed in a secondary contact between two relatively recently separated phylogeographical lineages (Pope *et al.* 2000). In contrast, substantial post-zygotic selection against hybrids was inferred from narrow clines and strong genetic disequilibrium in the contact zone between the deeply divergent lineages of *Carlia rubrigularis* (Phillips *et al.* 2004). Finally, hybrids appear rare at two contact zones between genetically divergent lineages of the hylid *Litoria genimaculata* (Hoskin *et al.* 2005). For this case, there is experimental evidence for strong, but asymmetric, post-zygotic isolation between northern and southern lineages and, in one contact zone, but not the other, evolution of strong assortative mating. Remarkably, this reinforcement-driven divergence in mate choice (and body size) has resulted in rapid speciation between closely related, allopatric populations of the southern lineage (Hoskin *et al.* 2005; Hoskin in press). More such studies are needed, but the evidence to date is consistent with the view that these phylogeographical lineages are incompletely isolated and should continue to be regarded as conspecific, even while their relevance to conservation is recognized (Moritz & McDonald 2005).

How do these observations accord with current thinking on the processes that generate species diversity, and why is recent (Pleistocene) speciation of vertebrates so rare within the Wet Tropics when compared to topographically complex regions such as the Andes and east African arc mountains (Moritz *et al.* 2000)? Both theory and evidence indicate that: (i) allopatric isolation will eventually result in speciation and is generally more probable than parapatric or sympatric

speciation; (ii) this is a slow process, taking of the order of millions of generations, unless there is divergent selection driving the evolution of post- or pre-zygotic isolation (which also is required for parapatric or sympatric speciation); and (iii) speciation by reinforcement (selection for assortative mating where hybrids are unfit) is more plausible than previously considered (Gavrilets 2003; Servedio & Noor 2003; Coyne & Orr 2004). Several aspects of the evidence from the Wet Tropics are consistent with these views. The distributions of sister taxa (within and outside of the Wet Tropics *per se*) are more consistent with allopatric than alternative modes of speciation (but see Schneider *et al.* 1999) and the varying levels of reproductive isolation observed in contact zones support the view that strongly differentiated phylogeographical lineages might represent 'species-in-waiting' (Avise & Walker 1998), but also that this is a very slow process. The enhancing effect of divergent selection on speciation can be inferred in some cases where morphologically distinct species are now recognized, e.g. a possible role for divergent sexual selection in the speciation of the blue-throated *Carlia rhomboidalis* from the southern lineage of the red-throated *C. rubrigularis* (Dolman & Moritz 2006) and ecologically mediated shifts in body size between sister taxa of *Mixophyes* and *Saproscincus*. Strong pre-zygotic selection has evidently caused rapid morphological divergence and speciation in *L. genimaculata* (Hoskin *et al.* 2005). How selection, genetic drift and isolation have interacted in diversification of *Cophixalus* frogs warrants close attention, as these include instances of closely related species in sympatry, parapatry and allopatry (Hoskin 2004). Similarly, the processes underlying the relatively recent divergence between allopatric sister taxa of *Dendrolagus* tree-kangaroos and of *Pseudochirulus* possums needs more exploration.

These and some other (e.g. *Acanthiza katherina*, *Sericornis keri*) examples aside, the overall impression is one of strong conservation of eco-morphology and of environmental niches throughout the climate-driven habitat fluctuations of the Quaternary. Several factors may be contributing to this conservatism. One is that the Wet Tropics rainforests may be too limited in geographical extent or environmental diversity (e.g. the lack of substantial lowland rainforests) to sustain conditions for speciation for all but the smallest vertebrates (e.g. the microhylids). A second factor could be that stabilizing, rather than diversifying, selection has predominated among refugial populations. In this context, it is notable that leaftail geckos (*Phyllurus*) have diverged morphologically (i.e. speciated) among isolates in the smaller and more marginal rainforest areas of central Queensland and the southernmost Wet Tropics (Couper *et al.* 2000; Hoskin *et al.* 2003), whereas the similarly divergent phylogeographical lineages of the widespread leaftail in the Wet Tropics (*Salturius cornutus*) have not. Perhaps ecological forcing, of the sort suggested by Vanzolini and Williams (1981), is more prevalent in these marginal settings than in the larger refuges of the Wet Tropics. Finally, compared to geologically younger settings such as the Andes and central New Guinea, it is likely that there has been very little by way of novel niche space, and thus potential adaptive radiations, in this ancient system.

The generally slow and conservative evolution of the Wet Tropics fauna has implications for assembly of ecological guilds. Much of the local diversity is the result of persistence of old taxa derived via gradual allopatric divergence, mostly at a spatial scale larger than the Wet Tropics, rather than *in situ* adaptive divergence (niche-filling). Overlying this process of long-term evolution are the effects of local extinctions and recolonizations during the late Pleistocene, with species responding at various spatial and temporal scales (see below). The arboreal mammal community, and the ringtail possums in particular, are a case in point. These include a phylogenetic relic (*H. lemuroides*), species with affinities to New Guinea (*P. archeri*) and recently diverged, allopatric sister-species (*P. herbertensis* and *P. cinereus*). On the basis of their ecological requirements and specialized physiology, Winter (1997) has argued that many of these species underwent substantial range fluctuations during the late Pleistocene and Holocene. Likewise, the arboreal gecko and litter skink guilds each include a phylogenetic relic and other species with affinities of various ages to congeners from southern rainforests (Table 10.1) with widely varying spatial scales of persistence through the Quaternary climate fluctuations. In contrast, the microhylids have achieved their local diversity primarily via *in situ* speciation, which may be related to their distinctive macroecology (Williams & Hero 2001; Graham *et al.* 2006). However, the evident disparity of phylogenetic and biogeographical history within guilds also cautions against attempts to fit a single spe-

ciation dynamic to evolutionary models of community assembly (e.g. Hubbell 2001).

Late Quaternary climate fluctuations and extinction–colonization dynamics

Ricklefs (1987) proposed that patterns of biological diversity should be interpreted in the light of both contemporary and historical influences. Schluter and Ricklefs (1993) expanded on this theme of the importance of historical biogeography in understanding contemporary diversity patterns. Research in the Wet Tropics completely supports the necessity of examining the combined effects of processes associated with speciation, extinction, historical biogeography and contemporary processes and environmental patterns. A combination of palaeoecological and biogeographical evidence indicates substantial climate-induced contractions and shifts of upland rainforest vegetation in the Wet Tropics during the late Quaternary period (reviewed in Moritz *et al.* 2000; Kershaw *et al.* 2005). Specifically, the rainforest contracted during restrictive cool-dry (e.g. Last Glacial Maximum, LGM) and warm-wet (mid-Holocene) periods and expanded during the cool-wet period of the early Holocene and again under current climate (Nix & Switzer 1991; Graham *et al.* 2006), and these contractions resulted in non-random patterns of localized extinctions and recolonizations (Williams 1997; Williams & Pearson 1997; Schneider & Williams 2005; Graham *et al.* 2006).

Williams and Pearson (1997) examined spatial patterns in the distribution of vertebrates in the Wet Tropics in order to determine the processes that shaped vertebrate assemblages and patterns of species richness and endemism. They concluded that the single most significant process that has influenced the current patterns of species richness in the region is the impact of contractions in rainforest area over the Quaternary. The contractions in rainforest area within each subregion resulted in non-random patterns of subregional extinctions (species sifting). Species were sifted out of the subregional species pool relative to their extinction proneness and the relative area the species requires to maintain a viable population. Subsequent wetter periods have allowed some species to recolonize, dependent on their degree of habitat specialization and dispersal ability. They made these conclusions on the basis of several macroecological analyses. Overall species richness of terrestrial rainforest vertebrates was positively related to rainforest area, but not influenced by rainforest shape at the subregional level. A positive relationship between species richness and area is commonly found and is attributed to the fact that larger areas have greater habitat and niche diversity. Larger areas also support larger populations of a species, which are less likely to go extinct, and may also present a larger 'target' to dispersing individuals, thus having lower extinction and higher recolonization probabilities (Gaston & Blackburn 2000). However, the species richness of regionally endemic species was highly correlated with both area **and** shape of the rainforest – with the degree of endemism being greater in rainforest blocks with a larger area that are closer to circular in shape. Areas with a convoluted, less spherical shape supported a higher proportion of less specialized species. Williams and Pearson (1997) interpreted the combined influence of area and shape as being an index of the relative severity of the reduction in rainforest area in each subregion over the Quaternary. This was supported by the high degree of assemblage nestedness among subregions exhibited by the vertebrates. High degrees of geographical nestedness in assemblage structure indicate a process of species-sifting via non-random patterns of extinction and/or colonization (Patterson & Atmar 1986). Graham et al. (2006) tested this hypothesis, and refined the analysis, by using palaeo-habitat modelling combined with cost-path analysis to examine the relative stability of rainforest over the Pleistocene and recolonization pathways during episodes of rainforest expansion. They demonstrated that the stability of suitable habitat areas through historical time is as important as, or in some cases more important than, current rainforest area in explaining current patterns of species richness, particularly for smaller, dispersal-limited taxa. Areas with high habitat stability (Figure 10.3) were found to have high species richness of regionally endemic rainforest vertebrates. Habitat stability was also found to influence patterns of species diversity independent of rainforest area. However, patterns of species richness for endemics with high dispersal capacity were best predicted by using current environmental parameters.

Further support for the profound effects of late Pleistocene and Holocene climate change on the distributions of rainforest-restricted vertebrates comes from the various phylogeographical studies discussed above and in Table 10.1. In general, the high stability areas predicted by spatial modelling (Figure 10.3) are also the foci of unique genetic lineages, especially in low dispersal taxa (Hugall et al. 2002; Graham et al. 2006). Conversely, regions thought to have been largely unsuitable for rainforest (especially during the cold-dry period of the LGM) harbour fewer distinct genetic lineages and show genetic affinities to geographically proximal refugial areas. In addition, many species show the genetic signature of Holocene range expansion (Schneider & Moritz 1999). However, it is also clear that species responded at varying spatial scales to this common history of rainforest fluctuation – some (e.g. S. cornutus, C. rubrigularis and L. genimaculata) have just two genetic lineages separated by the Black Mountain Corridor (BMC), whereas others (e.g. G. queenslandiae, C. laevis) have finer-scale geographical structuring, implying persistence in multiple refugia both north and south of the BMC. These somewhat idiosyncratic, climate-induced range shifts have interesting consequences for interpretation of current assemblage structure within subregions – some upland areas are clearly refugial (Figure 10.3), others have been recently assembled, while others represent a mix of persistent and recolonized populations. The lowlands, though of high species richness overall, have very few endemic rainforest-restricted species and also a lower density of birds than predicted (Williams 2006). Both modelling (Nix & Switzer 1991) and palaeoecological evidence (Hopkins et al. 1993; Kershaw et al. 2005) suggests that the expanded coastal platform supported relatively little rainforest at the LGM, such that warm-adapted species might have undergone extreme contractions. The almost complete loss of lowland rainforest may have resulted in the loss of lowland specialists and this hypothesis is being actively examined now. Another result of the Holocene rainforest expansion was the establishment of new connections between long isolated populations, forming the suture zone that extends across the BMC and Lamb Uplands (Figure 10.3).

Contemporary influences on biodiversity

It is evident that the historical biogeography of the region has had a massive influence on contemporary biodiversity in the Australian Wet Tropics. It has shaped the refugial dynamics and the subsequent patterns of extinction and speciation, as well as both within- and between-bioregion patterns of dispersal and colonization. However, the contemporary environment also has a very significant influence on patterns of abundance, species distributions and assemblage structure.

In almost any analysis, there is a positive species–area relationship. A number of studies have examined this relationship for different subsets of the vertebrate fauna (Williams *et al.* 1996; Williams 1997; Williams & Pearson 1997; Williams & Hero 2001) and, with the exception of microhylid frogs, area is a good correlate of species richness. Most of these analyses ascribed this relationship to habitat diversity; that is, a larger area has higher habitat diversity and hence more species. Williams *et al.* (2002) examined habitat diversity in more detail for mammals and concluded that habitat structure explained much of the local variation in assemblage structure but that it was the spatial heterogeneity, more than vertical complexity, of the vegetation that had the greatest influence on species richness.

For microhylid frogs, regional patterns of species richness are strongly related to consistent moisture levels throughout the year, driven by altitudinal gradients, and are limited by low rainfall in the dry season. However, in more generalist species of rainforest frog, species richness is primarily related to historical biogeographical processes, broad habitat diversity and current climatic variables (Williams & Hero 2001).

More recent studies have implicated a further factor proposed to explain much of the variation in local and regional species richness, the species–energy relationship (e.g. Currie 1991). This theory proposes that patterns of biodiversity can be explained by differences in environmental energy availability, with the most frequently cited pathway linking energy to species richness being that, in areas with high plant productivity, primary and secondary consumers are able to maintain larger populations that reduce their extinction risk, thus elevating species richness and local biodiversity (Wright 1983; Evans *et al.* 2005). Current research in the Wet Tropics suggests that, for avian species at least, the species–energy relationship does explain some of the variation in patterns of local species richness (Figure 10.4). While much of the variability in species richness of rainforest birds in the Wet

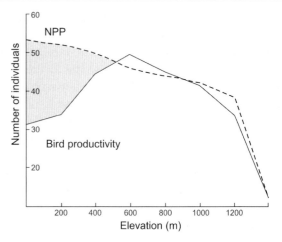

Figure 10.4 The relationship between net primary productivity per unit area (NPP, dashed line) and the realized productivity of the bird assemblage (mean total energy flux) per unit area (continuous line). The relationship is strong in uplands (> 500 m a.s.l.), $P = 0.000$, $R^2 = 0.919$, but breaks down in the lowlands. There is a significant portion of energy coming into the system that is not being realized in the bird assemblage (grey shaded area).

Tropics is explained by the species–area relationship and altitude, results indicate that ecosystem energy flow may be the most significant determinant of local abundance and species richness, at least in birds (Figure 10.4) (Williams *et al.*, unpublished data).

Life history and ecology can also shape local and regional patterns of biodiversity through their capacity to influence the ability of a particular species to disperse and recolonize suitable habitat, to utilize less specific habitat, to withstand environmental change and extinction events and to tolerate varying levels of intraspecific competition and predation. Endemic vertebrate species in the Wet Tropics are generally characterized by a common suite of ecological and life history traits: they tend to occur above 300 metres above sea level (m asl) in relatively isolated rainforest patches, have specialized diet (Williams *et al.* 2006) and/or micro- or macro-habitat requirements and a narrow range. Microhylid frogs are one group that have been less affected by patterns of extinction (Williams & Hero 2001), but the species restricted to single mountaintops have generalist diets while widespread species have specialized diets (Williams *et al.* 2006). This suggests ecological compensation to avoid extinction; that is, species confined to very small areas cannot afford to be specialist in other niche dimensions. These traits can influence the probability that a species will persist, disperse and colonize new habitats and undergo speciation. For example, Williams & Hero (2001) found that low dispersal ability predisposes Wet Tropics frog species to extinction. Range size in particular is thought to be an important determining factor in shaping patterns of biodiversity. One of the most prominent features in biogeography is that species differ markedly in the size of their geographical ranges (Brown 1996) and the range size–abundance relationship has been a controversial topic in macroecology for a number of years (Lawton 1993; Symonds & Johnson 2006). Generally, in a given assemblage of species, the majority tend to have relatively small ranges. However, those species that are widespread also tend to be locally abundant (Symonds & Johnson 2006), and this is probably because those species with a large range are typically generalists in terms of ecology and life history. Thus, local patterns of species diversity result from the complex spatial interactions of many species with small ranges and relatively few with large ranges (Kreft *et al.* 2006). It is argued that this is one of the most fundamental ways in which species share space within a community, influencing biodiversity and local species richness. However, while Murray and Hose (2005) examined the life history parameters of endemic Australian frog fauna and found that small range size was an important explanatory variable for frogs in decline, Symonds and Johnson (2006) found no evidence of a range size–abundance relationship in Australian passerine birds, including those found in the Wet Tropics.

Summary

Although there is now increasing information describing the ongoing evolutionary process within the region, most of the currently recognized species are very old. The current patterns of species richness and biodiversity for vertebrates in the Wet Tropics are probably a combined result of species-sifting via the historical processes of rainforest contraction and expansion together with the subsequent patterns of extinction and recolonization. Areas with high habitat stability over the late Quaternary acted as refuges for surviving species and currently act as 'hotspots' of biodiversity (Moritz *et al.* 2005; Graham *et al.* 2006). Other forces

shaping patterns of biodiversity probably include species-intrinsic factors such as life history, range size–abundance relationships, competition and other ecosystem interactions. The resulting species pool is then filtered by current environmental conditions, including climate and rainfall, habitat heterogeneity, vegetation complexity, primary productivity, altitude and latitude. Diversity is highest in those areas with higher habitat heterogeneity and higher net primary productivity. However, there is now considerable evidence suggesting that diversity is also limited by climatic seasonality; that is, in areas with a highly variable climate across the year, seasonal bottlenecks in resources limit species abundances and distributions. The concepts of limiting thresholds in these patterns of diversity and abundance are in need of further examination and are the topic of a number of ongoing studies.

Conservation of biodiversity in the Wet Tropics

Although the rainforests in the Australian Wet Tropics bioregion have been protected as a World Heritage Area since 1988, the region and biodiversity contained within it is still under threat from a number of factors, many of which similarly threaten biodiversity in other tropical regions. Major threats to biodiversity in the Wet Tropics include climate change, habitat fragmentation, introduced invasive species, disease and interactions among these. As discussed in detail in Williams *et al.* (Chapter 22, this volume) climate change is predicted to result primarily in a reduction in core habitat area, a reduction in overall diversity and particularly negative effects on specialized endemic fauna. Fragmentation of rainforest is relatively severe in some areas, especially the Atherton uplands and some lowland rainforest areas, and has been shown to affect the local composition of mammal (Laurance 1997) and bird (Laurance *et al.* 1993) communities. In addition, detailed studies of a rainforest skink revealed effects on age structure, local dispersal and geographical genetic variation (Sumner *et al.* 1999; Sumner 2005). Harrison and Congdon (2002) assessed the threats of introduced species to the biodiversity of the Wet Tropics and summarized the top worse invasive terrestrial vertebrate pests as pigs (*Sus scrofa*), feral cats (*Felix catus*), cane toads (*Bufo marinus*) and wild dogs (*Canis* spp.). Predation by feral dogs and cats can decimate populations of native small mammals and birds, pigs modify habitat and compete for resources, while cane toads are toxic and lethal to native predators such as quolls and snakes and can also compete with native frog fauna for resources. Finally, introduced diseases are also having a large impact on native fauna in the Wet Tropics. In particular, chytridiomycosis, caused by the fungus *Batrachochytrium dendrobatidis* or chytrid fungus, is thought to be responsible for recent dramatic declines of at least eight species of native frog in the Wet Tropics (Woodhams & Alford 2005). Evidence suggests that chytridomycosis is also responsible for widespread amphibian extinction worldwide and recent work by Pounds *et al.* (2006) indicates that climate change may be acting to increase the incidence of the fungus by optimizing its growth.

In the face of these threatening processes, protecting areas as reserves, and restoring habitat connections/corridors off reserves, is commonly thought of as the best strategy for conservation. Systematic conservation planning (Margules & Pressey 2001; Ferrier 2002) seeks to identify combinations of areas that most efficiently represent viable populations of species at the scale of concern. Extensive research, both in the Wet Tropics (e.g. Williams & Pearson 1997) and elsewhere, demonstrates a positive relationship between biodiversity and area. Thus, a crucial problem for biodiversity conservation is that protected areas should be large and effective enough to support viable populations of their original species (Marsden *et al.* 2005). The relationships between range size–abundance and energy–species richness should also be considered in conservation planning to maximize biodiversity in the Wet Tropics. For example, many range-restricted endemic species in the Wet Tropics are predicted to lose the majority of their core range under future climate change scenarios (e.g. Williams *et al.* 2003), while a potential increase in primary productivity at higher temperatures may ameliorate some of the detrimental effects of climate change on biodiversity predicted for the region (see Williams *et al.*, Chapter 22, this volume).

Consideration of the underlying evolutionary processes contributes to conservation planning in two ways: (i) by understanding the historical biogeography and processes that generate diversity, it should be possible to improve prediction of overall diversity (including poorly known taxa, and genetic and functional diversity); and (ii) by explicitly protecting generative processes, conservation will be more effective in the long term (Moritz 2002). At a global scale, patterns of species richness and endemism are highly correlated among amphibians, reptiles, birds and mammals (Lamoreux *et al.* 2006), but discordance is often seen at smaller spatial scales (e.g. Prendergast *et al.* 1993). For the Wet Tropics, the underlying history of climate-driven vicariance has resulted in spatial patterns of complementarity that are essentially shared across taxa and, for vertebrates, between species and genes. Given this, we would expect one taxonomic group to be an effective predictor of conservation priority for another. However, whereas insects and snails were effective as surrogates in conservation planning for vertebrate species, the converse was not true (Moritz *et al.* 2001). This is probably due to the smaller spatial scale of species turnover (and thus complementarity) for the former and is sobering given that patterns of avian and mammalian species are commonly used to develop strategies for overall biodiversity conservation. To capture species diversity of Wet Tropics rainforest vertebrates, the highest priority areas for conservation are the major coastal refugia – CU, BK, TU and geographical outliers, especially EU, FU and HU (Moritz *et al.* 2001, 2005). These areas also represent most of the phylogenetic diversity within the microhylids (Moritz *et al.* 2005). Remarkably, the species-rich Atherton Upland (AU) did not rank highly in any of these analyses because this area has little unique diversity for species or phylogeographical units. However, AU was highly ranked for snails and insects, and was the highest ranked overall, and also is important to sustain viable populations of vertebrates because of its large area. In relation to sustaining ongoing speciation, it is important to extend the priority area to represent phylogeographical units (as 'species-in-waiting') and the suture zone (hybridization and reinforcement, e.g. Hoskin *et al.* 2005). Under these criteria, the Malbon-Thompson Range, southern BMC and Lamb Uplands are of particular significance (Moritz 2002; Moritz & McDonald 2005). Within each priority area, it is important to protect habitats representing continuous environmental gradients to maintain the diversity that derives from the interaction between selection and gene flow (e.g. Schneider *et al.* 1999), as well as to allow migratory response to climate change.

Returning to the potential effects of ongoing climate change, it is especially alarming that current models predict marked reduction of montane rainforest habitats in the Wet Tropics. These areas, having retained populations of low-dispersal rainforest specialists, including those of narrowly endemic species, through the climatic oscillations of the Quaternary now appear severely threatened by effects of global warming (Williams *et al.* 2003; Thomas *et al.* 2004; Williams *et al.* Chapter 22, this volume). In this context, it is vital to retain maximum habitat connectivity across elevational gradients and, for montane endemics, to improve our knowledge of current physiological limitations of and the potential for adjustment of these limits via plasticity or evolutionary change.

References

Armstrong, L. A., Krajewski, C. & Westerman, M. (1998). Phylogeny of the dasyurid marsupial genus Antechinus based on cytochrome-b, 12S-rRNA, and protamine-P1 genes. *Journal of Mammalogy* **79**: 1379–89.

Avise, J. C. & Walker, D. (1998). Pleistocene phylogeographic effects on avian populations and the speciation process. *Proceedings of the Royal Society of London Series B Biological Sciences* **265**: 457–63.

Bouchard, P., Brooks, D. R. & Yeates, D. (2005). Mosaic macroevolution in Australian Wet Tropics arthropods: community assemblage by taxon pulses. In *Tropical Rainforests: Past, Present and Future*, Moritz, C.,Bermingham, E. & Dick, C. (eds). University of Chicago Press, Chicago.

Bowyer, J. C., Newell, G. R., Metcalf, C. J. & Eldridge, M. B. D. (2003). Tree-kangaroos *Dendrolagus* in Australia: are *D. lumholtzi* and *D. bennettianus* sister taxa? *Australian Zoologist* **32**: 207–13.

Brown, J. H. (1996). The geographic range: size, shape, boundaries, and internal structure. *Annual Review of Ecology and Systematics* **27**: 597.

Burk, A., Westerman, M. & Springer, M. (1998). The phylogenetic position of the musky rat-kangaroo and the evolution of bipedal hopping in kangaroos. Macropodidae: Diprotodontia. *Systematic Biology* **47**: 457–74.

Couper, P. J., Schneider, C. J. Hoskin, C. J. & Covacevich, J. A. (2000). Australian leaf-tailed geckos: Phylogeny, a new genus, two new species and other new data. *Memoirs of the Queensland Museum.* **45**: 253–65.

Couper, P. J., Wilmer, J. W., Roberts, L., Amery, A. P. & Zug, G. R. (2005). Skinks currently assigned to *Carlia aerata* (Scincidae: Lygosominae) of north-eastern Queensland: a preliminary study of cryptic diversity and two new species. *Australian Journal of Zoology* **53**: 35–49.

Coyne, J. A. & Orr, H. A. A. (2004). Phylogenetic analyses of the leaf-tailed geckos, based on DNA. In *Speciation*, Sunderland, M. A. (ed.), Sinauer Assoc.

Currie, D. J. (1991). Energy and large-scale patterns of animal-species and plant-species richness. *American Naturalist* **137**: 27

Dolman, G. & Moritz, C. (2006). A multilocus perspective on refugial isolation and divergence in rainforest skinks, Carlia. *Evolution* **60**: 573–82.

Endler, J. A., Westcott, D. A., Madden, J. R. & Robin, T. (2005). Animal visual systems and the evolution of color patterns: sensory processing illuminates signal evolution. *Evolution* **59**: 1795–818.

Evans, K. L., Greenwood, J. J. D. & Gaston, K. J. (2005). Relative contribution of abundant and rare species to species–energy relationships. *Biology Letters* **1**: 87–90.

Ferrier, S. (2002). Mapping spatial pattern in biodiversity for regional conservation planning: Where to from here? *Systematic Biology* **51**: 331–63.

Firestone, K. B., Elphinstone, M. S., Sherwin, W. B. & Houlden, B. A. (1999). Phylogeographical population structure of tiger quolls *Dasyurus maculatus* Dasyuridae: Marsupialia, an endangered carnivorous marsupial. *Molecular Ecology* **8**: 1613–25.

Gaston, K. J. (1996). *Biodiversity: A Biology of Numbers and Difference*. Blackwell Scientific Publishing, Carlton.

Gaston, K. J. & Blackburn, T. M. (2000). *Pattern and Process in Macroecology*. Blackwell Science, Oxford.

Gavrilets, S. (2003). Perspective. Models of speciation: what have we learned in 40 years? *Evolution* **57**: 2197–215.

Gentry, A. H. (1992). Tropical forest biodiversity: distributional patterns and their conservation significance. *Oikos* **63**: 19–28.

Goosem, S. (2002). *Update of Original Wet Tropics Nomination Dossier*. Cairns, Wet Tropics Management Authority.

Goudberg, N. J. & Bonell, M. (1991). *Some perspectives on management issues. In Tropical Rainforest Research in Australia: Present Status and Future Directions for the Institute for Tropical Rainforest Studies*, Goudberg, N., Bonell, M. & Benzarken, D. (eds). ITRS, Townsville. 210 pp.

Graham, C., Moritz C. & Williams, S. E. (2006). Habitat history improves prediction of biodiversity in rainforest fauna. *Proceedings of the National Academy of Science* **102**: 632–6.

Haffer, J. (1969). Speciation in Amazonian forest birds. *Science* **165**: 131–7.

Harrison, D. A. & Congdon, B. C. (2002). *Vertebrate Pest Risk Assessment Report*. Wet Tropics Management Authority Consultant Report, Cairns.

Hopkins, M. S., Ash, J., Graham, A. W., Head, J. & Hewett, R. K. (1993). Charcoal evidence of the spatial extent of the eucalyptus woodland expansions and rainforest contractions in north Queensland during the late Pleistocene. *Journal of Biogeography* **20**: 357–72.

Hoskin, C. J. (2004). Australian microhylid frogs (*Cophixalus* and *Austrochaperina*): phylogeny, taxonomy, calls, distributions and breeding biology. *Australian Journal of Zoology* **52**: 237–69.

Hoskin, C. J. (in press). Description, biology and conservation of a new species of Australian tree frog (Amphibia: Anura: Hylidae: *Litoria*) and an assessment of the remaining populations of *Litoria genimaculata* Horst (1883): systematic and conservation implications of an unusual speciation event. *Biological Journal of the Linnean Society.*

Hoskin, C.J. & Couper, P.J. (2004). A new species of Glaphyromorphus (Reptilia: Scincidae) from Mt Elliot, north-eastern Queensland. *Australian Journal of Zoology* **52**: 183–90.

Hoskin, C. J., Couper, P. J. & Schneider, C. J. (2003). A new species of *Phyllurus* (Lacertilia: Gekkonidae) and a revised phylogeny and key for the Australian leaf-tailed geckos. *Australian Journal of Zoology* **2**: 153–64.

Hoskin, C. J., Higgie, M., McDonald, K. R. & Moritz, C. (2005). Reinforcement drives rapid allopatric speciation. *Nature* **437**: 1353–6.

Hubbell, S. P. (2001). *The Unified Theory of Biodiversity and Biogeography.* Princeton University Press, Princeton, NJ.

Hugall, A., Moritz, C., Moussalli, A. & Stanisic, J. (2002). Reconciling paleodistribution models and comparative phylogeography in the Wet Tropics rainforest land snail *Gnarosophia bellendenkerensis* (Brazier 1875). *Proceedings of the National Academy of Sciences of the United States of America* **99**: 6112–17.

Joseph, L. & Moritz, C. (1993). Phylogeny and historical aspects of the ecology of eastern Australian scrubwrens *Sericornis* spp. – evidence from mitochondrial-DNA. *Molecular Ecology* **2**: 161–70.

Joseph, L. & Moritz, C. (1994). Mitochondrial DNA phylogeography of birds in eastern Australian rainforests: first fragments. *Australian Journal of Zoology* **42**: 385–403.

Joseph, L., Moritz, C. & Hugall, A. (1995). Molecular support for vicariance as a source of diversity in rainforest. *Proceedings of the Royal Society of London Series B Biological Sciences* **260**: 177–82.

Kershaw, A. P., Moss, P. T. & Wild, R. (2005). Patterns and causes of vegetation change in the Australian Wet Tropics region over the last 10 million years. In *Tropical Rainforests: Past, Present and Future*, Moritz, C., Bermingham, E. & Dick, C. (eds). University of Chicago Press, Chicago.

Keto, A. & Scott, K. (1986). *Tropical Rainforests of North Queensland: Their Conservation Significance.* AGPS. Canberra.

Kikkawa, J., Webb, L. J., Dale, M. B., Monteith, G. B. Tracey, J. G. & Williams, W. T. (1981). Gradients and boundaries of monsoon forests in Australia. *Proceedings of the Ecological Society of Australia* **11**: 39–52.

Körner, C. (2002). Mountain biodiversity, its causes and function: an overview. In *Mountain Biodiversity: A Global Assessment*, Körner, C. & Spehn, E. M. (eds). Parthenon Publishing, New York.

Kreft, H., Sommer, J. H. & Barthlott, W. (2006). The significance of geographic range size for spatial diversity patterns in Neotropical palms. *Ecography* **29**: 21–30.

Lamoreux, J. F., Morrison, J. C. & Ricketts, T.H. *et al.* (2006). Global tests of biodiversity concordance and the importance of endemism. *Nature* **440**: 212–14.

Laurance, W. F. (1997). Responses of mammals to rainforest fragmentation in tropical Queensland: a review and synthesis. *Wildlife Research* **24**: 603–12.

Laurance, W. F. Garesche, J. & Payne, C. W. (1993). Avian nest predation in modified and natural habitats in tropical Queensland: an experimental study. *Wildlife Research* **20**: 711–23.

Lawton, J. H. (1993). Range, population abundance and conservation. *Trends in Ecology and Evolution* **8**: 409–13.

McDonald, K. R. (1992). *Distribution Patterns and Conservation Status of North Queensland Rainforest Frogs.* Queensland Department of Environment and Heritage, Conservation Technical Report, pp. 1–51.

Mahoney, M., Donnellan S. C., Richards S. J. & McDonald, K. R. (2006). Species boundaries of barred river frogs, *Mixophyes* (Anura: Myobatrachidae) in north-eastern Australia, with descriptions of two new species. *Zootaxa* **1228**: 35–60.

Margules, C. R. & Pressey, R. L. (2000). Systematic conservation planning. *Nature* **405**: 243–53.

Marsden, S. J., Whiffin, M., Galetti, M., *et al.* (2005). How well will Brazil's system of Atlantic forest reserves maintain viable bird populations? *Biodiversity and Conservation* **14**: 2835–53.

Mayr, E. & O'Hara, R. J. (1986). The biogeographical evidence supporting the Pleistocene forest refuge hypothesis. *Evolution* **40**: 55–67.

Moritz, C. (2002). Strategies to protect biological diversity and the evolutionary processes that sustain it. *Systematic Biology* **51**: 238–54.

Moritz, C., Hoskin, C., Graham, C. H., Hugall, A. & Moussalli, A. (2005) Historical biogeography, diversity and conservation of Australia's tropical rainforest herpetofauna. In *Phylogeny and Conservation*, Purvis, A.,

Gittleman, J. L. & Brooks, T. (eds). Conservation Biology Series 8, Cambridge University Press, Cambridge, 243–64.

Moritz, C., Joseph, L. & Adams, M. (1993). Cryptic diversity in an endemic rainforest skink (*Gnypetoscincus queenslandiae*). *Biodiversity and Conservation* 2: 412–25.

Moritz, C., Joseph, L., Cunningham, M. & Schneider, C. J. (1997). Molecular perspectives on historical fragmentation of Australian tropical and subtropical rainforest: implications for conservation. In *Tropical Rainforest Remnants: Ecology, Management and Conservation of Fragmented Communities*, Laurance W. F. & Bieregard R. O. (eds). University of Chicago Press, Chicago, 442–54.

Moritz, C. & McDonald, K. R. (2005). Evolutionary approaches to the conservation of tropical rainforest vertebrates. In *Tropical Rainforests: Past, Present and Future*, Moritz, C., Bermingham, E. & Dick, C. (eds). University of Chicago Press. Chicago.

Moritz, C., Richardson, K. S., Ferrier, S. *et al.* (2001). Biogeographical concordance and efficiency of taxon indicators for establishing conservation priority in a tropical rainforest biota. *Proceedings of the Royal Society of London Series B Biological Sciences* 268: 1875–81.

Moritz, C., Patton, J. L., Schneider, C. J. & Smith, T. B. (2000). Diversification of rainforest faunas: An integrated molecular approach. *Annual Review of Ecology and Systematics* 31: 533–63.

Morley, R. J. (2000). *Origin and Evolution of Tropical Rainforests*. Wiley & Sons, Chichester.

Moussalli, A., Hugall, A. F. & Moritz, C. (2005). A mitochondrial phylogeny of the rainforest skink genus *Saproscincus*, Wells and Wellington (1984). *Molecular Phylogenetics and Evolution* 34: 190–202.

Murrary, B. R. & Hose, G. C. (2005). Life-history and ecological correlates of decline and extinction in the endemic Australian frog fauna. *Austral Ecology* 30: 564–71.

Myers, N. (1988). Tropical forests and their species: going, going … ? In *Biodiversity*, Wilson, E. O. & Peter, F. M. (eds). National Academy Press, Washington, DC, pp. 28–35.

Nicholls, J. A. (2001). Molecular systematics of the thornbills, *Acanthiza*. *Emu* 101: 33–7.

Nicholls, J. A. &. Austin, J. J. (2005). Phylogeography of an east Australian wet-forest bird, the satin bowerbird (*Ptilonorhynchus violaceus*), derived from mtDNA, and its relationship to morphology. *Molecular Ecology* 145: 1485–96.

Nicholls, J. A., Austin, J. J., Moritz, C. & Goldizen, A.W. (2006). Genetic population structure and call variation in a passerine bird, the satin bowerbird, *Ptilonorhynchus violaceus*. *Evolution* 60: 1279–90.

Nix, H. A. & Switzer, M. (1991). Rainforest animals. In *Atlas of Vertebrates Endemic to Australia's Wet Tropics*. Australian Nature Conservation Agency, Canberra.

Norman, J. A., Christidis, L., Joseph, L., Slikas, B. & Alpers, D. (2002). Unravelling a biogeographical knot: origin of the leapfrog distribution pattern of Australo-Papuan sooty owls (Strigiformes) and logrunners (Passeriformes). *Proceedings of the Royal Society Biological Sciences Series B* 269: 2127–33.

O'Connor, D. & Moritz, C. (2003). A molecular phylogeny of the Australian skink genera *Eulamprus*, *Gnypetoscincus* and *Nangura*. *Australian Journal of Zoology* 51: 317–30.

Patterson, B. D. & Atmar, W. (1986). Nested subsets and the structure of insular mammalian faunas and archipelagos. *Biological Journal of the Linnean Society* 28: 65–82.

Phillips, B. L., Baird, S. J. E. & Moritz, C. (2004). When vicars meet: a narrow contact zone between morphologically cryptic phylogeographic lineages of the rainforest skink, *Carlia rubrigularis*. *Evolution* 58: 1536–48.

Phillips, S. J., Anderson, R. P. & Schapire, R. E. (2006). Maximum entropy modeling of species geographic distributions. *Ecological Modelling* 190: 231–59.

Pope, L. C., Estoup, A. & Moritz, C. (2000). Phylogeography and population structure of an ecotonal marsupial, *Bettongia tropica*, determined using mtDNA and microsatellites. *Molecular Ecology* 9: 2041–53.

Pounds, J. A., Bustamante, M. R., Coloma, L. A. *et al.* (2006). Widespread amphibian extinctions from epidemic disease driven by global warming. *Nature* 439: 161–7.

Prendergast, J. R., Lawton, R. M., Quinn, J. H., Eversham, B. C. & Gibbons, D. W. (1993). Rare species, the coincidence of diversity hotspots and conservation strategies. *Nature* 365: 335–7.

Read, K., Keogh, J. S., Roberts, I. A. W. & Doughty, P. (2001). Molecular phylogeny of the Australian frog genera *Crinia*, *Geocrinia*, and allied taxa (Anura: Myobatrachidae). *Molecular Phylogenetics and Evolution* 21: 294–308.

Ricklefs, R. E. (1987). Community diversity: relative roles of local and regional processes, *Science* 235: 167–71.

Ricklefs, R. E. & Schluter, D. (1993). *Species Diversity in Ecological Communities: Historical and Geographical Perspectives*. University of Chicago Press, Chicago.

Rull, V. (2006). Quaternary speciation in the Neotropics. *Molecular Ecology* 15: 4257–9.

Schluter, D., & Ricklefs, R. E. (1993). Species diversity: an introduction to the problem. In *Species Diversity in Ecological Communities: Historical and Geographical Perspectives*, Schluter, D. & Ricklefs, R. (eds). University of Chicago Press, Chicago, pp. 1–10.

Schneider, C. J., Cunningham, M. & Moritz, C. (1998). Comparative phylogeography and the history of endemic vertebrates in the Wet Tropics rainforests of Australia. *Molecular Ecology* 7: 487–98.

Schneider, C. J., Smith, T. B., Larison, B. & Moritz, C. (1999). A test of alternative models of diversification in tropical rainforests: ecological gradients versus rainforest refugia. *Proceedings of the National Acadamy of Science* 96: 13869–73.

Schneider, C. J. & Moritz, C. (1999). Refugial isolation and evolution in the Wet Tropics rainforests of Australia. *Proceedings of the Royal Society of London B*. 266: 191–6.

Schneider, C. J & Williams, S. E. (2005). Quaternary climate change and rainforest diversity: insights from spatial analyses of species and genes in Australia's Wet Tropics. In *Tropical Rainforests: Past, Present and Future*, Moritz, C., Bermingham, E. & Dick, C. (eds). Chicago University Press, Chicago.

Schulte, J. A., Melville, J. & Larson, A. (2003). Molecular phylogenetic evidence for ancient divergence of lizard taxa on either side of Wallace's Line. *Proceedings of the Royal Society of London Series B – Biological Sciences* 270: 597–603.

Servedio, M. R. & Noor, M. A. F. (2003). The role of reinforcement in speciation: theory and data. *Annual Review of Ecology Evolution and Systematics* 34: 339–64.

Sites, J. W. Jr. & Marshall, J. C. (2003). Delimiting species: A Renaissance issue in systematic biology. *Trends in Ecology and Evolution* 18: 462–70.

Springer, M., McKay, G. Aplin, K. & Kirsch, J. A. W. (1992). Relations among ringtail possums (Marsupialia, Pseudocheiridae) Based on DNA–DNA Hybridization. *Australian Journal of Zoology* 40: 423–35.

Stuart-Fox, D. M., Hugall, A. F. & Moritz, C. (2002). A molecular phylogeny of rainbow skinks (Scincidae: *Carlia*): taxonomic and biogeographic implications. *Australian Journal of Zoology* 50: 39–51.

Sumner, J., Moritz, C. & Shine, R. (1999). Shrinking forest shrinks skink: morphological change in response to rainforest fragmentation in the prickly forest skink (*Gnypetoscincus queenslandiae*). *Biological Conservation* 91: 159–67.

Sumner, J. (2005). Decreased relatedness between male prickly forest skinks (*Gnypetoscincus queenslandiae*) in habitat fragments. *Conservation Genetics* 6: 333–40.

Symonds, M. R. E. & Johnson, C. N. (2006). Range size–abundance relationships in Australian passerines. *Global Ecology and Biogeography* 15: 143–52.

Thomas, C .D., Cameron, A., Green, R. E. *et al.* (2004). Extinction risk from climate change. *Nature* 427: 145–8.

Vanzolini, P. E. & Williams, E. E. (1981). The vanishing refuge: a mechanism for ecogeographic speciation. *Papeis Avulsos de Zoologia* 34: 251–5.

Webb, C. O., Ackerly, D. D., McPeek, M. A. & Donoghue, M. J. (2002). Phylogenies and community ecology. *Annual Review of Ecology and Systematics* 33, 475–505.

Webb, L. J. & Tracey, J. G. (1981). Australian rainforests: patterns and change. In *Ecological Biogeography of Australia*, Keast, A. (ed.). Dr W. Junk, The Hague, pp. 605–94.

Wiens, J. J. & Graham, C. H. (2005). Niche conservatism: integrating evolution, ecology, and conservation biology. *Annual Review of Ecology Evolution and Systematics* 36: 519–39.

Wiens, J. J. & Penkrot, T. A. (2002). Delimiting species using DNA and morphological variation and discordant species limits in spiny lizards (*Sceloporus*). *Systematic Biology* 51: 69–91.

Wilson, E. O. (1988). The current state of biological diversity In *Biodiversity*, Wilson, E. O. & Peter, F. M. (eds). National Academy Press, Washington, DC, pp. 3–20.

Williams, P. H., Gaston, K. J. & Humphries, C. J. (1994). Do conservationists and molecular biologists value differences between organisms in the same way? *Biodiversity Letters* 2: 67–78.

Williams, S. E. (1997). Patterns of mammalian species richness in the Australian tropical rainforests: are extinctions during historical contractions of the rainforest the primary determinant of current patterns in biodiversity? *Wildlife Research* 24: 513–30.

Williams, S. E. (2006). *Vertebrates of the Wet Tropics Rainforests of Australia: Species Distributions and Biodiversity*. Cooperative Research Centre for Tropical Rainforest Ecology and Management, Rainforest CRC, Cairns.

Williams, S. E., Bolitho, E. E. & Fox, S. (2003). Climate change in Australian tropical rainforests: an impending environmental catastrophe. *Proceedings of the Royal Society of London Series B – Biological Sciences* 270: 1887–92.

Williams, S. E. & Hero, J.-M. (2001). Multiple determinants of Australian tropical frog biodiversity. *Biological Conservation* 98: 1–10.

Williams, S. E., Marsh, H. & Winter, J. W. (2002). Spatial scale, species diversity, and habitat structure: small mammals in Australian tropical rain forest. *Ecology* 83: 1317–29.

Williams, S. E. & Pearson, R. G. (1997). Rainforest shape and endemism in Australia's Wet Tropics. *Proceedings of the Royal Society of London B* 264: 709–16.

Williams, S. E., Pearson, R. G. & Walsh, P. J. (1996). Distributions and biodiversity of the terrestrial vertebrates of Australia's Wet Tropics: a review of current knowledge. *Pacific Conservation Biology* 2: 327–62.

Williams, Y., Williams, S. E., Waycott, M., Alford, R. A. & Johnson, C. N. (2006). Niche breadth and geographical range: ecological compensation for geographical rarity in rainforest frogs. *Biology Letters* **2**: 532–5.

Winter, J. W. (1997). Responses of non-volant mammals to late Quaternary climatic changes in the Wet Tropics region of north-eastern Australia. *Wildlife Research* **24**: 493–511.

Winter, J. W., Bell, F. C., Pahl, L. I. & Atherton, R. G. (1984). *The Specific Habitats of Selected Northeastern Australian Rainforest Mammals*. Report to World Wildlife Fund, Sydney (unpublished).

Winter, J. W., Bell, F. C., Pahl, L. I. & Atherton, R. G. (1987). The distribution of rainforest in north-eastern Queensland. In *The Rainforest Legacy*, Kershaw, P. & Werren, G. (eds). AGPS, Canberra, pp. 223–6.

Woodhams, D. C. & Alford, R. A. (2005). Ecology of chytridiomycosis in rainforest stream frog assemblages of tropical Queensland. *Conservation Biology* **19**: 1449–59.

Wright, D. H. (1983). Species–energy theory, an extension of species–area theory. *Oikos* **41**: 496–506.

11 ORIGINS AND MAINTENANCE OF FRESHWATER FISH BIODIVERSITY IN THE WET TROPICS REGION

Brad Pusey[1]*, Mark Kennard[1]* and Angela Arthington[2]*

[1]Australian Rivers Institute, Griffith University, Nathan, Queensland, Australia
[2]Griffith University, Nathan, Queensland, Australia
*The authors were participants of Cooperative Research Centre for Tropical Ecology and Management.

Introduction

The streams and rivers of the Australian Wet Tropics bioregion are a prominent feature and formative agent of the landscape. Typically their flows are highly seasonal, reflecting higher rainfall in the summer wet season and lower rainfall in the drier season. Discharge patterns in rivers of the Wet Tropics region are also highly predictable, with much of the predictability being derived from the high constancy of flow from month to month. Coupled with distinctive landscape features such as waterfalls and other barriers to dispersal, as well as historical factors, these predictable, seasonal flow patterns have major implications for the evolution and maintenance of aquatic biodiversity, especially the richness and composition of the fish fauna and the structure of fish assemblages. This chapter describes the extraordinary diversity of freshwater fishes, their distribution patterns and adaptations to climate, discharge and landform, ending with a brief commentary on threats to this unique fauna.

Freshwater fish biodiversity

The freshwater fish fauna of the Wet Tropics region of north-eastern Australia is highly diverse and distinctive (Unmack 2001; Pusey & Kennard 1996; Pusey et al. 2004b). To date, a total of 104 native and six alien species has been recorded from freshwaters (<2000 µS cm⁻¹) of the region (Figure 11.1). These 110 species are contained within 37 families with almost half (52 of 110) within six families: Eleotridae (gudgeons, 15 species (spp.)), Gobiidae (gobies, 13 spp.), Chandidae (glassfishes, 9 spp.), Mugilidae (mullets, 6 spp.), Terapontidae (grunters, 5 spp.) and Plotosidae (eel-tailed catfish, 4 spp). These families typically contribute the majority of freshwater fish biodiversity in northern Australian rivers (Bishop & Forbes 1991; Allen et al. 2002). The fork tailed catfishes (Ariidae) are an important family in rivers to the north and south of the Wet Tropics region but are notably absent from freshwaters of the Wet Tropics region (Pusey et al. 2004b). Of the 110 species, 49% are entirely restricted to freshwater or have a limited estuarine phase in the life history, 34% have an obligate estuarine or marine life history interval, with the remainder being composed of species within marine families that occasionally may be found in freshwater rivers (e.g. Sillaginidae, Leiognathidae, Platycephalidae and Mugilidae).

As many as 36 alien and native species have been introduced to streams, farm dams and impoundments of the region over the past 100 years (Burrows 2004; Pusey et al. 2004b). Self-sustaining populations of

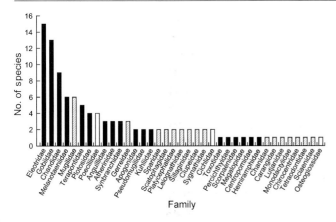

Figure 11.1 Species diversity within families recorded from natural aquatic habitats of the Wet Tropics region. Species present only in farm dams and impoundments are not included. Solid bars are predominantly freshwater families or families in which Wet Tropics species are predominantly freshwater inhabitants as adults. Hatched bars denote families composed primarily of either marine or estuarine species occasionally found in freshwaters of the Wet Tropics region. Open bars represent alien or translocated families known to have become established in freshwaters of the region. For species within each family and sources of information see Pusey *et al.* (2004b) and references therein.

at least six alien species – *Oreochromis mossambicus* (Cichlidae), *Tilapia mariae* (Cichlidae), *Gambusia holbrooki* (Poeciliidae), *Poecilia reticulata* (Poeciliidae), *Xiphophorus helleri* (Poeciliidae) and *Xiphophorus maculatus* (Poeciliidae) – have been recorded from rivers of the region (Figure 11.1). In addition, the native osteoglossid *Scleropages jardinii* has been recorded from Liverpool Creek; its presence due to translocation (Russell & Hales 1997). The threat to native fishes posed by introduced species and other human activities is discussed in Box 11.1.

Freshwater fish biogeography

The following discussion focuses on the native freshwater fish of the Wet Tropics region. We define freshwater fish in a relatively broad sense to include those fish species that can reproduce in freshwater and diadromous species that spend the majority of their lives in freshwater (following Allen *et al.* 2002; Pusey *et al.* 2004b). We excluded numerous species with strong marine or estuarine affinities that may enter freshwater for short periods.

Coastal drainages of north-eastern Australia can be divided into five regions based on landform, climate, the distribution of aquatic habitat types and aquatic biota (Australian Water Resources Commission 1976; Unmack 2001, Allen *et al.* 2002, Pusey *et al.* 2004b). The five regions are:

- Gulf of Carpentaria and Western Cape York Peninsula (GULF);
- eastern Cape York Peninsula (ECYP);
- the Wet Tropics region (WT);
- central Queensland (CQ);
- south-eastern Queensland (SEQ).

The biogeography of Australia's freshwater fishes tends to be highly congruent with these regions (Unmack 2001; Allen *et al.* 2002; Pusey *et al.* 2004b). Multivariate ordination of the freshwater fish species composition of Queensland's coastal river basins clearly depicts the separation of the five regions (Figure 11.2). Widespread species are common (29 species from 18 families were recorded in four or five regions), yet each region is distinctive.

Two major gradients are identifiable in the pattern depicted in Figure 11.2. Rivers within the GULF region (i.e. those located west of the Great Dividing Range (GDR)) are located to the left on axis 1, whereas all remaining drainages are located to the right (i.e. a gradient of longitude). Several species commonly recorded in ECYP rivers (e.g. *Anguilla reinhardtii, A. obscura, Kuhlia* spp., *Giurus margaritacea* and *Oxyeleotris aruensis*) were recorded only from the Jardine, Ducie and Wenlock river basins of the northernmost GULF region. These species either have an estuarine larval phase or are estuarine or marine spawners and, in the case of the eels, there is obviously some transmission of larvae into these two rivers despite the bulk of leptocephali (larvae) being transported down the east coast in the East Australia Current (Pusey *et al.* 2004b). Species characteristic of western rivers may occur in eastern rivers too. Unmack (2001) also noted substantial similarity between the most northern drainages of both ECYP and GULF regions. The GDR is of low relief <500 m above sea level (m a.s.l.) in this part of Cape York Peninsula, as opposed to further south, and may not therefore be an effective barrier to the transfer of species between regions. In addition, movement between basins at times of high discharge from rivers of the region and from the Fly River of southern Papua New Guinea may also facilitate interbasin transfer.

Box 11.1 Threats to the freshwater fishes of the Wet Tropics region

Changes in riparian and aquatic habitat, due largely to agricultural land-use and hydrologic alteration, are an important threat to the fishes of the Wet Tropics region. Foremost among these changes is the loss and degradation of riparian forests and loss of wetland/floodplain habitats (Russell & Hales 1993; Russell et al. 1996a, b; Pusey & Arthington 2003). Several fish species of high conservation significance such as the Cairns rainbowfish (*Cairnsichthys rhombosomoides*) and jungle perch (*Kuhlia rupestris*) are highly reliant on terrestrial insects derived from the riparian zone. Others such as the khaki grunter (*Hephaestus tulliensis*), an endemic species, feed extensively on riparian fruits. In the Wet Tropics region, loss of the riparian zone is nearly always accompanied by the invasion and proliferation of the introduced ponded pasture grass *Urochloa mutica* (para grass), a significant threat to the maintenance of biodiversity in northern Australia (Arthington et al. 1997). It has been estimated (Russell & Hales 1993; Russell et al. 1996a, b) that between one-half and two-thirds of all wetlands in the Johnstone, Moresby and Russell/Mulgrave River systems had been destroyed in the period 1951–92. The extent of damage is probably now much greater given that there has been little abatement in the rate of reclamation in the past decade. At present, there are no major dams or weirs present on the lowland main stem of rivers in the Wet Tropics region and thus migratory species such barramundi, jungle perch and eels can make use of all available riverine habitats necessary to complete their life-cycle. However, many smaller barriers to movement, such as road crossings and culverts, have been constructed in smaller streams of the region and especially limit movement between the main river channel and wetland/floodplain habitats. In the Wet Tropics region, major water infrastructure exists on Crystal Creek and the Barron (Tinaroo Falls Dam, Kuranda Weir and Copperlode Dam) and Tully (Koombaloomba Dam) rivers. Major flow regime changes and consequent ecological impacts have been reported with all such storages. Numerous small dams and weirs and points of abstraction also exist throughout Wet Tropics region to service irrigation and urban needs.

Introduced or alien species such as the tilapias *Oreochromis mossambicus* and *Tilapia mariae* and the topminnows *Poecilia reticulata* and *Xiphophorus maculatus* may threaten about one-third of the fishes of the region (Burrows 2004).

The tilapia species are declared 'noxious' species in Queensland and in the Wet Tropics region are most abundant and widely distributed in the lowlands of rivers. Alien species may impact on native fishes in a variety of ways, including predation on native species, direct competition for food and spawning areas via aggressive interactions, which reduce individual fitness, and the transmission of novel parasites. The translocation of native species outside of their natural range is also a problem in the Wet Tropics region (Burrows 2004). The most well-known example is the introduction of mouth almighty (*Glossamia aprion*) into Lake Eacham and the subsequent local extinction of the Lake Eacham rainbowfish (*Melanotaenia eachamensis*). This species had the dubious distinction of being the first Australian freshwater fish to be declared extinct in the wild, although it is now known to occur in a small number of surrounding rainforest streams. Translocated native species may have the same impacts on resident native species as alien species, especially when the receiving streams contain communities that have evolved in the absence of predators (Pusey et al. 2006). Translocation of native species has an additional impact, in that hybridization between native and introduced stocks may reduce the genetic distinctiveness of native stocks. Two species widely translocated throughout the Wet Tropics region are the sooty grunter and the eel-tailed catfish (*Tandanus* spp.) (Burrows 2004). Both species are known to consist of a complex of genetically distinctive and geographically isolated forms; the extent of genetic dissimilarity among populations is equivalent to that observed between species (see Pusey et al. 2004b), yet stocks from as far afield as the Murray-Darling River in the case of the catfish, and the Burdekin and Walsh Rivers in the case of the grunter, have been stocked in streams of the Wet Tropics region.

Additional threats are posed by global climate changes but are difficult to quantify. Rising sea levels are likely to reduce greatly the extent of lowland freshwater fish habitat and, although the endemic element of the fish fauna is typical of upland sections, it is the downstream lowland reaches that hold the greatest diversity of fishes. Changes in rainfall and cloud capture are likely to lead to decreased constancy of discharge, particularly in upland streams, and changes in habitat may impact on many endemic species and, particularly, riffle specialists.

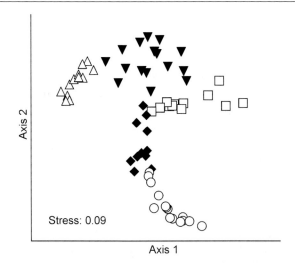

Figure 11.2 Spatial variation in freshwater fish community composition in rivers of north-eastern Australia. Multidimensional scaling ordination plot of individual north-eastern rivers based on Bray-Curtis dissimilarity measure using species presence/absence data. △, Gulf of Carpentaria; ▼, eastern Cape York Peninsula; □, Wet Tropics; ◆, central Queensland; ○ south-eastern Queensland. Species lists are derived from Pusey *et al.* (2004b) and do not contain estuarine vagrants.

Many species found in these rivers also occur in the Fly River (Allen 1991; Allen *et al.* 2002; Pusey *et al.* 2004b). Clearly, some marine dispersing larvae (i.e. in Anguillidae, Gobiidae and Eleotridae) are transported through the Torres Strait.

The second major gradient in Figure 11.2 is one of latitude. Northern regions (GULF, ECYP and WT) are located high on axis 2, whereas more southern regions are located low on this axis. Similar distinctions between the regions are evident when ordinations are based on the presence or absence of either genera or families. Multivariate statistical comparisons (using Mantel's tests) reveal that the association matrices based on river-by-genera or river-by-family matrices are significantly correlated with that for river-by-species, indicating that the underlying spatial pattern is consistent across higher taxonomic levels (unpublished data). When comparisons are undertaken at the generic or family level, but each is weighted by the number of species within each taxon, the associations are stronger than just presence or absence alone. This effect greatly reflects the regional differences in the diversity of species within the Gobiidae, Eleotridae, Terapontidae,

Chandidae, Melanotaeniidae and Plotosidae. For example, northern rivers (GULF, ECYP and WT) contain many species within the goby genus *Glossogobius* (seven spp.) and the gudgeon genus *Oxyeleotris* (five spp.), whereas both are absent in SEQ and represented by a single species in each genus in CQ. The WT region is particularly rich in both Gobiidae and Eleotridae but the diversity of gobies declines dramatically further to the south (Pusey *et al.* 2004b). Notably, the diversity of gudgeons does not decline with increasing latitude, partly due to the diversity of *Hypseleotris* in SEQ rivers but also due to the appearance of species within *Philypnodon* (two spp.) and *Gobiomorphus* (two spp.). The presence of *Retropinna semoni, Myxus petardi* and *Neoceratodus forsteri* within other less speciose families further distinguishes SEQ from the remaining regions.

This latitudinal gradient evident for community composition is also evident in comparisons of regional species richness. Regional richness is similarly high in the GULF, ECYP and WT regions (65, 59 and 59 species, respectively), intermediate in CQ (52 species) and low in SEQ (44 species) (Figure 11.3e), echoing a general global trend for freshwater biodiversity to be greater at low latitudes and presumably reflecting latitudinal differences in temperature, rainfall and hence productivity (Oberdorff *et al.* 1995). In addition to these gradients, other climate-related differences occur between regions. Mean annual runoff is greater in the WT region than elsewhere (Figure 11.3a) and markedly less variable (Figure 11.3b). In northern Queensland, most rainfall is derived from cyclones and low-pressure systems associated with the monsoonal trough. This leads to very well-defined wet and dry seasons in drainages of ECYP and GULF, with over 90% of rainfall and discharge occurring in the wet season from December to May. Rivers of the WT region are also strongly influenced by cyclones and the monsoonal trough, although significant groundwater inputs and the pronounced orographic effect of the GDR on moisture-laden south-easterly winds result in high dry season inputs such that up to an average of 25% of the mean annual flow occurs during the dry season. The southern regions of CQ and SEQ are less frequently and more unpredictably influenced by cyclones. Very little dry season rain occurs in the CQ region and as a consequence dry season flows contribute less than 10% of the mean annual flow.

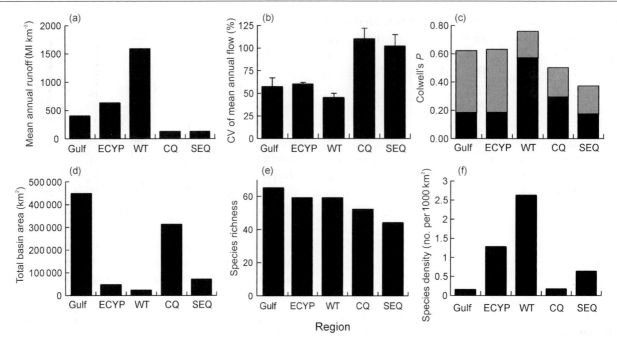

Figure 11.3 Regional variation in: (a) mean annual runoff (mean annual flow/catchment area) (±1 SE); (b) mean coefficient of variation of mean annual flow ([SD/mean] × 100) (±1 SE); (c) predictability of monthly flows; (d) region size (sum of basin catchment areas in km²); (e) species pool (total number of species across all rivers within each region); and (f) species density (regional pool/total catchment area). Biodiversity estimates based on data in Pusey *et al.* (2004b) and catchment and discharge information derived from the Queensland Department of Natural Resources, Mines and Energy. Predictability analysis based on Colwell's (1974) analysis of monthly flows over 20 years (Kennard & Pusey, unpublished data). The solid component of the bar refers to the contribution to predictability from constancy of monthly flows (C) and the shaded component refers to the contribution of seasonality (M) to predictability.

However, the low predictability of rainfall in this region frequently results in the absence of a pronounced wet season, and consequently interannual flow variability is high. The SEQ region may also experience infrequent and relatively unpredictable intrusion of temperate winter rainfall events originating from the south, which can lead to early dry season flooding (i.e. in June or July). Combined with unpredictable wet season rainfall, this leads to substantial year to year variation in mean annual flow in this region. These differences in the incidence and regularity of rainfall lead to spatial variation in discharge predictability and the contribution of seasonality to predictability (Figure 11.3c). Riverine discharge in GULF and ECYP rivers is moderately predictable, with most of the predictability being due to the high seasonal signal. Rivers of the Wet Tropics region are highly predictable, with much of the predictability being derived from the high constancy of flow from month to month. Less predictable discharge regimes are

present in CQ and SEQ (Figure 11.3c). For an expanded discussion of spatial variation in discharge in northeastern Australia refer to Pusey *et al.* (2004b). It is worth noting that these data pertain to large lowland river systems and may not apply to small basins or upstream tributaries.

These factors combine to create a diversity of riverine habitats across north-eastern Australia. Rivers of the Wet Tropics region tend to be short and very steep, often with significant discontinuities in profile in their middle sections, culminating in relatively short, low gradient, lowland reaches. High gradient tributaries typically discharge into the main channel within less than 10 km from the river mouth. Riffle and rapid habitats are abundant, even in the main channel of Wet Tropics rivers, and by virtue of the perennial flow regime, these habitats are available to fishes yearround. Elsewhere in north-eastern Australia, rivers tend to be of much lower gradient, contain few extensive riffle or rapid habitats, have ephemeral tributary habitats and have dry season habitats that tend to be limited to long deep reaches with little flow (Pusey *et al.* 2004b). Differences in region size are profound, with the Wet Tropics being less than one-tenth of the area of the GULF and CQ regions, for example (Figure 11.3d). When species richness is scaled by region size (i.e. species density), it is apparent that this small region (WT) contains comparatively many more species than do other regions of north-eastern Australia (Figure 11.3f).

Pusey *et al.* (2004a) presented an analysis of regional variation in biodiversity in which interbasin and interregional comparisons of biodiversity were scaled for variation in drainage catchment area. They noted a significant relationship between species richness (*S*) and catchment area (*CA*) (log $S = 1.096 + 0.999CA$; $r^2 = 0.297$, $p < 0.001$) but also noted that species richness was comparatively more strongly related to mean annual flow (MAF) (log $S = 0.197 + 0.203MAF$; $r^2 = 0.537$, $p < 0.001$). However, the WT and GULF regions consistently demonstrated greater diversity than predicted by these relationships (i.e. significantly greater mean residual scores: $F_{4,32} = 7.280$, $p < 0.001$ and $F_{4,32} = 4.612$, $p < 0.001$, respectively). A multiple regression model incorporating mean annual flow and a measure of flow perenniality (the proportional contribution to mean annual flow by the six driest months) accounted for over 62% of the variation in species richness ($F_{reg} = 21.702$, $p < 0.001$) with no significant regional variation in residual scores. WT rivers, although relatively small in size, have comparatively high discharges delivered more evenly throughout the year than do other rivers of north-eastern Australia and consequently may be able to support more species.

The WT region also differs from other north-eastern Australian regions with respect to the relative prevalence of the species within the region (Figure 11.4). The climatically variable regions of CQ and SEQ contain proportionally more widespread species and fewer restricted species than the northern, climatically predictable GULF, ECYP or WT regions (Figure 11.4). The WT region is distinctive, as it contains proportionally fewer species that are either very widespread or highly restricted, and more species that occur in between 25% and 75% of rivers in the region. The 29 species present in four or five of the regions discussed here are from a variety of families (18 in total), of varying body size (50–600 mm size range), and although 12 species have a marine or estuarine life history phase that may

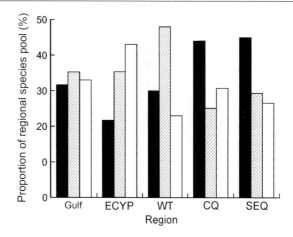

Figure 11.4 The proportional contribution of widespread (closed bars, present in ≥75% of rivers within the region), moderately widespread (hatched bars, present in 26–75% of rivers within the region) and restricted species (open bars, present in ≤ 25% of rivers within the region) to the regional species pool. Data on species distributions sourced from Pusey *et al.* (2004b).

partially account for their wide distributions, the remainder are exclusively freshwater in habitat use. With the exception of *Mogurnda adspersa*, all of the widespread species in north-eastern Australia are habitat generalists distributed widely across a range of habitats including lowland sections of the riverine landscape. Their presence in lowland rivers enables them to persist over periods of both seasonal and prolonged aseasonal low flow that may result in the desiccation or severe reduction in extent of upstream and tributary habitats.

Spatial patterns within the Wet Tropics region

Despite the presence of a large number of moderately widespread species within the WT region, spatial patterns in distribution within it are also evident. In contrast to the general pattern of decreasing species richness with increasing latitude, species richness in individual basins of the Wet Tropics increases until a maximum of about 50 species per basin is reached in the Mulgrave/Russell and Johnstone River basins. Richness then declines again further south. This pattern is probably due to a similar trend in rainfall and discharge. The GDR reaches its north-eastern

Australian maximum in the headwaters of these two river systems (>1750 m asl) and orographic forcing of rainfall from the interception of moisture laden south-easterly winds and from cloud capture contribute greatly to dry season flows (approx 25% of discharge occurs in the dry season). Despite their relatively small catchment areas (<3000 km^2), the Mulgrave/Russell and Johnstone River basins contain more species than any other north-eastern Australian river with the exception of the Jardine River.

Pusey and Kennard (1996) suggested that the high diversity recorded in these rivers, and of the region as a whole, is related to the reliability of discharge on a year-to-year basis, which may result in low rates of extinction over contemporary time scales. In addition, low rates of extinction may have been a more persistent feature of this region. The WT region contains more endemic species than any other region of north-eastern Australia (Unmack 2001) and many of these endemic species are restricted to single river basins. For example, *Guyu wujalwujalensis* is restricted to the upper Bloomfield River and its phylogenetic distinctiveness suggests it has persisted in this river for many millions of years (Pusey & Kennard 2001). Although complex changes in drainage alignment and landscape evolution have created the circumstances necessary for vicariant speciation and complex phylogeographical patterning (Hurwood & Hughes 1998, 2001; McGlashan & Hughes 2000, 2001; McGuigan *et al.* 2000), it is difficult to see how the outcomes of these processes persist in the landscape in the absence of low rates of extinction over both historical and prehistoric time scales. Not surprisingly, high rates of endemism and persistence of primitive taxa in the region are evident for many plant and animal groups, suggesting that although dramatic shifts in climate have occurred, especially during the Pleistocene, such changes have been insufficient to result in widespread and total extinction events. Modelling of climate changes and its impact on rainforests during the Pleistocene suggests that, in addition to rainforest contraction to the highest mountain tops, gallery rainforest remained intact in the river valleys of the Mulgrave and Johnstone River basins (D. Hilbert, pers. comm.). River flows and aquatic riverine habitats were clearly maintained during the driest phases of the Pleistocene. Further support for this hypothesis can be found from assessing

Table 11.1 Habitat use by the endemic freshwater fishes of the Wet Tropics region. Information on habitat use

Species	Altitudinal limits (m a.s.l.)	Gradient (%)	Preferred mesohabitat	Mesohabitat function
Melanotaenia eachamensis	400–790	1.1	Riffles, runs	Feeding, breeding
Melanotaenia utcheensis	55–80	11.9	Riffles, runs	Feeding, breeding
Craterocephalus spp.	>720	n.a.	Riffles, runs	Feeding, breeding
Cairnsichthys rhombosomoides	5–100	1.0	Cascades, riffles	Feeding, breeding
Guyu wujalwujalensis	>200	n.a.	Riffles, runs, pools	n.a.
Schismatogobius spp.	<50	n.a.	Riffles	Feeding, breeding
Glossogobius spp. 4	20–70	0.7	Rapids, riffles, runs	Feeding, breeding
Stiphodon alleni	<20	n.a.	Riffles	n.a.
Stiphodon spp.	<20	n.a.	Rapids, riffles	n.a.
Tandanus spp.	10–722	0.8	Riffles, runs	Feeding esp. juveniles
Hephaestus tulliensis	10–60	0.8	Riffle, runs, pools	Breeding, feeding as juvenile

n.a., information is currently not available.

Source: Allen *et al.* (2002), Pusey *et al.* (2004a), Theusen (pers. comm.).

the habitat requirements of the endemic species (Table 11.1), all of which are reliant to a great extent on shallow, fast-flowing habitats such as riffles. Riffle habitats are the most vulnerable to changes from reduced flow of all habitats within the riverine landscape. In addition, many of the endemic species occur in small streams at high altitude and again, it is likely that such systems would be vulnerable to changes in discharge. While it could be argued that prior increases in sea level (McGlashan & Hughes 2000, 2001) are likely to have affected only those species at low elevation and, as a consequence, the only endemic species that might persist during such phases are those occurring at elevations greater than marine incursions, the fact remains that the endemic species are predominantly riffle dwellers and that this mesohabitat type is vulnerable to temporal changes in discharge over a variety of temporal scales.

Recent and contemporary environmental drivers

Pusey and Kennard (1996) suggested that the predictable and relatively constant discharge in rivers of the Wet Tropics region may have contributed to increased

diversity within individual rivers by fostering greater habitat specialization and increased species packing at the mesohabitat scale. Discharge in rivers of the Wet Tropics region is much less variable than in south-eastern Queensland rivers (coefficients of variation of annual flow <50% and >100%, respectively), yet comparison of species richness at the mesohabitat scale (i.e. individual riffles, runs or pools) in rivers of these two regions suggests little difference (Figure 11.5a). However, although there is little difference across all

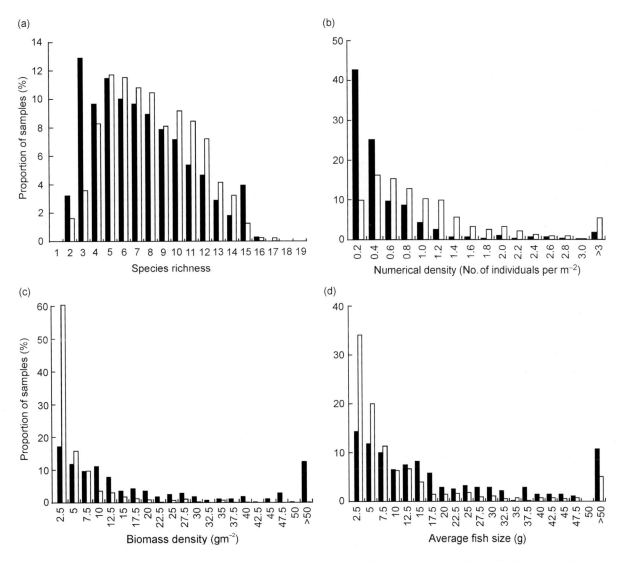

Figure 11.5 Comparison of emergent properties of freshwater fish assemblages in streams of the Wet Tropics (solid bars) and south-eastern Queensland (open bars). Data based on multipass electrofishing and supplementary seine netting in 279 and 555 individual mesohabitat units in the Wet Tropics and south-eastern Queensland, respectively, over the period 1994–7. Refer to Pusey *et al.* (2004b) for details.

mesohabitat types, riffle and rapid habitats of the Wet Tropics region tend to be slightly more speciose (by about two species) than similar habitats in rivers of SEQ (Pusey *et al.*, unpublished data). Riffle habitats in SEQ rivers vary greatly in depth and velocity through time owing to highly unpredictable variation in discharge and, as a consequence, there may be less opportunity for fishes of this region to specialize in this habitat type. There are also clear differences in other structural characteristics of the stream fish assemblages of these two regions that can be ascribed to differences in stream flow variability. For example, fish density is frequently higher in SEQ streams (Figure 11.5b) as a result of habitat contraction during the dry season and concentration of fish in reduced habitat space. Habitat structure is less variable in WT streams because of the perennial nature of stream flow. In contrast, biomass densities are much higher in WT streams and average fish size is greater also (Figure 11.5c, d). Greater longevity and body size may be expected in perennial streams, whereas the most appropriate life history strategy in variable systems may be one of small size and early maturity in order to take advantage of conditions appropriate for breeding when they become available. Finally, although greater diversity at the mesohabitat or local scale is not apparent, Pusey *et al.* (2000) demonstrated that greater beta diversity at the macrohabitat scale occurred in the WT region compared to SEQ. Attributes such as high vagility and the ability to move between habitats as they change in quality, which may confer benefit in variable environments, are likely to obscure any deeper impression of macrohabitat selectivity or spatial organization.

Very clear spatial organization of freshwater fish assemblages is evident in rivers of the WT region (Pusey *et al.* 1995; Pusey & Kennard 1996). Landscape-scale features are important in determining the distribution and abundance of fishes in WT rivers and streams. Waterfalls are a common feature of the WT landscape and, depending on their size, gradient and location in the river network, tend to depress upstream diversity by limiting upstream colonization by species with an estuarine phase in their life history (Pusey & Kennard 1996). Such barriers to movement are important in maintaining biodiversity through the isolation of endemic species and distinctive phylogenetic lineages (Hurwood & Hughes 2001; McGuigan 2001; Pusey & Kennard 2001). Other landscape-scale features, such

as stream elevation, distance from the river mouth and stream channel width, are also important in determining the distribution of species within the stream network in a highly predictable way. Pusey *et al.* (2000) developed predictive models of fish assemblage structure for two rivers of the WT region, the Mulgrave and Johnstone Rivers, based on these landscape features. Notably, the structure of fish assemblages based on the presence or absence of particular species was determined by landscape-scale features, whereas accurate prediction of assemblage structure based on the density of species required the inclusion of additional local-scale habitat variables such as stream depth, velocity and substrate composition.

Summary

The origin and maintenance of fish assemblages in streams and rivers of the Wet Tropics region involves a hierarchical set of processes. First, the regional species pool is determined by large-scale biogeographical factors. Second, the local species pool is determined by finer scale biogeographical processes such as within-drainage vicariance events and also by contemporary ecological processes related to catchment size and discharge regime. Third, landscape-scale features constrain which species within the local species pool are able to colonize various parts of the riverine landscape. This is especially important in the WT region given that about half of the local species pool must either move downstream to the lower reaches or estuary to breed or must colonize the river as larvae or small juveniles. In addition, factors such as altitudinal variation in water temperature may also limit the distribution of some species. Finally, local-scale features of habitat structure related to stream geomorphology (i.e. depth, gradient, water velocity and substrate composition) determine the abundance of individual species at different points in the stream network. Poff (1997) likened such hierarchical organization to a series of environmental 'filters', which in the present case include processes that have operated over both prehistorical and contemporary time frames. Over these temporal and spatial scales, the reliability of rainfall, and hence river discharge, has contributed greatly to the maintenance of the region's biodiversity and the distinctive nature of fish assemblages.

References

Allen, G. R. (1991). *Field Guide to the Freshwater Fishes of New Guinea*. Christensen Research Institute, Madang, Papua New Guinea.

Allen, G. R., Midgley, S. H. & Allen, M. (2002). *Field Guide to the Freshwater Fishes of Australia*. Western Australian Museum, Perth.

Arthington, A. H., Marshall, J., Rayment, G., Hunter, H. & Bunn, S. (1997). Potential impacts of sugarcane production on the riparian and freshwater environment. In *Intensive Sugar Cane Production: Meeting the Challenges Beyond 2000*, Keating, B. A. & Wilson, J. R. (eds). CAB International, Wallingford, UK, pp. 403–21.

Australian Water Resources Council (1976). *Review of Australia's Water Resources 1975*. Australian Government Publishing Service, Canberra. 170 pp.

Bishop, K. A. & Forbes, M. A. (1991). The freshwater fishes of northern Australia. In *Monsoonal Australia: Landscape, Ecology and Man in the Northern Lowlands*, Haynes, C. D., Ridpath, M. G. & Williams, M. A. J. (eds). A. A. Balkema, Rotterdam, pp. 79–107.

Burrows, D. W. (2004). *Translocated Fishes in Streams of the Wet Tropics Region: Distribution and Potential Impact*. Cooperative Research Centre for Tropical Rainforest Ecology and Management, Cairns. 83 pp.

Colwell, R. K. (1974). Predictability, constancy and contingency of periodic phenomena. *Ecology* **55**:1148–53.

Hurwood, D. A. & Hughes, J. M. (1998). Phylogeography of the freshwater fish *Mogurnda adspersa*, in streams of north-eastern Queensland, Australia: evidence for altered drainage patterns. *Molecular Ecology* **7**: 1507–17.

Hurwood, D. A. & Hughes, J. M. (2001). Historical interdrainage dispersal of eastern rainbowfish from the Atherton tablelands, north-eastern Australia. *Journal of Fish Biology* **58**: 1125–36.

McGlashan, D. J. & Hughes, J. M. (2000). Reconciling patterns of genetic variation with stream structure, Earth history and biology in the Australian freshwater fish *Craterocephalus stercusmuscarum* (Atherinidae). *Molecular Ecology* **9**: 1737–51.

McGlashan, D. J. & Hughes, J. M. (2001). Low levels of genetic differentiation among populations of the freshwater fish *Hypseleotris compressa* (Gobiidae: Eleotridinae): implications for its biology, population connectivity and history. *Heredity* **86**: 222–33.

McGuigan, K. L. (2001). An addition to the rainbowfish (Melanotaeniidae) fauna of north Queensland. *Memoirs of the Queensland Museum* **46**: 647–55.

McGuigan, K., Zhu, D., Allen, G. R. & Moritz, C. (2000). Phylogenetic relationships and historical biogeography of melanotaeniid rainbowfishes in Australia and New Guinea. *Marine and Freshwater Research* **51**: 713–23.

Oberdorff, T., Guégan, J. F. & Hugueny, B. (1995). Global scale patterns of fish species richness in rivers. *Ecography* **18**: 345–52.

Poff, N. L. (1997). Landscape filters and species traits: towards mechanistic understanding and prediction in stream ecology. *Journal of the North American Benthological Society* **16**: 391–409.

Pusey, B. J. & Arthington, A. H. (2003). Importance of the riparian zone to the conservation and management of freshwater fishes: a review with special emphasis on tropical Australia. *Marine and Freshwater Research* **54**: 1–16.

Pusey, B. J., Arthington, A. H. & Kennard, M. J. (2004a). Hydrological regime and its influence on broadscale patterns of fish diversity in north-eastern Australian rivers. In *Proceedings of the Fifth International Symposium on Ecohydraulics. Aquatic Habitats, Analysis and Restoration*. Madrid, September, 75–81.

Pusey, B. J., Arthington, A. H. & Kennard, M. J. (2005). Threats to freshwater fishes of the Wet Tropics Region. In *Proceedings of the Ozwater 2005 Watershed Symposium*. Australian Water Association, Townsville, May.

Pusey, B. J., Arthington, A. H. & Read, M. G. (1995). Species richness and spatial variation in assemblage structure in two rivers of the Wet Tropics of north Queensland. *Environmental Biology of Fishes* **42**: 181–99.

Pusey, B. J., Burrows, D., Arthington, A. H. & Kennard, M. J. (2006). Translocation and spread of piscivorous fishes in the Burdekin River, north-eastern Australia. *Biological Invasions* **8**: 965–77.

Pusey, B. J. & Kennard, M. J. (1996). Species richness and geographical variation in assemblage structure of the freshwater fish fauna of the Wet Tropics region of north Queensland. *Marine and Freshwater Research* **47**: 563–73.

Pusey, B. J. & Kennard, M. J. (2001). *Guyu wujalwujalensis*, a new genus and species (Pisces: Percichthyidae), from north-eastern Queensland, Australia. *Ichthyological Exploration of Freshwaters* **12**: 17–28.

Pusey, B. J., Kennard, M. J. & Arthington, A. H. (2000). Discharge variability and the development of predictive models relating stream fish assemblage structure to habitat in north-eastern Australia. *Ecology of Freshwater Fish* **9**: 30–50.

Pusey, B. J., Kennard, M. J. & Arthington, A. H. (2004b). *Freshwater Fishes of North-eastern Australia*. CSIRO Publishing, Collingwood. 684 pp.

Russell, D. J. & Hales, P. W. (1993). *Stream Habitat and Fisheries Resources of the Johnstone River Catchment*. DPI Queensland, No. QI93056.

Russell, D. J. & Hales, P. W. (1997). *Stream Habitat and Fisheries Resources of the Liverpool, Maria and North Hull Catchments.* DPI Queensland, No. QI97039.

Russell, D. J., Hales, P. W. & Helmke, S. A. (1996a). *Fish Resources and Stream Habitat of the Moresby River Catchment.* QDPI, Brisbane, QI96061.

Russell, D. J., Hales, P. W. & Helmke, S. A. (1996b). *Stream Habitat and Fish Resources in the Russell and Mulgrave Rivers Catchment.* QDPI, Brisbane. Report No. QI96008.

United Nations Educational, Scientific and Cultural Organisation (UNESCO) (2005). *Operational Guidelines for the Implementation of the World Heritage Convention.* UNESCO World Heritage Centre, Paris.

Unmack, P. J. (2001). Biogeography of Australian freshwater fishes. *Journal of Biogeography* **28**: 1054–89.

12 DIVERSITY OF INVERTEBRATES IN WET TROPICS STREAMS: PATTERNS AND PROCESSES

Niall M. Connolly[1]*, Faye Christidis[2], Brendan McKie[3], Luz Boyero[4] and Richard Pearson[2]*

[1]School of Marine and Tropical Biology, James Cook University, Cairns, Queensland, Australia
[2]School of Marine and Tropical Biology, James Cook University, Townsville, Queensland, Australia
[3]Department of Ecology and Environmental Science, Umeá University, Sweden
[4]School of Tropical Biology, James Cook University, Cairns, Queensland, Australia
*The authors were participants of Cooperative Research Centre for Tropical Ecology and Management

Introduction

As the name infers, the Wet Tropics is a unique region within Australia, characterized by high rainfall. As a consequence, streams and wetlands are prominent and conspicuous features in the landscape. The region is topographically varied, with mountain ranges rising sharply near the coast catching the easterly winds that carry moisture-laden air from the warm waters of the Coral Sea. The headwaters of countless streams rise in these rainforest-clad mountains then rapidly descend to a narrow floodplain. The climate is dominated by seasonal patterns of rainfall, with a summer wet season and winter dry season. Most streams, although seasonal in flow, are perennial and have probably been so for millions of years, in contrast to those in many other parts of Australia, including most of the tropical region. In an Australian context, therefore, Wet Tropics streams are exceptional because they sustain a unique and diverse freshwater fauna, including species-rich invertebrate communities (e.g. Pearson *et al.* 1986; Walker *et al.* 1995), distinctive fish fauna with many endemic species (Pusey & Kennard 1996; Pusey *et al.*, Chapter 11, this volume) and a diverse frog fauna (e.g. Williams *et al.* 1996). Here we describe the diversity and ecology of the stream invertebrate fauna of the Wet Tropics in relation to the region's biogeographical history and present-day environment.

The invertebrates are an important component of the stream biota. Most species are aquatic larval stages of aquatic and terrestrial insects, but representatives from many other invertebrate phyla occur too, including worms, molluscs, crustaceans, sponges and cnidarians. Invertebrates typically make the first linkages in aquatic food webs, connecting the primary producers (or microbial producers) to higher consumers, which include other invertebrates (e.g. dragonfly larvae), fish, amphibians, reptiles, birds and mammals. In shaded headwater streams, productivity is heterotrophic and supported by the allochthonous input of terrestrial organic matter, mostly leaves from the surrounding forest trees. Certain invertebrates (the shredders) play a vital role in transforming this detritus by scraping, gouging and shredding it, making it available to other invertebrates and facilitating microbial colonization and decomposition (Pearson *et al.* 1989; Nolen & Pearson 1992; Pearson & Connolly 2000).

Freshwater invertebrates have been shown to be the most important component of the diet of freshwater fish in Australia (Arthington 1992; Pusey *et al.* 1995; Kennard *et al.* 2001) and represent about 80% of the food of the platypus (Faragher *et al.* 1979). Emerged aquatic insects also represent an important food source for terrestrial predators, particularly insectivorous birds, thereby linking terrestrial and aquatic food webs (e.g. Likens & Bormann 1974; Jackson & Fisher 1986; Gray 1993). Invertebrates are thus crucial to the transfer of energy through the food web and their functional importance has made them the focus of investigations into relationships between species diversity and ecological process rates. Stream invertebrates, mostly being relatively sedentary, also offer a time-integrated sample of environmental conditions over their lifetime (weeks to years) and consequently have been regularly used as indicators of water quality and ecosystem health (Rosenberg & Resh 1993).

Despite the invertebrates' ecological importance and utility as biomonitors, knowledge of their patterns of diversity and distribution at most scales is generally deficient, especially outside the temperate northern hemisphere. Our research within the Wet Tropics,

including experiments examining small-scale processes and interactions (e.g. Pearson & Connolly 2000; Connolly *et al.* 2004; Cheshire *et al.* 2005; McKie & Pearson 2006; Connolly & Pearson 2007) and large-scale surveys crossing broad latitudinal and altitudinal gradients (e.g. Christidis 2003; McKie *et al.* 2005), has given us a good basis for commenting on small- and large-scale diversity patterns of stream invertebrates, particularly for three important groups: the Ephemeroptera (mayflies), Trichoptera (caddisflies) and Chironomidae (non-biting midges). These three taxa typically dominate the composition of invertebrate assemblages in Wet Tropics streams, in both abundance and numbers of species (e.g. Pearson *et al.* 1986; Pearson & Connolly 2000). However, their contrasting life histories, dispersal abilities and feeding ecologies facilitate an assessment of the factors causing or sustaining current patterns of distribution for a range of stream invertebrates in the Wet Tropics (see Box 12.1).

Box 12.1 Ephemeroptera, Trichoptera and Chironomidae

Ephemeroptera (Figure 12.1)

The mayfly fauna of the Wet Tropics comprises 28 genera and 58 species from five families (Table 12.1). For the size of the region, species richness appears to be comparable to that of other Australian regions: about 70 species of mayfly are known from Victoria (Dean & Suter 1996), 30 from Tasmania, 12 from south-western Australia (Dean & Suter 1996), 14 from South Australia (Suter 1986; Alba-Tercedor & Suter 1990), 9 from Cape York Peninsula (Wells & Cartwright 1993) and 24 from the Alligator Rivers region of the Northern Territory (Suter 1992). There are two major components of the Wet Tropics fauna: (i) a Gondwanan element, including the family Amelotopsidae and almost all of the Leptophlebiidae, which have phylogenetic affinities with taxa present on other Gondwanan landmasses, particularly southern South America and New Zealand; and (ii) an oriental element, including the family Prosopistomatidae, the leptophlebiid genus *Thraulus* and some Baetidae, which probably entered from the north and are restricted to

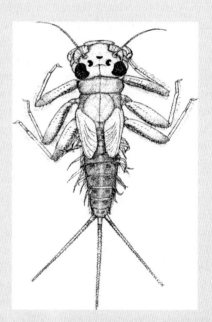

Figure 12.1 Leptophlebiidae (Ephemeroptera) nymph. Line drawing by Andi Cairns.

(Continued)

Box 12.1 (Continued)

northern Australia. The majority of mayflies in the Wet Tropics have Gondwanan affinities.

A striking feature of the mayfly fauna of the Wet Tropics is the high level of endemism. Among the Leptophlebiidae, 21 of the 29 species recorded from the Wet Tropics are endemic to the region, and some of these species are also subregional endemics.

Trichoptera (Figure 12.2)
Walker et al. (1995) provide a detailed examination of the species records of Trichoptera from a large number of sites in

the Wet Tropics. They found high species richness at site and regional scales, confirming previous reports (Pearson et al. 1986; Vinson & Hawkins 1998, 2003). They found that species richness of Trichoptera in the Wet Tropics was greater than in the Tasmanian World Heritage Area (Neboiss et al. 1989). The highest diversity recorded at any Wet Tropics site was 78 species at Yuccabine Creek (Pearson et al. 1986), exceeding the richest sites in Tasmania (Franklin River, Roaring Creek Junction, 45 species; Walker et al. 1995), Victoria (O'Shannassy River, 44 species; Dean & Cartwright 1987) and Cape York Peninsula (Gunshot Creek, 47 species; Wells & Cartwright 1993). The species lists for Yuccabine Creek and O'Shannassy River resulted from extensive collecting, and lower diversity at other sites may reflect a smaller collecting effort. However, there is strong evidence that species richness of Trichoptera is greater in the Wet Tropics than in other areas of Australia at site and regional scales.

Chironomidae (Figure 12.3)
McKie et al. (2005) identified 87 chironomid species in 49 genera in a survey of small streams across the Wet Tropics. Estimation of the total Australian species pool is difficult, since new species are frequently discovered, but the Wet Tropics count evidently represents a substantial proportion of the currently recognized pool of 160–200 species. On the basis of appropriate extrapolations, Cranston (2000) estimates the number of species in the Wet Tropics to be approximately 110–158; however, he points out that this total is much lower than for tropical regions of Asia and the Americas and that the chironomid fauna is relatively depauperate compared to that of other continents. For example, 216 species were recorded from two sites 3 km apart in Costa Rica, and

Table 12.1 The number of known genera and species of the nine families of Ephemeroptera that occur in Australia

Family	Wet Tropics		Australia	
	Genera	Species	Genera	Species
Ameletopsidae	1	1	1	3
Baetidae	9	15	9	31
Caenidae	3	12	3	25
Coloburiscidae	–	–	1	3
Leptophlebiidae	14	29	25	117
Oniscigastridae	–	–	1	3
Prosopistomatidae	1	1	1	1
Nesameletidae	–	–	1	1
Teloganodidae	–	–	1	1

Data include described and known undescribed taxa.

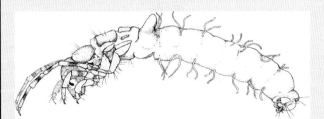

Figure 12.2 Leptoceridae (Trichoptera) larva. Line drawing by Andi Cairns.

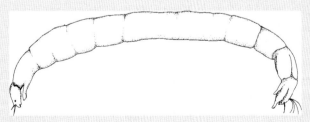

Figure 12.3 Chironomidae (Diptera) larva. Line drawing by Andi Cairns.

(Continued)

Box 12.1 *(Continued)*

174 species were collected from only one drift net deployed for 12 hours in Guinea (Cranston 2000). However, endemism in the Wet Tropics is high, with 15 of the genera being novel, and several novel species identified within previously recognized genera. Over half the genera occur worldwide and about 25% of the fauna has Gondwanan affinities, while some species are clearly of northern (Asian) origin.

Recent molecular studies of *Echinocladius martini* have found that dispersal of chironomids may be much more restricted than previously thought, and that the presupposed extensive distribution of *E. martini* along the east coast of Australia may actually constitute a broader species group (Krosch 2006).

Regional comparisons

The east coast of Australia lies on a latitudinal gradient, covering more than 24 degrees of latitude, with an almost continuous mountain chain that has a consistent geological history and therefore offers a unique opportunity to explore the effect of latitude on biodiversity. The diversity of stream invertebrates in the Wet Tropics has been reported to be high in comparison with similar streams elsewhere in Australia (Lake *et al.* 1994; Walker *et al.* 1995) and globally (Pearson *et al.* 1986; Vinson & Hawkins 2003). However, Cranston (2000), in a broad review of the Chironomidae along the Australian eastern states, reports no compelling evidence for a larger regional species pool in the tropics than in south-eastern Australian streams. These contrasting observations are interesting because there are conflicting accounts of the latitudinal patterns of lotic invertebrates. Stout and Vandermeer (1975) and Boyero (2002) found greater diversity of stream invertebrates at lower latitudes in the Americas. However, other authors have suggested that lotic invertebrates are an exception to the rule that biodiversity decreases with latitude (Patrick 1966; Coffman 1989; Coffman & De La Rosa 1998; Flowers 1991).

Most studies of lotic invertebrates rely on small-scale samples taken at a limited number of sites, and data on regional species pools are consequently deficient. The lack of comprehensive descriptions of lotic faunas at regional scales, particularly in the tropics, means that broad comparisons along latitudinal gradients are problematic. For example, Lake *et al.* (1994) compared the species richness of benthic invertebrates on stones in streams in the Wet Tropics and in Victoria, in temperate Australia. They estimated the regional species pool from within-site samples from two Victorian and two Wet Tropics sites, following procedures used by Stout and Vandermeer (1975) in a similar latitudinal comparison. These sites had been studied over long periods of time and so the species pools at these sites were well described. However, their estimates of species richness were actually at the site rather than the regional scale, raising the question of how well the assemblages at two sites (per region) could represent the regional species pool even if all species were recorded for each site.

Nevertheless, the comparisons by Lake *et al.* (1994) and Stout and Vandermeer (1975) highlight some interesting differences between sites at different latitudes, with both concluding that species richness on stones was greater at tropical sites than in temperate sites. In both studies species accumulation curves were less steep at tropical sites but species ultimately accumulated to a higher level. That is, species accumulated more gradually in tropical than in temperate sites, but continued to accumulate for longer as additional stone samples were added to the data. This observation reflects the distribution of species among the stones sampled. If the assemblages on individual stones are randomly assembled (see Box 12.2) then a lower slope could occur if there were fewer individuals found per stone at tropical sites, or if the distribution of species among stones was more heterogeneous. Stout and Vandermeer (1975) concluded that tropical invertebrates were more spatially heterogeneous than their higher latitude counterparts and observed lower densities on stones in the tropics. It is not clear if the same was observed by Lake *et al.* (1994), although Pearson (1994) does comment that fauna at Yuccabine Creek, one of the sites used in the comparison by

Box 12.2 The relationship between species richness and abundance at very small scales

The diversity of aquatic invertebrates within micro-habitats, such as on individual stones or small leaf packs, is strongly correlated with abundance (Pearson 2005). This relationship is important because it is at the scale of individual stones or small quadrats that conventional sampling is undertaken. The correlation is due to the assemblage being a subset of the pool of individuals available at any one time. In effect, these assemblages conform to a 'Type 1' community, which is unsaturated (Cornell & Lawton 1992), where the assemblage occurring on, say, a stone is essentially a sample of the larger site assemblage, and this stone sample is a random subset of animals. This is corroborated by Pearson (2005), who found that the assemblages colonizing bricks in streams at Paluma were a random subset of the surrounding riffle community, with a positive relationship between species richness and abundance at the single brick scale.

Similarly, recent experiments have shown that the densities of invertebrates inhabiting individual leaf packs at any point in time are the result of equilibrium dynamics involving the immigration and emigration of individuals from the leaf packs (N. Connolly, unpublished data). The equilibrium density determines the species richness by setting the sample size on a leaf pack. Thus, the fine-scale species richness is determined by both the larger regional pool and the number of individuals present (the sample size), with a greater abundance increasing the probability of more species being included.

Lake *et al.* (1994), 'was not abundant compared to streams elsewhere'. Thus, because the curves at tropical sites rise more gradually but plateau at a higher level, there appear more species in the site species pool in the tropics and a greater number of rare species with spatially patchy distributions. Interestingly, these within-site patterns may be mirrored in multi-site studies from the Wet Tropics, in which the rate of Trichoptera and Chironomidae species accumulation showed little sign of declining over samples collected

from 20 and 33 sites respectively (McKie 2002; Pearson 2005), but no corresponding data from temperate regions are available for comparison.

The differences in distribution of aquatic invertebrate diversity with latitude will, of course, have implications for any sampling for comparisons of species richness, with greater effort required at tropical sites to avoid underestimating species richness. Both Stout and Vandermeer (1975) and Lake *et al.* (1994) recognize that differences between sites at different latitudes were not detectable until a large number of stone samples were collected, because of the relatively slower rate of occurrence of taxa at the tropical sites. Stout and Vandermeer (1975) suggest that the lack of difference detected in other studies (e.g. Patrick 1966) was due to inadequate sampling, with further problems arising from inconsistent methodology.

At present, our best estimates of regional species pools are usually derived not from quantitative surveys but from accumulated knowledge of species records by taxonomists and systematists with intimate knowledge of their particular group (e.g. Walker *et al.* 1995; Cranston 2000). The review of Walker *et al.* (1995), which compared the diversity of Trichoptera in the Wet Tropics with that of the temperate Tasmanian World Heritage Area, is rare in considering the full suite of species records at a large number of sites. This study was not limited by taxonomy, as it was undertaken by recognized Australian Trichoptera systematists and sampling effort appears to have been high, including repeated sampling of some sites at multiple times, leaving less chance for overlooking taxa. They found that although richness at the family and genus levels does not appear to differ between the Wet Tropics and the Tasmanian WHA, species richness diverges markedly (Walker *et al.* 1995). Thus, Trichoptera in the Wet Tropics have high species richness at the scale of the site, the region and the subregion (defined on the basis of consistent topography), with a higher proportion of the total Australian species pool occurring in the Wet Tropics (36.7% of the Australian fauna) than in the Tasmanian WHA (22.7%). The Wet Tropics also had the highest number of species recorded at a single site, with 78 species collected from Yuccabine Creek, although this tally was derived from an extensive sampling programme and included several species with only a single record each (Benson & Pearson 1988). The average number of species for the ten most diverse sites in the two World Heritage Areas was 41.8 species for the Wet Tropics and 37.0 species for the Tasmanian World Heritage Area. This is not a large difference and given the higher regional difference, suggests that regional richness is greater but local richness is constrained in the Wet Tropics.

Cranston (2000) has similarly suggested that individual stream richness is constrained for the chironomid fauna, and argued that there was no overall difference in regional species pools between tropical and temperate regions. This contrast with observations for Trichoptera and some vertebrates (especially frogs) may relate to differences in the degree of ecological specialization characteristic of these different groups (McKie *et al.* 2004). More generally, recent assessment of regional diversity patterns of stream faunas based on published species lists (R. G. Pearson & L. Boyero, pers. comm.) suggests that those taxa that have a significant terrestrial phase (e.g. Odonata) are more species rich in the tropics (Boulton *et al.* in press). Inconsistency in the responses of different invertebrate groups may explain Hillebrand's (2004) observation that while a latitudinal gradient in species richness seems apparent for freshwater systems, it appears weaker than in marine or terrestrial environments, and differs between continents and habitat types.

Distributional patterns within the Wet Tropics bioregion

Large-scale patterns of distribution of fauna and flora are typically associated with regional or subregional differences in environmental conditions. However, as conditions are not constant over time, current distributions may reflect past rather than present climate and geology. The distribution of rainforest within the Wet Tropics has fluctuated with climate during the Quaternary (Quilty 1994). During the late Pleistocene (13 000–8000 years ago), drier sclerophyll forests displaced most rainforest, with only isolated moist upland refugia remaining (Nix 1991). Subsequently, increased rainfall allowed rainforest to expand to its current extent (Hopkins *et al.* 1996), but current diversity and distributional patterns of several groups of terrestrial organisms bear the signature of the past rainforest contractions; for example, diversity and endemism of

terrestrial vertebrates are greatest in the largest refugial areas (Williams & Pearson 1997; Winter 1997). In contrast, current diversity patterns of the Wet Tropics freshwater fish, Chironomidae and Trichoptera seem little affected by the Pleistocene contractions, with most species homogeneously distributed throughout the region's latitudinal range (Walker *et al.* 1995; Pusey & Kennard 1996; McKie *et al.* 2005; Pearson 2005), although there is some indication of loss of lowland species of Trichoptera, as seen for the terrestrial vertebrates (Pearson 2005), and some invertebrate species appear to have restricted distributions – for example, some mayflies (Christidis 2003) and *Euastacus* crayfish species restricted to mountain tops in the Wet Tropics (Morgan 1988; Short & Davie 1993).

McKie *et al.* (2005) found no evidence for an enduring effect of historical rainforest contractions on current distribution patterns for Chironomidae, with most species present at all latitudes within the Wet Tropics. Further, their surveys showed no 'hotspots' of species richness associated with rainforest persistence during dry glacial periods, as observed for the terrestrial vertebrate fauna (Williams 1997), though abundance (but not richness) of certain Gondwanan species tended to increase with altitude. They suggest that stream habitats may have been buffered from the effects of climate change where stream flow and shade were maintained. Tall ranges may have continued to capture enough precipitation to maintain flows and narrow riparian strips of rainforest vegetation, or even drier sclerophyllous vegetation, may have maintained shade and kept water temperatures cool. In locations where streams became ephemeral or dried completely, they suggested that the vagility of the adult chironomids would have allowed recolonization as the climate ameliorated and flow returned, obscuring any effects on contemporary distributional patterns. The widespread distribution across broad latitudinal and altitudinal bands of Australian lotic chironomids show their relatively unconstrained ecology (Brundin 1966), substantiated by ecophysiological studies demonstrating broad temperature tolerances, even for cool-Gondwanan species (McKie *et al.* 2004). However, recent molecular research highlighted substantial genetic differentiation among populations of one chironomid species, *Echinocladius martini*, inhabiting different rainforest blocks, and even adjacent streams

(Krosch 2006). This indicates that the forest environment may constitute a substantial barrier to dispersal between streams for adults of weakly flying taxa such as the Chironomidae, and demonstrates that the biogeography of the Wet Tropics can have a substantial influence on the distribution of freshwater diversity at the genetic level.

The Trichoptera appear to be homogeneously distributed throughout the latitudinal extent of the Wet Tropics, with no subregional species distribution patterns (Walker *et al*. 1995; Pearson 2005). In contrast, distributional patterns among leptophlebiid mayflies within the Wet Tropics suggest that the biogeographical history of the region has influenced the present-day distributions of species (Christidis 2003). Although some leptophlebiid species occur throughout the Wet Tropics (e.g. *Jappa serrata*, *Atalophlebia* sp. AV13, *Nousia* sp. NQ1), others appear to have far more restricted distributions. For example, *Austrophlebioides wooroonooran* and *A. porphyrobranchus* are known only from the Atherton subregion. Extensive collecting from the Paluma Range has failed to find these species, suggesting that their absence from the southern end of the Wet Tropics is genuine. These species do not appear to be present in the lower reaches of streams in the Daintree area. The genus *Austrophlebioides* is represented in Paluma by an undescribed species and in the Daintree region by *A. rieki*. The distribution of *A. rieki* may extend south of the Daintree area as nymphs belonging to this or morphologically similar species have been collected from the lower sections of streams in the Atherton subregion.

Several other leptophlebiid species also have restricted distributions. For example, WT sp. 4, an undescribed species of a new genus, is presently known only from the Daintree region, whereas its sister species WT sp. 2 occurs in the Daintree as well as the Cardwell/Ingham area north of the Herbert River, but is absent from the Paluma Range to the south of the Herbert River. Molecular data may show whether the presence of WT sp. 2 in the Cardwell/Ingham area represents recent dispersal into the area from populations further north. The absence of several species from the Paluma Range suggests that the Herbert River may be an effective barrier to the dispersal of some mayfly species. Interestingly, the distributions of a number of the endemic rainforest vertebrates also

do not extend south of the Herbert River (Nix 1991; Williams *et al*. 1996).

The restricted geographical distribution of many of the leptophlebiid species is not surprising given the limited dispersal abilities of mayflies. The nymphs of many leptophlebiid species are confined to cool forest streams and the potentially dispersive adults are short lived (two to three days) and prone to desiccation. It is noteworthy that widely distributed species tend to have broader ecological tolerances occurring in a range of flow regimes including pools with reduced flow, whereas species with more restricted distributions are found predominantly in fast flowing waters (Christidis 2003).

Overall, past climatic fluctuations in the region appear to have had an enduring effect on the present-day distributions of leptophlebiid species, probably because of their limited dispersal abilities and narrow ecological tolerances. In contrast, the possibly more vagile and tolerant Chironomidae and Trichoptera have long since overcome any restrictions imposed in the past and are now widespread in suitable habitats across the Wet Tropics.

Altitudinal and longitudinal gradients

It should be remembered that although streams are distributed throughout the Wet Tropics region, they are elongate linear water bodies that cross environmental gradients. Within-stream gradients are important determinants of aquatic invertebrate distributions and need to be accounted for in any intra- or interregional comparisons of stream sites. Wet Tropics streams descend quickly from high ranges, passing over tablelands and through steep gorges, before flowing across a narrow coastal floodplain to wetlands and estuaries. The rapid changes in altitude coupled with a relatively narrow floodplain have resulted in a sharp geomorphic gradient in these systems. Such gradients are usually reflected in the distributions of invertebrates (e.g. Allan 1975; Vannote *et al*. 1980; Bapista *et al*. 2001) that follow the physical changes that occur along the stream continuum, including gradients of temperature, stream size, slope, discharge, current velocity and substratum.

Typically, two main geomorphic zones occur along the stream length: the steeper upstream zone crossing a steep altitudinal gradient with substantial changes in temperature and vegetation; and the low altitude, low gradient zone on the floodplain with habitats largely governed by the distribution of bed materials resulting from variation in stream power.

Altitudinal gradient

McKie *et al.* (2005) found consistent trends in chironomid distributions in Wet Tropics streams with altitude, with cool Gondwanan taxa (originating in cooler regions of the former Gondwanan super continent and with distributions now centred on Australia's southeast) occurring predominantly in cooler upland streams (e.g. *Echinocladius martini*, *Botryocladius grapeth*), while species from tropical and cosmopolitan genera were more typical of lowland sites (e.g. *Rheocricotopus* spp. and *Nanocladius* spp.). However, cool Gondwanan species were also found in well shaded lowland streams (e.g. Gap Creek, north of Bloomfield) characterized by cooler conditions arising from the mountain mass effect (Nix 1991), whereby streams drop rapidly down steep escarpments, limiting the scope for warming. Conversely, poorly shaded upland streams (e.g. Yuccabine Creek), which can be two to three degrees warmer than predicted for their altitude, may support no cool Gondwanan species. Thus, there is no distinct zonation of chironomid faunas with altitude, as observed elsewhere in both tropical (e.g. Jacobsen *et al.* 1997) and temperate regions (e.g. Rossaro 1991) of the world, and also no systematic relationship with species diversity. Nevertheless, cool upland streams appear to have favoured the persistence of a cool temperate chironomid fauna in the otherwise warm Wet Tropics region, as the Australian continent drifted northwards. Furthermore, altitude is likely to prove an important determinant for distributions of other taxa; for example, distributions of Trichoptera are correlated with altitude, with more species in upland than lowland streams and with a greater number of exclusive trichopteran species above 700 m asl than at lower altitudes (Pearson 2005).

As the streams cross the escarpment, they typically create waterfalls, which have patchy occurrence, depending on topography and geology. Clayton (1995) demonstrated that waterfalls in the Wet Tropics have specialized faunas that are essentially isolated from each other by the lack of suitable intervening habitat. Special characteristics included high-velocity regions, dominated by species of Simuliidae, moderate velocity regions with Hydropsychidae and Blepharicidae, and low velocity and splash zones with Pyralidae and various beetles. As waterfalls often create a gap in the canopy, algal growth is possible and several of the specialist species are algal grazers (e.g., the blepharicids and pyralids). Others, such as the simuliids and hydropsychids, are filter feeders, benefiting from the food supplied in the strong currents.

Lowland longitudinal gradient

There are no published reports of the longitudinal distribution of invertebrates for streams in the Wet Tropics. However, Connolly *et al.* (2007) have recently sampled four streams at approximately 1.5 kilometre intervals from the base of the range to their confluence with the Russell or Mulgrave Rivers. The physical character of the streams changed gradually along their length, with reducing stream power and sediment size. Invertebrate distributions reflected this longitudinal gradient, with the number of species declining downstream, although abundance peaked mid-way across the floodplain (Figure 12.4a, b). Species richness was strongly correlated with sediment size, with more species present in riffles composed of mixed sediments dominated by cobbles (Figure 12.5). This pattern reflects the greater habitat stability and complexity provided by cobble riffles with large surface areas for attachment of invertebrates and growth of biofilms and macroscopic plants such a bryophytes and algae. As the physical structure of the habitat is a strong determinant of species distributions (e.g. Minshall 1984), it clearly needs to be considered when comparing patterns of diversity of benthic invertebrates. This is important not only when making comparisons between sites but also within sites, where the substratum is a complex matrix of various particles.

These streams had a consistent longitudinal pattern of species turnover with different assemblages in the upper, middle and lower parts of the stream. High densities of a few grazing or filter-feeding taxa (mayflies, chironomid and simuliid fly larvae, and hydropsychid and philopotamid caddisflies) were responsible for the peak in abundance midway across the floodplain, probably reflecting a combination of ideal substratum and high productivity resulting from more open streams, and nutrient supplements from agricultural activity on the floodplain. The abundances of

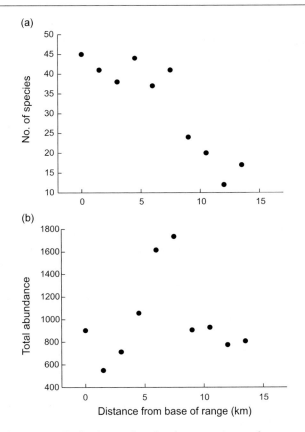

Figure 12.4 Preliminary data for the macroinvertebrates sampled from riffles at sites distributed along a longitudinal gradient in Behana Creek from the base of the range to the confluence with the Mulgrave River: (a) number of species; (b) total abundance.

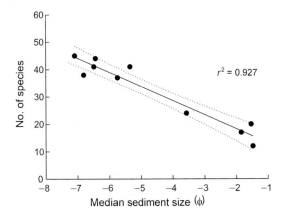

Figure 12.5 Preliminary data for macroinvertebrates sampled from riffles at sites distributed along a longitudinal gradient in Behana Creek from the base of the range to the confluence with the Mulgrave River, describing the relationship between taxon richness and the median sediment particle size. φ is a $\log_2$ scale where $\varphi = -\log_2$ (size mm). φ increases as sediment size decreases and sizes greater than 1 mm have a negative φ.

some taxa, including larval psephenid beetles and gripopterygid stoneflies, gradually declined at sites furthest from the range, possibly reflecting changes in flow and water temperature. Other groups, including leptophlebiid mayflies and hydropsychid caddisflies, were abundant where cobble substratum was available, but absent when sediments were reduced to fine gravels and sands in the most downstream sites. In contrast, caenid mayflies, although present at most sites across the floodplain, were most abundant at the most downstream sites, reflecting their preference for fine substrata, in which they burrow. Similarly, the larvae of the leptocerid caddisfly, *Oecetis* spp., occurred only in the downstream sites and made their cases from the fine sand grains.

It is clear that longitudinal gradients within a stream are very strong determinants of large-scale distributions of lotic invertebrates. The vagile adult stages of the aquatic insects appear to enable many of these taxa to disperse and access most locations within the Wet Tropics, but recent genetic evidence for the chironomid *E. martini* indicates that water-courses may provide the predominant means of dispersal through the landscape for weak flying species, since connectivity across intact forest was minimal even between adjacent streams (Krosch 2006). Furthermore, changes in the physical structure of habitats and in light and temperature regimes along channel gradients are important for the larval and nymphal stages, resulting in assemblage shifts along the length of the stream.

Local patterns of diversity

While patterns of biodiversity are predominantly structured by broad biogeographical and climatological histories and through broad environmental gradients, the composition of the invertebrate assemblage at any given location is further influenced by the outcomes of ecological interactions and processes, and by chance, through processes of immigration, emigration and disturbance.

Disturbance

Physical disturbance is a characteristic of Wet Tropics stream environments, caused by high flows that shift substrata and abrade everything in their path. High flows can cause catastrophic local mortality, but they also promote diversity by providing continually renewed habitat patches and preventing dominant species from displacing others. Flooding creates a somewhat unpredictable stream environment and resistance to disturbance and recolonization capacity are necessary characteristics of stream faunas world-wide. The occurrence of spates, associated with widespread density-independent mortality, may dilute the importance of factors such as predation and competition as structuring forces in streams, at least in the wet season, although these factors may be important during the dry season, when habitats are reduced in extent and animal densities are higher (e.g. Dudgeon 1993; Pearson 2005).

Rosser and Pearson (1995) showed that riffle faunas were generally well adapted to unpredictable high-flow events, such that recolonization of denuded areas of stream bed took place remarkably rapidly. On smaller scales, two chironomid species showed great resilience to physical disturbance in the laboratory, though sublethal effects of the disturbance on growth and fecundity were apparent (McKie 2004). Recolonization of denuded substrata following disturbance may be from several sources and depends on the scale of the disturbance and the source of colonists – local movements from nearby undisturbed patches, immigration from other sections of the stream or aerial immigration and egg-laying by adults of aquatic insects. Typically, many aquatic invertebrates drift with the current over short distances and this gives them the ability to recolonize available space quickly. For example, Benson and Pearson (1987a, b) demonstrated the capacity of the stream fauna to recolonize denuded substrata in experiments in Yuccabine Creek, particularly by drifting but also by means of upstream movements.

A series of recent experiments in Birthday Creek showed that at small scales (individual leaf litter packs) the assemblage of invertebrates at any moment in time was the result of equilibrium dynamics involving the immigration and emigration of individuals from the drift or surrounding substrata (N. Connolly, unpublished data) with approximately a 10% turnover of individuals on a leaf pack per day. Therefore, these animals not only have a strong capacity for recolonization after disturbance, they are also continually dispersing within the riffle substrata, and with around 10% of the riffle fauna available to colonize available

substrata every day, it is not surprising that small areas of experimentally denuded substrata were rapidly reoccupied.

Biotic interactions

Effects of competition and predation have been demonstrated in streams (Kohler 1992; Peckarsky *et al.* 1993), and succession was apparent in the colonization of the small units of substratum used in the experiments described above, suggesting that biotic interactions have some role at these small scales. However, it is difficult to demonstrate that changes in habitat use and distribution are the consequences of such interactions (Giller & Malmqvist 1998).

Some small-scale biotic interactions may be specific and subtle. For example, McKie and Pearson (2006) revealed remarkably specific developmental and behavioural responses by chironomid prey to different predator species. The presence of a net-spinning ambush predator (a polycentropodid trichopteran) reduced adult body and oocyte size for the riffle-inhabiting chironomid *Echinocladius martini* but not for the pool-inhabiting *Polypedilum australotropicus*, whereas the presence of the actively hunting chironomid predator *Australopelopia* had no effect on *E. martini* development, but caused *P. australotropicus* to shorten its pupal duration and to emerge with smaller oocytes. These results are attributable to observed effects of the predators on different aspects of prey behaviour and correspond with differences in the threat posed by the predators to the prey species *in situ* (McKie 2002). In another example, Boyero, Rincón and Pearson (2007) showed that fish predators had substantial indirect effects on detritus-based food webs in streams, mediated by changes in consumer behaviour. They identified reductions in both short-term overall activity and medium-term leaf breakdown rates in response to chemical cues from fish predators in laboratory conditions. Sublethal effects of predation on the behaviour and development of individuals have the potential to influence individual fitness and habitat use at small scales, but even at larger scales these effects have the potential to influence the demographics of prey populations, and hence patterns of species distribution, especially where responses of prey to predators are well differentiated, like those discussed here (McPeek & Peckarsky 1998; McKie & Pearson 2006).

Recent experiments have also shed some light on the interactions of detritivores (shredders) in Wet Tropics streams. For example, leaf processing by shredders decreased exponentially when other individuals of the same species were present (Boyero & Pearson 2006), although overall breakdown increased with greater density of shredders. Further, shredders were able to detect chemical cues from conspecifics and responded to them by reducing their immediate activity (Boyero, Allan & Pearson, unpublished data). However, this reduction in activity was not reflected in individual rates of shredding, so it is probable that visual or physical interactions are more important, or that individuals get used to the presence of the chemical cues.

Habitat partitioning

Although riffles may be regarded as well mixed habitats, habitat partitioning by invertebrates among various microhabitats does occur. It has long been accepted that species-specific responses to environmental variables related to current velocity, physical substratum and food sources affect maintenance of diversity and distribution at the riffle scale (Cummins & Lauff 1969; Rabeni & Minshall 1977; Minshall 1984). Therefore, the fact that partitioning occurs within the riffle is very interesting, given the continuous movement of individuals. In the Wet Tropics, McKie (2002) found clear patterns in the microdistribution of chironomids across five habitats in Birthday Creek. Even within a single genus, *Polypedilum*, that is otherwise morphologically uniform, there were distinct differences between pools and riffles, and between leaf litter and stones: thus, *P. australotropicus* was found in pool leaf packs, *P. vespertinus* was found only on pool rocks, *P. 'alpha'* was found in all leaf and rock riffle habitats, while *P. oresitrophuus* was found only in riffle leaf packs.

Likewise, fine divisions of habitat by mayflies were evident in Yuccabine Creek and were confirmed by manipulative experiments in the stream (Hearnden & Pearson 1991). Twelve species occurred in all the microhabitats examined, but each species had significant peaks in abundance in particular microhabitats; where two species preferred the same habitat, they were very different in size. Like the chironomids, the mayfly species showed distinct preferences for either pools or riffles, and further discriminated among habitats on the basis of substratum size (sand to large rock) and

presence of leaf litter. Such habitat preferences have been shown for entire assemblages in Birthday Creek (Benson 1999). Habitat partitioning is also a characteristic of odonates, again relating to the same habitat variables (Charlton 1989), but neither the mayflies (mostly detritivores) nor the odonates (predators) showed much partitioning of food resources, except perhaps by particle size. However, there is partitioning of food resources elsewhere in the aquatic food web (Cheshire *et al.* 2005), and within some invertebrate groups. For example, trichopteran shredders apparently have distinct preferences for different resources: some eat leaves in riffles, some eat leaves in pools and others eat mostly woody material (Boyero *et al.* 2007).

Phylogenetic influence on patterns of habitat use

Many studies on stream fauna have emphasized the importance of physical factors in structuring communities and in determining the distribution of species among habitats (e.g. Cummins & Lauff 1969; Hynes 1970; Rabeni & Minshall 1977; Hart & Finelli 1999), and it has often been assumed or inferred that this is due to competitive or other ecological pressures. Until recently there has been little consideration of the influence of evolutionary history in determining the distribution of species among the substrata in stream habitats. Christidis (2003) showed that substratum and habitat use by leptophlebiid mayflies in the Wet Tropics reflected phylogenetic relationships, with closely related species tending to occur on similar substrata (e.g. *Austrophlebioides* species on stones), and in some instances in similar habitats (e.g. *Koorrnonga* species in pools). High levels of partitioning based on substratum type resulted from differences among phylogenetic lineages in the use of substrata: the litter fauna was dominated by species of the *Nousia* lineage, whereas the stone fauna was dominated by species of the *Meridialaris* lineage. Likewise, association with a particular habitat type of species within some lineages contributed to the distinctiveness of mayfly assemblages of pool, run and riffle habitats. Thus phylogenetic history is important in determining some ecological traits and present-day species' distributions among substratum and habitat types. It is unknown how broadly such conclusions might apply, but it is likely that other elements of the fauna are similarly influenced by their phylogenetic history.

Influence of biodiversity on ecological processes

While attention has been paid to the historical and current factors that influence biodiversity patterns, there is increasing interest in how biodiversity influences ecosystem function because of concerns about the potential consequences of species loss (Loreau 2000; Loreau *et al.* 2001). It is possible that loss of a few key species will have a disproportionate effect on key processes. Research on this issue is limited and focuses mostly on terrestrial vegetation, with none reported for the tropics. Recent work in streams (Jonsson & Malmqvist 2003) has focused on the influence of biodiversity on the decomposition of terrestrially derived leaf litter, which drives the food web in forest streams (Vannote *et al.* 1980), including tropical rainforest streams (Benson & Pearson 1993; Dobson *et al.* 2002; Cheshire *et al.* 2005). Temperate streams harbour a numerically important shredder guild but shredders are reported to be scarce in tropical streams (Dudgeon & Wu 1999; Dobson *et al.* 2002; Mathuriau & Chauvet 2002). This is not true for Wet Tropics streams where insect shredders constitute around 20% of the total species richness (cf. world average of 11%) (Cheshire *et al.* 2005). Our previous research has demonstrated the importance of shredders in the community (Pearson & Tobin 1989; Pearson *et al.* 1989; Coughlan 1990; Nolen & Pearson 1992) and current experiments, outlined below, are investigating the relationship between the diversity of shredders and the decomposition process.

Activity of shredders (including several species of caddisfly, a leptophlebiid mayfly, a crayfish, a chironomid and a beetle) is high in Birthday and Camp Creeks. Colonization of experimental leaf litter occurs within two weeks and competitive interactions are likely to take place mostly among individuals of the most abundant species. Intraspecific interactions in these species also occur in laboratory conditions, as individual leaf breakdown activity decreases as density increases over a wide range (from two to twelve individuals per leaf) (Boyero & Pearson 2006). Although abundance of leaves is not usually a limiting factor, shredders show strong preferences for particular leaf species, which are patchily distributed (Bastian *et al.* 2007).

The number of shredder species, their individual patterns of behaviour and their intraspecific and interspecific interactions may all influence breakdown rates. It is possible that there is redundancy in the species complement, but the nature of the redundancy cannot be assumed and must be determined at the species level; for example, Bastian *et al.* (2007) showed that species identity was the main factor governing leaf breakdown rates in Wet Tropics streams, but species richness also had some effect. Leaf breakdown rates were higher than expected when the three most common shredder species were present together, suggesting the existence of at least one of three mechanisms (niche complementarity, interspecific facilitation or release from intraspecific interference). However, there is no evidence of different feeding abilities of the different species or facilitation, while release from intraspecific interference has been previously found in these species (Boyero & Pearson 2006) and other shredders (Jonsson & Malmqvist 2003).

Bastian *et al.* (2008) manipulated species richness and composition simultaneously across two trophic levels (leaves and shredders) and found that breakdown of leaf litter was affected by temperature, the composition of shredder communities and the diversity and identity of leaf species present, due largely to a so-called sampling effect. Different shredder species had different processing capabilities, but leaf preferences were similar for all shredder species. Increasing shredder diversity had no effect on leaf breakdown rates, but the identity of shredder species present was significant in determining those rates, as the different insect species process litter at different rates. These results corroborate findings from other studies that have shown the important role that taxonomic identity can play in the dynamics and functioning of ecosystems (Huston 1997; Vanni *et al.* 2002).

Summary

Our understanding of Wet Tropics streams is important at a regional scale for their inherent water resource, water quality and biodiversity values, and at a national scale because of their high biodiversity compared with the rest of the continent (Pearson *et al.* 1986; Lake *et al.* 1994). At a global scale, there are few other tropical regions where extensive tropical stream research is being undertaken (some examples are: Latin America

(Pringle 2000; Pringle *et al*. 2000); Hong Kong (Dudgeon 1995, 1999); and Kenya (Dobson *et al*. 2002)). Our research in the Wet Tropics, therefore, can make a substantial contribution to understanding of stream ecology generally, and to appropriate management of tropical streams, which are among the most threatened of ecosystems globally (Dudgeon 1992; Boyero 2000; Connolly & Pearson 2004).

As for any regional biota, the diversity of invertebrates in Wet Tropic streams is influenced by a range of interlinked biogeographical, evolutionary, ecological and behavioural processes, operating at a range of scales. Thus the species assemblage on a particular unit of habitat (e.g. a stone) is determined by the species pool, individual species' requirements, colonization dynamics, small-scale interactions and chance. Understanding of such influences, i.e. the generation and maintenance of species diversity, underpins further studies and, especially, appropriate management activities of ecosystems.

The invertebrate fauna of Wet Tropics streams is largely Gondwanan in origin, including several cooler-adapted elements, which might otherwise struggle to persist in the region, but includes some Asian-derived elements also. The overlap between major biogeographical regions contributes to the diversity of Wet Tropics streams, like other habitats (Willig *et al*. 2003). Diversity has also been increased by speciation in the region, or by the region acting as a refuge for particular species, as there is a high level of endemism for some groups (e.g. Trichoptera, Leptophlebiidae, Odonata), and subregional endemism for non-vagile taxa such as Leptophlebiidae and Parastacidae. Consequently, species richness for some taxa is higher than elsewhere in Australia and most other places that have been studied in detail globally (Vinson & Hawkins 2003; Boulton *et al*. in press). At subregional scales, there are clear patterns of distribution of some taxa according to altitude and position along the stream continuum, relating to physical factors such as temperature, current and substratum. At local scales there are differences in assemblages between pools and riffles, and between stone and leaf pack habitats, reflecting species' habitat preferences that stem from their phylogenetic origins or subsequent evolution. At a fine scale, interactions may determine which species immediately cohabit, and impacts of biotic interactions on individuals are of significance for population demographic parameters (growth, fecundity) and have the potential to influence

larger-scale patterns of diversity and distribution, although the importance of such effects in the frequently disturbed streams of the Wet Tropics is presently difficult to assess. Between-latitude differences in spatial distribution, relative abundances and rarity are interesting and warrant further investigation by means of standardized studies across regions and continents. Understanding the links between diversity and ecosystem function is becoming particularly important in the face of rapid global change and species loss, and more extensive research is required to understand the processes and to apply that understanding to management responses. Research in the Wet Tropics can play a major role because of its research base, and its special place in terms of Australian and global biodiversity.

References

Alba-Tercedor, J. & Suter, P. J. (1990). A new species of Caenidae from Australia: *Tasmanocoensis arcuata* spp. n (Insecta, Ephemeroptera). *Aquatic Insects* **12**: 85–94.

Allan, J. D. (1975). Faunal replacement and longitudinal zonation in an alpine stream. *Verh Int Verein Limnol* **19**: 1646–52.

Arthington, A. H. (1992). Diets and trophic guild structure of freshwater fishes in Brisbane streams. *Proceedings of the Royal Society of Queensland* **12**: 31–47.

Bapista, D. F., Dorville, L. F. M., Buss, D. F. & Nessiamian, J. L. (2001). Spatial and temporal organisation of aquatic insects assemblages in the longitudinal gradient of a tropical river. Rev. *Brasil Biol* **61**: 295–304.

Bastian, M., Boyero, L., Jackes, B. R., & Pearson, R. G. (2007). Leaf litter diversity and shredder preferences in an Australian tropical rainforest stream. *Journal of Tropical Ecology* **23**: 219–29.

Bastian, M., Pearson, R. G. & Boyero, L. (2008). The influence of biodiversity on ecosystem processes: shredder and leaf species interactions in tropical streams. *Austral Ecology* (in press).

Benson, L. J. (1999). Allochthonous input and dynamics in and upland tropical stream. PhD thesis, James Cook University.

Benson, L. J. & Pearson, R. G. (1987a). Drift and upstream movements of macro invertebrates in a tropical Australian stream. *Hydrobiologia* **153**: 225–39.

Benson, L. J. & Pearson, R. G. (1987b). The role of drift and effect of season in colonization of implanted substrata in a tropical Australian stream. *Freshwater Biology* **18**: 109–16.

Benson, L. J. & Pearson, R. G. (1988). Diversity and seasonality of adult Trichoptera captured in a light-trap at Yuccabine

Creek, a tropical Australian rainforest stream. *Australian Journal of Ecology* **13**: 337–44.

Benson, L. J. & Pearson, R. G. (1993). Litter inputs to a tropical Australian upland rainforest stream. *Australian Journal of Ecology* **18**: 377–83.

Boulton, A. J., Boyero, L., Covich, A. P., Dobson, M., Lake, P. S. &. Pearson, R. G. (in press). Are tropical streams ecologically different from temperate streams? In *Tropical Stream Ecology*, Dudgeon, D. & Cressa, C. (eds). Academic Press, San Diego, CA.

Boyero, L. (2000). Towards a global stream ecology. *Trends in Ecology and Evolution* **15**: 390–1.

Boyero, L. (2002). Insect biodiversity in freshwater ecosystems: is there any latitudinal gradient? *Marine and Freshwater Research* **53**: 753–5.

Boyero, L. & Pearson, R. G. (2006). Intraspecific interference in a tropical stream shredder guild. *Marine and Freshwater Research* **57**: 201–6.

Boyero, L., Pearson, R. G. & Bastian, M. (2007). How biological diversity influences ecosystem function: the separate role of species richness and evenness. *Ecological Research* **22**: 551–8.

Boyero, L., Pearson, R. G. & Camacho, R. (2006). Leaf processing in Australian tropical streams: the role of different species in ecosystem functioning. *Archiv für Hydrobiologie* **166**: 453–66.

Boyero, L., Rincon, P. A. & Pearson, R. G. (2007). Effects of a predatory fish on a tropical detritus based food web. *Ecological Research* (in press).

Brundin, L. (1966). Transantarctic relationships and their significance as evidenced by chironomid midges. Gleneagles Publishing, Glen Osmond, South Australia.

Charlton, L. (1989). The ecology of dragonflies (Odonata) in a tropical Australian stream. Honours thesis, James Cook University, Townsville.

Cheshire, K., Boyero, L. & Pearson, R. G. (2005). Food webs in tropical Australian streams: shredders are not scarce. *Freshwater Biology* **50**: 748–69.

Christidis, F. (2003). Systematics, phylogeny and ecology of Australian Leptophlebiidae (Ephemeroptera). PhD Thesis, James Cook University, Townsville.

Clayton, P. D. (1995). The ecology of waterfalls in the Australian Wet Tropics. PhD thesis, James Cook University, Townsville.

Coffman, W. P. (1989). Factors that determine the species richness of lotic communities of Chironomidae. *Acta Biol Debr Oecol Hung* **3**: 95–100.

Coffman, W. P. & De La Rosa, C. (1998). Taxonomic composition and temporal organization of tropical and temperate assemblages of lotic communities. *Kansas Entomol Soc* **71**: 388–406.

Connolly, N. M., Crossland, M. R. & Pearson, R. G. (2004). Effect of low dissolved oxygen on survival, emergence, and drift of tropical stream macroinvertebrates. *Journal of the North American Benthological Society* **23**: 251–70.

Connolly N. M. & Pearson R. G. (2004). Impacts of forest conversion on the ecology of streams in the humid tropics. In *Forests, Water and People in the Humid Tropics: Past, Present and Future Hydrological Research for Integrated Land and Water Management*, Bonnell, M. & Bruijnzeel, L. A. (eds). Cambridge University Press, Cambridge, pp. 811–35.

Connolly N. M. & Pearson R. G. (2007). The effect of fine sedimentation on a tropical stream macroinvertebrate assemblage: a comparison using flow-through artificial stream channels and recirculating mesocosms. *Hydrobiologia* **592**: 423–38.

Cornell, H. V. & Lawton, J. H. (1992). Species interactions, local and regional processes, and limits to the richness of ecological communities: a theoretical perspective. *Journal of Animal Ecology* **61**: 1–12.

Coughlan, J. F. (1990). Population and trophic ecology of the freshwater crayfish Cherax depressus (Crustacea: Parastacidae) in a North Queensland tropical rainforest stream. Honours thesis, James Cook University, Townsville.

Cranston, P. S. (2000). August Thienemann's influence on modern chironomidology – an Australian perspective. *Verhandlungen Internationale Vereinigung der Limnologie* **27**: 278–83.

Cummins, K. W. & Lauff, G. F. (1969). The influence of substratum particle size on the distribution of stream benthos. *Hydrobiologia* **34**: 145–81.

Dean, J. & Cartwright, D. (1987). Trichoptera of a Victorian forest stream: species composition and life histories. *Australian Journal of Marine and Freshwater Research* **38**: 845–60.

Dean, J. C. & Suter, P. J. (1996). *Mayfly Nymphs of Australia. A Guide to Genera*. CRC Freshwater Research Identification Guide No. 7.

Dobson, M., Mathooko, J. M., Magana, A. & Ndegwa, F. K. (2002). Macroinvertebrate assemblages and detritus processing in Kenyan highland streams: more evidence for the paucity of shredders in the tropics? *Freshwater Biology* **47**: 909–19.

Dudgeon, D. (1992). Endangered ecosystems: a review of the conservation status of tropical Asian rivers. *Hydrobiologia* **248**: 167–91.

Dudgeon, D. (1993). The effects of spate-induced disturbance, predation and environmental complexity on macroinvertebrates in a tropical stream. *Freshwater Biology* **30**: 189–97.

Dudgeon, D. (1995). The ecology of rivers and streams in tropical Asia. In *Ecosystems of the World 22. River and Stream Ecosystems*, Cushing, C. E., Cummins, K. W. & Minshall, G. W. (eds). Elsevier, Amsterdam, pp. 615–57.

Dudgeon, D. (1999). *Tropical Asian Streams*. Hong Kong University Press, Hong Kong.

Dudgeon, D. & Wu, K. K. Y. (1999). Leaf litter in a tropical stream: food or substratum for macroinvertebrates? *Archiv für Hydrobiologie* **146**: 65–82.

Faragher, R. A., Grant, T. R. & Carrick, F. N. (1979). Food of the platypus (*Ornithorhynchus anatinus*) with notes on the food of Brown Trout (*Salmo trutta*) in the Shoalhaven River, NSW. *Australian Journal of Ecology* **4**: 171–9.

Flowers, R. W. (1991). Diversity of stream-living insects in north western panama. *Journal of the North American Benthological Society* **10**: 322–34.

Giller, P. S. & Malmqvist. B. (1998). *The Biology of Streams and Rivers*. Oxford University Press, New York. 296 pp.

Gray, L. J. (1993). Response of insectivorous birds of emerging aquatic insects in riparian habitats of a tall grass prairie stream. *American Midland Naturalist* **129**: 288–300.

Hart, D. D. & Finelli, C. M. (1999). Physical-biological coupling in streams: the pervasive effects of flow on benthic organisms. *Annual Review of Ecology and Systematics* **30**: 363–95.

Hearnden, M. N. & Pearson, R. G. (1991). Habitat partitioning among the mayfly species (Ephemereoptera) of Yuccabine Creek, a tropical Australian stream. *Oecologia* **87**: 91–101.

Hillebrand, H. (2004). On the generality of the latitudinal diversity gradient. *The American Naturalist* **163**: 192–211.

Hopkins, M. S., Head, J., Ash, J., Hewett, R. & Graham, A. (1996). Evidence of a Holocene and continuing recent expansion of lowland rain forest in humid, tropical North Queensland. *Journal of Biogeography* **23**: 737–45.

Huston, M. A. (1997). Hidden treatments in ecological experiments: re-evaluating the ecosystem function of biodiversity. *Oecologia* **110**: 449–60.

Hynes, H. B. N. (1970). *The Ecology of Running Waters*. University of Toronto Press, Toronto.

Jackson, J. K. & Fisher, S. G. (1986). Secondary production, emergence, and export of aquatic insects of a Sonoran desert stream. *Ecology* **67**: 629–38.

Jacobsen, D., Schultz, R. & Encalada, A. (1997). Structure and diversity of stream invertebrate assemblages: the influence of temperature with altitude and latitude. *Freshwater Biology* **38**: 247–61.

Jonsson, M. & Malmqvist, B. (2003). Mechanisms behind positive diversity effects on ecosystem functioning: testing the facilitation and interference hypothesis. *Oecologia* **134**: 554–9.

Kennard, M. J., Pusey, B. J. & Arthington, A. H. (2001). *Trophic Ecology of Freshwater Fishes in Australia*. Summary Report, CRC for Freshwater Ecology Scoping Study ScD6. Centre for Catchment and In-stream Research, Griffith University, Brisbane.

Kohler, S. L. (1992). Competition and the structure of a benthic stream community. *Ecological Monographs* **62**: 165–88.

Krosch, M. N. (2006). Phylogeography of *Echinocladius martini* (Chironomidae: Orthocladiinae) in the Australian Wet Tropics. Honours thesis, Queensland University of Technology, Brisbane.

Lake, P. S., Schreiber, E. S. G., Milne, B. J. & Pearson, R. G. (1994). Species richness in streams: Patterns over time, with stream size and with latitude. *Verh Internat Verein Limnol* **25**: 1822–26.

Likens, G. E. & Bormann, F. H. (1974). Linkages between terrestrial and aquatic ecosystems. *Bioscience* **24**: 447–56.

Loreau, M., (2000). Biodiversity and ecosystem functioning: recent theoretical advances. *Oikos* **91**: 3–17.

Loreau, M., Naeem, S., Inchausti, P. *et al.* (2001). Biodiversity and ecosystem functioning: current knowledge and future challenges. *Science* **294**: 804–8.

McKie, B. G. L. (2002). Multiscale abiotic, biotic and biogeographic influences on the ecology and distribution of lotic Chironomidae (Diptera) in the Australian Wet Tropics. PhD thesis, School of Tropical Biology, James Cook University, Townsville.

McKie, B. G. (2004). Disturbance and investment: developmental responses of tropical lotic midges to repeated tube destruction in the juvenile stages. *Ecological Entomology* **29**: 457–66.

McKie, B. G., Cranston, P. S. & Pearson, R. G. (2004). Gondwanan mesotherms and cosmopolitan eurytherms: effect of temperature on the development and survival of Australian Chironomidae (Diptera) from tropical and temperate populations. *Marine and Freshwater Research* **55**: 759–68.

McKie, B. G. & Pearson, R. G. (2006). Effects of temperature and two contrasting predators on the survival and development of Chironomidae (Diptera): species-specific phenotypic plasticity in tropical rainforest streams. *Oecologia* DOI 10. 1007/s00442-006-0454-8.

McKie, B. G., Pearson, R. G. & Cranston, P. S. (2005). Does biogeographical history matter? Diversity and distribution of lotic midges (Diptera: Chironomidae) in the Australian Wet Tropics. *Austral Ecology* **30**: 1–13.

McPeek, M. A. & Peckarsky, B. L. (1998). Life histories and the strengths of species interactions: combining mortality, growth and fecundity effects. *Ecology* **79**: 867–79.

Mathuriau, C. & Chauvet, E. (2002). Breakdown of leaf litter in a neotropical stream. *Journal of the North American Benthological Society* **21**: 384–96.

Minshall, G. W. (1984). Aquatic insect–substratum relationships. In *The Ecology of Aquatic Insects*, Resh, V. H. and Rosenberg, D. M. (eds). Praeger, New York, pp. 358–400.

Morgan, G. J. (1988). Freshwater crayfish of the genus *Euastacus* Clark (Decapoda: Paratacidae) from Queensland. *Memoirs of the Museum of Victoria* **49**: 1–49.

Neboiss, A., Jackson, J. & Walker, K. (1989). Caddis-flies (Insecta: Trichoptera) of the World Heritage Area in Tasmania – species composition and distribution. *Occasional Papers of the Museum of Victoria* **4**: 1–41.

Nix, H. A. (1991). Biogeography: pattern and process. In *Rainforest Animals: Atlas of Vertebrates Endemic to Australia's Wet Tropics*, Nix, H. A. & Switzer M. A. (eds). Australian National Parks and Wildlife Service, Canberra, pp. 10–40.

Nolen, J. A. & Pearson, R. G. (1992). Life history studies of *Anisocentropus kirramus*, Neboiss (Trichoptera: Calamoceratidae) in a tropical Australian rainforest stream. *Aquatic Insects* **14**: 213–21.

Patrick, R. (1966). The Catherwood Foundation Peruvian Amazon Expedition: limnological and systematic studies. *Mongr Acad Nat Sci Philadelphia* **14**: 1–495.

Pearson, R. G. (1994). Limnology in the northeastern tropics of Australia, the wettest part of the driest continent. *Mitt Internat Verein Limnol* **24**: 155–63.

Pearson, R. G. (2005). Biodiversity of the freshwater fauna of the Wet Tropics region of north-eastern Australia: patterns and possible determinants. Pages 470–85. In *Tropical Rain Forests: Past, Present and Future*, Bermingham, E., Dick, C. W. & Moritz, C. (eds). University of Chicago Press, Chicago. 1004 pp.

Pearson, R. G., Benson, L. J. & Smith, R. E. W. (1986). Diversity and abundance of the fauna in Yuccabine Creek, a tropical rainforest stream. In *Limnology in Australia*, de Deckker, P. & Williams, W. D. (eds). CSIRO, Melbourne, pp. 329–42.

Pearson, R. G. & Connolly, N. M. (2000). Nutrient enhancement, food quality and community dynamics in a tropical rainforest stream. *Freshwater Biology* **43**: 31–42.

Pearson, R. G. & Tobin, R. K. (1989). Litter consumption by invertebrates from an Australian tropical rainforest stream. *Archiv für Hydrobiologie* **116**: 71–80.

Pearson, R. G., Tobin, R. K L., Benson, J. & Smith, R. E. W. (1989). Standing crop and processing of rainforest litter in a tropical Australian stream. *Archiv für Hydrobiologie* **115**: 481–98.

Peckarsky, B. L., Cowan, C. A., Penton, M. A. & Anderson, C. (1993). Subethal consequences of stream-dwelling predatory stoneflies on mayfly growth and fecundity. *Ecology* 74: 1836–46.

Pringle, C. M. (2000). Riverine conservation in tropical versus temperate regions: Ecological and socioeconomic considerations. In *Global Perspectives on River Conservation: Science, Policy and Practice*, Boon, P. J. & Davies, B. (eds). John Wiley & Sons, Chichester, pp. 367–78.

Pringle C. M., Scatena, F., Paaby, P. & Nunez, M. (2000). Conservation of riverine ecosystems of Latin America and the Caribbean. In *Global Perspectives on River Conservation: Science, Policy and Practice*, Boon, P. J. & Davies, B. (eds). John Wiley & Sons, Chichester, pp. 39–75.

Pusey, B. J. & Kennard, M. J. (1996). Species richness and geographical variation in assemblage structure of the freshwater fish fauna of the Wet Tropics Region of north Queensland. *Marine and Freshwater Research* **47**: 563–73.

Pusey, B. J., Read, M. G. & Arthington, A. H. (1995). The feeding ecology of freshwater fishes in two rivers of the Australian Wet Tropics. *Environmental Biology of Fishes* **43**: 85–318.

Quilty, P. G. (1994). The background: 144 million years of Australian palaeoclimate and palaeogeography. In *History of the Australian Vegetation: Cretaceous to Recent*, Hill, R. S. (ed.). Cambridge University Press, Cambridge, pp. 14–43.

Rabeni, C. F. & Minshall, G. W. (1977). Factors affecting the microdistribution of stream benthic insects. *Oikos* **29**: 33–43.

Rosenberg, D. M. & Resh, V. H. (eds) (1993). *Freshwater Biomonitoring and Benthic Macroinvertebrates*. Chapman & Hall, London. 488 pp.

Rossaro, B. (1991). Chironomids and water temperature. *Aquat Ins* **13**: 87–98.

Rosser, Z. C. & Pearson, R. G. (1995). Responses of rock fauna to physical disturbance in two Australian tropical rainforest streams. *Journal of the North American Benthological Society* **14**: 183–96.

Short, J. W. & Davie, P. J. F. (1993). Two new species of freshwater crayfish (Crustacea: Decapoda: Parastacidae) from northeastern Queensland rainforest. *Memoirs of the Queensland Museum* **34**: 69–80.

Stout, J. & Vandermeer, J. (1975). Comparison of species richness for stream-inhabiting insects in tropical and mid-latitude streams. *The American Naturalist* **109**: 263–80.

Suter, P. J. (1986). The Ephemeroptera (mayflies) of South Australia. *Records of the South Australian Museum* **19**: 339–97.

Suter, P. J. (1992). *Taxonomic Key to the Ephemeroptera (Mayflies) of the Alligator Rivers Region, Northern Territory*. Open File Record No. 96, Supervising Scientist for the Alligator Rivers Region.

Vanni, M. J., Flecker, A. S., Hood, J. M. & Headworth, J. L. (2002). Stoichiometry of nutrient recycling by vertebrates in a tropical stream: linking species identity and ecosystem processes. *Ecology Letters* **5**: 285–93.

Vannote, R. L., Minshall, G. W., Cummins, K. W., Sedell, J. R. & Cushing, C. E. (1980). The River Continuum Concept. *Canadian Journal of Fisheries and Aquatic Sciences* **37**: 130–7.

Vinson, M. R. & Hawkins, C. P. (1998). Biodiversity of stream insects: variation at local, basin, and regional scales. *Annu Rev Entomol* **43**: 271–93.

Vinson, M. R. & Hawkins, C. P. (2003). Broad-scale geographical patterns in local stream insect enera richness. *Ecography* **26**: 751–67.

Walker, K., Neboiss, A., Dean, J. & Cartwright, D. (1995). A preliminary investigation of the caddis-flies (Insecta: Trichoptera) of the Queensland Wet Tropics. *Australian Entomology* **22**: 19–31.

Wells, A. & Cartwright, D. I. (1993). Trichoptera, Ephemeroptera, Plecoptera and Odonata of the Jardine River area, Cape York Peninsula, north Queensland. In *Cape York Scientific Expedition Wet Season 1992.* The Royal Geographic Society of Queensland, pp. 221–30.

Williams, S. E. (1997). Patterns of mammalian species richness in the Australian tropical rainforests: are extinctions during historical contractions of the rainforest the primary determinant of current regional patterns in biodiversity. *Wildlife Research* **59**: 1211–29.

Williams S. E. & Pearson R. G. (1997). Historical rainforest contractions, localized extinctions and patterns of vertebrate endemism in the rainforests of Australia's Wet Tropics. *Proc R Soc Lond B* **264**: 709–16.

Williams, S. E., Pearson, R. G. &. Walsh, P. J. (1996). Distributions and biodiversity of the terrestrial vertebrates of Australia's Wet Tropics: a review of current knowledge. *Pacific Conservation Biology* **2**: 327–62.

Willig, M. R., Kaufman, D. M. & Stevens, R. D. (2003). Latitudinal gradients of biodiversity: pattern, process, scale, and synthesis. *Annual Review. of Ecology, Evolution and Systematics* **34**: 273–309.

Winter J. W. (1997). Responses of non-volant mammals to the late Quarternary climatic changes in the Wet Tropics region of north-eastern Australia. *Wildlife Research* **24**: 493–511.

13 THE INVERTEBRATE FAUNA OF THE WET TROPICS: DIVERSITY, ENDEMISM AND RELATIONSHIPS

David Yeates[1]* and Geoff B. Monteith[2]*

[1]The Australian National Insct Collection, CSIRO Entomology, Canberra, ACT, Australia
[2]Queensland Museum, South Brisbane, Queensland, Australia
*The author was participant of Cooperative Research Centre for Tropical Ecology and Management

Introduction

Tropical forests cover only 8% of the Earth's surface but their importance looms much larger because estimates suggest they are home to as much as 50–80% of the Earth's species and are therefore intense hotspots of biological diversity (Myers 1986; Stork 1988). Australia's tropical rainforests cover a much smaller proportion of Australia (only 6300 km^2, less than 1% of the land mass), but constitute a biome that is critical for understanding the evolution of Australian ecosystems (Crisp et al. 2004). They include a very important northern component of the mesotherm archipelago found along the east coast of Australia (Nix 1991; Adam 1994). Like rainforests in other parts of the world, the forests of the Australian Wet Tropics are very diverse, containing, for example, 25% of all plant genera found in Australia, 30% of all frog species, 62% of butterflies and 60% of bats (Keto & Scott 1986; Moritz 2005).

In this chapter we focus on recent work that has addressed the diversity, altitudinal zonation, faunal turnover and biogeography of Wet Tropics insects and other invertebrates at local, regional and global scales. Insects and other invertebrates comprise around 75% of biological diversity at species level (Wilson 1988), hence the scope of this chapter is broad and our knowledge fragmentary. What is certain from this review is that we are only just beginning along an exciting path of discovering the nature and roles of the Wet Tropics invertebrates and their importance in understanding the evolution of the Australasian biota.

Scope

This chapter largely reviews work that has been conducted on the fauna of the forest floor – the leaf litter, fallen logs and tree trunks. Readers should refer to Boulter et al. (Chapter 17, this volume) for reviews of pollination landscapes and to Connelly et al. (Chapter 12, this volume) on aquatic invertebrates for a more complete picture of the invertebrate fauna of the region. In addition, phylogeographical comparisons have used some invertebrate models, most notably the rainforest land snail Gnarosophia bellendenkerensis reviewed in Williams et al. (Chapter 10, this volume). The forest floor fauna has been the focus of much of the invertebrate faunal survey work in the region, and is likely to be where most of the forest diversity lies. The ratio of species found at the forest floor versus forest canopy is probably 5 : 1 or as much as

10 : 1 (Hammond 1992; Hammond *et al.* 1997) in Sulawesi, and comparable dominance of forest floor diversity over canopy diversity has been found in rainforests of Seram, Indonesia (Stork 1988, Stork & Brendell 1990). Almost four years of combined malaise and flight intercept trapping in the canopy and on the ground at the Australian Canopy Crane site in the Wet Tropics found 1158 species of beetle in the canopy and 895 on the ground (Stork & Grimbacher 2006), but this sampling would have missed flightless species only found in the leaf litter and upper soil layers, the subject of much of this chapter. This distribution of species in different structural components of the forest is likely to be the same in the Wet Tropics.

Most of the literature on the invertebrates of the Wet Tropics addresses the terrestrial insect fauna, but there are notable recent exceptions in non-insect groups such as spiders (Davies & Lambkin 2000; Monteith & Davies 1991; Raven 2000), Uropygi (Harvey 2000), Amblypygi (Harvey & West 1998), scorpions (Monod & Volschenk 2004), centipedes (Edgecombe 2001), Onychophora (Reid 1996), molluscs (Stanisic 2000, and references therein) and earthworms (Dyne 1985, and references therein). Arthropods represent 96% of individuals in invertebrate litter samples from the Wet Tropics (Frith & Frith 1990) and the few comparisons that we can make at this stage suggest that in general terms the insects are a reasonable proxy for macroinvertebrate pattern and processes in general.

We have also focused on the results of three major altitudinal transect studies in the Wet Tropics conducted by G. B. Monteith in the 1980s on the seaward faces of the Bellenden Ker Range, the Carbine Tableland and the Cape Tribulation area (Monteith 1985, 1989; Monteith & Davies 1991), and the results of faunal survey work in the upland subregions of the Wet Tropics (Yeates *et al.* 2002, and see below). Many invertebrate taxa have been described from the Wet Tropics over the past 20 years, often in the context of larger revisionary works (e.g., Baehr 1995, 2003; Monteith 1997; Matthews 1998; Hamilton 1999; Ballantyne & Lambkin 2000; Bouchard 2002) scattered in many publications. One journal issue devoted to the invertebrates of the Wet Tropics brought together many taxonomic revisions in a single volume (Monteith 2000). Our focus here is on the work that includes either quantitative or qualitative biogeographical or phylogeographical analyses, especially syntheses such as Bouchard *et al.* (2005).

History of discovery of the Wet Tropics ground fauna

European settlement of the Wet Tropics coast commenced in the late 1860s and from that time the more conspicuous insects, such as the forest-edge butterflies and the larger insects attracted out of the forest to lights, were steadily documented (e.g. Figure 13.1). However, the rate of discovery of the more cryptic ground fauna inside the rainforest, particularly the rich fauna at high elevations, was extremely slow until recent times. The experienced tropical biologist P. J. Darlington Jr made a landmark visit to the region in 1957–8. He collected systematically inside the forest, climbed several high peaks and discovered 25 new flightless carabid species (Darlington 1961). This sparked local interest, and in 1971 a telecommunications cablecar to the summit of Bellenden Ker gave easy access to the region's highest environment (1560 m) and revealed the presence of such unsuspected Gondwanan novelties as peloridiid moss bugs (Evans 1972) and chinamyersiine bark bugs (Monteith 1980).

The Queensland Museum commenced a systematic survey of high elevation ground fauna of the Wet Tropics in 1980. Some 50 individual mountains

Figure 13.1 *Sphaenognathus australicus* Moore (Lucanidae), a rare stag beetle known only from higher parts of the Carbine Tableland, one of two Australian members of an otherwise Neotropical genus.

were sampled using modern mass-collection methods of long-term pitfall and flight intercept traps, litter berlesates and pyrethrin spraying of ground level substrates. A suite of low-vagility insect taxa, known to be reliably sampled by these methods, was selected for critical sorting, databasing and mapping, in an attempt to reveal the real patterns of insect distribution across these complex mountain landscapes (Monteith 1995).

This survey produced a spectacular increase in species numbers during the 1980s as previously unsampled massifs were surveyed, now divided into 17 upland zones (Figure 13.2). From 1990 onwards, despite increased field effort, new species discovery levelled off, indicating that this component of the fauna is now well collected. Due to scores of papers by the global taxonomic community it is also now moderately well described. The resulting database, comprising 39 138 specimens from 457 localities, includes 406 species, with target groups including aradid bugs, flightless carabids, tenebrionids, dung beetles and other groups (Figures 13.1, 13.3, 13.4). Species diversity is concentrated in two upland zones (Figure 13.2a), the Carbine Tableland to the north of the Black Mountain Barrier (BMB) and Bellenden Ker to the south. Species diversity gradually decreases away from these foci, with upland zones in the south such as Mt Elliot and the Paluma/Bluewater Range, and those to the west such as the Hann Tableland, having fewest species.

This database has been used for Wet Tropics overview studies including surrogacy and concordance (Moritz *et al.* 2001), endemism (Yeates *et al.* 2002), mosaic macroevolution (Bouchard *et al.* 2005) and phylogenetic biodiversity (Crozier *et al.* 2005) and underpins much of the discussion in this chapter.

Biological and biogeographical highlights of the Wet Tropics insect fauna

The following examples indicate some aspects of the uniqueness of the region's insects. The primitive Gondwanan family of large crickets, Anostostomatidae, is well known in New Zealand under the Maori name 'weta', but the family is also diverse in Australia, reaching its global maximum in the Wet Tropics with seven genera and 28 species (Monteith & Field 2001). These include the only macropterous genera in the eastern hemisphere (*Exogryllacris* (Figure 13.4), *Gryllotaurus* and *Transaevum*), all restricted to the Wet Tropics uplands.

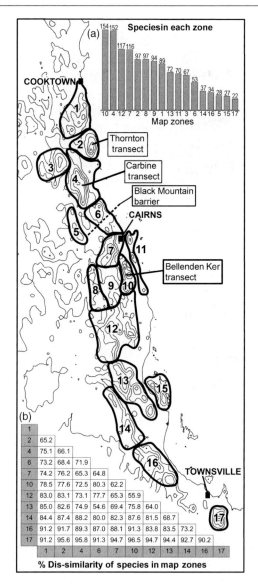

Figure 13.2 The main map shows the 17 upland insect endemism areas of the Wet Tropics used by the Queensland Museum. 1 Mt Finnigan; 2 Thornton Peak; 3 Windsor Tbld; 4 Carbine Tbld; 5 Hann Tbld; 6 Black Mountain; 7 Lamb Range; 8 Walsh/Hugh Nelson Range; 9 Atherton Tbld; 10 Bellenden Ker; 11 Malbon Thompson Range; 12 Walter Hill Range; 13 Kirrama/Cardwell Ranges; 14 Seaview Range; 15 Hinchinbrook Island; 16 Paluma/Bluewater Ranges; 17 Mt Elliot. These differ from the upland areas used by vertebrate researchers in that the Paluma and Bluewater Ranges are merged as one and the large Atherton area is divided into three. The location of three altitudinal transect studies discussed in the text is shown. Insert (a): relative species richness of the 17 areas based on 406 species of surveyed ground fauna insects. Insert (b): turnover values (percentage dissimilarity) of the same insect fauna of a north–south contiguous series of ten of the areas. Value for each pair of zones is the number of unshared species expressed as a percentage of the combined fauna of the two sites.

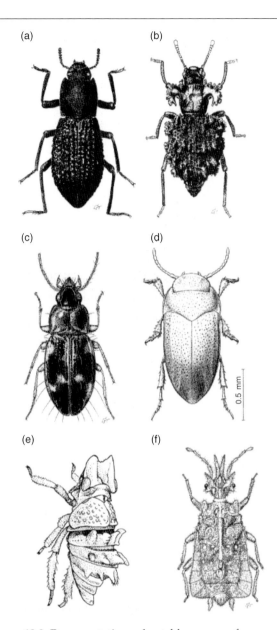

Figure 13.4 *Exogryllacris ornata* Willemse (Anostostomatidae), a monotypic, macropterous genus of king crickets, known only from the Lamb Range and mountains adjacent to the Atherton Tableland.

Figure 13.3 Representatives of notable groups of Wet Tropics insects. **Coleoptera**. (a) Tenebrionidae: *Apterotheca pustulosa* (Carter), one of more than 40 flightless species of this genus on Wet Tropics mountain tops. (b) Tenebrionidae: *Monteithium ascetum* Matthews, a bizarre, flightless member of the Gondwanan tribe Adeliini that is diverse in the Wet Tropics uplands. (c) Carabidae: *Philipis bicolor* Baehr, a genus with Neotropical affinities that has 37 high altitude species in the Wet Tropics. (d) Dytiscidae: *Terradessus caecus* Watts, a blind, wingless, terrestrial water beetle known only from the Thornton massif. **Hemiptera**. (e) Cicadellidae: *Myerslopella spinata* Hamilton, one of six species of this wingless, moss-feeding genus in the Wet Tropics. (f) Aradidae: *Granulaptera cooki* Monteith, one of 22 species of wingless Mezirinae restricted to dead wood of the forest floor of the Wet Tropics.

A rich fauna of mostly newly discovered, highly modified ant-inquiline Coleoptera in the Histeridae: Chlamydopsinae (Caterino 2003), Aphodiine Scarabaeidae (Howden & Storey 2000), Staphylinidae (*Clavigeropsis*), Carabidae (*Megalopaussus*) and Salpingidae (*Tretothorax*) indicates a long evolutionary history in the Wet Tropics.

On certain high-rainfall summits in the Thornton massif there has been movement of normally aquatic taxa into terrestrial forest floor habitats, including the world's only truly terrestrial anisopteran dragonfly nymphs (*Pseudocordulia*: Watson 1982), terrestrial water bugs (*Austrovelia*: Malipatil & Monteith 1983) and blind flightless water beetles (*Terradessus*: Brancucci & Monteith 1996) (Figure 13.3d).

Hamilton (1999) erected a new tribe, the Sagmatiini, for a curious, primitive, flightless group of litter/moss inhabiting leafhoppers with species known from Madagascar, Australia and New Caledonia. Its major radiation is the genus *Myerslopella* (Figure 13.3e) with six high altitude species restricted to the Wet Tropics. *Sagmation* has a species on the extreme summit of Bellenden Ker with congeners in New Caledonia and south Queensland.

Boulder caves, eroded by water, extend deep beneath the rainforest root zone on Mt Bartle Frere and the Walter Hill Range. These have a distinctive fauna including a giant, primitive spider (*Macrogradungula*: Forster *et al.* 1987), an undescribed genus of king crickets that browses on algal film of caves walls (Monteith & Field 2001) and a dung beetle that feeds on swiftlet guano (*Aptenocanthon winyar*: Storey & Monteith 2000).

Substantial allopatric speciation has occurred in high altitude flightless taxa within the Wet Tropics. Beetle examples include 37 species of the carabid genus *Philipis* (Baehr 1995) (Figure 13.3c), 41 species of the tenebrionid genus *Apterotheca* (Bouchard 2002) (Figure 13.3a), many adeliine tenebrionids (Matthews 1998) (Figure 13.3b) and 18 species of the dung beetle genus *Pseudignambia* (T. A. Weir, unpublished), all restricted to altitudes above about 800 m and many restricted to one or a few summits. Many wingless Aradidae (Hemiptera) (Figure 13.3f) are endemic to the Wet Tropics, with a large percentage restricted to the uplands (Russell 1997). These flightless faunas earned the Wet Tropics distinction as 'a centre of evolution, especially for low-vagility animals' (Bouchard 2002: 538).

Species associated with wet sclerophyll forests

The wet sclerophyll forests are a eucalypt-dominated, fire-maintained ecotonal vegetation type that occurs as a narrow band along the western (rain shadow) margin of rainforest on many of the mountain massifs of the Wet Tropics. It occupies a total area of only about 800 km^2 and appears to be rapidly declining in area in recent times due to replacement by rainforest in reduced fire regimes (Harrington & Sanderson 1994; Harrington *et al.* 1999). Certain mammals and birds are restricted to this environment and are considered at risk due to this habitat reduction (Winter *et al.* 1984). A suite of insect species is also limited to this habitat, including the dung beetles *Temnoplectron cooki* (Reid & Storey 2000) and *Onthophagus pinaroo* (Storey & Weir 1990), the dynastine scarab *Anomalomorpha monteithi* (Allsopp 2000) and the rare carabid beetle *Pamborus elegans* (Sota *et al.* 2005). Other species are long range disjunct populations of species that otherwise occur much further south in Australia, e.g. the dung beetle *Onthophagus incornutus* and the aradid bug *Carventus brachypterus* (Monteith 1997).

Transect surveys

Alpha diversity

We have no real way of addressing how many invertebrate or insect species inhabit the Wet Tropics, but there are a few intensive surveys of local areas that can give us some general impressions. A three-week survey of forest floor arthropods along an altitudinal transect on the Bellenden Ker Range (Figure 13.2), using a variety of collecting techniques, yielded 4385 species, with 4029 species of insects (Monteith & Davies 1991). This represents a staggering 2.5% of the estimated Australian terrestrial arthropod fauna (Yeates *et al.* 2003). Fully 40% of the beetles were represented by single individuals, suggesting that many more species existed in the fauna and only 13.5% were named to species level. The number of species at each altitudinal site (100 m, 500 m, 1054 m, 1560 m) was constant at just above 1000 species, except for the 100 m altitudinal survey that yielded almost 1800 species. The Coleoptera (beetles) and Lepidoptera (moths and butterflies) from these surveys were well sorted and yielded 1514 species and 1345 species respectively.

In comparison, a year-long survey of saproxylic beetles at a lowland rainforest site in the Wet Tropics yielded 336 species, with an estimated 500–600 species making up the total fauna. Single individuals made up 43% of species (Grove 2000). The Bellenden Ker transect collected 533 beetles at the lowland 100 m site, but this total includes both saproxylic and non-saproxylic species. While diverse, the Wet Tropics beetle fauna is not as rich as Asian tropical faunas. A 500 hectare lowland site in Sulawesi, Indonesia, yielded 4532 beetle species alone, with an estimated fauna of 6200 species (Hammond 1990). Reliable comparative data are available for butterflies where the total fauna for north-east Queensland of 271 species (Braby 2004) is only one-third the fauna of Papua New Guinea (820 spp.) and close to one-quarter that of the Malay Peninsula (1031 spp.) (Parsons 1999).

While these surveys are not directly comparable, it is likely that the total species diversity of the Wet Tropics fauna is less than tropical Asian rainforests and probably less than rainforests in other regions. It may be that the invertebrate fauna has suffered the same loss of species diversity as the flora has over the past few million years (e.g. Greenwood & Cristophel 2005; Kershaw *et al.* 2005). In addition, the fauna may never have been as large as other rainforests, as it is basically a temperate cool-adapted fauna that has rafted into the tropics, with processes causing extinction dominating over recent speciation and immigration.

Two other less intensive altitudinal transect surveys in the Wet Tropics, one on the eastern face of the Carbine Tableland at Mossman Bluff in 1989/90 (Monteith 1989) and another on the eastern face of the Thornton Uplands at Cape Tribulation in 1982/3 (Monteith 1985) (Figure 13.2) can be used to place the Bellenden Ker transect in regional context. The Carbine transect is about 100 km north of the Bellenden Ker transect site with the Thornton transect site a further 40 km north from there. Sampling and sorting was not directly comparable to the Bellenden Ker survey, but numbers of species collected in the Hemiptera (bugs), and selected families of Coleoptera were comparable, except that there were 40% fewer weevils (Curculionidae) collected on the Thornton Uplands transect (86 versus 133). This is probably due to the remarkably diverse fauna of weevils at the very summit of Bellenden Ker. In the two families compared, only around 30% of species were shared between the Thornton and Bellenden Ker transects (Monteith 1985). A comparison of Scarabaeine beetles at the Carbine and Thornton transects shows that about 50% of species are shared, ranging from 60% in the lowland sites to 33% in the upland sites, with no wingless species shared between transects.

We lack the information required to convert site and transect diversity data into regional diversity estimates. Some insights can be gained from examining the species overlap between rainforest subregions (Figure 13.2b), and examining the pattern of species turnover with altitude within transect surveys.

Species turnover between upland blocks

We calculated species turnover values (Figure 13.2b) for the upland low-vagility insect guild for a subset of the upland subregions arrayed in a contiguous sequence from the north (Mt Finnigan) to the south (Paluma/Bluewater Range). Turnover is uniformly high, and over 90% for the most distant upland regions at the end of the latitudinal sequence. Turnover is least (over 55%) for the geographically adjacent Bellenden Ker and Walter Hill Uplands, and for the lower altitude, relatively species-poor Black Mountain and Kirrama/Cardwell Ranges (54%). The faunas of most pairs of upland regions are between 60 and 80% dissimilar, even when geographically adjacent. This illustrates the highly fragmented nature of diversity in the upland regions, with all areas harbouring a unique fauna of endemic or otherwise highly restricted low-vagility litter invertebrates.

Altitudinal turnover and zonation

Results of the Bellenden Ker transect study show that at each altitude, around 50% of the Coleoptera species were only found at that site, with the summit site having 57% endemicity influenced by a large proportion of weevils (Curculionidae) not found at other sites. Turnover between sites was large, from 80 to 95%, with only 0.5% of species common to all altitudes. Of the 314 spiders collected, 43% were site endemics, a slightly lower proportion than for the insects. Site endemicity for selected Hemiptera and Coleoptera on the Thornton transect varied between 8% and 45%, and was highest in the uplands (Monteith 1985). This transect had sites spaced by about 100 m in altitude, so we would expect site endemicity to be lower than Bellenden Ker, with 500 m altitude between sites. The faunal similarity (shared species) between adjacent sites varied between 32 and 58% for the Carbine transect, which also had 100 m altitude site spacing (Monteith 1989).

Two faunal changes occurred on each transect, a less distinctive transition from a lowland to a mid-altitude fauna, and a more marked transition from the mid-altitude fauna to a high altitude community. The precise altitude of the transitions varied depending on the overall altitude of the mountain massif, but the lower transition occurs between 200 and 300 m and the upper transition occurs between 500 and 700 m (Monteith 1985, 1989). On Mt Bellenden Ker, part of the highest range in the region at just over 1560 m, the lower transition occurred at about 500 m, and the upper transition at 1000 m (Monteith & Davies 1991), perhaps an example of the 'massenerhebung' or 'mountain mass' effect. The marked upper transition could be detected in the earthworm and march fly (Diptera: Tabanidae) fauna of the Thornton transect (Dyne 1985; Yeates 1985).

These transitions correspond to the change in rainforest types from complex mesophyll to simple mesophyll and to simple notophyll vine-fern thicket respectively (Webb & Tracey 1981). The higher transition corresponds with the transition from macrotherm to mesotherm bioclimates of Nix (1991). The high altitude fauna associated with the simple notophyll forest also corresponds with the lower limit of the cloud base. These upland areas are often shrouded in clouds and receive a considerable portion of their precipitation from cloud stripping (McJannet & Reddell 2004). An altitude between the lower and upper transition points usually has the highest diversity of insects, reflecting an overlap between the lowland and upland faunas, and the proportion of flightless species increases at the higher altitudes (Monteith 1985, 1989; Monteith & Davies 1991).

Endemism

Invertebrate endemism in rainforest subregions of the Wet Tropics was studied using a system of subregionalization developed for studying vertebrate distributions (Winter *et al.* 1984; Williams *et al.* 1996). This system divides the Wet Tropics into 17 subregions, with 14 of these restricted to the upland rainforest regions above 300 m altitude. These 14 upland subregions are slightly different to the 17 shown in Figure 13.2. Hann Tableland (Area 5) and Hinchinbrook Island (Area 15) are excluded; Walsh/Hugh Nelson Range (Area 8), the Atherton Tableland (Area 9) and Walter Hill Range (Area 12) are united into a single block called the Atherton Uplands; and the Paluma/Bluewater Ranges (Area 16) are divided into separate areas, called the Halifax Uplands and the Spec Uplands respectively.

Of the 274 flightless insect taxa restricted to the 14 Wet Tropics upland subregions, a total of 50% (137 spp.) were found to be restricted to a single subregion (Yeates *et al.* 2002; Figure 13.2). In comparison, only 15% of vertebrate species found in the Wet Tropics are subregional endemics (Williams *et al.* 1996; Williams & Pearson 1997). By far the largest numbers of endemics were found in the Bellenden Ker Upland (28 endemics) and associated Atherton Upland (17 endemics) south of the BMB, and in the Carbine Uplands (21 endemics) north of the BMB. These are likely to be the longest standing refugia, and are predicted to have been covered in rainforest even in the last glacial maximum 18 000 years BP (LGM), when the temperature was 3–4°C cooler and precipitation was half what it is today (Nix 1991; Hugall *et al.* 2002). Multiple regression analysis showed that the combination of rainforest area and shape explained the most variance in the

numbers of species of regional endemic insects, suggesting that these numbers were controlled by the size and persistence of refugial rainforest patches in the subregions. The numbers of subregional endemic insects were not explained by subregional area and shape, suggesting that these species may have evolved *in situ* (Yeates *et al.* 2002).

Refugia determined by the numbers of endemic invertebrates and palaeoclimatic modelling largely agree (Hugall *et al.* 2002; Yeates *et al.* 2002). Significant refugia are to be found in the Finnigan, Thornton and Carbine Uplands north of the BMB. South of the BMB, major refugial areas through the Pleistocene were predicted to occur in the Lamb Uplands, Malbon Thompson Uplands, Bellenden Ker and Atherton. South of these rainforest blocks, climate modelling predicts either very small or no refugia in the Kirrama uplands, Lee Uplands, Spec and Halifax Uplands. Of these areas, only the Kirrama Uplands has a significant number (five) of endemic invertebrates and modelling predicts a total of only 27 ha of refugium at LGM, compared to 59 000 ha today. It seems likely that the small dissected refugia in the Kirrama Uplands could support invertebrates but not vertebrates.

Recolonization history determined from vertebrate population genetic data largely agrees with that determined from the size and proximity of refugia, and shared invertebrate distributions (Schneider *et al.* 1998; Schneider & Moritz 1999; Yeates *et al.* 2002). The Black Mountain area was probably recolonized largely from the Carbine Uplands in the north, while the Lee, Spec and Elliot Uplands in the south were probably recolonized from Atherton and Kirrama to the north and the Elliot Uplands near Townsville in the south. However, one intriguing anomaly exists north of the BMB. Insect distribution data predict Windsor to have been recolonized since the LGM from Carbine (Yeates *et al.* 2002) but population genetic data on vertebrates and snails predict Windsor to have been recolonized from refugia on the Thornton Uplands (Schneider *et al.* 1998; Hugall *et al.* 2002). The Carbine and Finnigan Uplands are thought to have been more dissected refugia than Thornton, and less likely to be refugial areas for vertebrates and the snail (Hugall *et al.* 2002). However, based on the numbers of endemic insects in the Carbine Uplands (21) and Finnigan Uplands (15), these areas, as well as Thornton, were significant refugia for invertebrates, if not for vertebrates. It seems likely that insects were able to survive in these smaller rainforest patches through the LGM and the similar periods of cool dry climate throughout the later Tertiary and the Quaternary. The snail population genetic results are based on two individuals (Hugall *et al.* 2002). Perhaps further sampling in the Windsor Uplands will reveal haplotypes linking it with the Carbine Uplands. In any case, stochastic events probably play a major role in the assembly of recolonizing faunas in areas such as the Windsor Uplands, and the communities are independently and probably differently assembled in non-refugial areas after each climatic fluctuation.

Relationships and biogeography

Based on a synthesis of formal phylogenetic treatments, Bouchard *et al.* (2005) found the South Pacific (Gondwanan) track of Crisp *et al.* (1999) to be common, as well as the equatorial track (Younger Northern, Indo Malayan and Asian Tertiary components of other authors).

We have tabulated the biogeographical relationships of a number of genera of the low vagility guild (Table 13.1) in order to highlight some of the distributional disjunctions evident in the fauna. These relationships are based on the distributions of species in

Table 13.1 Some taxa of Wet Tropics insects showing occurrence in other regions.

Insect taxa	MAL	NG	CYP	WT	CQ	SQ-VIC	TAS	LHI	NC	NZ	SA	MAD
Coleopt: Carabid: *Feronista*				H	X				X			
Coleopt: Carabid: Migadopini				H		X	X			X	X	
Coleopt: Carabid: *Prosopogmus*				H	X	X			X			
Coleopt: Carabid: *Notonomus*				H	X	X	X		X			
Coleopt: Carabid: Pamborini				H	X	X				X		
Coleopt: Carabid: *Trichosternus*				H	X	X						
Coleopt: Carabid: *Mecyclothorax*	X	X		H	X	X	X	X	X	X		
Coleopt: Carabid: *Colasidia*	X	X		L								
Coleopt: Lucanid: *Sphaenognathus*				H	X						X	
Coleopt: Lucanid: *Lissapterus*				H		X						
Coleopt: Tenebrionid: *Licinoma*				H	X	X					X	
Coleopt: Tenebrionid: *Nolicima*				H		X	X					
Coleopt: Tenebrionid: *Pseudobyrsax gp.*				H					X			
Hemipt: Peloridiidae				H		X	X	X	X	X	X	
Hemipt: Cicadellid: Sagmatiini				H	X	X			X			X
Hemipt: Delphacid: *Notuchus*				H				X	X			
Hemipt: Aradid: Tretocorini				H		X				X		
Hemipt: Aradid: *Chinessa*		X	X	L								
Hemipt: Mirid: *Schizopteromiris*				H		X		X	X			
Hemipt: Mirid: *Austrovannius*				H/L					X			
Hemipt: Mirid: *Vanniusoides*		X		L					X			
Hemipt: Tessaratomid: *Oncomeris*		X	X	L								
Hemipt: Mesoveliid: *Austrovelia*				H					X			
Blattod: Blattid: *Tryonicus*				H	X	X			X			

MAL, Malesia; NG, New Guinea; CYP, Cape York Peninsula; WT, Wet Tropics: CQ, Central Queensland; SQ - VIC, South Queensland to Victoria; TAS, Tasmania; NC, New Caledonia; NZ, New Zealand; SA, South America; MAD, Madagascar. H, High altitude; L, Low altitude.

well-defined genera or other higher taxa, and formal quantitative biogeographical analyses have been completed on only a subset of these groups. It is clear that most high altitude rainforest groups from the Wet Tropics have relatives elsewhere in the mesotherm archipelago of Australia (see below). Relatively few have relationships with Cape York Peninsula rainforests or land masses further north. Wet Tropics relatives are also commonly found in Gondwanan land masses still supporting rainforests, such as Lord Howe Island, New Caledonia, New Zealand and southern South America. There is a strong relationship between insects of the Wet Tropics and New Caledonia, evident in analyses of the distributions of both Coleoptera and Hemiptera (Monteith 1980; Matthews 1998; Hamilton 1999). Both areas were adjacent in the geological past and both have suffered similar latitudinal changes from temperate to tropical climates. The cicadellid tribe Sagmatiini shows a presumably ancient relationship with New Caledonia and Madagascar. These groups are ripe for treatment using modern biogeographical methods.

Regional patterns within Australia

The vast majority of invertebrate phylogenies for genera containing species in the Wet Tropics suggest that their relatives are in rainforests further south on the Queensland or New South Wales coasts (Russell 1997; Bouchard 2002; Bell et al. 2003; Sota et al. 2005), and this is consistent with findings for vertebrates (Schneider et al. 1998). The Wet Tropics fauna shows little influence of species from New Guinea or further north in Australia, unlike the forests of the McIlwraith and Iron Ranges, which show much greater faunal overlap with New Guinea (Kikkawa et al. 1981; Monteith 1997).

Age and mode of speciation

Using a combination of phylogenetic reconstruction, a detailed knowledge of species distributions and molecular sequence divergences, it is possible to estimate the timing and mode of speciation in some cases, as has been done for vertebrates (Schneider et al. 1999; Moritz et al. 2000; Schneider & Williams 2005). *Pamborus* species (Coleoptera: Carabidae) diverged in the late Miocene and the species have differentiated and

expanded repeatedly in eastern Australia, while the Wet Tropics species continued to diverge through to the late Pliocene (Sota et al. 2005). Geographically proximal species of *Temnoplectron* dung beetles (Coleoptera: Scarabaeidae) are often sister species, suggesting that allopatric speciation has been common (Reid & Storey 2000; Bell et al. 2003) and that molecular divergence between species suggested that species were older than Pleistocene (Bell et al. 2003). Most speciation events in *Aellocoris* (Hemiptera: Aradidae) occurred in the mid-Miocene and vicariant allopatric speciation was common (Russell 1997).

Using a modified form of Brooks Parsimony Analysis on insect phylogenies, Bouchard et al. (2005) found that the biogeographical history of the Wet Tropics was complex, reflected in multiple area cladograms and sympatric, parapatric and allopatric speciation modes all having about the same frequency in Wet Tropics insects. Field and experimental evidence suggests that speciation by sympatry or parapatry in the Wet Tropics can occur quickly (Higgie et al. 2000; Hoskin et al. 2005), but few molecular divergences suggest Pleistocene speciation, whatever the mechanism. This suggests that the 100 000-year duration of the Pleistocene glacial periods (Kershaw et al. 2005) is probably too short to cause speciation by genetic drift of allopatric populations of widespread species, but it can cause marked geographical structuring of populations within species (Hugall et al. 2002).

Major threats to Wet Tropics invertebrates

Much lowland rainforest habitat and its associated invertebrates has been lost from the Wet Tropics due to former clearing for sugarcane cultivation. Similarly, upland rainforest has been cleared for cultivation and grazing, especially on the Atherton Tableland, but this has practically ceased since establishment of the Wet Tropics of Queensland World Heritage Area (WHA) in the region. Considering the large number of narrowly endemic invertebrates now known from individual mountain summits, the frequent disturbance of summit vegetation for construction of communication, navigation and survey structures that took place in the 1970s and 1980s posed a potential threat but this has declined since satellite-based systems have largely supplanted the need for ground-based installations.

Competition and predation by invasive animals are the clearest ongoing threats. Two insectivorous vertebrates, the cane toad, *Bufo marinus* (Linnaeus), and the Asian house gecko, *Hemidactylus frenatus* Dumáril & Bibron, are now established, though only the toad invades natural habitats, reaching pristine rainforest habitats at maximum elevations. Data for their effect on invertebrates in the region are not available.

Invasive ants pose probably the greatest real threat to invertebrate biodiversity in the Wet Tropics. Species already established in Australia that have been shown to inhibit indigenous invertebrates include: the coastal brown ant, *Pheidole megacephala* (Fabr.), which devastates other invertebrates in monsoon forests near Darwin (Hoffman *et al.* 1999); the yellow crazy ant, *Anoplolepis gracilipes* (F. Smith), which affects terrestrial crabs on Christmas Island (O'Dowd *et al.* 2003), and the red imported fire ant, *Solenopsis invicta* Buren; which impacts invertebrate biodiversity in the USA (Wojcik *et al.* 2001). Of these three, *P. megacephala* has been established in the Wet Tropics for many years but has not significantly invaded rainforest habitats, *A. gracilis* is not yet established but there have been several short-term colonizations and *S. invicta* exists in Australia to date only in the Brisbane area where eradication is being attempted. The greatest threat is perhaps posed by an ant that was discovered in Cairns in May 2006, the little fire ant, *Wasmannia auropunctata* (Roger). This minute Neotropical species is currently spreading rapidly through the Pacific (Jourdan *et al.* 2002). It occurs up to 900 m in New Caledonia, which has similar climate, topography and vegetation to the Wet Tropics, and has been shown to be severely impacting lizards (Jourdan *et al.* 2001), native ants (Le Breton *et al.* 2003) and other invertebrates (Monteith *et al.* 2006). If not eradicated or controlled its effect on the biota of the Wet Tropics and other parts of northern Australia could be catastrophic.

The giant African snail, *Achatina fulica* Bowditch, which is established on many Pacific islands, including nearby New Guinea, competes with native snails where it occurs (Godan 1983). It is a periodic quarantine interception in Cairns and needs constant vigilance to prevent establishment in the Wet Tropics.

Perhaps the greatest long-term threat to the biota of the Wet Tropics is climate change. As we have seen, may Wet Tropics biotic elements represent the last remnant of cool-adapted mesic communities that have survived as northern Australia rafted north into tropical latitudes over the past few tens of millions of years and as global climate changed to restrict areas suitable for these groups. The increases in temperature predicted to be associated with climate change over the next 100 years will greatly restrict or extinguish the environmental envelopes of many vertebrate species in the Wet Tropics, especially those restricted to the upland mesotherm archipelago (Williams *et al.* 2003; Thomas *et al.* 2004). Results from surveys of rainforest fly (Diptera) communities suggest that approximately 25% of species are restricted to upland zones (Wilson *et al.* 2007) and will face a similar threat to upland vertebrate communities. Diptera represent perhaps 10–15% of animal diversity (Yeates & Wiegmann 1999, 2005), so these results can probably be extrapolated to all insects and other invertebrates. Results of experimental studies suggest that adaptation to the changed conditions is unlikely for many species (Hoffmann *et al.* 2003), and thus extinction may be the only possible outcome.

Summary

- The invertebrate fauna of the Wet Tropics is diverse, but probably not as diverse as tropical rainforest faunas elsewhere in the world.
- Alpha and beta diversity in the region is high. Levels of faunal turnover between rainforest upland blocks is high, varying from 55 to 90%, increasing with geographical distance (Figure 13.2b). There is about 50% faunal turnover in every 500 m of altitudinal change.
- There tends to be a major transition between lowland and upland faunas between 500 and 700 m, and flightless species increase from fewer than half the species to more than half the species as one ascends from lowlands to highlands. Faunas of intermediate altitudes (500–600 m) tend to be richest, having a mixture of lowland specialists and upland specialists.
- Species in the Wet Tropics tend to have closest relatives in rainforest blocks further south on the east coast of Australia.
- Some groups have closest relatives outside Australia, and when this occurs they are on other Gondwanan continental islands or continents such as New Caledonia, Lord Howe Island, New Zealand

and Chile. The faunal link with New Caledonia is quite strong and may reflect the fact that both New Caledonia and the Wet Tropics have been adjacent in the past, and harbour faunas that have undergone the same ecological sieving of mesic lineages that adapted to tropical climates.

- Refugia and recolonization patterns for vertebrates and invertebrates are largely concordant, with the Carbine upland and Bellenden Ker upland being the major refugial foci north and south of the Black Mountain Barrier respectively. These areas also harbour large numbers of subregional endemic invertebrate species.

- It is likely that invertebrates can survive in much smaller refugia than vertebrates; hence Kirrama uplands was probably an additional refugial area for invertebrates but not vertebrates.

- Major threats to the invertebrate faunas of the Wet Tropics are similar to the threats facing other components of the ecosystem. European agriculture has caused the elimination of much of the invertebrate biota at lower elevations. Invasive species such as fire ants and snails are probably the most threatening and immediate cause for concern. Over the next century, climate change predictions suggest that preferred habitats for large numbers of high altitude specialists will be extinct in the Wet Tropics.

References

Adam, P. (1994). *Australian Rainforests*. Oxford University Press, Oxford.

Allsopp, P. G. (2000). Revision of the Australian genus *Anomalomorpha* Arrow (Coleoptera: Scarabaeidae: Dynastinae) a new species from the Wet Tropics of Queensland. *Memoirs of the Queensland Museum* **47**: 1–7.

Baehr, M. (1995). Revision of *Philipis* (Coleoptera: Carabidae: Bembidiinae), a genus of arboreal tachyine beetles from the rainforests of eastern Australia: taxonomy, phylogeny and biogeography. *Memoirs of the Queensland Museum* **38**: 315–81.

Baehr, M. (2003). Psydrine ground beetles (Coleoptera: Carabidae: Psydrinae), excluding Amblytelini, of eastern Australian rainforests. *Memoirs of the Queensland Museum* **49**: 65–109.

Ballantyne, L. A. & Lambkin, C. (2000). Lampyridae of Australia (Coleoptera: Lampyridae: Luciolinae: Luciolini). *Memoirs of the Queensland Museum* **46**: 15–93.

Bell, K. L., Yeates, D. K., Moritz, C. & Monteith, G. B. (2003). Molecular phylogeny and biogeography of the dung beetle genus *Temnoplectron* Westwood (Scarabaeidae: Scarabaeinae) from Australia's Wet Tropics. *Molecular Phylogenetics and Evolution* **41**: 741–53.

Bouchard, P. (2002). Phylogenetic revision of the flightless Australian genus *Apterotheca* Gebien (Coleoptera: Tenebrionidae: Coelometopinae). *Invertebrate Systematics* **16**: 449–554.

Bouchard, P., Brooks, D. R. & Yeates, D. (2005). Mosaic macroevolution in Australian Wet Tropics arthropods: community assemblage by taxon pulses. In *Tropical Rainforests: Past, Present and Future*, Moritz, C., Bermingham, E. & Dick, C. (eds). University of Chicago Press, Chicago.

Braby, M. F. (2004). The Complete Field Guide to Butterflies of Australia. CSIRO, Melbourne. 229 pp.

Brancucci, M. & Monteith, G. B. (1996). A second *Terradessus* species from Australia (Coleoptera, Dytiscidae). *Entomologica Basiliensia* **19**: 585–91.

Caterino, M. S. (2003). New species of *Chlamydopsis* (Histeridae: Chlamydopsinae), with a review and phylogenetic analysis of all known species. *Memoirs of the Queensland Museum* **49**: 159–235.

Crisp, M. D., Cook, L. & Steane, D. (2004). Radiation of the Australian flora: what can comparisons of molecular phylogenies across multiple taxa tell us about the evolution of diversity in present-day communities? *Philosophical Transactions of the Royal Society of London B* **359**: 1551–71.

Crisp, M. D., West, J. G. & Linder, H. P. (1999). Biogeography of the terrestrial flora. In *Flora of Australia, Vol. 1, Introduction*, 2nd edn, Orchard, A. E. & Thompson, H. S. (eds). ABRS/CSIRO Australia, Melbourne, pp. 321–67.

Crozier, R. H., Dunnett, L. J. & Agapow, P.-M. (2005). Phylogenetic biodiversity assessment based on systematic nomenclature. *Evolutionary Bioinformatics Online* **1**: 11–36.

Darington, P. J. (1961). Australian carabid beetles. V. Transition of wet forest faunas from New Guinea to Tasmania. *Psyche* **68**: 1–24.

Davies, V. E. & Lambkin, C. (2000). *Wabua*, a new spider genus (Araneae: Amaurobioidea: Kababininae) from north Queensland, Australia. *Memoirs of the Queensland Museum* **46**: 129–47.

Dyne, G. (1985). Altitudinal transect studies at Cape Tribulation, north Queensland VIII. Oligochaeta. *Queensland Naturalist* **26**: 81–4.

Edgecombe, G. D. (2001). Revision of *Paralamyctes* (Chilopoda: Lithobiomorpha), with six new species from eastern Australia. *Records of the Australian Museum* **53**: 201–41.

Evans, J. W. (1972). A new species of Peloridiidae (Homoptera, Coleorrhyncha) from north Queensland. *Proceedings of the Royal Society of Queensland* **83**: 83–8.

Forster, R. R., Platnick, N. I. & Gray, M. R. (1987). A review of the spider superfamilies Hypochiloidea and Austrochiloidea (Araneae, Araneomorphae). *Bulletin of the American Museum of Natural History* **185**: 1–116.

Frith, D. & Frith, C. (1990). Seasonality of litter invertebrate populations in an Australian upland tropical rain forest. *Biotropica* **22**: 181–90.

Godan, D. (1983). *Pest Slugs and Snails. Biology and Control.* Springer-Verlag, Berlin, 445 pp.

Greenwood, D. R. & Christophel, D. C. (2005). The origins and Tertiary history of Australian tropical rainforests. In *Tropical Rainforests: Past, Present and Future*, Moritz, C., Bermingham, E. & Dick, C. (eds). Chicago University Press, Chicago, pp. 336–73.

Grove, S. J. (2000). Impacts of forest management on saproxylic beetles in the Australian lowland tropics and the development of appropriate indicators of sustainable forest management. PhD thesis, James Cook University.

Hamilton, K. G. A. (1999). The ground dwelling leafhoppers Myerslopiidae, new family, and Sagmatiini, new tribe (Homoptera: Membracoidea). *Invertebrate Taxonomy* **13**: 207–35.

Hammond, P.M. (1990). Insect abundance and diversity in the Dumoga-Bone National Park, North Sulawesi, with special reference to the beetle fauna of lowland rainforest in the Toraut region. In *Insects and the Rain Forests of South East Asia (Wallacea)*, Knight, W. J. & Holloway, J. D. (eds). Royal Entomological Society of London, London, pp. 197–254.

Hammond, P. M. (1992). Species inventory. In *Global Biodiversity, Status of the Earth's Living Resources*, Groombridge, B. (ed.). Chapman & Hall, London, pp. 17–39.

Hammond, P. M., Stork, N. E. & Brendell, M. J. D. (1997). Tree-crown beetles in context: a comparison of canopy and other ecotone assemblages in a lowland tropical forest in Sulawesi. In *Canopy Arthropods*, Stork, N. E., Addis, J. & Didham, R. K. (eds). Chapman & Hall, London, pp. 184–223.

Harrington, G. N. & Sanderson, K. D. (1994). Recent contraction of wet sclerophyll forest in the wet tropics of Queensland due to invasion by rainforest. *Pacific Conservation Biology* **1**: 319–27.

Harrington, G. N., Thomas, M. R., Bradford, M. G., Sanderson, K. D. & Irvine, A. K. (1999). Structure and plant species dominance in North Queensland wet sclerophyll forests. *Proceedings of the Royal Society of Queensland* **109**: 59–74.

Harvey, M. S. (2000). A review of the Australian schizomid genus *Notozomus* (Hubbardiidae). *Memoirs of the Queensland Museum* **46**: 161–74.

Harvey, M. S. & West, P. J. (1998). New species of *Charon* (Amblypygi, Charontidae) from northern Australia and Christmas Island. *Journal of Arachnology* **26**: 273–84.

Higgie, M., Chenoweth, S. &. Blows, M. W. (2000). Natural selection and the reinforcement of mate recognition. *Science* **290**: 519–21.

Hoffman, B. D., Anderson, A. N. & Hill, G .J. E. (1999). Impact of an introduced ant on native rain forest invertebrates: *Pheidole megacephala* in monsoonal Australia. *Oecologia* **120**: 595–604.

Hoffmann, A. A., Hallas, R. J., Dean, J. A. & Schiffer, M. (2003). Low potential for climatic stress adaptation in a rainforest *Drosophila* species. *Science* **301**: 100–2.

Hoskin, C. J., Higgie, M., McDonald K. R. &. Moritz, C. (2005). Reinforcement drives rapid allopatric speciation. *Nature* **437**: 1353–5.

Howden, H. F. & Storey, R. I. (2000). New Stereomerini and Rhyparini from Australia, Borneo and Fiji (Coleoptera: Scarabaeidae: Aphodiinae). *Memoirs of the Queensland Museum* **46**: 175–82.

Hugall, A., Moritz, C., Moussalli, A. & Stanisic, J. (2002). Reconciling paleodistribution models and comparative phylogeography in the Wet Tropics rainforest land snail *Gnarosophia bellendenkerensis* (Brazier 1975). *Proceedings of the National Academy of Science USA* **99**: 6112–17.

Jourdan, H., Bonnet de Larbogne, L. & Chazeau, J. (2002). The recent introduction of the Neotropical tramp ant *Wasmannia auropunctata* (Hymenoptera: Formicidae) into Vanuatu Archipelago (southwest Pacific). *Sociobiology* **40**: 483–509.

Jourdan, H., Sadlier, R. A. & Bauer, A. M. (2001). Little fire ant invasion (*Wasmannia auropunctata*) as a threat to New Caledonian lizards: evidences from a sclerophyll forest (Hymenoptera: Formicidae). *Sociobiology* **38**: 283–301.

Kershaw, A. P., Moss, P.T. & Wild, R. (2005). Patterns and causes of vegetation change in the Australian Wet Tropics Region over the last 10 million years. In *Tropical Rainforests: Past, Present & Future*, Moritz, C., Bermingham, E. & Dick, C. (eds). Chicago University Press, Chicago, pp. 374–400.

Keto, A. & Scott, K. (1986). *Tropical Rainforests of North Queensland: Their Conservation Significance.* Australian Government Publishing Service, Canberra.

Kikkawa, J., Monteith, G. B. & Ingram, G. (1981). Cape York Peninsula – the major region of faunal interchange. In *Ecological Biogeography in Australia*, Keast, A. (ed.). Junk, The Hague, pp. 1695–742.

Le Breton, J., Chazeau, J. & Jourdan, H. (2003). Immediate impacts of invasion by *Wasmannia auropunctata* (Hymenoptera: Formicidae) on native litter ant fauna in a New Caledonian rainforest. *Austral Ecology* **28**: 204–9.

McJannet, D. L. & Reddell, P. (2004). Water balance of mountain and coastal rainforests in the Wet Tropics of northern Australia and effects on conversion to pasture. In *Mountains in the Mist: Science for Conserving and Managing*

Tropical Mountain Cloud Forest. University of Hawaii Press, Honolulu.

Malipatil, M. B. & Monteith, G. B. (1983). One new genus and four new species of terrestrial Mesoveliidae (Hemiptera: Gerromorpha) from Australia and New Caledonia. *Australian Journal of Zoology* **31**: 943–55.

Matthews, E. G. (1998). Classification, phylogeny and biogeography of the genera of Adeliini (Coleoptera: Tenebrionidae). *Invertebrate Taxonomy* **12**: 685–824.

Monod, L. & Volschenk, E. S. (2004). *Liocheles litodactylus* (Scorpiones: Liochelidae): an unusual new *Liocheles* species from the Australian Wet Tropics (Queensland). *Memoirs of the Queensland Museum* **49**: 675–90.

Monteith, G. B. (1980). Relationships of the genera of Chinamyersiinae (Hemiptera: Aradidae), with description of a relict species from the mountains of North Queensland. *Pacific Insects* **21**: 275–85.

Monteith, G. B. (1985). Altitudinal transect studies at Cape Tribulation, north Queensland VII. Coleoptera and Hemiptera (Insecta). *Queensland Naturalist* **26**: 70–80.

Monteith, G. B. (1989). Entomology report. In *Expedition Devils Thumb, Mount Windsor and Mount Carbine Tableland, Queensland*, Burch, G. J. (ed.). Australia and New Zealand Scientific Exploration Society, Melbourne, pp. 82–96.

Monteith, G. B. (1995). *Distribution and Altitudinal Zonation of Low Vagility Insects of the Queensland Wet Tropics.* Report to the Wet Tropics Management Authority (Part 4). Queensland Museum, Brisbane. 120 pp.

Monteith, G. B. (1997). Revision of the Australian flat bugs of the subfamily Mezirinae (Insecta: Hemiptera: Aradidae). *Memoirs of the Queensland Museum* **41**: 1–169.

Monteith, G. B. (Ed) (2000). Invertebrates of the Australian Wet Tropics. *Memoirs of the Queensland Museum* 46(1), 358 pp.

Monteith, G. B, Burwell, C. J. & Wright, S. (2006). *Inventaire de l'entomofaune de la forêt humide de quatre ráserves spáciales botaniques du Grand Sud de la Nouvelle Caládonie.* Report to the Direction des Ressources Naturelles, Noumea. Queensland Museum, Brisbane.

Monteith, G. B. & Davies, V. T. (1991). Preliminary account of a survey of arthropods (insects and spiders) along an altitudinal rainforest transect in tropical Queensland. In *The Rainforest Legacy: Australian National Rainforest Study. Volume 2*, Special Australian Heritage Publication Series, Werren, G. & Kershaw, P. (eds). Australian Government Publication Service, Canberra, pp. 345–62.

Monteith, G. B. & Field, L. H. (2001). Australian king crickets: distribution, habitats and biology (Orthoptera: Anostostomatidae) pp. 79–94. In *The Biology of Wetas, King Crickets and Their Allies*, Field, L. H. (ed.). CABI Publishing, Walingford, UK. 540 pp.

Moritz, C. (2005). Overview: rainforest history and dynamics in the Australian Wet Tropics. In *Tropical Rainforests: Past, Present & Future*, Moritz, C., Bermingham, E. & Dick, C. (eds). Chicago University Press, Chicago, pp. 313–21.

Moritz, C., Patton, J. L., Schneider, C. J., & Smith, T. B. (2000). Diversification of rainforest faunas: an integrated molecular approach. *Annual Review of Ecology and Systematics* **31**: 533–63.

Moritz, C., Richardson, K. S., Ferrier, S., Monteith, G. B., Williams, S. E. & Whiffen, T. (2001). Biogeographic concordance and efficiency of taxon indicators for establishing conservation priority in a tropical rainforest biota. *Proceedings of the Royal Society of London B* **268**: 1–7.

Myers, N. (1986). Tropical deforestation and a megaextinction spasm. In *Conservation Biology: The Science of Scarcity and Diversity*, Soule, M. E. (ed.). Sinauer, Sunderland, pp. 394–409.

Nix, H. A. (1991). Biogeography: patterns and process. In *Rainforest Animals: Atlas of Vertebrates Endemic to Australia's Wet Tropics*, Nix, H. A. & Switzer, M. A. (eds). Australian National Parks and Wildlife Service, Canberra.

O'Dowd, D. J., Green, P. T. & Lake, P. S. (2003). Invasional meltdown on an oceanic island. *Ecology Letters* **6**: 812–17.

Parsons, M. J. (1999). *The Butterflies of Papua New Guinea. Their Systematics and Biology.* Academic Press, London. 736 pp. plus 136 plates.

Raven, R. J. (2000). A new species of funnel-web spider (*Hadronyche*: Hexathelidae: Mygalomorphae) from north Queensland. *Memoirs of the Queensland Museum* **46**: 225–30.

Reid, A. (1996). Review of the Peripatopsidae (Onychophora) in Australia, with comments on peripatopsid relationships. *Invertebrate Taxonomy* **10**: 663–936.

Reid, C. A. M. & Storey, R. I. (2000). Revision of the dung beetle genus *Temnoplectron* (Coleoptera: Scarabaeidae: Scarabaeini). *Memoirs of the Queensland Museum* **46**: 253–97.

Russell, B. L. (1997). Systematics of an aradid flat bug, *Aellocoris kormilev*, inhabiting rainforest blocks in the Wet Tropics. Honours thesis, University of Queensland.

Schneider, C. J. & Williams, S. E. (2005). Effects of Quaternary climate change on rainforest diversity: insights from spatial analyses of species and genes in Australia's Wet Tropics. In *Tropical Rainforests: Past, Present & Future*, Moritz, C., Bermingham, E. & Dick, C. (eds). Chicago University Press, Chicago, pp. 401–24.

Schneider, C. J., Cunningham, M. & Moritz, C. (1998). Comparative phylogeography and the history of endemic vertebrates in the Wet Tropics rainforests of Australia. *Molecular Ecology* **7**: 487–98.

Schneider, C. and Moritz, C. (1999). Rainforest refugia and evolution in Australia's Wet Tropics. *Proceedings of the Royal Society of London B* **266**: 191–96.

Schneider, C. J., Smith, T. B., Larison, B. & Moritz, C. (1999). A test of alternative models of diversification in tropical rainforests: ecological gradients vs. rainforest refugia. *Proceedings of the National Academy of Sciences USA* **96**: 13869–73.

Sota, T., Takami, Y., Monteith, G. B. & Moore, B. P. (2005). Phylogeny and character evolution of endemic Australian carabid beetles of the genus *Pamborus* based on mitochondrial and nuclear gene sequences. *Molecular Phylogenetics and Evolution* **36**: 392–404.

Stanisic, J. (2000). Taxonomy of the Australian rainforest snail, *Helix bellendenkerensis* Brazier, 1975 (Mollusca: Eupulmonata: Camaenidae). *Memoirs of the Queensland Museum* **46**: 337–48.

Storey, R. I. & Monteith, G. B. (2000). Five new species of *Aptenocanthon* Matthews (Coleoptera: Scarabaeidae: Scarabaeinae) from tropical Australia, with notes on distribution. *Memoirs of the Queensland Museum* **46**(1): 349–58.

Storey, R. I. & Weir, T.A. (1990). New species of *Onthophagus* Latreille (Coleoptera: Scarabaeidae) from Australia. *Invertebrate Taxonomy* **3**: 783–815.

Stork, N. E. (1988). Insect diversity – facts, fiction and speculation. *Biological Journal of the Linnean Society* **35**: 321–37.

Stork, N. E. & Brendell, M. J. D. (1990). Variation in the insect fauna of Sulawesi trees with season, altitude and forest type. In *Insects and the Rain Forests of South East Asia (Wallacea)*, Knight, W. J. & Holloway, J. D. (eds). Royal Entomological Society of London, London, pp. 173–90.

Stork, N. E. & Grimbacher, P. S. (2006). Beetle assemblages from an Australian tropical rainforest show that the canopy and the ground strata contribute equally to biodiversity. *Proceedings of the Royal Society of London B* **273**: 1969–75.

Thomas, C. D., Cameron, A., Green, R. E. *et al.* (2004). Extinction risk from climate change. *Nature* **427**: 145–8.

Watson, J. A. L. (1982). A truly terrestrial dragonfly larva from Australia (Odonata: Corduliidae). *Journal of the Australian Entomological Society* **21**: 309–11.

Webb, L. J. & Tracey, J. G. (1981). Australian rainforests; pattern and change. In *Ecological Biogeography of Australia*, Keast, J. A. (ed.). W. Junk, The Hague, pp. 606–94.

Williams, S. E., Bolitho, E. E. &. Fox, S. (2003). Climate change in Australian tropical rainforests: an impending environmental catastrophe. *Proceedings of the Royal Society of London B* **270**: 1887–92.

Williams, S. E. & Pearson, R. G. (1997). Historical rainforest contractions, localized extinctions and patterns of vertebrate endemism in the rainforests of Australia's Wet Tropics. *Proceedings of the Royal Society of London. Series B* **264**: 709–16.

Williams, S. E., Pearson, R. G. & Walsh, P.J. (1996). Distribution and biodiversity of the terrestrial vertebrates of Australia's Wet Tropics: a review of current knowledge. *Pacific Conservation Biology* **2**: 327–62.

Wilson, E. O. (1988). *Biodiversity*. National Academy Press, Washington, DC.

Wilson, R., Williams, S. E., Trueman, J. W. H. & Yeates, D. K. (2007). Communities of diptera in the Wet Tropics of North Queensland are vulnerable to climate Change. *Biodiversity and Conservation* **16**: 3163–77.

Winter, J. W., Bell, F. C., Pahl, L. I. & Atherton, R. G. (1984). *The Specific Habitats of Selected Northeastern Australian Rainforest Mammals*. Report to the World Wide Fund for Nature, Sydney.

Wojcik, D. P., Allen, C. R., Brenner, R. J., Forys, E. A., Jouvenaz, D. P. & Lutz, R. S. (2001). Red imported fire ants: impact on biodiversity. *American Entomologist* **47**: 16–23.

Yeates, D. K. (1985). Altitudinal transect studies at Cape Tribulation, North Queensland IV. The march flies (Diptera: Tabanidae). *Queensland Naturalist* **26**: 58–61.

Yeates, D. K., Bouchard, P. & Monteith, G. B. (2002). Patterns and levels of endemism in the Australian Wet Tropics rainforest: evidence from flightless insects. *Invertebrate Systematics* **16**: 605–19.

Yeates, D. K., Harvey, M. S. & Austin, A. D. (2003). New estimates for terrestrial arthropod species-richness in Australia. *Records of the South Australian Museum Monograph Series* **7**: 231–41.

Yeates, D. K. & Wiegmann, B. M. (1999). Congruence and Controversy: towards a higher-level phylogeny of Diptera. *Annual Review of Entomology* **44**: 397–428.

Yeates, D. K. & Wiegmann, B. M. (2005). *The Evolutionary Biology of Flies*. New York: Columbia University Press.

14 INTERNATIONAL PERSPECTIVE: THE FUTURE OF BIODIVERSITY IN THE WET TROPICS

*Jiro Kikkawa**

School of Integrative Biology, University of Queensland, Australia
*The author was participant of Cooperative Research Centre for Tropical Ecology and Management.

The Australian continent, since its separation from Antarctica in the early Tertiary, carried an isolated and diversifying Gondwanan flora and fauna during its northward drift. The collision of the Australian plate with the Southeast Asian plate in the Miocene created opportunities for the exchange of the biota between the Oriental Region and the Australian Region across Wallace's Line. However, for many groups of plants and animals there was little interchange between the two regions. For example, there are no dipterocarps, monkeys or woodpeckers in the Australian rainforests, while the Southeast Asian rainforests have no araucaria, marsupials or birds of paradise. There are several reasons for this, but the most important factor that has kept the two biotas separate is the climate. The region of Wallacea and northern Australia became semi-arid with only limited precipitation during the monsoon season, which was not suitable for the dispersal of the wet-adapted tropical Indo-Malesian flora and fauna to Australasia.

The Australian continent, originally covered with humid forests, started to lose rainforests of Gondwanan origin with increasing periods of aridity since the late Tertiary. A lowering of the mean annual temperature and leaching of the soil has also occurred. The consequence of all this was the isolation of tropical rainforests in refugia and a loss of many taxa adapted to humid tropical regions with nutrient-rich soil. Many of them had slow dispersal rates and were vulnerable to changes of environmental conditions.

In the wet periods large refugia may have acted as a minor epicentre of evolution permitting evolutionary divergence of wet-adapted forms in isolation. In the increasingly severe and prolonged dry periods, however, many wet-adapted forms would have become extinct, while dry-adapted forms in the wet periods would have survived and actively speciated by dispersing to suitable habitats over a much greater area than the refugia of the wet-adapted forms in the dry periods.

Aboriginal use of fire may have prevented the shifting of rainforest refugia by defining the boundaries, but the clearing of the forest since European settlement produced by far the greatest impact in the Wet Tropics, almost entirely eliminating the lowland rainforests on fertile soil and further fragmenting other rainforests.

Despite the great damage the rainforests of north Queensland suffered in the past 150 years, their remnants in the Wet Tropics are still the largest rainforest refugia in Australia. They carry ancient flowering plants, characteristic of the Australasian part of the forest of Gondwanaland. In just 0.1% of Australia's land area the Wet Tropics contain 56 endemic species of vertebrate animals. In fact, the vertebrate fauna of the Wet Tropics not only includes this extraordinary diversity of unique species but also one-third of the Australian species of marsupials and frogs, a quarter of Australian reptiles, a fifth of Australian birds and two-thirds of Australian bats. Thus the Wet Tropics of north Queensland had sufficient reason to be listed as a World Heritage area and was designated as such in 1988 (see Valentine & Hill, Chapter 6, this volume).

The work of the Rainforest CRC has greatly enhanced the World Heritage value of the region through the detailed study of flora and fauna and the ecosystems of the Wet Tropics. As summarized by Metcalfe and Ford (Chapter 9, this volume), we now know, for example, that the Wet Tropics rainforests contain 38 monotypic plant genera endemic to the region and a further 24 monotypic plant genera with a broader distribution. The survival of some primitive angiosperm families with many species is staggering. Of the 28 primitive dicotyledon families known in the world, 13, including one endemic family, are represented in the region. Here, beetles are implicated as the pollinators of some of the primitive flowering plants. The fact that many rainforest refugia acted as epicentres of evolution is evident in the chapter by Yeates and Monteith (Chapter 13, this volume), which shows 50% of 274 flightless insect taxa, with low-vagility ancestors of Gondwanan origin, to be restricted to single refugia within the Wet Tropics. The freshwater fauna of the Wet Tropics was little known until recently, but extensive surveys and ecological studies since the World Heritage listing (see Pusey *et al.*, Chapter 11, and Connolly *et al.*, Chapter 12, this volume) have demonstrated highly diverse and distinctive freshwater fish and stream invertebrate fauna in the region. Many of these groups are also of Gondwanan origin, and their high species diversity is largely due to speciation within the region rather than reflecting the relict status of wide ranging species.

It is interesting that the studies of indigenous culture in the Wet Tropics have also revealed relatively isolated human cultures within the region. This is reported by Pannell (Chapter 4, this volume) in her studies of different language groups and material culture. This shows that, unlike other environments of Australia, the rainforest environment could provide a stable supply of subsistence commodities locally in normal

years and permitted isolation of populations in relatively small areas for both people and animals. This would encourage the development of more or less closed ecosystems. Within the biotic community strong interdependent interactions include pollination, seed dispersal, parasitic relations and dynamic adjustments following disturbances. The influences of invasive exotic species on communities are better understood today in this context.

Of the natural disturbances tropical cyclones exert the greatest damage to the rainforest, but their impact is usually localized as shown by Turton and Stork (Chapter 3, this volume) and are unlikely to cause extinction of even the most endangered species in the rainforest. The rainfall patterns associated with a summer monsoon and tropical cyclones, as described by Bonell and Callaghan (Chapter 2, this volume), are responsible for the meteorological characteristics of the Wet Tropics, with strong run-off and forest landscapes featuring water in the rugged terrain. The sustainable nature-based tourism is therefore difficult to develop in this region. Pearce's research target (Chapter 7, this volume) is to develop community support and market analysis that will keep the impact on the rainforest flora and fauna minimal.

Thus we know today much more about the heritage values of the rainforest of the Wet Tropics and the management tools for their continued existence. What we do not know is the likely response of the diverse flora and fauna and ecosystems of the Wet Tropics to the rapid climatic change expected in the twenty-first century.

The ancient rainforest that has survived the climatic changes of the post-Miocene Australia and the post-Pleistocene sea level changes is facing an unprecedented threat today. Many groups of plants and animals during this period differentiated within the refugia, but the evolutionary setting, which once permitted migration, settlement and speciation for these organisms, no longer exists. This is because European settlement produced much greater fragmentation of the remaining patches of rainforest than any natural fragmentation in the past has ever produced (Turton, Chapter 5, this volume), making today's biota much more vulnerable to climate change than the worst scenario in the past. The past climate change eliminated large carnivores, large seed predators and dispersers in many parts of the tropics. The survivors in the rainforest today were obviously resilient to Pleistocene climate change. However, the rainforest community is still adjusting to the most recent climatic episode. It tends to have species with narrowly restricted niches, narrow altitudinal ranges, specific moisture requirements and, in general, specialists with specific food-web interactions.

Climate change biology, recording phenological shift of plants and animals in response to climate change, will uncover changes in the distribution and migratory habits of animals as well as the flowering and fruiting seasons of plants in response to global warming in temperate regions. In the tropics, however, new adaptive changes in the rainforest flora and fauna may be quite different (Lovejoy & Hannah 2005). In the Wet Tropics where the rainforest fragments are now very small and much more isolated than in the past,

many species will face greater chances of extinction. Hilbert's model shows that each major rainforest type would respond uniquely to climate change (Chapter 8, this volume). This natural experiment will also impact directly or indirectly on the genetic material of the organisms involved (Williams *et al.*, Chapter 10, this volume). Hence biodiversity research in the Wet Tropics inevitably involves molecular aspects of evolution. The chapters in this part point to the need for an integrated approach to the conservation of biodiversity in the Wet Tropics under climate change.

Reference

Lovejoy, T. E. and Hannah, L. (eds) (2005). *Climate Change and Biodiversity*. Yale University Press, New Haven, CT. 418 pp.

Part 2

Ecological Processes and Other Ecosystem Services

INTRODUCTION

Nigel E. Stork[1]* and Stephen M. Turton[2]*

[1]School of Resource Management and Geography, Faculty of Land and Food Resources, University of Melbourne, Burnley Campus, Richmond, Victoria, Australia
[2]Australian Tropical Forest Institute, School of Earth and Environmental Sciences, James Cook University, Cairns, Queensland, Australia
*The authors were participants of Cooperative Research Centre for Tropical Ecology and Management

In recent years there has been greater attention on the ecological processes and ecosystem services and values that tropical rainforests provide and how essential these are to the sustainability of this and other ecosystems, to climate and to people. In this part we examine some of these with particular attention to the Wet Tropics of Australia. McJannet *et al.* (Chapter 15) examine the importance of hydrological processes in tropical forests, with new information for both upland and lowland forests. The diversity and structure of plant communities are maintained by a number of processes, two of which are seed dispersal and pollination, and these are addressed at the landscape level by Westcott (Chapter 16) and Boulter *et al.* (Chapter 17), respectively. Both provide overview models of these systems. Insects are particularly important for pollination and in Chapter 18 Cunningham and Blanche look at the services and disservices that insects from tropical rainforests provide to surrounding agricultural landscapes. A few years ago Costanza *et al.* (1997) shocked the scientific community with their assessment that the world's ecosystem services were valued at an average of US\$33 trillion per year, almost twice the global gross national product (around US\$18 trillion per year). Curtis (Chapter 19) provides an assessment of different economic approaches to the value of tropical forests, again with a focus on the Wet Tropics. In the final chapter (Chapter 20), Joe Wright uses his knowledge and experience of forest processes to provide a perspective on this section.

Reference

Costanza, R., D'Arge, R., De Groot, R. *et al.* (1997). The value of the world's ecosystem services and natural capital. *Nature* **387**: 253–60.

15 HYDROLOGICAL PROCESSES IN THE TROPICAL RAINFORESTS OF AUSTRALIA

David McJannet[1]*, Jim Wallace[2]*, Peter Fitch[3]*,
Mark Disher[2]* and Paul Reddell[4]*

[1]CSIRO Land and Water, 120 Meiers Rd, Indooroopilly, Queensland, Australia
[2]CSIRO Land and Water, Davies Laboratory, Townsville, Queensland, Australia
[3]CSIRO Land and Water, Christian Laboratory, Black Mountain, ACT, Australia
[4]CSIRO Land and Water, Tropical Forest Research Centre, Atherton, Queensland, Australia
*The authors were participants of Cooperative Research Centre for Tropical Ecology and Management.

Introduction

The Wet Tropics bioregion of north Queensland covers an area of 9000 km[2] on the coastal strip adjacent to the Great Barrier Reef and contains the vast majority of the tropical rainforests of Australia. This region has high economic importance and exceptional environmental value, both of which need to be protected through well informed natural resource management and planning. An understanding of the hydrology of different rainforest types and the effects of changes in vegetation type on catchment response, water quality and water yield is vital in developing a more predictive understanding of the effects of possible land use and climate changes on water resources in the region (see Bonell & Callaghan, Chapter 2, this volume, for an overview of climate and hydrology).

In order to understand the hydrology of rainforests it is essential to understand the processes controlling their water balance. The water balance of a rainforest can be expressed by

$$(P_{ga} + P_c) = I + T + E_s + R + D + \delta\theta \qquad (15.1)$$

where P_{ga} and P_c are rainfall and cloud interception inputs to the forest, I is canopy interception, which is calculated as the difference between precipitation and the sum of throughfall (T_f) and stemflow (S_f). T is transpiration by the forest and E_s is forest floor evaporation. R and D represent the processes of runoff (i.e. lateral flows whether overland or subsurface) and recharge. Over annual periods, when the change in soil water storage ($\delta\theta$) is small, the combined sum of R and D is given by the difference between the amount of water entering the soil ($S_f + T_f$) and ($T + E_s$). In most hydrological studies in the tropics, precipitation inputs are generally derived from rainfall measurements (P_g) alone; however, at higher altitude exposed sites rainforests can receive significant contributions of water from three additional sources:

1 Wind blown rain being intercepted on sloping land (e.g. Sharon 1980; de Lima 1990; Holwerda 2005);

2 rain gauge wind-losses due to turbulence around the gauge (Førland et al. 1996);

3 cloud or fog droplets collecting on vegetation surfaces (e.g. Clark et al., 1998; Holder 2003; Bruijnzeel 2005).

P_{ga} represents rainfall adjusted for wind-losses and slope effects, while P_c refers to cloud interception. Over annual periods when the change in soil water storage ($\delta\theta$) is small, ($R + D$) is given by the difference between ($P_{ga} + P_c$) and evaporative losses through I, T

and E_s, which are collectively referred to as evapotranspiration (*ET*). Figure 15.1 provides a graphical overview of the water balance for a rainforest based on the processes described in Equation 1.

Hydrological research in Australian tropical rainforests is limited to a few studies: that of Gilmour (1975), who undertook water balance measurement and modelling at a coastal rainforest location; a study by Hutley *et al.* (1997), who measured the water balance in a subtropical forest in south-east Queensland; and the investigations of Herwitz (1985, 1986, 1987), who investigated interception storage capacities of rainforest trees, water flow on vegetative surfaces and surface runoff in relation to stemflow. These studies form a good basis for understanding the water balance of tropical rainforests in Australia, but research from around the world suggests that different rainforest types exhibit different water balances owing to rainfall variation and forest structural differences (Tracey 1982). To improve our understanding of the factors influencing the water

balance of different rainforest types we have implemented water balance monitoring projects at six locations across the Wet Tropics. The aim of this chapter is to summarize the findings from this recent research and to place the findings in an international context.

In this chapter we summarize data and methods used to measure the water balance components of rainforest sites from the coastal lowlands to the highest mountains in the region. Further details of the sites, methods and water balance results are given by McJannet *et al.* (2006 a–d).

Field site characteristics

Water balance components reported here are from four field sites that were selected to represent different forest types, geologies and altitudes found in this region. The locations of these four field sites in the Wet Tropics of northern Queensland are shown in Figure 15.2. Full characteristics of each of these sites are shown in Table 15.1. The four forest types are classified using the Australian tropical rainforest classification technique of Tracey (1982).

The site at Oliver Creek (OC) is a lowland rainforest with an easterly aspect located 1 kilometre from the coast. It is within the Daintree National Park at an altitude of 30 metres. This site is classified as a complex

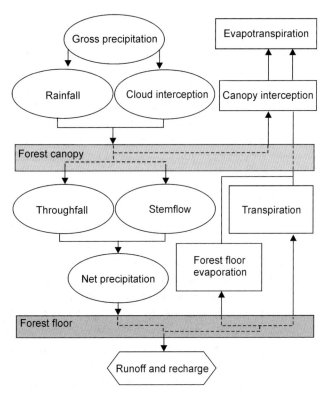

Figure 15.1 Forest water balance flow chart. Precipitation input pathways are shown in ellipses, evaporative losses are in rectangles and runoff/recharge losses are shown in a hexagon.

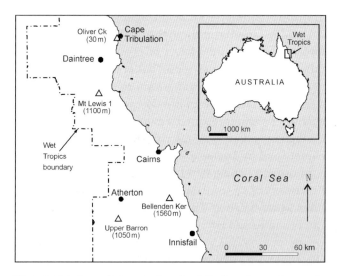

Figure 15.2 Location map showing Wet Tropics region and location of rainforest field sites (Δ). Altitudes of field sites are given in brackets. *Source*: after McJannet *et al.* (2006b).

Table 15.1 Location, altitude, aspect, slope, geology, forest type, plant species count (within 600 m^2 plots), LAI, canopy gap, stem density, basal area, average canopy height and long-term average annual rainfall from Bureau of Meteorology rain gauges (where available) (after McJannet *et al.* (2006b))

Site code	Lat. (S)	Long. (E)	Alt. (m)	Aspect	Slope (%)	Geology	Forest type	Plant species count	LAI (m^2 m^{-2})	Canopy gap (%)	Stem density (stems ha^{-1})	Basal area (m^2 ha^{-1})	Average canopy height (m)	Long-term average annual rain (mm)
OC	16°08.3	145°2 6.4	30	SE	4	Mud-stone	Complex mesophyll vine forest	141	4.2	3.5	644	64	27	3952
UB	17° 27.1	145° 29.7	1050	S	15	Basalt	Complex notophyll vine forest	128	4.1	4.0	925	69	25	n.a.
ML1	16° 31.7	145° 16.7	1100	SE	20	Granite	Simple notophyll vine forest	112	4.5	2.8	650	62	32	n.a.
BK	17° 16.0	145° 51.0	1560	E	10	Granite	Simple microphyll vine fern thicket	49	3.3	6.4	2019	74	8	8100

Source: McJannet et al. (2006).

mesophyll vine forest and has a leaf area index (LAI) of 4.2. The Upper Barron (UB) is a lower montane rainforest with a southerly aspect located 60 km from the coast on the Atherton Tablelands in Longlands Gap State Forest (1050 m). The forest type is complex notophyll vine forest with a LAI of 4.1. The Mount Lewis (ML1) site, which is also a lower montane rainforest, is one of two sites located on the Carbine Tablelands 25 km from the coast in the World Heritage listed Mt Lewis State Forest at an altitude of 1100 m. This site has a south-easterly aspect and is located in a sheltered valley. This simple notophyll vine forest has a LAI of 4.5. The highest altitude site is located within the Wooroonooran National Park at Bellenden Ker (BK) at 1560 m. This montane cloud forest is characterized by stunted trees (average canopy height of 8 m) with very few large diameter trees. This simple microphyll vine fern thicket forest has an easterly aspect and a LAI of 3.3, and is located 15 km from the coast. This site has the highest recorded average annual rainfall in Australia (>8000 mm). All sites are subject to a distinct wet and dry season. The dry season covers the months from April to October, while the wet season generally occurs from November to March. The predominant wind direction in this region is from the south-east.

Measurement methods

A concise description of the water balance measurements methods used is given below; full details can be found in McJannet *et al.* (2006a, c).

Rainfall and cloud interception

Total precipitation inputs at each site were determined from the combination of rainfall adjusted for slope effects and wind losses (P_{ga}) and cloud interception (P_c). Adjustments to rainfall were based on the methodology described by Holwerda (2005), which utilizes measurements of rainfall rate, hill slope angle, and wind speed and direction. Rainfall was measured using tipping bucket rain gauges located in clearings adjacent to the forests. P_c estimates were made using the canopy water balance methodology described in McJannet *et al.* (2006c). This method uses measurements of P_{ga}, throughfall (T_f) and stemflow (S_f) on days with and without cloud interception. T_f and S_f measurements are described in the canopy interception section below. The occurrence (not the amount) of cloud interception was determined using a gauge as shown in Figure 15.3. The gauge collects passing cloud droplets on a mesh screen, which then drains to a tipping

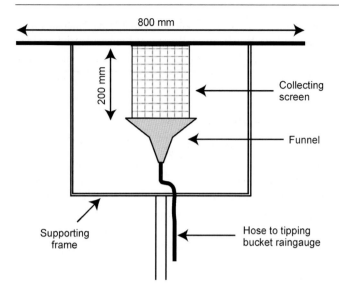

Figure 15.3 Gauge used for detecting the occurrence of cloud interception. The collecting screen is louvred screen (Kaiser Shadescreen, Alabama) with seven louvres per centimetre, each 23.1 mm long by 1.4 mm wide, angled at 17°. *Source*: after McJannet *et al.* (2006c).

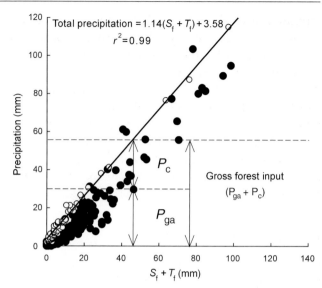

Figure 15.4 Canopy water balance methodology used to calculate cloud interception at Upper Barron. The regression line has been developed from days with no cloud interception. Days with cloud interception are shown as black circles. *Source*: after McJannet *et al.* (2006c).

bucket rain gauge. On days when no cloud interception was recorded by this gauge, a linear regression relationship was developed at each site that predicts the amount of precipitation required to produce the measured $(S_f + T_f)$. On days when cloud interception occurred, P_c was then estimated by subtracting the measured precipitation, P_{ga}, from the precipitation $(P_{ga} + P_c)$ predicted by this linear regression equation using the measured value of $(S_f + T_f)$. This process is illustrated in Figure 15.4. This indirect use of the cloud interception gauge overcomes any problems associated with how well the gauge mimics the cloud interception catching capacity of the forest. This is demonstrated in McJannet *et al.* (2006c) with a comparison of cloud interception calculated using direct and indirect methods.

Canopy interception

McJannet *et al.* (2006d) estimated canopy interception (*I*) at the six sites for periods ranging between 2 and 3.5 years. Data from four of these sites are included in this water balance analysis. Interception in this study was determined by calculating net precipitation; that is, a combination of throughfall (T_f) and stemflow (S_f).

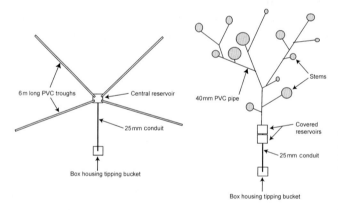

Figure 15.5 Throughfall (left) and stemflow (right) designs utilized at water balance sites. *Source*: after McJannet *et al.* (2006c).

Measurements of T_f were made using throughfall troughs, while (S_f) was determined by using a network of stemflow collars (Figure 15.5). Full technique descriptions are given in McJannet *et al.* (2006c).

Transpiration

Full descriptions of transpiration measurements at each site, including sampling and scaling methods, are presented by McJannet *et al.* (2006a). Measurements of sapflow at each site were made on sample trees using the heat pulse methodology. This method uses heat as a tracer to detect the rate of ascent of sap through the trunk of the tree. Strong relationships between the diameter at breast height (DBH; 1.3 metres) of trees and total daily water use enabled development of a robust method for scaling from sample tree water use to stand transpiration (*T*) (Figure 15.6). Strong relationships between tree size and daily water use were demonstrated despite the diversity of tree species found at all sites.

Forest floor evaporation

Evaporation from the forest floor (E_s) was not measured at any of the field sites in this study; therefore, estimates of this component of the water balance have been drawn from the studies reported in the literature. Average daily forest floor evaporation of 0.3 mm has been reported for mature temperate rainforest (McJannet *et al.* 1996) and broad leaved forest in New Zealand (Kelliher *et al.* 1992), while average daily rates of 0.1 mm are reported for Amazonian rainforest (Jordan & Heuveldop 1981). From these low rates it can be seen that E_s is likely to be a small component of

the water balance. Taking the average of values reported in the literature for temperate and tropical rainforests, daily forest floor evaporation was assumed to be 0.2 mm at all sites discussed in this chapter.

Runoff and drainage

Runoff and recharge ($R + D$) and soil moisture ($\delta\theta$) were not measured in this study, so the sum of these water balance components ($R + D + \delta\theta$) was determined from the difference between precipitation inputs ($P_{ga} + P_c$) and evaporative losses ($I + T + E_s$). Over the duration of this study the change in soil moisture storage is assumed to be negligible, so it is possible to get an estimate of long-term water yield ($R + D$) of each of the forest sites.

Results and discussion

Climatic characteristics during the study period

Measurements at each site were made over different periods between 2002 and 2005 (Table 15.2). During

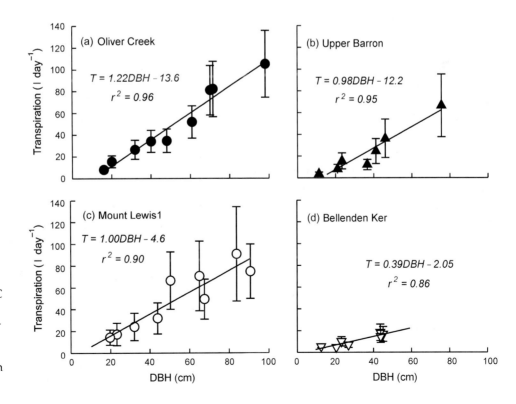

Figure 15.6 Relationship between DBH of individual sample trees and average daily transpiration (*T*) at OC (a), UB (b), ML1 (c) and BK (d). Average daily transpiration is determined from the full data set. Error bars are one standard deviation from the mean.

Table 15.2 Comparison period, total precipitation (rainfall and cloud interception), number of days with precipitation, maximum daily precipitation and average daily precipitation for rainfall days, average daily potential evapotranspiration and annual precipitation equivalent at each of the study sites

Site Name	Study period	Duration of study (days)	Total precipitation (mm)	Days with precipitation	Maximum daily precipitation (mm)	Average daily precipitation (mm)	Average daily ET_o (mm)	Annual precipitation equivalent (mm)
OC	17/08/02 to 31/12/03	502	3017	226 (45%)	417	17.3	3.3	2484
UB	13/09/03 to 30/06/05	657	5699	322 (48%)	454	16.1	3.0	2983
ML1	09/08/02 to 23/01/04	533	3877	281 (53%)	329	17.8	2.7	3040
BK	05/06/04 to 30/06/05	391	7898	271 (70%)	560	33.2	2.2	7471

Source: after McJannet *et al.* (2006b).

the years 2002 and 2003, rainfall in the region was well below average as a result of the development of strong El Niño conditions. This resulted in the driest rainy seasons on record at many Bureau of Meteorology weather stations in the region. Precipitation characteristics during the measurement period at each site are shown in Table 15.2. Total precipitation for the study period is a combination of P_{ga} and P_c and average annual precipitation equivalent is calculated from these totals for inter-site comparison purposes. The site with the highest precipitation input recorded in any single day was BK with 560 mm. BK has the highest average daily precipitation for rain days (33.2 mm) and the largest percentage of days with precipitation (70%). It should be noted that the annual precipitation equivalent calculated for each site in Table 15.2 is affected by the timing of the study: sites measured during the dry conditions of 2002 and 2003 had much lower precipitation inputs compared to those encompassing 2004 and 2005.

Seasonal water balance variations

Figure 15.7 shows the monthly water balance variations at each site. This figure demonstrates how evaporative losses via T, I and E_s affect the seasonality of the net water balance ($R + D + \delta\theta$) at each site. At all sites there is a distinct seasonal variation in water balance components, which reflects the variation in precipitation inputs between wet and dry seasons. Negative net

water balance around the end of the dry season at some sites indicates that evaporative losses can exceed precipitation inputs. Therefore, the forest must rely on soil moisture stores or have access to groundwater. While T is reduced during the wettest months, I is at its greatest at this time, therefore maintaining high evaporative losses. Data from the UB site can be used to illustrate how inter-annual differences in wet season rainfall affect the water balance (Figure 15.7b). Precipitation during the 2004/5 wet season was only 42% (1168 mm) of that during the 2003/4 wet season (2750 mm). However, in 2004/5 ($R + D + \delta\theta$) decreased by an even greater proportion than rainfall, being only 28% (521 mm) of its value in 2003/4 (1877 mm). This illustrates how there can be a twofold effect of lower precipitation on forest water yield: first, precipitation inputs are reduced; second, evaporative losses (as a percentage of inputs) increase. This finding has important implications for water yield from Wet Tropics catchments given the potential for climate change to alter precipitation inputs. With the huge precipitation inputs at BK, all months showed a net production of ($R + D + \delta\theta$) (Figure 15.7d). The wet conditions at this site (Table 15.2) and low energy inputs kept evaporative losses at this site very low.

Study period water balance components

Figure 15.8 provides an overview of the water balance at each site for the entire duration of measurements.

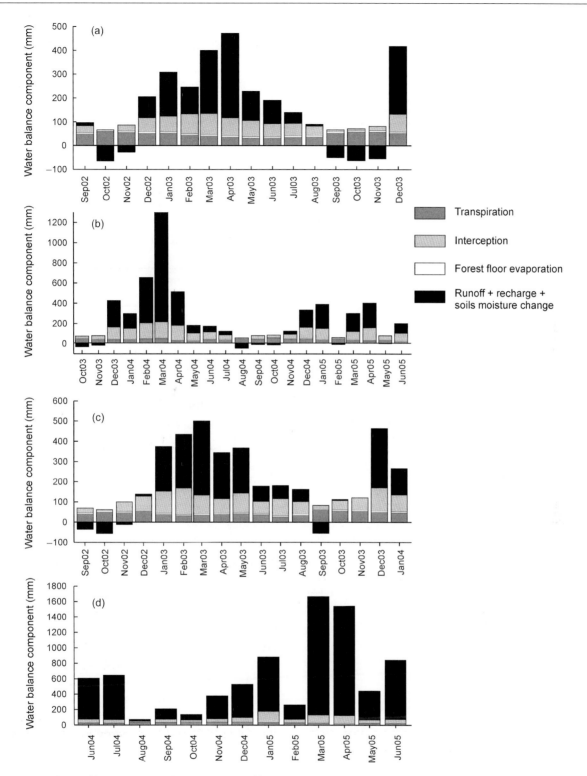

Figure 15.7 Total monthly evaporative losses and runoff, recharge and soils moisture change $(R + D + \delta\theta)$ for forest at OC (a), UB (b), ML1 (c) and BK (d). Negative $(R + D + \delta\theta)$ indicates that evapotranspiration is in excess of precipitation inputs. Shaded areas indicate wet season months. *Source*: after McJannet *et al.* (2006b).

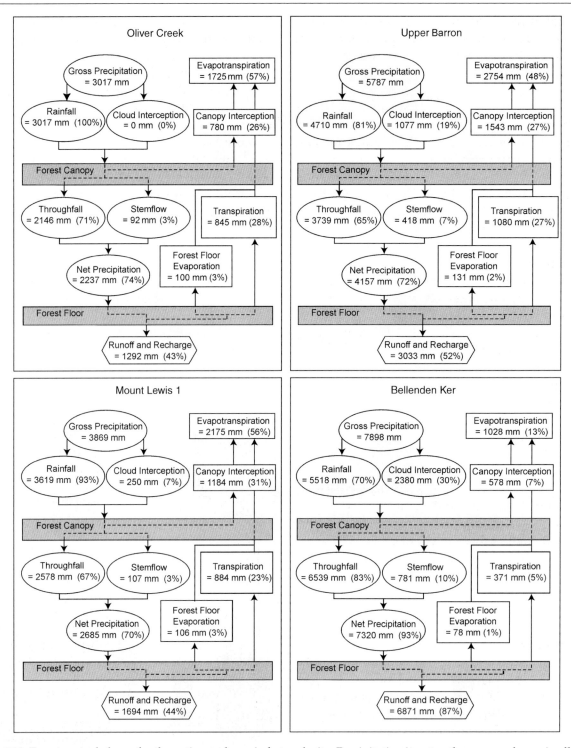

Figure 15.8 Forest water balance for the entire study period at each site. Precipitation input pathways are shown in ellipses, evaporative losses are in rectangles and runoff/recharge losses are shown in a hexagon. Percentages of gross precipitation are shown in parentheses.

These figures are based on the processes described in equation 1. Percentages of gross precipitation are shown in brackets. A summary of the water balance components at each site is given as follows.

Oliver Creek

This site is the only one where precipitation inputs are from P_{ga} alone. At this site T_f accounts for 71% (2146 mm) of gross rainfall and S_f accounts for 3% (92 mm). Evaporative losses through T were 845 mm and those through I were 780 mm. Overall, ET accounts for 57% (1725 mm) of gross precipitation, leaving 43% (1292 mm) for $(R + D)$.

Upper Barron

At UB, significant additional water input through P_c (19%, 1077 mm) was observed. S_f was 7% (418 mm) of gross precipitation and T_f was 65% (3739 mm), the lowest of any site. Evaporative losses through I (27%, 1543 mm) were about 50% greater than those through T (19%, 1080 mm), indicating that evaporation from the canopy dominates ET. This is an interesting observation considering that I and T were much more similar at the OC site. $(R + D)$ was equivalent to 52% (3033 mm) of gross precipitation.

Mount Lewis 1

Although ML1 is very similar in altitude to UB, the precipitation inputs through P_c at ML1 are much less (7%, 250 mm). This observation is a result of topographic factors that protect this site from direct exposure to prevailing winds. Net precipitation was about 70% (2685 mm) of gross precipitation, resulting in I of 31% (1184 mm). T was 23% (884 mm) of gross precipitation and $(R + D)$ was 44% (1694 mm).

Bellenden Ker

Precipitation inputs at this site were by far the largest (7898 mm) despite below average rainfall and the shortest observation period. Water inputs through P_c were very large (2380 mm), amounting to 30% of gross precipitation. This clearly indicates the importance of including this water balance component in hydrological analysis of high altitude rainforests. At 10% of gross precipitation, S_f at BK is the greatest of all sites. T_f was also much greater than at other sites (83%, 6539 mm) as a result of the lower LAI at this site (3.3). Despite the availability of very large amounts of precipitation, both

T and I are relatively small at 5% and 7% of gross precipitation, respectively. The reasons for this are related to a combination of forest structural characteristics and limited availability of energy to drive evaporation. The large inputs and small evaporative losses at BK result in the availability of 87% (6871 mm) of gross precipitation for $(R + D)$.

Water balance measurements at each of the sites were undertaken over different periods, thus, complicating comparisons between sites. In an attempt to overcome this, water balance components have been normalized to annual equivalents. Normalized site water balances are shown in Figure 15.9, where the size of pie charts for each site is equivalent to gross precipitation inputs. The total pie is then divided up into slices equivalent in size to the proportion of T, E_s, I and $(R + D)$. This figure shows that on an annual basis all of these forests have significant water yield. The three lower altitude sites, OC, UB and ML1, show quite similar water yields, with $(R + D)$ values ranging

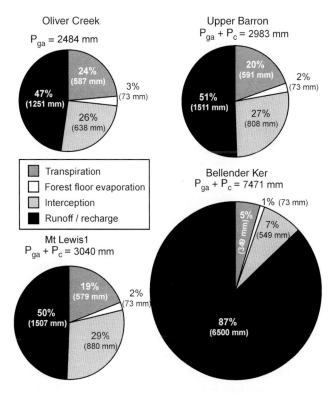

Figure 15.9 Annual water balance for each field site expressed as a percentage of total precipitation input $(P_{ga} + P_c)$. The size of each pie chart is proportional to the normalized annual precipitation input. *Source*: after McJannet *et al.* (2006b).

between 47 and 51% of gross precipitation. *I* at each of these sites is also similar (26–29%); therefore, differences in $(R + D)$ are largely due to differences in *T*. At BK *T* was just 5% (349 mm) of gross precipitation compared to 20% (591 mm) at UB, 24% (587 mm) at OC and 19% (579 mm) at ML1. McJannet *et al.* (2006a) concluded that the reason for these differences is likely to be the result of a combination of varying stem size distribution, sapwood area and climatic and soil moisture conditions at each site.

The international context

Annual transpiration rates for our Wet Tropics study sites (349–591 mm yr^{-1}) are at the lower end of annual transpiration estimates for tropical and warm temperate forests, which are reported in a review by Schellekens *et al.* (2000) (335–1375 mm yr^{-1}). In contrast, comparison of our results with transpiration rates from water balance studies in Australia reveals much more similar results: Gilmour (1975) undertook water balance estimates for coastal rainforest in the Wet Tropics of north Queensland and estimated annual transpiration of 420 mm (10% of precipitation), while Hutley *et al.* (1997), working in the sub-tropical forest of south-east Queensland, reported total annual transpiration of 513 mm (39% of precipitation). This finding suggests that Australian tropical rainforests could be transpiring at slower rates than those reported in other parts of the world and the reasons for this could be related to forest structural differences (i.e. fewer large trees), climatic differences or differences in tree physiology. The annual transpiration rate at BK was just 349 mm, far less than other sites but very similar to the annual transpiration of 310–390 mm reported by Bruijnzeel (2001) and Holwerda (2005) for lower montane forests subjected to frequent immersion in cloud.

Measurements of both S_f and T_f from all sites are within the range of values for rainforests found in the literature (see reviews in Cavalier *et al.* 1997; Fujieda *et al.* 1997; Tobón Marin *et al.* 2000). S_f at OC and ML1 was less than half that at UB (7%) and BK (10%), where cloud immersion was a major water input and stem density was high; similar observations have been made for other montane rainforests (Baynton 1969;

Weaver 1972; Mamanteo & Veracion 1985; Hafkenscheid 2000). As has been found in a number of other studies, a combination of altitude, exposure and canopy characteristics, rather than altitude alone, was found to control the amount of P_c measured at each site in the Wet Tropics. Again, similar findings are reported for the slopes of Mauna Loa, Hawaii (Juvik & Ekern 1978), the Central Cordillera in western Panama (Cavalier *et al.* 1996) and the Caribbean coast of South America (Cavalier & Goldstein 1989). Interception studies from rainforests around the world (Schellekens *et al.* 2000) show that the amount of water evaporated from the canopy can exceed that lost through *T*. This was certainly the case for three of our four sites where evaporative losses from the canopy were as much as 56% greater than losses through *T*.

Normalized annual *ET* for all sites ranged between 971 and 1532 mm, values very similar to *ET* rates reported for Australian rainforest by Hutley *et al.* (1997) (1259 mm) and Gilmour (1975) (1420 mm). These figures are at the lower end of values reported in the literature (865–2000 mm; see review in Schellekens *et al.* 2000), mainly as a result of lower rates of *T*. From the combined results of Australian studies it can be seen that *ET* of Australian rainforests tend to range between 1000 and 1500 mm; this suggests that the annual rate of water yield for a given rainforest site will increase with precipitation inputs beyond this range.

A most striking feature of water balance comparisons for the Wet Tropics is the difference between the montane forest at BK and all other sites. At this site, precipitation inputs are much greater and evaporative losses are much less. Of the 7471 mm of gross precipitation just 13% is lost through *ET*, leaving a huge 87% of gross precipitation for $(R + D)$. Similarly high yields have been reported for other montane forests (Bruijnzeel 2001; Holwerda 2005), illustrating the importance of these forests in generating streamflow. In north Queensland most of these high altitude forest locations are preserved within national parks, but the hydrological services provided by these forests are still subject to threats such as climate change and/or changes to large-scale circulation patterns such as El Niño.

There is increasing evidence that the water balance of tropical montane cloud forests may be threatened by global warming (Bruijnzeel 2001). With warmer atmospheric temperature, the average cloud condensation layer will tend to increase in elevation (Still *et al.* 1999), removing the capacity for intercepting cloud water and potentially increasing evaporation, the net effect of which is reduced water yield. The effects of such changes go beyond water supply. Ecologically such changes in moisture conditions could also have a major impact on plant and animal species dependent on wet, humid conditions. For example, Williams *et al.* (2003) undertook a study of the potential impacts of climate change on the spatial distribution of endemic rainforest vertebrates in the Wet Tropics and concluded that increases in temperature, raised orographic cloud base and reduced dry season precipitation have the potential to result in the extinction of many species currently found in mountain areas due to significant reduction or complete loss of their environment (see Williams *et al.*, Chapter 22, this volume).

Summary

• The information gained through this research programme has resulted in the most comprehensive study of the hydrology of tropical rainforests in north Queensland.

• Inter-site comparisons have revealed large differences in the water balance of different rainforest types, mainly as a result of precipitation inputs but also due to forest structural differences, altitude, site exposure and radiation climate. Such differences need to be considered during assessment of the impacts of climate and land use change.

• High altitude exposed rainforests have been shown to receive significant inputs of water from cloud interception. With as much as 30% of annual gross precipitation occurring in this way, this component of the water balance clearly needs to be properly accounted for in any hydrological study of high altitude rainforests.

• Annual transpiration rates from rainforest in other parts of the world can be as much as double those reported for Australian studies. This suggests that Australian tropical rainforests transpire less than those found in other parts of the world, with the reasons for this being related to forest structural differences (i.e. fewer large trees), climatic differences or differences in tree physiology.

• Combining the results from the four sites in this study and other studies in Australian rainforest reveals that annual *ET* is typically between 1000 and 1500 mm, despite large variations in precipitation input. This observation suggests that the annual water yield for a given rainforest site will increase with precipitation inputs beyond this range, a conclusion supported by the very large water yield from BK.

• Observations made during the dry El Niño conditions of 2002 and 2003 illustrate how there can be a double effect of lower precipitation on forest water yield: first, precipitation inputs are reduced; second, evaporative losses increase. This finding has important implications for reduction in water yield during the onset of El Niño conditions and/or climate change.

• Large water yields from the cloud forest at BK illustrate the importance of these forest types to stream flow and water supply in the region. However, such forests are potentially under threat from climate change, which could result in reduced cloud water interception and improved conditions for evaporation. The net effect of these changes is reduced yield, which is important hydrologically, and altered moisture conditions, which could also be disastrous for plant and animal species dependent on wet, humid conditions.

• The potential impacts of climate change and El Niño events on the water yield from tropical rainforest in the Wet Tropics needs to be considered in the management of the region's water resources. The hydrology of Wet Tropics rainforests has been shown to vary greatly between locations, seasons and years; hence, the development of a predictive understanding and rainforest water balance modelling tools is essential for effective natural resource management in the region.

Acknowledgements

The authors would like to thank Andrew Ford, Peter Richardson, Trevor Parker, Adam McKeown, Matt Bradford, Trudi Prideaux, Jenny Holmes and Pepper Brown for their help with field installation and maintenance. Funding for this research was provided by the Cooperative Research Centre for Tropical Rainforest

Ecology and Management. David Post and Albert van Dijk provided valuable comments on this chapter.

References

Baynton, H. W. (1969). The ecology of an elfin forest in Puerto Rico, 3. Hilltop and forest influences on the microclimate of the Pico Del Oeste. *Journal of the Arnold Arboretum* **50**: 80–92.

Bruijnzeel, L. A. (2001). Hydrology of tropical montane cloud forests: a reassessment. *Land Use and Water Resources Research* **1**: 1.1–1.18.

Bruijnzeel, L. A. (2005). Tropical montane cloud forest: a unique hydrological case. In *Forests, Water and People in the Humid Tropics*, Bonell, M. & Bruijnzeel, L. A (eds). Cambridge University Press, Cambridge, pp. 462–83.

Cavelier, J. & Goldstein, G. (1989). Mist and fog interception in elfin cloud forests in Colombia and Venezuela. *Journal of Tropical Ecology* **5**: 309–22.

Cavelier, J., Solis, D. & Jaramillo, M. A. (1996). Fog interception in montane forests across the Central Cordillera of Panama. *Journal of Tropical Ecology* **12**: 357–69.

Cavelier, J., Jaramillo, M., Solis, D. & de Leon, D. (1997). Water balance and nutrient inputs in bulk precipitation in tropical montane cloud forest in Panama. *Journal of Hydrology* **193**: 83–96.

Clark, K. L., Nadkarni, N. M., Schaefer, D. & Gholz, H. L. (1998). Atmospheric deposition and net retention of ions by the canopy in a tropical montane forest, Monteverde, Costa Rica. *Journal of Tropical Ecology* **14**: 27–45.

de Lima, J. (1990). The effect of oblique rain on inclined surfaces: a nomograph for the rain-gauge correction factor. *Journal of Hydrology* **115**: 407–12.

Førland, E. J., Allerup, P., Dahlström, B., *et al.* (1996). *Manual for Operational Correction of Nordic Precipitation Data*. Norwegian Meteorological Institute, Oslo, Norway. 66 pp.

Fujieda, M., Kudoh, T., Cicco, de V., *et al.* (1997). Hydrological processes at two subtropical forest catchments: the Serra do Mar, São Paulo, Brazil. *Journal of Hydrology* **196**: 26–46.

Gilmour, D. A. (1975). Catchment water balance studies on the wet tropical coast of north Queensland. PhD Thesis, James Cook University, Townsville. 129 pp.

Hafkenscheid, R. (2000). Hydrology and biogeochemistry of tropical montane rain forests of contrasting stature in the Blue Mountains, Jamaica. Academisch Proefschrift, Vrije Universiteit, Amsterdam. 302 pp.

Herwitz, S. R. (1985). Interception storage capacities of tropical rainforest canopy trees. *Journal of Hydrology* **77**: 237–52.

Herwitz, S. R. (1986). Infiltration-excess caused by stemflow in a cyclone-prone tropical rainforest. *Earth Surface Processes and Landforms* **11**: 401–12.

Herwitz, S. R. (1987). Raindrop impact and water flow on the vegetative surfaces of trees and the effects on stemflow and throughfall generation. *Earth Surface Processes and Landforms* **12**: 425–32.

Holder, C. D. (2003). Rainfall interception and fog precipitation in a tropical montane cloud forest of Guatemala. *Forest Ecology and Management* **190**: 373–84.

Holwerda, F. (2005). Water and energy budgets of rain forests along an elevational gradient under maritime tropical conditions. Academisch Proefschrift, Vrije Universiteit, Amsterdam. 168 pp.

Hutley, L. B., Doley, D., Yates, D. J. & Boonsaner, A. (1997). Water balance of an Australian subtropical rainforest at altitude: the ecological and physiological significance of intercepted cloud and fog. *Australian Journal of Botany* **45**: 311–29.

Jordan, C. F. & Heuveldop, J. (1981). The water budget of an Amazonian rain forest. *Acta Amazonica* **11**: 87–92.

Juvik, J. O. & Ekern, P. C. (1978). *A Climatology of Mountain Fog on Mauna Loa Hawaii Island*. Water Resources Research Centre, University of Hawaii, Honolulu. Technical Report 118; 63 pp.

Kelliher, F. M., Kostner, B. M. M., Hollinger, D. Y. *et al.* (1992). Evaporation, xylem sapflow, and tree transpiration in a New Zealand broad leaved forest. *Agricultural and Forest Meteorology* **62**: 53–73.

McJannet, D., Fitch, P., Disher, M. & Wallace, J. (2007a). Measurements of transpiration in four tropical rainforest types of north Queensland, Australia. *Hydrological Processes* DOI: 10.1002/hyp.6576 (In press).

McJannet, D. L., Vertessy, R. A., Tapper, N. J., O'Sullivan, S. K., Beringer, J. & Cleugh, H. (1996). *Soil and Litter Evaporation Beneath Re-growth and Old-growth Mountain Ash Forest*. Cooperative Research Centre for Catchment Hydrology, Melbourne.

McJannet, D., Wallace, J, Fitch, P., Disher, M. & Reddell, P. (2007b). Water balance of tropical rainforest canopies in north Queensland, Australia. *Hydrological Processes* DOI: 10.1002/hyp.6618 (In press).

McJannet, D., Wallace, J. & Reddell, P. (2007c). Precipitation interception in Australian tropical rainforests: I. Measurement of stemflow, throughfall and cloud interception. *Hydrological Processes* 21: 1692–1702.

McJannet, D, Wallace, J. & Reddell, P. (2007d). Precipitation interception in Australian tropical rainforests: II. Altitudinal gradients of cloud interception, stemflow, throughfall and interception. *Hydrological Processes* 21: 1703–1718.

Mamanteo, B. P. & Veracion, V. P. (1985). Measurements of fog drip, throughfall and stemflow in the mossy and Benguet pine (*Pinus kesiya* Royle ex Gordon) forests in the upper Agno river basin. *Sylvatrop (Philippines For. Res. J.)* **10**: 271–82.

Schellekens, J., Bruijnzeel, L. A., Scatena, F. N., Bink, N. J. & Holwerda, F. (2000). Evaporation from a tropical rain forest, Luquillo experimental forest, Eastern Puerto Rico. *Water Resources Research* **36**: 2183–96.

Sharon, D. (1980). The distribution of hydrologically effective rainfall incident on sloping ground. *Journal of Hydrology* **46**: 165–88.

Still, C. J., Foster, P. N. & Schneider, S. H. (1999). Simulating the effects of climate change on tropical montane cloud forests. *Nature* **398**: 608–10.

Tobón Marin, C., Bouten, W. & Sevink, J. (2000). Gross rainfall and its partitioning into throughfall, stemflow and evaporation of intercepted water in four forest ecosystems in western Amazonia. *Journal of Hydrology* **237**: 40–57.

Tracey, J. G. (1982). The vegetation of the humid tropical region of North Queensland. CSIRO, Melbourne.

Weaver, P. L. (1972). Cloud moisture interception in the Luquillo mountains of Puerto Rico. *Caribbean Journal of Science* **12**: 129–44.

Williams, S., Bolitho, E. & Fox, S. (2003). Climate change in Australian tropical rainforests: an impending environmental catastrophe. *Proceedings of the Royal Society of London Series B* **270**: 1887–93.

16 SEED DISPERSAL PROCESSES IN AUSTRALIA'S TROPICAL RAINFORESTS

David A. Westcott[1]*, Andrew J. Dennis[2]*, Matt G. Bradford[1]*, Graham N. Harrington[1]* and Adam McKeown[1]*

[1]CSIRO Sustainable Ecosystems, Atherton, Queensland, Australia
[2]CSIRO Sustainable Ecosystems, Herberton, Queensland 4887, Australia
*The authors were participants of Cooperative Research Centre for Tropical Ecology and Management

Introduction

A conspicuous feature of tropical rainforests is the interaction between the fruits of woody plants and the vertebrates that consume them. Plant species adapted for dispersal by vertebrates generally represent between 75% and 90% of woody plants in tropical rainforests (Willson *et al.* 1989; Jordano 1992), while their vertebrate dispersers can also represent significant proportions of local animal communities (Terborgh 1986a). Australia's rainforests are no different in this respect, with as many as 95% of woody plants being adapted for vertebrate dispersal and some 65 species of vertebrate acting as dispersers (Dennis & Westcott, in press). Interest in seed dispersal and frugivory by vertebrates has a long history (Aristotle 350 BCE; Theophrastus 350 BCE) and the subject has maintained its currency through to modern times (Levey *et al.* 2002) due to dispersal's broad influence, for example on organismal evolution and patterns of speciation (Schondube *et al.* 2001; Lucas *et al.* 2003; McNab 2005) and on the structuring and dynamics of plant and animal populations and communities (Nathan & Muller-Landau 2000; Wang & Smith 2000; Levin *et al.* 2003; Levine & Murrell 2003).

For frugivores, fruits are an important if not the major component of their diets and this has a variety of consequences. Patterns of fruiting determine resource availability, influencing not only the timing of events such as breeding but also population densities and distributions (Snow 1971, 1976). Poor fruiting seasons can result in population collapses or low reproductive success (Wright *et al.* 1999). Being easily harvested and nutrient rich, a frugivorous diet allows some consumers to allocate significant time and energy to pursuits beyond mere survival and reproduction. Many species, particularly birds, show extreme elaboration of secondary sexual characteristics or the evolution of complex social systems such as lek mating systems in association with frugivory (Snow 1971, 1976; Beehler & Pruett-Jones 1983).

For tropical forest fruits, frugivores represent agents of dispersal. This generally occurs as a result of the animal feeding on the fruit or seed and transporting it either internally or externally. Seed dispersal has implications for plants at all levels of ecological organization. For the individual seed, dispersal results in arrival at the germination site and thus determines the physical and biological conditions under which it must survive (Wenny & Levey 1998). Because dispersal determines the distribution and abundance of potential recruits it can be an important influence on population and meta-population dynamics (Levine & Murrell 2003), as well as influencing population genetic structure (Loiselle *et al.* 1995). At the community level, dispersal determines which species arrive at a site and are eligible to be part of the local community (Levin *et al.* 2003). Phenomenological models of seed dispersal indicate that its outcomes for plant population, species and community dynamics are fundamentally linked to the shape and scale of the dispersal kernel, i.e. the frequency distribution of dispersed seeds relative to the parent plant (Levin *et al.* 2003). Unfortunately, few dispersal kernels have been described for tropical rainforest plants.

In this chapter we provide an overview of frugivory and seed dispersal in the Wet Tropics of Queensland. We start by describing the plants and animals that participate in this process. We go on to consider similarities between the fruits of different plants and the dispersal services provided by different animals and identify groups with common features. We then outline how their combined interactions result in the dispersal of seeds and describe the dispersal kernel for a common Wet Tropics plant genus, *Cissus*.

Fruits of Australia's rainforests

The Wet Tropics contain about 1300 species of plants that are adapted for dispersal by vertebrates in that their fruits either contain a nutritive component that attracts dispersers or have seeds that act in this role (Westcott & Dennis, unpublished data). These species come from 132 families and 469 genera. The most speciose vertebrate-dispersed families are the Myrtaceae (102 species), including genera such as *Syzygium*, *Gossia* (*Austromyrtus*), *Acmena* and *Rhodomyrtus*, and the Lauraceae (100 species), including genera such as *Cryptocarya, Endiandra, Litsea, Beilschmedia, Cinnamomum* and *Neolitsea*. Other important vertebrate-dispersed families are the Sapindaceae (78 spp.), Rubiaceae (68), Euphorbiaceae (50), Moraceae (43), Annonaceae (40), Rutaceae (38), Eleaocarpaceae (36) and Meliaceae (31). At the genus level, diversity in vertebrate-dispersed species is concentrated in *Syzygium* (46 spp.), *Cryptocarya* (37), *Ficus* (35, Moraceae) and *Endiandra* (32).

In terms of general morphology and form, Wet Tropics fruits are similar to those of rainforests elsewhere.

Fleshy-fruited species, i.e. those with a fleshy pericarp or aril, represent the vast majority of vertebrate-dispersed species (about 1195 spp.), and include 37 species whose fruits are syconia, 28 with aggregate fruits, 930 drupes, berries and pomes, 192 dehiscent capsules or follicles with fleshy arils and 8 with indehiscent pods. Dry fruits, i.e. those with a dry pericarp at maturity, are represented by 33 dehiscent capsules, 23 indehiscent pods and follicles and 96 woody species (Westcott & Dennis, unpublished data). Most Wet Tropics vertebrate-dispersed fruits are small in terms of both their mass ($\bar{x}$ = 6.7 g, 16.2 SD, n = 1251) and their dimensions (volume $\bar{x}$ = 6.9 ml, 17.8 SD, n = 1250). A comparison of the family means of fruit length for species in the families Anacardiaceae, Burseraceae, Lauraceae, Meliaceae, Moraceae, Myrsisticaeae, Palmae and Simaroubiaceae from the Neotropics, Paleotropics (excluding tropical Australia) (Mack 1993) and Wet Tropics indicates that Wet Tropics fruits are similar in length ($\bar{x}$ = 24.7 mm) to the Neotropics ($\bar{x}$ = 24.9 mm) but smaller than those of the Paleotropics ($\bar{x}$ = 39.4 mm) (region $F_{2,14}$ = 5.5, $p < 0.02$; family $F_{7,14}$ = 1.5, $p = 0.24$).

Pan-tropical comparisons also suggest that differences occur between Wet Tropics rainforest seeds and those of rainforests elsewhere (Forget *et al.*, 2007). Wet Tropics seeds have a mean length of 12.8 mm (11.6 SD) and width of 9.1 mm (8.6 SD). In phylogenetically controlled comparisons, Wet Tropics seeds are smaller than African seeds, the same size as Neotropical seeds but larger than Asian seeds. Overall, Wet Tropics seeds are rounder than both African and Neotropical seeds but the largest Wet Tropics seeds are more elongate, similar to the pattern seen in Asia. One explanation for this is that seed width has been constrained by the size of dispersers in the Wet Tropics (Mazer & Wheelwright 1993). The largest Wet Tropics dispersers, for example the cassowary, are indeed smaller than those currently found elsewhere, for example elephants, rhinoceros and tapir, and there are fewer varieties of larger dispersers in Australia. Thus, the smaller Wet Tropics dispersers may have selected for more elongate large seeds than elsewhere in the tropics (Forget *et al.*, 2007).

Fruiting phenology

An important trait of rainforest fruit resources is their year round availability. This reliability allows for dietary specialization and is postulated to be a primary determinant of the degree to which frugivory is represented in regional faunas (Primack & Corlett 2005). However, most tropical rainforests show marked annual patterns of fruiting phenology (Dew & Boubli 2005). The Wet Tropics enjoys a single, but marked, wet and dry season each year, a pattern that is reflected in the flowering (Boulter *et al.* 2006) and fruiting (Westcott *et al.* 2005a) phenologies of the region.

There have been several studies of fruiting phenology in the Wet Tropics. Westcott *et al.* (2005a) combined the data for the upland studies (Paluma, Moore 1991; Atherton Tablelands, Dennis & Marsh 1997; Westcott *et al.* 2005a; Mt Windsor, Crome *et al.* unpublished) to provide a region-wide picture of patterns of fruiting. Across these sites and the years studied they recorded fruiting by 197 species from 123 genera and 57 families. On average, 78 species (range 59–100), 53 genera (range 41–64) and 33 families (range 24–45) were recorded fruiting per month. Species diversity was high, with 76% of genera and 44% of families represented by just a single species. Only six genera were represented by five or more species: *Endiandra* (13), *Syzygium* (11), *Ficus* (10), *Elaeocarpus* (9), *Cryptocarya* (9) and *Acronychia* (5). Seven families were represented by ten or more species: Lauraceae (30), Myrtaceae (16), Elaeocarpaceae (15), Proteaceae (12), Rutaceae (11), Sapindaceae (11) and Moraceae (10). Different locations showed marked differences in the diversity of species fruiting in any given month. An average of 20.45 (±3.82 SE) species were recorded fruiting per month at each site. On the Windsor Tableland a total of 47 species, 36 genera and 22 families with an average of 18 species (±1.06 SE) fruited each month. On the Atherton Tableland, the combined records show 139 species, 93 genera and 48 families fruited, with an average of 34.08 species (±2.85 SE) per month. At Paluma 38 species, 31 genera and 23 families were recorded, with just 6.48 (±7.2 SE) fruiting per month on average.

Fruiting phenology shows a clear and consistent seasonal pattern (Figure 16.1). At mid to high elevations, fruit is most abundant and diverse through the late dry and early wet seasons, roughly from October to January. Subsequently, there is a decrease in both biomass and diversity as the wet season continues, leading into a lean period between April and July. The average species fruits for 4.22 months (±0.21 SE) in a year, while genera and families fruit for 4.73 (±0.30 SE) and 6.43 (±0.50 SE) months a year, respectively.

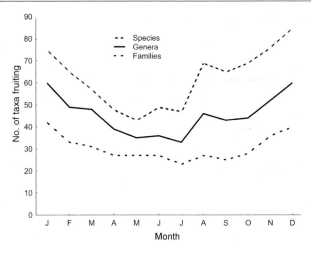

Figure 16.1 Annual phenology of fruiting at the species, genus and family level for mid-montane and upland Wet Tropics forests. *Source*: after Westcott *et al.* (2005a).

While productivity is seasonal, periods of scarcity and abundance differ between the ground and the canopy. In the uplands, the biomass of fruit available to frugivores on the ground averages 10.7 kg ha^{-1} month^{-1} (±9.4 SD), while in the canopy the average is 111 kg ha^{-1} month^{-1} (±38 SD). Not surprisingly, the phenology of fruit on the ground lags behind that of the canopy. Total biomass of canopy fruit peaks early, in August, and then declines through the dry and into the wet season, in January. In contrast, biomass on the ground increases through the dry season to a peak in December.

Although there is marked annual variation in the biomass and diversity of fruit production, this general pattern in the uplands remains relatively consistent across the region, rainforest vegetation types and years (Westcott *et al.* 2005a) and is reflected in other patterns within the forests. At lower elevations the patterns of fruiting are similar, though there is a shift in the timing of the peaks and troughs. Crome (1975) found that in the Mission Beach area the lean season extended through the January to May period, while fruit abundance increased from June/July, peaking between August and September and declining through to December. The period of resource scarcity between April and July corresponds with the departure of migratory species such as Torresian imperial pigeons (*Ducula bicolor*) and metallic starlings (*Aplonis metallica*), reports of starving cassowaries (P. Latch, pers. comm.) and high levels of mortality and poor condition in frugivores (Dennis & Marsh 1997).

Are there keystone fruits?

Early tropical rainforest seed dispersal research was characterized by attempts to describe strong, co-adaptive relationships between fruits and dispersers, and to identify species that played keystone roles in rainforest ecosystems (McKey 1975; Terborgh 1986b). Such efforts have had little success. Exclusive, coevolved dispersal relationships are uncommon and to date no species has been identified unequivocally as a keystone (Dew & Boubli 2005). It appears unlikely that Australia's rainforests will be any different. Dennis and Marsh (1997) found that musky rat-kangaroo (*Hypsiprymnodon moschatus*) populations were sustained during a period of exceptionally low fruit availability by *Ficus pleurocarpa*. They suggested that if this was a general phenomenon then *F. pleurocarpa* might play a keystone role in these forests. Subsequently, Westcott *et al.* (2005a) compared fruit availability and reliability in space and time across upland Wet Tropics sites in an attempt to identify potential keystone species. They reasoned that keystone species were most likely to be those that underpinned resource availability during periods of scarcity. Surprisingly few species provided a significant proportion of lean season fruit biomass, for example *Cryptocarya angulata* and *Ficus pleurocarpa*. However, even the most common and abundant of these species were not sufficiently reliable in space or time to allow them to act as reliable keystone species at a community level.

Frugivores of Australia's rainforests

The major frugivore taxa

In the Wet Tropics frugivory by vertebrates is common, particularly among mammals and birds. Of the 65 vertebrates for which fruit can be considered a regular dietary component, 17 are mammals and 48 are birds, a proportional breakdown that is roughly similar to that of other tropical forests. These include species that span the range from specialist frugivores through to those for which fruit is only a small component of the diet. Both the birds and mammals include species that process seeds gently and provide high quality dispersal, as well as species whose processing results in significant levels of seed damage or death. However, the relative proportions of gentle and predatory handling differ between the birds and mammals, with 64% of mammal

dispersers frequently causing seed damage or death as compared with just 23% of species in birds.

Among the mammals, five species of rat in the genera *Melomys*, *Rattus* and *Uromys* are frequent consumers of fruits and seeds; all are primarily seed predators but also consume some flesh. Their dispersal occurs as a product of moving fruits or seeds to feeding or caching sites (Theimer 2001; Dennis *et al.* 2005). One species of bandicoot in the genus *Perameles* is known to consume fruit occasionally and may provide some dispersal for small seeds. Three species of possum in the genera *Pseudochierops* and *Trichosurus* are recorded as occasional frugivores. The outcome of their fruit handling can vary from the seeds being discarded intact at the feeding site, to seeds being chewed and killed, and seeds being swallowed and passed intact (D. A. Westcott, pers. obs.). Three species of kangaroo, *Hypsiprymnodon moschatus*, *Thylogale stigmatica* and *Dendrolagus lumholtzii* consume fruit, though only *H. moschatus* provides significant quantities of high quality dispersal (Dennis 2003). The feral pig, *Sus scrofa*, consumes both flesh and seeds from entire fruit, although its handling is rough, crushing and killing most seeds it consumes. Five species of flying-fox in the genera *Pteropus*, *Syconicterus* and *Nyctimene* feed to differing extents on fruits and provide some dispersal, particularly for small-seeded species but also less frequently for larger fruits.

Frugivory is most widely distributed among the Wet Tropic's birds. Forty-eight species from 14 families are known to be regularly frugivorous, including some of the most iconic taxa. Australia's one representative of the family Casuaridae, *Casuarius casuarius*, is almost entirely frugivorous and consumes a large range of fruit types (Stocker & Irvine 1983; Westcott *et al.* 2005b). The most speciose group of frugivores are the honeyeaters (Meliphagidae), with 11 species regularly consuming fruit. These tend to feed on small-fruited species, generally swallowing them whole and passing seeds intact, along with insects and nectar. The pigeons and doves (Columbidae, 8 spp.) are a diverse group of frugivores and granivores. These birds fall into two general groups: those that feed on fruit flesh and pass seeds intact, for example fruit doves *Ptilonopus* spp., and those that are primarily granivorous, for example *Chalcophaps indica* and *Columba leucomela*. Seven species of parrot (Psittacidae and Cacatuaidae) consume fruit but provide only occasional dispersal, either because they are primarily seed predators, for example cockatoos and fig parrots, or because when they consume fruit flesh they ignore the seeds. A small but highly frugivorous group are the bower-birds (Ptilonorhynchidae), with four rainforest species. These birds feed on small to medium-sized fruits (<25 mm), providing significant quantities of high quality dispersal over distances usually less than 200 metres. Each of the families Dicaeidae, Zosteropidae and Sturnidae contains a single highly frugivorous species. The former two are small birds that move across habitat boundaries and disperse seeds over distances of hundreds of metres, while the latter is a medium-sized species that disperses seeds over longer distances, also across habitat boundaries. Two species of cuckoo (Cuculidae), the koel (*Eudynamis scolopacea*) and the channel-billed cuckoo (*Scythrops novaeholland-iae*), are highly frugivorous. However, both occur at low densities and probably play a limited dispersal role. One species each from the Dicruridae and the Paradisaeidae and three species from the Oriolidae are medium-sized birds that consume fruits and provide relatively short distance dispersal. The drongo, *Dicrurus hottentotus*, is the least frugivorous of these. The Campephagidae include four frugivores, with *Coracina lineata* being the most frugivorous. The Artamidae include two species that, though primarily carnivorous, regularly consume fruit. The Megapodiidae include one species that is recorded as frugivorous, though it is primarily a seed predator.

This list represents the fruit consumption records that are known either from our records or from those available to us from the literature or personal communication (see Dennis & Westcott 2007). While it represents current knowledge in terms of the most important frugivores it is certain not to be definitive. Other taxa such as reptiles and fish are known to disperse fruits elsewhere (Desouzastevaux *et al.* 1994; Godinez-Alvarez 2004) and in Australia (T. Lamb, pers. comm.; A. Dennis & D. A. Westcott, pers. obs.), and are likely to do so in the Wet Tropics region as well.

Is Australia different?

Many of the frugivores in the Wet Tropics are derived from Gondwanic taxa. These include the Meliphagidae, Ptilonorhynchidae, Paradisaeidae, Casuaridae and Hypsiprimnodontidae. While these groups have

evolved to fill niches that are elsewhere occupied by other taxa, several roles found elsewhere in the tropics do not have a direct Wet Tropics analogue. An example is the total absence of seed dispersing primates, squirrels, ungulates, porcupines and carnivores. In the Neotropics and most of the Paleotropics, fruit-eating primates are an important component of intact rainforest communities and can disperse tens of thousands of seeds per hectare (Stevenson 2000). While some primates kill a proportion of the seeds they ingest, many either remove the flesh and spit out seeds or swallow seeds intact. This group is completely lacking in Australia and New Guinea and no single group fills their role, despite the presence of fruit morphologies that elsewhere are considered to be adapted for primate dispersal, for example the tough dehiscent capsules of the Myristicaceae. Instead, our observations suggest that in Australia other frugivores are able to consume these fruits, either simply waiting until the pericarp can be peeled open or has opened fully, or using other strategies to access the fruit (Beehler & Durmbacher 1996). In contrast, the family Sciuridae, which is so diverse across the rest of the tropics, does have a partial analogue in the Wet Tropics. Squirrels elsewhere can be important seed predators and dispersers; in the Wet Tropics they are replaced by a small group of rodents, some of which, like squirrels, are highly scansorial, for example *Uromys caudimaculatus*.

Two additional groups find no clear analogues in the extant Wet Tropics frugivore fauna. In all but heavily hunted forests elsewhere the terrestrial mammalian fauna includes small (<25 kg) frugivorous deer (Artiodactylidae). In the Australian context this role is most closely assumed by the red-legged pademelon *T. stigmatica*. However, like most medium to large rain-forest Artiodactyls (Primack & Corlett 2005), these medium-sized kangaroos are primarily grazers and browsers that chew or digest most of the seeds from fruits they consume. Similarly, elsewhere in the world carnivores are regular consumers of fruit (Nowak 1999; Kitamura *et al.* 2002; Sunquist & Sunquist 2002; Gatti *et al.* 2006). No Wet Tropics carnivores are known to include significant quantities of fruit in their diet.

Among the birds perhaps the most notably absent groups are analogues of the large-billed Neotropical toucans (Ramphostidae) and Paleotropical hornbills (Bucerocidae). These birds' huge bills allow them to reach out and pluck fruits from terminal twigs that would not otherwise be accessible given their weight. In Australia, this beak morphology is most closely approximated by the channel-billed cuckoo, *Scythrops novaehollandiae*. However, in comparison to toucans and hornbills, *S. novaehollandiae* occurs at very low abundances, even compared to elsewhere in its range (D. A. Westcott, pers. obs.). This suggests that in the Wet Tropics context other groups, for example the agile fruit-doves, may overlap with the role usually provided by this group.

Most tropical forests contain one or more large-bodied vertebrates that ingest and disperse large fruits and seeds. Because of the constraints that consuming large fruits and seeds place on their mode of movement these species are usually terrestrial and mammalian. In the Paleotropics elephants, rhinoceros and hippopotamus, and in both the Paleotropics and Neotropics tapir, take on this role. Large-bodied primates may also contribute. In Australia this role is filled not by a mammal but by a bird, the southern cassowary, *Casuarius casuarius*. Although smaller in size than its mammalian equivalents it selects fruits of a similar size and morphology. For example, the largest seed sizes dispersed by elephants in Africa and tapir in the Neotropics are 65 × 35 mm (length and width) and 55 × 28 mm, respectively (Forget *et al.*, 2007). The largest seeds dispersed by cassowaries are those of *Endiandra microneura* (62 × 33 mm). Like elephants and tapir, cassowaries probably contribute to the persistence of large-sized fruits by reducing selection that might otherwise be imposed by smaller-bodied dispersers.

Seed dispersal in Australia's rainforests

Identifying the species participating in seed dispersal is an important but preliminary step in understanding seed dispersal, its outcomes and its consequences for population and community processes (Levin *et al.* 2003; Levine & Murrell 2003). For a variety of reasons, however, tropical ecologists have failed to describe dispersal adequately. There are several reasons why. First, there are many potential fruit-by-frugivore interactions, 65 × 1268 in the case of the Wet Tropics, so describing them all is simply not feasible. Second, documenting the outcome of dispersal in terms of survival and destination is complicated by the difficulties

of tracking both frugivores and seeds. This has seen ecologists focus on just one or a few fruit – frugivore interactions (for example Murray 1988; Westcott & Graham 2000; Dennis 2003) or examining dispersal over only a small part of its spatial scale. This failure to consider dispersal as a complete process has meant that the scale over which the process occurs, and therefore must be managed, has not been identified. Nor have realistic dispersal kernels been incorporated into attempts to predict the outcomes of population- or community-level processes.

In the following sections we derive an example of a complete dispersal kernel, that is, one that includes the relative contributions of all dispersers, for a Wet Tropics genus, *Cissus*. The approach we take assumes that higher level descriptions require the loss of some detail and has several steps. First, reduce the complexity introduced by high species diversity by deriving functional classifications of fruits and frugivores (fruit functional groups (FFGs) and disperser functional groups (DFGs)). Second, document the patterns and frequency of interaction between FFGs and DFGs. Third, determine the outcome of these interactions by describing dispersal kernels produced for a FFG by each DFG. Finally, combine these dispersal kernels to produce the complete dispersal kernel for an FFG.

Fruit functional groups

Our classification of fruits reduces 1347 vertebrate-dispersed plants to just nine FFGs (Table 16.1) (Dennis & Westcott, in preparation) and is based on characters that have the potential to restrict physically fruit processing and dispersal of seeds by particular frugivores and thereby determine which frugivores are capable of providing dispersal. The most fundamental division in the classification is between fruits whose flesh is the key attraction to dispersers and those where a dry covering or the seed itself is the attraction, i.e. those generally associated with frugivory and granivory. Despite some overlap, the suites of consumers attracted by fleshy and dry fruits are very different. Fleshy fruited species form the bulk of the diet of most DFGs, with only a few incorporating woody-fruited species into their diets (Figure 16.2). Those that do consume woody fruits also disperse a small proportion of them (Theimer 2001; Dennis *et al.* 2005; Dennis & Westcott, 2007).

Table 16.1 Functional classification of fruits

Group	Size-class	Fruit width (mm)	No. of seeds	Seed width (mm)	Fleshy/ dry
Fig	Small	<11	>50	<4	Fleshy
	Large	≥11	>50	<4	Fleshy
Multi-seeded	Small	<11	≥4	<4	Fleshy
	Large	≥11	≥4	<4	Fleshy
Few-seeded	Small	<11	<4		Fleshy
	Medium	11–24	<4		Fleshy
	Large	≥24	<4		Fleshy
Woody	Small	<11			Dry
	Large	≥11			Dry

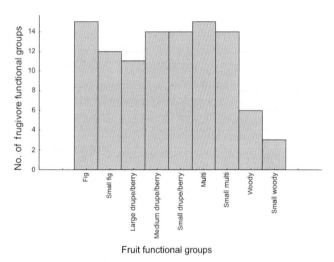

Figure 16.2 The number of frugivore functional groups visiting trees of the different fruit functional groups.

The next division in the classification is between fleshy fruits with multiple small seeds (width ≤4 mm; >3 seeds) and those with one or more large seeds (width >4 mm; ≤3 seeds). This division is important because fruits with few, large seeds can only be dispersed by species that are capable of swallowing or carrying those seeds. In contrast, fruits with many, small seeds tend to have their seeds accessed and dispersed by a wide range of dispersers. Even when their fruits are large, the seeds of these species can often be ingested by small dispersers if these can tear off sections of the fruit, for example *Ficus pleurocarpa*. We chose 4 mm as the seed size cut-off to distinguish between these two categories because this is the

largest seed ingested by the two smallest birds, the mistletoebird, *Dicaeum hirundinaceum*, and the silvereye, *Zosterops lateralis*, and is the largest sized seed swallowed by fruit bats (*Pteropus conspicillatus*).

Within the multiple-and small-seeded fruits we distinguish between figs and non-figs because figs have been repeatedly identified as particularly important (Shanahan *et al.* 2001; Westcott *et al.* 2005a), keystone (Terborgh 1986b) or attractive to a very wide arrays of potential dispersers (Coates-Estrada & Estrada 1986; Lambert 1989; Kitamura *et al.* 2002). This is potentially due to their spatial and temporal ubiquity as a fruit crop (Terborgh 1986b; Westcott *et al.* 2005a) and their high nutritive value (O'Brien *et al.* 1998). In the Wet Tropics, figs attract a much wider array of dispersers than any other plant group and have particularly high visitation rates (Dennis & Westcott, in preparation).

Fruit size defines the final categories within each of the fruit types. Fruit size has been shown to be one of the most significant factors determining selection and dispersal of fruits by frugivores and granivores (for example Herrera 1985; Levey 1987; Jordano 1992; Wheelwright 1993; Peres & van Roosmalen 2002) and there is commonly a negative relationship between the size of fruits and the size of disperser assemblages consuming them (for example Pratt & Stiles 1985; Hamman & Curio 1999; Githiru *et al.* 2002). In general, fruits smaller than 11 mm can be swallowed whole by the entire Wet Tropics frugivore community, with the exception of flying-foxes, which chew fruits, spitting out the majority of seeds. Larger fruits are consumed by an ever decreasing proportion of dispersers (Figure 16.2). For example, drupes and berries <11 mm are consumed by all dispersers, while those between 11 and 24 mm and those >24 mm are consumed by 84 and 27% of dispersers respectively.

Frugivore functional groups

Overlap in the dispersal service provided by dispersers means that a functional classification of dispersers based on the nature of their dispersal service can further simplify the description. Dennis and Westcott (2006) have done this for the Wet Tropics based on traits that influence dispersal outcomes (Schupp 1993): the quantity and diversity of seeds moved and the quality of handling. The most basic division is between species that are primarily granivorous and specifically

Table 16.2 Functional classification of dispersers

Division 1	Division 2	Division 3	Division 4	Service provided	Australian Wet Tropics representatives
Granivores	Poor dispersers	Generally not significant	Digestive predators	Kill most seeds, some long distance dispersal. Sometimes secondary dispersers	Columbidae – *Columba leucomela, Chalcophaps indica, macropygia amboinensis*; Megapodiidae – *Alectura lathami*
			Chewing predators	Kill most seeds, can disperse over short distances	Psittacidae – *Platycercus elegans, Alisterus scapularis, Cyclopsitta diopthalma, trichoglossus haematodus, T. chlorolepidotus* (rarely *Glossopsitta pusilla*); Cacatuidae – *Cacatua galerita*; Macropodidae – (rarely *Thylogale stigmatica, Dendrolagus lumholtzii, D. bennetti*); Psuedocheiridae – (rarely *Psuedochirops archeri, Psuedochirulus herbertensis*); Phalangeridae – (rarely *Trichosurus johnsonii*)
	Occasional dispersers	Significant for some species	Mega predators	Kill most seeds, can disperse hard seeds, potentially long distances	Suidae – *Sus scrofa*
			Predatory rodents	Kill most seeds but also scatter or larder-hoard. Sometimes secondary dispersers.	Muridae – *Melomys cervinipes, Rattus leucopus, R. fuscipes, Uromys caudimaculata, U. hadrourus*
Frugivores	Short distance	Within-forest	Small	Scatter or small clump-disperse over short distances	Ptilonorhynchidae – *Prionodura newtoniana, Ptilonorynchus violaceus, Scenopoeetes dentirostris, Ailuroedus melanotus*; Paradisaeidae – *Ptiloris victoriae*
			Large	Scatter or small clump-disperse wide range of fruit sizes over short to long distances	Columbidae – *Ptilinopus regina, P. superba, P. magnificus*; Oriolidae – *Oriolus sagittatus, O. flavocinctus*
			Terrestrial	Short distance dispersers and/or scatter-hoarders; kill few seeds	Hypsiprymnodontidae – *Hypsiprymnodon moschatus*
		Throughout landscape	Frugivores	Short to moderate distance dispersal of primarily small fruits and seeds throughout the landscape	Meliphagidae – *Meliphaga lewinii*; Dicaeidae – *Dicaeum hirundinaceum*; Zosteropidae – *Zosterops lateralis*
			Facultative	Regular, short distance dispersal of few species with small fruit	Meliophagidae – *Lichenostomus frenatus,Meliphaga notata, Xanthotis macleayana*; Campephagidae – *Lalage leucomela*
			Opportunists	Rare, short distance dispersal of very few species	Meliphagidae – *Philemon buceroides, Philemon corniculatus, Myzomela obscura, Meliphaga gracilis, Lichenostomus flavus, L. versicolor, L. Chrysops*; Cracticidae – *Cracticus quoyi, Coracina papuensis*; Dicruridae – *Dicrurus bracteatus*; Pteropodidae – *Syconycteris australis*
	Long distance	Cross landscape	Wide-ranging long-retention	Very long distance scatter or small clump-dispersal, primarily from canopy	Cuculidae – *Scythrops novaehollandiae*; Columbidae – *Lopholaimus antarcticus*; Cracticidae – *Coracina novaehollandiae, C. lineata*
			Wide-ranging short-retention	Moderate to long distance, scatter or small clump-dispersal from canopy	Oriolidae – *Sphecotheres viridis*; Sturnidae – *Aplonis metallica*
			Wide-ranging large fruit	Moderate to very long distance, scatter or small clump-dispersal of high wide range of fruits (nomadic or migratory species)	Cracticidae – *Strepura graculina*; Cuculidae – *Eudynamys scolopacea*; Columbidae – *Ducula bicolor*
			Wide-ranging small seeds	Long dispersal distances for seeds <5mm, Short distances for larger seeds, high levels of seed wastage and frequent clumping of some seeds	Pteropodidae – *Pteropus conspicillatus, Nyctimene robinsoni*, (rarely *P. alecto, P. scapulatus*)
		Terrestrial	Mega terrestrial frugivores	Moderate to long distance, clump disperseral	Casuaridae – *Casuarius casuarius*

Source: Dennis & Westcott (2006).

feed on seeds and those that are primarily frugivorous and feed on fruit flesh (Table 16.2).

Granivorous species generally kill most seeds but disperse a small number intact; granivores such as *U. caudimaculatus* are the major dispersers of some species (Theimer 2001). The Wet Tropics granivores include digestive predators (Columbidae, Megapodidae), chewing predators (Psittacidae, Cacatuidae, Cacropodidae, Pseudocheirodae, Phalangeridae), large terrestrial predators (Suiidae) and predatory rodents (Muridae).

Among the frugivores we distinguish six functional groups that share a characteristic of relatively short distance dispersal and five that disperse over longer distances (Table 16.2). Three of the short-distance groups include species that move predominantly under the forest canopy, and, in fragmented forests, tend to remain within fragments. These are further broken into one group of volant species that feed on small- to medium-sized fruits, provide short to medium distance dispersal and can produce both scattered and clumped patterns of deposition (Ptilonorhynicidae, Paradisaeidae), and another volant group that provide very similar dispersal but over a broader range of distances and for small to large fruits (Columbidae, Oriolidae). The third group includes a single species, *H. moschatus*, which provides short distance dispersal to generally medium to large fruits.

The next set of DFGs are species that provide short-distance dispersal but frequently cross habitat boundaries (Table 16.2). One contains primarily frugivorous birds (Meliphagidae, Dicaeidae, Zosteropidae) that generally disperse small- to medium-sized fruits. The next two groups are characterized by lower levels of frugivory. One is similar to the previous group but its members, though they feed on fruits regularly, eat only a few species and in small quantities (Meliphagidae, Campephagidae). For the remaining group, fruit represents a highly variable proportion of the diet (Meliphagidae, Artamidae, Dicruridae, Pteropidae).

The final set of DFGs comprise frugivores that provide regular, long distance dispersal (Table 16.2). All of these disperse a broad range of fruit types and sizes, and move widely across the landscape, most moving easily through both continuous forest and fragmented landscapes. These groupings include wide-ranging volant species that have a slow gut-passage rate (Cuculidae, Columbidae, Cracticidae), have a rapid gut-passage rate (Oriolidae, Sturnidae), consume large-seeded fruits (Cracticidae, Cuculidae, Columbidae) or disperse only

small seeds through the gut (Pteropidae). The final functional group is comprised of a single species, *C. casuarius* (Casuaridae), which consumes small- to large-seeded fruits, is terrestrial and provides long distance dispersal.

Visitation to fruiting trees

Having identified the FFGs and DFGs, we now consider how these functional groups interact and what the outcomes of these interactions are in terms of how far seeds move from their parents. Of particular interest is the identification of the relative contributions of each DFG to crop removal. This is done by observing fruiting trees and recording visitation and fruit removal. Once these relative contributions are known, the proportions of the crop can be assigned to each DFG contribution to an FFG complete dispersal kernel.

Different FFGs are visited by different DFGs (Figure 16.2). Large figs and large multi-seed fruits, because they can be torn apart and eaten piecemeal, are the only FFGs to be fed upon by all Wet Tropics DFGs. Small figs and multi-seeded fruits are fed on by fewer DFGs and in each case it is terrestrial DFGs that are missing. A similar pattern of decrease in the number of DFGs consuming a fruit with a decrease in the size of the fruit also occurs for woody fruits. This repeated pattern may be a function of the smaller reward small fruits provide to larger dispersers or for those that invest significant energy in handling. For few-seeded FFGs the patterns of fruit removal are very different. When small, these fruits are fed on by all DFGs with the exception of large terrestrial predators. In contrast, large few-seeded fruits are fed on by just 11 DFGs; those not represented have gape sizes that are small relative to the fruit.

Importantly, rates of visitation differ substantially across the different FFGs. Visitation rates, measured as the number of animals recorded in focal trees at 5-minute intervals, were greatest for figs, then few-seeded fruits and finally multi-seeded and woody species (Figure 16.3). Medium and small FFGs have higher rates of visitation by frugivores (size category $F_{1,2} = 4.36$, $p = 0.037$; fruit type $F_{1,2} = 8.46$, $p = 0.01$; interaction not significant). This is probably a function of their greater use by all species, and particularly by smaller species that tend to have frequent, short visits. Rates of fruit removal showed an almost identical pattern, with highest removal rates recorded in figs,

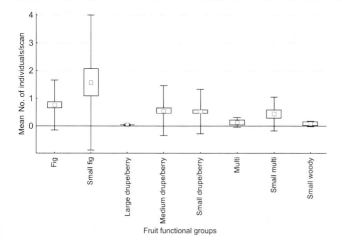

Figure 16.3 Mean visitation rate by frugivores to fruiting trees of the different fruit functional groups. Visitation rate is measured as the mean number of individual frugivores present in the tree during scans conducted every 5 minutes during observation periods.

Table 16.3 Relative contribution as percentage of total fruit crop removed by each disperser functional group for species in the medium-sized, few-seeded fruit functional group, which includes *Cissus* spp.

Disperser functional group	Medium drupes and berries
Chewing predators	6.8
Gristmill predators	6.4
Predatory rodents	8.8
Small within forest	10
Large within forest	11.2
Opportunists	1.6
Facultative	2.4
Terrestrial within forest	7.2
Wide-ranging, small	5.6
Wide-ranging, small seeds	6.5
Wide-ranging, large seeds	9.6
Wide-ranging, slow gut	7.7
Wide-ranging, rapid gut	6.4
Large terrestrial	9.6

followed by few-seeded, multi-seeded and woody species. Small figs and few-seeded fruit had higher rates of removal than large fruits of similar morphology. An example of the relative contribution to fruit removal of *Cissus* is given for the medium few-seeded FFGs in Table 16.3.

Estimating dispersal curves

Having documented the relative contribution of each DFG to fruit removal, the next step in estimating an FFG's complete dispersal kernel is to document how far each DFG moves the seeds it handles. Dispersers can be divided into two groups: (i) those that transport a seed or fruit to a new location for consumption or storage; and (ii) those that consume a fruit or seed *in situ* and transport it internally before depositing it. For the former group it is the factors influencing their choice of feeding location that determine dispersal distances, a choice that has proven difficult to predict (Theimer 2001; Dennis 2003). For these species dispersal distance is documented using spool lines or flags attached to seeds (Harrington *et al.* 1997; Theimer 2001; Dennis 2003; Dennis *et al.* 2005). Their dispersal tends to be short distance, for example < 70 m with a mean of just 13 metres for *U. caudimaculatus* (Theimer 2001) and 17 metres for *H. moschatus* (Dennis 2003).

Dispersers that consume fruits and seeds *in situ* tend to be species that range more widely and rapidly (for example birds), making direct documentation of dispersal distance difficult. For these species we assume that dispersal distances are a function of how long seeds are retained and how far dispersers move during this time. Dispersal distances can then be estimated as the product of the frequency distribution of gut passage times and displacement rates of the dispersers (Westcott & Graham 2000; Westcott *et al.* 2005b). The former we document using captive feeding trials in which representatives of each DFG are fed representatives from each FFG (Figure 16.4). Displacement rates are determined using continuous radiotelemetry. Representatives of each DFG are fitted with radio transmitters and locations along their movement pathway are determined as frequently as possible. Estimation of dispersal distances requires determination of the dispersers' straight-line displacement during each seed's retention time. A set of starting times in each telemetry session, chosen according to the temporal pattern of foraging, is assumed to be analogous to the time of ingestion. The displacement distance for each gut passage time observed in the captive feeding trials is then calculated from each of these starting points (Westcott *et al.* 2005b). The DFG dispersal kernel is the frequency distribution of these dispersal distances (Figure 16.5a).

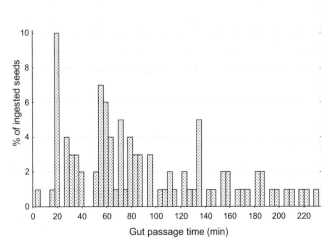

Figure 16.4 Gut passage rates of 197 seeds in the medium, few-seeded fruits fruit functional group through the guts of individuals of the wide-ranging disperser functional group.

Since all FFGs are fed upon by multiple DFGs, a single DFG dispersal kernel represents just a fragment of the dispersal the FFG receives. Complete dispersal kernels for a FFG incorporate the contribution of all DFGs. To estimate complete dispersal kernels it is necessary to scale each DFG contribution to removal relative to that of all other DFGs. These scaled kernels are then combined to give the complete dispersal kernel. In Figure 16.5b we provide an example of a complete dispersal kernel, one that incorporates the dispersal of all DFGs and is scaled according to their contribution to fruit removal, for species in the genus *Cissus* (Vitaceae), a medium-sized, few-seeded fruit. This figure shows the large proportion of dispersal that occurs within 200 metres of the vine. What is not clear is its very long tail; with dispersal occurring regularly to 1200 metres and infrequently to twice that distance (not shown). In addition, it highlights how the dispersal kernel is the cumulative contribution of a number of DFGs.

Several points need to be made about this kernel. First, it is clear that different dispersers provide very different dispersal in terms of their relative contribution to fruit removal and dispersal distance. Loss of particular DFGs will result in loss of particular types of dispersal. For example, cassowaries are responsible for much long distance dispersal for this genus; loss of cassowaries would dramatically truncate the tail of the kernel and consequently gene flow between populations. Second, dispersal occurs over

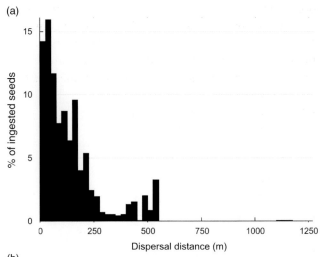

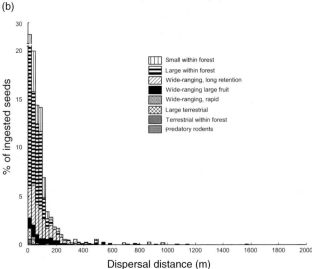

Figure 16.5 Dispersal curves produced for species in the genus *Cissus* (Vitaceae) by: (a) the wide-ranging, large-fruit disperser functional group, calculated using the total fruits consumed by that group; and (b) all disperser functional groups, i.e. the complete dispersal curve, calculated using the total number of fruits in the crop. The wide-ranging, large-fruit disperser functional group is solid black in both plots.

very long distances. We have shown dispersal distances of up to 1200 metres but estimate small amounts of dispersal by wide ranging DFGs to 2400 m. This suggests that dispersal occurs over distances that would not previously have been guessed at. Importantly, the species that are performing this long distance dispersal are all species that are, to varying degrees, willing to cross habitat boundaries, and thus we might expect that dispersal for *Cissus* and similar medium-sized fruits will occur across both continuous and fragmented landscapes. However, most dispersal occurs over comparatively short distances, less than 200 metres, and therefore within local populations. Since individual spacing in most tree species is less than 100 metres it would appear that much dispersal will be to locations where the nearest conspecific will not be the parent tree (Westcott *et al.* 2005b).

Summary

• Vertebrate seed dispersal in the Wet Tropics incorporates the interactions of at least 1268 plants and 65 dispersers, both frugivores and granivores. Birds provide the bulk of dispersal in most cases; however, mammals provide unique forms of dispersal and are particularly important for some plant species.

• Dispersal as a process operates similarly in the Wet Tropics as it does elsewhere in the tropics, despite differences in the taxa providing the services. Fruiting trees are visited by a wide range of dispersers and although they provide a broad range of dispersal services, there is marked overlap between species in the service provided in terms of how and which seeds are processed and the distances they are dispersed.

• By combining detailed data on each step of the dispersal process we have described complete dispersal curves for species in the genus *Cissus*. These dispersal curves show dispersal processes as operating at landscape scales most commonly up to 200 m but regularly over distances of up to 1200 m.

• Long-distance dispersal, up to 2400 m, is provided by a handful of species that to varying degrees will cross habitat boundaries. Thus, dispersal processes may be resilient in the face of fragmentation. However, long distance dispersal is probably the most critical component for the long-term persistence of plant metapopulations and the loss of threatened long distance dispersers may have significant impacts on plant populations.

References

Aristotle (*c.*350 BCE). *Historia Animalium, Vols IV–VI*, Peck, A. L. (translator). Loeb Classical Library, Harvard University Press, Cambridge, MA.

Beehler, B. M. & Dumbacher, J. P. (1996). More examples of fruiting trees visited predominantly by birds of paradise. *Emu* **96**: 81–8.

Beehler, B. M. & Pruett-Jones, S. G. (1983). Display dispersion and diet of birds of paradise: a comparison of nine species. *Behavioral Ecology and Sociobiology* **13**: 229–38.

Boulter, S. L., Kitching, R. L. & Howlett, B. G. (2006). Family, visitors and the weather: patterns of flowering in tropical rain forests of northern Australia. *Journal of Ecology* **94**: 369–82.

Coates-Estrada, R. & Estrada, A. (1986). Fruiting and frugivores at a strangler fig in the tropical rain forest of Los Tuxtlas, Mexico. *Journal of Tropical Ecology* **2**: 349–57.

Crome, F. H. J. (1975). The foraging ecology of fruit pigeons in tropical Northern Queensland. *Australian Wildlife Research* **2**: 155–85.

Dennis, A. J. (2003). Scatter-hoarding by musky rat-kangaroos, *Hypsiprymnodon moschatus*, a tropical rain-forest marsupial from Australia: implications for seed dispersal. *Journal of Tropical Ecology* **19**: 619–27.

Dennis, A. J., Lipsett-Moore, G., Harrington G. N., Collins, E. A. & Westcott, D. A. (2005). Secondary seed dispersal, predation and landscape structure: does context make a difference in tropical Australia? In *Secondary Dispersal and Seed Fate*, Forget, P.-M. & Vanderwaal, M. (eds). CABI Publishing, Wallingford, UK, pp. 117–35.

Dennis, A. J. & Marsh, H. (1997). Seasonal reproduction in musky rat-kangaroos, *Hypsiprymnodon moschatus*, a response to changes in resource availability. *Wildlife Research* **24**: 561–78.

Dennis, A. J. & Westcott D. A. (2006). Reducing complexity in the study of seed dispersal: a functional classification of seed dispersers in Australia's tropical rain forests. *Oecologia* **149**: 620–34.

Dennis, A. J. & Westcott, D.A. (2007). Estimating dispersal kernels produced by a diverse community of vertebrates. In *Frugivory and Seed Dispersal: Theory and Its Application in a Changing World*, Dennis, A. J., Green, R., Schupp, E., & Westcott, D. A. (eds). CABI Publishing, Wallingford, UK, pp. 201–228.

Desouzastevaux, M. C., Negrelle, R. R. B. & Citadinizanette, V. (1994). Seed dispersal by the fish *Pterodoras granulosus* in the Parana River basin, Brazil. *Journal of Tropical Ecology* **10**: 621–6.

Dew, J. L. & Boubli, J. P. (2005). *Fruits and Frugivores: The Search for Strong Interactors*. Kluwer Academic Publishers, New York.

Forget, P.-M., Dennis, A. J., Mazer, S., Jansen, P. A., Lambert, J. A. & Westcott, D. A. (2007). Seed allometry and frugivore size: a global comparison of patterns in tropical rainforests. In *Frugivory and Seed Dispersal: Theory and Its Application in a Changing World*, Dennis, A. J., Green, R.,

Schupp, E., & Westcott, D. A. (eds). CABI Publishing, Wallingford, UK, pp. 5–36.

Gatti, A., Bianchi, R., Rosa, C. R. X. & Mendes, S. L. (2006). Diet of two sympatric carnivores, *Cerdocyon thous* and *Procyon cancrivorus*, in a restinga area of Espirito Santo State, Brazil. *Journal of Tropical Ecology* **22**: 227–30.

Githiru, M., Lens, L., Bennur, L. A. & Ogol, C. P. K. O. (2002). Effects of site and fruit size on the composition of avian frugivore assemblages in a fragmented Afrotropical forest. *Oikos* **96**: 320–30.

Godinez-Alvarez, H. (2004). Pollination and seed dispersal by lizards: a review. *Revista Chilena de Historia Natural* **77**: 569–77.

Hamann, A. & Curio, E. (1999). Interactions among frugivores and fleshy fruit trees in a Philippine submontane rainforest. *Conservation Biology* **13**: 766–73.

Harrington, G. N., Irvine, A. K., Crome, F. H. J. & Moore, L. A. (1997). Regeneration of large-seeded trees in Australian rainforest fragments: a study of higher order interactions. In *Tropical Forest Remnants: Ecology, Management, and Conservation of Fragmented Communities*, Laurance, W. F. & Bierregaard, R. O. (eds). University of Chicago Press, Chicago, pp. 292–303.

Herrera, C. M. (1985). Determinants of plant–animal coevolution: the case of mutualistic dispersal of seeds by vertebrates. *Oikos* **44**: 132–41.

Jordano, P. (1992). Fruits and frugivory. In *Seeds: The Ecology of Regeneration in Plant Communities*, M. Fenner (ed.). CABI, Wallingford, UK, pp. 105–56.

Kitamura, S., Yumoto, T., Poonswad, P. *et al.* (2002). Interactions between fleshy fruits and frugivores in a tropical seasonal forest in Thailand. *Oecologia* **133**: 559–72.

Lambert, F. (1989). Fig-eating by birds in a Malaysian lowland rain-forest *Journal of Tropical Ecology* **5**(4): 401–12.

Levey, D, J. (1987). Seed size and fruit-handling techniques of avian frugivores. *American Naturalist* **129**: 471–80.

Levey, D. J., Silva, W. R. & Galetti, M. (eds) (2002). *Seed Dispersal and Frigovory: Ecology, Evolution, and Conservation*. CABI Publishing, Wallingford, UK.

Levin, S. A., Muller-Landau, H. C., Nathan, R. & Chave, J. (2003). The ecology and evolution of seed dispersal: a theoretical perspective. *Annual Review of Ecology Evolution and Systematics* **34**: 575–604.

Levine, J. M. & Murrell, D. J. (2003). The community-level consequences of seed dispersal patterns. *Annual Review of Ecology Evolution and Systematics* **34**: 549–74.

Loiselle, B. A., Sork, V. L. & Graham, C. (1995). Comparison of genetic variation in bird-dispersed shrubs of a tropical wet forest. *Biotropica* **27**: 487–94.

Lucas, P. W., Dominy, N. J., Riba-Hernandez, P. *et al.* (2003). Evolution and function of routine trichromatic vision in primates. *Evolution.* **57**: 2636–43.

Mack, A. L. (1993). The sizes of vertebrate-dispersed fruits: a neotropical–paleotropical comparison. *The American Naturalist* **142**: 840–56.

McKey, D. (1975). The ecology of coevolved seed dispersal systems. In *Coevolution of Animals and Plants*, Gilbert, L. E & Raven, P. H. (eds). University of Texas Press, Austin, pp. 159–91.

McNab, B. K. (2005). Food habits and the evolution of energetics in birds of paradise (Paradisaeidae). *Journal of Comparative Physiology B – Biochemical, Systematic and Environmental Physiology* **175**: 117–32.

Mazer, S. J. & Wheelwright, N. T. (1993). Fruit size and shape: allometry at different taxonomic levels in bird-dispersed plants. *Evolutionary Ecology* **7**: 556–75.

Moore, G. J. M (1991). Seed dispersal by male tooth-billed bowerbirds, Scenopoetes dentirostris (Ptilonorhynchidae), in north-east Queensland rain forests: processes and consequences. Unpublished doctoral dissertation, James Cook University, Townsville.

Murray, K. G. (1988). Avian seed dispersal of three neotropical gapdependent plants. *Ecological Monographs* **58**: 271–98.

Nathan, R. & Muller-Landau, H. C. (2000). Spatial patterns of seed dispersal, their determinants and consequences for recruitment. *Trends in Ecology and Evolution* **15**: 278–85.

Nowak, R. M. (1999). *Walker's Mammals of the World*, 6th edn. Johns Hopkins University Press, Baltimore, MD.

O'Brien, T. G. O., Kinnaird, M. F., Dierenfeld, E. S., Conklin-Brittain, N. L., Wrangham, R. W. & Silver, S. C. (1998). What's so special about Figures *Nature* **392**: 668.

Peres, C. A. & van Roosmalen, M. G. M. (2002). Patterns of primate frugivory in Amazonia and the Guianan shield: implications to the demography of large-seeded plants in overhunted forests. In *Seed Dispersal and Frugivory: Ecology, Evolution and Conservation*, Levey, D. J., Silva, W. R. & Galetti, M. (eds). CABI, Wallingford, UK.

Pratt, T. K. & Stiles, E. W. (1985). The influence of fruit size and structure on composition of frugivore assemblages in New Guinea. *Biotropica* **17**: 314–21.

Primack, R. & Corlett, R. (2005). *Tropical Rain Forests: An Ecological and Biogeographical Comparison*. Blackwell Publishing, Oxford.

Schondube, J. E., Herrera, M. L. G. & del Rio, C. M. (2001). Diet and the evolution of digestion and renal function in phyllostomid bats. *Zoology – Analysis of Complex Systems* **104**: 59–73.

Schupp, E.W. (1993). Quantity, quality and effectiveness of seed dispersal by animals. *Vegetatio* **107/108**: 15–29.

Shanahan, M. S, Compton S. G. & Corlett, R. (2001). Fig-eating by vertebrate frugivores: a global review. *Biological Reviews* **76**: 529–72.

Snow, D. W. (1971). Evolutionary aspects of fruit eating by birds. *Ibis* **113**: 194–202.

Snow, D. W. (1976). *The Web of Adaptation: Bird Studies in the American Tropics*. Demeter Press, New York City.

Stevenson, P. R. (2000). Seed dispersal by woolly monkeys (*Lagothrix lagothricha*) at Tinigua National Park, Colombia: dispersal distance, germination rates, and dispersal quantity. *American Journal of Primatology* **50**(4): 275–89.

Stocker, G. C. & Irvine, A. K. (1983). Seed dispersal by cassowaries (*Casuarius casuarius*) in north Queensland rainforest. *Biotropica* **15**: 170–6.

Sunquist, M. & Sunquist, F. (2002). *Wild Cats of the World*. University of Chicago Press, Chicago.

Terborgh, J. (1986a). Community aspects of frugivory in tropical forests. In *Frugivores and Seed Dispersal*, Estrada, A & Fleming, T. H. (eds). Dr W. Junk Publishers, Dordrecht, pp. 371–84.

Terborgh, J. (1986b). Keystone plant resources in the tropical forests. In *Conservation Biology: The Science of Scarcity and Diversity*, Soulá, M. E. (ed.). Sinauer Associates, Sunderland, MA, pp. 330–4.

Theimer, T. C. (2001). Seed scatter-hoarding by white-tailed rats: consequences for seedling recruitment by an Australian rain forest tree. *Journal of Tropical Ecology* **17**: 177–89.

Theophrastus (*c*.350 BCE). *De causis plantarum Vols I, II, & III*, Einarson, B. & Link, G. K. K. (translators). Loeb Classical Library, Harvard University Press, Cambridge, MA.

Wang, B. C. & Smith, T. B. O. (2002). Closing the seed dispersal loop. *Trends in Ecology and Evolution* **17**: 379–85.

Wenny, D. G. & Levey, D. J. (1998). Directed seed dispersal by bellbirds in a tropical cloud forest. *Proceedings of the National Academy of Sciences USA* **95**(11): 6204–7.

Westcott, D. A. & Graham, D. L. (2000). Patterns of movement and seed dispersal by a tropical frugivorous bird. *Oecologia* **122**: 249–57.

Westcott, D. A., Bentrupperbäumer, J., Bradford, M. G. & McKeown, A. (2005b). Incorporating disperser movement and behaviour patterns into models of seed dispersal. *Oecologia* **146**: 57–67.

Westcott, D. A., Bradford, M. G., Dennis, A. J. & Lipsett-Moore, G. (2005a). Key fruit resources in Australia's tropical rain forests. In *Fruits and Frugivores: The Search for Strong Interactors*, Dew, J. L. & Boubli, J. P. (eds). Kluwer Academic Publishers, New York, pp. 236–60.

Wheelwright, N. T. (1993). Fruit size in a tropical tree species, variation, preference by birds and heritability. *Vegetatio* **108**: 163–74.

Willson, M. F., Irvine, A. K. & Walsh, N. G. (1989). Vertebrate dispersal syndromes in some Australian and New Zealand plant-communities, with geographic comparisons. *Biotropica* **21**: 133–47.

Wright, S. J., Carrasco, C., Calderon, O. & Paton, S. (1999). The El Niño Southern Oscillation, variable fruit production, and famine in a tropical forest. *Ecology* **80**: 1632–47.

17 FLORAL MORPHOLOGY, PHENOLOGY AND POLLINATION IN THE WET TROPICS

Sarah L. Boulter[1]*, Roger L. Kitching[1]*, Caroline L. Gross[2],
Kylie L. Goodall[3]* and Bradley G. Howlett[3,4]*

[1]Griffith School of Environment, Griffith University, Nathan, Queensland, Australia
[2]Ecosystem Management, University of New England, Armidale, New South Wales, Australia
[3]Griffith University, Nathan, Queensland, Australia
[4]New Zealand Institute for Crop and Food Research Ltd, Christchurch, New Zealand
*The authors were participants of Cooperative Research Centre for Tropical Ecology and Management

Introduction

The reproductive ecology of most flowering plant species is a complex response to evolutionary processes that exist between a plant and its visitor array (Wyatt 1983; Dukas 2001), phylogenetic constraints (Johnson & Steiner 2000), phenotypic plasticity (Rathcke and Lacey 1985; Miller & Diggle 2003) and how these factors may have changed over its evolutionary history (Feinsinger 1983). Ultimately these are expressed in the form of flower morphology, flowering phenology and the attraction of associated flower visitors. A plant's fitness depends upon its reproductive success, particularly in allogamous matings; thus pollination becomes of paramount importance within the wider field of floral biology.

Several features of the Wet Tropics flora make it a potentially unique case study of reproductive ecology. First, rainforest dominates the bioregion's 1.8 million hectares (Sattler & Williams 1999). Second, the area has a high floristic diversity and a high species-to-area ratio (Myers *et al.* 2000; Metcalfe & Ford, Chapter 9, this volume). Third, the area has a high level of plant endemism (>30%) (Metcalfe & Ford, Chapter 9, this volume). In addition, there is a high generic diversity, a high incidence of monotypic genera (Gross 2005; Metcalfe & Ford, Chapter 9, this volume) and a high diversity of woody, phylogenetically basal, angiosperms (Worboys & Jackes 2005). The forests are also notable for their distinctive Gondwanan taxa (Webb & Tracey 1994), especially in the uplands. Unlike many equatorial areas of rainforest, the Wet Tropics region is properly categorized as seasonally dry (van Schaik *et al.* 1993), with at least five months of the year receiving less than 60 mm average rainfall in most parts (Gross 2005), with the exception of some small areas that receive higher rainfall (Metcalfe & Ford, Chapter 9, this volume).

Our understanding of the reproductive systems of the plants of the Wet Tropics remains *ad hoc* at best, with the pollination system of fewer than a handful of species known. This is also true of many other tropical floras worldwide, despite intensified efforts (see reviews in Bawa 1990; Williams & Adam 1994) and improved technologies, such as canopy access systems (i.e. canopy walkways, tree towers and canopy cranes) over the past four decades or so. This increased effort has highlighted the diversity of pollen vectors within tropical systems. Increased access to different layers of the rainforest, including the canopy, has been facilitated by the use of canopy walkways (van Dulmen 2001), canopy cranes (Boulter *et al.* 2005) and trapping techniques using devices that can be hauled into the canopy (House 1989, 1992), leading to an improvement in our understanding of the higher strata of the forest.

For example, the installation of the Australian Canopy Crane Research Facility at Cape Tribulation in 1999 has allowed increased opportunities to understand canopy animal–plant interactions in the Wet Tropics, including pollination.

This chapter reviews some of the key features of the reproductive ecology of plants and plant–pollinator systems of the rainforests of North Queensland. Four basic sources of data have been used. Two databases have been compiled by Boulter, Kitching, Gross and Howlett (Boulter *et al.* 2006). The first of these databases records key morphological features (e.g. habit, inflorescence form and position, flower size and colour, flower symmetry and sexual system) of the same species using information from existing floras (Cronin 2000; Hyland *et al.* 2003; Cooper & Cooper 2004). The second records collection month, altitude and latitude of individual plant species, using herbarium specimen sheets held at the North Queensland Herbarium and the Queensland Herbarium, Brisbane, that had flowers or buds (i.e. were reproductive at the time of collection). Both were compiled using the list of northern Australian trees, shrubs and vines recorded in Hyland *et al.* (2003), excluding any species not found in the Wet Tropics and any naturalized species. This information was compiled on a total of 1752 plant species. Information explicitly on breeding systems, fruit-types and floral visitors has been collated by Gross (Gross 2005; Sjöström & Gross 2006). In addition, we compile here a literature review on flower visitors and pollination interactions in the Wet Tropics.

Floral morphology

Flower morphology can provide clues to the reproductive ecology of a flora. Floral diversity is linked to the diversity of flower-visiting animals (Waser 1983). We summarize here some key floral features from the trees, shrubs and vines of the Wet Tropics.

Habit

Of the 1752 plant species we surveyed approximately 29% are trees, 23% are shrubs and a further 29% are trees/shrubs (i.e. those plants that can take the form of either a tree or a shrub). The vines made up almost 17% of the total, with a further 2% that could be described as either vines or shrubs (vines/shrubs).

Flower size

Hyland *et al.* (2003) categorize all flower corollas as either <10 mm or >10 mm in diameter. Nearly 75% of the plants we surveyed have flowers <10 mm in diameter. Where available, a more accurate estimate of flower size was entered into our database for 1459 species. For those species we see that the number of flowers with diameters larger than 10 mm falls rapidly over 10 mm (Figure 17.1) and very large flowers are a rare occurrence. The division of flower size appears fairly consistent across all habits (Figure 17.2) and a chi-square

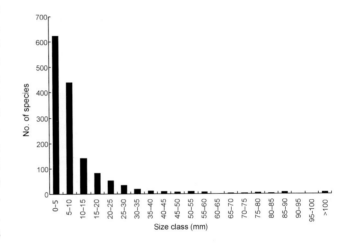

Figure 17.1 Total number of plant species with flower diameters in each 5 mm size class.

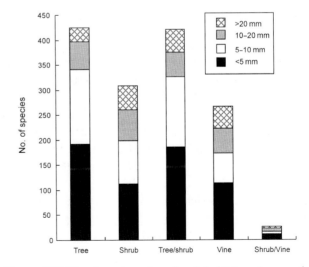

Figure 17.2 Number of species of each habit type in one of four flower diameter size classes.

test confirms that the two variables are independent ($\chi^2 = 1.537$, d.f. = 12, $P<0.05$).

Flower colour

Determining flower colour is a subjective task, and flowers are often the sum of their parts, which can be in a variety of colours. We assessed photographs and descriptions (Hyland *et al.* 2003), where available, to assign a dominant colour to the 1533 target species. These were then further grouped into the colour categories white/green, yellow/orange, pink/red, blue/purple and brown, based on colour groups generally associated with pollination syndromes (e.g. red or pink associated with bird pollination syndromes; Faegri & van der Pijl 1979). The overwhelming majority of flowers were white/green (72%), with 12% yellow/orange, 8% pink/red, 6% blue/purple, 1% brown and 1% having no corolla. Of course, we are aware that plant visitors will perceive colours in a non-human way (Kevan *et al.* 1996) but there is a clear understanding that colour, along with other elements of the flower's appearance, attracts pollinators and allows them to discriminate among flowers (Dafni *et al.* 1997). When considering the proportion of flowers representing each colour group within each habit type (Figure 17.3), not surprisingly we found white/green flowers dominant in all habit types. However, they do make up a greater proportion of tree and tree/shrub species

than vines and vines/shrubs. Using a chi-square test of independence, we found that colour and habit were dependent variables ($\chi^2 = 107.99$, d.f. = 20, $P<0.05$), suggesting that vines have colourful flowers more often than trees or shrubs. The proportion of white/green flowers appears to decrease with increasing flower size (Figure 17.4). A chi-square test showed that these variables were indeed dependent ($\chi^2 = 93.54$, d.f. = 15, $P < 0.05$), and that small flowers are more often a dull white or green colour than larger flowers are.

Sexual systems

The rainforest plants of northern Australia display the usual range of sexual systems found in tropical ecosystems (Gross 2005). As in rainforest communities elsewhere, most tree species in the Australian tropics have hermaphroditic flowers (approx. 60%, $n = 1100$). Significant differences, however, do occur in the proportions of unisexual systems at the population and landscape levels compared with tropical rainforest communities found elsewhere (see Table 2 in Gross 2005). Australia has significantly more monoecious taxa than expected when comparing the Australian data with other Old World or with New World tropical systems, where dioecy is the predominant unisexual system (see Table 2 in Gross 2005). In the main these differences arise because of a skew in the abundance

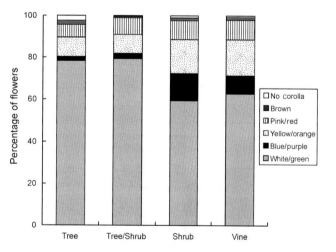

Figure 17.3 Proportion of dominant flower colour displayed by all Wet Tropics plant species surveyed by their habit type.

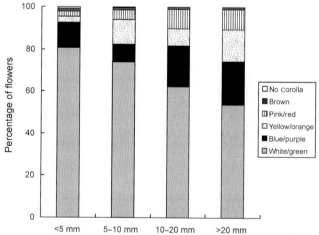

Figure 17.4 Proportion of dominant flower colour displayed by all Wet Tropics plant species surveyed by flower diameter size class.

of monoecious sex systems in the following families: Euphorbiaceae, Phyllanthaceae, Sapindaceae, Malvaceae, Moraceae and Combretaceae (Gross 2005). It is interesting to note that all of these families (and many of their genera) occur in Southeast Asia, where they are more often dioecious and in some cases less species-rich. In addition, monoecy is the most common unisexual breeding system for the Australian flora, which incorporates many diverse ecosystems (Sjöström and Gross 2006). Gross (2005) hypothesizes that monoecy is prevalent in the Australian tropics because most species are insect pollinated but concurrently pollinator-limited. This does not preclude specialization in plant–pollinator relationships as it has been shown that an oligolectic pollinator is not necessarily an efficient pollinator (e.g. *Myristica insipida* and beetle pollinators) (Armstrong and Irvine 1989a, b). Renner and Feil (1993) suggest that dioecy will not evolve when unreliable pollination systems are entrenched. Comparative studies are now required of dioecious and monoecious congeners to understand the complex relationship between pollinators and breeding systems.

Flowering phenology

The timing, duration and frequency of flowering are pivotal to plant reproductive success and, in turn, the success of animals that rely on floral resources. Driving these phenological patterns is a complex interaction between, and balance of, environmental factors, biotic factors and evolutionary factors (Wright & van Schaik 1994; Bawa *et al.* 2003; Bolmgren *et al.* 2003), although it is generally accepted that environmental or climatic factors are proximate cues to flowering (Wright 1996).

Individual plant species demonstrate a range of flowering patterns, from the most familiar, a simple and predictable annual rhythm, to more complex patterns such as supra-annual or mass flowering events. Flowering patterns can vary not only in timing but also in frequency, duration and amplitude. For example, flowering frequency may be continual (throughout the year), subannual (flowering in more than one cycle in a year), annual or supra-annual (one cycle over more than one year) (Newstrom & Frankie 1994). Variation in flowering patterns also occurs at the level of individual plants, individual species and the community.

Traditionally, phenological information requires long-term (up to ten years) regular observation of flowering behaviour in a number of individuals. In the absence of such data, surrogates for phenology data can be gleaned from extensive herbarium data sets (e.g. Borchert 1996; Boulter *et al.* 2006) to make an approximation of flowering patterns. No long-term flowering phenology studies have been published for the Wet Tropics bioregion, although limited short-term (two to three year) studies have been made (Frith & Frith 1985; Hopkins & Graham 1989; Geyer *et al.* in review). Since analysing the flowering shrubs and trees of the Wet Tropics using herbarium data (Boulter *et al.* 2006), we have expanded the database to include the vines of the Wet Tropics and present here an updated summary of flowering patterns in the Wet Tropics at the community and species level.

There is an *a priori* expectation that flowering in tropical floras will follow a seasonal rhythm, at the community level at least. This expectation arises from a link between flowering and the seasonal variation in limiting factors such as rainfall, solar irradiance and the presence of seasonally available pollinators, predators and frugivores or seed dispersers. Using the number of species recorded flowering in any given month, we see that such an annual rhythm in flowering exists for the Wet Tropics flora, in this case centred on the beginning of the wet season (October and November) (Figure 17.5). This pattern is equally represented in the vines, the trees and the shrubs. This analysis quite simply scores any flowering activity recorded for each species. Refining this analysis further, we have calculated a peak of flowering activity (see Boulter *et al.* 2006 for the method of calculation) for each species by employing circular vector statistics to estimate the mean or flowering midpoint for each individual. Using mean flowering midpoint or peak flowering (Figure 17.6) gives a similar seasonal trend as described above (Figure 17.5), although it is interesting to note that the vines appear to have a very strong seasonal trend, with low flowering midpoints in May through to August, the dry season. We conclude from these data that while flowering activity is continuous throughout the year, greater numbers of species, and probably higher intensity flowering, is experienced from the end of the dry season through the wet season across the Wet Tropics flora.

Determining the ultimate cause of flowering patterns, however, is problematic and a number of

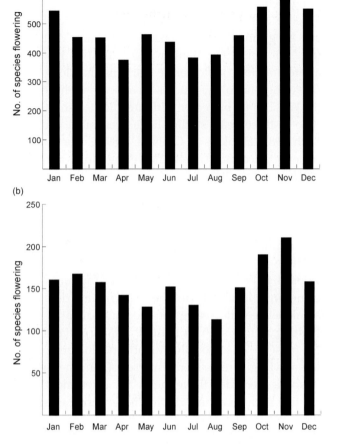

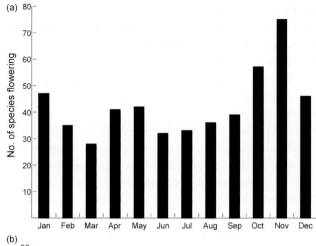

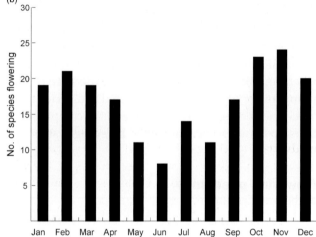

Figure 17.5 Total number of: (a) tree and shrub (reproduced from Boulter *et al.* 2006); and (b) vine species recorded flowering in any given month in the Wet Tropics bioregion.

Figure 17.6 The distribution of mean flowering times ('peak' flowering) for: (a) all tree and shrub (reproduced from Boulter *et al.* 2006); and (b) all vine species calculated as an algebraic vector.

hypotheses have evolved (see Boulter *et al.* 2006 for a review of the various hypotheses) to address this question. For convenience, we group them into three broad categories. The **climatic hypotheses**, alluded to above, link patterns of flowering and fruiting to variation in seasonally limited resources (Waser 1979; Wright & van Schaik 1994; Ramírez 2002). The **biotic hypotheses** link the activities of pollinators, predators and dispersers to the synchrony of phenological activity among plants. For example, the pollinator competition hypothesis suggests that to avoid competing with other individuals for a limited set of pollinators, flowering events should spread evenly through time (Pleasants 1980; Bolmgren *et al.* 2003). In the contrary view, the mass action hypothesis suggests that facilitation is more important than competition for plants that share

pollinators or predators and this will synchronize their flowering phenology (Rathcke 1983; Sakai 2002). The **phylogenetic hypothesis** proposes that phenological patterns are in some way constrained by phylogeny (Kochmer and Handel 1986). For example, taxonomically related groups of plants would flower at similar times regardless of geographical location. We note that the outcomes of many of these competing hypotheses, in terms of flowering pattern, may be indistinguishable, and may co-occur at different temporal or spatial scales. In the context of the Wet Tropics, we have concluded that, at the community level, there is evidence of coincidence of flowering among species of some families but we were unable to determine if this is the result of phenological or biotic factors (Boulter *et al.* 2006). Coincidence of flowering is probably limited to the family level. For example, some Wet Tropics families are clearly dry season flowerers, while others are wet season (see Table 4 in Boulter *et al.* 2006).

Flower associates and pollinators

Animals visit flowers for a number of reasons: in search of food resources (nectar, pollen, plant tissue), in search of mates (either using the inflorescence as a mating site or when deceived by the plant into pseudo-copulation) and in search of shelter and brood sites. The benefit for the visited flower can be successful pollination, but this outcome isn't always guaranteed. To qualify as a pollinator, a visitor must transfer pollen to the stigmatic surface of the flower, the pollen must germinate and a seed must set. The physiology of the visitor and the flower will play a significant role in this process, as does visitor behaviour. Few visitors actively pollinate a flower (e.g. fig-wasp) and most pollination relies on inadvertent touching of body parts to the reproductive structure. In addition some visitors remain in the host flower for their entire life and may have a limited role in the transfer of pollen among flowers.

There are few published records of visitors associated with flowers in the Wet Tropics and even fewer quantifiable assessments of successful pollinators (Table 17.1). Here we examine these few Wet Tropics studies and compare the findings with other tropical floras.

Wind pollination

Only 2–3% of tropical plants are wind pollinated (Regal 1982; Bawa 1990; Bush & Rivera 2001). Wind pollination is unreliable over large distances (Regal 1982) and is unlikely to be selected for in spatially isolated rainforest trees, although facultative wind pollination may play a greater role in rainforest pollination than is currently identified (Williams & Adam 1999). No Wet Tropics study has identified wind as a pollination mechanism (Table 17.1). Wind pollinated species are most commonly conifers and their near relatives. The Wet Tropics has 11 species of the conifer families Araucariaceae and Podocarpaceae. We would expect these all to be candidates for wind pollination.

Vertebrate pollination

Among vertebrate pollinators, birds and bats are readily identified, particularly in the New World. In Central America, hummingbirds constitute a major pollinator group (Bawa 1990). In other regions passerine birds (Sazima *et al.* 1993) and spiderhunters (Sakai *et al.* 1999a) have also been recognized as successful pollinators. Birds are persistent flower visitors in Australia (Ford *et al.* 1979) and the Australian bird fauna possess a large number of flower-adapted families (Armstrong 1979). Of the eight bird families listed by Proctor and Yeo (1972), five are known from Australia, and species from all five families occur in the Wet Tropics (Table 17.2). Species from the family Meliphagidae (particularly honeyeaters) are considered the principal nectar feeders and pollinators in Australia (Ford *et al.* 1979). Certainly those pollination studies from the Wet Tropics that have recorded bird visitors have all observed honeyeaters (*Syzygium sayeri*, Boulter *et al.* 2005; *S. cormiflorum*, Crome & Irvine 1986; *S. tieryanum*, Hopper 1980; *Hornstedia scottiana*, Ippolito & Armstrong 1993). Bird pollination systems are frequently associated with the flowers of one of four blossom types (Armstrong 1979; Faegri & van der Pijl 1979): brush-like inflorescences (e.g. *Callistemon*, *Banksia*), deep tubular corollas (e.g. *Correa* Rutaceae), gullet blossoms (e.g. *Calothamnus*, Myrtaceae) or flag-shaped blossoms (e.g. Fabaceae). They are often brightly coloured, have little odour and produce copious quantities of nectar (Ford *et al.* 1979). While characteristics such as appropriate flower shape and colour are well represented in the Wet Tropics flora, this does not mean that those flowers are bird pollinated. In an example where structure did seem to correlate with observed visitors, *Hornstedia scottiana*, a Wet Tropics ginger that is positioned close to the ground, seemed an unlikely candidate despite its flower having a scarlet corolla, lack of odour and copious nectar production consistent with a bird-pollination syndrome (Ippolito & Armstong 1993). Three species of honeyeater were identified as the only visitors, and the authors in that case concluded that honeyeaters were the pollinators of this species.

Bat pollination is also well documented in the tropics (Crome & Irvine 1986; Sazima *et al.* 1999), although it appears to be a fairly recent phenomenon in the Australian flora (Armstrong 1979). All Australian flower-visiting bats are from the family Pteropodidae,

Table 17.1 Known flower visitors and pollinators to plant species of the Wet Tropics region.

Species	Habit	Sexual system	Location/ Distribution	Flower visitors	Pollinator/(s)	Reference
Annonaceae						
Pseuduvaria froggattii[†]	Tree	Dioecious	Cape Tribulation	Diptera (Drosophilidae plus others),Coleoptera (Nitidulidae, Elateridae)	Diptera?	Silberbauer-Gottsberger et al. 2003
Pseuduvaria hylandii[†]	Tree	Dioecious/ monoecious	Cape Kimberley, Daintree National Park	Diptera	Diptera?	Silberbauer-Gottsberger et al. 2003
Goniothalamus australias[†]	Tree	Hermaphrodite	Atherton Tablelands	Coleoptera (Nitidulidae)		Silberbauer-Gottsberger et al. 2003
Melodorum spp. (Stone Crossing LWJ 814)[†]	Vine	Hermaphrodite	Atherton Tableland (cultivated)	Coleoptera (Curculionidae)		Silberbauer-Gottsberger et al. 2003
Meiogyne spp. (Henrietta Creek LWJ 512)[†]	Tree	Hermaphrodite	Atherton Tableland (cultivated)	Coleoptera (Nitidulidae)		Silberbauer-Gottsberger et al. 2003
Haplostichanthus spp. 1 (Topaz LWJ 520)[†]	Tree	Hermaphrodite	Atherton Tablelands	Coleoptera (Curculionidae)		Silberbauer-Gottsberger et al. 2003
Uvaria concava[†]	Vine/shrub	Hermaphrodite	Atherton Tablelands	Hymenoptera (Meliponinae, *Apis mellifera*)	Meliponinae	Silberbauer-Gottsberger et al. 2003
Arecaceae						
Normanbya normanbyi[*]	Palm	Dioecious	Cape Tribulation	Curculionidae, other Coleoptera, Diptera, Lepidoptera, Thysanoptera, Acari, other insects	Not identified	Kitching et al. 2007
Calamus radicalis[*]	Palm	Monoecious	Cape Tribulation	Diptera, Coleoptera, other insects	Not identified	Kitching et al. 2007
Licuala ramsayi[*]	Palm	Bisexual	Cape Tribulation	Diptera, Coleoptera, Hymenoptera (including ants), other insects	Not identified	Kitching et al. 2007
Balanophoraceae						
Balanophora fungosa[†]	Herb, root parasite	Monoecious	Nth 27 deg South	Nitidulid beetle, Tipulid fly, ants, rats		Irvine & Armstong 1990
Bignoniaceae						
Neosepicaea jucunda[*]	Vine	Bisexual	Cape Tribulation	Diptera, Hymenoptera, other insects	Not identified	Kitching et al. 2007
Calycanthaceae						
Idiospermum australiense[‡]	Canopy tree	Protogynous/ andromonoecious/ hermophrodite	Daintree, south of Cairns	Thysanoptera 2 spp., Coleoptera 14 spp., mites, spiders, Blattodea, Diptera, Hemiptera, Hymenoptera (including Formicidae), Lepidoptera (including Pyralidae), Orthoptera, Psocoptera	Generalist?	Worboys & Jackes 2005; Worboys 1998
Convolulaceae						
Merrimiapeltata[*]	Vine	Bisexual	Cape Tribulation	Diptera, Coleoptera, Hymenoptera, spiders, other insects	Not identified	Kitching et al. 2007
Ebenaceae						
Diospyros pentamera[‡]	Subcanopy/ Canopy tree	Dioecious	Nth NSW, south Qld & Nth Qld	Coleoptera 19 spp., Diptera 15 spp., Hymenoptera 11 spp.		Irvine & Armstrong 1990; House 1989
Eleocarpaceae						
Elaeocarpus largiflorens largiflorens	Subspp.	Tree	Hermaphrodite	Coleoptera 3 spp.; Diptera 2 spp.; Hymenoptera 4 spp.; Lepidopter 4 spp.		Weber 1994
Eupomatiaceae						
Eupomatia laurina[†]	Understorey tree, 3–10 m	Hermaphrodite	Sth NSW to Cairns area	Weevil		Irvine & Armstong 1990; Armstrong & Irvine

	Growth form	Breeding system	Location	Flower visitors	Effective pollinators	Reference
Flindersiaceae						
Flindersia brayleyana†	Canopy tree	Hermaphrodite	Cairns region	Beetles, flies & Lepidoptera		Irvine & Armstong 1990
Lauraceae						
*Neolitsea dealbata**	Canopy tree	Dioecious		Coleoptera 14 spp., Hymenoptera 11 spp., Diptera 17 spp.		House 1989
*Litsea leefeana**	Canopy tree	Dioecious		Coleoptera 14 spp., Diptera 15 spp., Hymenoptera 9 spp.		House 1989
Loganiaceae						
*Fagrea cambagei**	Tree	Bisexual	Cape Tribulation	Diptera, Coleoptera, Hymenoptera, other insects	Not identified	Kitching et al. 2007
Melastomaceae						
Melastoma affine	Shrub	Hermaphrodite		Hymenoptera 9 spp.	4 primary pollinators	Gross 1993a; Gross & Mackay 1998
Mimosaceae						
*Entada phaseoloides**	Vine	Bisexual	Cape Tribulation	Coleoptera, Diptera, Hymenoptera, spiders, other insects	Not identified	Kitching et al. 2007
Myristicaceae						
Myristica insipida†	Subcanopy/canopy tree	Dioecious	North of 20 deg south	Beetles		Irvine & Armstrong 1990; Armstrong & Irvin 1989a
Myrsinaceae						
Rapanea subsessilis†	Understory shrub	Gynodioecious			*Trigona* spp.	Harrison 1987
Maesa dependens		Androdioecious			Thysanoptera	Harrison 1987
Aegicera corniculatum†	Mangrove	Hermaphrodite/ Protandrous			Megachilidae spp.	Harrison 1987
Myrtaceae						
Syzygium tierneyanum†	Tree	Hermaphrodite		12 moth spp., 9 butterfly spp., 4 moths?, 4 flies, 1 wasp, 2 bees, 2 ants, 1 weevil, 1 bat		Hopper 1980
Syzygium cormiflorum‡	Tree	Hermaphrodite		Meliphagidae 4 spp., Pteropodidae 2 spp., Coleoptera 4 spp., Diptera 3 spp., Blattellidae 1 spp., Moths 2 spp., Formicidae 1 spp., Mite 1	Pteropodidae	Crome and Irvine 1986
Syzygium sayeri‡	Canopy tree	Hermaphrodite	Cape Tribulation	Honeyeaters ? spp., Bat 2/3 spp., Hymenoptera, Lepidoptera (Butterflies & Spingid moths), Coleoptera (? Families, ? spp.)	Day and night visitors	Boulter et al. 2005; Boulter 2003
Syzygium gustavioides	Canopy tree	Hermaphrodite	Cape Tribulation	Coleoptera (? Families, ? spp.), Diptera, Thysanoptera	Most insect visitors	Boulter 2003
*Acmena graveolens**	Tree	Hermaphrodite	Cape Tribulation	Diptera, Coleoptera, Hymenoptera, Thysanoptera, spiders, other insects	Not identified	Kitching et al. 2007
Orchidaceae						
Cymbidium madidum†	Epiphyte	Hermaphrodite	Herberton, Atherton Tablelands		Trigona	Barteau 1993
Cymbidium suave†	Epiphyte	Hermaphrodite	Herberton, Atherton Tablelands		Trigona	Barteau 1993
Dendrobium iichenastrum†	Epiphyte		Herberton, Atherton Tablelands		Trigona	Barteau 1993
Dendrobium monophyllum†	Epiphyte		Herberton, Atherton Tablelands		Trigona	Barteau 1993

(Continued)

Table 17.1 (*Continued*)

Species	Habit	Sexual system	Location/ Distribution	Flower visitors	Pollinator/s	Reference
Orchidaceae						
Dendrobium toresse[†]	Epiphyte		Herberton, Atherton Tablelands		Trigona	Barteau 1993
Dipodium ensifolium[†]	Herb		Herberton, Atherton Tablelands		Trigona	Barteau 1993
Diuris oporina[†]	Herb		Herberton, Atherton Tablelands		Hymenoptera	Barteau 1993
Pterostylis procera[†]	Herb		Herberton, Atherton Tablelands		Diptera	Barteau 1993
Proteaceae						
Banksia plagiocarpa[†]			Hinchinbrook Island & adjacent mainland	*Pteropus* spp., moth, *Trigona* spp., *Apis mellifera*	Fruit bats, moth	Robertson 1999
Rhamnaceae						
Alphitonia petriei[†]	Pioneer tree, 5–25 m	Hermaphrodite	Cairns region	Flies, beetles, wasps and bees		Irvine & Armstrong 1990
Rubiaceae						
Gardenia actinocarpa[‡]	Shrub to small tree	Dioecious	Noah Creek, Cape Tribulation	Lepidoptera 2 spp., Tiphiidae at least 2 spp., Hymenoptera ? spp., Diptera ? spp.		Osunkoya 1998
Rutaceae						
Flindersia brayleyana[†]	Tree	Hermaphrodite				Irvine & Armstrong 1990
Stangeriaceae						
Bowenia spectabilis[‡]	Cycad	Dioecious	Tarzali, Atherton Tablelands	weevils	*Milotranes prosternalis* (Lea) (Curculionidae: Curculioninae: Molytini)	Wilson 2002
Zingiberaceae						
Hornstedia scottiana[†]	Large herb	Hermaphrodite		3 spp. Honeyeaters		Ippolito & Armstrong 1993

Methods: *Insect trapping only.

[†] Observation only.

[‡] Combination trapping/observation methods and/or exclusion experiments.

Table 17.2 The main flower visiting families of birds (after Armstrong 1979), and the genera and total number of species (in parentheses) occurring in the Wet Tropics bioregion

Bird family	Genera
Dicaeidae (flowerpeckers)	Dicaeum* (1)
Meliphagidae (sugar birds, honeyeaters)	Acanthagenys (1), Acanthorhynchus (1), Certhionyx (1), Conopophila (2), Entomyzon (1), Grantiella (1), Lichenostomus (11), Manorina (2), Meliphaga (3), Melithreptus (3), Myzomela (2), Philemon (4), Phylidonyris (1), Plectorhyncha (1), Ramsayornis (2), Trichodere (1), Xanthotis (1).
Nectariniidae (Sunbirds)	Nectarinia (1)
Psittacidae (brush-tongued parakeets)	Alisterus* (1), Aprosmictus* (1), Cyclopsitta* (1), Glossopsitta (1), Psitteuteles* (1), Trichoglossus (2)
Zosteropidae (white-eyes)	Zosterops (2)

*Unlikely to be pollinators.
Source: Higgins and Peter (2002), T. Reis (pers. comm).

Table 17.3 Flower feeding bats of the Wet Tropics and genera identified as a food source. Number of species recorded in the Wet Tropics from each genus in parentheses

Species of the family Pteropodidae	Genera of identified food Sources
Macroglossus minimus (northern blossom bat)	Nectar: Syzygium (37), Sonneratia (3)
Syconycteris australis common blossom bat	Nectar: Syzygium (37), Melaleuca (5), Banksia (1), Grevillea (3), Eucalyptus (5).
Nyctimene robinsoni (eastern tubenosed bat)	Nectar: Banksia (1)
Pteropus alecto (black flying fox)	Blossom: Eucalyptus (5), Melaleuca (5), Syncarpia glomulifera
Pteropus conspicillatus (spectacled flying-fox)	Blossom: Eucalyptus (5)
Pteropus scapulatus (little red flying fox)	Blossom: various

Source: Churchill (1998), Hyland et al. (2003), Cooper and Cooper (2004).

a predominantly herbivorous group of bats that feed on fruits, flower structures and nectar (Armstrong 1979). Six species from this family are recorded in the Wet Tropics region and all have been recorded drinking nectar or feeding on blossom of various plant genera (Table 17.3). Bats are generally associated with tube- or brush-shaped flowers that provide copious nectar (Sazima et al. 1999). In the Wet Tropics, blossom bats visit Syzygium sayeri and S. cormiflorum (Crome and Irvine 1986; Boulter et al. 2005) and may be an important pollinator in both these species. We would expect this might be the case for other species from the families Myrtaceae and Proteaceae as well as some mangrove species (Table 17.3). Von Helversen and von Helversen (1999) note the evolution of acoustically conspicuous structures in bat-pollinated flowers that make locating the flowers easier and emphasize the importance of flower shape in bat attraction.

Bats and birds can be associated with the same flowers, no doubt due to similar feeding requirements (Ollerton & Watts 2000). Crome and Irvine (1986), in their study of Syzygium cormiflorum, showed that although bats appear to be the dominant pollinator (or at least nighttime visitors based on exclusion experiments), birds also make successful daytime visits. Bats may act as an effective pollinator in spatially isolated species. For example, in a subsequent study of S. cormiflorum, common blossom bats on the Atherton Tablelands were found to carry greater quantities of pollen, have shorter foraging times and engage in more frequent movements of greater than 200 metres than birds (Law & Lean 1999). Similarly, both birds and bats visited Syzygium sayeri, but there was some evidence, although inconclusive, that bats might be more efficient pollinators than birds or other daytime visitors (Boulter et al. 2005).

Other, non-flying mammals will exploit nectar-rich flowers or those flowers with nocturnal anthesis. Few non-flying mammals have been identified as pollinators in tropical floras (Bawa 1990). In the Wet Tropics, Irvine and Armstrong (1990) observed rats visiting and feeding on male flowers of the parasitic understorey herb Balanophora fungosa, brushing pollen on to their faces in the process.

Invertebrate pollination

Invertebrates represent the majority of flower visitors in the tropics in general (Bawa 1990) and this trend is reflected in what we know of the Wet Tropics flora (Table 17.1). Among observed insect visitors, Hymenoptera, Coleoptera and Diptera are most commonly identified. Bees have been identified as important and active pollinators in tropical systems (Dressler 1982; Frankie et al. 1990; Gross 1993a, b; Gross & Mackay 1998; Momose et al. 1998). Their role, diversity, foraging behaviour, spatial abundance or importance do, however, vary in response to biogeography, climate, latitude, altitude, vegetation structure and the distribution and phenology of plants (Bawa 1990; Williams & Adam 1994; van Dulmen 2001). For example, it has been suggested that bee pollination may be less important or entirely absent in Australian rainforests at higher latitudes (House 1989; Irvine & Armstrong 1990). We know that this is certainly not an absolute rule in the Australian tropics. The vast complement of orchids are likely to be bee pollinated (and possibly by some syrphid flies) (Bartareau 1993). It is likely that native bees have been overlooked as important pollinators perhaps because they are less likely to be detected using traditional sampling methods even when they can be seen in the canopy (C. Gross, unpublished data). We note that Hymenoptera are the only visitors to some Wet Tropics species. For example, the pioneer species *Melastoma affine* cannot set seed without bee visitors (Gross 1993a), and in one study, flowers were pollinated by five species of native bees (Figure 17.7) in the Paluma district of the southern Wet Tropics area (Gross & Mackay 1998; Gross & Kukuk 2001). Native bees are also frequent visitors to flowers of *Macadamia whelanii*, *Helicia nortoniana* (Proteaceae) (Figure 17.8), *Gevuina bleasdalei* (Proteaceae), *Sloanea macbrydei* (Elaeocarpaceae), *Cardwellia sublimi*, *Rhodomyrtus trineura* (Myrtaceae) (C. Gross, unpublished data) and ground dwelling flora that includes *Dianella* spp. and *Hibbertia* spp. as well as species in the Asteraceae (Gross and Mackay, unpublished data).

Beetles are an important and diverse group of rainforest pollinators world wide (Bawa 1990; Irvine & Armstrong 1990, Sakai *et al.* 1999b). In Australia,

pollination by beetles is claimed to occur in up to one quarter of rain forest plants and throughout the vertical profile of the forest (Irvine & Armstrong 1990). This assessment was based on rating individual plant species as beetle-pollinated using personal observations, case histories and extrapolations from species with similar morphologies. We note that beetles have been identified visiting the flowers of 26 of the 46 plant species listed in Table 17.1. In their study of members of the family Annonaceae, Silberbauer-Gottsberger *et al.* (2003) observed beetle visitors to the flowers of five out of the seven species studied in the Wet Tropics. The myrtaceous species *Syzygium gustavioides* was visited exclusively by a diverse array of insects, including an impressive 97 morphospecies of beetles from 26 families (Boulter 2003).

Diptera are also a conspicuous component of the visitor fauna of tropical forest flowers, although little is known about fly pollination (Bawa 1990; Kitching *et al.* 2007). Fly pollination is known in both Old and New World species of the archaic family Annonaceae (Silberbauer-Gottsberger *et al.* 2003). In the Wet Tropics, for three species of the genus *Pseudavaria*, *P. frogattii*, *P. hylandii* and *P. villosa*, flies were flower visitors and suggested to be the pollinators (Silberbauer-Gottsberger *et al.* 2003). In the case of *P. frogatti*, a study at Cape Tribulation indicated that flies visited this cauliflorous, dioecious tree during anthesis, licked nectar and touched the reproductive organs. Flies were also identified as the pollinator of

Figure 17.7 A female *Xylocopa disconota* buzz pollinating a flower of *Melastoma affine*. Note the pollen deposition on the bee from the antisepalous stamens. Image © C. L.Gross.

Figure 17.8 Flowers of *Helicia nortoniana* in the Mt Spec rainforests, September 2005 with a female *Hylaeus* species foraging at flowers. Image © C. L.Gross.

the herbaceous orchid *Pterostylis procera* on the Atherton Tablelands (Bartareau 1993). Diptera were the most abundant flower visitors to three Wet Tropics rainforest species studied extensively by House (1989, 1992, 1993), and for at least two of the tree species, Diptera carried pollen more often to pistillate trees than other visitors and were likely to be the most effective pollinators (House 1989).

Specialization versus generalization

The vast majority of rainforest taxa have long been thought to have specialized pollination systems, in that one or a few species belonging to the same taxonomic group are often identified as pollinators (Bawa 1990). This is no longer considered accurate and new thinking is that the majority of angiosperms can be described as generalists, in that they are pollinated by a range of taxonomically diverse organisms (Waser 1983; House 1989, 1993; Bronstein 1995; Waser *et al.* 1996), often across orders (e.g. birds and moths, Hopper 1980). For example, Williams and Adam (1994) suggest that the spatial and temporal distribution of pollinator

resources plus floral traits of Australian rainforest taxa indicate that generalist pollinators dominate this system. This is thought to be especially true in the canopy. They support this theory by suggesting that flexible pollination may have been necessary for survival during rapid and massive changes during the Quaternary. There is certainly evidence of generalized pollination systems in Australian rainforests (House 1989; Boulter 2003; Boulter *et al.* 2005). Certainly for those species studied in the Wet Tropics a wide range of insects visit their flowers (Table 17.1), many of which are likely to be carrying pollen (House 1989). In the case of *Syzygium gustavioides* many were shown to be capable of pollination (Boulter 2003). We have focused attention on the diversity of trappable and resident flower visitors in the canopy of the Wet Tropics. This work has shown not only the diversity of insects visiting flowers, but that the suite visiting a given set of co-flowering plant species can be quite different (Kitching *et al.* 2007). For example, the insects resident in a canopy tree (*Acmena graveolens*) and a vine (*Entada phaseoloides*) and the flying visitors to the same species show some differences (Figure 17.9). Using budding inflorescences as a

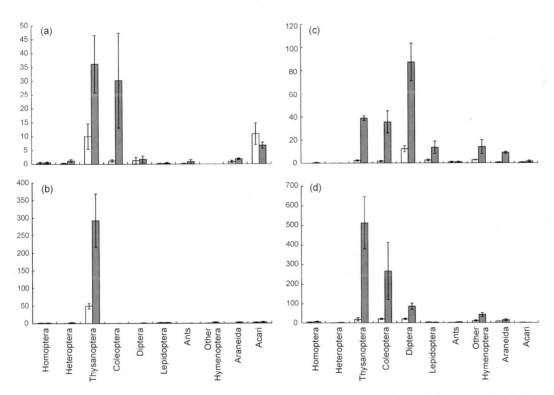

Figure 17.9 The resident fauna collected by ethanol washing from (a) *Acmena graveolens* and (b) *Entada phaseoloides*, and the visitor fauna (collected using interception and sticky traps) of (c) *Acmena graveolens* and (d) *Entada phaseoloides*.

control, we demonstrate that each set of insect visitors is not simply a random subset of the available insect fauna, but that each set of visitors is attracted to each species. We conclude that for many species, a generalist pollination system will still include some selection of host plants by insect visitors.

Conservation

The collapse of pollinator mutualisms has been identified as one potential consequence of anthropogenic land use change (Kearns & Inouye 1997; Allen-Wardell *et al.* 1998; Kearns *et al.* 1998; Wilcock & Neiland 2002). Declines in pollinators have been reported from most continents (Kearns *et al.* 1998; Kevan & Phillips 2001). Land clearance, fragmentation, agricultural practices, herbicides, pesticides and the introduction of exotic plant and pollinator species have all been implicated in a serious decline in pollinators that has been referred to as a pollination crisis (Buchman & Nabhan 1996). Although the level of impact these disturbances actually have on pollination systems worldwide is a matter for debate (Ghazoul 2005a, b; Steffan-Dewenter *et al.* 2005).

Loss of or interruption to pollinator services may have several outcomes. The most obvious result is a loss or reduction in seed set; however, impacts may also extend to reduced offspring vigour as a result of self-pollination, decreasing heterozygosity and the increased expression of deleterious traits resulting from inbreeding (Kearns & Inouye 1997). Ultimately, loss of seeds, fruits or plants will affect animals that rely on these resources.

Like most tropical landscapes, the Wet Tropics have been subjected to processes of fragmentation over the past 200 years. Plant species marooned in these fragments may or may not be part of viable populations, and it may take much longer than 100 years before this becomes evident. Pollination and the subsequent reproductive performance of plants in fragments become crucial issues. Understanding the changes that occur to pollination processes and outcomes in rainforest fragments is an essential first step in managing these changes and attempting to ensure the long-term future of our forests, though there are almost no data available on this topic. The study by Law and Lean (1999) on *Sygygium cormiflorum* did demonstrate that visits by vertebrates to the flowers were skewed in favour of bats over birds in fragmented situations. There is no question that much more work of this nature is required and, indeed, is crucial to future management decisions.

Summary

Considering the large number of plant species of the Wet Tropics, we know little about their breeding systems and the realized effect of pollen flow or, indeed, of plant mating systems. We see challenges for plant reproductive ecologists working in the region. They include recognizing:
- that a floral visitor may not necessarily be a pollinator (see Gross & Mackay 1998; Navarro 1999) and careful observations at flowers or manipulative techniques are required to verify pollinator status;
- that many populations must be included in the determination of plant breeding systems.

The latter is difficult in most terrains but particularly so where access to the canopy is restricted. Variation in reproductive systems can come about through differential selection pressures among pollinators (Morgan & Wilson 2005) through to complex interactions with the environment (Gross & Caddy 2006), particularly at the extremes of a species' distribution (e.g. Dorken & Eckert 2001). However, this variation is often only considered in single population studies (reviewed in Gross & Caddy, in review). For example, in tropical systems elsewhere self-incompatibility among tree species is a dominant feature (reviewed by Schatz in Bawa and Hadley 1990), but again much of this work has been conducted at one location. Much more work is required in northern Australia before we can contribute to a global discussion on the importance of breeding systems in structuring rainforest systems. Access to canopies through facilities such as canopy cranes and portable cherry pickers will greatly facilitate the collection of such valuable data.

References

Allen-Wardell, G., Bernhardt, P., Bitner, R. *et al.* (1998). The potential consequences of pollinator declines on the conservation of biodiversity and stability of food crop yields. *Conservation Biology* **12**: 8–17.

Armstrong, J. A. (1979). Biotic pollination mechanisms in the Australian flora – a review. *New Zealand Journal of Botany* **17**: 467–508.

Armstrong, J. E. & Irvine, A. K. (1989a). Floral biology of *Myristica insipida* (Myristicaeae) a distinctive beetle pollination syndrome. *American Journal of Botany* **76**: 86–94.

Armstrong, J. E. & Irvine, A. K. (1990). Functions of staminodia in the beetle-pollinated flowers of *Eupomatia laurina*. *Biotropica* **22**: 429–31.

Armstrong, J. E. & Irvine, A. K. (1989b). Flowering, sex ratios, pollen–ovule ratios, fruit set and reproductive effort of a dioecious tree, *Myristica insipida* (Myristicaceae), in two different rain forest communities. *American Journal of Botany* **76**: 74–85.

Bartareau, T. M. (1993). *The reproductive biology and ecology of eight orchid species endemic in North-east Queensland*. MSc Thesis, James Cook University of North Queensland.

Bawa, K. S. (1990). Plant–pollinator interactions in tropical rain forests. *Annual Review of Ecology and Systematics* **21**: 399–422.

Bawa, K. & Hadley, M. (eds) (1990). *Reproductive Ecology of Tropical Forest Plants*. UNESCO Parthenon, Park Ridge, NJ.

Bawa, K. S., Kang, H. & Grayum, M. H. (2003). Relationships among time, frequency and duration of flowering among tropical rain forest trees. *American Journal of Botany* **90**: 877–87.

Bolmgren, K., Eriksson, O. & Linder, H. P. (2003). Contrasting flowering phenology and species richness in abiotically and biotically pollinated angiosperms. *Evolution* **57**: 2001–11.

Borchert, R. (1996). Phenology and flowering periodicity of Neotropical dry forest species: evidence from herbarium collections. *Journal of Tropical Ecology* **12**: 65–80.

Boulter, S. L. (2003). *Reproductive ecology of two Myrtaceous canopy plant species:* Syzygium gustavioides (F. M. Bail.) B. Hyland *and* S. sayeri (F. Muell.) B. Hyland. BSc (Hons) thesis, Griffith University.

Boulter, S. L., Kitching, R. L. & Howlett, B. G. (2006). Family, visitors and the weather – patterns of flowering in tropical rainforests of northern Australia. *Journal of Ecology* **94**: 369–82

Boulter, S. L., Kitching, R .L., Howlett, B. G. & Goodall, K. (2005). Any which way will do – the pollination biology of a northern Australian rainforest canopy tree (*Syzygium sayeri*; Myrtaceae). *Botanical Journal of the Linnean Society* **149**: 69–84.

Bronstein, J. L. (1995). The plant–pollinator landscape. In *Mosaic Landscapes and Ecological Processes*, Hansson, L., Fahrig, L. & Merriam, G. (eds). Chapman & Hall, London.

Buchman, S. L. & Nabhan, G. P. (1996). *The Forgotten Pollinators*. Island Press, Washington, DC.

Bush, M. B. & Rivera, R. (2001). Reproductive ecology and representation among neotropical trees. *Global Ecology and Biogeography* **10**: 359–67.

Churchill, S. (1998). *Australian Bats*. New Holland Publishers, Amsterdam.

Cooper, W. & Cooper, W. (2004). *Fruits of the Australian Tropical Rainforest*. Nokomis Editions, Victoria.

Crome, F. H. J. & Irvine, A. K. (1986). Two bob each way: the pollination and breeding system of the Australian rain forest tree *Syzygium cormiflorum* (Myrtaceae). *Biotropica* **18**: 115–25.

Cronin, L. (2000). *Key Guide to Australian Palms, Ferns, Cycads and Pandans*. Envirobooks, Australia.

Dafni, A., Lehrer, M. & Kevan, P.G. (1997). Spatial flower parameters and insect spatial vision. *Biological Reviews* **72**: 239–82.

Dorken, M. E. & Eckert, C. G. (2001). Severely reduced sexual reproduction in northern populations of a clonal plant, *Decodon verticillatus* (Lythraceae). *Journal of Ecology* **89**: 339–50.

Dressler, R. L. (1982) Biology of the orchid bees (Euglossini). *Annual Review of Ecology and Systematics* **13**: 373–94.

Dukas, R. (2001). Effects of perceived danger on flower choice by bees. *Ecology Letters* **4**: 327–33.

Faegri, K. & van der Pijl, L. (1979). *The Principles of Pollination Ecology*, 3rd edn. Pergamon Press, Oxford.

Feinsinger, P. (1983). Coevolution and pollination. In *Coevolution*, Futuyma, D. J. & Slatkin, M. (eds). Sinauer Associates, Sunderland, MA.

Ford, H. A., Paton, D. C. & Forde, N. (1979). Birds as pollinators of Australian plants. *New Zealand Journal of Botany* **17**: 509–19.

Frankie, G. W., Vinson, S. B, Newstrom, L. E., Barthell, J. F., Haber, W. A. & Frankie, J. K. (1990). Plant phenology, pollination ecology, pollinator behaviour and conservation of pollinators in neotropical dry forest. In *Reproductive Ecology of Tropical Forest Plants*, Bawa, K. & Hadley, M. (eds). UNESCO Parthenon, Park Ridge, NJ.

Frith, C.B. & Frith, D.W. (1985). Seasonality of insect abundance in an Australian upland tropical rainforest. *Australian Journal of Ecology* **10**: 237–48.

Ghazoul, J. (2005a). Buzziness as usual? Questioning the global pollination crisis. *Trends in Ecology and Evolution* **20**: 367–73.

Ghazoul, J. (2005b). Response to Steffan-Dewenter *et al.*: Questioning the global pollination crisis. *Trends in Ecology and Evolution* **20**: 652–3.

Gross, C. L. (1993a). The breeding system and pollinators of *Melastoma affine* (Melastomataceae): a pioneer shrub in tropical Australia. *Biotropica* **25**: 468–74.

Gross, C. L. (1993b). The reproductive ecology of *Canavalia-rosea* (Fabaceae) on Anak Krakatau, Indonesia. *Australian Journal of Botany* **41**: 591–9.

Gross, C. L. (2005). A comparison of the sexual systems in the trees from the Australian tropics with other tropical biomes – more monoecy but why? *American Journal of Botany* **92**: 907–19.

Gross, C. L. & Caddy, H. A. R. (2006). Are differences in breeding mechanisms and fertility among populations contributing to rarity in *Grevillea rhizomatosa* (Proteaceae)? *American Journal of Botany* **93**: 1791–1799.

Gross, C. L. & Kukuk, P. F. (2001). Foraging strategies of *Amegilla anomola* at flowers of *Melastoma affine* – no evidence for separate feeding and pollinating anthers. *Acta Horticulturae* **561**: 171–8.

Gross, C. L. & Mackay, D. (1998). Honeybees reduce fitness in the pioneer shrub *Melastoma affine* (Melasomataceae). *Biological Conservation* **86**: 169–78.

Harrison, K. J. (1987). *Floral biology of three tropical Australian species:* Aegiceras corniculatum, Maesa dependens *and* Rapanea subsessilis *(Myrsinaceae).* BSc (Hons) thesis, James Cook University.

Higgins, P. J. & Peter, J. M. (Eds.) (2002). *Handbook of Australian, New Zealand and Antarctic Birds. Volume 6: Pardalotes to Shrike-thrushes.* Oxford University Press, Melbourne.

Hopkins, M. S. & Graham, A. W. (1989). Community phenological patterns of a lowland tropical rainforest in northeastern Australia. *Australian Journal of Ecology* **14**: 399–413.

Hopper, S. D. (1980). Pollination of the rainforest tree *Syzygium tierneyanum* (Myrtaceae) at Kuranda, North Queensland. *Australian Journal of Botany* **28**: 223–37.

House, S. M. (1989). Pollen movement to flowering canopies of pistillate individuals of three rainforest tree species in tropical Australia. *Australian Journal of Ecology* **14**: 77–94.

House, S. M. (1992). Population density and fruit set in three dioecious tree species in Australian tropical rain forest. *Journal of Ecology* **80**: 57–69.

House, S. M. (1993). Pollination success in a population of dioecious rain forest trees. *Oecologia* **96**: 555–61.

Hyland, B. P. M., Whiffin, T., Christophel, D. C., Gray, B. & Elick, R. W. (2003). *Australian Tropical Rain Forest Plants.* CSIRO Publishing, Melbourne.

Ippolito, A. & Armstrong, J. E. (1993). Floral biology of *Hornstedtia scottiana* (Zingiberaceae) in a lowland rain forest of Australia. *Biotropica* **25**: 281–9.

Irvine, A. K. & Armstrong, J. E. (1990). Beetle pollination in tropical forests of Australia. In *Reproductive Ecology of Tropical Forest Plants*, Bawa, K. & Hadley, M. (eds). UNESCO Parthenon, Park Ridge, NJ, 135–49.

Johnson, S. D. & Steiner K. E. (2000). Generalization versus specialization in plant pollination systems. *Tree* **15**: 140–3.

Kearns, C. A. & Inouye, D. W. (1997). Pollinators, flowering plants and conservation biology. *Bioscience* **47**: 297–307.

Kearns, C. A., Inouye, D. W. & Waser, N. M. (1998). Endangered mutualisms: The conservation of plant–pollinator interactions. *Annual Review of Ecological Systematics* **29**: 83–112.

Kevan, P., Giurfa, M. & Chittka, L. (1996). Why are there so many and so few white flowers? *Trends in Ecology and Evolution* **1**: 280–4.

Kevan, P. G. & Phillips, T. P. (2001). The economic impacts of pollinator declines: an approach to assessing the consequences. *Conservation Ecology* **5**: 8.

Kitching, R. L., Boulter, S. L., Howlett, B. G. & Goodall, K. L. (2007). Visitor assemblages at flowers in a tropical rainforest canopy, *Austral Ecology* **32**: 29–42.

Kochmer, J .P. & Handel, S. N. (1986). Constraints and competition in the evolution of flowering phenology. *Ecological Monographs* **56**: 303–25.

Law, B. S. & Lean, M. (1999). Common blossom bats (*Syconycteris australis*) as pollinators in fragmented Australian tropical rainforests. *Biological Conservation* **91**: 201–12.

Miller, J. S. & Diggle, P. K. (2003). Diversification of andromonoecy in *Solanum* section *Lasiocarpa* (Solanaceae): the roles of phenotypic plasticity and architecture. *American Journal of Botany* **90**: 707–15.

Momose, K., Ishii, R., Sakai, S. & Inoue, T. (1998). Plant reproductive intervals and pollinators in the aseasonal tropics: a new model. *Proceedings of the Royal Society of London B* **265**: 2333–9.

Morgan, M. T. & Wilson, W. G. (2005). Self-fertilization and the escape from pollen limitation in variable pollination environments. *Evolution* **59**: 1143–8.

Myers, N., Mittermeier, R. A., Mittermeier, C. G., da Fonseca, G. A. B. & Kent, J. (2000). Biodiversity hotspots for conservation priorities. *Nature* **403**: 853–8.

Navarro, L. (1999). Pollination ecology and effect of nectar removal in *Macleania bullata* (Ericaceae). *Biotropica* **31**: 618–25.

Newstrom, L. E. & Frankie, G. W. (1994). A new classification for plant phenology based on flowering patterns in lowland tropical rain forest trees at La Selva, Costa Rica. *Biotropica* **26**: 141–59.

Ollerton, J. & Watts, S. (2000). Phenotype space and floral typology: towards an objective assessment of pollination syndromes *Det Norske Videnskaps-Akademi. I. Matematisk Naturvidenskapelige Klasse, Skrifter, Ny Serie* **39**: 149–59.

Osunkoya, O. O. (1999). Population structure and breeding biology in relation to conservation in the dioescious *Gardenia actinocarpa* (Rubiaceae) – a rare shrub of North Queensland rainforest. *Biological Conservation* **88**: 347–59.

Pleasants, J. M. (1980). Competition for bumblebee pollinators in Rocky Mountain plant communities. *Ecology* **61**: 1446–59.

Proctor, M. & Yeo, P. (1972). *The Pollination of Flowers.* Taplinger Publishing Company, New York.

Ramírez, N. (2002). Reproductive phenology, life-forms, and habitats of the central plain. *American Journal of Botany* **89**: 836–42.

Rathcke, B. (1983). Competition and facilitation among plants for pollination. In *Pollination Biology*, Real, L. (ed.). Academic Press, Orlando, FL, pp. 305–29.

Rathcke, B. J. & Lacey, E. P. (1985). Phenological patterns in terrestrial plants. *Annual Review of Ecology and Systematics* **16**: 179–214.

Regal, P. J. (1982). Pollination by wind and animals: ecology of geographic patterns. *Annual Review of Ecology and Systematics* **13**: 497–524.

Renner, S. & Feil, J. (1993). Pollinators of tropical dioecious angiosperms. *American Journal of Botany* **80**: 1100–7.

Robertson, J. (1999). *The reproductive ecology of Banksia plagiocarpa (Proteaceae) A. S. George (1981): a rare species restricted to Hinchinbrook Is. and adjacent mainland*. BSc (Hons) Thesis, James Cook University of Northern Australia.

Sakai, S. (2002). General flowering in lowland mixed dipterocarp forests of south-east Asia. *Biological Journal of the Linnean Society* **75**: 233–47.

Sakai, S., Kato, M. & Inoue, T. (1999a). Three pollination guilds and variation in floral characteristics of Bornean gingers (Zingiberaceae and Costaceae). *Australian Journal of Botany* **86**: 646–58.

Sakai, S., Momose, K., Yumoto, T., Kato, M. & Inoue, T. (1999b). Beetle pollination of *Shorea parvifolia* (section Mutica, Dipterocarpaceae) in a general flowering period in Sarawak, Malaysia. *American Journal of Botany* **86**: 62–9.

Sattler, P. S. & Williams, R. D. (1999). *The Conservation Status of Queensland's Bioregional Ecosystems*. Environmental Protection Agency, Brisbane.

Sazima, I., Buzato, S. & Sazima, M. (1993). The bizarre inflorescence of *Norantea brasiliensis* (Marcgraviaceae): visits of hovering and perching birds. *Botanical Acta* **106**: 507–13.

Sazima, M., Buzato, S. & Sazima, I. (1999). Bat-pollinated flower assemblages and bat visitors at two Atlantic forest sites in Brazil. *Annals of Botany* **83**: 705–12.

Silberbauer-Gottsberger, I., Gottsberger, G. & Webber, A. C. (2003). Morphological and functional flower characteristics of New and Old World Annonaceae with respect to their mode of pollination. *Taxon* **52**: 701–18.

Sjostrom, A. & Gross, C. L. (2006) Life-history characters and phylogeny are correlated with extinction risk in the Australian angiosperms. *Journal of Biogeography* **33**: 271–90.

Steffan-Dewenter, I., Potts, S.G. & Packer, L. (2005). Pollinator diversity and crop pollination services are at risk. *Trends in Ecology and Evolution* **20**: 654–2.

van Dulmen, A. (2001). Pollination and phenology of flowers in the canopy of two contrasting rain forest types in Amazonia, Colombia. *Plant Ecology* **153**: 73–85.

van Shaik, C. P., Terborgh, J. W. & Wright, S. J. (1993). The phenology of tropical forests: adaptive significance and consequences for primary consumers. *Annual Review of Ecology and Systematics* **24**: 353–77.

von Helversen, D. & von Helversen, O. (1999). Acoustic guide in bat-pollinated flower. *Nature* **398**: 759–60.

Waser, N. M. (1979). Pollinator availability as a determinant of flowering time in Ocotillo (*Fouquieria splendens*). *Oecologia* **39**: 107–21.

Waser, N. M. (1983). Competition for pollination and floral character differences among sympatric plant species: a review of evidence. In *Handbook of Experimental Pollination Biology*, Jones, C. E. & Little, R. J. (eds). Van Nostrand Reinhold, New York, pp. 277–93.

Waser, N. M., Chittka, L., Price, M. V., Williams, N. M. & Ollerton, J. (1996). Generalization in pollination systems, and why it matters. *Ecology* **77**: 1043.

Webb, L. J. & Tracey, J. G. (1994). The rainforests of northern Australia, In *Australian Vegetation*, 2nd edn, Groves, R. H. (ed.). Cambridge University Press, Cambridge, pp. 87–129.

Weber, E. T. (1994). *The reproductive biology of (Elaeocarpus largiflorens) C. White subsp. largiflorens (Elaeocarpaceae)*. Honours thesis, James Cook University of Northern Australia.

Wilcock, C. & Neiland, R. (2002). Pollination failure in plants: why it happens and when it matters. *Trends in Plant Science* **7**: 270–7.

Williams, G. & Adam, P. (1994). A review of rainforest pollination and plant–pollinator interactions with particular reference to Australian subtropical rainforests. *Australian Zoologist* **29**: 177–212.

Williams, G. & Adam, P. (1999). Pollen sculpture in subtropical rain forest plants: is wind pollination more common than previously suspected? *Biotropica* **31**: 520–4.

Wilson, G. W. (2002). Insect pollination in the cycad genus *Bowenia* Hook ex Hook. f. (Stangeriaceae). *Biotropica* **38**: 438–41.

Worboys, S. J. (1998). *Pollination processes and population structure of Idiospermum australiense (Diels) S.T. Blake, a primitive tree of the Queensland wet tropics*. MSc thesis, James Cook University of Northern Australia.

Worboys, S. J. & Jackes, B. R. (2005). Pollination processes in *Idiospermum australiense* (Calycanthaceae), an arborescent basal angiosperm of Australia's tropical rain forests. *Plant Systematics and Evolution* **251**: 107–17.

Wright, S. J. (1996). Phenological responses to seasonality in tropical forest plants. In *Tropical Forest Plant Ecophysiology*, Mulkey, S. S., Chazdon, R. L. & Smith, A. P. (eds). Chapman & Hall, New York, pp. 440–60.

Wright, S. J. & van Schaik, C. P. (1994). Light and phenology of tropical trees. *The American Naturalist* **143**: 192–9.

Wyatt, R. (1983). Pollinator–plant interactions and the evolution of breeding systems. In *Pollination Biology*, Real, L. (ed.). Academic Press, Orlando, FL.

18 SERVICES AND DISSERVICES FROM INSECTS IN AGRICULTURAL LANDSCAPES OF THE ATHERTON TABLELAND

Saul A. Cunningham[1]* and K. Rosalind Blanche[2]*

[1]CSIRO Entomology, Canberra, ACT, Australia
[2]CSIRO Entomology, Tropical Forest Research Centre, Atherton, Queensland, Australia
*The authors were participants of Cooperative Research Centre for Tropical Ecology and Management

Introduction

Around the world tropical landscapes often support diverse mixed agriculture interspersed with remnant forest vegetation. Whereas the temperate agricultural landscape in Australia is dominated by broad acre plantings of cereal crops, landholdings in the tropics are typically smaller and the range of agricultural products is greater. The Atherton Tableland typifies this pattern, with a diverse range of agricultural crops being produced since European settlement began in the late 1880s (Table 18.1) (Turton, Chapter 5, this volume). Some of the original crops, like tobacco and navy beans, have disappeared or declined in importance but many others, including macadamias, longans, lychees, custard apple, cut flowers and rare fruits, are rapidly increasing in suitable parts of the region (Department of Primary Industries Queensland 1999). All this is happening in a landscape that abuts extensive rainforest and contains within it many remnant rainforest patches. This kind of heterogeneous landscape allows insects that use rainforest to disperse into agriculture and interact with crops.

Agriculturalists most often view insects as a source of problems. There is no doubt that insect pests can be a serious problem because of the production losses they cause. Tropical agriculture is particularly vulnerable to pest and disease problems because, unlike in the temperate zone where winter limits the life cycle of many organisms, conditions in the tropics can be suitable for pests year round (Janzen 1973). There is increasing interest, however, in the various benefits that insects can bring. These benefits from nature can collectively be thought of as ecosystem services (Daily 1997). Among the key beneficial services that insects might deliver are natural population control of pest insects and pollination of crops.

Crop pollination: an ecosystem service

Most plant species require pollination to set fruit and produce seeds (Regal 1982). In this way pollination is important for maintaining the production of economically significant crop species (Klein *et al.* 2007). Many of the staple foods for humans are descended from grasses that, like their forebears, are wind pollinated. Although we would not face famine if insect pollination declined (Ghazoul 2005), insect pollination of crops is critical in providing food diversity and in supporting the economies of communities that produce these crops for trade. Crop pollination has been

Table 18.1 Relative importance of the top 30 Atherton Tableland crops

Crop	Value ($ million)	Volume (t)	Area (ha)
Sugarcane	33	1 119 000	9 664
Mango	30	19 000	3 213
Avocado	20	14 400	1 300
Tobacco	17.4	3 000	1 130
Potato	9.5	25 000	1 056
Peanut	8.9	13 095	3 200
Banana	~6.3	n.a.	~180
Macadamia	5.1	16 670	667
Maize	4.5	32 000	6 800
Papaya	4.2	2 690	63
Flowers	3	–	33
Sweet potato	3	3 500	160
Tea tree	3	100	500
lychee	2.5	400	215
Navy bean	2.4	3 000	1 200
Citrus	2	1 725	230
Pumpkin	1.9	6 400	320
Aquaculture	1.5	n.a.	n.a.
Coffee	1.5	200	100
Farm forestry timber	1.25	n.a.	n.a.
Longan	1.2	600	138
Custard apple	1.05	300	30
Grape	1	188	40
Passionfruit	0.75	53	n.a.
Cashew	0.16	100	240
Tomato	0.12	n.a.	20
Stonefruit	0.05	15	5
Onion	0.04	600	20
Rare fruits	n.a.	n.a.	n.a.
Tea	n.a.	n.a.	600

Complied from Campbell (1998), Chudleigh (1999), Department of Primary Industries Queensland (2000b).

recognized as one of the key ecosystem services that nature provides for the human economy, and one for which we rarely pay directly (Daily 1997). There have been a number of studies aimed at understanding the economic impact of crop pollination. Most have

targeted coffee production because it is one of the most globally significant export crops and of great significance to the economies of a number of tropical countries (Klein *et al.* 2002, 2003a, b; Ricketts 2004; Ricketts *et al.* 2004).

Recognition of the benefits of crop pollination has a long history, famously captured in the early bas-relief depiction of hand pollination of date palm, *c*.1750 BCE (Griffith 2004). The provision of managed honeybee colonies to enhance crop pollination also has a long history, but in recent years there has been a developing interest in crop pollination by wild insects. For some time it has been recognized that although honeybees are easy to manage, they are not the most effective pollinators of some crops and are unable to pollinate some crop species (Westerkamp & Gottsberger 2000). The need for better management of pollinators also grew from concerns that insecticide application was harming wild insects, leading to problems, for example, with blueberry production in Canada (Kevan *et al.* 1997). Most recently the concern has been that populations of wild and managed honeybees are at risk from the parasitic mite *Varroa destructor* and other diseases. Bee populations in North America in particular were greatly reduced by a *Varroa* outbreak in the 1990s (Watanabe 1994). Being on the second last continent free of *Varroa destructor* (the other being Antarctica) and harbouring extensive natural forests, the Australian Wet Tropics might seem less vulnerable to crop pollination problems. However, the agricultural areas of this region have been extensively modified, pesticides are widely used and eventual incursion of *Varroa destructor* remains a high likelihood (Cunningham *et al.* 2002). Horticulture in this region could benefit greatly from pollinator research and management in advance of possible pollinator declines.

There have been a number of recent studies examining the degree to which different crops benefit from pollination services provided by wild pollinators (e.g. Kremen *et al.* 2002, 2004; Shuler *et al.* 2005). The hope is that this knowledge can be used to create landscapes in which cropping and natural remnants can coexist and provide agriculture with an economically significant and ecologically resilient pollination service. To allow for this kind of management we need to know which insects are significant crop pollinators and how these insects use both the agricultural and non-agricultural vegetation in the landscape. In Australia, as elsewhere, honey bees (*Apis mellifera*) have traditionally been considered the main crop pollinators. In addition to managed colonies, naturalized feral honeybees are widespread and are thought to be economically significant to pollination in many environments (Cunningham *et al.* 2002; Cook *et al.* 2007). However, Australian native bees, wasps, beetles and flies may be just as useful or even better pollinators for some plant species. For example, stingless bees, especially *Melipona* and *Trigona* species (spp.), are known to be effective pollinators of at least nine crops and contribute to the pollination of about 60 other crop species (Heard 1999). Papaya are known to be pollinated by hawk moths (*Macroglossum* spp.), which also pollinate the rainforest tree *Syzygium tierneyanum* (F. Muell.) (Hopper 1980) and use several rainforest plant species as a larval food source (Morrisen 1995).

Case study: five crops

We examined the pollination of five crops (atemoya, longan, macadamia, lychee and peanut) (Table 18.2) in a series of experiments designed to uncover the identity of flower visitors, the significance of insect pollination to fruit set and the possibility that proximity to rainforest remnants was associated with more flower visits by potential pollinators. Prior research in other regions indicated possible pollinators for the crops in question: for four of the crops bees are thought to be the most important pollinators. Experiments and observations on early commercial peanut varieties suggest that large bees (>8 mm) increase self-pollination and improve peanut yields (Girardeau & Leuck 1967; Rashad *et al.* 1979). Lychee flowers are thought to be pollinated by day-flying insects including honeybees (Pandey & Yadava 1970; Batten 1986) and perhaps other bees such as *Trigona* spp. (King *et al.* 1989) and carpenter bees (du Toit 1994), although early research also suggested a role for screw-worm flies and soldier beetles (Butcher 1957). Macadamia flowers in Hawaii are reported to be mainly pollinated by honeybees (Urata 1954; Shigeura *et al.* 1970) but in northern New South Wales, the native location for macadamia, stingless bees (native *Trigona* spp.) are thought to be major pollinators (Heard 1994). In other areas of Australia, outside the natural range of macadamia, the lycid beetle, *Metriorrhynchus rhipidius* (MacLeay), the flower wasp, *Campsomeris tasmaniensis* Saussure, and the halictid bee, *Homalictus* spp., are capable of contributing to macadamia pollination (Vithanage & Ironside 1986). Longan flowers are thought to be pollinated by *Apis* and *Trigona* bee species and possibly also by hoverflies (Syrphidae) (Boonithee *et al.* 1991), but the relationship between bee visitation and longan pollination remains unverified.

In contrast to the four beetle-pollinated crops, atemoya are known to be pollinated by small beetles, usually species of the family Nitidulidae. Lack of pollinators, however, is one of the main limitations to productivity of atemoya crops worldwide (Nagel *et al.* 1989). Commercial atemoya growers on the Atherton Tableland need to carry out labour-intensive hand pollination to produce enough high quality fruit. Atemoya is a member of the family Annonaceae, which includes numerous other species native to the region's tropical rainforest, some of which have flowers similar to those of atemoya. Native insect pollinators of these native tropical rainforest species might also pollinate atemoya.

Table 18.2 Crop species, orchards sampled for pollinators, and their distance from rainforest

Crop species	Near orchards	Distance from rainforest (m)	Far orchards	Distance from Rainforest (km)
Lychee (*Litchi chinensis* Sonn.)	4	≤200	4	11–35
Peanut (*Arachis hypogaea* L.)	3	≤400	4	6–15
Longan (*Dimocarpus longan* Lour.)	2	≤500	4	4–30
Atemoya (*Annona squamosa* L.× *A. cherimola* Mill. hybrid)	3	≤500	6	5–24
Macadamia (*Macadamia integrifolia* Maiden and Betche)	3	≤500	2	4–30

Case study: flower visitors

We selected orchards both near and far from the rainforest in which to observe flower visitors and examine the effect of pollinator exclusions (Table 18.2). We used small flight intercept traps placed near open flowers to collect flower-associated insects passively in lychee and peanut crops. For macadamia and longan we observed visitors and identified them visually, collecting samples where necessary for verification. The unusual pollination system of atemoya required a different approach. Atemoya flowers open only slightly at first, allowing small insects inside the chamber where they can contact the female flower parts. About 24 hours later the flower opens fully and the pollen-producing male phase begins. At this stage insects can leave freely, potentially carrying pollen to the next flower. To document potential pollinators of atemoya we collected flowers in the early female stage and surveyed the visitors inside the chamber.

For macadamia, longan and atemoya the profile of flower visitors we observed was broadly in line with predictions from the existing literature. The main visitors to atemoya flowers were nitidulid beetles. In addition to the single nitidulid species already recorded from Australian atemoya (*Carpophilus hemipterus*) (George *et al.* 1989), we collected four other exotic species, three of which are known to be atemoya pollinators elsewhere in the world (Blanche & Cunningham 2005). Importantly we also found five native beetle species that have never been recorded from atemoya flowers. The most common visitors to longan and macadamia flowers were bees, including the introduced honeybee and native bees (especially *Trigona*). During our pilot observations of macadamia we saw both honeybees and *Trigona* visiting flowers, but during the main study only honeybees were abundant. It is possible that this change in the visitor profile was associated with pesticide applications that occurred in the intervening period.

Flight intercept traps near lychee flowers trapped many flies, moths, wasps and beetles, but relatively few bees. It may be that the lychee flowers are generalized and open to a wide range of potential pollinators. In contrast, very few flower visitors of any kind were seen to visit peanut flowers in any of the orchards. To test if this was due to availability of visitors rather than the attractiveness of flowers we placed honeybee hives near the peanuts and repeated our observations. Even in these circumstances there were very few flower visits.

Lychee, longan, macadamia and atemoya all experienced patterns of flower visitation that suggested benefits of association with rainforest. For lychee visitors there was a significant negative relationship between distance from the nearest patch of rainforest (log transformed) and the total number of insect visitors ($r^2 = 68\%$, d.f. = 7, $P = 0.007$), but no relationship with diversity of visitors. Similarly, longan flowers in orchards near rainforest received more visits from the native *Trigona* bees than those far from rainforest, but honeybee visitation was not significantly different (Blanche *et al.* 2006b). Macadamia flowers received more honeybee visits in orchards near rainforest than in orchards far from rainforest, even though this species is not a native of rainforests in this region.

For atemoya, proximity to rainforest was associated with both greater abundance and species richness of flower-visiting beetles. Atemoya orchards less than 500 metres from rainforest were predominantly visited by five previously unrecognized native beetle pollinators that are likely to originate in tropical rainforest. The abundance and number of native, but not exotic, species decreased significantly as distance from tropical rainforest increased. Distance of atemoya orchard from rainforest was also correlated with a gradient of decreasing rainfall, but we found that the distance effect had greater explanatory power than the rainfall effect, supporting the inference that a larger part of the effect was driven by rainforest patches acting as a source of potential pollinators (Blanche & Cunningham 2005).

Case study: pollination as a determinant of fruit set

Variation in availability of flower visitors is only relevant to production if it is linked to variation in fruit set. Our research demonstrated this kind of linkage for longan, macadamia and atemoya. In line with observations of few flower visitors, however, we found that peanut pollinator exclusion did not diminish fruit set. It seems likely that selection for other desirable peanut traits has resulted in the development of varieties that are no longer attractive to flower-tripping bees and that there is no advantage to be gained by promoting bees in crops of these varieties (Blanche *et al.* 2006a). Unfortunately we were unable to attain suitable data on lychee pollination.

For atemoya the complete temporal separation of the male and female phases of flowering ensures that autogamy (self-pollination without visitors) is unlikely, and the nearly closed form of the female phase makes wind pollination unlikely too. To examine the effect of pollinator exclusion on atemoya fruit set we matched six open-pollination flowers with six pollinator exclusion flowers on each of ten trees. No fruit were set in the pollinator exclusion treatment, but almost 12% of open-pollinated flowers set fruit (Blanche & Cunningham 2005). This responsiveness to pollinator access means that atemoya is likely to benefit from the greater abundance of flower visitors in the orchards near rainforest. Further benefits might be gained by management approaches that further enhance pollinator availability, particularly in the orchards distant from rainforest experience the greatest shortages. Whether or not the diversity of flower-visiting species also confers benefits in terms of fruit set by atemoya is unknown but such a relationship has been shown for other crops (e.g. Klein *et al.* 2003a).

Fruit set in longan and macadamia was also reduced by pollinator exclusion. Our differential pollinator exclusion on longan established for the first time that pollen transfer is by a combination of self pollination (7%), wind (31%) and bees (62%) (Blanche *et al.* 2006b). We also determined that there was a significant relationship between the number of fruit initiated and the observed frequency of *Trigona* visits to longan flowers (Blanche *et al.* 2006b). For macadamia we used fewer levels of exclusion and found that fruit set from self-pollination was 17% of the open pollination level (Blanche *et al.* 2006b). On their own these results do not necessarily indicate a role for rainforest-using insects as pollinators. In this regard, the most informative observation was that there was a significant interaction between the effect of pollinator exclusion and the distance of the orchard from the nearest rainforest remnant, such that for both longan and macadamia the highest level of fruit set was only achieved when flowers were open to all insect pollinators and the orchard was within 500 metres of a rainforest remnant. These results, paired with observations of greater pollinator availability in orchards near rainforest, indicate that rainforest remnants help to support populations of economically significant crop pollinators, including native species such as *Trigona* bees, but also the introduced honeybee.

Pests and pest control

Insects have probably always competed with humans for food plants and this is particularly apparent in horticulture today where much time, effort and money is directed at reducing the proportion of crops consumed or damaged by insects. Tropical rainforest in north Queensland is perceived by many growers to be a significant source of such insects. To assess the potential impact of rainforest-using insects as pests we surveyed the literature on economically significant pests in the Atherton Tableland region (Table 18.3).

Only 18 of the 49 economically important arthropod pests of major crops (Table 18.3) are of Australian origin, the majority (63%) being introduced species. At least seven of these pests (the native avocado fruit borer, cane grubs, fruit-piercing moths, fruit-spotting bugs, macadamia flower caterpillar, macadamia nut borers and Queensland fruit flies, plus two exotic species, the cane weevil borer and pink wax scale) use rainforest plants for some part of their life cycle (Table 18.3). It is likely that at least some others, especially generalist species like the yellow peach moth and green coffee scale, are able to utilize rainforest plants and thus might be harboured by rainforest. Some of the pests associated with rainforest plants, such as fruit flies (Drew *et al.* 1984), pink wax scale (Smith *et al.* 1997), fruit-spotting bugs (Brimblecombe 1948) and fruit-piercing moths (Fay 1996), are among the most difficult to control and damage the greatest number of crops. For example, fruit piercing moths can cause up to 50% crop loss (Waterhouse & Norris 1987; Fay & Halfpapp 1991) and Queensland fruit flies can be responsible for up to 100% damage to unprotected fruit (Botha *et al.* 2000). However, rainforests also have the potential to provide predatory and parasitic insects that could help to reduce the abundance of pest insects and mitigate this negative aspect of cropping near rainforest.

Control of insect pests

Consumers are becoming increasingly concerned about the possible effects of pesticides on human health and on the environment, so there is a growing market for clean, green food produced without the use of synthetic pesticides (Department of Primary Industries Queensland 2000a; Cosic 2002). At the same time, we

Table 18.3 Economically important arthropod pests in the Atherton Tableland region, their origin, relationship with rainforest and the crops they damage

Common name	Species name	Origin	Found in rain forest	Crops affected by pest
Avocado fruit borer	*Thaumatotibia zophophanes*	Australia	yes	avocado
Avocado leaf roller	*Homona spargotis*	? Australia	?	avocado, atemoya, tea, coffee, lychee & others
Banana rust thrips	*Chaetanaphothrips signipennis*	exotic	?	banana
Banana scab moth	*Nacoleia octasema*	part Asia/Oceana	?	banana
Banana spider mite	*Tetranychus lambi*	? Australia	?	numerous crops
Banana weevil borer	*Cosmopolites sordidus*	Indo-Malaysia	?	banana
Bean fly	*Ophiomyia phaseoli*	exotic	?	bean
Cane weevil borer	*Rhabdoscelus obscurus*	New Guinea	yes	sugarcane, banana
Canegrubs	*Lepidiota* spp.	Australia	yes	peanut, sugarcane, pasture
Citrus bud mite	*Eriophyes sheldoni*	exotic	?	citrus
Citrus gall wasp	*Bruchophagus fellis*	Australia	?	citrus
Citrus mealybug	*Planococcus citri*	exotic	?	most fruits
Citrus rust mite	*Phyllocoptruta oleivora*	E & SE Asia	?	citrus
Citrus snow scale	*Unaspis citri*	SE Asia	no	citrus
Common mango scale	*Aulacaspis tubercularis*	exotic	?	mango
Corn earworm	*Helicoverpa armigera*	S Europ. & Medit.	?	corn, bean, tobacco, & others
Ectropis looper	*Ectropis sabulosa*	? Australia	?	avocado
Erinose mite	*Eriophyes litchii*	?	?	lychee
Fruit-piercing moths	*Eudocima* spp.	Aust., SE Asia, Africa	yes	most fruits
Fruit-spotting bug	*Amblypelta lutescens lutescens*	Australia	yes	most fruits
Green coffee scale	*Coccus viridis*	tropics	?	citrus, coffee, lychee, longan, mango
Hemispherical scale	*Saissetia coffeae*	Africa	?	citrus, avocado, coffee
Lacebug	*Ulonemia concava*	Australia	?	macadamia
latania scale	*Hemiberlesia lataniae*	exotic	?	avocado, macadamia
Macadamia flower caterpillar	*Cryptoblabes hemigypsa*	Australia	yes	macadamia
Macadamia nut borers	*Cryptophlebia* spp.	Australia	yes	macadamia, lychee, longan
Macadamia weevil	*Sigastus?* spp.	? Australia	?	macadamia
Mango planthopper	*Colgaroides acuminata*	Australia	?	mango
Mango scale	*Phenacaspis dilatata*	exotic	?	mango
Mango seed weevil	*Sternochaetus mangiferae*	Asia, Pacific, Africa	?	mango
Mussel scale	*Lepidosaphes beckii*	oriental	?	citrus, avocado
Nigra scale	*Parasaissetia nigra*	SE Asia or Africa	?	atemoya, avocado, citrus
Oriental scale	*Aonidiella orientalis*	? China	?	papaya, mango
Pink wax scale	*Ceroplastes rubens*	? Africa	yes	citrus, mango, avocado, atemoya, longan
Potato moth	*Phthorimaea operculella*	S America	?	potato
Queensland fruit fly	*Bactrocera tryoni*	Australia	yes	most fruits except strawberry & pineapple

(Continued)

Table 18.3 (*Continued*)

Common name	Species name	Origin	Found in rain forest	Crops affected by pest
Red scale	*Aonidiella aurantii*	? China	?	citrus, passionfruit
Redbanded thrips	*Selenothrips rubrocinctus*	tropics, subtropics	?	avocado, mango, macadamia, cashew
Redshouldered leaf beetle	*Monolepta australis*	? Australia	?	avocado, macadamia, mango, citrus, cashew, corn
Rhinoceros beetle	*Xylotrupes gideon*	SE Asia	?	lychee
Spherical mealybug	*Nipaecoccus viridis*	?Africa, SE Asia	?	citrus
Spined citrus bug	*Biprorulus bibax*	Australia	?	citrus
Sugarcane budmoth	*Opogona glycyphaga*	Australia	?	banana, sugarcane
Sugarcane soldier fly	*Inopus rubriceps*	Australia	no	sugarcane, pasture
Swarming leaf Beetles	*Rhyparida* spp.	Australia	no	numerous tree crops
Sweet potato weevil	*Cylas formicarius elegantulus*	exotic	?	sweet potato, corn
Two-spotted mite	*Tetranychus urticae*	? Europe	?	numerous crops
White fringed weevil	*Naupactus leucoloma*	S America	?	peanut, sugarcane, potato, maize, & others
Yellow peach moth	*Conogethes punctiferalis*	Asia	?	atemoya, macadamia, corn, papaya, lychee, & others

Source: Compiled from Sinclair (1979), Waite (1988), Ridgeway (1989), Broadley (1991), Fay and Huwer (1993), Fay *et al.* (1993), Crosthwaite (1994), Pinese and Piper (1994), Robertson and Webster (1995), Robertson *et al.* (1995), Brough *et al.* (1996), Ironside (1995), Agnew (1997), Smith *et al.* (1997), Peña *et al.* (1998), Waite and Huwer (1998), Pinese *et al.* (2001), Sallam and Garrad (2001), Waterhouse and Sands (2001).

know that pesticides are not always as effective as we would like, because pests can become resistant to them, they sometimes degrade rapidly or are washed away by rain, and might be applied during an unreceptive stage in the life cycle of the pest or when the pest is inside a plant organ or protected by a thick canopy. Integrated pest management (IPM) reduces the use of pesticides and improves their effectiveness (e.g. Pinese & Piper 1994; Smith *et al.* 1997). IPM practices include managing the existing communities of natural enemies for more effective pest control. The effectiveness of insect parasitoids and predators is influenced by many factors, especially the microclimate in the crop, the proximity of alternative hosts and the nature and timing of pesticide use (Waterhouse & Sands 2001).

To maximize their effectiveness we need to learn where natural enemies of crop pests come from so that we can provide conditions that attract and maintain them near the crop. Most native predators and parasitoids have not been studied in detail and their origins and habitat needs are unknown. Some, especially those targeting pest species associated with rainforest plants, may have followed their hosts from rainforest into crops. Other rainforest predators and parasitoids may

have extended their host ranges to include non-rainforest pest arthropods. For example, in Uganda, dragonflies have been observed sheltering in rainforest by day and searching for prey in nearby banana plantations at dusk and dawn (Corbet 1961). Similarly, carabid beetles are known to move from woodland remnants into arable fields near York, UK (Bedford & Usher 1994), and carabid and staphylinid beetles move between wetlands, pasture, wheat fields and dry meadows in the Rhine Valley, Switzerland (Duelli *et al.* 1990).

Australian tropical rainforest harbours numerous generalist predatory insects, such as species of robber flies and assassin bugs, with the potential to act as biocontrol agents in crops (Queensland Museum 2000). Presumably rainforest arthropods are also attacked by numerous native parasitoids that could be useful in controlling crop pests but few of these have been documented.

Case study: predatory insects in crops

To determine whether tropical rainforest vegetation on the Atherton Tableland is a source of predatory insects

Table 18.4 Diversity of predators (Simpson's reciprocal = 1/D) in flight intercept samples collected in four different crops. Sampling was evenly divided between the wet and dry season, except for peanut farms which were sampled in the wet season only. Larger values denote greater diversity

	Near rainforest		Far from rainforest	
Crop species	Samples	1/D	Samples	1/D
Macadamia (*Macadamia integrifolia* Maiden and Betche)	6	27	6	17
Atemoya (*Annona squamosa* L.× *A. cherimola* Mill. hybrid)	4	3	4	12
Avocado (*Persea americana*) Mill.	6	13	6	10
Peanut (*Arachis hypogaea* L.)	4	2	4	4

in crops we sampled insects in macadamia, atemoya, avocado and peanut crops near (up to 500 m away) and far (3–35 km away) from rainforest. We used flight intercept traps to sample in the three tree crops for two weeks in the dry season and two weeks in the wet season. Small intercept traps, cleared and reset every two weeks, were used to sample peanut crops for 12 weeks during the wet season (Table 18.4).

We trapped a total of 223 insect species known to be predators. These were from 22 insect families and seven orders. Hymenoptera, Diptera and Coleoptera were the most speciose groups. Samples from the near rainforest sites contained fewer insects (of any species) than those collected far from rainforest. This probably indicates that the far sites have greater overall population densities. However, the potential for a beneficial pest control service might be determined by the diversity of species available. To describe differences in diversity (considering both species richness and the evenness of species) we calculated Simpson's diversity (*D*), pooling samples from the same crops in the same proximity class. The highest diversity was achieved in macadamia orchards near to rainforest. Samples from avocado orchards near rainforest also had higher diversity than samples from far orchards. Among the samples from atemoya orchards, however, the difference was in the other direction, with low diversity near rainforest. Samples collected in peanut crops had very low diversity, in part because no samples were collected in peanut crops during the dry season (Table 18.4).

We also examined the similarities and differences in the samples using non-metric multidimensional scaling to produce two-dimensional ordinations (Clarke 1993) using PCOrd (McCune & Mefford 1999). This analysis indicated that samples were significantly grouped according to crop type (multi-response permutation procedure, *P*<0.001), and that this pattern was stronger than the structuring of communities explained by proximity to rainforest (i.e. near versus far). Consequently, management to enhance predator communities is likely to be quite specific to the crop in question.

We used indicator species analysis (Dufrene & Legendre 1997) to detect species that were characteristic of samples collected near rainforest (by contrast to far from rainforest) considering samples from all crops. We also examined the data with the specimens organized into subfamily to create fewer groups with more individuals. To be a statistically significant indicator a taxon must occur repeatedly in the category of interest (e.g. near rainforest sites) and be rare or missing from the other category (i.e. far sites). We found that a wasp species in the subfamily Polistinae was significantly associated with near sites (*P* = 0.024, Monte Carlo test, 1000 permutations), as was the dipteran subfamily Asilinae (*P* = 0.02, Monte Carlo test, 1000 permutations). Because they were consistently associated with near rainforest sites, even in the absence of management to enhance their populations, these taxa might offer the best potential for enhanced natural pest control service.

Summary

- Rainforest-using insects interact with agricultural crops of the Atherton Tableland in ways that can be of substantial economic significance. These interactions include negative effects, such as damage to plants by pests, and benefits, such as the pollination of crops and natural control of pest populations.

- We found that production by atemoya, macadamia and longan crops is enhanced by insect pollination, and this service is greater in orchards near rainforests. Strategic location of orchards could help to maximize this service in the future.

- Most pest species affecting crops in the region are not native, but a number of significant pests do use rainforests. There is also a rich fauna of native predatory insects about which little is known. We found that crop type has a major influence on the insect predator community, suggesting that management for pest control benefits will need to be very crop specific. Although the abundance of predators was often greater in sites far from rainforest, the pattern of predator diversity was less clear. The most diverse predator community was in macadamia crops near rainforest.

- Current management practices do not attract and maintain sufficient numbers of beneficial insects in crops to take full advantage of the ecosystem services available. An understanding of the value of rainforest as a source of beneficial insects can lead to management changes that improve both crop yields and rainforest conservation. Some beneficial insects could be retained in crops without literally creating more rainforest patches, but to achieve this we need to know more about the biology of the organisms concerned.

- There are exciting prospects for managing this mixed landscape of agriculture and rainforest for environmental, economic and social benefits. To do so we need to understand the benefits that can flow from rainforest to agriculture, the scale over which these processes occur and how land can be managed to maximize benefits and minimize negative effects.

Acknowledgements

We thank the many growers who allowed us to use their farms for our field studies, Rob Bauer for his help in the field and laboratory, John Ludwig and Simon Blomberg for their help with statistical analyses and Rob Floyd who was important in the development of the research project. This research was funded by the Rainforest CRC, the Myer Foundation and CSIRO.

References

Agnew, J. (ed.) (1997). *Australian Sugarcane Pests*. BSES, Indooroopilly, Queensland.

Batten, D. J. (1986). Towards an understanding of reproductive failure in lychee (*Litchi chinensis* Sonn.). *Acta Horticulturae* **175**: 79–83.

Bedford, S. E. & Usher, M. B. (1994). Distribution of arthropod species across the margins of farm woodlands. *Agriculture, Ecosystems and Environment* **48**: 295–305.

Blanche, K. R & Cunningham, S. A. (2005). Rain forest provides pollinating beetles for atemoya crops. *Journal of Economic Entomology* **98**: 1193–201.

Blanche, K. R., Hughes, M., Ludwig, J. A. & Cunningham, S. A. (2006a). Do flower-tripping bees enhance yields in peanut varieties grown in north Queensland. *Australian Journal of Experimental Agriculture* **46**: 1529–34.

Blanche, K. R., Ludwig, J. A. & Cunningham, S. A. (2006b). Proximity to rainforest enhances pollination and fruit set in orchards. *Journal of Applied Ecology* **43**: 1182–7.

Boonithee, A., Juntawong, N., Pechhacker, H. & Hüttinger, E. (1991). Floral visits to selected crops by four *Apis* species and *Trigona* sp. in Thailand. *Acta Horticulturae* **288**: 74–80.

Botha, J., Hardie, D. & Power, G. (2000). Queensland fruit fly *Bactrocera tryoni*. Exotic threat to Western Australia. Agriculture Western Australia, Fact Sheet 43.

Brimblecombe, A. R. (1948). Fruit-spotting bug as a pest of the macadamia or Queensland nut. *Queensland Agricultural Journal* 1 October: 206–11.

Broadley, R. H. (ed.) (1991). *Avocado Pests and Diseases*. Information Series Q190013. QDPI, Brisbane.

Brough, E. J., Elder, R. J. & Beavis, C. H. S. (eds). (1996). *Managing Insects and Mites in Horticultural Crops*. QDPI, Brisbane.

Batten, D. J. (1986). Towards an understanding of reproductive failure in lychee (*Litchi chinensis* Sonn.). *Acta Horticulturae* **175**: 79–83.

Butcher, F. G. (1957). Pollinating insects on lychee blossoms. *Proceedings of the Florida State Horticultural Society* **70**: 326–8.

Campbell, T. (1998). *Rural Development on the Atherton Tablelands: Change, Choices and Challenges* (www.agric. wa.gov.au/programs/new/newrural/T_Campbell.htm).

Chudleigh, P. (1999). *Review of the Prospects for the Australian Black Tea Industry*. RIRDC Publication No. 99/28 Project No. AGT-5A.

Clarke, K. R. (1993). Non-parametric multivariate analyses of changes in community structure. *Australian Journal of Ecology* **18**: 117–43.

Cook, D. C., Thomas, M. B., Cunningham, S. A., Anderson, D. L. & DeBarro, P. J. (2007). Predicting the economic impact of an invasive species on an ecosystem service. *Ecological Applications* **17**: 1832–1840.

Corbet, P. S. (1961). Entomological studies from a high tower in Mpanga forest, Uganda. XII Observations on Ephemeroptera, Odonata and some other orders. *Transactions of the Royal Entomological Society of London* **113**: 356–61.

Cosic, M. (2002). Eat, drink and be wary. *The Weekend Australian Magazine* 12–13 January: 23–6.

Crosthwaite, I. (1994). *Peanut Growing in Australia*. QDPI, Brisbane.

Cunningham, S. A., FitzGibbon, F. & Heard, T. A. (2002). The future of pollinators for Australian agriculture. *Australian Journal of Agricultural Research* **53**: 893–900.

Daily, G. C. (1997). *Nature's services*. Island Press, Washington, DC.

Department of Primary Industries Queensland (1999) *Centre for Tropical Agriculture, Mareeba*. (Booklet) QDPI, Mareeba.

Department of Primary Industries Queensland (2000a). *The Market for Organic Products* (www2.dpi.qld.gov.au/dpinotes/economics/rib00009.html).

Department of Primary Industries Queensland (2000b). *An Agricultural Profile of the Atherton Tablelands*. QDPI, Mareeba.

Drew, R. A. I., Zalucki, M. P. & Hooper, G. H. S. (1984). Ecological studies of eastern Australian fruit flies (Diptera: Tephritidae) in their endemic habitat. *Oecologia* **64**: 267–72.

Duelli, P., Studer, M., & Kakob, S. (1990). Population movements of arthropods between natural and cultivated areas. *Biological Conservation* **54**: 193–207.

Dufrene, M. & Legendre, P. (1997). Species assemblages and indicator species: the need for a flexible asymmetrical approach. *Ecological Monographs* **67**: 345–66.

du Toit, A. P. (1994). Pollination of avocados, mangoes and litchis. *Inligtingsbulletin Institut vir Tropiese en Subtropiese Gewasse* **262**: 7–8.

Fay, H. A. C. (1996). Evolutionary and taxonomic relationships between fruit-piercing moths and the Menispermaceae. *Australian Systematic Botany* **9**: 227–33.

Fay, H. A. C. & Halfpapp, K. H. (1991). Potential methods for the control of fruit piercing moths (Lepidoptera: Noctuidae) in tropical Australia. In *Proceedings of the First Asia-Pacific Conference on Entomology*, Chiang Mai, Entomological Zoology Association Thailand, Bangkok, pp. 204–8.

Fay, H. A. C. & Huwer, R. K. (1993). Egg parasitoids collected from *Amblypelta lutescens lutescens* (Distant) (Hemiptera: Coreidae) in north Queensland. *Journal of the Entomological Society of Australia* **32**: 365–7.

Fay, H. A. C., Storey, R. I., DeFaveri, S. G. & Brown J. D. (1993). Suppression of reproductive development and longevity in the redshouldered leaf beetle, *Monolepta australis* (Col. Chrysomelidae), by the tachinid, *Monoleptophaga caldwelli* (Dipt.). *Entomophaga* **38**: 335–42.

George, A. P., Nissen, R. J., Ironside, D. A. & Anderson, P. (1989). Effects of nitidulid beetles on pollination and fruit set of *Annona* spp. hybrids. *Scientia Horticulturae* **39**: 289–99.

Ghazoul, J. (2005). Buzziness as usual? Questioning the global pollination crisis. *Trends in Ecology and Evolution* **20**: 367–73.

Girardeau, J. H. & Leuck, D. B. (1967). Effect of mechanical and bee tripping on yield of the peanut. *Journal of Economic Entomology* **60**: 1454–5.

Griffith, M. P. (2004). The origins of an important cactus crop, *Opuntia ficus-indica* (Cactaceae): new molecular evidence. *American Journal of Botany* **91**: 1915–21.

Heard, T. A. (1994). Behaviour and pollinator efficiency of stingless bees and honey bees on macadamia flowers. *Journal of Apicultural Research* **33**: 190–8.

Heard, T. A. (1999). The role of stingless bees in crop pollination. *Annual Review of Entomology* **44**: 183–206.

Hopper, S. D. (1980). Pollination of the rainforest tree *Syzygium tierneyanum* (Myrtaceae) at Kuranda, North Queensland. *Australian Journal of Botany* **28**: 223–37.

Ironside, D. A. (1995). *Insect Pests of Macadamia*. QDPI, Brisbane, Queensland.

Janzen, D. H. (1973). Tropical agroecosystems, *Science* **182**: 1212–19.

Kevan, P. G., Greco, C. F. & Belaousoff, S. (1997). Log-normality of biodiversity and abundance in diagnosis and measuring of ecosystemic health: pesticide stress on pollinators on blueberry heaths. *Journal of Applied Ecology* **34**: 1122–36.

King, J., Exley, E. M. & Vithanage, V. (1989). Insect pollination for yield increases in lychee. *Proceedings of the 4th Australian Conference on Tree and Nut Crops*. Lismore, Australia, pp. 142–5.

Klein, A. M., Steffan-Dewenter, I., Buchori, D. & Tscharntke, T. (2002). Effects of land use intensity in tropical agroforestry systems on coffee flower-visiting and trap nesting bees and wasps. *Conservation Biology* **16**: 1003–14.

Klein, A. M., Steffan-Dewenter, I., Tscharntke, T. (2003a). Fruit set of highland coffee increases with the diversity of pollinating bees. *Proceedings of the Royal Society of London B* **270**: 955–61.

Klein, A. M., Steffan-Dewenter, I., Tscharntke, T. (2003b). Pollination of *Coffea canephora* in relation to local and regional agroforestry management. *Journal of Applied Ecology* **40**: 837–45.

Klein, A.-M., Vaissiere, B. & Cane, J. H. *et al.* (2007). Importance of crop pollinators in changing landscapes for world crops. *Proceedings of the Royal Society B* **274**: 303–13.

Kremen, C., Williams, N. M., Bugg, R. L., Fay, J. P. & Thorp, R. W. (2004). The area requirements of an ecosystem service: crop pollination by native bee communities in California. *Ecology Letters* **7**: 1109–19.

Kremen, C., Williams, N. M. & Thorp, R. W. (2002). Crop pollination from native bees at risk from agricultural intensification. *Proceedings of the National Academy of Sciences* **99**: 16812–16.

McCune, B. & Mefford, M. J. (1999). *PC-Ord: Multivariate Analysis of Ecological Data, Version 4.2.0.* MjM Software, Gleneden Beach, USA.

Morrisen, A. (1995). The pollination biology of papaw (*Carica papaya* L.) in central Queensland. Unpublished PhD thesis, Central Queensland University.

Nagel, J., Peña, J. E. & Habeck, D. (1989). Insect pollination of atemoya in Florida. *Florida Entomologist* **72**: 207–11.

Pandey, R. S. & Yadava, R. P. S. (1970). Pollination of litchi (*Litchi chinensis*) by insects with special reference to honeybees. *Journal of Apicultural Research* **9**: 103–105.

Peña, J. E., Mohyuddin, A. I. & Wysoki, M. (1998). A review of the pest management situation in mango agroecosystems. *Phytoparasitica* **26**: 1–20.

Pinese, B., Bauer, R. & Horak, M. (2001). *Avocado Pest Monitoring on the Atherton Tableland.* Final Report HRDC Project AV99006. QDPI, Mareeba, Queensland.

Pinese, B. & Piper, R. (1994). *Bananas, Insect and Mite Management.* QDPI, Brisbane.

Queensland Museum (2000). *Wildlife of Tropical North Queensland.* Queensland Museum, Brisbane.

Rashad, S. E., Ewies, M. A. & El Rabie, H. G. (1979). Pollinators of peanut (*Arachis hypogaea* L.) and the effect of honeybees on its yield. In *Proceedings of the IVth International Symposium on Pollination.* Maryland Agricultural Experiment Station Special Miscellaneous Publication 1, pp. 227–30.

Regal, P. J. (1982). Pollination by wind and animals: ecology of geographic patterns. *Annual Review of Ecology and Systematics.* **13**: 497–524.

Ricketts, T. (2004). Tropical forest fragments enhance pollinator activity in nearby coffee crops. *Conservation Biology* **18**: 1–10.

Ricketts, T. H., Daily, G. C., Ehrlich, P. R. & Michener, C. D. (2004). Economic value of tropical forest to coffee production. *Proceedings of the National Academy of Sciences* **101**: 12579–82.

Ridgeway, R. (ed.) (1989). *Mango Pests and Disorders.* Information Series Q189007. QDPI, Brisbane.

Robertson, L. N., Allsopp, P. G., Chandler, K. J. & Mullins, R. T. (1995). Integrated management of canegrubs in Australia: current situation and future research directions. *Australian Journal of Agricultural Research* **46**: 1–16.

Robertson, L. N. & Webster, D. E. (1995). Strategies for managing cane weevil borer. *Proceedings of the Australian Society of Sugarcane Technologists* **17**: 83–7.

Sallam, M. N. & Garrad, S. W. (2001). Distribution and sampling of adults of *Rhabdoscelus obscurus* (Boisduval) (Coleoptera: Curculionidae) and their damage in sugarcane. *Australian Journal of Entomology* **40**: 281–5.

Shigeura, G. T., Lee, J. & Silva, J. A. (1970). The role of honey bees in macadamia nut (*Macadamia integrifolia* Maiden & Betche) production in Hawaii. *Journal of the American Society of Horticultural Science* **95**: 544–6.

Shuler, R. F., Roulston, T. H. & Farris, G. E. (2005). Farming practices influence wild pollinator populations on squash and pumpkin. *Journal of Economic Entomology* **98**: 790–5.

Sinclair, E. R. (1979). Parasites of *Cryptophlebia ombrodelta* (Lower) (Lepidoptera: Tortricidae) in southeastern Queensland. *Journal of the Australian Entomological Society* **18**: 329–35.

Smith, D., Beattie, G. A. C., Broadley, R. (eds) (1997). *Citrus Pests and Their Natural Enemies.* QDPI & HRDC, Brisbane.

Urata, U. (1954). Pollination requirements of macadamia. *Hawaii Agricultural Experiment Station Technical Bulletin* **22**: 1–40.

Vithanage, V. & Ironside D. A. (1986). The insect pollinators of macadamia and their relative importance. *The Journal of the Australian Institute of Agricultural Science* **52**: 155–60.

Waite, G. K. (1988). Biological control of latania scale on avocados in south-east Queensland. *Queensland Journal of Agricultural and Animal Sciences* **45**: 165–7.

Waite, G. K. & Huwer, R. K. (1998). Host plants and their role in the ecology of the fruitspotting bugs *Amblypelta nitida* Stal and *Amblypelta lutescens lutescens* (Distant) (Hemiptera: Coreidae). *Australian Journal of Entomology* **37**: 340–9.

Watanabe, M. E. (1994). Pollination worries rise as honey bees decline. *Science* **265**: 1170.

Waterhouse, D. F. & Norris, K. R. (1987). *Biological Control: Pacific Prospects.* Inkata, Melbourne.

Waterhouse, D. F. & Sands, D. P. A. (2001). *Classical Biological Control of Arthropods in Australia.* ACIAR Monograph No. 77, Canberra, ACT.

Westerkamp, C. & Gottsberger, G. (2000). Diversity pays in crop pollination. *Crop Science* **40**: 1209–22.

19 ECONOMIC APPROACHES TO THE VALUE OF TROPICAL RAINFORESTS

*Ian Curtis**

Nature Conservation Trust of New South Wales, SMEC Australia Pty Limited, Townsville, Queensland, Australia; and School of Earth Environmental Sciences, James Cook University, Queensland, Australia
*The author was participant of Cooperative Research Centre for Tropical Ecology and Management

Introduction

The various branches of economics have endeavoured for decades to derive appropriate and effective methods to value the environment, or more particularly the goods and services provided by ecosystems for the public good. These attempts have often been, and still are, the subject of acrimonious debate, but with refinements they have survived and are still used today. The problem has been the public good nature of the environment, and the concept of market failure, where the consumption of these ecosystem goods and services goes uncompensated. In other words, they are non-market or unpriced goods and services. Moreover, the global nature of these externalities has frequently been neglected owing to the main purpose of environmental economic studies being to measure the national welfare (Aronsson & Lofgren 2001).

Fundamental limits on natural capital lead to unsustainable development, and because economic functions are embedded in nature, sustainability requires handing down to future generations local and global ecosystems that largely resemble our own (Goodstein 2002). As ecological economists do not accept that natural and created capital are substitutes, they reject the net national welfare approach

to measuring sustainability. Neoclassical economists have resource rents fully reinvested in created capital, which is described as weak sustainability, where strong sustainability is an intact stock of natural capital. If environmental quality and natural resources are not, in general, capable of restoration or substitution, it does not make sense to subtract off the reductions in natural capital from the increase in created capital to arrive at a measure of welfare (Goodstein 2002). The Daly Rule provides for protection of the stock of natural capital that does not have viable, reliable current substitutes (Goodstein 2002), and this goal should be regardless of cost to the current generation. Moreover, the stock of natural capital should be limited to the yield, that is, the flow of services from that capital, now commonly described as ecosystem goods and services (Costanza *et al.* 1997; Cork & Shelton 2000; Curtis 2003, 2004).

Closed canopy wet tropical forests have been high on the agenda for application of these economic valuation procedures owing to their species richness. Vegetation cover has long been regarded as a surrogate for habitat values and ecosystem integrity, with Mooney (1988) developing a metric that related canopy cover to species richness in a ratio of 3 : 2, except for Mediterranean ecosystems where it was 1 : 1.

Table 19.1 The now commonly accepted suite of ecosystem goods and services

Group	Type
Stabilization services	Gas regulation (atmospheric composition)
	Climate regulation (temperature, rainfall)
	Disturbance regulation (ecosystem resilience)
	Water regulation (hydrological cycle)
	Erosion control and soil/sediment retention
	Biological control (populations, pest/disease control)
	Refugia (habitats for resident and transient populations)
Regeneration services	Soil formation
	Nutrient cycling and storage (including carbon sequestration)
	Assimilation of waste and attenuation, detoxification
	Purification (clean water, air)
	Pollination (movement of floral gametes)
	Biodiversity
Production of goods	Water supply (catchment)
	Food production (that sustainable portion of GPP)
	Raw materials (that sustainable portion of GPP, timber, fibre etc.)
	Genetic resources (medicines, scientific and technological resources)
Life fulfilling services	Recreation opportunities (nature-based tourism)
	Aesthetic, cultural and spiritual (existence values)
	Other non-use values (bequest and quasi option values)

Sources: Curtis (2003, 2004), modified after Costanza *et al*. (1997) and Cork and Shelton (2000).

Accordingly, wet tropical forests with a high crown cover projection density are among the highest producers of ecosystem goods and services on the planet. The now commonly accepted suite of ecosystem goods and services is provided in Table 19.1.

Overconsumption of ecosystem goods and services leads to unsustainable resource use, which is akin to the tragedy of the commons (Hardin 1968). More recently the problem has been identified as intertemporal resource misallocation (Folmer *et al*. 2001). Ecological economists, however, take a different view. As indicated by the first law of thermodynamics, all of the energy and matter used by the economic system must come from the environment. Accordingly, ecological economists use material and energy flows to measure the impact of economic activities on the environment (Kaufmann 2001). In order to do so, ecological economists generate empirical models using appropriate quantitative techniques and use these results to

generate feasible policies that will succeed where existing policies have been ineffective.

Spatial aspects of resource use can be as important as the temporal dimension (Gerking & List 2001). Ecosystem goods and services provided by tropical rainforests and other landscape types exist at the biosphere/atmosphere interface on a variety of scales – global, regional, bioregional or landscape – and can be assessed down to the individual landholding, whether privately held or in the public domain (Curtis 2003, 2004). While mainstream economics describes some property rights in ecosystem goods and services as incomplete, this can occur in instances where landholders may not be rewarded for ecosystem service provision because they cannot legally or cost-effectively preclude potential downstream consumers. However, the unique nature of the goods and services in question and the benefits enjoyed by the landholder of the natural resource imply a qualified rather than an absolute right, which is reinforced by the power of the owner to limit or severely curtail ecosystem service provision, by, for example, deforestation (Sheehan & Small 2002).

Economic valuation procedures

Mainstream economic valuation procedures are based in the discipline known as microeconomics, which deals with human preferences and choice amidst scarcity. The procedures include methods such as willingness to pay (WTP), an explicit stated preference as to how much an individual is willing to pay for an environmental outcome, which is often then incorporated into the most commonly applied method, the contingent valuation method (CVM), where a contingent market is described to the respondent (Mitchell & Carson 1989; Johansson 1993; Hanley & Spash 1993; Bateman & Turner 1995; Fisher 1996; Judez *et al*. 2000). However, despite the popularity of the CVM, it is possibly the least theoretically rigorous of the economic valuation methods (Allison *et al*. 1996). One reason for its popularity is probably government preference for the democratic choice-making nature of the WTP process. Other methods include: the travel-cost method, where an individuals' cost of accessing a protected area is assessed as a measure of the individuals' revealed and implicit willingness to pay to visit the area; and hedonic pricing, where the value of a view or proximity to, say, a national park can be determined by analysing housing property prices with or without the benefit (Hanley & Spash 1993). The latter method is commonly regarded as the most rigorous of the valuation procedures because it is based on an empirical database of property prices, thus avoiding the vagaries and potential bias of stated preference surveys, and the possibility of, say, multiple destinations in the instance of travel-cost (Allison *et al.* 1996). The choice modelling technique has removed some of the inherent bias in stated preference surveys, but the choice of respondents and bid levels remains contentious.

Notably, none of the neoclassical or environmental economics valuation procedures include an ecological assessment of the integrity of the ecosystem being valued. Surprisingly, this is left up to the respondents in the case of WTP surveys and choice modelling, who most probably have no experience in this field whatsoever, nor of trading in markets for ecosystem services. Travel cost from a common point of origin is the same whether accessing the Great Barrier Reef or the Daintree rainforest, which says nothing about the intrinsic or habitat value of the respective ecosystems. The economic value of recreation is not a proxy for the value of the ecosystems in question, because respondents could hardly be expected to make a conscious choice to allocate money without knowing what ecosystem goods and services are included in their allocation for recreation. In other words, are the respondents to include just direct use values or are respondents supposed also to include indirect use and non-use values, bequest, option, quasi-option and existence values?

The term value is judgement-laden and despite 50 years of development of valuation techniques by the various sub-disciplines of economics, no widely accepted method has appeared to exist that unambiguously identifies the value of a whole ecosystem, or a component of it (Lally 1999). However, this author has recently developed a new approach that has been well received by land and ecological economists, because it builds on the rigour of hedonic pricing. The methodology is based in land valuation procedures that have evolved from English common law, which in Australia have been greatly elaborated and replaced by legislation, most of which was first enacted in the late nineteenth and early twentieth centuries.

This new approach also includes an ecological assessment of the whole suite of ecosystem goods and services as to the extent of their contribution, loosely based on the work of Holdridge (1967), Holdridge *et al.* (1971), Lugo (1988), Mooney (1988) and Brown and Lugo (1982). Lugo (1988: 61) postulated that 'statistically significant relationships suggest that life zone conditions relate to characteristic numbers of tree species, biomass and rate of primary productivity, and capacity to resist and recover from disturbance'. Individual ecosystem goods and services were also weighted in a pecuniary valuation table by using the collective opinions of a group of experts in economics and the natural sciences as to their relative contribution (Curtis 2003, 2004). Criteria were anthropocentric and utilitarian, which were then sensitized to criteria such as threats, risk, uncertainty and precaution, and the resistance and resilience of ecosystems. Bias was eliminated by the use of the Delphi technique, the features of which include iteration and the anonymity of respondents.

Ecosystem goods and services are hosted in the biosphere or the land/atmosphere interface (Figure 19.1), and if property rights were to be assigned, they would be assigned to the proprietor of the estate in fee simple of the land that hosts them. Ecosystem goods and services are thus a production function of land (the *usus fructus per annum*), and can be valued in much the same way as other more traditional products, such as sugar cane or tropical fruit (Roll 1961). Most recently, the NSW Department of Environment and Conservation (DEC) published a paper dealing with biodiversity certification and banking, and proposing historic reforms, stating that it sought 'to correct market failure to recognize important biodiversity values in land prices' (DEC 2005: 3). Clearly, too, as in traditional farm products, the ecological contribution needs to be quantified by ecologists, not economists, and the accounts prepared by an accountant, prior to the landholding being valued by a valuer based on its extended productivity. Economists are not accountants, nor are they valuers. For public land, that is, National Parks, where there are no other products other than ecosystem goods and services, which includes recreation, only the *usus fructus per annum* needs to be capitalized to produce a capital value, or vice versa (Curtis 2006).

Every use of land has an opportunity cost, that being the existing use or other uses to which the land could

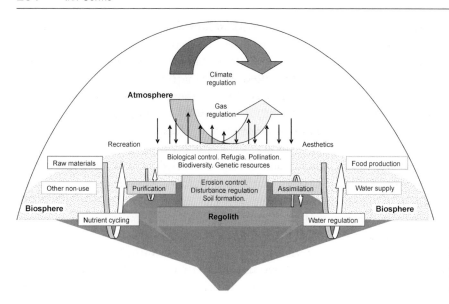

Figure 19.1 Atmosphere; atmosphere land/water interface; biosphere; biosphere/regolith interface; regolith: stratification of ecosystem services.

be put (the use forgone) (Edwards 1987; McNeeley 1988; Frank 1991). The value of a conservation area should be at least as much as the cost of preserving it, or measured by the cost of the forgone opportunities, as the area cannot be developed or redeveloped (Allison *et al.* 1996). McNeeley (1988: 33) described marginal opportunity cost as a 'very useful tool in making decisions about allocation of resources'. Moreover, McNeeley (1988: 33) argued that marginal opportunity cost 'can be used as a means by which those who will lose from having restrictions placed on their use of biological resources can be compensated to recover the value of their lost opportunity'.

Marginal opportunity cost can be expressed in terms of the annual net revenue forgone, in which case it would be capitalized, resulting in a land value in restricted and unrestricted use (McNeeley 1988, 1994). These concepts clearly link the natural production function of land with land valuation procedures. As ecosystem goods and services are the production function of land in its natural state (the *usus fructus per annum*), and as ecosystem goods and services are essential for planetary life support (Ke Chung & Weaver 1994), it could be argued that the provision of ecosystem goods and services is the highest and best use of land. It follows that the value of non-market environmental attributes can be derived indirectly by using prices from a related market that does exist (Allison *et al.* 1996), namely, the property market.

Property rights

The recent real and/or potential commodification of property rights in natural resources such as carbon, water and biota, confers three things:

- management power;
- the ability to receive income or benefits;
- an ability to sell or alienate the interest (Sheehan & Small 2002: 1).

However, the concepts of property rights of a more familiar strain are more concrete than the esoteric, incomplete or partial visions of property rights in natural resources other than land. These new forms of property rights are land property *sui generis*. Permits and licences have in a number of cases been held to be property (*Dovey* 1993; *Western Mining Corporation* 1994; *Newcrest Mining (WA) Ltd* 1997). The recognition of Native Title has also been instructive in identifying other types of property rights and quite different values ascribed to them, albeit they are still all expressions of worth. Clearly, Native Title is synonymous with the concept of usufructuary rights, which have been operational for centuries.

Sheehan and Small (2002: 11) claim that 'economic rights depend on, and are subsidiary to, the capacity of legal rights to permit and allow the holder to enjoy as a benefit … the natural resource in question'. A strong definition of property rights can ameliorate socio-economic and environmental impacts of natural resource allocation. Markets will act to provide the best allocation of property rights in natural resources, provided the rights are clarified and enforced. The crucial test of what defines a property right is fundamental to their widespread acceptance. Case history provides these insights:

- The definition, or notion of, property 'extends to every species of valuable right and interest including real and personal property, incorporeal hereditaments such as rent and services, rights of way, rights of profit or use of land of another, and chooses in action', and 'To acquire any such right is rightly described as an acquisition of property' (Starke 1944).

- 'The word "property" is often used to refer to a something that belongs to another ... "property" does not refer to a thing; it is a description of a legal relationship with a thing. It refers to a degree of power that is recognized in law as power permissibly exercised over the thing. The concept of "property" may be elusive. Usually it is treated as a bundle of rights', and 'at common law there could be no "absolute property" but only "qualified property" in fire, light, air, water and wild animals' (*Yanner* v. *Eaton* 1999).

The degree to which any of the three qualities of a property right in a natural resource are conferred depends on the mix of fundamental characteristics of the particular property, and warrants closer investigation, because many variations to property rights emerged only during the past century, including such novel ideas as strata title, where a nexus was created between air space and the Crown guarantee of title residing in land (Sheehan & Small 2002: 23). However, the key determinant in any pursuit of the creation of property rights is need, and the need for market incentives for property rights in biodiversity protection is now well established in the literature, and in reality.

Recent work in the Wet Tropics of Queensland

The total value of ecosystem goods and services provided by public and private land in the Wet Tropics of Queensland World Heritage Area (WHA) was found by this author to be in the range Australian $188–211 million year^{-1} or $210–237 ha^{-1} yr^{-1} in 2002 (Curtis 2003, 2004). Biodiversity was ranked most highly, followed by refugia, erosion control and soil/sediment retention, genetic resources, gas regulation and climate

regulation. As at 30 June 2005, the value of ecosystem goods and services in the WHA was estimated to have increased by about 94% due to population and development pressures, which while not affecting this land directly, as it is mostly protected by the World Heritage listing, is still significant as a measure of alternative land-use potential and commensurability. However, tropical rainforest on private land in the Wet Tropics bioregion, while less intact than public land, is more at risk and attracts a higher capitalization rate. As at 30 June 2005, ecosystem goods and services provided by tropical rainforest on private land were calculated by this author to be in the range $373–446 ha^{-1} yr^{-1}.

In a paper for the International Task Force on Global Public Goods, Clemencon (2005) used this author's paper in the journal *Ecological Economics* (Curtis 2004) to demonstrate the methodological complexity in deriving individual values for ecosystem goods and services, and commented that the annual worth derived for the WHA was a dramatic return, considering the estimated cost of managing protected areas in developed countries (US$10 ha^{-1} yr^{-1}). However, in valuation parlance, the return is the range of values given as a function of the capital value of the WHA. The cost of maintaining the WHA is an expense, and thus represents the difference between the gross and net production, and hence the yield. One does not capitalize the annual expense of maintaining a protected area over the annual value to society of the protected area, one deducts it. Clemencon's (2005) global estimate of the cost of managing protected areas in developed countries (i.e. US$10 ha^{-1} yr^{-1}) amounts to about US$8.94 million year^{-1} or about 0.06% of the current annual worth of ecosystem goods and services in the WHA, while James *et al.* (1999) put the cost at US$10.90 ha^{-1} yr^{-1} for developed countries and US$2.77 ha^{-1} yr^{-1} for developing nations. Moreover, prior to 1999, the amount spent in managing protected areas in tropical nations was just US$0.93 ha^{-1} yr^{-1}.

The travel cost method was used by Driml (1996) to estimate consumer surplus generated by Australian tourists to the WHA and also to address criticisms of the method as not providing an absolute measure of welfare by using different ways of estimating travel cost, which led to different but statistically acceptable measures of consumer surplus. The value of recreation was found to be in the range Au$83–166 million year^{-1}, which translated to about $100–200 million year^{-1}

in 2002 prices, or $112–224 ha^{-1} yr^{-1} (Driml 2002). Multipliers were used based on the results of this study to extrapolate the economic value of recreation to the local economy in the region. While the values ascribed by Curtis (2004) and Driml (2002) are not dissimilar (i.e. within the same order of magnitude), they are for different things. This author's work provided estimates for the whole suite of services, while Driml (1996, 2002) estimated only the value of recreation. Clearly, this one direct use, recreation, is not worth as much as all of the others. The travel cost method uses travel cost to access a natural area as a measure of consumer surplus, that is, a surrogate for value. However, this begs the question: the value of what? In the contingency valuation method a hypothetical market is described to respondents to elicit their WTP response to a scenario that may impact on a natural area, but are they being asked to value a specific attribute of the environment that is being impacted or the whole basket of goods and services? Psychologically it is difficult for respondents to separate out the recreational value, for example, and nominate a bid level, when in fact they have absolutely no idea what other attributes there are, what attributes are valuable and what values apply to them. As a result, the imputed price derived from studies of this kind is not just for, say, recreation, but for everything the respondent consciously or subconsciously perceives as being part of the natural environment in question, and as such it must include some indirect use and non-use values and perhaps even some option values and existence values. The same logic can be applied to the travel cost method. The economic values of the whole suite of ecosystem goods and services are constrained within measures that are consistent with all other uses to which land is put and other avenues of investment in the economic system. The values of individual ecosystem services are constrained within this overall basket of goods and services on a landscape or bioregional scale, but in some ecosystems certain goods and services may be worth more than others based on scarcity or limiting factors.

Other work in Australia and overseas

Costanza et al.'s (1997) seminal study of the value of ecosystem services in the global biomes has recently been updated and used by Blackwell (2005) to derive values for the coastal regions in Australia including up to 3 kilometres inland. Temperate and boreal forests were again within the same order of magnitude as other studies (Au$543 ha^{-1} yr^{-1}), but the mean for global forests was Au$1743 ha^{-1} yr^{-1} and the value of tropical forests a full order of magnitude above all other preceding and later studies (Au$3609 ha^{-1} yr^{-1}).

Rolfe and Bennett (2002) used the choice modelling technique to assess people's value preferences for conservation of rainforests in Queensland, New South Wales and overseas. A random sample of Brisbane residents was asked to reveal their preferences and to address the issues of location, features and qualities of the choices. It was found to be important to be able to distinguish between different components of value and to prioritize between a set of alternatives. The respondents chose location as possibly the most important attribute to them, and being Brisbane residents the results showed that they were parochial in ranking the choices as Queensland, then Australia, and overseas.

A contingency valuation study was undertaken by Duthy (2002) to determine community support for dedication of the Whian Whian State Forest in northeastern NSW as a new National Park. Whian Whian is one of the largest remaining sub-tropical rainforests in New South Wales. The two most important uses of the forest were found to be water catchment protection and habitat for endangered species. Respondents placed strongest values on bequest, existence and non-consumptive use values, and weakest on the productive functions of Whian Whian. The mean willingness to pay for the non-consumptive use and non-use values was Au$18.89 year^{-1} for three years (median $10). The population from which the sample was drawn was 119 148; thus the value of non-consumptive uses and non-uses was put at $1.19 million year^{-1} using the median bid, to $2.25 million year^{-1} using the mean bid, excluding the recreation value ascribed to potential visitors from outside the area. Whian Whian comprises 5567 hectares of State Forest, so these WTP estimates equate to a range of from $214 to $404 ha^{-1} yr^{-1} for non-consumptive uses and non-uses. Moreover, Duthy (2002) argued that by transferring recreational values estimated for Dorrigo National Park and Gibraltar National Park, also in northern NSW, these could add another $264–298 ha^{-1} yr^{-1} to the total for Whian Whian. In respect of this possible benefit transfer to add to the gross hectare values for Whian Whian, this author's same comments apply as to the value of recreation in the WHA of Queensland. The values of recreation elicited for Dorrigo and Gibraltar National Parks must include a component of indirect and non-use values and possibly also some option, bequest and existence values, so to use benefit transfer in such an unqualified way could amount to double counting.

Perhaps the best, but not so recent, example of financing environmental services on both public and private land is the Costa Rican experience (Chomitz *et al.* 1998). Costa Rica's forest cover decreased from more than 50% in 1950 to 29% in 1986, and thence reduced overall by 1.1% per year, with a much lower rate for the areas under World Heritage protection by 1997. Secondary forest, including plantation forest, covers about three-quarters of the deforested area. Economic values were estimated by Kishor and Constantino in 1993 for some use and non-use services at US$162–214 $ha^{-1} yr^{-1}$, the majority ascribed to carbon sequestration (US$120) (Chomitz *et al.* 1998). In 1996, Costa Rica passed a new forestry law that permitted landholders to be compensated for providing some environmental services. The law (no. 7575) explicitly recognized four environmental services of forests: carbon fixation, hydrological services, biodiversity protection and provision of scenic beauty. The implementation rules for the new law were adopted in 1997. A unique set of institutional arrangements was being put in place contemporaneously to enable the creation of markets for the forest's environmental services. Some of these novel arrangements revolved around the joint implementation (JI) and clean development mechanism (CDM) provisions under the Kyoto Protocol. There is no link between the provision of

services and financing as the government acts as an intermediary to sell the services, and the funds realized are used to finance the services, including those provided by national parks and other public land. Payments to landholders under the programme currently reimburse them for four types of actions over a 5-year period, after which time they are free to renegotiate or deal direct, but they commit to manage or protect the forest for 20 years, which is recorded on the public land register (Table 19.2).

At the start of the programme earlier incentive programmes already covered 145 000 hectares. In 1997, a further 79 000 hectares were placed under forest protection and 10 000 hectares under forest management, and 6500 hectares were destined for reforestation, for a gross payment to landholders of US$14 million. In 1998, the waiting list or excess demand, was estimated to be of the order of 70 000 hectares.

With the emergence of market-based instruments for environmental outcomes, financial incentives and rewards for environmental stewardship on private land and resource rent as it applies to access to national parks by commercial operators, markets are emerging for land set aside for conservation and other environmental benefits. Clearly for conservation to be a viable alternative land use it must be competitive with other uses to which land could be put, or no one will pay for it. Despite the use of many environmental valuation procedures, including this author's new approach based on opportunity cost, comparison of a range of work and programmes does evidence a surprising synergy in the valuation of ecosystem goods and services provided by rainforests. This synergy could be attributed to the similarity of people's expressed and

Table 19.2 Payment schedule to landholders for conservation contracts in Costa Rica

Activity	Min. Area (ha)	Max. Area (ha)	Total payment (US$ ha^{-1} over 5 years)	Yr 1	Yr 2	Yr 3	Yr 4	Yr 5
Reforestation	1	any	480	240	96	72	48	24
Natural forest management	2	300	321	161	64	32	32	32
Regeneration	2	300	200	40	40	40	40	40
Protection	2	300	200	40	40	40	40	40

Reforestation by organizations of small producers is limited to a maximum area of 10 ha. Exchange rate approx 250 colones/US$1, March 1998.
Source: modified after Chomitz *et al.* (1998).

Table 19.3 Selected valuation studies or payments made for environmental stewardship of forests on a dollar per hectare basis

Researcher/author	Subject of the research	Lower range Au$ h^{-1}yr^{-1}	Upper range Au$ h^{-1}yr^{-1}
Bennett 1995	Dorrigo National Park, NE NSW Australia (economic value of recreation)		1500
Bennett 1995	Gibraltar Range National Park, NE NSW Australia (economic value of recreation)		46
Blackwell 2005	Boreal and temperate forests (ecosystem services)		543
Blackwell 2005	Global forests (ecosystem services)		1743
Blackwell 2005	Tropical rainforests (ecosystem services)		3609
Castro 1994	Costa Rica Wildlands (all services)	170	357
Chomitz et al. 1998*	Costa Rica (various environmental stewardship practices)	40	96
Costanza et al. 1997	Global biomes (all services)		1343
Curtis 2004	Wet Tropics Queensland, Australia (all ecosystem goods and services within tenures)	210	236
Curtis 2004	Wet Tropics Queensland, Australia (all ecosystem goods and services across tenures)	149	342
Curtis (this publication)	Wet Tropics Queensland, Australia (ecosystem services, rainforest on private land)	373	446
Davis et al., in Duthy 2002	Gibraltar Range and Dorrigo National Parks, NE NSW Australia (recreation)	264	298
de Groot 1994	Panama's forests (use and non-use values)		835
Driml 2002	Wet Tropics Queensland, Australia (Tourism)	112	224
Duthy 2002	Whian Whian National Park, NE NSW Australia (use and non-use values)	214	404
Flatley & Bennett 1996	Vanuata tropical rainforest on the islands of Erromango and Malakula (conservation)		87
Gillespie 1997	Budderoo National Park, SE NSW Australia (economic value of recreation)		809
Kishor & Constantino in Chomitz et al. 1998	Costa Rican forests (use and non-use services)	162	214
Lockwood & Carberry 1998	Southern Riverina, Victoria, Australia (preserve remnants)	38	87
Lockwood & Carberry 1998	NE Victoria, Australia (preserve remnants)	43	98
Pimental in Myers et al. 1997	Global rainforests (sustainable use value)		367
Tobias & Mendelson 1991	Monte Verde Cloud Forest Reserve, Costa Rica (domestic recreational value)		20

*Denotes environmental payment scheme.
 Studies more than 10 years old have been adjusted to 2002 values.

revealed preferences, and perhaps subliminal awareness of scarcity and risk factors that are demonstrated by higher values being attributed to regions most at risk due to population density and development pressures.

Summary

- The methodologies and examples described in this chapter evidence a range of economic, valuation, ecological and statistical procedures used in order to develop an objective approach to valuing ecosystem goods and services provided by tropical rainforests.

- Property rights in ecosystem goods and services are not incomplete as postulated by mainstream economists, but new forms of unique property rights in natural resources, which are supported by case law in Australia.

- In the most part, estimates of the pecuniary value of the ecological benefits provided by tropical forests internationally fall well within one order of magnitude, but the amount spent on management in developed and developing nations is a full order of magnitude different.

- Tropical forest managers should be proactive in the development of initiatives to develop financial incentives for the protection of tropical forests on both public and private land, à *la* Costa Rican experience. These may not be limited to resource rents by way of entry fees by commercial operators to protected areas, but extended to include biodiversity credits, water resources, assimilation services, pollination services and soforth on both private and public land, with funding from external organizations that may be affected downstream to institute best practice environmental management of tropical rainforests and upstream catchments.

References

Allison, G., Ball, S., Cheshire, P., Evans, A. & Stabler, M. (1996). *The Value of Conservation? A Literature Review of the Economic and Social Value of the Cultural Built Environment for the Department of National Heritage, and the Royal Institution of Chartered Surveyors*. Royal Institution of Chartered Surveyors, London.

Aronsson, T. & Lofgren, K.-G. (2001). Green accounting and green taxes in the global economy. In *Frontiers of Environmental Economics*, Folmer, H., Landis Gabel, H., Gerking, S. & Rose, A. (eds). Edward Elgar, Cheltenham, UK.

Bateman, I. J. & Turner, R. K. (1995). Valuation of the environment, methods and techniques: The contingent valuation method. In *Sustainable Environmental Economics and Management*, Turner, R. K. (ed.). John Wiley & Sons, London.

Bennett, J. (1995). *Economic Value of Recreational Use: Gibraltar Range and Dorrigo National Parks*. A report prepared for the Environmental Economics Policy Unit, Environmental Policy Division, NSW National Parks and Wildlife Service.

Brown, S. & Lugo, A. E. (1982). The storage and production of organic matter in tropical forests and their role in the global carbon cycle. *Biotropica* **14**: 161–87.

Blackwell, B. (2005). *The Economic Value of Australia's Natural Coastal Assets: A Preliminary Estimate of the Ecoservices of Some of Australia's Natural Coastal Assets*. Coastal CRC, Indooroopilly, Brisbane.

Castro, R. (1994). *The Economics Opportunity Costs of Wildlands Conservation Areas: The Case of Costa Rica*. Department of Economics, Harvard University.

Chomitz, K. M., Brenes, E. & Constantino, L. (1998). *Financing Environmental Services: The Costa Rican Experience and its Implications*. Environmentally and Socially Sustainable Development, Latin America and Caribbean Region, World Bank, Washington, DC.

Clemencon, R. (2005). *Costs and Benefits of Global Environmental Public Goods Provision*. International Task Force on Global Public Goods, University of California, San Diego.

Cork, S. J. & Shelton, D. (2000). The nature and value of Australia's ecosystem services: a framework for sustainable environmental solutions. In *Sustainable Environmental Solutions for Industry and Government*, Proceedings of the 3rd Queensland Environmental Conference, May 2000. Environmental Engineering Society, Queensland Chapter, The Institution of Engineers, Australia, Queensland Division, and the Queensland Chamber of Commerce and Industry, pp. 151–9.

Costanza, R., d'Arge, R., de groot, R. *et al.* (1997). The value of the world's ecosystem services and natural capital. *Nature* **387**: 253–60.

Curtis, I. A. (2003). Valuing ecosystem services in a green economy. PhD Thesis, James Cook University, Cairns Campus.

Curtis, I. A. (2004). Valuing ecosystem services: a new approach using a surrogate market and the combination of a multiple criteria analysis and a Delphi panel to assign weights to the attributes. *Ecological Economics* **50**(3/4): 163–94.

Curtis, I. A. (2006). Valuing the environmental impact of a power line corridor through a State Forest in Queensland: a heuristic exercise in environmental valuation for the property profession. *Australian Property Journal* **39**(2): 87–96.

de Groot, R. S. (1994). Environmental functions and the economic value of natural ecosystems. In *Investing in Natural Capital: The Ecological Economics Approach to Sustainability*, Jansson, A. N., Hammer, M. & Costanza, R. A. (eds). Island Press, Washington, DC, pp. 151–68.

Department of Environment and Conservation NSW (DEC) (2005). *Biodiversity Certification and Banking in Coastal and Growth Areas*. Dept of Environment and Conservation (NSW), Sydney South, July.

Dovey v. *The Minister for Primary Industries* (1993). 119 *ALR* 108.

Driml, S. (1996). Sustainable tourism in protected areas? An ecological economics case study of the Wet Tropics World Heritage Area. PhD thesis, Australian National University, Canberra.

Driml, S. (2002). Travel cost analysis of recreation value in the Wet Tropics World Heritage Area. *Economic Analysis and Policy* **32**(2) special issue: 11–26.

Duthy, S. (2002). Whian Whian – state forest or national park? Community attitudes and economic values. *Economic Analysis and Policy* **32**(2) special issue: 91–111.

Edwards, S. F. (1987). *An Introduction to Coastal Zone Economics: Concepts, Methods and Case Studies*. Taylor & Francis, New York.

Fisher, A. C. (1996). The conceptual underpinnings of the contingent valuation method. In *The Contingent Valuation of Environmental Resources*, Bjornstad, D. J. and Kahn, J. R. (eds). Edward Elgar, Cheltenham, UK.

Flatley, G. W. & Bennett, J. W. (1996). Using contingent valuation to determine Australian tourists' values for forest conservation in Vanuatu. *Economic Analysis and Policy* **26**(2): 111–27.

Folmer, H., Landis Gabel, H., Gerking, S. & Rose, A. (eds) (2001). *Introduction. Frontiers of Environmental Economics*. Edward Elgar, Cheltenham, UK.

Frank, R. H. (1991). *Microeconomics and Behaviour*. McGraw-Hill, New York.

Gerking, S. & List, J. (2001). Spatial economic aspects of the environment and environmental policy. In *Frontiers of Environmental Economics*, Folmer, H., Landis Gabel, H., Gerking, S. & Rose, A. (eds). Edward Elgar, UK.

Gillespie, R. (1997). *Economic Value and Regional Economic Impact of Minnamurra Rainforest Centre, Budderoo National Park*. NSW National Parks and Wildlife Service.

Goodstein, E. S. (2002). *Economics and the Environment*, 3rd edn. John Wiley & Sons, New York.

Hanley, N. & Spash, C. L. (1993). *Cost Benefit Analysis and the Environment*. Edward Elgar, Cheltenham, UK.

Hardin, G. (1968). The tragedy of the commons. *Science* **162**: 1243–8.

Holdridge, L. R. (1967). *Life Zone Ecology*. Tropical Science Centre, San Jose, Costa Rica.

Holdridge, L. R., Grencke, W. C., Hatheway, W. H., Liang, T. & Tosi, J. A. Jr (1971). *Forest Environments in Tropical Life Zones: A Pilot Study*. Pergamon Press, Oxford, UK.

James, A. N., Green, M. J. B. & Paine, J. R. (1999). *Global Review of Protected Area Budgets and Staff*. WCMC, Cambridge UK.

Johansson, P.-O. (1993). *Cost – Benefit Analysis of Environmental Change*. Cambridge University Press, Cambridge, UK.

Judez, L., de Andres, R., Perez Hugalde, C., Urzainqui, E. & Ibanez, M. (2000). Influence of bid and subsample vectors on the welfare measure estimate in dichotomous choice contingent valuation: Evidence from a case study. *Journal of Environmental Management* **60**: 253–65.

Kaufmann, R. (2001). The environment and economic well-being. In *Frontiers of Environmental Economics*, Folmer, H., Landis Gabel, H., Gerking, S. & Rose, A. (eds). Edward Elgar, Cheltenham, UK.

Ke Chung, K. & Weaver, R. D. (1994). Biodiversity and humanity: paradox and challenge. In *Biodiversity and Landscapes: A Paradox of Humanity*, Ke Chung, K. and Weaver, R. D. (eds). Cambridge University Press, Cambridge, UK.

Lally, P. (1999). Identifying non-market public amenity value using a values jury. *Australian Property Journal* February: pp. 436–42.

Lockwood, M. & Carberry, D. (1998). *State Preference Surveys of Remnant Native Vegetation Surveys*. Johnstone Centre Report No. 104. Charles Sturt University, Albury, 24.

Lugo, A. E. (1988). Estimating reductions in the diversity of tropical forest species. In *Biodiversity*, Wilson, E. O. (ed.). National Academy of Sciences Press, Washington, DC.

McNeeley, J. A. (1988). *Economics and Biological Diversity: Developing and Using Economic Incentives to Conserve Biological Resources*. IUCN. Gland, Switzerland.

McNeeley, J. A. (1994). Lessons from the past: forests and biodiversity. *Biodiversity and Conservation* **3**: 3–20.

Mitchell, R. C. & Carson, R. (1989). *Using Surveys to Value Public Goods: The Contingent Valuation Method*. Resources for the Future, Washington, DC.

Mooney, H. A. (1988). Lessons from Mediterranean-climate regions. In *Biodiversity*, Wilson, E. O. (ed.). National Academy of Sciences Press, Washington, DC.

Myers, N. (1997). The world's forests and their ecosystem services. In *Nature's Services: Societal Dependence on Natural Ecosystems*, Daily, G. C. (ed.). Island Press, Washington, DC.

Newcrest Mining (WA) Ltd v. *Commonwealth* (1997). 147 *ALR* 42.

Rolfe, J. & Bennett, J. (2002). Assessing rainforest conservation demands. *Economic Analysis and Policy* **32**(2) special issue: 51–67.

Roll, E. (1961). *A History of Economic Thought*. Faber and Faber, London.

Sheehan, J. & Small, G. (2002). Towards a definition of a property right. Paper presented at the Pacific Rim Real Estate Society (PRRES) Conference, Christchurch, New Zealand,

Starke, J. In *The Minister for the Army* v. *Dalziel* (1944). 68 *CLR* 290 (Dalziel).

Tobias, D. & Mendelsohn, R. (1991). Valuing ecotourism in a tropical rain-forest reserve. *Ambio* **20**(2): 91–3.

Western Mining Corporation v. *Commonwealth* (1994). *121 ALR 661*.

Yanner v. *Eaton* (1999). HCA 53, at 8 per Gleeson CJ, Gaudron Kirby & Hayne JJ.

20 INTERNATIONAL PERSPECTIVE: ECOLOGICAL PROCESSES AND ECOSYSTEM SERVICES IN THE WET TROPICS

S. Joseph Wright

Smithsonian Tropical Research Institute, Balboa, Panama, Republic of Panama

Humans and the natural world affect one another wherever they come together. The natural world is structured by interactions among wild species and between wild species and their physical environment. Humans benefit from many of these interactions, suffer from others and potentially alter them all. This is especially true wherever humans create a mosaic landscape where fragments of the original, natural vegetation are interspersed among crops, pastures and towns. This describes the Wet Tropics World Heritage Area, and the chapters in the present part explore interdependencies among humans, wild species and the physical environment in the Wet Tropics.

The natural world provides a wide range of benefits free of cost that sustain all human societies (Costanza *et al.* 1997). The clearest example is our water supply. In the first chapter of this section, McJannet *et al.* (Chapter 15, this volume) contrast hydrological processes – essentially what happens to precipitation after it reaches the vegetation – and forest water yields for four contrasting rainforests and cloud forests from the Wet Tropics. They find that El Niño events reduce forest water yields in two ways, with greater

evaporative losses as well as lower precipitation inputs. Their precise measurements are essential to develop a mechanistic understanding of forest water balance. This mechanistic understanding will, in turn, permit the development of predictive models for effective water management throughout the region. McJannet and colleagues are well on the way towards this goal.

A perhaps less obvious benefit of the natural world is provided by the insects, birds, bats and other animals that pollinate economically important crops. The native pollinators in neighbouring rainforest fragments are likely to play an important role in the pollination of nearby crops. This is particularly true for the many crops that have been introduced throughout the world without their original pollinators. Boulter and her colleagues (Chapter 17, this volume) describe the diversity of types of flowers, when they are produced and what is known of their pollinators for the rainforests of the Wet Tropics. This information sets the stage for the pollination of agricultural crops throughout the Wet Tropics. Boulter *et al.* conclude that the state of knowledge is limited – pollinators have been identified for just a handful of Wet Tropics plant species – and that much more needs to be learned.

The movement of seeds, like the movement of pollen, is crucial for the conservation of natural plant populations because seed and pollen movement offers the only opportunity for genetic material to move among plant populations, and, in addition, seed dispersal offers the only opportunity for plants to colonize new habitats. Westcott *et al.* (Chapter 16, this volume) describe a powerful approach to understand the movement of seeds throughout the rainforests of the Wet Tropics. They focus on the animals that move seeds. They have accumulated a remarkable database recording how often and how far each of 65 species of birds, bats and other mammals carries the seeds of 1268 rainforest plant species. They then integrate seed movements over the entire community of fruit and seed-eating animals to model the distances seeds are dispersed for each plant species. Many of the more important animal species that disperse seeds are absent from forest fragments in the mosaic landscape of the Wet Tropics, and the database accumulated by Westcott and his colleagues offers a powerful tool to identify plant species that will require assistance by man to maintain the movement of genetic material among fragmented populations and to colonize newly opened habitats. Their knowledge base for the management and conservation of tropical forest plants is unrivalled anywhere else in the world.

Cunningham and Blanche (Chapter 18, this volume) take a remarkably different approach in the fourth chapter of this part. Earlier studies of the benefits human societies reap from the natural world have ignored the potential costs associated with close proximity to the natural world (Costanza *et al.* 1997). Following this tradition, Cunningham and Blanche document benefits that accrue from pollination services and from pest control by predatory insects when crops are located near forest fragments where native pollinators and predatory insects live. Cunningham and Blanche also depart from tradition, however, and examine the potential costs associated with insect pests that might move to crops from the same nearby forest fragments. Their approach heralds a new type of analysis where both the benefits and costs of proximity to wild nature are tallied to determine the net impact of wild nature on our lives. The Wet Tropics offers an ideal setting for these studies. The results of Cunningham and Blanche indicate that the net benefits of proximity to forest vary among the most important crops grown in the region. Their approach has the potential to increase crop yields through matching crops to the mosaic landscape of forest fragments interspersed among human land uses.

In the final chapter of this part, Curtis (Chapter 19, this volume) provides an overview of the economist's approach to the valuation of goods and services received from wild nature and then summarizes recent work, much of it his own, on the development of a globally accepted set of standards for such valuations. This is crucially important work because the conservation of rainforest will only be able to compete with agriculture and other human land uses after the commercial market recognizes the value of the goods and services provided by the original rainforest. Natural ecosystems provide 20 widely recognized goods and services ranging from a clean water supply to the opportunity for recreation and tourism, to aesthetic and spiritual values. The problem is how to place a dollar value on these goods and services so that society will recognize the contribution of natural ecosystems and so that the loss of the goods and services provided by natural ecosystems is included when calculating the net benefit of conversion of land to alternative human uses. As an example, Curtis calculated the value of ecosystem goods and services provided by rainforest on private land in the Wet Tropics to be in the range of Au\$370–446 ha^{-1} yr^{-1} in June 2005. This is sharply lower than the value of US\$2007 ha^{-1} yr^{-1} for tropical forests globally (Costanza *et al.* 1997). This large discrepancy points towards the need for a set of standards for such valuations. Again, the work of Curtis is an important step forward towards these standards, and his valuations for the Wet Tropics represent an important tool to be incorporated into regional development schemes.

To summarize, the five chapters in this part represent five significant advances towards the realization that the human occupants of the Wet Tropics benefit in diverse ways from their proximity to tropical rainforests. Several of the chapters provide concrete tools for the management and conservation of those rainforests or for the management of regional resources. All five chapters point towards the true net benefits realized by society through the preservation of the Wet Tropics of Queensland World Heritage Area.

This realization will place conservation on a firm financial footing, and the example provided by the Wet Tropics has the potential to lead the way for less studied tropical forests located in developing countries around the world.

Reference

Costanza, R., d'Arge, R., de groot, R. *et al*. (1997). The value of the world's ecosystem services and natural capital. *Nature* **387**: 253–60.

Part 3

Threats to the Environmental Values of the Wet Tropics

INTRODUCTION

Stephen M. Turton[1]* and Nigel E. Stork[2]*

[1]Australian Tropical Forest Institute, School of Earth and Environmental Sciences, James Cook University, Cairns, Queensland, Australia
[2]School of Resource Management and Geography, Faculty of Land and Food Resources, University of Melbourne, Burnley Campus, Richmond, Victoria, Australia
*The authors were participants of Cooperative Research Centre for Tropical Ecology and Management

This section looks at ongoing and emerging threats to the environmental values and integrity of the Wet Tropics. As discussed in Part 1, the Wet Tropics forest ecosystems are largely driven by the region's climatic and hydrological systems (Bonell & Callaghan, Chapter 2). Hence, climate change is considered to be perhaps the greatest single threat to the Wet Tropics. Balston (Chapter 21) provides an overview of factors impacting on climate variability in the region, followed by a discussion of the likely impacts of climate change on rainfall and temperature regimes across the Wet Tropics. Highland forest ecosystems and their biota are most at risk as suitable habitat is expected to be greatly reduced in area with only a modest amount of warming and a slight decrease in rainfall (Williams *et al.*, Chapter 22). Perhaps as much as 74% of the Wet Tropics vertebrates will become threatened due to climate change this century.

Fragmentation of forests on the coast and Atherton Tablelands over the past 100 years has created mosaics of relict forest fragments within a sea of agricultural landscapes, with numerous ongoing threats to the integrity of the forest fragments. While large-scale clearing is no longer a threat to the Wet Tropics forests, internal fragmentation due to roads and powerline clearings continues to create a suite of impacts, including road kill and barrier effects for susceptible fauna (Laurance & Goosem, Chapter 23). Like many other tropical forest landscapes, the Wet Tropics have had their fair share of invasive species, notably weeds (Goosem, Chapter 24) and feral animals (Congdon & Harrison, Chapter 25). Invasive species continue to be high on the list of management priorities for the Wet Tropics World Heritage Area.

In recent years, we have witnessed increasing use of remote sensing technologies to improve detection of threatening processes in forested landscapes. Gillieson *et al.* (Chapter 26) provide an overview of applications of high-resolution remote sensing in rainforest ecology and management, illustrating their work with Wet Tropics case studies. As discussed earlier (Pearce, Chapter 7), rainforest-based tourism has increased significantly in recent decades, together with demand for outdoor recreation by residents of the Wet Tropics. Turton and Stork (Chapter 27) discuss environmental impacts of visitation in the Wet Tropics and put forward management strategies to mitigate some of the more serious impacts. Bill Laurance provides a personal perspective to this section, based on his earlier experiences in the Wet Tropics and more recent research in the neotropics (Chapter 28).

21 IMPACTS OF CLIMATE VARIABILITY AND CLIMATE CHANGE ON THE WET TROPICS OF NORTH-EASTERN AUSTRALIA

Jacqueline Balston

Department of Primary Indusrtries and Fisheries, Cairns, Queensland, Australia

Introduction

The Wet Tropics region of north-eastern Queensland, and the diverse ecological systems it supports, are the result of millennium scale processes of continental and geological movement, biological evolution and climate changes over the past 120 million years. It is a region where the climate is variable enough to encourage diversity, yet stable enough to maintain habitat for species evolving over evolutionary time scales. Compared with other rainforest regions on Earth, such as those in the Amazon and Indonesia, the Wet Tropics is uniquely subtropical, exhibiting distinct seasonality and relatively high variability in climate and hydrological processes over time scales from months to decades and beyond. Natural climate variability in the region is driven by a number of mechanisms, including the seasons, the Indo-Asian monsoon, the Madden Julian Oscillation (Bonell & Callaghan, Chapter 2, this volume) and inter-annual mechanisms such as the El Niño Southern Oscillation (ENSO) as well as decadal and inter-decadal systems. As a result, the rainfall of north-eastern Australia is one of the most variable on Earth. Added to this, climate change as a result of anthropogenic greenhouse gas emissions is likely to alter existing relationships between the atmosphere and the ocean, modifying climate variability on a number of time scales.

Inter-annual variability

El Niño Southern Oscillation

The El Niño Southern Oscillation (ENSO) is a complex ocean–atmosphere system second only to the seasons in generating large-scale variability within the climate (Allan 2000). It is responsible for up to 40% of the rainfall variability in eastern Australia (Cordery 1998). The atmospheric circulation consists of an area of low pressure and ascending air in the western Pacific, upper level westerlies, a region of descending air in the eastern Pacific and low level easterly winds (Figure 21.1). These surface winds push warm surface water along the equator to the western equatorial Pacific, leaving behind an area of cool sea surface temperatures in the eastern equatorial Pacific (Graham & White 1988). During a La Niña event this atmospheric circulation is enhanced, producing stronger than average low level easterly winds and increased probabilities of above average rainfall across northern and eastern parts of Australia, Indonesia and other western Pacific regions.

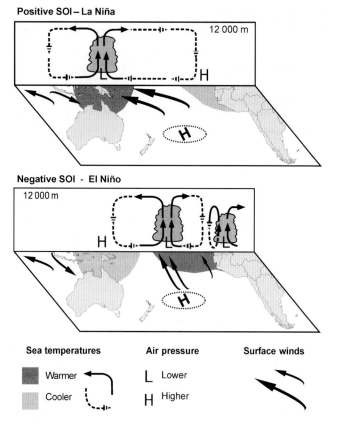

Figure 21.1 Diagram showing the two extremes of the ENSO. During El Niño events the Southern Oscillation Index (SOI) is negative, and anomalously warm water and increased convection is centred over the central equatorial Pacific, often resulting in anomalously cool water and subsidence over the western Pacific and northern Australia. This pattern is near reversed in La Niña years. *Source*: Clewett *et al.* (1994).

El Niño, the other extreme of the Southern Oscillation, is characterized by a migration of warm sea surface temperatures eastward along the equator and associated atmospheric changes. During an El Niño event, warm, moist air rises in the east, while cooler, drier air descends over the western Pacific. The trade winds slacken and may revert to westerlies with upper level winds blowing towards the east. Rainfall across northern and eastern Australia tends to be below average, with an increased probability of drought (e.g. Allen 1988).

El Niño events typically occur every 2–7 years, during which the monsoon trough is weakened and displaced further north. Maximum rainfall and peak freshwater flows in north-east Queensland occur slightly earlier than average (McBride 1987;

Evans & Allan 1992; Lough 2001). Regions like the Wet Tropics that are affected by ENSO exhibit greater inter-annual rainfall variability than those that are not, with the strongest and most spatially coherent effects in the southern hemisphere winter and spring, and weaker effects in summer and autumn (McBride & Nicholls 1983; Zhang & Casey 1992; Nicholls & Kariko 1993; Simmonds & Hope 1997; Lau & Nath 2000).

Long-term variability

In addition to intra-annual variability and that driven by ENSO, there are a number of longer cycles in the climate that impact on the Wet Tropics region, including biennial, decadal and inter-decadal cycles (Allan 2000). The Quasi-biennial Oscillation in the equatorial stratospheric winds (McGregor & Nieuwolt 1998) appears to have an impact on cyclone development, reducing the frequency of intense cyclones in the western Pacific during easterly phases (Gray & Scheaffer 1991). The Interdecadal Pacific Oscillation has been shown to correlate significantly with decadal rainfall anomalies in north-eastern Australia (Latif *et al.* 1997; Folland *et al.* 1998; Power *et al.* 1999), although it is considered by some to be part of the ENSO signal (Zhang *et al.* 1997). Variations to these longer scale cycles in response to climate change will possibly alter current relationships and increase the degree of climate variability (Allan 1988; Salinger *et al.* 2000; IPCC 2001a).

Cyclones

Tropical cyclones occur every year in the Australian region between October and May and add to the climatic variability of the region, bringing storm surges, high winds, intense rain and local flooding as they cross the coast (see Turton & Stork, Chapter 3, this volume for ecological effects of cyclones). The majority of cyclones affecting north-eastern Queensland from 1967 to 1997 occurred in January and February and were category 1 or 2 (average winds 17–33 ms^{-1}; central pressure 1000–970 hPa), with only about 10% reaching category 3 (33–44 ms^{-1}; 945–970 hPa) (Lough 2001). Recent analysis of detrital coral and shell deposits along the north-east coast suggests that super cyclones (category 5) occur in the region every two to three centuries, more frequently than previously estimated (Nott & Hayne 2001).

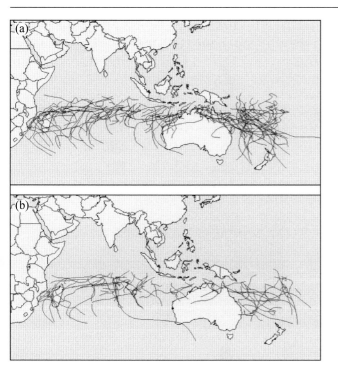

Figure 21.2 Influence of ENSO on cyclone tracks in the Australasian region. Lines show cyclone tracks for January from 1951 to 1995 when the September SOI was (a) consistently positive (indicative of a La Niña event) and (b) consistently negative (indicative of an El Niño event). *Source*: Jason Keys, © the State of Queensland Department of Primary Industries and Fisheries 2006.

The incidence and location of cyclone genesis is influenced by ENSO, although relationships have changed through time (Nicholls 1992). During El Niño events, tropical cyclone activity in the Coral Sea is reduced, with cyclones tending instead to be distributed along the equator away from the Queensland coastline (Dong 1988; Hastings 1990; Evans & Allan 1992; Basher & Zheng 1995) (Figure 21.2).

Climate change

Climate change in this context refers to that component of climate variability (including trends) now attributable to human activities and the release of greenhouse gases into the atmosphere. The Intergovernmental Panel for Climate Change Fourth Assessment Report (IPCC AR4) (IPCC 2007) represents a summary of hundreds of scientific studies from around the world and states that there is now 'very high confidence that the globally averaged net effect of human activities since 1750 has been one of warming'. The atmospheric concentration of all greenhouse gases has risen since the beginning of the industrial revolution, with current levels of CO_2 (379 p.p.m.) and methane (1774 p.p.b.) the highest in 6,500,000 years and concentrations of nitrous oxide 49 p.p.b. higher (IPCC 2007). Increased concentrations of greenhouse gases in the atmosphere enhance the naturally occurring greenhouse effect, further heating the troposphere by allowing short wave radiation from the sun to pass through and heat the Earth, but preventing long wave thermal radiation from escaping back out into space. This process is now validated by surface, balloon and satellite data (IPCC 2007).

Temperature

According to the IPCC AR4 (IPCC 2007) the global surface temperature has increased by 0.76 ± 0.2°C since 1850. Eleven of the past 12 years (1995–2006) rank among the 12 warmest years of the instrument record. Recent warming has been greater over land than the sea, with sea surface temperatures increasing at roughly half the rate of terrestrial surfaces. The rate of warming over the past 50 years is almost double that for the past 100 years (IPCC 2007), with some sources suggesting it has been up to 50 times faster than during the lead up to the last ice age 10 000 years ago (Flannery 2005).

Over the past century the Australian continental mean annual temperature has also increased (by about 0.8°C), with most warming in winter and spring. Minimum temperatures are increasing up to 30% faster than maximum temperatures, particularly in the northern half of the continent (Hughes 2003) (Figure 21.3). Sea surface temperatures on the Great Barrier Reef (GBR) have increased by 0.46°C in the north and by up to 2.56°C near Townsville (Lough 2000).

Rainfall

Since the beginning of the twentieth century atmospheric water vapour has increased several per cent per decade and cloud cover by some 2% (IPCC 2001a). Rainfall across Australia has increased slightly, although on a continent-wide basis the trend is not statistically significant due to the high inter-annual

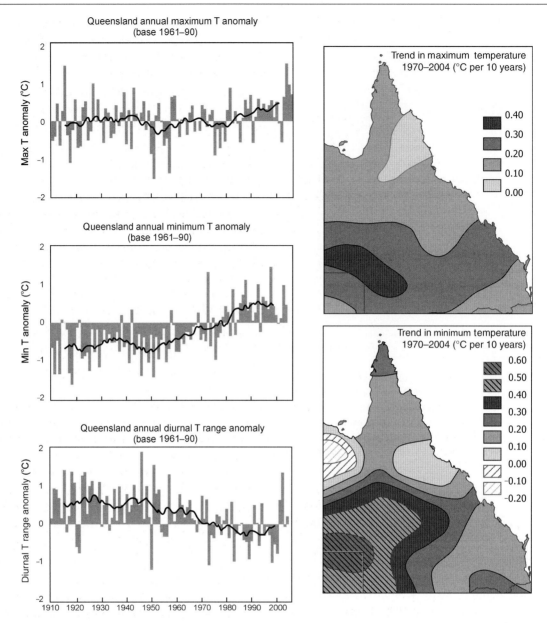

Figure 21.3 Trends in Queensland maximum and minimum temperature, and diurnal temperature range. Graphs show annual maximum and minimum temperature anomalies and annual diurnal temperature anomaly range from 1961 to 1990. Maps illustrate trends in maximum and minimum temperature in degrees Celsius per decade from 1970 to 2004. *Source*: adapted from Bureau of Meterology (2005).

variability (Hughes 2003). For regions where precipitation has increased (northern Western Australia and central Northern Territory) it has occurred more so in summer than winter and appears likely to be a result of increases in heavy rainfall events and the number of rain days (Hughes 2003). Since 1976 the frequency and intensity of El Niño events has increased, resulting in a rainfall decrease along the east coast, mostly in the summer and autumn months (Salinger *et al.* 2000; IPCC 2001a; Hughes 2003) (Figure 21.4). Historical rainfall records for Cairns highlight the degree of inter-annual variability and show a decreasing trend in rainfall over the past century, mostly in the April to June period (Clewett *et al.* 1994).

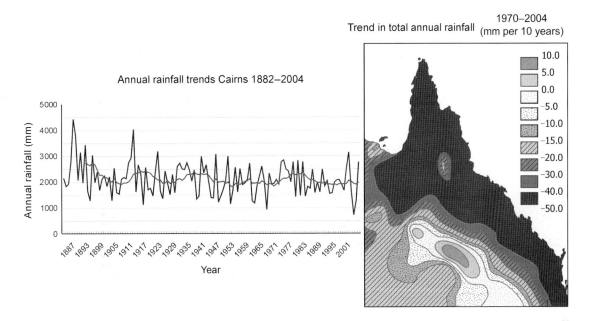

Figure 21.4 Total annual rainfall for Cairns (1882–2004) and trends across Queensland (1970–2004). *Source*: adapted from Bureau of Meterology (2005).

Sea levels

Global sea levels as recorded by tidal gauges show an increase of between 0.12 and 0.22 m during the twentieth century (IPCC 2007). The average rate of sea level rise around Australia is about 1.2 millimetres per year over the period from 1920 to 2000 (Church *et al.* 2004). Along the east coast of Queensland measurements vary. Townsville and Mackay have recorded a significant increase over the past 25 years (>1 mm yr^{-1}), while Cairns (−0.02 mm yr^{-1}), Bundaberg and Brisbane have recorded slight reductions (Mitchell *et al.* 2000).

Future climate change

In general, a warming of the Earth will reduce the temperature difference between the equator and the poles and decrease the kinetic energy of the atmosphere. The Earth will become more humid with more frequent and intense storms in the lower latitudes, while drier conditions will prevail in the mid-latitudes. The warming oceans will expand, with increasing sea levels contributed to by melting polar and continental ice sheets and glaciers (IPCC 2007). Future climatic conditions

are generated by global climate models (GCM) and usually provide a range of estimates based on the different scenarios of future emissions.

Temperature

The mean global temperature is expected to increase by 1.1–6.4°C over the 1990–2100 period (IPCC 2007). Globally averaged sea surface temperatures are also expected to increase, with a trend towards a more El Niño like state in the tropical western Pacific (IPCC 2001a; Cai 2003). It is highly likely that hot days and heat waves will be more frequent, with increases highest over land masses (particularly in the northern hemisphere) and areas where soil moisture decreases occur. Cold events and frosts are likely to decrease (IPCC 2007). An increase in average temperature of between 0.3°C and 2°C by 2030 and 0.8°C and 6°C by 2070 is predicted for Queensland. Spatial patterns of warming are expected to be consistent with current observations, with greater warming inland and less along the coastal strip (Cai *et al.* 2003) (Figure 21.5).

Local scale projections for climate change in the Wet Tropics region have been generated as a subset of the scenario data for Queensland and show an increase in

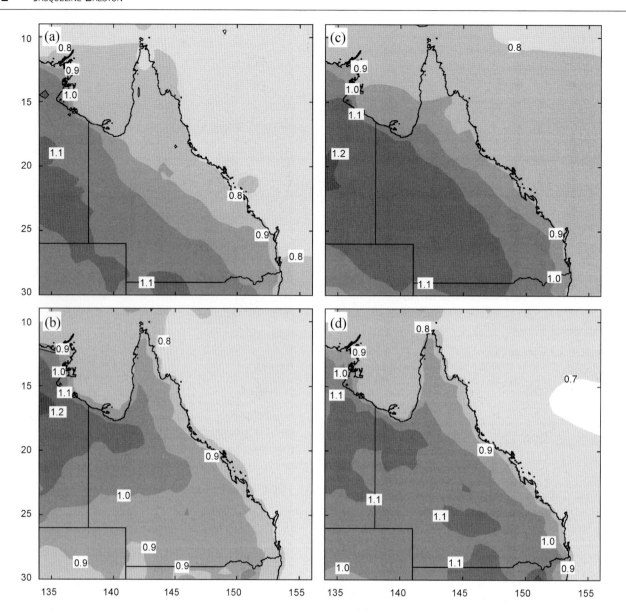

Figure 21.5 Change in temperature for Queensland per degree of warming interpolated to a 1° grid, for (a) summer (December–February), (b) winter (June–August), (c) autumn (March–May) and (d) spring (September–November). Contour interval is 0.1°C. *Source*: Walsh *et al.* (2001).

maximum temperature of 0.3–5.2°C for the period 2030–70 (Table 21.1) (Walsh *et al.* 2002; Whetton & Jones 2003).

Rainfall

Globally averaged atmospheric water vapour, evaporation and precipitation are all projected to increase under climate change. Increases in precipitation are very likely in high latitudes, while decreases are likely in most subtropical land regions (IPCC 2007). Rainfall extremes and intensities are projected to increase almost everywhere, while an increase in the frequency of El Niño events will continue the drying trend across northern and eastern Australia (IPCC 2001a). Increased temperatures and evaporation/evapotranspiration will lead to hotter and more severe droughts, such as that in 2002, even if rainfall totals remain largely unaffected (Rosenzweig & Hillel 1998; IPCC 2001a; Nicholls 2003; Risbey *et al.* 2003).

As on the global scale, rainfall projections for Queensland are varied depending on location and influence from systems such as the ENSO (Figure 21.6).

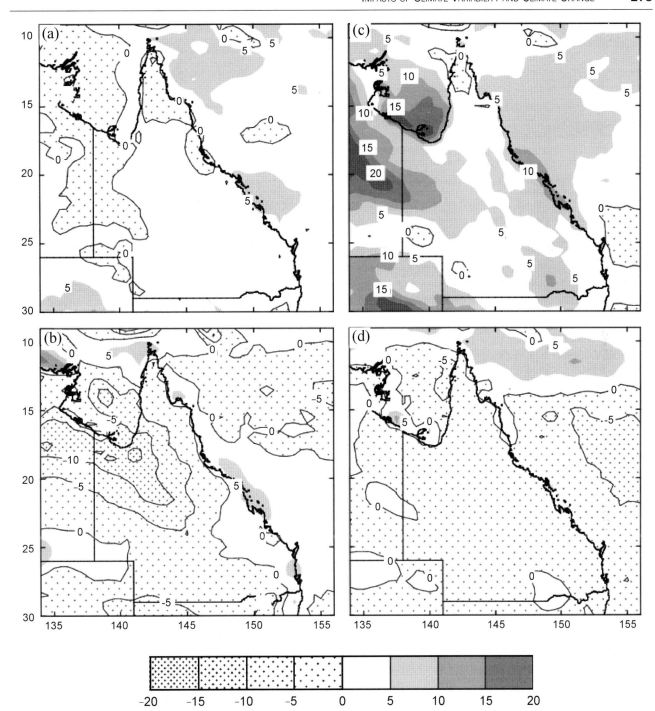

Figure 21.6 Percentage change of rainfall per degree of warming for Queensland interpolated to a 1° grid, for (a) summer (December–February), (b) winter (June–August), (c) autumn (March–May) and (d) spring (September–November). *Source*: Walsh *et al*. (2001).

Table 21.1 Projected climate changes for the Cairns region

Cairns	Now	2030	2070
Annual average maximum temperature (°C)	28.9	29.9 ± 0.7	31.9 ± 2.2
December–February days over 35°C	3	5.5 ± 2.5	41 ± 35
Annual rainfall (mm)	2028	1945 ± 160	1785 ± 485

Source: Whetton and Jones (2003).

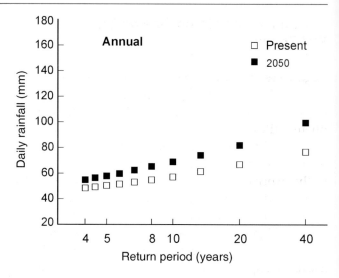

Figure 21.7 Return periods for extreme rainfall events for north Queensland. *Source*: Walsh *et al.* (2001).

Increases in the intensity and frequency of extreme rainfall events have been modelled across Australia and found to be highest in mountainous terrain, with an average increase in intensity of between 20% and 40% by 2040 (Abbs 2004).

Projections for the Wet Tropics region (Whetton & Jones 2003) suggest that by 2030 rainfall is expected to change by –5% to +15% from December to February, and –15% to +5% for March to November (Table 21.1). This would result in drier conditions and reduced soil moisture content when increases in evaporative demand are considered (Walsh *et al.* 2002; Crimp *et al.* 2003). The frequency of extreme rainfall events is expected to increase (CSIRO 2001; IPCC 2001a), with the return period for an 80 mm day^{-1} event dropping down from 40 to 20 years (Figure 21.7).

Sea levels

Projections for global sea level rise by the IPCC FAR lie between 0.20 and 0.59 m by 2090–9 across the range of scenarios (IPCC 2007). This is the result of thermal expansion of the water and contributions from glaciers, and the Greenland and Antarctica icecaps, but does not include uncertainties pertaining to changes in ice sheet flow.

Hydrology

Changes to the frequency and intensity of rainfall events will impact on the quality and volume of freshwater in the region (see Bonell & Callaghan, Chapter 2, this volume). Modelling these effects across Queensland indicates reduced flows for much of the state, with fewer, more intense events (Walsh *et al.* 2002; Chiew *et al.* 2003).

Cyclones

Records from the Australian basin (although limited) show a slight increase in the intensity of tropical cyclones (Plummer *et al.* 1999) due to changes in ENSO, but a decrease in the frequency of moderate to strong cyclones since 1970 (IPCC 2003). Modelling of cyclone activity globally has given mixed results, with the expected frequency of future cyclones varying by ±50% (IPCC 2003). Recent studies (e.g. Knutson & Tuleva 2004) are reinforcing the findings that cyclones will become more intense, with an expected increase in wind speeds of 5–10%, peak rainfall increases of 15–30% and an increase in the number of category 5 cyclones (see Turton & Stork, Chapter 3, this volume). Future tropical cyclones in the Queensland region are projected to increase in maximum intensity but show little change in their region of formation or number (Walsh *et al.* 2002).

Modelling the extent of cyclonic storm surges for the Cairns city area under climate change shows an area twice the size of that currently affected (McInnes *et al.* 2000). Other low lying areas and riverine tidal areas in the region may be affected up to 100 km inland (Waterman 1996).

Impacts of climate change on the Wet Tropics

Regardless of efforts to reduce emissions in the future, much of the climate change over the coming years will be the result of greenhouse gases already in the atmosphere (IPCC 2001a; Climate Change Risk and Vulnerability 2005). Australian ecosystems are considered particularly vulnerable to climate change owing to the expected increase in the frequency of El Niño events and resulting decreases in rainfall, as well as increases in cyclone and rainfall intensity. Those ecosystems highlighted as most at risk include coral reefs, arid and semi-arid regions, alpine systems (IPCC 2001b), mangroves and coastal wetlands, heathlands and the World Heritage reef and rainforest areas (Climate Change Risk and Vulnerability 2005). World Heritage regions in particular represent unique, endemic ecosystems of biological diversity, which in many cases are fragile or remnant systems affected by numerous other pressures, including tourism and pollution (see Williams *et al.*, Chapter 22, and Hilbert, Chapter 8, this volume). In addition, it is expected that projected changes will occur too quickly for the evolutionary adaptation of many species (Climate Change Risk and Vulnerability 2005).

A number of other publications have considered the impacts of climate change in the Australasian region in more detail than can be covered here (e.g. Pittock & Wratt 2001; Hughes 2003; Pittock 2003). This section aims instead to give a synopsis of climate change impacts relevant to the Wet Tropics region, by highlighting the risks to significant industries and ecosystems, and discussing possible management considerations for ameliorating negative effects.

Rainforest ecosystems

Rainforests thrive in the tropical regions on Earth where annual rainfall exceeds 2000 millimetres and the dry season is short or non-existent (Primack & Corlett 2005). The large mammal fauna and extensive rainforests that covered much of northern Australia during the early and mid-Miocene are today restricted to the Wet Tropics region due to a reduction in rainfall (see Hilbert, Chapter 8, this volume). Concerns for the future survival of these ecosystems are valid, with

numerous studies already recording changes to the distribution, migration, reproduction and abundance of both plants and animals in response to climate change (e.g. IPCC 2001b; Walther *et al.* 2002; Parmesan & Yohe 2003; Primack & Corlett 2005). Malhi and Wright (2004) reported that all the tropical rainforest regions have experienced a strong warming at a mean rate of $0.26 \pm 0.05°C$ per decade since the mid-1970s, and a decline in precipitation at a rate of 1% per decade (±0.8%). Future climate changes are expected to continue the observed poleward and altitudinal shift of species, altering species abundance, composition and dominance, and reducing favourable habitats to the point of extinction in some cases (Thomas *et al.* 2004; Williams *et al.*, Chapter 22, this volume).

As the result of previous extinctions and recolonizations in response to climatic changes, the Wet Tropics rainforest is perhaps more adaptive than other rainforest regions on Earth (Schneider *et al.* 1998), but will none the less be impacted upon by climate change in a number of ways (Ostendorf *et al.* 2001; Williams 2002; Williams *et al.* 2003; Krokenberger *et al.* 2004). An overview of these changes is presented in Box 21.1, with more detail on the impacts to biodiversity and rainforest ecosystem covered by Williams *et al.* (Chapter 22, this volume) and Hilbert (Chapter 8, this volume).

Aquatic ecosystems

Aquatic ecosystems are expected to be influenced by changes to freshwater flows, water temperature, wind, nutrient and salinity levels (Frye 1983). Changes to wind regimes will probably drive changes in ocean currents and water temperature, affecting areas of upwelling and associated primary production, nutrient transfer and salinity. These changes will impact on the spawning and transport of larvae and juvenile fish, adult migrations, growth rates, food supplies and perhaps even the predator–prey relationships for certain species (Frye 1983; Kapetsky 2000; Putten & Rassam 2001).

Freshwater and estuarine fisheries that rely on freshwater flows to bring nutrients and provide other services to the ecosystem will be affected by increased water temperatures, changes to flow regimes and an increase in the number or duration of drought events (Frye 1983; Kapetsky 2000). In the Wet Tropics,

Box 21.1 How might the rainforest change?

A collation of research that has examined the possible impacts of climate change on the rainforests of north-eastern Queensland suggests, in some cases, that dramatic changes are likely. Increased levels of atmospheric CO_2 are expected to increase forest productivity by increasing photosynthesis, but will reduce the digestibility and quality of vegetation (Ceulemans 1996) as leaf toughness and leaf defence compounds (such as tannins) increase and nitrogen : phosphorus ratios and proteins decrease (Wang & Polglase 1995; O'Neill 1999; Kanowski 2001; Williams et al. 2003). This effect may be exacerbated by the Wet Tropics rainforest acting as a net emitter of CO_2, as it does during El Niño years when temperatures are higher and rainfall is reduced (Liddell & Turton, unpublished), a response that conflicts with some other studies (e.g. Clark 2004; Fearnside 2004).

Temperature increases will force cool adapted species further up the mountain. Modelling shows that 1°C of warming will reduce montane areas in the Wet Tropics by 66%, and 2°C will reduce them by 95% (O'Neill 1999). Increased temperature will also lift the cloud base, and deprive upland rainforest of water stripped from clouds (Foster 2001).

Williams et al. (2003) predict a complete loss of all 65 species analysed in response to a 7°C temperature increase, and species distribution decreased by more than 50% on average by 2100 (Meynecke 2004). Additionally, 74% of rainforest birds in the region are predicted to become threatened (including 26 critically endangered species) as a result of mid-range warming over the same period (Shoo et al. 2005). Although many lowland forest species are already near their maximum temperature tolerance (Primack & Corlett 2005), lowland forests are expected to increase in area, while highland rainforest will decrease by up to 50% with 1°C of warming (Hilbert et al. 2001).

Projected increases in cyclone intensity are expected to impact heavily on the Wet Tropics rainforest communities (Nott & Hayne 2001; Turton & Stork, Chapter 3, this volume), as the shallow rooted, tall rainforest trees are susceptible to uprooting, breakage and defoliation (Primack & Corlett 2005).

Changes to the frequency and intensity of rainfall events will lead to increased erosion and sedimentation, which will impact on species that have evolved in tandem with past flow regimes (IPCC 2001b). A possible reduction in the area of wetlands essential for the survival of various estuarine and bird species is also possible (Crimp et al. 2003).

An expected increase in the frequency of El Niño events and the probable reduction in available moisture will possibly lead to an increase in the frequency of fire, and a shift from fire-sensitive vegetation to fire-tolerant species (Kershaw 1985; Mayle et al. 2004). Surface fires have recently been shown (Laurence 2003) to be disastrous in tropical forests, where they kill up to 48% of trees and 49% of the living tree liana biomass over a period of 3 years, and reduce the number of understorey birds. What is more, once burned even by a light surface fire, rainforests are prone to further fires. Research in other regions suggests that the limits of tropical rainforests are set by drought (Primack & Corlett 2005), and models of the Amazon rainforest show that the rainforest converts into dry savannah when more than 33 years of drought occur in a century, making extremes of fire and drought more important than changes to the average climate (Angelo 2005).

the catch of iconic species such as the barramundi, a popular sport and table fish, has been shown to relate closely to rainfall and sea surface temperatures (Balston 2005).

Estuarine and coastal habitats

Sea level rise and heightened storm surges will have significant impacts on coastal belt habitats and communities (IPCC 2001b). The coastal belt of the Wet Tropics region and the city of Cairns are at particular risk owing to low elevation and the number of large rivers (Crimp et al. 2003). Coastal ecosystems such as wetlands and mangroves are highly sensitive to sea level rise, with research showing that mangroves have moved inland in response to rising levels in the past. However, restraints in the form of urbanization and farming will limit this migration in the future, reducing available habitat (Woodroffe 1993). Incursions of salt water into fresh surface and ground water along the coastal strip, combined with a possible decrease in freshwater flows, are expected to affect the quality of freshwater sources currently tapped for urban and agricultural purposes (Crimp *et al.* 2003).

Tourism

With an increasing reputation as a nature-based destination, the tourism industry in the Wet Tropics generates approximately Au$1.1 billion annually and is considered to be highly dependent on the natural ecosystem (Crimp *et al.* 2003; Pearce, Chapter 7, this volume). Impacts from climate change on the Great Barrier Reef (GBR) and the rainforest, which are expected to reduce biodiversity and ecosystem health, will also reduce tourist appeal, no doubt affecting the economy of the region (Crimp *et al.* 2003; Climate Change Risk and Vulnerability 2005). Other changes, such as an increase in the incidence of mosquito-borne diseases and extreme events such as cyclones, may also reduce tourist visitation (Crimp *et al.* 2003).

Agriculture

With inter-annual variability in rainfall already accounting for huge fluctuations in agriculture production (Rosenzweig & Hillel 1998), amplification of this variability and a possible reduction in the predictability of events will increase agricultural risk. A poleward shift in the optimal conditions for crops and livestock will mean that competition against tropical producers will increase, leaving them to develop new and untested options, while increased rainfall and cyclone intensities will create additional risks to agriculture in the region (Crimp *et al.* 2003).

Current agricultural enterprises in the Wet Tropics region include the production of sugarcane, dairy and beef cattle, horticulture and tree crops including avocadoes, mangoes and tropical fruits (Crimp *et al.* 2003; Climate Change Risk and Vulnerability 2005). As with other systems, the impacts of climate change are the result of a complex interplay of factors. Vegetation responds to increases in CO_2 in different ways, with C3 plants (temperate grasses and dicotyledons, including many crops and weeds) exhibiting an increase in yields of up to 35% (doubled CO_2), while C4 plants (many tropical grasses including sugarcane) show increases of only 10%. The water use efficiency of both plants is enhanced under higher atmospheric CO_2 concentrations (Crimp *et al.* 2003), but increases in air temperature and vapour pressure deficit will boost evapotranspiration considerably and possibly offset any water efficiencies achieved (Rosenzweig & Hillel 1998).

As with native species, higher CO_2 levels also increase the carbon to nitrogen ratio in fodder plants, decreasing quality and digestibility. Some crops and tropical pastures are expected to become less competitive than the weeds, which will also change in their geographical distribution (Rosenzweig & Hillel 1998; Crimp *et al.* 2003). Changes to the climate will also affect crop and livestock pests and diseases, with an expected increase in tropical parasites such as buffalo fly, and disease vectors including mosquitoes that affect both domesticated and native wild animals (Rosenzweig & Hillel 1998; IPCC 2001b; Crimp *et al.* 2003). Those pests of most concern for the Wet Tropics region include fruit flies, locusts, soil breeding beetles, heliothis moths and fruit piercing moths (Crimp *et al.* 2003).

Increased temperatures will reduce the number of chilling days for plant species reliant on cold events (e.g. mangoes) and increase the incidence of heat stress in animals such as beef and dairy cattle (Crimp *et al.* 2003). Increased demands on freshwater for irrigation will put further pressure on the existing supplies, which might be reduced in both quantity and quality as increased rainfall intensity exacerbates erosion and freshwater sedimentation (Crimp *et al.* 2003). Current ground water quality ranges from poor to moderate (Freeman 2003).

Conclusions, implications and management considerations

The climate of the Wet Tropics of north-eastern Queensland is highly variable and influenced by a number of synchronous and asynchronous natural cycles. Anthropogenic climate change will lead to further increases in the variability and intensity of extreme events. Research that has considered the degree of species diversity in the Amazon (e.g. Colinvaux 1989) suggests that the highest species richness occurs where environmental disturbances are frequent but not excessive (intermediate-disturbance hypothesis). Climate variability, then, is beneficial to the continued evolution of a species-rich ecosystem. However, projections of future climate change, and the rate of this change, consider it more akin to the catastrophic changes responsible for mass species extinction (Colinvaux 1989). Global climate model projections for the Wet Tropics region indicate:

- A warmer, probably drier future that is increasingly affected by El Niño, more intense rainfall events and higher cyclone intensities.
- Impacts for the region will be complex and varied and will depend on the ability of the systems to adapt. Projections suggest a decrease in upland rainforest habitat and rainforest biodiversity, an increase in species extinction and increased inundation of coastal mangrove and wetland communities due to sea level rise, all of which are irreversible.
- The ramifications of expected changes are widespread and will impact on the biodiversity, lifestyle and economic sustainability of the region.

Many actions that will provide our best opportunities for adapting to climate change are already known to be environmentally and economically sustainable, and are currently used in the management of climate variability and drought preparedness. In many cases, the technology and processes to implement them already exist. Tropical land clearing, for example, currently accounts for 10–30% of global anthropogenic CO_2 emissions (Rosenzweig & Hillel 1998) and so retaining the current carbon stores in existing forested areas along with habitat and biodiversity should be a priority. More than half of the existing Wet Tropics rainforest areas are not protected by National Park or World Heritage listing (forestry reserves or private land tenure) and so are not protected from clearing. Additionally, minimizing habitat fragmentation and creating wildlife corridors is essential, as is reducing the impact from other detrimental sources such as pests, exotic weeds and disease in order to maintain ecosystem resilience.

Although human systems are considered better able to adapt to the impacts of climate change (IPCC 2001b), success will depend on a number of factors, including the levels of education, wealth and age structure of the community. Recommendations for adaptation include:

- the provision of buffer zones along the coast to provide protection to urban development from storm surge events and to allow for inland migration of littoral habitats from rising sea levels;

- development of slopes should be reconsidered as the risk of landslide in response to higher rainfall and cyclone intensities increases in an area already considered to be at high risk from earthquakes;
- adoption of agricultural processes that stabilize soil structure and organic carbon need to be encouraged so as to reduce both CO_2 emissions and the nutrient and sediment contamination of freshwater and coral reef areas during flood events;
- development of strategic plans to address the risks associated with climate change, including impacts on water resources, disease and weed dynamics, socioeconomic and infrastructure planning, extreme events and public health;
- knowledge gaps in other sectors have been identified (e.g. Crimp *et al*. 2003) for the region and need to be addressed as soon as possible.

With the full extent of climate change from current emissions expected to take up to 100 years, both urgent and significant reductions in greenhouse gas emissions are needed now to ameliorate future impacts on the unique ecosystems of the Wet Tropics. In addition, the implementation of strong public education and policy mechanisms to address what we already know is required.

References

Abbs, D. J. (2004). The effect of climate change on the intensity of extreme rainfall events. In *Climate and Water: 16th Australian New Zealand Climate Forum*. ANZCF 2004 Conference Committee, Lorne, Victoria.

Allan, R. J. (1988). El Niño Southern Oscillation influences in the Australasian region. *Progress in Physical Geography* **12**: 313–49.

Allan, R. J. (2000). ENSO and climatic variability in the last 150 years. El Niño and the Southern Oscillation: multiscale variability, global and regional impacts. Diaz, H. F. & Markgraf, V. (eds). Cambridge University Press, Cambridge, pp. 3–55.

Angelo, C. (2005). Punctuated disequilibrium: occasional but extreme climate could turn parts of the Amazon rainforest into dry savannas. *Scientific American* **292**: 22–3.

Basher, R. E. & Zheng, X. (1995). Tropical cyclones in the southwest Pacific: spatial patterns and relations to southern oscillation and sea surface temperature. *Journal of Climate* **8**(5): 1249–60.

Bureau of Meteorology (BOM) (2005). Bureau of Meteorology Website 2005 (www.bom.gov.au).

Balston, J. M. (2005). Modelling impacts of climate variability on the commercial barramundi (*Lates calcarifer*) fishery of north-eastern Queensland. In *International Congress on Modeling and Simulation (MODSIM)*, Melbourne.

Cai, W. (2003). Australian droughts, climate variability and climate change: insights from CSIRO's climate models. Science for Drought, Queensland Department of Primary Industries, Brisbane.

Cai, W., Crimp S, Cai, W. *et al.* (2003). *Climate Change in Queensland under Enhanced Greenhouse Conditions. Final Report 2002–2003*. CSIRO, Aspendale, p. 79.

Ceulemans, R. (1996). Direct impacts of CO_2 and temperature on physiological processes in trees. In *Impacts of Global Change on Tree Physiology and Forest Ecosystems*. Kluwer Academic Publishers, Wageningen, The Netherlands.

Chiew, F. H. S., Harrold, T. I., Siriwardena, L., Jones, R. N. & Srikanthan, R. (2003). Simulation of climate change impact on runoff using rainfall scenarios that consider daily patterns of change from GCMs. In *International Congress on Modelling and Simulation (MODSIM)*, Modelling and Simulation Society of Australia and New Zealand, Townsville, Queensland.

Church, J. A., Hunter, J., Mc Innes, K. & White, N. J. (2004). *Sea-level Rise and the Frequency of Extreme Events around the Australian Coastline*. National Coastal Conference – Coast to Coast, Hobart.

Clark, D. A. (2004). Sources or sinks? The responses of tropical forests to current and future climate and atmospheric composition. *Philosophical Transactions of the Royal Society of London. Series B, Biological Sciences* 359(1443): 477–91.

Clewett, J. F., Clarkson, N. N., Owens, D. T. & Abrecht, D. G. (1994). *Australian Rainman: Rainfall information for better management*. Australia, Queensland Department of Primary Industries, p. 45.

Climate Change Risk and Vulnerability (2005). Allen Consulting Group, Canberra, p. 159.

Colinvaux, P. A. (1989). The past and future Amazon. *Scientific American* May: 68–74.

Cordery, I. (1998). Forecasting precipitation from atmospheric circulation and SOI. *Hydrology in a Changing Environment* 1: 63–70.

Crimp, S., Balston, J. M. & Ash, A. (2003). *A Study to Determine the Scope and Focus of an Integrated Assessment of Climate Change Impacts and Options for Adaptation in the Cairns and Great Barrier Reef Region*. Australian Greenhouse Office, Canberra, p. 117.

CSIRO (2001). *Climate Change Projections for Australia*. CSIRO Atmospheric Research, p. 8.

Dong, K. (1988). El Niño and tropical cyclone frequency in the Australian region and the north-west Pacific. *Australian Meteorological Magazine* 36: 219–25.

Evans, J. C. & Allan R. J. (1992). El Niño/Southern Oscillation modification to the structure of the monsoon and tropical cyclone activity in the Australasian region. *International Journal of Climatology* 12: 611–23.

Fearnside, P. M. (2004). Are climate change impacts already affecting tropical forest biomass? *Global Environmental Change* 14: 299–302.

Flannery, T. (2005). *The Weather Makers: The History and Future Impact of Climate Change*. Text Publishing, Melbourne.

Folland, C., Parker, D. E., Colman, A. W. & Washington, R. (1998). Large scale modes of ocean surface temperature since the late ninteenth century. *Climate Research Technical Note*, No. 81, 37. Hadley Centre, London.

Foster, P. (2001). The potential negative impacts of global climate change on tropical montane cloud forests. *Earth-Science Reviews* 55: 73–106.

Freeman, J. (ed.) (2003). *State of the Environment*. Environmental Protection Agency Queensland, Brisbane.

Frye, R. (1983). Climatic change and fisheries management. *Natural Resources Journal* 23(1): 77–96.

Graham, N. E. & White, W. B. (1988). The El Niño cycle: A natural oscillator of the Pacific Ocean-atmosphere system. *Science* 240: 1293–302.

Gray, W. M. & Scheaffer, J. D. (1991). El Niño and QBO influences on tropical cyclone activity. In *Teleconnections Linking Worldwide Climate Anomolies: Scientific Basis and Societal Impact*, Glantz, M. H., Katz, R. W. & Nicholls, N. (eds). Cambridge University Press, Cambridge, pp. 257–84.

Hastings, P. A. (1990). Southern Oscillation influences on tropical cyclone activity in the Australian/south-west Pacific region. *International Journal of Climatology* 10: 291–8.

Hilbert, D. W., Ostendorf, B. & Hopkins, M. (2001). Sensitivity of tropical forests to climate change in the humid tropics of north Queensland. *Austral Ecology* 26: 590–603.

Hughes, L. (2003). Climate change and Australia: Trends, projections and impacts. *Austral Ecology* 28: 423–43.

Intergovernmental Panel on Climate Change (IPCC) (2007). *Climate Change 2007: The Physical Science Basis*. IPCC, Geneva, 21.

Intergovernmental Panel on Climate Change (IPCC) (2001a). *Climate Change: The Scientific Basis*. IPCC, Geneva, 20.

Intergovernmental Panel on Climate Change (IPCC) (2001b). *Climate Change: Impacts, Adaptation and Vulnerability*. IPCC, Geneva, 18.

Intergovernmental Panel on Climate Change (IPCC) (2003). *IPCC Workshop on Changes in Extreme Weather and Climate Events*. IPCC, Beijing, p. 107.

Kanowski, J. (2001). Effects of elevated CO_2 on the foliar chemistry of seedlings of two rainforest trees from northeast Australia: Implications for folivorours marsupials. *Austral Ecology* 26: 165–72.

Kapetsky, J. M. (2000). Present applications and future needs of meteorological and climatological data in inland fisheries and aquaculture. *Agricultural and Forest Meteorology* **103**: 109–17.

Kershaw, A. P. (1985). An extended late Quaternary vegetation record from north-eastern Queensland and its implications for the seasonal tropics of Australia. *Proceedings of the Ecological Society of Australia* **13**: 179–89.

Knutson, T. R. & Tuleya R. E. (2004). Impact of CO_2-induced warming on simulated hurricane intensity and precipitation: sensitivity to the choice of climate model and convective parameterization. *Journal of Climate* **17**(18): 3477–95.

Krockenberger, A. K., Kitching, R. & Turton, S. (2003). *Environmental Crisis: Climate Change and Terrestrial Biodiversity in Queensland*. Cooperative Research Centre for Tropical Rainforest Ecology and Management, Cairns.

Latif, M., Kleeman, R. & Eckert, C. (1997). Greenhouse warming, decadal variability or El Niño? An attempt to understand the anomalous 1990s. *Journal of Climate* **10**: 2221–39.

Lau, N. & Nath, M. J. (2000). Impact of ENSO on the variability of the Asian-Australian monsoons as simulated in GCM experiments. *Journal of Climate* **13**(12): 4287–309.

Laurence, W. F. (2003). Slow burn: the insidious effects of surface fires on tropical forests. *Trends in Ecology and Evolution* **18**(5): 209–12.

Lough, J. M. (2000). 1997–98: Unprecedented thermal stress to coral reefs? *Geophysical Research Letters* **27**: 3901–4.

Lough, J. M. (2001). Climate variability and change on the Great Barrier Reef. In *Oceanographic Processes of Coral Reefs: Physical and Biological Links in the Great Barrier Reef*, Wolanski, E. J. (ed.). CRC Press, Boca Raton, FL, pp. 269–300.

McBride, J. L. & Nicholls N. (1983). Seasonal relationships between Australian rainfall and the Southern Oscillation. *Monthly Weather Review* **111**: 1998–2004.

McBride, J. L. (1987). The Australian summer monsoon. In *Monsoon Meteorology*, Chang, C. P. & Krishnamurti, T. (eds). Oxford University Press, Oxford, pp. 203–31.

McGregor, G. R. & Nieuwolt, S. (1998). *Tropical Climatology: An Introduction to the Climates of the Low Latitudes*, 2nd edn. John Wiley & Sons, Chichester, UK.

McInnes, K. L., Walsh, K. J. E. & Pittock, B. (2000). *Impact of Sea Level Rise and Storm Surges on Coastal Resorts: Final Report*. CSIRO, p. 42.

Malhi, Y. & Wright, J. (2004). Spatial patterns and recent trends in the climate of tropical rainforest regions. *Philosophical Transactions of the Royal Society of London. Series B, Biological Sciences* **359**(1443): 311–29.

Mayle, F. E., Beerling, D. J., Gosling, W. D. & Bush, M. B. (2004). Responses of Amazonian ecosystems to climatic and atmospheric carbon dioxide changes since the last glacial maximum. *Philosophical Transactions of the Royal Society of London. Series B, Biological Sciences* **359**(1443): 499–514.

Meynecke, J. O. (2004). Effects of global climate change on geographic distributions of vertebrates in North Queensland. *Ecological Modelling* **174**(4): 347–57.

Mitchell, W., Chittleborough, J., Ronai, B. & Lennon, G. W. (2000). Sea level rise in Australia and the Pacific. *The South Pacific Sea Level and Climate Change Newsletter Quarterly Newsletter* **5**: 10–19.

Nicholls, N. & Kariko A. (1993). East Australian rainfall events: interannual variations, trends and relationships with the Southern Oscillation. *Journal of Climate* **6** (June): 1141–52.

Nicholls, N. (1992). Recent performance of a method for forecasting Australian seasonal tropical cyclone activity. *Australian Meteorological Magazine* **40**(2): 105–10.

Nicholls, N. (2003). *Climate Change: Are Droughts Becoming Drier and More Frequent?* Science for Drought, Queensland Department of Primary Industries, Brisbane.

Nott, J. & Hayne, M. (2001). High frequency of super-cyclones along the Great Barrier Reef over the past 5,000 years. *Nature* **413**: 508–12.

O'Neill, G. (1999). Treetop tremors. *Ecos* April: 15.

Ostendorf, B., Hilbert, D. W. & Hopkins, M. S. (2001). The effect of climate change on tropical rainforest vegetation pattern. *Ecological Modelling* **145**: 211–24.

Parmesan, C. & Yohe, G. (2003). A globally coherent fingerprint of climate change impacts across natural systems. *Nature* **421**(1): 37–42.

Pittock, B. (2003). *Climate Change: An Australian Guide to the Science and Potential Impacts*. Australian Greenhouse Office, Canberra.

Pittock, B. & Wratt, D. (2001). *Australia and New Zealand. Third Assessment Report of the Intergovernmental Panel on Climate Change*. Cambridge University Press, Cambridge.

Plummer, N., Salinger, M. J., Nicholls, N. *et al.* (1999). Changes in climate extremes over the Australian region and New Zealand during the twentieth century. *Climatic Change* **42**: 183–202.

Power, S., Tseitkin, F., Mehta, V., Lavery, B., Torok, S. & Holbrook, N. (1999). Decadal climate variability in Australia during the twentieth century. *International Journal of Climatology* **19**: 169–84.

Primack, R. & Corlett, R. (2005). *Tropical Rain Forests: An Ecological and Biogeographical Comparison*. Blackwell Publishing, Oxford.

Putten, M. V. & Rassam, G. (2001). Fisheries, climate change and working together. *Fisheries* **26**: 4.

Risbey, J., Karoly, D., Reynolds, A. & Braganza, K. (2003). *Drought and Climate Change. Science for Drought*. Queensland Department of Primary Industries, Brisbane.

Rosenzweig, C. & Hillel, D. (1998). *Climate Change and the Global Harvest*. Oxford University Press, New York.

Salinger, M. J., Stigter, C. J. & Das, H. P. (2000). Agrometeorological adaptation strategies to increasing climate variability and climate change. *Agricultural and Forest Meteorology* **103**: 167–84.

Schneider, C. J., Cunningham, M. & Moritz, C. (1998). Comparative phylogeography and the history of endemic vertebrates in the Wet Tropics rainforests of Australia. *Molecular Ecology* **7**: 487–98.

Shoo, L. P., Williams, S. E. & Hero, J. (2005). Climate warming and the rainforest birds of the Australian Wet Tropics: using abundance data as a sensitive predictor of change in total population size. *Biological Conservation* **125**: 335–43.

Simmonds, I. & Hope, P. (1997). Persistence characteristics of Australian rainfall anomalies. *International Journal of Climatology* **17**: 597–613.

Thomas, C. D., Cameron, A., Green, R. E. *et al.* (2004). Extinction risk from climate change. *Nature* **427**: 145–8.

Walsh, K. J. E., Cai, W., Hennessy, K. J. *et al.* (2002). *Climate Change in Queensland under Enhanced Greenhouse Conditions*. CSIRO.

Walsh, K. J. E., Hennessy K. J., Jones, R. *et al.* (2001). *Climate Change in Queensland under Enhanced Greenhouse Conditions: Third Annual Report, 1999–2000*. CSIRO, p. 90.

Walther, G. R., Post, E., Convey, P. *et al.* (2002). Ecological responses to recent climate change. *Nature* **416** (28 March): 389–95.

Wang, Y. P. & Polglase, P. J. (1995). Carbon balance in the tundra, boreal forest and humid tropical forest during climate change: scaling up from leaf physiology and soil carbon dynamics. *Plant, Cell and Environment* **18**: 1226–44.

Waterman, P. (1996). *Australian Coastal Vulnerability Project Report*. Department of the Environment, Sport and Territories, Canberra, p. 75.

Whetton, P. & Jones, P. D. (2003). Impacts of climate change on the Cairns Great Barrier Reef region. In *A Study to Determine the Scope and Focus of an Integrated Assessment of Climate Change Impacts and Options for Adaptation in the Cairns and Great Barrier Reef Region*. Australian Greenhouse Office, Canberra, p. 117.

Williams, S. E. (2002). *The Future of Biodiversity in the Rainforests of the Australian Wet Tropics*. Ecology 2002 Conference, Cairns.

Williams, S. E., Bolitho, E. E. & Fox, S. (2003). Climate change in Australian tropical rainforests: an impending environmental catastrophe. *Proceedings of the Royal Society of London B* **270**: 1887–92.

Woodroffe, C. D. (1993). Late Quaternary evolution of coastal and lowland riverine plains of southeast Asia and northern Australia: an overview. *Sedimentary Geology* **83**: 163–75.

Zhang, X. G. & Casey T. M. (1992). Long-term variations in the Southern Oscillation and relationships with Australian rainfall. *Australian Meteorological Magazine* **40**: 211–25.

Zhang, Y., Wallace J. M. & Battisti, D. S. (1997). ENSO-like interdecadal variability: 1900–93. *Journal of Climate* **10**: 1004–20.

22 THE IMPACT OF CLIMATE CHANGE ON THE BIODIVERSITY AND ECOSYSTEM FUNCTIONS OF THE WET TROPICS

Stephen E. Williams*, Joanne L. Isaac and Luke P. Shoo*

Centre for Tropical Biodiversity and Climate Change, School of Marine and Tropical Biology, James Cook University, Townsville, Queensland, Australia
*The authors were participants of Cooperative Research Centre for Tropical Ecology and Management

Introduction

There is no doubt that the climate is changing across the globe, primarily due to increases in anthropogenic greenhouse gas emissions. Emissions of carbon dioxide are 12 times greater today than 100 years ago and average temperatures have already risen approximately 0.6 °C and are continuing to increase (Houghton 2001). The Intergovernmental Panel on Climate Change (IPCC) predicts further increases of between 1.4 and 5.8°C by the end of the century, and the Australian Bureau of Meteorology has announced that 2005 was the hottest year on record. As well as increasing temperatures, rainfall patterns are changing, sea levels are rising and severe weather events, such as El Niño, drought and floods, are becoming more common (Easterling et al. 2000; Walsh & Ryan 2000; Milly et al. 2002; Palmer & Raianen 2002).

Assessing the impact of climatic change on global biodiversity is a research issue of global importance and a great challenge to ecologists. It is now widely recognized that climate change is one of the most significant current threats to biodiversity worldwide and contemporary research indicates that significant effects of climate change are already apparent in a variety of taxa and ecosystems (Hughes 2000; Parmesan & Yohe 2003; Root et al. 2003, 2005; Pounds et al. 2006). Research indicates that the most common effects are shifts in distribution and range (Parmesan et al. 1999; Thomas & Lennon 1999; Hill et al. 2002; Konvicka et al. 2003; Brommer 2004; Hickling et al. 2005; Wilson et al. 2005), earlier migration and breeding in many avian species (e.g. Visser & Both 2005) and variation in life history traits such as clutch size (e.g. Both & Visser 2005) and adult body size (Smith et al. 1998) in a number of vertebrate species. More recently, studies have shown that climatic changes have the potential to disrupt predator-prey relationships (Durant et al. 2005), increase the susceptibility of species to disease (Burrowes et al. 2004; Pounds et al. 2006), interfere with community dynamics and ecosystem function (Sinclair & Byrom 2006) and alter the genetic constitution of local populations (Umina et al. 2005). However, there is also limited evidence to suggest that some species can show evolutionary and behavioural compensation via mechanisms that may mitigate the negative effects of climate change (e.g. Wichmann et al. 2005).

There is likely to be huge variation in the vulnerability of different species to the impacts of climate change. More generalist species, such as those with wide geographical ranges and temperature tolerances, may be able to adapt to changing conditions. However, the risk of extinction will increase for many species that are already vulnerable, such as those with limited climatic ranges and/or restricted habitat requirements, or those already under threat from additional factors, such as habitat fragmentation. It is therefore crucial to determine which species and areas are most at risk in order to take appropriate action to mitigate impacts. In particular, species in isolated areas, such as islands and mountains, and populations and/or species with limited dispersal ability are thought to be at especially high risk, as suitable habitat will be reduced (Williams *et al.* 2003; Chamaille-Jammes *et al.* 2006).

Initial research on climate change suggested that temperate regions and their associated biodiversity would be most affected owing to the greater increases in mean annual temperatures at middle and higher latitudes. However, more recently it has come to light that biodiversity in tropical systems may be more at risk for a variety of reasons and, of course, there are much higher absolute levels of biodiversity in the tropics. For example, Malcolm *et al.* (2002) have demonstrated that migration rates are often significantly lower for tropical species, while Coley (1998) suggests that tropical areas often have more specific plant–herbivore relationships, making them more vulnerable to environmental disturbance. Tropical mountain systems, and tropical montane cloud forests in particular, are expected to be extremely vulnerable to climatic change (Foster 2001). The fragmentary nature of montane cloud forests and the compression of climatic zones over the elevation gradient promote explosive speciation and high rates of endemism, making them hotspots of biological biodiversity (Körner 2002). For example, 30% of Peru's endemic birds, mammals and anurans are found in cloud forest and globally it is estimated that over 20% of restricted range birds use the cloud forest extensively, and many of these are endangered species (Foster 2001). However, their unique and fragmentary nature, and association with elevation and temperature gradients, means tropical montane forests are extremely sensitive to climate change and also have a slower recovery rate than most other forest types (Foster 2001). The possible effects of climate change on tropical mountain areas include a general increase in temperature, increased water loss through evapotranspiration, a rise in the average basal altitude of the orographic cloud layer (McJannet *et al.* Chapter 15, this volume), changing rainfall input

(Balston, Chapter 21, this volume), decreased soil moisture, increases in fire and drought (Pounds *et al.* 1999; Foster 2001) and increased invasion potential for introduced pest species. The implications of these effects are likely to include deforestation, flooding and reduced stream flow (McJannet *et al.* Chapter 15, this volume) and a decrease in plant biomass and floristic diversity, all of which will severely affect other components of biodiversity.

Climate change is thus one of the most significant current threats to global biodiversity, predicted to have a profound effect on individuals and populations in animal and plant communities and on human society. At regional, state, national and international levels, understanding the nature of possible climate change and its impact on the environment as well as identifying and developing appropriate adaptive management and mitigation strategies is recognized as a pressing research issue. Policy at all levels emphasizes the need for research to increase our understanding of impacts, predict capability better, understand the potential for adaptation, identify vulnerable species, areas and ecosystem processes and instigate monitoring. The need for further research into the impacts of climate change on Australian biodiversity is recognized in *National Research Priority 1: An Environmentally Sustainable Australia*, and specifically within priority goals 5: *Sustainable Use of Australia's Biodiversity* and 7: *Responding to Climate Change and Variability*, and also in the *National Biodiversity and Climate Change Action Plan* (Natural Resource Management Ministerial Council 2004).

Research on climate change impacts in the Wet Tropics

The Wet Tropics bioregion is a major biodiversity hotspot of global significance, listed as a World Heritage Area in 1988, primarily because of the high biodiversity value of a unique regional biota: the area contains 84 species of regionally endemic rainforest vertebrates (Graham *et al.* 2006). The area is dominated by mountain ranges varying from sea level to over 1600 metres and altitude is the most influential environmental gradient determining species composition and patterns of biodiversity in the region (Williams *et al.* 1996; Williams 1997; Williams & Pearson 1997; Williams & Hero 2001).

Research on patterns of biodiversity has been ongoing in the Wet Tropics for several decades; however, the impact of climate change on biodiversity is a relatively recent development. Studies have centred on tropical rainforest within the region (an area of approximately 10 000 km²) and extensive, systematic data have been collected on the majority of rainforest vertebrates occurring in the area. This sustained research effort has enabled the development of one of the most comprehensive ecological data sets so far amassed for tropical rainforest fauna. In the context of contemporary climate warming, the data set has provided valuable opportunities to apply contemporary modelling techniques and develop novel approaches to predict impacts including change in the extent of species' distributions and population size. Further, detailed studies on individual species examining thermal tolerance, microhabitat requirements and dispersal capability have provided valuable first evidence to validate important assumptions of predictive models. These have been complemented by community analyses examining the potential for additional factors, such as ecosystem processes, to ameliorate or exacerbate impacts.

The spatial pattern of long-term habitat stability in the cool, wet uplands of the region has been a major factor determining the current spatial patterns of species richness via the processes of non-random local extinction and colonization (Williams & Pearson 1997; Winter 1997; Schneider & Williams 2005; Graham *et al.* 2006). The biogeography, both past and present, of the region predisposes the fauna to being vulnerable to climate change as regionally endemic fauna are mostly adapted to a cool, wet and relatively aseasonal environment.

Predicting the effects of climate change on biodiversity, with the ultimate aim of managing and conserving the ecosystem in the face of predicted threats, can only be achieved through a comprehensive understanding of both the historical and current factors that determine species distributions and diversity. Early research in the Wet Tropics focused on the roles of historical biogeography, climate fluctuations, current environment and biogeographical barriers in governing spatial patterns of vertebrate fauna diversity within the region (Hilbert, Chapter 8, this volume). More recently, however, research effort has been directed towards the challenge of predicting and monitoring the future impacts of contemporary climate change on biodiversity. Each of these issues is considered in turn below.

Historical climate fluctuations and present biodiversity

Biodiversity on a regional scale in the Wet Tropics bioregion has been shaped by contractions and expansions in the extent of rainforest area during the Quaternary period (reviewed by Williams & Pearson 1997). Current rainforest area and shape metrics index the degree to which each subregion (mountain range) has been affected by these historical climate and habitat changes. These relationships suggest that climate-driven habitat fluctuations have resulted in non-random patterns of localized extinctions in the past and are important in determining current patterns of biodiversity within the Wet Tropics (Williams & Pearson 1997). This study also determined that habitat diversity, including rainfall, vegetation and altitude, consistently explained much of the variability in current species richness in the Wet Tropics. For example, for microhylid frogs, regional patterns of diversity are strongly related to consistent moisture levels throughout the year and are limited by low rainfall in the dry season (Williams & Hero 2001). Thus climate change resulting in increased rainfall variation and a lifting cloud base (McJannet, Chapter 15, and Balston, Chapter 21, this volume) has serious implications for the persistence of most microhylid frog species. More recent studies using continuous bioclimatic paleomodels of rainforest extent and long-term stability support the original findings of Williams and Pearson (1997) and suggest that long-term habitat stability has been a major factor determining current patterns of species richness, particularly in low-vagility taxa such as litter skinks and microhylid frogs (Graham *et al.* 2006).

In summary, our early research provides evidence that previous climatic changes in the area resulted in significant levels of sub-regional species extinction, determined by the relative ecological plasticity of each species with little evidence for rapid evolutionary adaptation (Williams & Pearson 1997; Moritz *et al.* 2000; Schneider & Williams 2005; Graham *et al.* 2006; Williams *et al.*, Chapter 10, this volume). This was followed by periods of rainforest expansion when many species were able to recolonize the re-established rainforest.

Recolonization is dependent on many species-specific and biogeographical factors that determined which species are now present in a given area (Williams & Pearson 1997; Schneider & Williams 2005; Graham *et al.* 2006). Importantly, however, reconstructions of global temperatures for the past two millennia provide no evidence for warmer conditions than those observed during the post-1990 period (Moberg *et al.* 2005), suggesting that species are likely to be exposed to novel conditions under contemporary climate warming.

Contemporary climate change and future biodiversity

The effect of climate change on the distribution of rainforest habitats is undoubtedly one of the key factors in predicting how biodiversity will be affected overall, as any reduction in suitable habitat will impact on all levels of biodiversity. The effect of climatic changes, incorporating a conservative 1°C increase in temperature and several rainfall scenarios for the Wet Tropics, was modelled by Hilbert *et al.* (2001). Results demonstrate that most at risk is the cool-wet rainforest that occurs at the highest elevations (>600 m above sea level (m a.s.l.)). Highland rainforests are predicted to become greatly reduced in area, more fragmented and even eliminated from some parts of the Wet Tropics under a conservative scenario of 1°C temperature increase and a small reduction in rainfall (Hilbert *et al.* 2001).

In addition to these direct effects on vegetation distribution, climate change also has the potential to affect the nutritional content and digestibility of food plants for rainforest herbivores. For example, increased CO_2 levels reduce the nutritional value and increase toughness of foliage, and combined effects of increased CO_2 and temperature can also influence the concentration of plant secondary metabolites (Lawler *et al.* 1997; Kanowski *et al.* 2001; Moore *et al.* 2004). These factors will make plants less digestible for folivores such as tree kangaroos, ringtail possums and invertebrates, and are likely to influence the distribution of arboreal folivores in the Wet Tropics. The predicted climate-induced range shifts are also expected to shift species' distributions on to less productive granite soils at higher altitudes that currently support folivores at comparably lower densities (Kanowski *et al.* 2001).

Williams *et al.* (2003) developed bioclimate models of the spatial distribution for regionally endemic rainforests vertebrates in the Wet Tropics and used these models to predict the effects of climate warming on their distribution. The results of this study suggest that the impacts of climate change could be catastrophic for biodiversity in the region. Areas of core habitat were found to decline rapidly with increasing temperature, resulting in an amplification of extinction rates and significantly reducing overall biodiversity in the region. A relatively small increase of 1°C in temperature was found to reduce the area of core habitat of each species to about two thirds (~63%) of their existing distribution area and of the 65 species modelled, the area of core environment significantly declined for 63 species (97%). Under a 3.5°C temperature increase, all 65 endemic species are predicted to undergo dramatic declines in distribution, while 30 species lose their core environment completely. In a further analysis, which included data from the Wet Tropics species and also a variety of other data from species worldwide, Thomas *et al.* (2004) used climatic envelope models to assess future distributions and risk of extinction due to climate change. Their analysis indicates that even under a moderate climate change scenario (a 0.8–1.7°C increase), approximately 18% of species considered would be committed to extinction by 2050. Furthermore, this study indicates that climate change dominates extinction risk for species in the Wet Tropics, with a range of 7–58% of species extinction predicted, depending upon the climate change scenario used.

In a recent study focusing on Wet Tropics birds, Shoo *et al.* (2005a) predict that 74% of rainforest species will become threatened as a result of projected mid-range warming in the next one hundred years (Figure 22.1). However, extinction risk in rainforest birds varied according to where along the altitudinal gradient a species is currently most abundant. Upland birds are expected to be the most affected and are likely to be immediately threatened by small increases in temperature. However, there is a capacity for the population size of lowland species to increase, at least in the short term (Figure 22.2a). A recent study on the effects of a severe weather event on cloud forest birds in Mexico found a similar picture (Tejeda-Cruz & Sutherland 2005). Following a hurricane that resulted in large gaps in the vegetation in high altitude forest, high sensitivity

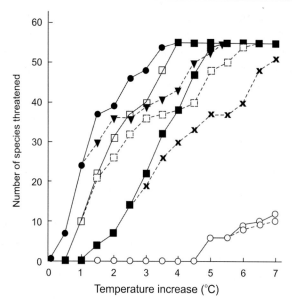

Figure 22.1 Accumulated number of rainforest bird species expected to qualify for threatened status as a consequence of a decline in population size with increasing temperature. Thresholds of population decline follow IUCN Red List Categories and Criteria (Criterion A, IUCN 2001) and correspond to the following threat categories: vulnerable (filled circles), endangered (open squares), critically endangered (filled squares) and extinct (open circles). Results from two alternative model scenarios are shown: (a) no occupation of newly created climatic habitat at low altitudes was inferred (scenario 1, continuous lines, $n = 55$ species); or (b) an abundance pattern symmetrically distributed about the abundance maxima was adopted to infer abundance of the species in newly created climatic habitat (scenario 2, broken lines, $n = 55$ species). Source: after Shoo *et al.* (2005b).

and specialist avian species declined, while generalist, low sensitivity species were more abundant after the event. Many microhylid frog species are also predicted to suffer large declines in population size as climates that currently support high density populations of species on mountaintops are likely to disappear under moderate levels of climate warming (Shoo 2005) (Figure 22.2b).

Shoo *et al.* (2005b) also incorporated spatial patterns of abundance into a predictive model and demonstrated that for three-quarters of the regionally endemic Wet Tropics bird species, as temperature increases population size is likely to decline more rapidly than distribution area (Figure 22.3a). This indicates that for these species, extinction risk associated with climate

change will be more severe than expected from a decline in distribution area alone. A similar scenario is also predicted for microhylid frogs (Luke Shoo, unpublished data) (Figure 22.3b). It is therefore important to consider both the overall range of a species and the spatial patterns of abundance of that species within its distribution when estimating extinction risk. Shoo *et al.* (2006) further point out that the effects of climate change on a species' distribution may go unnoticed if insufficient baseline data is collected to allow impacts, of the magnitude predicted under early climate warming, to be detected with conventional statistical analyses. They identify the need to equalize the distribution of sampling effort between contemporary and historical data sets and also the need for the statistical power of range shift analyses to be ascertained before discounting impacts.

Finally, species richness and density of Wet Tropics rainforest birds is highest at elevations of 600–800 metres and is positively related to net primary productivity and energy flux (Figure 22.4). Results suggest that total energy flux and species richness is a direct function of net primary productivity in the upland rainforests, but not in lowland areas. The authors suggest that an increase in temperature, due to global warming, may result in an increase in net primary productivity that could ameliorate some of the predicted negative effects of climate change on upland rainforest birds (Williams *et al.*, unpublished data).

It is not only the vertebrates that exhibit distributions strongly related to elevation. Studies of the invertebrate fauna have found assemblages that are strongly structured across the elevational gradient, with many species restricted to high altitudes including low vagility arthropods (Monteith 1985, 1995; Monteith & Davies 1991), schizophoran flies (Wilson *et al.*, in review) and ants (Yek, unpublished data). These results suggest that the impacts in the invertebrate assemblages will be similar to those previously predicted for regionally endemic vertebrates by Williams *et al.* (2003).

Relative species vulnerability and extinction risk

As mentioned previously, the relative effect of climate changes will differ considerably among species, dependent upon their current distributions, ecological requirements and life history. Traits that can predispose

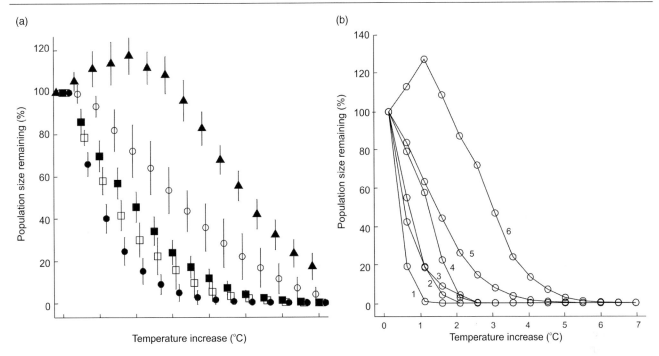

Figure 22.2 (a) Across-species mean decline in population size of rainforest birds within the region with increasing temperature. An abundance pattern symmetrically distributed about the abundance maxima was adopted to infer abundance of the species in newly created climatic habitat (*n* = 55 species). Population size is expressed as a percentage of current population size. Species were classified into altitudinal groups based on the altitudinal position of their abundance maxima: 0–299 m (filled triangles, *n* = 12); 300–599 m (open circles, *n* = 12); 600–899 metres (filled squares, *n* = 15); 900–1199 m (open squares, *n* = 6); 1200–1499 m (filled circles, *n* = 10). Data points were staggered to reveal overlapping values. Bars represent 95% confidence intervals of the mean. (b) Predicted decline in population size of microhylid species within the region with increasing temperature. Population size is expressed as a percentage of current population size. Number labels refer to individual species of microhylid frogs: (1) *Cophixalus monticola*; (2) *C. neglectus*; (3) *C. hosmeri*; (4) *C. aenigma*; (5) *C. ornatus*; (6) *C. infacetus*. *Source*: after Shoo (2005b).

species to a high degree of ecological specialization and high extinction risk include:

• low dispersal potential;
• low fecundity rates;
• a slow life history;
• small, fragmented or restricted populations;
• a narrow range of ecological requirements, including specific temperature, habitat, moisture and foraging requirements (McKinney 1998).

Endemic rainforest species of the Wet Tropics region display many of these characteristics, having small, restricted populations, with most species being distributed over areas with a very narrow range in annual mean temperature. For example, within their overall distribution, annual mean temperature varies spatially by as little as 1.1°C for the Tapping Nursery frog (*Cophixalus concinnus*) and even the most widely distributed, regionally endemic species, the leaf-tailed

gecko (*Saltuarius cornutus*), still only has a 9°C range in annual mean temperature (S. E. Williams, unpublished data). Laboratory studies show that a Wet Tropics mammal, the green ringtail possum (*Pseudochirops archeri*), is intolerant of temperatures over 30°C, with body temperature increasing linearly above this temperature and probably leading to lethal effects after only 4–5 hours (A. Krockenberger, unpublished data). Endemic species also have low dispersal ability. For example, evidence from genetic studies of microhylid frog species (Yvette Williams, unpublished data) indicates that dispersal events over distances of 1 kilometre are likely to be rare. This suggests that the ability of such species to avoid climate changes by dispersing upward in altitude will be potentially limited by differential dispersal and rate of change. Indeed, modelling predicts that even a 1°C increase will be likely to result in the extinction of *Cophixalus concinnus*, a microhylid

(a)

(b)

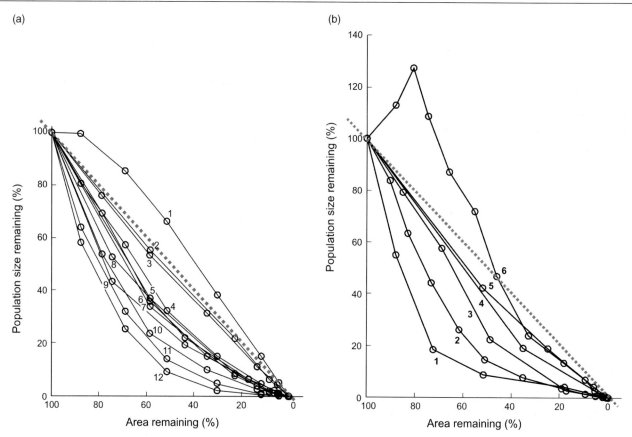

Figure 22.3 (a) Projected change in the relative population size and distribution area of rainforest birds following altitudinal range shifts associated with climate warming. The relationship is linear where population size declines in direct proportion to distribution area (i.e. dotted grey line) and curvilinear where population size declines more rapidly or more slowly than distribution area with increasing temperature. Number labels refer to individual species of rainforest birds: (1) Macleay's honeyeater, *Xanthotis macleayana*; (2) tooth-billed bowerbird, *Scenopoeetes dentirostris*; (3) pied monarch, *Arses kaupi*; (4) Victoria's riflebird, *Ptiloris victoriae*; (5) golden bowerbird, *Prionodura newtoniana*; (6) chowchilla, *Orthonyx spaldingii*; (7) mountain thornbill, *Acanthiza katherina*; (8) fernwren, *Oreoscopus gutturalis*; (9) Atherton scrubwren, *Sericornis keri*; (10) bridled honeyeater, *Lichenostomus frenatus*; (11) grey-headed robin, *Heteromyias albispecularis*; (12) Bower's shrike-thrush, *Colluricincla boweri*. *Source*: modified from Shoo *et al.* (2005a). (b) Projected change in the relative population size and distribution area of microhylid species following altitudinal range shifts associated with climate warming. The relationship is linear where population size declines in direct proportion to distribution area (i.e. dotted grey line) and curvilinear where population size declines more rapidly or more slowly than distribution area with increasing temperature. Number labels refer to individual species of microhylid frogs: (1) *Cophixalus hosmeri*; (2) *C. ornatus*; (3) *C. aenigma*; (4) *C. monticola*; (5) *C. neglectus*; (6) *C. infacetus*. *Source*: after Shoo (2005a).

frog restricted to Thornton's Peak, due to total loss of its core environment (Williams *et al.* 2003).

A summary of current research findings on the impacts of climate change on Wet Tropics biodiversity

• Climate fluctuations during the Quaternary have resulted in the localized extinction of endemic rainforest vertebrates, with recolonization under subsequent favourable climates dependent on the ecological characteristics of individual species (Williams & Pearson 1997).

• The distribution of highland rainforest vegetation and habitat (>600 m elevation) is predicted to become significantly reduced in area and to become more fragmented as the climate changes (Hilbert *et al.* 2001).

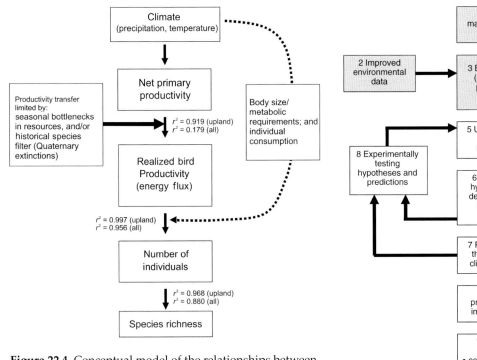

Figure 22.4 Conceptual model of the relationships between energy flow, limiting variables and bird productivity and species richness. Since energy flow explains a much higher proportion of variance in these relationships within the uplands (600 m and above), the R^2 values are provided separately for all elevations and for uplands only. Net primary productivity explains 89% of the variability in total species richness in the uplands: $R^2 = 0.887 = 0.919 \times 0.997 \times 0.968$. *Source*: modified from Williams *et al.* (2003).

• Predictions based on distribution area alone may be conservative. Analyses based on empirical density estimates indicate that, for many species, population size will decline faster than that predicted by a decline in distribution area alone (Shoo *et al.* 2005a, b).

• Ecosystem processes are intrinsically linked with assemblage dynamics and there is a possibility that increasing productivity with increasing temperature may exacerbate some of the impacts predicted from distribution models (Williams *et al.*, unpublished data).

Directions for future research

The development of a strategic approach to the climate change research conducted in the Wet Tropics bioregion will enable the knowledge gained here to be

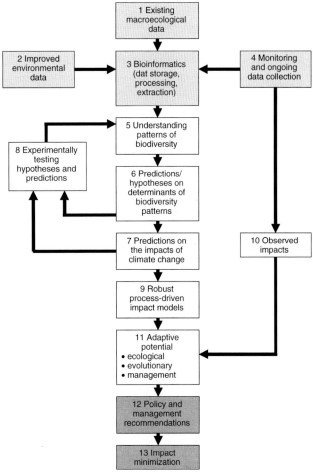

Figure 22.5 A flow diagram demonstrating a general model for the development of future research examining climate change impacts on biodiversity.

utilized nationally and internationally for management and conservation purposes. Summarized below, and in the corresponding flow diagram (Figure 22.5), is a general model for the development of future research examining climate change impacts on biodiversity.

• Existing macro-ecological data will enable meta-analyses that will produce working hypotheses driven by data to ensure efficient, targeted research with maximum efficiency in the use of funding.

• Spatial analysis will allow the latest vegetation and disturbance data to be combined with climate layers and detailed microclimate data to produce spatial microclimate maps.

• There is a need to combine existing ecological data with new environmental data and to utilize sophisticated data mining and extraction techniques to maximize data storage and pattern analysis.

• Ongoing data collection/monitoring is vital to provide improvement of the geographical and taxonomic scope as well as greater statistical power to detect impacts.

• Improving our baseline understanding of the spatial and temporal patterns of biodiversity is crucial for both research and monitoring.

• Biodiversity and environmental data should be combined to generate informed hypotheses about the deterministic processes driving the patterns of biodiversity.

• Knowledge about the processes affecting biodiversity can be used to make more robust predictions about the likely impacts of a changing climate.

• Hypotheses and predictions should be put through feedback loops of experimental testing via spatial modelling and simulation, natural field experiments and manipulative experiments.

• Continued improvement of patterns and processes allows the development of more robust, process-related models on the factors that will influence the impacts of climate change.

• By combining mechanistic models with extensive empirical data and ongoing monitoring, predictions on climate change impacts can be vastly improved.

• The impacts of global climate change will be determined by the interactions between changing ecosystem processes, existing ecological plasticity, evolutionary adaptation and active management. The potential for adaptive responses is a high priority research area both nationally and internationally.

• The outputs from this research will lead to informed policy on environmental sustainability, best management practice and impact mitigation procedures.

• The final goal is to minimize the impacts of global climate change on natural ecosystem processes and values using a combination of robust predictions, targeted management (specific species, geographical areas, ecosystem processes) and the maximization of adaptive resilience.

It is essential that the future focus of research in the Wet Tropics aims to understand the ecological and biological responses of rainforest species to climate change with a view to understanding:

1 whether these responses can partially rescue some species from the negative effects of climate change;

2 how differential responses between species can influence community dynamics and ecosystem processes.

These factors are explained in more detail below.

Behavioural and evolutionary responses to climate change

The research summarized previously indicates that, in general, climate change will have a catastrophic effect on the biodiversity of the Wet Tropics. However, more recent research indicates that the negative impacts of climate change on individuals, populations and communities may be ameliorated through a number of mechanisms. The recent research of Williams *et al.* (unpublished data) indicates that density and species richness of rainforest birds in the Wet Tropics is related to net primary productivity and energy flux (see Williams *et al.*, Chapter 10, this volume). The authors suggest that some of the predicted impacts of global warming on biodiversity in the region may be partially moderated by increases in net primary productivity owing to increasing temperatures, although more work is needed to incorporate these relationships into predictive models. Owing to the relatively small altitudinal range in the region, there is only limited scope for species to shift their distribution up in altitude before they get pushed off the top of the mountains. Evidence for altitudinal range shifts is currently scant (Shoo *et al.* 2006), and for species already restricted to mountaintops, such as *Cophixalus concinnus*, this is not a viable option.

There is some potential for species to use available microhabitat refuges, such as boulder piles, vegetation and logs, to shelter from more extreme temperatures. There is evidence from a number of species, including frogs, lizards and mammals, that vertebrates do use microhabitat refugia as thermal buffers and shelter from extreme temperatures (e.g. Pavey *et al.* 2003). Kerr *et al.* (2003) also demonstrate that sleepy lizards (*Tiliqua rugosa*) can modify their refuge site use as ambient temperatures increase, implying an ability to discriminate among refugia sites. Preliminary data from Thornton's Peak suggest that microhabitat buffering potential may be significant in boulder piles (Williams, unpublished data). Temperature within boulders was found to be extremely stable, even with a 30°C temperature differential between the surface air and the bottom of the boulder pile, a depth of only 5 metres.

The ability of a species to adapt, in terms of its ecology and life history, may also make it less susceptible to changing environmental conditions. For example, plasticity in niche specialization, behavioural or micro-habitat adjustments and physiological tolerance ranges can all make a species less vulnerable to climatic changes. Currently, there is scant empirical evidence demonstrating ecological adaptation to climatic changes. Previous work on the fly *Drosophila birchii*, a relatively widespread species occurring largely in tropical rainforests of Australia and New Guinea, has highlighted low potential for the evolution of desiccation resistance even after intense selection for over 30 generations (Hoffmann *et al.* 2003). However, wood frogs (*Rana sylvatica*) have shown rapid, behaviourally mediated responses to changing thermal environments (Friedenberg & Skelly 2004). Wichmann *et al.* (2005) also modelled extinction risk under climate change in African raptors and found that extinction risk was significantly lowered when the model assumed that birds adapt behaviourally by enlarging their territory in response to decreasing precipitation. These findings clearly have important implications for understanding population responses to climate change. Furthermore, Grant and Grant (1993) documented genetic evolution in Darwin's finches following El Niño events that resulted in selection for smaller beak sizes under changing food conditions. This occurred within a 5-year period, leading the authors to predict that micro-evolutionary changes might be possible under conditions of global warming. Most recently, Chamaille-Jammes *et al.* (2006) report that increasing temperatures over the past 18 years have resulted in an increase in body size in common lizards (*Lacerta vivipara*) in France, associated with an increase in clutch size, reproductive output and survival. This study is one of few to demonstrate a positive response to climate change in a vertebrate, but it should also be noted that *L. vivipara* is a relatively generalist species, with a wide geographical range.

Currently, apart from the studies outlined above, there has been little research effort focused on the potential for behavioural and genetic adaptations to mitigate the negative effects of rapid climatic changes on animal populations. Future research in the Wet Tropics needs to focus on these issues, incorporating long-term species-specific data encompassing the ecology and physiology of the species, contemporary modelling techniques and controlled experimental manipulations.

Impacts of climate change on community dynamics and ecosystem function

Climatic changes have the potential to disrupt food chains and interactions within animal communities, with unpredictable consequences for ecosystem function and stability. For example, climate-induced changes in the abundance of prey, themselves caused by changes in the phenology of primary production, could cause cascading effects through the food chain and influence the reproductive success and abundance of predators. Co-occurrence in food requirements of offspring and food availability is a key factor determining breeding success in most species and the trophic mismatch hypothesis proposes that climate change can create a mismatch between timing of peak breeding of predator and prey (Durant *et al.* 2005). Thus, if climate change influences season-specific life history traits in cyclic prey species, this may disrupt the coupling between species, with cascading effects on predators.

While there is mounting evidence that climate change can cause trophic mismatch, and associated changes in community dynamics and ecosystem function, in marine (e.g. Mouritsen *et al.* 2005) and temperate terrestrial systems (Durant *et al.* 2005), there has been little research focused on these effects in tropical ecosystems. In particular, it is currently unknown how variation among species in response to climatic changes affects community and ecosystem dynamics. However, Hooper *et al.* (2005) suggest that differential responses between species may stabilize ecosystem process rates in response to environmental fluctuations. Thus, if this is the case for tropical ecosystems, maintaining a diversity of organisms of different functional effect and response types will help to preserve a variety of management options in the long term.

Summary

It is imperative that we act now in order to minimize the impacts of climate change on the biodiversity of the Wet Tropics bioregion. The final, realized degree of change can only be addressed at a global level. Above all, Australia and the rest of the world must act to reduce emissions of the greenhouse gases CO_2 and methane. Without a significant reduction in the output of these gases, regional and species-specific management plans are doomed in the long term because temperatures will continue to rise. A reduction in habitat clearance is also necessary because vegetation clearance exacerbates the build-up of greenhouse gases and reduces the resilience of the ecosystem to environmental disturbance.

At a regional level, the impact of global climate change will also depend upon the resilience of the ecosystem in question. In order to maximize the resilience of the Wet Tropics bioregion, management must focus on maintaining healthy ecological processes and function, and minimize actions that might damage the inherent ability of the ecosystem to recover, including habitat loss, fragmentation and invasion by introduced animals, plants and diseases.

Management of reserves and surrounding areas

As the climate changes, it is possible that suitable habitat for Wet Tropics species will occur thousands of miles away from where it currently occurs. However, due to the biogeography of the majority of Wet Tropics species, there is almost no capacity for long distance movement to compensate for climate change. This means that we must take action to maintain the resilience of the area *in situ*. Such actions should include land acquisition in areas suitable in future climates, conservation of the land that surrounds reserves as well as the reserves themselves, the removal of invasive species, the establishment of additional wildlife corridors and the identification of buffer zones and possible areas for species translocation.

Adaptive management and species-specific plans

The nature of climate change and global warming means that we must take an adaptive management approach to conserving biodiversity by incorporating research into conservation action. This approach and the continuation of research in the Wet Tropics will enable us to undertake active management of those species that appear to be most at risk from climate change. Collection of baseline data and ongoing monitoring

programmes are also vital in this feedback system (Figure 22. 5). Although the best form of conservation is *in situ* conservation, the possibility of having to translocate the most vulnerable species to distant suitable areas should also be assessed, along with identification of suitable translocation sites. It is vital to obtain as much knowledge as possible now, if we are to make the most efficient use of scare resources. We must avoid wasting resources where they are not needed or failing to act in cases where active management is necessary.

References

Both, C. & Visser, M. E (2005). The effect of climate change on the correlation between avian life-history traits. *Global Change Biology* **11**: 1606–13.

Brommer, J. E. (2004). The range margins of northern birds shift polewards. *Annales Zoologici Fennici* **41**: 391–7.

Burrowes, P. A, Joglar, R. L. & Green, D. E. (2004). Potential causes for amphibian declines in Puerto Rico. *Herpetologica* **60**: 141–54.

Chamaille-Jammes, S., Massot, M., Aragon, P. & Clobert, J. (2006). Global warming and positive fitness response in mountain populations of common lizards *Lacerta vivipara*. *Global Change Biology* **12**: 392–402.

Coley, P. D. (1998). Possible effects of climate change on plant/herbivore interactions in moist tropical forests. *Climatic Change* **39**: 455–72.

Durant, J. M., Hjermann, D. O., Anker-Nilssen, T. *et al.* (2005). Timing and abundance as key mechanisms affecting trophic interactions in variable environments. *Ecology Letters* **8**: 952–8.

Easterling, D. R., Meehl, G. A., Parmesan, C, Changnon, S. A., Karl, T. R. & Mearns, L. O. (2000). Climate extremes: observations, modeling, and impacts. *Science* **289**: 2068–74.

Foster, P. (2001). The potential negative impacts of global climate change on tropical montane cloud forests. *Earth-Science Reviews* **55**: 73–106.

Friedenburg, L. K. & Skelly, D. K. (2004). Microgeographical variation in thermal preference by an amphibian. *Ecology Letters* **7**: 369–73.

Grant, P. R. & Grant, B. R. (1993). Evolution of Darwin finches caused by a rare climatic event. *Proceedings of the Royal Society of London Series B* **251**: 111–17.

Graham, C. H., Moritz, C. & Williams, S. E. (2006). Habitat history improves prediction of biodiversity in rainforest fauna. *Proceedings of the National Academy of Sciences* **103**: 632–6.

Hickling, R., Roy, D. B., Hill, J. K. & Thomas, C. D. (2005). A northward shift of range margins in British Odonata. *Global Change Biology* **11**: 502–6.

Hill, J. K., Thomas, C. D., Fox, R., *et al.* (2002). Responses of butterflies to twentieth century climate warming: implications for future ranges. *Proceedings of the Royal Society of London B* **296**: 2163–71.

Hilbert, D. W., Ostendorf, B. & Hopkins, M. S. (2001). Sensitivity of tropical forests to climate change in the humid tropics of north Queensland. *Austral Ecology* **26**: 590–603.

Hoffmann, A. A., Hallas, R. J., Dean, J. A. & Schiffer M. (2003). Low potential for climatic stress adaptation in a rainforest *Drosophila* species. *Science* **301**: 100–2.

Hooper, D. U., Chapin, F. S., Ewel, J. J. *et al.* (2005). Effects of biodiversity on ecosystem functioning: A consensus of current knowledge. *Ecological Monographs* **75**: 3–35.

Houghton, J. (2001). The science of global warming. *Interdisciplinary Science Reviews* **26**: 247–57.

Hughes, L. (2000). Biological consequences of global warming: is the signal already apparent? *Trends in Ecology and Evolution* **15**: 56–61.

International Union for Conservation of Nature (IUCN) (2001). *IUCN Red List of Threatened Species*. IUCN, Gland, Switzerland.

Kanowski, J., Hopkins, M. S., Marsh, H. & Winter, J. W. (2001). Ecological correlates of folivore abundance in north Queensland rainforests. *Wildlife Research* **28**: 1–8.

Kerr, G. D., Bull, C. M. & Burzacott, D. (2003). Refuge sites used by the scincid lizard *Tiliqua rugosa*. *Austral Ecology* **28**: 152–60.

Konvicka, M., Maradova, M., Benes, J., Fric, Z. & Kepka, P. (2003). Uphill shifts in distribution of butterflies in the Czech Republic: effects of changing climate detected on a regional scale. *Global Ecology and Biogeography* **12**: 403–10.

Körner, C. (2002). Mountain biodiversity, its causes and function: an overview. In *Mountain Biodiversity: A Global Assessment*, Körner, C. and Spehn, E. M. (eds). Parthenon Publishing, New York.

Lawler, I. R., Foley, W. J., Woodrow, I. E. & Cork, S. J. (1997). The effects of elevated CO_2 atmospheres on the nutritional quality of Eucalyptus foliage and its interaction with soil nutrient and light availability. *Oecologia* **109**: 59–68.

Mckinney, M. L. (1998). Branching models predict loss of many bird and mammal orders within centuries. *Animal Conservation* **1**: 159–164.

Malcolm, J. R., Markham, A., Neilson, R. P. & Garaci, M. (2002). Estimated migration rates under scenarios of global climate change. *Journal of Biogeography* **29**: 835–49.

Milly, P. C. D., Wetherald, R. T., Dunne, K. A. & Delworth, D. L. (2002). Increasing risk of great floods in a changing climate. *Nature* **415**: 514–17.

Moberg, A., Sonechkin, D. M., Holmgren, K., Datsenko, N. M. & Karlén, W. (2005). Highly variable Northern Hemisphere temperatures reconstructed from low- and high-resolution proxy data. *Nature* **433**: 613–17.

Monteith, G. B. (1985). Altitudinal transect studies at Cape Tribulation, north Queensland VII. Coleoptera and Hemiptera (Insecta). *Queensland Naturalist* **26**: 70–8.

Monteith G. B. (1995). *Distribution and Altitudinal Zonation of Low Vagility Insects of the Queensland Wet Tropics (part 4)*. P.120. Queensland Museum, Brisbane.

Monteith, G. B. & Davies, V. T. (1991). Preliminary account of a survey of arthropods (insects and spiders) along an altitudinal transect in tropical Queensland. In *The Rainforest Legacy Volume 2*, Werren, G. and Kershaw, P. (eds). Australian Government Publishing Service. Canberra, pp. 345–62.

Moore, B. D., Wallis, I. R., Wood, J. T. & Foley, W. J. (2004). Foliar nutrition, site quality, and temperature influence foliar chemistry of tallowwood (*Eucalyptus microcorys*). *Ecological Monographs* **74**: 553–68.

Moritz, C., Patton, J. L., Schneider, C. J. & Smith, T. B. (2000). Diversification of rainforest faunas: an integrated molecular approach. *Annual Review of Ecology and Systematics* **31**: 533–63.

Mouritsen, K. N., Tompkins, D. M. & Poulin, R. (2005). Climate warming may cause a parasite-induced collapse in coastal amphipod populations. *Oecologia* **146**: 476–83.

Natural Resource Management Ministerial Council (2004). *National Biodiversity and Climate Change Action Plan 2004–2007*. Australian Government, Department of the Environment and Heritage, Canberra.

Nix, H. A. & Switzer, M. A (1991). *Rainforest Animals: Atlas of Vertebrates Endemic to Australia's Wet Tropics*. Australian National Parks and Wildlife Service, Canberra.

Palmer, T. N. & Raianen, J. (2002). Quantifying the risk of extreme seasonal precipitation events in a changing climate. *Nature* **415**: 512–14.

Parmesan, C. & Yohe, G. (2003). A globally coherent fingerprint of climate change impacts across natural systems. *Nature* **421**: 37–42.

Parmesan, C., Ryrholm, N., Stefanescu, C. *et al.* (1999). Poleward shifts in geographical ranges of butterfly species associated with regional warming. *Nature* **399**: 579–83.

Pavey, C. R., Goodship, N., Geiser, F. (2003). Home range and spatial organisation of rock-dwelling carnivorous marsupial, *Pseudantechinus macdonnellensis*. *Wildlife Research* **30**: 135–42.

Pounds, J. A., Fogden, M. P. L. and Campbell, J. H. (1999). Biological response to climate change on a tropical mountain. *Nature* **398**: 611–15

Pounds, J. A., Bustamante, M. R., Coloma, L. A. *et al.* (2006). Widespread amphibian extinctions from epidemic disease driven by global warming. *Nature* **439**: 161–7.

Root, T. L., Price, J. T., Hall, K. R., Schneider, S. H., Rosenzweig, C. & Pounds, J. A. (2003). Fingerprints of global warming on wild animals and plants. *Nature* **421**: 57–60.

Root, T. L., MacMynowski, D. P., Mastrandrea, M. D. & Schneider, S. H. (2005). Human-modified temperatures induce species changes: Joint attribution. *Proceedings of the National Academy of Sciences* **102**: 746–59.

Schneider, C. & Williams, S. E. (2005). Quaternary Climate Change and Rainforest Diversity: Insights from Spatial Analyses of Species and Genes in Australia's Wet Tropics In *Tropical Rainforests: Past, Present and Future*, Moritz, C., Bermingham, E. & Dick, C. (eds). Chicago University Press, Chicago.

Shoo, L. P. (2005). Predicting and detecting the impacts of climate change on montane fauna in Australian tropical rainforests. PhD thesis, James Cook University.

Shoo, L. P., Williams, S. E. & Hero, J.-M. (2005a). Potential decoupling of trends in distribution area and population size of species with climate change. *Global Change Biology* **11**: 1469–76.

Shoo, L. P., Williams, S. E. & Hero, J.-M. (2005b). Climate warming and the rainforest birds of the Australian Wet Tropics: Using abundance data as a sensitive predictor of change in total population size. *Biological Conservation* **125**: 335–43.

Shoo, L. P., Williams, S. E. & Hero, J.-M. (2006). Detecting climate change induced range shifts: Where and how should we be looking? *Austral Ecology* **31**: 22–9.

Sinclair, A. R. E. & Byrom, A. E. (2006). Understanding ecosystem dynamics for conservation of biota. *Journal of Animal Ecology* **75**: 64–79.

Smith, F. A., Browning, H. & Shepherd, U. L. (1998). The influence of climate change on the body mass of woodrats *Neotoma* in an arid region of New Mexico, USA. *Ecography* **21**: 140–8.

Tejeda-Cruz, C. & Sutherland, W. J. (2005). Cloud forest bird responses to unusually severe storm damage. *Biotropica* **37**: 88–95.

Thomas, C. D., Cameron, A., Green, R. E. *et al.* (2004). Extinction risk from climate change. *Nature* **427**: 145–8.

Thomas, C. D. & Lennon, J. J. (1999). Birds extend their ranges northward. *Nature* **399**: 213.

Umina, P. A., Weeks, A. R., Kearney, M. R., McKechnie, S. W. & Hoffmann, A. A. (2005). A rapid shift in a classic clinal pattern in *Drosophila* reflecting climate change. *Nature* **308**: 691–3.

Visser, M. E. & Both, C. (2005). Shifts in phenology due to global climate change: the need for a yardstick. *Proceedings of the Royal Society of London Series B* **272**: 2561–9.

Walsh, K. J. E. & Ryan, B. F. (2000). Tropical cyclone intensity increase near Australia as a result of climate change. *Journal of Climate* **13**: 3029–36.

Wichmann, M. C., Groeneveld, J., Jeltsch, F. & Grimm, V. (2005). Mitigation of climate change impacts on raptors by behavioural adaptation: ecological buffering mechanisms. *Global and Planetary Change* **47**: 273–81.

Williams, S. E. (1997). Patterns of mammalian species richness in the Australian tropical rainforests: Are extinctions during historical contractions of the rainforest the primary determinants of current regional patterns in biodiversity? *Wildlife Research* **24**: 513–30.

Williams, S. E., Bolitho, E. E. & Fox, S. (2003). Climate change in Australian tropical rainforests: an impending environmental catastrophe. *Proceedings of the Royal Society of London Series B* **270**: 1887–92.

Williams, S. E. & Hero, J.-M. (2001). Multiple determinants of Australian tropical frog biodiversity. *Biological Conservation* **98**: 1–10.

Williams, S. E. & Pearson, R. G. (1997). Historical rainforest contractions, localized extinctions and patterns of vertebrate endemism in the rainforests of Australia's Wet Tropics. *Proceedings of the Royal Society of London Series B* **264**: 709–16.

Williams, S. E., Pearson, R. G. & Walsh, P. J. (1996). Distributions and biodiversity of the terrestrial vertebrates of Australia's Wet Tropics: a review of current knowledge. *Pacific Conservation Biology* **2**: 327–62.

Wilson, R. D., Trueman, J. W. H., Williams, S. E. & Yeates, D. K. (in review). Altitudinally restricted communities of Schizophoran flies in Queensland's Wet Tropics: vulnerability to climate change.

Wilson, R. J., Gutiérrez, D., Gutiérrez, J., Martinez, D., Agudo, R. & Monserrat, V. J. (2005). Changes to the elevational limits and extent of species ranges associated with climate change. *Ecology Letters* **8**: 1138–46.

Winter, J. W. (1997). Responses of non-volant mammals to late Quaternary climatic changes in the wet tropics region of north-eastern Australia. *Wildlife Research* **24**: 493–511.

23 IMPACTS OF HABITAT FRAGMENTATION AND LINEAR CLEARINGS ON AUSTRALIAN RAINFOREST BIOTA

William F. Laurance[1] and Miriam Goosem[2]*

[1]Smithsonian Tropical Research Institute, Balboa, Ancon, Panama, Republic of Panama; and Biological Dynamics of Forest Fragments Project, National Institute for Amazonian Research (INPA), Manaus, Amazonas, Brazil

[2]School of Earth and Environmental Sciences, James Cook University, Cairns, Queensland, Australia

*The author was participant of Cooperative Research Centre for Tropical Ecology and Management

Introduction

Tropical forests are being destroyed and degraded at alarming rates (Achard *et al.* 2002; Hansen & DeFries 2004; Laurance & Peres 2006). The most common aftermath of large-scale forest conversion is a mosaic of relict forest fragments encircled by modified habitats, such as cattle pastures, soya and sugarcane farms, oil-palm plantations, slash-and-burn farming plots or scrubby regrowth. In addition, many internal clearings, such as highways, roads, power lines and gas lines, perforate surviving forest tracts. Hence, the tropical world is becoming ever smaller, more subdivided and further degraded by a range of external and internal disturbances (Goosem 1997; Laurance & Bierregaard 1997; Peres *et al.* 2006).

Although tropical rainforests comprise but a tiny fraction (<0.2%) of the total land area in Australia, they are enormously important reservoirs of biological diversity and endemism, with a remarkably long evolutionary history that harkens back to the time when Australia was still part of Gondwana. As such, these forests are of exceptional international significance (Keto & Scott 1986; WTMA 2004). This, in concert with the development of world-leading tropical researchers in Australia, has led to impressive advances in conservation-related research, especially within the Wet Tropics of Queensland World Heritage Area (WHA) under the aegis of the Cooperative Research Centre for Tropical Rainforest Ecology and Management.

Here we synthesize selected research in the Australian Wet Tropics on the two principal kinds of habitat subdivision: typical habitat fragmentation, in which forest fragments are isolated by relatively wide swaths of modified land; and internal fragmentation, where forests are insidiously dissected by small, often-linear clearings. Both types have been, and continue to be, important threats to the biota and ecosystems of the Australian Wet Tropics. We briefly highlight the most important types of ecological changes in fragmented habitats and place research in tropical Queensland in a broader context by comparing it to findings from the Amazon and elsewhere in the tropics.

The scope of the problem

Forest loss and fragmentation

In the Australian Wet Tropics, as elsewhere, the processes of forest loss and fragmentation are inextricably

linked. In north Queensland, destruction of rainforests has been most severe in the coastal lowlands (>300 metres elevation), where large expanses of forest have been cleared for sugarcane farming, residential development and other human activities (WTMA 2004; Hilbert, Chapter 8, this volume). Among upland rainforests, deforestation has been most extensive on the Atherton Tableland, with over 76 000 ha of forest having been cleared for cattle pasture and croplands (Winter *et al.* 1987). Large-scale deforestation ended with designation of most surviving forest in the Townsville-Cooktown region as a WHA in 1988, although some forest clearing and selective logging still occurs on private land.

Deforestation is a highly non-random process. In the Wet Tropics, farmers cleared forests overlaying richer basaltic and alluvial soils far more extensively than those overlaying poorer metamorphic or granitic substrates. In addition, flat and moderately steep lands were cleared more frequently than were steep or sharply dissected areas (Laurance 1997a). Deforestation also tends to spread contagiously, such that areas near highways, roads and towns are cleared sooner than those located further from human settlements. This trend is evident in north Queensland and is even more apparent in expanding frontiers such as the Brazilian Amazon, where over 90% of all deforestation occurs within 50 km of roads or highways (Laurance *et al.* 2001a).

As a consequence of non-random deforestation, forest remnants in the Wet Tropics are a highly biased subset of the original forest cover. Remnants frequently persist in steep and dissected areas, on poorer soils, in scenic reserves adjoining waterfalls and crater lakes and on partially inundated lands such as paperbark swamps and mangrove forests. In addition to being poorly representative of the original forest cover, forest fragments near roads and townships may often be older, more isolated and possibly smaller than those located further afield, where deforestation is more recent (Laurance 1997a). Hilbert (Chapter 8, this volume) provides an impressive quantitative estimate of deforestation across recognized forest structural types in the Wet Tropics region, by comparing current forest cover with estimates of past forest distributions based on biophysical modelling.

The influence of non-random deforestation on fragmented plant and animal communities has been little studied (e.g. Laurance *et al.* 1999). It is, however, important to recognize that the biota of forest fragments are likely to have been influenced by non-random deforestation long before the effects of fragmentation *per se* are manifested.

Linear clearings

Despite its limited extent (approx. 900 000 ha), the WHA is dissected by extensive linear infrastructure (Figure 23.1), much of which is used by a growing regional population. This includes 1217 km of paved highways and smaller roads, 320 km of power lines and many unpaved forest roads used for management and access by tourists and landholders. In addition, another 6535 km of former forestry roads and skidder (snig) tracks ramify throughout the region, many of which have been abandoned (WTMA 2004).

The linear clearings themselves occupy areas that would otherwise be forest habitat. This includes 772 ha of deforestation for power lines, 3679 ha for highways and roads and 24 ha for railways and cableways (WTMA 2004). Disused forestry tracks comprise 2070 ha in varying stages of regeneration. Altogether, deforestation for linear clearings (4475 ha) comprises only 0.5% of the WHA, with half of this supposedly regenerating on old forestry tracks. As described below, however, the ecological impacts of linear infrastructure are far greater than just the limited forest loss associated with their clearings.

Area effects

In fragmented landscapes, the species richness of any particular taxonomic group, such as trees or mammals, tends to be positively correlated with fragment size, with intact forests containing more species per unit area than forest fragments. For example, the logarithm of fragment area explains much of the total variability in species richness of rainforest birds (Warburton 1997) and arboreal mammals (Laurance 1990) in north Queensland. Such correlations tend to be stronger when non-rainforest species that sometimes invade forest fragments are excluded.

Species–area correlations arise in part because of sample effects: small samples contain fewer species than do large samples simply because they include

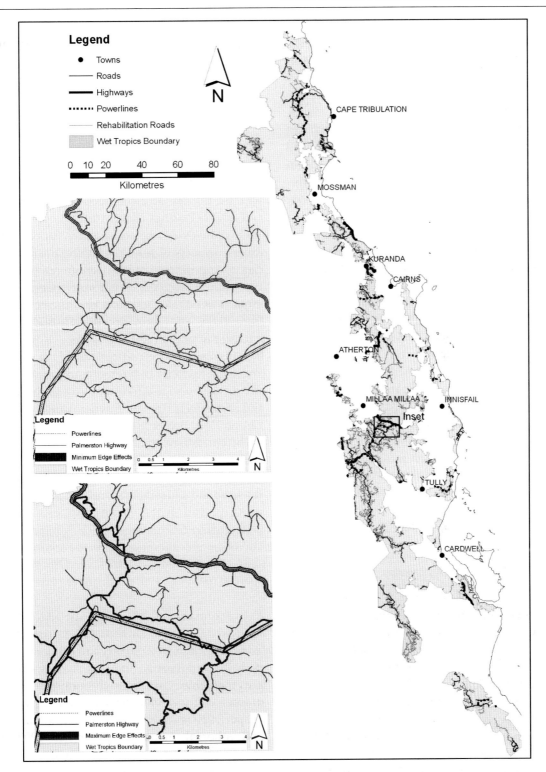

Figure 23.1 Major linear infrastructure in the Wet Tropics of Queensland World Heritage Area. Both insets show how the Palmerston Highway, Palmerston powerline clearing and minor road networks divide up the World Heritage Area into much smaller blocks of forest. The top inset demonstrates the extent of edge effects arising from linear clearings, using the minimum edge-penetration distance from microclimate, flora and fauna studies. The lower inset is similar but shows affected forest area based on an estimated maximum edge-penetration distance.

fewer total individuals. However, they also arise because certain species simply cannot persist in the limited universe of a forest fragment. Especially vulnerable in Australian rainforest fragments are large-bodied species, such as cassowaries (*Casuarius casuarius*); predators, such as spotted-tailed quolls (*Dasyurus maculatus*) and rufous owls (*Ninox rufa*); and old-growth rainforest specialists, such as the lemuroid ringtail possum (*Hemibelideus lemuroides*), Australian fernwren (*Crateroscelis gutteralis*) and golden bowerbird (*Prionodura newtoniana*) (Laurance 1991b, 1997a; Warburton 1997). Some invertebrate species also disappear from forest fragments; in the Amazon, a number of beetle, ant, bee, termite and butterfly species are highly sensitive to fragment area (Klein 1989; Souza & Brown 1994; Brown & Hutchings 1997; Didham 1997).

Fragment size strongly affects the rate of local extinctions, with species disappearing faster from small than large fragments (MacArthur & Wilson 1967; Ferraz *et al.* 2003). In fragmented rainforests on the southern Atherton Tableland, all of which were <600 ha in area, the most vulnerable species have vanished rapidly. For example, the lemuroid ringtail possum disappeared from a small (1.4 ha) fragment in just 3–9 years, and from two larger (43 and 75 ha) fragments in 35–60 years (Laurance 1990). Spotted-tailed quolls (Figure 23.2) were common on the Tableland prior to the Second World War but were never encountered during an intensive trapping and spotlighting survey in 1986–7 (Laurance 1994). Several other mammal species, such

as the Atherton antechinus (*Antechinus godmani*), brown antechinus (*A. stuartii*) and musky rat-kangaroo (*Hypsiprimnodon moschatus*), also disappear or decline sharply in fragmented forests in a few decades or less (Laurance 1997a).

Edge effects

Edge effects are physical and biotic changes associated with the abrupt, artificial margins of forest fragments and linear clearings (Janzen 1983; Lovejoy *et al.* 1986). Such changes can be remarkably varied, influencing diverse aspects of forest structure, microclimate, species composition and functioning (Murcia 1995; Laurance *et al.* 2002). For forest species, the overall impacts of edge effects may be difficult to distinguish from those of fragment-area effects, because both intensify in small fragments. Indeed, it is now believed that numerous local extinctions of species that were originally ascribed to fragment-area effects were actually driven by edge effects, or by a combination of edge and area effects operating synergistically (Laurance & Yensen 1991; Ewers & Didham 2006).

The creation of an abrupt forest edge has important impacts on forest microclimate and structure. Varied microclimatic alterations, such as increased light, elevated air and soil temperatures and increased moisture stress, prevail in the vicinity of forest edges (Kapos 1989; Turton & Freiberger 1997; Pohlman *et al.* 2007). Microclimatic alterations are especially strong near newly created edges, which tend to be structurally open (Kapos *et al.* 1997; Didham & Lawton 1999), near edges suffering periodic disturbance (Gascon *et al.* 2000; Pohlman *et al.* 2007) and near edges bordered by open environments such as cattle pastures or croplands (Mesquita *et al.* 1999). In addition, rates of tree mortality and damage often rise sharply near fragment margins. These are caused both by microclimatic stresses, which can exceed the physiological capacities of many tree species, and because wind shear and turbulence increase markedly near edges, causing elevated windthrow and trunk snapping (Laurance 1997b; Laurance *et al.* 1997, 1998). Chronically elevated tree mortality damages the protective forest canopy, allowing light and wind to penetrate to the forest floor and increasing the overall intensity of edge-related microclimatic changes (Malcolm 1998).

Figure 23.2 The spotted-tailed quoll (*Dasyurus maculatus*), a medium-sized predator (up to 7 kg in weight) that is highly vulnerable in fragmented landscapes of north Queensland (photo by Stephen Williams).

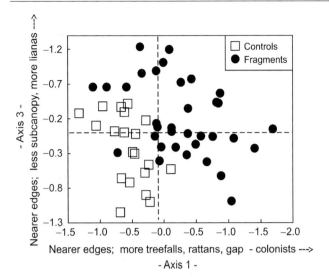

Figure 23.3 Ordination of forest structural and floristic data from 52 900-m² study plots in intact-forest controls and in ten forest fragments ranging from 1 to 600 ha in area on the Atherton Tableland in north Queensland. Data were collected 5–7 months after the study area was disturbed by cyclone Winifred in 1986. *Source*: after Laurance (1997).

Studies in north Queensland suggest that edge effects can have important impacts on forests, and that these effects are exacerbated by cyclones and seasonal windstorms. On the Atherton Tableland, forest fragments affected by periodic cyclones and storms are often heavily disturbed (Figure 23.3), as shown by reduced canopy cover, increased treefalls and elevated abundances of rattans (*Calamus* spp.), lianas and other disturbance-adapted plants (e.g. *Dendrocnide* spp., *Solanum* spp.; Laurance 1991a, 1997b). Many exotic weed species have been found in forest fragments, mostly along forest edges and in treefall gaps (Jenkins 1993). Comparably striking edge effects occur in some other fragmented regions, such as in the Amazon (Ferreira & Laurance 1997; Laurance *et al.* 1998, 2001b) and Brazilian Atlantic forest (Viana *et al.* 1997), but may be less apparent in Southeast Asian forests (Thomas 2004).

Roads and other linear clearings also experience significant edge effects in microclimate, vegetation structure, floristics and faunal species composition (Goosem & Marsh 1997; Siegenthaler *et al.* 2000; Goosem 2002a; S. G. Laurance 2004). These changes seem comparable in diversity to those in typical forest fragments, although their magnitude is probably lower because

narrow linear clearings have less 'fetch' than larger clearings, i.e. by virtue of their limited width they experience smaller changes in microclimate, wind-speed and evapotranspiration (Laurance 2004). In this sense there is certainly an important distinction between the narrowest linear clearings like small forest roads (<8 m width including verges), where the overhead canopy cover is substantially intact, and relatively wider openings, such as highways and powerline clearings (often 20–100 m wide). In such larger clearings, canopy cover is largely absent, microclimatic changes are far greater and light-loving grasses, weeds and exotic animal species commonly proliferate (Goosem & Marsh 1997; Siegenthaler & Turton 2000; Goosem 2002a; Maver 2002; Larsson 2003; Pohlman 2006). The nature of the ground surface within linear clearings also has an effect; paved highways cause greater surface-temperature increases than do weedy powerline clearings, resulting in greater desiccation stress (Pohlman *et al.* 2007).

Because of its highly fragmented nature, the Wet Tropics region is particularly vulnerable to edge effects. For example, even if one considers only narrow internal clearings (which currently total 16 144 km, of which more than 3074 km are actively being maintained), edge effects from lateral light penetration alter at least 8000 ha of forest (Siegenthaler & Turton 2000; Pohlman *et al.* 2007) and edge-related changes in vegetation and small-mammal communities affect 13 000–16 000 ha of forest (Goosem & Marsh 1997; Goosem 2000a, c, 2002b, c; Siegenthaler *et al.* 2000; Pohlman 2006). Such edge effects would be even greater if one considers the large areas affected by increased traffic noise (see below). Edge-related changes are likely to occur over even larger spatial scales for forests adjoining large clearings such as cattle pastures, sugarcane fields, banana plantations and residential areas (e.g. Laurance 1991a).

Species invasions

In addition to supporting some generalist forest species, the modified habitats surrounding rainforest fragments also contain many non-forest and exotic species (see Goosem, Chapter 24, and Congdon & Harrison, Chapter 25, this volume for overviews), some of which invade forest fragments. Unless heavily disturbed, species-rich rainforests are relatively resistant to invasions (Laurance & Bierregaard 1997), but linear clearings greatly facilitate invasions by allowing weeds and feral animals to penetrate deep into forest tracts (Goosem, Chapter 24, this volume). In the Amazon, for example, road verges provide corridors that allow weeds and non-forest insects, frogs and other species to invade remote frontier areas of the basin (Gascon *et al.* 1999).

In the Wet Tropics, linear clearings favour the ingress of a wide variety of weeds, feral animals and native fauna alien to rainforests. Weed invasions are facilitated by clearing-maintenance practices that reduce the cover and vitality of native competitors (Werren 2001). These measures include burning within powerline clearings (a practice recently discontinued), mowing, grading, removing trees and overhanging branches and spraying herbicides along road verges. Partly as a consequence, exotic grasses, including Guinea grass (*Urochloa maxima*) and molasses grass (*Melinis minutiflora*), dominate powerline clearings and road verges, along with exotic shrubs such as lantana (*Lantana camara*) and bramble (*Rubus alceifolius*) (Goosem 2002a, b). Forest functioning and regeneration can be greatly impaired by these 'transformer species', which alter the character and condition of large areas (Werren 2001). Exotic grasses can promote fire regimes that result in self-perpetuating grassland swathes, whereas monospecific stands of woody weeds can prevent recruitment of native trees and shrubs (Reynolds 1994; Goosem 2002a).

Linear clearings also provide habitat and movement corridors for many animal species, including cane toads (*Bufo marinus*), feral pigs (*Sus scrofa*), house mice (*Mus musculus*) and domestic cats (*Felis catus*) and dogs (*Canis familiaris*) (Goosem 2000c; Byrnes 2002; Larsson 2003). Although occurring in low numbers throughout much of the rainforest, cane toads are especially common on rainforest highways near open forest or pastures (Goosem 2000b), and appear to prefer forest edges with an open understorey (Larsson 2003). Sand and camera traps have shown that feral predators use narrow forest tracks, whereas pig activity increases near roads and powerline clearings (Byrnes 2002). Grassland small mammals dominate in powerline clearings (Goosem & Marsh 1997) and in grassy verges adjoining roads (Goosem *et al.* 2001), even where verges are quite narrow (Goosem 2000a, c, 2002b). The most effective way to reduce faunal invasions is to revegetate linear clearings (Larsson 2003).

Isolation and barrier effects

Among the most important insights from recent tropical research is that even remarkably small clearings can halt or impede the movements of many species. In the Amazon, clearings of just 15–100 m width are apparently insurmountable barriers for certain understorey birds, arboreal mammals, dung and carrion beetles and euglossine bees. Likewise, peccaries and specialized insect-gleaning bats are highly reluctant to enter clearings (Laurance *et al.* 2002 and references therein). Other species, such as some woodcreepers (Dendrocolaptidae) and antbirds (Formicariidae), will cross small clearings but are inhibited by larger expanses of degraded land (Harper 1989; Laurance & Gomez 2005).

Not only do many rainforest species avoid clearings, but they also tend to avoid forest edges abutting clearings. Laurance *et al.* (2004) demonstrated that, among Amazonian understorey birds, those that tend to avoid forest edges also have the lowest frequencies of road crossings. Such edge- and clearing-avoiding species are especially vulnerable to habitat fragmentation, suffering both from reduced habitat quality in forest fragments, which depresses their population size, and from being poorly adapted for using the degraded lands surrounding fragments. Populations of such species in fragments are strongly isolated demographically and genetically, sharply increasing the likelihood of local extinction (Laurance 1991b; Gascon *et al.* 1999; S. G. Laurance 2004).

In the Wet Tropics, clearings dramatically affect the movements of some fauna. Certain forest-interior birds tend to avoid the vicinity of highway clearings, with bird species richness and abundance declining within 3–100 m of the highway and with several highly specialized species not being recorded until 100–200 m inside the forest (Dawe & Goosem 2006). Small-mammal communities also change near road edges, even where roads are narrow (<10 m width) and have low traffic volumes (about 500 vehicles per day; Goosem 2002b), although wider roads, highways and powerline clearings have greater barrier effects (Goosem & Marsh 1997; Goosem 2000c). Several rainforest skink species are edge-avoiders, whereas others prefer the warmer, drier microclimate at powerline-clearing edges (Larsson 2003). In fragmented landscapes, non-flying mammal species that use or exploit the matrix of modified habitats surrounding rainforest remnants are much less vulnerable to local extinction than are those that avoid the matrix (Laurance 1991b).

For some fauna, the avoidance of roads and highways is probably increased by vehicle noise, vibrations, headlights or pollutants (Dawe 2005; Dawe & Goosem 2007; Wilson & Goosem 2007). Near Wet Tropics highways, vehicle noise is detectable 200 m or more from the source, at both canopy and ground level (Dawe 2005; Dawe & Goosem 2007). For nocturnal species such as ringtail possums, disturbance from traffic noise or headlights could potentially inhibit their activity near roads (Wilson 2000). Some species are willing and able to cross roads (or forage, breed or thermoregulate along roads) but are highly susceptible to road mortality (Figure 23.4), which could potentially create a population sink near high-volume roads and highways for vulnerable species (Goosem 2004). Collectively, road

Figure 23.4 Roadkill can be an important source of mortality for slow-moving animals and for those that forage, breed or thermoregulate on roads (photo by William Laurance).

mortality, in concert with the tendency of forest-interior species to avoid clearings and edge-related disturbances, and the presence of generalist competitors or predators in clearings, can create a formidable barrier to faunal movements.

Conservation corridors

Can the deleterious effects of forest fragmentation be mitigated? A considerable body of evidence indicates that movement corridors can increase biodiversity in fragmented landscapes, by partially countering the deleterious effects of population isolation on forest-dependent wildlife (e.g. Bennett 1990; Saunders & Hobbs 1991; Laurance & Laurance 2003). On the Atherton Tableland and coastal lowlands, for example, linear strips of secondary and relict primary vegetation along streams (typically ranging from 10 to 80 m in width) can sustain a number of forest bird and mammal species (Laurance 1991b; Crome *et al.* 1994; Laurance & Laurance 1999; Lawson 2004). Small (<20 ha) forest fragments linked to large forest tracts by such corridors sustain a greater diversity of arboreal marsupials than do those of nearby, isolated fragments of comparable area (Laurance 1990). The reconnection of Lake Barrine National Park, a 500-ha rainforest fragment on the Atherton Tableland, to nearby Gadgarra State Forest via an 80-m wide, 1.5-km long corridor that was planted along a stream has probably increased demographic movements and gene flow for some forest vertebrates and plants (Tucker 2000; Jansen 2005). Breeding activity and stable home ranges of many birds, mammals and amphibians in tropical corridors suggest that they can also provide useful wildlife habitat (e.g. Crome *et al.* 1994; Laurance & Laurance 1999; Lima & Gascon 1999). Conservation practitioners in many parts of the world have found corridors to have similar benefits for many species of forest-dependent wildlife (e.g. Saunders & Hobbs 2001; Hilty *et al.* 2006, and references therein).

It would, however, be a mistake to assume that corridors are a panacea. All but the widest corridors suffer significant edge effects and are likely to be selective filters, facilitating movements of some species but not forest-interior specialists, which are often the most vulnerable to fragmentation. For example, only the widest (>200 m) corridors composed of continuous, old-growth rainforest can sustain the lemuroid ringtail possum, the most vulnerable of all arboreal folivores to fragmentation (Laurance & Laurance 1999). For species that experience significant edge-related predation (Wilcove 1985) or hunting (Peres 2001) in fragmented landscapes, corridors might become demographic sinks that increase the vulnerability of nearby populations in fragments. It is also conceivable that corridors could increase the spread of pathogens among fragments (Simberloff & Cox 1987). These limitations should not reduce our enthusiasm for establishing conservation corridors, but they should inform our decisions and reinforce the notion that there is often no simple way to counter the negative effects of fragmentation on highly vulnerable species.

Environmental synergisms

Despite suffering seriously from fragmentation, the forests of the Australian Wet Tropics are less likely to sustain additional, direct human disturbances than are those elsewhere in the tropics. In many tropical regions, fragmented forests are logged and overhunted (Peres & Michalski 2006). Fragmented forests are also far more likely than intact forests to be degraded by destructive surface fires (Cochrane & Laurance 2002). Such changes can interact synergistically with fragmentation, sharply increasing the likelihood of local species extinctions (Laurance & Cochrane 2001). In addition, forests in developing nations are often subjected to spontaneous, unplanned colonization and exploitation when new roads and highways are created (e.g. Laurance *et al.* 2001a, 2006), a grave challenge that rarely plagues forests in tropical Australia.

The Wet Tropics region is, nonetheless, far from secure from future threats. Area-demanding species such as cassowaries are already highly vulnerable because extensive forest loss and fragmentation often preclude their normal foraging movements between lowland and highland areas, while mortality risk from roadkill rises as highways and traffic volume expand (Bentrüpperbaumer 1988; Crome 1990). Exotic pathogens, such as the chytrid fungus that has devastated many stream-dwelling amphibians worldwide, can imperil species even in large protected areas (Laurance *et al.* 1996; Daszak *et al.* 2003). Perhaps most alarming of all is that, by virtue of its limited extent, fragmented

nature and striking diversity of upland-endemic species, the Wet Tropics region appears exceptionally vulnerable to projected global warming (Williams *et al.* 2003; Williams & Hilbert 2006). Collectively, the synergistic effects of such environmental traumas could critically threaten one of the world's most special tropical biotas.

Summary

• Research in the Queensland Wet Tropics has provided important insights into the impacts of habitat fragmentation and linear clearings on rainforest biota.
• Among the most important conclusions are that edge effects have major impacts on forest fragments, that fragmented forests are increasingly vulnerable to windstorm damage and that even narrow clearings can greatly depress movements of some rainforest-dependent species.
• Because of its greatly limited extent, the Wet Tropics region and its biota are probably highly vulnerable to fragmentation and edge effects. These threats may increasingly interact with other ongoing alterations, such as anthropogenic climate change, in the future.

Acknowledgements

We thank Susan Laurance for commenting on the chapter. Partial support was provided by the Smithsonian Tropical Research Institute, Rainforest CRC, James Cook University, Queensland Department of Main Roads, Wet Tropics Management Authority, Queensland Electricity Commission, Powerlink, Queensland Parks and Wildlife Service, Douglas Shire Council and Queensland Department of Natural Resources.

References

Achard, F., Eva, H., Stibig, H. *et al.* (2002). Determination of deforestation rates of the world's humid tropical forests. *Science* **297**: 999–1002.

Bennett, A. F. (1990). *Habitat Corridors: Their Role in Wildlife Management and Conservation*. Department of Conservation and Environment, Melbourne.

Bentrupperbäaumer, J. (1988). *Numbers and Conservation Status of Cassowaries in the Mission Beach area following*

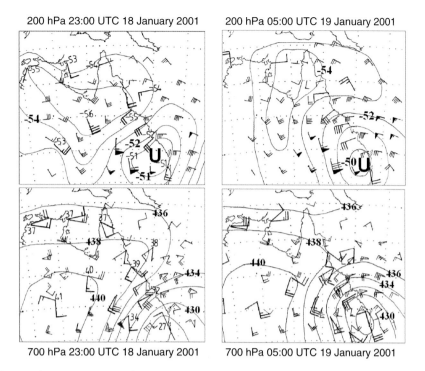

200 hPa 23:00 UTC 18 January 2001   200 hPa 05:00 UTC 19 January 2001

700 hPa 23:00 UTC 18 January 2001   700 hPa 05:00 UTC 19 January 2001

Plate 2.2 Corresponding to the same times as for Figure 2.1, the 200 hPa wind observations (large plots) and the NCEP/ NCAR reanalysis winds (smaller plots). The blue lines are the 200 hPa isotherms with the tropopause undulation marked by the 'U'. The lower frames are the 700 hPa winds and the 850–500 hPa shears, with the blue contours the 850–500 hPa thickness in decametres.

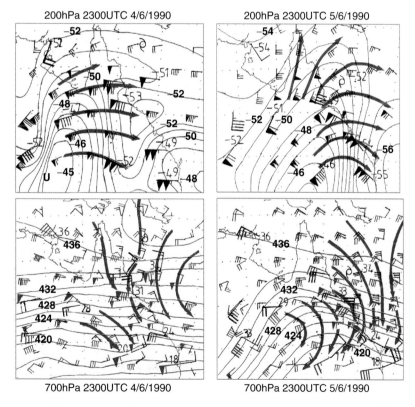

200hPa 2300UTC 4/6/1990   200hPa 2300UTC 5/6/1990

700hPa 2300UTC 4/6/1990   700hPa 2300UTC 5/6/1990

Plate 2.4 At the corresponding times shown in Figure 2.3 (top frames), the 200 hPa charts showing wind observations (large plots) and NCEP/NCAR reanalyses winds (smaller plots); the lines are the 200 hPa isotherms with the tropopause undulation marked by the 'U'. The lower frames are the 700 hPa winds and 850–500 hPa shears with the contours the 850–500 hPa thickness in decametres. Large streamlines highlight zones of warm (cold) air advection.

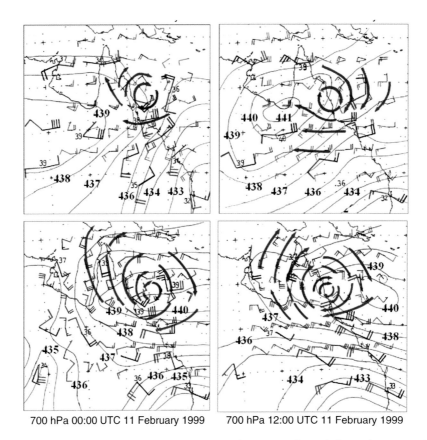

700 hPa 00:00 UTC 11 February 1999   700 hPa 12:00 UTC 11 February 1999

Plate 2.6 The NCEP/NCAR reanalyses 700 hPa winds and 850–500 hPa vertical wind shears for tropical cyclone Steve (top two frames) and tropical cyclone Rona (lower two frames). The contours are the 850–500 hPa thickness in decametres. The larger wind and shear plots are actual measurements taken from radiosonde (upper air sounding) stations. Large streamlines highlight zones of warm (cold) air advection.

700 hPa 23:00 UTC 29 January 1994   700 hPa 23:00 UTC 30 January 1994

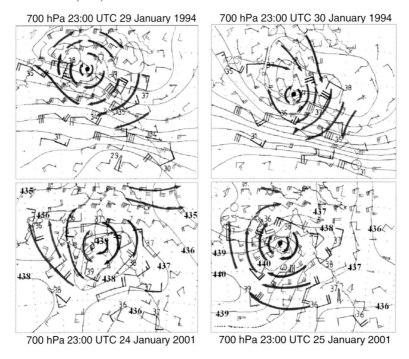

700 hPa 23:00 UTC 24 January 2001   700 hPa 23:00 UTC 25 January 2001

Plate 2.8 As in Figure 2.6, except for tropical cyclone Sadie (top two frames) and tropical cyclone Abigail (lower two frames).

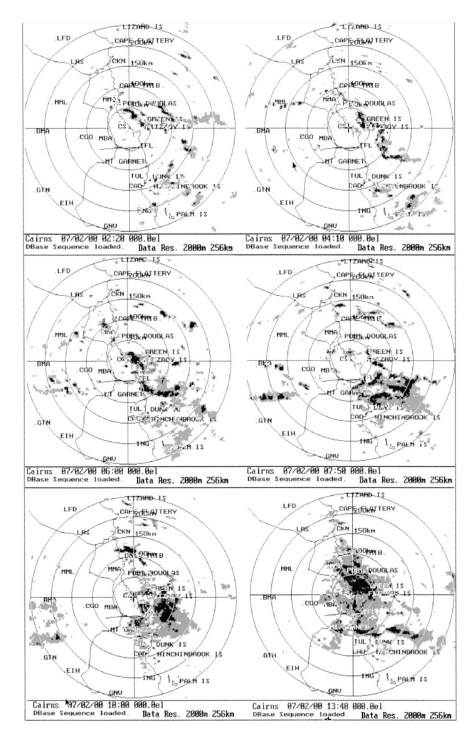

Plate 2.10 Sequence of rain echoes on Cairns radar from 02:20 UTC to 13:40 UTC 7 February 2000.

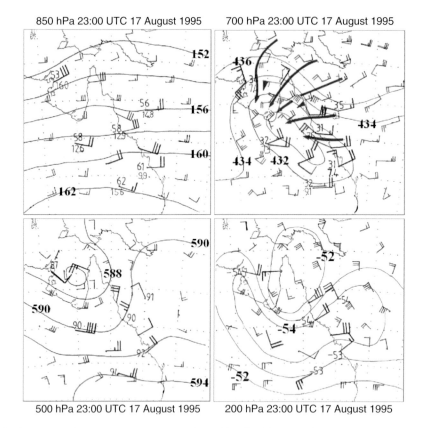

850 hPa 23:00 UTC 17 August 1995   700 hPa 23:00 UTC 17 August 1995

500 hPa 23:00 UTC 17 August 1995   200 hPa 23:00 UTC 17 August 1995

Plate 2.12 The NCEP/NCAR reanalyses at 850, 700, 500 and 200 hPa for 23:00 UTC 17 August 1995. Winds are plotted to 850–500 hPa vertical wind shears. Contours are the 850 and 500 hPa heights (decametres), 850–500 hPa thickness in decametres and 200 hPa isotherms. Larger wind and shear plots are actual from radiosonde stations. The large streamlines highlight zones of warm (cold) air advection.

Plate 26.1 Classified images (isodata algorithm for the Smithfield study area) (a) pre-fire January 2002 and (b) post-fire June 2004. Fire scars shown in purple.

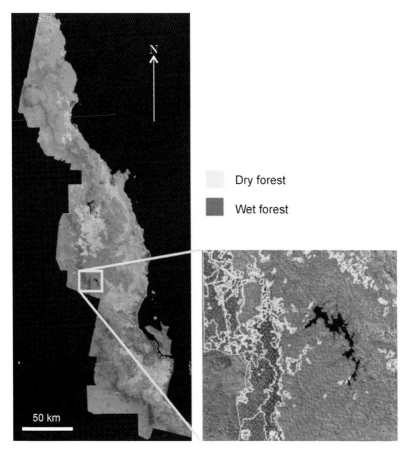

Plate 47.1 Landsat 7 Enhanced Thematic Mapper mosaic for the Wet Tropics bioregion with an enlarged subset showing the boundary between sclerophyll (yellow) and mesophyll (red) communities on the image and air-photo derived vegetation community boundaries superimposed.

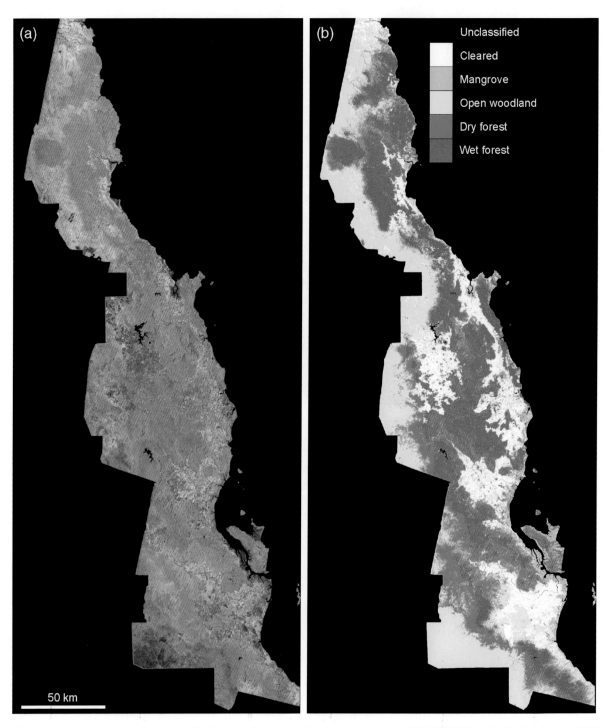

Plate 47.2 (a) A colour composite of the Wet Tropics based on: display red, Landsat ETM SWIR band; display green, JERS backscatter; display blue, Landsat ETM NIR band. (b) Vegetation community classification using combined Landsat ETM and JERS.

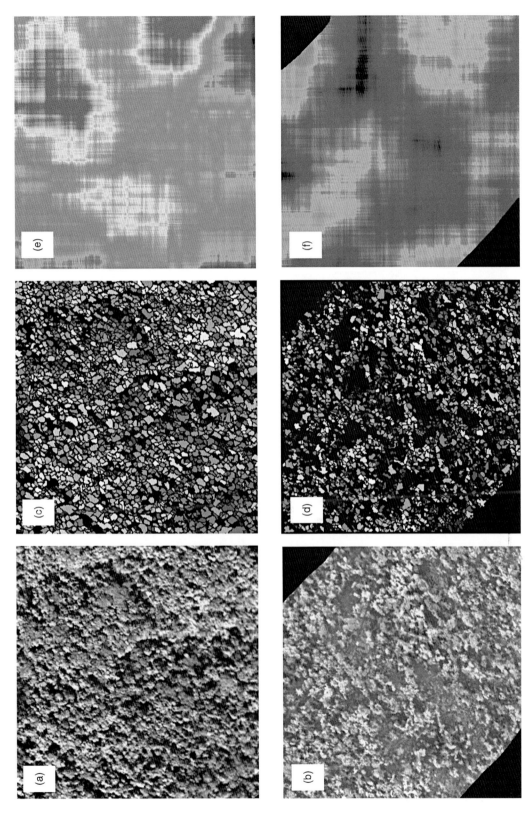

Plate 47.3 (a) Airborne hyperspectral image of undisturbed rainforest in the Cape Tribulation region of the Wet Tropics. (b) Airborne hyperspectral image of cyclone-damaged rainforest in the Cape Tribulation region of the Wet Tropics. (c) and (d) Tree crown map derived from airborne hyperspectral image. (e) and (f) Tree density maps derived from airborne hyperspectral image and tree crown map.

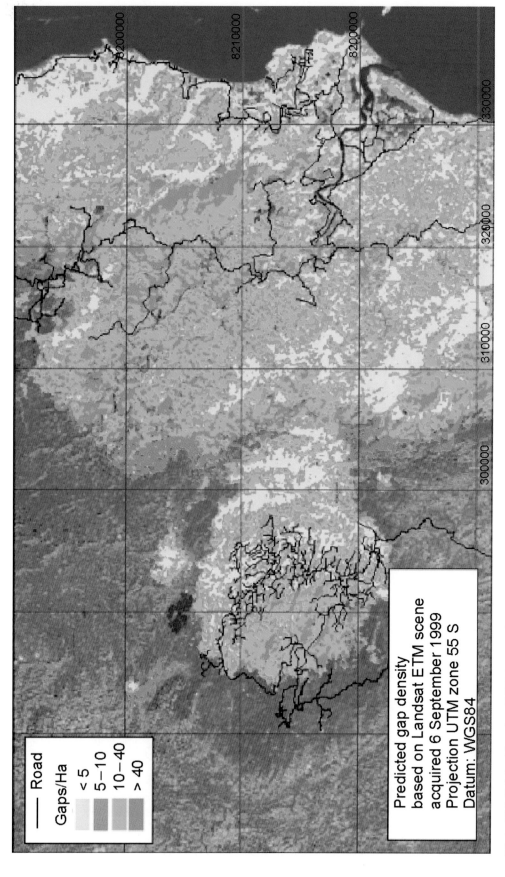

Plate 47.4 Original Landsat 7 ETM image for the Daintree–Cape Tribulation section of the Wet Tropics overlaid with estimated gap fraction.

Predicted gap density
based on Landsat ETM scene
acquired 6 September 1999
Projection UTM zone 55 S
Datum: WGS84

Road

Gaps/Ha

< 5
5–10
10–40
> 40

Plate 47.5 Example of preliminary biomass estimation product derived from radar interferometry based on the Wet Tropics AIRSAR data.

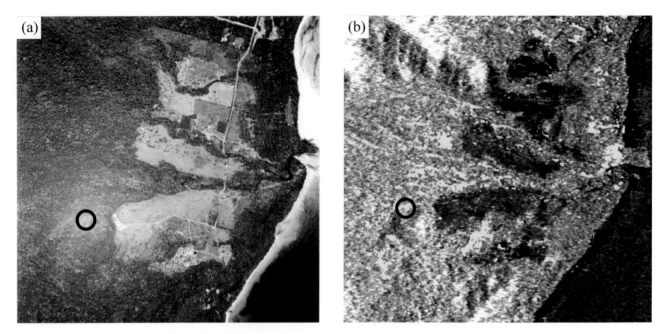

Plate 47.6 (a) Aerial photography (courtesy of Wet Tropics Management Authority) and (b) canopy height difference map produced by subtracting pre-cyclone from post-tropical cyclone Rona canopy heights derived from imaging radar data. Area shown is centred on the canopy crane site (black circle), Coconut Beach Resort, Cape Tribulation.

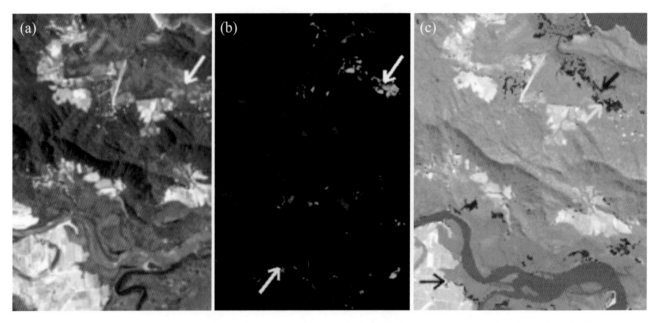

Plate 47.7 (a) Hyperion image in true colour of an area containing pond apple (arrows). Example of pond apple detection maps generated from processing (b) Hyperion image data and (c) Landsat 7 ETM image data and a constraining model. The coloured areas indicate where there is a spectral match with field measurements of pond apple.

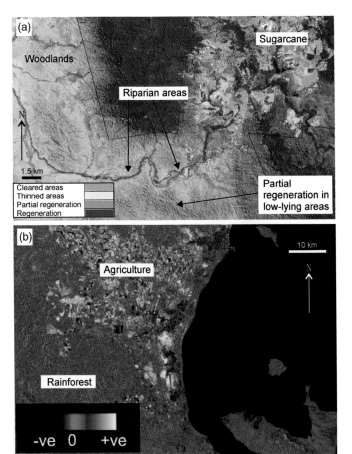

Plate 47.8 An example of changes in vegetation cover in the Wet Tropics mapped from (a) Landsat 7 ETM Enhanced Vegetation Index difference image 1988–90 and (b) JERS-1 difference image of January 1994 to July 1996.

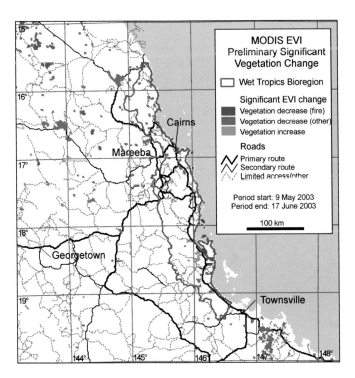

Plate 47.9 Example of a vegetation cover trend image produced from an integration of 48 daily enhanced vegetation index images and fire images for the period 19 May to 17 July 2003.

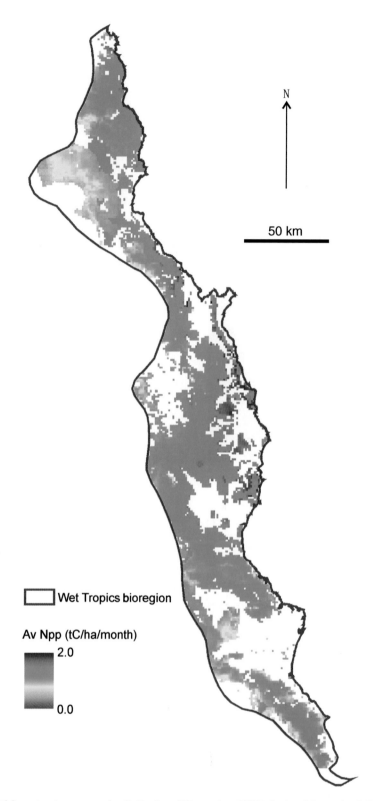

N

50 km

Wet Tropics bioregion

Av Npp (tC/ha/month)

2.0

0.0

Plate 47.10 Average monthly net primary productivity (as of December 2001) derived from the 3-PGS forest growth model for rainforest classed as national park, remnant native and production forests across the Wet Tropics bioregion of north Queensland.

Cyclone Winifred. Report to Queensland National Parks and Wildlife Service, Brisbane.

Brown, K. S. Jr & Hutchings, R. W. (1997). Disturbance, fragmentation and the dynamics of diversity in Amazonian forest butterflies. In *Tropical Forest Remnants: Ecology, Management and Conservation of Fragmented Communities*, Laurance, W. F. & Bierregaard, R. O. (eds). University of Chicago Press, Chicago, pp. 91–110.

Byrnes, P. (2002). Activity of feral pigs and cats associated with roads and powerline corridors within the Wet Tropics of Queensland World Heritage Area. BSc (Hons) thesis, James Cook University, Cairns.

Cochrane, M.A. & Laurance, W. F. (2002). Fire as a large-scale edge effect in Amazonian forests. *Journal of Tropical Ecology* **18**: 311–25.

Crome, F. H. J. (1990). Cassowaries in north-eastern Queensland: report of a survey and review of their status and conservation and management needs. *Australian Wildlife Research* **17**: 369–85.

Crome, F. H. J., Isaacs, J. & Moore, L. (1994). The utility to birds and mammals of remnant riparian vegetation and associated windbreaks in the tropical Queensland uplands. *Pacific Conservation Biology* **1**: 328–43.

Daszak, P., Cunningham, A. A. & Hyatt, A. D. (2003). Infectious disease and amphibian population declines. *Diversity and Distributions* **9**: 141–50.

Dawe, G. (2005). Traffic noise and its influence on the song of tropical rainforest birds. BSc (Hons) thesis, James Cook University, Cairns.

Dawe, G. & Goosem, M. (2007). *Noise Disturbance along Highways – Kuranda Range Upgrade Project*. Report to Queensland Department of Main Roads, Cairns.

Didham, R. K. (1997). The influence of edge effects and forest fragmentation on leaf-litter invertebrates in central Amazonia. In *Tropical Forest Remnants: Ecology, Management, and Conservation of Fragmented Communities*, Laurance, W. F & Bierregaard, R. O. Jr (eds). University of Chicago Press, Chicago, pp. 55–70.

Didham, R. K. & Lawton, J. H. (1999). Edge structure determines the magnitude of changes in microclimate and vegetation structure in tropical forest fragments. *Biotropica* **31**: 17–30.

Ewers, R. M. & Didham, R. K. (2006). Confounding factors in the detection of species responses to habitat fragmentation. *Biological Reviews* **81**: 117–42.

Ferraz, G., Russell, G., Stouffer, P., Bierregaard, R., Pimm, S. & Lovejoy, T. (2003). Rates of species loss from Amazonian forest fragments. *Proceedings of the National Academy of Sciences* **100**: 14069–73.

Ferreira, L. V. & Laurance, W. F. (1997). Effects of forest fragmentation on mortality and damage of selected trees in central Amazonia. *Conservation Biology* **11**: 797–801.

Gascon, C., Lovejoy, T., Bierregaard, R. *et al.* (1999). Matrix habitat and species persistence in tropical forest remnants. *Biological Conservation* **91**: 223–9.

Gascon, C., Williamson, G. B, & Fonseca, G. (2000). Receding edges and vanishing reserves. *Science* **288**: 1356–8.

Goosem, M. W. (1997). Internal fragmentation: the effects of roads, highways and powerline clearings on movements and mortality of rainforest vertebrates. In *Tropical Forest Remnants: Ecology, Managemen, and Conservation of Fragmented Communities*, Laurance, W. F & Bierregaard, R. O. Jr (eds). University of Chicago Press, Chicago, 2 pp. 41–55.

Goosem, M. W. (2000a). Effects of tropical rainforest roads on small mammals: edge changes in community composition. *Wildlife Research* **27**: 151–63.

Goosem, M. W. (2000b). Impacts of roads and powerline clearings on rainforest vertebrates with emphasis on ground-dwelling small mammals. PhD thesis, James Cook University, Cairns.

Goosem, M. W. (2000c). The effect of canopy closure and road verge habitat on small mammal community composition and movements. In *Impacts of Roads and Powerlines on the Wet Tropics World Heritage Area II*, Goosem, M. & Turton, S. M. (eds). Rainforest CRC, Cairns, pp. 2–18.

Goosem, M. W. (2002a). Weed penetration, edge effects and rehabilitation success in weedy swathes of the Palmerston powerline clearing. In *Weed Incursions Along Roads and Powerlines in the Wet Tropics of Queensland World Heritage Area: Potential of Remote Sensing as an Indicator of Weed Infestations*. Goosem, M. & Turton, S. M. (eds). Rainforest CRC, Cairns, pp. 52–67.

Goosem, M. W. (2002b). Weed surveys along highways, roads and powerline clearings traversing the Wet Tropics World Heritage Area. In *Weed Incursions Along Roads and Powerlines in the Wet Tropics of Queensland World Heritage Area: Potential of Remote Sensing as an Indicator of Weed Infestations*, Goosem, M. & Turton, S.M., Rainforest CRC, Cairns, pp. 1–25.

Goosem, M. W. (2002c). Weed penetration, edge effects and rehabilitation success in weedy swathes of the Palmerston powerline clearing. In *Weed Incursions Along Roads and Powerlines in the Wet Tropics of Queensland World Heritage Area: Potential of Remote Sensing as an Indicator of Weed Infestations*, Goosem, M. & Turton, S. M. (eds). Rainforest CRC, Cairns, pp. 52–67.

Goosem, M. W. (2004). Linear infrastructure in tropical rainforests: mitigating impacts on fauna of roads and powerline clearings. In *Conservation of Australia's Forest Fauna*, Lunney, D. (ed.). Royal Zoological Society of New South Wales, Mosman, pp. 418–34.

Goosem, M., Izumi, Y. & Turton, S. (2001). Efforts to restore habitat connectivity for an upland tropical rainforest fauna:

a trial of underpasses below roads. *Ecological Management and Restoration* **2**(3): 196–202.

Goosem, M. W. & Marsh, H. (1997). Fragmentation of a small-mammal community by a powerline corridor through tropical rainforest. *Wildlife Research* **24**: 613–29.

Hansen, M. C. & De Fries, R. (2004). Detecting long-term global forest change using continuous fields of tree-cover maps from 8-km Advanced Very High Resolution Radiometer (AVHRR) data for the years 1982–99. *Ecosystems* **7**: 695–716.

Harper, L. H. (1989). The persistence of ant-following birds in small Amazonian forest fragments. *Acta Amazônica* **19**: 249–63.

Hilty, J. A., Lidicker, W. Z. Jr & Merenlender, A. (eds) (2006). *Corridor Ecology: The Science and Practice of Linking Landscapes for Biodiversity Conservation*. Island Press, Washington, DC.

Jansen, A. (2005). Avian use of restoration plantings along a creek linking rainforest paches on the Atherton Tablelands, North Queensland. *Restoration Ecology* **13**: 275–83.

Janzen, D. H. (1983). No park is an island: Increase in interference from outside as park size decreases. *Oikos* **41**: 402–10.

Jenkins, S. I. (1993). Exotic plants in the rainforest fragments of the Atherton and Evelyn Tablelands, north Queensland. BSc (Hons) thesis, James Cook University, Townsville, Queensland.

Kapos, V. (1989). Effects of isolation on the water status of forest patches in the Brazilian Amazon. *Journal of Tropical Ecology* **5**: 173–85.

Kapos, V., Wandelli, E., Camargo, J. & Ganade, G. (1997). Edge-related changes in environment and plant responses due to forest fragmentation in central Amazonia. In *Tropical Forest Remnants: Ecology, Management, and Conservation of Fragmented Communities*, Laurance, W. F. & Bierregaard, R. O. (eds). University of Chicago Press, Chicago, pp. 33–44.

Keto, A. & Scott, K. (1986). *Tropical Rainforests of North Queensland: Their Conservation Significance*. Australian Government Publishing Service, Canberra.

Klein, B. C. (1989). Effects of forest fragmentation on dung and carrion beetle communities in central Amazonia. *Ecology* **70**: 1715–25.

Larsson, U. (2003). Edge- and linear-barrier effects on rainforest fauna, and the success of reducing those effects by revegetation. BApplSc (Hons) thesis, James Cook University, Cairns.

Laurance, S. G. (2004). Responses of understory rain forest birds to road edges in central Amazonia. *Ecological Applications* **14**: 1344–57.

Laurance, S. G. & Gomez, M. (2005). Clearing width and movements of understory rainforest birds. *Biotropica* **37**: 149–52.

Laurance, S. G. & Laurance, W. F. (1999). Tropical wildlife corridors: use of linear rainforest remnants in arboreal mammals. *Biological Conservation* **91**: 231–9.

Laurance, S. G. & Laurance, W. F. (2003). Bandages for wounded landscapes: Faunal corridors and their roles in wildlife conservation in the Americas. In *How Landscapes Change: Human Disturbance and Ecosystem Fragmentation in the Americas*, Bradshaw, G. & Marquet, P. (eds). Springer, New York, pp. 313–25.

Laurance, S. G., Stouffer, P. & Laurance, W. (2004). Effects of road clearings on movement patterns of understory rainforest birds in central Amazonia. *Conservation Biology* **18**: 1099–109.

Laurance, W. F. (1990). Comparative responses of five arboreal marsupials to tropical forest fragmentation. *Journal of Mammalogy* **71**: 641–53.

Laurance, W. F. (1991a). Edge effects in tropical forest fragments: application of a model for the design of nature reserves. *Biological Conservation* **57**: 205–19.

Laurance, W. F. (1991b). Ecological correlates of extinction proneness in Australian tropical rainforest mammals. *Conservation Biology* **5**: 79–89.

Laurance, W. F. (1994). Rainforest fragmentation and the structure of small mammal communities in tropical Queensland. *Biological Conservation* **69**: 23–32.

Laurance, W. F. (1997a). Responses of mammals to rainforest fragmentation in tropical Queensland: A review and synthesis. *Wildlife Research* **24**: 603–12.

Laurance, W. F. (1997b). Hyper-disturbed parks: edge effects and ecology of isolated rainforest reserves in tropical Australia. In *Tropical Forest Remnants: Ecology, Management and Conservation of Fragmented Communities*, Laurance, W. F. & Bierregaard, R. O. (eds). University of Chicago Press, Chicago, pp. 71–83.

Laurance, W. F. (2004). Forest-climate interactions in fragmented tropical landscapes. *Philosophical Transactions of the Royal Society, Series B* **359**: 345–52.

Laurance, W. F. & Bierregaard, R. O. (eds) (1997). *Tropical Forest Remnants: Ecology, Management and Conservation of Fragmented Communities*. University of Chicago Press, Chicago.

Laurance, W. F., Alonso, A., Lee, M. & Campbell, P. (2006). Challenges for forest conservation in Gabon, central Africa. *Futures* **38**: 454–70.

Laurance, W. F. & Cochrane, M. A. (2001). Synergistic effects in fragmented landscapes. *Conservation Biology* **15**: 1488–9.

Laurance, W. F., Cochrane, M., Bergen, S. *et al.* (2001a). The future of the Brazilian Amazon. *Science* **291**: 438–9.

Laurance, W. F., Ferreira, L., Rankin-de Merona, J. & Laurance, S. G. (1998). Rain forest fragmentation and the dynamics of Amazonian tree communities. *Ecology* **79**: 2032–40.

Laurance, W. F., Gascon, C. & Rankin-de Merona, J. (1999). Predicting effects of habitat destruction on plant communities: a test of a model using Amazonian trees. *Ecological Applications* 9: 548–54.

Laurance, W. F., Laurance, S.G., Ferreira, L., Rankin-de Merona, J., Gascon, C. and Lovejoy, T. (1997). Biomass collapse in Amazonian forest fragments. *Science* 78: 1117–18.

Laurance, W. F., Lovejoy, T., Vasconcelos, H. *et al.* (2002). Ecosystem decay of Amazonian forest fragments: a 22-year investigation. *Conservation Biology* 16: 605–18.

Laurance, W. F., McDonald, K. R & Speare, R. (1996). Epidemic disease and the catastrophic decline of Australian rain forest frogs. *Conservation Biology* 10: 406–13.

Laurance, W. F. & Peres, C. A. (eds) (2006). *Emerging Threats to Tropical Forests*. University of Chicago Press, Chicago.

Laurance, W. F., Perez-Salicrup, D., Delamonica, P. *et al.* (2001b). Rain forest fragmentation and the structure of Amazonian liana communities. *Ecology* 82: 105–16.

Laurance, W. F. & Yensen, E. (1991). Predicting the impacts of edge effects in fragmented habitats. *Biological Conservation* 55: 77–92.

Lawson, T. (2004). Landscape ecology of riparian vegetation in the Mossman Catchment. BApplSc (Hons) thesis, James Cook University, Cairns.

Lima, M. G.& Gascon, C. (1999). The conservation value of linear forest remnants in central Amazonia. *Biological Conservation* 91: 241–7.

Lovejoy, T. E., Bierregaard, R.O., Rylands, A. *et al.* (1986). Edge and other effects of isolation on Amazon forest fragments. In *Conservation Biology: The Science of Scarcity and Diversity*, M. E. Soulé (ed.). Sinauer, Sunderland, MA, pp. 257–85.

MacArthur, R. O.& Wilson, E .O. (1967). *The Theory of Island Biogeography*. Princeton University Press, Princeton, NJ.

Malcolm, J. R. (1998). A model of conductive heat flow in forest edges and fragmented landscapes. *Climatic Change* 39: 487–502.

Maver, N. (2002). The success of ecological rehabilitation in reducing edge effects on rainforest vegetation and microclimate. BApplSc (Hons) thesis, James Cook University, Cairns.

Mesquita, R., Delamonica, P. & Laurance, W. F. (1999). Effects of surrounding vegetation on edge-related tree mortality in Amazonian forest fragments. *Biological Conservation* 91: 129–34.

Murcia, C. (1995). Edge effects in fragmented forests: Implications for conservation. *Trends in Ecology and Evolution* 10: 58–62.

Peres, C. A. (2001). Synergistic effects of subsistence hunting and habitat fragmentation on Amazonian forest vertebrates. *Conservation Biology* 15: 1490–505.

Peres, C. A., Barlow, J. & Laurance, W. F. (2006). Detecting anthropogenic disturbance in tropical forests. *Trends in Ecology and Evolution* 21: 227–9.

Peres, C. A. & Michalski, A. (2006). Synergistic effects of habitat disturbance and hunting on Amazonian forest fragments. In *Emerging Threats to Tropical Forests*, Laurance, W. F. & Peres, C. A. (eds). University of Chicago Press, Chicago, pp. 105–26.

Pohlman, C. (2006). Internal fragmentation in the rainforest: edge effects of highways, powerlines and watercourses on tropical rainforest understorey microclimate, vegetation structure and composition, physical disturbance and seedling regeneration. PhD thesis, James Cook University, Cairns.

Pohlman, C., Turton, S. & Goosem, M. (2007). Edge effects of linear canopy openings on tropical rainforest understory microclimate. *Biotropica* 39: 62–71.

Reynolds, B. E. (1994). A study of weed (*Lantana camara*) infestation of tracks through rainforest on the Atherton Tableland, north Queensland. BA (Hons) thesis, James Cook University, Cairns, Queensland.

Saunders, D. S. & Hobbs, R. J. (eds) (1991). *Nature Conservation 2: The Role of Corridors*. Surrey Beatty, Chipping Norton, NSW.

Siegenthaler, S., Jackes, B., Turton, S. & Goosem, M. (2000). Edge effects of roads and powerline clearings on rainforest vegetation. In *Impacts of Roads and Powerlines on the Wet Tropics World Heritage Area II*, Goosem, M. W. & Turton, S. M. (Eds.). Rainforest CRC, Cairns, pp. 46–64.

Siegenthaler, S. & Turton, S. (2000). Edge effects of roads and powerline clearings on microclimate. In *Impacts of Roads and Powerlines on the Wet Tropics World Heritage Area II*, Goosem, M. & Turton, S. (eds). Rainforest CRC, Cairns, pp. 20–43.

Simberloff, D. & Cox, J. (1987). Consequences and costs of conservation corridors. *Conservation Biology* 1: 63–71.

Souza, de, O. F. F. & Brown, V. K. (1994). Effects of habitat fragmentation on Amazonian termite communities. *Journal of Tropical Ecology* 10: 197–206.

Thomas, S. C. (2004). Ecological correlates of tree species persistence in tropical forest fragments. In *Tropical Forest Diversity and Dynamism: Findings From a Large-scale Plot Network*, Losos, E. & Leigh, E. (eds). University of Chicago Press, Chicago, pp. 279–313.

Tucker, N. I. J. (2000). Linkage restoration: Interpreting fragmentation theory for the design of a rainforest linkage in the humid Wet Tropics of north-eastern Queensland. *Ecological Management and Restoration* 1: 35–41.

Turton, S. M. & Freiburger, H. (1997). Edge and aspect effects on the microclimate of a small tropical forest remnant on the Atherton Tableland, Northeastern Australia. In *Tropical Forest Remnants: Ecology, Management, and Conservation of Fragmented Communities*, Laurance, W. F. & Bierregaard, R. O. (eds). University of Chicago Press, Chicago, pp. 45–54.

Viana, V. M., Tabanez, A. & Batista, J. (1997). Dynamics and restoration of forest fragments in the Brazilian Atlantic Moist Forest. In *Tropical Forest Remnants: Ecology, Management, and Conservation of Fragmented Communities*, Laurance, W. F. & Bierregaard, R. O. (eds). University of Chicago Press, Chicago, pp. 351–65.

Warburton, N. H. (1997). Structure and conservation of forest avifauna in isolated rainforest remnants in tropical Australia. In *Tropical Forest Remnants: Ecology, Management, and Conservation of Fragmented Communities*, Laurance, W. F. & Bierregaard, R. O. (eds). University of Chicago Press, Chicago, pp. 190–206.

Werren, G. L. (2001). *Environmental Weeds of the Wet Tropics Bioregion: Risk Assessment and Priority Ranking*. Rainforest CRC, Cairns.

Wilcove, D. S. (1985). Nest predation in forest tracts and the decline of migratory songbirds. *Ecology* **66**: 1211–14.

Williams, S. E. & Hilbert, D. (2006). Climate change as a threat to the biodiversity of tropical rainforests in Australia. In *Tropical Forest Remnants: Ecology, Management, and Conservation of Fragmented Communities*, Laurance, W. F. & Bierregaard, R. O. (eds). University of Chicago Press, Chicago, pp. 33–52.

Williams, S. E., Bolitho, E. & Fox, S. (2003). Climate change in Australian tropical rainforests: an impending environmental catastrophe. *Proceedings of the Royal Society of London, Series B* **270**: 1887–92.

Wilson, R. (2000). The impact of anthropogenic disturbance on four species of arboreal, folivorous possums in the rainforests of north-east Queensland, Australia. PhD thesis, James Cook University, Cairns.

Wilson, R. & Goosem, M. (2007). *Vehicle Headlight and Streetlight Disturbance to Wildlife – Kuranda Range Road Upgrade Project*. Report to Queensland Department of Main Roads, Rainforest CRC, Cairns. 66 pp.

Winter, J. W., Bell, F. C., Pahl, L. I. & Atherton, R. G. (1987). Rainforest clearfelling in northeastern Australia. *Proceedings of the Royal Society of Queensland* **98**: 41–57.

WTMA (2004). *Wet Tropics Management Authority Annual Report and State of the Wet Tropics 2003–2004*. Wet Tropics Management Authority, Cairns.

24 INVASIVE WEEDS IN THE WET TROPICS

*Stephen Goosem**

Wet Tropics Management Authority, Cairns, Queensland, Australia
*The author was participant of Cooperative Research Centre for Tropical Ecology and Management

Introduction

This chapter deals with invasive plants, defined as introduced species capable of establishing, or considered to have a high probability of establishing, self-sustaining populations by invading native communities or ecosystems, and capable of causing major modification to species richness, abundance or ecosystem function. Consistent with this definition, invasive plants can be distinguished by the following five criteria (see e.g. McDonald *et al*. 1989; Cronk & Fuller 1995; Pysek 1995). They are:

- alien (non-native or exotic) species;
- introduced intentionally or accidentally by humans to areas where they never previously occurred;
- naturalized in this new area, whereby they are capable of creating self-sustaining populations;
- expanding their distribution and/or increasing their abundance;
- occurring in natural and semi-natural habitats.

Although some plant invasions can be attributed to natural dispersal processes and do occur without human assistance, it is the combination of the rate and magnitude, as well as the distances and agency involved, that separates human-driven invasion processes from self-perpetuated colonization events. While many of the exotic plants that have been introduced by intention or accident have not naturalized to any extent and can be considered ecologically benign, a significant proportion have become rampant environmental weeds to the extent that it is now widely accepted that biological invasions are the second most important threat to biodiversity, after direct habitat transformation (Vitousek *et al*. 1997; McNeely *et al*. 2001). Although the magnitudes of impacts of most invasive species are difficult to quantify (Blossey *et al*. 2001) and our ability to contrast impacts attributable to invasions with those due to other causes is problematic, it is becoming increasingly important to set objective priorities for managing invasive species based on their environmental impacts.

The loss and degradation of natural habitats that has occurred throughout the world has made it easier for alien species to establish and become invasive because many successful invading plants are colonizing species that benefit from the reduced competition that follows habitat degradation. The high rates of global habitat and species loss and degradation, in conjunction with biological invasions, put ecosystems under enormous stress, making it imperative that we understand how the loss of native species or the addition of a novel species influences the stability and function of ecosystems.

Table 24.1 Numbers of invasive and native plant species in various countries/areas

Country/area	Number of native species	Number of invasive species	Invasives as a percentage of wild plant species	Source
South Georgia	26	54	207.7	McNeely 2001
Tristan de Cunha	70	97	138.6	McNeely 2001
Hawaii	956	861	90.1	Wagner et al. 1990
New Zealand	1790	1570	87.7	Heywood 1989
Campbell Island	128	81	63.3	McNeely 2001
Australia	15638	2700	17.3	Low 1999
Queensland	9603	1298	13.5	EPA 2002
Wet Tropics of Queensland	4664	530	11.4	Werren 2001

Another more subtle danger from the ubiquity and magnitude of alien species naturalizations (e.g. Table 24.1) is their contribution to global species homogenization, where the unique is replaced by the commonplace. This process has been termed biological pollution (Luken & Thieret 1996) or the McDonaldization of world ecology (Low 1999). We have created a situation in which a tremendous mixing of plant species is occurring to the extent that it has become an important component of human-induced global change that decreases the local distinctiveness of floras and breaks down the geographical isolation that promotes and maintains global biodiversity.

More than 2700 weeds have become established in Australia so far (Table 24.1) at a cost to the economy of more than AUD$3 billion, and each year another ten take root (Weeds CRC 2005). Weeds now make up 17% of Australia's wild plant species (Low 1999). It needs to be emphasized that the history of European settlement in Australia is shorter than the lifetime of many trees, so Australia is therefore still in the first stages of invasion. The highly restricted nature of many endemic plant species within the Wet Tropics biogeographical region (Werren et al. 1995; Goosem et al. 1999) renders this regional flora particularly vulnerable to the threat posed by alien plant invasion.

Characteristics of invasive plants

An important distinction between types of invasive plants needs to be made. Some may only be significant in areas that are already modified and will not be important in, for example, continuous rainforest, whereas others may be a problem in both modified and unmodified habitat where disturbance may simply accelerate the rate of colonization.

Invasive plants do not constitute a separate biological/ecological category of organisms. However, they do have certain characteristics that allow them to invade new areas rapidly and compete with native plants for available light, water and nutrients (e.g. Binggeli et al. 1998; Westbrooks 1998; Werren 2001). Pointers to invasiveness in tropical environments include:
- rapid growth rate;
- early sexual maturity;
- long flowering and fruiting periods;
- high reproductive capacity by seeds and/or vegetative structures;
- effective dispersal;
- high numbers of invading individuals;
- large soil seed bank reservoirs;
- long viable seed persistence in the soil;
- seed dormancy that ensures periodic germination and prevents seedlings emerging during unfavourable conditions;
- allelopathic suppression of the germination and growth of other plant species;
- tolerance to a wide range of ecological conditions, especially for germination and growth;
- no special seed germination requirements.

Characteristics of Wet Tropics rainforest

Undisturbed rainforest is generally considered to be resistant to weed invasion and large intact stands of rainforest are, by and large, devoid of introduced plants except around their perimeter (e.g. Humphries & Stanton 1992; Werren 2001). Unfortunately, rainforest regeneration cycles are driven by disturbance. At any point in time, rainforests are a mosaic of patches at different stages of maturity. The nature of the stage in the cycle (and the inherent susceptibility of rainforest to weed invasion) depends on the size of the gap and the environment within it, which in turn is related to the intensity, scale and frequency of disturbance. In the absence of exotic invasive species, native ecosystems benefit from natural disturbances that provide regenerative diversity as well as temporal and spatial mosaics of successional communities.

Although recovery of rainforest following disturbance is usually treated as an example of secondary succession and although various categorizations of species into successional response groups have been proposed, there are basically only two qualitatively distinct groups, pioneer and non-pioneer species (Goosem 2003). Pioneers have seeds that only germinate in gaps large enough for sunlight to reach the ground for at least part of the day and require high irradiance levels for seedling establishment and growth. Seedlings and young plants of these species are never found under a closed canopy. This behaviour contrasts dramatically with non-pioneer species, which have seeds that can germinate under shade and whose seedlings can establish in shade. Some are capable of persisting in shade for prolonged periods, although most require increased light conditions through gap formation for their longer-term survival. Non-pioneer species are not restricted to below a closed canopy and may also exist with pioneers in large gaps. Within both groups there is variation in the response of species to a number of environmental factors, but pioneer and non-pioneer species are distinguished absolutely on the basis of their germination and seedling characteristics. Another distinguishing characteristic is that pioneer species form seed banks, whereas non-pioneer rainforest species generally form seedling banks having no, or a very limited, seed-dormancy phase (Goosem 2003). Similarly, there are two basic ecological categories of rainforest weeds: those that behave like rainforest pioneers and a smaller group that act like non-pioneers.

There are only about 20 species of native rainforest pioneer trees in the Wet Tropics, most of which have very wide geographical distributions. The relatively low proportion of pioneer tree species is a general feature of tropical rainforests, which are generally very resilient to small-scale natural disturbances but are not well equipped to handle large-scale artificial disturbances, having a generally impoverished pioneer flora with which to respond (Goosem 2003).

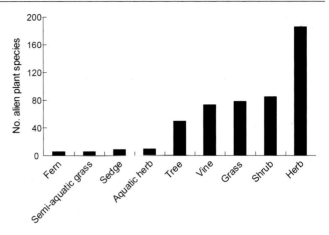

Figure 24.1. Numbers and life-forms of naturalized Wet Tropics alien plant species.

The rainforests of the Wet Tropics may also be vulnerable to invasion because they have relatively low numbers of native species in certain distinctive plant groups (Werren 2003). The rainforest flora of the Wet Tropics is relatively depauperate in shrubs and ground-cover herbs, for example. This characteristic may result in empty niches that new arrivals can exploit. It is interesting that these two life-form groups are the two most well-represented groups in the naturalized alien plant flora of the Wet Tropics (Figure 24.1).

Two other quintessential characteristics of rainforests that are relevant to the movement of invasive weed species across rainforest landscapes are: (i) the high proportion of native fleshy-fruited plant species reliant on animals for dispersal; and (ii) the general lack of wind below intact rainforest canopies (Goosem 2003). These features have important implications concerning both the mechanisms and pattern of potential weed invasion of rainforest landscapes. An understanding of the behaviour of the native and feral seed dispersal guilds, and their spatial and temporal interactions with the disturbed and intact parts of the landscape, is important (Westcott & Dennis 2003). Similarly, it is important to understand how air movement patterns and intensities are influenced by the configuration and intensity of clearing. In particular, the altered air movement patterns along cleared linear infrastructure corridors through rainforests may influence the spread of wind dispersed weeds and the flight patterns and behaviour of seed-dispersing birds and bats (Goosem 2003). The relative susceptibility of rainforest landscapes to invasion by weed species is therefore closely aligned not only to the pattern, extent, frequency and history of rainforest disturbance and land use but also to the demographics and behaviour of rainforest seed dispersers.

Although the Wet Tropics is a very small biogeographical region, it is characterized by a wide range of climates (encompassing tropical, subtropical, temperate and monsoon climatic zones), a wide range of geologies and soil types and a long and favourable growing season, making the region susceptible to invasion by a very broad range of alien plant species originating from a large range of climatic zones and environmental conditions.

Land use, disturbance and climate are driving factors in weed invasion, with most weeds currently confined to human-disturbed parts of the landscape (Humphries & Stanton 1992) even though pristine rainforest environments are not immune from the threat of weed invasion. Community infrastructure corridors such as powerline clearings and roads represent the primary conduits for the introduction of weed species into large rainforested areas and for weed propagule flow through these landscapes. Fortunately, most generalist weed species are incapable of colonizing less disturbed rainforest environments and confine themselves to the gross disturbance footprint of the clearing. Even so, this network of linear community infrastructure may still serve as sources of propagules that progressively percolate into disturbed rainforest edges or are liberated into interiors of pristine rainforest following natural disturbance events.

Werren (2001) compiled a comprehensive list of over 500 naturalized alien plant species for the Wet Tropics of Queensland. The naturalized flora includes a great variety of growth forms, including ferns, grasses, sedges, herbs, aquatic and semi-aquatic plants, vines, shrubs and trees (Goosem 2003). In terms of the total numbers of species represented in the region's naturalized flora, herbs, shrubs, grasses, vines and trees are the most important groups (Figure 24.1).

Wet Tropics invasive plant species priorities

The number of known naturalized alien plant species in the Wet Tropics region has grown rapidly over the past 50 years, with 200 new weed species having been identified in the past decade alone (Figure 24.2).

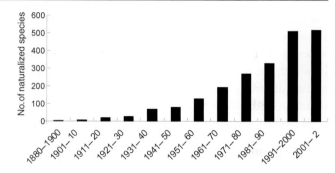

Figure 24.2 Cumulative total naturalized plant species in the Wet Tropics per decade from 1880 to 2002.

The majority of plant species that have become weeds have been intentionally introduced into the region for agricultural, horticultural or domestic purposes.

Werren (2001) devised a risk assessment and ranking system for prioritizing these naturalized species, tailored specifically for wet tropical environments. Assessment criteria included gross invasiveness (ecesis/establishment in native vegetation, including to what extent a species is advantaged by disturbance), a range of intrinsic attributes predisposing species to invasiveness (i.e. reproductive capacity/mode, dispersal capacity, competitive ability and ecesis needs) and extrinsic factors (including ecosystems/species at risk). The Wet Tropics Management Authority (WTMA 2004) subsequently modified the Werren (2001) priority ranking system by incorporating management feasibility criteria with an emphasis on the prevention or eradication of newly emerging, highly invasive species and the eradication of new outbreaks of certain longer-established species, in conjunction with containment of their larger established outbreaks. Table 24.2 is a shortlist of WTMA's priority invasive plant species. Some have already demonstrated their invasive potential in the Wet Tropics, while others are major problems in other tropical areas and are considered to have a high potential to spread here.

The combination of similarities in traits among invasive plant species, the ubiquity of species movements across the globe and the general commonalities in characteristics of receiving environments should lead to common suites of invasive species in areas of the globe experiencing similar climates. This is indeed the case when comparisons are made between Australia's

Table 24.2 Wet Tropics invasive environmental weed priorities

Priority	Common name	Species	Life-form	Origins and uses
Newly emerging weeds to be eradicated completely				
1	Miconia	Miconia calvescens	Tree	Central American ornamental
1	Koster's curse	Clidemia hirta	Shrub	Tropical American ornamental
1	Siam weed	Chromolaena odorata	Shrub, vine	Tropical American, accidental arrival
1	Hiptage	Hiptage benghalensis	Vine	SE Asian & Indian ornamental
1	Mikania	Mikania micrantha	Vine	Tropical American accidental invader
2	Miconia	Miconia racemosa	Shrub	Tropical American ornamental
2	Alligator weed	Alternanthera philoxeroides	Aquatic herb	South American
2	Brillantaisia	Brillantaisia lamium	Herb, shrub	African ornamental
2	Mistflower	Ageratina riparia	Herb	Mexican and West Indian
2	Limnocharis	Limnocharis flava	Aquatic herb	South American ornamental & food
2	White trumpet	Thunbergia laurifolia	Vine	Indian ornamental
2	Venezuelan pokeweed	Phytolacca rivinoides	Shrub	Tropical American ornamental
2	Coffee	Coffea arabica & Coffea liberica	Shrub, tree	Ethiopian horticultural
3	Gamba grass	Andropogon gayanus	Grass	Tropical African pasture grass
3	Aleman grass	Echinochloa polystachya	Grass	Tropical American ponded pasture
3	Panama rubber	Castilla elastica	Tree	Central American rubber tree
3	Indian plum	Flacourtia jangomas	Tree	Indian fruit tree
3	Cucumber tree	Parmentiera aculeata	Tree, vine	Asian food plant
3	Ceara rubber tree	Manihot glaziovii	Tree	Brazilian rubber tree
Potential weeds: prevent introduction and eradicate if introduced				
3	Mimosa	Mimosa pigra	Shrub	Tropical American
3	Ivy gourd	Coccinia grandis	Vine	Asian & African ornamental
3	Spiked pepper	Piper aduncum	Shrub, tree	Tropical American ornamental
3	Pagoda plant	Clerodendrum paniculatum	Shrub	Asian ornamental
3	Bronze leaved clerodendrum	Clerodendrum quadriloculare	Shrub	Philippine ornamental
3	Chandelier bush	Medinilla magnifica	Shrub	Philippine ornamental
3	Eared pepper	Piper auritum	Shrub	Tropical American ornamental & drug
New infestations to be eradicated and larger, established outbreaks to be contained				
4	Hymenachne	Hymenachne amplexicaulis	Aquatic grass	Tropical American ponded pasture
4	Pond apple	Annona glabra	Tree	American & African horticultural
4	Lantana	Lantana camara	Shrub	Tropical American garden plant
4	Leucaena,	Leucaena leucocephala	Tree, shrub	Central American fodder crop
5	Ardisia	Ardisia solanacea	Shrub	Sri Lankan
5	Para grass	Brachiaria mutica	Grass	American, African ponded pasture
5	Cabomba	Cabomba caroliniana	Aquatic herb	North American aquarium plant
5	Water hyacinth	Eichhornia crassipes	Aquatic herb	Tropical America ornamental

(Continued)

Table 24.2 (*Continued*)

Priority	Common name	Species	Life-form	Origins and uses
5	Harungana	*Harungana madagascariensis*	Tree	Tropical African ornamental
5	Woolly morning glory	*Argyreia nervosa*	Vine	Indian ornamental and medicinal
5	Salvinia	*Salvinia molesta*	Aquatic fern	South American aquarium plant
5	African tulip	*Spathodea campanulata*	Tree	West African, ornamental
5	Singapore daisy	*Sphagneticola trilobata*	Herb	Central American garden plant
5	Blue trumpet vine	*Thunbergia grandiflora*	Vine	Indian ornamental
5	Molasses grass	*Melinis minutiflora*	Grass	Tropical African pasture grass
6	Guinea grass	*Megathyrsus maximus*	Grass	African pasture grass
6	Water lettuce	*Pistia stratiotes*	Aquatic herb	World Tropics
6	Guava	*Psidium guajava*	Tree, shrub	Tropical American fruit tree
6	Turbina	*Turbina corymbosa*	Vine	Tropical American
7	Arrowhead	*Sagittaria graminea*	Aquatic herb	North American aquatic herb
7	Broad-leaved pepper tree	*Schinus terebinthifolius*	Tree	Brazilian ornamental
7	Elodea	*Elodea canadensis*	Aquatic herb	Canadian
7	Balloon vine	*Cardiospermum halicacabum*	Vine	North & tropical American
7	Brazilian nightshade	*Solanum seaforthianum*	Vine	Tropical American
7	Arrowhead vine	*Syngonium podophyllum*	Vine	South American garden plant
8	Yellow allamanda	*Allamanda cathartica*	Vine, shrub	South American ornamental
8	Glowvine	*Saritaea magnifica*	Vine	South American ornamental
8	Coral tree	*Erythrina x sykesii*	Tree	Indian fodder tree
8	Camphor laurel	*Cinnamomum camphora*	Tree	Asian, ornamental and windbreak
8	East Indian mahogany	*Chukrasia velutina*	Tree	Asian, commercial timber
8	White teak	*Gmelina arborea*	Tree	Asian commercial timber
8	Elephant grass	*Pennisetum purpureum*	Grass	African pasture grass
8	Coral berry	*Rivina humilis*	Herb	Tropical and south American

Source: WTMA (2004).

Wet Tropics region and other islands and areas within the South Pacific region.

Invasive plants of the South Pacific region

The greater susceptibility of island habitats to alien invasions is well documented (Heywood 1989; Rejmánek 1989; Lõvei 1997; Sax & Brown 2000). Various authors have commented on the natural insularity of Australian rainforests, employing phrases such as 'islands in a sea of sclerophyll vegetation' (Webb & Tracey 1981: 653) or 'islands of green in a land of fire' (Bowman 2000). Increased susceptibility to plant invasion appears to be a recurring feature associated with such a spatial configuration.

Meyer (2000) undertook a detailed review of invasive plant species across 15 Pacific islands or groups of islands that included:
- Polynesia: French Polynesia, Cook Islands, Niue, Pitcairn, Tonga, American Samoa and Samoa, Wallis and Futuna.
- Micronesia: Federated States of Micronesia, Guam, Nauru, Northern Mariana Islands and Palau.

- Melanesia: Fiji, New Caledonia and Vanuatu.

More than 30 invasive plant species considered to be serious threats to the native habitats of these Pacific islands were identified (Meyer 2000). Some of the key aggressive invasive species identified as occurring in many of these South Pacific countries include:

- Trees and shrubs: African tulip tree (*Spathodea campanulata*), leucaena (*Leucaena leucocephala*), guava (*Psidium guajava*), velvet tree (*Miconia calvescens*), Koster's curse (*Clidemia hirta*), lantana (*Lantana camara*), giant sensitive plant (*Mimosa diplotricha*).
- Vines: mile-a-minute (*Mikania micrantha*), passion-fruit species (*Passiflora* spp.), Cook's glory (*Merremia peltata*).
- Grasses: elephant grass (Pennisetum purpureum), paspalum (Paspalum urvillei).
- Aquatic plants: water hyacinth (*Eichhornia crassipes*).

Other significant aggressive species found on most of the islands included the trees *Albizia* spp. (especially *A. lebbeck*), the herbs *Stachytarpheta urticifolia* and *S. jamaicensis* and the tree *Acacia farnesiana*. Although many others exist, the above species were considered to be significant threats to conservation values in at least three, and often many more, Pacific island countries.

Among the dominant invasive taxa that are found in some Pacific islands only *Clidemia hirta* is highly invasive in the Mascarenes and widely naturalized in Malaysia; *Melinis minutiflora* is invasive in the Galapagos and the Ascension Islands; and *Ardisia elliptica* is an aggressive invader in the Mascarenes and the Seychelles.

Meyer (2000) also identified several serious invasive plants that penetrate montane rainforests on several islands, including the shrubs *Clidemia hirta* and *Cestrum* spp., and a number of very aggressive species presently found only on one or a few islands such as *Castilla elastica* in the Samoas, *Cordia alliodora* in Vanuatu, *Myrica faya* in Hawaii, *Piper aduncum* in Fiji, *Timonius timon* in Palau, *Cardiospermum grandiflorum* in the Cook Islands and *Miconia calvescens* in French Polynesia and Hawaii. *Miconia calvescens*, also called the purple plague in Hawaii or the green cancer in Tahiti, is now considered to be 'by far the most effective and destructive competitor of all native and established wet forest plants' in the Pacific islands

(Mueller-Dombois & Fosberg 1998: 412) and 'the worst of all exotic escapees' (Whittaker 1998: 244). *Miconia calvescens* is one of many significant invasive plant species that occur both in the Wet Tropics and many South Pacific islands (Table 24.3).

Ecological impact of invasive plants

Two types of plant invaders are particularly likely to have major environmental impacts: (i) species that constitute novel habitats; and (ii) species that alter ecosystem processes. Of course, some plant species both create novel habitats and alter ecosystem properties. Because physical habitat and ecosystem processes are fundamental to all species in a natural community, the cascade of impacts caused by such invaders can affect entire suites of plants and animals. Ecological changes resulting from plant invasions can be on composition, structure or ecosystem processes and commonly include:

- reductions in biodiversity;
- increases in dominance (e.g. monotypic stands);
- decreases in overall species richness;
- decreases in structural diversity;
- decreases in spatial heterogeneity;
- aggressive competition with native species;
- replacement (or displacement) of native species leading to the loss of one, several or all native populations;
- prevention of seedling establishment of native species;
- reduction in the amount of space, water, sunlight and nutrients available to native species;
- increases in erosion along stream banks and roadsides;
- changes to the characteristics of soil structure and chemistry;
- alterations to hydrological flows and conditions;
- loss of habitat for native wildlife;
- loss of food sources for native wildlife;
- changes to natural ecological processes such as plant community succession;
- disruptions to native plant–animal associations (e.g. pollination and seed dispersal);
- alterations to ecosystem-level processes (e.g. water or fire regimes, nutrient cycling patterns, light quantity and quality).

Table 24.3 Significant invasive plant species shared between South Pacific islands and the Wet Tropics

Family	Species	Common name	Pacific class[*]	Wet Tropics class[†]
Acanthaceae	*Brillantaisia lamium*	brillantaisia		2
Acanthaceae	*Sanchezia parvibracteata*		P	
Acanthaceae	*Thunbergia grandiflora*	blue clock vine	P	5
Acanthaceae	*Thunbergia laurifolia*	clock laurel vine		2
Alismataceae	*Limnocharis flava*	Limnocharis		2
Alismataceae	*Sagittaria graminea*	arrowhead		8
Amaranthaceae	*Alternanthera philoxeroides*	alligator weed		2
Anacardiaceae	*Schinus terebinthifolius*	broad-leaved pepper tree	S	8
Annonaceae	*Annona glabra*	pond apple	S	4
Apocynaceae	*Allamanda cathartica*	yellow allamanda		9
Araceae	*Pistia stratiotes*	water lettuce		7
Araceae	*Syngonium podophyllum*	syngonium		8
Asteraceae	*Ageratina riparia*	mistflower		2
Asteraceae	*Chromolaena odorata*	siam weed	D	1
Asteraceae	*Mikania micrantha*	mile a minute	D	1
Asteraceae	*Sphagneticola trilobata*	singapore daisy	D	5
Asteraceae	*Tithonia diversifolia*	japanese sunflower	P	
Bignoniaceae	*Parmentiera aculeata*	cucumber tree		2a
Bignoniaceae	*Saritaea magnifica*	glow vine		9
Bignoniaceae	*Spathodea campanulata*	african tulip tree	D	5
Cabombaceae	*Cabomba caroliniana*	cabomba		5
Caesalpiniaceae	*Caesalpinia decapetala*	mysore thorn	P	
Cannaceae	*Canna indica*	indian shot	S	
Caprifoliaceae	*Lonicera japonica*	japanese honeysuckle	P	
Clusiaceae	*Harungana madagascariensis*	harungana		5
Convolvulaceae	*Argyreia nervosa*	woolly morning-glory		5
Convolvulaceae	*Turbina corymbose*	turbine vine		7
Cucurbitaceae	*Coccinia grandis*	ivy gourd	D	3
Euphorbiaceae	*Manihot glaziovii*	ceara rubbertree		2a
Fabaceae	*Erythrina x sykesii*	coral tree		9
Fabaceae	*Pueraria lobata*	Kudzu	P	
Flacourtiaceae	*Flacourtia jangomas*	Indian plum		2a
Hydrocharitaceae	*Elodea Canadensis*	elodea		8
Lamiaceae	*Clerodendrum paniculatum*	pagoda flower	D	3
Lamiaceae	*Clerodendrum quadriloculare*	Bronze leaved clerodendrum	D	3
Lamiaceae	*Gmelina arborea*	white teak		9
Lauraceae	*Cinnamomum camphora*	camphor laurel	P	9
Malpighiaceae	*Hiptage benghalensis*	hiptage	P	1

(Continued)

Table 24.3 *(Continued)*

Family	Species	Common name	Pacific class*	Wet Tropics class†
Melastomataceae	*Clidemia hirta*	Koster's curse	D	1
Melastomataceae	*Medinilla magnifica*	chandelier bush		3
Melastomataceae	*Miconia calvescens*	miconia	D	1
Melastomataceae	*Miconia racemosa*	miconia		2
Meliaceae	*Chukrasia velutina*	East Indian mahogany		9
Mimosaceae	*Acacia farnesiana*	cassie flower	D	
Mimosaceae	*Leucaena leucocephala*	leucaena	D	4
Mimosaceae	*Mimosa diplotricha*	giant sensitive plant	D	
Mimosaceae	*Mimosa pigra*	mimosa		3
Moraceae	*Castilla elastica*	panama rubber	S	2a
Myrsinaceae	*Ardisia solanacea*	shoe button ardisia	D	5
Myrtaceae	*Psidium guajava*	yellow guava	D	7
Myrtaceae	*Syzygium cumini*	java plum	D	
Oleaceae	*Ligustrum sinense*	chinese privet	P	
Passifloraceae	*Passiflora foetida*	stinking passion flower	D	
Passifloraceae	*Passiflora laurifolia*	yellow granadilla	D	
Passifloraceae	*Passiflora quadrangularis*	granadilla	D	
Phytolaccaceae	*Phytolacca rivinoides*	Venezuelan pokeweed		2
Phytolaccaceae	*Rivina humilis*	coral berry		9
Piperaceae	*Piper aduncum*	spiked pepper	S	3
Piperaceae	*Piper auritum*	eared pepper		3
Poaceae	*Andropogon gayanus*	gamba grass		2a
Poaceae	*Arundo donax var. donax*	giant reed	P	
Poaceae	*Brachiaria mutica*	para grass		5
Poaceae	*Echinochloa polystachya*	Aleman grass		2a
Poaceae	*Hymenachne amplexicaulis*	hymenachne		4
Poaceae	*Megathyrsus maximus*	guinea grass	D	7
Poaceae	*Melinis minutiflora*	molasses grass	D	5
Poaceae	*Paspalum conjugatum*	sourgrass	D	
Poaceae	*Paspalum urvillei*	paspalum	D	
Poaceae	*Pennisetum clandestinum*	kikuyu grass	D	
Poaceae	*Pennisetum purpureum*	elephant grass	D	9
Poaceae	*Sorghum halepense*	johnson grass	D	
Pontederiaceae	*Eichhornia crassipes*	water hyacinth	D	5
Rhamnaceae	*Ziziphus mauritiana*	chinee apple	P	
Rosaceae	*Rubus alceifolius*	giant bramble	D	
Rubiaceae	*Coffea arabica*	coffee		2
Rubiaceae	*Coffea liberica*	liberian coffee		2

(Continued)

Table 24.3 (*Continued*)

Family	Species	Common name	Pacific class[*]	Wet Tropics class[†]
Salviniaceae	*Salvinia molesta*	salvinia	S	5
Sapindaceae	*Cardiospermum halicacabum*	balloon vine	S	8
Solanaceae	*Solanum mauritianum*	wild tobacco bush	P	
Solanaceae	*Solanum seaforthianum*	brazilian nightshade		8
Verbenaceae	*Lantana camara*	lantana	D	4
Verbenaceae	*Stachytarpheta jamaicensis*	blue snakeweed	D	
Zingiberaceae	*Hedychium coronarium*	butterfly flower	D	

[*]Pacific class: D, dominant plant invader in Pacific Islands; S, significant invader of some Pacific Islands; P, potential or moderate invader in the Pacific Islands.
[†]Wet Tropics class: 1, 2, 2a, high priority weeds (newly emerging weeds to be eradicated completely); 3, high priority potential weeds (prevent introduction and eradicate if introduced); 4, 5, high priority weeds (new infestations to be eradicated and larger, established outbreaks to be contained); 7, 8, 9, priority weeds, new infestations to be eradicated and larger, established outbreaks to be contained.

Invasive species are often noted for their tendency to alter species richness within invaded communities (Rejmánek & Rosén 1992; Hager & McCoy 1998; Parker *et al.* 1999; Meiners *et al.* 2001). Proposed mechanisms explaining invader effects on community diversity can be divided according to the demographic stage that is impacted by the invasion. The invader can affect species already established within the community (native species displacement), suppress the establishment of new native individuals (establishment limitation) or do both.

Native species displacement may occur through resource competition where the invading species is more efficient in securing limiting resources (Schoener 1983; Tilman 1997) or exhibits allelopathic effects that suppress germination or growth of other species (Callaway & Aschehoug 2000).

Establishment limitation can affect a community by reducing the colonization success of native species (Seabloom *et al.* 2003; Hager 2004; Yurkonis & Meiners 2004). This may be a consequence of site saturation by invader propagules (Brown & Fridley 2003) curtailing the rate of establishment by other native species as a result of their occupation of the available germination sites. Where competitive interactions shape community structure, the invasion process may more strongly inhibit colonization of species within the same functional group as the invader (Prieur-Richards *et al.* 2000; Symstad 2000; Fargione *et al.* 2003) due to their similarity in resource requirements.

Heywood (1989) argues that invasion of tropical forests is triggered by widespread disruption or conversion of the primary forest to secondary successional communities. The past history of widespread logging of the region's rainforests, combined with frequent cyclone disturbance, has brought about such a situation in the Wet Tropics. In addition, powerline easements and road networks represent major conduits for the introduction of most non-native species into the region's intact rainforests. Fortunately, most ruderal invasive plant species dominating disturbed roadsides and powerline corridors are incapable of colonizing less-disturbed natural environments. Even so, these disturbance corridors potentially serve as starting points for some invasive species to percolate from the edges to the interiors of pristine or naturally disturbed environments. Roadsides and powerline clearings also act as reservoirs of weed propagules that can be liberated in disturbance events.

High intensity winds associated with cyclones are a frequent natural phenomenon across tropical Australia. Webb (1958), for example, suggested that in north Queensland few areas of rainforest would escape some major cyclone damage for more than 40 years, so that cyclones are a major widespread and intensive natural disturbance agent that could aid in the spread and establishment of weeds. The invasion of parts of the steep inaccessible eastern slopes of the Bellenden Ker Range by *Harungana madagascariensis*, for example, has been attributed to structural damage to the rainforests caused by cyclone Winifred in 1986.

As cyclones move across the coast and decline, they may become intense rain-bearing depressions responsible for astonishingly high rainfall intensities; for example, Cyclone Peter dumped 1947 mm on Mt Bellenden Ker over a 48-hour period on 4 and 5 January 1979 (Adam 1994). Such high intensity rainfall also results in high velocity flows, erosion and flooding disturbance to stream and riparian communities, which is another common entry point for the establishment and conduit for the spread of many weeds (Werren 2003).

Invasions by exotic species are often studied and discussed as if they were a distinct ecological phenomenon (Elton 1958; Dukes & Mooney 1999). However, the basic processes that admit exotic plant species are essentially the same as those that facilitate colonizations by native species. If exotic invaders exhibited unique functional attributes compared to native colonizers, one might also conclude that establishment by the two groups involved different processes. However, a detailed comparison of the attributes of exotic invaders and native colonizers unsurprisingly concluded that the two groups were functionally indistinguishable (Thompson et al. 1995).

Much research effort has been spent exploring the relationship between the species richness of native communities and their susceptibility to invasion by alien species. A commonly espoused hypothesis, which was derived from the pioneering work of Elton (1958), is that species-rich communities resist invasion better than species-poor communities because the higher number of biotic interactions in species-rich communities exclude or restrict the recruitment and/or persistence of new arrivals including alien species. This hypothesis predicts, therefore, that native and alien species richness is negatively correlated. This view has now been challenged with increasing evidence of strong positive relationships between native and alien species richness at large spatial scales. Large-scale correlations studies have found that high diversity is associated with high environmental heterogeneity, which, in turn, does not discriminate between native or alien species and explains the positive correlation between native and alien species richness at larger scales (Rejmánek 1996; Levine & D'Antonio 1999; Lonsdale 1999; Deutschewitz et al. 2003; Stohlgren et al. 2003). Habitat heterogeneity also increases with human activities that in turn interact with the natural environmental variability, resulting in increased heterogeneity at

the same time that native, non-weed species would be removed, opening space for exotic invasive weeds.

The diversity–stability hypothesis suggests that diversity provides a general insurance policy that minimizes the chance of large ecosystem changes in response to environmental change. The larger the number of functionally similar species in a community, the greater is the probability that at least some of these species will survive changes in environment and maintain the current properties of the ecosystem. Chesson and Huntley (1997) showed that diversity cannot be maintained by variation alone. Instead, the maintenance of diversity requires the two following components: (i) the existence of flux or variability in ecosystems; and (ii) populations capable of differentially exploiting this flux or variability. Davis et al. (2000) suggest that invasion is influenced by three major factors:

1 propagule pressure (the number of seeds/fragments, repeated introductions, etc.);

2 characteristics of the introduced species;

3 the invasibility of the new (host) environment.

While the second factor pertains to intrinsic or biological traits of the prospective invader, the third is an emergent property that derives from climate, disturbance regimes and the competitive abilities of the resident native plants. Davis et al. (2000) argue that a plant community becomes more susceptible to invasion whenever there is an increase in the amount of unused resources. Their general theory of invasibility rests on the simple assumption that invading species must have access to resources such as light, nutrients and water, with species enjoying greater success in invading a community if they do not encounter intense competition for these resources from resident species (Davis et al. 1998).

Although many aspects of the fluctuating resource availability theory have been presented previously (e.g. Grime 1974, 1988; Huston 1994; Stohlgren et al. 1999), this is the first theory of invasibility to integrate resource availability, disturbance and fluctuating environmental conditions. Importantly, the theory is mechanistic, invoking a specific ecological process (fluctuating resource availability) to account for observed variation in invasibility due to a wide variety of causes.

An increase in resource availability can occur in two ways: (i) a decline in the use of resources by the resident vegetation; or (ii) an increase in resource supply at a rate faster than the resident vegetation can sequester it. Resource use could decline due to a disturbance that damages or destroys an area of vegetation, reducing light, water and nutrient uptake. A widespread disease, such as *phytophthora* dieback, would also reduce resource uptake. An increase in resource supply could arise in a particularly wet period (increased water supply) or a particularly dry period where drought conditions, if severe enough, cause a pulse of partial community leaf loss or patches of mortality, both of which create gaps in previously closed vegetation; the resulting increased light may increase invasibility, if not during the drought itself then once the drought is over. Whether resource uptake goes down, or supply goes up, there are more resources available and this is when a community is more vulnerable to invasion. This also means that successful species invasions are likely to occur as episodic or irregular events.

Disturbance, whether regular or sporadic, is a natural feature of all ecosystems but is also a feature that facilitates the invasion process by reducing the cover or vigour of native competitors and/or by increasing resource levels. This is especially so when it coincides with ready availability of the invading species' propagules. Consequently, Davis *et al.'s* (2000) fluctuating resource availability hypothesis would predict that:

• environments subject to pronounced fluctuations in resource supply are more susceptible to invasion than comparable systems with more stable resource supplies;

• environments are more susceptible to invasion immediately following abrupt disturbances that cause either an increases or a decrease in resource availability (e.g. following cyclone damage, fires);

• invasibility will be greater when there is a prolonged interval between an increase in resource supply and its eventual recapture by a site's native vegetation;

• a relationship between the species diversity of a plant community and its resilience to invasion does not necessarily exist;

• a relationship between a community's primary productivity and its susceptibility to invasion does not necessarily exist;

• whether or not invasion by an alien species actually occurs in a particular environment also depends on propagule pressure and the attributes of the invading species;

• the intense competition for resources characteristic of productive communities such as an intact rainforest may keep invasibility low most of the time, although these communities may become susceptible to invasion if there are periodic disturbances.

It is interesting that this general theory of invasibility emphasizes the importance of windows of opportunity, which periodically open and close. The range and magnitude of global change drivers that the Earth is presently experiencing is likely to mean that such windows of opportunity are going to be opened wider and for much longer periods than in the past, and in some cases the window may not be able to be closed again.

Global change

Global environmental changes, as the consequence of human activities, affect the environment directly and cumulatively. Climate change is the most notorious of three interrelated components of global environmental change. The others are atmospheric environmental change and land cover change. These changes can be detected as large shifts in average annual temperature, the distribution of land cover and forest types and other variables. Some forms of global change, such as urbanization and forest clearing, can happen within weeks or months, while others are measurable over a span of decades or centuries.

It has been argued that global environmental changes may accelerate species invasions (Dukes & Mooney 1999). Preliminary analyses by Hilbert *et al.* (2001) suggest that the rainforests of the Wet Tropics are extremely sensitive to climate change. Some authors have predicted that the increase in atmospheric CO_2 will favour invasions by certain species by increasing soil water availability owing to more efficient use of water by the resident plants (Idso 1992; Johnson *et al.* 1993; Dukes & Mooney 1999). Alternatively, when CO_2 increases, seedlings in forest gaps may grow more quickly, so turnover rates in tropical forests may increase. This is an example of increased resource availability owing to reduced uptake by the resident vegetation. Others have argued that invasions may be facilitated by increases in precipitation (Dukes & Mooney 1999), an example of increased resource availability owing to increased resource supply. Still others have argued that the global nitrogen eutrophication resulting from anthropogenic activities is already facilitating invasions (Wedin & Tilman 1996), another example of enhanced supply increasing resource availability. In most instances, the impacts of global change and the observed ecological responses can be subsumed under the theory of fluctuating resource availability.

A range of global change factors are likely to interact to favour greatly the introduction, naturalization, spread, establishment and impact of alien invasive plant species within rainforest environments. These include:

- the range and seasonality of rapidly changing environmental factors such as temperature and precipitation;
- the intensity and frequency of severe episodic events such as cyclones, storms, floods, drought and fires;
- the group of demographic, economic and social pressures related to human activities influencing disturbance patterns, alien species movements and spatial and temporal patterns of propagule pressure.

Climate change could open up new opportunities for invasive species that could devastate native flora and fauna. For example, if the species that are dominant in the native vegetation are no longer adapted to the environmental conditions of their habitat, what species will replace them? It may well be that exotic species will find these new habitats especially attractive, and the increasing presence of new species and the decline of old ones will drastically change successional patterns, ecosystem function and the distribution of resources. Consequently, concepts of global change need to include consideration of the behaviour and distribution of invasive species. It seems highly likely that invasive species are going to have far more opportunities in the future than they have at present.

References

Adam, P. (1994). *Australian Rainforests.* Oxford University Press, Oxford.

Binggeli, P., Hall, J. B. & Healey, R. (1998). *An Overview of Invasive Woody Plants in the Tropics.* School of Agriculture and Forest Science, University of Wales, Bangor Publication No. 13.

Blossey, B., Skinner, L. C. & Taylor, J. (2001). Impact and management of purple loosestrife (*Lythrum salicaria*) in North America. *Biodiversity and Conservation* **10**: 1787–807.

Bowman, D. M. J. S. (2000). *Australian Rainforests: Islands of Green in a Land of Fire.* Cambridge University Press, Cambridge.

Brown, R. L. & Fridley, J. D. (2003). Control of plant species diversity and community invasibility by species immigration: seed richness versus seed density. *Oikos* **102**: 15.

Callaway, R. M. & Aschehoug, E. T. (2000). Invasive plants versus their new and old neighbors: A mechanism for exotic invasion. *Science* **290**: 521–3.

Chesson, P. & Huntley, N. (1997). The roles of harsh and fluctuating conditions in the dynamics of ecological communities. *American Naturalist* **150**: 519–53.

Cronk, Q. C. B. & Fuller, J. L. (1995). *Plant Invaders. The Threat to Natural Ecosystems.* Chapman & Hall, London.

Davis, M. A., Grime, J. P. & Thompson, K. (2000). Fluctuating resources in plant communities: a general theory of invasibility. *Journal of Ecology* **88**: 528–34.

Davis, M. A., Wrage, K. J. & Reich, P. B. (1998). Competition between tree seedlings and herbaceous vegetation: support for a theory of resource supply and demand. *Journal of Ecology* **86**: 652–61.

Deutschewitz, K., Lausch, A., Kühn, I. & Klotz, S. (2003). Native and alien plant species richness in relation to spatial heterogeneity on a regional scale in Germany. *Global Ecology and Biogeography* **12**: 299–311.

Dukes, J. S. & Mooney, H. A. (1999). Does global change increase the success of biological invaders? *Trends in Ecology and Evolution* **14**: 135–9.

Elton, C. S. (1958). *The Ecology of Invasions by Animals and Plants.* Methuen, London.

Environmental Protection Agency (EPA) (2002). *Queensland Herbarium Achievements 2000–2001.* Environmental Protection Agency, Brisbane.

Fargione, J., Brown, C. S. & Tilman, D. (2003). Community assembly and invasion: An experimental test of neutral versus niche processes. *Proceedings of the National Academy of Science* **100**: 8916–20.

Goosem, S. P. (2003). Landscape processes relevant to weed invasion in Australian rainforests and associated ecosystems. In *Weeds of Rainforests and Associated Ecosystems*, Grice, A. C. & Setter, M. J. (eds). Cooperative Research Centre for Tropical Rainforest Ecology and Management, Rainforest CRC, Cairns.

Goosem, S., Morgan, G. & Kemp, J. E. (1999). Wet Tropics. In *The Conservation Status of Queensland's Bioregional Ecosystems*, Sattler, P. S. & Williams, R. D (eds). Environmental Protection Agency, Brisbane.

Grime, J. P. (1974). Vegetation classification by reference to strategies. *Nature* 250: 26–231.

Grime, J. P. (1988). The C-S-R model of primary plant strategies: origins, implications, and tests. In *Plant Evolutionary Biology*, Gottlieb, L. D. & Jain, S. K. (eds). Chapman & Hall, London.

Hager, H. A. & McCoy, K. D. (1998). The implications of accepting untested hypotheses: a review of the effects of purple loosestrife (Lythrum salicaria) in North America. Biodiversity and Conservation 7: 1069–79.

Hager, H. A. (2004). Competitive effect versus competitive response of invasive and native wetland plant species. *Oecologia* **139**: 140–9.

Heywood, V. (1989). Patterns, extents and modes of invasions by terrestrial plants. In *Biological Invasions: A Global Perspective*, Drake, J. A., Mooney, H. A., di Castri, F. *et al.* (eds). John Wiley & Sons, New York, pp. 31–51.

Hilbert, D. W., Ostendorf, B. & Hopkins, M. (2001). Sensitivity of tropical forests to climate change in the humid tropics of North Queensland. *Austral Ecology* **26**: 590–603.

Humphries, S. E. & Stanton, J. P. (1992). *Weed Assessment in the Wet Tropics World Heritage Area of North Queensland.* Report to the Wet Tropics Management Agency, Cairns.

Huston, M. A. (1994). *Biological Diversity.* Cambridge University Press, Cambridge.

Idso, S. B. (1992). Shrubland expansion in the American southwest. *Climate Change* **22**: 85–6.

Johnson, H. B., Polley, H. W. & Mayeux, H. S. (1993). Increasing CO_2 and plant–plant interactions: effects on natural vegetation. *Vegetatio* **104**: 157–70.

Levine, J. M. & D'Antonio, C. M. (1999). Elton revisited: a review of evidence linking diversity and invasibility. *Oikos* **87**: 15–26.

Lonsdale, W. M. (1999). Global patterns of plant invasion and the concept of invasibility. *Ecology* **80**: 1522–36.

Lõvei, G. L. (1997). Global change through invasion. *Nature* **388**: 627.

Low, T. (1999). *Feral Future*. Viking Books, Melbourne.

Luken, J, O. & Thieret. J. W. (1996). Amour honeysuckle, its fall from grace. *BioScience* **46**: 18–24.

Macdonald, I. A. W., Lloyd, L., Loope, M., Usher, B. &. Hamann, O. (1989). Wildlife conservation and the invasion of nature reserves by introduced species: a global perspective. In *Biological Invasions: A Global Perspective*, Drake, J. A., Mooney, H. A., di Castri, F. *et al.* (eds). John Wiley & Sons, New York.

McNeeley, J. A. (2001). *The Great Reshuffling: How Alien Species Help Feed the Global Economy*. Report of Workshop on Alien Invasive Species. Colombo, October 1999. IUCN Regional Biodiversity Programme, Asia, Colombo, Sri Lanka.

McNeely, J. A., Mooney, H. A., Neville, L. E., Schei, P. J. & Waage, J. K. (eds). (2001). *A Global Strategy on Invasive Alien Species.* IUCN, Gland, Switzerland.

Meiners, S. J., Pickett, S. T. A. & Cadenasso, M. L (2001). Effects of plant invasions on the species richness of abandoned agricultural land. *Ecography* **24**: 633–44.

Meyer J. (2000). Preliminary review of the invasive plants in the Pacific islands. In *Invasive Species in the Pacific: A Technical Review and Draft Regional Strategy*, G. Sherley, G. (ed.). South Pacific Regional Environment Programme (SPREP), Apia, Samoa.

Mueller-Dombois, D. & Fosberg, F. R. (1998). *Vegetation of the Tropical Pacific Islands*. Springer Verlag, New York.

Parker, I. M., Simberloff, D., Lonsdale, W. M., *et al.* (1999). Impact: toward a framework for understanding the ecological effects of invaders. *Biological Invasions* **1**: 3–19.

Prieur-Richards, A. H., Lavorel, S., Grigulis, K. & Dos Santos, A. (2000). Plant community diversity and invasibility by exotics: invasion of Mediterranean old fields by Conza bonariensis and Conza canadensis. *Ecology Letters* **3**: 412–22.

Pysek, P. (1995). On the terminology used in plant invasion studies. In *Plant Invasions. General Aspects and Special Problems*, Pysek, P., Prach, K., Rejmanek, M. & Wade, M. (eds). SPB Academic Publishers, Amsterdam.

Rejmánek, M. (1989). Invasibility of plant communities. In *Biological Invasions: A Global Perspective*, Drake, J. A., Mooney, H. A., di Castri, F. *et al.* (eds). John Wiley & Sons, New York.

Rejmánek, M. (1996). Species richness and resistance to invasions. In *Biodiversity and Ecosystem Processes in Tropical Forests*, Orians, G. H., Dirzo, R. & Cushman, J. H. (eds). Wiley & Sons, New York.

Rejmánek, M. & Rosén, E. (1992). Influence of colonizing shrubs on species–area relationships in alvar plant communities. *Journal of Vegetation Science* **3**: 625–30.

Sax, D. F. & Brown, J. H. (2000). The paradox of invasion. *Global Ecology and Biogeography* **9**: 363–71.

Schoener, T. W. (1983). Field experiments on interspecific competition. *American Naturalist* **122**: 240–85.

Seabloom, E. W., Harpole, W. S., Reichman, O. J. & Tilman, D. (2003). Invasion, competitive dominance, and resource use by exotic and native California grassland species. *Proceedings of the National Academy of Science* **100**: 13384–9.

Stohlgren, T. J., Barnett, D. T. & Kartesz, J. T. (2003). The rich get richer: patterns of plant invasions in the United States. *Frontiers in Ecology and Environment* **1**: 11–14.

Stohlgren, T. J., Binkley, D., Chong, G. W. *et al.* (1999). Exotic plant species invade hot spots of native plant diversity. *Ecological Monographs* **69**: 25–46.

Symstad, A. J. (2000). A test of the effects of functional group richness and composition on grassland invasibility. *Ecology* **81**: 99–109.

Thompson, K., Hodgson, J. G. & Rich, T. C. G. (1995). Native and alien invasive plants: more of the same? *Ecography* **18**: 390–402.

Tilman, D. (1997). Community invasibility, recruitment limitation, and grassland biodiversity. *Ecology* **78**: 81–92.

Vitousek, P. M., D'antonio, C. M., Loope, L. L., Rejmánek, M. & Westbrooks, R. (1997). Introduced species: a significant component of human-caused global change. *New Zealand Journal of Ecology* **21**: 1–16.

Wagner, W. L., Herbst, D. R. &. Sohmer, S.H. (1990). *Manual of the Flowering Plants of Hawaii*. University of Hawaii Press, Honolulu.

Webb, L. J. (1958). Cyclones as an ecological factor in tropical lowland rainforest, North Queensland. *Australian Journal of Botany* **6**: 220–8

Webb, L. J. & Tracey, J. G (1981). Australian rainforests: patterns and change. In *Ecological Biogeography of Australia*, Keast, A. (ed.). W. Junk, The Hague.

Wedin, D. A. & Tilman, D. (1996). Influence of nitrogen loading and species composition on the carbon balance of grasslands. *Science* **274**: 1720–3.

Weeds CRC. (2005). Weeding out the bad guys. *Cairns Post* 24 June.

Werren, G. L. (2001). *Environmental Weeds of the Wet Tropics: Risk Assessment and Priority Ranking*. Report to WTMA, Rainforest CRC, Cairns.

Werren, G. L. (2003). A bioregional perspective of weed invasion of rainforests and associated ecosystems: focus on the Wet Tropics of north Queensland. In *Weeds of Rainforests and Associated Ecosystems*, Grice, A. C. & Setter, M. J. (eds). Cooperative Research Centre for Tropical Rainforest Ecology and Management, Rainforest CRC, Cairns.

Werren, G. L., Goosem, S., Tracey, J. & Stanton, J. P. (1995). The Australian Wet Tropics centre of plant diversity. In *World Centres of Plant Diversity Volume 2*, Davies, S. D., Heywood, V. H. & Hamilton, A. C. (eds). WWF/IUCN, Oxford Information Press.

Westbrooks, R. (1998). *Invasive Plants, Changing the Landscape of America: Fact Book*. The Federal Interagency Committee for the Management of Noxious and Exotic Weeds (FICMNEW), Washington, DC.

Westcott, D. A. & Dennis, A. J. (2003). The ecology of seed dispersal in rainforests: implications for weed spread and a framework for weed management. In Weeds of Rainforests and Associated Ecosystems, Grice, A. C. & Setter, M. J. (eds). Cooperative Research Centre for Tropical Rainforest Ecology and Management, Rainforest CRC, Cairns.

Whittaker, R. J. (1998). *Island Biogeography. Ecology, Evolution, and Conservation*. Oxford University Press, Oxford.

WTMA (2004). *Wet Tropics Conservation Strategy*. Wet Tropics Management Authority, Cairns.

Yurkonis, K. A. & Meiners, S. J. (2004). Invasion impacts local species turnover in a successional system. *Ecological Letters* **7**: 764–9.

25 VERTEBRATE PESTS OF THE WET TROPICS BIOREGION: CURRENT STATUS AND FUTURE TRENDS

Bradley C. Congdon[1]* and Debra A. Harrison[2]*

[1]School of Marine and Tropical Biology, James Cook University, Cairns, Queensland, Australia
[2]Griffith School of Environment, Griffith University, Nathan, Queensland, Australia
*The authors were participants of Cooperative Research Centre for Tropical Ecology and Management

Introduction

The invasion of natural ecosystems by exotic species is considered one of the major processes threatening global biodiversity (Species Survival Commission 2000). Disruptions to ecosystem function and changes to patterns of native species richness and abundance are common forms of degradation attributable to exotic or pest species invasions (Sax & Brown 2000; Species Survival Commission 2000). To date, all continents have been negatively affected by the introduction of exotic species. However, islands and other isolated habitats have been significantly more susceptible to biodiversity loss (Mack *et al.* 2000; Sax & Brown 2000). This susceptibility is attributed to higher levels of endemism in these isolated regions (Species Survival Commission 2000).

By removing natural barriers to dispersal, globalization and the expansion of trade and tourism have facilitated many exotic species invasions (Mooney & Drake 1989; Mack *et al.* 2000; Species Survival Commission 2000). Similarly, human-induced habitat modification continues to allow invasive species into previously immune environments (Levin 1989; Mooney & Drake 1989). Of significance is that many invasions have succeeded only because of their association or dependence on human-modified environments (Sax & Brown 2000). The implications of these phenomena are obvious. Isolated areas of high endemism that have been, or are being, fragmented by human disturbance must be viewed as highly susceptible to the threat of exotic species invasion. The Wet Tropics bioregion of northern Australia is one such region.

Exotic species invasions (flora and fauna) have been identified as one of the most significant threats to the 41 ecosystems within the Wet Tropics bioregion that are considered endangered or of concern (Werren *et al.* 1995). Vertebrate pests constitute a substantial component of this threat (WTMA 1998, 2004). Fifteen exotic mammals, one amphibian, one reptile, five birds and at least six fish species are currently listed as present and/or undesirable in the Wet Tropics (WTMA 1998) (Table 25.1). Many of these pests are colonizing species that have benefited from human activities and habitat disturbance, particularly expanding agriculture and the establishment of linear service corridors such as roads and powerline easements (Weston & Goosem 2004). Although the total number of vertebrate pests has remained stable over recent years, their population numbers, distribution and ecological impacts are still very poorly understood (WTMA 2004).

Table 25.1 Identified vertebrate pests of the Wet Tropics bioregion

Family	Common name	Scientific name
Poeciliidae	Gambusia	*Gambuisa holbrooki*
Poeciliidae	Guppies	*Poecilia reticulata*
Poeciliidae	Sword tails	*Xiphorphorus hellerii*
Poeciliidae	Platys	*Xiphophorus macularta*
Cichlidae	Tilapia	*Tilapia mariae*
Cichlidae	Tilapia	*Oreochromis mossambicus*
Bufonidae	Cane toad	*Bufo marinus*
Gekkonidae	Asian house gecko	*Hemidactylus frenatus*
Columbidae	Rock dove	*Columba livia*
Columbidae	Spotted turtle-dove	*Streptopelia chinensis*
Passeridae	Nutmeg manikin	*Lonchura punctulata*
Passeridae	House sparrow	*Passer domesticus*
Sturnidae	Common myna	*Acridotheres tristis*
Muridae	House mouse	*Mus musculus*
Muridae	Brown rat	*Rattus norvegicus*
Muridae	Black rat	*Rattus rattus*
Canidae	Dog	*Canis familiaris*
Canidae	Dingo	*Canis familaris dingo*
Canidae	Red fox	*Vulpes vulpes*
Felidae	Cat	*Felis catus*
Leporidae	Rabbit	*Oryctolagus cuniculus*
Leporidae	Brown hare	*Lepus capensis*
Equidae	Horse	*Equus caballus*
Suidae	Pig	*Sus scrofa*
Cervidae	Rusa deer	*Cervus elaphus*
Cervidae	Fallow deer	*Dama dama*
Cervidae	Chital deer	*Cervus axis*
Bovidae	Goat	*Capra hircus*

Source. WTMA (1998).

The global significance of the Wet Tropics bioregion and increasing tourism and travel between northern Australia, New Guinea and areas of Southeast Asia with similar climate mean that an assessment of the current or potential pest status of invasive species within the region has become a priority. To this end the Wet Tropics Vertebrate Pest Risk Assessment Scheme (WTVPRAS) was developed as an ongoing research

and management tool for assessing the status of existing pests of the Wet Tropics bioregion, and as a framework for the development of similar dedicated risk assessment schemes elsewhere (Harrison & Congdon 2002). This chapter provides a brief overview of the pest status, species-specific pest characteristics and potential impacts of exotic vertebrates within the Wet Tropics bioregion based on the WTVPRAS process.

Assessing pest status and potential

Identifying life-history attributes that adequately define pest status (or potential) is a difficult problem that has received considerable attention in the ecological literature (e.g. Ehrlich 1986, 1989; Daehler & Strong 1993; Ricciardi & Rasmussen 1998; Davis *et al.* 2000; Mack *et al.* 2000). This debate clearly identifies a number of potential pest characteristics that provide some measure of pest potential (summarized in Ehrlich 1989). However, despite life-history similarities among some pest taxa, identifying pests prior to invasion still remains difficult. This is because general pest characteristics do not adequately define all pest species. For example, invasive species may develop from small but reproductively successful populations that originally have restricted distributions. The apparent non-detrimental effects attributable to these species while population numbers are low mean that they are often not recognized as potential pests. However, these sleeper species can cause severe ecosystem degradation and native species losses when climatic and/or resource fluctuations allow population expansion (Davis *et al.* 2000). Similarly, while some established exotic species may not display pest characteristics, and appear benign in their direct impact, they may vector other exotic organisms with devastating consequences on ecosystem integrity (Stone & Stone 1989; Pavlov 1992; Choquenot *et al.* 1993, 1996).

For invasive species to be successful they must also be able to adjust to the physical and biological characteristics of the habitat being invaded. Therefore, the characteristics of the target ecosystem greatly influence the potential pest status of any invader (Ehrlich 1989; Lodge 1993; Davis *et al.* 2000). Some general habitat characteristics appear to make ecosystems more susceptible; these include relative isolation, the prolonged absence of predators and previous human disturbance (Ehrlich 1989; Lodge 1993). However, the direct relationship between target ecosystem character and pest status makes identifying pest potential difficult without direct reference to the ecosystem under threat (e.g. Petren & Case 1998), and where possible, the impact of the potential pest in similar environments elsewhere (Pimm 1989).

Clearly, the ecological complexity of species–environment interactions make it difficult to generalize about the pest potential of invasive species based solely on broad life-history characteristics. Because of this, the possible impacts of invasive species in areas such as the Australian Wet Tropics can only be estimated through studies that also assess pest potential relative to defined habitat criteria, and consider the history of the potential pest in similar environments elsewhere.

A review of pest and invasive species literature establishes five generic pest species characteristics upon which existing risk/impact assessment schemes are generally based. These are

1 previous pest history;
2 reproductive and dispersal potential;
3 an ability to capitalize on variation in climatic and/or biological events;
4 ability to vector diseases or parasites;
5 threat to existing species via predation, competition and/or habitat degradation.
(e.g. Fox & Adamson 1979; Simberloff 1981; Groves & Burdon 1986; Norton & Pech 1988; Brown 1989; Ehrlich 1989; Hobbs 1989; Levin 1989; Pimm 1989; Hiebert & Stubbendieck 1993; Lodge 1993; Hone 1994; Arthington et al. 1999; Richardson et al. 2000; Sax & Brown 2000).

The Wet Tropics Vertebrate Pest Risk Assessment process (Harrison & Congdon 2002) evaluates published ecological data against these generic pest characteristics and generates four impact indices for each exotic species. These are

1 current impact;
2 potential future impact;
3 feasibility of control;
4 detrimental impacts of control.

The four indices are then used to produce a combined assessment of overall pest status. The relative independence of each value also allows comparisons of impact types to be undertaken in specific combinations for particular management purposes. The WTVPRAS is designed to gauge threats to the World Heritage values of the Wet Tropics region and to prioritize the eradication, containment and control of vertebrate pests in the region. It achieves this by assessing invasive species impacts relative to the Wet Tropics biodiversity values identified by Goosem et al. (1999).

High impact vertebrate pests of the Wet Tropics

Figure 25.1 provides a visual summary of the relative pest status of each exotic species within the Wet Tropics bioregion. This figure plots current and future potential impacts combined against control difficulty plus negative control effect. The pest space described by these axes comprises four regions, each containing a cluster of species having similar pest status. Species in the upper right hand region of this figure are difficult to control and have high overall impact. Species in the lower left region are significantly more controllable and have lower impacts. Relative pest status increases towards the upper right hand side of the figure.

The current major vertebrate pests of the Wet Tropics are identified as the pig and cat, closely followed by the cane toad and dog/dingo. These species rank highly primarily because of their extensive current impacts, but also because of their lack of controllability. These four species have wide distributions and directly quantifiable negative effects on endangered and/or threatened species and habitats (Pavlov et al. 1992; Werren 1993; Choquenot et al. 1996; Burnett 1997; Environment Australia 1999a; Schmidt et al. 2000).

Pigs (Sus scofa)

Sus scofa is the highest profile pest of the Wet Tropics bioregion. It is widely distributed and occurs in highly sensitive areas (Pavlov 1992; Pavlov et al. 1992; Olsen 1998). It modifies habitats, competes directly with endangered fauna for resources and transmits many diseases and parasites (Stone & Loope 1987; Bowman & McDonough 1991; Pavlov 1992; Pavlov et al. 1992; Choquenot et al. 1996; Laurance & Harrington 1997; Gadek 1999). It is a pervasive pest in similar environments elsewhere, has many generic pest species characteristics and is difficult to control from both a technical and cultural perspective (Stone & Loope 1987; Pavlov 1992; Pavlov et al. 1992; Caley 1994; Choquenot et al. 1996; Cape York Regional Advisory Group 1999).

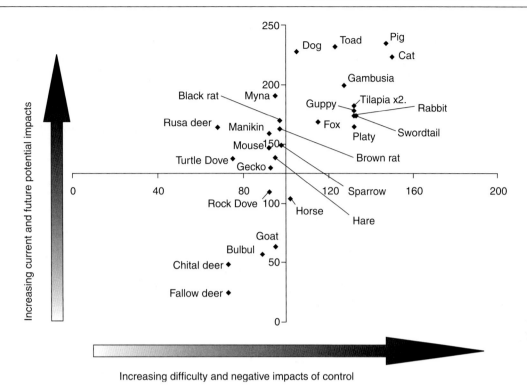

Figure 25.1 Assessment of current impacts plus potential impacts versus control difficulty plus negative impact of control. The pest space comprises four regions each containing species having similar pest status. Relative pest status increases towards the upper right hand corner of the graph.

Sufficient information is available to assess comprehensively the current and future pest potential of this species, but a current inability to establish accurately the effects of control on population size and demography significantly hampers control efforts.

Cats (Felis cattus)

Noted for their ability to decimate ground dwelling bird and small mammal populations, especially on islands and in isolated areas, cats pose a significant threat to species in the Wet Tropics bioregion. As with pigs, cats are widely distributed (Strahan 1993) and impact directly on vulnerable species in highly sensitive areas (Environment Australia 1999a; Goosem *et al.* 1999). They are highly dispersive and vector a range of other potential pests (Environment Australia 1999a). Currently populations are small but control methods are not very effective (Environment Australia 1999a). Where control is undertaken numbers are quickly replaced from stable populations in fringing

urban areas (Environment Australia 1999a). Cats are closely associated with humans and can be expected to spread with increasing urbanization. The extremely wet areas of the bioregion may offer some resistance to colonization (Environment Australia 1999a). Assessment information is relatively complete, although few studies have been undertaken specifically on feral cat impacts within the Wet Tropics. Further, the costs of control in heavily forested areas are unknown.

Cane toads (Bufo marinus)

The cane toad has been present throughout the Wet Tropics region for many years (Department of Natural Resources 1997) and the pest potential of this species is well known (Burnett 1997; Department of Natural Resources 1997). Cane toads are thought to be significant threats to quoll, monitor and native frog populations (Burnett 1997), but the long-term potential impacts on these threatened species have not been quantified. Studies have shown that available controls have limited effect (Tyler 1996; Zupanovic *et al.* 1998a, b). Presently widespread control of toads seems unrealistic. This suggests that toad research and control in the Wet Tropics may be best targeted at areas where they could be directly impacting vulnerable or endangered species.

Dogs/dingoes (Canis familiaris)

In general, dingoes occur in drier habitats than the Wet Tropics, suggesting that rainforest is not their preferred habitat (Strahan 1993). Dingoes/dogs are known to take other pests such as rabbits, foxes and feral cats, as well as native species (Usher 1989; Olsen 1998; Environment Australia 1999a). Currently, dingoes may limit numbers of these other vertebrate species (Usher 1989; Olsen 1998; Environment Australia 1999a). Any management strategy aimed at reducing dingo/dog numbers must consider the potential effects on these other pest species, particularly foxes. Identification of the principal perpetrators of dog kills, particularly of tree kangaroos (*Dendrolagus lumholtzi*, *D. bennettianus*) and cassowary (*Casuarius casuarius*), is required to focus management options. Wild dog control is extremely difficult. However, if many dog attacks are attributable to unsupervised domestic, rather than wild dogs, overall dog control may be more feasible. Attacks by unsupervised domestic dogs require substantially different control strategies to those used to reduce feral dog or dingo numbers.

Exotic fish

Exotic fish (gambusia, tilapia, guppies, swordtails, platys) and rabbits are also identified as current major pests. They have lower impact scores but are equally difficult to control. This is especially true for exotic fish, where it is extremely difficult to target pest species selectively using existing technology. Current impacts identified for most exotic fish species may also be underestimates. This is because of a lack of studies quantifying their detrimental effects on native wildlife either in Australia or overseas (also see the section on sleeper pests below).

Rabbits (Oryctolagus cunculus)

Rabbits are a slowly advancing species. They appear to be a greater threat to the region than the brown hare because they occur in upland sites (Williams *et al.* 1996) and have less selective feeding habitats (Strahan 1993). Rabbit spread may, in part, be contained by increased parasite load and mortality associated with high rainfall (Environment Australia 1999b). Similar mortality has been noted elsewhere (Strahan 1993; Environment Australia 1999b). The introduction of calcivirus may also significantly affect rabbit numbers (Environment Australia 1999b). The full impact of these potential controlling agents is not yet known. The impact of rabbits on the bioregion is complicated by their relationship with other exotic pests. Rabbit predation may buffer native species from the impacts of a number of other exotic carnivores such as cats, foxes and dogs/dingoes.

Sleeper pests of the Wet Tropics

Figure 25.2 is a plot of current impacts against potential future impacts that can be used to identify pest species not currently realizing their maximum pest potential. The species that occur in the upper right hand area of this plot are those previously identified as having both high current and future potential impact (pigs, cane toads, cats, dogs). Also clearly identified in this figure are a group of sleeper species (indicated by the boxed area) that have limited current impacts but substantial potential future impacts. These species in order of perceived threat are gambusia, the two tilapia species, the fox, rusa deer, swordtail, guppy, platy, black rat, Indian myna, rabbit, brown rat and nutmeg manikin.

Clearly, exotic fish constitute a significant unrealized threat to the Australian Wet Tropics. Six species have been identified as major or potential pests of the region. All species occur in degraded habitats and/or urban streams. Their present distributions probably still reflect human point source introductions and translocations (Arthington & Bluhdorn 1994; Pusey & Kennard 1994; Russell *et al.* 1996; North Queensland Joint Board 1997). Almost certainly, most of these species have not yet reached their full impact

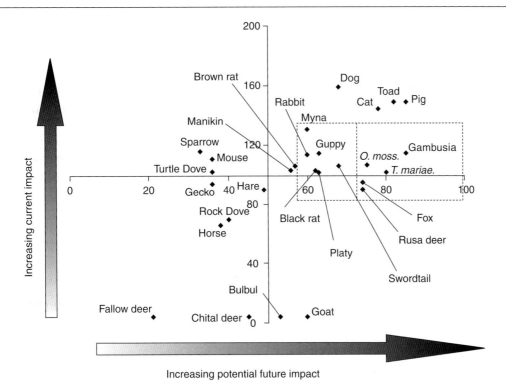

Figure 25.2 Assessment of current versus potential future impact. This plot identifies pest species having both high current and high future potential impacts (pigs, cane toads, cats, dogs) and also identifies sleeper species (indicated by the boxed area) with moderate current impacts but moderate to high future potential. Note also the goat, which has no current impact but moderate potential future impacts.

potential. Existing assessments probably substantially underestimate both their current and possible future affects. This is primarily because of limited historical knowledge of native stream fauna assemblages prior to introduction and the uncertainty regarding taxonomic status and levels of local endemism in the bioregion (Arthington & McKenzie 1997). The two tilapia species pose the more immediate threat. This is because of their large body mass and ability to adjust metabolism and age/size at maturity to limited resources (Arthington et al. 1999). In comparison, the smaller poeciliids (gambusia, swordtails, guppies, platies) do not overwhelm and modify habitat as quickly as tilapia are known to do (Arthington & Bluhdorn 1994; Pusey & Kennard 1994). The limited thermal tolerances of tilapia (Cnaani et al. 2000) may offer limited protection to catchment headwaters in the region.

It should also be noted that populations of other exotic fish occur along the central Queensland coast, principally in the Aplin Weir on the Ross River. These species include the green terror (*Aequidens rivulatus*), oscars (*Astronotus ocellatus*), red devils (*Amphilophus citrinellus*), banded cichlid (*Cichlasoma severum*), firemouth cichlid (*Cichlasoma meeki*), convict cichlid (*Cichlasoma nigrofasciatum*) and pearl cichlid (*Geophagus brasiliensis*) (Arthington et al. 1999). The jewel cichlid (*Hemichromis bimaculaus*) and oscars have also been reported, but not documented, in the Cairns area. These species exhibit very similar biology to the cichlid species already in the Wet Tropics bioregion. All have the potential to become major pests of the region. That similar potential impacts can be attributed to most exotic fish reflects these shared pest life-history characteristics and a general lack of detailed information on the extent of current impacts. Both these factors make species-specific descriptions difficult. Therefore, a description is provided for two of the principal pest species that represent the two major life-history strategies present in the Wet Tropics.

Mosquitofish (Gambusia holbrooki)

The distribution of gambusia within the bioregion is not well known. Various studies (Pusey & Kennard 1994; Herbert & Peeters 1995; Russell *et al.* 1996) and anecdotal evidence (Congdon, unpublished data) suggest they occur in limited numbers in degraded areas in most provinces of the Wet Tropics, especially irrigation channels and drainages. The negative impacts of gambusia on native fish and frog species (Arthington & Mitchell 1986; Lund 1994; Goodsell & Kats 1999; Knapp & Matthews 2000), as well as on water quality in natural areas (Lund 1994), are well documented. Given the difficulty in controlling fish numbers, the fecundity and dispersal capabilities of this species, and a reluctance to use current control methods because of negative impacts, it is expected that gambusia distributions and impacts will continue to expand.

Tilapia (Oreochromis mossambicus and Tilapia mariae)

Tilapia populations are currently correlated with their release points (Arthington & Bluhdorn 1994; Pusey & Kennard 1994; Arthington *et al.* 1999). At these points they dominate riverine communities by utilizing all resources and modifying habitat to their advantage (Arthington & Bluhdorn 1994; Froese & Pauly 2001). Human-induced removal of riparian vegetation alters river ecology by removing native fish habitat and increasing water temperatures (Cnaani *et al.* 2000). These processes further aid tilapia expansion. The extensive control required and the highly negative impacts of control, such as poisoning, suggest that control of tilapia species will be expensive and damaging (Arthington & McKenzie 1997). The exact limits of current distributions need to be established, as do the long-term effects of control measures on native fish. This is to determine if native populations are resilient against the broad-scale impacts associated with current controls. Tilapia populations in the Wet Tropics are also considered to be threatening the Gulf River Catchments (North Queensland Joint Board 1997). This issue further highlights the prioritizing of control for these species.

Fox (Vulpes vulpes)

Along with exotic fish, the fox is identified as a major future threat to the Wet Tropics. The fox's relatively low current impact value is due only to its limited distribution in the Atherton and Kirrama-Hinchinbrook provinces (North Queensland Joint Board 1997; Queensland Museum 2000; Earthworks 2001). The fox is an intelligent predator that will take other pest species such as mice and rabbits in open areas (Strahan 1993). However, it prefers native prey in forested areas (Strahan 1993). Foxes are currently thought to threaten the tropical bettong (Environment Australia 1999c). If fox populations continue to spread through forested areas they are likely to threaten many other native species (WTMA 1998; Environment Australia 1999c). As with dingoes, foxes may currently contribute significantly to rabbit control along the edges of the bioregion (Environment Australia 1999c). The impact of fox control on other pest species needs be considered in any major control programme. Fox risk assessments would be aided by detailed information on current distributions, including non-presence data, and native species interactions.

Rusa deer (Cervus elaphus)

A recent review of the distribution, abundance and pest status of deer, incorporating updated assessments using the WTVPRAS (Hudson 2006), clearly places rusa deer in the sleeper pest category. The current impacts of rusa deer are relatively low only because of their perceived small and patchy population distribution. However, within the Wet Tropics basic ecological information on population sizes, demography and level of interconnectedness is lacking. To date, the majority of feral deer sightings have been confined to the central part of the bioregion and are concentrated around known or former deer farms (Moriarty 2004; NRM & E 2004; Weston & Goosem 2004). However, other occasional sightings suggest that deer may be considerably more widespread (Hudson 2006). The potential for rusa to form significant pest populations in the Wet Tropics bioregion is high. Incidental observations and anecdotal evidence suggest that impacts from deer on vegetation and soils due to browsing and soil disturbance may be significant where populations are large (Hudson 2006). One priority for future research aimed at clarifying risk potential should be to quantify rusa dietary preferences and the effects of browsing on impacted rainforest plants (Hudson 2006).

Black Rat (Rattus rattus)

This species does not currently occur in uplands other than the Atherton Upland region (Williams *et al.* 1996; Queensland Museum 2000). Its presence in this region can be attributed to increased urbanization. The black rat has not previously displaced native rats in intact forest areas (Stone & Loope 1987), but may place pressure on native species in degraded natural areas (Strahan 1993). The black rat may establish feral populations independent of human settlements and will feed on a variety of food items (Strahan 1993). This species may require close attention, particularly given its high degree of adaptability, and the assistance increasing urbanization will provide for further establishment.

Indian myna (Acridotheres tristis)

The impact of this species on native taxa within the bioregion has not been accurately assessed. The myna is known to outcompete and exclude parrots and small mammals from nesting hollows in woodlands and southern eucalypt forests (Pell & Tidemann 1997; Tidemann & Pell 1997). If myna impacts in the Wet Tropics are comparable then this species poses a major threat to the region. Long (1981) suggests that mynas may be deterred by denser rainforest types. If so, this may restrict mynas to fringing zones and mediate impacts in some upland areas. This species is closely associated with human settlements and there seems little doubt that it impacts on native species with similar associations (Long 1981). Until information detailing myna impacts is available the effects of reducing myna numbers on biodiversity values in the bioregion cannot be estimated.

Goats (Capra hircus)

Importantly, other sleeper species that rank lower because of a lack of identified current impacts, such as goats, also have considerable potential to cause major problems should larger populations become established due to human error or translocation. Goats are not currently located within the bioregion but given their pest history elsewhere there is concern that they may become established (Environment Australia 1999d). Of the no-current-impact species, goats have the greatest potential to cause serious future problems. The goat is a generalist herbivore that modifies its diet according to available foliage (Environment Australia 1999d). Predation by dingoes or dogs and increased mortality from higher parasite loads in wetter climates (Environment Australia 1999d) may restrict goats in some areas surrounding the bioregion. Control of goat populations is labour intensive and is restricted to mustering, shooting and/or trapping. (Baiting with 1080 is not very successful.) The extensive foliage cover offered in the bioregion would also substantially decrease the effectiveness of control measures. The status of goat populations in adjacent areas needs to be very closely monitored.

Other taxa

The brown rat, nutmeg manikin, house mouse, house sparrow, brown hare and spotted turtledove have intermediate impact scores but these species appear somewhat easier to control. Many of these species are primarily associated with urban or disturbed habitats and appear to have limited potential to spread to natural areas. The overall impact of these species will increase as development increases throughout the region. Many may not yet have reached their full impact potential.

Impacts of control

Figure 25.3 is a plot of difficulty of control against the negative impact of control measures. Again, both the difficulty of control and the negative impacts of control increase towards the upper right hand corner of this figure. This figure clearly demonstrates that all control measures are perceived to impact native species to varying degrees. These effects can be significant. Figure 25.3 enables these effects to be examined in isolation for consideration during the development of

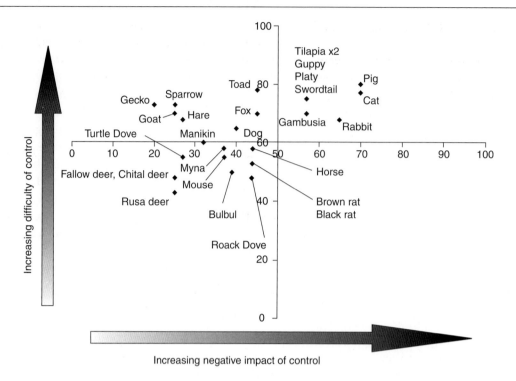

Figure 25.3 Difficulty of control plotted against negative impacts of control on non-target species. Overall controllability decreases as you move towards the upper right corner of the graph. Note that all control measures are perceived to have some negative effects on non-target native species or habitats.

future management objectives. Pig control poses the greatest perceived threat. This is because it is known to impact on high-profile vulnerable species such as the cassowary, or requires incursions by domestic animals into otherwise pristine habitat, with unknown consequences (McIlroy & Saillard 1989; Caley & Ottley 1995). Control methods for dogs, cats, cane toads, gambusia, tilapia, guppies, rabbits, swordtails and platys also have the potential to exhibit significant detrimental effects on native fauna. This is because in general they are not taxa-specific but affect any native species with similar ecological associations. This problem also gives these species their low controllability. For all other species, control measures have low impacts on non-target species but may not be particularly effective. This is particularly true for taxa where control options are restricted to mustering, trapping and/or shooting.

Generic problems of assessing pest status

There are a number of generic problems associated with assessing the current and future potential impacts

of vertebrate pests in the Wet Tropics bioregion. These are: (i) the lack of baseline data on historic or current native species distributions and abundance against which to gauge exotic species impacts (this is a general problem but is particularly relevant to native fish and other freshwater taxa); (ii) a lack of quantified exotic impacts on native species (this information is required not only to establish the extent and intensity of current impacts but also to determine the level and focus of future research that needs to be undertaken); (iii) the lack of basic data on pest population demography, including population sizes, ranges and levels of interconnectedness. Our current inability to census adequately most of the pest species within the Wet Tropics means that the effectiveness of control methods cannot be accurately assessed. Data on control effectiveness, where they exist, consist primarily of catch rates per technique per unit time. Unfortunately, these data do not provide information on the proportion of each pest population being removed per control event, or the relative impact of control on reproductive individuals. Obtaining these data requires accurate post-control censusing of known populations. These are the data that are needed to establish control effectiveness.

Many of these problems are applicable to determining pest status in general as well as to risk assessments in other systems. Examining them here serves to highlight both caveats in the assessments presented and the dependence of any assessment process on an extensive and robust understanding of both pest species biology and ecological dynamics of the impacted system.

Summary

Detailed assessments of pest status have confirmed the generally accepted belief that pigs, cats, dogs and toads are the high-impact pest species within the Wet Tropics bioregion, while exotic fish, foxes and rusa deer constitute the principal unrealized threat. Eradication of these species from the region does not seem feasible in the short term given current controls and costs. Ultimately, the effort required to control these high profile pests will depend on the management goals specified. Therefore, identifying and costing specific management objectives to mediate impacts in sensitive areas will ultimately determine their future impacts. To facilitate the collection of appropriate information, it will be necessary to establish management criteria on what are acceptable levels of infestation for specific areas, and what control impacts on native species are acceptable to achieve these goals.

As a general trend, most current and potential vertebrate pest species of the Wet Tropics are closely associated with human settlements or human disturbance. It is logical to conclude that an increase in human disturbance or expansion will bring with it increases in both high- and low-risk pest species. This finding must be considered during the development of any management aimed at containing vertebrate pests. This may be particularly important for high-diversity, high-risk areas such as the Daintree-Bloomfield province, where many species with strong human associations (e.g. indian mynas, mice, sparrows, turtledoves) have not yet been recorded.

References

Arthington, A. H. & Bluhdorn, D. (1994). Distribution, genetics, ecology and status of the introduced cichlid, Oreochromis mossambicus, in Australia. *International Association of Theoretical and Applied Limnology* **24**: 53–62.

Arthington, A. H. & McKenzie S. (1997). *Review of Impacts of Displaced/Introduced Fauna associated with Inland Waters*. Technical Paper Series (Inland Waters) Environment Australia, Community Information Unit Department of Environment, Canberra.

Arthington, A. H. & Mitchell, D.S. (1986). Aquatic invading species In *Ecology of Biological Invasions*, Groves, R. H. & Burdon, J. J. (eds). Cambridge University Press, Cambridge.

Arthington, A. H., Kailola, P. J., Woodland, D. J. & Zalucki, J. M. (1999). *Baseline Environmental Data Relevant to an Evaluation of Quarantine Risk Potentially Associated with the Importation to Australia of Ornamental Finfish*. Report to the Australian Quarantine and Inspection Service, Department of Agriculture, Fisheries and Forestry, Canberra.

Bowman D. M. J. S. & McDonough, L. (1991). Feral pig (*Sus scrofa*) rooting in a monsoon forest–wetland transition, Northern Australia. *Wildlife Research* **18**: 761–5.

Brown, J. H. (1989). Patterns, modes and extents of invasions by vertebrates. In *Biological Invasions: A Global Perspective*, Drake, J. A., Mooney, H. A., di Castri, F. *et al*. (eds). John Wiley & Sons, New York, pp. 85–110.

Burnett, S. (1997). Colonising cane toads cause population declines in native predators: reliable anecdotal information and management implications. *Pacific Conservation Biology* **3**: 65–72.

Caley, P. (1994). Factors affecting the success rate of traps for catching feral pigs in a tropical habitat. *Wildlife Research* **21**: 287–92.

Caley, P. & Ottley, B. (1995). The effectiveness of hunting dogs for removing feral pigs (Sus scrofa). *Wildlife Research* **22**: 147–154.

Cape York Regional Advisory Group (1999). *Animal and Weed Pests of Cape York Peninsula* (www.environ.gov. au/states/cyp_on_l/).

Choquenot, D., Kilgour, R. J. & Lukins, B. S.(1993). An evaluation of feral pig trapping. *Wildlife Research* **20**: 15–22.

Choquenot, D., McIlory, J. & Korn, T. (1996). *Managing Vertebrate Pests: Feral Pigs*. Bureau of Resource Sciences, Australian Government Publishing Service, Canberra.

Cnaani, A., Gall G. A. E. & Hulata, G. (2000). Cold tolerance of tilapia species and hybrids. *Aquaculture International* **8**: 289–298.

Daehler, C. C. & Strong, D. R. (1993). Prediction and biological invasions. *Tree* **8**(10): 380.

Davis, M. A., Grime, J. P. & Thompson, K. (2000). Fluctuating resources in plant communities: a general theory of invisibility. *Journal of Ecology* **88**: 528–534.

Department of Natural Resources (1997). *The Cane Toad DNR Pest Facts*. Queensland Government Printing Office,

Brisbane.

Earthworks (2001). The mammals of the Mount Molloy Stock Route Reserves and Spear Creek. *Earthworks Report 99c23*. Earthworks Environmental Services Pty Ltd, Townsville.

Ehrlich, P. R. (1986). Which animal will invade? In *Ecology of Biological Invasions of North America and Hawaii*, Mooney, H. A. & Drake, J. A. (eds). Springer-Verlag, New York, pp. 79–95.

Ehrlich, P. R. (1989). Attributes of invaders and the invading processes: vertebrates In *Biological Invasions: A Global Perspective*, Drake J. A., Mooney, H. A., di Castri, F. *et al.* (eds). John Wiley & Sons, New York.

Environment Australia (1999a). *Threat Abatement Plan for Predation by Feral Cats*. Australian Government Report, Canberra.

Environment Australia (1999b). *Threat Abatement Plan for Competition and Land Degradation by Feral Rabbits*. Commonwealth Government of Australia, Canberra.

Environment Australia (1999c). *Threat Abatement Plan for Foxes*. Commonwealth Government of Australia, Canberra.

Environment Australia (1999d). *Threat Abatement Plan for Competition and Land Degradation by Feral Goats*. Commonwealth Government of Australia, Canberra.

Fox, M. D. & Adamson D. (1985). The ecology of invasions. In *A Natural Legacy*, Recher, H. F., Lunney, D. & Dunn, I. (eds). Pergamon Press, New York.

Froese, R. & Pauly. D. (eds). (2001). *FishBase* (www.fishbase.org).

Gadek, P. A. (1999). *Patch Deaths in Tropical Queensland Rainforests: Association and Impacts of* Phytopthora cinnamomi *and Other Soil Borne Organisms*. Cooperative Research Centre for Tropical Rainforest Ecology and Management, Cairns.

Goodsell, J. A. & Kats, L. B. (1999). Effect of introduced mosquitofish on Pacific treefrogs and the role of alternative prey. *Conservation Biology* **13**(4): 921–4.

Goosem, S., Morgan, G. & Kemp, J. E. (1999). Wet Tropics. In *The Conservation Status of Queensland's Bioregional Ecosystems*, Sattler, P. S. & Williams, R. D. (eds). Environmental Protection Agency, Brisbane.

Groves, R. H. & Burdon, J. J. (eds). (1986). *Ecology of Biological Invasions*. Cambridge University Press, Cambridge.

Harrison, D. A. & Congdon, B. C. (2002). *Wet Tropics Vertebrate Pest Risk Assessment Scheme*. Cooperative Research Centre for Tropical Rainforest Ecology and Management, Cairns.

Herbert, B. & Peeters, J. (1995). *Freshwater Fishes of Far North Queensland*. Department of Primary Industries, Queensland Government, Brisbane.

Hiebert, R. D. & Stubbendieck, J. (1993). *Handbook for Ranking Exotic Plants for Management and Control*. Natural Resources Report NPS/NRMWRO/NRR-93/08. US Department of the Interior, Natural Resources Publication, Denver, CO.

Hobbs, R. J. (1989). The nature and effects of disturbance relative to invasions In *Biological Invasions: A Global Perspective*, Drake J. A., Mooney, H. A., di Castri, F. *et al.* (eds). John Wiley & Sons, New York.

Hone, J. (1994). *Analysis of Vertebrate Pest Control*. Cambridge University Press, Cambridge.

Hudson, S. (2006). *Feral deer in the Wet Tropics Bioregion: Distribution, Abundance and Management*. Cooperative Research Centre for Tropical Rainforest Ecology and Management, Cairns.

Knapp, R. A. & Matthews, K. R. (2000). Non-native fish introductions and the decline of the mountain yellow-legged frog from within protected areas. *Conservation Biology* **14**(2): 428–38.

Laurance, W. F. & Harrington, G. N. (1997). Ecological associations of feeding sites of feral pigs in Queensland Wet Tropics. *Wildlife Research* 24: 379–90.

Levin, S. A. (1989). Analysis of risk for invasions and control programs. In *Biological Invasions: A Global Perspective*, Drake J. A., Mooney, H. A., di Castri, F. *et al.* (eds). John Wiley & Sons, New York.

Lodge, D. M (1993). Biological invasions: lessons for ecology *Tree* **8**(4): 133–7.

Long, J. L. (1981). *Introduced Birds of the World*. A. H. & A. W. Reed Publishing, Sydney.

Lund, M. (1994). Interactions between riparian vegetation, macroinvertebrates and fish (*Gambusia holbrooki*) in Lake Monger, Western Australia. *Proceedings 33rd Congress of Australian Society for Limnology*, Rottenest Island, Perth.

Mack, R. N., Simberloff, D., Lonsdale, W. M., Evan, H., Clout, M. & Bazzaz, F. (2000). Biotic invasions: cause, epidemiology. *Global Consequences and Control Issues in Ecology* **5**.

McIlroy, J. C. & Saillard, R. J. (1989). The effect of hunting with dogs on the numbers and movements of feral pigs, *Sus scrofa*, and subsequent success of poisoning exercises in Namadgi National Park, A.C.T. *Australian Wildlife Research* **16**: 353–63.

Mooney, H. A. & Drake, J. A. (1989). Biological invasions: a SCOPE Program overview In *Biological Invasions: A Global Perspective*, Drake J. A., Mooney, H. A., di Castri, F. *et al.* (eds). John Wiley & Sons, New York.

Moriarty, A. (2004). The liberation, distribution, abundance and management of wild deer in Australia. *Wildlife Research* **31**: 291–9.

Natural Resources, Mines and Energy (NRM & E) (2004). *Annual Pest Assessment 2004*. Queensland Department of Natural Resources, Mines and Energy.

North Queensland Joint Board (1997). *Barron River Catchment Rehabilitation Plan*. North Queensland Joint Board, Cairns.

Norton, G. A. & Pech, R. P. (1988). *Vertebrate Pest Management in Australia – A Decision Analysis/Systems Analysis Approach.* CSIRO Division of Wildlife and Ecology, Canberra.

Olsen, P. (1998). *Australia's Pest Animals: New Solutions to Old Problems.* Bureau of Resource Sciences and Kangaroo Press.

Pavlov, P. M. (1992). *Investigation of Feral Pig Populations and Control Measures – Cape Tribulation Section of the Wet Tropics World Heritage Area.* Internal Report for Wet Tropics Management Authority, Cairns.

Pavlov, P. M., Crome, F. H. J. & Moore, L. A. (1992). Feral pigs, rainforest conservation and exotic disease in North Queensland. *Wildlife Research* **19**: 179–93.

Pell, A. S. & Tidemann, C. R. (1997). The impact of two exotic hollow-nesting birds on two native parrots in savannah and woodland in eastern Australia. *Biological Conservation* **79**: 145–53.

Petren, K. & Case, T. J. (1998). Habitat structure determines competition intensity and invasion success in gecko lizards. *Proceedings of the National Academy of Sciences* **95**(20): 11739–44.

Pimm, S. L. (1989). Theories of predicting success and impact of introduced species In *Biological Invasions: A Global Perspective*, Drake J. A., Mooney, H. A., di Castri, F. *et al.* (eds). John Wiley & Sons, New York.

Pusey, B. J. & Kennard, M. J. (1994). *The Freshwater Fauna of the Wet Tropics Region of Northern Queensland.* Consultancy Report for the Wet Tropics Management Authority, Cairns.

Queensland Museum (2000). *Wildlife of Tropical North Queensland.* Queensland Museum, Brisbane.

Ricciardi, A. & Rasmussen, J. B. (1998). Predicting the identity and impact of future biological invaders: a priority for aquatic resource management. *Canadian Journal of Aquatic Science* **55**: 1759–65.

Richardson, D. M., Pysek, P. Rejmanek M., Barbour, M.G., Panetta, R. D. & West, C. J. (2000). Naturalisation and invasion of alien plants: concepts and definitions. *Diversity and Distributions* **6**: 93–107.

Russell, D. J., Hales, P. W. & Helmke, S. A. (1996). *Stream Habitat and Fish Resources in the Russell and Mulgrave Rivers Catchment.* QDPI, Brisbane.

Sax, D. F. & Brown, J. H. (2000). The paradox of invasion. *Global Ecology and Biogeography* **9**: 363–71.

Schmidt, C., Felderhof, L., Kanowski, J., Stirn, B., Wilson, R. & Winter, J. W. (2000). *Tree-Kangaroos on the Atherton Tablelands: Rainforest Fragments as Wildlife Habitat. Information for Shire Councils, Land Managers and the Local Community.* Tree-Kangaroo and Mammal Group, Inc., Atherton. 36 pp.

Simberloff, D. (1981). Community effects of introduced species In *Biotic Crises in Ecological and Evolutionary Time*, Nitecki, M. H. (ed.). Academic Press, New York, pp. 53–83.

Species Survival Commission (2000). *Guidelines for the Prevention of Biodiversity Loss Caused by Invasive Species* (www.icun.org/themes/ssc/pubs/policy/invasivesEng. htm).

Stone, C. P. & Loope, L. L. (1987). Reducing negative effects of introduced animals on native biotas in Hawaii: what is being done, what needs doing, and the role of national parks. *Environmental Conservation Biology* **14**: 235–58.

Stone, C. P. & Stone, D. B. (1989). *Conservation Biology in Hawaii.* University of Hawaii Press, Honolulu.

Strahan, R. (1993). *The Complete Book of Australian Mammals.* Angus & Robertson Publishers, Sydney.

Tidemann, C. R & Pell, A. S. (1997a). The ecology of the common myna in urban nature reserves in the Australian Capital Territory. *Emu* **97**(2): 141–9.

Tyler, M. (1996). Cane toad control. *Australian Biologist* **9**(4): 120–1.

Usher, M. B. (1989). Ecological Effects of controlling invasive terrestrial vertebrates In *Biological Invasions: A Global Perspective*, Drake J. A., Mooney, H. A., di Castri, F. *et al.* (eds). John Wiley & Sons, New York.

Werren, G. L. (1993). Size and diet of *Bufo marinus* in rainforest of north eastern Queensland. *Memoirs of the Queensland Museum* **34**(1): 240.

Werren, G. L., Goosem, S., Tracey, J. & Stanton, J. P. (1995). The Australian Wet Tropics: centre of plant diversity. In *World Centres of Plant Diversity 2*, Davis, S. D., Heywood, V. H. & Hamilton, A. C. (eds). Oxford Information Press, WWF/ICUN, pp. 500–6.

Weston, N. & Goosem, S. (2004). *Sustaining the Wet Tropics: A Regional Plan for Natural Resource Management. Volume 2A Condition Report: Biodiversity Conservation.* Rainforest CRC and FNQ NRM Ltd, Cairns.

Williams, S. E., Pearson, R .G. & Walsh, P .J. (1996). Distributions and biodiversity of the terrestrial vertebrates of Australia's Wet Tropics: a review of current knowledge. *Pacific Conservation Biology* **2**: 327–62.

Wet Tropics Management Authority (WTMA) (1998). *Wet Tropics Management Plan.* Wet Tropics Management Authority, Cairns.

WTMA (2004). *Annual Report and State of the Wet Tropics Report 2003–2004.* Wet Tropics Management Authority, Cairns.

Zupanovic, Z., Lopez, G., Hyatt, A. D. *et al.* (1998a). Giant toads *Bufo marinus* in Australia and Venezuela have antibodies against ranaviruses. *Diseases of Aquatic Organisms* **32**: 1–8.

Zupanovic, Z., Musso, C., Lopez, G. *et al.* (1998b). Isolation and characterisation of iridoviruses from the giant toads *Bufo marinus* in Venezuela. *Diseases of Aquatic Organisms* **33**(1): 1–9.

26 APPLICATIONS OF HIGH RESOLUTION REMOTE SENSING IN RAINFOREST ECOLOGY AND MANAGEMENT

David Gillieson[1]*, Tina Lawson[2]* and Les Searle[1]

[1]School of Earth and Environmental Sciences, James Cook University, Cairns, Queensland, Australia
[2]CSIRO Sustainable Ecosystems, Tropical Forest Research Centre, Atherton, Queensland, Australia
*The author was participant of Cooperative Research Centre for Tropical Ecology and Management

Introduction

Remote sensing science has provided new, synoptic views of ecological processes and land use impacts since the 1970s and the development of digital image analysis in the late 1980s added a wide range of potential techniques for research applications. Remotely sensed data can be used to generate a wide range of estimates of forest structure and tree physiology that are of value to ecologists, including rainforest extent, canopy closure and connectivity, tree mortality, leaf area index (LAI), photosynthetic pigments and other indices of primary productivity.

The increasing availability of high resolution satellite and airborne sensors (here defined as having an instantaneous field of view (IFOV) of less than 16 m^2) has prompted new studies at local scales. These studies have permitted a better appreciation of the accuracy of modelled parameters and indices gained from low resolution sensors such as LANDSAT, AVHRR and MODIS (Roberts *et al.* 1998). In addition, they have allowed estimation of within-stand and individual tree canopy structure and physiology. The issue of scaling up from ground observations to satellite data is now more easily addressed, especially when local scale plot measurements are complemented by spectral radiometry of individual plants, plots and stands.

Implicit in all of these methodologies and their applications is the premise that there is a direct relationship between the forest structural and physiological parameters and the spectral reflectance as measured by the satellite or airborne sensor. It is assumed that the spectral signatures of individual trees or forest types are unique and easily separable. A final assumption is that each pixel in the image is uniform and contains only one vegetation type or cover class. If this last assumption is invalid, linear pixel un-mixing techniques (Adams *et al.* 1995) can resolve a limited number of components given adequate spectral resolution.

Methodologies for remote sensing of local scale processes and impacts in rainforests

Improvements in sensor technology over the past decade have made it possible to reduce the pixel size in images to sub-metre dimensions. Thus the latest generation of high resolution satellite sensors (Table 26.1) have IFOV dimensions of 0.6 m in panchromatic (greyscale) and 2–4 m in multispectral (multiple band) modes.

Table 26.1 High resolution satellite sensors for rainforest applications

Satellite/Launch	Operator	Instrument IFOV (m)	Revisit time (days)	Status
ALOS/2006	Japan	2.5 m Pan,10 m MS, 7-89 m SAR	2	Operational
CartoSat-1/2005	India	2.5 m PAN	5	Operational
EROS-A/2000	Israel	0.8–1.8 m PAN	2–3	Operational
EROS-B/2006	Israel	0.7 m PAN	2–3	Planned
EROS-C/2009	Israel	0.7 m PAN, 2.8 MS	2–3	Planned
IKONOS-2/1999	USA	0.82–1 m PAN, 4 m MS	3–5	Operational
IRS-1D/1997	India	5 m PAN, 23 m MS	5–24	Operational
Orbview-3/2003	USA	1 m PAN, 4 m MS	3	Operational
Orbview-5/2007	USA	0.41 m PAN, 1.64 m MS	< 3	Planned
QuickBird-2/2001	USA	0.61 m PAN, 2.4 m MS	1–3.5	Operational
RadarSat-2 2006	Canada	3 m SAR	3–24	Planned
ResourceSat-1 2003	India	5 m PAN, 23 m MS	5–24	Operational
SPOT-5 2002	France	2.5 m PAN, 10 m MS	1	Operational
Worldview-I 2006	USA	0.45 m PAN	1.7	Planned
Worldview-II 2008	USA	0.46 m Pan, 1.8 m MS	1.1	Planned

PAN, panchromatic (grey scale); MS, multispectral; HS, hyperspectral; SAR, synthetic aperture radar.

The new generation of satellite sensors have vastly improved spatial and spectral resolution, with additional radiometric resolution to 11-bit data (2048 grey levels). Repeat coverage is possible at 3–5 day intervals, making them ideal for assessing rapid changes such as fires and floods. IKONOS and Quickbird sensors have spatial resolutions of 0.6–1 m in panchromatic mode and 2.5–4 m in multispectral mode. Spectrally four bands from visible blue to near infrared are available, with good separation and narrow bandwidth. For the first time we are able to resolve features down to the size of individual tree canopies.

Airborne multispectral scanners have similar spectral characteristics to the satellite sensors, with three or four bands in visible and near infra-red (NIR) light. Spatial resolution can range from 10 cm to 1 m depending on aircraft height above the terrain, and linked differential global positioning system (GPS) makes precise location of flight lines possible. These scanners find ready application in smart agriculture and are now routinely used for monitoring of crop productivity and health in many areas of intensive, irrigated agriculture worldwide. They have the advantage that recording of images over a given area can be virtually on demand given good weather and positioning of the aircraft and its sensor in the region. They are very useful for gaining high resolution images, similar to aerial photographs, to be used in forest extent mapping or as backdrops in geographical information systems (GIS) analysis.

Irrespective of whether a satellite or airborne high resolution sensor is used, there are some technical issues that have to be addressed if digital image analysis is contemplated. The first issue relates to the need to orthorectify the image accurately to a topographic map base, allowing integration with GIS and extraction of thematic features such as roads, buildings and forest margins. This requires a good network of accurately located ground control points (GCPs) and a digital elevation model (DEM) with appropriate cell resolution. For many rainforest areas, we only have DEMs with cell resolutions of 30–80 m, which is insufficient precision for the task. Laser scanning techniques (LIDAR) and stereo models from aerial photography provide solutions to this problem, albeit at increased cost. The second issue relates to radiometric calibration, where the digital numbers (brightness values) in the image must be correlated with actual surface reflectance values. Most satellite sensors are well calibrated and maintain linearity throughout the full range of brightness values. For airborne scanners, this can be achieved using standard white and black reflectors placed in the field and recorded by the sensor at the same time as the rest of the scene. For some sensors, the relationship is non-linear at high reflectance values, limiting their utility. Finally, most of the airborne sensors use modified video cameras, in which lens curvature produces a vignetting effect. This variation in image brightness can be corrected given precise knowledge of the lens shape parameters and imaging geometry.

High resolution imagery is typically used in a wide range of applications for image interpretation, cartography and soft photogrammetry. Image interpretation can be either human visual interpretation or automated machine interpretation. Visual interpretation of orthorectified images is typically used in intelligence and disaster relief applications and can also produce high quality results when used in combination with ancillary GIS data such as forest mensuration and detailed vegetation mapping. Machine interpretation is commonly used for land use classification and feature extraction. Machine classification favours 11-bit products, such as IKONOS, that maintain absolute radiometric calibration of the multispectral imagery. There has been a recent move away from traditional, per-pixel classifiers towards object-oriented classifiers that rely on a hierarchy of image segmentation producing polygons of similar spectral response, shape and area. Thus contextual concepts derived from landscape ecology and photo interpretation can now be integrated with image data. Derived image products such as vegetation indices can be used, with appropriate ground measurements, to develop robust relationships capable of predicting above-ground biomass and ultimately carbon accounting.

Orthorectified images are the fundamental products for image maps, GIS image base and cartographic extraction of infrastructure such as roads, buildings and drainage. GIS systems commonly show hydrographic, transportation and other information as vector layers. Displaying those vectors on top of a base image adds useful context to the vector information.

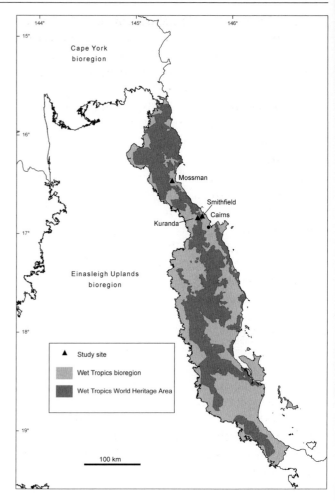

Figure 26.1 The location of three study sites in the Wet Tropics World Heritage Area of north Queensland.

Case studies

In this section we provide a number of examples using high resolution imagery to illustrate these various methods of digital image analysis. All are drawn from research carried out since 2000 in the Wet Tropics World Heritage Area (WTWHA) under the auspices of the Rainforest CRC in Cairns, north Queensland (Figure 26.1).

Assessing canopy connectivity across roads in rainforest areas

Road ecology is a rapidly developing branch of landscape ecology that focuses attention on the roles of roads as barriers to wildlife movement and as agents of fragmentation in forested landscapes (Forman & Alexander 1998). Few studies have been carried out on the ecology of roads in the humid tropics (Laurance *et al.* 2004; Goosem 2000) but there is a rapidly growing literature on the ecological effects of roads in temperate landscapes (Forman *et al.* 2003).

Roads affect wildlife in several ways at a variety of spatial scales. At a local scale, individual stretches of road with traffic form behavioural barriers to the movement of wildlife; there may be mortality due to animals being run over or the combination of road verge and pavement may inhibit or preclude animals from attempting to cross the road. The condition of the road verge may influence the risk: densely forested or shrub covered verges may be conducive to movement but the rapid emergence of animals on to the road pavement may increase mortality at black spots (Goosem 2002). For arboreal mammals, canopy connectivity is critical. Tree canopies must interfinger over a sufficient distance to allow free animal movement, although research has found that canopy bridges may provide a viable alternative.

In this study we estimated the forest canopy coverage and connectivity across a major road that runs through the Wet Tropics World Heritage Area of tropical North Queensland. The Kuranda Range Road (Figure 26.2) is the major arterial link between Cairns and the nearby towns of Kuranda, Mareeba and

Atherton on the northern Atherton Tablelands. It carries high volumes of traffic comprised of commuting residents, trucks and tourists. The road passes through tropical rainforest communities as well as cleared land, regrowth and residential areas.

The production of a detailed coverage of the road corridor was achieved by the integration of spatial data in different formats in ArcGIS 8.3. The imagery used was pan-sharpened IKONOS multispectral data, with a resolution of 1 m. A vector coverage in AutoCAD format provided road centreline and pavement edge data, based on ground surveys carried out by the Queensland Main Roads Department. This had millimetric precision.

The IKONOS imagery was classified using an ISODATA unsupervised classification in ER Mapper 6.4 to identify spectral classes of information. There was very good separability between forest canopy and road pavement owing to the greater data range in IKONOS (11-bit). In addition, spectral properties of individual image bands were used to differentiate between road and woody vegetation, and a series of spectral transects across the road used NIR/Red thresholding (Figure 26.3) to confirm separability. The binary image was then converted to vector.

The vector road data were cleaned and extraneous linework was removed, leaving only the pavement edges and centreline. Missing sections of centreline were reconstructed using an intersection of arcs pivoted on the pavement edge. The road pavement vectors were then intersected with the classified image to yield a linear vector data set that was bounded by the

Figure 26.2 The Kuranda Range road overlain on a visualization of terrain and rainforest vegetation.

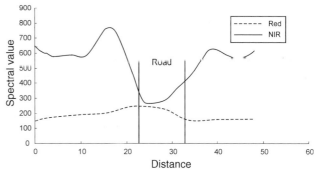

Figure 26.3 A spectral transect across Kuranda Range road at 9 km from the start, showing the decline in NIR reflectance across the road base. Vertical lines indicate the surveyed edge of the road base.

Figure 26.4 Canopy overhang polygons overlain on a visu-alization of terrain and rainforest canopy image (IKONOS infra-red false colour composite).

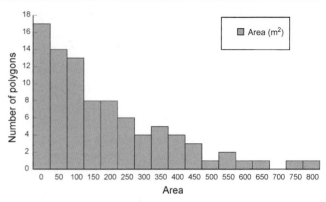

Figure 26.5 The distribution of connected canopy polygons along 19 km of the Kuranda Range road.

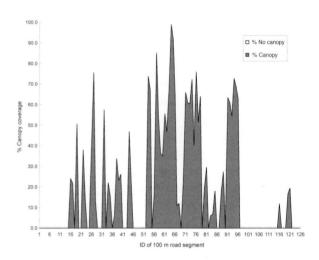

Figure 26.6 Proportional areas of canopy overhang to the road along the Kuranda Range road.

road edge and contained two categories, road and forest. This vector data set was subdivided into individual numbered polygons (forest canopy and road) and used in all subsequent spatial analysis, overlain on the IKONOS image (Figure 26.4) to allow checking of accuracy.

Canopy coverage was calculated as the percentage area of forest canopy in each 100-m segment of the road (total length 13 km). Canopy connectivity was defined by identifying all groups of contiguous polygons that intersected both pavement edges. A total of 55 connections were identified. The area (in square metres) and perimeter (in metres) of each connected polygon cluster were then calcu-lated. Areas of individual polygons ranged from 2 to 823 m^2 with a mean of 211 m^2. The distribution of connected polygons reveals a large number of small area connections and a few large area connections (Figure 26.5).

A shape index was derived using a vector interpola-tion method (triangulated irregular network or TIN). Each polygon cluster was subdivided into the mini-mum number of triangles that infilled the space. Two statistics were calculated:
• average area of the triangles in a polygon = poly-gon area ÷ number of triangles;
• shape index = polygon perimeter ÷ average area of triangle.

Percentage canopy cover on 100-m road segments, ranged from 0 to 99% with a mean of 27.9% and a standard deviation of 23.4%. Canopy coverage is low at the extremities of the road within approximately 2 km of the population centres of Kuranda and Smithfield Heights (Figure 26.6). Areas of consistently higher canopy coverage are found on the steep Kuranda Range section, around Streets Creek and near the Rainforestation tourist complex. These locations all fall within the WTWHA, where road operations such as vegetation management are subject to Best Practice Guidelines produced for the Wet Tropics Management Agency (Chester *et al.* 2006).

The degree and quality of canopy connectivity across the road can be assessed using the derived sta-tistics (Figure 26.7). Connectivity has to be assessed in

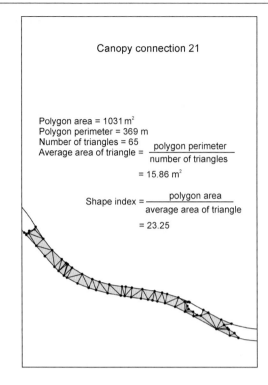

Canopy connection 21

Polygon area = 1031 m²
Polygon perimeter = 369 m
Number of triangles = 65
$$\text{Average area of triangle} = \frac{\text{polygon perimeter}}{\text{number of triangles}}$$
$$= 15.86 \text{ m}^2$$

$$\text{Shape index} = \frac{\text{polygon area}}{\text{average area of triangle}}$$
$$= 23.25$$

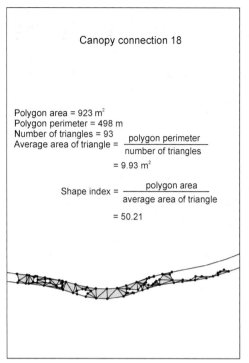

Canopy connection 18

Polygon area = 923 m²
Polygon perimeter = 498 m
Number of triangles = 93
$$\text{Average area of triangle} = \frac{\text{polygon perimeter}}{\text{number of triangles}}$$
$$= 9.93 \text{ m}^2$$

$$\text{Shape index} = \frac{\text{polygon area}}{\text{average area of triangle}}$$
$$= 50.21$$

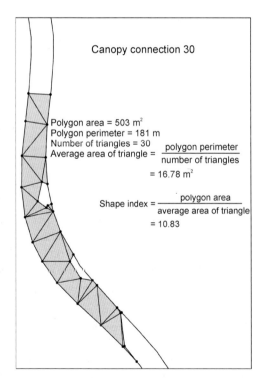

Canopy connection 30

Polygon area = 503 m²
Polygon perimeter = 181 m
Number of triangles = 30
$$\text{Average area of triangle} = \frac{\text{polygon perimeter}}{\text{number of triangles}}$$
$$= 16.78 \text{ m}^2$$

$$\text{Shape index} = \frac{\text{polygon area}}{\text{average area of triangle}}$$
$$= 10.83$$

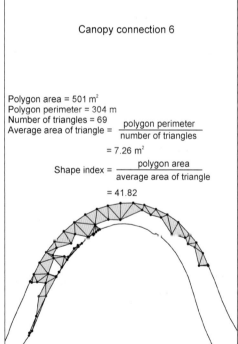

Canopy connection 6

Polygon area = 501 m²
Polygon perimeter = 304 m
Number of triangles = 69
$$\text{Average area of triangle} = \frac{\text{polygon perimeter}}{\text{number of triangles}}$$
$$= 7.26 \text{ m}^2$$

$$\text{Shape index} = \frac{\text{polygon area}}{\text{average area of triangle}}$$
$$= 41.82$$

Figure 26.7 Analysis of individual canopy overhang polygons providing shape statistics.

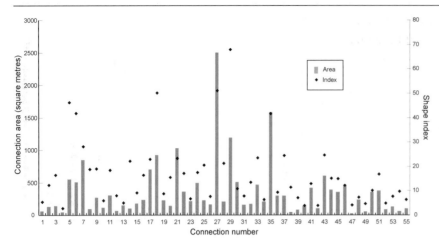

Figure 26.8 Canopy connectivity measures based on an analysis of canopy overhang polygons.

terms of both the area of the canopy connection and the shape index. A high area and low shape index indicate a connection that presents many points of contact and tends to a rectangular shape (e.g. connection 21 in Figure 26.7), rather than connections that are tangential and consist of linear strips along each side of the road.

There are a number of connections that have moderate area (500–1000 m²) and low shape index (15–20) (Figure 26.8), which may provide opportunities for marsupials and birds to move across the road corridor. Connections with an area of less than 200 m², though having a low shape index, may only provide one point for crossing and are thus vulnerable to disturbance. The road is used by many semi-trailers and their canopies may come within a few metres of the forest vegetation. Areas with high quality canopy connectivity may need to be flagged to avoid disturbance and disruption.

The accuracy of the canopy connectivity estimates was evaluated by walking the road and visiting each indicated connection (located using a GPS receiver). Indicated connections were rated into one of four categories (Table 26.2) as follows:
• unconnected – overlapping but vertically separated by more than 50 cm, or horizontally by a gap of more than 1 m;
• poor – canopies overlapping but touching at one point only, over a distance of less than 5 m, with little interfingering of branches;
• good – canopies overlapping over a distance of at least 5 m, with good interfingering of branches;

Table 26.2 Field evaluation of canopy connectivity, Kuranda Range Road

Quality	Number	%
Unconnected	33	49.25
Poor	19	28.36
Good	11	16.42
Very good	4	5.97

• very good – canopies overlapping over a distance of more than 10 m, with good interfingering of branches, and numerous connection points obvious.

An additional two canopy connections not indicated in the spatial analysis were noted and their positions determined. Where variation was noted in the connection quality, additional connections were noted for that polygon. A total of 51% of estimated canopy connections were found to be actually connected, but of these four were very good, 11 good and 19 poor. Of the estimated connections that were in fact unconnected, about half (*n* = 16) were the result of tree canopies overlapping in plan but not in elevation; there was thus a vertical separation of 1 and 5 m between the tree canopies. It is not possible to distinguish these in the spatial analysis, as the IKONOS imagery only provides a vertical view of the canopy.

This study has provided an integrative methodology for assessing canopy cover and connectivity across road and other corridors in tropical rainforests.

Its main advantage is that it provides a rapid and objective means of measuring connectivity and of targeting subsequent fieldwork to evaluate the quality of individual connections. Other applications could be found in assessing the quality and quantity of revegetation patches across powerline corridors or, given suitable stream channel definition, connectivity across tropical streams.

Tropical rainforest fires mapped using high resolution IKONOS imagery

Forty years ago tropical rainforests were considered by many ecologists to be immune to fire, due to the moisture content of the understorey (relative humidity usually over 90%) (Ginsberg 1998). However, El Niño events cause extended periods of drought, allowing the leaf litter on the forest floor to dry, while selective logging creates gaps in the canopy, allowing the fuel in the understorey to dry to potentially combustible levels (Uhl *et al.* 1988; Nepstad *et al.* 1999; Siegert *et al.* 2001). As these events are predicted to become more frequent and severe with increasing global temperatures, tropical rainforest susceptibility to fire and fire occurrence has the potential to increase dramatically (Timmerman *et al.* 1999; Cochrane 2003; Kershaw *et al.* 2003).

Forest fragmentation increases the perimeter to area ratio of the patches and leads to changes in forest structure and microclimate variables (Margules & Pressey 2000; Cochrane 2001). This increases the likelihood of fires in damaged or logged rainforests owing to the close proximity of fire-maintained land uses such as grazing. This is evident from frequent fire events in the Amazon and the 1997–8 Indonesian fires (Uhl & Kauffman 1990; Cochrane & Schulze 1998, 1999; Kinnaird & O'Brien 1998; Siegert & Hoffmann 2000).

Fires in tropical rainforests generally move slowly as creeping ribbons of flame. Fire intensity is low and fuel moisture largely controls fire propagation. Fires in previously burnt or logged rainforests are more severe. This is because of lower humidity levels and greater fuel loads (Cochrane 2003). Monitoring fires by remote sensing techniques is an important tool in inaccessible tropical rainforests because up to 80% of fires occur in the subtropics and tropics (Dwyer *et al.* 1998).

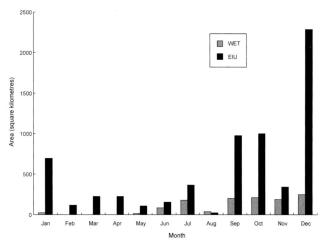

Figure 26.9 Monthly distribution of bushfire scars by area (in hectares) for the Wet Tropics (WET) and Einasleigh Uplands (EIU) bioregions of Queensland. Data supplied by Cape York Peninsula Development Association.

Much of the fire mapping in Australia uses NOAA and MODIS products. Due to the pixel size of this imagery (1 km^2) small fires (>20 hectares) are not detected. Many of the fires that occur in the Wet Tropics bioregion are small compared to those in the Einasleigh Uplands bioregions (Figure 26.9), so this type of imagery is unsuitable.

With the launch of the IKONOS satellite in September 1999, a new era of satellite imagery with resolution comparable to aerial photographs has become available (Dial & Grodecki 2003; Goetz *et al.* 2003), with 4 m multispectral and 1 m panchromatic resolution. IKONOS imagery has been previously used to differentiate between forest stand age classes (Kayitakire *et al.* 2002), to estimate tree crown diameter (Asner *et al.* 2002), to detect mountain pine beetle infestation (White *et al.* 2005), to quantify tropical rainforest tree mortality (Clark *et al.* 2004) and to map forest degradation (Souza & Roberts 2005). This study reports on the use of multi-temporal IKONOS imagery for fire scar mapping in a tropical rainforest in Far North Queensland.

This study was undertaken in the Smithfield Conservation Park (SCP), 17 km north of Cairns (Figure 26.10). The Smithfield Conservation Park has an area of approximately 270 ha, of which about 180 ha is rainforest vegetation. The rough terrain supports rainforest vegetation types of mesophyll vine forest, complex notophyll vine forest and simple notophyll vine forest (Stanton & Stanton 2003).

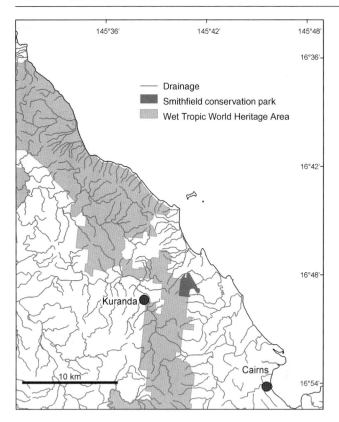

Figure 26.10 The location of the study area in the Cairns region of the Wet Tropics bioregion.

Table 26.3 Drought duration and rainfall data for Kuranda railway station from 1898 to 2005

Period	Duration (months)	Driest 12 months	Average SOI
May 2001 to Mar 2003	23	831	−4.8
Mar 1991 to Nov 1992	21	877	−11
Feb 1989 to Jan 1990	12	1015	5.6
Mar 1981 to Feb 1982	12	1029	2.7
Oct 1965 to Jan 1967	16	864	−4
Mar 1960 to Dec 1961	22	847	2.6
Jan 1951 to Dec 1951	12	1036	−0.7
Feb 1946 to Mar 1947	14	1016	−5.4
Jul 1925 to Aug 1926	14	1037	−8.5

Australia and tropical Queensland are no strangers to extreme El Niño events. Table 26.3 highlights drought duration and rainfall for the Kuranda railway station from 1898 to 2005, while Figure 26.11 shows the

occurrence of moderate and severe droughts in the Cairns region. The most severe drought was experienced in 2001–2, resulting in only 721 mm annual rainfall at Cairns, the lowest annual rainfall recorded for the region. This is well below the annual average of 2100 mm. This extended dry period (23 months in total) had serious consequences for the Wet Tropics. The fire that swept through the Smithfield Conservation Park in November 2002 burnt approximately 150 ha of the 180 ha of rainforest vegetation (Marrinan *et al.* 2005). This was a one in 100 year event that patchily burnt the steep rainforested ridges of the McAllister Range over a 3-week period. An extremely dry period in 2002 led to a large amount of dry leaf litter on the forest floor and this fuelled the fire (Marrinan *et al.* 2005). The fire burnt significant remaining examples of lowland complex mesophyll vine forest and mesophyll vine forest. Fire intensity for this event was thought to be low, as canopy scorching was absent and fire scars were restricted to the bases of trees (Marrinan *et al.* 2005).

IKONOS images of Smithfield, dated 29 January 2002 and 17 June 2004, were orthorectified using the Orthowarp extension for ER Mapper. Topographic correction of this imagery was necessary because in this area of rugged terrain similar vegetation types may vary highly in their spectral response owing to the shape of the terrain and low solar illumination angle, creating a shadowing effect. Data used in this study were topographically corrected using the C-correction (a modification of the cosine correction). Original means were preserved and the standard deviations were only slightly different to those of the original.

Once the 2002 and 2004 data were topographically corrected, the bands of the 2004 imagery were histogram matched to those of the 2002 imagery (i.e. red band 2004 matched to red band 2002, blue band 2004 matched to blue band 2002 etc.). This has the effect of giving the same bands in both images similar contrast and brightness. The histogram-matched image values were then converted from digital numbers (DN) to reflectance values.

The Enhanced Vegetation Index (EVI) was developed to optimize the vegetation signal with improved sensitivity in high biomass regions and improved vegetation monitoring through a decoupling of the canopy background signal and a reduction in atmosphere influences. The EVI formula was applied to the 2002

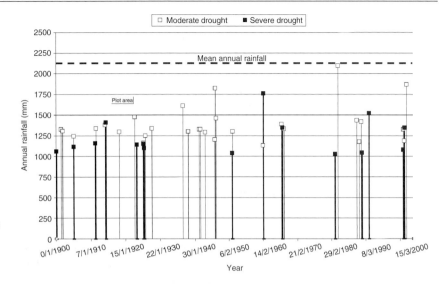

Figure 26.11 The occurrence of moderate and severe droughts in the Cairns region. Data from Kuranda railway station, processed in Rainman-Streamflow software.

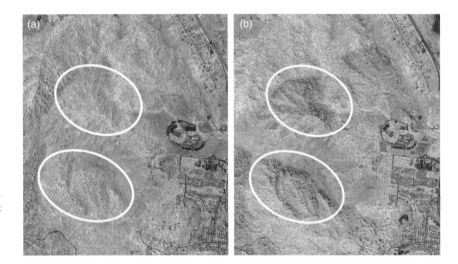

Figure 26.12 Enhanced vegetation index (EVI) images of the Smithfield study area: (a) pre-fire January 2002; (b) post-fire June 2004. No cloud-free images were available between these two dates.

and 2004 images. The ISODATA unsupervised classification used in this study identifies spectrally similar groups of pixels without any information on class land cover.

The results of the EVI on the 2002 and 2004 IKONOS imagery clearly show areas of burnt terrain on the latter image (Figure 26.12). These results are consistent with field observations that were carried out in April 2003. Through a reduction in atmospheric noise and a decoupling of the canopy background signal, the EVI has highlighted areas with significantly different vegetation signals (i.e. areas of green vegetation and burnt areas).

The 2004 image classification clearly delineates the fire scars (Plate 26.1). Analysis of the scattergrams of the classification revealed that the new class (fire scars) occurred on the soil line where dry vegetation or soil would occur. The areas mapped in the classification as fire scars match up visually with the areas delineated as fire scars by the EVI. During April 2003 the perimeters of the larger fire scars indicated in Plate 26.1 were mapped using GPS; other fire scars were visually checked from the Saddle Mountain fire-trail to the north of the study area. There is a very good correspondence between the areas classified in Plate 26.1 and burnt areas on the ground. However, the internal patchiness of burnt areas makes precise spatial corre-spondence difficult. This is a common problem with per pixel classifiers.

Multitemporal IKONOS imagery, with its high reso-lution, is very useful in detecting fire scars in the tropi-cal rainforests of far north Queensland. The high resolution (4 m multispectral and 1 m spatial cells) of the imagery allows for the discrimination of fire scars on a much finer scale than NOAA or MODIS. However, a number of image processing procedures must be employed before the imagery can be used to discrimi-nate successfully between burnt and unburnt land.

Assessing the health of riparian rainforests in the Mossman catchment

In dynamic tropical landscapes with high natural vari-ability in disturbance regimes, human impacts must be evaluated against the long-term record of environmen-tal change to assess their ecological effects reliably (Landres *et al.* 1999). GIS and remote sensing tech-niques have been used in many areas to assess vegeta-tion change over time (Mosugelo *et al.* 2002; Rhemtulla *et al.* 2002; Narumalani *et al.* 2004; Bouma & Kobryn 2004). The role of historical data is increasingly becom-ing an important measure in the fields of ecosystem management and restoration ecology (Swetnam *et al.* 1999; Landres *et al.* 1999).

This study was undertaken on the coastal lowlands of Douglas Shire. The area has been heavily trans-formed for agricultural purposes, predominantly sug-arcane production. A large proportion of the remaining vegetation consists of riparian rainforest, the focus of the study. The Mossman River (Figure 26.13) is the major watercourse in this catchment, with the South Mossman River and Cassowary Creek being its two major tributaries (Burrows 1998).

Traditional remote sensing products such as aerial photography can also be analysed using the new soft-ware tools developed for digital image analysis and GIS. This is advantageous when historical images pre-date the introduction of digital formats. Conversion to digital images, rectification to a map base and integra-tion within GIS allow detailed analysis of landscape change. In this study, changes in the extent of riparian vegetation along the Mossman River, South Mossman River and Cassowary Creek catchments were assessed using aerial photography from 1944 and 2000. The oldest

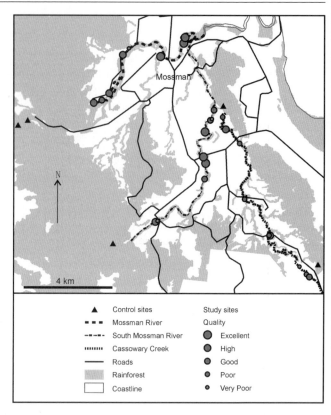

Figure 26.13 The location of study sites around the town of Mossman, north Queensland.

aerial photography available was flown in September 1944 by the Royal Australian Air Force (RAAF) at a scale of approximately 1 : 24 600 (greyscale). The more recent set of colour photos were flown in August 1998 and 2000 at a scale of approximately 1 : 25 000. Both sets of photos were orthorectified and mosaiced using ER Mapper 6.4 and OrthoWarp 2.3 software. The extent of the riparian vegetation along the Mossman River, South Mossman River and Cassowary Creeks was dig-itized on-screen on the 1944 data set, working at a scale of approximately 1 : 3500. This digitized layer was then overlain on the 2000 mosaic and areas of vegetation loss or gain were identified and digitized. Areas of vegetation thickening (regrowth) were also digitized together with areas that showed no difference in the extent of the riparian vegetation (within error limits) between the two dates. Consequently, four new GIS layers were created and were used to produce a final map (Figure 26.14) highlighting the change in extent of the riparian vegetation along the Mossman River, the South Mossman River and Cassowary Creek between 1944 and 2000.

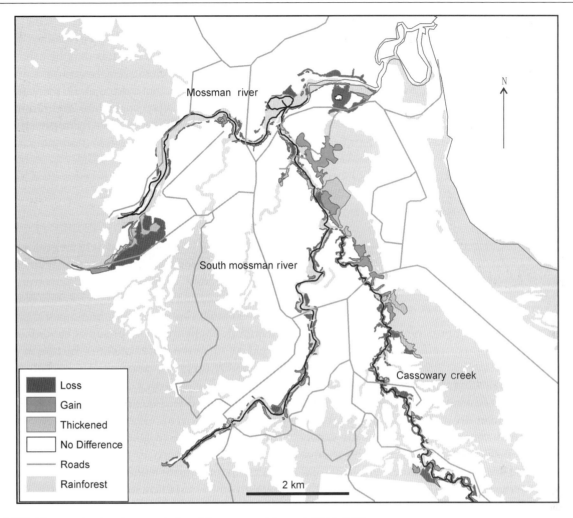

Figure 26.14 Changes in riparian vegetation, Mossman catchment, 1944–2000.

The use of repeat aerial photography in this study showed that during the 56 year period 1944–2000, a total of 285 ha of riparian forest changed in its extent (Table 26.4). Of this change approximately 124 ha were gained and 102 ha were lost. This resulted in a net gain of 22 ha of riparian rainforest along the three watercourses. A further 59 ha of forest became obviously thicker in its cover over this time period. Differences in riparian vegetation extent were identified as either a result of clearing, changes in farm management practices or stream channel movement.

The coastal lowlands are highly valued for agriculture owing to reliable rainfall and high soil fertility. Subsequently large tracts of rainforest have been cleared to facilitate these activities (Figure 26.14)

Table 26.4 Total change in extent of riparian rainforest in the Mossman Catchment between 1944 and 2000

	Area (ha)	% of total change
Loss	102	30
Thickening	59	21
Gain	124	43
Total	285	100

(Winter *et al.* 1987). Where forests existed in the mid-1940s but in 2000 had been developed for agricultural production, clearing was identified as the major cause of change in riparian forest extent. Within the study area, most of this type of vegetation change has

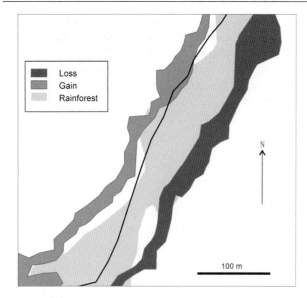

Figure 26.15 An example of riparian vegetation change due to natural stream channel movement, South Mossman River and Cassowary Creek.

occurred in two large areas, one towards the headwaters of the Mossman River and the other towards the mouth of the Mossman River. There are many smaller patches of riparian vegetation loss along all watercourses that may be attributed to clearing. This change has resulted in the loss of approximately 92 ha of riparian vegetation in the study area between 1944 and 2000, while changes in farm management result in a gain of approximately 110 ha of riparian forest and the thickening of approximately 55 ha of hill slope and open riparian vegetation.

Stream channel movement can be identified in GIS layers by loss of vegetation on one side of the river and consequent gain of vegetation on the other side of the river (Figure 26.15). Stream channel movement occurs in several locations along the Mossman River, the South Mossman River and Cassowary Creek. This type of change has resulted in the loss of approximately 10 ha and subsequent gain of 14 ha along with the thickening of approximately 3 ha of riparian vegetation over the study area.

The change in extent of riparian vegetation along the Mossman, South Mossman and Cassowary riparian tracts showed that both anthropogenic activities (clearing and changing land management practices) and natural processes (stream channel movement) have influenced the extent of riparian vegetation in this area.

There is now a clear and objective view of where changes have occurred in the riparian vegetation between the two dates and this may be used to model natural vegetation change in the future. Knowing the types of vegetation that existed along the watercourses in the past may help revegetation programmes in the future by giving them a base from which to work, when combined with vegetation descriptions currently available (Tracey 1982).

Summary

The choice of remotely sensed data for ecological applications in rainforest areas and for management studies will ultimately depend on precise formulation of research questions and careful evaluation of the necessary spatial, spectral and radiometric resolution required. There are many potential candidates among a galaxy of sensors and satellites. Factors of price and availability cannot be discounted, especially in tropical regions where persistent cloud cover can thwart efforts to obtain cloud-free imagery for particular dates and months of any year. There has also been a shift from low cost, government sponsored sensors to higher cost, commercial sensors over the past decade. Maintaining access to the data for a particular sensor may be a problem given the volatility of the commercial scene, with many company mergers and an evolving trend to restrictions on data access for strategic reasons.

It is now possible to sense remotely many forest structural parameters and some plant physiological parameters using high resolution remote sensing. All of these require detailed ground verification of both deterministic and empirical models, adding to the cost of any project. It may be that these models are specific for a particular rainforest type, but robust methodologies should be adaptable to different settings.

The methodologies for remote sensing of land use change, forest clearance and fragmentation are now well established at a variety of spatial scales and ome are becoming routine operational tools for land managers. But there is a real need for adequate training for land managers in remote sensing techniques, especially in developing countries and also in more remote regions of developed countries.

The challenges remaining for researchers are to evaluate new image classification methodologies, such as object oriented classifiers, for rainforest environments and to refine further the models that relate forest structural and physiological parameters to remotely sensed data. For managers, there needs to be an acceptance of what remote sensing can and cannot do in evaluating environmental impacts and land use change. Thus researchers and managers need to form strategic alliances to guide research and to inform government policy on remote sensing and its applications.

References

Adams, J. B., Sabol, D. E., Kapos, V *et al.* (1995). Classification of multispectral images based on fractions of endmembers: Application to land-cover change in the Brazilian Amazon. *Remote Sensing of the Environment* **52**: 137–54.

Asner, G. P., Palace, M., Keller, M., Pereirs, R. Jr, Silva, J. N. M. & Zweede, J. C. (2002). Estimating canopy structure in an Amazon forest from laser range finder and IKONOS satellite observations. *Biotropica* **34**: 483–92.

Bouma, G. A. & Kobryn, H. T. (2004). Change in vegetation cover in East Timor, 1989–1999. *Natural Resources Forum* **28**: 1–12.

Burrows, D. W. (1998). *FNQ 2010 Regional Environmental Strategy – Key Waterways Report*. Report for the Department of Environment, Cairns. Australian Centre for Tropical Freshwater Research, James Cook University, Townsville.

Chester, G., Goosem, M., Cowan, J., Harriss, C. & Tucker, N. (2006). *Roads in Tropical Forests – Best Practice Guidelines, Planning, Design and Management*. Rainforest CRC, Cairns.

Clark, D. B., Castro, C. S., Alvarado, L. D. A. & Read, J. M. (2004). Quantifying mortality of tropical rainforest trees using high-spatial-resolution satellite data. *Ecology Letters* **7**: 52–9.

Cochrane, M. A. (2001). Synergistic interactions between habitat fragmentation and fire in evergreen tropical forests. *Conservation Biology* **15**: 1515–21.

Cochrane, M. A. (2003). Fire science for rainforests. *Nature* **421**: 919.

Cochrane, M. A. & Schulze, M. D. (1998). Forest fires in the Brazilian Amazon. *Conservation Biology* **12**: 948–50.

Cochrane, M. A & Schulze, M. D. (1999). Fire as a recurrent event in tropical forests of the eastern Amazon: Effects on forest structure, biomass, and species composition. *Biotropica* **31**: 2–16.

Dial, G. & Grodecki, J. (2003). *Applications of IKONOS Imagery*. ASPRS Annual Conference Proceedings, Anchorage, Alaska.

Dwyer, E., Gregoire, J. M. & Malingreau, J. P. (1998). A global analysis of vegetation fires using satellite images: spatial and temporal dynamics. *Ambio* **27**: 175–81.

Forman, R. T. T. & Alexander, L. E. (1998). Roads and their major ecological effects. *Annual Review of Ecology and Systematics* **29**: 207–31.

Forman, R. T., Sperling, D., Bissonette, J. A., *et al.* (2003). *Road Ecology – Science and Solutions*. Island Press, Washington, DC. 481 pp.

Ginsberg, J. R. (1998). Perspectives on wildfire in the humid tropics. *Conservation Biology* **12**: 942–3.

Goetz, S. J., Wright, R. K., Smith, A. J., Zinecker, E. & Schaub, E. (2003). IKONOS imagery for resource management: tree cover, impervious surfaces, and riparian buffer analysis in the mid-Atlantic region. *Remote Sensing of Environment* **88**: 195–208.

Goosem, M. W. (2000). Effects of tropical rainforest roads on small mammals: edge changes in community composition. *Wildlife Research* **27**: 151–63.

Goosem, M. W. (2002). Effects of tropical rainforest roads on small mammals: fragmentation, edge effects and traffic disturbance. *Wildlife Research* **29**: 277–89.

Kayitakire, F., Farcy, C. & Defourny, P. (2002). IKONOS-2 imagery potential for forest stands mapping. Presented at ForestSAT Symposium, Heriot Watt University, Edinburgh.

Kershaw, A. P., Moss, P. & Van der Kaars, S. (2003). Causes and consequences of long-term climatic variability on the Australian continent. *Freshwater Biology* **48**: 1274–83.

Kinnaird, M. F. & O'Brien, T. G. (1998). Ecological effects of wildfire on lowland rainforest in Sumatra. *Conservation Biology* **12**: 954–6.

Landres, P. B., Morgan, P. and Swanson, F. J. (1999). Overview of the use of natural variability concepts in managing ecological systems. *Ecological Applications* **9**: 1179–88.

Laurance, S. G., Stouffer, P.C. & Laurance, W. F. (2004). Effects of road clearings on movement patterns of understory rainforest birds in central Amazonia. *Conservation Biology* **18**: 1099–109.

Margules, C. R. & Pressey, R. L. (2000). Systematic conservation planning. *Nature* **405**: 243–53.

Marrinan, M. J., Edwards, W. & Landsberg, J. (2005). Resprouting of saplings following a tropical rainforest fire in north-east Queensland, Australia. *Austral Ecology* **30**: 817–26.

Mosugelo, D. K., Moe, S. R., Ringrose, S. & Nellemann, C. (2002). Vegetation changes during a 36-year period in northern Chobe National park, Botswana. *African Journal of Ecology* **40**: 232–40.

Narumalani, S., Mishra, D. R. & Rothwell, R. G. (2004). Analysing landscape structural change using image interpretation and spatial pattern techniques. *Mapping Sciences and Remote Sensing* **41**: 25–44.

Nepstad, D. C., Verissimo, A., Alencar, A., *et al.* (1999). Large-scale impoverishment of Amazonian forests by logging and fire. *Nature* **398**: 505–8.

Rhemtulla, J. M., Hall, R. J., Higgs, E. S. & Macdonald, S. E. (2002). Eighty years of change: vegetation in the montane ecoregion of Jasper National Park, Alberta, Canada. *Canadian Journal of Forest Research* **32**: 2010–21.

Roberts, D. A., Batista, G., Pereira, J., Waller, E. & Nelson, B. (1998). Change identification using multitemporal spectral mixture analysis: applications in eastern Amazonia. In *Remote Sensing Change Detection: Environmental Monitoring Applications and Methods*, Elvidge, C. & Lunetta R. (eds). Ann Arbor Press, Ann Arbor, MI, pp. 137–61.

Siegert, F. & Hoffmann, A. A. (2000). The 1998 forest fires in East Kalimantan (Indonesia): a quantitative evaluation using high resolution, multitemporal ERS-2 SAR images and NOAA-AVHRR hotspot data. *Remote Sensing of Environment* **72**: 64–77.

Siegert, F., Ruecker, G., Hinriches, A. & Hoffmann, A. A. (2001). Increased damage from fires in logged forests during droughts caused by El Niño. *Nature* **414**: 437–40.

Souza, C. M. Jr & Roberts, D. (2005). Mapping forest degradation in the Amazon region with IKONOS images. *International Journal of Remote Sensing* **26**: 425–9.

Stanton, P. & Stanton, D. (2003). *Various Map Sheet Reports.* Unpublished reports to the Wet Tropics Management Authority, Cairns.

Swetnam, T. W., Allen, C. D. & Betancourt, J. L. (1999). Applied historical ecology: using the past to manage for the future. *Ecological Applications* **9**: 1189–206.

Timmerman, A., Oberhuber, J., Bacher, A., Esch, M., Latiff, M. & Roeckner, E. (1999). Increased El Niño frequency in a climate model forced by future greenhouse warming. *Nature* **398**: 694–7.

Tracey, J. G. (1982). *The Vegetation of the Humid Tropical Region of North Queensland.* CSIRO, Melbourne.

Uhl, C. & Kauffman, J. B. (1990). Deforestation, fire susceptibility, and potential tree responses to fire in the eastern Amazon. *Ecology* **71**: 437–49.

Uhl, C., Kauffman, J. B. & Cummings, D. L. (1988). Fire in the Venezuelan Amazon 2: environmental conditions necessary for forest fires in the evergreen rainforest of Venezuela. *Oikos* **53**: 176–84.

White, J. C., Wulder, M. A., Brooks, D., Reich, R. & Wheate, R. D. (2005). Detection of red attack stage mountain pine beetle infestation with high spatial resolution satellite imagery. *Remote Sensing of Environment* **96**: 340–51.

Winter, J. W., Bell, F. C., Pahl, L. I. & Atherton, R. G. (1987). Rainforest clearfelling in northeastern Australia. *Proceedings of the Royal Society of Queensland* **98**: 41–57.

27 ENVIRONMENTAL IMPACTS OF TOURISM AND RECREATION IN THE WET TROPICS

Stephen M. Turton[1]* and Nigel E. Stork[2]*

[1]Australian Tropical Forest Institute, School of Earth and Environmental Sciences, James Cook University, Cairns, Queensland, Australia
[2]School of Resource Management and Geography, Faculty of Land and Food Resources, University of Melbourne, Burnley Campus, Richmond, Victoria, Australia
*The authors were participants of Cooperative Research Centre for Tropical Ecology and Management

Introduction

Tourists and local residents are continually seeking new recreation and leisure experiences, particularly in biologically complex and pristine environments such as coral reefs and tropical rainforests (Turton & Stork 2006). By the very nature of these complex and sensitive environments, there is a significant risk that over-visitation or inappropriate visitor use of these unique resources, or the inappropriate construction of resort and associated facilities, may lead to their decline and hence threaten their long-term sustainability (Buckley 2004). These potential dangers are exacerbated by the rapid expansion of international tourism and the steady growth of domestic tourism and recreation in many tropical countries (Christ *et al.* 2003).

There are conflicting signs that nature-based tourism or ecotourism, in some locations, may be responsible for causing ongoing damage to some of the most critical tropical forest regions around the world but in other places tourism is helping to promote and fund rainforest conservation (Christ *et al.* 2003; Turton & Stork 2006). Like any industry that relies on natural assets, tourism requires appropriate management strategies and systems to ensure

both visitor enjoyment and resource protection for current and future generations (Buckley 2004; Turton 2005). Nowhere is this more challenging a task than in tropical forests where natural resource exploitation has led to a decline in the extent of native forests and the distribution of biodiversity, which in turn is resulting in severe socio-economic impacts on the host (developing) countries. One of the critical negative impacts is on local (indigenous) communities in developing countries with expanding rainforest tourism (Christ *et al.* 2003).

Despite these negative trends, a number of international organizations, including the World Tourism Organization, United Nations Environment Programme and Conservation International (Christ *et al.* 2003), have recognized that tourism and recreation, if combined with effective protected area management, have the potential to bring much needed ecological, socio-cultural and economic benefits to communities in developed and developing countries (Buckley 2004; Turton & Stork 2006). Tourism potentially is a much more sustainable use of tropical forest environments than other current uses of forest landscapes, such as logging and conversion to agriculture or plantations, because large numbers of visitors can be channelled through a few small well designed and managed access routes and visitor sites (Turton 2005).

World travel and tourism is expected to generate US$6477.2 billion of economic activity (total demand) in 2006, growing (in nominal terms) to US$12 118.6 billion by 2016 (World Travel and Tourism Council 2006), making tourism one of the world's biggest and fastest growing industries. In the late 1980s, nature-based tourism accounted for only 2% of all tourism revenue, rising to about 20% (about US$20 billion year^{-1}) by the late 1990s (Newsome *et al.* 2002). Since tourism has the potential to improve substantially the economies of many developing countries, bringing in large revenues, many countries have been prepared to allow the industry to grow very rapidly without putting in place safeguards and management protocols (Christ *et al.* 2003). This has led to some commentators considering the growth of tourism in tropical forest regions to be both an opportunity and a threat (Turton & Stork 2006).

Despite the massive growth of nature-baed tourism at the global scale (Newsome *et al.* 2002), and with the exception of some detailed work undertaken in north-east Australia and central America, there have been few studies of the impacts and sustainable management of tourism in tropical rainforests (Turton 2005; Turton & Stork 2006; Reser & Bentrupperbäumer, Chapter 34, and Pearce, Chapter 7, this volume). Australia is the only developed country in the world with tropical rainforest on its mainland, and therefore the standards for protection, sustainable use and management of this resource are very high (Stork *et al.*, Chapter 48, this volume). The nature of rainforest tourism and recreation and how they are managed in north-east Australia is therefore an example of how rainforest visitation may be managed and developed elsewhere in the world.

Here, and elsewhere in this book (Pearce, Chapter 7, Reser & Bentrupperbäumer, Chapter 34), we show how research, guided by informed and willing industry and management agency partners, can help to improve the way tourism can be managed within a sustainable framework. This chapter focuses on environmental impacts of tourism and recreation in tropical rainforests, and provides examples of how negative impacts may be mitigated.

Tourism and tropical rainforests: a global perspective

In the past 30 years there has been growing concern about the fate of the world's tropical forests and the biodiversity they contain. Smith *et al.* (1993), for example, examined the rate of change of threat status of species on global Red Data Book lists and concluded that 50% of all mammals and birds would disappear in the next 250 and 350 years respectively. Increasingly, attention has been focused on a remarkably few unique biodiversity 'hotspots' (Myers 1988). Myers and colleagues (Myers *et al.* 2000) suggest that 40% of the world's biodiversity is located in 20 tropical forest hot spots around the world. Many of these locations are prime targets for tourism and recreation as international and domestic visitors want to see their unique biodiversity, possibly before these species become extinct (Turton & Stork 2006).

Along with tropical reef ecosystems, tropical forests are the most biologically diverse biome on Earth and this biodiversity is the primary draw card for tourists. While tropical forests have impressive structural and floristic diversity, it is the highly enigmatic animal life that dominates tourism interest in tropical forests at the international level (Newsome *et al.* 2002). Examples include gorillas in Central Africa, orangutans in Indonesia and bird watching in the Neotropics, Australia and New Guinea.

In a very timely globally focused study of tourism and biodiversity, Christ *et al.* (2003) show that in the past 10 years tourism has increased by more than 100% in 23 of the world's biodiversity hotspots located in developing countries, and that more than half of these receive over 1 million international tourists each year. More than half of the world's 15 poorest countries fall within these biodiversity hotspots and in all of these, tourism is already nationally significant or is forecast to increase.

Christ *et al.* (2003) also recognize that tourism development is a complex interaction among many players, with the private sector usually driving the process. The establishment of facilities is heavily dependent on multi- and bilateral development agencies. They recognize that the effective management of tourism and protection of the environment must be a partnership between the private sector, public sector and society at large. Our experience of the sustainable development of rainforest in north Australia concurs with that conclusion.

The benefits of tourism to developing countries are enormous. Tourism can be confined to relatively small areas, unlike less sustainable industries such as agriculture and forestry. Most visitors are happy to be guided to a few locations where they are guaranteed the desirable views of nature or adventure tourism that they are seeking. The careful design of access to these sites and the design of on-site infrastructure such as roads, walking tracks, parking areas, toilets and other facilities means that tourists can be directed and carefully managed, while their impacts can be monitored and restricted. In this way visitors may only access much less than 1% of key rainforest sites but still be highly satisfied with their visits. Most tourists have a limited time to explore rainforests and are therefore very willing to be guided to some of the most spectacular sites and places where they are more likely to see some of their target animals or plants, or impressive natural features such as waterfalls and streams. What have we learnt about the impacts of tourism and recreation in the Wet Tropics and how we may mitigate any adverse impacts?

Environmental impacts of tourism and recreation in Australia's tropical rainforests

The Wet Tropics of Queensland World Heritage Area (WHA) is an outstanding tourist destination, and rainforest-based tourism alone is estimated to generate over Au$750 million each year (Driml 1997). The Wet Tropics region has experienced significant increases in domestic and international tourism over the past 20 years, with some 2 million visitors per year in 1995 and an estimated 3 million in 2003 (WTMA 2003a). Recent projections suggest that tourist numbers will be about 4 million per year by 2016, with an increase in international visitors being a major contributing factor (WTMA 2003b).

Bentrupperbäumer and Reser (2002) have estimated about 4.4 million visits per year to recognized Wet Tropics sites, with 60% of these visits being domestic and international tourists. The remaining 40% were local residents engaging in rainforest-based recreation activities. Local residents were more likely to visit the region in the wet season (December–May), while domestic tourists were more likely to visit during the dry season. Both types of visitors tended to favour specific sites in the WHA, with many of the most adverse ecological impacts being attributed to local residents' inappropriate behaviour and use of such sites (Turton 2005).

The main recreation and tourism activities associated with visitor use of the Wet Tropics (Bentrupperbäumer & Reser 2000, 2002; Reser & Bentrupperbäumer, Chapter 34, this volume) involve about 180 regularly used visitor nodes and sites in the region (WTMA 2000). While this is a significant number of sites, the area reserved for tourism and recreation use is concentrated and relatively small. Site use is unevenly distributed across the region, with a few sites receiving most of the visitation (WTMA 2000). In a survey of 2780 visitors to the WHA (Bentrupperbäumer & Reser 2002), the most popular activities included observing scenery (83%), undertaking short walks (63%), relaxing (47%) and observing wildlife (32%), while less popular activities included bird watching (24%), swimming (21%), picnics/barbeques (17%), long walks (7%) and camping (6%).

The *Wet Tropics Conservation Strategy* (WTMA 2003b) identified a number of direct and underlying threats to the world heritage values of the Wet Tropics region, including internal fragmentation by community infrastructure, climate change, the introduction and spread of weeds, feral animals and pathogens, as well as altered fire regimes, water quality, flow regimes and drainage patterns. Tourism and recreation activities have the potential to exacerbate many of these threats as a consequence of visitor activities in the WHA, combined with increasing demand for infrastructure to service the growing tourist industry and increasing demand by local residents for recreation activities in the protected areas (Turton & Stork 2006).

The Wet Tropics bioregion contains several of the fastest growing local government areas in Australia. In June 1999, about 200 000 people lived in the region, with an annual growth rate of 2.4% (WTMA 2003a), compared with the Australian average growth of 1.0%. It is estimated that some 270 000 people will live in the region by 2016 (WTMA 2003a). The population growth rate is largely driven by the tourist industry, the largest employer in the region. As the population continues to grow, there will be a corresponding increase in demand for outdoor recreation activities in the rainforests of the adjacent WHA. In addition, increasing numbers of people are drawn to live in the region because of the unique environment and the climate. In this sense, the region is part of the 'seachange' and 'treechange' phenomena that typify many of the eastern Australian coastal and near-coastal communities (see Dale *et al.*, Chapter 32, this volume).

Tourism and recreation activities, and their associated environmental impacts in the Wet Tropics, have been largely associated with visitor use of walking tracks and trails, old forestry roads and tracks, day use areas, camping areas, water holes and rivers (Table 27.1). While most visitor activities result in highly localized impacts at a small number of sites, threats such as the spread of weeds, feral animals and soil pathogens as a consequence of tourism and recreation activities potentially affect much larger areas, largely because of the extensive network of old forestry (logging) roads in the region (Table 27.1). While demand for recreational use of forestry roads and long-distance walking tracks is much lower than demands for use of other visitor settings in the region, the threats to world heritage values are considerably greater owing to the more dispersed nature of these activities (see Goosem *et al.*, Chapter 36, this volume).

Implications for visitor management in the Wet Tropics

The primary management goal of the Wet Tropics Management Authority (WTMA) is to protect, conserve, present, rehabilitate and transmit to future generations the Wet Tropics WHA within the meaning of the World Heritage Convention (WTMA 2003a). Tourism and recreation represent an effective mechanism for achieving this mandate, provided visitation and use of the WHA are effectively managed within a sustainable framework. Strategies and on-ground actions for managing tourism and recreation activities and their associated adverse impacts in the region are shown in Table 27.2.

Hammitt and Cole (1998) advocate a combination of site and visitor management techniques for managing recreation impacts in protected areas in temperate ecosystems, and a combination of these techniques is also recommended for the Wet Tropics. Site management

Table 27.1 Principal environmental impacts of tourism and recreation in the Wet Tropics of Queensland World Heritage Area

Recreational activity (location in landscape)	Environmental impacts
Bushwalking, hiking, mountain bike riding, horse-riding (walking tracks and trails)	• Trampling of vegetation • Changes to plant composition • Introduction of weeds • Increased social (undesignated) tracks • Changes in soil conditions and increased soil erosion • Spread of soil pathogens
Off-road driving with 4WDs and trail bikes (old forestry roads and off-road tracks)	• Vegetation damage • Alterations to plant composition • Alterations to faunal composition • Barrier effects on fauna • Increased soil erosion • Rainforest vertebrate mortality • Introduction of weeds • Introduction of feral animals • Spread of soil pathogens • Increased road noise
Picnicking, barbeques (day use areas)	• Trampling of vegetation • Changes in plant composition • Introduction of weeds • Changes in soil conditions • Collection and burning of fire wood • Habituation of native fauna
Camping (camping areas)	• Trampling of vegetation • Changes in plant composition • Introduction of weeds • Changes in soil conditions • Increased social (undesignated) tracks • Site manipulation and mutilation • Collection and burning of firewood • Development of social trails • Habituation of native fauna
Swimming, white water rafting, kayaking (water holes and rivers)	• Bank side erosion/sedimentation • Trampling of vegetation • Increased social tracks • Changes in water quality

Source: Turton (2005).

techniques should be applied at heavily used sites, such as day-use areas, camping areas and short-distance rainforest walks (Table 27.2). Examples include the construction of board walks, artificial surfacing and compaction of walking tracks with durable materials, rotation of heavily used camp sites to allow for recovery and closure of some water holes to swimming in the dry season to prevent the introduction of water borne pathogens. In extreme cases, it may be necessary

Table 27.2 Management strategies and on-ground actions for reducing visitor impacts in the Wet Tropics of Queensland World Heritage Area

Recreational Activity (Location in Landscape)	Impact Management Strategies for Sustainable Use
Bushwalking, hiking, mountain bike riding, horse-riding (walking tracks and trails)	• Wet season closure of some tracks and trails • Design new tracks to minimise development of social (undesignated) tracks • Permanent closure of severely eroded tracks and trails • Avoidance of basalt soil series during track construction • Avoidance of die-back and erosion prone areas during construction of long-distance walking tracks • Designate separate track systems for mountain bikes, walkers and horse-riding • Canopy cover should be maintained • Apply best practice (minimal impact bushwalking) to reduce spread of soil pathogens • Construction of board walks at high use sites
Off-road driving with 4WDs and trail bikes (old forestry roads andoff-road tracks)	• Wet season closure of roads in areas susceptible to die-back and erosion • Retention of canopy cover to reduce weeds and spread of feral animals • Revegetation of road verges • Apply best practice to reduce spread of soil pathogens in infected and uninfected areas • Breeding season closure of some roads • Traffic calming to reduce road noise and road kills
Picnicking, barbeques (day use areas)	• Concentration of use at a small number of hardened sites • Canopy cover should be maintained
Camping (camp areas)	• Concentration of visitor use at a small number of hardened sites • Rotation of camp sites every 18–24 months, depending on vegetation type • Wet season closure of some camp areas • Canopy cover should be maintained over camp sites • Minimal impact techniques for isolated camp areas (situated on long distance walking trails)
Swimming, white water rafting, kayaking (water holes and rivers)	• Prevent development of social trails from accessing water holes • Revegetation of river banks • Apply best practice to reduce spread of water borne pathogens • Dry season closure of some water holes for swimming

Source: Turton (2005).

to close some forestry roads and long-distance walking tracks during the wet season to prevent soil and the spread of soil pathogens.

It may also be necessary to close some roads during the breeding season for rare and threatened fauna or to introduce traffic-calming measures to reduce penetration of road noise into the forest (Table 27.2). At heavily used camping areas every attempt should be made to ensure retention of canopy cover and ground-level ecological processes, as these have been shown to be an effective means of reducing the spread of weeds and soil nutrient loss.

Visitor management techniques should be applied to long-distance walking tracks and forestry roads, and include educating visitors about best-practice techniques to prevent the spread of soil pathogens along walking tracks and forestry roads as well as minimal-impact camping techniques for remote camp sites (Table 27.2). Site management techniques for forestry roads and long-distance walking tracks include retention of canopy cover over tracks and roads to minimize the spread of weeds, feral animals and non-rainforest fauna and to reduce the penetration of edge effects into the forest interior (Table 27.2). Maintenance of canopy cover as part of management policy for old forestry roads will also reduce linear barrier effects on arboreal and ground-dwelling rainforest fauna and road (see Goosem *et al.*, Chapter 36, this volume).

Towards sustainable tourism and recreation in Australia's tropical rainforests

The Wet Tropics Nature-Based Tourism Strategy was developed to provide a framework for visitor management and development in the WHA (WTMA 2000). The Strategy incorporates a three-tiered planning and management framework for the WHA at three spatial scales: regional, precinct and site levels. The WHA has been divided into 12 sub-regions that recognize specific and unique visitor opportunities in various precincts. At the site level, individual places have been classified according to their management intentions, as well as emphasizing the management of potential and actual impacts from visitation and use (WTMA 2000). Visitor nodes and sites across the WHA range from 'Core Natural' sites that provide opportunities for small groups of visitors to experience the region in its natural state in a self-reliant manner, to 'Icon' sites that provide opportunities for large numbers of visitors and groups to experience outstanding natural features and values.

The Wet Tropics Walking Track Strategy was released to provide a framework for management of over 170 walks in the WHA, while recognizing that research on visitor satisfaction as well as biophysical impacts is essential for successful management of walks (WTMA 2001). The Strategy employs a track classification system that aims to provide consistent management across a diversity of walking track types. These range from formed pathways, boardwalks and graded tracks to rough tracks and marked routes through remote areas (long-distance trails).

Both strategies address tourism and recreation issues in the WHA and the need to develop a visitor monitoring system for ongoing evaluation of the environmental condition of visitor nodes and sites in the area. Successful implementation of these strategies requires sound scientific advice on the environmental impacts of visitation and use in the region, and effective management strategies to ameliorate any significant adverse impacts. The Visitor Monitoring System developed recently by Wilson *et al.* (2004a, b) has been designed to provide such advice to managers of the WHA on the basis of a hierarchical monitoring system

involving tour operators, park rangers and researchers (Figure 27.1). It is hoped that the newly adopted system will address the critical need for environmental agencies to base land management decisions on sound scientific advice (Buckley 2003). The monitoring system considers whether management objectives are being met and assists in understanding visitor patterns, demands and behaviours (see Reser & Bentrupperbäumer, Chapter 34, this volume). Management response includes developing visitor management strategies that will result in positive trends (Figure 27.1). The system has been successfully applied to four contrasting visitor sites in the Wet Tropics (Wilson *et al.* 2004a, b), and is sufficiently generic to be of value to other tropical forest protected areas elsewhere.

Summary

- There is currently an enormous expansion of nature-based tourism in many of the world's biodiversity hotspots and in particular tropical forests.
- Globally, there have been few studies of the impact and management of tourism and recreation in tropical forests. Work undertaken in rainforests of north-east Australia provides a sound theoretical and practical

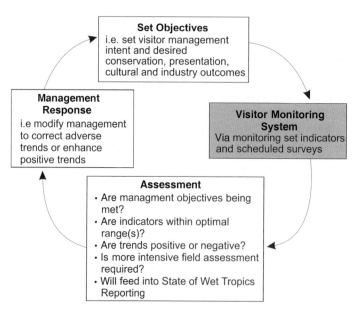

Figure 27.1 A model of the Visitor Monitoring System (VMS) and its relationship to nature-based visitor management in the Wet Tropics of Queensland World Heritage Area. *Source*: Wilson *et al.* (2004a).

framework for application to tropical forest tourist destinations around the world.

• The benefits of tourism to developing countries are enormous, as it can be confined to relatively small areas, unlike less sustainable industries such as agriculture and forestry.

• Tourism and recreation activities and their associated environmental impacts in the Wet Tropics have been largely associated with visitor use of walking tracks and trails, old forestry roads and tracks, day-use areas, camping areas, water holes and rivers.

• While most visitor activities result in highly localized impacts at a small number of sites, threats such as the spread of weeds, feral animals and soil pathogens as a consequence of tourism and recreation activities potentially affect much larger areas, largely because of the extensive network of old forestry (logging) roads in the region.

• The most appropriate management strategy for sustainable visitor use of rainforests is to concentrate visitor activities at a small number of heavily used (hardened) sites.

• While use-concentration is the most desirable method for controlling visitor impacts associated with the majority of visitor activities, it must be acknowledged that other activities are dispersed across the region and associated with low visitor numbers. Management techniques, such as site hardening and shielding, are not appropriate for such low-use sites, mainly due to aesthetic and cost considerations.

• Application of best practice methods by the tourist industry and recreational users, such as removal of soil from vehicles and hiking boots before traversing areas susceptible to forest dieback, is the preferred management strategy for these dispersed visitor activities. Wet season closure of parts of the Wet Tropics may also be required in areas susceptible to soil erosion and the spread and activation of soil pathogens.

• Retention of canopy cover at camp areas and day-use areas, as well as along walking tracks and forestry roads, is a simple yet effective management strategy to reduce numerous adverse impacts, including dispersal of weeds and feral animals, edge effects, soil erosion and nutrient loss, road kill and linear barrier effects on rare and threatened rainforest fauna.

References

Bentrupperbäumer, J. M. & Reser, J. P. (2000). *Impacts of Visitation and Use in the Wet Tropics of Queensland World Heritage Area, Stage 2*. Wet Tropics Management Authority and Rainforest CRC, Cairns.

Bentrupperbäumer, J. M. & Reser, J. P. (2002). *Measuring and Monitoring Impacts of Visitation and Use in the Wet Tropics World Heritage Area 2001/2002: A Site-based Regional Perspective*. Wet Tropics Management Authority and Rainforest CRC, Cairns.

Buckley, R. (2003). Conclusions. In *Nature-based Tourism, Environment and Land Management*, Buckley, R., Pickering, C. & Weaver, D. B. (eds). CABI Publishing, Wellingford, UK, pp. 109–10.

Buckley, R. (ed.) (2004). *Environmental Impacts of Ecotourism*. CABI Publishing, Wellingford, UK. 389 pp.

Christ, C., Hittel, O., Matus, S. & Sweeting, J. (2003). *Tourism and Biodiversity: Mapping Tourism's Global Footprint*. United Nations Environment Programme and Conservation International, Washington DC.

Driml, S. (1997). *Towards Sustainable Tourism in the Wet Tropics World Heritage Area*. Report to the Wet Tropics Management Authority, Cairns.

Hammitt, W. E. & Cole, D. N. (1998). *Wildland Recreation: Ecology and Management*, 2nd edn. John Wiley & Sons, New York.

Myers, N. (1988). Threatened biotas: hotspots in tropical forests. *Environmentalist* **8**: 187–208.

Myers, N., Mittermeier, R. A., Mittermeier, C. G., da Fonseca, G. A. B. & Kent, J. (2000). Biodiversity hotspots for conservation priorities. *Nature* **403**: 853–8.

Newsome, D., Moore, S. A. & Dowling, R. K. (2002). *Natural Area Tourism: Ecology, Impacts and Management*. Channel View Publications, Clevedon, UK.

Smith, F. D. M., May, R. M., Pellew, R., Johnson, T. H. & Walter, K. S. (1993). Estimating extinction rates. *Nature* **364**: 494-6.

Turton, S. M. (2005). Managing environmental impacts of recreation and tourism in rainforests of the Wet Tropics of Queensland World Heritage Area. *Geographical Research* **43**(2): 140–51.

Turton, S. M. & Stork, N. E. (2006). Tourism and tropical rainforests: opportunity or threat? In *Emerging Threats to Tropical Forests*, Laurance, W. F. & Peres, C. A. (eds). University of Chicago Press, Chicago, pp. 377–92.

Wet Tropics Management Authority (WTMA) (2000). *Wet Tropics Nature-based Tourism Strategy*. Wet Tropics Management Authority, Cairns.

Wet Tropics Management Authority (WTMA) (2001). *Wet Tropics Walking Strategy*. Wet Tropics Management Authority, Cairns.

Wet Tropics Management Authority (WTMA) (2003a). *Wet Tropics Management Authority Annual Report 2002–03*. Wet Tropics Management Authority, Cairns.

Wet Tropics Management Authority (WTMA) (2003b). *Draft Wet Tropics Conservation Strategy*. Wet Tropics Management Authority, Cairns.

Wilson, R. F., Turton, S. M., Bentrupperbäumer, J. M. & Reser, J. P. (2004a). *Visitor Monitoring System for the Wet Tropics World Heritage Area, Volume 1: Procedural Manual*. Wet Tropics Management Authority and Rainforest CRC, Cairns

Wilson, R. F., Turton, S. M., Bentrupperbäumer, J. M. & Reser, J. P. (2004b). *Visitor Monitoring System for the Wet Tropics World Heritage Area, Volume 3: Case Studies – Biophysical Assessment*. Wet Tropics Management Authority and Rainforest CRC, Cairns.

World Travel and Tourism Council (2006). (www.wttc.org/ 2006TSA/pdf/Executive%20Summary%202006.pdf).

28 INTERNATIONAL PERSPECTIVE: CONSERVATION RESEARCH IN THE AUSTRALIAN WET TROPICS

William F. Laurance

Smithsonian Tropical Research Institute, Balboa, Ancon, Panama, Republic of Panama; and Biological Dynamics of Forest Fragments Project, National Institute for Amazonian Research (INPA), Manaus, Amazonas, Brazil.

Environmental promise and peril

At first glance, one might be tempted to conclude that the Australian Wet Tropics is among the best-protected natural areas in the world. With the bulk of its forest sheltered within a World Heritage Area since 1988, the region suffers few of the pervasive problems – rampant deforestation, logging, overhunting, widespread surface fires and land speculation, among others – that plague tropical regions in many developing countries. By and large, the tropical forests of north Queensland are effectively managed and protected, and the region has been a hotspot for applied and basic research, as the many chapters in this volume attest.

Yet it would be a mistake simply to assume that all is well. For all its biogeographical uniqueness, the Wet Tropics region is perilously small, less than a million hectares in area, if one considers forests under World Heritage status. It is also highly fragmented, from both deforestation during the past century and the pronounced natural patchiness of its vegetation. This renders it inherently vulnerable to edge effects and a variety of external development pressures: from new roads and infrastructure, expanding cities and suburbs, air pollution, water harvesting and even ecotourism, which are mounting as the population of north Queensland continues to grow apace.

These forests, moreover, seem especially susceptible to invading species, with a veritable zoo of exotic plants and feral animals already assaulting the region. Like oceanic islands, which are notoriously invasible, the Wet Tropics region has long been insular, persisting as a wet archipelago of forest in a sea of arid land. It is also battered by cyclones with some regularity, and has been logged extensively in the past; such recurring disturbances can facilitate the incursions of foreign species into rainforests. One particularly insidious invader is the chytrid fungus, an emerging pathogen that has already devastated entire assemblages of stream-dwelling frogs in the Wet Tropics and elsewhere in the world.

Perhaps the most vexing threat to these forests comes not from within but from without. If the Wet Tropics has a single remarkable feature, it is its striking concentration of endemic species, many of them archaic in nature, some virtually living fossils. Nearly all of these endemics have tiny geographical ranges and most are confined to upland or montane areas, being specialized for cool conditions or high elevation habitats. With the expected onset of substantial global warming in the coming century, the geographical ranges of many species are likely to be squeezed and their populations beaten down, greatly increasing the risks of extinction.

These, then, are the array of threats facing the Wet Tropics. None are unique to the region, and in this sense the Wet Tropics might serve as a microcosm for the study of environmental perils facing other protected areas throughout the tropical world.

The leading role of Australian tropical science

As the chapters in this part illustrate, scientists working in tropical Australia have established themselves among the world leaders in a number of conservation-related fields. In my view, there are at least a half-dozen such areas in which Australian science, led by the Rainforest CRC, is arguably among the elite worldwide:

- tropical restoration ecology;
- bioclimatic modelling of vegetation and faunal distributions;
- studies of habitat fragmentation (including internal fragmentation);
- applications of molecular genetics and phylogeography to conservation;
- assessing the impacts of tropical ecotourism;
- integrating applied research into natural-resource management.

To focus on a single example, I know of no tropical region anywhere in the world in which the geographical distributions of numerous species, particularly endemic vertebrates, vegetation types and certain endemic invertebrate and plant species, have been so exhaustively studied and modelled. The bioclimatic models that have been devised to predict species distributions in the past, present and future have built upon the pioneering efforts of Henry Nix, and have been informed by detailed palynological and charcoal studies of late Pleistocene forest distributions led by Peter Kershaw and Michael Hopkins. Also essential to this effort has been a massive input of biogeographical and ecological data, such as the fine-scale typology of tropical forest types devised by Len Webb and Geoff Tracey, and remarkably comprehensive databases on endemic vertebrate distributions compiled by Stephen Williams and Richard Pearson.

Armed with these impressive data sets and an array of cutting-edge modelling tools, Australian tropical scientists are now attempting to predict changing species distributions in response to impending climate change, with a level of precision that I believe is unparalleled anywhere in the tropics. Of course, many uncertainties still remain, particularly when one endeavours to project how global warming will affect precipitation, which along with rising temperatures will have a critical impact on forests. Despite such inherent doubts, the ongoing work on vertebrate and vegetation responses to future climate change, led respectively by Stephen Williams and David Hilbert, is arguably the best in the tropical world.

Where now?

For all their achievements and sophistication, tropical scientists in Australia have largely failed, in my view, to attain their real potential, not just as world-class researchers in the Australian Wet Tropics, but as international leaders in tropical conservation, research and training. Australia sits just on the doorstep of some of the most biologically important and imperilled forests in the world, in New Guinea, the Solomon Islands, New Caledonia, the Sundaland region of Indonesia and Malaysia and the scattered archipelagos of Polynesia and Micronesia. As a leading industrial nation in the Asia-Pacific region, Australia is ideally situated to play a far more active and collaborative role in conservation research and training.

Why has this not happened? The problem can be traced, I think, to a kind of parochialism that has pervaded the thinking of those in Australian funding agencies and research bodies. Living in a vast and sparsely populated land, many have long felt that Australian science had its hands full simply addressing Australian issues, despite the growing urgency of environmental threats in the Asia-Pacific region. Scientists in Australia must also accept some blame, for far too few have attempted to stretch the bounds of their science to incorporate projects internationally. There are also daunting logistical challenges, language barriers and physical dangers involved in working in places like the New Guinea Highlands or the interior of Borneo.

For all the practical challenges involved, I believe that embracing an international leadership role would have enormous benefits not just for the Asia-Pacific region, but also for tropical science in Australia. James Cook University in Cairns would be an ideal headquarters for such a body, building directly upon the foundations of the Rainforest CRC. How better to energize conservation science in the Wet Tropics than to use it as an international model for rainforest management? How better to attract international funding than to develop cooperative projects that parlay Australian expertise into real-life conservation initiatives in nearby developing nations? How better to galvanize Australian scientists than to offer them a whole world of new challenges?

Part 4

Living in a World Heritage Area

INTRODUCTION

Stephen M. Turton[1]* and Nigel E. Stork[2]*

[1]Australian Tropical Forest Institute, School of Earth and Environmental Sciences, James Cook University, Cairns, Queensland, Australia
[2]School of Resource Management and Geography, Faculty of Land and Food Resources, University of Melbourne, Burnley Campus, Richmond, Victoria, Australia
*The authors were participants of Cooperative Research Centre for Tropical Ecology and Management

The Wet Tropics is a dynamic living landscape and the purpose of this part is to emphasize the importance of linking social and ecological systems in a landscape context. The underlying philosophy of the Wet Tropics Management Authority is that conservation of the World Heritage Area can only be achieved through community partnerships (Clarke, Chapter 29), requiring cooperation within and outside the protected estate to build resilience to threats such as climate change. The Wet Tropics has been the home to Rainforest Aboriginal people for tens of millennia, and is considered to be a cultural landscape by the first inhabitants (Pannell, Chapter 30). Bentrupperbäumer and Reser (Chapter 31) explore interactions between people and the World Heritage landscape, distinguishing the local community and visitors to the region.

The Wet Tropics have provided a living laboratory for researchers for many decades, with much of this science and knowledge being the basis for natural resource management (NRM) planning for the region. Dale *et al.* (Chapter 32) evaluate the Wet Tropics experience in the NRM arena, and the highly successful partnership between the regional NRM Body and the research community, resulting in the preparation and delivery of a regional plan that directs investment towards outcomes for biodiversity conservation, sustainable use of natural resources and community

capacity building and institutional change. As part of the wider NRM process, an Aboriginal natural and cultural resource management plan was developed for the Wet Tropics, the first of its kind in Australia. Pannell (Chapter 33) discusses the process of including Aboriginal interests and aspirations into the regional NRM plan.

Tourism and recreation are important activities in the Wet Tropics World Heritage Area, and Reser and Bentrupperbäumer (Chapter 34) provide a new approach for framing and researching the impacts of visitation and use in protected areas, emphasizing the importance of linking social and biophysical systems for sustainability outcomes. Hill (Chapter 35) discusses how we might link the natural and cultural diversity of the Wet Tropics to vibrant regional economies, in particular for the traditional owners of the land. Miriam Goosem provides an overview of the extensive road ecology research undertaken within the Wet Tropics under the auspices of the Rainforest CRC (Chapter 36). Much of this research has focused on mitigation of negative impacts of roads and other linear infrastructure towards sustainable outcomes. Jeff McNeely (Chapter 37) brings an international perspective to this section, drawing heavily on his experiences in contested, multiple-use landscapes across the tropics.

29 THE WET TROPICS CONSERVATION STRATEGY: CONSERVATION IN A COMMUNITY CONTEXT

Campbell Clarke*

Wet Tropics Management Authority, Cairns, Queensland, Australia
*The author was participant of Cooperative Research Centre for Tropical Ecology and Management

Unless we enlarge our thinking about land, I fear that in the 21st century a park will become the equivalent of something like a bonsai tree. A bonsai tree is living and green. It's beautiful and elegant. But its roots are confined, its branches pruned. A bonsai tree is an artificial creation of human more than nature. … In our new concept, other lands are not outside a park but connected and complementary to these fully conserved areas. In our new concept, people and their uses of land are not regarded as hostile but are an inextricable component in the integrated landscape. (McCormick 2003: 3)

Introduction

The Wet Tropics Conservation Strategy (WTMA 2004) was developed by the Wet Tropics Management Authority (WTMA) to promote coordinated conservation of the Wet Tropics of Queensland World Heritage Area (hereafter WHA). The strategy promotes biodiversity conservation that incorporates different cultural and social perspectives on resource management as well as some low impact sustainable use of natural resources in the WHA.

The long-term integrity of the WHA depends on community conservation both within and outside the Area and across all tenures including private lands. The WHA must be managed as the core component of the Wet Tropics bioregion, which has many of the same natural and cultural values. The majority of threats to the WHA originate directly or indirectly from human activities outside its boundaries (see Part 3, this volume).

Economic systems are needed that provide financial incentives for biodiversity conservation and better incorporate the environmental cost of threatening processes outside the WHA. Funding and resources for conservation management should reflect the true social, environmental and economic value of the WHA to the community.

The Conservation Strategy promotes community education about World Heritage values to help to create a culture of environmental appreciation. The development of the strategy has allowed the WTMA to learn about the community's attitudes and has created ongoing relationships with a broad range of community sectors. The willingness and capacity of the community to provide their expertise and support is essential to conserve the integrity of the WHA and surrounds.

This chapter outlines the context for the development of the Conservation Strategy and emphasizes the importance of incorporating community values and participation in all aspects of World Heritage conservation.

World Heritage listing for natural values

The WHA forms the core of the Wet Tropics bioregion (IBRA 2003). The bioregion (1 849 725 ha) is about twice the size of the WHA (894 420 ha) and comprises remnant vegetation over 77% of its area. The bioregion, including coastal sections of the WHA, adjoins the Great Barrier Reef World Heritage Area.

The WHA meets all four natural criteria specified by the World Heritage Committee for inclusion as a property on the World Heritage List (see Valentine & Hill, Chapter 6, this volume). The World Conservation Union's summary of the World Heritage nomination (IUCN 1988) described some of the qualities that justified its listing:

• The WHA contains one of the most complete and diverse living records of the major stages in the evolution of land plants, from the very first land plants to higher plants. It also has one of the most important living records of the history of marsupials and songbirds.

• Levels of species diversity and endemism in the Wet Tropics are exceptionally high, reflecting the long isolated, ancient biota of the Australian Wet Tropics.

• The Wet Tropics provides the only habitat for numerous rare or threatened species of plants and animals.

• The Wet Tropics contains some of the most significant regional ecosystems in the world, with outstanding features of natural beauty and magnificent sweeping landscapes. The coastal scenery is exceptional and contains tropical rainforest, white sandy beaches and fringing reefs.

A complete, updated list of World Heritage values can be found in the Periodic Report for the WHA (WTMA 2002). However, there is still much to understand about the complex biodiversity values of the WHA and their role in the life of the community. It should be noted that, while the WHA is only listed for its natural values, this does not diminish its cultural significance to Aboriginal Traditional Owners and the broader community. The WTMA has given in principle support for a cultural renomination. Legislation for the WHA focuses on protection of its natural values, but also provides specifically for participation in management by Rainforest Aboriginal people and the broader community.

Legislation and community participation

The primary goal of World Heritage management is to fulfil Australia's international duty to protect, conserve,

present and rehabilitate the WHA for future generations. The WHA was listed in December 1988 amid much community controversy and opposition from the Queensland government of the day (for the history see Valentine & Hill, Chapter 6, this volume). In 1993 the Wet Tropics World Heritage Protection and Management Act 1993 (Qld) (the WTWHPM Act) established the management regime for the WHA as agreed by the Australian and Queensland governments. The Act provides for an independent Board of Management, which reports to a Ministerial Council. It also provides for the appointment of statutory and advisory committees. The Board has established a Community Consultative Committee, a Rainforest Aboriginal Advisory Committee and a Scientific Advisory Committee. It has also established formal liaison groups to ensure healthy interactions with the conservation sector and the tourism industry. As a result, the community plays an integral role in the management of the WHA (see Valentine & Hill, Chapter 6, this volume).

In 1998, after prolonged public consultation and conflict, the Wet Tropics Management Plan 1998 (Qld) (the Plan) was completed as required under the Act. The Plan provides legislative protection for the natural values within the WHA through a zoning scheme that provides a framework for activities that are prohibited, allowed or allowed only with a permit. Any subsequent changes to the Plan or its zoning maps require formal public consultation.

Two other Acts have a particularly important role in the management of the WHA. The Commonwealth Environment Protection and Biodiversity Conservation Act 1999 (EPBC Act) regulates matters of national environmental significance, which include the protection of World Heritage Areas and threatened species. Importantly the EPBC Act has extended legislative protection to activities occurring outside the Area that have, or are likely to have, a significant impact on World Heritage values. Section 475 of the EPBC Act also grants standing to interested individuals to seek injunctions, in the public interest, to prevent contraventions of the Act.

The Queensland Nature Conservation Act 1992 provides for the conservation and ecologically sustainable use of biological diversity in national parks and other protected areas (now about 80% of the WHA). Under this Act, the Nature Conservation (Wildlife)

Regulation 1994 regulates wildlife management throughout Queensland, including the conservation of rare and threatened plants and animals. The Wet Tropics Management Plan 1998 protects forest products, but does not regulate animal wildlife.

Conservation and land management must also take into account Aboriginal rights and interests as defined under legislation such as the Commonwealth Native Title Act 1993 and the Queensland Aboriginal Cultural Heritage Act 2003.

A range of other laws have been introduced at all levels of government that help to achieve conservation of the WHA and surroundings. Legislation and regulations cover such issues as vegetation clearing, water quality, agriculture, wildlife, weeds and feral animals, fisheries, fire management, endangered species, development assessment, coastal zones and wetlands.

Management in a community context

Legislation alone cannot achieve positive management to protect and enhance conservation values. Article 5 of the World Heritage Convention states that cultural and natural heritage should be given a function in the life of the community. The Australian Committee for the World Conservation Union (ACIUCN 2000) recommended that a high priority be placed on the development of a Wet Tropics Conservation Strategy to ensure greater community involvement in the management of the WHA.

The diverse legislation and tenures that apply across the WHA and its surrounds mean that land management requires cooperative partnerships between government agencies, Aboriginal Rainforest people, landholders and the broader community. A few facts and figures demonstrate how complex cooperative management of the WHA can be

- The WTMA has the principal role in setting policy and coordinating management across the WHA under the Wet Tropics Ministerial Council, which consists of two ministers from both the Queensland and Australian governments.
- The Australian government administers referrals and assessments under the EPBC Act.
- The Queensland Parks and Wildlife Service (QPWS) is the on-ground land manager for the national parks, forest reserves and timber reserves that make up almost 80% of the area.

- Unallocated state land (7% of the Area) is managed by the Department of Natural Resources and Water (DNRW), as are numerous leases within the WHA. The Department of Primary Industries and Fisheries (DPIF) manages fisheries resources and fish habitats such as mangrove systems.
- A number of state agencies, local governments and authorities manage service infrastructure such as roads, powerlines, water supplies and telecommunications.
- There are 14 local governments that have part of their areas within the WHA. They are responsible for pest control, development assessment and approvals. This is particularly important adjacent to the WHA.
- There are 18 Aboriginal Rainforest tribal groups who assert Native Title rights to live on and manage their traditional lands throughout the WHA.
- The WHA includes about 96 freehold blocks or parts thereof (2% of the Area) and over 100 leases (10% of the Area, mainly due to some large grazing leases).
- Neighbouring land is managed for a multitude of purposes, including conservation, timber production, grazing, sugar cane and other agriculture, beekeeping, tourism and private residences. There are more than 2500 individual blocks of land neighbouring the WHA's 3000 km boundary and many more in the catchment areas.
- The Wet Tropics NRM Body (FNQ NRM Ltd) was established in 2003 as a community-based organization to allocate Natural Heritage Trust (NHT) funding and other conservation investment in the Wet Tropics region and to assist in the monitoring and evaluation of resource condition (see Dale *et al.*, Chapter 32, this volume).
- There are numerous organizations that undertake research in the WHA and the surrounding bioregion. To date researchers have primarily been based at the Rainforest Cooperative Research Centre (CRC), James Cook University and CSIRO. Relevant fields of study include ecology, zoology, botany, genetics, taxonomy, geology, geography, environmental studies, economics, anthropology, engineering, sociology and psychology.
- The Wet Tropics region has numerous volunteer community groups that promote environmental education and participate in rehabilitation activities such as tree planting and weeding.

The role of the World Heritage Area in the life of the community

The principal aim of the Conservation Strategy is to maintain biological diversity in the form of intact, fully functioning ecosystems for future generations. While the biodiversity of the WHA has inherent natural value *per se*, effective conservation will only be achieved if the WHA plays a significant role in the life of the community. Conservation of biodiversity, both within and outside the WHA, is inseparable from conservation of the WHA's cultural, social, economic and spiritual values. Biodiversity conservation and community well-being are interdependent.

Since World Heritage listing there has been a significant change in community attitudes to the WHA: the majority of the Wet Tropics community now support and identify with the Area (Bentrupperbäumer & Reser 2002; Bentrupperbäumer & Reser, Chapter 31, this volume). The WHA has a special place in the hearts of our regional community, being central to our sense of place, identity and spirituality. Its spectacular scenery provides a backdrop for our urban and rural lifestyles, with many locals relishing opportunities to experience the beauty and grandeur of its rainforests, mountains, rivers, waterfalls and wildlife. The WHA provides socio-economic benefits to the community through services such as tourism, recreation, water supply and genetic resources (see Part 2, this volume). Aboriginal tribal groups continue to live in and around the WHA and sustain their traditional cultural knowledge and connections to the country.

Socio-economic values

The WHA supports a variety of natural processes through which ecosystems sustain and fulfil human needs. In economic terms, the natural capital of the WHA's resources provides ecosystem goods and services that have a wide range of social and economic benefits for the Wet Tropics community (see Curtis, Chapter 19, this volume). These benefits may range from direct products to vicarious pleasures, from the personal to the international. Benefits may be environmental, economic, cultural, spiritual, educational, recreational or medicinal. By far the most important of these goods and services are those that provide support for life. There are also numerous others that enhance the quality of life in the Wet Tropics. Table 29.1 lists some of the ecosystem goods and services provided by the WHA.

The use of the term ecosystem goods and services implies that the WHA has both intrinsic and flow-on economic benefits. The true economic benefits of most ecosystem goods and services remain unquantified. However, there have been some attempts to quantify the economic benefits of tourism in the Wet Tropics. Driml (1997) has estimated the direct and indirect

Table 29.1 Ecosystem goods and services provided by the WTQWHA

Environmental values and processes	Environmental regulation	Community provision	Community enrichment
• Habitats and refugia	• Regulation of regional and micro climates	• Food	• Tourism
• Soil formation and fertility carbon sequestration	• Flood mitigation	• Clean water supply	• Recreation and leisure activities
• Conversion of solar Energy	• Water purification	• Energy (hydro, solar and wind)	• Spiritual values and enjoyment
• Biomass production	• Erosion control	• Shade and shelter	• Natural values
• Pollination	• Pest control	• Soils	• Scenic and aesthetic values
• Nutrient recycling	• Groundwater recharge	• Pharmaceutical and biological products	• Cultural and historical values
• Nitrogen fixing	• Waste treatment	• Horticultural products	• Education
• Water cycles	• Energy conversion	• Art and craft materials	• Scientific discovery
• Genetic resources			• Sense of place and identity
• Fire regimes			• Maintaining options for the future

Source: WTMA (2004).

benefits of tourism and recreation in the Wet Tropics to be over Au$750 million per annum. Similarly, Kleinhardt –FGI (2002) has estimated Daintree tourism and recreation to be worth Au$395 million per annum. More recently Prideaux and Falco-Mammone (2007) have estimated the direct benefits of tourism in the Wet Tropics World Heritage Area to be Au$426 million per annum – about 21 percent of the region's annual tourism expenditure of Au$2 billion. The estimation is considered to be conservative. It did not include any multiplier to calculate flow-on economic benefits, nor did it include the expenditure of local residents. The overall value of the WHA's ecosystem goods and services has yet to be calculated, but would be significantly higher than the estimates for tourism alone (see Curtis, Chapter 19, this volume).

The spiritual and aesthetic qualities of the WHA are also highly valued by the local community. Results from a survey conducted by Bentrupperbäumer and Reser (2002) showed that the most important benefits of the WHA for the regional community were its contribution to quality of life and social and recreational opportunities (see Bentrupperbäumer & Reser, Chapter 31, this volume). The WHA's economic benefits were considered less important. The community viewed the WHA as an integral and cherished part of their natural and cultural landscape, with just knowing it is there rating as the most important benefit overall. The WHA and its attributes contribute significantly to the community's sense of place and identity.

There is evidence that conservation of the WHA has greater economic benefit than any exploitation of its resources. Balmford *et al.* (2002) calculated that on average, about one-half of an ecosystem's total economic value is lost when that ecosystem is replaced or exploited for economic uses such as logging, agriculture or aquaculture. The study also found that the annual cost of maintaining a system of nature reserves is very small when compared with the total economic loss of ecosystem goods and services if those habitats were lost.

Aboriginal cultural heritage values

At least 80% of the WHA is potentially claimable under the Native Title Act 1993 (Cwlth) and 32% is already

under claim. There are also legal requirements for the WTMA to consider Aboriginal interests in management. The WTWHPM Act states that, in protecting natural heritage values, the Authority must have regard to the Aboriginal tradition of, and liaise and cooperate with Aboriginal people particularly concerned with land in, the Wet Tropics area.

Rainforest Aboriginal people have occupied, used and enjoyed their lands in the Wet Tropics region since time immemorial. The WHA is part of a series of complex living cultural landscapes for 18 Rainforest Aboriginal tribes with connections to the land. This means that the country and its natural features and resources are central to Rainforest Aboriginal people's spirituality, culture, social organization and economic use (see Pannell, Chapter 30, this volume, for an overview of this cultural landscape).

Conservation of natural World Heritage values is inextricably linked with that of Aboriginal cultural and spiritual values. The ecosystems of the Wet Tropics region have evolved over thousands of years through active Aboriginal interaction with the land and management of its resources. The participation of Traditional Owners and their cultural knowledge and perspectives of plants, animals and ecological processes creates a special context for conservation management and research of the WHA. Activities such as fire management, hunting and gathering, and the harvesting of materials for shelter, tools, ceremony or art and craft are essential for the maintenance of Aboriginal culture and have always been integral to the ecology of the Area.

Table 29.2 lists some of these Aboriginal cultural values (also see Pannell, Chapter 30, this volume). It should be noted that many of these values could easily fit into several categories. For instance, hunting can also have spiritual, social and ecological values.

Community heritage values

The WHA has important historical, cultural and spiritual value for the general community. People have lived and worked in the rainforests and surrounds since the first explorations of the region and early pioneering days (see Turton, Chapter 5, this volume, for history of European settlement). They have developed strong

Table 29.2 Aboriginal cultural values relating to conservation of the WTQWHA

Spiritual	Ecological	Social	Economic
● Creation stories	● Bush tucker (animals and plants)	● Language	● Hunting and gathering
● Sacred sites	● Bush medicine	● Living areas and camps	● Tools
● Burial grounds	● Knowledge of ecological relationships	● Walking tracks	● Food preparation
● Bora grounds	● Fire management	● Kinship systems	● Shelter building
● Ceremony	● Seasonal calendar	● Clans	● Harvesting resources
● Responsibility for country		● Cultural identity	● Art and craft
● Totems		● Traditional law	

Source: WTMA (2004).

and varied relationships with the WHA. The WHA contains evidence of early explorers such as Edmund Kennedy and Christie Palmerston blazing trails and marking trees. This provides a context for records of their travels and meetings with Aboriginal people in the nineteenth century. Historically, much of the WHA and adjacent landscapes have been used as a resource base by the logging, mining and grazing industries. Railways, roads and powerlines have been constructed throughout the WHA and several rivers have been dammed. Farming, agriculture and settlements have become established all around the WHA. Despite the fact that many of these activities dramatically altered the landscape and diminished biodiversity values, they are of immense historical importance in the social development of the Wet Tropics region. Some of these industries also played a vital role in conservation of the WHA. For instance, in the late 1920s and early 1930s the forestry industry was active in conservation of Wet Tropics forests for sustainable use and opposed the broad-scale clearing that occurred under the settlement policies of the day (Gould 2000).

The fight to save the rainforests and achieve World Heritage listing also has a strong personal significance for many in the community (see Valentine & Hill, Chapter 6, this volume). Many current residents and neighbours of the WHA and those who have worked for its conservation have a spiritual attachment to the WHA and a sense of community pride in achieving World Heritage status to protect the remaining rainforest vegetation communities.

Table 29.3 Underlying pressures and direct threats to the WTQWHA

Underlying pressures	Direct threats
Regional population growth	Vegetation clearing and fragmentation
Urban development and pollution	Weed invasions
Climate change	Feral animals
Demand for community infrastructure (water, electricity, roads)	Introduced pathogens
Farming (agriculture, grazing and aquaculture)	Altered water flows and drainage of waterways and wetlands
Tourism and recreation	Altered fire regimes

Source: WTMA (2004).

Threats to biodiversity

The listing of the WHA was a major achievement, protecting the Wet Tropics forests and preventing their rapid deterioration due to extractive activities such as logging. However, the values of the WHA remain under persistent threat from biophysical and social processes, emanating mostly from outside the WHA (see Part 3, this volume). Threats to the WHA's biodiversity and scenic values arise primarily from the impacts of community activities and use of natural resources. Table 29.3 lists the major threats that must be addressed if the community wishes to continue to enjoy the economic, cultural, social and spiritual benefits outlined above.

The Conservation Strategy discusses the major threats to the WHA in detail (WTMA 2004; also see Part 3, this volume). Two condition reports (Armour *et al.* 2004; Weston & Goosem 2004) prepared for the Wet Tropics Natural Resource Management Plan (FNQ NRM Ltd & Rainforest CRC 2004) describe the effects of such pressures on the health of the Wet Tropics bioregion's biodiversity, water, soil and fisheries. Two brief examples are given here to illustrate the need for the community to work together outside the WHA to address threats to World Heritage values.

Mission Beach development

The cumulative impacts of numerous coastal urban developments in the Mission Beach region threaten the long-term survival of the southern cassowary in the local area. The lowland rainforests at Mission Beach are already isolated from the rest of the WHA by adjacent clearings for agriculture and major roads. Urban development and clearing threaten cassowary populations, primarily through increased habitat fragmentation, road kills and attacks by dogs. The impacts of urban development continue to increase despite a range of Commonwealth and state laws and policies that protect endangered species and the integrity of the WHA, prevent vegetation clearing and promote cassowary conservation and rehabilitation of wildlife corridors in the area. The willingness of developers, private landholders and the Mission Beach community to accommodate the needs of the cassowary will be vital for its long-term future.

Environmental weeds

There are now over 500 weed species in the WHA and their spread has been increasing at an alarming rate (see Goosem, Chapter 24, this volume). Over 200 new weed species have been identified in the past decade. The majority of plant species that have become environmental weeds were intentionally introduced into the region for agricultural, horticultural, grazing or domestic purposes. Community cooperation to prevent new weed introductions is the most cost-efficient means of limiting future weed proliferation. The work of private landholders to identify weeds and control

incursions on their land is essential to prevent further weed spread into the WHA.

Biodiversity conservation on private lands outside the WHA

The Conservation Strategy promotes cooperative conservation on *private* lands outside the WHA as essential to maintain its integrity. Significant areas of high biodiversity value and other areas vital for habitat connectivity remain outside the WHA. The majority of threats to World Heritage values arise from outside the WHA. There are a variety of land uses, management techniques and conservation agreements that actively promote biodiversity conservation. Voluntary conservation mechanisms will often achieve better and more equitable conservation outcomes than regulation, which is often limited in scope and is difficult and expensive to enforce. However, some conservation mechanisms are underpinned by supporting legislation. The Conservation Strategy summarizes a range of conservation mechanisms available for private lands:

• Voluntary conservation agreements allocate duties and responsibilities to both parties, have rights and obligations that are enforceable under law and, in return, offer incentives to the landholder. Some offer long-term security and are listed on the title.

• Ecotourism ventures promote biodiversity conservation and educate visitors about World Heritage values (see Pearce, Chapter 7, Turton & Stork, Chapter 27, and Reser & Bentrupperbäumer, Chapter 34, this volume). Some actively incorporate forest rehabilitation as a significant part of their tourism activities.

• Farming activities such as selective agroforestry with mixed endemic species are compatible with World Heritage conservation (see Part 5, this volume). Less compatible farming activities can still provide conservation benefits through appropriate stocking levels and distribution, weed control, fire management, protection of remnant vegetation and organic farming practices.

• Farmers may benefit from retaining or restoring areas of native forests through ecologically sustainable harvesting, increased crop diversity, improved water quality, erosion control, flood mitigation, shade and windbreaks, improved pollination, pest reduction and aesthetic landscapes.

• A range of voluntary accreditation schemes, environmental management systems, improved farming techniques and alternative crops are now being developed by primary industries.

• The introduction of a range of tradeable rights and financial incentives can encourage landholders to conserve biodiversity.

Tradeable rights and financial incentives

Voluntary conservation on private lands will be dependent on incorporating the true value of biodiversity and the natural environmental into our economic systems so that conservation-minded landholders are encouraged and rewarded rather than penalized. The important roles of natural ecosystem services and their significant socio-economic benefits are not properly recognized in economic markets, government policies or land management funding and practices. 'Degradation of natural systems occurs because our economy makes it cheaper to degrade Australia than to look after it. The market signals are back to front' (Prime Minister's Science, Engineering and Innovation Council 2002).

Current regulatory and economic systems often provide conflicting incentives to the community and sometimes encourage wasteful practices. For instance, the cost of rehabilitation of cleared lands in the Wet Tropics has been estimated to be Au\$18 000–26 000 ha^{-1} (Catterall & Harrison, 2006; Catterall *et al.*, Chapter 41, this volume). Yet clearing and habitat fragmentation are still taking place in areas where land managers are funding rehabilitation works on neighbouring lands. Graziers are still planting pasture crops such as *hymen-achne* and *leucaena*, which have fast become invasive and costly environmental weeds (Goosem, Chapter 24, this volume).

Across Queensland and the Wet Tropics a variety of tradeable rights and incentives are being investigated to help to place a realistic economic value on our environmental resources and to promote environmentally sound farming and development practices. They include:

• purely financial benefits such as grants, tax concessions and rate rebates;

• incentives such as free labour and trees for rehabilitation, expert advice and awards for environmentally sustainable farm management;

• tenders and auctions for the opportunity to undertake biodiversity management and rehabilitation schemes;

• tradeable rights in water, vegetation clearing, development and carbon credits, often with a cap on the resource use;

• profit *à prendre* agreements that allow a landholder a commercial ownership or interest in the natural resource products of their plantation that is separate from the tenure and registered as an encumbrance on the tenure (such agreements can be used for carbon sequestration, timber harvesting, flowers, fruits and other vegetation products);

• mitigation banking of development impacts through rehabilitation of other sites;

• economic offsets for biodiversity conservation;

• increased security of tenure as a reward for implementing improved land management planning and monitoring;

• participation in accreditation schemes, environmental management systems (EMS) and eco-marketing.

Funding and other resources for conservation

The incorporation of true environmental values and costs into economic systems would significantly reduce the dire need for conservation funding to maintain protected areas (see Curtis, Chapter 19, this volume). The Conservation Strategy emphasizes that available resources are insufficient to undertake the full range of conservation works required in the WHA and surrounds. Long-term funding commitments are required to ensure effective conservation. Available funding is too often tied to particular projects and provided over relatively short periods. This can mean that the benefits of projects such as weed or feral animal control can be quickly undone through an inability to follow through on initial work. Land management agencies have increasing demands on their limited resources. Out of necessity, visitor management and safety often predominate in operational resource allocation. Limited staffing and resources may mean that monitoring and enforcement of environmental legislation and regulations become compromised. A loss of corporate knowledge, particularly local expertise acquired over time, can also result if resource allocation restricts the numbers and continuity of on-ground staff.

Economic and accounting systems can also influence land management funding. State government and local government land managers need to demonstrate the long-term economic benefits of maintaining healthy ecosystems and preventing cumulative deterioration of our natural capital. Some state and local governments have begun to assess the value, maintenance and depreciation of their infrastructure and facilities, but they have yet to incorporate any effective measure to value their natural assets and the costs of degradation. In the past there has been a false assumption that, once protected, the natural values and ecosystem services of the area will automatically be maintained. However, resources are required to manage these natural resource assets actively and mitigate the many threatening processes. These resources should be viewed as a future investment to maintain an economic asset.

Community capacity and education

The value of conservation funding can be magnified many times by the willingness and capacity of the community to maintain biodiversity, mitigate threats and participate in rehabilitation activities. It is extremely important to foster the capacity of community groups to continue to operate in an era of reduced and uncertain funding. In addition to achieving on ground outcomes, community conservation activities create a culture of caring for the environment and a heightened awareness and appreciation within the general community of the values of biodiversity.

Protected area managers also benefit from learning about the needs and desires of private land managers and the broader community. The process of developing the Conservation Strategy has been integral to building strong ongoing relationships between the Wet Tropics Management Authority and various sectors of the community. The development of the strategy has helped to educate both the WTMA and the community about biodiversity conservation throughout the Wet Tropics.

The future

Community conservation will become increasingly important as climate change and urban development reduce the ability of biodiversity to survive in isolated protected areas (Krockenberger *et al.* 2003; Williams *et al.* 2003; Williams *et al.*, Chapter 22, this volume). Conservation on private lands will be vital for habitat connectivity across the landscape to help fauna and flora to adapt to such threats.

It is predicted that climate change and ecosystem fragmentation will also significantly affect the ecosystem services provided by the WHA. Clean and plentiful water supplies may be diminished (see McJannet *et al.*, Chapter 15, this volume). Rainforest tourism and recreation may attract lower numbers of visitors if the quality of the rainforests deteriorates. Community conservation will become increasingly important in order to conserve the WHA and help to maintain the region's socio-economic fabric.

The Wet Tropics Natural Resource Management Plan (FNQ NRM Ltd & Rainforest CRC 2004) has built on the Conservation Strategy to address conservation and land use across the Wet Tropics landscape, within and outside protected areas (see Dale *et al.*, Chapter 32, this volume). The Plan will help to guide future environmental investment in the Wet Tropics region. Most importantly, investment in conservation activities, research and education will be guided and undertaken by the local community, a step towards a new community land ethic.

Ideally, values like ecological integrity would be sustained throughout the landscape without the need for centralised bureaucracies, fenced-off ecosystems and enforcement personnel. Moving towards such a goal means forging new socio-economic relations with the land and creating new institutions and systems outside the industrial models that make protected areas necessary in the first place. Widespread public awareness of the nature and importance of the land ethic both inside and outside protected areas will be fundamental to realise these broad and sweeping changes. One of the main roles of the park system is to build and nourish support for a land ethic. (Dearden & Dempsey 2004: 225).

References

Armour, J., Cogle, L., Rasiah, V. & Russell, J. (2004). *Sustaining the Wet Tropics: A Regional Plan for Natural Resource Management, Volume 2B Condition Report: Sustainable Use.* Rainforest CRC and FNQ NRM Ltd, Cairns.

Balmford, A., Bruner, A., Cooper, P. *et al.* (2002). Economic reasons for conserving wild nature, *Science* **297**: 950–3.

Bentrupperbäumer, J. M. & Reser J. P. (2002). *The Role of the Wet Tropics in the Life of the Community: A Wet Tropics World Heritage Area Community Survey*, Rainforest CRC, Cairns.

Australian Committee for the World Conservation Union (ACIUCN) (2000). *Wet Tropics of Queensland World Heritage Area – Condition, Management and Threats.* Australian Committee for International Union of Conservation of Nature Inc., Sydney.

Catterall, C. P. & Harrison D. A. (2006). *Rainforest Restoration Activities in Australia's Tropics and Subtropics.* Cooperative Research Centre for Tropical Rainforest Ecology and Management, Rainforest CRC, Cairns, Australia.

Dearden, P. & Dempsey, J. (2004). Protected areas in Canada: decade of change. *The Canadian Geographer* **48**(2): 225.

Driml, S. (1997). *Towards Sustainable Tourism in the Wet Tropics World Heritage Area.* Report to the Wet Tropics Management Authority, Cairns.

FNQ NRM Ltd & Rainforest CRC. (2004). *Sustaining the Wet Tropics: A Regional Plan for Natural Resource Management 2004–2008.* FNQ NRM Ltd, Innisfail.

Gould, K. (2000). A historical perspective on forestry management, In *Securing the Wet Tropics*, McDonald, G. & Lane, M. (eds). The Federation Press, Sydney.

Interim Biogeographic Regionalisation of Australia (IBRA) (2003): *The Interim Biogeographic Regionalisation of Australia* (http://ea.gov.au/parks/nrs/ibra/).

International Union for the Conservation of Nature (IUCN) (1988). *World Heritage Nomination – IUCN Summary, Wet Tropical RainForests (North East Australia).* IUCN Gland, Switzerland.

Kleinhardt – FGI (2002). *Tourism and Recreation Values of the Daintree and Fraser Island.* Australian Tropical Research Foundation, Kleinhardt – FGI (Corporate Advisors).

Krockenberger, A. K., Kitching, R. & Turton, S. (2003). *Environmental Crisis: Climate Change and Terrestrial Biodiversity in Queensland.* Cooperative Research Centre for Tropical Rainforest Ecology and Management, Cairns.

McCormick, S. (2003). *Towards Sustainable Parks in the 21st Century.* Speech to the World Parks Congress, Durban, South Africa (http://nature.org/event/wpc/files/mccormick_speech.pdf).

Prideau, B. & Falco-Mammone, F. (2007). *Economic Values of Tourism in the Wet Tropics World Heritage Area.* Cooperative Research Centre for Tropical Rainforest Ecology and Management, James Cook University, Cairns, Australia.

Prime Minister's Science, Engineering and Innovation Council (2002). *Sustaining our Natural Systems and Biodiversity.* PMSIEC, Canberra.

Weston N. & Goosem, S. (2004). *Sustaining the Wet Tropics: A Regional Plan for Natural resource Management, Volume 2A Condition Report: Biodiversity Conservation.* Rainforest CRC and FNQ NRM Ltd, Cairns.

Williams, S. E., Bolitho, E. E. & Fox, S. (2003). Climate change in Australian tropical rainforests: an impending environmental catastrophe. *Proceedings of the Royal Society of London B* **1527**: 1887–92.

Wet Tropics Management Authority (WTMA) (2002). *Periodic Report on the Application of the World Heritage Convention.* Wet Tropics Management Authority, Cairns (www.wettropics.gov.au/mwha/mwha_wettropicsreport.html).

Wet Tropics Management Authority (WTMA) (2004). *Wet Tropics Conservation Strategy.* Cairns (www.wettropics.gov.au/mwha/mwha_conservation.html).

30 CULTURAL LANDSCAPES IN THE WET TROPICS

Sandra Pannell*

Discipline of Anthropology and Archaeology, School of Arts and Social Sciences, James Cook University, Townsville, Queensland, Australia
*The author was participant of Cooperative Research Centre for Tropical Ecology and Management

Introduction

In the recent State of Environment report card for heritage (Lennon *et al.* 2001), Australia is identified as a land of unique heritage value: 'where for Aboriginal and Torres Strait Islander people the ancestral beings inscribed the law on the lands and waters as they created the landscape' (Lennon *et al.* 2001: 1). For many non-Indigenous Australians, this landscape is regarded as something quite natural, the product of evolutionary processes stretching back into the dim recesses of time. Perhaps nowhere else in Australia is this sense of deep evolutionary history more apparent than in the Wet Tropics of North Queensland. Nominated in 1988 for inscription on the World Heritage List as one of the 'most complete and diverse living records of the major stages of the earth's evolutionary history' (DASETT 1987: 19), the Wet Tropics is also valued for its 'exceptional natural beauty' (DASETT 1987: 29). While the 'magnificent sweeping landscapes' (IUCN 1988: 5) of the region resonate with a European cultural aesthetic about pleasing spaces, they also stand as physical monuments to a different cultural tradition.

Throughout the Wet Tropics of Queensland World Heritage Area (WHA) this different landscape tradition is gradually being revealed to non-Indigenous eyes. In some places, it reveals itself in the form of national park signage declaring the cultural significance of a particular geographical feature (e.g. *Din Din* or the Barron Falls) or an area (e.g. Wooroonooran National Park). For visitors and tourists, it declares itself in glossy brochures and posters proclaiming the existence of an ancient culture, or on the handful of Aboriginal-guided tours that operate in the Wet Tropics, and perhaps most obviously, at the Tjapukai Aboriginal Cultural Park near Cairns.

Although the Wet Tropics was originally nominated for its natural values, as the 2005 signing of the Wet Tropics Regional Agreement attests, Indigenous heritage places and cultural values are increasingly recognized as an integral part of environmental management of the region. This long overdue acknowledgement of Aboriginal cultural values reflects some of the changes that have occurred at an international level in the way that the environment and heritage are conceptualized. This is particularly apparent in the 1992 and 1994 revisions of World Heritage assessment criteria, the adoption of new cultural landscape categories (Titchen 1995) and the 2003 international convention for the safeguarding of intangible cultural heritage (UNESCO 2003). As this suggests, regimes for the protection of cultural heritage, and the values ascribed to it, change over time. As Lennon and her colleagues point out, these transformations reflect the dynamics of cultural identity and our evolving sense of place (Lennon *et al.* 2001), both of which come together in the concept of cultural landscapes.

What are cultural landscapes?

Since 1992, the World Heritage Convention Operational Guidelines have recognized the category of cultural landscapes for inscription on the World Heritage List. By the end of 2002, 30 of the 563 World Heritage cultural properties had been listed as official cultural landscapes. However, in a report on World Heritage cultural landscapes, Fowler (2003: 7) argues that this number does not accurately reflect the fact that more than 'one hundred cultural landscapes actually exist on the current World Heritage List'. These cultural landscapes haven't been formally recognized or renominated to date.[1]

Broadly speaking, the Convention defines cultural landscapes as properties that 'represent the combined work of nature and of man' (UNESCO 2005: 83). More specifically, cultural landscapes are defined in terms of three main categories, namely

1 'a landscape designed and created intentionally by man';
2 an 'organically evolved landscape', which may be a 'relict (or fossil) landscape' or a 'continuing landscape';
3 an 'associative cultural landscape', which may be valued because of the 'religious, artistic or cultural associations of the natural element' (UNESCO 2005: 84).

Examples of World Heritage cultural landscapes include the religious landscape of Vat Phou in Laos, the pilgrimage landscape of the Kii Mountain Range, Japan, the artistic and aesthetic landscape of Val d'Orcia, Italy, the vineyard cultural landscape of the Alto Douro Wine Region, Portugal, the archaeological landscape of the First Coffee Plantations in the southeast of Cuba, the regal landscape of Ambohimanga, Madagascar, and the industrial landscape of Blaenavon in the United Kingdom. As these few examples indicate, the associations of a cultural landscape can be quite diverse, and many of the listed landscapes possess characteristics of more than one World Heritage landscape category. Furthermore, the diverse nature of the cultural landscapes listed suggests that the concept

itself, as defined in the operational guidelines for the World Heritage Convention, can accommodate a range of cultural understandings and values, and not just European ones at that. This said, of the 30 cultural landscapes listed by the end of 2002, more than half of these (18) were listed as category 2b landscapes: continuing landscapes that retain 'an active social role in contemporary society closely associated with the traditional way of life, and in which the evolutionary process is still in progress' (UNESCO 2005: 84). While properties are nominated in terms of one of the three main World Heritage landscape categories, cultural landscapes are assessed and inscribed on the basis of the six cultural heritage criteria[2] that apply to all potential World Heritage sites.

In Australia, only one World Heritage property, Uluṟu Kata Tjuṯa National Park, is listed as an official World Heritage cultural landscape. Uluṟu Kata Tjuṯa National Park was renominated in 1994 as an associative cultural landscape with attributes that comply with World Heritage criteria (v) and (vi) for the assessment of outstanding universal value.[3] Central to the significance of this associative cultural landscape is the Indigenous philosophy of *Tjukurpa* or the Law, and the ancestral heroes who are regarded by *Aṉangu* (Aboriginal Traditional Owners) as giving form and meaning to the natural and social environment.

Associative cultural landscapes, like Uluṟu Kata Tjuṯa, make up a small percentage of the overall number of listed cultural landscapes. For example, as of 2002, only seven properties were what could be considered numinous inscriptions (Fowler 2003) or category 3 (iii) associative cultural landscapes. Fowler (2003: 28) observes that category 3 landscapes, specifically based upon criteria (vi) (see note 2), are 'used only rarely' by nominating state parties. He further states that the already inscribed cultural landscape of Tongariro in New Zealand 'set such a high standard … that extreme care is being taken with further claimants'. As Fowler (2003) points out using examples of failed or deferred cases, not all nominated cultural landscapes are able to meet the World Heritage Committee's criterion of outstanding universal value and the additional requirements of authenticity, integrity and distinctiveness. Since 1994, nominated properties are also required to take into account UNESCO's Global Strategy for a Balanced, Representative and Credible World Heritage List. In part, this strategy represents an attempt to address the underrepresentation of living cultures and traditional cultures on the World Heritage List. Indeed, Lennon points out that of the 690 properties on the list at the beginning of 2001, only seven are specifically listed for their Indigenous values, with four of these being in Australia (Lennon 2003).[4] While Lennon's point about underrepresentation is well taken, she raises the complex issue of defining Indigenous in a global cultural heritage framework. This is an issue that the World Heritage Convention largely sidesteps by adopting broad definitions of cultural heritage and landscapes, and in confining its administrative interactions to member states only. As Sullivan (2004) points out, First Nation peoples and local communities have no direct dealings with the World Heritage Committee.

It could also be argued that the underrepresentation noted in the Global Strategy reflects the way in which culture and heritage are regarded by different state parties, as well as indicating something about the politics and economics of heritage nomination and listing.[5] Indeed, Fowler (2003: 45) concludes that the concept of cultural landscape, as a mechanism for inscribing World Heritage sites of a non-monumental nature, 'has not in fact so far been realized'. This said, Aboriginal Traditional Owners of the WHA have long advocated that the area in question be renominated as a cultural landscape or a series of integrated cultural landscapes.

Preliminary research undertaken by Titchen (1995) and Horsfall (2003) suggests that the Wet Tropics could be renominated as either a continuing (category 2b) or an associative cultural landscape (category 3). It may also be possible to identify specific sites within this landscape(s) as having outstanding universal significance on the basis of the six cultural heritage criteria recognized under the convention. As Lennon *et al.* (2001: 14) point out, these sites should be seen as representing 'selected concentrations of meaning or significance in a wider intellectual and cultural landscape context'. In the following sections, I discuss some of the Indigenous values and conceptual themes associated with cultural landscapes in the Wet Tropics.

Tribal and linguistic landscapes

The WHA includes some, or all, of the traditional lands and waters of 18 Rainforest Aboriginal tribal groups

(WTMA 2005).[6] According to linguistic research, while the dialects and languages spoken by these 18 groups are part of the larger Pama-Nyungan family of languages, they are also associated with six different language groups, that is, the Yalandyic, Yidinyic, Dyirbalic, Nyawaygic, Wargamayic and Maric groups (Dixon 1972; Patz 1991, 2002; Wurm & Hattori 1981).[7] As such, the Wet Tropics constitutes one of the more linguistically diverse regions of Aboriginal Australia.[8] Dixon (1972: 351) hypothesizes that some of the linguistic similarities and divergences apparent today in the languages of the Cairns rainforest region are the result of changing patterns of vegetation, 'sclerophyll scrub giving way to tropical rain forest', and the expansion and schism of the proto-Dyirbal tribe into a number of new tribes. Dixon's (1972) proposition alerts us to the dynamic nature of both human populations and the environment, and the interdependent nature of the relationship between the two in this part of Aboriginal Australia.

Throughout the Wet Tropics, this relationship is mediated by language. Beneath the palimpsest of European names that feature on contemporary maps of the region lies a dense and interwoven Indigenous semantic landscape where names not only signify Storytime events and beings, but also refer to the flora and fauna of the locality. For instance, the Yidiny term for the place now known as Cairns is *Gimuy*, 'after the slippery blue fig tree (*Ficus albipila*) that grew there' (Dixon 1991: 1). As Bottoms (1999: 18) points out, *Gimuy* is also a 'Bama drinking well'. Some Indigenous place names have been anglicized and are now a part of the spatial lexicon of the region. For example, the towns of Babinda and Deeral respectively derive their names from the Yidiny words for waterfall (*binda*) and teeth (*dirra*) (Dixon 1991). The published sources, dating from Christie Palmerston's 1884–6 diaries to the present, reveal a more detailed linguistic mapping of the contours of country than current European charts indicate. For example, along a 6-km stretch of Toohey Creek, a tributary of the Mulgrave River, Dixon (1991) recorded 44 Yidiny-named places in the early 1980s. In a similar manner, Bottoms (1992) records hundreds of Djabugay and Yidinji places and campsites in the region from the Mowbray River in the north to the North Johnstone River in the south. As the work of Dixon, Bottoms and others (see Henry 1967; Field 1988; Hill *et al.* 2004) demonstrates, language remains an integral element in contemporary Aboriginal connections to land. For Aboriginal people of the Wet Tropics, named places also remind them of the ancestral beings who created the physical landscape we see today.

Cosmological landscapes

As Dixon discovered, Aboriginal people of the rainforest region have their own accounts of origins and movements that appear, in part, to correspond with linguistic reconstructions of the evolution of languages in the area. For example, Dixon (1991: 6) reports that Yidinji legends 'say the tribe came from the north' and certainly Yidiny is most similar to languages spoken in the north. Dyribal legends, on the other hand, 'describe a southerly origin' and, as Dixon (1991: 6) observes, 'Dyirbal is in fact most similar to languages spoken further south down the coast'.

The legends that Dixon (1977: 13) refers to are a part of a rich corpus of Aboriginal oral history that explains the 'origin of animals and birds, of geographical formations, and so on'. Indeed, throughout the Wet Tropics, the topographic and ecological features that contribute to the outstanding natural values of the World Heritage Area are regarded by Rainforest Aboriginal people as the creations and instantiations of a variety of ancestral beings. For some language groups, for example, Yidiny and Djabugay, these accounts of how 'country came to be' (Dixon 1991: 2) represent a 'common stock of dreamtime myths' (Patz 1991: 247). The two *Buluru* or Storytime brothers, Guyala and Damarri, are central ancestral figures in Yidinji and Djabugay origin accounts. These two brothers are responsible for providing the plant foods found in the rainforest and instituting the laws and customs of the two groups. They are also said to have founded the moiety system that regulates marriage and social interaction among some of the 18 tribal groups. For Djabugay people, prominent geographical features in their country are said to be the body of Damarri. For example, the supine body of Damarri can be seen in the contours of the Barron River and Redlynch valley, and the surrounding hills and outcrops. Damarri is just one of a number of form-shaping ancestral beings acknowledged by Aboriginal people in the region. In the southern regions of the Wet Tropics, Girrugar is

an important ancestral figure for Girramay, Gulngay, Djiru, Jirrbal and Bandjin people. Girrugar is said to have emerged from the sea and named mountains, lakes and islands in the region (Henry 1967; Dixon 1972). It has been speculated that Aboriginal accounts of the travels of Girrugar refer to the period at the end of the last ice age, some 8000–10 000 years ago, when the sea level was low enough that it would have been possible to walk to all of the islands in the Coral Sea (see Dixon 1972).

Water features, such as rivers, lakes and billabongs, are regarded as the creation and present-day homes of totemic snakes. For Dyirbal language speakers, *Yamani*, the Rainbow Serpent, is credited with the creation of regional water sources, as well as being the incarnation of the eponymously named meteorological phenomenon (Toohey 2001). In other parts of the Wet Tropics, the rainbow serpent is known by other names. For example, Bottoms (1999: 4) records that Gudju-Gudju, the Rainbow Serpent, 'made the rivers and creeks from Crystal Cascades [Yaln.giri] to Kuranda [Ngunbay]', travelled along the Mowbray River to the hill at present-day Port Douglas and finally came to rest at Wangal Djungay or Double Island. Whether identified as Yamani, Gudju-Gudju, Wungul, Kurriyala or Budaadji, as Dixon (1972: 29) points out, the Rainbow Serpent is 'a major spirit of the area', as it is throughout the rest of Aboriginal Australia (see McConnel 1930).

Throughout the Wet Tropics, the localized configuration of totemically imbued topographic features and ancestrally created named places, together with the Storytime-given flora and fauna of the area, is glossed in Aboriginal English as country. Djabugay and Yidiny speakers refer to their country as *bulmba* (Dixon 1977; Bottoms 1999), while for Kuku Yalanji speakers this sentient landscape is *bubu* (Anderson 1984).[9] As a living landscape, permeated with the presence of Storytime beings and the spirits of Aboriginal people's ancestors, country is 'interacted with, spoken to, acted upon and in turn it reacts, providing goods and resources or withholding them and bringing hardship' (Anderson 1984: 80).

Thus, while the origin of country is associated with the period variously known as *Buluru*, *Ngujakurra* and *Jujaba* throughout the Wet Tropics region, this creative epoch should not be regarded as a 'long dead and fixed point of reference' (Bell 1983: 90).[10] Similar to the

Anangu concept of *Tjukurpa* described in the renomination document for Uluṟu Kata Tjuṯa National Park, *Buluru*, or its other linguistic variants, is commonly glossed by Aboriginal people as the Law/Lore. It is incumbent upon Aboriginal people to follow the Law and, in doing so, maintain the interconnectedness of life, time and space. In this sense, the Law provides Aboriginal people with their 'plan of life' (Stanner 1963: 10). Like *Tjukurpa*, *Bulurru*, *Jujaba* and *Ngujakurra* denotes an ancient, yet living Indigenous religious heritage (see Box 30.1).

Box 30.1 All things come from the Law

Marra [cycad nuts] comes from the Ngujakura, from the beginning of time. Marra is a woman's thing. The two sisters, they made marra. Junbirr, the skink, taught them how to make marra. Kalkarubarr the snake chased those two sisters away up into the rainforest mountains. That's where they sit now, Ngalbabulal, the two sisters. (Dolly Yougie, Kuku Yalanji; Hill *et al.* 2004)

All things come from Bulurru, the sun, the moon and stars, the food we eat, the creatures of the world, the plants and trees, the rain, the very land itself. We ourselves come from Bulurru. Bulurru is the good spirit that protects life and Law. (Rhonda Duffin and Rosetta Brim, Djabugay; Duffin & Brim: 5)

Rainbow serpent lived at the bottom of the crater. He controlled fire. All the other animals wanted the fire so they sent *Badjandjila* the bird down to get it. The snake, *Yamani*, got angry and threw a stick at the bird. The stick hit the bird on the tail and created the split in its tail you can see today. *Yamani* left the crater and created a creek, the one that comes into Lake Barrine. *Yamani* took the water from the crater to Lake Barrine. Barrine is a living or moving water. That snake is restless and if tourists don't behave he'll take the water and leave. (Warren Canendo, Ngadjon-Jii)

There is a Girramay *Jujaba* story ... which tells how the Cassowary speared so many of the tribe that the people were frightened to go out hunting in case they met him and were attacked. So the Scrub Wallaby, helped by his daughters, cut off the Cassowary's arms as he lay sleeping. After that the Cassowary carried a *wuybali* spear with his toes. As a result, today all cassowaries have sharp toes which are as dangerous as a spear. (Desley Henry, Girramay; Pedley 1992a)

Archaeological landscapes

Aboriginal narratives and Storytime accounts point to an oral tradition that precedes the arrival of Europeans, in the form of Captain Cook, in 1770. Indeed, Ngadjon-Jii, Ma:Mu and Yidinji narratives recall the creation of the three volcanic crater lakes on the Atherton Tableland (Barrine, Eacham and Euramoo)[11] some 10 000 years ago, and describe the coast line of the region at the height of the last ice age (Dixon 1991; Toohey 2001). As a number of commentators have pointed out, archaeological and other research in the Wet Tropics provides supporting evidence for the veracity of local origin accounts (see Dixon 1977, 1991; Bottoms 1999).

The oldest known site in the region is Jiyer Cave, a rock shelter with Aboriginal art near the Russell River, in Ngadjon-Jii country (Horsfall 2003). Excavations at the cave since 1982 have revealed an array of terrestrial and marine artefactual material dating occupation of the site to about 6000 years ago (Horsfall 2003). To date, archaeological research in the region is quite limited and, as Horsfall (2003) notes, is confined to nine excavations sites in the central and southern sections of the Wet Tropics.[12] Notwithstanding this paucity, Horsfall (2003: 2.6) concludes that 'Aboriginal people have probably lived in the area for at least 40 000 years, through times of major environmental change'. Palaeoecological investigations at Lynch's Crater on the Atherton Tableland appear to confirm this suggestion. Turney *et al.* (2001) conclude that a rapid increase in the charcoal content of the sediments, dated to 46 000 years ago, indicates major fires in the area that were, most probably, caused by the arrival of humans. It appears that Aboriginal burning practices resulted in the 'replacement of fire-sensitive rainforest by fire-tolerant sclerophyll vegetation' (Turney *et al.* 2001: 770) over a period of some 20 000 years. As Kershaw (1974: 222) observes, 'at some time before 7,000 BP, rainforest again invaded the area, certainly as a result of increased precipitation'. The Lynch's Crater data support the 'view that human occupation of Australia occurred by at least 45 000–55 000 cal. yr BP [calculated year before present]' (Turney *et al.* 2001: 767) and point to the Aboriginal landscapes of the Wet Tropics as among the oldest in Australia.

As Hill (1988) and Hill *et al.* (2004) record for Kuku Yalanji people, Indigenous burning practices not only maintain areas of open forest within the rainforest, but also promote the growth of groves of carbohydrate-rich cycad species. There are further suggestions that Aboriginal aboricultural practices have, over the space of thousands of years, selected for certain floral species and hybrids. Indigenous burning and clearing practices were not just for subsistence purposes. Early explorers report that some Aboriginal groups used stone axes, or a combination of axes and fire, to prepare circular open ceremonial grounds in the rainforest (Jack 1888).[13] Horsfall (2003) reports that the discovery of considerable quantities of large stone artefacts (at least 20 000 axes), mainly edge-ground axes, grindstones and top-stones, is indicative of intensive occupation over a short time by a large population, points to continuous occupation over a longer period by smaller groups or suggests a combination of both. Certainly, David Harris's (1978) estimates of pre-contact population density in the central Wet Tropics region as 2.14 km^2 per person are high for hunter-gatherers and suggest a rich ecological environment.

Rainforest hunter-gatherer landscapes

More than 40 000 years of Indigenous burning and aboriculture, together with the performance of local increase ceremonies (Sharp 1938–9), point to the fact that Aboriginal tribal groups created a tropical forest landscape rather than just inhabited one (see Tsing 2003). Evidence of this creative and interactive relationship between humans and the environment is found in the way Aboriginal occupation and subsistence practices are inextricably linked to calendar faunal and floral species.[14] For example, the strident call of the green cicada (*gunyal*) announces the arrival of the story season to Yidinji people (Davis 2001). For Girramay and Jirrbal people, 'a rush grass, jindarigan (*Lomandra longifolia*) flowers when it is time to dig out the scrub-hen eggs and a plant commonly called "dog foot", gudamurran (*Balanophora fungosa*), grows when the carpet snakes are fat' (Pedley 1992a: 3). In the northern sections of the Wet Tropics, Kuku Yalanji people possess detailed knowledge of the relationship between the seasons and the availability of bush tucker.[15] In the words of Kuku Yalanji man Peter Fischer, 'we don't have a calendar. Bama story goes by the tree. The tree knows better that we do. That's why the flower comes

on. That tells us that's the time things are going to change' (Hill *et al.* 2004: 67).

The intimate relationship that Rainforest Aboriginal people have with country is further evident in their heavy reliance upon toxic plants, such as cycad nuts, black bean and cheeky yam, as staple food sources. As Horsfall (2003: 2.27) remarks, 'one of the most distinctive features of Rainforest Aboriginal cultures in the Wet Tropics, which distinguishes them from all others, is the dependence on several toxic food plants'. Horsfall (2003: 2.27) states that at least 15 toxic plant species eaten by Aboriginal people in the Wet Tropics require 'complex processing methods, involving washing or leaching in water'. As a number of researchers have commented, these plant species cause illness and, in some cases, death if not processed correctly (Pedley 1992a; Ashman 2001; Hill *et al.* 2004). For other less lethal species, cooking may be sufficient to destroy the toxins and make the plant more digestible. In conjunction with the evolution of sophisticated processing techniques, Rainforest Aboriginal groups developed a distinctive material culture in the form of lawyer cane baskets and bark string dilly bags, and also a range of species-specific hunting tools and traps.[16] Some of the material culture of the region, such as bicornual baskets, bark blankets, decorated tree buttress shields, huge wooden swords, slate grindstones with cut grooves, triangular or T-shaped stone implements known as 'ooyurkas' and large edge-ground axes, is unique to this region and found nowhere else in Australia (Horsfall 2003). Such is the diversity and array of Rainforest Aboriginal material culture that Dixon reports that he recorded over 30 pages of Dyirbal names for artefacts (Dixon & Koch 1996).

In her comparative discussion of hunter-gatherer occupation of tropical forests, Horsfall (2003: 2.28) concludes that 'the Wet Tropics of North Queensland are [universally] unique in the proportion of toxic plant species that were used as food staples'. The high proportion of toxic plant foods processed and eaten by Rainforest Aboriginal people, together with their exploitation of faunal and other floral species, made possible a hunter-gatherer subsistence lifestyle not reliant on access to cultivated carbohydrate staples. As Horsfall (2003: 2.29) observes, 'hunter-gatherer use of tropical rainforests in isolation from horticulture or agriculture is not represented by any other World Heritage site'.

While few, if any, Rainforest Aboriginal people today rely solely on hunting and gathering for subsistence, many people still utilize the skills and knowledge of their ancestors to supplement their diet and teach their children about their distinctive heritage.

Cultural itineraries

Critical to a rainforest hunter-gatherer lifestyle was the existence and maintenance of a series of walking tracks (Loos 1982). Djabugay people refer to these well defined and used paths as *djimburru*, while Yidinji people call them *gabay* (Dixon 1991).[17] Many of these tracks served to connect the rainforested uplands to the coastal plains and facilitated the movement of people to harvest seasonal resources, attend ceremonies, visit storyplaces, avoid tabu sites and maintain kin and affinal relations with other groups.[18] As Timothy Bottoms (1999) points out, some of these tracks re-enact the journeys of totemic beings; for example, the travels of the carpet snake, *Budaadji*, as it travelled up the Barron River to the tablelands to exchange nautilus shells for dilly bags. In his 1910 ethnographic bulletin, Walter Roth provides considerable detail about the routes used by Rainforest Aboriginal groups between Cardwell and Port Douglas to trade and barter all manner of material culture items, including grass necklaces, spears, shields, large swords, dilly bags, beeswax, nautilus shell necklaces and cockatoo feather headdresses. As Roth observes, some items, such as pearl-shell chest ornaments, were traded into the region from as far away as the Gulf of Carpentaria. As major thoroughfares throughout the region, Aboriginal tracks in the rainforest were commented upon and used by Europeans as early as 1886. For example, Christie Palmerston in his exploration for gold in the Russell River catchment regularly used 'native paths' (Savage 1992: 205). In a relatively short period of time, strategic Aboriginal paths were appropriated by Europeans and became the basis of early colonial tracks, such as the Clark's Track from Herberton to the coast via the Mulgrave River (May 1969: 4A), and the Bump Track from Thornborough to Port Douglas (Bottoms 1999).

While many pre-colonial Aboriginal walking tracks are entangled in European histories of the region, the extent, nature and significance of these Indigenous cultural itineraries continues to be acknowledged by rainforest tribal groups. In addition, these routes have been well documented by historians, ethnographers and archaeologists (see McCracken 1989; Bottoms *et al.* 1995; Lopatich 1996; Bottoms 1999; Toohey 2001), and are an integral part of walkways in a number of protected areas within the Wet Tropics (e.g. the Barron Gorge National Park Walking Track Strategy) (Lopatich 1996).

The Operational Guidelines for the Implementation of the World Heritage Convention have recently recognized the concept of heritage routes and cultural itineraries (UNESCO 2005: 89) as a 'specific, dynamic type of cultural landscape'.[19] In their definition of heritage routes, the Guidelines state that the cultural significance of such routes 'comes from exchanges and a multi-dimensional dialogue across countries or region' (UNESCO 2005: 88). Certainly, the network of Aboriginal heritage routes in the Wet Tropics highlights the multidimensional, interactive and plural nature of movement and interaction in this region in both pre- and post-contact times.

Colonized landscapes

> Mist which lies across the country
> A bulldozer nosing into Guymaynginbi
> It becoming just a cleared place
> The place really becoming cleared
> My father's father's country
> I had to sing about it
> That which was my home.
>
> (Paddy Birran and Jack Murray,
> *Girramayin*; Dixon & Koch 1996)[20]

The concerted arrival and presence of Europeans in the region around the 1860s initiated an invasion of the Aboriginal world, and the wholesale and irreversible transformation of the area's physical environment.[21] Within less than a decade of this arrival, large tracts of Aboriginal land were taken over, cleared of forest and used to graze cattle or for growing sugar and bananas. As Dixon and Koch report, 'many Aborigines died from white man's illnesses ... and those that dared protest against the theft of their land they were simply shot' (Dixon & Koch 1996: 4). Kuku Yalanji elder Peter Fischer recalls when the *Waybala* came: 'I've got to tell you how things have been happening, they shoot the black like a wild dingo you know, so bad … I seen my people go with the handcuffs, the women and all, in China Camp, taken away to Palm Island and Yarrabah' (Hill *et al*. 2004: 59).

Palm Island and Yarrabah were part of a reserve system instigated by colonial authorities in Queensland in 1897 ostensibly to protect Aborigines from European and Chinese settlers and the debilitating moral influences of these two societies (Meston 1895). The Queensland system was the first of its kind in Australia and variations of it were soon enacted in other states. So-called protection entailed the massed and forced removal of Aboriginal people from their traditional homelands to one of the newly established settlements. Prior to the establishment of the government's reserves, a number of church-run mission stations already existed, the first of these being set up in the 1870s in south-east Queensland (Department of Aboriginal and Islanders Advancement 1982). In the Wet Tropics region, Yarrabah[22] was established by the Church of England in 1892, Bloomfield River Mission was initially a government settlement established in 1886 but then taken over by the Lutheran Mission Council of South Australia in 1887, Mona Mona Mission was founded by the Seventh Day Adventist Church in 1913 and the government-run Hull River Aboriginal Settlement began operating in 1914 (Department of Aboriginal and Islanders Advancement 1982). By 1948, 18 Aboriginal and Islander reserves, missions and settlements had been established in North Queensland and on Cape York Peninsula (Department of Aboriginal and Islanders Advancement 1982).[23] In 1982, Deed of Grant legislation enabled elected Aboriginal councils to hold in trust Aboriginal reserve land in Queensland. Six years later, some of the recently received (1985) Yarrabah and Wujal Wujal (formerly Bloomfield River Mission) Deed of Grant in Trust (DOGIT) land, together with the Mona Mona reserve, was incorporated into the boundaries of the WHA (Commonwealth of Australia 1986). While no longer reserves or missions, today Yarrabah and Wujal Wujal continue to support large Aboriginal communities, while Djabugay people are redeveloping the former mission site of Mona Mona.

Most Aboriginal people living in North Queensland have some form of direct or mediated experience of the reserve system as part of what it meant to live under the Act in the twentieth century.[24] Today, the architectural remains of the mission days at Yarrabah, Wujal Wujal and Mona Mona are salient reminders of the injustices of the reserve system and the strength of Aboriginal people to survive this regime.[25] The Aboriginal memories and values accorded this experience and its physical expression are an integral dimension of the lived cultural landscapes of the Wet Tropics.

In 2000, the World Heritage Committee recommended for inscription on the List the Archaeological Landscape of the First Coffee Plantations in the southeast of Cuba. In doing so, the Committee recognized the 'role of Indigenous people in delaying the establishment of plantations systems' (cited in Fowler 2003: 89) and 'the sweat and blood of the African slaves who increased the wealth of their masters' (ICOMOS 2000: 73). The listing of this cultural property suggests that the physical remains and Aboriginal values of the Queensland reserve system for the protection, segregation and control of Indigenous people could equally be recognized as evidence of the creation of a unique cultural landscape illustrating a significant stage in Aboriginal – European relations and throwing considerable light upon the cross-cultural history of the region.

Cross-cultural landscapes

A number of official World Heritage cultural properties recognize the 'coming together of diverse cultural influences to create a cultural landscape' (Fowler 2003: 75). For example, the recently listed Historic Centre of Macau is said to be a 'unique testimony to the meeting of aesthetic, cultural, architectural and technological influences from East and West' (UNESCO 2006). What is interesting about this site, and the previously mentioned Coffee Plantations of Cuba property, is the identification of colonial encounters and the impact of colonialism as creating a form of cultural heritage worthy of recognition and protection.

The Wet Tropics bears testimony to one of the earliest and longest-lasting encounters between the forces of colonialism and Indigenous people in Australia. These encounters date from 1770, when Captain Cook sailed along the coast, went ashore in Gunggandji country in search of water and renamed *Djilibirri* (Barramundi Head), calling it Cape Grafton (Bottoms 1999). In the period since Cook's voyage, large swathes of rainforest have been transformed into a cross-cultural environmental patchwork bearing testimony to the arrival and activities of different ethnic groups, including English, Chinese, Japanese, Kanak, Italians and Hmong people. While not often recognized in European pioneering accounts, rainforest Aboriginal people also contributed to the production of this multicultural landscape, working on banana and coffee plantations, packing mules and digging races on the region's goldfields, labouring on dairy and potato farms, cutting sugar cane and timber, as well as assisting in the development of roads and railways in the region. Their sweat and blood is an integral dimension of the settled vistas and sweeping natural beauty we see today.

Legal and rhetorical landscapes

What is not seen in the exceptional natural beauty of the WHA is the jungle of legislation and legal tenures that form the lie of the land. Arguably, the creation of the Wet Tropics in 1988 represents one of the most recent waves of Aboriginal dispossession, as Aboriginal DOGIT and reserve land was included within the area without the informed consent of the elected Aboriginal councils concerned (Brennan 1992; Toyne 1994). As previously mentioned, the approximately 900 000 ha of the WHA also incorporated and cut across the traditional country of some 18 tribal groups. For some of the coastal groups, sections of their traditional estates, in the form of sea country, are also included within the 1981 listed Great Barrier Reef World Heritage Area. Since 1993, a new legal landscape has emerged as Aboriginal groups seek native title to country incorporated within both the Great Barrier Reef and Wet Tropics World Heritage Areas.

While rhetorically the Wet Tropics is presented as an outstanding example of rainforest, the area listed is 'fringed and to some extent dissected by sclerophyll forests, woodlands, swamps and mangrove forests' (IUCN 1988: 3). In a similar manner, the country of Aboriginal groups identified as Rainforest tribes contains areas of dry and wet sclerophyll forest, as well as wet tropical forest (Loos 1982). As fire research demonstrates (Hill *et al.* 2001; Hill & Baird 2003),

Aboriginal people in the region have a long tradition of maintaining sclerophyll areas and grasslands within their predominantly rainforested tribal areas. Indeed, as Hill and her colleagues point out, without Aboriginal intervention, in the form of traditional burning practices, rainforest rapidly invades these productive open areas (Hill *et al.* 2001).

The presence of dry sclerophyll forests and grasslands within the WHA, and the human creation and maintenance of these eco-zones, alerts us to the existence of rhetorical landscapes. These landscapes pivot as much upon the classification of the region's forests as natural, wet and tropical as they do upon the construction of Aboriginal people of the region as a unified cultural bloc of rainforest dwellers. As these comments suggest, landscapes are created within, and are the products of, specific cultural systems. They refer as much to particular ideas about people and place as they do to physical entities.

Landscapes of ideas: conclusions, implications and considerations for management

In Australia, Uluru Kata Tjuta, listed as an associative cultural landscape, is often cited as an exemplary model of the possibilities and practicalities of managing a landscape of ideas (Fowler 2003). If we look behind and beyond our notions about nature and culture, we can see that other World Heritage properties in Australia also bring together the 'tangibility of landscape – earth and rock and water – with the intangibility of an abstract idea' (Fowler 2003). For example, the values ascribed to the Wet Tropics, and other natural World Heritage properties in Australia, express particular Western ideas about space, its configuration, classification and meaning; though, as Tsing (2003) points out, all too often Western cultural frameworks are concealed by claims to scientific objectivity and neutrality. The history of protected area gazettal and management in Australia, whether national parks or World Heritage properties, certainly points to a perceptual (and obvious political) inability to see culture in nature (Smyth 2001). In situations like the wet tropical rainforests of north-east Queensland, the cultural presence of hunter-gatherer people, who 'exploit the natural environment in a sustainable way', have 'minimal material culture' and largely 'non-monumental life-styles' (Fowler 2003: 56), is not always apparent to Western ways of looking and thinking about forest landscapes. Yet, as Fowler remarks, the statement that 'there is nothing there' is in effect an inconceivable conclusion to expect from a serious examination of any area of land' (Fowler 2003: 56). As the struggle to secure the Wet Tropics illustrates (McDonald & Lane 2000), listing properties for their natural values alone often has serious ethical and wide-ranging political implications.

In an IUCN report on human use of World Heritage natural sites, the authors found that of the 126 natural and mixed sites listed, 51 have resident human populations (Thorsell & Sigaty 1998). In Australia, at least five natural or mixed World Heritage properties (i.e. Wet Tropics, Kakadu, Uluru Kata Tjuta, Great Barrier Reef and Willandra Lakes) are occupied by Aboriginal Traditional Owners (Thorsell & Sigaty 1998). While in the past, human occupation and use of protected areas was often viewed as an obstacle to the protection of the natural values of an area, in the past two decades there has been a 'huge increase in the attention to social factors in conservation' (Thorsell & Sigaty 1998: 9). The new generation of community-based approaches to protected area management has been strenuously pushed by Indigenous people as they attempt to secure recognition of their traditional rights and interests, which all too often have been ignored in the listing of World Heritage properties. It has also been driven by the realization that the way of life of many Indigenous people is 'now under severe threat' (Fowler 2003: 56). Indeed, one commentator has even gone so far as to ask whether 'preserving small, essentially non-Westernized indigenous populations in their natural' habitats is the proper business of those implementing the World Heritage Convention' (Fowler 2003: 56). While history shows that many of the state parties to the Convention have woefully neglected their fiduciary duty to their Indigenous citizens, Fowler's question brings home the realization that the protection of cultural landscapes, heritage and values pivots upon the sustainable involvement of, and support for, those groups and communities who are the traditional custodians of the cultural values expressed in the landscape. As Lennon (2003: 123) points out, 'World Heritage associative cultural landscapes have special needs for strategies and actions to maintain the traditional

associations which give the place its outstanding universal values'. Maintaining these associative values thus entails maintaining the cultural associations and cultural well-being of the groups whose values have been inscribed on the World Heritage List. For some groups and peoples this may involve creating new opportunities for the transmission of traditional skills and knowledge or revitalizing customary cultural activities. In either situation, effective management of an associative cultural landscape will need to address those social problems and economic pressures that impact upon the cultural viability of the group. As Lennon (2003) concludes, failure to recognize the very different management requirements of associative cultural landscapes may result in the placement of the landscape on the World Heritage in Danger List, or the reclassification of the property as a relict landscape. Lennon's (2003) comments beg the question as to whether the current regime for managing the natural values of the Wet Tropics has the capacity to address the challenges of protecting a cultural landscape. If the belated acknowledgement of Aboriginal cultural values and incidental involvement of Traditional Owners in the management of the Wet Tropics to date is any indication of this capacity, the answer would have to be a resounding no.[26]

Notes

1 Interestingly, Fowler (2003) does not include the Wet Tropics in his 'wider view' of cultural landscapes on the World Heritage List, even though the official nomination documents for this area stated that 'The wet tropics of North-East Australia preserves the only recognized extant Aboriginal rainforest culture and is therefore a major component of the cultural record of an Aboriginal society which has a long continuous history in the nominated area for at least 4000 years' (DASETT 1987: 19).

2 Previously these six criteria were identified as 'cultural heritage' criteria and formerly presented as a separate criteria set, distinct from the four natural heritage criteria set. The 2005 Operational Guidelines for the Implementation of the World Heritage Convention merge the ten criteria (UNESCO 2005: 19).

3 '(v) be an outstanding example of a traditional human settlement, land-use, or sea use which is representative of a culture (or cultures), or human interaction with the environment especially when it has become vulnerable under the impact of irreversible change; (vi) be directly or

tangibly associated with events or living traditions, with ideas, or with beliefs, with artistic and literary works of outstanding universal significance' (UNESCO 2005: 20).

4 The four Australian properties are Kakadu National Park, Tasmanian Wilderness, Willandra Lakes Region and Uluru Kata Tjuta National Park (Lennon *et al.* 2001).

5 While there are 122 nation-state signatories to the World Heritage Convention (Lennon *et al.* 2001: 31), it is apparent from looking at sessions of the World Heritage Committee that issues of state party autonomy and resourcing also inform nominations.

6 These self-identified 'tribal' groups are Banjin, Djabugay, Djiru, Girramay, Gugu Badhun, Gulngay, Gunggandji, Jirrbal, Koko Muluridji, Kuku Yalanji, Ma:Mu, Ngadjon-Jii, Nywaigi, Warrgamay, Warungu, Wulgurukaba, Yidinji and Yirrganydji (WTMA 2005). Of these 18 groups, only Djabugay, Djiru, Girramay, Gulngay, Gunggandji, Jirrbal, Ma:Mu, Ngadjon-Jii, Yidinji and Yirrganydji are identified as exclusively occupying rainforest. The traditional lands of the remaining groups are regarded as 'drier, more open country' (Dixon 1972: 27), and only contain a small proportion of wet tropical forest in its current distribution.

7 According to linguists, these language affiliations and groups are as follows: Koko Muluridji and Kuku Yalanji constitute two of the 12 dialects of the Kuku Yalanji language group; Djabugay and Yirrganydji people speak dialects of the Djabugay language group; Yidinji and Gunggandji people speak related dialects belonging to the Yidiny language group; Djiru, Girramay, Gulngay, Jirrbal, Ngadjon-Jii and Ma:Mu belong to the 'Dyirbal' language group; according to Dixon (1981), Wargamay and Bandjin are two dialects of a single language group, but Wurm and Hattori (1981) identify them as distinct languages within the Dyirbalic group; Nyawaigi and Wulugurukaba are dialects in the Nywaygic language group; and Warungu and Gugu Badhun are dialects belonging to the 'vast Maric group' of languages (Dixon 1972; Wurm & Hattori 1981; Patz 1991, 2002). I should point out here that linguistic classifications of Aboriginal languages into dialects and languages may not necessarily correspond to the way in which Aboriginal people themselves distinguish their own way of speaking from that of another group (see Patz 2002).

8 Dixon (1972: 2) states that 'the most divergent languages are found in and around Arnhem Land'.

9 In Wargamay, country is *ngardji* (Dixon 1981), while in Nyawaygi country can be referred to as *runggu* (Dixon 1983). Sutton (1973) records the Gugu Badhun term for 'ground' as *ngani*. To date, little linguistic research has been undertaken on Warungu and Wulgurukaba.

10 *Buluru* is the term used by members of the Yidiny and Djabugay language groups, while *Ngujakurra* denotes the concept of Storytime for Kuku Yalangi language group speakers. *Jujaba* is the term recognized by Dyirbal language group speakers (Pedley 1992a; Sharp 1938–9).

11 In the Ngadjon dialect, these three lakes are known as *Barany*, *Yidyam* and *Ngimun* (Dixon 1972).

12 Horsfall (2003: 2.20) also identities 'preservation factors' as contributing to the 'absence of old dates for the Wet Tropics region'.

13 In Dyirbal and Yidiny these ceremonial grounds are referred to as *buluba* (Dixon 1991; Dixon & Koch 1996). In Djabugay they are known as *djirrbibarra bulmba* (Patz 1991: 320), while in Nyawaygi and Wargamay they are called *burrun* (Dixon 1981: 129; Dixon 1983: 516). Throughout the region, Aboriginal ceremonial grounds are also known as 'bora grounds'. While not in use today, Indigenous oral histories indicate that regional ceremonies were conducted at these grounds until quite recently. These accounts point to the existence of a dense, interconnected, ceremonial landscape. To early Europeans, bora grounds and clearings in the rainforest for semi-permanent Aboriginal camps provided convenient house sites and thus many bora grounds and former camp sites are today known to Anglo-Australians as 'pockets' and are named after the first European to colonize these sites (Toohey 2001: 31).

14 R. M. W. Dixon (see 1972, 1977, 1991), assisted by biologists from the Queensland Museum and CSIRO, has recorded extensive inventories of Aboriginal knowledge of rainforest faunal and floral species in Yidiny, Dyirbal and Ngadjon languages and dialects. Further information on Aboriginal use of rainforest species is provided in Pedley (1992a, b, 1993a, b), Ashman (2001) and Smith *et al.* (2001).

15 Unlike the European view of the seasons in the Wet Tropics as either 'wet' or 'dry', Kuku Yalanji people recognize five seasons (Anderson 1984). Further south, Yidindji people acknowledge four seasons (Davis 2001).

16 As Pedley (1992a, b) documents, these tools and other items include barbed and three-pronged spears, turkey, wallaby, fish and eel nets and traps, goanna nooses, boomerangs, firesticks, bark and dug-out canoes.

17 In the Kuku Yalanji language these tracks are called *baaral* (Deeral 1999: 3).

18 A 1994 cultural heritage survey on Girramay land recorded two Aboriginal 'highways' – the *Juburriny* and *Gayjal* tracks. These two tracks are part of a wider network of paths that link living areas in Girramay country (Pentecost 1999).

19 The Sacred Sites and Pilgrimage Routes in the Kii Mountain Range in Japan (listed in 2004) is a recent

example of a cultural landscape envisaged as a 'heritage route'.

20 Dixon recorded the song poem *Destruction of Our Country* in 1964 at a time when the Queensland government had in the previous year leased a large tract of Girramay country to King Ranch, an American pastoral company. The company proceeded to clear the country with bulldozers and dynamite, in the process destroying people's traditional rainforest homeland and many sacred sites (Dixon & Koch 1996). Paddy Birran's and Jack Murray's eloquent commentary on the actions of King Ranch was subsequently published in *The Collins Book of Australian Poetry* (Hall 1981).

21 As Hill *et al.* (2000: 139) observe, 'settlement of tropical Queensland by non-Aboriginal people began only after 1860, following the opening of the Kennedy Pastoral District' in 1861 (Loos 1982: 29). Dixon and Koch (1996: 4) note that Cardwell was established in 1864.

22 Yarrabah was formerly known as Cape Grafton and also Bellenden Ker Mission.

23 In the 1960s, Aboriginal reserves were also established at Ravenshoe and Malanda in the Wet Tropics region.

24 In its various manifestations, the Aboriginals Protection and Restriction of the Sale of Opium Act of 1897 controlled Aboriginal lives in Queensland until 1984 (Bottoms 1999).

25 Upon its closure in 1962, a number of the Mona Mona mission houses were purchased by Aboriginal residents and relocated to the communities of Koah, Oak Forest, Kowrowa and Mantaka. Some of the mission buildings were sold to local farmers (Bottoms 1999).

26 In April 2005, some 17 years after the area was first listed, the Wet Tropics of Queensland World Heritage Area Regional Agreement for the involvement of Rainforest Aboriginal people in the management of the property was finalized. Signatories to the agreement include the 18 Rainforest Aboriginal tribal groups, the Wet Tropics Management Authority, Queensland's Environment Protection Agency, the Queensland Department of Natural Resources and Mines and the Federal Department of the Environment and Heritage (WTMA 2005).

References

Anderson, C. (1984). The political and economic basis of Kuku-Yalanji social history. Doctoral thesis, Department of Anthropology and Sociology, University of Queensland, St Lucia.

Ashman, D. E. (2001). *Sharing Culture: Rainforest*. Steve Parish Publishing, Brisbane.

Bell, D. (1983). *Daughters of the Dreaming*. McPhee Gribble Publishers, Melbourne.

Bottoms, T. (1992). *The Bama: People of the Rainforest*. Gadja Enterprises, Cairns.

Bottoms, T., Lee Long, D. & Verevis, R. (1995). *Djina:la Galing: Going on Foot: A Cultural Heritage Study and Archaeological Survey for the Barron Gorge Walking Tracks Strategy*. Report to the Queensland National Parks and Wildlife Service.

Brennan, F. (1992). *Land Rights Queensland Style*. University of Queensland Press, St Lucia, Queensland.

Bottoms, T. (1999). *Djabugay Country: An Aboriginal History of Tropical North Queensland*. Allen and Unwin, St Leonards, NSW.

Commonwealth of Australia (1986). *Tropical Rainforests of North Queensland: Their Conservation Significance*. A Report to the Australian Heritage Commission by the Rainforest Conservation Society of Queensland. Special Australian Heritage Publication Series No. 3. Australian Government Publishing Service, Canberra.

Davis, G. (2001). *The Mullunburra: People of the Mulgrave River*. Cassowary Publications, Cairns.

Deeral, E. (1999). Balga: the right Baaral track. *Rainforest Aboriginal News*, March, 3:3.

Department of Aboriginal and Islanders Advancement (1982). *Aboriginal and Islander Communities in Queensland*. Department of Aboriginal and Islanders Advancement, North Quay, Brisbane.

Department of Arts, Sports, the Environment, Tourism and Territories (DASETT) (1987). *Nomination of Wet Tropical Rainforests of North-East Australia by the Government of Australia for Inclusion in the World Heritage List*. Department of Arts, Sports, the Environment, Tourism and Territories, Canberra.

Dixon, R. M. W. (1972). *The Dyirbal Language of North Queensland*. Cambridge University Press, Cambridge.

Dixon, R. M. W. (1977). *A Grammar of Yidiny*. Cambridge University Press, Cambridge.

Dixon, R. M. W. (1981). Wargamay. In Dixon, R. M. W. & Blake, B. J. (eds) *Handbook of Australian Languages: Volume 2*. The Australian National University Press, Canberra, pp. 1–144.

Dixon, R. M. W. (1983). Nyawaygi. In Dixon, R. M. W. & Blake, B. J. (eds) *Handbook of Australian Languages: Volume 3*. The Australian National University Press, Canberra, pp. 431–525.

Dixon, R. M. W. (1991). *Words of Our Country: Stories, Place Names and Vocabulary in Yidiny, the Aboriginal Language of the Cairns-Yarrabah Region*. University of Queensland Press, St Lucia.

Dixon, R. M. W. & Koch. G. (1996). *Dyirbal Song Poetry: The Oral Literature of an Australian Rainforest People*. University of Queensland Press, St Lucia.

Duffin, R. and R. Brim (nd). *Ngapi Garrang Bulurru-m: All Things Come from Bulurru*. Kuranda.

Field, D. (1988). *Jakalbaku*. Douglas Shire Council, Mossman.

Fowler, P. J. (2003). *World Heritage Cultural Landscapes 1992–2002*. UNESCO World Heritage Centre, Paris.

Hall, R. (1981). The Collins Book of Australian Poetry. Collins, Sydney.

Harris, D. (1978). Adaptation to a tropical rain-forest environment, Aboriginal subsistence in northeastern Queensland. In *Human Behaviour and Adaptation*, Blurton-Jones, N. G. & Reynolds, V. (eds). Taylor and Francis, London, pp. 113–30.

Henry, G. J. (1967). *Girroo Gurrll the First Surveyor and other Aboriginal Legends*. W. R. Smith & Patterson Pty Ltd, Fortitude Valley, Brisbane.

Hill, R. (1998). Vegetation change and fire in Kuku-Yalanji country: implications for management of the Wet Tropics of Queensland World Heritage Area. Doctoral thesis, School of Tropical Environmental Studies and Geography, James Cook University, Townsville.

Hill R. & Baird, A. (2003). Kuku Yalanji Rainforest Aboriginal people and carbohydrate resource management in the Wet Tropics of Queensland, Australia. *Human Ecology* **30**: 27–52.

Hill, R., Baird, A., Buchanan, D. C. *et al.* (2004). *Yalanji-Warranga Kaban. Yalanji: People of the Rainforest Fire Management Book*. Little Ramsay Press, Cairns.

Hill, R., Griggs, P. & Bama Bamanga Ngadimunku Incorporated (2000). Rainforests, agriculture and Aboriginal fire-regimes in Wet Tropical Queensland, Australia. *Australian Geographical Studies* **39**(2): 138–57.

Hill, R., Smyth, D., Shipton, H. & Fischer, P. (2001). Cattle, mining or fire? The historical causes of recent contractions of open forest in the wet tropics of Queensland through invasion by rainforest. *Pacific Conservation Biology* **7**: 185–94.

Horsfall, N. (2003). *Aboriginal Cultural Values of the Wet Tropics Bio-Region*. Collaborative Cultural Heritage Research Report, Rainforest CRC, Cairns, unpublished.

International Council on Monuments and Sites (ICOMOS) (2000). *Coffee Plantations (Cuba), ICOMOS Advisory Body Evaluation*.

International Union for the Conservation of Nature (IUCN) (1988). *World Heritage Nomination – IUCN Summary. Wet Tropical Rainforests (North-East Australia)*. IUCN, Gland, Switzerland.

Jack, R. L. (1888). Report by Robert L. Jack, Government Geologist, on the Geology of the Russell River. *Votes and Proceedings of the Legislative Assembly*. Government Printer, Brisbane.

Kershaw, A. P. (1974). A long continuous pollen sequence from north-eastern Australia. *Nature* **251**: 222–3.

Lennon, J., Pearson, M., Marshall, D. *et al.* (2001). *Natural and Cultural Heritage, Australia State of the Environment Report 2001 Theme Report*. CSIRO Publishing, Canberra.

Lennon, J. (2003). Values as the basis for management of World Heritage cultural landscapes. In *Cultural Landscapes: The Challenges of Conservation*. UNESCO World Heritage Centre, Paris, pp. 120–7.

Loos, N. (1982). *Invasion and Resistance: Aboriginal – European Relations on the North Queensland Frontier, 1861–1897*. Australian National University Press, Canberra.

Lopatich, L. (1996). *Djina:la Galing, Going on Foot: Barron Gorge National Park Walking Track Strategy*. Queensland National Parks and Wildlife Service, Far North Region.

McConnel, U. H. (1930). The Rainbow Serpent in North Queensland. *Oceania* **1**: 347–9.

McCracken, C. R. (1989). Some Aboriginal walking tracks and camp sites in the Douglas Shire, North Queensland. *Queensland Archaeological Research* **6**: 103–14.

McDonald, G. & Lane, M. (2000). *Securing the Wet Tropics?* The Federation Press, Sydney.

May, J. (1969). *Eacham Shire Centenary Booklet of Historical Data*. Eacham Historical Society, Malanda.

Meston, A. (1895). *Queensland Aboriginals: Proposed System for their Improvement and Preservation*. Edmund Gregory, Government Printer. Brisbane.

Patz, E. (1991). Djabugay. In Dixon, R. M. W. & Blake, B. J. (eds) *Handbook of Australian Languages: Volume 4*. The Australian National University Press, Canberra, pp. 245–347.

Patz, E. (2002). *A Grammar of the Kuku Yalanji Language of North Queensland*. Pacific Linguistics, Research School of Pacific and Asian Studies, The Australian National University, Canberra.

Pedley, H. (1992a). *Aboriginal Life in the Rainforest*. Department of Education, Cairns.

Pedley, H. (1992b). *Aboriginal Tools of the Rainforest*. Department of Education, Cairns.

Pedley, H. (1993a). *Garrimal Wuju Wabungga: Summer Fruit of the Rainforest*. Innisfail.

Pedley, H. (1993b). *Jaban Buningga Nyajun Wabunnga: Eel Cooking in the Rainforest*. Innisfail.

Pentecost, P. (1999). Aboriginal walking tracks: linking the past to the present. *Rainforest Aboriginal News* March, 3:14.

Roth, W. E. (1910). Transport and trade. *North Queensland Ethnography* **14**: 1–19.

Savage, P. (1992). *Christie Palmerston, Explorer*. James Cook University Press, Townsville.

Sharp, L. R. (1938–9). Tribes and totemism in north-east Australia. *Oceania* **9**(4): 439–61.

Smith, S. and the Aboriginal People of Jumbum (2001). *Fruits of the Forest*. Department of Education, Cairns.

Smyth, D. (2001). Joint management of national parks. In *Working on Country: Contemporary Indigenous Management of Australia's Lands and Coastal Regions*, Baker, R., Davies, J. & Young, E. (eds). Oxford University Press, Melbourne, pp. 75–91.

Stanner, W. E. H. (1963). *On Aboriginal Religion*. Oceania Monographs No. 11, University of Sydney.

Sullivan, S. (2004). Local involvement and traditional practices in the World Heritage system. In *Linking Universal and Local Values: Managing a Sustainable Future for World Heritage*. UNESCO World Heritage Centre, Paris, pp. 49–55.

Sutton, P. (1973). Gugu-Badhun and its neighbours: a linguistic salvage study. MA thesis, Macquarie University, Sydney.

Thorsell, J. & Sigaty, T. (1998). *Human Use of World Heritage Natural Sites: A Global Overview*. IUCN, Gland, Switzerland.

Titchen, S. M. (1995). *A Strategy for the Potential Renomination of the Wet Tropics of Queensland Using the Newly Defined World Heritage Cultural Landscape Categories*. Unpublished report to the Wet Tropics Management Authority, Cairns.

Toohey, E. (2001). *From Bullock Team to Puffing Billy: The Settling of the Atherton Tableland and Its Hinterland*. Central Queensland University Press, Rockhampton.

Toyne, P. (1994). *The Reluctant Nation: Environment, Law and Politics in Australia*. ABC Books, Sydney.

Tsing, A. L. (2003). Cultivating the wild: honey-hunting and forest management in southeast Kalimantan. In *Culture and the Question of Rights: Forests, Coasts, and Seas in Southeast Asia*, Zerner, C. (ed.). Duke University Press, Durham, NC, pp. 24–56.

Turney, C. S. M., Kershaw, A. P., Moss, P. *et al.* (2001). Redating the onset of burning at Lynch's Crater (north Queensland): implications for human settlement in Australia. *Journal of Quaternary Science* **16**: 767–71.

United Nations Educational, Scientific and Cultural Organisation (UNESCO) (2003). *Report of Commission IV*. General Conference 32nd Session, Paris.

United Nations Educational, Scientific and Cultural Organisation (UNESCO) (2005). *Operational Guidelines for the Implementation of the World Heritage Convention*. World Heritage Centre, Paris.

United Nations Educational, Scientific and Cultural Organisation (UNESCO) (2006). World Heritage Centre (http://whc.unesco.org).

Wet Tropics Management Authority (2005). *The Wet Tropics of Queensland World Heritage Area Regional Agreement*. Wet Tropics Management Authority, Cairns.

Wurm, S. A. & Shiro Hattori (1981). *Language Atlas of the Pacific Area*. Australian Academy of the Humanities, Canberra.

31 ENCOUNTERING A WORLD HERITAGE LANDSCAPE: COMMUNITY AND VISITOR PERSPECTIVES AND EXPERIENCES

Joan Bentrupperbäumer[1]* and Joseph Reser[2]*

[1]School of Earth and Environmental Sciences, James Cook University, Cairns, Queensland, Australia
[2]School of Psychology, Griffith University, Gold Coast, Queensland, Australia
*The authors were participants of Cooperative Research Centre for Tropical Ecology and Management

Introduction

The focus of this chapter is on the nature of the encounter visitors and residents have with the Wet Tropics of Queensland World Heritage Area (WHA) landscape, including Indigenous and non-Indigenous community residents as well as local, domestic and international visitors.

This applied focus addresses those aspects of people's experience and behaviour that have implications for management, whether these relate to protection of the natural and human environments, achieving sustainable visitation practices or the presentation of the WHA and its values. The Wet Tropics WHA is a very powerful and meaningful place and a very prominent and salient natural and cultural landscape. Living within and visiting such a World Heritage landscape provides the opportunity for a range of very positive and powerful experiences. Close examination of these people–environment encounters has enabled us to understand better how transactions with this natural environment both constitute and impact on the experience that local residents and visitors have of these environments, their enjoyment and perceptions of the environment and their behaviour within such areas. In a protected area management context an informed understanding of the dynamics of people–environment encounters is particularly useful, because it alerts managers to the importance of including an individual or psychological level of analysis when considering critical issues linked to positive and negative impacts. These impacts include perceived environmental quality, community environmental concerns, experienced satisfactions or disappointments, development or visitation threats to privacy, a sense of place and community, quality of life or environment or World Heritage experience.

While World Heritage values, presentation, perception and appraisal, management agency practices, interpretation and setting design, and visitor motivations and expectations were all areas of research in what was a comprehensive research programme, this chapter's coverage reflects a strategic focus on a number of key topic areas only. This allows for a more substantive consideration of these topics from the differing perspectives of those Indigenous and non-Indigenous residents who live in, drive through, visit or experience this World Heritage landscape daily, and of those domestic and international visitors who physically encounter this same environment for a moment in time, while on holiday or travelling through the region. The approach taken substantially broadens the focus from tourist destination and image to that of an internationally recognized WHA, which is also a well known and very significant part of a living cultural and natural landscape and catchment, providing multiple non-economic and economic benefits to local residents.

The research

The research undertaken within the Rainforest CRC Project 4.1 (strategies for sustainable access and use) involved multiple and integrated social and natural science based projects that, in combination, provided a comprehensive and composite overview of the nature, quality and extent of the interactions Indigenous and non-Indigenous people have with a World Heritage landscape in northern Australia. The WHA core business of the research was

- to conceptualize adequately and address the nature of the impacting processes associated with people–environment interactions in this WHA;
- to examine more closely the nature and quality of these transactions between individuals and settings;
- to document and measure the reciprocal and interdependent influences and impacts occurring during these encounters;
- to translate these findings and insights into a research approach and plan that would allow for continued monitoring, assessment and evaluation of management strategies and practices over time.

An important research objective was to establish and maintain a database and provide a methodology and measures that would pragmatically inform management decision-making and on-the-ground management strategies. An additional and overarching objective was to provide an interdisciplinary framework that would allow for a genuine synthesis of social science and natural science in the context of managing impacts in World Heritage environments (Reser & Bentrupperbäumer, 2008; Chapter 34, this volume).

The role of a WHA in the life of a community

The role of a WHA in the life of a community living within and adjacent to such iconic protected areas can be framed and understood in a number of ways.

An immediate and conventional understanding relates to how local communities value the contribution of such a protected natural area and its associated natural and cultural heritage, to their well-being, to their quality of life and to their sense of place and belonging (Hartig *et al.* 1991; Hartig 1993; Kaplan 1995; Kaplan *et al.* 1998; Bushell *et al.* 2002). In addition, community perceptions of importance and salience, advantages and disadvantages, threats and concerns for such a landscape, as well as an understanding of their use of and attachment to favourite places, provide valuable insights into the contributions to the regional community of the WHA itself. In the case of the WHA it was clear, from the previous findings and past history, that the World Heritage listing and protracted debate ensured that this area was a particularly important feature of the political and historical landscape of the region, and an object collective lesson in the nature and politics of local communities and protected areas (Anderson 1987; McDonald & Lane 2000).

The importance and salience of the World Heritage Area

Consideration of the contribution a WHA makes to experienced quality of life and perceived environmental quality provides a sensitive and very meaningful way to assess the role of the WHA in the life of the local community (e.g. Hunter & Green 1995; Cantrill 1998). In the community survey research undertaken ($n = 788$), quality of life, quality of environment benefits were identified as the most important personal benefits derived from the WHA and were considered far more important to this local community than direct or indirect economic benefits (Bentrupperbäumer & Reser, 2003). It is particularly noteworthy that quality of life advantages, such as just knowing that it is there, that it exists, received the highest importance rating, indicating that virtual use and symbolic value are fundamentally important benefits of protected areas for regional residents, which are not often assessed or taken into serious account. The considerable difference between rated quality of life/environment benefits (60% very important) and economic benefits (15% very important) suggests that socio-economic assessments need to be strongly qualified and contextualized in considerations of community perceptions and priorities in the context of World Heritage landscapes, and that an exclusive or even predominant socio-economic focus may well miss factors that are actually most important to the community in question (e.g. Driml & Common 1995).

The most salient benefits for the whole community that could be derived from the WHA were the ecosystem and protection services afforded by this landscape, with the provision of clean water and air, the protection of rainforest plants and animals and the protection of scenic landscapes being rated by community residents as most important (see also Curtis, Chapter 19, this volume). Again, commercial and economic benefits to the community as a whole were deemed less important than all other nominated benefits, the proportion of respondents rating these benefits as very important being considerably lower for economic benefits (34.5%) compared to protection and ecosystem services (72.5–81.2%) (Bentrupperbäumer & Reser 2003). This type of valuation and expression of fundamental importance is once again missed in socio-economic assessments generally, as is the fact that everyone is a stakeholder when it comes to World Heritage assets. The existence and health of the WHA may be of considerable importance to the perceived environmental quality and the quality of life of individuals whose residence may be remote from north Queensland, or for those local residents who do not physically visit or otherwise directly use this environment. Such virtual use has become an important touchstone and consideration in environmental assessment (e.g. Gee 1994; Bentrupperbäumer & Reser 2000a).

Place meaning, place identity and place attachment

The local natural environment plays an important role not only in perceived environmental quality and quality of life but also in place meaning and self-identity (e.g. Altman & Low 1992; Groat 1995; Hirsch & O'Hanlon 1995; Canter 1997; Burnham 2000). When local residents of this World Heritage landscape characterized the place where they live, the rainforest, the national parks, ready access to peaceful and unspoilt natural environments, specific sites and places within the WHA, childhood memories and re-establishing connection to place were all invariably mentioned (Bentrupperbäumer & Reser 2003).

Importantly, local residents' own identity and sense of being a responsible community member and/or landholder became intimately tied up with local geography and region, such that specific environmental values, concerns and specific issue-related attitudes were of particular salience and personal significance (e.g. Bonaiuto *et al.* 1996; Knight & Landres 1998; Borschmann 1999; Birksted 2000; Cotter *et al.* 2001).

This sense of belonging and connection to the World Heritage landscape was also evident in research on place attachment undertaken by O'Farrell (2003). In this study O'Farrell found that the favourite places nominated by local residents that were located inside the WHA boundaries were all unique with respect to biophysical, social and cultural attributes, in addition to some having a considerable and often controversial political history. Nevertheless, the physical or natural features of the place were the most important reason given for nominating such places as being favourite places. While convenience and proximity to a place were considered relevant, some places within the WHA, but remote from location of residence and/or inaccessible were also nominated as favourite places. In spite of this remoteness such places can be experienced by all residents because of the very visible and dominant nature of this landscape in the everyday lives of community residents and the very visible media and marketing profile of some of these iconic places (e.g. Mossman Gorge, Daintree National Park, Cape Tribulation). The strong emotional attachment to place, and the inclusion of features of the landscape in references to home, is particularly evident among long-term residents and the elderly who are once again returning to their familiar landscape and connecting that landscape with knowing they are home. There are a number of Australian sources that make a powerful and cogent case for the central importance of place attachment, landscape meaning and identity for Australians of very diverse cultural backgrounds (e.g. Haynes 1998; Borschmann 1999). One possible reason why local appreciation of and attachment to the natural landscape has been underemphasized in the protected area context is that many surveys, reports and planning documents are generally regional planning and tourism industry oriented, for understandable reasons, and the rich multidisciplinary literature on place meaning and connection is not widely known by many working in natural resource management or regional planning contexts.

Loss of place, displacement

A non-Indigenous perspective

Given the importance of local resident attachment to this World Heritage landscape, loss of place or the sense of displacement from places within the WHA is of considerable concern (Andersen & Brown 1984), particularly when over 25% of community respondents noted feeling displaced from their favourite places (Bentrupperbäumer & Reser 2003). In both this research and research undertaken by Bryden (2001), residents who felt displaced from places in the WHA were the longer-term residents who had frequently visited such places in the past. As suggested by O'Farrell (2003), identity and attachment to the place is clearly developed through familiarity and knowledge gained from repeat visits.

Reasons why community residents felt displaced from places in the WHA varied, but were most frequently reported as due to being fenced off, denied access and crowding, too many tourists. Exclusion due to denial of access was further supported in the selection by local residents of rules, regulations and restrictions as the main disadvantage of living in this World Heritage landscape. The issue of crowding at certain locations in the WHA, in particular those icon sites in the Daintree National Park (e.g. Mossman Gorge, Cape Tribulation), has also led to the displacement of local residents. These two sites receive very high visitor numbers and, importantly, very high numbers at any one time within a relatively small recreational space. For example, Mossman Gorge can receive up to 500 000 visits per year with over 300 visitors at one time and 150 vehicles accessing the site, which has facilities for only 30 vehicles (Bentrupperbäumer & Reser 2000a, b; Bentrupperbäumer 2002a). Experienced crowding was also evident in a visitor survey undertaken at Mossman Gorge ($n = 717$), where 48% of respondents agreed that there were too many people at the site. Such site-level, carrying capacity considerations (e.g. Graefe *et al.* 1984; Lindberg *et al.* 1997) and their differential impacts on residents, visitors and particularly resident visitors well illustrate management complexities and challenges in conservation and recreational landscapes that are simultaneously protected natural environments,

popular outdoor recreation areas and behaviour settings and heritage places providing for multiple other benefits and needs, such as restoration, and place and cultural identity and affirmation (e.g. Hartig 1993; Twigger-Ross & Uzzell 1996; Cantrill 1998).

An Indigenous perspective

An Indigenous perspective on visitation, use and impacts in a protected area owned, used by and sacred to local Indigenous communities is particularly helpful as a comparative and reflective touchstone with respect to natural and cultural heritage management and the multiple advantages and critical need for working management partnerships (e.g. Kellert *et al.* 2000; Reser *et al.*, 2000). The real and perceived displacement from places within the WHA was particularly significant among Indigenous residents, as evident in survey research undertaken with Indigenous local residents from the Mossman Gorge community ($n = 43$), a community living adjacent to this icon WHA recreation site (Daintree National Park, Mossman Section) (Bentrupperbäumer & Reser 2000c, 2005). In addition to a 63% reduction in the number of Aboriginal community members visiting the park, of those who continued to visit, a shift had occurred in the frequency of their visits, with a 75% reduction in the number visiting on a regular basis (daily to once per week). Of those who continued to visit the park, the majority did so either early in the morning or late in the afternoon, so as to avoid the crowds, the traffic and the noise. Reasons given for not visiting at all provided a valuable insight into why this displacement had occurred. Disturbance due to crowding, traffic and/or noise was clearly the most important factor, with 59% of the Indigenous respondents providing this as their first response to an open-ended question addressing this issue, and with this being the major factor identified in their second and third responses. When specifically categorized, 'crowding, too many people' was identified by 93% of respondents and received the highest mean importance score (scale 1–6; mean = 5.28 ± 1.55).

For this Indigenous community, a loss of sense of place/belonging was found to be the second most important reason for displacement, with 90.2% and 92.5% of respondents agreeing that it was no longer a place for locals and that they feel they have been pushed out of the area (importance mean = 5.06 ± 1.68;

5.18 ± 1.47). Rules and regulations imposed as a consequence of the area being a national park were identified as the third most important reason for displacement. A total of 83% of respondents agreed with 'can't do what I want to do' (importance mean = 4.8 ± 2.19). The issue of lowered self-esteem, which was apparent in respondents feeling uncomfortable and/or embarrassed and/or under scrutiny when they visit the Mossman Gorge recreation site, was identified as the fourth most important reason for displacement, with a considerable percentage of respondents (68.3%), particularly young boys and men, experiencing this discomfort. While the remaining reasons did not feature as highly in responses to both sets of questions asked, their very mention establishes their importance in terms of a more comprehensive and in depth understanding of the displacement process. Over half of the Indigenous respondents experienced strange behaviour such as others staring at them, which they translate to mean that they are considered to be weird, different, a novelty, etc. Being the object of such looking or staring also carries other meanings and concerns in an Aboriginal cultural context (Lawrence 1985). Many are not happy with visitors taking photos of them and feel they have no privacy. What is also very disturbing is the bad behaviour that members of this Aboriginal community have been subjected to. Abusive language and racist taunts, sensing a lack of respect and believing that their very colour immediately leads to perceptions of illegitimate use and possible accusations of stealing all appear to be very important reasons for why displacement from the national park has occurred for members of this Aboriginal community.

Visitation and use of the World Heritage Landscape

While the significance of vicarious use of this World Heritage landscape for both local residents and others is acknowledged for the contribution it makes to heightening awareness, connection to place and place identity, the extent and pattern of actual use of the WHA provides another interesting insight into just how important this landscape is to local residents and to domestic and international visitors as a place to go, to spend time in, to enjoy and to experience. Over 4.65 million visits are made each year to 100 designated recreation sites in this World Heritage landscape, with over 3.5 million visits made to just 15 of these sites (Bentrupperbäumer & Reser 2003). While a considerable number of these visits include domestic and international visitors, the visitation and use figures evidenced a very high level of proportional use by local residents with respect to the actual number of visits made, how often and over what time period. It was found that 85% of community respondents surveyed had visited the WHA, with 66% of these reporting that they had visited the area within the past 6 months (Bentrupperbäumer & Reser 2003). That just over 8% of community respondents visit or pass through the WHA virtually every day, and another 12.3% at least once a week, is an important index of just how interwoven this WHA is with the lives of community residents.

This high local resident use of the WHA was also evident in the site-based survey results, which demonstrated that local residents not only say they are using but clearly do very regularly use these WHA sites, and that they are enjoying a range of highly valued benefits from such use. A total of 34% of respondents surveyed at WHA recreation sites were local residents (Bentrupperbäumer & Reser 2002). This high proportional representation of local residents is consistent with other survey findings in the WHA bioregion. An analysis of visitors to five of the most frequently visited 79 sites that were surveyed by Manidis Roberts in 1993 and 1994, for example (Moscardo 1999), with each site representing samples of more than 250 visitors, found that 45% of visitors were local residents. A comprehensive site level survey of visitation and use at WHA sites in 1999, including a number of icon sites, found that 34% of visitors were local residents (Bentrupperbäumer & Reser 2000a, d). These figures take on particular meaning when it is appreciated that local residents were actually the principal visitors and users of many WHA recreation sites. Local residents, for example, were the largest visitor group (as compared with international and domestic visitors) at seven of eleven key sites surveyed, with this strong local representation increasing at smaller and more remote locations (e.g. Goldsborough Valley and Big Crystal, 83% locals; Lake Eacham, 63% locals) (Bentrupperbäumer & Reser 2000a; Bentrupperbäumer 2002b, c). Furthermore, 50, 59 and 61% of these local residents, respectively, made repeat visits to these three sites.

Recreation, experience and learning opportunities

The site-based visitor survey findings regarding reasons for visiting particular WHA recreation sites relate to preference, destination choice and motivation, as well as to recreation and experience opportunities. Findings highlighted the central importance of experience itself when visiting this WHA, as evident in comparisons of experiential reasons with activity and educational reasons ($n = 2780$) (Bentrupperbäumer & Reser 2002). The five most important reasons reported for visiting the recreational sites within the WHA related to experience, to see natural features and scenery, to be close to and experience nature, to experience tranquility, to experience the Wet Tropics. Reasons relating to experiences and experience opportunities were rated as significantly more important than recreation opportunities relating to activities, or to educational and learning opportunities. This was a finding that held true across all groups and across all recreational sites. However, these findings are somewhat at odds with the implicit assumptions and emphases of the Recreation Opportunity Spectrum (ROS) planning models of outdoor recreation management (e.g. Driver & Brown 1978; Clark & Stankey 1997; Hammitt & Cole 1998) and with conventional wisdom generally about why people visit national parks and WHAs. Nonetheless, the findings are very congruent, with a diverse literature documenting the attraction, fascination and restorative benefit of natural environments and the very important role such environments play in experienced quality of life and perceived environmental quality (e.g. Hartig *et al.* 1991; Herzog & Bosley 1992; Hartig 1993; Kaplan 1995; Kaplan *et al.* 1998; Montgomery 2002).

An interesting and non-intuitive finding from the site-based visitor surveys was that local and domestic Australian visitors rated experiential reasons as significantly more important than overseas visitors (Bentrupperbäumer & Reser 2002). This may reflect place attachment and meaning for local Australian respondents, and the salience and symbolic importance of the WHA following its controversial listing and legacy. On the other hand, learning about nature and culture were important considerations for international visitors, who are visiting a relatively unknown and very impressive environment that is a declared WHA. The general perceived importance and status of learning and education reasons for visiting are important considerations given various government and agency commitments to improve the level of understanding of local communities and World Heritage visitors with respect to the nature and status of WHA (e.g. AHC 1998a, b, 2002; Hockings 2000).

Impacts and impact assessment

Impacts on a natural environment

While a preceding section addressed a number of the perceived benefits associated with this World Heritage landscape in the eyes of community residents, it is worth more closely considering the nature of perceived threats and impacting processes, as these both create and constitute salient areas of community and visitor concern, and the extent to which these threats are viewed as being adequately addressed relates very directly to perceptions of management effectiveness. The research findings suggest that the threats to the WHA that community residents see as particularly important relate to human activities both inside and outside the WHA, as well as introduced animals, plants and pests (Bentrupperbäumer & Reser 2003). The specificity of the introduced animal and plant threats mentioned, the frequency with which these threats were mentioned and the very low ratings with respect to their being adequately managed all suggest that these are real and important issues and matters of concern for the community, concerns that they feel are not entirely shared by management agencies. It is very important to distinguish between perceived impacts on the protected area and perceived impacts on adjacent human communities. These are quite separate impact considerations and domains for monitoring, but are of course ultimately interrelated in their consequences.

Much of natural resource management concerns the identification and management of adverse biophysical impacts. Increasingly we realize that effective management must also address very important and interactive social and psychological impacts, on and of both natural environments and human settings (see Reser & Bentrupperbäumer, 2008; Chapter 34, this volume). Such impacts are conventionally conceptualized and measured as threats or changes in the environment or landscape that have potential short- and long-term consequences. The diverse impacts of visitation and use on protected areas constitute a particularly important and complex set of biophysical and psychosocial impacts, including very different user and interest groups, a range of adverse and positive impacts (consequences, costs/benefits, changes, responses to changes, concerns). In our own research we have given particular consideration to the impacts of visitation and use on local resident communities, and human environments and settings, as not only are these a badly neglected part of the ecological equation, but these psychological and social impacts directly affect and interact with biophysical impacting processes. The impacts of visitation and use collectively cover not only what are conventionally thought of as social impacts on the structural, demographic and socio-economic aspects of local affected communities and stakeholders but more direct and immediate causal consequences of rapidly increasing World Heritage visitation for individual local residents and groups, with respect to their own use of, involvement with and concerns for this personal, social, cultural and out-standing natural place and landscape.

Impacts on a human environment

In a study undertaken with an Aboriginal community living immediately adjacent to a World Heritage designated recreational site, Mossman Gorge, a spectrum of important issues relating to impacts from recreational use of a WHA on an adjacent human community was examined (Bentrupperbäumer & Reser 2000c). Their individual and community perceptions and experiences, including issues of privacy, cultural commodification and trivialization, trespass, displacement and multiple losses relating to individual and community control, amenity value and economic benefit, were considered (Bentrupperbäumer & Reser 2000c, 2005; Bentrupperbäumer et al. 2001). The study location epitomized many of the current planning and management issues as well as decisions relating to sustainable use, carrying capacity and biophysical and psychosocial impacts of high level visitor use (e.g. Ryan & Huyton 2002). In addition, this World Heritage site was accessed by a road that traversed the community, exposing residents to multiple impacts from the recreational use of the site. In the management context the situation in many ways epitomized the classic dilemma for World Heritage management: finding a working balance between a very prescriptive management structure and regime, and the needs, aspirations and involvement of a local, resident community (Lane 1994; Beltran & Phillips 2000; Kellert et al. 2000). An aspect of Aboriginal community concern not often appreciated is that communities often feel a keen sense of responsibility for the well-being and safety of anyone passing through their country, with many protected areas that receive large numbers of visitors being viewed as particularly dangerous country, especially for those unfamiliar with its cultural and natural dangers (Brereton et al. 2005).

The Indigenous psychosocial impact assessment (Bentrupperbäumer & Reser 2000c, 2005) undertaken in concert with an assessment of the biophysical and psychosocial impacts of visitation and use at the Mossman Gorge visitation site itself (Bentrupperbäumer & Reser 2000b) suggests that impacts on natural and cultural World Heritage are interdependent and interwoven in the case of Aboriginal communities, raising a number of different and complex management issues (e.g. Lawrence 2000). The nature, salience and consequences of the impacts were in many ways different for an Aboriginal local community as compared with non-Aboriginal local communities. This is, in part, because the relationship, history and responsibilities of an Indigenous community to their traditional country are very different from those of non-Indigenous residents (Rose 1996; Read 2000). A brief example might suffice to illustrate the superficial nature of typical majority culture understandings of Aboriginal community issues and concerns. The issue of privacy provides a particularly apt example. Concerns about privacy were frequently expressed by Aboriginal community residents at Mossman Gorge in the context of the impacts of site visitation and use. We should first be clear that privacy is a very complex domain that has to date eluded any satisfactory social science analysis and clarity (Altman & Chambers 1980). Hence the specification and measurement of privacy is difficult. It is, however, possible to assess the extent to which people in a community feel that privacy issues are important and/or the extent to which privacy is threatened. In an Aboriginal community context reference to privacy includes a number of non-obvious concerns. In such communities, for example, there are matters that the community does not want to share with the outside world. Media coverage of social problems, poverty, financial inequalities or community conflict is the source of considerable embarrassment and shame. The more open and outdoor life of many Aboriginal communities means that a public road through a community puts personal and community life on show in an often invidious and unfair way. In addition, public and uncontrolled roads heighten local fears of travellers, who are often associated with magical powers and malevolent intent. Such traditional beliefs and fears are not matters to be lightly discussed with a sceptical audience and might well elicit a response relating to privacy, when the issue in many ways is about felt security and control. Looking and staring are also behaviours that have particular cultural meanings that differ from many non-Aboriginal social contexts in Australia, and can be the cause of anxiety, anger, discomfort and shame. This situation is different but in some ways exacerbated when individuals or communities are the objects of the tourist gaze, in the form of a curious look by an interested visitor. The cultural context of privacy in Aboriginal communities is only now receiving renewed cross-cultural research interest, but it is clear from past studies that privacy and perceived cultural issues are very important aspects of interpersonal behaviour and individual security in Aboriginal communities and very directly tied to social control.

Conclusions, implications and management considerations

As is evident in the discussion presented, this research has taken a more ecological, natural and human landscape perspective, in which different individuals, groups and users have differing functional, temporal and personal relationships and connections with the WHA, and perceive and appraise the Area, its management agencies and landscape alterations and changes, from their rather different perspectives. For protected area managers of World Heritage landscapes the challenge is to acknowledge, become adequately informed about and respond appropriately to the demands of the multiple landscapes that make up a WHA – biophysical, cultural, human, political, economic, etc. This chapter has highlighted the central importance of one of those landscapes, the human landscape.

The importance and salience of the WHA

Research findings highlighted the importance of the World Heritage landscape in the lives of community residents and for visitors to the WHA. Not only is this landscape providing the physical backdrop and view from the window for many community residents, it is integral to their quality of life and environment. The WHA is also seen as an amenity and resource that provides for community recreation, restoration and inspiration, as an important component of place identity, as well as a livelihood for some. The importance of non-monetary benefits highlights the need for management policy to reflect, address and otherwise acknowledge community expectations, concerns and associations with this World Heritage landscape.

Place meaning and attachment

With respect to place meaning and attachment the depth and strength of community attachment to, and emotional and lifestyle involvement with, the WHA provides a clear sense of just how important these individual and community connections are. In many ways the conferring of World Heritage status on the area has given formal and international recognition to what was already a community conferred special status and sacred trust.

Loss of place, displacement

For management agencies the principles of preservation and presentation undoubtedly require attention to be paid to both the protection of the natural environment and the promotion of community engagement, which in some instances involve strategies that are in conflict with either one of these principles. Local residents can feel excluded when management agencies restrict access to certain places for conservation or cultural purposes, put in place regulations with respect to removal of plants and animals or enforce rules regarding the presence of domestic animals.

Visitation and use

The high local resident visitation and use of this WHA landscape obviously reflects a very natural and familiar extension of residents' lifestyles and general involvement with the natural environment. Unlike the one-off tourist experience associated with visiting a distant national park or WHA, for the region's residents the WHA is an everyday, well-known extension of their backyard or neighbourhood park. It is just this type of natural area that many researchers feel plays an important role in everyday well-being and enjoyment (e.g. Hartig *et al.* 1991; Kaplan 1995; Kaplan *et al.* 1998).

Recreation, experience and learning opportunities

There is much to be said for broadening the ROS framework to include the notion of an Experience Opportunity Spectrum (EOS), which more directly acknowledges a spectrum of motivations and satisfactions that are unrelated to activity *per se* but closely tied to personal connections with significant natural places. Such a reframing of the reasons why people visit areas such as a WHA would allow for a more serious consideration of how site design, interpretation materials and management policies might more adequately address and support the seemingly fundamental needs of many visitors, both local and others, for a restorative experience and encounter in a powerful natural environment. Such experiences and the felt connection and at times identification with such natural and protected environments also foster more ecologically sensitive behaviours and respect, therefore advancing management objectives of minimizing adverse biophysical impacts while maximizing positive psychosocial benefits and impacts.

Impacts and impact assessment

The impacts on an Aboriginal community of visitation and use of an icon World Heritage site are substantial and largely adverse. These impacts are psychological and social, as well as socio-economic, impacting on individuals, families, organizations, the community as a whole and community relations with local non-indigenous communities. Specific impacts include privacy loss, cultural and community disrespect, trespass, racist behaviours, displacement, loss of individual and community control, as well as loss of amenity value and economic benefit.

The nature and magnitude of the cumulative impacts are such that immediate management attention must be given to a collaborative review of all management options with respect to numbers of visitors accessing and using this recreation site, some measure of jurisdiction and control of the road passing through the community and the management of associated adverse psychosocial impacts.

Summary

This chapter, and the conclusion of the Rainforest CRC, have provided an opportunity to think about what has been accomplished, our success in getting the research and findings out to both end users and research colleagues, and what outcomes and findings have been particularly striking and/or of potential importance to natural resource and protected area management generally. It is clear that the single most important accomplishment and contribution has been to integrate social science and environmental psychology more fully in this multidisciplinary enterprise. This has made the focus of this chapter particularly apropos, as this captures the reality before us. Environments such as the WHA are natural environments and human settings. They constitute the backdrops for people's lives in a myriad of important ways, and are invested with significance and meaning not only in World Heritage and tourist destination image terms, but in the context of local history, heritage and identification as well as personal experience and encounter (Birksted 2000; Read 2000). Of particular importance, from a monitoring and managing the impacts perspective, is that people are an important and integral part of protected areas, and that biophysical and psychosocial impacts, that is, impacts on and of the area in question, are interdependent, and are experienced nationally and internationally, as well as in local communities and at visitor sites. All of this means that social and behavioural science, which has strong roots in the natural sciences as well as the humanities, is critically necessary for effective World Heritage management.

References

Altman, I. & Chambers, M. (1980). *Culture and Environment.* Cambridge: Cambridge University Press.

Altman, I. & Low, S. (eds.) (1992). *Place Attachment.* Human Behaviour and Environment Series. New York, Plenum.

Anderson, C. (1987). Aborigines, ecology and conservation: The case of the Daintree–Bloomfield road. *Australian Journal of Social Issues* **24**: 214–27.

Anderson, D. H. & Brown, P. J. (1984). The displacement process in recreation. *Journal of Leisure Research* **16**: 61–73.

Australian Heritage Commission (AHC) (1998a) *Protecting Local Heritage Places: A Guide for Communities*. Australian Heritage Commission, Canberra.

Australian Heritage Commission (AHC) (1998b) *Natural heritage places handbook*. Australian Heritage Commission, Canberra.

Australian Heritage Commission (AHC) (2002). *World Heritage icon value*. Australian Heritage Commission, Canberra.

Beltran, J. & Phillips, A. (Eds) (2000) *Indigenous and Traditional Peoples and Protected Areas*. Island Press, Washington, DC.

Bentrupperbäumer, J. M. (2002a). *Mossman Gorge: Site Level Data Report 2001/2002*. James Cook University and Rainforest Cooperative Research Centre Cairns.

Bentrupperbäumer, J. M. (2002b). *Goldsborough: Site Level Data Report 2001/2002*. James Cook University and Rainforest Cooperative Research Centre. Cairns.

Bentrupperbäumer, J. M. (2002c) *Big Crystal: Site Level Data Report 2001/2002*. James Cook University and Rainforest Cooperative Research Centre Cairns.

Bentrupperbäumer, J. M., Hill, R., Peacock, C. & Day, T. (2001). *Mossman Gorge Community-Based Planning Project. Bama Babu Sees the Future*. James Cook University and Rainforest Cooperative Research Centre, Cairns.

Bentrupperbäumer, J. M. & Reser, J. P. (2000a). *Impacts of visitation and use: Psychosocial and Biophysical Windows on Visitation and Use in the Wet Tropics World Heritage Area*. James Cook University and Rainforest (CRC) Co-operative Research Centre, Cairns.

Bentrupperbäumer, J. M. & Reser, J. P. (2000b). Roads and access, visitation and use. In *Impacts of Visitation and Use: Psychosocial and Biophysical Windows on Visitation and Use in the Wet Tropics World Heritage Area*, J. Bentrupperbäumer, J. & Reser J. (eds). James Cook University and Rainforest (CRC) Co-operative Research Centre, Cairns, pp. 61–74.

Bentrupperbäumer, J. M. & Reser, J. P. (2000c). Aboriginal perspectives. In *Impacts of Visitation and Use: Psychosocial and Biophysical Windows on Visitation and Use in the Wet Tropics World Heritage Area*, J. Bentrupperbäumer, J. & Reser J. (eds). James Cook University and Rainforest (CRC) Co-operative Research Centre, Cairns, pp. 23–40.

Bentrupperbäumer, J. M. & Reser, J. P. (2000d). Cultural background and prior experience. In *Impacts of Visitation and Use: Psychosocial and Biophysical Windows on Visitation and Use in the Wet Tropics World Heritage Area*, J. Bentrupperbäumer, J. & Reser J. (eds). James Cook University and Rainforest (CRC) Co-operative Research Centre, Cairns, pp. 102–10.

Bentrupperbäumer, J. M. & Reser, J. P. (2002). *Measuring and Monitoring Impacts of Visitation and Use in the Wet Tropics World Heritage Area 2001/2002: A Site Based Bioregional Perspective*. James Cook University and Rainforest CRC, Cairns.

Bentrupperbäumer, J. M. & Reser, J. P. (2003). *The Role of the Wet Tropics in the Life of the Community: A Wet Tropics World Heritage Area Survey*. James Cook University and Rainforest CRC, Cairns.

Bentrupperbäumer, J. M. & Reser, J. P. (2005). Psychosocial Impacts of Visitation and Use of a World Heritage Area on an Aboriginal Community. Paper presented to the 11th International Symposium on Society and Resource Management, Östersund, Sweden, 16–19 June.

Birksted, J. (ed.) (2000). *Landscapes of Memory and Experience*. Spon Press, London.

Bonaiuto, M., Breakwell, G. M. & Cano, I. (1996). Identity processes and environmental threat: the effects of nationalism and local identity upon perception of beach pollution. *Journal of Community and Applied Social Psychology* **6**: 157–75.

Borschmann, G. (1999). *The People's Forest. A Living History of the Australian Bush*. The People's Forest Press, Blackheath.

Brereton, D., Memmott, P., Reser, J. *et al.* (2005) *Mining and Indigenous Tourism in Northern Australia*. CRC Tourism, Griffith University, Gold Coast.

Bryden, N. H. (2001). Recreational displacement of local residents in the Wet Tropics of Queensland: Implications for planning and management. Unpublished masters thesis, School of Tropical Environmental Studies and Geography, James Cook University, Cairns.

Burnham, P. (2000) *Indian Country, God's Country: Native Americans and the National Parks*. Island Press, Washington, DC.

Bushell, R., Staiff, R. & Conner, N. (2002) The role of nature-based tourism in the contribution of protected areas to quality of life in rural and regional communities in Australia. *Journal of Hospitality and Tourism Management* **9**: 24–36.

Canter, D. (1997). The facets of place. In *Advances in Environment, Behavior and Design. Volume 4: Toward the Integration of Theory, Methods, Research, and Utilization*, Moore G. T. & Marans R. W. (eds). Plenum Press, New York, pp. 109–147.

Cantrill, J. G. (1998). The environmental self and a sense of place: Communication foundations for regional ecosystem management. *Journal of Applied Communication Research*, **26**: 301–18.

Clark, R. N. & Stankey, G. H. (1997). *The Recreation Opportunity Spectrum: A Framework for Planning, Management and*

Research. General Technical Report PNW-98. Department of Agriculture, Forestry Service, Pacific Northwest Forest and Range Experiment Station, Portland, OR.

Cotter, M., Boyd, B. & Gardiner, J. (eds) (2001) *Heritage Landscapes: Understanding Place and Communities*. Southern Cross University Press, Lismore.

Driml, S. & Common, M. (1995). Economic and financial benefits of tourism in major protected areas. *Australian Journal of Environmental Management* **2**: 19–39.

Driver, B. L. & Brown, P. J. (1978). The opportunity spectrum concept and behavioural information in outdoor recreation resource supply inventories: rationale. In *Integrated Inventories of Renewable Natural Resources*, Lund, G. H., La Bau, V. J., Folliott, P. F. and Robinson, D.W. (eds). General Technical Report RM-55. USDAFS, Rocky Mountain Forest and Range Experiment Station, Fort Collins, CO, pp. 24–31.

Gee, M. (1994), Questioning the concept of the user. *Journal of Environmental Psychology* **14**: 113–24.

Graef, A., Vaske, J. & Kuss, F. (1984). Social carrying capacity: An integration and synthesis of twenty years of research. *Leisure Sciences*, **6**: 395–431.

Groat, L. (ed.) (1995). *Giving Places Meaning*. Plenum, New York.

Hammitt, W. E. & Cole, D. N. (1998) *Wildland Recreation: Ecology and Management*, 2nd edn. John Wiley, & Sons New York.

Hartig, T. (1993). Nature experience in transactional perspective. *Landscape and Urban Planning* **25**: 17–36.

Hartig, T., Mang, M. & Evans, G. W. (1991). Restorative effects of natural environment experiences. *Environment and Behavior* **23**: 3–26.

Haynes, R. D. (1998). *Seeking the Centre: The Australian Desert in Literature, Art and Film*. Cambridge University Press, Melbourne.

Herzog, T. R. & Bosley, P. J. (1992). Tranquility and preference as affective qualities of natural environments. *Journal of Environmental Psychology* **12**: 115–127.

Hirsch, E. & O'Hanlon, M. (eds) (1995). *The Anthropology of Landscape: Perspectives on Place and Space*. Clarendon Press, Oxford.

Hockings, M. (2000). *Evaluating Effectiveness: A Framework for Assessing the Management of Protected Areas*. IUCN, Gland, Switzerland.

Hunter, C. & Green, H. (1995). *Tourism and Environment*. Rourtledge, London.

Kaplan, R., Kaplan, S. & Ryan, R. J. (1998) *With People in Mind: Design and Management of Everyday Nature*. Island Press, Washington, DC.

Kaplan, S. (1995). The restorative benefits of nature: toward an integrative framework. *Journal of Environmental Psychology* **15**: 169–182.

Kellert, S. R., Mehta, J. N., Ebbin, S. A. & Lichtenfeld, L. L. (2000). Community natural resource management: Promise, rhetoric, and reality. *Society and Natural Resources* **13**: 705–15.

Knight, R. L. & Landres, P. B. (eds) (1998) *Stewardship across Boundaries*. Island Press, Washington, DC.

Lane, M. B. (1994) *Public Involvement in the Wet Tropics: A Review*. Griffith University, Faculty of Environmental Sciences, Brisbane.

Lawrence, D. (2000). *Kakadu: The Making of a National Park*. Melbourne University Press, Carlton South.

Lawrence, R. (1985). The tourist impact and the Aboriginal response. In *Aborigines and Tourism: A Study of the Impact of Tourism on Aborigines in the Kakadu Region Northern Territory*, Palmer, K. (ed.). Northern Lands Council, Darwin.

Lindberg, K., McCool, S. & Stankey, G. (1997) Rethinking carrying capacity. *Annals of Tourism Research*, **24**: 461–5.

McDonald, G. & Lane, M. (2000) *Securing the Wet Tropics?* The Federation Press, Sydney.

Montgomery, C. A. (2002). Ranking the benefits of biodiversity: an exploration of relative values. *Journal of Environmental Management* **65**: 313–26.

Moscardo, G. (1999). *Making Visitors Mindful: Principles for Creating Sustainable Visitor Experiences through Effective Communication*. Sagamore Publishing, Champaign, IL.

O'Farrell, S (2003). Place attachment to the World Heritage Area: a Wet Tropics bioregional perspective. Hons thesis, School of Psychology, James Cook University, Cairns.

Read, P. (2000). *Belonging: Australians, Place and Aboriginal Ownership*. Cambridge University Press, Cambridge.

Reser, J. P., Bentrupperbäumer, J. & Pannell, S. (2000). Roads, tracks and World Heritage: natural and cultural values in the Wet Tropics. *People and Physical Environment Research* **55–6**: 38–53.

Reser, J. P. & Bentrupperbäumer, J. M. (2008) A new paradigm for conceptualizing and researching the impacts of visitation. In N. Stork & S. Turton (Eds) *Living in a dynamic tropical forest landscape* (pp x-x).

Rose, D. B. (1996). *Nourishing Terrains: Australian Aboriginal Views of Landscape and Wilderness*. Australian Heritage Commission, Canberra.

Ryan, C. & Huyton, J. (2002). Tourists and Aboriginal people. *Annals of Tourism Research* **29**: 631–47.

Twigger-Ross, C. L. & Uzzell, D. L. (1996). Place and identity processes. *Journal of Environmental Psychology* **16**: 205–20.

32 INTEGRATING EFFORT FOR REGIONAL NATURAL RESOURCE OUTCOMES: THE WET TROPICS EXPERIENCE

Allan Dale[1], Geoff McDonald[2]*[†] and Nigel Weston[3]*

[1]Terrain Natural Resource Management, Innisfail, Queensland, Australia
[2]CSIRO Sustainable Ecosystems, St Lucia, Queensland, Australia
[3]Griffith University, Nathan, Queensland, Australia
*The authors were participants of Cooperative Research Centre for Tropical Ecology and Management
[†]deceased

Introduction

World Heritage listing in the late 1980s delivered a significant slice of Australia's Wet Tropical landscape to biodiversity conservation and associated ecotourism activities. The wider region, however, comprises a complex social, cultural and natural landscape of multiple uses, values and resource types. The health of this landscape, in turn, contributes to the health of the Great Barrier Reef lagoon, another natural icon of international significance.

Following World Heritage listing, a flurry of fragmented planning for the region resulted in the regulatory frameworks that defined the broad limits for sustainable use of the region's resources. These approaches, however, were not designed to progress the integrated and sustainable management of these resources within those limits. Continued biodiversity loss, the ongoing loss of productive soils and poor water quality flowing into the reef lagoon were the result.

In 2002, the Australian government led a shift towards Queensland adopting a more integrated approach to natural resource management that had the potential to fill the remaining gaps in landscape planning and management. The new approach required a shift towards stronger community-based regionalism, including the development of a Regional Natural Resource Management (NRM) Plan and a consequent Regional Investment Strategy. These integrated plans were to be developed through a stronger alliance between the regional community and Australian and state governments through the formation of Regional Natural Resource Management Bodies.

This chapter explores the shift to integrated regionalism for sustainable natural resource management and reviews the establishment of the Wet Tropics Regional NRM Body and development of the region's first integrated Regional NRM Plan. It explores the strengths and weaknesses in the processes behind these new approaches. Finally, it proposes a positive future path for this bold experiment in integrated regionalism.

Why integrated regional planning?

Sustainable natural resource management (NRM) is now the stated goal of natural resource managers in many countries. It recognizes the importance of natural resource productivity, the prosperity of natural resource dependent communities, the elements of eco-system health and biodiversity, the need to maintain landscape productivity and the significance of the cultural, social and aesthetic values of natural resources. It takes a longer-term, target-focused, landscape-scale perspective on natural resource values in comparison to the narrow, fragmented resource use regimes of the past. In its most general form, this means taking account of triple bottom line factors in development and management decisions and measuring progress towards sustainability.

Regional planning systems have a high potential to contribute to sustainable development by integrating economic, social and environmental policies in a spatial context. Many of the issues of natural resource management, such as water quality, biodiversity and the sustainable use of natural resources, can be best measured and addressed at a regional scale. Regions provide a better stage for improved integration of politics and the administration needed to balance the economic, social and ecological dimensions of development. Finally, there is a need to decentralize decisions closer to the local community to enhance NRM activity at the property and local scales and to facilitate more open participatory decision-making processes.

Traditionally, there has been no single regional plan in any region in Australia, but there has been a diversity of plans addressing various NRM and other issues. National programmes such as the National Action Plan for Salinity and Water Quality (NAP) and the Natural Heritage Trust (NHT) are now the main drivers for more integrated regional resource use planning throughout Australia. The NAP and NHT focus on water quality, salinity and riparian zone management, sustainable agriculture, biodiversity and coastal management issues. All states and territories now produce regional natural resource management plans covering diverse themes such as rivers, forests and economic development.

Regional governance

Australia does not have formalized structures of government at the regional scale. Not only does each level of government typically adopt its own approach, but also state and federal governments continue to design and implement programme-specific arrangements that differ in scale, style, how they are resourced and

accountability standards. While we do not have formalized structures of government at the regional scale, it has become clear that many tasks of governance need to occur at this scale. These tasks also need to be integrated across local, state and federal jurisdictions.

Regional policy frameworks are now widely promoted in Australia at the federal level. They are aimed at dealing with global pressures, accelerating technological advances, increasing productivity growth from commodity sectors and ensuring sustainable development (e.g. AFFA 1999; DOTRS 1999). Such new policy frameworks are manifest in infrastructure development, telecommunications and networking and, more recently, NRM, and include:

● The National Action Plan on Salinity and Water Quality. The NAP is a partnership approach to natural resource management based on the development of regional plans by landholders and local communities. These plans fit within a national accreditation framework to address salinity and water quality problems. NAP targets the 21 regions across Australia most affected by salinity and poor water quality. It aims to help communities, in partnership with governments, to take responsibility for planning and implementing regional natural resource management strategies.

● The National Heritage Trust. The Trust is focused on investing money in large, strategic projects at the regional level. It has the capacity to provide funding on a 3-year rolling basis for the implementation of regional investment strategies. Regional NRM Bodies are responsible for developing and implementing integrated natural resource management plans in consultation with a range of community interests. The plans form the basis for investment in priority areas by the Commonwealth, the states and territories and industry and other community groups.

Importantly, the new regional governance frameworks reflect a global trend in government devolving specific decision-making closer to its source or context. The regional turn in public policy is not just an Australian phenomenon, but is also strong in Southeast Asia, Europe and elsewhere.

Regional integration

Integration in governance is essential if the principles of sustainable development are to be achieved (Productivity Commission 1999). Governments must make multi-objective decisions on policy priorities for budget allocations, development directions and community well-being. Conventionally, they divide their policy-making and administrative systems into a set of manageable ministries and agencies are usually divided along sectoral lines for efficiency. A recurring consequence is that ministries and agencies tend to focus on narrow sets of objectives and to operate within very defined limits.

As the fourth dimension of sustainability, integration is a critical principle. Integration, however it occurs, boosts the level of connectivity and the multi-objective efficiency of decisions. A useful integration diagnostic has been developed that defines attributes of integration (Morrison 2004; McDonald *et al.* 2003). The diagnostic builds on previous research (Margerum & Born 2000) and when applied using empirical interview and content analysis techniques can provide insights into the relationships between government policies and laws and the operational relationships between agencies of government and the community.

Community engagement at regional level

Significant areas of environmental policy, in both developed and developing countries, have been restructured and devolved. This action is based on the belief that decentralizing government agencies and institutions and devolving responsibility for the formation and implementation of development policies to local communities and non-state associations creates local participation in and control of planning. A more effective, context-sensitive mode of planning is said to result (Agrawal & Gibson 1999; Leach *et al.* 1999). Underpinning this approach is the assumption that local communities are better able to understand and intervene in environmental problems because they are closer to both the problem and the solution.

The community-based approach can enable sensitivity and deployment of local knowledge, greater understanding of the local context and more efficient implementation through cooperative partnerships. Important sources of weaknesses in top-down or centralist planning include the failure to recognize, respect and utilize local knowledge and the crude simplification of the local social and physical environment (Sandercock 1998; Scott 1998; Lane & McDonald 2005).

Sensitivity to local knowledge has been associated with minimizing the local social impacts of plans, reducing unintended consequences and providing equitable access to planning processes (Scott 1998). Furthermore, community-based models may generate context-sensitive plans by being more responsive to local priorities and imperatives. The community-based model is also thought to provide greater efficiency in plan implementation by recruiting local communities, thereby leveraging more resources to address problems. Addressing problems manifest at the local scale requires, among other things, a sophisticated and nuanced understanding of the local social and physical environment.

Building the capacity of local people and institutions to develop and implement improved natural resources management is central to sustainable resources management in Australia. This entails support for individuals, landholders, industry and communities with skills, knowledge, information and institutional frameworks to promote biodiversity conservation and sustainable resource use and management. Capacity building is necessary for diverse players such as Regional NRM Bodies, Landcare groups, Indigenous communities, industry sectors, local government and state/territory and Commonwealth government agencies.

The Wet Tropics: a case in integrated regionalism

Following World Heritage listing in the 1990s, a range of fragmented strategic and regulatory planning activities were carried out in the Wet Tropics region. This effort was particularly focused on the region's Growth Management Framework, the Wet Tropics of Queensland World Heritage Area (WHA) Management Plan, water resource plans, vegetation management plans and coastal management plans under various state government-led legislation. While these activities resulted in the regulatory frameworks that defined the broad limits for sustainable use of the region's resources, they were not designed to progress substantively the integrated management of those resources within those limits. Progress towards a more integrated system in the Wet Tropics region had its roots in the Australian government's responses to the social impacts of the World Heritage listing process. This progress was driven, further years, later by a national shift to more integrated approaches to natural resource management.

Emerging 'new' regionalism in the Wet Tropics

With the listing of the World Heritage Area in 1988, the Australian government announced it would implement a structural adjustment package (SAP) in order to address any negative impacts associated with the listing. As reported by Lynch (2000), the SAP was worth Au$75million and had three components: job creation, business compensation and financial assistance for forcibly displaced timber workers. The job creation component of the SAP comprised public works projects, tree planting projects, private sector initiatives and local community initiatives. Delivery of the SAP was managed by the Rainforest Unit in Cairns and was primarily the responsibility of two environmental scientists (Lynch 2000).

In some ways the Cairns Rainforest Unit could be thought of as the first Regional NRM Body in the Wet Tropics. It administered the Wet Tropics Tree Planting Scheme, among other things, and was the forerunner to the North Queensland Reforestation Joint Board, the North Queensland Afforestation Association, the NRM Board (Wet Tropics) and, ultimately, FNQ NRM Ltd, the region's new Regional NRM Body.

Wet Tropics Tree Planting Scheme

A significant component of the SAP, the Wet Tropics Tree Planting Scheme (WTTPS) was a shire-based operation first set up in four regional shires: Eacham, Hinchinbrook, Mareeba (November 1989) and the then Mulgrave Shire (January 1991). The Cairns Rainforest Unit negotiated the scheme and contracts were signed between participating shires and the then Department of Arts, Sport, Environment, Tourism and Territories (DASETT). The main aims and objectives of the WTTPS were the following:

- to employ ex-timber workers made redundant through the Wet Tropics listing;
- to retrain participants into a local government workforce that would be capable of carrying out environment-related works during the term of the SAP and beyond;

- to build within regional shires a professional level of expertise and sufficient resources to undertake ecosystem rebuilding, rehabilitation of degraded land, stream bank/riparian restoration, road impacts and fire management through revegetation works;
- to assist ecologically based community groups, for example, Trees for the Atherton and Evelyn Tablelands (TREAT), Trees for the Cairns Environment (TREEFORCE) and Landcare groups.

The national government provided each shire with a technical supervisor and a purpose-built nursery facility to conduct works. As a further commitment to longer-term environmental outcomes, ex-timber workers who left the programme were allowed to be replaced by long-term unemployed people on various labour market programmes. The SAP scheme ended in 1994, but the Australian government provided a $3 million grant to allow for the continuation of the shire-based environmental programme and specifically to maintain the nursery infrastructure.

North Queensland Reforestation Joint Board

In 1992, the North Queensland Reforestation Joint Board was set up to administer the Community Rainforest Revegetation Program (CRRP), funded by the Commonwealth from the SAP. All shires whose boundaries were common to the WHA were members. In July 1994, the Joint Board took on the additional role of administering the extended WTTPS programme, including the extension of the programme to an additional seven shires. The two programmes ran in parallel for several years, notwithstanding several efforts to share resources and complement each other. A major difficulty in ensuring integration was the fact that the Queensland Department of Primary Industries (DPI) ran the CRRP operation and the WTTPS remained with local government.

North Queensland Afforestation Association

When the CRRP scheme was nearing completion, a new organization was created in 1996–7 called the North Queensland Afforestation Association (NQAA). It was similar in function to the Joint Board but its corporate structure had changed from a quasi-government body to a private incorporated body. With the run down of Commonwealth funds for both the CRRP and WTTPS there were lean times, but in most cases the shires took an increasing responsibility for funding the units and merged them into mainstream shire activities. When the first round of NHT funding became available, NQAA was well prepared to coordinate applications and was successful in securing reasonable Commonwealth investment. However, funding was still below the old SAP levels and there was new competition from the emerging state government-backed catchment groups that were evolving to take a more integrated approach to catchment management. One of these groups, the Johnstone River Catchment Management Association, was a national leader in the development of such approaches.

NRM Board (Wet Tropics) Inc.

When the first round of Natural Heritage Trust Funding (NHT Mark 1) was introduced by the Australian government, there was a view within the Queensland Department of Primary Industries that the stream bank rehabilitation funds in particular should be sent to catchment groups. Some rural landowners also held the view that the catchment groups could provide a wider range of farm production support than an essentially environment/ecological focused group like NQAA. Consequently, another group was formed called the NRM Board (Wet Tropics) Inc. It was set up as a Regional Strategy Group and based in Innisfail, a rural part of the region. This group was essentially a confederation of catchment and Landcare groups that operated under a board structure that comprised representatives from government (Australian, state and local), indigenous and conservation interests. The board met regularly, with meetings linked to the timelines for the delivery of NHT. The NHT provided partial support for the board's operations and administrative staff.

Between 1997 and 2000, there were effectively two Regional NRM Bodies receiving funding applications as well as assessing and prioritizing projects in the Wet Tropics: one a local government-based group (NQAA) and the other state-based (NRM Board). By 2001 the situation had become both politically and economically unwieldy and the Australian and state governments pressured the two groups to merge. In August of that year, the two groups held a joint forum in Cairns to review the regional NRM arrangements. The meeting resolved to prepare a new regional plan that would give regional direction to future funding and prioritize investment from a range of sources,

especially the proposed second phase of NHT. A second resolution was that the NQAA and NRM Board should amalgamate. This new body would be responsible for the development of the new regional plan and for ensuring that all key stakeholders – including local government, Commonwealth and state agencies, land managers, industry, Indigenous people, the academic/scientific sector and environmental groups – were effectively involved (Weston *et al.* 2003).

Established in September 2003, FNQ NRM Ltd is now the designated body established to work with and represent the community in managing the region's natural resources. Bodies such as FNQ NRM Ltd are the Australian and state governments' primary mechanism to involve the community in natural resource decision-making and management. The organization's role is recognized under the Natural Heritage Trust of Australia Act (1997) as well as through a bilateral agreement signed by both state and Commonwealth governments.

FNQ NRM Ltd and a new integrated plan

The Regional NRM planning process

The first priority of the new FNQ NRM Ltd Board was to finalize the Regional NRM Plan and prepare an investment strategy to help in its implementation. The Cooperative Research Centre for Tropical Rainforest Ecology and Management (Rainforest CRC) was originally contracted to prepare the new plan by NQAA and the NRM Board after their joint forum in August 2001.

One of the first tasks of the original project team was the formation of a Plan Steering Group (or Key Stakeholder Reference Group). This group of government, community and industry came together in early 2002 to oversee the project. A Working Group of Traditional Owner Elders was also formed to progress negotiations and make decisions on behalf of the Indigenous community. Initially this group was assisted by an Indigenous Technical Support Group and together the two groups formed an Indigenous (Bama) Reference Group to oversee the development of the Aboriginal Plan (Weston *et al.* 2003). This group was superseded by the Traditional Owner Advisory Committee. More information on the Aboriginal planning process is contained in the Wet Tropics Aboriginal Cultural and Natural Resource Management Plan (Wet Tropics Aboriginal Plan Project Team 2005; see also Smyth 2004).

The Wet Tropics NRM Plan directs investment towards activities that deliver outcomes against the three overarching objectives of biodiversity conservation, sustainable use of natural resources and community capacity building and institutional change. These were the three areas of activity that would direct decision-making under NHT. To ensure an integrated and strategic approach to these activities, six Community Working Groups were established during the planning phase. In keeping with government terminology, four of these followed the simplified programme areas slated for NHT, namely:

- biodiversity or Bushcare (activities that contribute to conserving and restoring habitat for the native flora and fauna that underpin the health of the environment);
- coasts and marine or Coastcare (activities that contribute to protecting coastal catchments, ecosystems and the marine environment);
- land and water or Landcare (activities that contribute to reversing land degradation and promoting sustainable agriculture);
- rivers and wetlands or Rivercare (activities that contribute to improved water quality and environmental flows in river systems and wetlands).

A Community Capacity Building Working Group was also formed, while Indigenous interests were overseen by the Traditional Owner Advisory Committee. There was considerable overlap between these elements and they were integrated through the Plan accordingly. These working groups, as well as providing a direct link between the community and the planning process, served to organize research and consultation. Membership was non-exclusive and involved community members with both technical and practical skills. More than 70 people were involved.

A science panel provided high-level scientific support to the Regional Body to ensure that its decisions were based on the best available science. The panel met once in November 2002 and members subsequently contributed to the development of the Plan through participation in targeted workshops and the like. It was the science panel meeting that provided the impetus for the preparation of the technical reports that underpin the Regional NRM Plan.

The planning structure described above provided an excellent opportunity to engage the science and planning skills of the Rainforest CRC and other institutions with the community. This was important, because there were rigorous criteria for how plans are to be prepared and for the scientific information they need to have for determining priorities, target setting and implementation.

The Regional NRM Plan

The technical reports underpinning the Regional NRM Plan included a background report, two condition reports, a report on community capacity and capacity building, and the Aboriginal Plan. The background report provided a background to, and context for, the new NRM Plan and associated technical documents (McDonald & Weston 2004). It described regional processes and outlined the relationship of the Regional NRM Plan to other plans. It reported that over 100 plans and strategies had been prepared previously for coasts, catchments, endangered species, local government areas, the World Heritage Area and so on in the Wet Tropics. It noted that the challenge of the Regional NRM Plan was to integrate this previous work into a coordinated action plan to improve the condition of the region's natural resources.

The condition reports dealt with two of the three overarching objectives of NHT, namely, biodiversity conservation (Weston & Goosem 2004) and sustainable use (Armour et al. 2004). Adopting the condition – pressure – response framework used for Commonwealth and state-scale State of the Environment Reporting, the documents contained information about the condition of natural resources (state), about human activities that affect these resources (pressures) and about human efforts to address resource conservation issues (responses).

Complementing the condition reports was a consultancy report on community capacity and capacity building requirements in relation to NRM in the Wet Tropics (Fenton 2004). The aim of all three documents was to establish a continuing information base for developing sound environmental strategies and management, and for assessing the conservation of biodiversity and sustainable use of natural resources in the Wet Tropics, as well as community involvement in these activities. The Aboriginal Plan also contained much information that informed the Plan.

The Regional NRM Plan itself is presented in four parts. Part A describes the purpose and scope of the Plan. Part B provides an overview of the Wet Tropics region (drawn largely from the background report) and explains why and how the Plan was developed. Part C describes the assets, the threats or challenges to these assets and the current management arrangements (summarized from the condition reports, the capacity building consultancy report and the Aboriginal Plan). Part C also sets performance goals and intermediate targets for improvement in resource condition and identifies management options that, if undertaken, could effect the changes needed to reach the targets and, ultimately, the goal itself. For various reasons it is not possible to protect every asset from every challenge it faces, so Part D identifies regional priorities for the investment of NRM funds. It also provides details on the monitoring, evaluation and reporting arrangements linked to the achievement of regional targets and defines the roles and responsibilities of key stakeholders in delivering the Plan.

The Plan is based around the primary assets of biodiversity, climate, land and water as well as the human or social asset of community. Community is treated as an asset because it is recognized that the future health of the environment is dependent on the people of the region. Aboriginal cultural and natural resource management, incorporating traditional ecological knowledge and customary practices, is also recognized as a critical asset that can provide benefits to the management of natural resources in the region.

The Plan is intended to be dynamic and responsive to community and regional needs and will be improved as more information becomes available and feedback is obtained as on-ground change takes place. A Regional Investment Strategy (RIS) will also be developed and regularly reviewed to complement the Plan. The purpose of the RIS, essentially a business plan, is to attract funds from the state government, the Commonwealth government and other sources for priority actions that need to be undertaken.

Target setting: a key integrative mechanism

Targets and actions are at the core of the new Regional NRM Plans. The need to set targets for the natural

resources outcomes was the major new innovation in the integrated plan. Setting targets improved the focus of planning and action on resource conditions and guided the type and location of management actions and capacity building needed to achieve those conditions. The logic for this planning approach emerges from the rational view that the key questions are:

- Where do we want to be in the long term (aspirational targets or goals)?
- What are the long-term targets or statements about the desired condition of natural resources in the longer term (e.g. 50 plus years)? These targets guide regional planning and set a context for measuring achievement.
- How do we plan to get there? Management action targets are the short-term targets (1–5 years) for actions that will contribute to the longer-term resource condition targets and in turn the long-term goals.

Prior to the preparation of the new Wet Tropics Plan, not only were the existing regulatory plans fragmented but also few were based on or even mention targets or specifically quantifiable objectives. The two most significant existing plans on which this new plan was based, the FNQ2010 Regional Plan and the NRM Board's existing Regional Strategy, covered most of the topics addressed, and in both cases these older plans had a strong objective-based foundation (which could be interpreted as aspirational targets in many cases). Their implementation components had also set out many actions (if only generally) that might be used for management action targets in the new plan. The matters for targets for the Wet Tropics region are listed in Table 32.1. These take into account the requirements of Commonwealth and state guidelines.

There was a real integration challenge for the regional plan and the Regional Body. On the one hand, existing programmes took care of a considerable proportion of the matters for natural resource management and thereby set (or implied) targets within the scope of their focus. Statutory programmes, for example, provided strong protection for some assets and threats within their scope (e.g. national parks or licensed water pollution). These commitments needed to roll into any integrated plan. There were also, however, substantial gaps in coverage (e.g. off-reserve conservation) and the regional plan needed to address them, even if it meant promoting future statutory programmes (see Table 32.2).

The integrated plan needed to identify those targets and actions beyond regulation. While statutory plans

Table 32.1 Matters for Targets in the Wet Tropics

Theme	Scope
Biodiversity conservation	
Terrestrial ecosystems	Extent, diversity, distribution and condition of terrestrial ecosystems (including terrestrial vegetation, fauna, micro-organisms, and soils)
Inland aquatic ecosystems	Extent, diversity, condition and ecological values of inland aquatic ecosystems including rivers and other wetlands
Estuarine, coastal and marine ecosystems	Extent, diversity and condition of estuarine, coastal and marine habitats
Significant species and ecological communities	Extent and condition of significant native species and ecological communities
Environmental weeds, pests and diseases	
Weeds	Extent, impact and number of significant invasive plant species
Pest animals	Extent, impact and number of significant invasive animal species
Pathogens/diseases	Extent, impact and number of significant diseases/pathogens
Global Carbon	Stocks and flows of carbon in soil and plants
Land resources	
Physical and chemical quality of soils and landscapes	The condition of soils and landscapes including chemical processes (acidification, acid sulfate soils, salinity and fertility) and soil erosion
Pastures	The condition of native pasture resources used for grazing
Forests	The condition of the ecological resources used for production forestry including farm forestry
Water resources, waterways and fisheries	
Water quality	The physical and chemical quality of water including sediments, nutrients and pesticides
Water quantity	Water use, environmental flows, flooding and drainage
Waterways and wetlands	The condition of river channels, riparian zones and wetlands
Freshwater and marine fisheries	The size and condition of fish populations and fish habitats

Table 32.2 Statutory and non-statutory aspects of natural resource management

Asset	Statutory resource allocation	Non-statutory resource management
Water	*Queensland Water Act*: water resource plans *Queensland Environmental Protection Act*: pollution discharge licensing	Water use efficiency Research and development, best management adoption
Land	*Queensland Land Act*: land tenure and lease conditions *Queensland Integrated Planning Act*: land use controls	Best management practices, subsidies for on-ground works. Conservation works
Biodiversity	*Queensland Nature Conservation Act*: reserves *Queensland Environmental Protection Act*: species and habitat protection	Restoration programmes Capacity building
Climate	None exist	Energy programmes Design codes

are strong on development control of future actions (e.g. land use change), they are often largely unable to address liabilities of previous decisions and provide for restoration and rehabilitation. These actions often require community involvement and a more diverse set of actions, such as incentives or capacity building.

Regional NRM Plan delivery arrangements

Now that the Wet Tropics Regional NRM Plan has been completed and accredited by the state and Commonwealth governments, the core business of FNQ NRM Ltd has become the continued development and maintenance of the Plan on behalf of the region and the alignment of regional effort towards its achievement. This agenda defines the role of Regional NRM Bodies relative to the roles and functions of their critical delivery partners. Now that the Regional NRM Plan has been accredited, programme alignment and the monitoring of progress towards resource condition and management action targets are becoming increasingly important functions.

To help to deliver the Regional NRM Plan, however, FNQ NRM Ltd also manages a range of federally funded natural resource management programmes. It delivers these programmes through a range of sectoral, regional and catchment scale partnerships. Its partners include local government (via the FNQ Regional Organisation of Councils), industry (via an industry advisory group), the conservation sector (via Cairns and Far North Environment Centre), Indigenous communities (via the Aboriginal Rainforest Council) and the Queensland government (via the North Queensland Regional Coordination Committee).

At the catchment level, FNQ NRM Ltd works closely with catchment management associations, local government councils, river improvement trusts, Landcare and other community groups and community nurseries.

Beyond these core functions, the Board also sees itself as having emerging critical roles in preparing the institutional frameworks for delivering ecosystem services payments within the region in future years. The concept of ecosystems services advocates financial rewards for land managers in rural landscapes who carry out environmental services on their properties that are for the public good. It presents a national way forward and common agenda in the planning, funding and delivery of biodiversity conservation and sustainable land use. Valuing and purchasing ecosystem services uses established market systems to achieve the conservation and sustainable use of the Australian landscape.

FNQ NRM Ltd envisages the Wet Tropics region becoming nationally competitive in attracting government investment, environmental offsets and corporate and philanthropic investments in ecosystem services. Success will rely on the Board continuing to support its key delivery partners, establishing a sound research and development framework and reporting on the achievement of the Regional NRM Plan.

The future of integrative regionalism: key priorities for continuous improvement

The FNQ NRM Ltd integrated resource management process grew in the context of experimental policy implementation and political uncertainty. The Australian and Queensland governments successfully created unique non-statutory regional arrangements that now, after almost three years, are producing outcomes on the ground. The success of the system will depend on maintaining this momentum. The Queensland cabinet recently decided upon Option 1.2 from its publicly released Options Paper concerning the state's future regional arrangements for natural resource management (Department of Natural Resources and Mines 2005). This option was to maintain the current system with some specific improvements. Acceptance of this model illustrates the broad level of cross-sectoral support for the emerging integrative regionalism across Queensland.

Some of the possible improvements needed in the context of the Wet Tropics region follow.

Encouraging alignment of effort towards implementation of the Regional NRM Plan

Implicit within the intent of the bilateral agreement for the Natural Heritage Trust is the need for the alignment of regional effort towards achievement of the NRM Plan's management action targets. To date, effort alignment activity is widely recognized as being in its infancy. In relation to this issue, McDonald et al. (2005) consider that the highly siloed nature of programmes and funding currently impede regional planners' ability to design and deliver integrated outcomes.

As state and Commonwealth governments do not see a role for themselves in taking overall responsibility for coordinating effort alignment, this role is likely to fall squarely on Regional Bodies. A problem facing the Bodies is that government allocation of core operational costs does not explicitly recognize this role.

In the Wet Tropics, progress is being made to align regional efforts towards the achievement of plan targets among regional state government agencies, but this effort will require significant cultural changes within the various agencies. In some parts of government, a historical view exists that Regional NRM Bodies are seeking an explicit allocation of agency resources to match their Commonwealth funds expenditure. This perception is a hangover from the matching funding focus of NHT Mark I and has made some agencies reluctant to engage. Other agencies see additional external funds as a way to extend their capacity to deliver core mandates.

In the Wet Tropics, FNQ NRM Ltd and state agencies are considering the concept of an annual alignment process to forward project effort alignment on a larger scale. Efforts will also commence to involve other significant sectors in this approach, particularly local governments.

Better integrating Regional Plans and Regional NRM Plans

Regional NRM Plans are not statutory documents, but must take account of statutory limits to the use and management of natural resources (e.g. water resource plans under the Queensland Water Act). Equally, Regional NRM Plans need to work within the framework set by regional plans developed by Regional Planning Advisory Committees (RPACs) established under the Queensland Integrated Planning Act. At the same time, issues emerging in Regional NRM Plans need to be reflected in the regional planning process.

To draw this much stronger link between RPACs and the Regional NRM Planning responsibilities of Regional Bodies, two complementary actions are possible. One could be for relevant regional organizations of councils or regional alliances of local governments to accept formally Regional Bodies as Advisory Committees under Division 3, Section 452 (b) of the Queensland Local Government Act. The other could be for RPACs established under the Integrated Planning Act to adopt formally Regional NRM Plans as the natural resources component of their regional plans. Some plan adjustments may be required for this to occur and these could be negotiated between individual Regional Bodies and RPACs or regional organizations of councils.

Under these arrangements, it would be important for Regional Bodies to remain skills-based corporate entities, with a particular focus on community engagement, integrated NRM planning and corporate governance. Their core roles would remain the development and implementation of accredited Regional NRM Plans. Regional NRM Bodies would be expected to account for Regional Plans established by RPACs, and equally Regional Body chairs would be expected to participate as RPAC members, ensuring that critical regional NRM issues are dealt with in RPAC considerations. Significant disputes between RPACs and Regional Bodies could be resolved by agreement.

Strengthening the role for Regional NRM Bodies in engagement

The act of bilateral government designation of Regional Bodies remains a critical avenue for ensuring they have the three skills required to deliver an effective regional NRM programme. These designation criteria include:
- the capacity to engage critical sectors and groups within the regional community;
- the technical capacity to develop and implement a Regional NRM Plan;
- governance capacity to manage significant funds effectively.

One of the reasons why there is currently reasonable governance strength in Queensland's Regional Bodies is because the state and Commonwealth have run a rigorous designation process under the mandate of the NHT Bilateral Agreement. The designation criteria used by both governments, particularly those related to engagement, need to be continuously refined, and there needs to be regular review of the capacity of Regional Bodies against these criteria.

The capacity of regional communities has been strengthened through the planning process undertaken by the Regional Bodies. The Wet Tropics community, for example, has demonstrated its ability to build and share knowledge, resolve conflict and empower others, and has increased its capacity to plan for and manage change. There are now reasonably clear engagement mechanisms for critical sectors, though some sectors (e.g. tourism) remain underengaged. Continuous review of strategic engagement arrangements is always needed to keep a high level of engagement and to ensure these gaps don't remain.

Strengthening the programme delivery processes of state government agencies

Regional Bodies have experienced two key problems in dealing with the programme (funding) delivery processes of state government, and specific changes are needed. First, there are a number of complex steps in the investment approval process and numerous points at which significant delays can occur. This has often left Regional Bodies with major delays in receiving signed contracts and subsequent payments. On a number of occasions this has led Regional NRM Bodies to the brink of insolvency. Some of the biggest risks emerge from the need for four separate ministers (two state and two Commonwealth) to sign off on all investments. Additionally, Commonwealth ministers have to revisit any delivery delays in implementation over 6 months. The standard contracts are not partnership-oriented and need a major overhaul through negotiation with Regional Bodies.

Second, significant cultural changes are needed within both state and Commonwealth agencies and Regional NRM Bodies to make the investment and performance management process more efficient. A stronger risk management approach is required alongside more effective performance benchmarking and governance improvement processes.

Strengthening the functions of regional bodies to review and improve the quality of Regional NRM Plans

Continuous improvement in the understanding of the state and trend of Queensland's environment is essential. Regional NRM planning to date has allowed for increased knowledge, integration of information and analysis of gaps in data. This will continue to be a critical component in decision-making for future natural resource management.

McDonald *et al*. (2005) have shown that Queensland's high-quality Regional NRM plans are one of the key achievements of the new regional arrangements to date. They indicate that the plans are strategic and provide a sound basis for investment in regions for the next 3–5 years, finding that regional coordination between statutory and voluntary planning and management activities is improving in regions, with more purposeful interaction between regional bodies, agencies planners, managers and technical staff. McDonald *et al*. (2005) reported that Regional Body interviewees generally supported the role of statutory planning and government 'backing' to achieve the targets in regional NRM plans.

As part of their core operating costs, Regional Bodies need to be supported to effect continuous improvement in the plans and to monitor the delivery of management action targets. At this stage, there is an unfounded assumption that this will occur, except in cases where specific review work is identified in the NRM Plan and Regional Investment Strategy.

Importantly, the state government itself needs to take a more clearly defined role in monitoring resource condition and trend within regions. While Regional Bodies are well placed to facilitate state of the region type reporting processes, the state itself should be responsible for the resource condition monitoring that should underpin such reporting.

Providing greater recognition to the voluntary sector

Regional Bodies rely on a wide range of people taking voluntary roles in natural resource management. Partnering Landcare and sub-regional or catchment groups have made substantial contributions to planning and implementation. Indeed, at the local and catchment scale, many projects are reliant on a dedicated workforce of voluntary participants.

In the Wet Tropics, the Regional Body has actively engaged community groups, and many remain active participants that are successfully taking up the challenge of the new method of project delivery. Regardless of the strength of this relationship, Regional Bodies are required under their designation requirements to work closely with their community groups. Such groups are a key component in the roll-out of the plan. One of the challenges of the system is to maintain an enhanced level of support for such partner-based arrangements.

Improving the corporate governance of Regional NRM Bodies

Government funding for programmes such as the Natural Heritage Trust are not guaranteed in the long term. Government budget allocation processes, both state and federal, are very competitive. In order to maintain effective funding from governments, and to attract other funding opportunities, the governance and performance of Regional NRM Bodies in Queensland needs to demonstrate quantitative and justifiable benefits.

A programme of quality assurance that is able to quantify the benefits delivered by Regional Groups could be established in Queensland. Such a system would best encourage a self-development approach, but also ensure the public release of evidence of the continuous improvement of the strength of the regional arrangements. Such a programme may also benefit from an accreditation system that allows for governments to pay additional incentives for continuous improvement. In other words, an accreditation scheme would allow for recognition and additional benefits if regions achieve more than the minimum standard. Such a scheme may provide added confidence for a range of potential investors in regional NRM activities.

In the meantime, the current designation process under the NHT Bilateral Agreement in effect provides both governments with a functional quality assurance mechanism. To date, FNQ NRM Ltd has easily demonstrated its designation qualities. Indeed, under a recent governance review, external reviewers stated:

Overall, we found that FNQ NRM Ltd was well governed with a strong, skills based board led by an experienced and effective Chairperson. The Board is actively involved in the governance of the organisation through active involvement in board meetings and a range of other organisational activities. The Board is supported by an experienced CEO and senior management team. (Turnbull 2005: 3)

FNQ NRM Ltd takes the view that open review and discussion of governance strengths and weaknesses is critical, and that continuous improvements in governance based on regular monitoring is important. Some form of regular partnership-based review by the Bodies and the state and Australian governments would add value to this approach.

Establishing long-term financial security for Regional Bodies

Regional Bodies have been looking beyond the current funding from the Australian government programmes, but certainly rely on the state and Australian governments to provide core operational funds. Many Regional Bodies have developed into significant community-based organizations that are working towards financial independence and security to ensure their ongoing viability beyond government programme funding.

Consideration has to be given to registering Regional NRM Bodies as tax deductible organizations to attract philanthropic and other investment from both the private and public sectors. However, to achieve the significant landscape changes required in Australia, both the state and Australian governments should continue to contribute to major project funding streams.

Acknowledgements

This chapter recognizes the historical efforts of many within Far North Queensland to establish robust arrangements for the sustainability of the region. We particularly thank Mr Bill Sokolich for insights into the state of play in the Wet Tropics in the early stages of new regionalism.

References

Agrawal, A. & Gibson, C. C. (1999). Enchantment and disenchantment: the role of community in natural resource conservation. *World Development* **27**(4): 629–49.

Armour, J., Cogle, L., Rasiah, V. & Russell, J. (2004). *Sustaining the Wet Tropics: A Regional Plan for Natural Resource Management, Volume2B Condition Report – Sustainable Use.* Rainforest CRC and FNQ NRM Ltd, Cairns.

Department of Natural Resources and Mines (2005). *Options for Future Community Engagement in Regional Natural Resource Management.* Queensland Government, Brisbane.

Department of Transport and Regional Services (1999). *Proceedings of the Regional Australia Summit.* Australian Government, Canberra.

Fenton, M. (2004). *Capacity Building and Capacity Building Requirements in Relation to NRM in the Wet Tropics.* FNQ NRM Ltd Consultancy Report, Innisfail.

Lane M. B. & McDonald G. (2005). Community-based environmental planning: operational dilemmas, planning principles and possible remedies. *Journal of Environmental Planning and Management* **48**(5): 709–31.

Australian Forestry, Fisheries and Agriculture (AFFA) (1999). *Mid-term Review of the National Heritage Trust: Integrated Regional Summary Final Report.* Australian Government, Canberra.

Leach, M., Mearns, R. & Scoones, I. (1999). Environmental entitlements: dynamics and institutions in community-based natural resource management. *World Development* **27**(2): 225–47.

Lynch, M. (2000). The social impacts of conservation. In *Securing the Wet Tropics: A Retrospective on Managing Australia's Northern Tropical Rainforests*, McDonald, G. T. and Lane, M. B. (eds). Federation Press, Sydney.

McDonald, G., Morrison, T. & Lane, M. (2003). Integrating natural resource management systems for better environmental outcomes. *Proceedings of the Australian Water Summit*, Sydney, February.

McDonald, G. T., Taylor, B., Bellamy, J. *et al.* (2005). *Benchmarking Regional Planning Arrangements 2004–5*. CRC Tropical Savannas, Darwin.

McDonald, G. & Weston, N. (2004). *Sustaining the Wet Tropics: A Regional Plan for Natural Resource Management, Volume 1 – Background to the Plan*. Rainforest CRC and FNQ NRM Ltd, Cairns and Innisfail.

Margerum, R. D. & Born, S. M. (2000). A co-ordination diagnostic for improving integrated environmental management. *Journal of Environmental Planning and Management* **43**: 5–21.

Morrison, T. H. (2004). Institutional integration in complex environments: pursuing rural sustainability at the regional Level in Australia and the USA. Unpublished PhD thesis, School of Geography, Planning and Architecture, University of Queensland, Brisbane.

Productivity Commission (1999). *Implementation of Ecologically Sustainable Development by Commonwealth Departments and Agencies, Canberra*. Report No. 5. AusInfo, Canberra.

Sandercock, L. (1998). *Towards Cosmopolis: Planning for Multicultural Cities*. Wiley, Chichester, UK.

Scott, J. C. (1998). *Seeing Like a State: How Certain Schemes to Improve the Human Condition Have Failed*. Yale University Press, New Haven, CT.

Smyth, D. (2004). Case Study 4: Developing an Aboriginal Plan for the Wet Tropics NRM region in North Queensland. In *Case Studies in Indigenous Engagement in Natural Resource Management in Australia*, Smyth, D., Szabo, S. & George, M. (eds). Department of Environment and Heritage, Canberra.

Turnbull, W. (2005). *Evaluation of the Current Governance Arrangements to Support Regional Investment Under NHT2: Case Study*. FNQ NRM Ltd, DEH and DAFF, Canberra.

Weston, N. & Goosem, S. (2004). *Sustaining the Wet Tropics: A Regional Plan for Natural Resource Management, Volume 2A Condition Report Biodiversity Conservation*. Rainforest CRC and FNQ NRM Ltd, Cairns.

Weston, N., McDonald, G., Fenton, J. & Dorrington, B. (2003). Planning for sustainable natural resource management in the Wet Tropics. Paper presented to the Planning Institute of Australia National Congress, Adelaide, March/April 2003.

33 'GETTING THE MOB IN': INDIGENOUS INITIATIVES IN A NEW ERA OF NATURAL RESOURCE MANAGEMENT IN AUSTRALIA

*Sandra Pannell**

Discipline of Anthropology and Archaeology, School of Arts and Social Sciences, James Cook University, Townsville, Queensland, Australia
*The author was participant of Cooperative Research Centre for Tropical Ecology and Management

Introduction

A new approach to natural resource management (NRM) is sweeping across Australia. As an extension of the Commonwealth government's Natural Heritage Trust (NHT) programme, regional bodies and plans are being established to direct the management of rivers, coastlines, biodiversity and vegetation. In this multibillion dollar nationwide experiment in environmental management and social change, there is an increased emphasis upon integrated, asset-based responses to regional issues and pressures concerning natural values. In the first phase of the NHT programme, 1997–2001, Indigenous groups were largely excluded from the planning process and consequently only a handful of Aboriginal communities Australia-wide received funding support for NRM projects.[1] Notwithstanding the belated development of national and state Indigenous engagement protocols (NHT 2004a, b, c), this situation does not appear to have significantly improved in the second tranche of funding (referred to as NHT2), and operating from 2002 to the present. As one commentator concludes, 'Overall, the NHT … program has demonstrated little actual involvement with, or consultation of, Indigenous Australians' (Worth 2005: 4).

In the handful of reviews, commissioned by the NHT to identify why Aboriginal people are minor players in arguably the most significant intergovernmental environmental initiative undertaken in Australia, a raft of reasons are identified. These range from a lack of representation on NHT decision-making structures, a poor understanding within government of Indigenous perspectives on environmental issues, to the profound political and economic marginalization of Aboriginal people in Australian society. One mid-term review even goes so far as to suggest that the 'current overall guidelines, programs, community structures and administrative procedures may not be suitable for Aboriginal proponents' (Dames & Moore 1999: 49).

These unsuitable structures, processes and practices are an intrinsic dimension of contemporary environmental discourse in Australia. In current, acronym-based NRM parlance, the language of modernization, historically allied with resource development and Indigenous peoples, has been replaced by the nomenclature of the new green managerialism. Within this discursive framework, nature is depicted as an asset or capital infrastructure, environmental protection is regarded as an investment strategy, ecological knowledge is seen as capacity, while a diverse range of social practices are categorized as forms of management.

In the emerging eco-bureaucracies, Indigenous people are depicted as the subjects of engagement or their social forms are identified as targets for institutional change.

Based upon a case study of the Wet Tropics NRM region of north Queensland, this chapter examines Aboriginal people's engagement with some of the language, spaces and social formations associated with this new era of environmental planning and management. The Wet Tropics NRM region covers 22 000 km^2 and supports a population of 216 000 people, with 90% of people residing in the Cairns area. This region is also home to more than 20 000 Indigenous people (Australian Bureau of Statistics 2001).[2] Australia-wide, the Wet Tropics NRM region has been heralded as a 'case study of exemplary consultation with Indigenous communities in the NHT program' (Worth 2005: 4). Often overlooked in the bureaucratic peddling of a NHT Indigenous success story is the agency of Aboriginal people themselves, and the way in which key NRM concepts and structures are contested, subverted and denied in the intersection of government and Aboriginal ideoscapes (Appadurai 1996: 36). In this chapter, I argue that the indigenization of key elements of the NHT programme is one of the unintended outcomes of the latest state-sponsored initiative in social and environmental engineering.

'Getting the mob in': a brief history

While the NHT programme was officially launched by the Australian Prime Minister on 22 May 1997, it wasn't until March 2002 that Aboriginal people in the Wet Tropics became fully aware of this new funding framework. At a regional forum, attended by Traditional Owners, representatives from Aboriginal organizations and Indigenous and non-Indigenous employees of key land management agencies, the Aboriginal participants took the first steps towards 'getting the mob in' and redressing their marginalization in the NRM planning process. These steps included the establishment of an Indigenous Working Group (IWG) and the formulation of two position statements calling for (i) 'guaranteed 50% funding for Indigenous projects'; and (ii) '70% majority Indigenous representation' on the proposed regional NRM board (Smyth 2004: 118).

A second regional meeting was convened in August 2002 in which participants drafted a vision statement and set down guiding principles for Indigenous NRM in the Wet Tropics region. At this forum, a new IWG was established, comprised solely of Aboriginal elders, while the original IWG was renamed the Indigenous Technical Support Group (ITSG) to reflect better its advisory role and the NRM expertise of its Indigenous and non-Indigenous members.[3] The formation of both of these groups, not envisaged in any of the NHT guidelines, sent a clear statement to the regional NRM body and departmental technocrats that Aboriginal decision-making processes had 'taken charge' (Smyth 2004: 120) of the government's attempts at Indigenous engagement. These processes were authorized by the Aboriginal 'mob' at this and subsequent meetings convened to discuss NHT procedures. The ongoing support of the IWG, the body of Aboriginal elders from the region, further served to stamp Aboriginal decision-making with the authority of both tradition and culture.[4]

One of the key outcomes from the August 2002 forum was the decision to develop a separate Aboriginal Cultural and Natural Resource Management Plan, key elements of which would be integrated into the mainstream regional NRM Plan (see McDonald & Weston 2004; FNQ NRM Ltd & Rainforest CRC 2004).[5] The development of the mainstream plan had already begun a year earlier in August 2001. Notwithstanding the disadvantage created by this catch-up scenario, the Aboriginal Plan would become the first dedicated Indigenous NRM plan for a multitenured region in Australia.[6] In the following sections, I discuss how the formulation of an Aboriginal Plan (Wet Tropics Aboriginal Plan Project Team 2005) challenged and subverted existing NHT guidelines, objectives and categories.

From invisible subjects, to Indigenous stakeholders, to traditional owners

In many respects, the near absence of Indigenous people in current NRM planning processes reflects the official invisibility of Aboriginal and Torres Strait Islander peoples in Anglo-Australian society prior to 1967. Surprising as it may sound, the historic Referendum of 1967 initiated a period when Aboriginal people would be counted as part of the national census for the first time.[7] As Benedict Anderson (1991) observes, the counting and classification of populations is an integral element in the planning and mapping exercises undertaken by the state in the name of governance.

In the post Second World War environment, planning represents one of the key strategies in the project of modernity embarked upon by Western democracies. Under the auspices of planning and management, Western ideas about development and modernization have been imposed upon the non-Western world. For Indigenous peoples in Australia, and elsewhere in the world, planning signifies one of the techniques used by governments to order, to control and, too often, to intervene in their lives in ways that often appear passive and even beneficial. Indeed, planning as a basic tool for achieving social change and controlling people and space often goes unchallenged (see Howitt 2001). Yet, as many Indigenous Australians know from their own experiences, all too often state planning and development projects produce inequity and intolerance, and not just ordered and predictable environmental templates (Hobart 1993). This intolerance is apparent in the way of the Indigenous people's refusal to comply with state policies, their resistance to development efforts and their sluggish participation in a range of planning processes characterized as anti-modern. In this situation, Indigenous people are readily blamed for their apparent inability to embrace the ideas and benefits of development. In the twenty-first century, the old language of modernization and development has been replaced by new talk about management, capacity building, governance and institutional change for Aboriginal people. Nowadays, all manner of things are fed into the meat-grinder of management, including cultural and natural values, fauna and flora, even understandings and perceptions.

In the current NRM planning regime, Aboriginal people are no longer invisible subjects inhabiting a fourth world within a first. The NHT programme speaks of these former colonial subjects as Indigenous people and, more often, as Indigenous stakeholders. Similarly, Anglo-Australians are spoken of in broad, though more nuanced, categorical terms: landholders, natural resource managers, resource users and the community. As these labels suggest, the NHT programme attempts to create new landed identities and

affinities as part of its redrawing and compartmentalization of the social and environmental landscape of Australia. In many respects, NHT terms of identity articulate and give form to the new environmental tribes that are integral to the roll-out of the programme. Throughout Australia, hundreds of Landcare, Bushcare, Rivercare and Coastcare community groups constitute the embodied and on-ground realization of the government's environmental objectives.

These large-scale identities are essential to the government's attempts at sustainable democracy or, in the case of Aboriginal people, practical reconciliation. The notion of political participation promoted by the NHT programme is based upon the idea of an 'educated, postethnic, calculating individual subsisting in the workings of the free market and participating in a genuine civil society' (Appadurai 1996: 142). NHT, as the latest form of state-sponsored environmental governance, attempts to deterritorialize old identities and replace these with new ones, which are increasingly linked to the delivery of entitlements and benefits.

Aboriginal people in the Wet Tropics have resisted government attempts at environmental corporatism; there is no Countrycare or Indigicare yet. They have also defied attempts to homogenize their identities and erase cultural difference by adopting NHT-preferred terms of identity, such as Indigenous people or stakeholders. Instead, they assert that they are the Traditional Owners of the Wet Tropics NRM region and identify themselves in terms of their different language, tribal or clan group names. As a slap in the face of political correctness, in the context of regional meetings to discuss the NRM planning process, Traditional Owners also use Aboriginal Creole terms, such as mob, Murri and blackfella, to talk about themselves and others. Giving voice to their own primordial categories, Traditional Owners emphasize their shared claims to blood, soil and language, drawing 'their affective force from the sentiments that bind small groups' (Appadurai 1996: 140). In naturalizing their identity in local bodies and places, Traditional Owners mark out a history of cultural and racial differences. In the not so distant past, these and other essentialized indices of difference formed the basis of racist policies and practices in Australia. In the contemporary context, these primordial sentiments and essentialized differences serve as an effective political counterpoint to the cool rationalities of the modern nation-state. As demonstrated in the

Wet Tropics, the politics of affect can empower Aboriginal people and disenfranchise them as well.

In the context of NRM planning, the assertion of a Traditional Owner identity explicitly and intentionally excludes those Aboriginal and Torres Strait Islander people referred to as historical people or who are part of the stolen generation. Many of these people were forcibly removed from their traditional lands and relocated to one of the missions or settlements in the Wet Tropics region (e.g. Mona Mona and Yarrabah). The explicit exclusion of these people from the Aboriginal planning process certainly challenges the neoliberal discourse of national egalitarianism that underlies the government's delivery of NHT. At the same time, this ethos of egalitarianism, the level playing field of Australia, was invoked by some Anglo-Australian government employees in the early phases of the planning process as a critique of the proposal for a separate Aboriginal Plan. As Indigenous people in Australia well know, special treatment does not conform to European fantasies about privilege and position.

In many ways, the exclusion of historical people reflects the regional realignment of identities and affiliations associated with the belated recognition of native title in Australia, as well as giving some form to traditional ideas about strangers and the inherent dangers of being and talking out of place. As this comment suggests, Aboriginal responses to, and interpretation of, the NHT programme take place within a complex topography of new spaces and old places.

Modern colonies, new topographies

Planning projects can be seen as one of the ways in which modern states produce the space of nationness and, through a range of bureaucratic structures, simultaneously create the sites and apparatus for surveillance, discipline and mobilization. In the production of these sites, planning projects all too often impose linear models of time and produce new bounded spaces in which socio-economic activities are understood. Under the regime of NHT2, an emphasis upon regionalism defines these new spaces.

To realize the government's vision for coordinated and targeted action to address pressing environmental issues, the NHT programme has created a new,

nationwide administrative topography. The creation of the 56 NRM regions is supposedly based upon strategic natural features, such as catchments or bio-regions. For example, the Wet Tropics NRM region is defined by the amalgamation of six river catchment areas. In other parts of Australia, however, the logic of bio-physical parameters does not necessarily apply. As NHT maps indicate, some NRM regions are based on the boundaries of Aboriginal lands (north-west South Australia), some on the political borders of a territory (e.g. the Northern Territory), while others appear to be based upon the physical extent of a historically characteristic economic activity, such as pastoralism (e.g. the Rangelands NRM, which includes a large proportion of Western Australia). Regardless of their basis, the boundaries of the new NRM regions overlay a number of existing, regionally based designations. For example, the newly formed Wet Tropics NRM region does not conform to either the parameters of the World Heritage Area of the same name or the bio-region denoted by the geographical distribution of tropical rainforests.

Catchment units and bio-regions are based upon scientific notions about what constitutes an interacting system or integrated area. Aboriginal perceptions of country do not necessarily conform to these scientific spaces, even though Aboriginal ideas about country can be regarded as systematic and holistic. As reported recently, 'regional approaches to NRM do not adequately recognize [Aboriginal] cultural boundaries' (FNQ NRM Ltd and Rainforest CRC 2004: 145). As such, the new NRM regions cut across Aboriginal ideas about the extent of their traditional lands and waters. As Smyth et al. (2004: 21) point out, 'many Indigenous groups' countries lie within more than one NRM region'. In this respect, the Wet Tropics NRM region includes some, or all, of the country of 17 different Traditional Owner groups.[8]

The government's emphasis upon regionality has gained momentum in recent years and increasingly all manner of socio-economic services are delivered at this scale. Aboriginal notions of country and people's attachment to traditional localities appear at odds with the general push towards regionalism and the imposition of the new NRM administrative spaces. In this era of bureaucratic regionalism, it is easy to see how Indigenous people's focus upon the local could result in a range of exclusions. Aware of the encroaching reality of, and challenges posed by, regionalism, the Traditional Owners of the Wet Tropics have attempted to span the distances between locally based Aboriginal landscapes and neighbourhoods and the larger-scale spatial and social formations of NHT2.

In the modern NRM colonies established by NHT, Traditional Owners have annexed the concept of regionalism to their own practices of producing meaningful localities and local subjects (Appadurai 1996). So-called regional meetings are more than just mechanical techniques of social aggregation. For Aboriginal people, they function as public settings for showcasing and confirming the importance of local affinities and identities. For example, large-scale meetings served to direct the investigations of the Aboriginal Plan Project Team to local workshops, held on country, with each of the 17 Traditional Owner groups in the Wet Tropics NRM region. In this context, large-scale meetings provided a framework for thinking regionally but acting locally.

There is no suggestion here of creating a dedicated pan-Aboriginal regional body with the sole responsibility of addressing the issues of NRM planning. In tackling the lack of Aboriginal involvement in the management of the WHA since its inception in 1988, Traditional Owners in this part of Australia already have some experience of the issues of representation and authorization faced by such bodies (see Review Steering Committee 1998). This ongoing struggle for recognition in the WHA has certainly sensitized Aboriginal people to some of the inequities and politics associated with environmental protection. Moreover, it has also alerted Aboriginal people to the benefits of a coordinated approach in effecting change at the higher levels of government. In this respect, Traditional Owners have harnessed their demands for a regional agreement for the management of the WHA to the locally devised strategies and actions identified in the Aboriginal Plan.

Similar to other Aboriginal people in Australia, Traditional Owners have also had to deal with the plethora of regional forms (the most recent being the now defunct Aboriginal and Torres Strait Islander Commission (ATSIC)) promoted by the government as democratic avenues for Indigenous empowerment. Mindful of the problems with these entities, Traditional Owners have responded to the regional delivery of NRM funding by advocating country-based planning

and mapping projects. As indicated in the draft regional plan, the development and delivery of management plans at a country-based scale reflects 'both the cultural boundaries and customary laws of each group for their country' (FNQ NRM Ltd & Rainforest CRC 2004: 145). Side-stepping the rallying call of NHT2 for greater regional coordination and integration to address nationally identified environmental targets, Traditional Owners have thus opted instead to emphasize the importance of their own localized spatial forms and characteristics.

From science to sentiment: the assertion of a different ethos

Throughout the Wet Tropics, there are many ways of speaking about the environment and natural resources as there are different Aboriginal ways of talking about country and culture. Increasingly, the language we hear, particularly from government environmental agencies, is that of the new managerialism. In this discursive environment, the challenge for Aboriginal people is to uphold a holistic approach to caring for country and culture, when the language of environmental management speaks of bits and pieces.

This comment points to one of the fundamental epistemological challenges facing Aboriginal people and environmental technocrats alike, the reconciliation of an Aboriginal environmental ethos with the Western scientific paradigm underpinning the NHT programme. These contrasting world views amount to more than just a series of simplistic dichotomies, in which science is constructed as global, modern and theoretical, whereas traditional ecological knowledge (TEK) is portrayed as local, traditional and concrete. As a number of authors have commented, Aboriginal country denotes a sentient and sentimental spatial experience (Rose 1996; Baker 1999). Country is alive, but not just with a range of biological objects, the natural resources and biodiversity that are the focus of the NHT programme. The presence and, at times, unpredictable behaviour of a variety of powerful totemic, Dreaming and ancestral beings animates country as well. These non-human or post-human forces are part of a terrain that is simultaneously ecological, social and cosmological in nature. Accordingly, Aboriginal people tend to subscribe to a non-scientific theory of causation when accounting for what they regard as environmental change and continuity. The distinction between Aboriginal and scientific knowledge hinges not just upon differences in perception, but also upon differences in power. This is particularly evident in the realm of NRM, where Western scientific concepts and values regarding nature and conservation shape NHT priorities, actions and understandings.

Notably, absent from the discourse of environmental management that informs the NHT programme are references to affect and aesthetics, and the strong emotions and attachments embodied in the Aboriginal notion of country. Throughout the preparation of the Aboriginal Plan, Traditional Owners spoke of the affective force of country in shaping their sense of identity and their aspirations for themselves and their children. All too often the sentiments and feelings that Aboriginal people express in relation to their connections to land are regarded as irrational or inappropriate in terms of the predominant NRM models.

Painfully aware of the way language reflects, shapes and limits action, Traditional Owners challenged the bio-physical emphasis of the NHT programme. Instead, Traditional Owners advocated a more inclusive and relational approach to resource planning and management. In meetings and workshops, Aboriginal participants spoke of caring for country in terms of promoting individual and community well-being, maintaining linguistic diversity, recognizing Indigenous ways of knowing and securing economic certainty. To suggest that this approach involved putting culture into nature imposes an artificial distinction upon Aboriginal experiences and, I would add, attributes to Traditional Owners a self-conscious awareness of culture that is notably lacking in people's quotidian actions. In linking economic and social disadvantage to NRM planning, and calling for improvements in Aboriginal health, employment and education, Traditional Owners give voice to those values and qualities that characterize both their historical and contemporary experience of locality (see Appadurai 1996).

In attempting to reflect this experience of dwelling in place, the Aboriginal Plan introduces a new lexicon for talking about NRM. It refers to cultures and concerns, landscapes and values, sentiments and senses. The use of these words points to the need to develop different kinds of literacy and skills with respect to

NRM. In retaining some of the language identified in NHT regional planning guidelines, the Aboriginal Plan also recognizes the multilingual nature of environmental experiences and attempts to build linkages between the various ways Indigenous people and Anglo-Australians talk about their worlds. Among Aboriginal people in the Wet Tropics, some people are more familiar with this latter way of speaking, situated as they are within a range of government organizations and agencies. These individuals play an important role in their community as cross-cultural interpreters and knowledge brokers. As members of the Aboriginal Plan Project Team, some of these Aboriginal people have also been instrumental in building and bridging Indigenous and Anglo-Australian relations in the development of the Aboriginal Plan.

The actions of the Wet Tropics Traditional Owners in questioning the language and logic of the NHT programme can be seen as part of a broader Indigenous movement sweeping across northern Australia.[9] Based upon an alliance between Traditional Owner groups, Aboriginal organizations and Native Title Representative Bodies, the aim of this political and geographical bloc is to effect a paradigm shift in existing protected area management and conservation efforts on Aboriginal country.

The future of the Aboriginal Plan?

Aware of the real possibility that a distinctively Aboriginal response to the NHT programme could result in further ghetttoization, Traditional Owners have successfully argued for the inclusion of the key elements of the Aboriginal Plan in the mainstream NRM plan for the Wet Tropics. This action highlights the contradictory and ambiguous nature of Aboriginal people's experience of society and the state, whereby they are both marginalized and mainstreamed. Certainly, with the coalition government's policy of practical reconciliation, there has been a trend towards mainstreaming Indigenous programmes and services. In requiring Aboriginal organizations and individuals to compete in the market place and engage with agencies with little, if any, experience of Indigenous issues, mainstreaming effects the opposite of what it sets out to achieve. In this respect, the paradoxical results of mainstreaming are no different from the inverted outcomes of modernization theory (see Collmann 1988; Hobart 1993). At the same time, the devolution of environmental governance to regional NRM bodies raises a real concern among Indigenous people that the tense nature of race relations in rural areas will result in their further marginalization (see Smyth *et al.* 2004).

To combat the possibility of future sidelining, Traditional Owners, together with members of the Aboriginal Plan Project Team, have successfully lobbied the regional NRM body, FNQ NRM Ltd, to allocate specific funding targets for Indigenous projects in the Regional Investment Strategy (RIS). At the same time, the identification of key Aboriginal aspirations and strategies in the mainstream plan ensures a broader-based funding commitment. While the RIS details those 'actions, costs and timeframes required to implement the regional plan' (Worth 2005: 26), as Dermot Smyth points out, it does not count the following:

Emotional cost of living through a process in which core values of identity, belief and inherited responsibility to country become matters for public scrutiny, to be translated into planning jargon and compete for attention with the interests of other people and organisations for whom history had delivered greater control over that same country. (Smyth 2004: 141)

Mindful of this history, and the general exclusion of Aboriginal people in NRM planning to date, Traditional Owners have also secured a skills-based, rather than a tokenistic representative, presence on NHT-related decision-making structures, including the Regional NRM Board and the advisory science panel. In April 2005, Traditional Owners launched the Aboriginal Plan, in conjunction with the signing of the Wet Tropics Regional Agreement, to an audience of more than 350 people. Members of the Aboriginal Plan Project Team arranged for senior government employees, as well as state and federal politicians, not only to share this historic moment, but also to play a central role in the future implementation of the Plan.

While regional NRM plans are promoted as the key to the strategic delivery of Natural Heritage Trust funds, the Aboriginal Plan has its limits. It in no way replaces customary means of caring for country, which Traditional Owners may enact on a day-to-day basis;

for example, those cultural responsibilities associated with speaking to and about country. Nor does the Aboriginal Plan adequately address the trauma, stress and frustration experienced by Traditional Owners when country is used in inappropriate ways, at times resulting in misfortune, injury and death to others. The Plan is not a quick fix for current environmental ailments and previous social injustices. Nor does it spell out some magical formula for Indigenous engagement, which can be replicated, regardless of context, in other regions of Australia. While there are lessons to be learnt, the post hoc success of the Plan can be attributed to a combination of historically specific factors, including the somewhat idiosyncratic input of a number of strategically situated and politically motivated individuals. Ironically, it was the 'failure to provide for comprehensive Indigenous engagement' (Smyth *et al.* 2004: 137) in the mainstream Wet Tropics NRM Plan that 'resulted in Aboriginal people of the region taking control and ownership of their own planning process' (Smyth *et al.* 2004). In this respect, the Aboriginal Plan does not present a set of abstract facts that can be readily exploited by government agendas or harnessed to the needs of environmental management in the form of a one size fits all, best practice manual. Indeed, although plans have yet to be finalized and accredited for many NRM regions, it is apparent that throughout Australia Aboriginal people have engaged with, and responded to, the planning process in a myriad of ways (see Smyth *et al.* 2004). For example, Traditional Owners in the Burdekin Dry Tropics NRM region, which adjoins the Wet Tropics in the south, have focused upon improving their involvement in NRM decision-making structures. Other Traditional Owners have worked towards establishing protocol agreements for indigenous involvement in NRM with regional authorities (see Smyth *et al.* 2004). The varied and circumscribed nature of indigenous responses to contemporary NRM planning processes reflects the historical conditions and experience of marginalization throughout Aboriginal Australia. It also points to the rhetorical limits of both science and the state in addressing and effecting social, let alone environmental, change in the concentrated time-frames identified for the NHT programme. In many ways, the circumspect approach adopted by Aboriginal people to current planning initiatives reveals the hurdles that still remain regarding cross-cultural understanding of the lie of the land (Carter 1996) throughout settled Australia.

Acknowledgements

As the Coordinator of the Indigenous Research and Facilitation Program at the Rainforest CRC, the programme that hosted and supported the development of the Aboriginal Plan, I was afforded the unique opportunity of transforming my employment into a participant-observation research experience. For this privilege, I am primarily indebted to Professor Nigel Stork, Chief Executive Officer, Rainforest CRC. I would also like to acknowledge the input of Libby Larsen, Planning Officer for the Aboriginal Plan, and her determination in working together with Traditional Owners, members of the ITSG, staff at FNQ NRN Ltd, various government employees and the general public to create a 'recognition space' for the ideas and aspirations outlined in the Plan. Finally, as a member of the Aboriginal Plan Project Team, I would like to acknowledge and pay my respects to the Traditional Owners of the Wet Tropics region. Their support and ongoing involvement has been critical in breaking down the barriers that have, until now, excluded meaningful Indigenous involvement in natural resource management.

Notes

1 For example, in the first phase of NHT, Aboriginal people in Queensland received less than 1% of funding, outside of the Cape York Peninsula (Wet Tropics Aboriginal Plan Project Team. 2005: 27). In the latest round of NHT funding (Round 5, 2004–5), Australia-wide Indigenous projects received 4% of the overall amount of $8 493 851 allocated for projects (Worth 2005). As David Worth (2005: 36) points out, one of the problems with assessing NHT funding of Indigenous projects is that 'there is no central collection of data for Indigenous projects that have been funded under the various rounds of NHT'.

2 According to Australian Bureau of Statistics figures for 2001, of the Indigenous population of 20 483, 14 338 people identify as Aboriginal, 3761 as Torres Strait Islanders and 2384 people identify as both Aboriginal and Torres Strait Islander.

3 In mid-2004, a new group, the Traditional Owner Advisory Committee (TOAC), was formed. The nine-member TOAC replaced the ITSG and thus saw the Aboriginal Plan through to its completion in April 2005. With the launch of the Aboriginal Plan, TOAC's primary role is to advise the Wet Tropics NRM Board members on the implementation of both the Aboriginal Plan and the mainstream Wet Tropics NRM Plan (Wet Tropics Aboriginal Plan Project Team 2005: 28).

4 Further background information and specific details about the development of the Aboriginal Plan are provided in the Plan itself (Wet Tropics Aboriginal Plan Project Team 2005) and in Smyth et al. (2004).

5 Traditional Owners and members of the IWG and ITSG were supported in the development of the Aboriginal Plan by a number of organizations, including the Rainforest CRC. Among other things, this research institution funded and provided a neutral hosting environment for the planning officer selected by the Traditional Owners to prepare the Plan (see Smyth et al. 2004 for more details of the support provided by the Rainforest CRC and other organizations).

6 A separate indigenous plan has also been developed for the Aboriginal Lands Integrated NRM Region of South Australia (Aboriginal Lands Integrated Natural Resource Management Group 2004). In contrast to the Wet Tropics NRM region, with the exception of land reserved as national park, this NRM region includes areas gazetted, or held in trust, as Aboriginal Land.

7 Prior to 27 May 1967, Section 127 of The Commonwealth of Australia Constitution Act 1900 stated that: 'In reckoning the numbers of the people of the Commonwealth, or of a State or other part of the Commonwealth, aboriginal natives shall not be counted.' As a result of the 1967 Referendum, Section 127 of the Australian Constitution was repealed.

8 Of the 17 Traditional Owner groups included in the Wet Tropics NRM region, the country of Kuku Yalanji people lies within two NRM regions, Cape York and Northern Gulf. Much of the country of Bar-Barrum people also falls within the Northern Gulf NRM region, while Traditional Owner groups in the southern part of the Wet Tropics NRM region, such as Warrgamay, Nywaigi and Warungu, are also part of the Burdekin Dry Tropics NRM region (Wet Tropics Aboriginal Plan Project Team 2005: 34).

9 Established to 'support indigenous land managers ranging from Broome in Western Australia to Cairns in Queensland', the North Australian Indigenous Land and Sea Management Alliance (NAILSMA) is a unique collaboration involving the Northern Land Council, Kimberley Land Council, Balkanu Cape York

Development Corporation and Tropical Savannas Cooperative Research Centre.

References

Aboriginal Lands Integrated Natural Resource Management Group (2004). Integrated Natural Resource Management Plan for the Aboriginal Lands Integrated Natural Resource Management Region of South Australia. Aboriginal Lands Integrated Natural Resource Management Group, Adelaide.

Anderson, B. (1991). *Imagined Communities: Reflections on the Origin and Spread of Nationalism.* Verso. London,

Appadurai, A. (1996). *Modernity at Large: Cultural Dimensions of Globalization.* University of Minnesota Press, Minneapolis.

Australian Bureau of Statistics (2001). *Census of Population and Housing, Basic Community Profile.* Australian Bureau of Statistics, Canberra.

Bzaker, R. (1999). Land Is Life: From Bush to Town – The Story of the Yanyuwa People. Allen & Unwin, St Leonards.

Carter, P. (1996). *The Lie of the Land.* Faber, London.

Collmann, J. (1988). *Fringe-Dwellers and Welfare: The Aboriginal Response to Bureaucracy.* University of Queensland Press, St Lucia.

Dames and Moore NRM (1999). *Mid-term Review of the Natural Heritage Trust: Gascoyne-Murchison Region Final Report.* Dames and Moore, Adelaide.

FNQ NRM Ltd. & Rainforest CRC (2004). *Sustaining the Wet Tropics: A Regional Plan for Natural Resource Management 2004–2008.* FNQ NRM Ltd, Innisfail.

Hobart, M. (ed.) (1993). *An Anthropological Critique of Development: The Growth of Ignorance.* Routledge, London.

Howitt, R. (2001). *Rethinking Resource Management: Justice, Sustainability and Indigenous Peoples.* Routledge, New York.

McDonald, G. & Weston, N, (2004). *Sustaining the Wet Tropics: A Regional Plan for Natural Resource Management, Volume 1 – Background to the Plan.* Rainforest CRC and FNQ NRM Ltd, Cairns and Innisfail.

Natural Heritage Trust (2004a). *Working with Indigenous Knowledge in Natural Resource Management – Ways to Improve Community Engagement.* Natural Heritage Trust, Canberra.

Natural Heritage Trust (2004b). *Working with Indigenous Knowledge in Natural Resource Management – Guidelines for Regional Bodies.* Natural Heritage Trust, Canberra.

Natural Heritage Trust (2004c). *Working with Indigenous Knowledge in Natural Resource Management – Recommendations for Commonwealth Agencies.* Natural Heritage Trust, Canberra.

Review Steering Committee (1998). *Which Way Our Cultural Survival. The Review of Aboriginal Involvement in the Management of the Wet Tropics World Heritage Area. Volume 1 Report.* Wet Tropics Management Authority, Cairns.

Rose, B. (1995). *Land Management Issues: Attitudes and Perceptions amongst Aboriginal People of Central Australia.* Central Land Council, Alice Springs.

Rose, D. B. (1996). *Nourishing Terrains: Australian Aboriginal Views of Landscape and Wilderness.* Australian Heritage Commission, Canberra

Smyth, D. (2004). Case study 4: Wet Tropics region, Queensland. In *Case Studies in Indigenous Engagement in Natural Resource Management in Australia,* Smyth, D., Szabo, S. & George M. (eds). Department of Environment and Heritage, Canberra, pp. 109–44.

Smyth, D., Szabo, S. & George, M. (2004). *Case Studies in Indigenous Engagement in Natural Resource Management in Australia.* Department of Environment and Heritage, Canberra.

Wet Tropics Aboriginal Plan Project Team (2005). *Caring for Country and Culture: The Wet Tropics Aboriginal Cultural and Natural Resource Management Plan.* Rainforest CRC and FNQ NRM Ltd, Cairns and Innisfail.

Worth, D. (2005). *The Natural Heritage Trust and Indigenous Engagement in Natural Resource Management.* National Native Title Tribunal, Perth.

34 FRAMING AND RESEARCHING THE IMPACTS OF VISITATION AND USE IN PROTECTED AREAS

Joseph Reser[1]* and Joan Bentrupperbäumer[2]*

[1]School of Psychology, Griffith University, Gold Coast, Queensland, Australia
[2]School of Earth and Environmental Sciences, James Cook University, Cairns, Queensland, Australia
*The authors were participants of Cooperative Research Centre for Tropical Ecology and Management

Introduction

This chapter provides an overview of the thinking behind and the implementation of an environmental psychological, transactional, approach to the impacts of visitation and use in protected area management and Word Heritage Area contexts. This more social science-based and interdisciplinary approach to natural resource management incorporates both environmental impact assessment (EIA) and monitoring and social impact assessment (SIA), and integrates and synthesizes the logic of these approaches with an environmental psychological perspective and methodology, allowing for a consideration and integration of individual experience and behaviour within the context of a World Heritage Area encounter. Importantly, this approach simultaneously considers the impacts of World Heritage Areas on visitors as well as the impacts of visitors on the biophysical environment and on the local social environment. The resulting psychosocial impact assessment (PSIA) model and framework (e.g. Reser & Bentrupperbäumer 2001a) provides a more accessible and cross-disciplinary sampling and decision-making framework for strategic research consideration and the selection of variables, relationships, processes and measures. This approach also directly addresses the needs of managers when considering the people side of natural resource and protected area management, as it takes into account not only visitor perceptions, expectations, needs and concerns, but also the nature of their experience, encounters and impacts within a World Heritage or other protected area, and direct natural resource planning and management implications.

The research undertaken within the Rainforest CRC Project 4.1 (strategies for sustainable rainforest access and use) provided an excellent opportunity, in a context of very real and pressing management needs, to incorporate a social science and environmental psychology approach in what has now been an in-depth examination of the impacts of visitation and use, broadly defined, on and of the Wet Tropics World Heritage Area (WTWHA). The research developed, trialled, refined and implemented a series of visitor site-based assessments and surveys that were complemented by community catchment surveys and other focused studies. (Comprehensive research reports and more specific coverage of theoretical and methodological matters are found in Bentrupperbäumer & Reser 2000, 2002, 2003, 2006, b; Reser & Bentrupperbäumer 2001a, 2005a; Bentrupperbäumer & Reser, Chapter 31, this volume.) Together these studies allowed the authors to field test and adjust models and methods developed in other environmental and sustainability contexts, and apply them to the management and monitoring needs of protected area managers and catchment communities in the WTWHA catchment. The sequenced, site-based surveys undertaken were of particular value in examining and documenting individual experiences and transactions *in situ*, at a point in time when experiences were immediate, and provided for a representative sampling of tourists and local resident visitors as well as substantial ecological validity (Reser & Bentrupperbäumer 2002).

Understanding the nature and quality of human transactions in natural and human-designed settings

A consistent focus in this research has been the development of a more adequate way of understanding the nature and quality of human transactions in natural and human-designed settings. This has necessitated an exploration and reframing of alternative ways of thinking about environmental encounters so as to retain the experiential and transactional integrity and dynamic of the interface between behaving person and setting. The objective has been the development of a robust conceptual model and approach that would lend itself to the applied, problem-solving, interdisciplinary needs of natural resource management (NRM). A concomitant aim has been the development of approaches that can more effectively address the negative and unsustainable impacts of problematic and incongruent people-setting transactions, both in terms of biophysical impacts in natural settings and in terms of both diminished and stressful human experience and functioning. This work looks closely at what is actually happening at the interface of the transaction, both in terms of objective transaction elements and with respect to subjective experience, awareness and psychological impact. The objective has been to achieve a more holistic and multidisciplinary analysis that leads to clear problem-solving interventions as well as planning and design improvements that enhance the nature and quality of people – environment interactions, and mitigate the negative psychosocial and setting impacts. This approach gives equal weight and status to the nature and characteristics of positive and rewarding environmental interactions and encounters, and the value of conceptualizing the planning and management focus as the design of experiences as well as settings. We have been working on the development of conceptual frameworks and measures that are more transactional and transparent, and that make sense to architects, designers, urban and regional planners, and natural resource managers, as well as to resident communities and conservation groups.

A more environmental psychological approach

The more psychological approach adopted in this research is informed by the experience, methods and insights of environmental psychology, an area of applied psychology that focuses on the nature and behavioural implications of the physical and social settings in which people live and behave (e.g. Ittelson *et al.* 1974; Bonnes & Secchiarolli 1995; Bell *et al.* 2001; Gifford 2002). Environmental psychology draws from both the natural and social sciences, as well as from the arts and humanities, in its interest in aesthetics, design, place, and the symbolic as well as the functional nature of human buildings and settings. The nature of human interactions with the natural environment as well as attitudes, values, and understandings of the natural environment are important areas of interest within environmental psychology (e.g. Altman & Wohlwill 1983; Bell *et al.* 2001; Kaplan & Kaplan 1989; Kaplan *et al.* 1998; Reser 2002, 2003). Environmental psychology itself is reasonably interdisciplinary and has strong linkages with leisure studies, visitor studies, natural resource management, human ecology and heritage management (e.g. Proshansky *et al.* 1970; Stokols & Altman 1987; Uzzell & Ballantyne 1998; Gobster 1999, 2000; Bechtel & Churchman 2002). Nonetheless, and for a variety of reasons (e.g. Reser & Bentrupperbäumer 2001b), environmental psychological approaches have been markedly absent in natural resource and protected area management research and practice in Australia, and the social science component of much World Heritage Area research has been very modest, selective, and tourism market focused.

While our WTWHA research initiative articulates with international social science and natural resource management literature (e.g. Cordell & Bergstrom 1999;

Manfredo *et al.* 2004), and the more generic natural resource and protected area management literature (e.g. Hammitt & Cole 1998; Worboys *et al.* 2001), it more fully incorporates social and environmental psychology and environment-behaviour studies (e.g. Wapner *et al.* 2000; Bell *et al.* 2001; Bechtel & Churchman 2002; Gifford 2002; Schmuck & Schultz 2002), as distinct from social ecology, rural, recreation or interpretation sociology, or non-psychologically based leisure studies approaches (e.g. Machlis & Field 1992; Manning 1999; Dunlap & Michelson 2002). This environmental psychological perspective is characterized by a multi-level and often interdisciplinary approach that includes individual experience and the environmental encounter, as well as community and cultural understandings and representations, as critical and neglected elements. In this World Heritage context, such an approach seeks to understand better the interests and motivations of those who visit, the connections and concerns of those who live in the catchment, the multiple impacts of this powerful and majestic environment on visitors and their attitudes and behaviours, and the public understandings of environmental pressures and management responses and practices (e.g. Reser *et al.* 2000; Bentrupperbäumer & Reser 2006b). There exists a widespread misunderstanding of what constitutes social science and social science input into natural resource and protected area management (e.g. Reser & Bentrupperbäumer 2001b). Socio-economic considerations, for example, typically examine a very narrow spectrum of human behaviour and experience, and this paradigm and its associated methods and measures, applied in natural resource management contexts, can be limiting and distorting (Bazerman *et al.* 1997; Becker & Jahn 1999; Bell & Morse 1999). Anthropological and sociological approaches also differ markedly from environmental psychological approaches in that their theoretical models privilege cultural and societal processes and parameters, and methods are not well suited to the systematic analysis of individual – environment transactions, or many aspects of individual experience (e.g. Little 1999; Reser & Bentrupperbäumer 2001b; Dunlap & Michelson 2002).

Of particular importance to the current World Heritage context in which we are working is that taking such an environmental psychological approach has enabled us to move from individual experience at a visitor site to within- and between-group comparisons;

and, over time, from concerned residential community to national environmental protection legislation and international agreements, from on the ground manager need to methodological and measurement meaning, transparency and credibility. This very applied perspective has kept us focused both on the needs of management (and with respect to their formal mandate to protect, preserve and present) and on the motivations, needs and experiences of all visitor and user groups. The more psychological character of the approach also underscores the fact that what needs to be planned and managed is more often than not human behaviour and aspects of the psychosocial and built environment, not natural environments *per se*, and that effective management and monitoring of the people side of things is often the most effective and strategic way to manage and minimize adverse biophysical impacts, while maximizing positive psychosocial impacts and benefits.

A transactional framework

Given the inadequacy of existing frameworks in addressing the issues and challenges associated with our research, and in order to investigate adequately and consequently manage the complex issue of impacts of visitation and use in the WTWHA, it was necessary to develop a conceptual framework that would guide us in the exploration of, and help managers to understand, the critical components of the impact system. While any impact assessment is concerned with how changes introduced into the natural system affect ecosystem integrity and functioning, it is clear that visitation and use constitutes a multifaceted and dynamic set of transactions and impacts, the nature and magnitude of which are very consequential for both the visitor/user and the setting. This research therefore considered impacts of visitation and use in a more comprehensive and interactional way, underscoring the fact that environment includes the human environment, and impact includes impacts on visitor experience. It has been the case that impact assessments in protected area management contexts have had a somewhat loaded meaning for the tourism industry, as they can be seen to imply the adverse and unidirectional biophysical impacts of visitation and use. In our own research and in the context of a transactional approach, impacts are assumed to be reciprocal and potentially positive as well as adverse. Clearly the superlative World Heritage Areas of Australia and elsewhere have very powerful and lasting positive impacts on most of those individuals who visit them, and, of course, these areas confer multiple and very positive benefits to those who are fortunate enough to live in their catchments.

This transactional model of physical and psychosocial environmental impacts differentiates between four types of impact domains: those relating to conventional social impacts, psychological impacts, biophysical impacts and impacts on the human infrastructure or built environment (Figure 34.1). Of these four domains, the initial two relate to individual and community functioning and response, while the latter two relate to the physical environment, both natural and human-made. The model also conveys the reality of multiple feedback loops and interacting processes in environmental transactions, such that all impacts are potentially interactive and synergistic. Both physical and psychological impact domains mediate and moderate impacting processes.

The transactional approach that informs this framework has been articulated and used by many theorists in environmental psychology (e.g. Proshansky *et al.* 1970; Moos 1976; Stokols 1976; Stokols & Schumaker 1981; Wapner 1981; Altman & Rogoff 1987; Reser & Scherl 1988; Demick & Wapner 1990; Werner & Altman 2000; Werner *et al.* 2002), and other psychologists (e.g. Lewin 1951; Gibson 1979; Magnusson 1981; Pervin 1984) and design and system theorists working in the environment–behaviour domain (e.g. Moore & Marans 1997; Wapner *et al.* 2000). A core premise of the transactional framework is that dynamic and reciprocal impacts characterize all people – setting interactions, with the processes underlying and mediating these effects being typically both interdependent and interactive. The emphasis is on the ongoing flow and patterning of behaviour – environment interactions, with phenomena being viewed holistically and comprised of mutually defining aspects that change with experience (Werner & Altman 2000). A social science perspective acknowledges and underscores the existence and influence of the social as well as the biophysical environment, with each mediating transactions with the other. This social environment is also seen as separate and distinct from the internal, psychological environment of the individual, although this latter is powerfully and constantly influenced by the immediate biophysical and social environment. Environmental psychology in particular tends to examine the psychological processes (including perceptions, appraisals, experienced stress and adjustments) involved in environmental transactions and negotiations, along with the actual and perceived characteristics of the situation and the setting, with overt behaviour being an important adaptive, expressive and feedback-generating domain.

Just as visitors' presence and behaviour in protected environments impact on the existing physical, natural and social environment, these coalesced and interacting environments impact on individual visitor behaviour and experience. An important, obvious, but overlooked impact and sustainability consideration is that natural settings impact on people, often in very positive and beneficial ways. This is particularly true of dramatic, powerful, natural environments that have the additional conferred status and respect of a national park or a World Heritage site. Effective presentation and experience opportunities in these venues constitute powerful vehicles for behaviour and attitude change, as well as formative environmental encounters. What is often lost in current social impact analyses is that planned intervention changes the very nature and quality of ongoing people – setting transactions and impacts on people are typically mediated by impacts on settings (e.g. Evans & Lepore 1997).

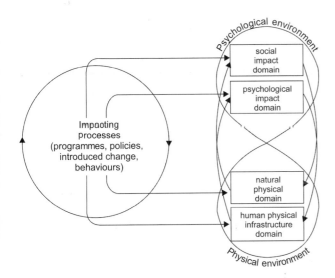

Figure 34.1 A transactional model of physical and psychosocial environmental impacts. *Source*: Reser and Bentrupperbäumer (2001).

This model differs from the widely used Organisation for Economic Co-operation and Development (OECD) reporting model (e.g. OECD 1991, 1994; Australian State of the Environment Committee 1996, 2001) in that it has been necessary to flag explicitly the condition of the biophysical environment as perceived and experienced by those visiting and using this environment, and the nature and quality of the experience itself that visitors have in the WTWHA. This is arguably critical to any assessment of the biophysical and psychosocial impacts of visitation and use and management. The mandate and management objective to present as well as protect the WTWHA makes this emphasis very sensible, quite apart from the integral links between impacts on people and impacts on setting. The OECD model does not treat individual and institutional responses to biophysical condition as impacts in their own right, nor is the matter of psychosocial indicators addressed. The model presented in this report is more sympathetic to the widely used Visitor Impact Management/Planning Process (VIM) model (Graefe *et al.* 1990), which incorporates field assessment of social impact indicators and strongly emphasizes the interrelationships of impacting processes, the experience of the visitor and the central role of human judgement in visitor impact assessment. Nevertheless, the VIM model, in our view, does not adequately address the actual nature and quality of individual experience, the nature of the transaction itself or the reciprocal and interdependent nature of biophysical and psychosocial impacts.

EIA, SIA, PSIA and transactional EIA

Clearly natural resource management has a strong mandate to ensure that adverse actual or potential human and natural impacts on the environment or ecosystem in question are routinely and strategically assessed, monitored, and managed. The impacts of visitation and use in protected areas and in the context of EIA/EIS are conventionally understood as the biophysical impacts of visitation, with these at times extending to wear and tear on human-built infrastructure and services (Hammitt & Cole 1998; Worboys *et al.* 2001). Social impact assessment (SIA) in the context of protected area management can encompass a number of considerations, ranging from the impacts of visitation

and use of a protected area on proximate residential communities, to the adverse impacts of crowding or pollution on visitor experience, to the role of a World Heritage Area in the life of a community (e.g. Commonwealth EPA 1994; Burdge & Vanclay 1995; Thomas & Elliott 2005). Social impact assessment underscores the fact that there is a social environment at issue (i.e. a psychological, social, community, cultural, socio-economic, socio-political environment), as well as a biophysical environment of concern, and that the impacts of visitation and use affect, interactively, both of these environments (Dale *et al.* 2001; Pol 2002; Berkes *et al.* 2003).

Psychosocial impact assessment (PSIA) is a more inclusive framing of how biophysical environments, human-designed environments and social environments impact on people and individual experience and behaviour. An important consideration is that, in a natural resource management or outdoor recreation context, many management considerations (such as visitor satisfaction, visitor site design, involvement, behaviour) relate to the impacts of park setting, venues, and environments on people. While this more psychosocial perspective and framing of impact assessment is unfamiliar to many, an integral consideration for an environmental psychological perspective is that environmental encounters are invariably interactive and dynamic, and need to be seen, understood and managed as a transactional, reciprocal, set of processes. Simply put, this means that the powerful experience that visitors have of a World Heritage status protected area is not only an important and typically positive impact consideration, it is also integral to the presentation and other management of impacts responsibilities. An important role of management agencies is to respond to and utilize more fully this presentation, interpretation and management of visitor behaviour opportunity.

More recent emphases on state of the environment reporting, management-community partnerships and meaningful measures (e.g. ANZECC 2000; ASEC 2001; CSEE 2003; Millennium Ecosystem Assessment 2003) have required frameworks and indicators that more fully embrace and reflect the psychosocial and the full spectrum of impacts on and of protected areas and adjacent catchments. To date, however, this has not eventuated, and sensitive and informative social and psychosocial measures are virtually absent in current suites of state of the environment reporting indicators, purportedly encompassing the social environment (e.g. ANZECC 2000).

How does taking a more transactional and environmental psychological approach change things?

Taking a more transactional and environmental psychological approach, in effect, changes most considerations relating to NRM in protected areas. Initially, and importantly, it changes how we think about the objectives and tasks of management, and the research and monitoring side of NRM. Presentation and legibility considerations, for example, become more sensitized to the prior expectations and motivational and sense-making needs and interests of visitors. Planning and design are articulated more closely with actual experiences and experience options, as well as actual behaviours. Impact considerations immediately include the impacts of both the immediate social and natural environment on visitors, and a necessary rethinking of just what is being managed with respect to impacts and environments. The perspectives of psychosocial and reciprocal impacts and transactional encounters also encompass perceived changes in the landscape and how local residents as well as visitors might view and understand particular management interventions that might impact on visual amenity or community concern levels. Each of these considerations has its own often surprising and often challenging management implications. A prior focus on the Recreation Opportunity Spectrum (ROS) (Clark & Stankey 1979) as a planning instrument and touchstone, for example, alters to encompass an Experience Opportunity Spectrum (EOS) as an equally important planning and site design consideration. From a managing the impacts perspective, multiple possibilities present themselves for managing impacts through presentation and interpretation as well as more conventional approaches, and not just adverse biophysical impacts, but typically very positive psychosocial impacts. Monitoring and managing World Heritage values encompasses how perceptions, values and concerns might be changing in the local, regional or more encompassing political catchment with respect to priority issues, ecosystem well-being or perceived management effectiveness.

Natural science and social science differences here also make this matter of monitoring and managing World Heritage values a particular challenge (Reser & Bentrupperbäumer 2005b; Bentrupperbäumer & Reser 2006a).

This transactional, psychosocial, impact assessment approach also substantially broadens our understanding of impacts to include the very positive impacts of these powerful World Heritage environments on people, as well as the adverse biophysical impacts of dramatically increasing levels of visitation and use, which impact on both the environment itself and the perceptions and experiences of visitors. This broadened notion of reciprocal impacts has multiple implications for visitor presentation, interpretation, and management, with impacting processes and management response strategies becoming a logical and integral part of interpretation and the management of visitor impacts. Further, the monitoring and reporting of the impacts of visitation and use in these protected area public spaces become more balanced, incorporating both positive and adverse consequences of visitation.

Summary

What has driven and informed this research and framework development was the need for a holistic framework and methodology that could encompass the scope of the research, which was grounded in relevant theory and research, and which integrated existing natural science-based, social science-based and impact assessment frameworks. This framework also needed to be one that made sense in the protected area, outdoor recreation, World Heritage context in which we were working, which informed and guided decisions with respect to the identification and selection of representative and sensitive variables and measures, which was practical and operationalizable in this research context, and which could be clearly communicated to partner organizations and end users, most particularly management agencies and adjacent communities. While the research squarely addressed the substantive impacts of visitation and use in the Wet Tropics World Heritage Area, and the measuring and monitoring undertaking this implied, the emergent and ultimately prerequisite research outcome was undoubtedly the development of a model and framework that made the larger undertaking possible.

Our principal research finding has been that this transactional, biophysical, and psychosocial impact assessment framework has worked very well. It is grounded in theory, best practice guidelines, the multidisciplinary context of environmental research, and the needs of management agencies and local communities. It is pragmatic, flexible and adaptive, and was readily operationalized in the field, in the face of multiple and differing needs and expectations. Importantly, it was also a framework that was transparent and made sense to very different stakeholder groups as well as other researchers. It was also a framework that educated all parties with respect to what could and should be done when attempting to research, monitor and manage the psychosocial as well as the biophysical impacts of visitation and use in a protected area, outdoor recreation venue, and larger backdrop and catchment area to a thriving and growing region.

The model we have developed and applied in our own work in protected area management in Australia confers the following advantages:

- it provides a more holistic and dynamic transactional model that both situates and locates differing levels of analysis and impacting processes, such that an integrated, system-based, multidisciplinarity is possible;
- it enables a more balanced and representative consideration of differing variables or indicator domains from which to select nominated indicators for monitoring both biophysical and psychosocial impacts over time and location;
- it allows for a more reasonable consideration of what is happening with respect to the individual, and with respect to individual experience, motivation, concerns and outcomes;
- it allows for more strategic and comparable monitoring approaches and a data collection focus on those points of transaction where problems are apparent;
- it provides for a better understanding of the reciprocal nature of biophysical and psychosocial impacts, of wear and tear on people and settings when there is poor fit between human need and setting provision;
- it provides an encompassing perspective that is equally meaningful and transparent to natural and social scientists, as well as to managers, stakeholder groups and the public;

- this framework fosters and promotes sustainability and sustainable practice by more clearly identifying and specifying what we need to consider, bring together, and systematically research and monitor to ensure the human transactions and encounters within a protected area are maximizing positive psychosocial impacts and themselves promoting ecologically responsible behaviour and minimization of adverse biophysical impacts.

The research undertaken within Rainforest CRC Project 4.1 (Strategies for Sustainable Access and Use) has been diverse and applied, as this research effort has been as much in response to expressed and/or prioritized management agency need as it has been a reflection of stated research objectives. At the same time, it has been both strategic and convergent with respect to its more general mission, which has been to present, implement, and develop a pragmatic, social science informed framework and set of procedures, methodologies and measures for researching the impacts of visitation and use in the WHA bioregion of the Wet Tropics.

Acknowledgement

The authors acknowledge the use of the transactional model figure, its description and the summary comment presented in this chapter as initially appearing in the authors' chapter in Filho (2005), courtesy of Peter Lang Publishers.

References

Altman, I. & Rogoff, B. (1987). World views in psychology: Trait, interactional, organismic and transactional perspectives. In Stokols, D. & Altman, I. (eds) *Handbook of Environmental Psychology, Volume I*. Wiley, New York, pp. 1–40.

Altman, I. & Wohlwill, J. (1983). *Human Behaviour and Environment: Volume 6, Behavior and the Natural Environment*. New York: Plenum Press.

Australia State of the Environment Committee (ASEC) (1996). *Australia State of the Environment 1996*. Independent report to the Commonwealth Minister for the Environment and Heritage. CSIRO Publishing on behalf of the Department of the Environment and Heritage, Canberra.

Australia State of the Environment Committee (ASEC) (2001). *Australia State of the Environment 2001*. Independent report to the Commonwealth Minister for the Environment and Heritage. CSIRO Publishing, Canberra.

Australian and New Zealand Environment and Conservation Council (ANZECC) (2000). *Core Environmental Indicators for Reporting on the State of the Environment*. Department of the Environment and Heritage, Canberra.

Bazerman, H., Messick, D. M., Tenbrunsel, A. E. & Wade-Benzoni, K. A. (eds) (1997). *Environment, Ethics and Behavior*. New Lexington Press, San Francisco.

Bechtel, R. B. & Churchman, A. (eds) (2002). *Handbook of Environmental Psychology*. John Wiley, New York.

Becker, E. & Jahn, T. (eds) (1999). *Sustainability and the Social Sciences: A Cross-disciplinary Approach to Integrating Environmental Considerations into Theoretical Reorientation*. Zed Books, London.

Bell, P. A., Greene, T. C., Fisher, J. D. & Baum, A. (2001). *Environmental Psychology*, 5th edn. Harcourt College, New York.

Bell, S. & Morse, S. (1999). *Sustainability Indicators: Measuring the Immeasurable*. Earthscan, London.

Bentrupperbäumer, J. M. & Reser, J. P. (2000). *Impacts of Visitation and Use: Psychosocial and Biophysical Windows on Visitation and Use in the Wet Tropics World Heritage Area*. James Cook University and Rainforest (CRC) Centre for Co-operative Research, Cairns.

Bentrupperbäumer, J. M. & Reser, J. P. (2002). *Measuring and Monitoring Impacts of Visitation and Use in the Wet Tropics World Heritage Area 2001/2002: A Site Based Bioregional Perspective*. James Cook University and Rainforest CRC, Cairns.

Bentrupperbäumer, J. M. & Reser, J. P. (2003). *The Role of the Wet Tropics in the Life of the Community: A Wet Tropics World Heritage Area Community Survey 2001/2002*. James Cook University and Rainforest CRC, Cairns.

Bentrupperbäumer, J. M. & Reser, J. P. (2006a). Uses, meanings and understandings of values in the environmental and protected area arena: a consideration of 'World Heritage' values. *Society and Natural Resources*, **19**: 723–41.

Bentrupperbäumer, J. M. & Reser, J. P. (2006b). *The Role of the Wet Tropics World Heritage Area in the Life of the Community*, rev edn. Rainforest CRC, James Cook University, Cairns.

Berkes, F., Colding, J. & Folke, C. (eds). (2003). *Navigating Social-ecological Systems: Building Resilience For Complexity and Change*. Cambridge University Press, New York.

Berkes, F., Folke, C. & Colding J. (2000). *Linking Social and Ecological Systems*. Cambridge University Press, New York.

Bonnes, M. & Secchiaroli, G. (1995). *Environmental Psychology: A Psycho-social Introduction*. Sage, London.

Burdge, R. J. & Vanclay, F. (1995) Social impact assessment. In *Environmental and Social Impact Assessment*, Vanclay, F. & Bronstein, D. A. (eds). John Wiley, New York, pp. 31–66.

Center for Science, Economics and the Environment (CSEE) (2003). *The State of the Nation's Ecosystems: Measuring the Lands, Waters, and Living Resources of the United States*. New York: Cambridge University Press.

Clark, R. & Stankey, G. (1979). *The Recreation Opportunity Spectrum: A Framework for Planning, Management and Research*. USDA Forest Service Research paper PNW-98.

Commonwealth Environment Protection Agency (1994). *National Pollutant inventory*. Public discussion paper. CEPA, Canberra.

Cordell, H. K. & Bergstrom, J. C. (eds) (1999.) *Integrating Social Sciences with Ecosystem Management*. Sagamore Publishing, Chapaign, IL.

Dale, A., Taylor, N. & Lane, M. (eds) (2001). *Integrating Social Assessment in Resource Management Institutions*. CSIRO Publications, Canberra.

Demick, J. & Wapner, S. (1990). Role of psychological science in promoting environmental quality. *American Psychologist* **45**(5): 631–2.

Dunlap, R. E. & Michelson, W. (eds) (2002). *Handbook of Environmental Sociology*. Greenwood Press, Westport, CT.

Evans, G. W. & Lepore, S. J. (1997). Moderating and mediating processes in environment-behavior research. In *Advances in Environment, Behaviour and Design, Volume 4*, Moore, G. T. & Marans, R. W. (eds). Plenum, New York, pp. 255–85.

Filho, W. L. (ed.) (2005). *Handbook of Sustainability Research* Peter Long, New York.

Gibson, J. J. (1979). *The Ecological Approach to Visual Perception*. Houghton-Mifflin, Boston.

Gifford, R. (2002). *Environmental Psychology: Principles and Practice*, 3rd edn. Optimal Books, Colville, WA.

Gobster, P. (1999). An ecological aesthetic for forest landscape management. *Landscape Journal* **18**: 54–63.

Gobster, P. (2000). Foreword. In *Forests and Landscapes: Linking Ecology, Sustainability and Aesthetics*, Sheppard, S. R. J. & Harshaw, H. W. (eds). CABI Publishing, New York, pp. xxi–xxviii.

Graefe, A. R., Kuss, F. R. & Vaske, J. J. (1990). *Visitor Impact Management: The Planning Framework*. National Parks and Conservation Association, Washington, DC.

Hammitt, W. E. & Cole, D. N. (1998). *Wildland Recreation: Ecology and Management*, 2nd edn. John Wiley, New York.

Ittelson, W. H., Proshansky, H. M., Rivlin, L. G. & Winkel, G. H. (1974). *An Introduction to Environmental Psychology*. Holt Rinehart and Winston, New York.

Kaplan, R. & Kaplan, S. (1989). *The Experience of Nature: A Psychological Perspective*. Cambridge University Press, New York.

Kaplan, R., Kaplan, S. & Ryan, R. J. (1998). *With People in Mind: Design and Management of Everyday Nature*. Island Press, Washington, DC.

Leal F. W. (ed.) (2005). *Handbook of Sustainability Research*. Peter Lang, Frankfurt am Main.

Lewin, K. (1951). *Field Theory in Social Science*. Harper, New York.

Little, P. E. (1999). Environments and environmentalisms in anthropological research: facing a new millennium. *Annual Review of Anthropology* **28**: 253–84.

Machlis, G. E. & Field, D. R. (eds) (1992). *On Interpretation: Sociology for Interpreters of Natural and Cultural History*, rev. edn. Oregon State University Press, Corvallis, OR.

Magnusson, D. (1981). *Toward a Psychology of Situations*. Erlbaum, Hillsdale, NJ.

Manfredo, M. J., Vaske, J. J. Bruyere, B. L., Field, D. R. & Brown, P. J. (eds) (2004). *Society and Natural Resources: A Summary of Knowledge*. Modern Litho, Jefferson, MS.

Manning, R. E. (1999). *Studies in Outdoor Recreation: Search and Research for Satisfaction*, 2nd edn. Oregon State University Press, Corvallis, OR.

Millennium Ecosystem Assessment (2003). *Ecosystems and Human Well-being: Millennium Ecosystem Asessment*. Island Press, Washington, DC.

Moore, G. T. & Marans, R. W. (eds) (1997). *Advances in Environment, Behavior and Design. Volume 4: Toward the Integration of Theory, Methods, Research, and Utilization*. Plenum Press, New York.

Moos, R. (1976). *The Human Context: Environmental Determinants of Behavior*. John Wiley, New York.

Organisation for Economic Co-operation and Development (OECD) (1991). *State of the Environment*. OECD, Paris.

Organisation for Economic Co-operation and Development (OECD) (1994). *Environmental Indicators – OECD Core Set*. OECD, Paris.

Pervin, L. A. (1984). *Current Controversies and Issues in Personality*, 2nd edn. John Wiley, New York.

Pol, E. (2002). Environmental management: a perspective from environmental psychology. *Handbook of Environmental Psychology*, In Bechtel, R. B. & Churchman, A. (eds). John Wiley, New York, pp. 55–84.

Proshansky, H. M., Ittelson, W. & Rivlin, L. G. (eds) (1970) *Environmental Psychology: Man and His Physical Setting*. Holt Rinehart & Winston, New York.

Reser, J. P. (2002). Environmental psychology and cultural and natural environmental heritage interpretation and management. *Journal of Environmental Psychology* **22**: 307–17.

Reser, J. P. (2003). Thinking through conservation psychology: prospects and challenges. *Human Ecology Review* **10**: 167–74.

Reser, J. P. & Bentrupperbäumer, J. M. (2001a). Reframing the nature and scope of social impact assessment: a modest proposal relating to psychological and social (psychosocial) impacts. *Integrating Social Assessment in Resource Management Institutions*, In Dale, A., Taylor, N. & Lane, M. (eds). CSIRO Publications, Canberra, pp. 106–22.

Reser, J. P. & Bentrupperbäumer, J. M. (2001b). Social science in the environmental studies and natural science arena: misconceptions, misrepresentations and missed opportunities. In *Social Sciences in Natural Resource Management: Theoretical Perspectives*, Higgins, V., Lawrence, G. & Lockie, S. (eds). Edward Elgar, London, pp. 35–49.

Reser, J. P. & Bentrupperbäumer, J. M. (2002). Capturing visitation and use: the relative merits of site-level sampling and monitoring in protected area management survey assessments. Paper presented to the Ecological Society of Australia and the New Zealand Ecological Society, Ecology 2002 Conference, Cairns, 2–6 December.

Reser, J. P. & Bentrupperbäumer, J. M. (2005a). The psychosocial impacts of visitation and use in world heritage areas: researching and monitoring sustainable environments and encounters. In *Handbook for Sustainability Research*, Filho, W. L. (ed.). Peter Lang Scientific Publishers, New York, pp. 235–63.

Reser, J. P. & Bentrupperbäumer, J. M. (2005b). What and where are environmental values? Assessing the impacts of current diversity of use of environmental and World Heritage values. *Journal of Environmental Psychology* **25**: 125–46.

Reser, J. P, Bentrupperbäumer, J. & Pannell, S. (2000). Roads, tracks and World Heritage: natural and cultural values in the Wet Tropics. *People and Physical Environment Research* **55–6**: 38–53.

Reser, J. P. & Scherl, L. M. (1988). Clear and unambiguous feedback: a transactional and motivational analysis of environmental challenge and self encounter. *Journal of Environmental Psychology* **8**(3): 269–86.

Schmuck, P. & Schultz, W. P. (eds) (2002). *Psychology of Sustainable Development*. Kluwer Academic Publishers, Boston.

Stokols, D. (1976) *Psychological Perspectives on Environment and Behavior: Conceptual and Empirical Trends*. Plenum Press, New York.

Stokols, D. & Altman, I. (eds) (1987). *Handbook of Environmental Psychology*. Wiley, New York.

Stokols, D. & Schumaker, S. A. (1981). People in places: a transactional view of setting. In *Cognition, Social Behavior, and the Environment*, Harvey, J. (ed.). Lawrence Erlbaum, Hillsdale, NJ, pp. 441–88.

Thomas, I. & Elliott, M. (2005). *Environmental Impact Assessment: Theory and Practice*. The Federation Press, Annandale, NSW.

Uzzell, D. & Ballantyne, R. (eds) (1998). *Environmental Psychology and Cultural and Natural Heritage Interpretation and Management: Contemporary Issues in Heritage and Environmental Interpretation*. The Stationary Office, London.

Wapner, S. (1981). Transactions of persons in environments: some critical transitions. *Journal of Environmental Psychology* **1**: 223–39.

Wapner, S., Demick, J., Yamamoto, T. & Minami, H. (eds) (2000). *Theoretical Perspectives in Environment-Behavior Research: Underlying Assumptions, Research Problems, and Methodologies*. Plenum, New York.

Werner, C. M. & Altman, I. (2000). Humans and nature: insights from a transactional view, In *Theoretical Perspectives in Environment-Behavior Research: Underlying Assumptions, Research Problems, and Methodologies*, Wapner, S., Demick, J., Yamamoto, T. & Minami, H. (eds). Plenum, New York, pp. 21–38.

Werner, C. M., Brown, B. B. & Altman, I. (2002). Transactionally oriented research: Examples and strategies. In *Handbook of Environmental Psychology*, Bechtel, R. B. & Churchman, A. (eds). John Wiley, New York, pp. 203–21.

Worboys, G., Lockwood, M. & De Lacy, T. (2001). *Protected Area Management: Principles and Practice*. Oxford University Press, South Melbourne.

35 LINKING CULTURAL AND NATURAL DIVERSITY OF GLOBAL SIGNIFICANCE TO VIBRANT ECONOMIES

*Rosemary Hill**

School of Earth and Environmental Sciences, James Cook University, Cairns, Queensland, Australia; and CSIRO Sustainable Ecosystems, Cairns, Queensland, Australia
*The author was participant of Cooperative Research Centre for Tropical Ecology and Management

Cultural and natural diversity

Cultural diversity influences both natural resource management and natural diversity throughout the world. A wide variety of cultural landscapes with high biodiversity values are the joint works of people and nature; examples include traditional coffee production forests in Mexico, highly diverse aynoka potato cultivation in the Andes and coltura mista farming in the Northern Appenines (Farina 1995; Sotomayor 1995; Moguel & Toledo 1999). Many of these systems are now undergoing replacement by industrial mono-cultures, or abandonment, resulting in rapid biodiversity decline (Michon & de Forestra 1995; Pandey 2003). Tropical rainforests, however, are places where the biophysical influence of Indigenous peoples is more difficult to discern, despite their millennia of occupation. Biodiversity in old fallows of Amazonian tropical forest agricultural systems approaches that of primary forests (Fujisaka *et al.* 1998). Although Kayapó resource strategies in the Amazon clearly concentrate non-domesticated resources near campsites, the distribution of Brazil nut trees, linked to Indigenous resource strategies, can also be explained by the scatter hoarding of the large rodent, the red-rumped agouti (*Dasyprocta agouti*) (Posey 1988, 1999; Peres & Baider 1997). Nevertheless, tropical rainforests and their plants and animals are integral to Indigenous rainforest peoples' cultures; for example, yams are of great religious significance to the Baka people of Cameroon (Joiris 1993), while in West Kalimantan, the Kantu people's augury is based on omens from several birds, notably the *Nenak* (white-rumped shama, *Copsychus malabaricus*) (Dove 1999). Alliances based on recognizing and respecting this deep significance of tropical rainforest to Indigenous peoples through territorial rights are also proving effective for protection of natural diversity (Schwartzman & Zimmerman 2005).

Rainforest Aboriginal peoples in the Australian Wet Tropics have deep cultural and spiritual links to the lands and waters of the region, and ongoing custodial rights and responsibility for management. The research presented in this chapter initially aimed to elucidate better these links between Indigenous cultural diversity and natural diversity, both biophysical and less tangible, and to develop policies, protocols and practical mechanisms for recognizing Aboriginal peoples' values, rights and responsibilities associated with rainforest environments, as a means of providing for culturally sustainable management

(Rainforest CRC 2000). More recent research has focused on the role of appropriate economies in enhancing the sustainability of natural and cultural diversity.

Elucidating the links between natural and cultural diversity in the Wet Tropics

In the Australian tropics, the biophysical influence of Indigenous societies on biodiversity is linked to fire management practices, and to domiculture, the concentration of non-domesticated resources, particularly fruit trees (Horsfall 1987; Chase 1989; Bowman 2000; Yibarbuk *et al.* 2001; Hill & Baird 2003). For the Indigenous peoples, the intangible aspects of such interactions with nature, imbued with spirituality, inherited rights, responsibilities, knowledge and the whole range of human emotions appear to have primacy (Bright 1995; Rose 1996; Bradley 2001; Kendall 2005). The Wet Tropics of Queensland World Heritage Area (WHA) is the home of 18 Rainforest Aboriginal tribal groups, who view natural and cultural values as inseparable, although the area is currently only recognized as a natural World Heritage site (Wet Tropics Regional Agreement 2005). Collaborative research projects with the Kuku-Yalanji, Ma:Mu, Girramay, Gulngay, Jirrbal and Djabugay groups revealed that a holistic appreciation of Indigenous cultures, including customary law, ongoing custodial responsibility, language, traditional knowledge and spiritual beliefs, is central to illuminating the links with natural diversity (Hill 1998; Smyth 2002; Worboys 2002; Talbot 2005).

Kuku-Yalanji fire management

The collaboration with Kuku-Yalanji people, traditional owners of the northern third of the WHA, focused on elucidating the biophysical imprint and cultural basis of their fire management practices. Kuku-Yalanji fire management is extraordinarily fine-grained, targeted towards particular cultural values and resources, both plant and animal, determined by seasonal indicators (Figure 35.1) (Hill *et al.* 2004). Fire knowledge is encapsulated in the landscape of stories signifying the ongoing power and meaning of the ancestral beings

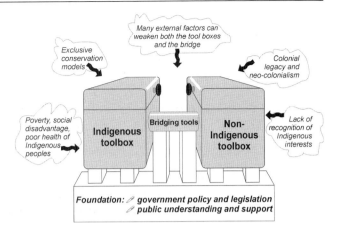

Figure 35.1 Kuku-Yalanji fire management, plant resources and seasons. Printed with permission of Eastern Kuku-Yalanji people, based on material included in Hill *et al.* (2004).

(Box 35.1). Fire management protects fire-sensitive rainforest, as well as promotes fire-prone open forest. Kuku-Yalanji people in their traditional subsistence economy are socially mediated fire managers who optimize resource availability. They suppress and control fire to protect edible fruits, yams and seeds, and the animals that depend on these resources; they also promote fire to sustain certain seeds, grasses and other plants, and to hunt game (Table 35.1).

Although culturally profound, this fine-grained Kuku-Yalanji fire management practice has a subtle effect on the biodiversity, producing a finer grain to the heterogeneity of vegetation communities, with little overall impact on the distribution of any one community at the regional scale (Hill 2003). In particular, small patches of fire-prone forest were maintained to enable access to critical resources, including the dietary staple carbohydrate *Cycas media*, on each traditional clan estate of some 2000–5000 ha (Hill & Baird 2003). Management of carbohydrate resources has emerged as a critical factor in hunter-gatherer rainforest occupation globally (Sponsel *et al.* 1996). The *Cycas media* stands were essential for the Kuku-Yalanji traditional subsistence strategy: the productivity was potentially 104 kJ m^{-2} yr^{-1}, somewhat less than from slash and burn agriculture of 200–500 kJ m^{-2} yr^{-1}, but nevertheless able to produce some 28% of the annual energy requirement for inhabitants of a clan estate (Hill & Baird 2003).

Box 35.1 The living cultural landscape of stories signifies the ongoing power of the ancestral beings

Alma Kerry describes how *Dabu*, the rainforest honey bee, and *Walarr*, the open forest bee, fought over fire.

In the beginning, at *Buru*, *Kija* the moon had started a fire. *Dabu* are those little bees that make wild honey in the rainforest. *Dabu* didn't want the fire to spread out. That fire was too hot, they were frightened *Kadar* the wallaby would burn his feet. *Dabu* cut some branches and leaves to put out the fire, by beating on it. *Dabu* was singing out 'don't make too much fire'. But the fire didn't stop. So he ran away from the fire, he flew away, and ended up near the mangroves down there. There's *yirmbal*, a spirit now, at the place where *Dabu* went. No-one can go near it or touch it. You're not to go near the mangroves, or eat anything from that area, shell, mussel, or walk around there. (13 November 1995, Wujal Wujal, translated from Yalanji by Adelaide Baird)

Permission of Eastern Kuku-Yalanji people was obtained to reprint this extract from Hill *et al.* (2004).

The rainforests are a rich source of plant foods for Kuku-Yalanji people, who use at least 60 different species of roots, seeds, fruits and greens. The seeds from rainforest tree nuts, which provided carbohydrate during the wet season in the traditional economy, included a much higher proportion of endemic Wet Tropics species than in the tree flora as a whole (Table 35.2). Many of the tree nuts, including *Cycas media*, require processing to remove toxic substances. Knowledge of processing techniques has been passed down to generations of Kuku-Yalanji women from the ancestral beings, the two sisters, who were active in the *Ngujakura*, the dreaming. The seasonally critical use of rainforest tree nuts appears to be a unique cultural adaptation that occurred in the Wet Topics bioregion, perhaps stimulated by the availability of the substantial number of large-seeded rainforest trees that are found in the wet tropics (Hill & Baird 2003). Although carbohydrate today is supplied primarily from wheat and rice, Kuku-Yalanji people place high value on maintaining the cultural knowledge and practices associated with these resources, and on the transmission of the knowledge to younger women (Hill *et al.* 2004).

The Kuku-Yalanji research collaboration revealed specific physical links between their culture and a fine-scale patterning on the heterogeneity of vegetation. However, this physical link is profoundly embedded in Kuku-Yalanji culture as a whole and forms an essentially inextricable strand within their Indigenous knowledge, practice and cultural system (ICSU 2002).

Table 35.1 Associations between Kuku-Yalanji fire management activities, habitat responses and staple carbohydrate resource utilisation in their traditional subsistence economy

Season (months indicate approx correspondence only as the timing varies each year)	Important fire management activities	Habitat and other responses to the fires	Staple carbohydrate resource utilization
Kambar: wet time, late December to March	No fires: too wet	Not applicable	Peak period for seeds from rainforest tree nuts; mangrove fruits
Kabakada: winter rain time, April to May	Small enclosed and other hunting fires	These fires ensure regions of low fuel as breaks for later fires	Beginning of season for yams; some seeds from rainforest tree nuts
Bulur: cold time, June to August/September	Fires along rainforest margins for hunting wallabies	These fires protect the rainforest and particularly the yams in the rainforest margins from later fires	Peak period for some yams; beginning of the period for cycad seeds
Wungariji: hot time, September/October to early November	Fires lit for ease of access to cycad stands	These fires enhance the productivity of cycads, and ensure maintenance of the open forest habitat	Peak period for cycad seeds
Jarramali: storm time, late November to mid-December	Fewer fires; fires are lit for hunting; green pick encouraged by storm rains attracts animals	The potential destructive effects of fires at this hot time is ameliorated by fire break effect of earlier fires	Some cycad seeds available; beginning of period for seeds from rainforest tree nuts

Source: Hill and Baird (2003). Reproduced with kind permission of Springer Science and Business Media.

Table 35.2 Rainforest seeds used as food by Kuku-Yalanji people

Species	Distribution and usage
Aleurites moluccana L. Willd.	Species also eaten in Malesia, Lockhart
Aleurites rockinghamensis (Baill.) P.I. Forst.	Species occurs only in the Wet Tropics in Australia and may also occur in New Guinea*
Beilschmiedia bancroftii (F.M. Bailey) C.T. White	Australian endemic species restricted to the wet tropics; genus occurs Malesia and Lockhart*
Bowenia spectabilis Hook. ex Hook. f.	Australian endemic species restricted to the wet tropics; genus Australian endemic[†]
Castanospermum australe A. Cunn. & Fraser ex Hook.	Species occurs in Cape York, Vanuatu and New Caledonia*, but is not eaten in the Lockhart region; genus does not occur in Malesia
Elaeocarpus bancroftii F. Muell. & F.M. Baill	Australian endemic species restricted to the wet tropics*; genus occurs but the seed is not used in Malesia or the Lockhart region
Elaeocarpus stellaris L.S. Sm.	Australian endemic species restricted to the wet tropics*
Entada phaseoloides (L.) Merr.	Species eaten in Malesia and Lockhart region
Lepidozamia hopei Regel	Australian endemic species restricted to the wet tropics region; Australian endemic genus, does not occur in Lockhart region[†]
Prunus turnerana (F.M. Bail.) Kalkman	Species occurs in wet tropics and New Guinea, genus in Malesia where it is eaten. Genus does not occur in Lockhart region*
Sterculia quadrifida R. Br.	Species occurs in New Guinea and northern Australia and is eaten at Lockhart; genus is eaten in Malesia*

Sources: Hill and Baird (2003); Data on usage at Lockhart from Harris 1975; data on Malesian usage from Burkill (1935);
*Hyland and Whiffin (1993); Clifford & Constantine (1980). Reproduced with kind permission of Springer Science and Business Media.

For Kuku-Yalanji people, the link between their culture and their country lies in their Yalanji language, the names of all the plants, animals, features, the locations of the human stories, the burial and birthing sites, the camping places, the sacred sites where the activities of the ancestral beings can still be seen in the landscape today. The link is also about Kuku-Yalanji customary law, the history of the lives of the Senior law men and women in the contact era, the stories women pass on of the *Ngujakura* as they gather bush tucker with their children, in talking to country, in the knowing that when a certain tree flowers, the river mullet will be fat (Hill *et al.* 2004). Maintaining the link between Kuku-Yalanji culture and the tropical rainforest diversity is then about maintaining Kuku-Yalanji culture in a holistic way, including Kuku-Yalanji law, language, connection to country and knowledge system.

Holistic Indigenous cultures

This holistic view of the links between Indigenous culture and the rainforest is reinforced by the results

of other collaborative research in the Wet Tropics. Ma:Mu people, traditional owners of the coastal region in the central Wet Tropics, identify themselves as true rainforest people (Worboys 2002). Ma:Mu people's world view encapsulates the rainforest as the medium through which all values are developed. The signification of rainforest within cultural identity has been highlighted by the convergence of the importance of rainforest and rainforest plants in traditional cultural practices, in ongoing subsistence and the reduced extent of rainforest. As rainforest people, the Ma:Mu value highly both ongoing access to rainforest resources and cultural practices associated with these resources, in some cases incorporating invasive exotic species now dominating the wild resources, in order to maintain these cultural practices (Worboys 2002). Smyth's (2002) research with Girramay, Gulngay and Jirrbal people in Jumbun Community identified a holistic set of indicators as vital for considering Indigenous cultural links with the Wet Tropics environment: rights to country, looking after country, cultural knowledge (including species names, breeding patterns, life-cycles, food requirements, medicinal qualities, seasonal patterns of growth, languages) and an understanding of history. The concept of a holistic Indigenous world view emerges as critical to understanding Djabugay people's connection with and management of their country (Talbot 2005).

Linking management of globally significant natural and cultural heritage: a paradigm shift under way?

A living cultural landscape of world heritage significance

These holistic links between Indigenous peoples, their cultures and natural diversity in the Wet Tropics are clearly important and deeply significant to the Indigenous people concerned, but are also of significance to the global human community. The Wet Tropics is recognized as a living cultural landscape with the potential for inclusion on the World Heritage list for its cultural values. Both Rainforest Aboriginal peoples and the relevant governments have made a commitment to work towards future listing (Wet Tropics Regional Agreement 2005). This Rainforest Aboriginal cultural landscape has outstanding scientific, social, aesthetic and historic significance (Titchen 1995). The unique cultural adaptation of utilizing fire and processing restricted endemic toxic seeds, to manage the carbohydrate limitations of tropical rainforest, is of great scientific interest. Archaeological sites within the region have the capacity to contribute to further understanding of the occupation of rainforest during the Holocene, and perhaps earlier (Horsfall & Hall 1990; Cosgrove 1996). The high aesthetic and social significance derives in part from the cultural sites, including the story places, tracks and dreaming sites, signifying the ongoing power and meaning of the ancestral beings, the occupation sites, stone quarries, carved and scarred trees, art sites, burial places, ceremonial grounds, walking tracks and waterways (Wet Tropics Aboriginal Plan Project Team 2005). Other aspects of social significance include the ongoing use of the area for collection of bush foods and medicines, for ceremonies, for language use (naming of unique places, animals, plants, stories) and as an ongoing source of material for expression through dance and music. Historical significance arises in sites of accommodation, where Aboriginal people developed economic relationships with miners and pastoralists that enabled survival during the era of frontier violence, and sites of resistance and conflict between the cultures. Together, these and other Aboriginal cultural values in the Wet Tropics make essential contributions to the global presentation of several cultural heritage sub-themes already identified as of outstanding universal value; for example. complex persistence of a hunting-and-gathering society on a single continent and dreaming sites (Titchen 1995; Hill 1998). Preliminary research indicates that there is evidence to support the renomination of the WHA as a series of Aboriginal cultural landscapes that complements its current listing as a natural World Heritage site (Horsfall 2003).

Cooperative management

Recognition globally of such connections between cultural and natural diversity, encapsulated in the term biocultural diversity, is leading to a paradigm shift in natural resource management globally towards a more inclusive approach (Brechin *et al.* 2002; Schwartzman & Zimmerman 2005). In protected area management, the shift is characterized as a shift from protected areas as islands, managed reactively with a narrow focus on conservation and control of the activities of local people, to protected areas as elements of a network, managed adaptively with a broader focus on socio-economic and cultural objectives, and established and run together with local peoples (Borrini-Feyerabend *et al.* 2004a).

Within the WHA, government policies generally support cooperation with Indigenous peoples in management. The *Wet Tropics World Heritage Protection and Management Act 1993 (Queensland)* requires the government to have regard to Aboriginal peoples' tradition and involve them in management of the WHA. The Queensland government now recognizes that Aboriginal people may hold native title rights over many national parks, and that management will be enhanced by respecting and supporting the laws, customs, knowledge, responsibilities and interests of Indigenous people (EPA 2001). The *Native Title Act Reprint 1 1993 (Commonwealth)* provides for

Indigenous Land Use Agreements (ILUAs) as a means of reaching negotiated agreement about accommodation of native title rights; several agreements are in progress in the Wet Tropics region. Development of the Wet Tropics Aboriginal Cultural and Natural Resource Management Plan (see Pannell, Chapter 33, this volume) is a real acknowledgement of the positive contribution that Traditional Owners make to natural resource management (WTAPPT 2005). The Wet Tropics World Heritage Area Regional Agreement, signed in April 2005 between Rainforest Aboriginal peoples and the Queensland and Australian governments, develops a cooperative framework, underpinned by the principles of recognition, involvement, ensuring the right people for the right country, engagement in policy development, strategic and park planning, permitting and the protection of intellectual and cultural property. These principles will be achieved through commitments to the recognition of cultural values, participation in decision-making, engagement protocols and the establishment of an Aboriginal Rainforest Council (Wet Tropics Regional Agreement 2005). Traditional Owners regard the Regional Agreement as an important step forward, although falling short of their joint management aspirations (WTAPPT 2005).

Establishing a cooperative approach to management across cultures is a complex undertaking. Analysis in the Kuku-Yalanji case study highlighted the cross-cultural aspects of six dimensions of ecosystem management in the WHA, and identified areas of commonality and conflict (Hill *et al.* 1999) (Table 35.3). From this initial analysis, a conceptual framework was developed of an Indigenous and non-Indigenous management toolbox with a number of bridging tools linking the two approaches, thereby providing a handle by which others could pick up and use the toolboxes (Figure 35.2) (Hill 2003, 2006; see also Kendall 2005). These bridging tools are in essence new policy instruments that are aiming to transform public policy direction. Analysis of two such tools developed with Kuku-Yalanji people, an ILUA, and a Fire Protocol, considered Dovers's (2003) core principles for successful public policy innovation within a framework that recognizes that change occurs at three levels:

- first order change in policy instruments;
- second order change in both instruments and policy goals;
- third order change in policy paradigm or the hierarchy of goals behind the policy setting (Eckersley 2003).

A central feature of the ILUA, proposed by Kuku-Yalanji people to the Queensland government, is the division of the area under consideration into three zones:

- areas of state-owned land as national park;
- areas of inalienable Aboriginal-owned land managed for conservation under binding agreements;

Table 35.3 Comparative ecosystem management framework

Ecosystem management dimensions	Kuku-Yalanji ecosystem management (examples)	WTWHA ecosystem management (examples)
Desired outcome	Integrity of the cultural landscape	Integrity of the natural landscape
Political and legal control	Elders and councils, customary law	WTMA, Queensland legislation
Economic base	Subsistence, cultural tourism, government welfare and other payments	Protection of biodiversity in the context of the national economy
Technology of management	Story places, taboo sites and totems, clan ownership and management rights	Management plans, weed and pest control
Social aspects	Education in the field	Institutionally based education
Ethical and spiritual basis	*Ngujakura*, the dreaming	Land and conservation ethic
Knowledge base	Traditional knowledge systems	Biology, ecology

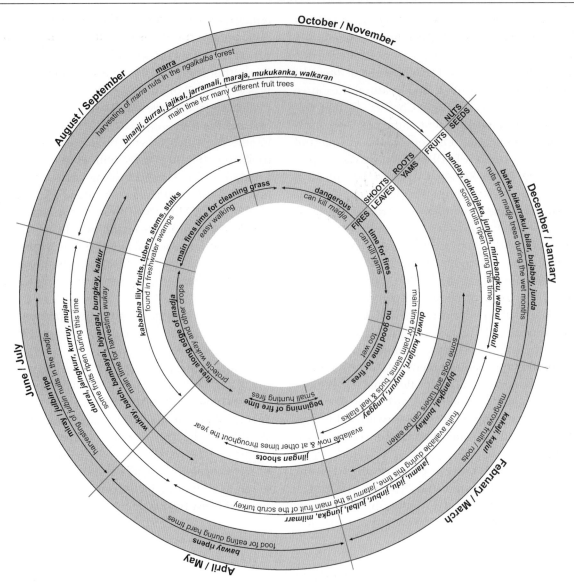

Figure 35.2 Conceptual framework for bridging between two different management frameworks. © 2006 from *Society and Natural Resources* by Hill. Reproduced by permission of Taylor & Francis Group, LLC (http://www.taylorandfrancis.com).

• much smaller areas of inalienable Aboriginal-owned land for community development.

Kuku-Yalanji people saw the national parks as providing primarily for WHA managers' approaches (but sought the future development of co-management arrangements), and the provision of conservation agreement zones providing primarily for Kuku-Yalanji management approaches, while the recognition of Aboriginal ownership compensated partially for the

loss of territoriality within the parks (Table 35.2). The Kuku-Yalanji Fire Protocol was promoted by Kuku-Yalanji people as potentially an effective tool to bridge the mutual rights and responsibilities of both parties in relation to fire management; government managers agreed to apply the Protocol to vegetation burns conducted in the upper Daintree National Park during 2001 (Hill 2006). The Protocol begins with identification of the appropriate people with custodial responsibilities, and includes steps of establishing a collaborative team between Traditional Owners and protected area managers, identifying natural and cultural values, agreeing on and subsequently implementing and evaluating a Fire Plan (Hill *et al.* 1999).

The analysis of the effectiveness of the ILUA and the Protocol as bridging tools demonstrated that both instruments adhere to many of Dovers's (2003) core principles for successful policy change (Hill 2006). However, consideration of Eckersley's (2003) three levels shows that change occurred most effectively at the second order level of change in policy goals. Both Kuku-Yalanji people and the WHA managers share goals associated with achieving lasting policy change to accommodate better a bicultural approach. At the first order level of change through application of new instruments, both the ILUA and the Protocol were hampered by inadequate resources. At the third order level of change in policy paradigm, both instruments were limited by a failure of government to support policy commitments by appropriate legal mechanisms. Traditional Owners believe that their rights and management approaches require them to be in a decision-making role, and able to exercise the broader dimensions of Indigenous management, founded on customary law and Indigenous knowledge, and encapsulated by Kuku-Yalanji in the concept of Bama National Parks. Although the *Native Title Act Reprint 1 1993 (Commonwealth)* provides a framework for negotiating agreements, the Queensland land and protected area management laws provide primarily for consultation with the Traditional Owners, without effective means of supporting traditional law, despite the commitment to do so noted above.

Analysis of a Heads of Agreement (HoA) signed between the Queensland government and Ma:Mu people concerning the development of a canopy walkway in the Wooroonooran National Park also showed important differences in the perspectives of the non-Indigenous and Indigenous parties (van der Zijden 2002). While the process of developing the HoA created a much better working relationship between the parties, for the Ma:Mu people, protection and maintenance of culture emerged as foremost, whereas the non-Indigenous parties emphasized the aspects of management, tourism and cultural interpretation. Van der Zijden (2002) also noted that Ma:Mu people explic-

itly promoted the concept of joint management through the HoA, a term not used by the non-Indigenous parties, again reflecting the lack of enabling and binding legal mechanisms. The recently signed Wet Tropic World Heritage Area Regional Agreement, although an important step forward in recognition and changed policy goals, is similarly limited by clauses that commit the parties to abide by its spirit and intent, but deny that the agreement creates any legal relationship between them. The Queensland government policy on protected area partnership with Indigenous peoples appears to be an example of the common approach of acknowledging a paradigm shift through rhetorical means, but not backing it up with substantive binding mechanisms (Hill 2006). The conceptual framework of bridging between these two toolboxes for management has therefore been revised to take account of the impact of this political and social context (Figure 35.3).

Joint management and community conserved areas

The government managers of the WHA have recently strengthened mechanisms for Indigenous access, benefit-sharing and advice through the Regional Agreement, HoA and ILUAs. However, these arrangements fall short of the joint management sought by

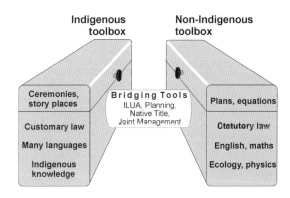

Figure 35.3 Revised conceptual framework to recognize the effect of the policy environment on the Indigenous toolbox, attempts to create a bridge and the necessity for a strong foundation in government policy and public support. © 2006 from *Society and Natural Resources* by Hill. Reproduced by permission of Taylor & Francis Group, LLC (http://www.taylorandfrancis.com).

Rainforest Aboriginal people. In Australia, the term joint management is generally used to refer to situations where authority is shared between governments and Aboriginal people through the establishment of legal partnership and management structures that reflect the rights and responsibilities of both parties (Smyth 2001). Joint management arrangements for national parks are in place in the Northern Territory (NT), South Australia and New South Wales (NSW). The features of joint management that provide for shared authority often include: Aboriginal ownership through inalienable freehold; a lease to the government for the national park; a board of management with Aboriginal majority as the primary means of shared decision-making; a shared process for preparation of the management plan; a management agreement addressing issues including recognition of traditional knowledge and customs; arrangements for employment and training; reservations of rights of use and access; and payment of rent and a share in revenue from the park. The NT recently enacted a major reform to provide for joint management over 27 national parks that eliminates the requirement for a board of management, reflecting a view that such arrangements were too inflexible and expensive, and did not provide a good avenue for cross-cultural decision-making. A key driver for this reform was a recognized need for Indigenous socio-economic development through engagement in the parks-related tourism economy,

the biggest employer in the NT (D. Ritchie, Director NT Parks Service, pers. comm. 1 June 2005). Legislative options are available in Queensland for joint management through Aboriginal ownership of some existing parks, but are accompanied by requirements for perpetual lease-back and government management, reflecting their enactment prior to the recognition by Australia's High Court of native title in 1992. Governments and Traditional Owners have not been able to reach satisfactory arrangements for sharing authority through these mechanisms and no jointly managed national parks in Queensland have been finalized.

Numerous joint management regimes globally have proven an effective means of strengthening management, protection, connectivity at the landscape and seascape level and public support for conservation, while increasing the flexibility and responsiveness of the protected areas system, improving sustainability and strengthening the relationship between people and nature (Borrini-Feyerabend *et al.* 2004b). As a result of these positive outcomes, IUCN now recognizes a diversity of government types for protected areas, including both co-managed and community conserved areas (CCAs) (Table 35.4). The option for community governance is proving an important tool for conservation, as co-management regimes are not effective when offered as a token resolution of land grievances (Coombes & Hill 2005). For example, the Colombian

Table 35.4 IUCN governance types and categories for protected areas

	Governance types			
Protected area categories	A. Government Managed Protected Areas	B. Co-managed Protected Areas	C. Private Protected Areas	D. Community Conserved Areas
za - Strict Nature Reserve Ib Wilderness Area				
II National Park				
III National Monument				
IV Habitat/Species Management				
V Protected Landscape/Seascape				
VI Managed Resource Protected Area				

government in 2003 approved the creation of the Matavén Forest Indigenous Territory, bringing together 1.8 million hectares of land into management as a strict conservation area by the 16 Indigenous peoples, based on authority powers granted to them by the Colombian constitution. The option for a CCA-delivered conservation outcome with co-management through a national park was viewed as imposing unacceptable limitations on territorial rights and management authority (Luque 2003).

In Australia, recognition and support for Indigenous community governance has primarily emerged through the Australian government's Indigenous Protected Areas (IPA) programme, which formally recognizes and financially supports protected areas declared and managed by Indigenous landowners (Smyth 2001; Szabo & Smyth 2003). However, no Australian state or territory government has yet established legal mechanisms to recognize IPAs formally as part of their respective protected area estates, though some state/territory conservation agencies provide operational and technical support to Indigenous managers of IPAs. The model advanced in the Queensland Wet Tropics through the Kuku-Yalanji ILUA attempts to overcome the limitations on territorial rights imposed by government-managed national parks through ensuring that substantial areas of land with high conservation values are recognized as inalienable Aboriginal tenure, unencumbered by national park status, but with protection of natural and cultural values assured through conservation agreements under the *Nature Conservation Act 1992*. However, such agreements are more in keeping with a private governance category for protected area management (Table 35.4). Legislative reform to provide for all four of the governance types recognized by IUCN within the WHA, including joint management and CCAs, would bring the management more closely into alignment with international best practice, as well as provide a better means of linking the outstanding natural and cultural diversity.

Frameworks to support natural and cultural heritage as an economic driver

A key strategic direction identified for sustainable economic development in the Wet Tropics is the positioning of the region as a world leader in tropical reef and rainforest preservation and management. Our key competitive advantage lies in the outstanding heritage values, both natural and cultural, and management expertise (Blackwood 2003). However, while the reef and rainforest currently generate enormous wealth for non-Aboriginal people in Far North Queensland, many Traditional Owners live in poverty (Foley 2005). The socio-economic disadvantage of Australian Indigenous people is pervasive: the life expectancy of Indigenous people is around 17 years lower than the Australian population; the age standardized unemployment rate in 2002 was 3.2 times higher for Indigenous than non-Indigenous people; and Indigenous household and individual incomes are on average lower (Steering Committee for the Review of Government Service Provision 2005). Nevertheless, the data show that participation in the customary and cultural activities are very significant components of the economy for Indigenous people in northern Australia, within a hybrid economy, where state, market and customary activities are all important (Altman 2004). Traditional Owners in northern Australia have developed some innovative businesses based on harvesting native cycads and bush honey that are compatible with maintaining the customary economy without altering the landscape thorough forestry, agriculture and grazing (Cooke 2005; Maningrida Wildlife Centre 2005). Altman (2004) argues that there are significant national benefits generated by these hybrid economies, including biodiversity conservation, and they deserve more active support, not in the form of the welfare state, but with the state as the underwriter and guiding hand of development.

In the Wet Tropics, Rainforest Aboriginal people have articulated a strong desire to continue their customary economic activities through hunting and gathering, while developing market opportunities in forestry, fishing, aquaculture and tourism (WTAPPT 2005). A number of Aboriginal tourism businesses that directly link cultural and natural heritage are now operating in the Wet Tropics, including Kuku-Yalanji Dreamtime Tours, the Walker Family Tours, Native Guide Safari Tours, Djabugay Country Tours and Yindili tours (Zeppell 2002; WTAPPT 2005). At a workshop in 2003, Aboriginal people from the Wet Tropics and Cape York Peninsula, together with scientists, economists, small business operators and environment groups, identified a number of culturally and environmentally appropriate business opportunities. Such businesses link the protection of cultural and natural heritage with industries like land and sea management, art, traditional medicines, nature-based and cultural tourism, protected areas, education and training, communications and information technology, eco-commodities including carbon credits, nondestructive research, feral animal and weed management, language renewal, market gardens, seed collection, bush foods, small-scale novel crops and nurseries (Hill & Turton 2004). However, there are currently many barriers to realizing these opportunities fully, including poor access to capital and expertise, lack of respect for Indigenous rights and interests, unresolved land use issues, marginalization of Indigenous people in decision-making and poor understanding between Indigenous and environment interests. The primary measures identified at the workshop to overcome these barriers were: reform to land tenure and conservation management arrangements through acquisition of lands of high natural and cultural value and return to Indigenous ownership; reform of the conservation, protected area, land tenure and management arrangements to take better account of Traditional Owner rights and interests; and funding for conservation and land management undertaken by Traditional Owners (Hill & Turton 2004). The Queensland government has since acted on these recommendations, although only within the Cape York Peninsula region, by funding an ongoing programme of acquisitions of lands of high natural and cultural conservation value, and return of ownership of parts to Traditional Owners, with the remainder gazetted as national parks. Negotiations are ongoing about arrangements for joint management of these parks and for the recognition of community conservation approaches. Other recommendations from the meeting included pilot projects, exchange programmes and the development of a private capital stream directed towards environmental and social ends, based on the Ecotrust model from Canada. WTAPPT (2005) similarly identified that the structures to support both the customary and nascent market activities need improvement, and that current frameworks can block projects based on native biota; for example, the Ma:Mu bush tucker industry initiative has been restricted by their inability to obtain a permit under the *Nature Conservation Act 1992 (Queensland)* to collect seed from Ma:Mu country for propagation and development into cultivars.

Summary

The links between Indigenous peoples and natural diversity in the Wet Tropics are embedded holistically within the Indigenous cultures, and are reflected in their influence on the patterns of biodiversity, on the cultural landscapes imbued with deeply significant spiritual meaning, useful plant and animal resources, languages, traditional ecological knowledge, cultural sites and human history. The Wet Tropics is recognized as a living cultural landscape with the potential for inclusion on the World Heritage list for its cultural values, and both Rainforest Aboriginal peoples and the relevant governments have made a commitment to work towards future listing (Wet Tropics Regional Agreement 2005). Recognition globally of such connections between cultural and natural diversity is leading to a paradigm shift in natural resource management towards a more inclusive approach (Brechin *et al.* 2002; Schwartzman & Zimmerman 2005). Although the management approaches of Indigenous and non-Indigenous peoples to the Wet Tropics can be characterized as two toolboxes, potentially able to be bridged through policy instruments including ILUAs, protocols and agreements, the current political and social context makes such bridging difficult. Available government mechanisms to support Indigenous partnerships within the WHA are expressed in aspirational terms and provide for cooperation, rather than the substantive and binding legal instruments required for lasting paradigmatic shift and real progress in collaboration and shared decision-making between Indigenous peoples and governments. Legislative reform to provide for all four of the protected area governance types recognized by IUCN within the WHA, including joint management and community conserved areas, would bring the management more closely into alignment with international best practice, as well as provide a better means of linking the outstanding natural and cultural diversity, and bridging the two management toolboxes (Hill 2006).

Involvement by Rainforest Aboriginal people in customary and market economies that link the outstanding natural and cultural heritage clearly has the potential to build on the unique comparative advantage in the Wet Tropics, and drive a vibrant economy for all Australians. However, a more supportive policy framework would assist in the realization of this potential. I have argued previously for broad frameworks in the national interest to support Indigenous culture and land management, encompassing legislation, regulations, functional governance institutions, together with financial resourcing and career structures; in short a package of measures that enable Indigenous people to manage the land they currently own, and to participate in management of land where their rights coexist with others (Hill 2003). Subsequent research has identified the potential for corporate and non-government collaborations to play a central role in providing such support and in linking nature conservation to Indigenous peoples' role on country as the means of sustainable economic development (Hill & Turton 2004; McCaul 2005).

Australia is the developed nation with the most substantial area of tropical rainforest. Norman Myers, a world leader in rainforest conservation, in his address to the World Heritage Tropical Forests Conference in Cairns in 1996, highlighted the unprecedented opportunity for global leadership in rainforest protection and management. However, the claim to global leadership cannot be maintained without greater attention to the fundamental issues of Indigenous peoples in the rainforest. Throughout the world, rainforests are the homelands of Indigenous societies. For tropical forests globally, the critical decisions about conservation are also about the future of the Indigenous inhabitants (Colchester 2000; Romero & Andrade 2004). Research has demonstrated the extensive links between the natural and cultural diversity of global significance in the Wet Tropics. More supportive frameworks that provide for Aboriginal tenure, joint management and CCAs, funding and support for management and business development, and that that recognize and respect Indigenous leadership and community as central, will allow Rainforest Aboriginal peoples' expertise in cultural and natural heritage management to drive a vibrant future.

References

Altman, J. C. (2004). Economic development and Indigenous Australia: contestations over property, institutions and ideology. *Australian Journal of Agricultural and Resource Economics* **48**(3): 513-34.

Blackwood, B. (2003). *Far North Queensland Sustainable Industries: A Study*. Institute for Sustainable Regional Development, Rockhampton.

Borrini-Feyerabend, G., Kothari, A. & Oviedo, G. (2004a). *Indigenous and Local Communities and Protected Areas: Towards Equity and Enhanced Conservation*. IUCN, Gland, Switzerland.

Borrini-Feyerabend, G., Pimbert, M., Farvar M. T., Kothari A. & Renard, Y. (2004b). *Sharing Power. Learning by Doing in Co-management of Natural Resources throughout the World*. IIED and IUCN/CEESP/CMWG, Cenesta, Tehran.

Bowman, D. M. J. S. (2000). *Australian Rainforests: Islands of Green in a Land of Fire*. Cambridge University Press, Cambridge.

Bradley, J. J. (2001). Landscapes of the mind, landscapes of the spirit: negotiating a sentient landscape. In *Working on Country: Contemporary Indigenous Management of Australia's Land and Coastal Regions*, Baker, R., Davies, J. & Young, E. (eds). Oxford University Press, Oxford, pp. 295–307.

Brechin, S. R., Wilshusen, P. R., Fortwangler C. L. & West P. C. (2002). Beyond the square wheel: toward a more comprehensive understanding of biodiversity conservation as a social and political process. *Society and Natural Resources* **15**: 41–64.

Bright, A. (1995). Burn grass. In *Country in Flames. Proceedings of the 1994 Symposium on Biodiversity and Fire in North Australia*, Rose, D. B. (ed.). Biodiversity Unit, Department of the Environment, Sport and Territories and the North Australia Research Unit, Australian National University, Canberra and Darwin, pp. 59–62.

Burkill, I. H. (1935). *A Dictionary of the Economic Products of the Malay Peninsula in Two Volumes*. Crown Agents for the Colonies on Behalf of the Governments of the Straits Settlements and Federated Malay States, London.

Chase, A. K. (1989). Domestication and domiculture in northern Australia: a social perspective. In *Foraging and Farming. The Evolution of Plant Exploitation*, Harris R. & Hillman, G. C. (eds). Hyman and Unwin, London.

Clifford, H. T. &. Constantine J. (1980). *Ferns, Fern Allies and Conifers of Australia*. University of Queensland Press, Brisbane.

Colchester, M. (2000). Self-determination or environmental determinism for Indigenous peoples in tropical forest conservation. *Conservation Biology* **14**(5): 1365–7.

Cooke, P. (2005). Sugarbag dreaming: natural heritage. *Journal of the Natural Heritage Trust* **24**(Winter): 11.

Coombes, B. L. &. Hill, S. (2005). Na whenua, na Tuhoe. Ko D.o.C. te partner. Prospects for comanagement of the Te Urewera National Park. *Society and Natural Resources* **18**: 135–52.

Cosgrove, R. (1996). Origin and development of Australian Aboriginal tropical rainforest culture: a reconsideration. *Antiquity* **70**(270): 900–13.

Dove, M. R. (1999). Forest augury in Borneo: indigenous environmental knowledge about the limits to knowledge of the environment. In *Cultural and Spiritual Values of Biodiversity*, Posey, D. & Dutfield, G. (eds). UNEP, London, pp. 376–80.

Dovers, S. (2003). Processes and institutions for resource and environmental management: why and how to analyse. In *Managing Australia's Environment*, Dovers, S. & Wild-River, S. (eds). The Federation Press, Sydney, pp. 3–14.

Eckersley, R. (2003). Politics and policy. In *Managing Australia's Environment*, Dovers, S. & Wild-River, S. (eds). The Federation Press, Sydney, pp. 485–500.

Environmental Protection Agency (EPA) (2001) *Master Plan for Queensland's Parks System*. Queensland Government, Environment Protection Agency, Brisbane.

Farina, A. (1995). Upland farming systems of the Northern Appenines. The conservation of biological diversity. In *Conserving Biodiversity Outside Protected Areas: The Role of Traditional Agro-ecosystems*, Halladay, P. & Gilmour, D. A. (eds). IUCN, Gland, Switzerland, pp. 123–35.

Foley, R. (2005). Statement from the Indigenous Technical Support Group. In *Caring for Country and Culture: The Wet Tropics Aboriginal Cultural and Natural Resource Management Plan*. Rainforest CRC and FNQ NRM Ltd, Cairns.

Fujisaka, S., Escobar, G. & Veneklaas E. (1998). Plant community diversity relative to human land uses in an Amazon forest colony. *Biodiversity and Conservation* **7**(1): 41–57.

Harris, D. R. (1975). Traditional patterns of plant-food procurement in the Cape York Peninsula and Torres Strait Islands. Report on field work carried out August–November 1974. Unpublished manuscript, Department of Geography, University College London.

Hill, R. (1998). Vegetation change and fire in Kuku-Yalanji country: implications for management of the Wet Tropics of Queensland World Heritage area. Doctoral thesis, School of Tropical Environment Studies and Geography, James Cook University, Townsville.

Hill, R. (2003). Frameworks to support Indigenous managers: the key to fire futures. In *Australia Burning: Fire Ecology,*

Policy and Management Issues, Cary, G., Lindenmayer, D. & Dovers, S. (eds). CSIRO Publishing: Canberra, Australia, pp. 175–86.

Hill, R. (2006). The effectiveness of agreements and protocols to bridge between Indigenous and non-Indigenous tool-boxes for protected area management:. a case study from the Wet Tropics of Queensland. *Society and Natural Resources* **19**(7): 577–90.

Hill, R. &. Baird, A. (2003). Kuku-Yalanji Rainforest Aboriginal people and carbohydrate resource management in the Wet Tropics of Queensland, Australia. *Human Ecology* **30**: 27–52.

Hill, R. &. Turton. S. M. (eds.) (2004). *Culturally and Environmentally Appropriate Economies for Cape York Peninsula*. Rainforest CRC, Cairns.

Hill, R., &. Baird, A. & Buchanan, D. (1999). Aborigines and fire in the Wet Tropics of Queensland, Australia: ecosystem management across cultures. *Society and Natural Resources* **12**: 205–23.

Hill, R., Baird, A., Buchanan, D. C. *et al.* (2004). *Yalanji-Warranga Kaban. Yalanji: People of the Rainforest Fire Management Book*. Little Ramsay Press, Cairns.

Horsfall, N. (1987). Living in rainforest: the prehistoric occupation of North Queensland's humid tropics. Doctoral thesis, James Cook University of North Queensland, Townsville.

Horsfall, N. (2003). *Aboriginal Cultural Values of the Wet Tropics Bio-Region. Collaborative Cultural Heritage Research Report*. Rainforest CRC, Cairns, unpublished

Horsfall, N. &. Hall, J. (1990). People and the rainforest: an archaeological perspective. In *Australian Tropical Rainforests*, Webb L. J. & Kikkawa J. (eds). CSIRO, Melbourne, pp. 33–9.

Hyland, B. P. M. & Whiffin, T. (1993). *Australian Tropical Rainforest Trees*. CSIRO, Melbourne.

International Council for Science (ICSU) (2002). *Science, Traditional Knowledge and Sustainable Development*. ICSU and UNESCO, Paris.

Joiris, D. V. (1993). The mask that is hungry for yams: ethno-ecology of *Dioscoria mangenotiana* among the Baka, Cameroon. In *Tropical Forests, People and Food*, Hladik, C. M., Hladik, A., Linares, O. F., Pagezy, H., Semple, A. & Hadley M. (eds). UNESCO, Paris, pp. 633–42.

Kendall, G. (2005). A burgeoning role for Aboriginal knowledge. *Ecos* **125**(June/July): 10–13.

Luque, A. (2003). The people of the Matavén Forest and the national park system: allies in the creation of a Community Conserved Area in Colombia. *Policy Matters* **12**(September): 145–151.

McCaul, J. (2005). A planned cultural economy for Cape York. *Ecos* **126** (August/September): 21–3.

Maningrida Wildlife Centre (2005). Arnhem entrepreneur: natural heritage. *Journal of the Natural Heritage Trust* **24**(Winter): 9.

Michon, G. & de Forestra, H. (1995). The Indonesian agro-forest model. Forest resource management and biodiversity conservation. In *Conserving Biodiversity Outside Protected Areas: The Role of Traditional Agro-ecosystems*, Halladay, P. and Gilmour, D. A. (eds). IUCN Gland, Switzerland, pp. 90–106.

Moguel, P. & Toledo. V. M. (1999). Biodiversity conservation in traditional coffee systems of Mexico. *Conservation Biology* **13**(1): 11–21.

Pandey, D. N. (2003). Cultural resources for conservation science. *Conservation Biology* **17**(2): 633–35.

Peres, C. A. & Baider, C. (1997). Seed dispersal, spatial distribution and population structure of Brazilnut trees (*Bertholletia excelsa*) in southeastern Amazonia. *Journal of Tropical Ecology* **13**: 595–616.

Posey, D. (1988). Kayapo Indian natural-resource management. In *People of the Tropical Rain Forest*, Denslow J. S. & Padoch C. (eds). University of California Press in association with Smithsonian Institution Travelling Exhibition Service, Berkeley, pp. 89–90.

Posey, D. A. (1999). Cultural landscapes, chronological ecotones and Kayapó resource management. In *Cultural and Spiritual Values of Biodiversity*, Posey, D. & Dutfield, G. (eds). UNEP, London, pp. 363–4.

Rainforest CRC (2000). *Annual Report 1999–2000*. Cooperative Research Centre for Tropical Rainforest Ecology and Management, Cairns.

Romero, C. &. Andrade, G. I. (2004). International conservation organizations and the fate of local tropical forest conservation initiatives. *Conservation Biology* **18**(2): 578–80.

Rose, D. B. (1996). *Nourishing Terrains: Australian Aboriginal Views of Landscape and Wilderness*. Australian Heritage Commission, Canberra.

Schwartzman, S. &. Zimmerman, B. (2005). Conservation alliances with indigenous peoples of the Amazon. *Conservation Biology* **19**(3): 721–7.

Smyth, D. (2001). Joint management of national parks. In *Working on Country: Contemporary Indigenous Management of Australia's Land and Coastal Regions*, Young, E., Davies, J. & Baker, R. M. (eds). Oxford University Press, Oxford.

Smyth, D. (2002). *Indicating Culture. Development of Cultural Indicators for the Management of the Wet Tropics World Heritage Area*. Rainforest CRC, Cairns.

Sotomayor, M. (1995). Traditional farming systems and biodiversity in the High Andes of Bolivia. The case of Ayllu Mujlli. In *Conserving Biodiversity Outside Protected Areas. The Role of Traditional Agro-ecosystem,*

Halladay, P. & Gilmour, D. A. (eds). IUCN, Gland, Switzerland, pp. 50–62.

Sponsel, L. E.,. Bailey R. C. & Headland T. N. (1996). Anthropological perspectives on the causes, consequences, and solutions of deforestation. In *Tropical Deforestation The Human Dimension*, Sponsel, L. E., Headland, T. N. & Bailey, R. C. (eds). Columbia University Press, New York, pp. 3–52.

Steering Committee for the Review of Government Service Provision (2005). *Overcoming Indigenous Disadvantage Key Indicators 2005 Overview*. Productivity Commission, Canberra.

Szabo, S. & Smyth, D. (2003). Indigenous protected areas in Australia: incorporating Indigenous owned land into Australia's national system of protected areas. In *Innovative Governance: Indigenous Peoples, Local Communities and Protected Areas*, Jaireth, H. & Smyth, D. (eds). Ane Books, New Delhi, pp. 145–64.

Talbot, L. (2005). Indigenous land management techniques of the Djabugay People. Masters thesis, School of Tropical Environment Studies and Geography, James Cook University, Cairns.

Titchen, S. M. (1995). A strategy for the potential renomination of the Wet Tropics of Queensland using the newly defined World Heritage cultural landscape categories. Unpublished report to the Wet Tropics Management Authority, Cairns.

van der Zijden, R. (2002). Ma:Mu Canopy Walkway: the integration of Indigenous and non-Indigenous approaches to rainforest management. Honours thesis, School of Tropical Environment Studies and Geography, James Cook University, Cairns.

Wet Tropics Aboriginal Plan Project Team (WTAPPT) (2005). *Caring for Country and Culture: The Wet Tropics Aboriginal Cultural and Natural Resource Management Plan*. Rainforest CRC and FNQ NRM Ltd, Cairns.

Wet Tropics Regional Agreement (2005). *Wet Tropics of Queensland World Heritage Area Regional Agreement between Rainforest Aboriginal People and the Wet Tropics Management Authority, Queensland Environmental Protection Agency, Queensland Parks and Wildlife Service, Queensland Department of Natural Resources and Mines and The Commonwealth of Australia Department of Environment and Heritage for Management of Wet Tropics World Heritage Area 2005*. Wet Tropics Management Authority, Cairns.

Worboys, P. M. (2002). Ma:Mu and cultural values of plants: an environmental and oral history of culturally important rainforest plants. Honours thesis, School of Tropical Environment Studies and Geography, James Cook University, Cairns.

Wet Tropics Aboriginal Plan Project Team. (WTAPPT) (2005). *Caring for Country and Culture – The Wet Tropics*

Aboriginal Cultural and Natural Resource Management Plan. Cairns, Australia: Rainforest CRC and FNQ NRM Ltd.

Yibarbuk, D., Whitehead P. J., Russell-Smith J. *et al.* (2001). Fire ecology and Aboriginal land management in central Arnhem Land, northern Australia: a tradition of ecosystem management. *Journal of Biogeography* **28:** 325–43.

Zeppel, H. (2002). Indigenous tourism in the Wet Tropics World Heritage Area, north Queensland. *Australian Aboriginal Studies* Autumn: 65.

36 RETHINKING ROAD ECOLOGY

Miriam Goosem*

School of Earth and Environmental Sciences, James Cook University, Cairns, Queensland, Australia
*The author was participant of Cooperative Research Centre for Tropical Ecology and Management

Introduction

In the past 17 years, increasing knowledge in the fields of road ecology and internal fragmentation, together with strict legislative requirements to reduce environmental impacts, have caused infrastructure organizations to alter their attitudes with regard to linear clearing construction and upgrade. In the Wet Tropics, the Rainforest CRC (Cooperative Research Centre) has been at the forefront of global research regarding linear infrastructure impacts in rainforest and, more recently, the application of mitigatory strategies for rainforest infrastructure. These research projects involve collaboration with government environmental and infrastructure agencies and community conservation groups. This chapter discusses the background to this research, the impacts investigated, suggested mitigation strategies and case studies of collaborative projects between researchers and government road, power and environmental agencies, as well as the conservation community. This research is aimed at producing the best possible environmental outcomes.

Background

During the late 1980s when the impacts of roads and other linear infrastructure on wildlife were first being considered and data collected within the Wet Tropics (Burnett 1992; Goosem 1997, 2000b), there was a paucity of information in the fields of road ecology and internal fragmentation throughout the world, and no data whatsoever regarding tropical rainforests. The impetus for impact studies to be undertaken in rainforest was provided by the illegal construction of a road through state forest. The road was built using the simple expedient of bulldozing through ridge tops and using that soil and rock to fill adjoining gullies. At that time, relatively few articles had been published concerning road impacts, the majority of these being descriptive studies in temperate zones that listed road mortality on sections of road. In many cases data had been collected opportunistically while driving an often-travelled route (e.g. McClure 1951; Hodson 1960, 1966, Vestjens 1973). A few studies analysed potential underlying factors associated with such mortality, focusing on human safety issues such as deer-vehicle accidents in the northern hemisphere (Bellis & Graves 1971; Allen & McCullough 1976; Case 1978; Coulson 1982, 1989; Bashore et al. 1985). Seasonal, diurnal, topographic, visibility and habitat factors were implicated.

A second group of studies, also in temperate habitats, considered the potential barrier effects to wildlife of roads and linear clearings (see also Laurance & Goosem, Chapter 23, this volume). Roads were found to inhibit movements of large mammals, including deer, mountain goat, antelope, bear and wolf (Pienaar 1968; Klein 1971; Singer 1975; Thiel 1985; McLellan & Shackleton 1988; Mech *et al*. 1988; Brody & Pelton 1989), and also to reduce crossings by small mammals of many species (King 1978; Kozel & Fleharty 1979; Wilkins 1982; Garland & Bradley 1984; Mader 1984; Swihart & Slade 1984; Bakowski & Kozakiewicz 1988). One study in temperate coniferous forests examined elements of both impacts (Oxley *et al*. 1974), concluding that clearing width was the most important factor causing barrier effects for small mammals. Wider clearings formed greater barriers, the result being few or no movements across wide highways and also little or no road mortality on wide roads. The influence of traffic volume and road surface on crossings appeared negligible, but habitat specialization played a large role in species mobility across roads (Oxley *et al*. 1974). Similarly, Barnett *et al*. (1978) found that the mobility of small mammal species in subtropical Australian rainforests influenced road crossing and that wider clearing widths reduced crossing rates.

Given this background, the new studies in the Wet Tropics aimed to determine the extent to which these impacts also occurred in tropical rainforests and, where possible, to establish factors that influenced the impacts. It was postulated that a wide variety of impacts would be observed, including habitat loss, disturbance in the form of emissions of energy or pollutants, or from maintenance regimes, together with invasions of weeds and feral animals, edge effects in microclimate, vegetation structure, floristics and faunal composition, as well as physical and behavioural barriers to movements. Linear barrier impacts were expected to be greater for wildlife adapted to the high structural complexity of the rainforest than for animals from open habitats, and especially strong for species found only in rainforest interior habitats and/or for species with low mobility or low levels of exploratory behaviour.

Impact research in rainforests of the Wet Tropics

Initial Wet Tropics research was aimed at examining impacts on wildlife. Road mortality and its influencing factors were studied on the Kuranda Range section of the Kennedy Highway. This road divided the important connection between upland rainforest areas formed by the Black Mountain corridor, in a similar fashion to the newly constructed illegal road 20 km to the north (Goosem 1997, 2000b). Road mortality was substantial in many vertebrate groups (extrapolated estimates 1100–3200 vertebrates per kilometre annually) and did not occur randomly, but was influenced by the surrounding landscape. More animals were killed near gullies that form a natural conduit for movements through the landscape than in areas adjacent to road cuttings and embankments, features that might create a physical barrier for some species (Goosem 2000b). Clearing width was also found to have a large impact on most groups of rainforest-dependent vertebrates (Goosem 2000b). Road mortality decreased as clearing width increased, suggesting that fewer crossing attempts occurred at wider clearings. Trapping and radio-tracking studies of small and arboreal mammal movements in relation to roads and powerline clearings also established inhibition of movements (Burnett 1992; Goosem & Marsh 1997; Goosem 2000a, c, 2001; Wilson 2000; Laurance & Goosem, Chapter 23, this volume), with some clearings forming a complete barrier (Goosem & Marsh 1997). Greater width of clearing was implicated in barrier creation, but altered habitat within the clearing and the degree to which habitat structure differed from that of rainforest also increased barrier effects. Rainforest specialists, including arboreal mammals that do not venture to the ground, and forest-dependent species of the understorey and ground layer were found to be most seriously affected (Wilson 2000; Goosem 2000b, c). Barrier effects were reduced where canopy could be maintained across a narrow road clearing (Goosem 2000a, c).

Faunal community composition also differed near the edges of linear clearings compared with the forest interior, with several small mammal species appearing to avoid edges (Goosem 2000a, c, 2001), while others became more abundant. These faunal edge effects were exacerbated by the presence of habitat within the clearing (matrix) that was structurally and floristically greatly dissimilar from rainforest (Goosem & Marsh 1997; Goosem 2000c). Avoidance of highway edges by certain rainforest birds might be related to traffic noise levels as well as these edge habitat alterations (Dawe & Goosem 2007). The ingress of species alien to the rainforest habitat along linear clearings (including weeds and feral animals) might also be a factor in faunal edge effects, with the altered habitat within the clearing potentially allowing aliens to compete with or predate on native rainforest species at the edge (Goosem & Marsh 1997; Mitchell & Mayer 1997; Byrnes 2002; Goosem 2000c, 2001, 2002a, b; Larsson 2003).

To consider factors causing edge effects further, we examined the degree of edge alterations occurring to microclimate and vegetation (Siegenthaler & Turton 2000; Siegenthaler et al. 2000; Goosem 2002b; Maver 2002; Pohlman 2006; Pohlman et al. 2007). Clearing width and clearing habitat were again found to influence the extent of penetration of edge effects. Microclimatic edge effects and alterations in vegetation structure and floristics were less noticeable where the canopy was closed above narrow roads, or where the clearing or road verge had been revegetated (Siegenthaler & Turton 2000; Siegenthaler et al. 2000; Goosem 2002b; Maver 2002). In contrast, these edge changes were greatest for highways where bitumen surfaces store heat during the day and where there was no additional evaporative cooling provided by creeks or by moisture retained in the tall dense grasses and weeds found in powerline clearings (Pohlman et al. 2007). Overall, linear infrastructure causes significant habitat loss and alteration within the Wet Tropics.

Recently, disturbance impacts have been examined. Traffic noise from highways penetrated to distances greater than 200 m (Marks & Turton 2000; Dawe 2005; Dawe & Goosem 2007), while headlights can still be detected 75 m into the forest in some areas (Wilson & Goosem 2007). The degree of road noise suffered adjacent to busy highways may be sufficient to cause alterations in the dominant frequencies that certain birds call near the highway edge, possibly increasing the energy expended in communication (Dawe 2005; Dawe & Goosem 2007). In contrast to noise penetration, heavy metal pollutants appeared to be screened by the sealed rainforest edges and did not penetrate far into the forest (Diprose et al. 2000), but instead collected in roadside sediments, eventually being washed by run-off into streams, resulting in unsafe levels of several heavy metals in road run-off (Pratt and Lottermoser 2007). Unsealed roads tend to erode more seriously when there is no canopy above them (Bacon 1998), with eroded sediment expected to enter watercourses during high rainfall events, causing sedimentation of streams, while the concentrated run-off flowing from impervious road surfaces can cause

stream channelization. Together these two alterations in stream habitats could potentially result in major alterations to stream flora and fauna.

Conclusions regarding impacts

Many of the earlier hypotheses with regard to the impacts of linear infrastructure in rainforest areas were demonstrated to be true:

- habitat loss from linear infrastructure clearings is substantial and greatly exacerbated by edge and disturbance effects penetrating to different distances;
- disturbance impacts emanate from linear clearings in the form of noise, headlights, vehicle-borne pollutants and erosion (and possibly others such as herbicide sprays);
- invasions of feral animals, weeds and flora and fauna alien to the surrounding environment occur along linear clearings, possibly forming a base for further incursions into the adjacent habitat;
- edge effects occur in microclimate, floristics, vegetation structure and faunal community composition, and are exacerbated by several factors, including greater clearing width, loss of canopy above the clearing and greater alteration of habitat within the clearing;
- road mortality is common for many terrestrial and understorey rainforest species, although rainforest interior specialists and species mobile enough to avoid vehicles are less common than expected; with season, traffic speed, microtopography, clearing width and possibly traffic volume being influential factors;
- linear barrier effects or inhibition of crossing movements occur for some faunal species, notably small mammals and arboreal mammals, and are exacerbated by greater clearing width and greater degree of alteration of habitat within the clearing, and also by alterations to microtopography adjacent to the clearing and any potential physical barriers such as fences and highway barriers within it.

Strategies for mitigation

Several recurring themes arising from this research suggested strategies for mitigation of the impacts of linear clearings. First, edge effects, erosion, invasions, barrier effects and possibly pollution from herbicides are reduced if canopy can be maintained above the linear clearing. Second, many impacts, including habitat loss, edge effects, linear barrier effects and invasions, are minimized if the width of the clearing is minimized, although road mortality may be increased. Third, impacts in edge effects, invasions and linear barrier effects are reduced if the habitat within the clearing is maintained to be similar to the adjoining rainforest, rather than completely cleared or allowed to be invaded by grassland or woody weeds. Finally, reducing the volume or speed of traffic along roads should reduce disturbance within the road corridor and adjacent forest from noise, pollutants and other emissions, and also reduce wildlife mortality.

Although the most successful strategy to minimize impacts in rainforest would be to avoid this habitat altogether, unfortunately this is generally not feasible for the major roads in the Wet Tropics. These roads traverse the mountain ranges and associated rainforest between the growing population centres on the coast and the rural regions of the Tablelands (Goosem 2004). However, the rehabilitation of many minor tracks through the forest, arising from log removal during past forestry operations, could reduce impacts of habitat loss, invasions, edge and consequent linear barrier effects. Ideally, rehabilitation would involve removal of any drainage structures and ripping of track surfaces. In practice, due to lack of funding to relevant management agencies, tracks have been left to revegetate without assistance, emphasizing the current need to monitor whether such natural rehabilitation is successful or whether weed invasions are self-perpetuating. Similarly, the removal of redundant powerline clearings or those that are increasingly expensive to maintain could reduce impacts substantially, provided the clearings were either completely revegetated or, as a more economically feasible alternative, planted systematically to fragment the clearings gradually with revegetated strips. Natural regeneration could then expand from each new set of plantings.

A second strategy that would minimize many impacts of linear clearings is to maintain canopy closure above road clearings, thus reducing the severity of edge effects in microclimate, vegetation structure, floristics and faunal community composition, and reducing or almost eliminating weed invasions and their associated feral or alien animal communities. The synergistic linear barrier impact caused by the combination of these effects would be concomitantly reduced,

although not completely eliminated, as the alignment of home ranges along the anthropogenic edge and avoidance of the area devoid of understorey and ground cover would still cause at least partial inhibition of animal movements. This is a very feasible alternative for narrow and unsealed tourist roads within the Wet Tropics (Box 36.1). An economic benefit of this strategy is reduced funding for maintenance of unsealed roads, resulting in both less road grading after erosion and less spraying of weeds on verges. Sediments and herbicide spray pollutants entering aquatic habitats would also decrease. Along roads in protected areas, canopy closure also creates a green tunnel effect attractive to tourists. In particular, the absence of roadside weeds allows the visitor to see into and enjoy the forest environment, providing a far more pleasant experience than viewing an impenetrable expanse of weeds.

Unfortunately, where roads carry high levels of traffic and are aimed at providing fast transit, safety issues dictate that tree species with the potential to drop limbs cannot be left to overhang the road. Such roads tend to have wider clearings, but clearing width should be minimized by maintaining trees as close to the constructed road surface as safety allows. Widths of cuttings and embankments should also be minimized. An alternative group of native tree species (e.g. figs) with root systems that stabilize banks and reduce the prospect of uprooting, together with stable branch networks with a wide spread that may encompass a wide clearing, should be considered for the revegetation of verges along major roads (Biotropica Australia 2006). Such species could also provide canopy cover, while visibility for drivers around road curves could be maintained by the removal of understorey, with shading from the spreading canopy preventing the growth of tall grasses. Driver visibility provided in this way would also allow recognition of dangers that include large animals such as cassowaries, but where such frugivores occur, plant species without attractive fruit must be used for revegetation. However, to reduce road mortality of these species effectively, a concomitant reduction in traffic speeds is also required.

The third strategy to reduce impacts requires alternative engineering solutions for improving connectivity across roads and reducing road mortality. These engineering solutions include road underpasses and bridges that provide dry animal passage and preferably vegetative cover for terrestrial rainforest

Box 36.1 Canopy closure over unsealed and sealed tourist roads

In the Wet Tropics of Queensland World Heritage Area (WHA) there are about 960 km of unsealed roads used by tourists, landholders and managers. The majority of these can be managed to maintain canopy closure, creating a green tunnel effect and mitigating many of the edge, barrier and invasive impacts of roads (Figure 36.1a). Removal of canopy above the road allows weeds to dominate and feral and grassland fauna to invade these unnatural grassland habitats (Figure 36.1b). Where weeds dominate the road verge,

herbicide control in 1999 and 2000 by the Queensland Department of Primary Industries was successful (Figure 36.1c). When continued over several years, these methods encourage native regrowth on verges and, eventually, canopy closure. Similarly, closed canopy above sealed (Figure 36.1d) and unsealed minor roads that traverse rainforest (Figure 36.1e) provides a very pleasant driving experience for tourist and locals alike, while reducing impacts.

Figure 36.1 (a) Closed canopy roads restrict weed growth; (b) open canopy roads allow road verges to become overgrown with grass; (c) open canopy road after herbicide weed control; (d) closed canopy sealed tourist highway with no weeds; (e) closed canopy unsealed tourist road with few weeds.

species, together with canopy bridges for arboreal species (Weston *et al.* 2008). These structures have been trialled in the Wet Tropics, with designs being updated using best practice engineering methodologies (see Boxes 36.2–36.4). Unfortunately, these do not have the added benefits of reducing edge effects, weed and feral animal invasions and erosion impacts, and are therefore not as successful in conservation outcomes as

complete avoidance of habitat or maximization of canopy closure and minimization of clearing width.

A fourth strategy is to build infrastructure above the canopy, in the form of bridges high above the vegetation, or powerlines that swing from ridge top to ridge top (see Boxes 36.5 and 36.6). This can be extremely successful for the reduction of many impacts, provided the construction methods ensure minimal disturbance

Box 36.2 East Evelyn underpasses

In 2000, QDMR widened and straightened the East Evelyn road, which passed through an important north–south connectivity gap in the World Heritage rainforests of the southern Atherton Tablelands. Prior to the upgrade, the single-lane road traversed a strip of abandoned weedy pasture, which, together with the road, created a barrier to movements of rainforest fauna (Goosem et al. 2001). The adjacent rainforest provides habitat for rare rainforest animals, including Lumholtz's tree kangaroo, *Dendrolagus lumholtzi*, the southern cassowary, *Casuarius casuarius johnsonii*, and three rainforest ringtail possums (see Box 36.3), all of which are negatively impacted by roads, through either road mortality or barrier effects (Bentrupperbäumer 1988; Moore & Moore 1999; Wilson 2000; Izumi 2001; Kanowski et al. 2001).

The best location and design of underpasses was decided in discussions between researchers, QDMR engineers and environmental personnel, Queensland Environmental Protection Agency (EPA) personnel from the Centre for Tropical Restoration and Tablelands community conservation groups, including the Tree Kangaroo and Mammal Group and Wildlife Rescue. The focus of this collaboration was the inclusion of four large underpasses that allow large species such as the cassowary to see attractive habitat from either underpass entrance, thereby encouraging passage (Figure 36.2a). To facilitate use of underpasses by smaller rainforest fauna that might be intimidated by a large empty space, conditions on the rainforest floor were simulated using soil, leaf and branch litter. Underpass furniture comprised rocks and logs that

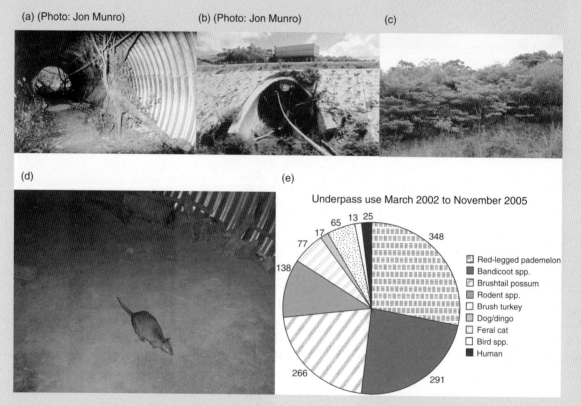

(a) (Photo: Jon Munro)   (b) (Photo: Jon Munro)   (c)

(d)   (e)

Underpass use March 2002 to November 2005

- Red-legged pademelon
- Bandicoot spp.
- Brushtail possum
- Rodent spp.
- Brush turkey
- Dog/dingo
- Feral cat
- Bird spp.
- Human

Figure 36.2 (a) East Evelyn underpass; (b) underpass furniture with soil, leaf litter, rocks, logs, erect escape poles and ropes, showing visibility of attractive habitat at far entrance; (c) corridors of rainforest trees leading to underpasses showing growth after 3 years; (d) red-legged pademelon, *Thylogale thetis*, photographed by infra-red triggered camera while traversing an underpass; (e) graph showing species or species groups using the underpass up to November 2005.

(Continued)

Box 36.2 *(Continued)*

provide cover for smaller species from predators such as dogs, cats and owls, together with vertical tree branches erected to provide an escape route from predators for arboreal species (Kanowski & Tucker 2002), especially tree kangaroos, which are threatened by dog attacks (Figure 36.2b). Thick ropes were swung from the roof and tied to external trees for possible use by obligate arboreal species (see Box 36.3). The conservation group TREAT (Trees for the Evelyn and Atherton Tablelands) collaborated with EPA personnel to plant corridors of rainforest trees from the entrances of the underpasses to rainforest to the north and south of the road upgrade (Figure 36.2c).

Monitoring of underpasses using sand-tracking and infra-red triggered cameras has demonstrated use by a variety of generalist rainforest species (Figure 36.2d, e), while a slight decrease in road mortality has been recorded. Road mortality remains substantially lower than occurs on a highway further to the north that carries similar traffic levels through the same rainforest corridor (Goosem et al. 2006).

Tree kangaroos have occasionally moved through the underpasses, although this is relatively uncommon, while one of the few remaining cassowaries in the area has been observed attempting to enter an underpass (Jon Munro, pers. comm.). Feral predators have rarely been recorded in underpasses, with only one incidence of predation recorded near an underpass.

The rainforest corridors leading to the underpasses have grown substantially, reaching 3–6 m in height after 3 years (Figure 36.2c) and creating a closed canopy attracting rainforest-dependent birds so that bird community composition has approached that found on the edges of rainforest nearby. However, colonization of corridors by rainforest-dependent small mammals has been slower (Goosem et al. 2006). It is expected that, as the corridors grow and a complex understorey forms below the canopy, rainforest-dependent species will dominate and movements through the underpasses by rainforest specialists will be further encouraged.

Box 36.3 Canopy bridges for possums

Collaboration between government environmental agencies, Rainforest CRC researchers and infrastructure agencies has been a feature of the erection and research into the effectiveness of canopy bridges above roads. Target species are the rare rainforest ringtail possums. One of these possums, the lemuroid ringtail, *Hemibelideus lemuroides*, will not descend to the ground, resulting in roads constituting a severe barrier to movements, while the other two ringtails, the Herbert River, *Pseudochirulus herbertensis*, and green, *Pseudochirops archeri*, rarely move at ground level, preferring to remain in the safety of the treetops (Weston 2000, 2003; Wilson 2000). Canopy connections via branches above the road are the preferred means of crossing (Weston 2003), but where clearings are too wide, or are maintained without branches over the road, arboreal overpasses that use simple engineering designs to provide connections may provide a solution to the road barrier. Such overpasses also provide a safe route for less obligate arboreal species often killed on the road. These include the coppery brushtail

possum, *Trichosurus vulpecula johnsoni*, the striped possum, *Dactylopsila trivirgata*, and the fawn-footed melomys, *Melomys cervinipes*, while several less well known species could also benefit, including the long-tailed pygmy possum, *Cercartetus caudatus*, and an arboreal rodent, the prehensile-tailed rat, *Pogonomys mollipilosus*.

In 1995, a trial canopy bridge in the form of a rope tunnel (Figure 36.3a) was erected above a 7 m wide unsealed road with almost no traffic, as a result of collaboration between Rupert Russell of Queensland National Parks and Wildlife Service (QPWS), the Wet Tropics Management Authority (WTMA), which funded the structure, and the Far North Queensland Electricity Board (FNQEB), which provided equipment and personnel for construction. Monitoring from 2000 to 2002 by scat collection, hair analysis, spotlighting and remote photography revealed use of the structure by all target ringtail possum species (Figure 36.3b) as well as the striped possum and fawn-footed melomys (Weston 2000).

(Continued)

Box 36.3 (*Continued*)

In 2001, collaboration between Rainforest CRC researcher Nigel Weston and QDMR resulted in the erection of a simpler, rope ladder design (Figure 36.3c). This structure above a tourist road was 14 m long and was attached at either end to trees in an area where no branch connections occurred across the road and thus lemuroid ringtail possums could not cross. Lemuroid ringtails are now regular users of the canopy bridge (Figure 36.3d), with multiple individuals crossing, while many different Herbert River ringtail possums also cross via the structure and green ringtail, coppery brushtail and striped possums have also been observed (Weston 2003).

In late 2005, 40 m long structures were erected over a wide two-lane highway (Figure 36.3e). QDMR funded the bridges and monitoring, and also helped in installation, together with Rainforest CRC researchers and Ergon Energy, which provided personnel and equipment to erect the four bridges (Figure 36.3f). Both tunnel and rope ladder designs are being tried. Several possums in the vicinity of the overpasses have been radio-collared and tracked to examine their home ranges and follow any crossing movements. Monitoring of the tunnel using an infra-red triggered digital camera revealed that the first animal to use the structure did so in May 2006, when a giant white-tailed rat, *Uromys caudimaculatus*, moved from one side to the other, then returned (Figure 36.3f). Occasional use by two rainforest ringtail species was observed in 2007.

(a) (Photo: Nigel Weston)   (b) (Photo: Nigel Weston)

(c) (Photo: Nigel Weston)   (d) (Photo: Wild about Australia)

(e) (Photo: Birgit Kuehn)   (f)

Figure 36.3 (a) Tunnel-style canopy bridge erected in 1995 over a narrow unsealed road by Rupert Russell (QPWS) and FNQEB, funded by WTMA; (b) green ringtail possum, *Pseudochirops archeri*, using rope tunnel; (c) rope ladder-style canopy bridge erected in 2001 across a 14 m wide canopy gap by Nigel Weston and QDMR; (d) lemuroid ringtail possum, *Hemibelideus lemuroides*, using rope ladder; (e) three rope ladders and one rope tunnel, all 40 m long, being erected over the Palmerston Highway in 2005; (f) the first animal (giant white-tailed rat, *Uromys caudimaculatus*) photographed using the long rope tunnel.

to the forest below. Means of achieving this minimal disturbance include: the construction of powerline and cablecar towers by helicopter, rather than building tracks to import construction materials, thus ensuring that only the footprint of the tower needs to be cleared (Box 36.5); top-down construction of bridges, so that only the footprint of the bridge pylons needs to be cleared (Box 36.6); and by ensuring that sufficient light and moisture can penetrate below the bridges to maintain growth of the forest beneath (Box 36.6).

Box 36.4 Inexpensive bridges provide faunal connectivity in riparian corridors

Rainforest vegetation remaining along creeks and rivers in agricultural areas protects stream banks against erosion, but can also provide a corridor function for rainforest fauna that otherwise cannot move around the cleared landscape (Lawson 2004; Lawson et al. 2007). Although rainforest-dependent birds are found to inhabit and, presumably, move along these riparian rainforest strips (Lawson 2004; Lawson et al. 2007), trapping has revealed that small rainforest mammals may be less mobile. Where wide sections of grass divide the connectivity of rainforest riparian vegetation, small rainforest mammals are very seldom captured.

However, where dry passage occurs under bridges adjacent to rainforest (Figure 36.4a, b), small mammals can cross under the road easily. Such inexpensive structures can play a role in the maintenance of connectivity for terrestrial fauna. The important factor is to maintain areas under either end of the bridge that provide dry passage for the majority of the year and that connect rainforest habitat on either side of the bridge, while still being able to carry flood waters in extreme rainfall events. Inexpensive connectivity improvements that do not affect water carrying capacity could include understorey plants or rocks that provide cover from predators.

(a)   (b)

Figure 36.4 (a) A bridge dividing rainforest riparian vegetation; (b) a passage under the bridge by which small mammals move underneath the bridge for the majority of the year.

Box 36.5 Powerlines strung above the canopy

In 1998, the Chalumbin – Woree powerline through the WHA was upgraded. Following Rainforest CRC research information provided to Powerlink and WTMA, Powerlink built this line by clearing only for the tower footprint and swinging the line over the canopy from ridge top to ridge top (Figure 36.5a), thereby avoiding the habitat loss, barrier, edge and invasion impacts caused by swathe clearing (Figure 36.5b) (Goosem & Marsh 1997). Helicopters were used to install the line, and continue to maintain it by landing on helicopter pads, on each tower. Thus, no tracks are needed for powerline maintenance and the only clearing undertaken was for the base of the tower, an area subsequently revegetated with rainforest plants.

(Continued)

Box 36.5 (*Continued*)

(a)

(b)

Figure 36.5 (a) The new Chalumbin–Woree powerline swings above the canopy between towers on ridge tops; (b) the grassland swathe of the Palmerston powerline clearing.

Box 36.6 Kuranda Range highway upgrade design

The culmination to date of the collaborative approach between researchers, infrastructure institutions and environmental agencies was demonstrated when the Rainforest CRC provided advice to environmental managers and road planners during the design of the Kuranda Range upgrade project. The Kuranda Range section of the Kennedy Highway travels through World Heritage rainforest on the range west of Cairns and is currently a two-lane, winding road with traffic in 2005 averaging 7000 vehicles a day. QDMR included numerous bridges within the design of the four-lane upgrade after consideration of the importance of different habitats and species found adjacent to the road. Four of these bridges were of sufficient height to retain rainforest canopy underneath, while others were high enough to maintain subcanopy or understorey connectivity. Initially, the Rainforest CRC provided advice regarding environmental best practice methodology that could be used to improve the design of the upgrade (Goosem et al. 2004). The focus was habitat con-

nectivity, edge effects and road mortality issues, together with maintaining excellent stream water quality and catchment integrity, minimizing habitat loss and maximizing the success of rehabilitation of road cuttings, embankments and disused sections of road. Second, the net adverse impacts of the design provided by QDMR were examined (Chester et al. 2004). One recommendation was that all four bridges above the canopy should be built using top-down construction, an expensive technique that causes habitat disturbance only where the bridge pylons enter the ground, avoiding the use of tracks and clearing of vegetation. Areas of high conservation value are thereby preserved (Figure 36.6a, b). Where gaps in knowledge existed, QDMR agreed to fund further research that examined a series of issues.

• Road mortality data from 1989 to 1992 were updated in 2005/6 to examine any changes in spatial and temporal patterns owing to increased traffic or recent road widening. Mortality data can be an aid in the siting of bridges designed

(*Continued*)

Box 36.6 (Continued)

(a)   (b)

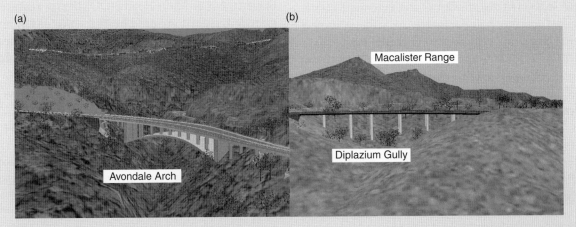

Figure 36.6 The design of bridges in the Kuranda Range upgrade that allow retention of rainforest canopy underneath in areas of high biodiversity and conservation values. (a) Avondale Creek arch design; (b) Diplazium Gully.

to provide faunal connectivity and reduce road mortality (Goosem 2006; Goosem & Weston 2007).

• Traffic noise impacts were examined, building on earlier studies (Marks & Turton 2000; Dawe 2005). In several areas, topographical features adjacent to the road create quiet refuges protected from the traffic noise that penetrates elsewhere to distances of more than 200 m (Dawe & Goosem 2007). These refuges may be important to birds and amphibians that communicate by sound, so retention of refuges or creation of similar areas is a priority for the upgrade design. Certain bird species avoid the noisy road edge and others alter their dominant song frequency adjacent to noisy roads, thereby possibly expending more energy on communication (Dawe & Goosem 2007).

• Possible faunal disturbance by headlights and streetlights was examined by establishing the distance of light penetration into adjacent forest (Wilson & Goosem 2007) and the influence of topography and slope on this area affected. Little information on this potential impact is available worldwide.

• Whether arboreal fauna might use the long canopy bridges that would be required to provide connectivity over the four-lane highway was another topic of research (see Box 36.3).

• The problem of pollutants that enter waterways after being washed from the road by runoff was addressed. Potential pollutant scrubbers were identified in a PhD project (Pratt 2006)

and feasible alternatives for introducing these pollutant removal agents into the stream of runoff directed along road gutters are being trialled.

• One problem identified when bridges above the canopy were designed was whether such bridges would allow the penetration of sufficient light and moisture to sustain the rainforest beneath them. This question was examined in a modelling exercise that examined the light requirements of various tree and understorey species and whether the orientation of the bridges would allow these requirements to be met (Turton & Pohlman 2006). Designs of certain bridges have been modified to provide a gap between the lanes, wide enough to allow sufficient light to penetrate to maintain rainforest canopy growth underneath, also allowing greater rainfall penetration. A second recommendation is to spread runoff water under bridges subsequent to pollutant removal so that moisture levels are maintained (Turton & Pohlman 2006).

• Best practice means of rehabilitating the steep cuttings, embankments and disused road sections through the planting of rainforest species sourced from the Kuranda Range area was another knowledge gap identified and researched (Biotropica Australia 2006).

Together, these research topics have increased knowledge on impacts, while providing best practice information to QDMR to improve the design of this large infrastructure project.

Conclusions

Research on the impacts of roads and powerline clearings has informed the design of management strategies to reduce those impacts, but to achieve impact reduction the strategies must be implemented. This is impossible without meaningful collaborations between rainforest ecologists, environmental and infrastructure management agencies, and the conservation community. Boxes 36.1–36.6 demonstrate a variety of collaborations between researchers, research users and the community in the interests of better conservation outcomes. Similarly, research must be incorporated in manuals of best practice for infrastructure design, construction and operation. The *Roads in the Wet Tropics Best Practice Manual* (QDMR 1997) and the recently revised *Roads in Tropical Forests Best Practice Guidelines* (Chester *et al.* 2006) demonstrate the practical application of Rainforest CRC research to on-ground problems faced by the infrastructure management agency and their contractors.

Summary

Over the past 17 years, sound ecologically based research into impacts has provided the information necessary, while changes in legislative policy have provided the impetus, to employ strategies for mitigation of the impacts of roads and highways through rainforest ecosystems. Such strategies include:

- road designs that avoid rainforest habitat or rehabilitation of habitat previously disturbed by redundant linear infrastructure;
- the maintenance or re-establishment of rainforest tree canopy above narrow roads and the identification and planting of tree species that create canopy over wider roads without dropping branches;
- the minimization of clearing width;
- engineering solutions for connectivity of habitat that include canopy bridges for arboreal animals and underpasses with furniture for cover or bridges that retain habitat underneath for terrestrial wildlife;
- the erection of powerlines or road bridges above the canopy.

Collaboration between researchers, road authorities, government agencies and conservation groups has applied this research to upgrades of several roads and powerlines through Wet Tropics rainforest. Serial modifications of road designs and funding from Queensland Department of Main Roads (QDMR) for research into knowledge gaps ensured best practice environmental outcomes based on the latest research. The accumulated knowledge base is applicable to roads through rainforests throughout the world.

References

Allen, R. E. & McCullough, D. R. (1976). Deer – car accidents in southern Michigan. *Journal of Wildlife Management* **40**: 317–25.

Bacon, T. A. (1998). Erosion on unsealed roads within the Wet Tropics World Heritage Area: A comparison of road conditions under varying amounts of canopy cover. MApplSci thesis, James Cook University, Townsville.

Bakowski, C. & Kozakiewicz, M. (1988). The effect of a forest road on bank vole and yellow-necked mouse populations. *Acta Theriologica* **33**: 345–53.

Barnett, J. L., How, R. A. & Humphreys, W. F. (1978). The use of habitat components by small mammals in eastern Australia. *Australian Journal of Ecology* **3**: 277–85.

Bashore, T. L., Tzilkowski, W. M. & Bellis, E. D. (1985). Analysis of deer – vehicle collision sites in Pennsylvania. *Journal of Wildlife Management* **49**: 769–74.

Bellis, E. D. & Graves, H. B. (1971). Deer mortality on a Pennsylvania interstate highway. *Journal of Wildlife Management* **35**: 232–7.

Bentrupperbaumer, J. (1988). *Numbers and Conservation Status of Cassowaries in the Mission Beach Area Following Cyclone Winifred*. Report to Queensland National Parks and Wildlife Service, Brisbane.

Biotropica Australia P/L (2006). *The Kuranda Range Road Upgrade – A Preliminary Restoration Design*. Report to Queensland Dept of Main Roads, Cairns.

Brody, A. J. & Pelton, M. R. (1989). Effects of roads on black bear movements in western North Carolina. *Wildlife Society Bulletin* **17**: 5–10.

Burnett, S. (1992). Effects of a rainforest road on movements of small mammals: mechanisms and implications. *Wildlife Research* **19**: 95–104.

Byrnes, P. (2002). Activity of feral pigs and cats associated with roads and powerline corridors within the Wet Tropics of Queensland World Heritage area. BSc (Hons) thesis, James Cook University, Cairns.

Case, R. M. (1978). Interstate highway road-killed animals: a data source for biologists. *Wildlife Society Bulletin* **6**: 8–13.

Chester, G., Goosem, M., Cowan, J., Harriss, C. & Tucker, N. (2006). *Roads in Tropical Forests – Best Practice Guidelines, Planning, Design and Management*. Rainforest CRC, Cairns.

Chester, G., Harriss, C. & Goosem, M. (2004). *Kennedy Highway, Kuranda Range Proposed Upgrade Project. Assessment of Net Adverse Impacts*. Report to the Queensland Department of Main Roads, Rainforest CRC, Cairns.

Coulson, G. (1982). Road-kills of macropods on a section of highway in central Victoria. *Australian Wildlife Research* **9**: 21–6.

Coulson, G. M. (1989). The effect of drought on road mortality of macropods. *Australian Wildlife Research* **16**: 79–83.

Dawe, G. (2005). Traffic noise and its influence on the song of tropical rainforest birds. BSc (Hons) thesis, James Cook University, Cairns.

Dawe, G. & Goosem, M. (2007). *Noise Disturbance along Highways – Kuranda Range Upgrade Project*. Report to Queensland Department of Main Roads, Cairns.

Diprose, G., Lottermoser, B., Marks, S. & Day, T. (2000). Geochemical impacts on roadside soils in the Wet Tropics of Queensland World Heritage Area as a result of transport activities. In *Impacts of Roads and Powerlines on the Wet Tropics World Heritage Area II*, Goosem, M. W. & Turton, S. M. (eds). Rainforest CRC, Cairns, pp. 66–82.

Garland, T. Jr & Bradley, W. G. (1984). Effects of a highway on Mojave Desert rodent populations. *American Midland Naturalist* **111**: 47–56.

Goosem, M. W. (1997). Internal fragmentation: the effects of roads, highways and powerline clearings on movements and mortality of rainforest vertebrates. In *Tropical Forest Remnants: Ecology, Management and Conservation of Fragmented Communities*, Laurance, W. F. & Bierregaard, R. O. Jr (eds). University of Chicago Press, Chicago, pp. 241–55.

Goosem, M. W. (2000a). Effects of tropical rainforest roads on small mammals: edge changes in community composition. *Wildlife Research* **27**: 151–63.

Goosem, M. W. (2000b). Impacts of roads and powerline clearings on rainforest vertebrates with emphasis on ground-dwelling small mammals. PhD thesis, James Cook University, Cairns.

Goosem, M. W. (2000c). The effect of canopy closure and road verge habitat on small mammal community composition and movements. In *Impacts of Roads and Powerlines on the Wet Tropics World Heritage Area II*, Goosem, M. W. & Turton, S. M. (eds). Rainforest CRC, Cairns, pp. 2–18.

Goosem, M. W. (2001). Effects of tropical rainforest roads on small mammals: inhibition of crossing movements. *Wildlife Research* **28**: 351–64.

Goosem, M. W. (2002a). Weed surveys along highways, roads and powerline clearings traversing the Wet Tropics World Heritage Area. In *Weed Incursions Along Roads and Powerlines in the Wet Tropics of Queensland World Heritage Area: Potential of Remote Sensing as an Indicator of Weed Infestations*, Goosem, M. & Turton, S. M. (eds). Rainforest CRC, Cairns, pp. 2–25.

Goosem, M. W. (2002b). Weed penetration, edge effects and rehabilitation success in weedy swathes of the Palmerston powerline clearing. In *Weed Incursions Along Roads and Powerlines in the Wet Tropics of Queensland World Heritage Area: Potential of Remote Sensing as an Indicator of Weed Infestations*, Goosem, M. & Turton, S. M. (eds). Rainforest CRC, Cairns, pp. 52–67.

Goosem, M. W. (2004). Linear infrastructure in tropical rainforests: mitigating impacts on fauna of roads and powerline clearings. In *Conservation of Australia's Forest Fauna*, Lunney, D. (ed.). Royal Zoological Society of New South Wales, Mosman, pp. 418–34.

Goosem, M. W. (2006). *Frog Status, Threats and Mitigation of Highway Impacts – Kuranda Range Upgrade Project*. Report to Queensland Department of Main Roads, Cairns.

Goosem, M. W. & Marsh, H. (1997). Fragmentation of a small-mammal community by a powerline corridor through tropical rainforest. *Wildlife Research* **24**: 613–29.

Goosem, M. W. & Weston, N. (2007). *Roadkill on the Kennedy Highway, Kuranda Range Section, 2005–2006 and Its Application to Connectivity Issues for the Proposed Road Upgrade. July 2006*. Report to Queensland Department of Main Roads.

Goosem, M., Izumi, Y. & Turton, S. (2001). Efforts to restore habitat connectivity for an upland tropical rainforest fauna: A trial of underpasses below roads. *Ecological Management and Restoration* **2**(3): 196–202.

Goosem, M., Harriss, C., Chester, G. & Tucker, N. (2004). *Kuranda Range: Applying Research to Planning and Design Review*. Report to the Queensland Department of Main Roads. Rainforest CRC, Cairns.

Goosem, M., Weston, N. & Bushnell, S. (2006). Effectiveness of arboreal overpasses and faunal underpasses in providing connectivity for rainforest fauna. In *Proceedings of the 2005 International Conference on Ecology and Transportation*, Irwin, C. L., Garrett, P., & McDermott, K. P. (eds). Center for Transportation and the Environment, North Carolina State University, Raleigh.

Hodson, N. L. (1960). A survey of vertebrate road mortality. *Bird Study* **7**: 224–31.

Hodson, N. L. (1966). A survey of road mortality in mammals (and including data for the Grass snake and Common frog). *Journal of Zoology (London)* **148**: 576–9.

Izumi, Y. (2001). Impacts of roads and fragmentation on fauna on the Atherton and Evelyn Tablelands, north Queensland. MApplSci thesis, James Cook University, Cairns.

Kanowski, J. & Tucker, N. I. J. (2002). Trial of shelter poles to aid the dispersal of tree-kangaroos on the Atherton Tablelands, north Queensland. *Ecological Management and Restoration* **3**: 137–8.

Kanowski, J., Felderhof, L., Newell, G. *et al.* (2001). Community survey of the distribution of Lumholtz's Tree-kangaroo on the Atherton Tablelands, north-east Queensland. *Pacific Conservation Biology* **7**: 79–86.

King, D. (1978). The effects of roads and open space on the movements of small mammals. BSc (Hons) thesis, Australian National University.

Klein, D. R. (1971). Reaction of reindeer to obstructions and disturbances. *Science* **173**: 393–8.

Kozel, R. M. & Fleharty, E. D. (1979). Movements of rodents across roads. *Southwestern Naturalist* **24**: 239–48.

Larsson, U. (2003). Edge- and linear-barrier effects on rainforest fauna, and the success of reducing those effects by revegetation. BApplSc (Hons) thesis, James Cook University, Cairns.

Lawson, T. (2004). Landscape ecology of riparian vegetation in the Mossman Catchment. BApplSc (Hons) thesis, James Cook University, Cairns.

Lawson, T., Goosem, M. & Gillieson, D. (2007). Rapid assessment of habitat quality in riparian rainforest vegetation. *Pacific Conservation Biology* **13** (in Press).

McClure, H. E. (1951). An analysis of animal victims on Nebraska's highways. *Journal of Wildlife Management* **15**: 410–20.

McLellan, B. N. & Shackleton, D. M. (1988). Grizzly bears and resource extraction industries: effects of roads on behaviour, habitat use and demography. *Journal of Applied Ecology* **25**: 451–60.

Mader, H. J. (1984). Animal habitat isolation by roads and agricultural fields. *Biological Conservation* **29**: 81–96.

Marks, S. & Turton, S. (2000). Vehicular noise disturbance in rainforest. In *Impacts of Roads and Powerlines on the Wet Tropics World Heritage Area II*, Goosem, M. W. & Turton, S. M. (eds). Rainforest CRC, Cairns, pp. 84–94.

Maver, N. (2002). The success of ecological rehabilitation in reducing edge effects on rainforest vegetation and microclimate. BApplSc (Hons) thesis, James Cook University, Cairns.

Mech, L. D., Fritts, S. H., Radde, G. L. & Paul, W. J. (1988). Wolf distribution and road density in Minnesota. *Wildlife Society Bulletin* **16**: 85–7.

Mitchell, J. & Mayer, R. (1997). Diggings by feral pigs within the Wet Tropics World Heritage Area of North Queensland. *Wildlife Research* **24**: 591–601.

Moore, L. A. & Moore, N. J. (1999). *Cassowary Management Project for the Wet Tropics Management Authority. 1. Interim Report for the Kuranda Survey (Nov–Dec, 1998). 2. Streets Creek Cassowary Road Crossing Points and Management Strategies.* Report to the Wet Tropics Management Authority, Cairns, Australia.

Oxley, D. J., Fenton, M. B. & Carmody, G. R. (1974). The effects of roads on populations of small mammals. *Journal of Applied Ecology* **11**: 51–9.

Pienaar, U. de V. (1968). The ecological significance of roads in a nature park. *Koedoe* **11**: 169–74.

Pohlman, C. (2006). Internal fragmentation in the rainforest: edge effects of highways, powerlines and watercourses on tropical rainforest understorey microclimate, vegetation structure and composition, physical disturbance and seedling regeneration. PhD thesis, James Cook University, Cairns.

Pohlman, C., Turton, S. & Goosem, M. (2007). Edge effects of linear canopy openings on tropical rainforest understory microclimate. *Biotropica* **39**, 62–71.

Pratt, C. and Lottermoser (2007). *Remediation of Heavy Metal Impacts in Roadside Corridors.* Wet Tropics World Heritage Area, Townsville.

Queensland Department of Main Roads (QDMR) (1997). *Roads in the Wet Tropics: Planning, Design, Construction, Maintenance and Operation Best Practice Manual.* Queensland Department of Main Roads Technology and Environment Division, Cairns.

Siegenthaler, S., Jackes, B., Turton, S. & Goosem, M. (2000). Edge effects of roads and powerline clearings on rainforest vegetation. In *Impacts of Roads and Powerlines on the Wet Tropics World Heritage Area II*, Goosem, M. W. & Turton, S. M. (eds). Rainforest CRC, Cairns, pp. 46–64.

Siegenthaler, S. & Turton, S. (2000). Edge effects of roads and powerline clearings on microclimate. In *Impacts of Roads and Powerlines on the Wet Tropics World Heritage Area II*, Goosem, M. & Turton, S. (eds). Rainforest CRC, Cairns, pp. 20–43.

Singer, F. (1975). Behaviour of mountain goats in relation to US Highway 2, Glacier National Park, Montana. *Journal of Wildlife Management* **42**: 591–7.

Swihart, R. K. & Slade, N. A. (1984). Road crossing in *Sigmodon hispidus* and *Microtus ochrogaster*. *Journal of Mammalogy* **65**: 357–60.

Thiel, R. T. (1985). Relationship between road densities and wolf habitat suitability in Wisconsin. *American Midland Naturalist* **113**: 404–7.

Turton, S. & Pohlman, C. (2006). *Vegetation and Connectivity under the Proposed Kuranda Range Road Bridges: Impacts of Existing Bridges on Microclimate with Implications for the Proposed Road Bridges.* Report to Queensland Department of Main Roads, January. Rainforest CRC, Cairns.

Vestjens, W. J. M. (1973). Wildlife mortality on a road in New South Wales. *Emu* **73**: 107–12.

Weston, N. (2000). Bridging the rainforest gap. *Wildlife Australia* **37**: 16–19.

Weston, N. G. (2003). The provision of canopy bridges for arboreal mammals: a technique for reducing the adverse effects of linear barriers, with case studies from the Wet Tropics region of north-east Queensland. MSc thesis, James Cook University, Cairns.

Weston, N., Goosem, M., Marsh, H. &. Russell, R. (2008). A review of technologies aimed at reducing roadkill and restoring habitat connectivity for arboreal mammals. *Landscape and Urban Planning* (in Press).

Wilkins, K. T. (1982). Highways as barriers to rodent dispersal. *Southwestern Naturalist* **27**: 459–60.

Wilson, R. (2000). The impact of anthropogenic disturbance on four species of arboreal, folivorous possums in the rainforests of north-east Queensland, Australia. PhD thesis, James Cook University, Cairns.

Wilson, R. & Goosem, M. (2007). *Headlight and Streetlight Disturbance – Kuranda Range Upgrade Project*. Report to Queensland Department of Main Roads, June.

37 LIVING IN A WORLD HERITAGE LANDSCAPE: AN INTERNATIONAL PERSPECTIVE

Jeffrey A. McNeely

International Union for Conservation of Nature and Natural Resources (IUCN),
Gland, Switzerland

Introduction: nature and people

One of the great successes of the World Heritage Convention has been to draw the attention of international tourists to areas of outstanding universal value, both cultural and natural. Many World Heritage sites have benefited from substantial increases in tourism and tourism earnings since they were added to the World Heritage List (Thorsell 1992). International tourists often appreciate both natural and cultural sites, and the majority of these visitors come from cities where such amenities may be a distant dream, a virtual visit via video or a once-in-a-lifetime experience. So the economic values of the sites are important for the host country and the experience of visiting a World Heritage site is important for the visitor – a perfectly reasonable and fair exchange of value.

Having said that, the World Heritage Convention has not been totally successful in linking cultural and natural heritage, although a handful of the sites are listed as mixed (as of 2006, the 830 World Heritage sites include 644 cultural, 162 natural and 24 mixed). Some visitors to natural sites may have the impression that these sites are 'wilderness', essentially unmodified by human action. But anthropologists are very well aware that humans have had a profound influence on nature

throughout the world (with the exception of Antarctica and the sub-Antarctic islands, and even those have been indirectly affected by atmospheric transport of pollutants and invasive alien species carried by ships, and even by tourists).

In Western culture, nature is often considered to be that which operates independently of people (Hoerr 1993), and a major focus of development has been to bring nature under greater human control. In fact, progress is often measured by technological innovations that have enabled humans to gain a greater share of the planet's productivity. Conservation, on the other hand, traditionally has been based on the idea of sequestering the largest possible tracts of nature in a state of imagined innocence as national parks and other kinds of protected areas that exclude people, at least officially. Forests that are pristine or virgin or primary are thus given particularly high value for conservation and considered likely to have high biological diversity (Gomez-Pompa & Kaus 1992).

Despite the dominance of this view of nature, work in ecology (Sprugel 1991), palaeontology (Martin & Klein 1984), forestry (Poffenberger 1990), history (Boyden & Dovers 1992; Ponting 1992; Flannery 2001), archaeology (Audric 1972; Raven-Hart 1981), anthropology (Denevan 1992; Roosevelt 1994), and ethics

(Taylor 1986) is calling into question the separation of people from nature, supporting instead the age-old view of many cultures that people are part of nature and that the biodiversity – that is, the variety of genes, species and ecosystems – found in today's forests result from a combination of cyclical ecological and climatic processes and past human action. Evidence is building to support the view that very few of today's forests anywhere in the world can be considered pristine, virgin or even primary (Clark 1996; Williams 2003), and that conserving biological diversity in tropical forests requires a far more subtle appreciation of both human and natural influences.

An iconic wilderness World Heritage site is Yellowstone National Park, often considered the first of the new generation of national parks (established in 1872). But Yellowstone had long been considered a sacred site by several of the Native American ethnic groups (including Crow, Blackfeet, Bannock, Nez Perce and Shoshone) occupying that part of the American West, and it was merely a matter of military history that Yellowstone was relatively empty at the moment it was established as a national park (Burnham 2000). Other wilderness World Heritage sites, such as Serengeti National Park (Tanzania), were intentionally emptied of resident peoples, in that case focused on the Maasai, in the belief that people and nature were incompatible. The irony of this is that the 20 or so ethnic groups of local people living in the greater Serengeti ecosystem were largely responsible for making Serengeti such a paradise for large mammals, through regular burning and other compatible forms of habitat management that maintained the savannah in productive grassland habitats. Now that humans have been excluded, the savannah is changing its character from a grassland to a shrubland, with an associated change in the assemblage of large mammals (Sinclair & Arcese 1995, Kideghesho et al. 2006).

Tropical forests, too, have been significantly influenced by humans, even though some are today considered wilderness by some people, and some tropical forest landscapes appear to be empty. But increasing archaeological and anthropological evidence is indicating that tropical forests have been profoundly influenced by people for thousands, even tens of thousands, of years. The wilderness of New Guinea, for example, is thoroughly settled, even in the most remote areas (Flannery 1998). The wilderness of the Amazon is now

thought to have contained a fairly dense human population in pre-Columbian times, before diseases brought by European explorers led to a substantial population crash (Denevan 1992; Mann 2006). The forests of Sumatra have grown on the foundation of towns from an ancient civilization called Srivijaya (Schnitger 1964), and the forest was taking over the remains of the Cambodian civilization centred at Angkor Wat (now a World Heritage site) before archaeologists fought back against the forest and restored at least some of the temples to a semblance of their former glory (Audric 1972; Higham 2004). The Australian tropics are no exception and it is pleasing to see that this volume recognizes both the dynamism of tropical ecosystems and the role of aboriginal peoples in creating or at least influencing the natural ecosystems that we see today.

McNeely (1994) drew from recent studies to suggest four general conclusions about the history of forests and biodiversity:

- humans have been a dominant force in the evolution of today's forests;
- as humans develop more sophisticated technology, their impact on forests increases until forests are degraded to the long-term detriment of the over-exploiting society;
- over-exploitation is usually followed by a culture change that might reduce human pressure, after which some forests may return to a highly productive and diverse, albeit altered, condition, and others may be permanently altered to much less productive and diverse conditions;
- the best approach to conserving forests and their biodiversity is through a variety of management approaches over a large landscape ranging from strict protection to intensive use, with a careful consideration of the distribution of costs and benefits of each.

This progression is not inevitable, and recognizing the Australian Wet Tropics as part of the world's natural heritage may have helped avoid the worst symptoms of the second stage. While considerable forest area has been lost, much still remains. Different systems of management may enhance or reduce the remaining forest diversity. Completely excluding human intervention from species-rich communities found in hilly, high rainfall areas of the tropics, for example, may reduce genetic and species diversity by changing the mix of successional stages, or it may help to conserve species that are confined to old-growth stages of succession. Further, the sheer number of species is not necessarily a useful measure. Australia, like California, Hawaii and New Zealand, may have more species now than ever before. Many of these are non-native and maintained by human action, but some are invasive alien species that threaten the native species (McNeely *et al.* 2001). The notions of natural vegetation or ecosystem processes, therefore, are still useful goals in forest management, but they should be revised to recognize that a range of ecosystems can legitimately be considered natural (Sprugel 1991) and nearly all of them have been significantly influenced by people.

The role of myth in conserving biodiversity

The idea that nature exists as something separate from people has become part of the mythology of industrial society. The *Oxford Dictionary* (Pollard & Lieback 1994) defines myth as a story arising from an unknown source and containing ideas or beliefs that purport to explain natural events without a basis in fact. It appears that this mythic image of nature is essential to the psychological well-being of modern humanity (Jung 1964; Campbell 1985), whose industrial approach to conquering nature can be so destructive. Non-industrial societies have different myths, often treating what industrial society calls nature as a set of very real threats to human existence or as alternative expressions of the human spirit. Campbell (1985: 10) suggests that societies that cherish and keep their myths alive 'will be nourished from the soundest, richest strata of the human spirit', and that myths often contain many elements that ring true to the people who believe them. Bringing Aboriginal concepts like Dreamtime into the management of the Wet Tropics will enrich the value of the World Heritage site and enhance options for interpreting the region for visitors.

The myth of the virgin forest and nature untouched by humans began with European colonizers seeking new empty land, and today is a myth of urban-dwelling people who are well separated from the reality of the forest. Those who actually live in the forest are faced with a rich diversity that is mostly hidden far above or in the heavy cover of the ground vegetation where various dangers lurk, so the rich forest-related mythology of the Aborigines should come as no surprise.

Rites and rituals are carried out by forest-dwelling peoples to help to reinforce the myths, while the myths provide the mental support for rites and rituals, helping to ensure that children are made well aware of the social and natural environment into which they must fit in order to become a competent member of society (Campbell 1985). The corresponding rituals of modern urban society in relation to the myth of nature may be watching nature programmes on television, giving money to organizations that claim to conserve nature and visiting protected areas, many of which assume almost a sacred character (Putney & Schaaf 2003). The myths of visitors to the Wet Tropics may be as exotic to the local people, especially the Aborigines, as the myths of the former are to the latter. Managed well, this potential conflict of myths can be a source of enlightenment for all concerned.

The Western mythical vision of an untouched wilderness has permeated global policies and politics in resource management (Gomez-Pompa & Kaus 1992). But this view of forests is based on an outmoded ecological perspective, misunderstanding of the historical relationship between people and forests, and ignorance of the role people have played in maintaining biodiversity in forested habitats.

Changing circumstances are bringing about new perceptions and new demands on forest managers. Deepened understanding of the roles of trees and forests in carbon, climatic and water cycles will stimulate new approaches to forest management and biodiversity conservation. A challenge inherent in a multiple-use approach is that products of forests that can be allocated by markets are relatively easy to quantify and exploit, while benefits that cannot easily be given a market value, such as biodiversity, tend to be undervalued and are therefore likely to be degraded over time.

Utilitarian values are often in conflict with the strongly held romantic and symbolic values that I have called myths. To many urban people today, clearing rare old-growth forests for their commodity values and subsequent conversion to plantations is as sensible as tearing down the Sydney Opera House and using the rubble to build new houses. Any money yielded by such an action would be inconsequential relative to the social value of the national symbol. As non-product benefits like biodiversity, or simply the existence of a place like the Daintree Rainforest, become more important to urban citizens, modern social and political systems inevitably will become more prominent in protected area management.

Another of Australia's World Heritage sites has set an international precedent in linking nature with culture. Originally listed in 1987 on the basis of its natural criteria, the Uluru-Kata Tjuta National Park was relisted in 1994, this time including cultural criteria based on the importance given to the site by its Aboriginal owners. The fact that the Wet Tropics now has an Aboriginal Plan, involving 18 Aboriginal tribal groups, provides a new entry point for enhancing the link between people and the rest of nature. A challenge is going to be managing sustainable uses, such as tourism, with the increasing pressure from globalization and the temptations dangled by foreign governments seeking access to the resources of the Wet Tropics. It might be more sustainable for the Wet Tropics to export knowledge and experiences rather than products.

The managers of the Wet Tropics are increasingly seeking diverse combinations of compatible forest uses. They are finding, for example, that conserving biodiversity, promoting tourism and storing carbon for reducing atmospheric carbon dioxide are highly compatible forest services, and that such uses can also yield non-timber forest products as well as conserve soil and water. These uses are incompatible with clear-felling, but may be compatible with well managed agroforestry. The trend away from single-product forestry is continuing, delivering more diversity and benefits for people living in and around the forests.

It appears that the best way to maintain biodiversity in the ecosystems of Australia's Wet Tropics in the twenty-first century is through a combination of: strategically selected, strictly protected areas as World Heritage sites; multiple-use agroforestry areas intensively managed by local people; natural forests extensively managed by forestry professionals for sustainable production of logs and other commodities; and forest plantations intensively managed for other wood products needed by society. This diversity of approaches and uses will provide humanity with the widest range of options, the greatest diversity of opportunities, for adapting to the cyclical changes that are certain to continue. The incredibly productive research carried out by the Rainforest CRC has provided the concepts, data and institutions to enable such a system to function to the benefit of both Australians and the rest of the world.

The interface between biodiversity conservation and sustainable forest management

An essential component of any effort at sustainable forest management is the economic viability of the various enterprises that are involved. While timber extraction is the most obvious money-earner, such a constrained area as Queensland's Wet Tropics cannot sustain heavy logging for very long. Increasing foreign demand for logs and woodchips may threaten the World Heritage site by reducing it to small islands of forested habitat. Fortunately, many other economic activities are possible, providing economic benefits to local people, especially Aborigines. Benefiting from enterprises that depend on the biodiversity of the forest within which they live encourages local people to support the conservation and sustainable use of the forest ecosystem.

This attractive idea has been extensively tested across 39 project sites in Asia and the Pacific, involving activities such as eco-tourism, distilling essential oils from wild plant roots, producing jams and jellies from forest fruits, collecting other forest products and sustainably harvesting timber. The conclusion was that a community-based enterprise strategy can indeed lead to conservation, but only under certain specified conditions that are critically dependent on external factors such as market access. Further, any such enterprise can be sustainable only if it is designed to be adaptable to changing conditions. Because many forested areas are subject to political or economic turmoil, fires, droughts and other external factors, this adaptability of the enterprise is essential to long-term sustainability of the forest (Salefsky *et al.* 2001). The complexity of factors affecting the forest also calls for multiple levels for protecting biodiversity, with actions at local, national and international levels providing the redundancy needed for ensuring that all genes, species and ecosystems are conserved.

Converting the potential benefits of forest biodiversity conservation into real and perceived goods and services for society at large (and especially for local people) requires a systems approach, as suggested above. Elements of this approach include:

• At the national level, an integrated set of protected areas encompassing various levels of management and administration, including the national, provincial and local governments, non-governmental organizations, local communities and indigenous peoples, the private sector and other stakeholders (McNeely 1999).

• Within the framework of the market-based economic systems that are becoming increasingly widespread, greater participation by the civil society in economic development that extends to the management of both production forests and protected areas, especially for tourism and the sustainable use of certain natural resources (Szaro & Johnston 1996).

• A fairly large geographical scale (sometimes called a bioregion) for resource management programmes, within which protected areas are considered as components in a diverse landscape, including farms, harvested forests, fishing grounds, human settlements and infrastructures (Miller 1996).

• Cooperation between private landowners, indigenous peoples, other local communities, industry and resource users; the use of economic incentives, tax arrangements, land exchanges and other mechanisms to promote biodiversity conservation; and the development of administrative and technical capacities that encourage local stakeholders, universities, research institutions and public agencies to harmonize their efforts.

A programme for sustainable forest management in a region like Queensland's Wet Tropics needs to include both firm governmental action and alliances with the other stakeholders. The national government cannot delegate its role of guarantors of the conservation of the country's natural World Heritage, so the appropriate authorities need to build the capacity to fulfil their regulatory and management duties and responsibilities. The Queensland government has the lead responsibility for overall forest management ensuring compatibility between the Wet Tropics World Heritage Site, the management of the other forests in the region and the equitable sharing of benefits with the Aboriginal rights holders. The broader civil society, too, can share certain rights and responsibilities regarding the management of living natural resources after careful preparations and an adequate definition of roles and responsibilities. Given the interests of non-governmental organizations, business, Aborigines and local communities who live within or close to protected areas and other forested regions, alliances should be created among stakeholders, which that enable each to play an appropriate role according to clear government policies and laws.

Summary

Australia's Wet Tropics are widely recognized as of 'outstanding universal value', which is why they have been included on the World Heritage List. Often neglected in discussion of the world's tropical rainforests, the Queensland forests are now one of the best studied tropical forests in the world, largely as a result of the Rainforest CRC. The Queensland Wet Tropics can now join other well-studied tropical forest areas, such as La Selva in Costa Rica, the island of Barro Colorado in the middle of the Panama Canal, Sinharaja World Heritage site in Sri Lanka, Pasoh Forest Reserve in Malaysia and Tambopata Research Centre in Peru, in providing essential information on how tropical forests can best be managed, and how they are responding to changing conditions.

The work that has been carried out in the Wet Tropics over the past decade or so will be especially valuable as a baseline against which further changes can be measured. It is highly likely that climate change will bring profound modifications to the ecosystems of the region, increasing rainfall, increasing droughts and making the entire region more vulnerable to external effects. The impact of invasive alien species will continue to expand, requiring vigorous management measures and continuing research; the Wet Tropics provides an ideal laboratory for studying the interaction of climate change and invasive alien species. The impacts of various forest management practices on biodiversity, the effects of differing types of tourism on the forest and on the economy, and increasing demands for forest goods and services all demand a continuing research effort.

Many people believe that protected areas are established to represent pristine nature. But protected areas are probably better seen as expressions of culture, an idea well captured by the World Heritage Convention. We establish such areas to demonstrate our concern about conserving samples of ecosystems that are important to us, along with the biodiversity they contain and the cultural diversity they support.

The active role Australia has played in the World Heritage Convention and other biodiversity-related Conventions, the investment made in better understanding the Queensland Wet Tropics through the Rainforest CRC and the popularity of nature-based tourism are clear reflections of the care for the environment that is such an important part of the character and culture of the Australian people.

References

Audric, J. (1972). *Ankor and the Khmer Empire*. Robert Hale, London.

Boyden, S. & Dovers, S. (1992). Natural-resource consumption and its environmental impacts in the western world: impacts of increasing per capita consumption. *Ambio* **21**(1): 63–9.

Burnham, P. (2000). *Indian Country, God's Country: Native Americans and the National Parks*. Island Press, Washington, DC.

Campbell, J. (1985). *Myths to Live By*. Grenada Publishing, London.

Clark, D. (1996). Abolishing virginity. *Journal of Tropical Ecology* **12**: 735–9.

Denevan, W. M. (1992). *The Native Population of the Americas in 1492*, 2nd edn. University of Wisconsin Press, Madison.

Flannery, T. (1998). *Throwim Way Leg: Tree-kangaroos, Possums, and Penis Gordes. On the Track of Unknown in Wildest New Guinea*. Atlantic Monthly Press, New York.

Flannery, T. (2001). *The Future Eaters: An Ecological History of the Australasian Lands and People*. George Braziller, New York.

Gomez-Pompa, A. and Kaus, A. (1992). Taming the wilderness myth. *BioScience* **42**(4): 271–9.

Higham, C. (2004). *Encyclopedia of Early Asian Civilizations*. Facts on File, New York.

Hoerr, W. (1993). The concept of naturalness in environmental discourse. *Natural Areas Journal* **13**(1): 29–32.

Jung, C. G. (1964). *Man and His Symbols*. Doubleday, New York.

Kideghesho, J. R., Nyahongo, J., Hassan, S., Tarimo, T. & Mbije, N. (2006). Factors and ecological impacts of wildlife habitat destruction in the Serengeti ecosystem in northern Tanzania. *AGEAM-RAGEE* **11**: 917–32.

McNeely, J. A. (1994). Lessons from the past: forests and biodiversity. *Biodiversity and Conservation* **3**. 3–20.

McNeely, J. A. (1999). *Mobilizing Broader Support for Asia's Biodiversity: How Civil Society Can Contribute to Protected Area Management*. Asian Development Bank, Manila.

McNeely, J. A., Mooney, H. A., Neville, L. E., Schei, P. & Waage, J. K. (eds) (2001). *A Global Strategy on Invasive Alien Species*. IUCN, Gland, Switzerland.

Mann, C. O. (2006). *1491: New Revelations of the Americas before Columbus*. Vintage, New York.

Martin, P. S. and Klein, R. G. (eds) (1984). *Quaternary Extinctions: A Prehistoric Revolution*. University of Arizona Press, Tucson.

Miller, K. R. (1996). *Balancing the Scales: Guidelines for Increasing Biodiversity's Chances Through Bioregional Management*. World Resources Institute, Washington, DC.

Poffenberger, M. (1990). *Keepers of the Forest: Land Management Alternatives in Southeast Asia*. Kumarian Press, West Hartford, CT.

Pollard, E. and Lieback, H. (1994). *The Oxford Paperback Dictionary*. Oxford University Press, Oxford.

Ponting, C. (1992). *A Green History of the World: The Environment and the Collapse of Great Civilizations*. St Martin's Press, New York.

Putney, A. & Schaaf, T. (2003). Guidelines for the management of sacred natural sites. Paper presented at the Vth World Parks Congress, Technical Sessions on Building Cultural Support for Protected Areas, 13 September, Durban, South Africa.

Raven-Hart, R. (1981). *Ceylon: History in Stone*. Lakehouse Investments, Colombo, Sri Lanka.

Roosevelt, A. C. (1994). Amazonian anthropology: strategy for a new synthesis. In *Amazonian Indians: From Prehistory to the Present*, Roosevelt, A. C. (ed.). University of Arizona Press, Tucson, pp. 1–32.

Salafsky, N. *et al.* (2001). A systematic test of an enterprise strategy for community-based biodiversity conservation. *Conservation Biology* **15**(6): 1585–95.

Schnitger, F. M. (1964). *Forgotten Kingdoms in Sumatra*. E. J. Brill, Leiden, The Netherlands.

Sinclair, A. R. E. and Arcese, P. (eds) (1995). Serengeti II: Dynamics, Management and Conservation of an Ecosystem. University of Chicago Press, Chicago.

Sprugel, D. G. (1991). Disturbance, equilibrium and environmental variability: what is 'natural' vegetation in a changing environment? *Biological Conservation* **58**: 1–18.

Szaro, R. and Johnston, D. W. (1996). *Biodiversity in Managed Landscapes: Theory and Practice*. Oxford University Press, Oxford

Taylor, P. W. (1986). *Respect for Nature: A Theory of Environmental Ethics*. Princeton University Press, Princeton, N.J.

Thorsell, J. (ed.) (1992). *World Heritage Twenty Years Later*. IUCN, Gland, Switzerland.

Williams, M. (2003). *Deforesting the Earth: From Pre-history to Global Crisis*. University of Chicago Press, Chicago.

Part 5

Restoring Tropical Forest Landscapes

INTRODUCTION

Nigel E. Stork[1]* and Stephen M. Turton[2]*

[1]School of Resource Management and Geography, Faculty of Land and Food Resources, University of Melbourne, Burnley Campus, Richmond, Victoria, Australia
[2]Australian Tropical Forest Institute, School of Earth and Environmental Sciences, James Cook University, Cairns, Queensland, Australia
*The authors were participants of Cooperative Research Centre for Tropical Ecology and Management

Restoration of tropical forest landscapes is highly controversial and difficult, with many examples of failures and few examples of successes. In this respect this is a new field of research with a very large potential market for the take up of this science. The Wet Tropics has been the focus of much concerted effort in this regard and this part summarizes the research and evaluation of tree plantings from riparian zone repair through to timber plantings. This field of research has benefited from the close collaboration of a group of researchers with practitioners, state government departments, community groups and others. Lamb and Erskine (Chapter 38) provide an overview of such forest restoration at the landscape level and the different approaches. Tucker (Chapter 39) brings the perspective of someone who has been practically involved in landscape restoration and has learnt much from that experience. He provides examples of some recent advances.

Forests are increasingly valued for more than just their timber values and Wardell-Johnson *et al.* (Chapter 40) discuss this issue. One of the major drivers for forest restoration is the enhancement of biodiversity values and Catterall *et al.* (Chapter 41) look at the different phases of biodiversity development through this process. One of the key elements that previously has been ignored or poorly addressed is the monitoring of restoration outcomes and the lessons that should be learnt (Kanowski *et al.* Chapter 42). We are at a period of rapid change for forested landscapes and Harrison and Herbohn (Chapter 43) discuss the future of forest-based industries in the Wet Tropics. As a former Director-General of the Centre for International Forest Research and with his role with WWF International, Jeff Sayer brings a unique perspective on the importance of research and learning on tropical forest landscape restoration with a hard hitting commentary (Chapter 44).

38 FOREST RESTORATION AT A LANDSCAPE SCALE

David Lamb* and Peter Erskine*

School of Integrative Biology, University of Queensland, St Lucia, Queensland, Australia
*The authors were participants of Cooperative Research Centre for Tropical Ecology and Management

Introduction

Restoring forests is a difficult task. The ecological knowledge about how this is done is limited and landholders usually face a host of social and economic constraints that can make it difficult to use the knowledge that is available, and trade-offs are usually required. These include deciding how much land should be devoted to agriculture or reforestation and whether to maximize productivity or to foster biodiversity. Restoring forests at a scale that matches the recent rates of deforestation is even more daunting. The UN Food and Agriculture Organisation estimates that primary forests are currently being lost at the rate of 6 million hectares a year, while plantations are only being established at a rate of 2.8 million hectares per year

The cost of treating large areas is high and the diversity of stakeholders present in these larger areas mean the social difficulties are also considerable. But, paradoxically, the heterogeneity of the landscape mosaic sometimes offers opportunities to manage some of the production–conservation trade-offs that are more difficult to make at individual sites. Here we review experiences in assembling forests in north Queensland and discuss some of the dilemmas involved in attempting to do this at a landscape scale.

Forest clearing has been taking place in north Queensland for well over 100 years but the cessation of rainforest logging in 1988, as a result of the World Heritage listing of the area, prompted significant interest in reforestation. This interest was enhanced by the financial package designed to compensate former timber industry workers. The landscape in which this reforestation was subsequently carried out is a complex mosaic of agricultural areas and residual forest. The agricultural areas are used for cropping and grazing, while the remaining forests include both large areas that were once part of the selectively logged State Forest network and smaller remnants scattered within the agricultural areas.

This complex ecological mosaic is matched by a similarly complicated social mosaic as the region contains a variety of landowners and stakeholders. Many hold sharply different views on the extent to which reforestation is desirable or even a necessary land use. However, recent economic circumstances are leading to changes in attitude about the ways land might be used. While much of the agriculture in the region is productive, some historically important activities have declined or become threatened in recent years (see Turton, Chapter 5, this volume). For example, the tobacco and dairy industries have both declined despite having been important land uses for more than 50 years. Perhaps even more significantly, sugarcane, one of the region's key crops, has had fluctuating economic fortunes because of changes in international prices. These changes may lead to reforestation being viewed as a more attractive land use.

Queensland government forestry agencies had undertaken research on reforestation over many years but most plantations were established using the exotic *Pinus caribaea* or the native *Araucaria cunninghamii* (Lamb *et al*. 2001). This was despite the commercial attractiveness of the so-called cabinet timbers (i.e. high value timber trees) harvested from the natural rainforests of north Queensland by selection logging. The primary reason for this apparent paradox was that it made little sense to use these mostly slow-growing species in plantations when the government owned no cleared land it could reforest and when there were apparently steady supplies of cabinet timbers coming on to the market from the natural forest.

The attractiveness of reforestation changed following the listing of the Wet Tropics on the World Heritage register that led to the cessation of logging. After this time many in the community were keen to restore the original species-rich forests on underutilized agricultural areas in the region. Other landowners were interested in growing high-value cabinet species in plantations for commercial reasons. Interestingly, however, many of these latter landowners also wished to foster conservation on their lands rather than use the more traditional monoculture plantations favoured until then by the government forestry agencies. These two approaches were supported by separate government funding programmes; one, the Wet Tropics Tree Planting Scheme (WTTPS), helped to support restoration plantings, while the other, the Community Rainforest Reforestation Program (CRRP) supported mixed species timber tree plantations (Vize *et al*. 2005).

This meant that there were two urgent research tasks in the early 1990s. One was to develop ways of restoring diversity to degraded sites, while the second was to develop silvicultural systems using native tree species that generated economic benefits for landowners but also produced some biodiversity benefits. Both tasks were difficult because there was only limited knowledge of the autecology of most tree species in the region and very limited knowledge about methods of assembling species-rich communities; there was even

less knowledge about ways of making trade-offs between production and biodiversity. The initial work focused largely on finding methods of assembling new forests on particular sites. Only later was it realized that some of the trade-offs might be more easily managed at a landscape level and that there might be other, more significant tasks that needed to be addressed at this larger scale.

Assembling forest stands at particular sites

In this section we briefly describe different forms of reforestation used in north Queensland, including methods developed simply to restore biodiversity as well as others directed at enhancing production and biodiversity. Much of the research into methods of reforestation and the forms of silviculture that were developed has been described in Goosem and Tucker (1995), Bristow *et al*. (2005) and Erskine *et al*. (2005b). Some general principles are outlined in Lamb (2000).

Restoration for biodiversity purposes

Three broad approaches have been developed to use in restoration projects. These are the Framework Species method, Direct Seeding and the Maximum Biodiversity method. The main elements of these approaches are outlined in Table 38.1. All three require weeds to be excluded from the site at the time of planting or sowing and all are initiated by a single planting or sowing event (although further species might be added as seeds or seedlings once a tree canopy has been created).

These three approaches differ in the rate at which biodiversity is restored to a site. The Framework Species method and the Direct Seeding method take some time because they use a very small number of species to initiate the natural successional processes. By contrast, the Maximum Diversity method generates a rapid increase in plant species richness over a short time but depends on managers being able to raise seedlings of a large number of species for a particular planting date.

Subsequent natural recruitment and successional development in all three methods depend on appropriate dispersal agents being present and these being able to carry seed to the site. This means that recruitment is constrained by a series of ecological filters (Table 38.2).

Table 38.1 Methods of restoration plantings tested in north Queensland

Method	Main elements of this approach
Framework Species (Goosem & Tucker 1995)	Plant a few fast growing hardy species with wide canopies and fruit rewards (30% of which should be short-lived pioneers) at around 4000 trees ha^{-1} to provide a closed cover and shade out weeds in a short time period; these forests then facilitate the colonization of the site by species from nearby natural forests
Direct Seeding (Doust 2005)	Use a small number of large-seeded species or readily available pioneer species to provide an initial tree cover. These then facilitate the colonization of the site by species from nearby natural forests
Maximum Diversity (Goosem & Tucker 1995)	Plant as many non-pioneer species as possible at densities of around 4000 tree ha^{-1} to occupy the site, exclude weeds and quickly create a species-rich forest communities. Although there is less dependence on subsequent colonization from nearby natural forests it can be useful to include some short-lived pioneer species that create canopy openings and allow this to occur

Table 38.2 Ecological filters affecting the ways additional species colonise newly planted forests

Species pool	Ecological filter	Amplification
Habitat pool	Regional environmental constraints on total species available	Abiotic environmental changes caused by previous land uses (e.g. decreases in soil fertility) may prevent some of the species originally present at a site from reoccupying it
Geographical pool	Species present in the region may not be able to reach the site	Restoration plantings or mixed species plantations may be too distant from natural forest remnants to be colonized by species from these forests. Species with small seed are likely to be more widely dispersed than species with larger seed and therefore arrive earlier at a particular site and in larger numbers
Ecological pool	Species shared by habitat and geographic species pool	The species able to tolerate site conditions and able to reach the site from natural forest remnants
Community	Species able to regenerate and grow in the community as a consequence of intra-specific interactions	Internal dynamics such as seed predation, herbivory and other biotic interactions may exclude certain species able to reach a site and tolerate the particular site conditions

The ecological pool potentially able to colonize the site is a subset of species from the habitat pool and the geographical pool. These are species able to tolerate the regional environment and able to reach the site from existing forest remnants. The actual species that finally do colonize the site are those able to tolerate the micro-environments and competitive interactions they are then exposed to at the site. These include above- and below-ground competition from plants already present as well as seed predation and seedling herbivory by wildlife. In the case of animals (including seed-dispersers), there is a feedback mechanism at play here, since wildlife will only visit a new forest when it acquires habitat features of interest such as structural complexity or provides a food reward (Willson & Crome 1989; Toh *et al.* 1999). In time, however, it appears that there can be a causal relationship between floral and faunal diversity (Kikkawa 2005).

There appears to be little specific information on how the arrival sequence of species makes a difference to successional processes, although some evidence suggests this may be significant. For example, the understoreys that colonize monoculture plantations of the native gymnosperm *Araucaria cunninghamii* acquire diversity at a much slower rate than in plantations of angiosperms such as *Flindersia brayleyana* (Keenan *et al.* 1997 Langi 2001; Firn 2005). This suggests that the inclusion of a particular species in a seed or seedling mix that is used to initiate restoration may have consequences for the subsequent successional sequence. It is not clear whether this difference persists or is eroded over time.

The costs of restoration can be high (Erskine 2002) and some limited field experience suggests that direct seeding could be a much cheaper method than the other two approaches under certain circumstances. These include being able to eradicate weeds cheaply, to have exposed mineral soils and to collect large numbers of seed of appropriate species (Sun *et al.* 1995; Doust 2004). However, the Maximum Diversity method has been more widely used in a number of plantings because of the rate at which diversity is created, even though it is a more costly approach.

Reforestation for production and biodiversity purposes

In contrast to restoration plantings, the large scale industrial plantations traditionally established for timber production in Australia have all been simple, even-aged monocultures. But many landowners in the wet tropics of north Queensland who contemplated reforestation for timber production were more interested in mixed species plantations than in monocultures. This was because they seemed to offer the possibility of increased biodiversity while also offering some commercial benefits.

Mixed plantations are difficult to design. The two most common ways of assembling mixed species timber plantations have been to use alternate row plantings of a small number of species or to use random plantings of a larger number of species. In both cases trees are usually established at densities of 1100 trees per hectare. Alternate row plantings offer the possibility of harvesting trees of a particular species, if necessary, without harming those left behind. Random mixtures create a more natural forest but are likely to be much more difficult to manage over time.

The composition of most of these mixed stands was initially determined by issues such as seedling availability, the commercial value, the presumed site requirements of individual species and then, later, the subjective assessment of the complementarity of various species mixes (Vize *et al.* 2005). This meant that many plantings were essentially ad hoc mixtures of easily acquired species. Tucker *et al.* (2004) suggest other species that, with hindsight, might have been given greater prominence in order to encourage wildlife.

A large amount of subsequent research has been undertaken to improve on the design of the mixtures used in these plantations. This has involved research into the site requirements of particular species (Bristow 1996; Merkel 1996; McNamara 2002), the physiology of these species (Swanborough *et al.* 1998), the nutritional requirements of species growing at sites degraded by agricultural practices (Webb *et al.* 2005) and work on the relative competitive abilities of different species

(Nhan 2001). Draft guidelines for identifying complementary species to use in mixtures are described in Lamb *et al.* (2005).

Besides identifying which species to use in particular mixtures, a key issue in designing mixtures is to determine the number of species that should be included in each plantation: just what is the relationship between the number of species (or species types) present in an ecosystem and the way that a system is able to function? This matter has been a contentious issue in ecology in recent years. In the case of the CRRP plantations in north Queensland the modal number of species used in each plantation was eight to ten species, although up to 38 species were used at one site (Figure 38.1). Erskine *et al.* (2006) have found evidence of increased production, with increases of up to eight species in these even-aged CRRP mixtures. On the other hand, Langi (2001) and Firn (2005) have both found a decrease in productivity when comparing old (>50 years) plantations with differing levels of tree diversity (caused by natural colonization by tree species from nearby intact forest). This difference probably represents the difference between two quite different assembly processes. In the CRRP plantings all species were chosen at random from a large pool of potential species. The increases in production are probably caused by additional species being able to use resources in unutilized niches. This process has been referred to as the niche-differentiation hypothesis

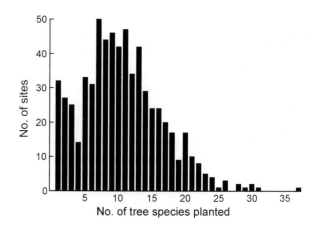

Figure 38.1 The frequency of Community Rainforest Reforestation Program sites (*n* = 658) at which plantations had differing numbers of species in the mixture. *Source*: CRRP database.

(Kinzig *et al.* 2001). In the second case, foresters had specifically chosen highly productive timber species such as *Araucaria cunninghamii* to establish the plantations; then subsequent, less productive colonists arrived and competed with, rather than complemented, the plantation species (i.e. the earliest arrival). This process has been referred to as the sampling effect hypothesis (Kinzig *et al.* 2001). The importance of the identities of species in tree mixtures was demonstrated by Nhan (2001), who studied various paired species combinations of north Queensland species and found significant increases in production with certain species groups but not others.

Unless all trees are equally valuable, large numbers of species in a mixture can have implications for the financial profitability of plantations when, at some point, the lower commercial value of additional species will cause a decline in the net financial value of the timber produced from a plantation. This occurs simply because the numbers of high value timber trees are being reduced to accommodate increased numbers of lower valued species.

It is difficult to manage the trade-off between achieving good commercial outcomes from timber production and fostering biodiversity at the same site. Different landowners will make different judgements depending on both their ecological and economic circumstances. Some of these site level trade-off issues are discussed by Erskine and Catterall (2004) and Catterall *et al.* (2005).

The legacy of more than a decade of reforestation using multi-species plantations

The reforestation programmes in north Queensland were conceived in some haste and took a while to develop methods and planning processes (Vize *et al.* 2005). Despite this early learning period, it is useful to reflect on just what has been the legacy of more than a decade of reforestation, restoration and research. There are several ways of doing this. These include an assessment of reforestation methodologies and silvicultural techniques that have been developed over that time, an assessment of the scale of the areas reforested and a review of the functional consequences of the planting programme.

Reforestation methodologies

Two distinct planting methodologies were developed. One of these was to restore plant diversity as a means of restoring biodiversity to cleared sites. The other was to devise ways of creating mixed species timber plantations. There is ample evidence that some successful assembly methods have now been developed to tackle both of these tasks. This is not to say that further research is not needed to refine both. For example, ways still need to be found to reduce establishment costs for restoration plantings. Likewise, ways of managing the mixed-species timber plantations in the longer term to generate reasonable economic outcomes and protect biodiversity are still to be explored (Brown *et al.* 2004; Erskine *et al.* 2005a).

Areas reforested

A second way of evaluating what has been achieved is to consider whether the reforestation that did occur was able to enlarge significantly the forest area and create a new timber resource or, given the relatively short time involved, whether it has fostered a planting programme that will eventually do so. In this case the answer is less positive. By 1998, in the 10 years following the cessation of logging in natural forests, around 3000 ha of reforestation had been carried out using native species (Vize *et al.* 2005). The majority of this (1850 ha) was done using CRRP funded mixed-species plantings, with the remainder (1250 ha) being restoration plantings established for biodiversity purposes. This area represents a trivial increase in overall forest area in these essentially agricultural landscapes. This was not because land deserving reforestation was not available, as estimates by Annandale *et al.* (2003) suggest that there are at least 86 000 ha of land within 200 km of Cairns that are inappropriate for sustainable agriculture and probably more suited for timber plantations. Instead, it was because those interested in reforestation face a number of significant hurdles.

One of these hurdles is that reforestation is expensive. Eono and Harrison (2002) suggest around Au$4.81 million were spent directly on reforestation by the CRRP between 1992–3 and 1994–5 (to plant 1117 ha), which represents around $4300 ha^{-1}. This expenditure excludes funds used for education and the retraining of displaced timber workers. Their analysis of the economic value of the programme concluded that it was only marginally justifiable in economic terms (generating internal rates of return of just 6.5% or 5.95% depending on whether non-timber benefits, conservation or training benefits were taken into account).

This high cost might have been acceptable if it had initiated an ongoing process of reforestation by landowners, but this was not the case. Funds for planting under the CRRP planting programme ceased after 1998 and very little additional planting has been carried out by private landowners since then. The major reason for this is probably the weak local market for cabinet timbers that exists at present. The previously more profitable market collapsed when timber supplies ceased followed the cessation of logging in natural forests. In the face of high initial costs and in the absence of a market for timber (or ecological services) most landowners have been reluctant to invest money or land in the long-term process of tree growing. Ironically, there is unlikely to be an increase in timber prices unless they do so and sufficient quantities of these timbers start appearing on the market once more. Herbohn *et al.* (2005) and Harrison *et al.* (2005) discuss some of the other socio-economic impediments to reforestation in the region, which include rights to harvest planted trees and taxation matters.

Despite these constraints, Eono and Harrison (2002) thought that many landowners were still interested in planting more trees on their properties. This is probably because they believed the timber market will eventually recover and because of the conservation benefits of tree planting. These authors suggested that a small but well targeted incentive scheme could be beneficial. This would allow a plantation resource to accumulate gradually and this might eventually begin to re-supply a market. It is worth noting that the original concept for the reforestation scheme that followed the cessation of logging in natural forest recognized that this process would need time to break the supply-price dilemma and had recommended a 30-year funding period (Shea 1992). Such long-term support had previously been needed to initiate the pine reforestation programme in southern Queensland and the sugarcane industry in northern Queensland (Lamb *et al.* 2001).

Like the timber plantations, the area reforested using restoration plantings was also small. Some of these were carried out with government funds such as the Wet Tropics Tree Planting Scheme and others were done by non-government organizations such as TREAT (Trees for Evelyn and Atherton Tablelands; website: www.treat.net.au). Estimates made of the cost of some of these plantings range from Au$15 000 to 25 000 ha^{-1} (Erskine 2002). These costs partly reflect the higher planting densities used at these plantings but they are also a consequence of the fact that WTTPS operations were dispersed over a number of shire councils. This precluded any economies of scale. Nonetheless, the high cost of restoration plantings and the absence of any direct financial return to the landowners will probably mean that these restoration plantings will always be carried out at a relatively small scale.

Functional outcomes

A third way of evaluating these past plantings is to consider their functional consequences and whether the areas planted were in locations likely to influence key ecological processes. In the case of biodiversity this might be assessed in terms of the extent to which plantings helped to enlarge small forest remnants or improved the connectivity between forest fragments. Quantitative data are limited, although Harrison *et al.* (2004) report that 55% of CRRP plantings in the Atherton and Eacham shires adjoined an existing forest area and another 20% were within 1 km of a forest area. This suggests that these plantings may have contributed to the capacity of the forests in the region to self-repair by reducing dispersal distances and fostering seed dispersal and species movement. On the other hand, it is not clear that these plantings were established in optimal locations and many large agricultural areas remain as relatively simple homogeneous landscapes. Several specific attempts to improve connectivity were made by creating corridors between areas of intact forest, although the number of these was small and their effectiveness is still being assessed (Table 38.3). Aerial photographs of one new corridor are shown in Figure 38.2.

A second important functional process concerns sedimentation and the extent to which tree plantings act as filters and reduce erosion and soil movement into streams. Both the CRRP and the Wet Tropics Tree Planting Scheme sought to overcome land degradation and address riverine erosion (e.g. Vize *et al.* 2005; North

Table 38.3 Examples of corridors established in north Queensland between 1990 and 2005

Name	Location	Design
Donaghy's Corridor	Linking Lake Barrine National Park and Gadgarra forest. Two landowners involved	Total length of 1.5 km and width of 100 m with buffer rows of *Araucaria cunninghamii* which has a deep crown and which 'seals' the edge; built along a riparian strip; a total of 101 species planted at densities of 2500–4400 tph. (Tucker 2000). Planting completed in 1999
Crater Lakes Corridor	Linking forests surrounding Lake Eacham and Lake Barrine. Eight landowners involved	A total length of 1.5 km and >25 m wide; planted at density of 2500 trees tph. A total of 42 species planted at densities of 1600–4500 tph. Completed 1998
Peterson Creek	Linking the Lake Eacham section of the Crater Lakes National Park and the Curtain Fig State Forest and Yungaburra National Park. Involves eight landowners	The project started in 1998 and some 38000 trees have been planted. The initial work planted trees in patches as stepping stones and over time the gaps are being filled in

Figure 38.2 Corridor at Petersons Creek on the Atherton Tableland. The trees are up to 8 years old (photograph: Kevin Mackay).

Queensland Joint Board undated). In the case of the CRRP, Harrison *et al.* (2004) reported that 65% of all plantings in the Atherton and Eacham Shires had a riparian component, although the absolute sizes of the areas treated were comparatively small and there are few quantitative data on whether these plantings were successful in preventing erosion or stabilizing stream banks. Nor is it clear that areas in the watersheds most in need of treatment were, in fact, treated.

Any reforestation of agricultural land has an opportunity cost for the individual landowner (e.g. farmland taken out of production) and there are contrasting views among landowners in the region on the merits of riparian reforestation. Some take the view that rivers and stream banks should be reforested for the benefit of the community, while others (probably only a vocal minority) argue that it is better to leave river banks bare of trees in order to allow floodwaters to drain away quickly. Rehabilitation treatments that benefit the community more than individual landowners probably need to be accompanied by some kind of compensation or incentive payments to increase their attractiveness to landowners.

The overriding issue influencing all of these considerations, however, is that most individual plantings were small in scale. For example, the total CRRP plantation area of 1850 ha was spread over 658 individual plantings, with 90% of these being less than 5 ha. The Wet Tropics Tree Planting Scheme plantings were also generally less than 1 ha. This occurred because in both cases there was a political need to disperse planting across each of the local government councils in the region irrespective of ecological circumstances or functional requirements

In summary, the legacy of these many site level interventions has been to increase the area of multispecies plantings across the region. Some very diverse plantings were established although at considerable expense. Some less expensive plantings were also established where biological diversity was traded off against increased levels of production in an attempt to facilitate an enlargement of the reforested area. Some of these plantings may contribute to a future commercial timber resource. All are likely to have also generated some functional benefits, although the magnitude of these benefits and the identity of the beneficiaries are quite unclear.

Assembling functionally effective landscapes

In most cases the restoration of species populations to degraded areas and changes to ecological processes such as hydrological cycles or erosion cannot be done by interventions at single sites. Instead, these require strategic reforestation at particular places across landscapes. It is also likely to be easier to make trade-offs between environmental and socio-economic goals across the mosaic of land uses in a landscape than it is at a single site. However, moving up from restoration at a particular site to restoration at a landscape scale prompts four questions:
- What is the role of the agricultural matrix surrounding forest remnants?
- How big an area of the landscape must be reforested before any functional benefits are achieved?
- Where in the landscape should these different types of reforestation be carried out?
- What type of reforestation should be done at particular locations?

The role of the agricultural matrix

The agricultural matrix surrounding forest remnants is often regarded as functionally degraded and an unsuitable habitat for any of the original species that once occupied forests in the area. The truth of the matter is often more complicated than this. Simple croplands may be quite unsuitable habitat for most wildlife but some particular species may be able to use or transit pastures that still contain scattered trees or shrubs. Likewise, regularly ploughed croplands may be highly susceptible to erosion but pastures, such as many of those used by dairy farmers on the Atherton Tablelands, might provide excellent watershed protection. That is, not all agricultural areas are identical and

the functional effectiveness of any particular part of the agricultural landscape will depend on the function in question.

The minimum area of reforestation needed

The minimum area needing to be reforested is also a matter about which it is difficult to be precise, since it too, almost certainly, depends on what function is being considered. Different prescriptions will be needed depending on whether it is soil erosion, water tables and salinity levels or biodiversity protection that is being addressed. Soil erosion may be limited, even when overall forest cover across the landscape is low, provided steep slopes and riverine areas are protected. But biodiversity conservation almost certainly requires a more substantial amount of forest cover. Some have argued that around 30% of an agricultural area should be covered by forest if biodiversity is to be protected (see reviews by Boulter *et al.* 2000; McIntyre *et al.* 2000). This may be a plausible target in situations where large areas of forest still remain and where the option of clearing or not clearing is still open. It may be a quite different situation when an extensively cleared landscape is being reforested. In these cases a 30% cover may be a very ambitious goal.

In any case, simple targets such as these can be deceptive because it is the vagility of the species concerned that is important. For example, easily dispersed wildlife species are able to tolerate much greater degrees of forest loss and fragmentation than poorly dispersed species (Turner *et al.* 2001). For most wildlife it is the organism's perception of landscape heterogeneity that is most critical rather than some supposed threshold proportion. The question then is which should be the driving factor: the requirements of a species with poor dispersal abilities or those of species with 'average' dispersal abilities?

Whatever the merits of benchmarks such as a 30% target, it is clear that further large scale reforestation in the north Queensland region is unlikely without some kind of additional payment to landowners for either the goods or services produced from reforestation. There is considerable interest in restoration planting by various non-government organizations such as TREAT or Eacham Landcare, but the areas such groups can reforest are small compared with what might be done if the market for timber or ecological services increased and trees became an attractive crop for landowners.

The spatial pattern of reforestation

The spatial pattern of residual forests and the location of new plantings are especially important in landscapes without much forest cover. Andren (1994) concluded that species were lost when habitat areas diminished but that the species loss was more dependent on landscape configuration than overall habitat area once deforestation exceeded 70% of the area. This suggests that it might sometimes be better to have five strategically located 20-ha plantations within a landscape than a single 100-ha plantation. For example, these small plantings might stabilize several spatially distinct point sources of erosion or they might allow several isolated remnants to be linked by new forest to form a wildlife corridor. The question is: which should be the priority? Should erosion sources be fixed first, should improved connectivity be the priority or should the target be to surround forest remnants with new buffer areas? Of course, there is no single prescription that suits all situations and every landscape is different. What needs to be done is to develop a range of possible reforestation scenarios and evaluate these. These evaluations might be done using various landscape metrics to assess connectivity or functioning (e.g. McGarigal & Marks 1994; Turner *et al.* 2001). The choice then depends on what spatial pattern gives the best outcome in that landscape.

However, the functional and biodiversity consequences of these spatial patterns also depend on the types of forest established at particular locations (Tucker *et al.* 2004; Kanowski *et al.* 2005; Wardell-Johnson *et al.* 2005). A monoculture plantation with a dense understorey may provide a stable watershed cover but may not be as useful at that particular site as a smaller restoration planting that creates a structurally complex and species-rich community able to sustain wildlife in an adjoining forest remnant. Of course, this will depend on the particular wildlife species, with some being more affected by structural complexity and others by spatial pattern.

Landscapes are commonly social mosaics and contain a variety of stakeholders. This usually means that any reforestation across a landscape depends on the aspirations of these landowners. Furthermore, the final outcome is then the result of a number of small decisions made in isolation by these individual landowners without reference to the likely final consequences across the landscape as a whole. There are occasions where landowners' goals and broader societal goals can coincide. For example, Ward *et al.* (2003) describe a situation where restoration plantings along grassy riparian strips have led to a decline in rodents and a substantial reduction in damage to nuts in an adjoining macadamia plantation. Similar benefits have been reported as a consequence of riparian plantings adjacent to sugar cane plantations (Qureshi & Harrison 2001; Tucker *et al.* 2004). More commonly, however, the final list of reforestation priorities and types of plantings may end up being a compromise between what is ecologically desirable and what is socially and economically possible. Of course, the latter might be modified if financial incentives or subsidies can be offered by planning agencies to the landowners to make certain choices more attractive.

Cyclones and landscape restoration

One final issue affecting the way forest landscape restoration is carried out in future in north Queensland concerns the role of cyclones. Large areas of forest, including both natural and planted forests, may be blown over during the strong winds that can occur in the region (e.g. Cyclone Larry in 2006). There is still insufficient information about the wind-firmness of trees and forests in this region to provide guidelines for restoration planning. However, some patterns are emerging from local observations and from studies in other regions. For example, some locations are more prone to windthrow than others, with trees on exposed ridges being more likely to suffer damage than those planted along rivers or in valleys. Likewise, thin, linear plantings such as those along fence lines are often more susceptible to damage than block plantings, although there appear to be some differences between species. In north Queensland, plantations of *Pinus caribaea* and *Eucalyptus pellita* were particularly susceptible to windthrow, while plantations of *Elaeocarpus*, *Agathis*, *Casuarina* and some of the *Flindersia* spp. appear to have been more stable after Cyclone Larry (M. Bristow & N. Tucker, pers. comm.). More detailed studies are needed to confirm these observations.

These differences may be caused by differences in tree shape and there is some evidence that a low tree height to diameter ratio is more stable than a high ratio. Simple ratios such as these can, of course, be manipulated by thinning (Wilson & Oliver 2000). A key question is whether the design of a plantation can affect its ability to withstand cyclones. That is, whether structurally complex mixed species plantings can withstand cyclones better than monocultures. Insufficient study has been carried out to give an unequivocal answer to this question but there is some limited evidence that they might (Mason 2002; Gardiner et al. 2005). It is likely to be some time before any of these relationships are sufficiently well understood to include them in landscape restoration planning.

Assembling landscapes and building resilience

An alternative way of approaching forest landscape restoration is to consider reforestation approaches that foster increases in ecological (and economic) resilience. Holling (1973) defined resilience as the amount of disturbance a particular ecosystem system can absorb and still remain within the same state or domain of attraction. In the present context, therefore, resilience describes the extent to which the newly reforested landscape is able to re-establish key ecological processes that maintain productivity, conserve resources and buffer the system against unexpected environmental changes. Holling's (1973) definition also includes the degree to which the system is capable of self-organization (versus a lack of organization or being a system that is largely dependent on external inputs such as fertilizers or weedicides) and how much it has a capacity for learning and adaptation.

Ecosystem resilience is usually fostered by enhanced biodiversity. However, it is not necessarily taxonomic diversity that is the most crucial element but the diversity of functional types present (Kinzig et al. 2001). Elmqvist et al. (2003) argue that these functional types should have the capacity to generate a diversity of responses to varying environmental changes. That is, there should be a capacity for back-ups and alternatives able to carry out crucial ecological functions if circumstances change. The same is likely to be true for economic resilience (Berkes & Folke 2003).

The significance of functional diversity is highlighted by the results reported by Walker et al. (1999). In this case they examined the abundance of species in a grassland where some species were abundant and others were less common. Several attributes of each species were measured to assess ecological functioning (e.g. height, biomass, specific leaf area, longevity) and the species present in the grassland were grouped into several functional types. The more dominant species were found to be functionally dissimilar (i.e. there was functional diversity) but each type had analogues among the less common species. When the grassland was heavily grazed some of the dominant species were lost but were replaced by the less common but analogous functional types. These increased in abundance and helped to maintain ecosystem function. That is, there was a response diversity (Elmqvist et al. 2003).

Most of the early CRRP plantations were established as mixed species plantations on the assumption that this would generate other conservation benefits as well as creating a timber resource. But the production–biodiversity trade-off is difficult to manage for the reasons described earlier. That is, the alpha diversity in such plantations is always likely to be constrained. On the other hand, the trade-off is much easier to manage at a landscape scale by using different species combinations at different sites in the landscape mosaic (or even monocultures of different species at different sites). Under these circumstances there is every chance of increasing the landscape (or gamma) diversity (Lamb 1998). A corollary of this is that it may be possible to enhance resilience across a landscape in a similar way. While the modal species richness (i.e. alpha diversity) in each of the CRRP planting sites was between eight and ten species, the total number of species planted in farm plantations across the region between 1992 and 1998 (i.e. the gamma diversity) was 175 species. The distribution showing the number of sites at which particular species were planted is shown in Figure 38.3. The most widely planted species were *Flindersia brayleyana* and *Elaeocarpus grandis*, which were present at 67% and 59% of the sites planted, respectively ($n = 658$ sites). On the other hand, there were 65 species found at only four or fewer sites.

There are several reasons why species such as *Flindersia brayleyana* or *Elaeocarpus grandis* were more commonly planted than others (see Table 38.4). While there is clearly scope to alter the rankings in the

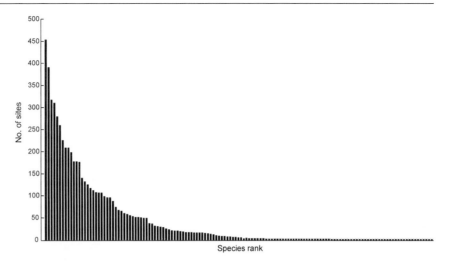

Figure 38.3 The number of individual Community Rainforest Reforestation Program sites ($n = 658$) in north Queensland at which various tree species were planted between 1992/3 and 1997/8. The total number of species used was 175. The most commonly planted species was found at more than 450 sites, while some species were present at only one or two sites.

Table 38.4 Reasons why certain species were more widely planted than others in the Community Rainforest Reforestation Program

Attribute	Specific example
It was easy to raise seedlings of these species	Species had regular, annual seeding
	Seed could be stored until required
	There were well developed nursery techniques to raise seedlings of these species
	Seedlings did not require long periods in nursery
The species were tolerant of a wide range of site conditions and were expected to grow rapidly	Species tolerated various soil conditions or altitudes
	Species was tolerant of occasional dry periods
The species were easy to manage	The ecological and silvicultural attributes were well known
	There were no significant insect problems
	There was no need for nurse crop or early shade
Timber of these species was assumed to have high market value	The species was commercially well known
	The species had desirable timber properties

frequency distribution of Figure 38.3, it is also clear that certain species will always be more attractive than others for either ecological or economic reasons. How functionally different are the more commonly planted species? The proportion of these with some simple functional properties is summarized in Table 38.5. Although most species tend to be longer lived large trees and many have dry, non-dehiscent fruit, there is considerable variation among the different species present in terms of growth, physiological attributes and reproduction mechanisms. A scan of the less widely planted species suggests that these same functional properties are well represented here as well. That is, the reason for their lower abundance is because they lack the attributes of Table 38.4 rather than because of a lack of the functional properties of Table 38.5.

Table 38.5 Proportion of the 15 most commonly planted species in the CRRP with particular physical or functional attributes

Species attribute	Unit	Proportion (%)
Tall trees when grown in plantations	>25 m	87
Fast growth	> 1 m yr^{-1}	20
Long lived	>100 yr	87
Shade tolerant		30
Able to enhance soil fertility (e.g. N-fixers)		7
Annual fruiting		100
Fleshy fruit		13
Widely dispersed seed	> 200 m	40
Dense timber	> 600 kg m^{-3}	27

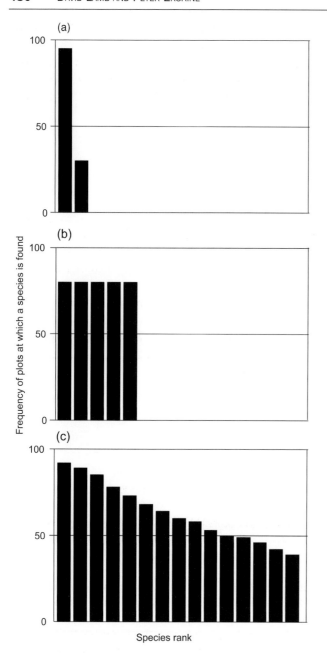

Figure 38.4 Alternative species distribution patterns for building resilience across landscapes. (a) A few species are used at most sites. This is the pattern currently used in Queensland where most of the large state-owned plantations use *Pinus caribaea* or *Araucaria cunninghamii*. (b) A mixed species plantation containing five complementary species established at a large number of separate sites. These plantations have a greater alpha diversity than (a) but the overall diversity is still low. (c) A still comparatively small number of species planted in mixtures. However, the composition of the mix used at any particular site varies with site conditions such that the alpha diversity is low but the gamma diversity is higher.

Collectively, therefore, the landscape mosaic now has a greater degree of resilience than traditional monoculture plantations would have and there has been an increase in regional biodiversity.

One must also ask if this is the most efficient way of generating a more resilient landscape? Several alternative arrangements of distributing species in the landscape are shown in Figure 38.4. The traditional timber monocultures that have been planted throughout Queensland in the past, such as *Pinus caribaea* or *Araucaria cunninghamii*, are shown as (a). In this case a small number of species are planted in most plantations and the resilience here would be low. Option (b) shows a theoretical example of a well designed mixed plantation system involving just five commercially attractive species that together form a stable and complementary mixture. This five-species design is widely replicated in many small plantations across the landscape because it is apparently successful. In this case the resilience is also low because there is little back-up in the system.

Option (c) represents a compromise between (b) and the much greater gamma diversity of the CRRP plantations shown in Figure 38.3. In this case a large number of species are used but the number at any particular site might still be only three to five species (or the plantation at this particular site might even be a monoculture), but the total diversity of species across the landscape comes from better matching of species to particular site conditions while excluding commercially unattractive species. Ecological resilience is higher than in (a) or (b) and the species used are primarily those that are commercially attractive, making it a better option than the current CRRP.

It is difficult to be precise about just how many species might be needed to achieve this balance. There are six attributes in Table 38.5 where the proportion of species represented is less than 50%. As a first approximation, therefore, perhaps the number to plant across the landscape might be two or three times the number of these functional categories (i.e. around 10–20 species).

Summary and conclusions

• A significant amount of research on reforestation with rainforest species has been carried out since logging in natural forests ceased in the late 1980s. Many of these experimental forests are comparatively young but there is already evidence that appropriate successional trajectories are developing in some of the restoration plantings. Methods have also been developed to grow a number of potentially useful timber tree species in monocultures and mixtures in farm forestry plantations.

● This is not to suggest that these reforestation technologies are settled. Instead, it is to argue that good progress has been made in a relatively short time. The task now is to monitor these new forest areas and document future developments as the stands mature. Despite over 100 years of tropical forest research there is surprisingly little useful knowledge on how to grow many native forest species and these recent plantings provide an outstanding opportunity to learn more. As these plantings age new problems will emerge and different managers will adopt contrasting solutions. All of these experiences need to be followed and recorded as part of an adaptive management process.

● While such plantings may be seen as successful from an individual landowner's perspective, they have been less successful from a landscape perspective. The two key issues are that most of the areas planted have been small and most have been located on land made available by landowners rather than on land that should have been reforested. This means that the overall impact of the reforestation effort on biodiversity conservation or watershed protection has probably been less than it might have been.

● There are general principles available for guiding how new forest areas might be spatially arranged across landscapes to optimize ecological benefits from reforestation. However, most of these assume a single benign landowner who is willing to change existing land uses to accommodate these changes. This is rarely the case. Most landscapes contain a variety of landowners who may have sharply contrasting objectives. Many landscapes also have external non-landowning stakeholders who have a legitimate interest in how the landscape is managed. This is certainly the case in north Queensland, where many people across the region are interested in conservation issues.

● This means that the real benefits to the community from reforestation will only occur when some form of collective landscape planning is carried out. This planning may allow synergistic benefits to emerge from many small plantings and should lead to the overall benefits of reforestation being greater than the sum of all the many small parts. Landscape planning on this scale has not yet been achieved in north Queensland but would have substantially increased the benefits from reforestation if it had.

Acknowledgements

We would like to acknowledge the assistance we have had over many years from a number of colleagues, including Mila Bristow, Rod Keenan, Nigel Tucker, Carla Catterall, Grant Wardell-Johnson, John Kanowski, Mark Annandale, David Yates and David Doley. We would also like to thank Susan Doust, Oliver Woldring, Tina Langi, Jennifer Firn, Sean McNamara, Bruce Dunn and all the other postgraduate students that have worked on different aspects of forest restoration. Finally, we thank Trees for the Evelyn and Atherton Tableland (TREAT) for the use of the photograph of Peterson Creek corridor.

References

Andren, H. (1994). Effects of habitat fragmentation on birds and mammals in landscapes with different proportions of suitable habitat – a review. *Oikos* **71**: 355–66.

Annandale, M., Bristow, M., Storey, R. & Williams, K. (2003). *Land Suitability for Plantation Establishment within 200 kilometres of Cairns*. Queensland Department of Primary Industries, Brisbane.

Berkes, F. & Folke, C. (eds). (2003). *Navigating Social–Ecological Systems: Building Resilience for Complexity and Change*. Cambridge University Press, Cambridge.

Boulter, S., Wilson, B., Westrup, J., Anderson, E., Turner, E. & Scanlan, J. (2000). *Native Vegetation Management in Queensland*. Department of Natural Resources, Queensland.

Bristow, M. (1996). Species–site relationships: early growth of plantation trees under varying climatic conditions in north Queensland, Honours thesis, University of Queensland, Brisbane.

Bristow, M., Annandale, M. & Bragg, A. (2005). *Growing Rainforest Timber Trees: A Farm Forestry Manual for North Queensland*. RIRDC Publication 03/010. Rural Industries Research and Development Corporation, Canberra.

Brown, P. L., Doley, D. & Keenan, R. J. (2004). Stem and crown dimensions as predictors of thinning responses in a crowded tropical rainforest plantation of *Flindersia brayleyana* R Muell. *Forest Ecology and Management* **196**: 379–92.

Catterall, C., Kanowski, J., Lamb, D., Killin, D., Erskine, P. & Wardell-Johnson, G. (2005). Trade-offs between timber production and biodiversity in rainforest plantations: emerging issues and an ecological perspective. In *Reforestation in the Tropics and Sub-tropics of Australia Using Rainforest Tree Species*, Erskine, P. D., Lamb, D. & Bristow, M. (eds). Rural Industries Research and Development Corporation, Canberra and Rainforest CRC, Cairns, pp. 206–21.

Doust, S. (2004). Seed and seedling ecology in the early stages of rainforest restoration. PhD thesis, University of Queensland, Brisbane.

Elmqvist, T., Folke, C., Nystrom, M. *et al.* (2003). Response diversity, ecosystem change, and resilience. *Frontiers in Ecology and Environment*. **1**: 488–94.

Eono, J.-C. &. Harrison, S. R. (2002). Estimation of costs and benefits of the Community Rainforest Reforestation program in north Queensland. *Economic Analysis and Policy* **32**: 69–89.

Erskine, P. D. (2002). Land clearing and forest rehabilitation in the Wet Tropics of north Queensland, Australia. *Ecological Management and Restoration* **3**: 135–7.

Erskine, P. D. & Catterall. C. (2004). *Production versus Rainforest Diversity: Trade-offs or Synergies in Farm Forestry Systems*. Cooperative Research Centre for Rainforest Ecology and Management, Rainforest CRC, Cairns.

Erskine, P. D., Lamb, D. & Borschmann, G. (2005a). Growth performance and management of a mixed rainforest tree plantation. *New Forests* **29**: 117–34.

Erskine, P. D., Lamb D. & Bristow, M. (eds) (2005b). *Reforestation in the Tropics and Sub-tropics of Australia Using Rainforest Tree Species*. RIRDC, Canberra and Rainforest CRC, Cairns.

Erskine, P. D., Lamb, D. & Bristow, M. (2006). Tree species diversity and ecosystem function: can tropical multi-species plantations generate greater productivity? *Forest Ecology and Management* **233**: 205–210.

Firn, J. (2005). What are the functional consequences of diversity in tropical plantations and secondary rainforest ecosystems? Honours thesis, University of Queensland, Brisbane.

Gardiner, B., Marshall, B., Achim, A., Belcher, R. & Wood, C. (2005). The stability of different silvicultural systems: a wind-tunnel investigation. *Forestry* **78**: 471–84.

Goosem, S. P. & Tucker, N. I. J. (1995). *Repairing the Rainforest. Theory and Practice of Rainforest Re-establishment in North Queensland's Wet Tropics*. Wet Tropics Management Authority, Cairns.

Harrison, J., Herbohn, J., Smorfitt, D. & Slaughter, G. (2005). Economic issues and lessons arising from the Community Rainforest Reforestation Program. In *Reforestation in the Tropics and Sub-tropics of Australia Using Rainforest Tree Species*, Erskine, P. D., Lamb, D. & Bristow, M. (eds). Rural Industries Research and Development Corporation, Canberra and Rainforest CRC, Cairns, pp. 245–61.

Harrison, R., Harrison, S. & Herbohn, J. (2004). An evaluation of the performance of a community rainforest reforestation program in north Queensland, Australia. In *Proceedings of Human Dimensions of Family, Farm and Community Forest International Symposium, March 29–April 1*, Baumgartner, D. (ed.). Washington State University, Pullman, WA.

Herbohn, J., Emtage, N., Harrison, S., Smorfitt, D. & Slaughter, G. (2005). The importance of considering social issues in reforestation programs. In *Reforestation in the Tropics and Sub-tropics of Australia Using Rainforest Tree Species*, Erskine, P. D., Lamb, D. & Bristow, M. (eds). Rural Industries Research and Development Corporation, Canberra and Rainforest CRC, Cairns, pp. 224–44.

Holling, C. S. (1973). Resilience and stability of ecological systems. *Annual Review of Ecology and Systematics* **4**: 1–23.

Kanowski, J., Catterall, C., Proctor, H., Reis, T., Tucker, N. & Wardell-Johnson, G. (2005). Biodiversity values of timber plantations and restoration plantings for rainforest fauna in tropical and sub-tropical Australia. In *Reforestation in the Tropics and Sub-tropics of Australia Using Rainforest Tree Species*, Erskine, P. D., Lamb, D. & Bristow, M. (eds). Rural Industries Research and Development Corporation, Canberra and Rainforest CRC, Cairns, pp. 183–205.

Keenan, R., Lamb, D., Woldring, O., Irvine, A. & Jensen, R. (1997). Restoration of plant biodiversity beneath tropical tree plantations in Northern Australia. *Forest Ecology and Management* **99**: 117–31.

Kikkawa, J. (2005). Reforestation and biodiversity in the Asia-Pacific region. In *Plantation Technology in Tropical Forest Science*, Suzucji, K., Ishii, K., Sakurai, S. & Sasaki, S. (eds). Springer-Verlag, Tokyo, pp. 247–63.

Kinzig, A. P., Pacal, S. W. & Tilman, D. (eds). (2001). *The Functional Consequences of Biodiversity*. Princeton University Press, Princeton, NJ.

Lamb, D. (1998). Large scale ecological restoration of degraded tropical forest lands: the potential role of timber plantations. *Restoration Ecology* **6**: 271–9.

Lamb, D. (2000). Some ecological principles for re-assembling forest ecosystems at degraded tropical sites. In *Forest Restoration for Wildlife Conservation*, Proceedings of a Workshop, 30 January to 4 February. University of Chiang Mai, Thailand, pp. 35–43.

Lamb, D., Huynh Duc Nhan & Erskine, P. (2005). Designing mixed-species plantations: progress to date. In *Reforestation in the Tropics and Sub-tropics of Australia Using Rainforest Tree Species*, Erskine, P. D., Lamb, D. & Bristow, M. (eds). Rural Industries Research and Development Corporation, Canberra and Rainforest CRC, Cairns, pp. 129–40.

Lamb, D. & Keenan, R. (2001). Silvicultural research and development of new plantation systems using rainforest tree species. In *Sustainable Farm Forestry in the Tropics: Social and Economic Analysis and Policy*, Harrison, S. R. & Herbohn, J. L. (eds). Edward Elgar, Cheltenham, pp. 21–34.

Lamb, D., Keenan, R. & Gould, K. (2001). Historical background to plantation development in the tropics: a north Queensland case study. In *Sustainable Farm Forestry in the Tropics: Social and Economic Analysis and Policy*, Harrison, S. & Herbohn, J. L. (eds). Edward Elgar, Cheltenham, pp. 9–20.

Langi, M. A. (2001). Nutrient cycling in tropical plantations and secondary rainforests: the functional role of biodiversity. PhD thesis, University of Queensland, Brisbane.

McGarigal, K. & Marks, B. (1994). *FRAGSTATS. Spatial Pattern Analysis Program for Quantifying Landscape Structure*. Forest Science Department, Oregon State University.

McIntyre, S., McIvor, J. & MacLeod, N. (2000). Principles for sustainable grazing in eucalypt woodlands: landscape scale indicators and the search for thresholds. In *Management for Sustainable Ecosystems*, Hale, P., Petrie, A., Malony, D. & Sattler, P. (eds). Center for Conservation Biology, University of Queensland, Brisbane, pp. 92–100.

McNamara, S. (2002). Growth and performance of native timber species in mixed species plantations in north Queensland. Honours thesis, University of Queensland, Brisbane.

Mason, W. L. (2002). Are irregular stands more windfirm? *Forestry* **75**: 347–55.

Merkel, C. (1996) The nutrition and water stress status of young rainforest plantations in north Queensland. Honours thesis, Botany Department, University of Queensland.

Nhan, H. D. (2001). The ecology of mixed species plantations of rainforest tree species. PhD thesis, University of Queensland, Brisbane.

North Queensland Joint Board (undated). *Barron River Catchment Rehabilitation Plan: Technical Report on Rehabilitation Needs*. North Queensland Joint Board, Cairns.

Qureshi, M. & Harrison, S. (2001). Economic evaluation of riparian revegetation options in north Queensland. In *Sustainable Farm Forestry in the Tropics: Social and Economic Analysis and Policy*, Harrison, S. & Herbohn, J. L. (eds). Edward Elgar, Cheltenham, pp. 147–60.

Shea, G. (1992). *New Timber Industry Based on Valuable Cabinetwoods and Hardwoods*. Consultancy Report for Councils of the Wet Tropics Region, Queensland Forest Service, Brisbane.

Sun, D., Dickinson, G. R. & Bragg, A. L. (1995). Direct seeding of *Alphitonia petrei* (Rhamnaceae) for gulley revegetation in tropical northern Australia. *Forest Ecology and Management* **73**: 249–57.

Swanborough, P. W, Doley, D., Keenan, R. J. &. Yates, D. (1998). Photosynthetic characteristics of *Flindersia brayleyana* and *Castanospermum australe* from tropical lowland and upland sites. *Tree Physiology* **18**: 341–7.

Toh, I., Gillespie, M., Lamb, D. (1999). The role of isolated trees in facilitating tree seedling recruitment at a degraded sub-tropical rainforest site. *Restoration Ecology* **7**: 288–97.

Tucker, N. I. J. (2000). Wildlife colonisation on restored tropical lands: What can it do, how can we hasten it and what can we expect? In *Forest Restoration for Wildlife Conservation*, Proceedings of a Workshop, 30 January to 4 February. University of Chiang Mai, Thailand, 279–95.

Tucker, N. I. J., Wardell-Johnson, G., Catterall, C. P. & Kanowski, J. (2004). Agroforestry and biodiversity: Improving the outcomes for conservation in tropical north-eastern Australia. In *Agroforestry and Biodiversity Conservation in Tropical Landscapes*, Schroth, G., da Fonseca, G., Harvey, C., Gascon, C., Vasconcelas, H. & Izac, A.-M. (eds). Island Press, Washington, DC, pp. 431–52.

Turner, M. G., Gardner, R. H. & O'Neill, R. V. (2001). *Landscape ecology in theory and practice: pattern and process*. Springer, New York.

Vize, S., Killin, D. &. Sexton, G. (2005). The Community Rainforest Reforestation Program and other farm forestry programs based around the utilization of rainforest and other species. In *Reforestation in the Tropics and Sub-tropics of Australia Using Rainforest Tree Species*, Erskine, P. D., Lamb, D. & Bristow, M. (eds). Rural Industries Research and Development Corporation, Canberra and Rainforest CRC, Cairns, pp. 7–22.

Walker, B., Kinzig, A. & Langridge, J. (1999). Plant attribute diversity, resilience and ecological function: the nature and significance of dominant and minor species. *Ecosystems* **2**: 95–113.

Ward, D., Tucker, N. I. J. & Wilson, J. (2003). Cost-effectiveness of revegetating degraded riparian habitats adjacent to macadamia plantations in reducing damage. *Crop Protection* **22**: 935–40.

Wardell-Johnson, G., Kanowski, J., Catterall, C., McKenna, S., Piper, S. & Lamb, D. (2005). Rainforest timber plantations and the restoration of plant biodiversity in tropical and subtropical Australia. In *Reforestation in the Tropics and Sub-tropics of Australia Using Rainforest Tree Species*, Erskine, P. D., Lamb, D. & Bristow, M. (eds). Rural Industries Research and Development Corporation, Canberra and Rainforest CRC, Cairns, pp. 162–82.

Webb, M., Bristow, M., Redell, P. & Sam, N. (2005). Nutritional limitations to the early growth of rainforest timber trees in north Queensland. In *Reforestation in the Tropics and Sub-tropics of Australia Using Rainforest Tree Species*, Erskine, P. D., Lamb, D. & Bristow, M. (eds). Rural Industries Research and Development Corporation, Canberra and Rainforest CRC, Cairns, pp. 69–82.

Willson, M. & Crome, F. (1989). Patterns of seed rain at the edge of a tropical Queensland rainforest. *Journal of Tropical Ecology* **5**: 301–8.

Wilson, J. S. &. Oliver, C. D (2000). Stability and density management in Douglas-fir plantations. *Canadian Journal of Forest Research* **30**: 910–20.

39 RESTORATION IN NORTH QUEENSLAND: RECENT ADVANCES IN THE SCIENCE AND PRACTICE OF TROPICAL RAINFOREST RESTORATION

Nigel Tucker

Biotropica Australia Pty Ltd, Malanda, Queensland, Australia

Introduction

The natural contractions of Australia's formerly widespread Gondwanan rainforest and recent anthropogenic fragmentation presents our contemporary tropical rainforest as fragments of a fragment. By the early 1980s, however, a growing awareness of the effects of this loss, and its impact on biodiversity, stimulated interest in the restoration of degraded lands, in both the tropics and sub-tropics. Led by the prescient ecologists Len Webb, Geoff Tracey, Joan Wright and Aila Keto, this change was marked by approaches that were significantly different to traditional forestry techniques. Row planted exotic species and native conifers were abandoned in favour of densely planted, local, native species and a strong emphasis was placed on the restoration of biodiversity. Specific restoration prescriptions were developed based on soil, landform and drainage, and by using existing vegetation fragments as a guide to the assemblages that could be established on a site.

These changes were accompanied by knowledge developed in other related fields, including the propagation of rainforest plants, the relationship between frugivory, dispersal and regeneration, and the process of recolonization within restored areas by vertebrates and invertebrates. The application of this knowledge to fieldwork, and the feedback resulting from monitoring activities, has resulted in the development of techniques that are highly successful in terms of both plant establishment and community recovery.

This chapter documents the development of restoration ecology expertise over the past two decades in the Wet Tropics of north-eastern Queensland. It relates the current methodologies in place to treat a range of degraded areas, the target species approach and its use in the restoration of habitat linkages, and discusses some of the recent research undertaken to improve biodiversity outcomes.

The restoration process

There are three phases involved in the restoration process: propagation and maintenance of plant stock; site preparation and planting; and the maintenance of established areas. This includes monitoring a range of variables including weed cover, plant growth and the recruitment of other plants and animals into the site.

Plant propagation and maintenance

While the propagation of plant stock for commercial wood and chip production is highly advanced, there has been comparatively little information available on the propagation of native rainforest plants. Interest in restoration provided a major impetus to the development of a range of techniques that are now applied to ensure the majority of desirable species are available for restoration at particular sites.

There are eight different treatments that are now used to induce germination depending on size, pericarp or mesocarp structure, and any known dormancy mechanisms. Table 39.1 summarizes these seed types and the techniques developed for their germination.

There are a number of species that remain difficult to germinate despite the fact that they are abundant in the areas where they occur, for example, yellow walnut (*Beilschmiedia bancroftii*). Recent research suggests that some plants may be dependent upon passage through the vertebrate gut. Webber and Woodrow (2004) found a significant increase in the rate of germi-

Table 39.1 Common seed types and germination treatments for each type

Seed type	Common family(ies)	Treatment
Seed enclosed in a hard shell or nut	Eleocarpaceae Proteaceae	Place nut end-to-end in a vice, turn slowly until cracked
Single seeds enclosed in a fleshy covering	Myrtaceae Lauraceae	Remove flesh and sow fresh
Aggregate fleshy fruits excluding *Ficus* spp.	Rubiaceae Rutaceae	Allow flesh to soften, remove seeds, sow fresh
Aggregate fleshy fruits including *Ficus* spp.	Moraceae	Allow flesh to dry, crumble material and sow
Arillate seeds in a leathery capsule	Sapindaceae	Remove capsule and aril, sow fresh
Conifers	Araucariaceae	Collect whole cones and allow seed to fall from scales, sow fresh
Pioneers	Euphorbiaceae Rhamnaceae	Soak *Alphitonia* spp. and *Acacia* spp. in boiling water. Often recalcitrant
Large (>60 mm) seeds usually with a leathery mesocarp	Lauraceae	Place individually in a single pot, sow fresh

nation within the rare plant *Ryparosa* spp. nov. 1 (Achariaceae) after seeds were fed to a southern cassowary (*Casuarius casuarius johnsonii*) at the Melbourne Zoo. *Ryparosa* germination in a nursery setting has previously been erratic. Decline in the local cassowary population, an important disperser of large fruits, may therefore be detrimental to the long-term persistence of stems located within isolated fragments no longer visited by key dispersal agents. Individual birds that are kept in zoos would be an obvious way to stimulate germination in species where gut passage may be important.

Other life forms (ferns and grasses) that might be suitable for specific surfaces are also deserving of further research. The scrambling fern jungle brake (*Dicranopteris linearis*) is common across the Asia Pacific and an ideal species for the restoration of highly degraded surfaces, including mine sites, landslips and road cuttings. Despite its ubiquity this plant has proved virtually impossible to germinate. Current research is now focusing on alternative techniques to mass produce this plant.

Plant stock maintenance

The past decade has seen a significant change in the production of plant stock for restoration projects. Much of this has been in response to the potential spread of weeds and soil pathogens, particularly into sensitive natural areas including national parks, other reserves and riparian zones. Higher standards of hygiene in relation to plants and plant production areas are now becoming more commonplace. The exotic cinnamon fungus (*Phytophthora cinnamomi*) is a federally listed threatening process for which a National Threat Abatement Plan exists. This organism affects both natural areas and some horticultural plantations (e.g. avocado) and its accidental introduction through restoration has been potentially reduced through an accreditation process managed by the Nursery Industry Association of Australia (Goosem & Tucker 1998).

Accreditation involves the adoption of a range of minimum standards. Treatment of water is required either by chemical or UV sterilization or by drawing water from depths where these organisms are not present. A soilless potting medium is required, and once plants have been potted they are maintained on racks above the ground to limit the potential for weed/pathogen introduction, and also to deter exotic flatworms and earthworms, which could also be accidentally introduced. Plants are potted into tubes with a volume of 700 ml of potting mix, and this is sufficient to maintain plants for a full 12 months until they are planted out. Regular monitoring is undertaken to check nursery run-off for nutrients, the pH of potting mixes and the sterility of the water, while regular use is made of lupins (*Lupinus* spp.) to bait for any pathogens that may be present in potting media or water. Many production facilities are now adopting these standards and some have attained formal accreditation in recognition of the benefits that accrue from improved plant health.

Provenance selection

There has been an increasing awareness of the role of provenance matching. Many species occur in a range of different plant communities (regional ecosystems). Previously there has been a tendency to collect seed from a limited number of stems or from a single individual in one geographical location then to replant progeny across a variety of locations. While distances may not be large, there are often significant differences in the altitude and substrate of disparate locations. Natural ecological barriers (rivers, mountains) have also contributed to the natural population division. There are a number of species with very disjunctive distributions resulting from historical climatic sifting, for example, the local endemic *Ristantia pachysperma* is known from two populations, one north of the Daintree River and the other south of Cairns (a distance of 150 km), and deliberate mixing of these discrete populations may not be appropriate. For this reason, most restoration practitioners now source propagation material from a minimum number of individuals, generally from a 2–5 km wide radius of the proposed restoration site. The local effects of using non-matched stock on both genetics (i.e. the potential for inappropriate mixing) and performance (i.e. whether unmatched provenances perform poorly or comparably) are, however, unknown.

Site preparation, planting and maintenance

There has been little change to the preparation and planting techniques that were developed in the mid 1980s. Preparation almost exclusively utilizes a non-residual herbicide (glyphosate) preparation, which is blanket sprayed over the site. This treatment is generally applied at least twice. The first application kills all weeds on site and generally provides a vegetation (mulch) cover over the site. A second spray is applied when seed from the soil seed bank germinates following the release from competition.

Holes are dug using a motorized auger, then 300–500 g of slow release organic fertilizer is placed at the base of the hole immediately before planting. Planting is concentrated in the wet season (December to May) to reduce the need for supplementary watering.

Because most sites have had a history of disturbance and/or are located adjacent to ongoing disturbance, weed invasion is inevitable. Post-planting maintenance of restored areas therefore focuses on the use of non-residual herbicide to control weeds that arise from the seed bank or are dispersed into the site. The grasses are the main target for control because of their competitive nature. Experience is required in the mixing and application of the herbicide to ensure spray drift does not harm planted stems. This maintenance technique requires effort over a 1–2 year period, by which time a closed canopy has formed and weeds are no longer exerting a competitive influence.

Restoration methodology

There are a number of methods used to restore forest in the Wet Tropics area. The factor that most dictates methodology is the distance from the restoration site to the nearest habitat patch. Distance plays a key role in determining the rate and nature of species that are likely to recolonize a restoration site, and hence the rate and direction of plant and animal succession within the site. Sites close to habitat patches receive more seed inputs from the frugivorous birds and mammals utilizing those patches, than sites that are distant from habitat (McKenna 2001; Tucker & Simmons 2004; White *et al.* 2004). Sites that are distant also receive seed inputs but these inputs are biased towards exotic species that are common in agricultural settings.

The effect of distance on dispersal is sufficient to separate two broad approaches to restoration: passive and active. Passive regeneration relies only on succession, and the presence of weeds in the early stage of this succession is tolerated because of their transitory nature. Active regeneration implies intervention and seeks to bypass the first stages of succession and accelerate the recruitment of a broad range of indigenous plants from a range of successional stages. The passive approach has shown promise (Tucker *et al.* 2004) and has in some cases forced reconsideration of long held beliefs regarding threats posed by plants previously considered serious environmental weeds, for example, camphor laurel (*Cinnamomum camphora*) (Neilan *et al.* 2006). Resource management goals have largely driven the active restoration approach and these include riparian zone restoration to improve water quality (Bunn *et al.* 1999) and bank stability, re-establishment of habitat linkages and reducing pest species habitat (Ward *et al.* 2003), none of which goals are mutually exclusive.

An appreciation of the role played by seed-dispersing wildlife in adding species and life form diversity has led to an increased awareness of the need to plant a year round supply of fruit resources, so that regular visitation by dispersers is encouraged. It has also led to the identification of plants whose propagules are especially attractive to frugivorous dispersers because of size or nutritional reward. This interaction has been a key influence on the type of plants selected for active restoration works. This appreciation of the role of wildlife has prompted more attention to other habitat needs in projects where a particular species may be the restoration target.

The Framework Species method

The Framework Species method (Goosem & Tucker 1995; Lamb *et al.* 1997) was developed in recognition of the potential for vertebrate-mediated dispersal to accelerate native vegetation recovery in a highly fragmented landscape. The method selects plant species for re-establishment based on a range of characteristics, including:

- the attractiveness of propagules to a wide range of vertebrate dispersers;
- regular and/or reliable production of fruit resources, some from an early age;
- the ability of their offspring to germinate and persist in other areas away from the restoration site;
- the ability to grow quickly and/or persistently in open and degraded areas;
- ease of germination and maintenance in a nursery setting;

- provision of fruit resources during periods when resources are scarce.

Pioneer plants comprise around 30% of the total species planted and most display all of the above characteristics with the exception of some species. Pioneer plants, however, are most useful because of their ability to accelerate structural habitat complexity, first through their very rapid growth and subsequent contribution to litterfall, second through their contribution as logs to the ground storey habitat following comparatively earlier mortality and finally because of the mortality-induced disturbance they create in a stand of even age and size class.

This method has been the most commonly used for closed forest restoration in the Wet Tropics area. Around 30 species are used, representing between 5% and 10% of the plant species that may be present within a particular assemblage. These are established at densities between 2500 and 3000 stems per ha. The results of this method have been documented (Tucker & Murphy 1997; Tucker *et al.* 2004). There is now an increasing trend towards the further simplification of this technique by reducing the numbers of species used. This simplification has also led to the development of the foster ecosystems approach, using pioneer monocultures to capture a site quickly and foster successional development by providing suitable germination for components of the seed rain.

The Maximum Diversity method

Non-random deforestation in north Queensland inevitably led to some areas being heavily cleared during the agricultural expansion between 1900 and 1950 (e.g. Birtles 1978; Winter *et al.* 1987). Rainforests occurring on basalt soils on the Atherton Tablelands and the well drained alluvial soils of the coastal plain were extensively cleared and, with the exception of mangrove systems, the majority of the original ecosystems occurring in these areas are either Endangered (i.e. <10% of pre-European extent) or Of Concern (10–30% of pre-European extent). Remaining fragments are usually very small (<300 ha) and are isolated from adjacent patches (Tucker 2002).

Size and isolation (i.e. surrounding matrix hostility) combine to limit the opportunities for effective dispersal into and from these fragments. Restoration plots in these areas have typically recruited very low numbers of native species, almost completely derived from the immediate surrounds, with higher numbers of exotic species (McKenna 2001; White *et al.* 2004). This outcome led to the development of the Maximum Diversity method, one implicitly accepting that in this instance diversity will not 'beget diversity', species diversity must be incorporated into the initial species matrix. In these plantings, up to 100 species may be used (around 30% of the total species pool within some assemblages), representing close to the maximum number that would be available for collection, propagation and planting in a typical season. This method is increasingly used as restoration practitioners have begun to counter this non-random loss. Much of this has been driven by the refinement and availability of vegetation mapping that has stemmed from the Vegetation Management Act (1999), showing practitioners the extent and distribution of assemblages that are naturally restricted by altitude, landform and parent material.

Direct seeding

Recent research (Doust *et al.*, 2006) has provided some insight into the role that direct seeding may play in future restoration works. While this technique has only been subject to small scale trials, the Doust *et al.* study suggests that with refinement direct seeding would be a cost-effective way to accelerate succession through the provision of woody cover over large areas. In a trial on basaltic kraznozems of the Pin Gin series, across three altitudinal gradients, Doust *et al.* demonstrated that germination of large seeded species buried below the soil surface was reliable across gradients.

More recent observations at these sites (Tucker, unpublished data) suggest that the wattles (*Acacia* spp.) are likely to be an important component of this approach. At all sites, *Acacia* has become the dominant genus. Stems have now begun to exert a competitive influence on adjacent grass communities that are being replaced by woody plants, both native and exotic. Natural regeneration patterns in direct seeded plots are now similar to those recorded in manually planted plots (e.g. Tucker & Murphy 1997; McKenna 2001; Tucker 2002; White *et al.* 2004). Differences appear to be mainly temporal; in manually planted plots natural regeneration generally commences after 18–24 months, whereas within direct seeded plots, the same patterns have taken over 4 years to establish.

Habitat linkages

An increased awareness of vegetation assemblages, their status and their degree of connectivity has been accompanied by an increased understanding of restored habitat linkages in enhancing the dispersal of plants and animals across a fragmented landscape. Connectivity has now become a central element of restoration planning and has become intertwined with the management of rare, threatened and/or iconic wildlife. The restoration of habitat linkages has emerged as a key aim of many restoration projects in the Wet Tropics area.

Plantings that form direct linkages between habitat fragments are increasingly common, with the majority of these linkages following riparian zones. In some cases, such as the Walter Hill Ranges, Donaghy's and Petersen Creek Corridors (Tucker 2000, 2001, 2002; Freeman & Seabrook 2006), restoration has been undertaken solely to meet a wildlife conservation outcome. At these locations, despite their relative proximity to native forest, over 100 species have been used in the planting matrix to maximize the availability of wildlife resources. Species that have limited effective dispersal (i.e. large fruited taxa), those that are rare or threatened and those that are known food plants of fragmentation-sensitive vertebrates (e.g. arboreal folivores, flying mammals and cassowaries) are also included to meet conservation outcomes within these linkage plantings.

Habitat linkages are designed to attract wildlife using a range of planting strategies. In addition, there is an increasing emphasis on providing nesting, resting and escape cover in the form of ground storey furniture. This takes the form of logs, rock piles and other debris. This material is placed randomly across the restoration site prior to planting. Restoration is also used adjacent to purpose built tunnels, which are designed to facilitate movement beneath major roads traversing the Wet Tropics. Linkages have also been replanted across powerline corridors to improve connectivity for forest-dependent invertebrates, birds and mammals. In these cases, restoration has focused on a range of plants that are attractive to the faunal target groups (e.g. arboreal folivores and the cassowary) and these are established to the very edge of the tunnel or cleared corridor. Tunnels are also fitted with internal furniture (e.g. ceiling ropes) to ensure that arboreal mammals do not have to traverse the ground (Goosem *et al.* 2001). In these cases, manipulating bird or animal behaviour is critical because the wrong choice when crossing roads is likely to result in mortality.

The other major advance in plot design has been the treatment of plot margins. Edge effects such as weed invasion, microclimatic change and species assemblages are well accepted and documented (Murcia 1995; Laurance & Bierregaard 1997). These effects are also present along the edges of restoration plots, commonly manifested by weed colonization and ingress. In recognition of this problem, margins are now treated by establishing a belt of lower growing shrubs that are known for a bushy habit. These species are used to seal the margin artificially and increase the survival of natural regeneration along the margin. This zone has typically shown the lowest numbers of naturally regenerated seedlings, despite the edge being the most productive zone once it has become sealed (Tucker 2002).

Indigenous food plants

Over the past decade there has been an expanding interest in the establishment of plantations of native species that have economic potential as foods, condiments, cut flowers and foliage, or other novel uses such as essential oils or didgeridoo sticks. While such plantations do not conform to restoration in the strictest sense, they offer particular benefits in the provision of structurally complex wildlife habitat and provide landholders with an alternative income source. The latter is a particularly important consideration in economically depressed rural areas.

Agroecological systems using a subset of local trees, shrubs and vines could also provide an income source for indigenous societies when established on traditional lands. Many tribal corporations have acquired former pastoral lands through the Commonwealth's Native Title Act (1993). Many of these areas contain degraded sites that would benefit from active restoration. Such systems would not require intensive management, and the subsequent invasion of the understorey by plants from the surrounding landscape can increase their structural complexity without significantly detracting from their original purpose.

There are a range of native species with commercial potential now being established in orchard configurations.

These include Atherton nut (*Athertonia diversifolia*), Davidson's plum (*Davidsonia pruriens*), lemon aspen (*Acronichya acidula*), water cherry (*Syzygium aqueum*), acid berry (*Melodorum* spp.) and the Boonjie tamarind (*Diploglottis bracteata*). This industry is very much in its infancy and the issues of continuity of supply, market acceptance, intellectual property rights and processing techniques will require significantly more investigation before a 'bush foods' industry becomes economically viable. However, as a restoration technique, indigenous food plantations offer some promise, in particular on Indigenous lands where there is an urgent need to reconnect people with country (Tucker 2005).

Hostile surfaces

There have also been significant advances in the restoration of very hostile surfaces that arise from gross disturbance events such as road construction. Cuts and embankments that arise from the construction of most linear infrastructure are problematic surfaces to restore because the surfaces are generally smooth and compacted, contain no organic matter, often contain high levels of aluminium ions that inhibit plant growth and exhibit an angle of repose that sheds both moisture and organic matter (O'Brien 2004). Where they are adjacent to ongoing disturbance, these surfaces are typically colonized by exotic species that arrest succession and spread into adjacent intact ecosystems. Some of these surfaces, particularly those on south facing slopes, remain unvegetated for many years and contribute significantly to sediment loads in adjacent streams. Such surfaces are not only sites for weed ingress; they also significantly detract from visual amenity, being an unattractive accompaniment along many roadsides through the Wet Tropics World Heritage Area.

A recent study (Biotropica Australia 2006) classified the surfaces associated with new highway construction and the decommissioning of an old road through closed forest, to enable restoration prescriptions to be developed for each. Each surface requires particular approaches based on slope, aspect and whether the area is being re-established primarily for habitat linkage purposes. The preparation, plant establishment and maintenance of each site type vary markedly, depending on the degree of modification and the

Table 39.2 Surfaces arising from road decommissioning and their restoration treatment

Surface	Site preparation and maintenance	Establishment technique
Old road surface	Remove tarmac, reform drainage and deep rip soil. Control weeds using glyphosate	Manually plant local provenance seedlings at 3000 stems per ha
Old road cuts	Selective control only on woody weeds Control grasses using glyphosate	Direct seeding and/or natural regeneration. Use woody weeds to encourage natural regeneration
Old road embankments	Selective control only on woody weeds Control grasses using glyphosate	Direct seeding and/or natural regeneration. Use woody weeds to encourage natural regeneration
Stream banks	Remove concrete and mechanically reform Secure banks above flood zone with geotextile	Manually plant with riparian species. Use 50% local *Ficus* species to rapidly secure banks
Stream beds	Remove concrete. Rebuild riprap using local gravels, rock and in-stream woody debris	Coffer dams required to maintain water quality during construction. Planting not required
New road cuts	Scarify surface to improve seed microsites	Hydromulch, using mung bean (*Munga vigna*) to add nitrogen/organic material, with sword grass (*Gahnia* spp.) and jungle brake (*Dicranopteris linearis*) as long-term cover
New road embankments	Cover bare surface with geotextile. Control grasses using glyphosate	Manually establish local provenance seedlings at 3000 stems per ha
Newly exposed margins	Control weeds using glyphosate	Fertilise emerging pioneer component of the soil seed bank
Key wildlife crossing point	Install furniture (logs, rock piles etc.) to provide nesting, resting and escape cover	Plant selected species depending on species/guild/community target
Temporary exposed surfaces	Install silt fences where appropriate	Hydromulch using sterile grasses, for example, Japanese millet (*Echinochloa esculentum*)

subsequent works required to attain a more natural restoration surface. Some surfaces, such as old road cuts and embankments, require minimal intervention, whereas stream banks and beds crossed by the old road had been concreted and the original surface must be completely rebuilt. Table 39.2 shows the basic elements of a strategy to restore each site type.

Hydromulching has typically been the most common technique to establish plants on these surfaces, although research has indicated that many hydromulched sites have little resilience to weed invasion (O'Brien 2004). This effect is especially noticeable on soils of higher fertility and on slopes with a northerly aspect where solar radiation is more concentrated. Because of their roadside position regular inputs of weeds are guaranteed at these sites, so building resilience to invasion is a key aim. The majority of hydromulching on roadside surfaces, within the Wet Tropics World Heritage Area, has involved the establishment of mostly exotic grasses. This treatment has been ineffective in directing sites towards a natural recovery path (O'Brien 2004).

Recent research (Biotropica Australia 2006) has examined the role of native species in hydromulch recipes to treat these surfaces. In a replicated glasshouse study, sub-surface soil from a new road cutting was hammer packed into pots to mimic the impermeable batter surface. Instead of the usual range of native and exotic grasses, a combination of mung bean (*Munga vigna*) and the native sword grasses (*Gahnia sieberiana, G. aspera*) were sown on to the surface and combined with the standard hydromulch components of organic matter, fertilizer and organic binders (glues). Results show that supplementary watering is sufficient to induce high rates of germination and persistence of sword grass, which germinates well in conjunction with mung bean as the cover crop. While this result has yet to be trialled in the field, it demonstrates that appropriate native plants can be found for most of the surfaces that require restoration.

Summary

Increasing sophistication in the restoration process is reflected in all phases of restoration in tropical north Queensland. The propagation and maintenance of native plants is significantly more efficient, as refinement of germination techniques and nursery hygiene continues. This has led to decreased weed spread and potential *Phytophthora cinnamomi* infection through poorly maintained plant stock. Provenance selection has been widened, though establishment techniques remain largely unchanged. With the exception of developing germination techniques for recalcitrant species, this component of restoration is at a very high standard.

An increased knowledge of the relationship between frugivory, dispersal and regeneration has focused practitioners on planting species matrices that can accelerate succession. These species are intended to accelerate succession at the restoration site and to provide a source of appropriate species to promote natural recovery in adjacent, degraded areas. Distance between fragments is a key determinant of the species composition at a site: adjacent sites require only a small subset of the original vegetation, while more distant sites require replanting as much as 30% of the pre-clearing species composition. Trials of direct seeding indicate that the technique has potential to recover large areas at a reduced cost, though establishment and weed control tasks require further study.

Improving landscape connectivity is now a key aim of most restoration works. Much of this work is based on a target species or guild approach, requiring evaluation of food plants and structural requirements to ensure that important niche requirements are met wherever possible. Restoration is also an important component of efforts to promote safe crossing through linear barriers such as roads and powerline corridors.

Hydromulch remains the favoured technique to recover sites such as road batters, temporary construction sites and minesites. Although the technique is greatly hampered by weed invasion at moister sites, refinement of species selection and fertilizing techniques should improve current outcomes.

References

Biotropica Australia P/L (2006). *The Kuranda Range Road Upgrade – A Preliminary Restoration Design*. Report to Queensland Dept of Main Roads, Cairns (peer reviewed).

Birtles, T. G. (1978). Changing perception and response to the Atherton–Evelyn rainforest environment 1880–1920. In *Proceedings of the 15th Annual Conference, Institute of Australian Geographers*, Canberra.

Bunn, S. E., Davies, P. M. & Mosisch, T. D. (1999). Ecosystem measures of river health and their response to riparian and catchment degradation. *Freshwater Biology* **41**: 333–45.

Doust, S. J., Erskine, P. D. & Lamb, D. (2006). Direct seeding to restore rainforest species: microsite effects on the early establishment and growth of rainforest tree seedlings on degraded land in the wet tropics of Australia. *Forest Ecology and Management* **234(1**–3): 333–43.

Freeman, A. D. & Seabrook, L. S. (2006). Increase in riparian vegetation along Petersen Creek, north Queensland 1938–2004. *Ecological Management and Restoration* **7**(1): 63–5.

Goosem, M., Izumi, Y. & Turton, S. (2001). Efforts to restore habitat connectivity for an upland tropical rainforest fauna: a trial of underpasses below roads. *Ecological Management and Restoration* **2**(3): 196–202.

Goosem, S. P. & Tucker, N. I. J. (1995). *Repairing the Rainforest. Theory and Practice of Rainforest Re-establishment in North Queensland's Wet Tropics*. Wet Tropics Management Authority, Cairns.

Goosem, S. P. & Tucker, N. I. J. (1998). Current concerns and management issues of *Phytophthora cinnamomi* in the rainforests of the Wet Tropics. In *Patch Deaths in Tropical Queensland Rainforests: Association and Impact of* Phytophthora cinnamomi *and Other Soil Borne Organisms*, Gadek, P. A. (ed.). CRC for Tropical Rainforest Ecology and Management, Technical Report, Cairns, pp. 9–16.

Lamb, D., Parrotta, J. A., Keenan, R. & Tucker, N. I. J. (1997). Rejoining habitat remnants: restoring degraded rainforest lands. In *Tropical Forest Remnants: Ecology, Management and Conservation of Fragmented Communities*, Laurance, W. F. & Bierregaard, R. (eds). University of Chicago Press, Chicago, pp. 366–85.

Laurance, W. F. & Bierregaard, R. O. (eds) (1997). *Tropical Forest Remnants: Ecology, Management and Conservation of Fragmented Communities*. University of Chicago Press, Chicago.

McKenna, S. (2001). Native plant species regeneration in restoration plantings on the Atherton Tablelands, North Queensland. BSc (Hons) thesis, James Cook University, Cairns.

Murcia, C. (1995). Edge effects in fragmented forests: Implications for conservation. *Trends in Ecology and Evolution* **10**: 58–62.

Neilan, W., Catterall, C. P., Kanowski, J. & McKenna, S. (2006). Do frugivorous birds accelerate rainforest succession in weed dominated old-field regrowth in sub-tropical Australia? *Biological Conservation* **129**: 393–407.

O'Brien, K. (2004). Evaluating the current design and effectiveness of remediation treatments on road cut slopes in the Wet Tropics, north Queensland. Honours thesis, James Cook University, Cairns.

Tucker, N. I. J. (2000). Linkage restoration: interpreting fragmentation theory for the design of a rainforest linkage in the humid Wet Tropics of north-eastern Queensland. *Ecological Management and Restoration* **1**(1): 35–41.

Tucker, N. I. J. (2001). Wildlife colonization on restored tropical lands: what can it do, how can we hasten it and what can we expect? In *Forest Restoration for Wildlife Conservation*, Elliott, S., Kerby, J., Blakesley, D., Hardwicke, K., Woods, K. & Anusarnsunthorn, V. (eds) International Tropical Timber Organisation and the Forest Restoration Research Unit, Chiang Mai University, Thailand, pp. 279–95.

Tucker, N. I .J. (2002). Vegetation recruitment in a restored habitat linkage in the Wet Tropics of north-eastern Queensland. Masters thesis, James Cook University, Cairns.

Tucker N. I. J. (2005). Healing country and healing relationships. *Ecological Management and Restoration* 6(2): 83–4.

Tucker, N. I. J. & Murphy, T. (1997). The effects of ecological rehabilitation on vegetation recruitment: some observations from the Wet Tropics of North Queensland. *Forest Ecology and Management* 99: 133–52.

Tucker, N. I .J. & Simmons, T. S. (2004). Animal–plant interactions in tropical restoration: observations and questions from north Queensland. In *Animal–Plant Interactions in Conservation and Restoration*, Kanowski, J., Catterall, C. P., Dennis, A. J. and Westcott, D. A. (eds). Cooperative Research Centre for Tropical Rainforest Ecology and Management, Cairns.

Tucker, N. I .J., Wardell-Johnson, G., Catterall, C. P. & Kanowski, J. (2004). Agroforestry and biodiversity: Improving the outcomes for conservation in tropical north-eastern Australia. In *Tropical Biodiversity and the Role of Agroforestry*, Scroth, G., Gascon, C., Izac, I. M. N. & Harvey, C. (eds). Island Press, Washington, DC.

Ward, D., Tucker, N. I .J. & Wilson, J. (2003). Cost effectiveness of revegetating degraded riparian habitats adjacent to macadamia orchards in reducing crop damage. *Crop Protection* 22: 935–40.

Webber, B. L. & Woodrow, I. E. (2004). Cassowary frugivory, seed defleshing and fruit fly infestation influence the transition from seed to seedling in the rare Australian rainforest tree, *Ryparosa* sp. nov. 1 (Achariaceae). *Functional Plant Biology* 31: 505–16.

White, E., Tucker, N. I. J., Wilson, J. & Meyers, N. (2004). Seedling recruitment in ten year old restoration plantings in far north Queensland. *Forest Ecology and Management* 192: 409–26.

Winter, J. W., Bell, F. C., Pahl L. I. & Atherton, R. G. (1987). Rainforest clearing in north-eastern Australia. *Proceedings of the Royal Society of Queensland* 98: 41–57.

40 RAINFOREST RESTORATION FOR BIODIVERSITY AND THE PRODUCTION OF TIMBER

Grant W. Wardell-Johnson[1,3]*, *John Kanowski*[2]*,
Carla P. Catterall[1]*, *Mandy Price*[3] and *David Lamb*[4]*

[1]Griffith School of Environment, Griffith University, Nathan, Queensland, Australia
[2]Centre for Innovative Conservation Strategies, School of Environment, Griffith University, Nathan, Queensland, Australia
[3]Natural and Rural Systems Management, University of Queensland, Gatton, Queensland, Australia
[4]School of Integrative Biology, University of Queensland, St Lucia, Queensland, Australia
*The authors were participants of Cooperative Research Centre for Tropical Ecology and Management

Introduction

Rainforest timber plantations in Australia

Australia's diverse, but fragmented, tropical and subtropical rainforests have long supported the cultural, spiritual and economic needs of their human occupants (Hill *et al.* 1999; Pannell, Chapter 4, this volume). For the first two centuries of European colonization, Australian rainforests were intensively exploited for their timbers and large areas were converted to pasture (Webb 1966; Dargavel 1995; Kanowski *et al.* 2005). The conversion of other areas of native rainforest to plantations commenced as the desired rainforest timber resource began to dwindle in the early part of the twentieth century. By the 1980s, about 50 000 ha of rainforest timber plantations had been established as monocultures in the subtropics and tropics of eastern Australia (Keenan *et al.* 1997; Kanowski *et al.* 2003, 2005). Species used were primarily hoop pine (*Araucaria cunninghamii*), with smaller areas of bunya pine (*A. bidwillii*), kauri pine (*Agathis robusta*), red cedar (*Toona ciliata*), Queensland maple (*Flindersia brayleyana*) and other cabinet timbers.

Recently, native forests and plantations have been more widely viewed as sources of multiple benefits such as timber, biodiversity, carbon sequestration, catchment protection, recreation and regional economic development (Catterall *et al.* 2005; Kanowski *et al.* 2005). In response to increasing awareness of these multiple benefits, rainforest timber plantations have, since the 1980s, mostly been established on cleared land that has proved marginal for agriculture (Lamb *et al.* 2001; Kanowski *et al.* 2005). These plantations have included monocultures of hoop pine and eucalypts, as well as mixed species plantations of rainforest cabinet timbers established by joint ventures between state forest agencies and private landholders. At the same time, other plantings to restore rainforest ecosystems and rainforest biodiversity (Tracey 1986; Kooyman 1996) have also been established in the subtropics and tropics of eastern Australia. Hence reasons for reforestation are many, with some plantings seeking to achieve multiple objectives in the one stand, and others through combination plantings at the wider or landscape scale.

Plantations for biodiversity and timber?

The multiple management objectives of rainforest reforestation may involve synergies, but in other cases may require trade-offs in the tropics as elsewhere

(Catterall *et al.* 2005). The wide array of reforestation styles implemented in recent decades allows some preliminary examination of this balance. Although the size of the plantation estate on cleared rainforest land in tropical and subtropical Australia remains relatively small, the decline in some agricultural industries in these areas has stimulated proposals to expand the resource base (Spencer *et al.* 1999; CRA/RFA Steering Committee 2000; Annandale *et al.* 2003; Kanowski *et al.* 2005). In addition, many rural landholders have planted trees since the late 1980s with an expectation of achieving multiple benefits, including both timber and biodiversity (Emtage *et al.* 2001; Harrison *et al.* 2003). While various plantation design and management options are able to be assessed against economic criteria (Herbohn & Harrison 2000; Hunt 2003), there has to date been limited scrutiny against criteria developed to assess biodiversity (Kanowski *et al.* 2005). However, such assessment deserves consideration because negative impacts will have a bearing on the public acceptance of plantation proposals, affect the willingness of landholders to participate in plantation schemes, affect the likelihood that such schemes will be supported by governments or large corporations (Emtage *et al.* 2001; Kanowski *et al.* 2005) and determine the attraction of environmentally linked funds (Binning *et al.* 2002; Lindenmayer & Hobbs 2004; Kanowski *et al.* 2005).

There may also be biodiversity (and other) benefits of an expanded plantation programme through the application of thoughtful design and management practices. Thus plantations may be a cost-effective means of restoring biodiversity to cleared rainforest landscapes (Lamb 1998), as well as increasing public acceptance for subsequent plantation proposals. In addition, the biodiversity value of plantations may soon be linked to financial benefits to landholders (Binning *et al.* 2002). However, there are currently few guidelines on the changes to plantation design, management or site configuration to enable judgements concerning biodiversity criteria (Kanowski *et al.* 2005). Without a better understanding of how to design plantations to meet multiple goals, there is a risk that investment in plantings will not achieve the outcomes desired by landholders, investors and the public (Catterall *et al.* 2004; Kanowski *et al.* 2005). Guidelines for improving biodiversity outcomes in rainforest timber plantations have been suggested (e.g. Keenan *et al.* 1997;

Lamb 1998; Lamb & Keenan 2001; Tucker *et al.* 2004; Kanowski *et al.* 2005), with systematic research on the development of such guidelines commencing in recent years. The measurement of biodiversity values in rainforest restoration has been carried out across a variety of sites representing the range of land cover types with a potential to provide timber in the tropics and subtropics of eastern Australia since 1999.

Catterall *et al.* (2005) examined synergies and trade-offs between biodiversity and production in plantations, and suggested that there has tended to be a negative relationship between the two. Kanowski *et al.* (2005) examined the on- and off-site costs and benefits of various plantation configurations for developing approaches to produce synergistic or compromise outcomes at a landscape scale. This chapter examines the interaction between biodiversity and timber production in different styles of reforestation (referred to hereafter as land cover types) with specific reference to the Wet Tropics of Australia, by building on frameworks developed by Catterall *et al.* (2005) and Kanowski *et al.* (2005). We discuss the potential for achieving both high biomass and high biodiversity through the application of the additive basal area phenomenon. We then discuss aspects of the design and management of plantations that are likely to affect the outcome of plantation programmes, including trade-offs associated with plantation design and management, those related to timber harvest cycles, those involving landscape contexts and those involving site configurations in landscapes such as the different allocations of different primary goals to different areas.

Background to trade-offs and synergies in biodiversity and timber

Biodiversity and the rainforests of the Wet Tropics

The rainforests of the Wet Tropics vary greatly in both their plant taxa and physical structure, associated with soil type, temperature, moisture, topography, elevation and latitude (Adam 1994; Metcalf & Ford, Chapter 9, this volume). Forest structure can be complex, and includes upper, middle and lower tree strata, as well as unusual features such as buttressing and strangler figs, and an array of special life forms such as epiphytes and lianas (Kanowski *et al.* 2003; Wardell-Johnson *et al.* 2005). The canopy generally reaches around 30 m in height, with some emergent species such as hoop pine reaching 50 m. These rainforests are also notable for being diverse, with nearly 90 tree species having been recorded in 5000 m² plots (Metcalf & Ford, Chapter 9, this volume). In addition, there are many species of herbs, epiphytes and vines (Wardell-Johnson *et al.* 2005; Metcalf & Ford, Chapter 9, this volume). The rainforests of the Wet Tropics cover an area of less than 1% of the nation's land mass, but include 25% of Australia's plant genera (Metcalf & Ford, Chapter 9, this volume), 30% of the marsupials and frogs, 25% of the reptiles and over 60% of the butterfly species (Williams *et al.*, Chapter 22, this volume). Of course, biodiversity is not just the number of species present in any habitat, land system or region. Instead, it can be described as species, genetic and ecosystem diversity (see Stork & Turton, Introduction, this volume, for a perspective from the Wet Tropics). An area's biodiversity may be greater if a species has multiple values in its function within an ecosystem.

Current means for assessing biodiversity attributes in replanted rainforest sites of the Wet Tropics (Wardell-Johnson *et al.* 2002, 2005; Catterall *et al.* 2004; Kanowski *et al.* 2003, 2005) have two components. First, a broad range of biodiversity attributes are measured in target plantations. These include forest structure (e.g. canopy height and cover, stem densities and diameters, special life forms), biota (e.g. birds, reptiles, particular invertebrate taxa and vascular plants) and ecological processes (e.g. seed predation patterns, litter decomposition). Second, the numerical values of attributes derived from these measurements are compared with those obtained from a set of reference sites, whose background environmental properties broadly match those of the replanted sites. These include both (F) forest and (P) pasture. The various sites being assessed against the reference sites have included:

- (OP) old monoculture plantations;
- (E) ecological plantings;
- (C) cabinet timber plantations;
- (YP) young monocultures;
- (R) unmanaged regrowth.

Of the 104 sites that include 100 m by 30 m quadrats established by Catterall *et al.* (2004) in the Wet Tropics of north-east Queensland, and in subtropical south-east Queensland and northern New South Wales, 50 were in the Wet Tropics, and covered the seven different land cover types. In this chapter, we provide some comparison between surrogates for biodiversity and for timber production derived from these sites. Further detail on biodiversity and restoration is provided by Catterall *et al.* (Chapter 41, this volume) and Kanowski *et al.* (Chapter 42, this volume).

Fragmentation, biodiversity loss and restoration

Much of the remaining natural forested or vegetated areas of rainforests on the Atherton Tablelands are fragmented (Kanowski *et al.* 2003; Catterall *et al.* 2004). This is where most of the current restoration plantings in the Wet Tropics are currently established, and where most research on biodiversity restoration in the Wet Tropics of Australia has been carried out. Fragmentation affects the ecological integrity of the fragments themselves, with impacts being determined by surrounding land use, species specialization, shape, size and history of the remnant patch (Catterall *et al.*, Chapter 41, this volume). Edge effects as well as introduced predators and invasive species also have an influence on the species composition of the patch. Proximity to other stands, particularly older stands, influences the species composition of a fragmented stand (Kanowski *et al.* 2003).

Linkages between these fragmented patches and larger stands of native vegetation are crucial for the movement of some wildlife and the distribution of some plant species (Laurance 1990, 1994; Tucker 2000; Kanowski *et al.* 2006). Reforestation not only aids in the dispersal and movement of species but also contributes to water quality, creek bed and bank stability, and creates a buffer for agricultural chemical runoff (Sweeney & Czapka 2004). Restoration programmes that have a biodiversity objective have become prominent in the landscape along creeks and in disused agricultural land, to accelerate and/or replicate natural succession (Chen *et al.* 2003; Abensperg-Traun *et al.* 2004; Sweeney & Czapka 2004). Restoration programmes recognize the importance of connectivity between remnant patches and larger stands of natural forest, and of slowing land degradation, protecting vulnerable species, preserving biodiversity and recreating suitable habitat for wildlife (Kanowski *et al.* 2003; Abensperg-Traun *et al.* 2004).

Biodiversity, biomass and productivity

Biomass plays an important role in timber volume and carbon stocks (Hunter 2001; Keller *et al.* 2001; Tickle *et al.* 2001; Coomes *et al.* 2002; Kauffman *et al.* 2003; Losi *et al.* 2003; Specht & West 2003). Biomass is an indicator of the most preferable silvicultural applications for achieving maximum potential production yield. For non-timber forest products, biomass can be an indicator of how best to design planting schemes (Nolte *et al.* 2003). There is demand for estimates of forest growth rates, prospective productivity and current forest biomass to satisfy the requirements of the United Nations Framework Convention on Climate Change (UNFCCC), the Montreal Process and the Kyoto Protocol (Tickle *et al.* 2001; Specht & West 2003). It is likely that both publicly owned native and plantation forests wishing to sell carbon credits will need to produce reliable estimates of their carbon stocks to prospective buyers in the future (Specht & West 2003).

The variation in biomass from site to site and forest type to forest type is quite substantial. For example, biomass density estimates in the Brazilian Amazon range from 155 to 464 Mg ha^{-1} (Keller *et al.* 2001), making it difficult to use a broad relationship to estimate biomass on large scales or from site to site. Kumar *et al.* (1998) found that species, management practices and stand age all impacted upon both biomass and productivity. Keenan *et al.* (1995) proposed that having a diverse range of species for timber production provides additional security from the loss of species from pathogen attack, as well as providing a greater range of products (including non-timber forest products). Cabinet timber monoculture stands are more disposed to pathogen attack than those planted in species diverse plantations (Montagnini *et al.* 1995).

Most of the biomass in rainforest is represented by trees. For example, Clark and Clark (2000) found that more than 90% of the biomass was represented by tree species in the humid lowlands of Costa Rica, while larger subcanopy vegetation such as palms accounted for about 25% of individuals but only about 6% of the total biomass. Liu *et al.* (2002) also found that forest trees accounted for 85–98% of the total biomass of the forest in montane evergreen rainforest in south-western China. The variation in biomass in the Liu *et al.* (2002) study was attributed to forest density, with the denser forest being younger. However, even in rainforest with considerable tree species richness (such as the Wet Tropics), most of the overall species richness of plants is represented by life forms other than trees. For example, Wardell-Johnson *et al.* (2005) examined the proportions of various life forms (specifically epiphytes, vines, ground plants, shrubs and canopy trees) in a range of land cover types including pasture and rainforest reference sites in the subtropics and tropics of Australia. They found that shrubs were the most species-rich group, followed by trees (30%), but with vine species also numerous.

Basal area as a surrogate for the measurement of biomass and productivity

Because of its ease of measurement and the information that it provides about stand density with respect to stem quantity and size, basal area (BA) has become a key variable in forest management (Jacobs 1953; Opie 1968) and is frequently used as a rapid indirect index of biomass (e.g. Wardell-Johnson *et al.* 2007). By comparison with neighbouring sites, BA is used as an indirect measurement of competition within the stand. This intra-stand competition may be for resources such as light, soil nutrients or water, but is usually considered as competition for space (Opie 1968). BA varies greatly between forest types (Malhi *et al.* 2006), with the highest BAs having been recorded in coastal redwood (*Sequoia sempervirens*) stands of north-west America, where they have reached up to 346 m^2 ha^{-1} (Fujimori 1977). Montagnini *et al.* (1995) found that mixed species plantations had a larger BA than did the monocultures in plantations of the humid neotropics, while Ball *et al.* (1985) also found that mixed plantations had more biomass in young plantations than monocultures. In the reference sites in the rainforests of the Wet Tropics, BA was found to average around 52 m^2 ha^{-1}, while BAs averaged over 72 m^2 ha^{-1} in monoculture hoop pine plantations (Figure 40.1).

The relationship between biodiversity and productivity in various land cover types in the Wet Topics was

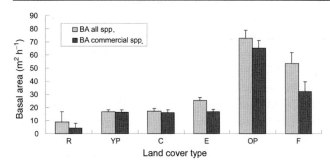

Figure 40.1 Site basal area for (a) all species, and (b) commercial species, at each land cover type.

determined by comparing various growth and form records of species in the same plots used for the recording of biodiversity measures. Forty-one of the fifty Wet Tropics sites that were sampled for taxon-based biota measures were also sampled to assess measures of biomass and productivity. The diameter of all individuals 5 cm or greater in diameter at 1.3 m height was measured in each of three 10 m radius plots in each of these sites. It was possible to analyse these data, both for all species and for all species currently considered commercial (Table 40.1).

Table 40.1 Commercial species used in commercial species analysis.

Species	Origin	Fruit type	Species	Origin	Fruit type
Acacia aulacocarpa	L	Fleshy	Eucalyptus grandis	L	Dry
Acacia melanoxylon	L	Fleshy	Eucalyptus pellita	L	Dry
Acmena smithii	L	Fleshy	Eucalytus resinifera	L	Dry
Agathis robusta	L	Dry	Flindersia acuminata	L	Dry
Alphitonia petriei	L	Fleshy	Flindersia bourjotiana	L	Dry
Alstonia muelleriana	L	Dry	Flindersia brayleyana	L	Dry
Alstonia scholaris	L	Dry	Flindersia pimenteliana	L	Dry
Aluerites moluccana	L	Fleshy	Flindersia schottiana	L	Dry
Anthocarapa nitidula	L	Fleshy	Geissois biagiana	L	Dry
Araucaria bidwillii	L	Dry	Grevillea robusta	A	Dry
Araucaria cunninghamii	L	Dry	Hylandia dockrillii	L	Fleshy
Argyrodendron peralatum	L	Dry	Khaya nyasica	E	Dry
Beilschmiedia bancroftii	L	Fleshy	Litsea bindoniana	L	Fleshy
Beilschmiedia obtusifolia	L	Fleshy	Litsea leefeana	L	Fleshy
Blepharocarya involucrigera	L	Dry	Melia azedarach	L	Fleshy
Canarium australasicum	L	Fleshy	Musgravea heterophylla	L	Dry
Cardwellia sublimis	L	Dry	Myristica insipida	L	Fleshy
Castanospermum australe	L	Dry	Nauclea orientalis	L	Fleshy
Cedrella odorata	E	Dry	Niemeyera prunifera	L	Fleshy
Celtis paniculata	L	Fleshy	Opisthiolepis heterophylla	L	Dry
Ceratopetalum succirubrum	L	Dry	Pararchidendron pruinosum	L	Dry
Chukrasia tabularis	E	Dry	Placospermum coriaceum	L	Dry
Cinnamomum camphora	E	Fleshy	Planchonella obovoidea	L	Fleshy
Cinnamomum laubatii	L	Fleshy	Pouteria obovoidea	L	Fleshy
Cinnamomum oliveri	A	Fleshy	Prumnopitys amara	L	Fleshy

(Continued)

Table 40.1 *(Continued)*

Species	Origin	Fruit type	Species	Origin	Fruit type
Cryptocarya oblata	L	Fleshy	*Prunus turneriana*	L	Fleshy
Darlingia darlingiana	L	Dry	*Pullea stutzeri*	L	Dry
Darlingia ferruginea	L	Dry	*Sloanea australis*	L	Fleshy
Doryphora aromatica	L	Dry	*Sloanea langii*	L	Fleshy
Dysoxylum mollissimum	L	Fleshy	*Sloanea macbrydei*	L	Fleshy
Dysoxylum oppositifolium	L	Fleshy	*Symplocos cochinchinensis*	L	Fleshy
Dysoxylum pettigrewianum	L	Fleshy	*Syncarpia glomulifera*	L	Dry
Elaeocarpus angustifolius	L	Fleshy	*Syzygium cormiflorum*	L	Fleshy
Elaeocarpus coorongaloo	L	Fleshy	*Syzygium gustavioides*	L	Fleshy
Elaeocarpus foveolatus	L	Fleshy	*Syzygium kuranda*	L	Fleshy
Elaeocarpus ruminatus	L	Fleshy	*Syzygium luehmannii*	L	Fleshy
Endiandra acuminata	L	Fleshy	*Syzygium papyraceum*	L	Fleshy
Endiandra cowleyana	L	Fleshy	*Terminalia sericocarpa*	L	Fleshy
Endiandra insignis	L	Fleshy	*Toona ciliata*	L	Dry
Endiandra palmerstonii	L	Fleshy	*Xanthostemon whitei*	L	Dry

The current list of species occurring within the 41 sites sampled for BA of species in the Wet Tropics considered by the Department of Primary Industries (DPI Forestry 1983) to be commercial. Information is also provided on their origin (local, Australian but non-local, exotic) and fruit type (fleshy, dispersed by vertebrates, e.g. drupes, berries, arillate seeds; dry, wind dispersed or otherwise not dispersed by vertebrates).

While there were many tree species in the 41 Wet Tropics plots, the BA tended to be dominated by relatively few species. When all species are considered (Figure 40.1), both old plantation and rainforest reference sites had high BAs, with old plantation sites the highest. Of the young plantings (YP, C and E), restoration plantings had the highest BA. When only commercial stems are considered, the old plantations have a much higher BA than other reforested sites or even rainforest reference sites (although most of the rainforest sites had been selectively logged on one to three occasions in the past 10–90 years). Young planted sites (YP, C and E sites) all show similar mean BAs for commercial species. The mean BA of commercial species in rainforest and ecological restoration sites was 60–70% of the total mean BA for those sites. However, total mean BA for timber plantations (YP, C and OP sites) was very similar to the mean BA for commercial species only.

The relationship between total plant species richness (see Wardell-Johnson *et al.* 2005) and BA across the 41 sites encompassing six land cover types (pasture is not included) in the Wet Tropics was investigated using scatter plots and correlation (Figures 40.2 and 40.3). A comparison of the total and commercial basal area against the number of stems counted at each site for each land cover type was also made (Figures 40.4 and 40.5).

In the subset of young reforested sites assessed in the Wet Tropics, total BA generally increased with species richness (YP, C and E sites, $r = 0.61$, $p = 0.006$) (Figure 40.2; see also Erskine *et al.* 2006). However, BA of commercial species in young reforested sites was independent of species richness ($r = 0.13$, $p = 0.60$) (Figure 40.3). Nevertheless, it should be noted that the list of commercial species used will have a large bearing on the outcome of such an analysis. We used the list of species occurring within the 41 sites sampled for basal area in the Wet Tropics currently considered by the Department of Primary Industries to be commercial (DPI Forestry 1983) (Table 40.1). However, we also acknowledge that there are many other uses besides timber, and that species other than trees have

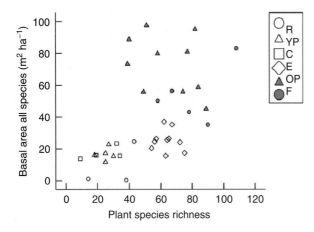

Figure 40.2 Site basal area plotted against species richness, by land cover type.

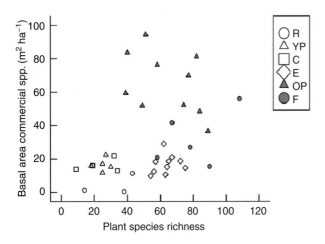

Figure 40.3 Site basal area of commercial species plotted against plant species richness, by land cover type.

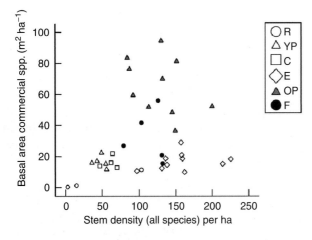

Figure 40.4 Site basal area plotted against stem density for different land cover types.

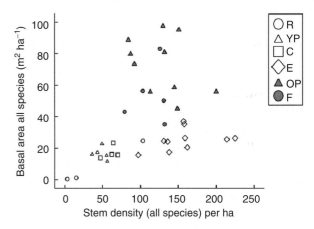

Figure 40.5 Site basal area of commercial species plotted against stem density for different land cover types.

important (including commercial) uses. Changes in technology and in economic valuation are likely to lead to an increasing proportion of total species being considered as commercial (including timber). This analysis suggests that the higher proportion of the species in an area that are considered useful or commercial, the closer productivity and biodiversity are likely to coincide.

A comparison of total BA against the number of stems counted at each site for each land cover type shows that the old plantation and rainforest sites have both large BA and high stem density (Figure 40.4). Although the BA of restoration plantings is relatively low (and also 20–50 years younger than OP sites), these plantings generally have higher stem densities than other plantings. However, while total BA generally increases with stem density for young reforested sites (YP, C and E sites) ($r = 0.67$, $p = 0.002$) (Figure 40.4), BA of commercial species in young reforested sites is independent of stem density ($r = 0.13$, $p = 0.58$) (Figure 40.5). In view of the finding that pruning, thinning and other stand management were closely associated with increased weed invasion and hence reduced biodiversity values in the Wet Tropics (Wardell-Johnson *et al.* 2005), the relationship between stem density, productivity and biodiversity is worthy of further exploration. For rainforest biodiversity, it is generally seen as an advantage to reach and maintain canopy closure early in the rotation (see Catterall *et al.* 2004, Chapter 41, this volume; Kanowski *et al.* 2005; Wardell-Johnson *et al.* 2005). This is usually achieved by close plantings and rapid early growth.

The additive BA theory

There is intense competition between a large stem and its near-neighbours in stands at full carrying capacity (Reineke 1933; Moller 1947; Smith 1962; Yoda *et al.* 1963; Drew & Felling 1977). Thus stands established with a wide range of initial stem density will, over time, produce similar volumes (Smith 1962). In practical terms the stand volume may consist of many small stems or fewer large stems so that as a fully stocked even-aged stand develops, competition between stems results in a decrease in stand density with the increase of stem volume. Thus large stems will become dominant and lead to the mortality of smaller or less tolerant stems. This loss of BA and volume of suppressed stems is substituted by the growth of the larger dominant stems. According to the law of constant yield this mortality from competition, or self-thinning, is independent of plant species (Reineke 1933; Yoda *et al.* 1963; Drew & Felling 1977).

However, Paijmans (1970) and Enright (1982) found that the BA of plots including *Araucaria hunsteinii*, in eastern Papua New Guinea, was much greater than that of those without it. These authors also found that there was minimal change in residual BA regardless as to whether or not the plot contained an emergent stem. *A. hunsteinii* is an emergent species and Paijmans (1970) proposed that emergents are extra, or additive, in terms of BA and biomass of the stand, due to this apparent contradiction to accepted theory. Enright (1982) proposed that in forests with emergents competition is reduced by minimizing the light and spatial competition between the emergent and its near-neighbours, allowing extra stacking of stand BA. Midgely *et al.* (2002) undertook the first extensive study of the additive BA theory, in a closed subtropical forest dominated by *Podocarpus latifolius*, but without emergents, in Knysna, South Africa. Midgely *et al.* (2002) proposed that the most likely cause of the additive basal area (ABA) phenomenon was the spatial efficiency of large trees. Lusk *et al.* (2003) proposed an alternative explanation for the ABA in forests without emergents based on data gathered in a closed-canopy old-growth temperate forest in the western foothills of the Andes, Chile. Lusk *et al.* (2003) suggested that the extra BA is created from large individuals having a proportionally smaller canopy than smaller stemmed individuals. He argued that this attribute enables them to pack space more efficiently and allow the BA of the rest of the forest matrix to be minimally affected.

While the cause of the ABA phenomenon has not been resolved, and may vary in different circumstances, its existence in various rainforest systems has not been questioned. If the ABA theory is generally applicable in rainforest, then restoration programmes may be able to be designed both to increase production and to provide positive biodiversity outcomes. Certainly, productivity is usually enhanced in well designed plantation programmes. The question now centres on whether it is possible to provide high levels of biodiversity, either at the same sites or in conjunction with plantations, in a landscape context.

Trade-offs and synergies in biodiversity and timber in practice

Plantation design and management

Choice of stand design and management regime has important consequences for the interaction of biodiversity and timber. There has been extensive review of the influence of plantation style by the Biodiversity Values in Reforestation team (e.g. Catterall 2000; Wardell-Johnson *et al.* 2002, 2005; Kanowski *et al.* 2003, 2005; Nakamura *et al.* 2003; Catterall *et al.* 2004, 2005; Proctor *et al.* 2004; Tucker *et al.* 2004). In particular, Catterall *et al.* (2005) provided an ecological perspective on emerging issues in rainforest reforestation associated with trade-offs and synergies in biodiversity and production. Catterall *et al.* (2005) showed that an increase in tree cover from pasture to forest is accompanied by an increase in both biodiversity and timber values. However, if only forested land is considered, then different plantation styles will have greater or lesser levels of biodiversity and/or production, depending on numerous factors.

At one end of the biodiversity – production spectrum, rainforests that are lightly harvested retain high rainforest biodiversity values, but low levels of timber production. At the other end, an intensively managed hoop pine plantation, established far from any other forest on a long-cleared site, could maximize the production of high value timber, but would support limited rainforest biodiversity. Hence, within a given area there is a continuum between styles of plantation providing extremes of high levels of biodiversity and low timber production, and low levels of biodiversity and high timber production. However, design and management compromises may provide intermediary levels of each, and in certain circumstances may be able to provide high (or low) levels of both (Catterall *et al.* 2005) (Figure 40.1). Compromises with respect to management, design and situation include the provision of non-timber trees that are particularly important to wildlife (Tucker *et al.* 2004), dense plantings with limited thinning or pruning (Wardell-Johnson *et al.* 2005), reduced levels of pruning and understorey removal, greater variety of life form and age class, and greater plantation area or patch size (Kanowski *et al.* 2003; Tucker *et al.* 2004; Catterall *et al.* 2005). These factors will all in turn be influenced by landscape-scale features such as isolation or proximity to forest and other contexts associated with the location in time and space of the plantation.

The measurement of various biodiversity components and BA (merchantable and non-merchantable) data in the same sites, in the various land cover types of the Wet Tropics, has allowed quantification of the relationship between these components of biodiversity and the timber production potential. In this study high BAs and high plant species richness suggest that there can be synergies between productivity and biodiversity in the same stand (Figures 40.2 and 40.3). However, considerable time (>50 years) is required to achieve high levels of both. Fortunately, the rainforest species currently most used in a plantation environment in the Wet Tropics (particularly hoop pine) are recognized as maintaining increment for much longer than most plantation-grown species. However, hoop pine has been well known as a plantation species for almost 100 years, while the silviculture of most native rainforest species is presently unknown. Thus there is potential to develop a forest assemblage resembling rainforest, and a forest structure that provides greater levels of productive timber than the rainforest reference sites. The maintenance of some sections of old plantation of hoop pine and other rainforest trees in the Wet Tropics would allow the ongoing study of the developmental relationship between biodiversity and production. It is likely that the ABA theory is applicable to forests that include a component of hoop pine, thus providing future advantages for both production and biodiversity. The testing of these ideas will require the retention

of sections of these old plantation forests long beyond their currently recognized optimal rotation length for timber. The retention of such old plantation forests (Figures 40.6 and 40.7) may also lead to their recognition as cultural icons, for their establishment is notable in the early history of Australian forestry (Dargavel 1995).

Timber harvest cycles

If plantation design and management only is considered, then a static view of the growth of trees and the development of a plantation and its situation in the wider spatial and temporal context of the region is assumed. Three other factors associated with biodiversity and production require consideration. These are the timber harvest cycles, the broader landscape context and allocation between various site configurations at a landscape scale.

The Biodiversity Values in Reforestation team has recorded high levels of biodiversity in plantation monocultures of rainforest timber species established

Figure 40.6 Old plantations of red cedar and hoop pine on Atherton Tablelands. A 61-year-old red cedar (*Toona ciliata*) plantation at Wongabel, north-east Queensland (photo: John Kanowski, 2001).

Figure 40.7 A 68-year-old hoop pine (*Araucaria cunning-hamii*) plantation at Gadgarra, north-east Queensland (photo: John Kanowski, 2001).

40–70 years previously in the Wet Tropics. These sites also have high BAs and high quantities of merchantable timber (see Figures 40.2 and 40.3), and presumably are also highly profitable (see Hunt 2003). These sites have biodiversity values that approach those of neighbouring rainforest, although they are unlikely to achieve the levels present in mature rainforest. These sites can be compared with other sites of the same species 10–15 years since establishment. Thus the circumstances for achieving both high levels of biodiversity and timber must be considered together with the overall harvest cycle.

Catterall *et al.* (2005) have presented a model of the development of biodiversity levels based on differential rates of biodiversity recovery or acquisition. The nature and context of a plantation will affect both its maximum biodiversity value and the rate at which biodiversity develops. Certain components of biodiversity take much longer to form or be provided than is achieved in standard rotation lengths, even within native forests where rotation lengths have tended to be longer than plantations (see Calver & Wardell-Johnson

2004; Wardell-Johnson & Calver 2005). Catterall *et al.* (2005) provided an example whereby high levels of biodiversity may be reached in a given area of plantation over 100 year rotation with fast or slow rates of biodiversity recovery. However, under a 30 year rotation with slow biodiversity development, any given area of the plantation would always have limited biodiversity value.

In this example the biodiversity value would vary greatly within any given area through the life of the plantation. However, if the plantation comprises stands of different ages, then the within-site variation would be buffered at the scale of the whole plantation, even though the long-term average would not differ greatly between different scenarios. However, a greater continuity of habitat availability could improve the overall biodiversity outcome. Since older forest stages would always be present in the plantation, it would not be necessary for species to recolonize the plantation repeatedly from suitable habitat elsewhere as each cycle progressed. However, it should be noted that for this approach to have high biodiversity value, the older stages would have to have biodiversity values that approach rainforest. Unfortunately, this is unlikely to be the case in typical timber plantations established away from forest. From a production and profitability viewpoint such a practice towards sustainable yield of a range of products, including biodiversity (see Bradshaw 1992), would also be desirable because it generates regular returns on investment, the usual practice in plantation management.

The harvesting design itself may have a large bearing on the biodiversity values in plantations designed to achieve the broader vision of a sustained supply of timber, profits and biodiversity values. For example, selectively harvesting individual stems may ensure that timber removal has less effect on biodiversity than if all stems in a stand were removed simultaneously (i.e. clear felled). The disturbance and damage to residual trees resulting from selective harvesting will be a function of the design, with trees planted in alternate rows being easier to harvest selectively than those planted (or growing) in a more intimate mix. Both the cost and efficiency of timber harvest are a consideration if the level of production is not to be compromised. Thus, individual tree removal results in the disturbance of a much larger area than clear felling of a patch for the removal of a given level of timber resource.

Of course, the scale of harvesting is critical, and for a variety of environmental reasons very large coupes can result in undesirable environmental outcomes such as erosion and microclimate changes. However, this requires consideration of patch and gap sizes rather than the less spatially defined term clear fell (see Bradshaw 1992). Where canopy gaps have been advocated (see Spellerberg & Sawyer 1997), many small gaps may be more desirable than a single large gap. This may be the situation where there are few interior dependent species. However, in rainforest environments of the Wet Tropics, the most sensitive rainforest elements that are least likely to occur in plantations tend to be interior-dependent rather than gap-dependent species. Actually, the large number of species and processes to consider, along with the limited information on biology, provide only limited opportunity for reliable predictions of the outcomes of differences in management. Thus, Catterall *et al.* (2005) advocate the empirical determination of biodiversity, timber and economic outcomes of differing forms of plantation design and management using long-term large-scale field trials.

Landscape issues in biodiversity and timber production

Both the absolute size (area and shape) of a plantation and its landscape context are important to its acquisition of rainforest biodiversity value. This is because size and context affect the ability of new species to colonize a site, as well as determining the species able to occur in particular habitat types based on their maximum home ranges. For example, while large natural areas nearby are likely to be a source of native colonists, small distant areas are not. In addition, many species are sensitive to the size of patches because small areas suffer the greatest edge effects. Small areas also suffer the greatest ongoing effects of the establishment of introduced weedy species. Shape becomes increasingly problematic in small areas because of edge effects (see Wardell-Johnson *et al.* 2005). Thus patches less than a few hectares in area are unlikely ever to support the full range of rainforest taxa, no matter what their quality, design and management, unless they are very close to larger areas of high quality rainforest habitat (Price *et al.* 1999; Catterall *et al.* 2004, 2005).

The development of species richness in a restoration planting, is accentuated by recruitment aids, such as stags (old dead trees), proximity to remnant forest and the utilisation of planted trees such as fruit bearing trees by fauna (Carnevale & Montagnini 2002; Dunmortier *et al.* 2002; Kanowski *et al.* 2003; Catterall *et al.* 2004). Kanowski *et al.* (2003) found that, apart from certain old-growth attributes such as strangler figs, old plantations and intact rainforest showed similar forest structure. In addition, the presence of species associated with intact rainforest was noted in old plantations. This is due to the recruitment of rainforest biota to the plantation from already present propagules, the proximity of the plantations to natural forest, and the connectivity properties of the plantations and natural forest (Wardell-Johnson *et al.* 2005). Wardell-Johnson *et al.* (2005) found that in younger plantings, the species richness of plants was greater in the restoration plantings than in cabinet timber and young monocultures. They attributed this to the diversity of species planted in the restoration plantings. In general, a plantation would be expected to acquire greater levels of rainforest biodiversity if its context included greater proximity to rainforest, fewer weedy species in the landscape and less exposure to adverse processes from adjacent areas (e.g. dry winds, insecticide drift or run-off, heavy grazing, fire).

Plantations can also provide off-site biodiversity benefits or negative impacts based on their landscape context, adding further complexity to the accounting of their biodiversity outcomes. Benefits include microhabitat effects and stepping stone or corridor effects, while negative affects include sources of invasive exotic species or potential sources of genetic introgression (Catterall *et al.* 2005; Kanowski *et al.* 2005).

Area also complicates the measurement of biodiversity value because species–area curves are non-linear. Thus the rate per unit area at which new species are recorded at a site declines as the surveyed area increases (Connor & McCoy 1979), affecting the contribution of the area of the plantation to its biodiversity (Catterall *et al.* 2005). Sites of the same area can be compared when they differ in other factors. Similarly, sites of the same type of planting can be compared when they differ in area. However, it is more difficult to compare quantitatively two plantations that differ in both area and type of planting. Resolving these issues will require better information on species – area accumulation curves for different types and ages of plantation (Catterall *et al.* 2005).

Biodiversity, production and the configuration of sites across landscapes

A fourth consideration in trade-offs or synergies between biodiversity and production lies in the configuration of the planted area (Catterall *et al.* 2005; Kanowski *et al.* 2005; Lamb & Erskine, Chapter 38, this volume). The consequences of altering the design, management and harvesting schedule over an area are relatively manageable if the design (e.g. tree spacing, species selection, early management) of the plantation is uniform. However, in practice, an alternative approach would be to consider the incorporation of spatial heterogeneity into the design of the plantation. This may be achieved by designing and managing some sections of the nominal plantation area for timber production with limited expected biodiversity benefit, and other sections as restoration plantings with limited expected timber production (Lamb 1998; Lamb *et al.* 2001; Lindenmayer & Franklin 2002; Catterall *et al.* 2005; Kanowski *et al.* 2005; Lamb & Erskine, Chapter 38, this volume). Kanowski *et al.* (2005) have also provided a detailed analysis of the expected on- and off-site benefits and negatives of such a scenario for the Wet Tropics in particular, by considering the range of plantation designs currently being considered. Without further empirical testing, it is not possible to say whether such area-based trade-offs could produce better production/biodiversity compromises than simply altering design and management within the plantation structure. The preferred type of trade-off is also likely to be affected by the context and nature of the plantation site (Kanowski *et al.* 2005). Area-based trade-offs could, however, be the most pragmatic from the viewpoint of balancing different primary objectives. However, spatially separating the area of production forest from ecological plantings could assist in project management and accounting procedures, especially in large-scale projects that involve both private and public sector funding (Catterall *et al.* 2005).

The rainforest landscapes of tropical and subtropical Australia offer an opportunity to compare the performance of a range of different plantation designs and approaches to management. Most of these are young (less than 20 years old), and there remains much to be learned about the rates and patterns of their biodiversity and production development. Thus the information base upon which to plan the various design,

management, context and configuration scenarios of future rainforest plantations remains limited. There is an urgent requirement for large-scale, long-term monitoring and research programmes in collaboration with broad-scale plantation and reforestation projects.

Summary

- During the past two centuries there have been major paradigm shifts in the management of Australian rainforests and the use of their timbers. This has been accompanied by similar shifts in the management and conservation of biodiversity. This chapter has discussed the ability of plantations to act as a source of both timber and rainforest biodiversity benefits.
- The additive basal area phenomenon suggests that it is possible to have high basal areas of some species, without altering the residual basal area. It may also be possible to have both high basal area and high levels of rainforest biodiversity in the same stand. Examples of high levels of both in the Wet Tropics include monocultures of hoop pine that are in close proximity to rainforest and that have been subject to limited recurrent disturbance.
- The trade-off between biodiversity and timber production can be managed by considering the problem in the broader context of plantation design and management, timber harvest cycles, the landscape context and site configurations in landscapes.
- The expansion of commercial forestry plantations can have positive biodiversity benefits where biodiversity is explicitly considered in the design, management, context and configuration of these plantations. Such consideration will facilitate planned expansion of the plantation estate.
- There are many landscape-scale plantation scenarios that can be ranked based on their positive and negative consequences for biodiversity, although in practice relative rankings may vary with landscape cover. While replanting for strictly biodiversity purposes is expensive and unlikely to be carried out over large areas, mixed purpose plantations offer the prospect of financial return and thereby increase the opportunity to reforest larger areas of cleared land.
- The information base upon which to design landscape-scale plantation scenarios remains limited, and will require large-scale, long-term monitoring and

research programmes in collaboration with broad-scale plantation and reforestation projects. Implementing such a management and research programme will require a new vision for biodiversity and trees in landscapes.

Acknowledgements

We thank the Rainforest CRC, the University of Queensland and Griffith University for supporting this research. We also thank Heather Proctor, Terry Reis, Nigel Tucker and Robert Kooyman for fieldwork, discussion and review.

References

Abensperg-Traun, M., Wrbka, T., Bieringer, G. *et al.* (2004). Ecological restoration in the slipstream of agricultural policy in the old and new world. *Agriculture, Ecosystems and Environment* **103**(3): 601–11.

Adam, P. (1994). *Australian Rainforests*. Oxford University Press: Oxford.

Annandale, M., Bristow, M., Storey, R. & Williams, K. (2003). *Land Suitability for Plantation Establishment within 200 Kilometres of Cairns*. Queensland Department of Primary Industries, Brisbane.

Ball, D. S., Whitesell, C. D. & Schubert, T. H. (1985). *Mixed Plantations of Eucalyptus and Leguminous Trees Enhance Biomass Production*. Research Paper–Pacific Southwest Forest and Range Experiment Station (EUA). PSW-175. EUA, Berkeley, Calif. 6p.

Binning, C., Baker, B., Meharg, S., Cork, S. & Kearns, A. (2002). *Making Farm Forestry Pay – Markets for Ecosystem Services*. RIRDC publication 02/005.

Bradshaw, F. J. (1992). Quantifying edge effect and patch size for multiple-use silviculture – a discussion paper. *Forest Ecology and Management* **48**: 249–64.

Calver, M. & Wardell-Johnson, G. (2004). Sustained unsustainability? An evaluation of evidence for a history of overcutting in the jarrah forests of Western Australia and its consequences for fauna conservation. In *Conservation of Australia's Forest Fauna*, Lunney, D. (ed.). Royal Zoological Society of New South Wales, Mosman, pp. 94–114.

Carnevale, N. J. & Montagnini, F. (2002). Facilitating regeneration of secondary forest with the use of mixed and pure plantations of indigenous tree species. *Forest Ecology and Management* **163**: 217–27.

Catterall, C. P. (2000). Wildlife biodiversity challenges for tropical rainforest plantations. In *Opportunities for the New Millennium. Proceedings of the Australian Forest Growers*, Snell, A. & Vize, S. (eds). Biennial Conference, pp. 191–5.

Catterall, C. P., Kanowski, J., Wardell-Johnson, G. *et al.* (2004). Quantifying the biodiversity values of reforestation: perspectives, design issues and outcomes in Australian rainforest landscapes. *Conservation of Australia's Forest Fauna*, 2nd edn. Lunney, D. (ed.). Royal Zoological Society of New South Wales, Sydney, pp. 359–93.

Catterall, C., Kanowski, J., Lamb, D., Killin, D., Erskine, P. & Wardell-Johnson, G. (2005). Trade-offs between timber production and biodiversity in rainforest plantations: emerging issues and an ecological perspective. In *Reforestation in the Tropics and Subtropics of Australia Using Rainforest Tree Species*, Erskine, P., Lamb, D. & Bristow, M. (Eds). Rural Industries Research and Development Corporation, Canberra and Rainforest CRC, Cairns, pp. 206–22.

Chen, X., Li, B.-L. & Lin, Z.-S. (2003). The acceleration of succession for the restoration of the mixed-broadleaved Korean pine forests in Northeast China. *Forest Ecology and Management* **177**: 503–14.

Clark, D. B. & Clark, D. A. (2000). Landscape-scale variation in forest structure and biomass in a tropical rain forest. *Forest Ecology and Management* **137**: 185–98.

Connor, E. F. & McCoy, E. D. (1979). Statistics and biology of the species – area relationship. *American Naturalist* **113**: 791–833.

Coomes, D. A., Allen, R. B., Scott, N. A., Goulding, C. & Beets, P. (2002). Designing systems to monitor carbon stocks in forests and shrublands. *Forest Ecology and Management* **164**: 89–108.

CRA/RFA Steering Committee (2000). Identification of Commercial Timber Plantation Expansion Opportunities in New South Wales – Upper and Lower North East CRA Region. Bureau of Rural Sciences, State Forests New South Wales and Australian Bureau of Agricultural and Resource Economics, Canberra.

Dargavel, J. (1995). *Fashioning Australia's Forests*. Oxford University Press, Melbourne.

Department of Primary Industries and Forestry (DPI) (1983). *Rainforest Research in North Queensland: A Position Paper*. Queensland Department of Forestry, Brisbane.

Drew, T. J. & Felling, J. W. (1977). Some recent Japanese theories of yield – density relationships and their applications to Monterey Pine plantations. *Forestry Science* **23**: 517–34.

Dunmortier, M., Butaye, J., Jacquemyn, H., Camp, N. V., Lust, N. & Hermy, M. (2002). Predicting vascular plant species richness of fragmented forests in agricultural landscapes in central Belgium. *Forest Ecology and Management* **158**: 85–102.

Emtage, N. F., Harrison, S. R. & Herbohn, J. L. (2001). Landholder attitudes to and participation in farm forestry activities in sub-tropical and tropical eastern Australia.

In *Sustainable Farm Forestry in the Tropics*, Harrison, S. R. & Herbohn, J. L. (eds). Edward Elgar, Cheltenham, pp. 195–210.

Enright, N. J. (1982). Does *Araucaria hunsteinii* compete with its neighbours? *Australian Journal of Ecology* 7: 97–9.

Erskine, P. D., Lamb, D. & Bristow, M. (2006). Tree species diversity and ecosystem function: can tropical multi-species plantations generate greater productivity? *Forest Ecology and Management* 233 (2/3): 205–210.

Fujimori, T.J. (1977). Stem biomass and structure of a mature *Sequoia sempervirens* stand on the Pacific coast of northern California. *Journal of the Japanese Forestry Society* 59: 435–41.

Harrison, R., Wardell-Johnson, G. & McAlpine, C. (2003). Rainforest reforestation and biodiversity benefits: a case study from the Australian Wet Tropics. *Annals of Tropical Research* 25: 65–76.

Herbohn, J. L. & Harrison, S.R. (2000). Assessing financial performance of small-scale forestry. In *Sustainable Farm Forestry in the Tropics*, Harrison, S. R. & Herbohn, J. L. (eds). Edward Elgar, Cheltenham, pp. 39–49.

Hill, R., Baird, A. & Buchanan, D. (1999). Aborigines and fire in the Wet Tropics of Queensland, Australia: ecosystem management across cultures. *Society and Natural Resources* 12: 205–23.

Hunter, I. (2001). Above ground biomass and nutrient uptake of three tree species (*Eucalyptus camaldulensis*, *Eucalyptus grandis* and *Dalbergia sissoo*) as affected by irrigation and fertiliser, at 3 years of age, in southern India. *Forest Ecology and Management* 144: 189–99.

Jacobs, M. R. (1953). *Growth Habits of the Eucalypts*. Commonwealth Government Printer, Canberra.

Kanowski, J., Catterall, C. P. & Wardell-Johnson, G. W. (2005). Consequences of broadscale timber plantations for biodiversity in cleared rainforest landscapes of tropical and subtropical Australia. *Forest Ecology and Management* 208: 359–72.

Kanowski, J., Reis, T., Catterall, C. P & Piper, S. D. (2006). Factors affecting the use of reforested sites by reptiles in cleared rainforest landscapes in tropical and subtropical Australia. *Restoration Ecology* 14: 67–76.

Kanowski, J., Catterall, C. P., Wardell-Johnson, G. W., Proctor, H. & Reis, T. (2003). Development of forest structure on cleared rainforest land in eastern Australia under different styles of reforestation. *Forest Ecology and Management* 183: 265–80.

Kauffman, J. B., Steel, M. D., Cummings, D. L. & Jaramillo, V. J. (2003). Biomass dynamics associated with deforestation, fire and conversion to cattle pasture in a Mexican tropical dry forest. *Forest Ecology and Management* 176: 1–12.

Keenan, R., Lamb, D. and Sexton, G. (1995). Experience with mixed species rainforest plantations in north Queensland. *Commonwealth Forestry Review* 74(4): 315–21.

Keenan, R., Lamb, D., Woldring, O., Irvine, A. & Jensen, R. (1997). Restoration of plant biodiversity beneath tropical tree plantations in Northern Australia. *Forest Ecology and Management* 99: 117–31.

Keller, M., Palace, M. and Hurtt, G. (2001). Biomass estimation in the Tapajos National Forest, Brazil: examination of sampling and allometric uncertainties. *Forest Ecology and Management* 154: 371–82.

Kooyman, R. M. (1996). *Growing Rainforest. Rainforest Restoration and Regeneration Recommendations for the Humid Sub-tropical Region of Northern New South Wales and South East Queensland*. Greening Australia, Brisbane.

Kumar, B. M., George, S. J., Jamaludheen, V. & Suresh, T. K. (1998). Comparison of biomass production, tree allometry and nutrient use efficiency of multipurpose trees grown in woodlot and silvopastoral experiments in Kerala, India. *Forest Ecology and Management* 112: 145–63.

Hunt, M. A. (2003). Biodiversity versus production trade-offs: a production perspective. *Production versus Rainforest Biodiversity: Trade-offs or Synergies in Farm Forestry Systems?* Rainforest CRC, Cairns.

Lamb, D. (1998). Large scale ecological restoration of degraded tropical forest lands: the potential role of timber plantations. *Restoration Ecology* 6: 271–9.

Lamb, D. & Keenan, R. (2001). Silvicultural research and development of new plantation systems using rainforest tree species. In *Sustainable Farm Forestry in the Tropics: Social and Economic Analysis and Policy*, Harrison, S. R. & Herbohn, J. L. (eds). Edward Elgar, Cheltenham, pp. 21–4.

Lamb, D., Keenan, R. & Gould, K. (2001). Historical background to plantation development in the tropics: a north Queensland case study. In *Sustainable Farm Forestry in the Tropics: Social and Economic Analysis and Policy*, Harrison, S. R. & Herbohn, J. L. (eds). Edward Elgar, Cheltenham, pp. 9–20.

Laurance, W. F. (1990). Comparative responses of five arboreal marsupials to tropical forest fragmentation. *Journal of Mammalogy* 71: 641–53.

Laurance, W. F. (1994). Rainforest fragmentation and the structure of small mammal communities in tropical Queensland. *Biological Conservation* 69: 23–32.

Lindenmayer, D. B. & Franklin, J. F. (2002). *Conserving Forest Biodiversity: a Comprehensive Multiscaled Approach*. Island Press, Washington, DC.

Lindenmayer, D. B. & Hobbs, R. J. (2004). Fauna conservation in Australian plantation forests – a review. *Biological Conservation* 119: 151–68.

Liu, W., Fox, J. E. D. & Xu, Z. (2002). Biomass and nutrient accumulation in montane evergreen broadleaved forest (*Lithocarpus xylocarpus* type) in Ailao Mountains, SW China. *Forest Ecology and Management* 158: 223–35.

Losi, C. J., Siccama, T. G., Condit, R. & Morales, J. E. (2003). Analysis of alternative methods for estimating carbon stock in young tropical plantations. *Forest Ecology and Management* **184**: 355–68.

Lusk, C. H., Jara, C. & Parada, T. (2003). Influence of canopy tree size on stand BA may reflect uncoupling of crown expansion and trunk diameter growth. *Austral Ecology* **28**(2): 216–18.

Malhi, Y., Wood, D., Baker, T. R. *et al.* (2006). The regional variation of aboveground live biomass in old-growth Amazonian Forests. *Global Change Biology* **12**: 1107–38.

Midgely, J. J., Parker, R., Laurie, H. & Seydack A. (2002). Competition among canopy tree in indigenous forests: an analysis of the additive BA phenomenon. *Austral Ecology* **27**(3): 269–72.

Moller, C. M. (1947). The effect of thinning, age and site on foliage, increment and loss of dry matter. *Journal of Forestry* **45**: 293–404.

Montagnini, F., Gonzalez, E., Porras, C. & Rheingans, R. (1995). Mixed and pure forest plantations in the humid neotropics: a comparison of early growth, pest damage and establishment costs. *Commonwealth Forestry Review* **74**(4): 306–14.

Nakamura, A., Proctor, H. & Catterall, C. (2003). Using soil and litter arthropods to assess the state of rainforest restoration. *Ecological Management and Restoration* **4**: 20–6.

Nolte, C., Tiki-Manga, T., Badjel-Badjel, S., Gockowski, J., Hauser, S. & Weise, S. F. (2003). Effects of calliandra planting pattern on biomass production and nutrient accumulation in planted fallows of southern Cameroon. *Forest Ecology and Management* **179**: 535–45.

Opie, T. E. (1968). Predictability of individual tree growth using various definitions of competing basal area. *Forest Science* **14**(3): 314–23.

Paijmans, K. (1970). An analysis of four tropical rainforest sites in New Guinea. *Journal of Ecology* **58**: 77–101.

Price, O. F., Woinarski, J. C. Z. & Robinson, D. (1999). Very large area requirements for frugivorous birds in monsoon rainforests of the Northern Territory, Australia. *Biological Conservation* **91**: 169–80.

Proctor, H. C., Kanowski, J., Wardell-Johnson, G., Reis, T. & Catterall, C. P. (2004). Does diversity beget diversity? A comparison between plant and leaf-litter invertebrate richness from pasture to rainforest. In *Proceedings 5th Invertebrate Biodiversity and Conservation Conference*. Records of the South Australian Museum, Supplement Series, 7, 267–74.

Reineke, L. H. (1933). Perfecting a stand-density index for even-aged forests. *Journal of Agricultural Reserves* **46**: 627–38.

Smith, M. J. (1962). *The Practice of Silviculture.* J. Wiley & Sons, New York. 578 pp.

Specht, A. & West, P. W. (2003). Estimation on biomass and sequesterting carbon on farm forest plantations in northern New South Wales, Australia. *Biomass and Bioenergy* **25**: 363–79.

Spellerberg, I. & Sawyer, J. (1997). Biological diversity in plantation forests. In *Conservation Outside Nature Reserves*, Hale, P. & Lamb, D. (eds). Centre for Conservation Biology, University of Queensland, Brisbane, pp. 516–21.

Spencer, R., Bums, D., Andrzejewski, K., Bugg, K. P., Lee, A. & Whitfield, A. D. (1999). *Opportunities for Hardwood Plantation Development in South East Queensland.* Bureau of Rural Sciences and Australian Bureau of Agricultural and Resource Economics, Canberra.

Sweeney, B. W. & Czapka, S. J. (2004). Riparian forest restoration: why each site needs an ecological prescription. *Forest Ecology and Management* **192**(2/3): 361–73.

Tickle, P. K., Coops, N. C., Hafner, S. D. and The Bago Science Team (2001). Assessing forest productivity at local scales across a native eucalypt forest using a process model, 3PG-SPATIAL. *Forest Ecology and Management* **152**: 275–91.

Tracey, J. G. (1986). *Trees on the Atherton Tableland: Remnants, Regrowth and Opportunities for Planting.* Centre for Resource and Environmental Studies, Australian National University, Canberra.

Tucker, N. I. J. (2000). Linkage restoration: interpreting fragmentation theory for the design of a rainforest linkage in the humid Wet Tropics of north-eastern Queensland. *Ecological Management and Restoration* **1**: 35–41.

Tucker, N. I. J., Wardell-Johnson, G., Catterall, C. P. & Kanowski, J. (2004). Agroforestry and biodiversity: improving the outcomes for conservation in tropical north-eastern Australia. In *Agroforestry and Biodiversity Conservation in Tropical Landscapes*, Schroth, G., Fonseca, G., Harvey, C. A., Gascon, C., Vasconcelos, H. & Izac, A. M. (eds). Island Press, Washington, DC, pp. 431–52.

Wardell-Johnson, G. & Calver, M. (2005). Toward sustainable management: southern Africa's Afromontane, and Western Australia's jarrah forests. In *A Forest Consciencesness*, Calver, M. C., Bigler-Cole, H., Bolton, G. C. *et al.* (eds). Millpress, Rotterdam, pp. 729–39.

Wardell-Johnson, G. Lawson, B. & Coutts, R. (2007). Are regional ecosystems compatible with floristic heterogeneity: a case study from Toohey Forest Brisbane, Australia. *Pacific Conservation Biology* **13**: 47–59.

Wardell-Johnson, G., Kanowski, J. Catterall, C. P., Proctor, H. & Reis, T. (2002). Measuring the restoration of rainforest biodiversity: a case study in research design, and its implications for establishing monitoring frameworks. *Biodiversity – The Big Picture. Proceedings of the South-east Queensland Biodiversity Recovery Conference,* November 2001.

Wardell-Johnson, G. W., Kanowski, J., Catterall, C., McKenna, S., Piper, S. & Lamb, D. (2005). Rainforest timber plantations and the restoration of plant biodiversity in tropical and subtropical Australia. In *Reforestation in the Tropics and Subtropics of Australia using Rainforest Tree Species,* Erskine, P., Lamb, D. & Bristow, M. (eds). Rural Industries Research and Development Corporation, Canberra and Rainforest Cooperative Research Centre, Cairns, pp. 162–82.

Webb, L. J. (1966). The rape of the forests. In *The Great Extermination. A Guide to Anglo-Australian Cupidity, Wickedness and Waste,* Marshall A. J. (ed.). Heinemann, London, pp. 156–205.

Yoda, K., Kira, T., Ogawa, H. & Hozumi, K. (1963). Self-thinning in over crowed pure stands under cultivated and natural conditions (intra-specific competition among higher plants). *Journal of Biology,* Osaka City University **14**: 107–29.

41 BIODIVERSITY AND NEW FORESTS: INTERACTING PROCESSES, PROSPECTS AND PITFALLS OF RAINFOREST RESTORATION

Carla P. Catterall[1]*, John Kanowski[2]* and Grant W. Wardell-Johnson[1,3]*

[1]Griffith School of Environment, Griffith University, Nathan, Queensland, Australia
[2]Centre for Innovative Conservation Strategies, School of Environment, Griffith University, Nathan, Queensland, Australia
[3]Natural and Rural Systems Management, University of Queensland, Gatton, Queensland, Austalia
*The authors were particulars of Cooperative Research for Tropical Ecology and Mangement.

The challenge of reforestation and biodiversity

Processes of deforestation and reforestation

To reverse the effects of the worldwide clearing of tropical rainforests, large-scale reforestation is required (Dobson *et al*. 1997; Chazdon 2003; Lamb *et al*. 2005). Widespread clearing has caused significant losses of biodiversity at local, landscape and global scales, as well as causing impacts on soils, climate and hydrological processes (Laurance & Bierregard 1997; Mooney *et al*. 2005). Australian tropical rainforest landscapes are no exception. Following European settlement in the late nineteenth century, most land that could yield a short-term profit from agricultural development was cleared and farmed, even when its real agricultural potential was marginal (Webb 1966; Winter *et al*. 1987; Lamb *et al*. 2001; Catterall *et al*. 2004; Turton, Chapter 5, this volume). Most of the remaining rainforest occurs on steep mountain slopes, and a number of vegetation types have been reduced to less than 10% of their former extent (Goosem *et al*. 1999). With growing scientific and public awareness of the consequences of deforestation, most of the remaining Wet Tropics rainforests are now either reserved or protected from clearing through government regulations (Erskine 2002; Tucker *et al*. 2004).

Simultaneously, there has been an increasing occurrence of reforestation (Erskine 2002; Tucker *et al*. 2004; Catterall *et al*. 2004, 2005), driven in part by increases in local community interest in biodiversity and conservation. Additionally, the national and global economic conditions of the early twenty-first century offer improved prospects for reforestation. It is more widely recognized that rainforest within production landscapes may provide economic benefits (e.g. through pest mitigation, crop pollination, tourism, future timber or carbon trading). Moreover, the previously dominant primary industries of dairy and sugarcane farming have become much less profitable. In some areas where land has been abandoned from production, woody regrowth is increasingly occurring, often with a novel tree species mix, which includes both native plants and invasive exotics (Erskine *et al*., 2007). The term 'new forests' has been applied to such vegetation in tropical America (Lugo & Helmer 2004). In Australia, we here apply the term to all forms of reforestation on land formerly used for production or human settlement, including those in which there has been active intervention to accelerate forest development.

The main method of active reforestation has been tree planting, encouraged since the late 1980s by a range of government funding schemes, and practised by a growing number of private landholders, community organizations and government agencies (Catterall *et al.* 2004; Tucker *et al.* 2004). The local goals of these efforts have varied widely, spanning biodiversity and wildlife habitat, riparian and stream condition, mitigation of erosion, timber production, farm windbreaks and aesthetic considerations. Goals of individual projects have often been mixed; for example, many landholders established timber plantations with a concurrent desire to improve biodiversity (Emtage *et al.* 2001). However, only a small area of land has been reforested. Satellite imagery analysis revealed around 3.7 km^2 of apparent new regrowth between 1991 and 1999 (Erskine 2002). The overall area of cleared land on which active replanting efforts had reinstated vegetation up to 2002 is around 10 km^2, which is equivalent to 0.5% of the 1800 km^2 of rainforest estimated to have been cleared from the region (Catterall & Harrison 2006).

Promises and pitfalls of reforestation

Developing a scientific understanding of reforestation processes, which is sufficient to provide a sound basis for biodiversity planning and management actions, poses considerable challenges. The consequences of deforestation are much better understood than those of reforestation (Young 2000). For example, it is well known that deforestation causes fragmentation, which leads to biodiversity loss (Laurance & Bierregaard 1997). However, there is unresolved debate concerning how much of this loss depends on each of four interrelated aspects of fragmentation (Figure 41.1). These involve decreases in total landscape-scale habitat area, remnant patch sizes, functional linkages among patches and habitat quality within patches (Harrison & Bruna 1999; Fahrig 2003). Reforestation may assist recovery of rainforest biodiversity by improving any of these factors, but the field of restoration ecology has yet to reach a stage where it can provide clear predictions of the consequences. Without such guidance, practical restoration priorities and plans must rely on educated guesswork, or use ecological theories (such as corridor concepts) whose effectiveness remains largely untested (Tucker 2000).

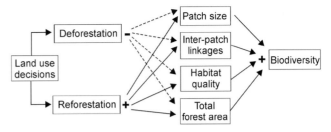

Figure 41.1 Processes whereby anthropogenic deforestation and reforestation affect rainforest biodiversity. The four central boxes refer to characteristics of forest cover (the top three relating to remnant patches, and the fourth to the whole landscape); + and − signs are directions of effects. Note that linkages may be functional rather than physical.

In the Wet Tropics, land clearing occurred over little more than a century, a timespan that is considerably less than the lifetime of many rainforest trees. Furthermore, other rapid changes have also occurred, including the introduction and spread of new species, changes in climate and the abandonment of Aboriginal burning practices (Catterall *et al.* 2004). It is therefore unlikely that conditions of ecological equilibrium can be assumed even for areas of retained intact forest. For example, even if clearing has stopped, remnant forest patches are likely to be currently experiencing ongoing biodiversity loss, due to the time-lagged consequences of past deforestation of surrounding or nearby areas (the so-called 'extinction debt') (Tilman *et al.* 1994). Furthermore, there is acute ecological disequilibrium in the modified parts of the landscape. Therefore, observed biodiversity outcomes in differently aged areas of past reforestation may provide a poor predictor of the likely future outcomes of current reforestation activities.

Nevertheless, management actions cannot be postponed until better scientific knowledge is available. Since 2000, research into the ecological processes associated with reforestation in the Wet Tropics and elsewhere has rapidly increased. This work has begun to reveal the factors that can improve or limit biodiversity outcomes in reforested areas. In the remainder of this chapter we first consider what factors are most likely to influence biodiversity outcomes in rainforest reforestation. Second, we look at the ways in which the developing biodiversity feeds back to influence the restoration process. Finally, we consider implications for future management and the need for a large-scale experimental approach to the practice of ecological restoration. While the focus is on the Wet Tropics, we also draw upon recent results from similar landscapes in the Australian subtropics and relevant research from other continents.

What drives the biodiversity benefits from reforestation?

Pathways to rainforest biodiversity

Any change in vegetation cover is likely to lead to a change in biodiversity, with some components being lost, while others are gained. In the context of rainforest restoration, 'rainforest biodiversity value' can be conceptualized as a rainforest-like set of biota, ecological processes and physical structures (Catterall *et al* 2004). Many of these are lost when rainforest is cleared, and achieving a biodiversity benefit from reforestation depends on regaining them. In Australia, uncleared rainforest forms a spatial mosaic with more open-canopied and biologically dissimilar sclerophyll (e.g. eucalypt-dominated) forest and woodlands. In this case, if a reforested site acquired a biota typical of eucalypt forest, then it would not have been restored as rainforest. To assess the development of rainforest biodiversity within reforested areas, it is necessary to place them in relation to specified reference conditions (Parker & Pickett 1997; Catterall *et al*. 2004), such

as pasture or cropland at one end of the impact spectrum and rainforest at the other (see Kanowski *et al.*, Chapter 42, this volume).

The reforestation pathways that can lead to the redevelopment of rainforest biodiversity are themselves diverse. They vary in features such as starting conditions (e.g. the time elapsed since the former rainforest was cleared and the type of intervening land use), method of initial forest re-establishment (e.g. planting trees, seeds, unassisted regrowth), details of establishment (e.g. numbers and types of plant species) and the subsequent management (e.g. whether exotic plants are removed) (see Ashton *et al.* 2001; Lamb *et al.* 2005). These interact to determine the rate and trajectory of biodiversity development that takes place during reforestation. The biodiversity outcomes are also influenced by factors related to spatial and landscape context, such as patch size, proximity to existing forest and surrounding land use. Further, the reforestation process may itself be either anthropogenic, where there is active human intervention, or autogenic, in which self-organizing biological processes lead to forest regrowth in the absence of human intervention (Erskine *et al.*, 2007).

For Wet Tropics rainforests, there are currently three broad types of potential reforestation pathways: ecological restoration plantings, timber plantations and regrowth management (Table 41.1 and Table 38.2 in Lamb & Erskine, Chapter 38, this volume). There is considerable variation within each, and they intergrade

Table 41.1 Types of reforestation pathway for Wet Tropics rainforests

Name	Description and selected references	Establishment characteristics and current extent in tropical and subtropical Australia
Ecological restoration plantings	Seedlings or seeds are planted in order to rapidly create native rainforest-like vegetation (2, 4, 6, 7, 9, 13, 14)	Typically there are high densities of seedlings (c. 3000–6000 per ha), including a diversity of rainforest plants (often 20–50+ species). Often in small patches (<5 ha) or linear strips, on steep banks of waterways. Total area around 10 km²
Timber plantations	Seedlings of timber trees are planted, with a primary goal of timber production; includes a variable proportion of rainforest elements (3, 4, 5, 6, 7, 8, 10, 11, 12, 13)	Tree seedlings (typically 1–10 species, including some non-rainforest or exotic species) are sufficiently spaced to assist rapid timber production (c. 400–1000 per ha). Plantation area varies, total area >500 km²
Regrowth management	Self-organizing recruitment and growth of woody plants follows land abandonment; this may be managed to produce a rainforest-like vegetation (1, 6, 7, 13, 15, 16, 17)	Seedling density and diversity vary greatly with context (duration of cleared phase and location in relation to remnant forest); may be dominated by exotic species. Total area <1 km², although unmanaged regrowth area is much larger, perhaps thousands of km²

Sources: Wet Tropics: 1, Gilmour and Riley (1970), 2, Goosem and Tucker (1995), 3, Keenan *et al.* (1997), 4, Lamb *et al.* (1997), 5, Lamb *et al.* (2001), 6, Kanowski *et al.* (2003a), 7, 8, Catterall *et al.* (2004, 2005), 9, Florentine and Westbrooke (2004), 10, Tucker *et al.* (2004), 11, Kanowski *et al.* (2005b), 12, Erskine *et al.* (2005), 13, Erskine *et al.* (2007). Subtropics: 14, Kooyman (1996), 15, Woodford (2000), 16, Big Scrub Rainforest Landcare Group (2005), 17, Neilan *et al.* (2006). General overviews are given in 6,7 and 13.

in practice. They differ in their main goals, method of initial establishment, nature and extent of human intervention and the financial costs and benefits associated with their establishment. They also differ in the types and rates of biological development that they undergo; for example, there are differences in the patterns of development of floristic composition and physical structure. All occur mainly on land of relatively low agricultural value. Timber plantations have the longest history and occupy the largest area in the region, with early plantations established on the Atherton Tableland from the early 1900s. Ecological restoration plantings have been undertaken since the 1980s, although they have mostly developed since the 1990s and cover a smaller area. To date, there has been little active management of the increasing area of regrowth in the region.

Developmental factors at the site scale

Irrespective of the pathway involved (Table 41.1), a set of common factors will be involved in biodiversity outcomes from reforestation at the site level (Figure 41.2; see Tucker 2001; Lindenmayer & Franklin 2002; Catterall *et al.* 2004; Tucker *et al.* 2004; Kanowski *et al.* 2005a). The specific values of these factors will be influenced by the interacting outcomes of the establishment method (Table 41.1), the age of a site and the nature of ongoing management. The development of a rainforest-like physical structure, especially the closure of the tree canopy, plays an important role in the early stages of site development.

Canopy closure is a significant threshold for a site's biodiversity development, because an open canopy allows light to penetrate to ground level, which enables a dense growth of grasses and herbs (including vigorous introduced species of grass genera such as *Imperata*, *Melinus* and *Panicum*). These grasses inhibit the survival and growth of rainforest seedlings, probably through some combination of competition, microclimatic modification and fire-enhancement (Toh *et al.* 1999; Florentine & Westbrooke 2004; Erskine *et al.*, unpublished). After canopy closure the ground layer becomes shady and litter-covered, which suppresses grass and herb growth but favours the survival and growth of rainforest seedlings. Furthermore, soils that have been grazed or farmed for decades have lost their store of rainforest seeds and seedlings (Hopkins &

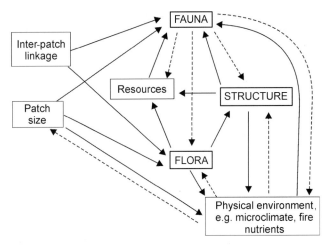

Figure 41.2 Major factors and pathways of cause and effect that contribute to outcomes for rainforest biodiversity from reforestation at the site level. Capital letters show attributes that undergo developmental changes as the site ages. Structure refers to the physical structure of the vegetation. Resources are materials needed by fauna (e.g. food, nest sites). Decisions concerning the method of vegetation establishment will affect the flora and structure. Decisions concerning site location affect linkage, patch size and the physical environment. Management could modify any factor. Feedbacks that are likely to occur during site development are shown with dotted lines.

Graham 1984), and dispersal of new seeds is needed to overcome this. A dense tree canopy provides suitable habitat for fruit-eating rainforest birds that carry in the seeds of rainforest plants, providing a positive feedback to the development of the site's flora.

Reforestation pathways in which canopy closure is achieved at a younger site age should therefore rapidly provide suitable conditions for the recruitment, survival and growth of rainforest plants. This principle has been the basis for a recommended high planting density (1–2 m between plants) in ecological restoration plantings, informed and tested by the field trials of restoration practitioners (Goosem & Tucker 1995; Kooyman 1996). Neither timber plantations nor regrowth reach canopy closure within the 3–5 years (or less) achieved by restoration plantings (Kanowski *et al.* 2003a; Catterall *et al.* 2004, 2005).

Canopy closure also contributes to the maintenance of a cool and humid microclimate that is required by many species which depend on rainforest habitat (Grimbacher *et al.* 2006; Kanowski *et al.* 2006b). Other aspects of physical structure, such as foliage density,

---

dead timber, leaf litter, trunk crevices and areas of vine tangle or high stem density also contribute important elements of the habitat required by many rainforest-dependent animals (Kikkawa 1990; Jansen 1997, 2005; Grove 2002; Nakamura *et al.* 2003; Proctor *et al.* 2003; Catterall *et al.* 2004; Kanowski *et al.* 2003a, 2006b). A high early planting density encourages the early development of many of these structural features. Fauna that feed on decaying matter, other fauna, nectar or fungi are likely to be influenced at least as strongly by the vegetation structure as by the plant species composition, whereas a smaller group will be most strongly influenced by plant species composition (in particular, some herbivores such as phytophagous insects and marsupial folivores, and some frugivores) (Jones & Crome 1990; Kitching *et al.* 2000; Kanowski *et al.* 2003b; Moran *et al.* 2004a, b).

Developmental rates

Recent studies of a range of reforestation pathways in the Wet Tropics have begun to provide information on the patterns and rates of biodiversity development (Box 41.1; Tucker & Murphy 1997; Keenan *et al.* 1997; Jansen 1997, 2005; Kanowski *et al.* 2003a, 2005a, 2006b; Nakamura *et al.* 2003; Proctor *et al.* 2003; Tucker &

Box 41.1 Bird species' responses to reforestation in the Wet Tropics

The QBVR project (quantifying the biodiversity values of reforestation) measured the occurrence of different organisms across a range of reforestation pathways (Kanowski *et al.* 2003a, 2005a, 2006a,b; Proctor *et al.* 2003; Catterall *et al.* 2004; Wardell-Johnson *et al.* 2005). In the Wet Tropics there were sites representing four types of young reforestation (5–22 years) – oldfield regrowth (RG), monospecific timber plantations (YP), mixed-species cabinet timber plantations (CT), ecological restoration plantings (ER) – and one type of older reforestation (mono-specific timber plantations, OP, 40–70 years). Similar measurements from reference sites (pasture, rainforest) provided a context. These different site-types showed a gradient of increasing vegetation complexity (Kanowski *et al.* 2003a). Sample sizes were: P, RG, YP, CT, each five sites; ER, OP, F, each ten sites.

Figure 41.3a shows that the total number of bird species per site did not vary greatly among the types of reforestation. However, rainforest-dependent species (the black section of each bar) occurred less frequently in young plantations (YP, CT, ER) and regrowth, while other species (grey sections) were more frequent. Figure 41.3b and c subdivide the rainforest-dependent birds into two groups; species endemic to the Wet Tropics bioregion show the strongest decrease in frequency in the simpler forms of reforestation.

Table 41.2 lists species recorded from at least 80% of sites of each type (endemic rainforest-dependent species are unshaded with an asterisk, other rainforest-dependent birds are shaded in dark grey and all other bird species are in light grey and italicized; species' vertical orders are by decreasing

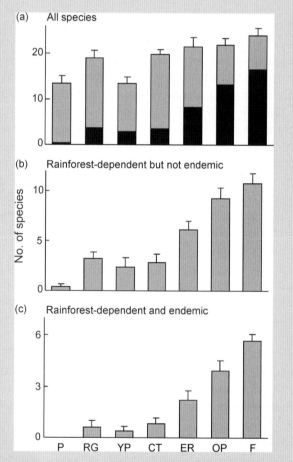

Figure 41.3 Bird assemblage responses to reforestation in the Wet Tropics. The black shading in (a) shows rainforest-dependent species.

(*Continued*)

Box 41.1 (Continued)

frequency). Rainforest-dependent endemics were uncommon in all younger reforested site types. Even in old plantations adjacent to rainforest they were underrepresented. Species labelled 'F' are fruit-eaters considered capable of regularly dispersing seeds (after Moran et al. 2004b). While the number and variety of common frugivores was greatest in rainforest, even pasture had one common seed disperser (the silvereye), and all reforested site-types had at least two.

The graphs (Figure 41.3) represent means and standard errors of the number of species within a transect 100 × 30 m at each site, accumulated across eight 30 minute visits that were repeated across a year. 'Rainforest-dependent' species are largely confined to, or apparently dependent on, rainforest, based on independent assessments of their distributions within uncleared landscapes (Catterall et al. 2004; Kanowski et al. 2005a). The remaining species include those that regularly use both rainforest and drier vegetation types or wetland. All patterns of variation are statistically significant (ANOVA, $P < 0.0001$).

Table 41.2 The most common bird species recorded from each site-type

Generally increasing complexity of vegetation ⟶

P	RG	YP	CT	ER	OP	F
White-rumped swiftlet	Brown gerygone	Brown cuckoo-dove	Silvereye F	Brown gerygone	Victoria's riflebird*F	Victoria's riflebird*F
Silvereye F	Black-faced monarch	Silvereye F	Lewin's honeyeater F	Black-faced monarch	Grey-headed robin*	Bower's shrike-thrush*
Chestnut-breasted mannikin	Silvereye F	Lewin's honeyeater F	Rainbow lorikeet	Silvereye F	Brown gerygone	Macleay's honeyeater*F
Cattle egret	Lewin's honeyeater F	Red-browed finch	Mistletoebird F	Lewin's honeyeater F	White-throated treecreeper	Grey-headed robin*
	Mistletoebird F	Mistletoebird F	White-throated gerygone	Golden whistler	Large-billed scrubwren	Brown gerygone
	Chestnut-breasted mannikin	Little shrike-thrush		Little shrike-thrush	Eastern whipbird	Eastern whipbird
	White-rumped swiftlet	Red-backed fairy-wren			Pale-yellow robin	White-throated treecreeper
	Tawny grassbird				Lewin's honeyeater F	Large-billed scrubwren
	Golden-headed cisticola				Golden whistler	Wompoo fruit-dove F
					Sulphur-crested cockatoo	Spectacled monarch
						Lewin's honeyeater F Grey Fantail

Unshaded and asterisked, endemic rainforest-dependent species; dark grey, other rainforest-dependent birds; light grey and italicized, all other birds.

Simmons 2004; Catterall et al. 2004; Hausmann 2004; Wardell-Johnson et al. 2005; Grimbacher & Catterall, 2007). These have shown that a range of rainforest-dependent plants and animals may begin to use ecological restoration plantings within 5 years. By 10 years these plantings can show a moderate similarity to rainforest in species composition (although rainforest-dependent species are underrepresented) and high similarity in some structural and functional measures. Timber plantations are colonized by a more limited range of rainforest animals within the first 10 years and are more likely to support habitat generalists or open-habitat fauna species and a weedy flora. Much less is known about the early development of regrowth, although limited work suggests that its biota develops at a somewhat slower rate than for timber plantations. However, older timber plantations (40–70 years) adjacent to rainforest, and on which reforestation began immediately after clearing, support a fauna species composition that is very similar to that of rainforest, and a high diversity of understorey and seedling rainforest flora.

Even in the best-developed available older plantations and regrowth, up to several decades old, there is a suite of rainforest specialist species that remain absent, or are present at substantially lower densities than in intact rainforest (Catterall *et al*. 2004; Hausmann 2004; Jansen 2005; Kanowski *et al*. 2005a, 2006b). The reasons underlying this have not been studied in detail, but it seems likely that a lack of particular structural attributes that occur in old-growth rainforest plays a role. For example, the skink *Eulamprus tigrinus*, which is moderately common in rainforest, typically occurring in crevices on the trunks of rainforest trees such as those formed by the braided roots of strangler figs, was not recorded from old timber plantations adjacent to tropical rainforest (Catterall *et al*. 2004; Kanowski *et al*. 2006a). Older restoration plantings have not been available for study, although it is clearly the hope of practitioners that improved restoration planting designs may somehow provide at least some of the important structural attributes that are missing from old timber plantations and regrowth.

The roles of landscape context

The flora and fauna of a reforested site can also be affected by its landscape context and by land-use history. For example, land areas near to uncleared rainforest, that were abandoned decades ago shortly after being cleared, are now likely to support regrowth that is dominated by native rainforest plants, whereas more recently abandoned areas of former pasture within extensively cleared regions that were grazed for decades after clearing now often support regrowth that is dominated by invasive exotic plants (Erskine *et al*. 2007). The proximity of a reforested patch to remnant rainforest affects its ability to acquire rainforest-dependent species of plants and animals, both for ecological restoration plantings (Tucker & Murphy 1997; McKenna 2001; Kanowski *et al*. 2005a; Grimbacher & Catterall, 2007) and for regrowth (Hausmann 2004; Neilan *et al*. 2006). However, the specific shape of the biodiversity–distance relationship remains poorly understood and is likely to vary between taxa and between regions, depending on factors such as landscape forest cover and history of clearing (Kanowski *et al*. 2005a, 2006b). In addition, the effects of patch size on the biota within reforested areas remain virtually unstudied.

At the broader landscape scale, different forms of reforestation may have a range of different types of on-site and off-site consequences for biodiversity. These include both positive and negative consequences for particular taxa or taxon groups (Box 41.3). Whether a particular consequence occurs will depend on the nature of the reforestation and its environmental context. These could be considered at

Table 41.3 Potential ways in which reforestation may produce positive and negative biodiversity outcomes, at local and landscape (off-site) scales

Scale	Possible positive outcomes	Possible negative outcomes
On-site	• Added intrinsic value of planted species • Facilitation of rainforest regeneration • Provision of habitat for rainforest biota	• Provision of habitat for weeds and undesirable native species • Replacement of existing rainforest remnants or regrowth
Off-site*	• Facilitation of dispersal of rainforest biota • Increased regional populations of rainforest species • Buffering of remnant forests to improve their habitat value • Habitat improvement for aquatic biota	• Invasion of native forests by non-indigenous planted species • Invasion of native forests by non-local genes from planted populations • Provision of a source of weeds for surrounding landscapes

* Further possible off-site effects may also occur through indirect pathways. For example: (i) contributions of reforested areas to reducing climate change may prevent loss of some rainforest species; (ii) if reforestation improves (or reduces) downstream and offshore water quality, coastal biotas will ultimately be affected.

the project planning stage. However, the real-world measurement of off-site consequences of reforestation is likely to be difficult, and they remain largely unstudied.

Wet Tropics endemics: a special case

The characteristics of different species also affect their response to reforestation. Many species that are endemic to the Wet Tropics region appear to be both particularly susceptible to the effects of habitat modification and fragmentation, and particularly slow to recolonize reforested areas. This pattern can be seen in the response of reptiles such as the prickly forest skink, *Gnypetoscincus queenslandiae* (Sumner *et al.* 1999; Hausmann 2004; Kanowski *et al.* 2006a), but is so far most evident in birds (Box 41.1; Hausmann 2004). There is logic to this phenomenon, since most of these species have a narrow habitat range and are strong upland rainforest specialists. They are considered to be relict survivors from the late Pleistocene, when rainforest disappeared from most areas, and the Wet Tropics upland rainforests were important refugia (Adam 1994; Williams & Pearson 1997). Thus, their contemporary endemism, conservation significance, current vulnerability to deforestation and limited response to reforestation appear to be all connected to their rainforest specialization, and they are least likely to benefit in the short term from reforestation.

How is biodiversity a driver of the reforestation process?

Important interactions

A fascinating aspect of rainforest biodiversity, and its ecological functioning, is the large number of interactions and interdependencies between species (Jones & Crome 1990). Pollination, seed dispersal and predation on seeds or seedlings are all strongly influenced by interactions with animals and are also crucial processes in the regeneration of rainforest plants (Kanowski *et al.* 2004; see also Bawa 1990; Levey *et al.* 2002; Wright 2002). Most of these plants rely on animals (mainly insects, but also birds and mammals) for pollination. Birds and mammals are important dispersers of the seeds. The feeding activities of seed

predators (including mammals such as rodents, and insects such as beetles and moths) and herbivores (such as various insects and mammals) set limits to seedling densities. Animals that disturb the leaf litter on the forest floor in search of food can also limit seedling regeneration.

Other interactions affect growth within the new generation of recruited plants. For example, many rainforest trees and shrubs are involved in root symbioses with fungi (e.g. mycorrhizae; Hopkins *et al.* 1996) or bacteria (e.g. the root nodules of leguminous species), which help them to acquire nutrients to assist growth. Other forms of complex interaction are involved in the decomposition of organic matter (involving plants, invertebrate animals, fungi and bacteria). Finally, competition between plant species, for factors such as light, water and nutrients, affects both the balance between trees and grass in the early stages of reforestation, and the relative abundances of different rainforest trees. Indeed, part of the influence of seed and seedling predation on patterns of plant regeneration may occur because selective predation on some types of plant alters the outcomes of competition with other plants by, for example, suppressing dominant competitors (Wright 2002).

Within the vegetation of rainforest remnants there is likely to be a progressive cascade of changes as decades elapse after deforestation, due in part to the effects of decreases or increases in abundances of animal species that contribute to pollination, seed dispersal, and to seed or seedling predation. For example, the cassowary, *Casuarius casuarius*, a declining bird of Wet Tropics rainforests, is also an important and wide-ranging seed-disperser (Westcott *et al.*, Chapter 16, this volume). Its local extinction would greatly reduce the dispersal of many large-seeded rainforest plants, with longer-term consequences for the composition of rainforest tree species (e.g. Harrington *et al.* 1997). Reforestation could restore seed dispersal processes within a remnant rainforest patch, if used to enlarge its area, to buffer it or to provide habitat connections with other remnants, and if these changes helped to sustain important disperser species.

Within the 'new forests' developing on formerly cleared land, the fauna will also exert strong influences on trajectories and rates of development (Kanowski *et al.* 2004). Direct human intervention, such as handpollinating, seeding, tree planting or weeding, can substitute for some local interactions between animals and plants. However, in the long term, the nature of future rainforest landscapes will depend mainly on the ability of rainforest plants to replace themselves within remnant forest, to recolonize and establish themselves in areas formerly cleared and to persist in the face of colonization by introduced plants. In some cases, management of the animal–plant interactions may be more effective than intervention that directly targets plants. Different animals can have positive or negative impacts on the restoration of rainforest biodiversity, depending on the establishment method, the intensity of particular interactions and the characteristics and age of individual sites. For example, suppression of the growth of some plants by marsupial browsers may cause an undesirable reduction in the density of rainforest seedlings (Woodford 2000), but alternatively it may assist reforestation if the browsers reduce competition from grasses and vines through selective feeding (Wahungu 2000).

Animals also drive rainforest restoration through their appeal to people. For example, practitioners have found that providing information about the value of specific rainforest plants as food for birds or butterflies generates an increased public demand for their seedlings (Russell 2004). Fauna that are both rainforest-dependent and threatened, such as the cassowary and Lumholtz's tree-kangaroo in the Wet Tropics and Coxen's fig-parrot in the Australian subtropics, have become flagships for increasing public interest in reforestation (Crome & Bentrupperbäumer 1993; Kanowski & Tucker 2002; Bower 2004; Russell 2004).

Phases of development

The nature and importance of specific interactions will vary over the course of reforestation at a site. It is useful to distinguish the following three phases: establishment, building and maintenance (see also Kanowski *et al.* 2004). These phases are relevant to all reforestation pathways, although the rate of passage from one phase to the next will vary among pathways, and with other factors (including the landscape context). Better knowledge of all phases will allow management intervention to accelerate or redirect the reforestation process. Important ecological interactions during each phase are as follows.

1 The establishment phase. During establishment, the developing canopy has not yet closed, and therefore the site has not yet been 'captured' from the competing grass and herb cover. For planted sites, the most important interactions determining the outcome of this phase will be between herbivores and plants. In regrowth sites, seed dispersal from a seed source, and pollination at that source, will also be important.

2 The building phase. Following canopy closure, the most important interactions will be those that facilitate the development of a diverse, rainforest-like flora and fauna, and the development of the ecological processes that would allow its maintenance. For this to occur, it may be necessary for a reforested site to acquire a full range of functional types of pollinators, dispersers and seed/seedling predators, even if the species mix differs from that in the original forest. The importance of species diversity as opposed to functional diversity in re-establishing ecological processes is presently unclear. Developing an understanding of the range of functional categories involved, and the relationship between species diversity and functional diversity, will enable better targeted management intervention.

3 The maintenance phase. If restoration is successful, at some stage the rate of species turnover, the functional composition and perhaps the species composition in the reforested area should stabilize at a level similar to that which characterizes intact rainforest. After this, further intervention would be unnecessary or even undesirable. Because the building phase is likely to take many decades, and possibly centuries (especially in the absence of intervention aimed to speed it up), these latter steps have yet to be researched.

The specific roles of plant–pollinator and plant–predator (or herbivore) interactions across these phases remain largely unknown (Kanowski *et al.* 2004). However, the experience of Wet Tropics revegetation practitioners suggests that reproduction in rainforest plants in relatively young reforested areas may not be pollinator limited (Tucker & Simmons 2004). It also seems that there may be a window of opportunity during the transition between the establishment and maintenance phases for effective plant species enrichment using direct seeding. At this time, canopy closure is sufficient to facilitate germination and growth, but most rodent seed predators are yet to colonize (Tucker & Simmons 2004). To develop such anecdotal observations into knowledge requires systematic research.

Fruit-eating fauna and seed dispersal outcomes

Seed dispersal is an especially important ecological process in rainforest restoration, and is better known because it affects the future trajectory of the plant communities in three different ways (Wunderle 1997; Levey et al. 2002; Moran et al. 2004a, b; White et al. 2004; Neilan et al. 2006; Westcott et al., Chapter 16, this volume). First, within intact rainforests (and new forests in the maintenance phase), dispersal is important in maintaining plant diversity because it moves seeds to locations where they are more likely to germinate and survive. Second, within fragmented landscapes, dispersal between remnant patches allows the replenishment of plant species that are otherwise likely to go locally extinct periodically because of fragmentation-related processes. Third, in long-cleared and reforested parts of the landscape, from which the bank of rainforest seeds and seedlings has been lost, dispersal allows a diversity of plant species to reaccumulate.

More than 70% of plant species in Australia's tropical rainforests are fleshy-fruited and dispersed by animals, especially birds and bats (Jones & Crome 1990), a pattern shared with rainforests worldwide (Willson et al. 1989). Consequently, many methods for active rainforest restoration depend on vertebrate-assisted seed dispersal for building up plant species richness (Goosem & Tucker 1995; Tucker & Murphy 1997; Toh et al. 1999; Florentine & Westbrooke 2004). Reforested areas typically attract frugivorous birds within the first few years of development (Box 41.1; Jansen 2005), but the quality and quantity of seed dispersal will be affected by the types and abundances of the frugivore species involved. Timber plantations are often dominated by wind-dispersed rather than fleshy-fruited plants, and rainforest frugivores tend to visit young timber plantations less frequently than they visit similar-aged restoration plantings (Kanowski et al 2005b; Erskine et al., in press). With respect to frugivore-dispersed plants, the dispersers of larger-seeded species may be the most likely to decline following rainforest fragmentation, and the slowest to recolonize following reforestation (Levey et al. 2002; Moran et al. 2004b; Westcott et al., Chapter 16, this volume). For example, in Box 41.1, the common frugivores in all reforested sites were silvereyes and Lewin's honeyeaters; both species have narrow beaks that would prevent them from swallowing larger (>1.5 cm diameter) seeds.

This would mean that wind-dispersed and larger-seeded plants may be less likely to be moved into areas undergoing reforestation, and therefore a number of practitioners have suggested that a focus of management during the building phase should be enrichment plantings of such species (Tucker & Murphy 1997; Russell 2004; Tucker et al. 2004; White et al. 2004). Large seeded plants are also likely to survive relatively well when introduced through direct seeding (Doust 2004). During the establishment phase, the most effective means of rapidly accumulating the majority of fleshy-fruited rainforest plants may be to achieve a rapidly closing canopy of fast-growing and early-maturing tree species that produce large crops of smaller-seeded fleshy fruits, in order to attract a diversity and abundance of frugivores (Goosem & Tucker 1995). Of the 240 tree, shrub and vine species recorded from the Wet Tropics rainforest sites represented in Box 41.1, 66% had small-diameter (< 1.5 cm) propagules likely to be dispersed by vertebrates, 17% had larger-diameter propagules that were also vertebrate-dispersed, while 11% were considered wind-dispersed (the authors, unpublished data). The remaining 6% are considered to be dispersed by neither wind nor vertebrates, or their dispersal mode is unknown (data from surveys of five 5 m radius quadrats on each of ten rainforest sites; the total survey area was approximately 0.4 ha) (Wardell-Johnson et al. 2005).

Most potential disperser species are birds; in the Wet Tropics there are over 20 bird and two bat genera likely to disperse the seeds of rainforest plants regularly (Erskine et al., 2007). Frugivorous bird species of Australian rainforest landscapes show large interspecific differences in ecological traits that determine their likely seed dispersal roles (Moran et al. 2004a, b; Westcott et al., Chapter 16, this volume). These traits include their abundance in highly cleared regions, their ability to consume or disperse fruits of different types and the frequency with which they undertake longer-distance movements. Management of developmental trajectories within the new forests requires a better understanding of these functional differences, their relationships with recruitment in plant species or plant functional groups and the extent to which the seed dispersal roles of rare or deforestation-intolerant species can be filled by more common or deforestation-tolerant species.

Exotic woody plants may rapidly colonize abandoned agricultural land, perhaps because they are more able to set fruits, disperse and compete with introduced pasture grasses and herbs than are their rainforest counterparts (at least within the cleared areas). Over time, birds can disperse rainforest seeds into this regrowth, and its canopy may then close, assisting the survival and growth of the rainforest seedlings (Box 41.2). A management approach that involves tolerance of the initial weedy regrowth, followed by actions targeted at reducing any suppression of the growing rainforest seedlings by the exotic overstorey, may prove to be a cost-effective restoration technique over large areas (Neilan *et al.* 2006). Such suggestions are controversial, because the control of exotic plants has been an important part of both the rehabilitation of remnant rainforest and the management of planted areas (e.g. Harden *et al.* 2004). Paradoxically, birds also play a large role in dispersing the seeds of invasive exotic plants across the landscape (Gosper *et al.* 2005; Buckley *et al.* 2006). During the building phase, bird dispersal can introduce an unwanted flora of aggres-

sive vines and other exotic plants, which in the absence of maintenance, may overgrow reforested areas (Catterall *et al.* 2004; Wardell-Johnson *et al.* 2005). However, in extensively cleared and long-modified landscapes, it may be necessary to accept that exotic plants (and animals) can play both undesirable and useful roles in successional processes, and that improving biodiversity development may require careful intervention to manage these roles rather than universal eradication (Buckley *et al.* 2006).

Managing new forests of the future

There is a need to undertake large-scale reforestation now, in order to avert further biodiversity decline and loss of ecosystem services. However, reliable prediction of outcomes, and prescriptions for best-practice management, are impeded by two factors. First, there are large gaps in the current knowledge about many of the ecological processes involved in reforestation, and this makes it unlikely that useful predictive models can be built. Second, rapidly changing physical conditions, coupled with species invasions

---

Box 41.2 A role for weedy regrowth in rainforest restoration?

The decline of the dairy and banana industries in the Australian subtropics in the mid to late twentieth century was followed by the rapid spread of the camphor laurel (*Cinnamomum camphora*), a fast-growing Asian species, which bears large crops of small (1 cm diameter) fruit that are consumed by many fruit-eating birds. By the year 2000, regrowth dominated by camphor laurel occupied around 25% of the former Big Scrub region, a 750 km^2 landscape in which more than 99% of the original rainforest had been cleared for many decades. This regrowth both provides fauna habitat and allows the regeneration of rainforest seedlings, which appear poorly adapted to recruiting within pasture.

Neilan *et al.* (2006) measured the abundance of bird and plant species across 24 well developed camphor laurel patches (around 20–40 years old). They recorded 16 fruit-eating bird species with a medium to high potential for dispersing the seeds of rainforest plants, including five species

present in >90% of the patches. Among the latter was the threatened rose-crowned fruit-dove *Ptilinopus regina*, for which the weedy regrowth probably provided important habitat. Native rainforest species comprised 25% of individual mature trees (>10 or 20 cm stem diameter, depending on species), but 47% of the next generation of the very young trees (<2.5 cm diameter). Statistical analyses showed that bird-dispersed rainforest trees, especially those of later-successional stages, were becoming increasingly prevalent among young trees. These findings suggest that the growth of camphor laurel may help rainforest plants to recruit in abandoned agricultural areas, although later removal of the older camphor trees may be needed to help the recruits to reach adult size more rapidly. Controlled trials would be needed to test these possibilities. We expect this type of situation to occur increasingly in other landscapes, as both invasion by exotic plants and abandonment of agricultural lands are increasing.

and extinctions, will make it impossible to forecast the future simply through observation of past pathways of development.

The uncertainty in current knowledge creates further practical dilemmas, both general and specific. Is it more important to increase the total area of rainforest habitat or to provide physical linkages between remnant patches? To create a single large patch or several small ones? For a given level of financing, is it more effective to reforest a larger area with moderate biodiversity outcomes per unit area, or a smaller area with better biodiversity outcomes per unit area (see also Wardell-Johnson, Chapter 40, this volume)? Would it be better to allocate funds to create a riparian buffer 10 m wide along 2 km of stream, or one that is 50 m wide along only 0.4 km of the stream? Should well established invasive exotic plants be tolerated and managed as an aid to reforestation or eliminated as unwanted aliens? Realistically, many situations may be best served by spatially explicit combinations of options (e.g. Kanowski *et al.* 2005b), but this depends on developing the knowledge to help to make the decisions that can link site-specific actions to landscape outcomes.

A practical solution to this problem is to approach large-scale reforestation as experimental management (see also Catterall *et al.* 2005). Instead of educated guesswork, which often requires arbitrary choices that are focused on a presumed (and possibly contentious) best method, experimental management of reforestation would involve the identification of alternative options. These options would then be trialled within a context where their planning and design are carefully controlled, and where monitoring of biodiversity (Kanowski *et al.*, Chapter 42, this volume) is used to assess, compare and interpret the outcomes. This can provide better knowledge, not only of the consequences of different approaches to reforestation, but also of the ecological processes involved. To achieve this, however, requires a shift in attitudes to reforestation management, from 'best practice' to 'experimental adaptive learning'. It also requires an institutional structure and context that enables such a large-scale planning enterprise to develop, provides the resources for its implementation and includes incentives for cooperation between managers, researchers and landholders. Within such a context, creative manipulation of autogenic regrowth deserves further

investigation as an opportunity for restoring rainforest biodiversity.

Summary

- Human efforts to increase rainforest land cover are expanding, driven by changing values, attitudes and economics. For Australian rainforests, there are currently three broad types of potential reforestation pathway: ecological restoration plantings, timber plantations and regrowth management.
- Three phases of biodiversity development during reforestation are defined: establishment, building and maintenance. Canopy closure heralds the transition from the establishment to the building phase as it is a key factor in providing environmental conditions for the recruitment, survival and growth of many rainforest plants and animals.
- By 10 years, some plantings can show a moderate similarity to rainforest in species composition, although rainforest-dependent species are underrepresented. The best early biodiversity outcomes occur in ecological restoration plantings that were initially established with a high diversity and density of rainforest seedlings. However, regionally endemic fauna species appear to be relatively unresponsive to all forms of reforestation in the short term.
- Vegetation development and plant regeneration are strongly influenced by interactions with animals (such as pollination, seed dispersal and predation on seeds or seedlings). Most rainforest plant species are fleshy-fruited, and therefore seed dispersal by birds and bats is especially important to the future trajectories of plant communities during reforestation.
- Reliable predictions about practical outcomes of reforestation efforts are currently impeded by large gaps in current knowledge and rapidly changing environmental conditions. A solution to this problem could be to approach reforestation as experimental management.

Acknowledgements

Much of the recent research discussed here was funded and stimulated by the Rainforest Cooperative Research Centre, 1999–2005, through its Rehabilitation and

Restoration programme. Thanks to the editors for their encouragement; to Stephen McKenna for compiling information on plant propagule traits; to Cath Moran, Nigel Tucker, Peter Erskine, David Lamb, Peter Grimbacher, Aki Nakamura, Wendy Neilan, Ralph Woodford, Hank Bower, Mark Dunphy, Marc Russell, Mark Heaton and many other researchers, reforestation practitioners and landholders for discussions about reforestation processes and practices; and to the Department of Wildlife Ecology and Conservation, University of Florida, for providing facilities to CC during manuscript preparation.

References

Adam, P. (1994). *Australian Rainforests*. Oxford University Press: Oxford.

Ashton, M. S., Gunatilleke, C. V., Singhakumara, B. M. & Gunatilleke, I .A. (2001). Restoration pathways for rainforest in southwest Sri Lanka: a review of concepts and models. *Forest Ecology and Management* **154**: 409–30.

Bawa, K. S. (1990). Plant–pollinator interactions in tropical rain forests. *Annual Review of Ecology and Systematics* **21**: 399–422.

Big Scrub Rainforest Landcare Group (1998, 2005). Subtropical Rainforest Restoration: A Practical Manual and Data Source for Landcare Groups, Land Managers and Rainforest Regenerators. Big Scrub Rainforest Landcare Group, Bangalow, NSW.

Bower, H. (2004). Animal–plant interactions: applying the theory on the ground in north-east New South Wales. In *Animal–Plant Interactions in Rainforest Conservation and Restoration*, Kanowski, J., Catterall, C. P., Dennis, A. J. & Westcott, D. A. (eds). Cooperative Research Centre for Tropical Rainforest Ecology and Management. Rainforest CRC, Cairns, pp. 27–9.

Buckley, Y. M., Anderson, S., Catterall, C. P. *et al.* (2006). Management of plant invasions mediated by frugivore interactions. *Journal of Applied Ecology* **43**: 848–57.

Catterall, C. P. & Harrison, D. A. (2006). *Rainforest Restoration Activities in Australia's Tropics and Subtropics*. Rainforest CRC (www.jcu.edu.au/rainforest/publications/restoration_activities.htm).

Catterall, C. P., Kanowski, J., Lamb, D., Killin, D., Erskine, P. & Wardell-Johnson, G. (2005). Trade-offs between timber production and biodiversity in rainforest tree plantations: emerging issues from an ecological perspective. In *What Have We Learnt from Growing Rainforest Timber Species in the Last 10 Years?* Erskine, P., Lamb, D. & Bristow, M. (eds). Rainforest CRC and RIRDC, Canberra, pp. 206–22.

Catterall, C. P., Kanowski, J., Wardell-Johnson, G. *et al.* (2004). Quantifying the biodiversity values of reforestation: perspectives, design issues and outcomes in Australian rainforest landscapes. In *Conservation of Australia's Forest Fauna*, 2nd edn, Lunney, D. (ed.). Royal Zoological Society of New South Wales, Sydney, pp. 359–93.

Chazdon, R. L. (2003). Tropical forest recovery: legacies of human impact and natural disturbances. *Perspectives in Plant Ecology Evolution and Systematics* **6**: 51–71.

Crome, F. H. J. & Bentrupperbaumer, J. (1993). Special people, a special animal and a special vision: the first steps to restoring a fragmented tropical landscape. In *Nature Conservation 3. The Reconstruction of Fragmented Ecosystems*, Saunders, D. A., Hobbs, R. J. & Ehrlich, P. R. (eds). Surrey Beatty and Sons. Chipping Norton, NSW, pp. 267–79.

Dobson, A. P., Bradshaw, A. D. & Baker, A. J. M. (1997). Hopes for the future: restoration ecology and conservation biology. *Science* **277**: 515–22.

Doust, S. (2004). Seed and seedling ecology in the early stages of rainforest restoration. PhD thesis, University of Queensland.

Emtage, N. F., Harrison, S. R. & Herbohn, J. L. (2001). Landholder attitudes to and participation in farm forestry activities in sub-tropical and tropical eastern Australia. In *Sustainable Farm Forestry in the Tropics*, Harrison, S. R & Herbohn, J. L. (eds). Edward Elgar, Cheltenham, pp. 195–210.

Erskine, P. D. (2002). Land clearing and forest rehabilitation in the Wet Tropics of north Queensland, Australia. *Ecological Management and Restoration* **3**: 136–8.

Erskine, P., Lamb, D., & Bristow, M. (Eds.) (2005). *What Have We Learnt from Growing Rainforest Timber Species in the Last 10 Years?* Rainforest CRC and RIRDC, Canberra.

Erskine, P. D., Catterall, C. P. Lamb, D & Kanawski, J. (2007). Patterns and processes of old field reforestation in Australian rainforest landscapes. In *Old Fields Dynamics and Restoration of Abandoned Farmland*, Camer, V. A. & Hobbs, R. J. (eds). Island Press, Washington, DC, pp. 119–43.

Fahrig, L. (2003). Effects of habitat fragmentation on biodiversity. *Annual Review of Ecology, Evolution and Systematics* **34**: 487–515.

Florentine S. K & Westbrooke, M. E. (2004). Evaluation of alternative approaches to rainforest restoration on abandoned pasturelands in tropical north Queensland, Australia. *Land Degradation and Development* **15**: 1–13.

Gilmour, D. A. & Riley, J. J. (1970). Productivity survey of the Atherton Tableland and suggested land use changes. *Journal of Australian Institute of Agricultural Science* **36**: 259–272.

Goosem., S. P & Tucker, N. I J. (1995). *Repairing the Rainforest. Theory and Practice of Rainforest Re-establishment in North Queenslands Wet Tropics.* Wet Tropics Management Authority, Cairns.

Goosem S. P., Morgan G. & Kemp J. E. (1999). Regional ecosystems of the wet tropics. In *The Conservation Status of Queensland's Bioregional Ecosystems*, Sattler, P. S. & Williams, R. D. (eds). Environmental Protection Agency, Brisbane, pp. 1–73.

Gosper, C. R., Stansbury, C. D. & Vivian-Smith, G. (2005). Seed dispersal of fleshy-fruited invasive plants by birds: contributing factors and management options. *Diversity and Distributions* **11**: 549–58.

Grimbacher, P. S. & Catterall, C. P. (2007). How much do site age, habitat structure and spatial isolation influence the restoration of rainforest beetle species assemblages? *Biological Conservation* **135**: 107–118.

Grimbacher, P. S., Catterall, C. P. & Kitching, R. L. (2006). Beetle species' responses suggest that microclimate mediates fragmentation effects in tropical Australian rainforest. *Austral Ecology* **31**: 458–470.

Grove, S. (2002). The influence of forest management history on the integrity of the saproxylic beetle fauna in an Australian lowland tropical rainforest. *Biological Conservation* **104**: 149–71.

Harden, G. J., Fox, M. D. & Fox, B. J. (2004). Monitoring and assessment of restoration of a rainforest remnant at Wingham Brush, NSW. *Austral Ecology* **29**: 489–507.

Harrington, G. N., Irvine, A. K., Crome, F. H. J. & Moore, L. A. (1997). Regeneration of large-seeded trees in Australian rainforest fragments: a study of higher order interactions. In *Tropical Forest Remnants. Ecology, Management, and Conservation of Fragmented Communities*, Laurance, W. F. & Bierregaard, R. O. (eds). University of Chicago Press, Chicago, pp. 292–303.

Harrison, S. & Bruna, E. (1999). Habitat fragmentation and large-scale conservation: what do we know for sure? *Ecography* **22**: 225–32.

Hausmann, F. (2004). The utility of linear riparian rainforest remnants as habitat and/or dispersal routes for ground dwelling vertebrates. MPhil thesis, Griffith University.

Hopkins, M. S. & Graham, A. W. (1984). Viable soil seed banks in disturbed lowland tropical rainforest sites in North Queensland. *Australian Journal of Ecology* **9**: 71–9.

Hopkins, M. S., Reddell, P., Hewett, R. K. & Graham, A. W. (1996). Comparison of root and mycorrhizal characteristics in primary and secondary rainforest on a metamorphic soil in North Queensland, Australia. *Journal of Tropical Ecology* **12**: 871–85.

Jansen, A. (1997). Terrestrial invertebrate community structure as an indicator of the success of a tropical rainforest restoration project. *Restoration Ecology* **5**: 115–24.

Jansen, A. (2005). Avian use of restoration plantings along a creek linking rainforest patches on the Atherton Tablelands, North Queensland. *Restoration Ecology* **13**: 275–83.

Jones, R. E. & Crome, F. H. (1990). The biological web: plant/animal interactions in the rainforest. In *Australian Tropical Rainforests: Science, Value, Meaning*, Webb, L. J. & Kikkawa, J. (eds). CSIRO Publications, Melbourne.

Kanowski, J. & Tucker, N. I. J. (2002). Trial of shelter poles to aid the dispersal of tree-kangaroos on the Atherton Tablelands, north Queensland. *Ecological Management and Restoration* **3**: 138–9.

Kanowski, J., Catterall, C. P., Wardell-Johnson, G. W., Proctor, H. & Reis, T. (2003a). Development of forest structure on cleared rainforest land in eastern Australia under different styles of reforestation. *Forest Ecology and Management* **183**: 265–80.

Kanowski, J., Irvine, A. K. & Winter, J. W. (2003b). The relationship between the floristic composition of rain forests and the abundance of folivorous marsupials in north-east Queensland. *Journal of Animal Ecology* **72**: 627–32.

Kanowski, J., Catterall, C., Reis, T. & Wardell-Johnson, G. (2004). Animal–plant interactions in rainforest restoration in tropical and subtropical Australia. In *Animal–Plant Interactions in Rainforest Conservation and Restoration*, Kanowski, J., Catterall, C. P., Dennis, A. J. & Westcott, D. A. (eds). Cooperative Research Centre for Tropical Rainforest Ecology and Management. Rainforest CRC, Cairns, pp. 20–3.

Kanowski, J., Catterall, C. P., Proctor, H., Reis, T., Tucker, N. I. J. & Wardell-Johnson, G. W. (2005a). Rainforest timber plantations and animal biodiversity in tropical and subtropical Australia. In *What Have We Learnt from Growing Rainforest Timber Species in the Last 10 Years?* Erskine, P., Lamb, D. & Bristow, M. (eds). Rainforest CRC and RIRDC, Canberra, pp. 183–205.

Kanowski, J., Catterall, C. P. & Wardell-Johnson, G. W. (2005b). Consequences of broadscale timber plantations for biodiversity in cleared rainforest landscapes of tropical and subtropical Australia. *Forest Ecology and Management* **208**: 359–72.

Kanowski, J., Reis, T., Catterall, C. P. & Piper, S. D. (2006a). Factors affecting the use of reforested sites by reptiles in cleared rainforest landscapes in tropical and subtropical Australia. *Restoration Ecology* **14**: 67–76.

Kanowski, J., Catterall, C. P. & Wardell-Johnson, G. W. (2006b). Impacts of broadscale timber plantations on biodiversity in cleared rainforest landscapes of North Queensland. In *Sustainable Forest Industry Development in Tropical North Queensland*, Harrison, S. R. *et al.* (eds). Rainforest CRC, Cairns.

Keenan, R., Lamb, D., Woldring, O, Irvine, A. & Jensen, R. (1997). Restoration of plant biodiversity beneath tropical tree plantations in Northern Australia. *Forest Ecology and Management* **99**: 117–31.

Kikkawa, J. (1990). Specialisation in the tropical rainforest. In *Australian Tropical Rainforests: Science, Value, Meaning*, Webb, L. J. & Kikkawa, J. (eds). CSIRO Publications, Melbourne, pp. 67–73.

Kitching, R. L., Orr, A. G., Thalib, L., Mitchell, H., Hopkins, M. S. & Graham, A. W. (2000). Moth assemblages as indicators of environmental quality in remnants of upland Australian rain forest. *Journal of Applied Ecology* **37**: 284–97.

Kooyman, R. M. (1996). *Growing Rainforest. Rainforest Restoration and Regeneration Recommendations for the Humid Sub-tropical Region of Northern New South Wales and South East Queensland*. Greening Australia, Brisbane.

Lamb, D., Erskine, P. D. & Parrotta, J. A. (2005). Restoration of degraded tropical forest landscapes. *Science* **310**: 1628–32.

Lamb, D. Keenan, R. & Gould, K. (2001). Historical background to plantation development in the tropics: a north Queensland case study. In *Sustainable Farm Forestry in the Tropics*, Harrison, S. R & Herbohn, J. L. (eds). Edward Elgar, Cheltenham, pp. 9–20.

Lamb, D., Parrotta, J. A., Keenan, R., Tucker, N. (1997). Rejoining habitat remnants: restoring degraded rainforest lands. In *Tropical Forest Remnants. Ecology, Management, and Conservation of Fragmented Communities*, Laurance, W. F. & Bierregaard, R. O. (eds). University of Chicago Press, Chicago, pp. 366–85.

Laurance, W. F. & Bierregaard, R. O. (eds) (1997). *Tropical Forest Remnants. Ecology, Management and Conservation of Fragmented Communities*. University of Chicago Press, Chicago.

Levey, D. J., Silva, W. R. & Galetti, M. (eds) (2002). *Seed Dispersal and Frugivory: Ecology, Evolution and Conservation*. CABI Publishing, Wallingford, UK.

Lindenmayer, D. B. & Franklin, J. F. (2002). *Conserving Forest Biodiversity: a Comprehensive Multiscaled Approach*. Island Press, Washington, DC.

Lugo, A. E. & Helmer, E. (2004). Emerging forests on abandoned land: Puerto Rico's new forests. *Forest Ecology and Management* **190**: 145–61.

McKenna, S. (2001). Native plant species regeneration in restoration plantings on the Atherton Tablelands, north Queensland. BSc (Hons) thesis, James Cook University.

Mooney, H., Cropper, A. & Reid. W. (2005). Confronting the human dilemma: how can ecosystems provide sustainable services to benefit society? *Nature* **434**: 561–2.

Moran, C., Catterall, C. P., Green, R. J. & Olsen, M. F. (2004a). Fates of feathered fruit eaters in fragmented forests. In *Conservation of Australia's Forest Fauna*, 2nd edn, Lunney, D. (ed.). Royal Zoological Society of New South Wales, Sydney, pp. 699–712.

Moran, C., Catterall, C. P., Green, R. J. & Olsen, M. F. (2004b). Functional variation among frugivorous birds: implications for rainforest seed dispersal in a fragmented sub-tropical landscape. *Oecologia* **141**: 584–95.

Nakamura, A., Proctor, H , & Catterall. C. (2003). Using soil and litter arthropods to assess the state of rainforest restoration. *Ecological Management and Restoration* **4**: S20–S26.

Neilan, W., Catterall, C. P., Kanowski, J. & McKenna, S. (2006). Frugivorous birds may facilitate rainforest succession in weed-dominated regrowth in subtropical Australia. *Biological Conservation* **129**: 393–407.

Parker, T. V. & Pickett, S. T. A. (1997). Restoration as an ecosystem process: implications of the modern ecological paradigm. In *Restoration Ecology and Sustainable Development*, Urbanska, K. M., Webb, N. R. & Edwards, P. J. (eds). Cambridge University Press, Cambridge, pp. 17–32.

Proctor, H. C., Kanowski, J., Wardell-Johnson, G., Reis, T. & Catterall, C. P. (2003). Does diversity beget diversity? A comparison between plant and leaf-litter invertebrate richness from pasture to rainforest. In *Proceedings of the 5th Invertebrate Biodiversity and Conservation Conference*. Records of the South Australian Museum, Supplementary Series **7**, 267–74.

Russell, M. (2004). Animal–plant interactions: their role in rainforest conservation and restoration at a community level in south-east Queensland. In *Animal–Plant Interactions in Rainforest Conservation and Restoration*, Kanowski, J., Catterall, C. P., Dennis, A. J. & Westcott, D. A. (eds). Cooperative Research Centre for Tropical Rainforest Ecology and Management. Rainforest CRC, Cairns, pp. 25–6.

Sumner, J., Moritz, C. &. Shine, R. (1999). Shrinking forest shrinks skink: morphological change in response to rainforest fragmentation in the prickly forest skink (*Gnypetoscincus queenslandiae*). *Biological Conservation* **91**: 159–67.

Tilman, D., May, R. M., Lehman, C. L. & Nowak, M. A. (1994). Habitat destruction and the extinction debt. *Nature* **371**: 65–6.

Toh, I., Gillespie, M. & Lamb, D. (1999). The role of isolated trees in facilitating tree seedling recruitment at a degraded sub-tropical rainforest site. *Restoration Ecology* **7**: 288–97.

Tucker, N. I. J. (2000). Linkage restoration: interpreting fragmentation theory for the design of a rainforest linkage in the humid Wet Tropics of north-eastern Queensland. *Ecological Management and Restoration* **1**: 35–41.

Tucker, N .I. J. (2001). Wildlife colonisation on restored tropical lands: What can it do, how can we hasten it and what can we expect? In *Forest Restoration for Wildlife Conservation*, Elliott, S., Kerby, J., Blakesley, D., Hardwicke, K., Woods, K. & Anusarnsunthorn, V. (eds). International Tropical Timber Organisation and the Forest Restoration Research Unit, Chiang Mai University, Thailand, pp. 279–95.

Tucker, N. I. J. & Murphy, T. (1997). The effect of ecological rehabilitation on vegetation recruitment: some observations from the Wet Tropics of north Queensland. *Forest Ecology and Management* **99**: 133–52.

Tucker, N. I. J. & Simmons, T. M. (2004). Animal–plant interactions in tropical restoration: observations and questions from north Queensland. In *Animal–Plant Interactions in Rainforest Conservation and Restoration*, Kanowski, J., Catterall, C. P., Dennis, A. J. & Westcott, D. A. (eds). Cooperative Research Centre for Tropical Rainforest Ecology and Management. Rainforest CRC, Cairns, pp. 30–3.

Tucker, N. I. J., Wardell-Johnson, G., Catterall, C. P. & Kanowski, J. (2004). Agroforestry and biodiversity: improving the outcomes for conservation in tropical north-eastern Australia. In *Agroforestry and Biodiversity Conservation in Tropical Landscapes*, Schroth, G., Fonseca, G., Harvey, C. A., Gascon, C., Vasconcelos, H. & Izac, A. M. (eds). Island Press, Washington, DC, pp. 431–52.

Wahungu, G. (2000). Selective herbivory at rainforest edges and its effects on regeneration: a case study with red-necked pademelon *Thylogale thetis*. PhD thesis, Griffith University.

Wardell-Johnson, G. W., Kanowski, J., Catterall, C. P., Piper, S. & Skelton, D. (2005). Rainforest timber plantations and plant biodiversity: the Community Rainforest Reafforestation Program in context. In *What Have We Learnt from Growing Rainforest Timber Species in the Last 10 Years?* Erskine, P., Lamb, D. & Bristow, M. (eds). Rainforest CRC and RIRDC, Canberra, pp. 162–82.

Webb, L. J. (1966). The rape of the forests. In *The Great Extermination. A Guide to Anglo-Australian Cupidity, Wickedness and Waste*, Marshall, A. J. (ed.). Heinemann, London, pp. 156–205.

White, E., Tucker, N., Meyers, N. & Wilson, J. (2004). Seed dispersal to revegetated isolated rainforest patches in North Queensland. *Forest Ecology and Management* **192**: 409–26.

Williams, S. E. & Pearson, R. G. (1997). Historical rainforest contractions, localised extinctions and patterns of vertebrate endemism in the rainforests of Australia's wet tropics. *Proceedings of the Royal Society of London Series B* **264**: 709–16.

Willson, M. F., Irvine, A. K. & Walsh, N. G. (1989). Vertebrate dispersal syndromes in some Australian and New Zealand plant communities, with geographic comparisons. *Biotropica* **21**: 133–47.

Winter, J. W., Bell, F. C., Pahl, L. I. & Atherton, R. G. (1987). Rainforest clearfelling in northeastern Australia. *Proceedings of the Royal Society Queensland* **98**: 41–57.

Woodford, R. W. (2000). Converting a dairy farm back to a rainforest water catchment. *Ecological Management and Restoration* **1**: 83–92.

Wright, S. J. (2002). Plant diversity in tropical forests: a review of mechanisms of species coexistence. *Oecologia* **130**: 1–14.

Wunderle, J. M. (1997). The role of animal seed dispersal in accelerating native forest regeneration on degraded tropical lands. *Forest Ecology and Management* **99**: 223–35.

Young, T. P. (2000). Restoration ecology and conservation biology. *Biological Conservation* **92**: 73–83.

42 MONITORING THE OUTCOMES OF REFORESTATION FOR BIODIVERSITY CONSERVATION

John Kanowski[1]*, Carla P. Catterall[2]* and Debra A. Harrison[2]*

[1]Centre for Innovative Conservation Strartegies, School of Environment, Griffith University, Nathan, Queensland, Australia
[2]Griffith School of Environment, Griffith University, Nathan, Queensland, Australia
*The authors were participants of Cooperative Research for Tropical Ecology and Management

Introduction

The reforestation of cleared rainforest land is a recent activity in Australia. In the Wet Tropics of north Queensland two major reforestation schemes, the Community Rainforest Revegetation Program (CRRP) and the Wet Tropics Tree Planting Scheme (WTTPS), were established after the listing of the Wet Tropics World Heritage Area in 1988. Since 1997, the Natural Heritage Trust (NHT) has been the main source of funds for conservation-oriented revegetation works in the Wet Tropics region and elsewhere in Australia (Catterall *et al.* 2004; Vize *et al.* 2005).

None of the major reforestation schemes in the Wet Tropics have incorporated a formal monitoring programme to assess their outcomes for biodiversity, even though the stated goals of the schemes included addressing the problems of land degradation (CRRP; Vize *et al.* 2005), environmental repair (WTTPS; Gleed 2002) and restoring habitat (NHT; Australian Government 2006a). Although projects funded by the NHT are required to report on administrative outcomes (e.g. area planted, volunteer hours expended), the monitoring of ecological outcomes is optional. Not surprisingly, few NHT-funded reforestation projects

in the Wet Tropics have conducted quantitative monitoring of vegetation or fauna (Catterall & Harrison 2006) (Box 42.1). Freeman (2004) found that most community groups in the Wet Tropics lacked the time, personnel, resources, motivation and expertise to monitor projects properly. A wider review of NHT projects in the Wet Tropics concluded that 'monitoring and evaluation is carried out randomly and not against set criteria … [there] is a distinct lack of knowledge regarding monitoring and evaluation methodology and practices, and the lack of capacity (knowledge, time, money) within many groups to undertake such work in the first place' (Greening Australia 2003). In fact, these problems are endemic to the NHT programme. A review of over 13 000 NHT-funded revegetation projects across Australia found that 'only a handful' of projects had monitored the outcomes of revegetation for biodiversity (Freudenberger & Harvey 2003).

A list of revegetation projects in the Wet Tropics that have been monitored for their biodiversity outcomes follows. The community group TREAT (Trees for the Evelyn and Atherton Tablelands), along with researchers from CTR (Centre for Tropical Restoration), a small team within the Queensland Parks and Wildlife Service with close links to TREAT, has monitored

Box 42.1 Documenting reforestation projects in the Wet Tropics 1997–2001

Catterall and Harrison (2006) compiled available records of revegetation projects funded by the Natural Heritage Trust (NHT) in the Wet Tropics between 1997 and 2001. During this period, there were 49 NHT-funded projects focused primarily on the reforestation of cleared land in the Wet Tropics. In aggregate, these projects comprised restoration works (mainly tree-planting) on many sites, with a total area of 644 ha, at a total cost of over Au$16 million ($25 600 ha^{-1}), of which 36% was funded by the NHT. Community groups comprised 29 of the 44 different proponents involved in these activities, and volunteer labour contributed around 25% of the total cost of the projects.

Monitoring of project outcomes was a requirement of NHT funding. However, the nature of any monitoring programme was determined entirely by proponents, and the conduct of proposed monitoring activities was not audited by the NHT.

Table 42.1 lists monitoring activities reportedly undertaken within 87 vegetation-related projects that included both reforestation and the protection or repair of existing remnants. The majority of the projects (83%) reported some form of monitoring, with up to four different monitoring activities conducted per project. However, considering just those monitoring activities aimed at assessing vegetation or fauna, 44% of reported activities consisted simply of taking photos, while an additional 31% were unspecified. Fewer than 20% of activities comprised surveys of plants or animals, and these surveys were not necessarily quantitative or robustly designed. This low return of information on the outcomes of previous reforestation works has provided the stimulus for developing a regionally consistent approach to documenting and monitoring of the outcomes of reforestation in the Wet Tropics (Kanowski & Catterall 2007).

Table 42.1 Monitoring activities reported by vegetation-related projects funded by the NHT in the Wet Tropics 1997–2001

Monitoring activities per project	Projects conducting specified number of monitoring activities	Number of projects that monitored attributes			
		Vegetation	Fauna	Pest species	Water quality
Attributes targeted for monitoring					
0	15	—	—	—	—
1	35	31	0	1	3
2	26	25	9	3	15
3	9	11	7	2	6
4	2	1	4	2	1
Total	87	68	20	8	25

Focus of monitoring activities	Total number of monitoring activities	Type of monitoring activity			
		Photo	Survey	Growth rates	Unspecified
Type of monitoring					
Vegetation	68	39	11	6	12
Fauna	20	0	5	0	15

Source: Catterall and Harrison (2006).

changes in vegetation structure, plant recruitment and assemblages of birds and/or small mammals on a number of revegetated sites (Tucker & Murphy 1997; Tucker 2000, Grundon *et al.* 2002; White *et al.* 2004). Some in addition, a number of these revegetation projects have also been surveyed by universities and other research institutions; for example, Jansen (1997, 2005) monitored leaf litter invertebrates and birds, and Grove and Tucker (2000) surveyed saproxylic beetles, in one corridor site revegetated by TREAT and CTR while Ward *et al.* (2003) monitored rodents in another a revegetated site. The most comprehensive monitoring of revegetation projects in the Wet Tropics to date, at least in terms of number of sites assessed and number of taxa surveyed, has been the collaborative surveys of birds, reptiles, invertebrates, plants and forest structure in different types of reforestation, pasture and rainforest reference sites (Kanowski *et al.* 2003, 2006; Proctor *et al.* 2003; Catterall *et al.* 2004; Tucker *et al.* 2004; Wardell-Johnson *et al.* 2005). While this project has provided robust comparisons of the biodiversity outcomes of different types of revegetation, it is nevertheless a snapshot survey, mostly of young reforested sites. The structure, composition and availability of resources in reforested sites are likely to change considerably as plantings mature (Kanowski *et al.* 2004, 2005a).

Consequences of the lack of monitoring of reforestation projects

Because most reforestation projects in the Wet Tropics have not been monitored, the significant investment in reforestation in the region has resulted in only modest advances in our understanding of the factors affecting the biodiversity value of reforested sites (Catterall *et al.* 2005; Catterall *et al.*, Chapter 41, this volume). The insights that have been obtained have generally been derived from post hoc studies that have often been poorly replicated or confounded across key factors (Catterall *et al.* 2004). Opportunities for examining trade-offs or synergies between biodiversity conservation and other objectives of reforestation (e.g. production, water quality), which are fundamental to rational land-use planning, have simply not been addressed in the design of the major reforestation schemes, such

that consideration of these issues for the Wet Tropics is still mainly speculative (Catterall *et al.* 2005; Kanowski *et al.* 2005b). Of course, this problem is not restricted to the Wet Tropics or even Australia. The failure of many restoration projects worldwide to include a monitoring component has been strongly criticized by ecologists because of lost opportunities for learning about and improving restoration practices (e.g. Block *et al.* 2001; Lake 2001; Gillilan *et al.* 2005). Nevertheless, the problem has been felt particularly keenly by Australian ecologists because the NHT programme does not fund research (Australian Government 2006a). Hence, the significant investment in restoration by the NHT programme has not been matched by a significant increase in knowledge about the consequences of restoration for biodiversity. Thus, Lake (2001) noted that,

in Australia, the practice of undertaking poorly designed restoration projects has been widespread; driven by political pressures to initiate community-based projects without interference from monitoring or research. In terms of ecology, very little of value has been learnt from such projects, and in terms of restoration, design inadequacy has meant that lessons learnt on the sites cannot be reliably evaluated to improve restoration on these sites or elsewhere.

More bluntly, Chapman and Underwood (2000) have stated: 'In Australia, much money and resources are wasted in on-the-ground projects which are poorly designed and seldom evaluated.'

Effective monitoring of community-based restoration projects

There is extensive recent literature on the monitoring of biodiversity in forests and other ecological communities (Boyle & Boontawee 1995; Stork & Samways 1995; Jermy *et al.* 1996; Krebs 1999; Noss 1999; Henderson & Southwood 2000; Elzinga *et al.* 2001; Downes *et al.* 2002; Nakashizuka & Stork 2002; Kitching *et al.* 2005; Magnusson *et al.* 2005; Spellerberg 2005), and the monitoring of restoration projects in particular (Michener 1997; Block *et al.* 2001; Holl & Cairns 2002; Society for Ecological Restoration 2002; Wardell-Johnson *et al.* 2002; Catterall *et al.* 2004). Readers seeking information on the design, implementation and analysis of monitoring programmes are referred to this literature.

Here we wish to address two issues pertinent to effective monitoring of community-based restoration projects, the prevailing type of restoration in the Wet Tropics and, indeed, much of Australia. These issues are to do with the specification of performance criteria for evaluating the success of restoration projects, and the rigour with which restoration projects should be monitored.

Specifying performance criteria for restoration projects

Monitoring can be defined as the repeated survey of target variables and their assessment against performance criteria (Stork & Samways 1995; Spellerberg 2005). Performance criteria for ecological restoration projects are typically derived from comparisons with a reference system; for example, remnant forest similar to the type that existed on a site prior to clearing (Society for Ecological Restoration 2002). A fundamental problem with assessing the success of rainforest restoration projects is that, in Australia at least, most reforested sites are young (years to decades), whereas it is likely to take decades to centuries for many of the attributes of restored sites to converge on mature rainforest (Hopkins 1990; Guariguata & Ostertag 2001).

An alternative to judging young restored sites against mature reference systems would be to assess changes in restored sites against a target trajectory, using criteria appropriate for particular states in that trajectory (Society for Ecological Restoration 2002; Walker & del Moral 2003). Unfortunately, knowledge of succession in reforested rainforest, and how succession might be affected by the particular characteristics of a site or deflected by various disturbances, is currently insufficient to assess whether restoration is on track or not (Hopkins 1990; Kooyman 1996). A large number of case studies of restored sites, covering a wide variety of ecological situations (e.g. size, composition, proximity to remnant forest), would need to be documented for restoration practitioners to be able confidently to assess sites against an ideal restoration trajectory (Catterall *et al.* 2004). Such data are available for few ecosystems. One exception is the jarrah (*Eucalyptus marginata*) forests that were mined for bauxite in Western Australia. Data collected over 15 years

from more than 400 plots were used by Grant (2006) to develop a model of successional trajectories in rehabilitated jarrah forests and to identify deviations from the target trajectory. This study demonstrates the scale of the effort required to determine successional trajectories even when, as in the jarrah forest, the course of succession is largely dependent on the initial plant species composition of the rehabilitated site. A greater effort may be required to document successional trajectories in rainforests, because succession in rainforest is not only strongly dependent on the dispersal of seeds to disturbed sites, a process that is spatially and temporally variable, but also influenced by a series of complex interactions between animals and plants (Hopkins 1990; Catterall *et al.* 2004; Kanowski *et al.* 2004).

Even if successional trajectories can be identified, performance criteria will need to be tailored to specific revegetation models. For example, *ex situ* recruitment is likely to be a critical measure of success in the so-called Framework Species model, where pioneer species are planted to provide rapid canopy closure and conditions suitable for the recruitment of mature phase species (Goosem & Tucker 1995; Kooyman 1996). In contrast, recruitment is likely to be less important, at least initially, in a Maximum Diversity model, where a wide range of species are planted at a site. In this case, survival and *in situ* recruitment may be a more informative measure of success during the early stages of restoration. Suitable performance criteria would also need to be developed for rainforest fauna and ecological processes, such as seed predation and decomposition, in reforested sites (Catterall *et al.* 2004). At present, the responses of animals to rainforest succession are poorly understood, even for well known taxa such as birds in well studied landscapes such as the Wet Tropics (Crome 1990), while basic data on rates of ecological processes in restored rainforest are almost entirely lacking.

How rigorously should restoration projects be monitored?

There is widespread agreement in the scientific literature, cited above, that ideally the monitoring of restoration projects should be conducted with similar rigour as ecological research, that is, with clearly defined objectives, adequate sampling and replication, appropriate control and reference sites, and valid statistical analysis. Some authors are adamant that this is the only worthwhile approach to restoration (Chapman & Underwood 2000), while others draw a distinction between what may be a successful restoration and what may be a successful monitoring programme (Palmer et al. 2005). Regardless of this distinction, it is widely agreed that there are very few examples of restoration projects where monitoring has met the ideal scientific criteria. The reasons for this are both technical (i.e. difficult experimental conditions, such as a lack of suitable replicate treatment, control or reference sites) and social/political/economic, whereby proponents may not have the resources or skills necessary for monitoring, or may simply lack interest in conducting monitoring, particularly if projects do not have to demonstrate success against ecological objectives.

These constraints are especially acute in the monitoring of community-based restoration projects. Community groups are largely reliant on volunteers and few have the expertise to design, implement and interpret a scientific monitoring programme (Freudenberger & Harvey 2003). According to Freeman (2004), volunteers in environmentally focused community groups in the Wet Tropics are typically motivated to do something for the environment and may not perceive monitoring activities to be as valuable as growing or planting trees. The high turnover of volunteers in community groups also militates against the long-term commitment necessary for tracking the progress of revegetation. Some quotes from representatives of community groups interviewed by Freeman (2004) illustrate these issues: 'We don't have time to monitor', 'We don't know what to do', 'Our funding is stretched as it is', 'Our membership changes through time', 'We can see the trees growing and funding agencies aren't enforcing monitoring anyway', 'Community groups don't know where these results are going' and 'Monitoring is not necessary'.

According to Freeman (2004), the provision of specific funding for monitoring activities, training in survey techniques for volunteers and the availability of simple, standardized and relevant monitoring protocols would overcome a number of the obstacles that currently prevent most community groups from monitoring their revegetation projects. However, these measures may not address fundamental concerns held by community groups about the value of monitoring or the lack of interest in monitoring by volunteers. Freeman (2004) argued that funding bodies may need to engage professional ecologists to monitor restoration projects established by community groups. Similar conclusions were reached by Greening Australia (2003) in their review of NHT projects in the Wet Tropics and by Freudenberger and Harvey (2003) in their review of the entire NHT programme.

Recent developments in the monitoring and evaluation of reforestation projects in the Wet Tropics

An important recent development in the planning and funding of reforestation projects in Australia has been the establishment of regional Natural Resource Management organizations, partly funded by NHT funds, that are responsible for managing investment in regional restoration projects over the medium to long term (Australian Government 2006b). These institutional arrangements are far more conducive to the development of monitoring and evaluation programs than the short-term, insecure funding associated with previous reforestation schemes. In the Wet Tropics, Far North Queensland Natural Resource Management Ltd (FNQ NRM) has funded a number of monitoring and evaluation projects (FNQ NRM Ltd & Rainforest CRC 2004).

Monitoring and evaluation of revegetation works

One of the projects funded by FNQ NRM specifically targets the monitoring of community-based reforestation works. The project aims to develop a toolkit to help community groups to document, monitor and report on their revegetation works. The initial version of the toolkit includes protocols for documenting on-ground works (establishment statistics) and for monitoring vegetation condition in revegetation projects (basic indicators) (Kanowski & Catterall 2007). Subsequent versions are planned to address the monitoring of more specialized targets (advanced indicators).

This project addresses many of the concerns raised by Freeman (2004) in relation to the effective monitoring of community-based revegetation projects. First, rather than placing the unrealistic obligation of developing scientifically credible monitoring protocols on to community groups, the project has been contracted to ecologists with specific expertise in biodiversity surveys in reforested sites, including in the Wet Tropics. Second, the project is being developed with the assistance of revegetation practitioners and staff from FNQ NRM and the community groups themselves. This collaboration has the dual purpose of exposing the ecologists responsible for developing the project to the ideas, skills and constraints of the revegetation practitioners and community groups responsible for the on-ground works, and of promoting the value of monitoring among practitioners and community groups. Third, the timeframe of the project allows testing of the protocols by their intended users and subsequent revision of the protocols. The major components of the toolkit are discussed below.

Establishment statistics

Interpreting the results of a monitoring programme generally requires knowledge of the condition of sites prior to the on-ground works, as well as detailed information on the on-ground works themselves (Society for Ecological Restoration 2002). Unfortunately, many revegetation projects in Australia have been poorly documented. The review by Freudenberger and Harvey (2003) of NHT-funded revegetation projects across Australia found that very few had adequately documented the on-ground works and that only a few appeared to have collected baseline data. Furthermore, data reported by NHT-funded projects are not readily accessible by the public.

The monitoring and evaluation toolkit being developed for FNQ NRM specifies the collection of a range of establishment statistics to enable subsequent monitoring of projects for biodiversity. Some of these statistics are already required for projects funded by the NHT (e.g. site location, area planted, number of stems planted, budget, timeline), while other statistics have been included to provide the background data needed to interpret the monitoring results, including a basic ecological description of the site (soils, rainfall, landscape context), a description of the revegetation approach used in a project (e.g. framework species, maximum diversity) and the identification of suitable control and reference sites, if relevant. It is intended that information collected on revegetation projects will be maintained by FNQ NRM and made more accessible to interested parties.

Basic indicators

The selection of target variables is an important element of a monitoring programme. Criteria used to select target variables include relevance to project objectives, sensitivity to management practices, the feasibility and cost-effectiveness of surveys and whether variables can act as surrogates for other components of the ecosystem (Stork & Samways 1995; Holl & Cairns 2002). Ideally a diverse range of variables relating to habitat structure, biota and ecological processes would be surveyed to assess biodiversity, but community groups are limited in their ecological expertise (Freeman 2004; Spellerberg 2005). Hence, monitoring protocols intended for use by community groups need to identify target variables that can be easily and rapidly assessed in the field, given the available expertise. Details of the approach used to identify these target variables, termed basic indicators, in the monitoring toolkit that is being developed for FNQ NRM, are given in Box 42.2.

Advanced indicators

In some cases, it may be better to monitor directly the value of revegetation projects for various components of biodiversity than to rely on surrogates. For example, evaluation of the success of a reforestation project aimed at conserving the cassowary (*Casuarius casuarius*) would probably require surveys of the use of the reforested sites by cassowaries, rather than an assessment of vegetation structure or condition. Although community groups could commission local naturalists to survey such advanced indicators, they might still need guidance on appropriate survey methodologies, particularly if the naturalists have little training in survey design (Spellerberg 2005).

Monitoring the different stages of revegetation

Regardless of whether basic or advanced indicators are used to monitor revegetation projects, it is important that target variables are relevant to the stage of development of the revegetation. A key distinction is between the establishment phase of revegetation (the period

Box 42.2 Selection of basic indicators for monitoring reforestation projects

The development of a monitoring toolkit to help community groups to monitor reforested sites (Kanowski & Catterall 2007) required the identification of target variables, or basic indicators, that would:

1 provide information on the development of revegetated sites;

2 be correlated with the use of revegetated sites by rainforest wildlife;

3 be readily applied in the field by non-specialists.

Potential indicators were selected from the literature on the vegetation structure of rainforests (Webb *et al.* 1976), revegetated rainforest sites (Kooyman 1996; Kanowski *et al.* 2003) and Australian forests and woodlands (McElhinny 2002; Parkes *et al.* 2003; Gibbons *et al.* 2005).

The potential for these attributes to act as surrogates for rainforest fauna was examined using data from reforested sites in eastern Australia (see Catterall *et al.* 2004 for overview of sites and methods). For these analyses, rainforest birds and reptiles were identified from natural history observations. The habitat affinities of mites were determined from analysis of the relative occurrence of mite taxa in pasture and rainforest reference sites, while the affinities of beetles were derived from the similarity of assemblages in revegetated sites to reference rainforests. The results of these analyses (Table 42.2) show that nearly all the structural attributes assessed were potentially useful surrogates of rainforest fauna when the full set of reforested sites was considered. However, a much smaller subset of variables was correlated with rainforest taxa when analyses were restricted to young replanted sites. In young sites, six of thirteen structural attributes assessed were correlated with rainforest bird richness, but only two were correlated with the richness of rainforest reptiles and beetles, and none with mites.

Table 42.2 Correlation (*rs*) of structural attributes with the richness of rainforest birds, reptiles and mites, and the assemblage composition of beetles, in reforested sites in tropical Australia. Reforested sites included young monoculture plantations, young cabinet timber plantations, young ecological restoration plantings, old monoculture plantations and regrowth

Attribute	Birds	Reptiles	Beetles	Mites
All reforested sites (*n* = 35)				
Tree species richness	0.71*	0.66*	0.65*	0.52*
Canopy cover	0.75*	0.71*	0.68*	0.49*
Canopy height	0.57*	0.54*	0.52*	0.53*
Tree density	0.54*	0.56*	0.51*	0.34*
Large tree density	0.61*	0.60*	0.48*	0.59*
Basal area	0.68*	0.69*	0.54*	0.48*
Tree height diversity	0.52*	0.51*	0.41*	0.54*
Tree diameter diversity	0.43*	0.44*	0.26	0.29
Special life forms	0.59*	0.51*	0.44*	0.59*
Shrub density	0.63*	0.53*	0.51*	0.49*
Woody debris	0.46*	0.50*	0.50*	0.52*
Leaf litter	−0.05	0.02	0.01	0.06
Grass cover	−0.71*	−0.64*	−0.56*	−0.56*
Young replanted sites only (*n* = 20)				
Tree species richness	0.75*	0.58*	0.57*	0.32
Canopy cover	0.59*	0.70*	0.32	0.04
Canopy height	0.03	−0.12	−0.07	0.01
Tree density	0.46*	0.43	0.16	−0.01
Large tree density	−0.06	0.26	−0.04	−0.19
Basal area	0.49*	0.40	0.08	−0.16
Tree height diversity	−0.06	0.01	−0.25	−0.22
Tree diameter diversity	−0.07	0.01	−0.35	−0.34
Special life forms	0.26	0.30	0.31	0.15
Shrub density	0.65*	0.39	0.58*	0.18
Woody debris	0.07	−0.10	0.18	0.14
Leaf litter	−0.20	−0.20	−0.06	−0.10
Grass cover	−0.47*	−0.33	−0.31	−0.10

*P < 0.05.

Sources: Kanowski *et al.* (2003, 2006), Proctor *et al.* (2003), Catterall *et al.* (2004), Wardell-Johnson *et al.* (2005), Grimbacher *et al.* (submitted 2007).

of several years following planting until trees form a relatively closed canopy that suppress grasses) and the subsequent building and maintenance phases (Catterall *et al.*, Chapter 41, this volume). Appropriate targets to monitor during the establishment phase would include the survival and growth of planted trees, the development of canopy closure and the proportion of the ground covered by grasses and weeds that might suppress the recruitment of rainforest plants. Such targets would be directly related to site capture and their assessment could provide useful feedback on the maintenance requirements of a site. In the subsequent building and maintenance phases, a typical objective for restoration projects might be the development of a floristically and structurally diverse forest that provides habitat for native wildlife. Appropriate monitoring targets during these phases would include vegetation structure, floristic composition, the recruitment of plants and the use of reforested sites by wildlife.

Summary

• With some important exceptions, most reforestation projects in the Wet Tropics of north Queensland have not been monitored for their outcomes for biodiversity. Consequently, despite a large investment of public funds, the value of reforestation projects for biodiversity remains poorly known.

• The monitoring and evaluation of reforestation projects requires the development of performance criteria that are appropriate for particular stages and models of reforestation.

• Developing suitable criteria will require data from a large number of reforested sites, representing a range of ecological situations.

• The effective monitoring of reforestation projects poses significant challenges for community groups. Volunteer members of community groups typically lack sufficient training in survey design and evaluation, and may not perceive the value of information obtained from monitoring.

• A monitoring and evaluation programme involving collaboration between ecologists, revegetation practitioners and community groups has recently been instigated in the Wet Tropics region. The project has the potential to improve our knowledge of the biodiversity outcomes of reforestation, although it will require long-term commitment to realize its potential.

• Rapid advances in knowledge of the outcomes of reforestation for biodiversity and other objectives are likely to require the incorporation of experimental research and monitoring programmes within reforestation projects.

Acknowledgements

This project was funded by the Rainforest CRC and FNQ NRM Ltd. Researchers from the Quantifying Biodiversity Values of Reforestation project of the Rainforest CRC, particularly Peter Grimbacher, Stephen McKenna, Cath Moran, Scott Piper, Heather Proctor, Terry Reis, Nigel Tucker and Grant Wardell-Johnson, contributed ideas and data that have been used in developing the monitoring and evaluation project. The collaboration of staff from FNQ NRM Ltd and the advice of numerous revegetation practitioners and community groups are gratefully acknowledged. Cath Moran kindly commented on a draft of the manuscript.

References

Australian Government (2006a). Natural Heritage Trust (www.nht.gov.au/).

Australian Government (2006b). Natural Resource Management (www.nrm.gov.au/).

Block, W. M., Franklin, A. B., Ward, J. P. Jr., Ganey, J. L. & White, G. C. (2001). Design and implementation of monitoring studies to elucidate the success of ecological restoration on wildlife. *Restoration Ecology* **9**: 293–303.

Boyle, T. J. B. & Boontawee, B. (eds) (1995). *Measuring and Monitoring Biodiversity in Tropical and Temperate Forests.* Center for International Forestry Research, Bogor, Indonesia.

Catterall, C. P. & Harrison, D. A. (2006) *Rainforest Restoration Activities in Australia's Tropics and Subtropics.* Rainforest CRC, Cairns.

Catterall, C. P., Kanowski, J., Wardell-Johnson, G. *et al.* (2004). Quantifying the biodiversity values of reforestation: perspectives, design issues and outcomes in Australian rainforest landscapes. In *Conservation of Australia's Forest Fauna*, 2nd edn, Lunney, D. (ed.). Royal Zoological Society of New South Wales, Sydney, pp. 359–93.

Catterall, C., Kanowski, J., Lamb, D., Killin D., Erskine, P. & Wardell-Johnson, G. (2005). Trade-offs between timber production and biodiversity in rainforest plantations:

emerging issues and an ecological perspective. In *Reforestation in the Tropics and Subtropics of Australia Using Rainforest Tree Species*, Erskine, P. D., Lamb, D. & Bristow, M. (eds). RIRDC and Rainforest CRC, Canberra and Cairns, pp. 206–21.

Chapman, M. G. & Underwood, A. J. (2000). The need for a practical scientific protocol to measure successful restoration. *Wetlands (Australia)* **19**: 28–49.

Crome, F. H. J. (1990). Vertebrates and succession. In *Australian Tropical Rainforests: Science, Value, Meaning*, Webb, L. J. & Kikkawa, J. (eds). CSIRO Publications, Melbourne, pp. 53–64.

Downes, B. J., Barmuta, L. A., Fairweather, P. G., *et al.* (2002). *Monitoring Ecological Impacts: Concepts and Practices in Flowing Waters.* Cambridge University Press, Cambridge.

Elzinga, C. L., Salzer, D. W., Willoughby, J. W. & Gibbs, J. P. (2001). *Monitoring Plant and Animal Populations.* Blackwell Science, Oxford.

FNQ NRM Ltd & Rainforest CRC (2004). *Sustaining the Wet Tropics: A Regional Plan for Natural Resource Management 2004–2008.* FNQ NRM Ltd, Innisfail.

Freeman, A. (2004). Constraints to community groups monitoring plants and animals in rainforest revegetation sites on the Atherton Tablelands of far north Queensland. *Ecological Management and Restoration* **5**: 199–204.

Freudenberger, D. & Harvey, J. (2003). *Assessing the Benefits of Vegetation Enhancement for Biodiversity: A Framework.* Report for Environment Australia (www.ea.gov.au).

Gibbons, P., Ayers, D., Seddon, J., Doyle, S. & Briggs, S. (2005). *BioMetric Version 1.8.* NSW Department of Environment and Conservation (www.nationalparks.nsw.gov.au/npws.nsf/Content/biometric_tool).

Gillilan, S., Boyd, K, Hoitsma, T. & Kauffman, M. (2005). Challenges in developing and implementing ecological standards for geomorphic river restoration projects. *Journal of Applied Ecology* **42**: 223–7.

Gleed, S. (2002). *A Report on the On-ground Activities by the Wet Tropics Tree Planting Scheme for the Wet Tropics Vegetation and Biodiversity Management Program.* North Queensland Afforestation Association, Cairns.

Goosem, S. & Tucker, N .I. J. (1995). *Repairing the Rainforest: Theory and Practice of Rainforest Re-establishment in North Queenslands Wet Tropics.* Wet Tropics Management Authority, Cairns.

Grant, C.D. (2006). State-and-transition successional model for bauxite mining rehabilitation in the Jarrah forest of Western Australia. *Restoration Ecology* **14**: 28–37.

Greening Australia. (2003). *Bushcare Support 2003. Native Vegetation Management: A Needs Analysis of Regional Service Delivery in Queensland – Wet Tropics* (www.deh.gov.au/land/publications/nvm-qld/pubs/qld-wet-tropics.pdf).

Grimbacher, P.S., Catterall, C. P., Kanowski, J. & Proctor, H. C. (2007). Responses of beetle assemblages to different styles of reforestation on cleared rainforest land. *Biodiversity and Conservation* **16**: 2167–84.

Grove, S. J. & Tucker, N. I. J. (2000). Importance of mature timber habitat in forest management and restoration: what can insects tell us? *Ecological Management and Restoration* **1**: 62–4.

Grundon, N., Wright, J. & Irvine, T. (2002). *Pelican Point Revegetation, Atherton Tableland: An Example of a Community Participatory Project – Establishment and Measuring Post-Development Success.* TREAT, Atherton.

Guariguata, M. R. & Ostertag, R. (2001). Neotropical secondary forest succession: changes in structural and functional characteristics. *Forest Ecology and Management* **148**: 185–206.

Henderson, P. A. & Southwood, T. R. E. (2000). *Ecological Methods*, 3rd edn. Blackwell Science, Oxford.

Holl, K. D. & Cairns, J. Jr. (2002). Monitoring and appraisal. In *Handbook of Ecological Restoration. Volume 1*, Perrow, M. R. & Davy, A. J. (eds). Cambridge University Press, Cambridge, pp. 411–32.

Hopkins, M. S. (1990). Disturbance – the forest transformer. In *Australian Tropical Rainforests: Science, Value, Meaning*, Webb, L. J. & Kikkawa, J. (eds). CSIRO Publications, Melbourne, pp. 40–52.

Jansen, A. (1997). Terrestrial invertebrate community structure as an indicator of the success of a tropical rainforest restoration project. *Restoration Ecology* **5**: 115–24.

Jansen, A. (2005). Avian use of restoration plantings along a creek linking rainforest patches on the Atherton Tablelands, north Queensland. *Restoration Ecology* **13**: 275–83.

Jermy, C., Long, D., Sands, M., Stork, N. & Winser, S. (eds) (1996). *Biodiversity Assessment: A Guide to Good Practice.* HMSO, London.

Kanowski, J. & Catterall, C. P. (2007). *Monitoring Revegetation Projects for Biodiversity in Rainforest Landscapes. Toolkit version 1, revision 1.* Marine and Tropical Sciences Research Facility, Cairns (www.rrrc.org.au/publications/research reports.html).

Kanowski, J., Catterall, C.P., Wardell-Johnson, G.W., Proctor, H. & Reis, T. (2003). Development of forest structure on cleared rainforest land in eastern Australia under different styles of reforestation. *Forest Ecology and Management* **183**: 265–80.

Kanowski, J., Catterall, C. P., Reis, T. & Wardell-Johnson, G.W. (2004). Animal–plant interactions in rainforest restoration in tropical and subtropical Australia. In *Animal–Plant Interactions in Rainforest Conservation and Restoration*, Kanowski, J., Catterall, C. P., Dennis, A. J. & Westcott, D. A. (eds). Cooperative Research Centre for

Tropical Rainforest Ecology and Management. Rainforest CRC, Cairns, 20–3.

Kanowski, J., Catterall, C. P. & Wardell-Johnson, G. W. (2005b). Consequences of broadscale timber plantations for biodiversity in cleared rainforest landscapes of tropical and subtropical Australia. *Forest Ecology and Management* **208**: 359–72.

Kanowski, J., Catterall, C. P., Proctor, H., Reis, T., Tucker, N. I. J. & Wardell-Johnson, G. W. (2005a). Rainforest timber plantations and animal biodiversity in tropical and subtropical Australia. In *Reforestation in the Tropics and Subtropics of Australia Using Rainforest Tree Species*, Erskine, P. D., Lamb, D. & Bristow, M. (eds). RIRDC and Rainforest CRC, Canberra and Cairns, pp. 183–205.

Kanowski, J., Reis, T., Catterall, C. P. & Piper, S. D. (2006). Factors affecting the use of reforested sites by reptiles in cleared rainforest landscapes in tropical and subtropical Australia. *Restoration Ecology* **14**: 67–76.

Kitching, R. L., Boulter, S. L., Vickerman, G., Laidlaw, M. J., Hurley, K. L. & Grimbacher, P. S. (2005). *The Comparative Assessment of Arthropod and Tree Biodiversity in Old-World Rainforests*. The Rainforest CRC/Earthwatch Protocol Manual. Rainforest CRC, Cairns (www.jcu.edu.au/rainforest/publications/earthwatch_manual.htm).

Kooyman, R. M. (1996). *Growing Rainforest. Rainforest Restoration and Regeneration – Recommendations for the Humid Subtropical Region of Northern New South Wales and South East Queensland*. Greening Australia, Brisbane.

Krebs, C.J. (1999). *Ecological Methodology*. Benjamin/Cummings, California.

Lake, P. S. (2001). On the maturing of restoration: linking ecological research and restoration. *Ecological Management and Restoration* **2**: 110–15.

McElhinny, C. (2002). *Forest and Woodland Structure as an Index of Biodiversity: A Review*. Report commissioned by NSW NPWS (www.nationalparks.nsw.gov.au/npws.nsf/content/biometric_tool).

Magnusson, W. E., Lima, A. P., Luizao, R. *et al.* (2005). RAPELD: a modification of the gentry method for biodiversity surveys in long-term ecological research sites *Biota Neotropica* **5**(2) (www.biotaneotropica.org.br/v5n2/pt/abstract?point-of-view+bn01005022005).

Michener, W. K. (1997). Quantitatively evaluating restoration experiments: research design, statistical analysis, and data management considerations. *Restoration Ecology* **5**: 324–37.

Nakashizuka, T. & Stork, N. (2002). *Biodiversity Research Methods: IBOY in Western Pacific and Asia*. Kyoto University Press, Kyoto and Trans Pacific Press, Melbourne.

Noss, R. F. (1999). Assessing and monitoring forest biodiversity: a suggested framework and indicators. *Forest Ecology and Management* **115**: 135–46.

Palmer, M. A., Bernhardt, E. S., Allan, J. D. *et al.* (2005). Standards for ecologically successful river restoration. *Journal of Applied Ecology* **42**: 208–17.

Parkes, D., Newell, G. & Cheal, D. (2003). Assessing the quality of native vegetation: the habitat hectares approach. *Ecological Management and Restoration* **4**: S29–S38.

Proctor, H. C., Kanowski, J., Wardell-Johnson, G., Reis, T. & Catterall, C. P. (2003). Does diversity beget diversity? A comparison between plant and leaf-litter invertebrate richness from pasture to rainforest. In *Proceedings 5th Invertebrate Biodiversity and Conservation Conference*. Records of the South Australian Museum, Supplementary Series **7**: 267–74.

Society for Ecological Restoration Science and Policy Working Group (2002). *SER Primer on Ecological Restoration* (www.ser.org/).

Spellerberg, I. F. (2005). *Monitoring Ecological Change*, 2nd edn. Cambridge University Press, Cambridge.

Stork, N. E. & Samways, M. J. (1995). Inventory and monitoring of biodiversity. In *Global Biodiversity Assessment*, Heywood, V. H. (eds). Cambridge University Press, Cambridge.

Tucker, N .I. J. (2000). Wildlife colonisation on restored tropical lands: what can it do, how can we hasten it and what can we expect? In *Forest Restoration for Wildlife Conservation*, Elliott, S., Kerby, J., Blakesley, D., Hardwicke, K., Woods, K. & Anusarnsunthorn, V. (eds). International Tropical Timber Organisation and the Forest Restoration Research Unit, Chiang Mai University, Thailand, pp. 279–95.

Tucker, N. I. J. & Murphy, T. (1997). The effect of ecological rehabilitation on vegetation recruitment: some observations from the Wet Tropics of north Queensland. *Forest Ecology and Management* **99**: 133–52.

Tucker, N. I. J, Wardell-Johnson, G., Catterall, C. P. & Kanowski, J. (2004). Agroforestry and biodiversity: improving the outcomes for conservation in tropical north-eastern Australia. In *Agroforestry and Biodiversity Conservation in Tropical Landscapes*, Schroth, G., Fonseca, G., Harvey, C. A., Gascon, C., Vasconcelos, H. & Izac, A. M. (eds). Island Press, Washington, DC, pp. 431–52.

Vize, S., Killin, D. & Sexton, G. (2005). The Community Rainforest Reforestation Program and other farm forestry programs based around the utilisation of rainforest and other species. In *Reforestation in the Tropics and Subtropics of Australia Using Rainforest Tree Species*, Erskine, P. D., Lamb, D. & Bristow, M. (eds). RIRDC and Rainforest CRC, Canberra and Cairns, pp. 7–22.

Walker, L. R. & del Moral, R. (2003). *Primary Succession and Ecosystem Rehabilitation*. Cambridge University Press, Cambridge.

Ward, D., Tucker, N. I .J. & Wilson, J. (2003). Cost-effectiveness of revegetating degraded riparian habitats adjacent to macadamia orchards in reducing rodent damage. *Crop Protection* **22**: 935–40.

Wardell-Johnson, G., Kanowski, J., Catterall, C., McKenna, S., Piper, S. & Lamb, D. (2005). Rainforest timber plantations and the restoration of plant biodiversity in tropical and subtropical Australia. In *Reforestation in the Tropics and Subtropics of Australia Using Rainforest Tree Species*, Erskine, P. D., Lamb, D. & Bristow, M. (eds). RIRDC and Rainforest CRC, Canberra and Cairns, pp. 162–82.

Wardell-Johnson, G. W., Kanowski, J., Catterall, C. P., Proctor, H. C. & Reis, T. (2002). Measuring the restoration of rainforest biodiversity: a case study in research design, and its implications for establishing monitoring frameworks. *Biodiversity – The Big Picture. Proceedings of Southern Queensland Biodiversity Recovery Conference*, pp. 72–81.

Webb, L. J., Tracey, J. G. & Williams, W. T. (1976). The value of structural features in tropical forest typology. *Australian Journal of Ecology* **1**: 3–28.

White, E., Tucker, N., Meyers, N. & Wilson, J. (2004). Seed dispersal to revegetated isolated rainforest patches in north Queensland. *Forest Ecology and Management* **192**: 409–26.

43 THE FUTURE FOR FOREST-BASED INDUSTRIES IN THE WET TROPICS

Steve Harrison* and John Herbohn*

School of Natural and Rural Systems Management, University of Queensland, Gatton, Queensland, Australia
*The authors were Particpants of Cooperative Research for Tropical Ecology and Management.

Introduction

Since 1992, various research projects conducted by members of the Cooperative Research Centre for Tropical Rainforest Ecology and Management (Rainforest CRC) have shed light on socio-economic issues of plantation forestry in the Queensland Wet Tropics. This chapter reviews research findings, particularly with respect to plantation options and organizational support measures relevant to the re-establishment of a forest industry in tropical north Queensland based on existing cleared farm land through non-industrial or farm forestry.

Within the chapter a number of questions are raised. Should plantation forestry be promoted in the Queensland Wet Tropics? If so, what form should it take in terms of ownership type, species choice, method of financing and target products and markets? What support measures would be most cost-effective? The following section reviews the timber industry in the Wet Tropics. Some of the particular industry characteristics are then outlined. The main industry constraints are examined, together with measures that may be adopted to overcome them. Next, various futures for a forestry industry are explored. This is followed by an assessment of the various forestry support infrastructure and measures.

Historical context to forest-based industries

There is a long tradition and rich history of forestry in the Wet Tropics of Queensland, based on logging of native rainforest cabinetwoods and eucalypts, particularly on the Atherton Tablelands. The early history of forestry in north Queensland has been described by Gould (2000) and Lamb *et al.* (2001).

A log quota system for timber mills was introduced in the late 1940s. The aggregate quota was 207 000 m^3 in 1979, but was progressively reduced to 60 000 m^3 (the estimated sustainable yield) in 1986. The timber industry suffered a further severe contraction with the listing of the Wet Tropics of Queensland World Heritage Area (WTWHA) in 1988.

The progressive reductions in log allocation and finally total logging ban over nearly 1 million hectares due to the World Heritage listing of the Wet Tropics rainforests led to a severe contraction of the north Queensland timber industry, with closure of many timber mills. From about 40 licensed mills in the 1970s, the number of mills declined spectacularly. Information complied by Smorfitt (1999) from sawmilling records indicated that shortly after World Heritage listing there were approximately 120 portable sawmills (eight registered[1]) and four fixed-site mills operating in North Queensland, the largest of which had annual

Table 43.1 Fixed-site hardwood sawmills in the Wet Tropics shortly after World Heritage listing

Owner	Location	Annual log allocation
G. J. Keough	Evelyn Central	590 m³ private timber only
L. J. and B. A. Johnston	Tarzali	790 m³ private and 960 m³ Crown
G. T. and J. W. Lawson, trading as AUZ Building Supplies Pty Ltd	Innisfail	2090 m³ private and 600 m³ Crown
Rankine Bros Enterprises	Kairi	1999 m³ unrestricted

log throughput volumes as indicated in Table 43.1. Portable sawmills are recognized as having a role in farm forestry where volumes are small and spatially dispersed (Smorfitt *et al.* 2003).

Development of plantation forestry

The first reforestation activity in north Queensland was the planting of red cedar between 1903 and 1906 (Gould 2000). More recent plantings of this icon rainforest species were made with southern silky oak as a cover crop, on the Atherton Tablelands. State government planting of Caribbean pine (*Pinus caribaea* var. *hondurensis*) in north Queensland commenced in 1966–7, with clearing of about 400 ha of native forest per year undertaken to establish the plantations (Kent 2004). Other small-scale plantings took place from the 1950s through to the 1980s, including school forestry plots and farm forestry, under various government subsidy schemes.

The strong national expression of support for natural values of the Wet Tropics at the time of the World Heritage listing of rainforests led to interest in replacing rainforest logging with a plantation resource, to ensure survival of the timber industry as a source of revenue generation and employment. However, little private investment in forestry has taken place, some possible explanations being the long payback period and low prices for farm-grown timber associated with 'the deliberate government policy of maintaining artificially low timber prices' (Lamb *et al.* 2001: 17), as well as lack of knowledge about the growing of native rainforest timber species in plantations.

According to the australian Academy of Technological Sciences and Engineering (AATSE 1988), hoop pine plantations date back to 1917, with some plantings of open-rooted nursery stock on clear-felled areas in Queensland. Survival rate was markedly improved after 1922, when Deputy Forester Weatherhead of the Queensland Forest Service developed and introduced a technique of planting from metal tubes that helped the young plants to withstand the strong weed competition.

A breeding programme was begun in 1951 and a seed orchard in 1957. Propagation proved difficult using conventional grafting methods but in 1959 a new method, bark patch grafting, was developed as a result of research that showed major differences in the biology of the species, particularly in its bud systems, when compared with *Pinus* species. Two clonal seed orchards were established to produce stock of improved growth rate and stem straightness and this facilitated further genetic improvement. Queensland now has about 40 000 ha of hoop pine plantations that currently account for one-quarter of all publicly owned plantations in Queensland (DPI Forestry 2004). NSW also started to establish hoop pine plantations, in the 1930s, but phased them out in the mid-1950s because of their high maintenance cost (AATSE 1988).

The current north Queensland timber industry

Stakeholders in the Wet Tropics timber industry

In examining the timber industry in the Wet Tropics, it is necessary to recognize the diversity of groups that have a legitimate interest in the industry. A list of these (which is not exhaustive) is presented as Table 43.2. The main focus in this section is on plantation owners and timber processors, as the most directly financially involved in the industry, but it is recognized that for a business case for forestry expansion to be adopted it must have wide stakeholder support.

A brief overview of the current timber industry

The timber industry in the Wet Tropics operates within the structure of the state's timber industry.

Table 43.2 Stakeholders in the Wet Tropics timber industry

Stakeholder type	Specific groups
Plantation owners or tree growers	Government, industrial, farm
Timber processors	Softwood and hardwood mills, joinery plants
Timber users	Builders, furniture makers, woodturners
Community groups	Tableland residents in general, TREAT, Landcare groups, tree kangaroo and mammal group
Regulatory agencies	DPI&F, NRMW, DSDI, LGAs, WTMA, NRMW
Researchers	UQ, Rainforest CRC, MTSRF, JCU and CSIRO

Three characteristics stand out in defining the timber industry in Queensland: (i) the dominance of the south-east region in terms of the location of the timber plantations, processors and markets; (ii) the dominance of softwoods in the plantation sector; and (iii) the rapid expansion of short-rotation hardwood plantations.

To place the timber industry in the Wet Tropics in context, the majority of timber plantations and processing occurring in Queensland is based in the south-east region (DPI Forestry 2004; MBAC 2005). Eighty-five per cent of the plantations in Queensland are located in the south-east region, and timber processing companies in south-east Queensland use 87% of the state's log supplies (MBAC 2005). Most of the total plantation estate (92%) is comprised of softwood plantations, and most of the plantations (90%) are owned by the state government (MBAC 2005). In the south-east region, most of the hoop pine timber is processed by the Hyne and Son mill at Imbil (DPI Forestry, undated).

There has been a change in the types of new plantations established in Australia over the past 10 years driven largely by private investment facilitated through managed investment schemes (MIS) (Herbohn & Harrison 2004). In the 1960s and 1970s the majority of plantations established involved state government agencies utilizing long-rotation softwood species. According to Studdert (2006: 35), 'Since 1997 another 700 000 ha have been planted, largely with private investment money, and almost entirely of hardwoods.' The new plantings have been primarily in Western Australia, Victoria, Tasmania and New South Wales (FWPRDC 2005). Nearly 80% of these new plantations

are short-rotation hardwood plantations designed to produce wood chip for export. The other new plantation areas consist of long rotation temperate hardwoods (12%) and long rotation softwood (10%), with less than 1% long rotation tropical hardwoods (FWPRDC 2005). In the past 5 years the companies that have driven the plantation establishment in Western Australia and the southern states under these schemes have begun to acquire land through purchase and lease in central Queensland for short rotation plantation development (Lewty 2005).

Industrial timber supply and processing in the Wet Tropics

The timber industry is sharply segmented in the Wet Tropics, particularly between:
- Plantation softwood species (mainly Caribbean and hoop pine, the main commercial forestry), and native rainforest and eucalypt species (generally grouped as hardwoods) from private land.
- Industrial growers and processors and the non-industrial timber industry sector.
- The Atherton Tablelands and coastal areas. Because of the steep winding roads between the coast and the Atherton Tablelands (the Kuranda Range section of the Kennedy Highway; the Gillies Highway Gordonvale–Yungaburra section; and the Palmerston Highway from near Innisfail to Milla Milla) these areas are, to a large degree, separate log markets.

Commercial or industrial timber supply
By far the main timber grower in the Wet Tropics at present and in the foreseeable future is Forestry Plantations Queensland (FPQ), a government-owned corporation that replaced DPI Forestry in May 2006, and is independent of the Department of Primary Industries and Fisheries (DPIF). Kent (2004) reported a DPI Forestry plantation area in North Queensland of 14 614 ha. The largest area of plantations is at Kennedy, near Cardwell, on the coastal plain. Planting with Caribbean pine commenced in 1966–7. On the Atherton Tablelands, the FPQ plantation estate includes 2367 ha of Caribbean pine and 1057 ha of hoop pine. Tablelands plantations are located at Kuranda (about 1600 ha), Danbulla (1100 ha), Wongabel (220 ha) and Gadgarra (350 ha). The latter two sites comprise mainly hoop pine first established in the 1940s.

Industrial timber processing

According to DPI Forestry (2004), 13% of the licensed primary processing plants in Queensland (of the total of 321) operate in north Queensland, processing 1% of the state's log timber. On the Atherton Tablelands, softwoods are processed at a single mill (Ravenshoe Timbers), which is the sole customer for the FPQ hoop pine and Caribbean pine plantations. Ravenshoe Timbers (operated by D. Simms and Sons) recommenced operation at a mill site at Ravenshoe that had been closed as a result of World Heritage listing of Wet Tropics rainforests in 1988, and utilizes almost exclusively timber from Crown plantations. The Ravenshoe mill has a capacity of about 30 000 m³ per year. Athough this is small by international standards,[2] the mill uses modern technology, including two finger-jointing plants, computer-controlled sawing patterns and laser-guided sawing, as well as bandsaws that have a high sawn-timber recovery rate (about 50%). The mill has a preference for hoop pine, as a timber having superior qualities to Caribbean pine, but over 50% of its timber resource is the latter species. Ravenshoe Timbers also has a plant in Cairns, and is constructing a softwood mill with annual roundlog capacity of 47 000 m³ near Mareeba (D. Simms 2006, personal communication), and at least one other small softwood mill operates on the north Queensland coastal strip.

The softwood timber resource on the Atherton Tablelands, of about 2400 ha of Caribbean pine and 1100 ha of hoop pine, much of which is now at a harvestable age, is sufficient to meet the requirements of Ravenshoe Timbers and support some log exports out of the area.

Hardwood timbers utilized in north Queensland include eucalypts (used for lumber and poles) and rainforest species (some of which are prized for furniture production). The Tarzali and Gadgarra hardwood mills also operate on the Atherton Tablelands, obtaining relatively small amounts of hardwood from private and Crown land – even though there are presently approximately 95 000 ha of rainforest on private landholdings (Annandale 2002, cited in Weston & Goosem 2004). These other sawmills do not mill hoop and Caribbean pine, and would have to spend of the order of Au$1 million to install steam treatment plant to handle the high resin content of these timbers. Hence the current Atherton Tablelands softwood log market approximates the form of a bilateral monopoly (single seller and buyer), for which prices are determined by negotiation rather than market forces.[3] Should private large-area eucalypt plantings take place, as has happened in central and southern Queensland, it is likely that new sawmills would be built and a pulpwood plant developed.

The timber activities of Pentarch Forest Products

A sale of 3 million m³ of plantation Caribbean pine over 15 years from FPQ softwood plantations at Ingham, Cardwell, Atherton and Cathu was recently made to Melbourne-based Pentarch Forest Products, which is expected to create 'additional employment in harvesting, haulage, log export, value-adding and plantation establishment activities' (Palaszczuk 2004). Palaszczuk further noted that Pentarch would initially export 80% of the timber it harvested, with the remainder going to Townsville Pallet and Crate Manufacturers and to Ravenshoe Timbers, but would have its own sawmill operating by 2007, and planned to establish 5000 ha of its own plantations, but this did not eventuate.

The non-industrial (farm) plantation area and species grown

Over about the past 50 years, various timber tree species have been planted on private land in north Queensland, and particularly on the Atherton Tablelands. As has typically been the case elsewhere in Australia, these programmes have been of short-term duration, unlike the more stable forestry support programmes in some other countries (Harrison et al. 2001). Hoop pine (Araucaria cunninghamii) has been commonly used in farm woodlots, fenceline plantings, silvopastoral systems (as evaluated by Hardman et al. 1985) and urban forestry in the region. Caribbean pine was also promoted by government and planted on farms.

In the 1990s, following World Heritage listing of the Wet Tropics, two schemes based on native tree species, the Community Rainforest Reforestation Scheme (CRRP) (described by Vize et al. 2005) and the Plantation Joint Venture Scheme (PJVS) (Anderson & Halpin 1998, Harrison et al. 1999), were promoted by government. The CRRP introduced a new era in tree planting, with a focus on native rainforest and eucalypt tree species, motivated in part as a means of replacing the timber resource lost by World Heritage listing of the Wet Tropics rainforests. The programme initially had a list of over 150 species, though this narrowed down quickly as experience was gained with establishing these species in plantations. Herbohn *et al.* (2001) reported that the most widely planted species in the CRRP by area planted were red mahogany (*Eucalyptus pellita*, 12.7% of plantings), hoop pine (10.3%), Queensland maple (*Flindersia brayleyana*, 7.7%) and kauri pine (*Agathis robusta*, 6.0%). A similar ranking in terms of total number of seedlings planted under the CRRP for all shires was reported by Vize *et al.* (2005), except that Gympie messmate (*Eucalyptus cloesiana*) ranks third after hoop pine.

The PJVS in north Queensland has involved equity sharing between landholders and FPQ using legally binding profit à prendre share-farming agreements. Eucalypt species and hoop pine were planted, with areas of about 10 ha (the stated minimum for participation in the scheme) on each farm. Eucalypt species and hoop pine were planted on 16 farms, with a total area of only 160 ha, over the three years 1997–9, after which no new plantings were made. FPQ has since expanded its joint-venture programme in the south-east of the state, with 5025 ha of hardwood plantations established by 2003 and plans to establish a further 5000 ha of hardwood plantations by 2009 (DPI Forestry 2004).

Vize *et al.* (2005: 10), drawing on unpublished sources, estimated that there currently exists an area of 3100–3200 ha of 'mainly mixtures of eucalypts, maples, hoop pine and other rainforest species such as silver quandong' in the Wet Tropics. This area includes 1780 ha on about 600 farms under the CRRP, about 850 ha under the Wet Tropics Tree Planting Scheme, 160 ha under the PJVS and less than 100 ha under the Treecare programme. Harrison *et al.* (2004) noted that landholders are applying silviculture to – and intend to harvest – a substantial proportion of their CRRP plantings.

Cooperation between sectors and expansion of private forestry

Some softwood production takes place on private land, but small growers are in general locked out of the softwood market. When Ravenshoe Timbers obtains a log allocation from FPQ, the performance clauses in the agreement require that Ravenshoe Timbers takes its annually allocated amount. The company is required to pay for this amount of timber, regardless of whether it uses it, which acts as a barrier for purchasing timber from private growers. Ravenshoe Timbers has recently purchased some hoop pine logs from private growers, including salvage timber following tropical cyclone Larry. The Caribbean pine resource on private land, grown with government encouragement, and generally of low quality due to lack of silviculture, now appears unsaleable despite the efforts of the North Queensland Timber Cooperative to find a market for the material (Skelton 2005).

As noted by Skelton (2004: 82), 'There are many cabinet-makers and small-scale timber users but they source their local cabinetwood timber somewhat ad hoc from local millers.' Smorfitt *et al.* (2004) examined various aspects of utilization of cabinetwoods in north and south Queensland, as well as the demand by consumers for products made out of cabinetwoods, eucalypts and composite timbers. They noted that some consumers are prepared to pay higher prices for items made out of rainforest timbers than those of eucalypts or composite timber, but that cabinetmakers hold only small stocks of rainforest timbers, operating on a just-in-time basis.

The tree planting schemes have so far made little contribution to expansion of the north Queensland timber industry. As noted by Skelton (2004: 80), although the CRRP established nearly 2000 ha of mainly mixed-species plantings on private land, it 'failed to re-stimulate the timber industry, but given the short timeframe and the intense anger in the community affected [by the World Heritage listing], any other outcome would have been a miracle!' Most local governments have been supportive of farm forestry, although, as noted by Lamb *et al.* (2001: 18), there has been some negativity to planting trees on agricultural land, because it is 'a source of pests (e.g. rats, foxes) and earlier farm plantings … were viewed as economic failures'.

Private Forestry North Queensland (PFNQ) is seeking to compile an inventory of the timber resources on private lands (Skelton 2004). In relation to the development issues confronting the industry in FNQ, Skelton (2004: 85) observed that 'the real issue is to establish a large enough new resource to re-establish a viable industry'.

Complementarities between industrial and small-scale forestry

It has sometimes been suggested that the fledgling farm-forestry movement could piggyback on established industrial forestry (see, e.g., Herbohn & Harrison 2004), to become established on a firmer footing. This raises the question of whether complementarities exist between industrial and farm forestry. Complementarities would be possible in a number of forestry activities, including:

- land-use planning and land-use approval;
- nursery establishment and supply of high-quality seedlings;
- silvicultural systems, including availability of contractors;
- log marketing;
- timber salvage after storm damage;
- timber harvesting, transport, milling, and value adding;
- timber exporting.

In practice, industrial and small-scale forestry appear to rely on separate nursery seedling sources. There are probably some complementarities with regard to the development of a pool of contractors for land preparation, tree planting, pruning and thinning, and timber logging and hauling. In terms of marketing, there appears to be some competitive (and anti-competitive) behaviour by the industrial forestry sector. The situation differs strikingly from that existing in some countries; for example, in the Black Forest area of south-west Germany the state forest service strongly supports private growers, particularly in timber marketing and in timber salvage, following severe storm events (Hartebrodt 2004). Cooperation and collaboration can potentially help both small-scale and industrial forestry holders.

The conditions for forest industries to survive and develop

In order for forest industries in the Wet Tropics to survive and develop further, a number of conditions must be met. The first is that plantation forestry must be profitable, preferably in its own right, or alternatively with the assistance of government through direct and indirect payments for public good benefits generated by plantations. The latter is the case in Europe, where subsidies can comprise up to 50% of the revenue from forests. In addition, there needs to be sufficient suitable land and appropriate species for forestry, along with

an appropriate social setting for forest-based industries to develop.

Profitability of plantation forestry

The most reliable plantation financial performance information for a native species in North Queensland relates to hoop pine. The generic hoop pine financial analysis of DPI Forestry (1996) estimates an internal rate of return of 6.25% for a 45 year rotation. An analysis of the potential returns from hoop pine plantations on the Atherton Tablelands suggests that returns are in the order of 5–6% (Herbohn & Harrison, in preparation). Herbohn and Harrison (2000) reported internal rates of return (IRRs) for 11 native species in north Queensland, estimated using the Australian Cabinet Timbers Financial Model, including 8.3% for rose gum (*Eucalyptus grandis*), 7% for Gympie messmate (*E. cloesiana*) and 6.3% for hoop pine. Little information is available about the profitability of mixed species plantings, although Herbohn and Harrison (2001) estimated a small positive net present value (NPV) at a discount rate of 8% for a Gympie messmate/Queensland maple mixture. There would appear to be moderately high financial risk in forestry investment (Emtage *et al.* 2001), although the risk may be countercyclical to other farm financial risk (Louis 2000).

There are many factors that determine stumpage prices that growers receive for their timber. In north Queensland, thin markets for rainforest cabinet timbers adversely affect stumpage prices. These thin markets (and the resulting lower prices associated with lack of demand) are largely owing to the uncertain reliability of supply. Many end users, such as cabinetmakers, choose readily available imported species and southern hardwoods in preference to rainforest cabinet timbers because of difficulties in obtaining ready supplies. For example, Herbohn *et al.* (2001: 101) noted that 'Brazilian oak, which is almost indistinguishable from locally sourced northern silky oak (*Cardwellia sublimis*), was recently selling for $2,200/m³. At the same time, *C. sublimis* was selling for $1,800/m³.' Uncertainty of regular supplies also means that processors are unwilling to invest in new equipment, with the resulting lower recovery rates also contributing to low stumpage prices. In addition, low and erratic volumes of timber mean that processors are unlikely to achieve economies of scale and are probably operating at inefficient throughput volumes, thus driving up costs and reducing the stumpage prices they can pay while still remaining profitable.

A question remains about whether forestry, particularly farm forestry based on cabinet timbers, is profitable. This is due to a combination of the low stumpage prices currently being received and long rotation periods. Many landholders who have established farm woodlots believe that in the short term, forestry will result in higher land values (Maczkowiack *et al.*, in press). This may not be the case. Interviews with land valuers and real estate agents in north Queensland suggest that the cost of plantation establishment is not fully factored into land values (Harrison *et al.* 2001).

Appropriate species and suitable land

A core of at least 20 rainforest species with timber of commercial value – including some 'diamond' or very high quality species such as Queensland maple and northern silky oak – have been demonstrated to grow in plantations in the Wet Tropics, as well as a variety of native and exotic pines, eucalypts and acacias, and several exotic tropical species (including African mahogany, teak and West Indian cedar). Growth rates for many species are impressive relative to those for the same species in temperate climates, though rotation length is typically 30 years or more.

Kent and Tanzer (1983a, b) identified an area of 36 780 ha of land on the Atherton Tablelands as more suited to forestry or catchment protection than cropping and pastures. Fullerton (1985) subsequently identified 40 000 ha of degraded land on the Atherton Tablelands suitable and potentially available for farm forestry. Keenan *et al.* (1998) using a geographical information systems (GIS) approach, estimated that there was 41 000 ha within 150 km of Cairns with rainfall, slope and soil conditions suitable for plantation forestry, and potentially available for forestry (taking into account current economic conditions for competing land-uses). In 2002, an updated estimate of 86 000 ha was obtained, for an extended area of up to 200 km from Cairns (Annandale 2002). Anderson and Halpin (1998: 414) reported, in relation to the planning of the Plantation Forestry Joint Venture Scheme and based on a 200 km radius of Cairns, that 'the area

of suitable land available for commercial timber production … in far north Queensland' was 134 000 ha. Astuti (2004) estimated that about 40 000 ha of land suitable for forestry are available on the central and southern Atherton Tablelands. Land available for forestry varies from very high rainfall sites (>4000 mm yr^{-1}) on the coast to cooler elevated tableland areas (much around 800–1100 m above sea level (m a.s.l.)), of both high (1200–2000 mm) and moderate (1000–1200 mm) annual rainfall. Much of the land has soils well suited to forestry, including basaltic soils.

Social dimensions

The importance of the social dimensions of forest industry development is becoming increasingly recognized. In the case of forestry in the Wet Tropics region the social dimensions are particularly important. In that the Wet Tropics region is recognized to have high environmental values, any form of plantation forestry must be compatible with community attitudes to the landscape. Environmental sensitivity favours the development of plantations of native tree species, especially as mixtures, but not the growing of exotics or clearfell logging, as clearly reflected in recent surveys on community attitudes to land use (e.g. Blumenfeld 2005; Lwanga 2005; Suh 2005). With rapid economic growth in Cairns, in other coastal areas and at Atherton, land within commuting distance of employment centres (typically about a 1 hour drive) is in high demand, and is becoming too expensive for forestry.

Tourism is the largest single industry in the Wet Tropics, with two World Heritage Areas listed on the basis of natural features, a pleasant winter climate and relatively safe visitor destination, and proximity to Asia (through Cairns international airport). Ecotourism, including farmstays, is of growing importance and reforestation of degraded farm land in particular can enhance the visitor attractiveness of the area (Harrison 2000). The experience in Europe, particularly in the Black Forest region in Germany, is that farms can develop a diversified income stream through a combination of agriculture and farm forestry activities combined with developing accommodation for farm stays. Farm woodlots are seen as being complementary to farmstays because they offer farm visitors further variety in the on-farm activities, as well as improving farm aesthetics.

Constraints to forest industry development

Finding markets is a major obstacle for small-scale tree growers. The variety of species grown (many of which are lesser-know species), variable silvicultural management and timber quality, small harvest quantities and infrequent sales of small growers prevent them from developing regular supply relationships with timber processors.

While there is strong landholder interest in the Wet Tropics in growing native trees, evidenced, for example, by additional and self-funded plantings by many CRRP participants (Harrison 2001), there is lingering distrust by landholders of government land administration. Survey evidence (e.g. Harrison *et al.* 2001; Emtage *et al.* 2001; Herbohn *et al.* 2005) reveals considerable uncertainty about harvest rights, stemming in part from the federal government's nomination of the Wet Tropics for World Heritage listing against strong opposition from the north Queensland community and Queensland government (e.g. see Winter 1991). Recent proposals in the draft private forestry Code of Practice

to exclude logging of rainforest and regrowth forest on private land have refuelled landholder anxiety.

Some of the main constraints to smallholder forestry development in the Wet Tropics are summarized in Table 43.3. In addition, industrial forestry companies face further issues, including the high price of purchasing or leasing land, the long distance to markets for some products, a lack of infrastructure (e.g. the lack of a pulp mill to take thinnings) and potential changes in environmental regulations, which may restrict harvesting in some areas.

Forest industry development options, visions and futures

A wide variety of options exist for expansion of the timber industry in the Wet Tropics. Questions that may be posed include:
- Which species should be grown: native softwood, rainforest hardwood, and monoculture or species mixture?

Table 43.3 Constraints on small-scale forestry in the Wet Tropics

Constraint	Contributing factors
Low timber profitability	Lack of competition between timber buyers, and difficulty selling timber Infrequent sales by small-scale growers Multiple species, with limited market recognition Low quality of some farm-grown timber due to inadequate silvicultural practices Physical damage to plantations from pests, diseases and tropical cyclones Current market arrangements locking out new sellers
Long wait for returns	Many native species have long rotations, and there is little opportunity to generate revenue prior to clearfell (little local market for thinnings, lack of annuity financing arrangements)
Lack of technical, institutional and financial support	Lack of information on growing lesser-known species, including establishment methods and site-species matching, Lack of information about species mixtures and interactions btween species Lack of forestry technical advice (extension) Limited support for forest industry cluster and tree grower cooperative Lack of joint venture, venture capital, forestry partnership and forestry land lease schemes Absence of markets for ecological services
Attitudinal Impediments	Negative attitudes to plantation forestry (High concern over sovereign risk following World Heritage listing of the Queensland Wet Tropics rainforests; historical community attitude that timber is available from natural forests; negative signals given by lack of financial performance of earlier farm forestry schemes; lack of a plantation forestry culture among landholders, though high interest in growing native trees; opposition to plantation forestry by environmental groups)
Regulatory	Local government constraints over tree planting
Impediments	Concern over possible new controls in the Vegetation Management Act (logging on private land, regrowth control)

Sources: Stork *et al.* (1998), Harrison *et al.* (2001), Vize and Creighton (2001), Sharp (2002), Herbohn *et al.* (2005)..

Table 43.4   Plantation options for timber production on the Atherton Tablelands

	Producer category						
Plantation type	Government	Industrial grower	Small-scale farm grower	Medium-scale farm grower	Farm grower – joint venture	Local government	Community group
Conifer – exotic	x	x					
Conifer – native	x	x		x	x	x	
Eucalypt		x	x	x	x		
Native cabinetwood – single species	?	x	x		x	x	
Native cabinetwood – mixed species	?	x	x				x
Exotic hardwood		x	x	x			

- Grown by whom – government, industrial foresters, farmers – and and what locations?
- What silvicultural systems should be adopted?
- How will investment in plantations be financed?
- How will sustainability of the system (economic, social, environmental, cultural) be ensured?
- What proportion of plantings will be logged versus permanent stands?
- What products will be produced, and where will they be marketed?

Some plantation options are set out in Table 43.4, with more promising combinations marked with a cross. It seems probable that softwoods will play an important part in any industry expansion, since they currently account for the main plantation area and timber processing. As well as plantings at an industrial scale, by timber companies or government, it can be expected that further small-scale plantings of rainforest species and eucalypts will take place, provided markets and technical advice are available.

Various forestry options have been identified by Jeffreys (in preparation). The potential role of venture capital to finance forestry development has been examined by Sharp (2002), who concluded that the rotation period and timber prices achievable for plantation forestry in the Wet Tropics do not in general support a rate of return of the order of 12% as required by joint-venture companies to entice them to invest in forestry.

Forestry successes in other countries provide a model that could assist in reforestation design in the Queensland Wet Tropics. The integration of forestry with other income-earning activities, such as cattle grazing, tourism including homestay accommodation and off-farm work in the Black Forest of southern Germany (Selter 2003), is such a model.

Support measures to expand forest-based industries in the Wet Tropics

Given the lack of forest industry development in the Wet Tropics to date, an argument can be advanced that support measures will be required if an industry is to develop. Support can take many different forms.

Forestry cooperatives and associations

Forestry cooperatives and associations have proved successful in other parts of Australia and overseas. They can bring together the knowledge, objectives and organizational capacities of the various stakeholder types into an organized group to formulate forestry promotion policies and actions and to pursue collective interests. The North Queensland Timber Co-op (NQTC) was formed in 1996, registered under the Co-operatives Act (1997) in 1997, and has approximately 15 paid-up members. The NQTC has conducted an inventory on the private pine resource, and attempted to market fragmented lots of farm forestry on behalf of its members, but has been unable to achieve sufficient scale to be a price-maker. The NQTC is struggling to survive in a region of low-priced imports from Papua New Guinea and the Asia-Pacific, and a culture of independence among small-scale forest growers and processors. Lack of start-up capital and the long lead time involved with farm forestry are other contributing factors to its lack of support by regional growers.

Compensation for provision of ecosystem services

Compensation for provision of ecosystem services through offering tree growers annual payments for the ecological services they provide, such as carbon sequestration, biodiversity protection, salinity mitigation and water quality protection, has frequently been raised (e.g. Binning et al. 2002). This could be through payments by government (which would probably be seen as an unacceptable subsidy scheme) or through trading credits using a market mechanism. Such trading requires the development of a central, efficient, financially stable and widely accepted market mechanism bringing buyers and sellers together. It also relies on small-scale forest growers achieving both individual and collective scale sufficient to ensure that the value of ecological services exceeds the transaction and monitoring costs involved with activities such as the collection of data required for auditing and verification purposes.

Forestry research and development

Forestry research and development has an important role in industry development. In that much is still to be understood about the growth characteristics and optimal silviculture for monoculture and mixed-species plantations of native species in the Wet Tropics, further research potentially can make a major contribution to forestry expansion. Further, there is great scope for more research into tree breeding and plant protection in relation to native species, for example, developing the capacity to grow red cedar successfully in plantations.

Forestry extension services

Forestry extension services are critical where there is no culture or history of private plantation forestry as is the case in the Wet Tropics. Forestry extension officers located within the lead forest agency or a university forestry department can play an important role in the promotion of forestry activities, promoting interest by landholders and providing technology transfer and capacity building with respect to the management of plantations and native forest. Through identification of landholder typologies, it is possible to target extension activities towards those landholders who are more

likely to establish plantations (Emtage et al. 2001, 2006, in press). A network of forestry extension officers was formed in Queensland (including north Queensland) in the 1990s. However, this programme has now ceased, which is hardly surprising giving the earlier disbandment of a large and free agricultural extension service. It would appear that the state government places a low priority on expenditure on rural extension, due to a perception that the private sector will take over this function.

Direct government involvement in marketing of privately grown timber

Direct government involvement in the marketing of privately grown timber has proven to be effective in a number of countries. A government forest agency can take a range of positions in relation to non-industrial private foresters. At the one extreme is that of an aggressive low-pricing competitor with comparative advantage in relation to production costs (such as land tax exemption) and with exclusive contracts with processors. At the other extreme is a government forest agency that provides (subsidized or at-cost) services to private growers to assist them in overcoming impediments of small-scale operation. This is the case, for example, in Baden-Württemberg in Germany, where following severe storm damage in 1999 the state government undertook the timber recovery, wet storage and marketing of windthrow timber for those growers who wished to adopt this option (Hartebrodt et al. 2005). In Queensland, the state government has commercialized the lead forestry agency and collects an annual dividend from it in the order of Au$2–10 million, thus making it very difficult for the forest agency to undertake any activities that are deemed non-profit.

Removing institutional impediments

In that sovereign risk has been found to be a major impediment to forestry in the Wet Tropics (Herbohn et al. 2005), measures such as introducing harvest rights legislation and uniform regulations across local governments, and improved taxation treatment, can provide a major stimulus for forestry. Similarly, allowing log exports (now approved for Pentarch Forest Products in north Queensland) can have a major effect on log prices and hence plantation profitability. The Queensland government is undertaking steps to address institutional impediments, although a code of practice for plantation forestry is not yet available.

Government-funded forestry subsidy arrangements

Government-funded forestry subsidy arrangements are commonplace in most European countries and are an option to facilitate industry development in the Wet Tropics. There is, however, a climate of opinion that the forest industry should stand on its own feet; given the various efforts to promote forestry over many years, the infant industry argument for subsidization is not a strong one.[4] Governments elsewhere often provide practical assistance for non-industrial forestry (such as planting grants or provision of free or subsidized seedlings), typically channelled through regional offices of the national or state government forest agency. A more complex arrangement is where more than one level of government (but usually including the lead government agency for forestry) cooperates to support a regional reforestation programme. The CRRP provided an example of an innovative public assistance programme, involving all three levels of government (including 14 local governments), for primarily production forestry, to replace the resource lost through World Heritage listing. In north Queensland, subsidy programmes such as the CRRP appear to have had limited success in the creation of a viable timber industry, despite considerable public investment. The prevailing federal and state government forestry policies are aimed more at enticing large players to become involved.

Financing industry development

Closely associated with support measures for forestry is the manner in which the expansion of the forest estate is financed. There are many different forms that this support can take. These financing options can be roughly divided into five main groups:
1 Government-led initiatives through the expansion of the existing softwood estate;
2 joint-venture arrangements between government and landholders in which landholders contribute land

and possibly some labour and the government agency contributes the silviculture expertise and other inputs, with profit sharing on the basis of relative inputs;
3 companies leasing land from landholders and with lease payments providing annual income;
4 direct planting by landholders;
5 purchase of land by companies that then establish and maintain plantations, usually in the form of managed investment schemes.

The advantages and disadvantages of each of the five options are summarized in Table 43.5.

The way forward

The current forest industry in the Wet Tropics is dominated by government softwood plantations. One option for forest industry development is through the expansion of these government plantations. However, it appears to be FPQ policy not to expand softwood plantations in north Queensland, particularly for the long-rotation hoop pine species. The lack of suitable land at a reasonable cost and the current focus of FPQ operations in south-east Queensland makes this option unlikely. Joint-venture arrangements between government and private landholders have proved popular in some areas in Queensland and elsewhere in Australia, and are an option for the Wet Tropics, particularly where landholders have underutilized land. However, the high transaction costs of having to deal with the issues of co-ownership and management with a large number of landowners with relatively small planting areas can make joint ventures unattractive to industry and government partners. Joint ventures are often attractive to 'lifestylers' who lack the time and skills to enter into intensive forestry operations.

While an argument may be advanced for additional government plantings, there are good reasons for advocating private sector investment also. Private sector investment, either in the form of industrial style development (e.g. companies and MIS) or through woodlots being established by existing landholders, will dilute the bilateral monopoly market that currently exists with one dominant buyer (Ravenshoe Timbers) and one dominant seller (FPQ).

In other areas in Australia (e.g. Tasmania, the Green Triangle, Western Australia) the purchase of land by companies involved in MIS has been common. In at

Table 43.5 Summary of options to finance the expansion of the plantation estate in the Wet Tropics

Option	Advantages	Disadvantages
1 DPI Forestry to expand existing estate	DPI Forestry has established a successful record in native and exotic pine silviculture	Reinforces the perception that forestry is a 'government' activity Maintains monopoly situation Not consistent with DPI Forestry focus of operations in SEQ New processors are unlikely to enter market Not conducive to export market developing Most suitable land may be too expensive
2 Joint Venture (JV) arrangements	Landholders have access to forestry skill base Can be low labour input use for land Access to markets through large organization Often have flexibility in terms inputs from landholders	High transaction costs for government/company partners dealing with many landholders Some loss of control by landholders Planting limited to small number of commercial species in standard planting configurations
3 Leasing arrangements	Same as JVs Income payments are made on a annual basis	Same as JVs but with substantial loss of control by landholders
4 Direct planting by landholders	Is often highest value land use Way of diversifying farm income Relatively low labour input activity Forestry integrated into traditional farming systems and the broader community can contribute to community resilience and retain families	Large capital requirements in early years Long wait for returns (30+ years) Lack of farm forestry culture The resource is likely to be fragmented, with many farmers
5 Purchase of land by companies	No transaction costs with dealing with landholders after initial purchase Complete control over operations on farm Economies of scale and centralized access to markets	High initial cashflows required to purchase the land (however companies involved in Managed Investment Schemes (MISs) have large cash reserves) Community resistance to companies moving into regions Most MISs focus on shorter rotation tree crops (radiata pine, *E. globulus*), which are not suited to many areas in the wet tropics High land costs may mean returns are lower than in other regions with cheaper land

least some instances, there has been community unrest about perceived negative impacts through loss of farming land and families. However, there appears to be limited scope for industrial scale forestry development in the Wet Tropics. For instance, on the Atherton Tablelands, the high value of land, especially in the highly productive agricultural areas and those within commuting distance to Cairns, may make purchase of land prohibitively expensive. MIS may need some form of subsidy as returns may be higher in other areas with lower land costs. Some scope may exist for leasing arrangements between companies and existing landholders. This type of arrangement would suit landholders who are not interested in the growing of trees and simply want a constant income stream. Leasing arrangements are particularly suited to retirees or agriculturalists looking to use surplus land for income diversification.

Direct planting activities by existing landholders probably offer the greatest potential for the future of timber-based industries in the Wet Tropics. In Europe there is a tradition of farm forestry as an integral part of the farm business; many farmers in Europe see their forests as a bank to draw upon in times of need. There are, however, large subsidies and most farms have mature forests that can be harvested to generate income. In Australia, however, there is no such culture and farm forestry is in a developmental phase. For farm forestry to develop quickly, as has been the case in Ireland, subsidies and other forms of assistance will be required.

There is ample evidence from surveys of landholders in north Queensland (see Harrison & Herbohn 2005 for a summary) and elsewhere in the world that small-scale foresters value highly the broader benefits that their forests produce, particularly the environmental benefits. This type of forestry development is highly suited to the Wet Tropics region. There is great potential for well planned small-scale forestry to complement the high conservation areas within the Wet Tropics.

Notes

1 Not all portable sawmills require licensing. An approximate guide is that if the saw blade moves over a stationary log then licensing is not mandatory.

2 Internationally, a throughput of about 100 000 m³ is now considered minimal in terms of achieving high economies of scale.

3 Economic theory postulates a range in which negotiated price may fall in a bilateral monopoly, given production costs and buyer demand (Quayle 2004). In that Ravenshoe Timbers is the largest single employer in the Herberton Shire (with about 75 staff), and timber is a perishable product, it is probable that the price would settle low in the negotiation range.

4 Notably, substantial financial support has been provided to industrial forestry in the Wet Tropics, for example subsidization of a finger-jointing plant at Ravenshoe Timbers and low pricing of Caribbean pine sold to Pentarch Forest Products.

References

Anderson, M. & Halpin, N. (1998). Private forestry plantation: joint venture scheme. In *Managing and Growing Trees: Farm Forestry and Vegetation Management*, Grodecki, A. (ed.). Department of Natural Resources and Greening Australia, Brisbane, pp. 413–16.

Annandale, M. (2002). *Ecologically Sustainable Forest Management of Private Native Forest in the Wet Tropics*. Queensland Forestry Research Institute, Forestry Research Institute, Atherton.

Astuti, I. E. (2004). A land capability study to identify the potential sites on the Atherton Tablelands for hoop pine plantation establishment using geographical information systems and expert opinion. Masters thesis, The University of Queensland, Gatton.

Australian Academy of Technological Sciences and Engineering (AATSE) (1988) Technology in Australia 1788–1988. Plantations – High Productivity Resources: Native Species (www.austehc.unimelb.edu.au/tia/215.html), accessed 22 February 2006.

Binning, C., Baker, B., Meharg, S., Cork, S. & Kearns, A. (2002). *Making Farm Forestry Pay: Markets for Ecosystem Services*. Rural Industries Research and Development Corporation, Report No.02/005, Canberra.

Blumenfeld, N. (2005). *Public Perceptions of Land Use Options on the Atherton Tablelands*. World Learning, School for International Training, Natural and Cultural Ecology, Cairns.

DPI Forestry (undated) *SE Hoop Region Notes*. Department of Primary Industries and Fisheries, Kenilworth, Queensland.

DPI Forestry (2004). *The Queensland Forest Industry: An Overview of the Commercial Growing, Management and Processing of Forest Products in Queensland*. DPI, Brisbane.

DPI Forestry (1996). Generic financial spreadsheet for hoop pine. Faxed sheet, DPI, Brisbane.

Emtage, N. F., Herbohn, J. L & Harrison, S.R. (2006). Landholder typologies used in the development of natural resource management programs in australia: a review. *Australian Journal of Environmental Management* **13**(2): 79–94.

Emtage, N. F., Herbohn, J. L. & Harrison, S.R. (in press). Landholder profiling and typologies for natural resource management policy and program support: potential and constraints. *Environmental Management*.

Emtage, N. F., Harrison, S. R. & Herbohn, J. L. (2001), Landholder attitudes to and participation in farm forestry activities in sub-tropical and tropical eastern Australia. In *Tropical Small-scale Forestry: Social and Economic Analysis and Policy*, Harrison, S. R. & Herbohn, J. L. (eds). Edward Elgar, Cheltenham, pp. 195–210.

Forest and Wood Products Research and Development Corporation (FWPRDC) (2005). *Impediments to Investment in Long-rotation Timber Plantations*. Forest and Wood Products Research and Development Corporation, Canberra.

Fullerton, L. W. (1985). *The Reforestation Potential of Parts of the Southern Atherton Tableland*. Queensland Department of Forestry, Brisbane.

Gould, K. (2000). An historical perspective on forestry management. In *Securing the Wet Tropics?* McDonald, G. & Lane, M. (eds). The Federation Press, Sydney, pp. 85–102.

Hardman, J. R., Nicholson, D. & Moore, P. P. (1985). The economics of agroforestry on the Atherton Tableland. *Queensland Agricultural Journal* **111**(4): 207–10.

Harrison, R. (2001). *An Evaluation of the Community Rainforest Reforestation Program*, B.Sc. Honours Thesis, The University of Queensland, Brisbane (unpublished).

Harrison, R., Harrison, S. & Herbohn, J. (2004). Evaluation of the performance of the Community Rainforest Reforestation Program in north Queensland. *Small-scale Forest Economics, Management and Policy*, **3**(3): 411–29.

Harrison, S. R. (2000). Landscape amenity and recreation values of farm forestry. In *Sustainable Small-scale Forestry: Socio-economic Analysis and Policy*, Harrison, S. R., Herbohn, J. L. & Herbohn, K. F. (eds). Edward Elgar: Cheltenham, pp. 77–88.

Harrison, S. R. & Herbohn, J. L. (2005). Relationship between farm size and reforestation activity: evidence from Queensland studies. *Small-scale Forest Economics, Management and Policy* **4**(4): 471–84.

Harrison, S., Hill, P. & Herbohn, K. (1998). Reforestation incentives in the UK and Australia: a comparative evaluation. Paper prepared for the First World Congress of Resource and Environmental Economists, Venice, 22–26 June.

Harrison, S. R., Miamo, J. & Anderson, M. W. (1999). Government and private sector joint venturing in natural resource development: the Queensland forestry joint venture scheme plantation, *Economic Analysis and Policy*, **29**(1): 15–29.

Harrison, S. R., Herbohn, J. L., Emtage, N. F. & Smorfitt, D. B. (2001). Landholder attitudes to forestry incentive schemes in North Queensland. In *Managing and Growing Trees: Farm Forestry and Vegetation Management*, Grodecki, A. (ed.). Kooralbyn, 19–21 October 1998. Department of Natural Resources and Greening Australia, Brisbane, pp. 406–11.

Hartebrodt, C. (2004). The impact of storm damage on small-scale forest enterprises in the south-west of Germany. *Small-scale Forest Economics, Management and Policy* **3**(2): 203–22.

Hartebrodt, C., Fillbrandt, T. & Brandl, H. (2005). Community forests in Baden-Württemberg (Germany): a case study for successful public–public-partnership. *Small-scale Forest Economics, Management and Policy* **4**(3): 229–50

Herbohn, J. & Harrison, S. (in preparation). Hoop pine – is it a good investment for farm foresters?

Herbohn, J. L. & Harrison, S. R. (2000). Assessing financial performance of small-scale forestry. In *Sustainable Small-scale Forestry: Socio-economic Analysis and Policy*, Harrison, S. R., Herbohn, J. L. & Herbohn, K. F. (eds). Edward Elgar: Cheltenham, pp. 39–49.

Herbohn, J. L. & Harrison, S. R. (2001). Financial analysis of a two-species farm forestry mixed stand. In *Sustainable Farm Forestry in the Tropics: Social and Economic Analysis and Policy*, Harrison, S. R. & Herbohn, J. L. (eds). Edward Elgar, Cheltenham, pp. 39–46.

Herbohn, J. L. & Harrison, S. R. (2004). The evolving nature of small-scale private forestry in Australia. *Journal of Forestry* **102**(1): 42–7.

Herbohn, J. L., Emtage, N. F., Harrison, S. R. & Smorfitt, D. B. (2005). Attitudes of landholders to farm forestry in tropical eastern Australia. *Australian Forestry* **68**(1): 50–8.

Herbohn, J. L., Smorfitt, D. B & Harrison, S. R. (2001). Choice of timber inputs by small to medium sized cabinet-making firms in Queensland and implications for marketing lesser-know tropical timbers. In *Sustainable Farm Forestry in the Tropics: Social and Economic Analysis and Policy*, Harrison, S. R. & Herbohn, J. L. (eds). Edward Elgar, Cheltenham, pp. 89–104.

Jeffreys, I. (in process). PhD thesis in progress, School of Natural and Rural Systems Management, University of Queensland, Gatton.

Keenan, R. J., Annandale, M. & Philip, S. (1998). North Queensland Private Forest Plantation Land Suitability Study, Internal Report, Queensland Forest Research Institute, Atherton.

Kent, G. (2004). DPI's role in forestry development in north Queensland. Paper presented to the North Queensland Forestry Industry Development workshop, Cairns Student Lodge, Cairns, April.

Kent, D. J. & Tanzer, J. M. (1983a). *Evaluation of Agricultural Land, Atherton Shire, North Queensland*. Queensland Department of Primary Industries, Brisbane.

Kent, D. J. & Tanzer, J. M. (1983b). *Evaluation of Agricultural Land, Eacham Shire, North Queensland*. Queensland Department of Primary Industries, Brisbane.

Lamb, D., Keenan, R. J. & Gould, K. (2001). Historical background to plantation development in the tropics: a north Queensland case study. In *Sustainable Farm Forestry in the Tropics: Social and Economic Analysis and Policy*, Harrison, S. R. & Herbohn, J. L. (eds). Edward Elgar, Cheltenham, pp. 9–20.

Lewty, M. (2005). Prospects for forestry on the canelands of south-east Queensland. Presentation to a workshop held at Maroochy Shire Council, 3 October 2004, Nambour.

Louis, H. (2000). The potential role of high value cabinet timber plantations to diversity farm business risk with specific application to the Community Rainforest Reforestation Program. BAdmin (Hons) thesis, James Cook University, Townsville.

Lwanda, R. (2005), Masters thesis, School of Natural and Rural Systems Management, The University of Queensland, Gatton.

Maczkowiack, R., Herbohn, J. L., Emtage, N. R., Slaughter, G. & Harrison, S. R. (in press). *Assessment of Agroforestry Projects in Northern Australia using the Australian Farm Forestry Financial Model*. Rural Industries Research and Development Corporation Research Report Series, Canberra

MBAC Consulting Group (2005). *A Socio-economic Assessment of the Plantation Processing Sector in Queensland*. Report prepared for Timber Queensland, Canberra.

Palaszczuk, H. (2004). Major boost for north Queensland's timber industry, Press Release (QldGovSpin), Department of Primary Industries and Fisheries (http://us.altnews.com.au/drop/node/view/504), accessed 25 August 2004.

Quayle, M. (2004). Some pricing issues of long-term log sales in tropical north Queensland, Australia. In *Marketing of Farm-Grown Timber in Tropical North Queensland*, Suh, J., Smorfitt, D. B., Harrison, S. R. & Herbohn, J. L. (eds). Rainforest CRC, Cairns, pp. 65–74.

Selter, A. (2003). Farm forestry in the Black Forest Southern Nature Park from a socio-political point of view. *Small-scale Forest Economics, Management and Policy* **2**(1): 37–48.

Sharp, B. (2002). Venture capital raising for small-scale forestry in the Queensland Wet Topics: a strategic alliance view. BEcon (Hons) thesis, School of Economics, University of Queensland, Brisbane.

Skelton, D. (2004). A forestry industry cluster view on timber marketing in north Queensland. In *Marketing of Farm-grown Timber in Tropical North Queensland*, Suh, J., Smorfitt, D. B., Harrison, S. R. & Herbohn, J. L. (eds). Rainforest CRC, Cairns, pp. 77–88.

Skelton, D. (2005). Executive Officer, Private Forestry North Queensland, Kairi, personal communication.

Smorfitt, D. B. (1999). *Unpublished Statistics on Timber Mills in the Wet Tropics*. Rainforest CRC, James Cook University, Townsville.

Smorfitt, D. B., Harrison, S. R. & Herbohn, J. L. (2003). Portable sawmills in a high value rainforest cabinet timber industry in North Queensland. *Small-Scale Forest Economics, Management and Policy* **2**(1): 21–36.

Smorfitt, D. B., Herbohn, J. L. & Harrison, S. R. (2004). The relevance of end-user perceptions and purchase habits for the current and future marketing of Australian rainforest cabinet timber products. In *Marketing of Farm-grown Timber in Tropical North Queensland*, Suh, J., Smorfitt, D. B., Harrison, S. R. & Herbohn, J. L. (eds). Rainforest CRC, Cairns, pp. 123–32.

Stork, N. E., Harrison, S. R., Herbohn, J. L. & Keenan, R. (1998). Biodiversity conservation and sustainable management of forests: socioeconomic problems with farm-forestry of rainforest timber production in North Queensland. Keynote address. In *Overcoming Impediments to Reforestation*, Proceedings of the Sixth Annual International Bio-Reforestation Workshop, Brisbane, 2–5 December 1997. BIO-REFOR, The University of Tokyo, Tokyo, pp. 159–68.

Studdert, J. (2006), Hardwood creates solid returns. *The Australian* 8 March.

Suh, J. (2005). Public attitudes to forest landscapes in the southern Atherton Tableland: exploring economic evaluation and valuation methods for the landscape. Presentation at postgraduate student seminar, School of Natural and Rural Systems Management, University of Queensland, Brisbane.

Suh, J., Lwanga, R., Harrison, S. and Herbohn, J. (2006). Visitor attitudes to a proposed loop pine plantation establishment on the southern Atherton Tablelands;. In *Sustainable Forest Industry Development in Tropical North Queensland,* Harison, S. R. & Herbohn, J. L. (eds). Rainforest CRC, Cairns, pp. 135–46.

Vize, S. & Creighton, C. (2001). Institutional impediments to farm forestry. In *Sustainable Farm Forestry in the Tropics: Social and Economic Analysis and Policy,* Harrison, S. R. & Herbohn, J. L. (eds). Edward Elgar, Cheltenham, pp. 241–55.

Vize, S., Killin, D. & Sexton, G. (2005). The Community Rainforest Reforestation Program and other farm forestry programs based around the utilisation of rainforest and other species. In *Reforestation in the Tropics and Subtropics of Australia Using Rainforest Tree Species*, Erskine, P. D., Lamb, D. & Bristow, M. (eds). RIRDC and Rainforest CRC, Canberra and Cairns, pp. 7–22.

Weston, N. & Goosem, S. (2004), Sustaining the *Wet Tropics: A Regional Plan for Natural Resource Management, Volume 2A Condition Report: Biodiversity Conservation*, Rainforest CRC and FNQ NRM Ltd, Cairns.

Winter, J. (1991). The Wet Tropics of North-East Queensland. In *Environmental Research in Australia: Case Studies*. Australian Science and Technology Council, AGPS, Canberra, pp. 151–72.

44 INTERNATIONAL PERSPECTIVE: RESTORING TROPICAL FOREST LANDSCAPES; RESTORING WHAT AND FOR WHOM?

Jeffrey Sayer

Forest Conservation Programme, International Union for Conservation of Nature and Natural Resources (IUCN), Gland, Switzerland

Vast sums are spent on reforestation and forest restoration throughout the world. Such programmes have often been controversial and have often failed to re-establish forest cover. They are frequently driven by simplistic political agendas and suffer from having vague objectives, or objectives that are not consistent with established scientific facts. Thus we see major investments in planting fast growing, thirsty trees in programmes to counter desertification in Northern China and Africa where other land cover types would be more ecologically appropriate. We see highly fire-sensitive conifers with low timber values planted to restore forest cover in burnt areas in southern Europe and the western USA where the native fire-tolerant species would protect against future fires and provide a variety of products and services of local value. We see trees planted on hillsides to augment water flows and prevent landslides in situations where empirical evidence suggests that they may have the opposite effects to those sought. Fast growing commodity timber trees are being widely planted for low value pulp in place of natural forest that provided multiple benefits for local people.

We are also seeing numerous initiatives by governments, forest departments, industry and private citizens to restore forest landscapes in a much more thoughtful and targeted way. These initiatives aim to increase the flows of forest goods and services in ways appropriate to the needs of society. The mismanagement of forests over recent decades means that we now have deficits of many of the benefits that forests provide and the demand for reforestation and restoration is likely to increase in coming decades. The experience of forest management and restoration in Queensland, and of scientists based in various Australian research institutions, that is summarized in the chapters in this section have much to offer to the rest of the world.

There are two divergent tendencies driving changes in the world's forests. Globalization, the deregulation of trade and the search for economic efficiency are favouring uniform products and economies of scale. Thus we see a consolidation of the world's commodity fibre industries, with most pulp production now in the hands of a small number of multinational corporations. Increasingly these are focusing on short rotation plantation-grown fibre. This tendency has not yet impacted extensively on timber industries but it seems inevitable that it will. Already traditional low value pulp species like *Acacia mangium* are being used for sawn timber and previously valueless early pioneer species are now being grown for plywood filler. This globalizing tendency is pushing us towards extremely homogeneous forests managed for a single function – fibre production.

At the same time there is a widespread movement to decentralized forest management from the national level to the level of local authorities and communities. When forests are under local control they tend to be managed for multiple functions. The home gardens and agroforestry systems in many parts of the tropics are good examples, as are the community forests of much of Europe. People exploit fruits and nuts, firewood, timber, wildlife for hunting, mushrooms etc. Local people tend to manage forests for multiple functions.

Restoration and reforestation have in the past, especially when government sponsored, been generally equated with the large scale uniform plantation model. Many of us now believe that the broader needs of society would be better met if reforestation and restoration focused more on the locally integrated multifunctional model (Sayer & Maginnis 2005). The World Conservation Union and the Worldwide Fund for Nature launched a Forest Landscape Restoration Initiative in 2001 to counter the tendency towards uniform forests and to promote restoration programmes that aimed to build forest assets to meet the full range of societies' needs. The word 'landscape' was chosen as the descriptor of the geographical scale of these initiatives to stress the idea that societies' needs could best be met when a range of different forest types were distributed in a way that combined with other land uses to cover the range of human needs – to be multifunctional. The word landscape was also used rather than, for example, ecosystem, because it implies that it is not just the biophysical components of the system that are being considered but also the legal, regulatory, social and aesthetic components of that system (Mansourian *et al.* 2005).

The concept of Forest Landscape Restoration (FLR) was also introduced as a counter to the prevalent 'threats-based' approach that is still pursued by many conservation organizations. Instead of simply opposing any change to the status quo ante, WWF and IUCN wanted to make a statement that they were not just concerned about threats and that they wanted their conservation programmes to be firmly rooted in visions of outcomes. Conservation programmes often impose opportunity costs on local people. They compete for land with other human enterprises. FLR seeks to address the issue of trade-offs at the landscape scale.

The FLR initiative has now captured the imagination of a number of governments and international organizations and has led to the establishment of a Global Partnership on FLR (www.unep-wcmc.org/forest/restoration/globalpartnership/) that promotes these integrative multifunctional ideas and facilitates learning among practitioners throughout the world who are trying to build forest assets for their local needs. A fundamental tenet of FLR is that programmes have to be built on multistakeholder processes and that they will therefore require the negotiation of trade-offs between the interests of people with conflicting expectations.

An interesting feature of FLR is that it requires that answers are given to the fundamental questions of how much forest is needed, what type of forest it should be and how it should be distributed through the landscape. These questions should be fundamental to all forest management and conservation activities and it is surprising how rarely they are explicitly answered. Many conservation organizations and some government forest agencies still see maximizing forest area as the ultimate objective. Even the intergovernmental processes, the Convention on Biological Diversity and the United Nations Forum on Forests implicitly endorse maximizing forest area when they focus on halting deforestation. But in a world where we are approaching seven billion people all land is going to have to be used efficiently. Optimizing the extent, nature and location of forests is clearly the major present challenge. The Global Partnership on FLR is innovative in exploring these difficult questions – if it succeeds in convincing decision-makers of the wisdom of this approach it will revolutionize thinking on forests and landscapes.

Recent decades have seen the most rapid reduction of forest cover that the world has ever experienced. Many countries have gone through what is known as the 'forest transition', as previously remote and inaccessible areas have been opened up for agriculture. We now have forest–agriculture mosaics that were arrived at largely in an unplanned and chaotic way. We also have vast areas that have suffered catastrophic forest destruction from fires, cyclones, the tsunami, drought etc. These landscapes are not ideal for biological diversity or hydrological functions; they fail in many ways to meet societies' needs for the goods and services of forests. So we have a major challenge to restore these landscapes so that they better meet human needs now but also so that they have the resilience to meet those needs in the future.

The challenge is considerable. Forests in our future landscapes will have to support biodiversity in greatly reduced areas, they will have to sequester and store carbon while retaining other environmental and productive values and they will have to withstand changed climates in ways that we still understand only poorly. Recent models portray a bleak future for the world's most extensive tropical forests in the Amazon (Giles 2006). The days are gone when we can just allocate land to forest, fence it off and hope for the best. Given the intensity of demand for land and for forest values there will be a need to intensify greatly the management of all forest systems – and a vastly expanded body of science will be needed to support this management. The international Tropical Timber Organisation has recently taken initiatives to place forest restoration and management on a broader multifunctional footing with the production of sets of guidelines (ITTO 1993a, b, 2002) and is in the process of developing Guidelines for the Conservation of Biological Diversity in Tropical Production Forests. The latter cover the needs of planted forests in addition to managed natural forests. Demand is expanding for research to support this growing need to re-establish multifunctional forests.

This provides a fascinating and vital challenge or forest ecologists but also for scientists and practitioners from a range of disciplines concerned with land use issues and human behaviour. The challenge is nothing less than that of helping to shape a world that is habitable for its growing population – providing them with fuel and fibre but also with amenity and aesthetic values, while retaining a functioning and resilient system. The work from the Rainforest CRC reported in this volume has much to contribute to the knowledge, skills and understanding that will be needed in this endeavour.

References

Giles, J. (2006). The outlook for Amazonia is dry. *Nature* **442** (7104): 17 August.

ITTO (1993a). *ITTO Guidelines for the Establishment and Sustainable Management of Planted Tropical Forests.* ITTO Policy Development series number 4, Yokohama, Japan.

ITTO (1993b). *ITTO Guidelines for the Conservation of Biological Diversity in Tropical Production Forests.* ITTO Policy Development series number 5, Yokohama, Japan.

ITTO (2002). *ITTO Guidelines for the Restoration, Management and Rehabilitation of Degraded and Secondary Tropical Forests.* ITTO Policy Development series number 13, Yokohama, Japan.

Mansourian, S., Dudley, N. & Vallauri, D. (2005). Restoring Forests in Landscapes. In *Forest Landscapes: Ecosystem Approaches to Sustainability*, J. A. Sayer & S. Maginnis (eds). Earthscan, London.

Sayer, J. A. & Maginnis, S. (eds) (2005). *Forests in Landscapes: Ecosystem Approaches to Sustainability.* Earthscan, London.

Part 6

Science Informing Policy and Conservation and Management of Tropical Forests

INTRODUCTION

Nigel E. Stork[1]* and Stephen M. Turton[2]*

[1]School of Resource Management and Geography, Faculty of Land and Food Resources, University of Melbourne, Burnley Campus, Richmond, Victoria, Australia
[2]Australian Tropical Forest Institute, School of Earth and Environmental Sciences, James Cook University, Cairns, Queensland, Australia
*The authors were participants of Cooperative Research Centre for Tropical Ecology and Management. participants of Cooperative Research Centre for Tropical Ecology and Management

In this final part of our book we bring together some of the more heavily applied aspects of research in the Wet Tropics and surrounding regions. The first of these chapters (Chapter 45), by Richard Pearson and Nigel Stork, discusses one of the most pressing issues for Australia: the importance of improving water quality and ecosystem health in tropical streams and the impact that current poor conditions is having on the Great Barrier Reef. This is an issue that is equally important in many parts of the world but is largely ignored because it is difficult to monitor and because it crosses two major biomes. The next chapter (Chapter 46), by Kristen Williams and colleagues, is the culmination of an enormous amount of data collection and verification and the development of highly sophisticated modelling tools for conservation prioritization – something this group is world renowned for. This work is the precursor to listing the Wet Tropics and associated systems as one of the world's biodiversity hotspots – something will surely follow in the

next year or so. Chapter 47 by Stuart Phinn and his team is an assessment of the use of remote sensing tools to assist the maintenance and management of world heritage values for tropical rainforests, and will be of use in many similar rainforests across the world. Those of us engaged in research and the use of research by land managers are aware of the misunderstandings and lack of dialogue that can frequently occur. In Chapter 48, Steve Goosem, Chief Scientist at the Wet Tropics Management Authority, joins us in assessing how we can improve that dialogue and improve understanding of the different roles of researchers and research users. So what have we learnt from our studies in the Wet Tropics and are there lessons for other parts of the world? In the final chapter of the book (Chapter 49) our international commentators join us in addressing these important questions. We hope these lessons and the ideas and information in the other chapters in this part will be of benefit to all those who read them.

45 CATCHMENT TO REEF: WATER QUALITY AND ECOSYSTEM HEALTH IN TROPICAL STREAMS

Richard Pearson[1]* and Nigel E. Stork[2]*

[1] School of Marine and Tropical Biology, James Cook University, Townsville, Australia
[2] School of Resource Management and Geography, Faculty of Land and Food Resources, University of Melbourne, Burnley Campus, Richmond, Victoria, Australia
*The authors were participants of Cooperative Research Centre for Tropical Ecology and Management.

Introduction

Around the world agricultural and industrial landscapes have massively affected terrestrial and aquatic habitats as well as the water quality of run-off from entire landscapes, resulting in dramatically altered ecosystem integrity in freshwaters, coastal wetlands, estuaries and coastal marine systems. Although these systems are reasonably well understood in temperate areas, in the tropics there is much less information on such environmental impacts and fewer examples of remediation. In Australia, the export of nutrients, sediments and contaminants into near coastal waters and the Great Barrier Reef (GBR) lagoon, the world's largest reef system, has increased substantially since European settlement, severely impacting on the viability and condition of these ecosystems and the industries that depend on them (Baker 2003). Approximately 200 near-shore reefs are under immediate direct pressure from declining water quality in the Queensland Wet Tropics and Whitsunday areas. Degradation of habitats (e.g. removal of riparian vegetation and invasion by weeds) and run-off of excess nutrients, sediments and agricultural chemicals are also causing substantial impacts on the rivers and wetlands that feed into the GBR lagoon and threaten the biodiversity and ecology of these systems and their important role in the greater GBR environment. Protection and restoration of these ecosystems are critical challenges to managers and policy-makers in Australia and elsewhere in the tropics.

Delays in controlling run-off and providing remediation have been due largely to the failure of governments to act and because of some dispute over the extent of the problem. The Commonwealth government has established long-term targets for improving water quality in the GBR lagoon (Reef Water Quality Protection Plan 2003) and now needs effective tools for monitoring the status and trends of water entering the GBR World Heritage Area. Community-led Regional Management Bodies have some of the resources needed to influence significantly and broker land-use changes, but need advice on options for catchment and riparian vegetation management, on methods of measuring the health of rivers, estuaries and coastal waters, and on methods of measuring status and trends in water quality for benchmarking and ongoing monitoring. While new institutional arrangements and policy frameworks are required to bring about effective change in catchments and marine ecosystems, it is imperative that tools be developed to gauge the effectiveness of improving land- and water-use practices.

Solving Australia's most urgent environmental issue

Improvement in water quality and mitigation of the ecosystem effects of degraded water supplies is arguably Australia's most important environmental issue. The Commonwealth government has responded to this challenge through the creation of the National Action Plan for Salinity and the Natural Heritage Trust. These funding programmes aim to support a community-led change in land use across the spectrum of human activities that affect the quality of the water running off the land and the ecological health of rivers and coastal ecosystems.

Recent work indicates that high loadings of nutrients, sediments and other contaminants, and invasion by exotic weeds, are major causes of declining ecological health of rivers and coastal habitats, with the likelihood of significant impact on Australia's tourist industry and external earnings (Baker 2003). Targets for reduced loads in catchments draining into the GBR have been identified to halt the decline in water quality (GBRMPA 2001), but the levels of these targets are disputed and seen as unrealistic by farmers. Land management practices to minimize loads have been researched and promoted for many years. However, there is limited observable improvement to date in water quality; in fact, conditions seem to be degrading further or not responding at anywhere near expected rates. This could be due to a range of factors, including the small spatial extent of changes in land management in comparison with the cumulative effect of past land-use practices, other drivers besides best resource management, institutional and industry collaboration issues and the time lag for ecological responses to become evident.

Over the past few years, several key events have brought these issues into sharp focus and resulted in intense public and media concern about deterioration of catchments and inshore reefs. For example

• an ABC television *Four Corners* programme (20 April 2002) devoted to the linked issues of land use, water quality and reef deterioration;

• the state government established a Reef Protection Interdepartmental Committee Science Panel with terms of reference that reflected the need for a scientific basis for assessing impacts, including assessment tools and protocols (Baker 2003);

• the sugar industry – one of the groups regarded as a major contributor to environmental problems – largely withdrew funding for environmental research (through changed priorities of sugar research organizations);

• water quality was identified as a major management issue by one far-sighted local government (Douglas Shire), which received federal funding as a pilot programme to develop mitigation protocols;

• the Great Barrier Reef Marine Park Authority (GBRMPA) produced end-of-system water quality targets, causing some controversy, particularly in agricultural circles;

• several studies pointed to deteriorating water quality – for example, the Burdekin Water Resources Plan (Queensland Department of Natural Resources and Mines/James Cook University; and the Herbert Integration Study (CSIRO/AIMS/JCU);

• fish kills in coastal waterways continued to be a major concern for professional and recreational fishing organizations, with agricultural contamination and habitat destruction implicated;

• concern was expressed forcefully by conservation bodies, such as the World Wide Fund for Nature and the Queensland Conservation Council.

Furthermore, during this period, climate change has become a major issue: reefs are suffering increasing coral bleaching events, compounding the effects of other anthropogenic impacts; and changing climate has been identified as a likely major influence on river flows and conservation values of rivers and wetlands. The lack of effective processes to provide integrated ecosystem-level science, incorporating social, economic and biophysical aspects, and transparent processes for decision-making, conflict resolution and trade-offs, has hindered progress towards long-term sustainability. Institutional and policy disincentives have also hindered the changes required to reverse the current levels of degradation of rangelands, rainforest, reefs and coastal ecosystems through the narrow focus of institutions and sectors.

A major problem in the past has been the lack of integrated response from commonwealth, state and local governments and a high level of institutional dysfunction. This is mirrored by the 'silo'-like nature of some of the major research organizations that had the skills and ability to undertake the necessary research and deal with the technological problems, and their lack of focus on what was needed by the user community. What was required was a bold, holistic and integrated approach: first, to overcome the impediments, constraints and barriers to achieving wide-scale, multi-issue, cost-effective and efficient improvements to land-use practice and ecosystem restoration; second, to meet explicit targets and build robust, sustainable communities and industries. The Catchment to Reef research programme aimed to address this challenge.

Protecting Australia's most valuable assets

The Catchment to Reef programme was novel in that it linked two Cooperative Research Centres – Rainforest CRC and CRC Reef, each a consortium of interested groups, that had a crucial role in research provision for the successful management of two World Heritage Areas. It addressed problems that linked the foci of the two CRCs through the medium of water and specifically water quality. The CRCs were the best-placed research bodies in the Wet Tropics/GBR regions to tackle these issues.

The problems addressed by the Catchment to Reef programme relate to water quality and ecosystem health in streams, rivers and coastal waters, and to the connectivity from the headwaters in the catchment to reefs and other marine habitats in the GBR lagoon. The focus of the research was on the Wet Tropics biogeographical region, and contiguous marine habitats, reflecting the areas of interest of the Reef and Rainforest CRCs. This region is remarkable for its natural habitats, for the greatest concentration of biodiversity in Australia and for the juxtaposition of two World Heritage Areas with extensive human use of the land and seascapes: for grazing, agriculture, mining, urban development and infrastructure in the catchments; for recreational and commercial fishing in the reef lagoon; and for intensive and extensive tourism across the whole region. It is also remarkable for its proportional contribution to Australia's total run-off: with about 0.2% of the continent's land area, the Wet Tropics contributes approximately 6.6% of its annual run-off (Figure 45.1). The links from catchment to reef are, therefore, of particular pertinence in management of the natural assets of the region. There are substantial management challenges for sustaining all industries while retaining the qualities that warrant World

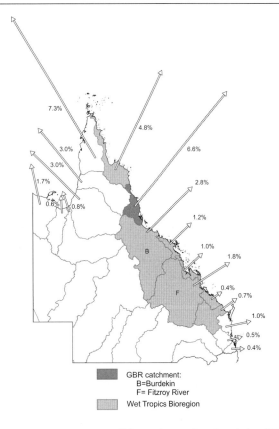

Figure 45.1 Annual run-off from Queensland and the GBR catchments, relative to that of the continent (total = 370 million megalitres per annum). Arrows indicate the proportion (%) of Australia's annual run-off from each major catchment. The total for Queensland = 40% of the continent's run-off, and for GBR = 19.2%.

Heritage listing. The Catchment to Reef programme was designed to provide the necessary science to underpin management of a crucial component of ecosystem health, namely water quality in its broadest sense.

The research programme recognized that water quality of aquatic ecosystems from the catchment to the reef is an issue that cannot be divorced from broader issues of natural ecosystem processes and human land use, land and water management, infrastructure development and changes over time. Some of these considerations are captured in Figure 45.2, which paints with a broad brush the characteristics of different parts of the catchment and the biophysical processes that occur there.

Water in the catchment is derived from precipitation over the whole area, and particularly in the uplands. Precipitation includes rainfall, but also cloud capture by montane forests, which can supplement water

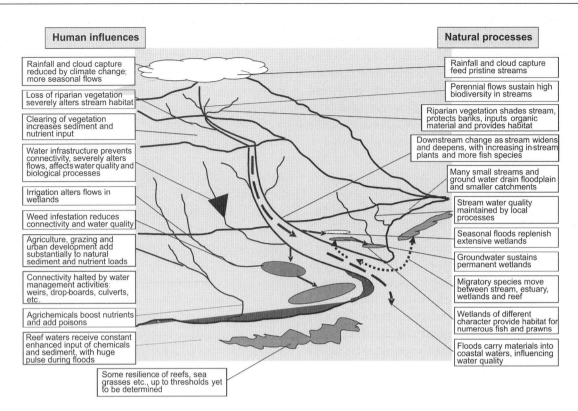

Figure 45.2 Natural processes and human influences in GBR catchments.

accession to the soil and streams by a substantial amount in the dry season (McJannet & Wallace 2006). During heavy rainfall, water runs over the surface and picks up sediments, suspended and dissolved organic matter, and nutrients, which are washed into streams. The quantities of these materials that reach streams depend on a number of factors, such as how well protected the soil is by vegetation and how disturbed the soil is, as well as the concentrations of nutrients at the soil's surface. Water also percolates through the soil, picking up dissolved organic matter and nutrients *en route* to the stream. Again, the amount of material dissolving in the water depends on the concentration in the soil, whether natural or supplemented by fertilizers. What then happens to the water and the materials it transports depends on the amount of water (e.g. wet season floods or dry season base flows) and the nature of the water body. Australia is typified by episodic rainfall events, which characterize much of the GBR catchment. The Wet Tropics has more predictable year-round rainfall than much of the continent, but nevertheless the largest volume of water with suspended and dissolved materials is delivered to coastal

waters by floods following unpredictable cyclonic events (Furnas 2004).

During large events, the capacity of streams to carry flood waters is exceeded and the floods spill out on to the floodplain, adding sediment to the land and replenishing networks of distributaries and wetlands. This flood pulse (Junk *et al.* 1989) is crucial to the ecology of the floodplain and its wetlands, although in many cases this natural process has been changed by human intervention through construction of drains, levee banks and the like. Flooding and deposition of sediments helps to reduce the amount of material reaching coastal waters, although the extent of this effect is not well documented.

Furnas (2004) has provided a broad description of the input of water and the dissolved and suspended materials to the GBR. A large proportion of the inputs apparently come from the large Burdekin and Fitzroy rivers to the south, and from the combined Wet Tropics rivers; however, inputs via floodplains, smaller surface drainages and ground water are not known. Furnas (2004) documents changes that have occurred in the past two centuries, including increases in sediments and nutrient loads due to broad-scale grazing and cropping. Much of the current information is derived from event-driven monitoring, which records the major loads exiting through river mouths. However, water quality in the catchments through the non-event periods is much less well documented. It is this ambient or chronic water quality that is of greatest importance to the ecology of the rivers and wetlands, as opposed to the short-term events that appear to drive water quality in coastal waters (Box 45.1). The relative importance of ambient inputs to coastal ecosystem processes is also not known.

Environmental problems that need addressing

The multitude of human impacts that might affect catchment and reef health is summarized in Table 45.1, and partly illustrated in Figure 45.4. The most important of these potential impacts is land clearing

(see Connolly & Pearson 2002). However, in the Wet Tropics, there is little available land left to clear, so the issue is more one of management of cleared land. Changing land use is a growing concern because, as climate changes and as the economics of particular crops changes, new land uses may bring problems that have not yet been experienced. Even the change in sugarcane harvesting, from the old method of burning first to remove trash, to the current approach of green cane harvesting, had unpredicted impacts: leaving the trash on the land had the major benefits of retaining organic material, removing smoke impacts and protecting the soil against erosion; but the interaction of rainfall with trash can produce organic pollution in streams, leading to fish kills. Fine-scale land and water management can alleviate such problems.

Nutrients

Nutrients are chemicals that occur naturally and are vital to plant production. Low availability of nutrients

---

Box 45.1 The contrast between ambient and flood conditions and their influences

In a typical Wet Tropics river, the annual hydrograph looks like Figure 45.3a, which is based on the Mulgrave River in 2004, showing the wet season peak and much lower flow during the dry season. Although there may be substantial variation in the loads of suspended and dissolved materials, they roughly follow this pattern. Therefore, the wet season is the period of maximum delivery of natural and human-supplemented loads to the marine environment, especially during major floods, and this is the period targeted for monitoring; however, delivery in the dry season months may provide above-natural loads of materials and influence marine systems during this period.

Daily flows during the dry season (Figure 45.3b) include two components – the base flow and minor peaks due to rain events. Base flow contributes small amounts of sediment and variable amounts of dissolved nutrients that percolate through the soil. The minor peaks may briefly contribute relatively large amounts of sediment and nutrients to the stream, especially from agricultural sources. The total amount of materials reaching the marine environment

during this period is generally low, but this pattern of events determines the water quality, productivity and ecology of the stream for much of the year, so monitoring during this period is of prime importance in understanding stream health.

In Figure 45.3c, which shows the day–night cycle (light and dark bars) over 4 days, the value of monitoring over a 24 hour cycle is illustrated. In slow-flowing streams and lagoons, dissolved oxygen can cycle dramatically, with abundant plants producing oxygen during the day, and plants, animals and microbes consuming it. At night, no oxygen is produced, so consumption causes the concentration to drop. On cloudy days, or in turbid water, less oxygen is produced during the day (e.g. day 2), potentially creating critical hypoxia conditions for the biota overnight, and sometimes leading to fish kills.

The type of monitoring required therefore depends on the questions being addressed, and contrasts particularly between the needs of land and water managers within catchments and the needs of managers of coastal waters.

(Continued)

Box 45.1 *(Continued)*

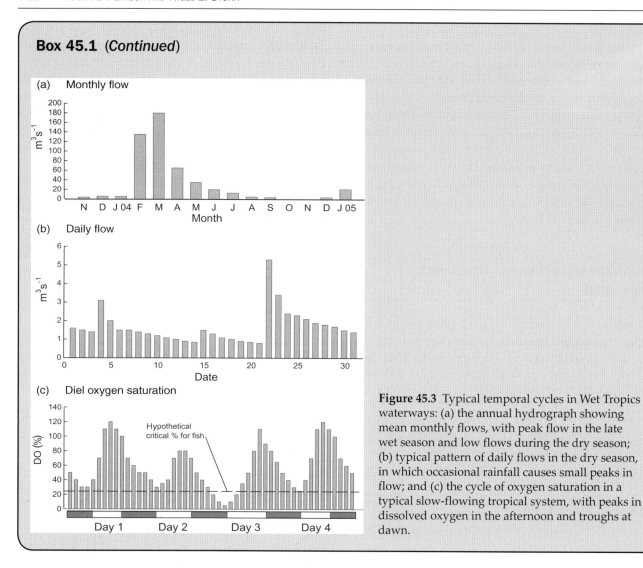

Figure 45.3 Typical temporal cycles in Wet Tropics waterways: (a) the annual hydrograph showing mean monthly flows, with peak flow in the late wet season and low flows during the dry season; (b) typical pattern of daily flows in the dry season, in which occasional rainfall causes small peaks in flow; and (c) the cycle of oxygen saturation in a typical slow-flowing tropical system, with peaks in dissolved oxygen in the afternoon and troughs at dawn.

relative to a plant's needs leads to poor growth and poor production of crops. Where soil nutrient status is perceived to be deficient, farmers add fertilizer supplements. While these added nutrients (especially nitrate and phosphate) are readily taken up by crop plants, varying amounts are lost, either to the atmosphere or into waterways. Phosphates attach to soil particles, which can be transported into waterways during heavy rain, where they are carried down-stream as suspended sediments, then deposited in waterways, on floodplains, in wetlands or in coastal waters. Nitrates readily dissolve in water and may be transported into the ground water to emerge much later some way downstream, or may be carried in flood waters overland into waterways (Rasiah *et al.* 2003).

Nutrients are normally not toxic to aquatic organisms, but they do promote aquatic plant growth. In streams, rivers and wetlands this typically results in enhanced growth of encrusting algae, phytoplankton and large aquatic plants. In many cases, where there has been removal of riparian vegetation and shade, and the invasion of exotic weeds, excess nutrients greatly enhance weed growth, with resultant clogging of waterways, blocking of fish passage and severe depletion of dissolved oxygen, making the habitat unavailable for many species. In coastal waters, nutrients can enhance phytoplankton blooms, some of which are toxic to other biota, and they can enhance benthic algal production, leading to smothering of reefs (Jompa & McCook 2002). It is also apparent that

Table 45.1 Examples of catchment impacts on ecological patterns and processes in rivers and coastal waters. All agents may interact with each other, with additive or multiplicative effects. These interactions are largely unknown; especially concerning is the potential interaction of these agents with changes in climate

Agents causing impacts	Sources	Impacts	Outcomes
Suspended sediments in excess of normal	Land clearing Grazing – soil and creek damage Agriculture and horticulture Mining Infrastructure Recreation	Smothering of stream habitats, plants and animals Smothering of reef systems Reduction of light penetration, productivity, and dissolved oxygen in waterways Reduction of light penetration and productivity, and possibly bleaching, of corals	
Nutrients in excess of normal	Agricultural fertilisers Breakdown of organic material Aerial input from industrial sources Point sources such as sewage treatment works, animal production units, dairies	Increased plant production and weed infestation in waterways Increased oxygen cycling, leading to fish kills in pools Increase algal production on reefs Plankton blooms Increased recruitment of crown-of-thorns starfish	
Organic material	Agricultural trash; decaying weed masses Stock on land or in intensive units; industry; sewage treatment works Replacement of natural (e.g. riparian) with exotic inputs	Increased oxygen demand by bacteria, leading to fish kills and barriers to movement in waterways Change in ecological community structure due to change in food supply in waterways	
Pesticides	Agriculture; gardens	Not well known; possibly toxicity to native plants, both freshwater and marine, and to instream fauna and coral communities	
Toxic minerals etc	Mining and industrial effluents	Toxicity and death to downstream plants and animals where effluents are not sufficiently diluted	
Salinity	Irrigated agriculture; cleared landscapes; deep drainage channels	Loss of intolerant plants and animals in waterways and riparian habitats	
Acidity	Deep drains penetrating acid sulphate soils	Toxicity to plants and animals in waterways	
Changed flow regimes	Climate change Water infrastructure – dams, weirs etc. Irrigation channels and drains Water harvesting Urbanization	Changed geomorphology and bank stability Sediment build up, or sediment scouring Changed water quality Weed invasion Altered triggers for animal breeding Unseasonable drying or inundation Changed habitat availability, loss of connectivity, etc. Water-logging or drying of riparian soils and dieback of vegetation Algal blooms	Loss of: • habitat • biodiversity • productivity • ecosystem services • amenity values
Changed temperature regimes	Altered flows Dam releases Loss of riparian shade	Effect on breeding triggers in waterways Exclusion of intolerant species Exacerbate processes leading to low dissolved oxygen levels	
Low dissolved oxygen (hypoxia)	Dam releases Nutrients, weeds, algal blooms, organic inputs Suspended sediments	Exclusion of intolerant species from waterways and change in community composition	
Barriers to movement of animals	Infrastructure – dams, weirs, culverts, flow control structures Weeds Poor water quality	Exclusion of species from important habitats in waterways and change in community composition	
Weed infestation	Especially through loss of riparian shade and increase in nutrient availability Intentional introduction of pasture grasses (e.g. para grass, hymenachne)	Effects in waterways on hydrogeomorphology, habitats, water quality, connectivity etc.	
Feral animals	Pigs and other large mammals Exotic fish escapes or willful introductions	Large mammals damage vegetation and benthic habitats; may cause local pollution through wastes and through mobilizing of sediments Exotic fish may displace native species from waterways	
Changes in light regime	Loss of riparian shade through clearing for agriculture etc.	Increase in stream temperature, possibility of increased hypoxia in waterways Increased production of weeds, leading to loss of habitat, barriers to connectivity, enhanced hypoxia etc. in waterways	

Figure 45.4 Examples of land-use impacts on catchments, waterways and the GBR: (a) reef bleaching owing to sediment deposition; (b) urbanization causes hydrological change and input of contaminants; (c) cattle damage banks and introduce nutrients and organic material to rivers; (d) broad-scale agriculture removes habitat, and can increase nutrient and sediment inputs to waterways; (e) loss of riparian vegetation causes increased temperature and light levels, invasion by weeds and changed channel characteristics; (f) riparian loss also reduces habitat values and connectivity across the landscape; (g) sediment from disturbed land and stream banks is deposited on stream beds and reefs, smothering habitats, plants and animals; (h) excess nutrients in agricultural and urban run-off cause increased production of algae and plants, blocking of waterways, hypoxia and fish kills.

outbreaks of crown-of-thorns starfish are linked to nutrient pulses into the GBR lagoon (Fabricius *et al.* 2005). Clearly it is in farmers' interests to retain as much as possible of their fertilizer for use by their crops, and best practice farming aims to do just that, by adding no more fertilizer than can be used by the crops, by adding it at an appropriate time and by adopting techniques that reduce run-off.

Sediments

Sediments occur naturally in aquatic systems as a result of normal erosion processes: the processes that shape landscapes, create floodplains and replenish beaches. Sediments become a problem when the rate of erosion, suspension and deposition is greatly increased by human activity. A major source of sediments is the vast cattle-grazing landscape in the GBR catchment, which in some places is nearly 1000 km from the GBR. Cattle grazing enhances sediment erosion and transport through damage to soils, removal of protective grasses and damage to stream banks. Cropping also increases sediment mobilization by exposing soils to heavy rain, by providing rapid drainage of fields along tractor tracks and by removal of riparian vegetation that might otherwise arrest water flow and sediment transport.

Sediments suspended in water reduce light penetration and so reduce the ability of plants to photosynthesize. When this happens during a flood, the effect is short-lived. But long-term turbidity in the water, such as occurs in the Burdekin River Dam (Lake Dalrymple) and in the river downstream, may have a long-term detrimental effect on normal processes. In coastal waters, suspended materials severely curtail the ability of the zooxanthellae (algae) living in coral tissues to photosynthesize, and so reduce the viability of their coral hosts. Moreover, if that suspended material settles to the substrate it can have major effects on benthic habitats and organisms, by clogging substrate interstices and micro-habitats, and smothering plants and small animals (e.g. Connolly & Pearson 2007).

Loss of sediments from farms represents loss of soil. It is in farmers' best interests to prevent this loss, via best practice management of the land and riparian strips, and careful construction and management of drainage channels.

Organic inputs

Organic inputs to aquatic systems, such as effluents from sewage works or dairies, typically cause oxygen depletion through bacterial respiration of organic materials, with subsequent loss of hypoxia-intolerant species of invertebrates and fish. In the Wet Tropics, sugar mill effluents were once the main source of problems (Pearson & Penridge 1987), but there has been substantial effort to remove or clean up discharges to waterways. Currently, major sources of organic materials are decaying vegetation such as sugar trash on fields, and dead aquatic weeds. Where trash sits in wet fields, fermentation can take place, deoxygenating the water and producing a toxic solution that can find its way into waterways and lead to fish kills. This typically happens under moderate rainfall, sufficient to wash the organic material into waterways but insufficient to flush it through. Floods are less of a problem because they dilute the contaminants. Organic material that is resident in waterways also causes oxygen depletion. The effect on the oxygen content of the waterway is largely dependent on the flow regime: fast-flowing streams are generally well flushed, whereas still lagoons are particularly vulnerable to hypoxia and fish kills. Good management of fields and riparian zones can greatly reduce the risk of these occurrences, but currently such management is challenging given that riparian vegetation has been almost completely removed in many sub-catchments.

Organic inputs are likely to be either absorbed or greatly diluted by the time they reach marine systems, so in themselves may not be implicated in contamination of reef waters. However, the organic snow that is found in enriched coastal waters is probably a culmination of the supplementation of terrestrially derived nutrients and organic material.

Pesticides

Pesticides include the chemicals that are applied to crops and stock to kill weeds, parasites etc. By their nature they can be damaging to non-target species, including native species. In the Wet Tropics, major pesticides include weed-control agents such as Diquat. Some monitoring of pesticide concentrations has been done in the Wet Tropics (e.g. Hunter *et al.* 1999) but there is very little information on the impacts of pesticides on native biota (exceptions include Kevan & Pearson 1993). Clearly our understanding of the fate and impacts of pesticides is a major knowledge gap that needs addressing.

Weeds and feral animals

Weeds and feral animals are species not native to the local region that become a nuisance. Typically they are introduced for their perceived values: the rabbit and the fox are two classic examples that have had devastating effects on Australian ecosystems. In the Wet Tropics there are many non-native species that are invading natural systems, with severe effects on biodiversity. In waterways some of the major problems are due to non-native grasses introduced as superior pastures (especially ponded pastures), out-performing native species substantially. In the Wet Tropics the major problem is para grass (*Brachiaria mutica*), which grows in profusion wherever there is sufficient light and appropriate substrate. Its growth is enhanced by nutrients in the water. It now infests most minor drainage channels, small streams and river banks. It is a severe impediment to normal drainage, and has substantial effects on the morphology of waterways (Bunn *et al.* 1998). It is a nuisance to farmers who have to manage it through mechanical or chemical means.

Water hyacinth (*Eichornia crassipes*) and salvinia (*Salvinia molesta*) are two plants introduced for their ornamental values, but which have become major weeds, especially in tropical Australia. They infest lagoons and slow-flowing waterways, eventually covering the whole water body in a thick mat that blocks out light and prevents gas exchange, rendering the waterway hypoxic and uninhabitable for native plants, fish and other animals.

As far as reef waters are concerned, weeds such as para grass in the catchment may have a positive effect in arresting the loss of sediment and nutrients to the marine environment. However, during major floods, much of this benefit is lost as weed mats and deposited sediments are mobilized and washed out of drainage systems. In blocking the movement of fishes, weeds prevent migratory species from reaching important habitat; for example, barramundi typically spend much of their lives in wetlands, but do not access those habitats that are blocked by weed.

Feral animals include pigs that can severely disturb the sediments and benthic fauna of shallow wetlands, and also several species of fish. In the Wet Tropics, of major concern are tilapia (*Oreochromis mossambicus*) and other related cichlid fishes of African origin, which, it is feared, might displace native species (Burrows 2004). Currently it appears that introduced fishes do especially well in disturbed habitats, but they are not yet implicated in displacement of natives in more pristine systems (Webb 2003).

Riparian vegetation

The natural riparian (riverbank) vegetation typically has different species composition from the catchment vegetation because of the more ready availability of water. It includes forest trees, shrubs and, with sufficient light penetration, some grasses and herbs. Where drainage is poor, species that are tolerant of waterlogging may dominate. The benefits of riparian vegetation to normal ecosystem function are well documented (e.g. Pusey & Arthington 2003). They include: habitat and habitat corridors for terrestrial animals and plants; habitat for semi-aquatic animals; shade; filtration mechanisms; organic inputs; bank stability; instream habitat via roots and snags and basking sites for reptiles. In the past, farmers were often encouraged to clear land right up to the river banks. It is now acknowledged that this policy was ill-conceived as the lost amenity values greatly outweighed the value of the land exposed. Despite broad acceptance of this assessment, restoration of riparian zones is only occurring very slowly.

Waterways that have undamaged riparian zones are characterized by clean and cool water, diverse habitat, armoured banks, rich instream and terrestrial wildlife and the absence of weeds. The richness of the terrestrial wildlife is proportional to the width of the riparian zone, as some species are very sensitive to edge effects; for example, Keir *et al.* (2007) estimate that a width of greater than 80 m (on each bank) is required to sustain most of the normal rainforest bird fauna.

Small breaks in the riparian vegetation allow opportunistic native species a chance to establish, and give saplings an opening for exploitation. Large gaps are prone to weed invasion, and the typical damaged Wet Tropics stream bank is infested with para grass, which has few of the values of native vegetation. It is clearly in the interests of efficient farming and of biodiversity values that riparian vegetation be protected and restored, but also that its values be demonstrated quantitatively.

Dissolved oxygen

Many of the impacts described above contribute to the oxygen status of waterways in the catchment. Riparian shade provides cooler water, with lower biological oxygen demand. Floating weed such as hyacinth prevents oxygen diffusion from the air. Conversely, submerged plants oxygenate the water by photosynthesis during the day, but consume oxygen at night. In situations where plant growth is excessive, oxygen consumption can lead to hypoxia and fish kills (often at dawn). Organic material feeds bacteria that deplete oxygen and block waterways, causing stagnation and reduced oxygenation, as well as contributing suspended sediments that reduce photosynthesis and re-oxygenation (paradoxically, this can be a benefit in open nutrient-rich systems as it reduces plant biomass, and so respiration etc.) The dynamics of oxygen (and, incidentally, pH) in catchment waterways are thus complex, and dependent on a range of natural and human-influenced variables (Pearson *et al.* 2003). Natural oxygen status can best be achieved: by maintaining riparian zones, curtailing weed growth; by preventing the input of nutrients; and by removing blockages to flow. While the tropical Australian invertebrate and fish fauna appear very resilient to low dissolved oxygen status (Pearson *et al.* 2003; Connolly *et al.* 2004), their tolerance thresholds can be breached, as evidenced by the occasional fish kills that occur in floodplain waterways.

This problem is essentially one of the catchment. Terrestrial run-off has not been reported to cause hypoxia in reef waters, presumably because of the very large dilution factor.

The special nature of the tropics

Much of the above discussion is based on knowledge gained over many years of working in the tropics on

extensive research programmes (e.g. Pearson *et al.* 2003), and from discussion with expert colleagues (see acknowledgements). While on face value the processes described above might occur in cooler zones, it is the year-round elevated temperatures and high light levels that make the tropics very different from the temperate zone. As long as there is water available, biological productivity can continue throughout the year in the tropics, and generally at a faster rate than elsewhere (temperate Australian systems may in summer match the tropics, but they slow down substantially in winter). This situation means that various processes are exaggerated in tropical waterways: in particular, enhanced photosynthesis leads to greater oxygen extremes through a 24-hour cycle, with a greater chance of fish kills in weedy slow-flowing waterways. The Wet Tropics has much greater diversity of fishes (Pusey *et al.* 2004) and invertebrates (Pearson *et al.* 1986; Pearson 2004) than other parts of Australia, so human activities are likely to have greater impact there. The high rate of processes is exemplified by the speed at which waterways can become covered by floating plants, making management of weed infestations particularly challenging. The tropics is also characterized by its hydrology – while Australia is a hydrological outlier compared with the rest of the world (Peel *et al.* 2001), the tropics is regarded as the most hydrologically active region of the continent (Bonnell 1998). Wet Tropics streams typically have perennial flows; however, some streams on the periphery of the Wet Tropics are intermittent, and some rivers that are partly fed by Wet Tropics streams (e.g. the Burdekin) cease to flow in some years.

Marine systems in the tropics are characterized by the richness and extent of their coastal plant assemblages (mangroves and sea grasses) and by their extensive and diverse coral reefs. It is in recognition of the unique nature of these marine systems that the Catchment to Reef programme was supported, although it is generally recognized that reef and catchment waters and habitats cannot be separated, either biophysically or in terms of their management.

Climate change

Over the past 100 years or so, recognition of the severe impact of human activity on the health of aquatic systems has developed from observations of point sources of organic pollution and their impacts (e.g. sewage in the Thames in the early 1900s; see Hynes 1960), through to the identification of pesticides as severe problems (Carson 1962) and to broad-scale land-use effects through non-point sources of contamination (e.g. Brodie & Furnas 1996). All of these problems have been related to burgeoning human population and demand for resources leading to land clearing for agriculture, infrastructure development, mining, heavy industry etc. While these problems are well recognized, it has taken a long time for land and water management policy and process to catch up, and there is still a long way to go, even in the more developed Western world. But over the past 10 years or so, the looming spectre of climate change has appeared, and it is clear that this is yet another problem that governments and on-ground managers must come to grips with. Assuming that some of the well documented effects will occur (IPCC 2001), it is a problem that will affect land availability and use, water availability and much human development, as well as ecosystem function and biodiversity. It is a problem that must be addressed in its own right, but not to the exclusion of the other issues discussed above, all of which interact with climate and so will themselves be affected by climate change. In some cases, some of the worst impacts of climate change may be felt through exacerbated effects of the processes described above. There is a major need to pursue appropriate research to address this issue.

Catchment to reef role

The above discussion points to our developing understanding of the nature of tropical waterways, and to the types of action required to improve their condition, which are to a great extent self-evident. However, it is not clear to what extent action needs to be taken, or the nature of the relationship between action and effect; for example, does doubling the width of the riparian zone double the improvement in stream condition? Tools are required to identify and monitor conditions and change in conditions resulting from management actions. While tools for river health monitoring have been developed around the world (e.g. Wright 1995), rarely have they explicitly linked levels of a particular contaminant and, say, biodiversity. Furthermore,

monitoring systems for temperate freshwaters are much more advanced than those for tropical systems; and marine monitoring systems lag way behind those for freshwaters. Therefore, the major goal of the Catchment to Reef programme was to develop tools, protocols and expertise to identify, monitor and mitigate riparian and water quality problems and to assess the functional health of aquatic ecosystems in the Wet Tropics and GBR World Heritage Areas. Previous research on water quality in the Wet Tropics focused on gaining a synoptic view of the fate and flux of nutrients and sediment. The Catchment to Reef programme focused on the links between water and the land (Figure 45.2), and on tools for measuring success or failure of improved land-use practices.

Specific research requirements

The issues outlined above need addressing by coordinated management and monitoring activities, using new tools that are appropriate to the questions being asked, as outlined below.

A major management tool for improving water quality is improved land management and manipulation of the riparian zone. While the importance of the riparian zone is well known, a pressing research need is to quantify its roles to facilitate management activities aimed at controlling water quality while simultaneously sustaining processes vital for river ecosystem health. The very special nature of GBR catchment hydrology, particularly the rainfall quantity and intensity, and stream flow extremes, creates an unusual circumstance in Australia that demands assessment at this bioregional scale. A clear need is tools to quantify the filtering role of the riparian ribbon, and the different effects of different land uses, including sugar and banana growing and forestry, on contributions to sediment and nutrient run-off and to provide guidelines towards enhancing performance.

New monitoring tools are required for water quality assessment against benchmarks. Water quality monitoring has frequently not been done well because:

• variables such as time of day, extent of instream vegetation and antecedent hydrological and other disturbances are of crucial importance but are ignored;
• end-of-river data do not reflect overall condition – for example, drainage from floodplain agriculture may

discharge not through the river but via wetland complexes or underground;

• end-of-river monitoring integrates within-river water quality improvements due to natural processes, but misses the stripping of contaminants by wetlands;

• flood-based monitoring quantifies material loads leaving the system but misses periods of steady contaminant transport; chronic inputs to the inshore marine system may be just as important as flood pulses because they supply readily accessible nutrients through the year;

• within rivers and wetlands, ambient conditions rather than flood events are the most important in determining the nature of those systems – that is, their water quality, habitat integrity, biodiversity and ecosystem function;

• benchmark or target development of necessity has been attempted without adequate reference to the above considerations and without the benefit of extended analysis;

• rarely does monitoring provide an explicit framework that addresses the issues of why, what, when, how and who to monitor water quality, so that there may be no sensible rationale for variable selection, sampling methods and guidelines for site selection through the catchment, that is at the sub-catchment scale, at the river mouth and across floodplain distributaries;

• there have been inadequate analysis and reporting frameworks in both scientific and management spheres.

Holistic assessment of system health is required because, in many cases, water quality is not the sole or most important issue; instead it is a combination of issues, such as habitat degradation, riparian condition and invasive species, as well as the physico-chemical properties of water, that determines the integrity of the system. River 'health' is a concept that describes the naturalness (and deviation from it) of river/stream ecosystems; it incorporates measures of physical condition and biological well-being, including key components of structure and function such as biodiversity, habitat integrity, food webs and community metabolism, water quality and response to natural disturbances (flood, drought etc.). River health is important because:

• it reflects catchment condition, especially the condition of the riparian zone;

• healthy river systems provide contaminant-stripping mechanisms;

• healthy river systems support rich biodiversity (including important fishery species) and provide for successful multiple use (e.g. drainage, water supply, recreation, fishery production);

• healthy river systems should sustain healthy downstream ecosystems of estuaries and coastal waters.

Box 45.2 Sources of contaminants: a case study

Figure 45.5 captures some views of drainage channels in the Russell River catchment during a period of heavy rain, but not during flood conditions, and demonstrates the need for protocols that are scale-sensitive so that management needs can be identified accurately. The main river (Figure 45.5a) was somewhat turbid, but its most obvious feature was a plume of suspended sediment that was tracking the left bank. This plume was emanating from a small creek, which was fed by a number of agricultural drains. Initial observation suggested that cane drainage was implicated (Figure 45.5b), but close inspection showed that the sediment was coming from upstream of the cane paddock and that water from the cane was clear (Figure 45.5c). Further upstream, the source of the sediment was

clearly a banana plantation (Figure 45.5d) in which sediment was being mobilized by water flowing along tractor tracks between the rows of bananas. In the cane paddock, the vegetation was much denser, preventing the erosion of the top soil.

The relative influence of two crops on suspended sediments in this particular example is clear, and points to the need to reduce the water velocity and its entrainment capacity in this local situation. However, there is no indication of the relative input of dissolved nutrients, which are invisible. This example points to the need for fine-scale understanding of water quality issues and mitigation methods, to improve both within-catchment water quality and end-of-river delivery to the GBR lagoon.

(Continued)

Box 45.2 (*Continued*)

(a)

(b)

(c)

(d)

Figure 45.5 Views of the Russell River and farm drains in its catchment. (a) The river with a sediment plume along the bank, emerging from a small tributary (arrow); (b) one of the main drains feeding the tributary, sediment-laden and passing through cane fields; (c) close-up showing sediment in drain, but clear water running from the cane; and (d) sediment entering the drain from a banana plantation.

A number of indices of river health have been developed that compare measures of different components of the system against measures expected from benchmark or reference sites. Different measures (e.g. dissolved nutrients, presence of fish) quantify different aspects of river health, and the preferred contemporary model (e.g. in south-east Queensland; Smith & Storey 2001) is to combine a suite of important variables. Given that the GBR catchments, especially the Wet Tropics bioregion, are very different from elsewhere (in terms of hydrology, biodiversity etc.), protocols need to be adapted or developed specifically for this region. Typically, river health measures are flawed because of the difficulty of finding suitable unimpacted (reference) sites. This necessitates intensive sampling of a number of streams to describe natural change along the stream, allowing for much more precise application of benchmarks, while also facilitating direct comparisons between more and less impacted systems.

Integrated catchment management is a major goal of natural resources management. Within sub-catchments there are usually several disparate forms of land use, tenure, protection and disturbance. Managing the quantity and quality of riparian and catchment forest cover within such landscapes has been linked to outcomes of improved water quality and river health, maintenance of water quality in estuaries and the GBR lagoon, and many terrestrial processes including biodiversity and microclimate regulation. Hence, the restoration of riparian vegetation cover is increasingly advocated as a key component of regional natural resource management. How best to assess and manage land use and disturbances to improve river condition and restore ecosystems is a major challenge. Rivers are characterized by longitudinal, lateral and surface to groundwater linkages, set in a matrix of spatial and temporal variability, including varying intensities of deforestation and land-use effects, often differing in different parts of catchments, as well as extreme hydrological variation due to the generally very high, but seasonal, rainfall. This includes, typically, a stream's progression from forested mountains, across agriculture-dominated tablelands, down the escarpment through steep forested gorges and to larger rivers and wetlands on the agriculture-dominated floodplain (Figure 45.2). There is also a lateral gradient on the floodplain, dependent on floods, flood distributaries

and connected wetlands through to the saline wetlands and the sea. These longitudinal and lateral progressions are subject to extreme hydrological variation owing to the generally high, but seasonal, rainfall. Typically, riparian vegetation has been extensively damaged in agricultural areas of the tablelands and the floodplain. In most disturbed reaches, weed invasion has further disrupted the normal functions of the riverine ecosystem.

Complex interactions among these factors must be considered in planning for development, conservation and rehabilitation. Frameworks that achieve integrated management are rare, and there is presently no clear set of guidelines on what might constitute best practice in integrated catchment management in the GBR catchments. Major issues are usually considered individually as isolated programmes of research or action (e.g. riparian restoration, water quality management, environmental flows) and outcomes for a catchment may be fragmented and inadequate to protect the component ecosystems. Clearly there is a major need for integrated catchment management frameworks to help to determine priorities for conservation of stream reaches that are performing well; to determine priorities for restoration works to ensure the most cost-effective use of resources allocated for restoration and rehabilitation; and to determine appropriate techniques for the restoration of river banks, riparian vegetation, instream habitat and connectivity, and water quality.

There is a need for coordinated condition and trend assessments for coastal marine communities. Coastal marine systems (especially coral reefs and seagrass beds within 20 km of the coast) are most at risk from the consequences of changing land use in adjacent catchments. However, despite their importance to fisheries, biodiversity and normal ecosystem functioning, there is little information about trends in the condition of these systems. Most monitoring effort in the GBR region has focused on offshore systems, which are less subject to land-based influences. Monitoring protocols used on outer reefs are not appropriate for inshore systems because of different conditions of reef extent, visibility etc. Coastal systems are naturally exposed to a range of disturbances, including coral bleaching, storm damage, fishing and crown-of-thorns starfish; however, run-off from human-influenced land uses (agriculture, urbanization etc.) is likely to be the most pervasive and chronic impact of human activity upon the quality of coastal coral reefs. Therefore, there is a need to establish a rigorous monitoring programme on coastal reefs. Once in place, the spatial and temporal history of changes observed on these reefs will provide critical context for interpreting change and assessing the relative importance of the various natural and anthropogenic threats to condition.

There is a need to develop tools for the detection of sublethal stress in aquatic organisms exposed to elevated levels of contaminates. Traditional monitoring techniques can be slow and ineffective ways of detecting environmental impacts on aquatic populations, especially when changes are subtle and/or mixed with other sources of variability. There is a need for more sensitive and unambiguous indicators of environmental quality, such as measurements of the physiological stress that develops in organisms long before conditions become so bad that populations or communities change substantially or crash. Some stress biomarkers have been proven for marine organisms, particularly fish from the temperate zone, but little comparable work has been done on tropical aquatic organisms even though they have great potential to indicate the presence of a wide range of contaminants. Potential new tools include semi-permeable membrane devices and diffusive gradients in thin films.

Some major unresolved issues

While research is proceeding to address the above issues, there are several major questions that require substantial attention, as outlined below.

Broadening the brief

While the issues outlined in this chapter are of global significance, it is at the regional scale that they are best understood and addressed. Previous work elsewhere in the world (e.g. south-east Queensland; Smith & Storey 2001) can provide models to be tested elsewhere, but the special nature of different biogeographical zones and different systems typically requires special responses, determined by regional case studies. The Catchment to Reef concept needs to build on its base in the Queensland Wet Tropics and contiguous coastal waters and encompass the whole GBR catchment, to provide for improved management of the

GBR region and also to provide a model for tropical systems elsewhere.

Quantifying connectivity

Understanding connectivity between the catchment and the reef (Figure 45.2) is a requirement for achieving successful outcomes. This connectivity is partly driven by the hydrological cycle whereby water is delivered from the sea to the land via the atmosphere, and then returns to the sea via surface and subsurface drainage. Major aspects of this cycle are the timing of events and the materials that the water picks up along its journey and delivers downstream. But other aspects of connectivity are also very important, particularly the biological links between the sea and freshwaters, and among the waterways of the catchment. While we know that many species use both marine and brackish or freshwater systems during their life cycles (Cappo 1998) (e.g. prawns use estuaries as juveniles and marine systems as adults; barramundi breed at sea but undertake much of their development in rivers and wetlands; mangrove jack, which is a marine and estuarine fish species, may venture a long way into freshwater streams), the quantitative importance of these connections, and impediments to them, is largely unknown. We do know that there are major impediments to fish passage in GBR catchments: for example, in the Burdekin River system there is a major dam without a fishway in the middle reaches of the river, there are weirs downstream and there are many barriers in the complex of waterways on the floodplain, such as culverts, drop-boards, weed infestation and zones of hypoxia (Perna & Burrows 2005). However, we have little detailed knowledge of the requirements of fishes through the region, and the management activities required to restore the values of these ecosystems.

Catchment versus coastal management

Major publications on water quality and the health of the GBR, such as the Baker Report (Baker 2003) and the Reef Water Quality Protection Plan (2003), were driven by perceptions of declining water quality and reef health. At the same time, these publications recognized the need to improve catchment health in order to halt and reverse the decline. What is less often clearly espoused is the inherent value of the catchments, reef or no reef. For example, the catchments house a large proportion of the Wet Tropics WHA and all its biodiversity and other values, including the highest diversity of aquatic invertebrates in Australia (Pearson *et al.* 1986; see also Connolly *et al.*, Chapter 12, this volume) and ranking very high on a global scale (Vinson & Hawkins 2003), and the highest diversity of Australia's unique fish fauna (Pusey *et al.* 2004). The waterways, and their contiguous riparian and terrestrial habitats, have inherent and high values. The waterways can maintain these values through their entire length, if not heavily disturbed, and the riparian ribbons that track the waterways provide habitat and terrestrial connectivity. The waterways are also popular for recreational use, they have high aesthetic values and are a major focus for recreational anglers, while the waterways and associated groundwaters are the only source of water to meet the increasing demand for irrigation and human consumption. These issues are of major concern for people involved in land and water management, and although their aims include good end-of-river water quality, the immediate needs are rather different. For example, weedy waterways are bad for catchment health, but they may enhance contaminant trapping, and so water quality, at the end of the river, at least in the short term. These issues need to be explicit in prioritizing the effort in improving land and water condition and management.

Therefore, despite the clear connections between the catchments and coastal waters, our approach to understanding the two parts of the system needs to be partitioned because the health of the catchments is largely driven by ambient conditions, while the health of coastal waters is mostly influenced by large events, as discussed previously. Furthermore, as much of the water quality monitoring effort over recent years has been driven by concerns about delivery of contaminants to the coast by rivers, monitoring effort has focused at the ends of rivers. So we now know a substantial amount about end-of-river contaminant loads, especially during wet-season events (Furnas 2004), but we know much less about water quality in the catchments, about loads delivered across floodplain distributaries and in groundwater, and about low-level but continuous inputs during the dry season.

For the plants and animals living in the catchments, wet-season floods are a mixed blessing: they can scour and remove organisms, causing severe population decline, especially of plants and invertebrates. But floods shape the waterways, they clean out potential weedy barriers and they feed the floodplain. Fish can largely avoid floods by taking refuge along banks and among vegetation, and in many cases use floods to provide access across or around otherwise impassable obstacles such as weirs and falls. Floods essentially reset freshwater systems and, given that they are part of the normal environment, the biota has evolved with them either to be resistant and withstand them or to be resilient and recover well. During the floods, water quality is affected dramatically: large loads of sediment and nutrient are transported, photosynthesis will be negligible and normal ecological processes (apart from dispersal) slow down or stop. However, floods dilute any potential toxicants (including hypoxic water), and flood conditions are typically short-lived, so that these events do not have a major water quality influence on the biota. Water quality in the catchment is driven more by the ambient natural processes of erosion, dissolution of chemicals and their transport; by continuous or occasional input of contaminants (Table 45.1); and by the interactions of physical, chemical and biological characteristics and processes leading to day-to-day and even hour-by-hour changes. End-of-river water quality monitoring tells us little about these processes and is of little use for managers wishing to improve the health of catchment waterways. On the other hand, monitoring of ambient conditions in the catchment gives no indication of what total loads of contaminants are reaching coastal waters. End-of-river monitoring does provide this sort of information, although, as indicated earlier, it does not allow estimates of loads across floodplains and via networks of distributaries, or through groundwater; and it gives no idea of whether chronic dry-season inputs are a crucial determinant of dry-season coastal conditions of turbidity, nutrient status, phytoplankton and benthic algal production etc. Greater understanding of the dynamics of nutrients, suspended sediments, productivity and the impact on reefs, especially during the non-event period of the year (i.e. most of it), is still required. This would involve not only monitoring of inputs throughout the catchments (including subsurface waters) but also studies on resuspension of deposited materials delivered previously by floods.

The need for models of ecosystem function

The above discussion involves conceptual models of how different parts of the catchment-to-reef ecosystems function. However, to facilitate our understanding of management needs, especially to enable responses to unexpected change, and to assist in the prioritization of efforts in restoration and rehabilitation, more explicit and quantitative models of stream, river, estuary and wetland function are required. Extending and quantifying the models discussed here will underpin management in all parts of the catchment-to-reef continuum.

The need to quantify roles of riparian zone and wetlands, beyond the popular rhetoric

As discussed above, the roles of the riparian zone have been well outlined in the literature. Similarly, there is much written about the role of the coastal wetlands: they support unique plant and animal communities, contribute substantially to biodiversity and provide breeding and nurturing habitats for key species (e.g. barramundi). These wetlands also store freshwaters over long periods of time, both arresting flows to the sea and enhancing the prospects for perennial aquatic habitat; and in slowing flows and promoting plant growth, they can strip sediments and nutrients from the water that is eventually delivered to the coast. However, few of these processes have been quantified, even at a small scale. We do not have the hydrological models to indicate how much flood run-off is captured, and we do not know what proportion of contaminants is retained temporarily or in the long term. These processes require further research to provide guidance to managers of riparian and wetland environments, especially with regard to where the major rehabilitation effort and cash should be expended.

Identifying and managing new impacts

While we have a good appreciation of the types of impacts affecting the catchment-to-reef continuum (Table 45.1), our lack of quantitative understanding of the mechanisms involved hampers our ability to predict the effects of novel impacts or novel combinations of impacts. For example, we do not have good models of how different climate change scenarios might directly affect pristine ecosystems, let alone interact with current impacts. Climate change may also lead to new interactions, as crops more suitable to the new climatic conditions in an area may raise a range of issues not previously encountered in a particular region. We do not have enough detailed information on particular crops and their impacts to predict the effects of change. Again, conceptual models of the processes that are involved in the impact of different land uses are required, followed by quantification of the models.

Summary

- The Catchment to Reef programme is focused on particular issues, particularly the development of appropriate monitoring tools for water quality and ecosystem health in the Wet Tropics and the adjacent GBR lagoon.
- It is restricted in time and scope, and cannot address many of the issues raised in this chapter. However, it is highlighting where further research is required to facilitate appropriate management (and perhaps to affect overarching policy) of aquatic systems and the activities that impact on them.
- Currently, the programme is drawing to a close, but some of these issues are being taken up by a new research initiative, the Marine and Tropical Science Research Facility. However, this collaboration between research providers, government and managers, like the Catchment to Reef programme, is limited in its scope, both geographical and conceptual, so to address the issues raised at the end of this chapter on a GBR-wide scale will require substantial partnerships with appropriate stakeholders.
- Clearly there has been political will to make a difference (halt and reverse the decline), as evidenced by investment in research and monitoring, in management through the NRM Boards, government agencies and GBRMPA and Wet Tropics Management Authority (WTMA), but this cannot be a short-term, flavour-of-the-month exigency: it needs to be maintained into perpetuity if the diverse values of the GBR, the WTWHA and their catchments are to be sustained.

Acknowledgements

This chapter reflects discussions with Catchment to Reef project participants, especially Angela Arthington, Jon Brodie, Barry Butler, Niall Connolly, Katharina Fabricius, John Faithful, Miles Furnas, Brad Pusey and the late Garry Werren, and other close co-workers – Nicole Flint, Colton Perna and Michael Crossland – as well as an understanding of tropical waterways developed over many years of working with Barry Butler and Niall Connolly. We are pleased to acknowledge the important inputs of these colleagues.

References

Baker, J. (chair) (2003). *A Report on the Study of Land-based Pollutants and their Impacts on Water Quality in and adjacent to the Great Barrier Reef.* Report to Intergovernmental Steering Committee, GBR Water Quality Action Plan, Brisbane.

Bonnell, M. (1998). Possible impacts of climate variability and change on tropical forest hydrology. *Climatic Change* **39**: 215–72.

Brodie, J. & Furnas, M. (1996). Cyclones, river flood plumes and natural water quality extremes in the central Great Barrier Reef. In *Downstream Effects of Land Use.* Hunter, H. M. *et al.* (eds). Queensland Department of Natural Resources, Brisbane, pp. 367–74.

Bunn, S. E., Davies, P. M., Kellaway, D. M. & Prosser, I. P. (1998). Influence of invasive macrophytes on channel morphology and hydrology in an open tropical lowland stream, and potential control by riparian shading. *Freshwater Biology* **39**: 171–8.

Burrows, D. W. (2004). *Translocated Fishes in Streams of the Wet Tropics Region, North Queensland: Distribution and Potential Impact.* Rainforest CRC Report No. 30. 83 pp.

Cappo, M. (1998). *A Review and Synthesis of Australian Fisheries Habitat Research* (www.aims.gov.au/pages/research/afhr/afhr-00.html).

Carson, R. (1962). *Silent Spring.* Houghton Mifflin, New York.

Connolly, N. M., Crossland, M. R. & Pearson R. G. (2004). Effect of low dissolved oxygen on survival, emergence and drift in tropical stream macroinvertebrate communities. *Journal of the North American Benthological Society* **23**: 251–70.

Connolly, N. M. & Pearson, R. G. (2002). Impacts of deforestation on stream ecology. In *Impacts of Deforestation in the Humid Tropics,* Bonell, M. (ed.). Cambridge University Press, Cambridge.

Connolly, N. M. & Pearson, R. G. (2007). Resistance to experimental sedimentation by a tropical stream macroinvertebrate assemblage. *Hydrobiologia* **592**: 423.

Fabricius, K., De'ath, G., McCook, L., Turak, E. & Williams, D. M. (2005). Changes in algal, coral and fish assemblages along water quality gradients on the inshore Great Barrier Reef. *Marine Pollution Bulletin* **51**: 384–98.

Furnas, M. (2004). *Catchments and Corals – Terrestrial Run-off to the Great Barrier Reef.* Australian Institute of Marine Science.

GBRMPA (2001). *Water Quality Action Plan: A Report to the Ministerial Council on Targets for Pollution Loads* (www. gbrmpa.gov.au/corp_site/key_issues/water_quality/action_plan).

Hunter, H. M., Hargreaves, P. A. & Raymnent, G. E. (1999). Pesticide residues in surface waters of two North Queensland catchments. In *Sources, Fates and Consequences of Pollutants in the Great Barrier Reef and Torres Strait.* Great Barrier Reef Marine Park Workshop Series No 25, Townsville, 24.

Hynes, H. B. N. (1960). *The Biology of Polluted Waters.* Liverpool University Press, Liverpool, 202 pp.

Intergovernmental Panel on Climate Change (IPCC) (2001). *Climate Change 2001: Synthesis Report.*

Jompa, J. & McCook, L. (2002). The effects of nutrients and herbivory on competition between a hard coral (*Porites cylindrica*) and a brown alga (*Lobophora variegata*). *Limnology and Oceanography* **47**: 527–34.

Junk, W. J., Bayley, P. B. & Sparks, R. E. (1989). The flood pulse concept in river-floodplain systems. *Special Publications of the Canadian Journal of Fisheries and Aquatic Science* **106**: 110–27.

Keir, A. F., Pearson, R. G. & Congdon, R. A. (2008). Determinants of bird assemblage composition in riparian vegetation on sugarcane farms in northern Queensland. *Wildlife Research*, in press.

Kevan, S. D. & Pearson, R. G. (1993). Toxicity of Diquat pulse exposure to the tropical freshwater shrimp *Caridina nilotica* (Atyidae). *Bulletin of Environmental Contamination and Toxicology* **51**: 564–7.

McJannet, D. & Wallace, J. (2006). *Methodology for Estimating Cloud Interception Inputs to Tropical Rainforest.* CSIRO Land and Water Science Report 6/06. Rainforest CRC.

Pearson, R. G. (2004). Biodiversity of the freshwater fauna of the Wet Tropics region of north-eastern Australia: patterns and possible determinants. In *Tropical Rain Forests: Past, Present and Future,* Bermingham, E. C., Dick, W. & Moritz, C. (eds). University of Chicago Press, Chicago, pp. 470–85.

Pearson, R. G. & Penridge, L. K. (1987). The effects of pollution by organic sugar mill effluent on the macro-invertebrates of a stream in tropical Queensland, Australia. *Journal of Environmental Management* **24**: 205–15.

Pearson, R. G., Benson, L. J. & Smith, R. E. W. (1986). Diversity and abundance of the fauna in Yuccabine Creek, a tropical rainforest stream. In *Limnology in Australia*, de Deckker, P. & Williams, W. D. (eds). CSIRO, Melbourne, pp. 329–42.

Pearson, R. G., Crossland, M., Butler, B. & Manwaring, S. (2003). *Effects of Cane-field Drainage on the Ecology of Tropical Waterways*, 3 volumes. Report on SRDC projects JCU016 & JCU024 Australian Centre for Tropical Freshwater Research Report No. 3/04-1-3.

Peel, M. C., McMahon, T. A., Finlayson, B. L. & Watson, F. G. R (2001). Identification and explanation of continental differences in the variability of annual run-off. *Journal of Hydrology* **250**: 224–40.

Perna, C. & Burrows, D. E. (2005). Improved dissolved oxygen status following removal of exotic weed mats in important fish habitat lagoons of the tropical Burdekin River floodplain, Australia. *Marine Pollution Bulletin* **51**: 138–48.

Pusey, B. J. & Arthington, A. H. (2003). Importance of the riparian zone to the conservation and management of freshwater fish: a review. *Marine and Freshwater Research* **54**: 1–16.

Pusey, B. J., Kennard, M. J. & Arthington, A. H. (2004). *Freshwater Fishes of North-Eastern Australia*. CSIRO Publishing, Melbourne.

Rasiah, V., Armour, J. D., Menzies, N. W., Heiner, D. H., Donn, M. J. & Mahendrarajah, S. (2003). Nitrate retention under sugarcane in wet tropical Queensland deep soil profiles. *Australian Journal of Soil Research* **41**: 1145–61.

Reef Water Quality Protection Plan (2003). Queensland and Commonwealth Governments publication. Queensland Department of Premier and Cabinet. 25 pp.

Smith, M. J. & Storey, A. W. (2001). *Design and Implementation of a Baseline Monitoring (DIBM3) Phase 1 Final Report*. Report to the SEQRWQMS, Brisbane, November 2000.

Vinson, M. R. & Hawkins, C. P. (2003). Broad-scale geographical patterns in local stream insect genera richness. *Ecography* **26**: 751–67.

Webb, A. C. (2003). The ecology of invasions of non-indigenous freshwater fishes in northern Queensland. PhD thesis, James Cook University.

Wright, J. F. (1995). Development and use of a system for predicting the macroinvertebrate fauna in flowing waters. *Australian Journal of Ecology* **20**: 181–97.

46 A PRELIMINARY ASSESSMENT OF PRIORITY AREAS FOR PLANT BIODIVERSITY CONSERVATION IN THE WET TROPICS BIOREGION

Kristen J. Williams*, Chris R. Margules*, Petina L. Pert*
and Tom Barrett*

CSIRO Sustainable Ecosystems; and Tropical Forest Research Centre, Atherton, Queensland, Australia
*The authors were participants of Cooperative Research Centre for Tropical Ecology and Management

Introduction

The Wet Tropics bioregion covers an area of approximately two million hectares. The Wet Tropics of Queensland World Heritage Area (WHA), covering 893 706 ha or 44.7% of the Wet Tropics bioregion, was inscribed in 1988, thus protecting a substantial proportion of the rainforests and associated habitats of the bioregion. The WHA is a large secure protected area and well serves the purpose for which it was established (natural heritage criteria i to iv, see www.deh. gov.au/heritage/worldheritage/criteria.html#natural). However, the WHA was not set up specifically to protect a representative sample of the biodiversity of the Wet Tropics bioregion, which extends beyond the WHA and includes species, habitats and ecosystems that are not covered by the WHA. In this chapter we identify those parts of the Wet Tropics bioregion outside the WHA that should have priority for the allocation of scarce biodiversity conservation resources, with the aim of increasing the proportion of regional biodiversity that is protected.

The Wet Tropics bioregion is considered remarkable for its high generic plant endemism and diversity, relatively low speciation within genera and abundance of gymnosperms and primitive angiosperms (Metcalfe &

Ford, Chapter 9, this volume). In this chapter, we focus on the identification of potential priority areas for the conservation of threatened or unprotected examples of extant vascular plant biodiversity in the Wet Tropics bioregion.

Biodiversity surrogates

Rather than considering all elements of biodiversity, we use locality records for vascular plants as surrogates for the known distribution of these taxa. The concept of total biodiversity is too vague to quantify or measure in the field because it includes diversity at every level of structural, taxonomic and functional organisation of biota (Sarkar & Margules 2002; Margules & Sarkar, in press). Even the simplifying proposal that biodiversity be considered at three levels of organization, genes, species and ecosystems (e.g. Meffe & Carroll 1994), while sensible in the face of intractable complexity, is unworkable for conservation action. Acquiring complete inventories by measuring and mapping all genes, species and ecosystems is not a practical option. Therefore some partial measure that can be quantified in the field is required and it has to be mapped. Such partial measures have come to be called biodiversity surrogates (Austin & Margules

1986; Sullivan & Chesson 1993; Sarkar & Margules 2002; Faith *et al.* 2004). Surrogates are attributes used to obtain information about biological diversity (however defined) in lieu of measuring biological diversity directly (Sullivan & Chesson 1993).

Taxa subsets, species assemblages and environmental variables or classes are commonly used as biodiversity surrogates in assessments of biodiversity conservation value. Taxa subsets include, for example, species at risk (Sarakinos *et al.* 2001), mammals (Kerley *et al.* 2003) and vascular plants (Pharo *et al.* 2000). Assemblages are species communities (e.g. McKenzie *et al.* 1989) and habitat types (e.g. Airamé *et al.* 2003). Examples of environmental classifications include land systems (Christian & Stewart 1968) and environmental domains (e.g. Mackey *et al.* 1989). While species are direct representations of biodiversity and would therefore usually be preferred as surrogates, assemblages and environmental classes have the advantage that they integrate ecological processes and, crucially, are normally easier to obtain at a consistent level of detail across planning regions. This is important because the identification of biodiversity priority areas is essentially a matter of comparing areas with one another across the landscape.

In an investigation of the efficiency of taxa as indicators for determining conservation priorities, Moritz *et al.* (2001) found the highly diverse and narrowly distributed taxa, such as plants and invertebrates, to be more efficient as predictors of other species than the less diverse and more broadly distributed taxa, such as vertebrates. More commonly, however, when identifying biodiversity priority areas, combinations of taxon and higher level surrogates are used. For example, Faith *et al.* (2001) used rare and threatened species, vegetation types and environmental domains to derive a conservation plan for Papua New Guinea. Habitat units, species and ecological process variables such as soil and altitude gradients and catchment hydrology were used to develop a conservation plan for the Cape Floristic Region in South Africa (for an overview see Cowling & Pressey 2003).

Another concern with taxa subsets is that records of where species occur in the field are often spatially biased. They are usually collected in a haphazard or opportunistic manner, contain records of species of interest to the collector but not necessarily other species that might have been there and are recorded from locations that were accessible, or represent places where the collectors expected to find the species they were looking for (Margules & Austin 1994). Museum and herbarium records are notorious in this regard.

Fortunately, in contrast with most regions, the biota of the Wet Tropics bioregion has been relatively well sampled. The plants in particular have been recorded from right across the range of environmental variation that exists in the bioregion. While some places have been sampled more densely than others, there are few actual gaps in environmental coverage. Consequently, the plant data have the crucial strength of spatial consistency, with the advantage of being direct components of the biota and not one or two places removed. Thus, for this preliminary analysis of biodiversity priority areas outside the WHA, plant records were used as surrogates for the known distribution of this taxa group in the Wet Tropics bioregion.

Identifying priority areas

We use the tools that have come to be associated with systematic conservation planning for identifying biodiversity priority areas. The principal tool is an algorithm that implements complementarity (Margules *et al.* 1988; Vane-Wright *et al.* 1991; Pressey *et al.* 1993; Margules & Pressey 2000). Complementarity is a measure of the contribution an area makes to the full complement of biodiversity features: species, assemblages, ecological processes etc. In order to implement complementarity it is necessary to set goals for the representation of biodiversity. Complementarity can then be measured by the contribution an area makes to the conservation goal. For example, if the conservation goal is to represent at least one population of all of the plant species then the first area we choose might be the one with the most species. The next area we choose will be the one with the most unrepresented species, and the next area, the one with the most remaining unrepresented species, and so on until all species are represented.

There are two important points to note about this process. The first is that as areas are added to the set, the contribution that remaining areas make to the biodiversity goal changes. This is because some of the species in those areas may have already been contributed by other areas selected previously. The second is that areas with the highest complementarity will not necessarily be those with the most species. In this example, complementarity is measured at each step as the number of unrepresented species. If an area has few species, but they do not occur widely in the planning region, it may have higher complementarity than an area with many species that are widespread throughout the region (depending on which species occur in areas that have already been selected). Thus, species richness *per se* cannot be used to measure complementarity.

There are many variations on the simple area selection process outlined above, including the incorporation of cost trade-offs and other spatial constraints (e.g. Faith *et al.* 2001). As it turns out, for example, it is more economical to begin the area selection process by selecting that area with the rarest species on it and then the one with the highest number of unrepresented species, and so on until all species are represented (Sarkar *et al.* 2002). In this chapter, we present the results of an analysis that prioritizes areas for the contribution they make to the full representation of the plants of the Wet Tropics bioregion, without explicitly considering social or economic trade-offs, which will be the next step.

Methods

Study area

The study area for our analysis was the Wet Tropics bioregion, a region of the Interim Biogeographic Regionalisation for Australia, Version 6.1 (Environment Australia 2005). The boundaries of the Wet Tropics bioregion, WHA and protected areas (national parks and forest reserves) are shown in Figure 46.1.

Vascular plant database

A database of vascular plant locations was compiled from a range of sources within CSIRO and the Queensland government (Table 46.1). The taxonomic nomenclature adopted by the Queensland Herbarium (Henderson 2002) was principally used to combine species names between databases. Records with collection dates older than about 1950 and spatial precision greater than about ±500 m were not used in this analysis. Records with locations in areas that have since been cleared were also excluded from analysis (as we only include records within areas identified as remnant vegetation current to 2003). This analysis is based on an extract of the compilation database, as at

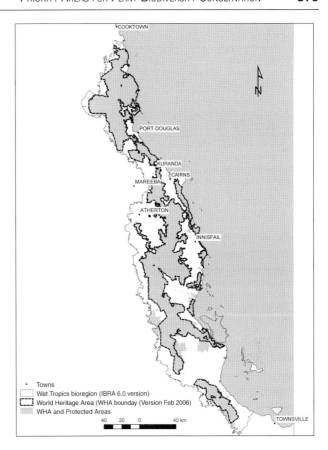

Figure 46.1 The Wet Tropics bioregion showing the World Heritage Area boundary and other protected areas (national parks and forest reserves).

Table 46.1 Source of plant records used in this analysis

Database	Source	Extract data
HERBRECS (specimen based collection, BRI)	Queensland Herbarium, Environmental Protection Agency, Brisbane	December 2004
CORVEG (systematic survey sites)	Queensland Herbarium, Environmental Protection Agency, Brisbane	December 2003
Andrew Ford collection (systematic surveys and incidental sightings)	CSIRO and Rainforest-CRC	July 2005
National Herbarium (specimen based collection, QRS)	CSIRO	July 2005

November 2005. Some errors are likely because the identification and correction of taxonomic and location errors is an ongoing, incremental process.

Plant species were designated rainforest endemic by A. Ford (unpublished data). Rainforest endemics were

Table 46.2 Categories of threat, in relation to risk of extinction, used for vascular plants in this analysis

Threat Level	Nature Conservation Act 1992*	Environmental Protection and Biodiversity Conservation Act 1999	IUCN Red List 2004
Endangered	Extinct in the wild, Endangered	Extinct, Critically Endangered, Endangered	Extinct, Extinct in the Wild, Critically Endangered, Endangered
Vulnerable	Vulnerable	Vulnerable	Vulnerable
Rare	Near threatened, Rare		Near threatened, Conservation dependent

* The Queensland Nature Conservation Act 1992 was reprinted (April 2006) with revised classes of wildlife (Part 5, Division 2): extinct in the wild, endangered, vulnerable, rare (to be phased out by 2010), near threatened, least concern.

defined as those species considered to be largely obligate to wet tropical rainforest assemblages, without limitation to the Wet Tropics bioregion.

Threat level

We adopted the threat levels for vascular plants reflected in state, national and international legislation or treaties. These were the Nature Conservation Act 1992 (NCA, Queensland), the Environmental Protection and Biodiversity Conservation Act 1999 (EPBC, Commonwealth of Australia) and the IUCN Red List 2004 (IUCN, the World Conservation Union). The different categories of threat are given in Table 46.2. 'Threat level' for our analysis was the highest listed status for a taxon. Some species regarded as presumed extinct or extinct in the wild have recently been rediscovered (A. Ford, unpublished data). Valid occurrences for previously presumed extinct species were therefore included in our analysis of priority areas.

Planning units

Planning units, the areas of land used for selecting priority areas, were derived from property boundaries (based on the digital cadastre database for Queensland; Department of Natural Resources and Mines 2004) and remnant vegetation (where a regional ecosystem type is mapped as at 2003; Environmental Protection Agency 2005a). Small cadastre units less than 10 ha often reflect urban and other intensive uses with fewer planning controls. These small cadastre units were merged where spatially adjacent and within the same tenure group and local government authority. This pre-processing step substantially reduced the number of planning units considered in the analysis. More recent information about tenure of national parks and reserves was attached to the planning unit data set from the Queensland Estates layer (Environmental Protection Agency 2005b) and the World Heritage Area boundaries (WTMA 2006). In addition, areas designated as nature refuges (Environmental Protection Agency 2005c) were included in the boundaries of the planning units. Nature refuges are privately owned properties with conservation covenants attached to all or part of the property.

The resulting base layer contains 153 368 planning units within the Wet Tropics bioregion. The number and extent of planning units by tenure and land category are shown in Table 46.3.

Targets for representing plant biodiversity

We used different sets of targets (i.e. quantitative goals) to explore three scenarios for representing unprotected examples of plant biodiversity in the Wet Tropics bioregion. These were: (i) all species, each taxon represented in at least one planning unit; (ii) rainforest endemics, each taxon represented in at least three planning units; and (iii) threatened species, all endangered taxa represented wherever they occur and vulnerable and rare taxa represented in at least three planning units. In the first target scenario we take the simplest case in which plant taxa are represented in at least one planning unit. In the second and third target scenarios we allow for some cases in which endemic or threatened plant taxa are represented across more than one planning unit.

Identifying priority areas

We used the conservation planning software C-Plan (Pressey *et al.* 1994, 1995; Ferrier *et al.* 2000; NSW NWPS 2001) to identify potential priority areas for vascular plant species. C-Plan iteratively selects planning units depending on their complementarity until the target is met using the *Minset* algorithm to identify an approximate 'minimum set' of planning units based on the selection rules described below. This results in a representative, repeatable, but not necessarily optimal, set of sites. The following selection rules describe the *Minset* algorithm in C-Plan.

Table 46.3 The number and areal extent of planning units for the entire Wet Tropics bioregion base layer (a) and the layer used for analysis (b) by tenure and land category

Tenure categories represented in the WHA*		Land type with area in hectares (and number of planning units)	
Code	Description	Non-remnant	Remnant
WHA	World Heritage Area	(a) 9704 (7 842) (b) 1117 (77)	a) 883 770 (5487) (b) 799 941 (563)
NP	National park (land reserved for a national park (including scientific), conservation park or resource reserve)	(a) 3619 (3624) (b) 229 (26)	(a) 383 193 (1442) (b) 344 064 (227)
FR	Forest reserve (tenure of an interim nature with associated conditions)	(a) 1877 (3057) (b) 56 (12)	(a) 447 586 (923) (b) 394 684 (251)
SF	State forest (land reserved for state forest purposes)	(a) 16 604 (2637) (b) 4 097 (54)	(a) 74 227 (896) (b) 50 464 (87)
TR	Timber reserve (land reserved for timber reserve purposes)	(a) 290 (326) (b) 10 (7)	(a) 108 039 (180) (b) 103 415 (61)
SL	State land (land held by the state of Queensland as unallocated state land and other areas vested in the state (or Crown) but not held in fee simple or as a lease issued under the lands Act 1994)	(a) 3276 (3 264) (b) 125 (18)	(a) 96 047 (1434) (b) 66 328 (110)
WR	Water resource (land vested in the department of natural resources and mines)	(a) 150 (57) (b) 0 (0)	(a) 124 (33) (b) 1 (1)
RE	Reserve (land reserved for community or public purposes)	(a) 5006 (3996) (b) 658 (20)	(a) 18 644 (1658) (b) 8 952 (75)
LL	Lands lease (leasehold land administered by the Department of Natural Resources and Mines excluding Mining Homestead Tenement Leases)	(a) 18 078 (5 426) (b) 2 712 (59)	(a) 201 307 (3291) (b) 81 020 (96)
F	Freehold (tenures: CV, covenant; EA, easement; AP, easement proposed; FD, below the depth plans; FH, freehold: predominantly land held by the State in fee simple (freehold title) which includes titles surrendered to the State of Queensland (or crown) in terms of section 358 of the Land Act 1994)	(a) 382 843 (53 034) (b) 32 714 (907)	(a) 162 416 (24 442) (b) 43 528 (525)
AP	Action pending (unallocated state land in the process of a reservation action or in the case of a dedication of a reserve proceeding on a registered survey plan)	(a) 3 (5) (b) 0 (0)	(a) 0.2 (5) (b) 0 (0)
CA	Commonwealth acquisition (land acquired by the Commonwealth of Australia and held prior to issue of a formal title, generally this land is used for military or government store purposes)	(a) 5 (2) (b) 0 (0)	(a) 1681 (2) (b) 1461 (1)
PP	Profit a *prendre* (a registered right or interest of use over the property of another that allows the holders to enter and take some natural produce: mineral deposits, timber)	(a) 140 (22) (b) 120 (3)	(a) 27 (8) (b) 0 (0)
R	Road (land vested in the Queensland Transport or the Department of Main Roads prior to issue of title or road dedication)	(a) 22 648 (12 602) (b) 6330 (127)	(a) 13 682 (10 860) (b) 1295 (139)
TP	Transferred property (land transferred to the Commonwealth of Australia on federation, usually for lighthouse or post and telegraph reserves)	(a) 6 (6) (b) 0 (0)	(a) 214 (8) (b) 63 (1)
RY	Railway (land vested for railway purposes in the Queensland Transport or Queensland Rail)	(a) 8 (21) (b) 0 (0)	(a) 1 (4) (b) 0 (0)
U	Crown urban infrastructure (tenures: HM – boat harbours, HL – housing land; id – industrial estates; PH – port and harbour boards)	(a) 15 (52) (b) 0 (0)	(a) 7 (3) (b) 0 (0)

(Continued)

Table 46.3 (*Continued*)

Tenure categories represented in the WHA*		Land type with area in hectares (and number of planning units)	
Code	Description	Non-remnant	Remnant
Other (excluded)			
Non-rem	Non-remnant	(a) 409 691 (64 886) (b) 40 033 (1029)	
Dist	Disturbed	(a) 26 325 (25 354) (b) 1 058 (157)	
Ocean	Ocean	(a) 9545 (6354) (b) 722 (23)	
Water	Water	(a) 5323 (1354) (b) 2618 (12)	
Plant	Plantation	(a) 14 861 (2707) (b) 2 988 (41)	

* DCDB tenure parcels with area ≤10 ha were merged where also in the same tenure group and same local government area.

- Rule 1: select the planning unit with the greatest single contribution to the goal. If there is a tie (more than oneplanning unit) then go to Rule 2.
- Rule 2: select the planning unit with the lowest area. If there is a tie then go to Rule 3.
- Rule 3: select the first planning unit in the list.

The stopping rule selected was 'until all available species that contribute to targets have been selected'. On the first iteration the planning unit with the greatest contribution to complementarity relative to the existing set of protected areas will be selected, and if there is more than one planning unit with equal highest contribution to the goal, the second rule acts as a tie-breaker. If there is still a tie at the second rule, i.e. two or more sites with equal area, then the next rule acts as a tie-breaker for those planning units. The higher the rule is in the algorithm the more influence it will have on the final set of planning units selected. After each selection C-Plan recalculates all complementarity values and a new planning unit is selected in the same way as the last.

Our analysis matrix contained 2064 plant species within 2871 planning units containing extant remnant vegetation. Regardless of the number of species records within a planning unit, each unit was coded in a binary fashion, with presence of a species indicated by a value of one. We found areas that are currently not protected, which added the most plant species to the list known

Table 46.4 Category of tenure and hierarchy used to define planning units available for selection or excluded in each target scenario. Tenure categories are described in Table 46.3

Tenure hierarchy	Tenure category available for selection	Tenure categories excluded from selection in steps 1 and 2	Area (ha)*
Step 1	RE, SF, TR, SL, WR	LL, F	79 981
Step 2	RE, SF, TR, SL, WR, LL	F	149 752
Step 3	RE, SF, TR, SL, WR, LL, F		188 744

* Non-remnant land types were excluded from area calculations.

to occur on the WHA and other protected areas. We then identified the area that added the most remaining unrepresented species and continued to find new areas in this way until all species were represented in the set of selected areas. We did this for the three different target scenarios – all species, rainforest endemics and threatened species.

We applied a tenure hierarchy (Table 46.4) to each scenario to reflect perceived preferences for choosing public land tenures before freehold when selecting areas that potentially contribute to the targets. In our analysis, therefore, some planning units were always excluded because they made little or no contribution to biodiversity (i.e. cleared areas; no plant taxa records; urban and other intensive use zones), and some were considered mandatory, always included at the beginning of each analysis and never exchanged (i.e. WHA and other protected areas) (see definitions in Table 46.5). Other tenures were temporarily excluded and then made available for selection according to the hierarchy in Table 46.5.

Results

We compiled records for 2064 vascular plant species occurring within the Wet Tropics bioregion (Figure 46.2). Threatened taxa included 35 endangered species, 75 vulnerable species and 242 rare species (Table 46.6).

Table 46.5 Definition of planning unit categories used in target scenarios

Planning unit groupings	Description (definitions of tenure categories are given in Table 46.3)
Mandatory	Planning units that fell inside the World Heritage Area, national parks and forest reserves were always selected first and could not be traded
Excluded	Planning units with land type non-remnant (i.e. disturbed, plantation, ocean, water and non-remnant) or tenure category (AP, CA, PP, TP, U, W, RY, R) were always excluded
Available	Planning units that are 'available' for selection
Selected	Planning units that contribute to satisfying targets

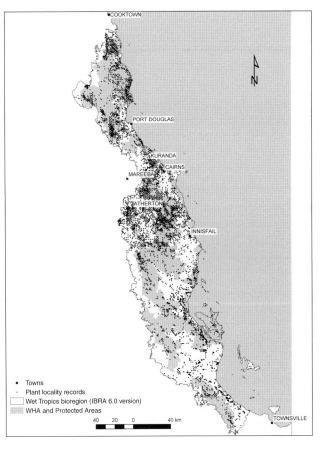

Figure 46.2 The distribution of reliable location records for vascular plants in the Wet Tropics bioregion.

Table 46.6 Summary of protection status of plants in the Wet Tropics bioregion. See Table 46.2 for explanation of EN, VU and Rare. We used a threshold to assign species to a land use. The number in other protected areas is the number that does not reach a threshold of 80% of all records in the WHA and the number in the unprotected category is the number that does not meet that threshold in either the WHA or other protected areas. The number in parentheses refers to all records. For threatened species, targets for representation include all populations of these species

Land use	Number of species (records)	Number of rainforest endemic species (records)	Threat level			
			Presumed extinct species (records)	EN species (records)	VU species (records)	Rare species (records)
World Heritage Area	1930 (60 468)	628 (21 927)	2 (5)	27 (429)	65 (851)	217 (4749)
Other protected areas (NP or FR) outside WHA	924 (3882)	207 (553)	0 (0)	8 (19)	12 (86)	48 (170)
Outside WHA and protected areas (unprotected)	1727 (27 983)	481 (5479)	0 (0)	22 (233)	45 (698)	159 (1178)
All areas	2064 (92 333)	637 (27 959)	2 (5)	35 (681)	75 (1635)	242 (6097)

Table 46.7 Summary of the number of plant species and locality records by land protection type and threat level. See Table 46.2 for explanation of EN, VU and Rare. Values for 'all species' and 'rainforest endemics' are number of species. Values for 'threatened species' also show number of records in brackets

		Species group								
		Frequency of records per species						Threatened species Threat level		
		All species			Rainforest endemic species					
		1–5	6–10	>10	1–5	6–10	>10	EN	VU	Rare
Land use	World Heritage area	427	259	1244	61	73	494	29 (434)	66 (851)	218 (4749)
	Other protected areas	467	284	1154	72	77	465	30 (392)	67 (781)	213 (4029)
	Outside protected areas	590	347	790	165	118	198	22 (233)	46 (698)	130 (1178)

Table 46.8 Minimum number of planning units additionally selected from each tenure group within the Wet Tropics bioregion for three target scenarios after first including World Heritage Area, national parks and forest reserves, the areas currently under protection. Area in hectares is given in parentheses

| Target scenario | (No. of available planning units that contribute to the target) | Number of planning units additionally selected | | | Total no. of planning units additionally selected |
		Tenure gp 1	+ Tenure gp 2	+ Tenure gp 3	
All species	800 (153 523)	23 (27 050)	11 (8170)	16 (447)	50 (35 667)
Rainforest endemics	831 (188 339)	3 (405)	0 (0)	9 (186)	12 (591)
Threatened species	741 (136 647)	64 (30 318)	29 (21 779)	87 (7754)	180 (59 850)

Of the 352 threatened taxa, 201 are considered endemic to Wet Tropic rainforest habitat (A. J. Ford, unpublished data). A summary of the number of plant species and frequency of locality records by land protection type and threat level is given in Table 46.7.

Table 46.8 presents the results of each target scenario for each successive land tenure group. In the 'All species' target scenario, 50 planning units were selected that contribute at least one occurrence of a plant species that is not currently recorded in the WHA, national parks or forest reserves, the areas currently protected. Only 12 planning units were needed to represent at least three occurrences of rainforest endemics in addition to areas currently protected. This contrasts with 180 planning units needed to represent all probable extant occurrences of endangered species and at least three occurrences of vulnerable or rare species in addition to areas currently protected.

Figures 46.3, 46.4 and 46.5 show the planning units selected in addition to the WHA to meet the three scenarios: all species; rainforest endemics; and threatened, vulnerable and rare species. Of the selected planning units, 17 are also nature refuges in all of the scenarios. Nature refuges (and coordinated conservation areas) are declared under the Nature Conservation Act 1992. A nature refuge can be declared over any land, state or freehold, to protect significant natural resources such as wildlife habitat and provide for controlled use of those natural resources, taking account of the landholder's interests. Areas are gazetted on the basis of an agreement between the government and private land owner. Of the 11 named nature refuges in the Wet Tropics bioregion, six were selected in one or more target scenario(s) (Table 46.9). Figure 46.6 shows the location of selected nature refuges.

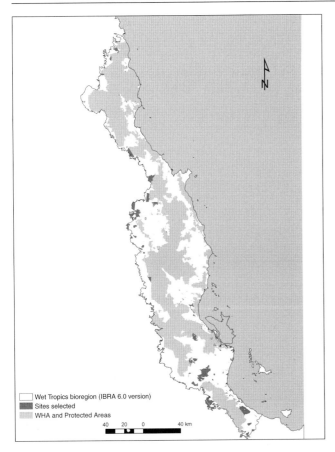

Figure 46.3 Planning units in addition to the WHA and other protected areas selected to represent all plant species at least once (the minimum set solution under target scenario 1, All Species).

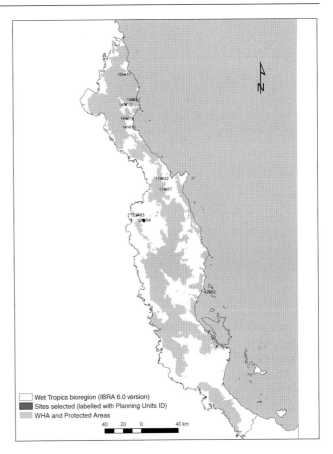

Figure 46.4 Planning units (labelled) in addition to the WHA and other protected areas selected to represent rainforest endemic plants in at least three planning units (the minimum set solution under target scenario 2, Rainforest Endemics).

Discussion

As might reasonably be expected, the WHA includes representative populations of most rainforest endemics, but does not necessarily represent all plant taxa known to occur in the Wet Tropics bioregion or those under the most serious threat in the fragmented coastal lowlands.

For this analysis, we used over 100 000 locality records for 2064 plant species, filtered by accuracy and reliability, as our biodiversity surrogate. It is well known, however, that taxa subsets, although reflecting the 'true surrogate' (*sensu* Sarkar & Margules 2002) under our objective, are limited because complete inventories across whole regions are neither practical nor feasible. Even though there are few gaps in the coverage of the Wet Tropics flora compared to other

data sets in other places and remaining gaps in environmental coverage have been systematically targeted for data collection (Ford *et al.* 2004, unpublished), georeferenced information on distributions remains incomplete. This has potentially serious implications for biodiversity priority areas selection. For example, gaps in our database for plant species occurring within the WHA and other protected areas may lead to areas being identified from among other tenures that are not actually required to fulfil biodiversity conservation targets. For this reason, Ferrier (2002) calls for new approaches to mapping spatial pattern in biodiversity that shifts the focus from individual to collective properties (*sensu* Austin 1999) of biodiversity and automatically fills gaps in a coverage through statistical estimation. One such promising approach is based on an extension of matrix regression to estimate compositional turnover in species using locality records as the response variable and spatial environmental data as the explanatory variables. This method is termed generalized dissimilarity modelling (Ferrier 2002; Ferrier *et al.* 2004). Research is proceeding, in collaboration with Simon Ferrier, to estimate compositional turnover in plant species for the Wet Tropics bioregion. In future work, we intend comparing the effectiveness of these different biodiversity surrogates in similar priority setting exercises (for example, see Sarkar *et al.* 2004).

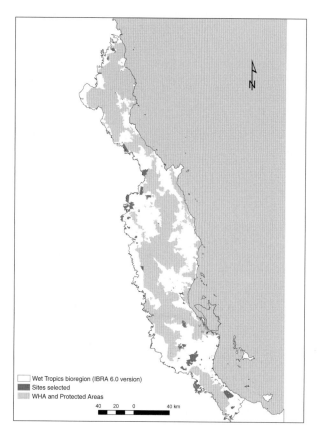

Figure 46.5 Planning units in addition to the WHA and other protected areas selected to represent all endangered taxa wherever they occur and vulnerable and rare taxa in at least three planning units (the minimum set solution under target scenario 3, Threatened Species).

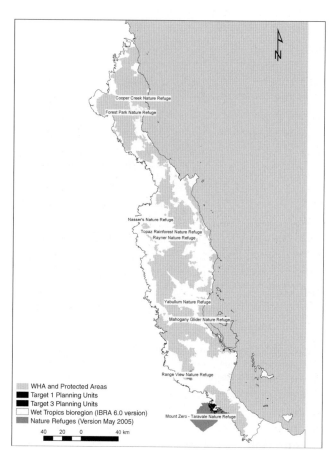

Figure 46.6 Nature refuges (named labels) selected in one or more target scenarios (1 = All Species; 3 = Threatened Species).

Table 46.9 Nature refuges (Nature Conservation Act 1992) selected in one or more target scenarios

Target scenario	Planning Unit ID	Area (ha)	Tenure	Nature refuge name	Nature refuge area (ha)	% of nature refuge
1	2508	877	LL	Mount Zero - Taravale Nature Refuge	60312	1.5
3	2508	3058	LL	Mount Zero - Taravale Nature Refuge	60312	5.1
2	3690	1451	LL	Mount Zero - Taravale Nature Refuge	60312	2.4
3	11991	249	LL	Range View Nature Refuge	4576	5.4
3	35609	69	LL	Mahogany Glider Nature Refuge	69	100.0
3	39785	27	F	Yabullum Nature Refuge	37	72.5
3	145383	19	F	Forest Park Nature Refuge	21	89.7
3	148979	6	F	Manani Nature Refuge	8	77.5

We will also be able to use the results to target any remaining gaps in environmental coverage of plant taxa records in this region (e.g. Funk *et al.* 2005).

We used the heuristic selection algorithm available with the C-Plan software to assist the identification of potential priority areas for vascular plant species. Heuristic algorithms, as their name suggests, do not necessarily select an optimal set of sites. Alternative selection algorithms are available in other conservation planning software. For example, the main optimization algorithm used in *MARXAN* (Ball & Possingham 2000; Possingham *et al.* 2000) is simulated annealing, which, for the larger and more complex conservation planning problems, is likely to produce a more efficient solution. In future work, we intend comparing the efficiency of our results with solutions derived using *MARXAN* and similar planning tools, such as *TARGET* (Faith & Walker 1998) and ResNet (Garson *et al.* 2002).

C-Plan uses the term 'irreplaceability' to describe the index of complementarity used to identify spatial priorities for conservation (Pressey *et al.* 1994; Pressey *et al.* 1995; Ferrier *et al.* 2000). The spatial pattern of the C-Plan irreplaceability index, like all measures of complementarity, changes in a region if the targets for biodiversity surrogates are changed and as planning units are added to a conserved set. Higher targets for all surrogate features will increase the irreplaceability of planning units, producing fewer spatial options for achieving those targets. Lowering targets will lower the irreplaceability values of most planning units and produce more spatial options for designing the system of priority areas. Clearly, the setting of conservation targets for biodiversity surrogates must reflect broad aims and decisions made at a policy level. In this sense, complementarity-based algorithms have sometimes been termed 'policy algorithms' (e.g. Faith *et al.* 2003). While consistent with the principal of complementarity, different priority-setting algorithms and tools have been developed that allow different spatial options or optimal solutions to be considered for practical planning in more complicated contexts (e.g. Leslie *et al.* 2003; Tsuji & Tsubaki 2004).

It would be desirable to distinguish geographically distinct populations that potentially represent broad scale genetic variation when selecting biodiversity priority areas to represent more than one occurrence of a taxon. In future work, we intend developing an approach to guide the selection of areas based on targets for biodiversity features that incorporate geographical stratification in their design. For example, we could simply introduce a stratification rule based on the subregions within the Wet Tropics bioregion or latitude and/or longitude to ensure that targets for capturing more than one occurrence of a plant taxon are optimally selected from among distinct populations across their geographical and/or environmental range.

Our analyses sought solutions from publicly owned and leasehold land before resorting to freehold tenure when no alternative was available. Included among freehold land tenure are areas already gazetted as nature refuges. These are areas in which landholders are already making contributions to the protection of biodiversity. We wished to focus attention on these areas to determine how much plant biodiversity they contribute that is not already captured within protected areas or other public tenures. Of 11 gazetted nature refuges in the Wet Tropics bioregion, one is needed to represent otherwise unprotected plant biodiversity under the 'all species' or 'rainforest endemics' scenario, and six capture threatened species.

In our scenarios, we did not impose a budget on the amount of area selected, but simply allowed each analysis to continue until the biodiversity targets were achieved. In most planning situations, it is not feasible to implement a set of priority areas without taking other constraints into account. These constraints include annual budgets for implementing conservation actions and other competing uses and interests over the land that must be negotiated. Social and economic trade-offs are inherent to decision-making for biodiversity conservation, necessitating several stages of analysis. In this example, we presented the first stage of a prioritization process; that is, without explicitly considering opportunity costs and potential trade-offs in area selection. In future work, we intend considering potential trade-offs linked to the costs of area acquisition for conservation and subsequent management to ameliorate, for example, threats to biodiversity persistence.

Similarly, we did not consider spatial configuration of remnants, minimum effective habitat area or condition, or other ecological processes and their probable effects on population viability and persistence of plant biodiversity. We note the existence of many small remnants in upland and lowland developed regions that are also important for representing plant biodiversity. For successful plant conservation, it will be necessary to evaluate the effective area and condition of remaining habitat, taking into account the spatial configuration and characteristics of remnant habitats, with implications for measures of biodiversity persistence.

Summary

The process of systematic conservation planning has its foundation in land allocation decision-making, with the special characteristic that the frequent multiple criteria used to describe the complexity of biodiversity be properly taken into consideration. Margules and Pressey (2000) reviewed and summarized the process of systematic conservation planning, which has been articulated and refined in conservation planning software tools developed by different research groups. Here we chose to use C-Plan (Pressey *et al.* 1994), but there are alternative tools, such as *TARGET* (Faith & Walker 1998; Faith *et al.* 2001), *MARXAN* (Possingham *et al.* 2000) and ResNet (Garson *et al.* 2002), that are currently undergoing rapid development, which provide potential alternative algorithms to explore more optimal solutions.

Conservation planning is a dynamic and iterative process in which software tools facilitate data integration and scenario development, support interactive learning by land managers and enable exploration of policy options affecting land allocation decisions where biodiversity is a consideration. These tools are decision-support tools, not decision-making tools. The whole process is ideally conducted with the participation of land management stakeholders during real planning exercises. Stakeholders are consulted on all database design elements and are notified of any necessary assumptions in configuring the spatial planning scenarios. There is a wealth of information about biodiversity and its management in the Wet Tropics from a wide range of biological, environmental and planning databases. The challenge is effective integration and use of these data with planning tools to improve policy formulation for strategic management of scarce resources.

The foundation goals of systematic conservation planning are to conserve: (i) a representative sample of all species and their habitats (the principle of representation); and (ii) the ecological and evolutionary processes that allow this biodiversity to persist over the long term (the principle of persistence). The design of biodiversity metrics that capture elements of both pattern and process is an area of continuing research and development (see, e.g. the review by Sarkar *et al.* 2006). A useful biodiversity surrogate needs to describe biophysical variation in the landscape consistently so that the contribution made by one area can be realistically compared with another, and be accepted by planners and policy-makers. We discussed models of species compositional turnover as one such possibility, but in the simplest case, the regional ecosystems (Environmental Protection Agency 2005a) could be used to identify bioregional gaps in the representation of vegetation pattern. Representations of plant taxa and threatened species could be used as preferences when alternative areas are approximately equal, or the data could be configured to incorporate phylogenetic diversity patterns (Faith 1992; Walker & Faith 1994). Assumptions about the effectiveness of particular surrogates to represent biodiversity variation, however, also need to be considered as part of the planning process.

While much attention has been given to biodiversity metrics that describe regional patterns, there is increasing emphasis on ecological processes. New metrics are being designed, such as those developed by Scholes and Biggs (2005), that capture the effects of habitat fragmentation, land management regimes (e.g. grazing, timber harvesting) and threatening processes (e.g. invasive species) on overall biodiversity persistence. These facets of biodiversity are now necessary considerations in conservation planning, increasing the complexity required of spatial planning databases and of the tools used to facilitate optimal prioritization of areas requiring changed management regimes.

Finally, the issues facing the maintenance of biodiversity, even within a region such as the Wet Tropics, which has extensive protected areas, require that decisions about conservation management go beyond strict reservation. Conservation planning has evolved with this shift in societal values. Tools and data can be configured to guide, through scenarios, a wide range of contextual decisions about conservation action for biodiversity, from complex arrangements such as incentive-based schemes operating through market instruments (e.g. Gole *et al.* 2005), to prioritizing management actions within existing protected areas.

References

Airamé, S., Dugan, J. E., Lafferty, K. D., Leslie, H., McArdle, D. A. & Warner, R. R. (2003). Applying ecological criteria to marine reserve design: A case study from the California Channel Islands. *Ecological Applications* **13**: S170–S184.

Austin, M. P. & Margules, C. R. (1986). Assessing representativeness. In *Wildlife Conservation Evaluation*, Usher, M. B. (ed.). Chapman & Hall, London, pp. 47–52.

Austin, M. P. (1999). The potential contribution of vegetation ecology to biodiversity research. *Ecography* **22**: 465–84.

Ball, I. R. & Possingham, H. P. (2000). *MARXAN (V1.8.2): Marine Reserve Design Using Spatially Explicit Annealing, a Manual.* (www.ecology.uq.edu.au/index.html?page=27710), accessed 8 May 2006.

Christian, C. S. & Stewart, G. A. (1968). Methodology of integrated surveys. In *Proceedings of the Toulouse Conference on Aerial Surveys and Integrated Studies*. UNESCO, Paris, pp. 233–80.

Cowling, R. M. & Pressey, R. L. (2003). Introduction to systematic conservation planning in the Cape Floristic Region. *Biological Conservation* **112**: 1–13.

Department of Natural Resources and Mines (2004). *Digital Cadastre Database* (digital data). Department of Natural Resources and Mines, Brisbane.

Environmental Australia (2005) *Revision of the Interim Biogeographic Regionalisation for Australia (IBRA) and Development of Version 5.1 – Summary report (2000). Updated, IBRA Version 6.1* (Digital Data, metadata) (www.deh.gov.au/parks/nrs/ibra/version6-1/index.html). accessed 3 May 2006.

Environmental Protection Agency. (2005a). *rev5_0/: Pre-clearing and 2003 Remnant Vegetation Communities and Regional Ecosystems of Queensland, version 5.0* (Digital Data, metadata). Environmental Protection Agency, Brisbane.

Environmental Protection Agency. (2005b). *estates_11_2, Queensland Estates (Protected Areas), Estates version 11.2, current to October 2005* (Digital Data, metadata). Environmental Protection Agency, Brisbane.

Environmental Protection Agency. (2005c). *Nature Refuges and Coordinated Conservation areas, version 9.0 and updates as at May 2006, Queensland* (Digital Data). Environmental Protection Agency, Brisbane.

Faith, D. P. (1992). Conservation evaluation and phylogenetic diversity. *Biological Conservation* **61**: 1–10.

Faith, D. P., Carter, G., Cassis, G., Ferrier, S. & Wilkie, L. (2003). Complementarity, biodiversity viability analysis, and policy-based algorithms for conservation. *Environmental Science and Policy* **6**: 311–28.

Faith, D., Ferrier, S. & Walker, P. (2004). The ED strategy: how species-level surrogates indicate general biodiversity patterns through an 'environmental diversity' perspective. *Journal of Biogeography* **31**: 1207–17.

Faith, D. P., Margules, C. R. & Walker, P. A. (2001). A biodiversity conservation plan for Papua New Guinea based on biodiversity trade-offs analysis. *Pacific Conservation Biology* **6**: 304–24.

Faith, D. P. & Walker, P. A. (1998). *TARGET: Software for the Analysis of Priority Protected Areas Representing Biodiversity. User Manual.* CSIRO, Canberra.

Ferrier, S. (2002). Mapping spatial pattern in biodiversity for regional conservation planning: where to from here? *Systematic Biology* **51**: 331–63.

Ferrier, S., Powell, G. V. N., Richardson, K. S. *et al.* (2004). Mapping more of terrestrial biodiversity for global conservation assessment. *Bioscience* **54**(12): 1101–9.

Ferrier, S., Pressey, R. L. & Barrett, T. W. (2000). A new predictor of the irreplaceability of areas for achieving a conservation goal, its application to real-world planning, and a research agenda for further refinements. *Biological Conservation* **93**: 303–25.

Ford, A. J., Richardson, K. S., Williams, K. J., Bruce, C. M. & Parker, T. A. (2004). *An Analysis and Discussion of Plant Data in the Wet Tropics NRM Planning Region*. A report prepared for FNQ NRM Ltd. Rainforest CRC Project 6.5. CSIRO, Atherton.

Funk, V. A., Richardson, K. S. & Ferrier, S. (2005). Survey-gap analysis in expeditionary research: where do we go from here? *Biological Journal of the Linnean Society* **85**: 549–67.

Garson J., Aggarwal A. & Sarkar S. (2002). *ResNet Ver 1.2 Manual*. University of Texas Biodiversity and Biocultural Conservation Laboratory, Austin (http://uts.cc.utexas.edu/~consbio/Cons/reports.html).

Gole, C., Burton, M., Williams, K. J. *et al.* (2005). *Auction for Landscape Recovery: ID 21 Final report, September 2005*. Commonwealth Market Based Instruments program, WWF Australia (www.napswq.gov.au/mbi/round1/project21.html).

Henderson, R. J. F. (ed.) (2002). *Queensland Plants, Algae and Lichens: Names and Distributions*. Queensland Herbarium Environmental Agency, Indooroopilly, Queensland.

Kerley, G. I. H., Pressey, R. L., Cowling, R. M., Boshoff, A. F. & Sims-Castley, R. (2003). Options for the conservation of large and medium-sized mammals in the Cape Floristic Region hotspot, South Africa. *Biological Conservation* **112**: 169–90.

Leslie, H., Ruckelshaus, M., Ball, I. R., Andelman, S. & Possingham, H. P. (2003). Using siting algorithms in the design of marine reserve networks. *Ecological Applications* **13**: 185–98.

McKenzie, N. L., Belbin, L., Margules, C. R. & Keighery, G. J. (1989). Selecting representative reserve systems in remote areas: a case study in the Nullarbor region, Australia. *Biological Conservation* **50**: 239–61.

Mackey, B. G., Nix, H. A., Stein, J. A., Cork, S. E. & Bullen, F. T. (1989). Assessing the representativeness of the Wet Tropics of Queensland World Heritage property. *Biological Conservation* **50**: 279–303.

Margules, C. R. & Austin, M. P. (1994). Biological models for monitoring species decline: the construction and use of data bases. *Philosophical Transactions of the Royal Society of London Series B* **343**: 69–75.

Margules, C. R. & Pressey, R. L. (2000). Systematic conservation planning. *Nature* **405**: 243–53.

Margules, C. R. & Sarkar, S. (in press). *Systematic Conservation Planning*. Cambridge University Press, Cambridge.

Margules, C. R., Nicholls, A. & Pressey, R. L. (1988). Selecting networks of reserves to maximise biological diversity. *Biological Conservation* **43**: 63–76.

Meffe, G. K. & Carroll, C. R. (1994). *Principles of Conservation Biology*. Sinauer Associates, Sunderland, MA.

Moritz, C., Richardson, K. S., Ferrier, S., Montieth, G. B., Williams, S. E. & Whiffen, T. (2001). Biogeographic concordance and efficiency of taxon indicators for establishing conservation priority for a tropical rainforest biota. *Proceedings of the Royal Society of London B* **268**: 1875–81.

NSW National Parks and Wildlife Service (1991). *C-Plan Conservation Planning Software. User Manual for C-Plan Version 3.06*. NSW National Parks and Wildlife Service, Armidale.

Pharo, E. J., Beattie, A. J. & Pressey, R. I. (2000). Effectiveness of using vascular plants to select reserves for bryophytes and lichens. *Biological Conservation* **96**: 371–8.

Possingham, H. P., Ball, I. & Andelman, S. (2000). Mathematical methods for identifying representative reserve networks. In *Quantitative Methods for Conservation Biology*, Ferson, S. & Burgman, M. (eds). Springer-Verlag, New York, pp. 291–305.

Pressey, R. L., Ferrier, S., Hutchinson, C. D., Sivertsen, D. P. & Manion, G. (1995). Planning for negotiation: using an interactive geographic information system to explore alternative protected area networks. In *Nature Conservation: The Role of Networks*, Saunders, D. A., Craig, J. L. & Mattiske, E. M. (eds). Surrey Beatty and Sons, Sydney, pp. 23–33.

Pressey, R. L., Humphries, C. J., Margules, C. R., Vane-Wright, R. I. & Williams, P. H. (1993). Beyond opportunism: key principles for systematic reserve selection. *Trends in Ecology and Evolution* **8**: 124–8.

Pressey, R. L., Johnson, I. R. & Wilson, P. D. (1994). Shades of irreplaceability: towards a measure of the contribution of sites to a reservation goal. *Biodiversity and Conservation* **3**: 242–62.

Sarikinos, H., Nicholls, A. O., Tubert, A., Aggarwal, A., Margules, C. R. & Sarkar, S. (2001). Area prioritization for biodiversity conservation in Quebec on the basis of species distributions: a preliminary analysis. *Biodiversity Conservation* **10**: 1419–72.

Sarkar, S., Aggarwal, A., Garson, J., Margules, C. R. & Zeidler, J. (2002). Place prioritization for biodiversity content. *Journal of Biosciences* **27**(Suppl. 2): 339–46.

Sarkar, S., Justus, J., Fuller, T., Kelley, C., Garson, J. & Mayfield, M. (2004). Effectiveness of environmental surrogate for the selection of conservation area networks. *Conservation Biology* **19**(3): 815–25.

Sarkar, S. & Margules, C. R. (2002). Operationalizing biodiversity for conservation planning. *Journal of Biosciences* **27**(Suppl. 2): 299–308.

Sarkar, S., Pressey, R. L., Faith, D. P. *et al.* (2006). Biodiversity conservation planning tools: present status and challenges for the future. *Annual Review of Environment and Resources* **31**: 123–59

Scholes, R. J. & Biggs, R. (2005). A biodiversity intactness index. *Nature* **434**: 45–9.

Sullivan, M. & Chesson, J. (1993). *The Use of Surrogate Measurements for Determining Patterns of Species Distribution and Abundance*. Resource Assessment Commission Research Paper No. 8. Australian Government Publishing Service, Canberra.

Tsuji, N. & Tsubaki, Y. (2004). Three new algorithms to calculate the irreplaceability index for presence/absence data. *Biological Conservation* **119**: 487–94.

Vane-Wright, R. I., Humphries, C. J. & Williams, P. H. (1991). What to protect? Systematics and the agony of choice. *Biological Conservation* **55**: 235–54.

Walker, P. A. & Faith, D. P. (1994). DIVERSITY-PD: Procedures for conservation evaluation based on phylogenetic diversity. *Biodiversity Letters* **2**: 132–9.

Wet Tropics Management Authority (2006). *World Heritage Area Boundary Version February 2006* (Digital Data, metadata). Wet Tropics Management Authority, Cairns.

47 NEW TOOLS FOR MONITORING WORLD HERITAGE VALUES

Stuart Phinn[1]*, Catherine Ticehurst[2]*, Alex Held[2]*,
Peter Scarth[1]*, Joanne Nightingale[1]* and Kasper Johansen[1]

[1]Biophysical Remote Sensing Group, School of Geography, Planning and Architecture, University of Queensland, Australia
[2]Division of Land and Water, CSIRO, Canberra, ACT, Australia
*The author was participant of Cooperative Research Centre for Tropical Ecology and Management

Tropical rainforests: information needs for science and management

One of the critical types of information required to understand an ecosystem, and to enable monitoring and modelling for management, is spatial information. In particular, these data are required for key components and processes, that is, the type of vegetation, its structural properties and their change over time. Spatial data are now applied extensively, in both rainforest science and monitoring and management activities. These activities range from baseline inventory and mapping of rainforest extent, its vegetation communities and invasive weeds, to mapping human induced changes and disturbance impacts, such as fire and tropical cyclones. As remotely sensed data sets and their derived products cover a range of spatial and temporal scales, they are increasingly being used to monitor environmental change and to drive models of ecosystem processes, such as vegetation growth, gap dynamics, hydrology and regional scale climate systems.

This chapter demonstrates how established tropical rainforest ecosystem health indicators can be mapped and monitored using current and proposed satellite image data sets in the context of Australian rainforests.

An extensive review of the types of remotely sensed data able to be used and their application in rainforests has been completed by the authors (see Lucas *et al.* 2004). A framework is first presented to demonstrate how remote sensing is able to map specific types of environmental information on tropical rainforests. The following sections then cover the state of the art techniques that have been successfully applied in tropical rainforests around the world to map a specific rainforest environmental variable or process. Specific examples are then presented for the application and validation of these techniques within the Wet Tropics rainforests of Australia. The first application area explains how rainforest vegetation communities can be mapped using aerial photography, multispectral Landsat Thematic Mapper and (JERS-1) L-band synthetic aperture imaging radar. These data sets are also used to demonstrate approaches for mapping rainforest vegetation structure (e.g. height, gap density, leaf area index or LAI) and changes in these features over time. Assessing the dynamics of tropical rainforests has been difficult due to problems of cloud cover and smoke. Several approaches are presented using imaging radar and high temporal frequency (daily) MODIS vegetation index images to map disturbance events and assess longer-term ecosystem dynamics. This leads into image-based modelling applications for assessing how tropical rainforests will change under specific climate and land-cover/land-use regimes, with an example of ecosystem productivity models. A final assessment is then made of the current and near-future applications of remote sensing for tropical rainforest mapping, monitoring, modelling and management.

Linking remote sensing to state of environment reporting for the Wet Tropics

Important environmental indicators applicable to remote sensing technologies have been defined for the Wet Tropics of Queensland World Heritage Area (WHA), to assist in the sustainable management of this region (Phinn et al. 2001). Phinn et al. (2001) investigated the feasibility of remote sensing for monitoring the State of the Wet Tropics (SoWT) indicators, by first establishing the link between the remote sensing variable and its possible environmental indicators (Figure 47.1 and Table 47.1), as well as indicating the likelihood of each indicator being retrieved from

Table 47.1 Wet Tropics Management Authority environmental indicators (state of Wet Tropics indicators) and matched remotely sensed variables (from Phinn et al. 2001)

Remotely sensed variable	SoWT indicator
Land cover	Land cover classes
	Extent of vegetation fragmentation
	Extent and severity of edge effects
	Extent of burnt area by spatial unit and assemblage
	Changes to drainage pattern
Land-cover change	Extent of clearing by stratification
	Extent of vegetation fragmentation
	Extent of burnt area by spatial unit and assemblage
	Changes to drainage pattern
Vegetation type	Extent of introduced environmental weed species by spatial unit and native plant assemblage
Vegetation index	Extent and severity of edge effects
	Structural modifications forest health
	Extent of burnt area by spatial unit and assemblage
Soil index	Erosion features
Structure/biomass index	Structural modifications forest health

Source: Phinn et al. (2001).

remote sensing based on the latest instruments, resources and techniques (Table 47.2). Indicators such as land cover, extent of clearing and fragmentation, and structural modifications were identified as being important variables, with remote sensing the only realistic and practical method for their assessment.

The following paragraphs are based on Phinn et al. (2002a) and provide a description of the process used successfully to communicate remote sensing requirements with the rainforest management agency of concern, Wet Tropics Management Authority (WTMA). The brief of the Phinn et al. (2001) report was to evaluate the feasibility of remote sensing techniques to monitor regional scale SoWT indicators as defined by the WTMA (Scientific Advisory Committee and Board). Wallace and Campbell (1998) provided a preliminary evaluation of the feasibility of remote sensing for monitoring an extensive set of national scale state of the environment indicators. This chapter builds on the results of Wallace and Campbell (1998)

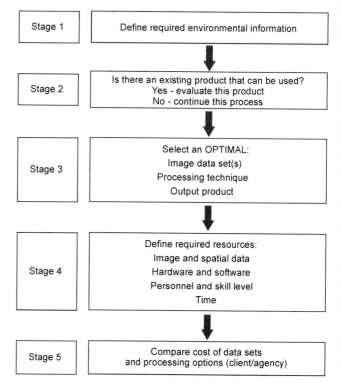

Figure 47.1 Schematic outline of the framework used to link remote sensing approaches to monitoring specific environmental indicators relevant to rainforest ecosystem condition. *Source*: modified from Phinn et al. (2001, 2005).

Table 47.2 Assessment of Wet Tropics Management Authority environmental indicators (state of Wet Tropics indicators) and their ability to be mapped and monitored by remotely sensed variables (from Phinn *et al.* 2001)

Indicator (surrogate)	Status
Extent of clearing by stratification (within land cover types: linear service corridors, inundation, spot clearings, boundary anomalies)	Operational
Extent of vegetation fragmentation (area of powerlines, roads)	Operational
Extent and severity of edge effects	Feasible
Structural modifications forest health	Feasible
Extent of burnt area by spatial unit and assemblage (within Webb–Tracy communities and land-cover types)	Operational
Extent of introduced environmental weed species by spatial unit and native plant assemblage	Likely/possible (dependent on scale of feature)
Erosion features (exposed soil)	Feasible (dependent on scale of feature)
Changes to drainage pattern (dams, stream geometry)	Feasible (dependent on scale of feature)

and represents a model for further assessment of remote sensing to monitor regional ecosystems or bioregions.

The approach taken was to define the characteristics or attributes of the SoWT indicators that could be used to select remotely sensed data (Table 47.1) and processing techniques capable of providing the required information at appropriate spatial, temporal, accuracy and cost levels. The end result is an evaluation of the feasibility of mapping set rainforest parameters from remotely sensed data (Table 47.2). This approach built on an internationally reviewed and recognized technique developed by Phinn (1998) that has been applied in a number of different environments to determine the feasibility of remote sensing for specific environmental monitoring and management problems (Phinn 1998; Phinn *et al.* 2000, 2005). The key to this approach is defining the spatial and temporal scale(s) of data and information required to address each indicator or its surrogate. The spatial and temporal scales of required indicator/surrogate information (and type of information) provide a direct link to remotely sensed data. Once the data set(s) and required processing technique are defined, details can be provided on the

cost of the work, based on required data and personnel, software, hardware and other resources. An example of this evaluation for mapping rainforest vegetation types is included in Table 47.3.

Mapping rainforest vegetation communities

Information goals

Owing to the high level of species and spatial biodiversity within tropical rainforest environments it is rare for mapping to be required at the level of individual species, unless the focus is on weeds mapping. The majority of baseline mapping of rainforest environments for vegetation information has been at the scale of vegetation communities, which in this context are defined as frequently recurring assemblages of species with mixed floristic and structural attributes. Within the Wet Tropics of Australia, vegetation communities were originally mapped at 1 : 100 000 scale in 1975 by Tracey and Webb (1975). The maps were compiled using their already extensive experience of the area, with field visits, air photos and helicopter flights over most of the region (Tracey 1982). Vegetation patterns were marked directly on to air photos (scale 1 : 80 000) and then transferred to contour maps (National Mapping 1 : 100 000 R631 series). Experience in working in the rainforest region enabled the classification of patterns into mapping units denoting recognizable vegetation types. These were established from forest structure, floristic dominants and major environmental features. Validity was confirmed by floristic classification based on numerical analysis of 740 species of trees and shrubs on 146 sites throughout the region. Most recently a similar approach has been taken by Stanton and Stanton using aerial photo interpretation, mapping to 1 : 50 000 scale and the Queensland Herbarium. The mapping system adopted in these programmes is based on extensive manual interpretation, fieldwork and additional spatial data. This section uses results from Ticehurst *et al.* (2003) and Ticehurst and Phinn (2005) to describe how commercially available satellite optical multispectral and imaging radar can be used to obtain similar information on rainforest vegetation communities, if the target communities have distinct structural attributes. The optical multispectral data rely on reflected sunlight, and are

Table 47.3 The remotely sensed variable vegetation type (applies to SoWT indicators: extent of introduced environmental weed species by spatial unit and native plant assemblage) and the listing of data types, processing requirements and costs for mapping and monitoring this variable using several suitable types of remotely sensed data

Vegetation type	Indicator attributes	Data type 1	Data type 2	Data type 3
Spatial scale extent MMU/GRE	Regional–local 10,000–100 km^2 1 ha to 100 m^2	Landsat ETM 185 ☐☐185 km per scene 15 m panchromatic 30 m multispectral 60 m thermal	Airborne hyperspectral Up to 100 km^2 0.5–10 m	Aerial photographs 1.3–33 km^2 5–250 m
Temporal	Annual eg by June for December delivery or event driven (WTMA) Baseline data collection for land cover	Approx 9.45 a.m. every 16 days (archive from CRC Rainforest and ACRES)	User controlled (subject to weather and aircraft availability)	User controlled (subject to weather and aircraft availability)
Variable	Land-cover class (refer to Table 2 with list of indicators addressed by land-cover classes)	Reflectance in up to 7 spectral bands	Reflectance in up to 126 spectral bands	Contact prints (23 ☐ 23 cm) requiring scanning and orthocorrection to produce a digital mosaic
Processing technique (output)		Image classification or feature detection (vegetation type map and target features). Note: the ability to map specific targets will depend on their growth form and extent	Image classification or (hyperspectral) feature detection (vegetation type map and target features) Note: the ability to map specific targets will depend on their growth form and extent.	Manual delineation of vegetation types either on hard-copy photographs or on-screen digitizing. (Vegetation type map)
Resource (equipment)		PC Image processing software GIS with image classification module (e.g. Arc-View Image Analyst)	PC Image processing software capable of hyperspectral data processing	PC A3 size or larger scanner Softcopy photogrammetry software Image processing software GIS with image classification module (e.g. Arc-View Image Analyst)
Resource (personnel)		Trained in image classification Experience with Landsat data Knowledge of area to be mapped	Trained in image classification and spectral unmixing or matching Experience with Hyperspectral data Knowledge of area to be mapped	Training in softcopy photogrammetry and image processing Extensive knowledge of area to be mapped
Estimated task and times		Image pre-processing (1 day) Image classification to, WTMA Broad Habitat Types Annual Report 1998–9 (p. 26) (15 days per scene) Field/Photo verification for a select number of sample sites (8 days) Map output production (2 days) Total = 26 days per scene	Note: this estimate is for a 10 x 10 km area Image pre-processing (2 days) Image analysis using classification, unmixing or matching to define WTMA Broad Habitat Types Annual Report 1998–9 (p. 26) and target features (8 days per area) Field/Photo verification for a select number of sample sites (3 days) Map output production (1 days) Total = 14 days per 10 x 10 km scene	Note: This estimate is for a 20 ☐ 20 km area (10 ☐ 10 photos) Aerial Photograph Scanning (1 day) Digital photographs ortho-correction (5 days) Photograph interpretation and digitizing boundaries (25 days) Build and clean up vegetation type layer Map output production; (2 days) Total = 33 days per 20 ☐ 20 km scene

(Continued)

Table 47.3 (*Continued*)

Vegetation type	Indicator attributes	Data type 1	Data type 2	Data type 3
Estimated cost (Note that these estimates are flexible)		Data acquisition: Image data = $1,950 Aerial photos (10) = $90 per frame to acquire or less to hire from Dept of Natural Resources Ancillary data (topo sheets) = $200 Processing = 28 days of technical officer @ $150 per day = $4,200 Total = $7,250 Note: this assumes software has been purchased	Data acquisition: Image data = $15,000 Aerial photos (10) = $90 per frame to acquire or less to hire from Dept of Natural Resources Ancillary data (topo sheets) = $200 Processing = 14 days of technical officer @ $150 per day = $1,700 Total = $2,100 Note: this assumes software has been purchased	Data acquisition: Aerial Photos (10) = $90 per frame to acquire or less to hire from Dept of Natural Resources = $9,000 Ancillary data (topo sheets) = $200 Processing = 33 days of technical officer @ $150 per day = $4,950 Total = $14,150 Note: this assumes software have been purchased
Evaluation result		**Operational**	**Feasible**	**Operational**

MMU, minimum mapping unit and GRE, ground resolution element.

distorted by cloud and smoke, while the radar can operate under all conditions.

Optical multispectral image data: Landsat 7 ETM

The importance of optical remote sensing as a tool for rainforest mapping and monitoring is well established. Landsat TM (Thematic Mapper) and ETM (Enhanced Thematic Mapper) and SPOT (Systeme pour l'Observation de la Terre) data have been commonly used for moderate-resolution (10–30 m pixel size) rainforest applications (Lucas *et al.* 2004). Landsat scenes have been applied to mapping tropical forest degradation (e.g. Sgrenzaroli *et al.* 2002), as well as the use of multitemporal imagery for mapping deforestation (e.g. Skole & Tucker 1993), forest clearing and regrowth (Hayes & Sader 2001) and secondary forest regeneration (Lucas *et al.* 1993).

For the mapping of the Wet Tropics, a radiometrically and geometrically corrected Landsat ETM mosaic (Plate 47.1) was provided by C. Bruce and D. Hilbert through the Rainforest Cooperative Research Centre, and their processing methods are described in detail in Bruce and Hilbert (2003). The acquisition dates of the six images used for producing this landcover map were in August and September 1999. The Stanton and Stanton vegetation maps were used to guide the selection of training and validation regions. A number of training sites were selected for each class to represent the variability within the different land-covers, and throughout the Wet Tropics region. The land-cover and aggregated vegetation community types were cleared (including urban, pasture and agriculture), open woodland, littoral zone, wet forest and dry forest. A supervised maximum likelihood classification was used to classify the image to an accuracy of 80.7% based on the validation sites.

Active radar image data: JERS-1

While data from the Landsat TM and ETM and other passive optical satellite sensor systems are widely used around the world for regional scale analyses (e.g. Skole & Tucker 1993; Archard & Estreguil 1995; Mayaux & Lambin 1997; Lucas *et al.* 2000; Boyd & Duane 2001; Castro *et al.* 2003), the limited availability of cloud-free scenes over tropical areas is a significant complication for land-cover change mapping. The applicability of synthetic aperture radar (SAR) for rainforest analysis has also been recognized owing to its cloud-free imagery, and radar backscatter's relationship to rainforest structure and biomass. SAR imaging sensors are able to penetrate clouds and haze, enabling frequent and/or routine data acquisition in regions of consistently high cloud cover and areas with extensive smoke and haze cover. The Japanese Earth Resource Satellite (JERS-1), an L-band radar satellite that operated between 1992 and 1998, has been used for rainforest mapping over extensive areas of South America, Central Africa and Southeast Asia (Rosenqvist *et al.* 2000). The long wavelength of this L-band radar system (approx. 25 cm in wavelength) is especially advantageous because it partially penetrates into dense vegetation and provides information on the structural characteristics and moisture content of different vegetation types (e.g. Lucas *et al.* 2004). L-band radar is also better at discriminating between forests and cleared areas in regions of high biomass compared to the more common 3–7 cm wavelength C-band SARs. For regional scale mapping of tropical rainforests using JERS-1 data, image texture has often been applied (e.g. Saatchi *et al.* 2000; Podest & Saatchi 2002) and used in conjunction with multitemporal scenes (Simard *et al.* 2000). L-band has been shown to perform better than C-band for detecting changes in woody biomass (Rignot *et al.* 1997), and although single band/frequency radar has limitations, the benefit of combining the single L-band JERS with C-band ERS-1 for tropical vegetation mapping has been shown (Mayaux *et al.* 2002; Simard *et al.* 2002).

Two important factors that influence the radar response to vegetation are the effects due to vegetation structure/biomass and moisture content in the soil and vegetation. An increase in moisture content in the vegetation, and especially soil, will reduce radar penetration, often resulting in an increase in backscatter (French *et al.* 1996; Griffiths & Wooding 1996). While flat open areas will have a low backscatter due to specular scattering of the microwaves away from the receiver (Figure 47.2), trees tend to act as strong reflectors, causing the microwaves to bounce off the ground and then their trunks, before returning a signal to the

receiver (double bounce). In many cases, the radar microwaves also get scattered inside the tree canopies before returning to the receiver (volume scattering) (Figure 47.2). Volume scattering tends to dominate in dense forests where less radar signal is able to reach the ground. Conversely, double bounce dominates in more open forests and cleared areas tend to show up in radar imagery as darker patches of reduced backscatter. Radar backscatter has thus been shown to correlate with above-ground biomass (e.g. Kasischke *et al.* 1995; Luckman *et al.* 1997; Castro *et al.* 2003).

Owing to the side-looking nature of SAR, the influence of terrain on radar return tends to be more extreme, and has been acknowledged as a problem that needs to be addressed for rainforest analysis (Castro *et al.* 2003). Methods are available to correct for these distortions (e.g. Curlander *et al.* 1987; Johnsen *et al.* 1995), but this relies on a digital elevation model (DEM) that is accurate and has suitable spatial resolution. While it is not ideal, these corrections have been avoided in the past by using a large pixel size to smooth out the effects (Simmard *et al.* 2000), using band ratios when more than one band is available (Ranson *et al.* 2001), or utilizing the topographic information to help in the vegetation discrimination (Ticehurst *et al.* 2004).

For the Wet Tropics, Ticehurst *et al.* (2004) developed a vegetation mapping method based on imaging radar. The method employed a topographic correction and mapping procedure that was developed to enable vegetation structural classes to be mapped from a geometrically corrected 1996 JERS-1 mosaic created for the Wet Tropics. A new technique for image smoothing was applied to the JERS-1 texture bands and a DEM, before a maximum likelihood classification was applied to identify major landcover and vegetation communities. Despite these efforts, dominant vegetation community classes could only be classified to low levels of accuracy (57.5%), which were partly explained by the significantly larger pixel size of the DEM (80 m) in comparison to the JERS image (12.5 m). In addition, the spatial and floristic detail contained in the classes of the original validation maps were much finer than the JERS classification product was able to distinguish. However, there are benefits in the combination of imaging radar with additional data sets to map vegetation communities accurately across the entire Wet Tropics.

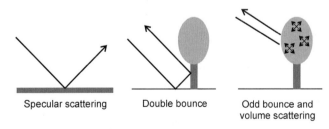

Specular scattering   Double bounce   Odd bounce and
volume scattering

Figure 47.2 The arrow indicates the three main interactions characteristic of the path of incident radiation from the radar transmitter and then scattered back to the radar sensor.

Integrated or fused image data: Landsat 7 ETM/JERS

The integration of optical and radar remote sensing utilizes the chemistry, colour and underlying moisture properties of the forest, along with the vegetation structure, density and dielectric characteristics. Combining optical with radar remote sensing data has been shown to improve classifications compared to their individual use; however, their combined use for tropical vegetation mapping is still not common. When applied together, radar has been shown to assist in the discrimination of different woody biomass levels, medium density vegetation and forest disturbance, while the optical imagery has proven useful for mapping deforested areas and secondary forest regrowth (Nezry *et al.* 1993; Rignot *et al.* 1997; Saatchi *et al.* 1997)

When a fully corrected JERS-1 and Landsat scene were combined to classify the Wet Tropics region into general landcover classes (cleared, littoral, woodland, wet forest, dry forest), a small but statistically significant improvement to a Landsat ETM classification was found for the combined data set (an accuracy of 83.4% based on validation regions extracted from Stanton and Stanton vegetation maps) (Plate 47.2). The main advantage in combining radar with optical imagery was the improvement in distinguishing between wet and dry forest, as well as for mapping agricultural areas.

Mapping rainforest vegetation structure

Vegetation structure definition

In the context of analysis of remote sensing data a number of physical properties of rainforest vegetation can be estimated from the scale of individual trees, to stands, gaps, patches, communities and ecosystems. These physical properties are often the structural elements of plants that control how sunlight or a beam of electromagnetic radiation (e.g. from an imaging radar or airborne laser scanner) is scattered, absorbed or transmitted. Detailed descriptions of radiative transfer theory and interactions from leaf to canopy, patch and ecosystem scale are provided in Tenhuenen and Kabat (1999) and Ustin (2004), and the reader is referred to

these references for the models used to map vegetation structural elements. In the context of work within the Wet Tropics rainforest in Australia, we have focused on estimation and mapping of LAI, tree density, canopy height, gap size and above-ground biomass. Passive and active sources of remotely sensed data are sensitive to different structural elements of rainforest vegetation, and extensive work has been completed on the use of imaging radar for mapping tropical rainforest biomass (see Lucas *et al.* 2004). This section presents examples of structural estimates from passive image data to extract tree density, canopy height and gap characteristics, and complementary approaches using active imaging radar to estimate biomass.

Tree density: high spatial resolution image data

The delineation of tree crowns is a useful tool for mapping and analysing forest environments through high-resolution remote sensing imagery. Delineated crowns are represented as a single polygon entity, which can then be used for deriving statistical information about a forest. The isolation of individual trees assists in canopy gap and vegetation structure analysis, while allometric relationships with canopy crown size lend themselves to forest inventory analysis and possible biomass estimation. Furthermore, the availability of high-resolution satellites (e.g. Ikonos, Quickbird) improves the accessibility of the data needed for delineating crowns.

Tree crown delineation has already been applied to monospecific plantations, and temperate and subtropical old-growth forests (e.g. Gougeon 1995; Culvenor 2000; Scarth 2000; Wulder *et al.* 2000). However, dense canopies of complex morphology (such as tropical rainforests) create another challenge in crown delineation. Canopy crowns of different species may be too close for a shadow to fall between them, which is a necessity of the traditional crown delineation process. Fortunately, when neighbouring crowns are of different species, which is a reasonable assumption in a tropical rainforest, their differences in spectral reflectance may be exploited in high-resolution hyperspectral imagery (Ticehurst *et al.* 2001) and may also have potential for species discrimination in tropical forests (Clark *et al.* 2005).

When applied to high resolution imagery over the Wet Tropics, crown delineation allows tree density, average tree size and canopy gaps to be mapped. (Plate 47.3) shows how the difference between untouched (top) and cyclone damaged (bottom) rainforest is easy to detect and is clearly visible in the density maps.

Canopy height, LAI and gap fractions: geometrical-optical models from Landsat ETM

Geometrical-optical models are a particular class of spectral models, where the bidirectional reflectance distribution function (BRDF) is modelled as a purely geometrical phenomenon resulting from scenes of discrete three-dimensional objects that are illuminated and viewed from different locations. The shape, density and patterns of objects when illuminated and viewed from different directions controls the proportions of sunlit and shaded object and background reflectance components visible to the observer. The overall reflectance is treated as an area weighted sum of the reflectance components (Strahler & Jupp 1990). To invert the geometrical-optical model requires the recovery of the proportions of the reflectance components and this entails the use of spectral mixture analysis. Previous work (e.g. Scarth & Phinn 2000) has demonstrated the utility of this approach.

Models of open forests typically use sunlit and shadowed, canopy and background (soil and litter) as the modelled objects. The measured reflectance of a forested scene then depends on the amount of each object visible, as well as the spectral reflectance of the components and the structure of the vegetation. To develop such a model for a tropical rainforest environment requires analysis of gaps rather than individual canopy elements. Since rainforests have a complex three-dimensional form, the effect of illumination by the sun gives rise to shadowing of the ground cover or soil, understorey and tree canopies. This shadowing becomes visible by a downward looking sensor when there are gaps in the rainforest canopy. The presence of wind-thrown gaps, primarily attributed to tropical cyclones, is a common component of the tropical rainforests of North Queensland.

These gaps range from the relatively small, caused by the effect of the multilayered leaf structure and the random orientation and spatial organization of individual leaves, to extremely large gaps, caused by wind-throw and tree fall events. The size of the gap, in conjunction with the solar zenith angle, will determine the proportion of the gap that is shaded. Thus, a simple geometrical-optical model of tropical rainforest environments can have three components:

- Sunlit canopy is the major scene component due to the dense interlocked foliage that is common in these environments.
- Sunlit background is apparent in those areas where the canopy is interrupted such as creeks and rivers, rocky outcrops, clearings, and gaps in the canopy.
- Shade is the third component. Unlike geometrical-optical models for woodland environments that generally have a shaded canopy and a shaded background component, in rainforest environments where there are no individual tree canopies there is no mutual shadowing component and therefore shade can be considered as a single constituent.

Scarth *et al.* (2002) developed a simple geometrical-optical model for closed tropical forest canopies that combined the mean gap size, the gap density, the canopy height and the solar zenith angle to derive the final pixel reflectance. This model was inverted by first unmixing a topographically corrected Landsat ETM scene into sunlit canopy, sunlit background and shade components, followed by analysis of the variance of the sunlit canopy fraction to separate the mean gap size from the gap density (Plate 47.4). The canopy height was then recovered from the mean gap size and the proportions of both sunlit canopy and sunlit background.

To assess the validity of the spectral mixture analysis and model inversion, field data were collected from 27 sites located across most of the image. Due to the difficulty in traversing the terrain, it was decided to use transect-based measurements of tree height (using a handheld laser range-finding clinometer) and canopy closeness (calculated from upward-looking canopy photos). This allowed the testing of both the unmixing operation and the model inversion. The slope of the line through the origin correlating the canopy closeness and the sunlit canopy fraction was estimated as 0.984 ± 0.041 with an R value of 0.978. The slope of the line through the origin correlating the measured and the predicted canopy height was estimated as 1.046 ± 0.056 with an R value of 0.978 (Figure 47.3).

It is notable that the slopes of the regression lines are both very close to 1, which indicates not only that there is a correlation between the field and modelled values,

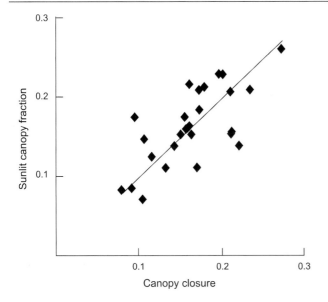

Figure 47.3 Comparison of field measured and image estimates of canopy gap fraction for the Daintree–Cape Tribulation section of the Wet Tropics.

but that the model definitions of canopy cover and canopy height are the same as the ground-based definitions. From the results of the validation, it is apparent that the inversion of the geometrical-optical model through the use of spectral mixture analysis and second order scene statistics works well in the tropical rainforest environments of North Queensland.

Biomass: imaging radar

Forest biomass is often used as a key indicator for biodiversity, forest structure and carbon content (Lucas et al. 2004). In particular, the sensitivity of imaging radar data to rainforest vegetation structural properties has already demonstrated its capacity for deriving empirical relationships between backscatter and biomass (Kasischke et al. 1997; Castro et al. 2003). However, at high biomass levels imaging radar is unable to differentiate changes in forest biomass, due to the saturation of radar backscatter. Past research has established C, L and P-band backscatter intensity saturation limits estimated to be approximately 20, 40 and 100 tons ha^{-1} respectively (Imhoff 1995). For P-band (approx. 60 cm wavelength) this saturation threshold has sometimes been extended to 200 tons ha^{-1} (Hoekman & Quinones 2000) or 270 tons ha^{-1} (Santos et al. 2003) using different band combinations. Even so, this threshold is particularly a problem since undisturbed tropical

forests typically range from 400 to 650 tons ha^{-1} (Lucas et al. 2004).

Radar interferometry has been developed and applied extensively for terrain and elevation mapping, and uses two radar images of the same area, with slightly different viewing geometry, to retrieve an estimate of surface elevation (Toutin & Gray 2000). It has also been investigated for its ability to derive vegetation structural information. In particular, the degree of correlation (or phase coherence) between the two radar images is influenced by vegetation in a different way to backscatter intensity (Askne et al. 1997) and has potential for improving the mapping and monitoring of tropical forests (Luckman et al. 2000) and the extraction of forest size parameters (Hyyppa et al. 2000), and for determining biomass in high-biomass forests (Fransson et al. 2001). Work to date has concentrated on the use of repeat-pass imagery, where the two images are acquired at different times. However, this temporal component leads to errors when trying to characterize a forest (Zebker & Villasenor 1992; Treuhaft et al. 1996) and tends to be worse for wetter vegetation environments, such as rainforests (Toutin & Gray 2000).

Ticehurst and Phinn (2005) examined the potential for radar backscatter and the coherence from single-pass radar interferometry (i.e. when images are acquired simultaneously), for estimating above-ground biomass of tropical rainforest in the Wet Tropics region (Plate 47.5). The radar data were acquired from the Airborne Synthetic Aperture Radar (AIRSAR) system flown in the interferometric (TOPSAR) mode in 1996 and processed to 10 m pixel size. Biomass data, allometrically derived from collected field data, were compared to AIRSARs backscatter intensity for L and P-band, as well as C-band interferometric coherence. While the backscatter intensity showed poor correlations and appeared to have saturated, the coherence band exhibited a negative correlation with biomass ($R^2 = 0.49$), up to the highest biomass values used (470 tons ha^{-1}). Since this preliminary examination of interferometric coherence is derived from a single-pass airborne instrument (AIRSAR), it indicates that the biomass saturation threshold may be higher when the temporal component is removed. However, single-pass interferometry is currently limited to airborne systems, so future work needs to focus on reducing the temporal effects of repeat-pass interferometry if the phase coherence is to become more operational.

Mapping rainforest disturbances: cyclones and weeds

Impacts of disturbance on rainforest structure and condition

Significant gaps in the canopy of tropical rainforests are associated with death or tree-fall produced by natural and artificial disturbances. Cyclones (or hurricanes and typhoons) are the principal cause of these gaps, after fires and human activities. Through changes in the microclimate and light conditions beneath the canopy, the size, shape and extent of these canopy gaps play an important role in the ecological processes within a tropical rainforest (e.g. Gray et al. 2002). For example, changes to the gap structure can increase a forest's susceptibility to fire (Nepstad et al. 1999) and influence forest species composition (Pinard et al. 1996). Weeds are often introduced in these gaps or at forest edges, and significantly impact the composition and condition of the forest. Tropical cyclones, strong wind gusts and lightning are relatively frequent natural disturbances in a number of coastal rainforest regions. Wind-thrown trees and defoliation resulting from these weather systems and their associated defoliation alter the composition, structure and function of tropical rainforests (Grove et al. 2000). Monitoring the condition of tropical forests has been based on relatively static assessment of their floristic composition, vegetation structure and gap dynamics. Hence, an ability to map accurately and repeatedly the impacts of disturbance, in terms of gap size and distribution, canopy height and extent of weed infestation, in a tropical forest will be highly useful for scientists and resource managers concerned with the condition of tropical forests (Scarth et al. 2002; Asner et al. 2004, 2005).

Tropical cyclone impact assessment: imaging radar for assessment of canopy condition

The sensitivity of imaging radar to forest structure properties such as stand density, tree height and differences in canopy structure has proven useful for mapping deforestation and fragmentation in tropical rainforests (Saatchi et al. 1997, 2001; Kuntz & Siegert 1999). Kuntz and Siegert (1999) used temporal and textural analysis of European Remote Sensing satellite (ERS) images to interpret regions of change visually, including selective logging, at scales of 1 : 100 000 in tropical rainforests of Borneo. Radar interferometry has also been shown to add information to standard SAR images for mapping disturbed tropical forests (Luckman et al. 2000).

In February 1999 tropical cyclone Rona crossed the coastline of far north Queensland, causing significant damage to the rainforest canopy in some areas. Ticehurst et al. (unpublished) examined the application of multitemporal canopy DEMs generated from the airborne TOPSAR interferometric radar data in 1996 and 2000, for detection of canopy gaps caused by cyclone damage. Canopy gaps were mapped here by identifying areas with significant decrease in canopy height estimated from the 1996 and 2000 TOPSAR images. Aerial photographs, flown shortly after the cyclone (March 1999), were used to validate the TOPSAR-based map of cyclone-damaged rainforest canopy. A detailed automated image registration was performed on the two TOPSAR DEMs. The results showed that the height difference between the two DEMs mapped where the cyclone damage occurred (Plate 47.6). The results demonstrate the potential for weather-independent mapping of forest canopy change that is not always possible from traditional sources of canopy elevation, for example, airborne laser data and stereo aerial photography. Naturally, the DEM resolution and accuracy, both horizontally and vertically, will determine the level of detail able to be detected. Furthermore, the instrument used here was airborne; hence the implications this has on the DEMs produced by satellite radar interferometry are yet to be tested, particularly in repeat-pass mode.

Mapping invasive weeds: passive multispectral and hyperspectral image data

The inaccessibility and extent of regions threatened by many weeds in the Wet Tropics makes remote sensing technologies a potential candidate for their detection and monitoring. The degree of success of mapping weeds using remote sensing imagery depends on the spatial resolution of the data with respect to weed extent, not only influencing the proportion of mixed pixels, but in some cases allowing textural discrimination (Lamb 1998; Lamb & Brown 2001; Phinn et al. 2002b). A unique spectral signature is also critical for allowing separation of the target species from other vegetation. This separation can sometimes be enhanced based on acquisition dates with respect to seasonal variation in the target species (McGowen 1998).

Pond apple (*Annona glabbra*) is a semi-deciduous invasive weed growing throughout the Wet Tropics region of far north Queensland. It has been declared a Weed of National Significance owing to its current and potential ability to spread and establish monocultural communities. Phinn *et al.* (2002b) assessed the ability of different remote sensing technologies for the detection and mapping of pond apple stands. A series of leaf and canopy field spectra were collected to capture the spectral reflectance variability of the pond apple tree canopy, as well as the vegetation it is likely to be associated with. These data were used to investigate the potential to separate pond apple from its surrounding environment using multispectral and hyperspectral airborne and satellite image data sets (Plate 47.7). Owing to pond apple's general association to brackish, low lying and wet locations, data from DEM, stream and rainfall surface layers were used to model the regions where it is most likely to occur. The predicted distribution was then used to constrain the regions identified in the hyperspectral and multispectral imagery. The results show that the known pond apple sites could be identified, but there was confusion with surrounding vegetation. Based on current and previous methods tested and used to identify and map pond apple in the Wet Tropics, visual interpretation of stereo colour aerial photography could possibly yield most effective results, if the photos are captured at a time when the pond apple canopy is yellow. High spatial resolution hyperspectral and regional-scale multispectral imagery with the inclusion of predictive models are feasible methods for predicting the distribution of pond apple.

Monitoring rainforest condition and dynamics

Rainforest condition: the role of environmental indicators

Monitoring requires the repetitive measurement of a defined environmental variable and the subsequent measuring and mapping of the type and amount of change. The majority of remote sensing monitoring applications in tropical forests have examined either bitemporal (between two dates) change at the scale of Landsat TM/ETM images for vegetation cover mapping or multi-date data analyses of coarse spatial resolution (>1 km pixels) for forest and non-forest areas. Based on our experience in the Wet Tropics we recommend that monitoring programmes be tied to established environmental indicators at specified spatial and temporal scales. The environmental or ecological indicators to use will be singular or integrative variables or indices (physical, chemical or biological) acknowledged to be strongly related to the structure, condition or functioning of specific environments (see Tables 47.1 and 47.2). The section below demonstrates how specific tropical rainforest indicators can be assessed in a bitemporal change analysis context, using both passive multispectral and active imaging radar data. In addition, a novel approach for identification of trends in environmental indicators, using image data collected on a daily basis, is demonstrated.

Rainforest vegetation change: multispectral (Landsat ETM) and imaging radar (JERS-1)

Landsat TM/ETM+ have already proved useful for monitoring clearing, thinning and regeneration of tropical rainforests (Lucas *et al.* 2004). Some studies have found that the use of a combination of TM/ETM+ bands 3, 4 and 5 has been effective for monitoring deforestation in tropical regions (Roy *et al.* 1991; Nelson 1994; Lillesand & Kiefer 2000). Even though much research has focused on developing image processing approaches related to mapping tropical rainforests using SAR, problems are often encountered (Conway 1997; Luckman *et al.* 1997; Grover *et al.* 1999; Kuntz & Siegert 1999; Van der Sanden & Hoekman, 1999; Ticehurst *et al.* 2003). Problems associated with the use of SAR are related to moisture content of vegetation and soils, landscape topography and poor signal-to-noise ratio (Kuntz & Siegert 1999; Ticehurst *et al.* 2003). The use of optical remote sensing systems provides a spectrally precise representation of features on Earth (Campbell 2002), even though cloud coverage hinders some areas from being optically mapped on a regular basis (Foody & Hill 1996).

Change in vegetation cover pre and post World Heritage listing: Landsat 7 ETM imagery

Johansen *et al.* (2005) mapped vegetation cover change using spectral vegetation index image mosaics of the Wet Tropics bioregion derived from Landsat 5 TM and Landsat 7 ETM+ images collected in 1988 and 1999. The image data sets had previously been geometrically, radiometrically and atmospherically corrected by Bruce and Hilbert (2003). Processing in this project conducted further checks of geometric and atmospheric correction accuracy before applying vegetation index transformations and classification. The use of spectral vegetation indices (SVI) provided maps for the entire Wet Tropics in 1988 and 1999 of a continuous biophysical variable directly related to vegetation cover, biomass and LAI.

Johansen *et al.* (2005) found that total clearing of vegetation was the smallest land-cover change class, with less than 5% of the Wet Tropics bioregion being classified as cleared and less than 1% of the WHA. Those areas of total clearing that were detected corresponded mainly to the expansion of agricultural fields in the Tully–Innisfail area and clearance of road and power line corridors throughout the region. The most extensive type of land-cover or vegetation cover change observed in both the bioregion and WHA was partial regeneration, which was attributed to the change from a drier year (owing to below average rainfall in 1988) to a more standard wet year in 1999. However, the estimates of aerial extent of change for each of the land cover change classes should be treated with caution because the images used to perform the mapping represent two snapshots at single points in time of a highly dynamic system.

The Enhanced Vegetation Index (EVI) spectral vegetation index was considered to provide the most accurate representation of vegetation cover change in the high biomass/LAIs environment of the Wet Tropics. The more traditional Normalized Difference Vegetation Index (NDVI) was subject to saturation effects, where small changes in vegetation cover at high biomass levels would not be able to be detected. With respect to the operational monitoring on an annual or more frequent basis, Johansen *et al.* (2005) recommended that the SVI used by SLATS (the Statewide Landcover and Trees Study) for producing woody vegetation cover could be requested and used for identifying change in the Wet Tropics. Alternatively, the freely available MODIS EVI image data, provided on a daily or weekly basis, could be used for mapping and monitoring vegetation cover and its change over time.

Change in Wet Tropics vegetation cover: imaging radar (JERS-1)

Owing to frequent cloud cover during parts of the year, radar can be used as a surrogate for detecting changes to the rainforest during periods when optical remote sensing imagery is unavailable. The two major types of change that are able to be mapped by both SAR and Landsat TM/ETM are clearing/regrowth and seasonal change. For forest clearing and regrowth, radar backscatter would be expected to decrease or increase in backscatter respectively.

Ticehurst *et al.* (2003) tested JERS-1 (SAR) imagery from different dates for its ability to detect rainforest clearing and compared it to similar products from optical Landsat TM imagery (Plate 47.8). One of the main advantages in using two radar scenes from the same area acquired at different acquisition dates but similar viewing geometry is that the effects on radar backscatter due to topography and viewing angle are the same in both scenes. Some of the key implications found when using SAR imagery for detecting rainforest clearing were that while optical satellite data has been shown to be effective in change detection, SAR images can be collected under any conditions at any time of the year; hence it does provide a true year-round regional-scale monitoring tool for tropical regions. The L-band radar systems appeared to offer a good option for the detection of large-scale forest clearing and the more subtle seasonal changes in forest cover in tropical forest and forest/woodland conditions. However, while SAR can be used as a surrogate for optical imagery for change detection, caution needs to be applied when interpreting the data. For JERS-1 data, regions as small as 10 pixels in size could be detected using visual interpretation, while more extensive regions (>100 m $\Box$ 100 m) could be automatically extracted. Attempts to extract thin linear regions of change were not recommended using JERS imagery, owing to residual geo-location inaccuracies combined with speckle, which reduces the interpretability of the imagery. However, radar data of finer resolution and shorter wavelengths were considered to be worth investigating.

The seasonal effects observed in the SAR imagery contribute additional information to optical image classification programmes designed to distinguish between different forest types. However, to avoid inter-annual change detection, the radar data need to be acquired from the same time of year and climatic conditions to avoid confusion with seasonal effects. Furthermore, rainforest clearing would only be detectable in SAR and optical imagery when the cleared land contains relatively lower levels of biomass, or when forest successional stages have differing optical reflectance and/or radar backscatter properties.

Rainforest vegetation dynamics: trends in vegetation cover mapped from MODIS images

A more recent development in remote sensing applications has been the analysis of environmental change from more than two successive images. Product-driven sensors, such as MODIS, deliver products on a daily basis, enabling trends in biophysical variables to be quantified from daily to decadal time scales. Phinn *et al*. (2005b) report on a project that aimed to demonstrate the utility and continuity of a freely available and continually updated archive of environmental indicator maps derived from satellite images for monitoring and managing tropical Australian environments, including the Wet Tropics and the Wet-Dry Tropics. The text below is taken from Phinn *et al*. (2005b). This report matched maps of biophysical properties produced from the MODIS sensor to the SOWT indicators required by the WTMA (Phinn *et al*. 2001), and indicators required by local councils and other state government agencies.

A summary of the match between a range of environmental indicators relevant to the Wet Tropics and image data sets that could be used to map and monitor these indicators was developed. MODIS products were placed in six groups, based on the type of information they provide:
- base-level data: reflectance, temperature and elevation model;
- land-cover: land-cover classes and continuous vegetation (cover) classes;
- vegetation: vegetation indices, leaf area index, fraction of absorbed photosynthetically active radiation, net photosynthesis and net primary productivity;
- processes: thermal anomalies;
- high-level science: albedo, bidirectional reflectance distribution;
- ocean products.

Three potential applications for the MODIS image-map products in regional scale monitoring and management were identified:
- static or snapshot image maps, showing the state of a selected indicator for a particular point in time, for example, the most recent land-cover or vegetation index image;
- change detection products, showing the change in state of a selected indicator between two particular points in time, for example, land-cover change from 2004 to 2003;
- trend detection products, showing the changes to a selected indicator over multiple points in time, for example, monthly trend in vegetation index values over the 2003–4 period (Plate 47.9).

Results presented by Phinn *et al*. (2005b) demonstrate that a sound link can be established between recognized indicators of environmental condition in the Wet Tropics and information products derived from satellite imaging systems. If appropriate and validated image-based products are selected for further use and the necessary infrastructure is put in place, a regional scale monitoring system capable of delivering spatial data on SoWT environmental indicators could be established. Collection of the MODIS products on a regular basis and further processing of these data with change or trend detection techniques will deliver maps capable of measuring the magnitude and spatial extent of changes to select environmental indicators. One limitation with the current set of image-based products is a lack of local field validation of the measured environmental parameters, such as land-cover, vegetation indices, LAI, amount of exposed ground and burnt areas.

Modelling rainforest ecosystem productivity

Tropical rainforests play a key role in the global carbon cycle, but there is considerable uncertainty about the quantity and distribution of carbon stored in these ecosystems. Regional scale analyses of forest growth and productivity are critical for the development of realistic global carbon budgets and for projecting how these ecosystems will be affected by climate change (Drake *et al*. 2002; Clark 2004). Studies have found that the tropical forests of north Queensland are highly sensitive to climate change within the range expected early in this century (Hilbert & Ostendorf 2001; Hilbert *et al.* 2001, 2004). The focus of the research conducted by Nightingale (2005) was the development of a model-based system for assessing forest growth and biomass accumulation dynamics within Australia's tropical rainforest bioregion.

The 3-PG (Physiological Principles Predicting Growth) forest growth model and its remote sensing driven variant (3-PGS) were applied to simulate forest growth and the productivity of regrowth plantations and mature old-growth components of the Australian tropical rainforest. Both models were effective at estimating growth and carbon dynamics with highly diverse species mixtures using generic parameterizations. Our research showed for the first time that statistically significant relationships can be obtained between 3-PG and 3-PGS modelled and field measured estimates of stand structural attributes in tropical forests including, basal area (BA), diameter at breast height (DBH) and above-ground biomass (AGB). Model estimates of site above-ground annual net primary production (NPP) for the relatively young, fast growing and highly productive restoration rainforest and plantation sites were within the range of values obtained from a study of similar seasonally wet rainforests throughout South America and Asia (Clark *et al.* 2001a, b). Annual NPP for the old-growth rainforest field sites related well to published field-based estimates of NPP from other rainforest sites throughout the world (Chambers *et al.* 2001; Clark *et al.* 2001a, b).

The Tropical Rainforest Growth (TRG) system developed by Nightingale *et al.* (2007) employs the 3-PG and 3-PGS models to account for both old-growth and forest regeneration from seedlings. The system was applied to assess AGB stocks and accumulation dynamics in response to human-induced and natural disturbances, including commercial plantation forestry and the impact of tropical cyclones. AGB stocks (December 2001) of the old-growth forest throughout the bioregion were estimated to be approximately 201.5 tC ha^{-1} (tonnes of carbon per hectare) (Plate 47.10). Replacement of areas of old-growth with commercial timber plantations decreased overall AGB stocks within the bioregion to approximately 198 tC ha^{-1}. Plantation carbon accumulation rates were higher than the mature rainforest, representing their potential to accumulate more biomass over a longer analysis time period. As tropical cyclones may significantly alter the carbon stocks of old-growth rainforests, the effect of tropical cyclone Rona on the Wet Tropics rainforest was assessed. The impact of the cyclone reduced AGB stocks within the region by only approximately 3.5 tC ha^{-1}, yet these systems are an important factor to be considered in regional carbon modelling activities in the tropics. The TRG system is an advanced modelling tool providing a rapid process-based assessment of biomass stocks and accumulation dynamics within Australia's tropical rainforest bioregion and has the potential for application in tropical forest ecosystems at both national and international levels.

Summary

- Several remote sensing technologies are at an operational level (based on commercially available images and software) for mapping and monitoring vegetation characteristics that have been established as indicators of rainforest ecosystem health or condition.
- Rainforest science and management agencies commonly use geographical information systems that are able to process and display information derived from remotely sensed images.
- Rainforest environmental indicators can to be mapped and monitored on an operational basis using commercially available remotely sensed data are
  - land cover;
  - vegetation communities;
  - extent of vegetation fragmentation;
  - extent and severity of edge effects;
  - extent of cleared area;
  - extent of burnt areas.
- Image data sets can be acquired from local to global spatial scales (from 0.5 m to 1 km pixels) at revisit frequencies of twice daily to monthly, and as a result a multiscale sampling and measurement tool for rainforest environments are provided.
- The use of active image data, such as imaging radar that is not subject to cloud or smoke effects, is strongly recommended as it provides an all weather mapping and monitoring capability when used in combination with optical image data. There are currently no L-band radar satellite systems operating, but this is expected to change within the near future.
- Scientific and government agencies working on rainforest environments should take advantage of the

significant global and national image data sets currently being collected from all earth resource monitoring satellites. These data sets contain current and archival data, and are often accessed for a minimal or no charge. Spatial data, including image data, are now regarded as basic infrastructure for a country to manage and maintain its natural and built resources. Rainforest science and management agencies should take a similar approach to understand the extent and composition of their forests and how they are changing over time.

• Significant work is still required to develop and verify mapping approaches suited to rainforest environmental indicators using high-spatial resolution images (e.g. Quickbird, Ikonos), high-temporal frequency images (e.g. MODIS, SPOT-VI, MERIS) and imaging radar (e.g. ALOS, ENVISAT).

Acknowledgements

Funding for the majority of this work was provided through Program 1 in the Rainforest Cooperative Research Centre 1999–2000. We would like to thank the ongoing support of Professors Nigel Stork and Geoff McDonald for maintaining the project and Dr Earl Saxon for initiating the work. The GIS staff at Wet Tropics Management Authority provided us with extensive data sets and guidance for completing our work. We thank Mr Michael Stanford and Ms Catherine Simpson for research assistance in the first phase of the projects reported here from 1999 to 2002 and Mr Tim Edmonds for assistance with the pond apple mapping and modelling. Peter Scarth's PhD was supported through CSIRO Land and Water, and Maths and Information Science top-up grants. Joanne Nightingale's PhD was supported by the Greenhouse Accounting CRC. Dr David Hilbert, Mr Trevor Parker and Ms Caroline Bruce provided access to extensive field and image data sets.

References

Archard, F. & Estreguil, C. (1995). Forest classification of southeast Asia using NOAA AVHRR data. *Remote Sensing of Environment* **54**: 198–208.

Askne, J. I. H., Dammert, P. B. G., Ulander, L. M. H. & Smith, G. (1997). C-band repeat-pass interferometric SAR observations of the forest. *IEEE Transactions on Geoscience and Remote Sensing* **35**: 25–35.

Asner, G. P., Keller, M., Pereira, R., Zweede, J. C. & Silva, J. N. M. (2004). Canopy damage and recovery following selective logging in an Amazon forest: integrating field and satellite studies. *Ecological Applications* **14**(4): 280–98.

Asner, G. P., Knapp, D. E., Broadbent, E. N., Oliveira, P. J. C., Keller, M. & Silva, J. N. (2005). Selective logging in the Brazilian Amazon. *Science* **310**: 480–2.

Boyd, D. S. and Duane, W. J. (2001). Exploring spatial and temporal variation in middle infrared reflectance (at 3.75 nm) measured from the tropical forest of west Africa. *International Journal of Remote Sensing* **22**(10): 1861–78.

Bruce, C. M. &. Hilbert. D. W (2003). *Pre-processing Methodology for Application to Landsat TM/ETM+ Imagery of the Wet Tropics*. Report prepared for the Rainforest CRC. CSIRO Tropical Forest Research Centre. Atherton

Campbell, J. B. (2002). *Introduction to Remote Sensing*, 2nd edn. Guilford Press, New York.

Castro, K. L., Sanchez-Azofeifa, G. A. & Rivard, B. (2003). Monitoring secondary tropical forests using space-borne data: implications for central America. *International Journal of Remote Sensing* **24**(9): 1853–94.

Chambers, J. Q., dos Santos, J., Ribeiro, R. J. & Higuchi, N. (2001). Tree damage, allometric relationships, and aboveground net primary production in central Amazon forest. *Forest Ecology and Management* **152**: 73–84.

Clark, D. A. (2004). Sources or sinks? The responses of tropical forests to current and future climate and atmospheric composition. *Philosophical Transactions of the Royal Society of London. Series B* **359**(1443): 477–91.

Clark, D. A., Brown, S., Kicklighter, D. W. et al. (2001a). Net primary production in tropical forests: An evaluation and synthesis of existing field data. *Ecological Applications* **11**(2): 371–84.

Clark, D. A., Brown, S., Kicklighter, D. W., Chambers, J. Q., Thomlinson, J. R. & Ni, J. (2001b). Measuring net primary production in forests: concepts and field methods. *Ecological Applications* **11**(2): 356–70.

Clark, M. L., Roberts, D. A. & Clarke, D. B. (2005). Hyperspectral discrimination of tropical rain forest tree species at leaf to crown scales. *Remote Sensing of Environment* **96**: 375–98.

Conway, J. (1997). Evaluating ERS-1 SAR data for the discrimination of tropical forest from other tropical vegetation types in Papua New Guinea. *International Journal of Remote Sensing* **18**(14): 2967–84.

Culvenor, D. S. (2000). Development of a tree delineation algorithm for application to high spatial resolution digital imagery of Australian native forest. PhD thesis, University of Melbourne.

Curlander, J. C., Kwok, R. &. Pan, S. S (1987). A post-processing system for automated rectification and registration of spaceborne SAR imagery. *International Journal of Remote Sensing* **8**(4): 621–38.

Drake, J. B., Dubayah, R. O., Clark, D. B. *et al.* (2002). Estimation of tropical forest structural characteristics using large-footprint lidar. *Remote Sensing of Environment* **79**: 305–19.

Foody, G. M. & Hill, R. A. (1996). Classification of tropical forest classes from Landsat TM data. *International Journal of Remote Sensing* **17**(12): 2353–67.

Fransson, J. E. S., Smith, G., Askne, J. & Olsson, H. (2001). Stem volume estimation in boreal forests using ERS-1/2 coherence and SPOT XS optical data. *International Journal of Remote Sensing* **22**(14): 2777–91.

French, N. H. F., Kasischke, E. S., Bourgeau-Chavez, L. L., & Harrell, P. A. (1996). Sensitivity of ERS-1 SAR to variations in soil water in fire-disturbed boreal forest ecosystems. *International Journal of Remote Sensing* **17**(15): 3037–53.

Gougeon, F. A. (1995). A crown-following approach to the automatic delineation of individual tree crowns in high spatial resolution aerial images. *Canadian Journal of Remote Sensing* **21**(3): 274–84.

Gray, A. N., Spies, T. A. & Easter, M. J. (2002). Microclimatic and soil moisture responses to gap formation in coastal Douglas-fir forests. *Canadian Journal of Forest Research* **32**: 332–43.

Griffiths, G.H. & Wooding, M.G. (1996). Temporal monitoring of soil moisture using ERS-1 SAR data. *Hydrological Processes* **10**(9): 1127–38.

Grove, S. J., Turton, S. M. & Siegenthaler, D. T. (2000). Mosaics of canopy openness induced by tropical cyclones in lowland rain forests with contrasting management histories in northeastern Australia. *Journal of Tropical Ecology* **16**: 883–94.

Grover, K., Quegan, S. & Freitas, C. (1999). Quantitative estimation of tropical forest cover by SAR. *IEEE Transactions on Geoscience and Remote Sensing* **1**: 479–90.

Hayes, D. J. & Sader, S. A. (2001). Comparison of change-detection techniques for monitoring tropical forest clearing and vegetation regrowth in a time series. *Photogrammetric Engineering and Remote Sensing* **67**(9): 1067–75.

Hilbert, D. W. & Ostendorf, B. (2001). The utility of artificial neural networks for modeling the distribution of vegetation in past, present and future climates. *Ecological Modelling* **146**: 311–27.

Hilbert, D .W., Ostendorf, B. & Hopkins, M. (2001). Sensitivity of tropical forests to climate change in the humid tropics of north Queensland. *Austral Ecology* **26**: 590–603.

Hilbert, D. W., Bradford, M., Parker, T. & Westcott, D. A. (2004). Golden bowerbird (*Prionodura newtonia*) habitat in past, present and future climates: predicted extinction of a vertebrate in tropical highlands due to global warming. *Biological Conservation* **116**: 367–77.

Hoekman, D. H. & Quinones, M. J. (2000). Land cover type and biomass classification using AirSAR data for evaluation of monitoring scenarios in the Colombian Amazon. *IEEE Transactions on Geoscience and Remote Sensing* **38**(2): 685–96.

Hyyppa, J., Hyyppa, H., Inkinen, M., Engdahl, M., Linko, S. & Zhu, Y.-H. (2000). Accuracy comparison of various remote sensing data sources in the retrieval of forest stand attributes. *Forest Ecology and Management* **128**: 109–20.

Imhoff, M. L. (1995). Radar backscatter and biomass saturation: Ramifications for global biomass inventory. *IEEE Transactions on Geoscience and Remote Sensing* **33**(2): 511–18.

Johansen, K., Phinn, S., Ticehurst, C. & Held, A. (2005). *Vegetation Cover Change Pre- & Post- World Heritage Listing for the Wet Tropics, Far North Queensland using Landsat TM/ETM+ Imagery*. Cooperative Research Centre for Tropical Rainforest Ecology and Management, Rainforest CRC, Cairns,

Johnsen, H., Lauknes, L. & Guneriussen, T. (1995). Geocoding of fast-delivery ERS-1 SAR image mode product using DEM data. *International Journal of Remote Sensing* **16**(11): 1957–68.

Kasischke, E. S., Christensen, N. L. & Bourgeau-Chavez, L. L. (1995). Correlating radar backscatter with components of biomass in Loblolly pine forests. *IEEE Transactions on Geoscience and Remote Sensing* **33**(3): 643–59.

Kasischke, E. S., Melack, J. M. & Dobson, M. C. (1997). The use of imaging radars for ecological applications – a review. *Remote Sensing of Environment* **59**: 141–56.

Kuntz, S. & Siegert, F. (1999). Monitoring of deforestation and land use in Indonesia with multitemporal ERS data. *International Journal of Remote Sensing* **20**(14): 2835–53.

Lamb, D. (1998). Opportunities for satellite and airborne remote sensing of weeds in Australian crops. In *Precision Weed Management in Crops and Pastures*, Medd, R. & Pratley, J. (eds). Proceedings of a workshop held at Charles Sturt University, Wagga Wagga, NSW, 5–6 May, 48–54.

Lamb, D. W. & Brown, R. B. (2001). Remote sensing and mapping of weeds in crops. *Journal of Agricultural Engineering Research* **78**(2): 117–25.

Lillesand, T. M. & Kiefer, R. W. (2000). *Remote Sensing and Image Interpretation*, 4th edn. John Wiley & Sons, New York.

Lucas, R., Held, A., Phinn, S. & Saatchi, S. (2004). Tropical forests. In *Manual of Remote Sensing*, Ustin, S. (ed.). American Society for Photogrammetry and Remote Sensing, Bethesda, MD, 4, 768.

Lucas, R. M., Honzak, M., Curran, P. J., Foody, G. M. & Nguele, D. T. (2000). Characterizing tropical forest regeneration in Cameroon using NOAA AVHRR data. *International Journal of Remote Sensing* **21**(15): 2831–54.

Lucas, R. M., Honzak, M. & Foody, G. M. (1993). Characterizing tropical secondary forests using multi-temporal Landsat sensor imagery. *International Journal of Remote Sensing* **14**(16): 3061–7.

Luckman, A., Baker, J., Kuplich, T. M., Yanasse, C. C. F. & Frery, A. C. (1997). A study of the relationship between radar backscatter and regenerating tropical forest biomass for spaceborne SAR instruments. *Remote Sensing of Environment* **60**: 1–13.

Luckman, A., Baker, J. & Wegmuller, U. (2000). Repeat-pass interferometric coherence measurements of disturbed tropical forest from JERS and ERS satellites. *Remote Sensing of Environment* **73**: 350–60.

McGowen, I. J. (1998). Remote sensing: background to the technology and opportunities for mapping pasture weeds. In: R. Medd & J. Pratley (Eds.) *Precision weed management in crops and pastures*. Proceedings of a workshop held at Charles Sturt University, Wagga Wagga, NSW, 5–6 May, 1998, 36–47.

Mayaux, P., De Grandi, G. F., Rauste, Y., Simard, M. & Saatchi, S. (2002). Large-scale vegetation maps derived form the combined L-band GRFM and C-band CAMP wide area radar mosaics of Central Africa. *International Journal of Remote Sensing* **23**(7): 1261–82.

Mayaux, P. & Lambin, E. F. (1997). Tropical forest area measured from global land-cover classifications: inverse calibration models based on spatial textures. *Remote Sensing of Environment* **59**: 29–43.

Nelson, B. W. (1994). Natural forest disturbance and change in the Brazilian Amazon. *Remote Sensing Reviews* **10**: 105–25.

Nepstad, D. C., Verissimo, A., Alencar, A. *et al*. (1999). Large-scale impoverishment of Amazonian forests by logging and fire. *Nature* **398**: 505–8.

Nezry, E., Mougin, E., Lopes, A. & Gastellu-Etchegorry, J. P. (1993). Tropical vegetation mapping with combined visible and SAR spaceborne data. *International Journal of Remote Sensing* **14**(11): 2165–84.

Nightingale, J. M. (2005) Carbon dynamics within tropical rainforest environments. PhD thesis, School of Geography, Planning and Architecture, University of Queensland.

Nightingale, J.M. Hill, M. Phinn, S.R., and. Held, A.A. (2007) Comparison of Australian Tropical Rainforest Productivity Derived from the 3-PG Forest Growth Model and MODIS Productivity Products. *Canadian Journal of Remote Sensing.* **33**(4): 278–88.

Phinn, S. (1998). A framework for selecting appropriate remotely sensed data dimensions for environmental monitoring and management. *International Journal of Remote Sensing* **19**(17): 3457–63.

Phinn, S., Edmonds, T. Ticehurst. C. & Held, A. (2002b). *Mapping Current Infestations: Developing Remote Sensing Procedures for Early Detection of New Pond Apple Infestations*. Report prepared for the Natural Heritage Trust Weeds of National Significance Program. Rainforest Cooperative Research Centre. 71 pp.

Phinn, S., Held, A., Stanford, M., Ticehurst, C. & Simpson, C. (2002a). Optimising state of environment monitoring at multiple scales using remotely sensed data. In *Proceedings of the 11th Australasian Remote Sensing and Photogrammetry Conference*, Causal Publications, Brisbane.

Phinn, S., Joyce, K., Scarth, P.& Roelfsema, C. (2005a) The Role of integrated information acquisition and management in the analysis of coastal ecosystem change. In: Le Drew, E. & Richardson, L. (eds.) *Remote Sensing of Coastal Aquatic Ecosystem Processes*. Remote Sensing and Digital Image Processing Series–Springer, Dordrecht, pp. 217–50.

Phinn, S. R., Menges, C., Hill, G. J. E. & Stanford, M. (2000). Optimising remotely sensed solutions for monitoring, modelling and managing coastal environments. *Remote Sensing of Environment* **73**(2): 117–32.

Phinn, S., Scarth, P., Johansen, K., Ticehurst, C. & Held, A. (2005b). *MODIS Products for Environmental Monitoring In Tropical Far-North Queensland*. Cooperative Research Centre for Tropical Rainforest Ecology and Management, Rainforest CRC, Cairns.

Phinn, S., Stanford, M., Held, A. & Ticehurst, C. (2001). *Evaluating the Feasibility of Remote Sensing for Monitoring State of the Wet Tropics Environmental Indicators*. Cairns, Cooperative Research Centre for Tropical Rainforest Ecology and Management. 71 pp.

Pinard, M., Howlett, B. & Davidson, D. (1996). Site conditions limit pioneer tree recruitment after logging of dipterocarp forests in Sabah, Malaysia. *Biotropica* **28**: 2–12.

Podest, E. &. Saatchi, S (2002). Application of multiscale texture in classifying JERS-1 radar data over tropical vegetation, *International Journal of Remote Sensing* **23**(7): 1487–506.

Ranson, K. J., Sun, G., Kharuk, V. I. & Kovacs, K. (2001). Characterization of forests in Western Sayani mountains, Siberia from SIR-C SAR data. *Remote Sensing of Environment* **75**: 188–200.

Rignot, E., Salas, W. A. & Skole, D. L. (1997). Mapping deforestation and secondary growth in Rondonia, Brazil, using imaging radar and thematic mapping data. *Remote Sensing of Environment* **59**: 167–79.

Rosenqvist, A., Shimada, M., Chapman, B. *et al*. (2000). The Global Rain Forest Mapping project – a review. *International Journal of Remote Sensing* **21**(6/7): 1375–87.

Roy, P. S., Ranganath, B. K., Diwakar, *et al*. (1991). Tropical forest type mapping and monitoring using remote sensing. *International Journal of Remote Sensing* **12**(11): 2205–25.

Saatchi, S. S., Agosti, D., Alger, K., Delabie, J. & Musinsky, J. (2001). Examining fragmentation and loss of primary forest in the southern Bahian Atlantic forest of Brazil with radar imagery. *Conservation Biology* **15**(4): 867–75.

Saatchi, S. S., Nelson, B., Podest, E. &. Holt, J. (2000). Mapping land-cover types in the Amazon Basin using 1km JERS-1 mosaic. *International Journal of Remote Sensing* **21**(6/7): 2101–23.

Saatchi, S. S., Soares, J. V. & Alves, D. S. (1997). Mapping deforestation and land use in Amazon rainforest by using SIR-C imagery. *Remote Sensing of Environment* **59**: 191–202.

Santos, J. R., Freitas, C. C., Araujo, L. S. *et al.* (2003). Airborne P-band SAR applied to the aboveground biomass studies in the Brazilian tropical rainforest. *Remote Sensing of Environment* **87**: 482–93.

Scarth, P., (2000). Mapping Koala habitat and eucalyptus trees: integration and scaling of field and airborne hyperspectral data. In *Proceedings of the 10th Australasian Remote Sensing and Photogrammetry Conference*, Adelaide, 21–25 August.

Scarth, P. &. Phinn, S. (2000). Determining forest structural attributes using an inverted geometricoptical model in mixed eucalypt forests, southeast Queensland, Australia. *Remote Sensing of Environment* **71**: 141–57.

Scarth, P., Phinn, S. & Held, A. (2002). Inversion of a tropical rainforest gap model for mapping disturbance at local to regional scales. In *Proceedings of 11th Australasian Remote Sensing and Photogrammetry Conference*, 2–6 September, Brisbane.

Sgrenzaroli, M., Baraldi, A., Eva, H., De Grandi, G. & Archard, F. (2002). Contextual clustering for image labeling: An application to degraded forest assessment in Landsat TM images of the Brazilian Amazon. *IEEE Transactions on Geoscience and Remote Sensing* **40**(8): 1833–48.

Simard, M., De Grandi, G., Saatchi, S. & Mayaux, P. (2002). Mapping tropical coastal vegetation using JERS-1 and ERS-1 radar data with a decision tree classifier. *International Journal of Remote Sensing* **23**(7): 1461–74.

Simard, M., Saatchi, S. S. & De Grandi, G. (2000). The use of decision tree and multiscale texture for classification of JERS-1 SAR data over tropical forest. *IEEE Transactions on Geoscience and Remote Sensing* **38**(5): 2310–21.

Skole, D. & Tucker, C. (1993). Tropical deforestation and habitat fragmentation in the Amazon: Satellite data from 1978 to 1988. *Science* **260**: 1905–10.

Strahler, A. H. & Jupp, D. L. B. (1990). Modeling bidirectional reflectance of forests and woodlands using Boolean models and geometric optics. *Remote Sensing of Environment* **34**: 153–66.

Tenhunen, J. D. & Kabat, P. (1999). *Integrating Hydrology, Ecosystem Dynamics and Biogeochemistry in Complex Landscapes.* John Wiley & Sons, London.

Ticehurst, C., Held, A. & Phinn, S. (2004). Integrating JERS-1 imaging radar and elevation models for mapping tropical vegetation communities in far north Queensland, Australia. *Photogrammetric Engineering and Remote Sensing* **70**(11): 1259–66.

Ticehurst, C., Lymburner, L., Held, A. *et al.* (2001). Mapping tree crowns using hyperspectral and high spatial resolution imagery. In *Proceedings of Third International Conference on Geospatial Information in Agriculture and Forestry*, Denver, CO, 5–7 November.

Ticehurst, C., Held A., & Phinn S. (2005) Integrating JERS-1 imaging radar and elevation models for mapping tropical rainforest communities in far north Queensland, Australia. *Photogrammetric Engineering and Remote Sensing*, **70**(11): 1259–1266.

Ticehurst, C., Phinn, S. & Held, A. (2003). *All-weather Land-cover Change Mapping System for the Wet Tropics* (a study on the feasibility of JERS-1 imaging radar for change detection analysis in the Wet Tropics Bioregion of Far North Queensland). Cooperative Research Centre for Tropical Rainforest Ecology and Management. Rainforest CRC, Cairns. 28 pp.

Toutin, T. & Gray, L. (2000). State-of-the-art of elevation extraction from satellite SAR data. *ISPRS Journal of Photogrammetry and Remote Sensing* **55**: 13–33.

Tracey, J. G. (1969). Edaphic differentiation of some forest types in eastern Australia. I. Soil physical factors. *Journal of Ecology* **57**: 805–16.

Tracey, J. G. & Webb, L. J. (1975). *Vegetation of the Humid Tropical Region of North Queensland* (15 maps at 1 : 100 000 scale + key). CSIRO Long Picket Labs, Indooroopilly, Queensland.

Treuhaft, R. N., Madsen, S. N., Moghaddam, M. & Van Zyl, J. J. (1996). Vegetation characteristics and underlying topography from interferometric radar. *Radio Science* **31**(6): 1449–85.

Tracey, J. G. (1982). *The Vegetation of the Humid Tropical Region of North Queensland.* CSIRO, Melbourne,

Ustin, S. (Ed.) (2004). *Manual of Remote Sensing, Volume 4, Remote Sensing for Natural Resource Assessment.* American Society for Photogrammetry and Remote Sensing, Bethesda, MD. 768 pp.

Van der Sanden, J. J. & Hoekman, D. H. (1999). Potential of airborne radar to support the assessment of land cover in a tropical rain forest environment. *Remote Sensing of Environment* **68**: 26–40.

Wallace, J. & Campbell, N. (1998). *Evaluation of the Feasibility of Remote Sensing for Monitoring National State of the Environment Indicators.* Australia: State of the

Environment Technical Paper Series (Environmental Indicators). Department of Environment, Canberra, pp. 1–24.

Wulder, M., Niemann, K. O. & Goodenough, D. G. (2000). Local maximum filtering for the extraction of tree loca-tions and basal area from high spatial resolution imagery. *Remote Sensing of Environment* **73**: 103–14.

Zebker, H. A. & Villasenor, J. (1992). Decorrelation in inter-ferometric radar echoes. *IEEE Transactions on Geoscience and Remote Sensing* **30**: 950–9.

48 RAINFOREST SCIENCE AND ITS APPLICATION

Stephen Goosem[1]*, Nigel E. Stork[2]* and Stephen M. Turton[3]*

[1]Wet Tropics Management Authority, Cairns, Queensland, Australia
[2]School of Resource Management and Geography, Faculty of Land and Food Resources, University of Melbourne, Burnley Campus, Richmond, Victoria, Australia
[3]Australian Tropical Forest Institute, School of Earth and Environmental Sciences, James Cook University, Cairns, Queensland, Australia
*The authors were participants of Cooperative Research Centre for Tropical Ecology and Management.

Introduction

The Wet Tropics of Queensland were inscribed on the World Heritage List on 9 December 1988 as a direct consequence of the accumulated scientific research and understanding of the region's rainforests up to that time. In 1992 the Wet Tropics Management Authority was established as the body responsible for the coordinated management of the World Heritage Area, while the Rainforest CRC (Cooperative Research Centre for Tropical Rainforest Ecology and Management) was established in 1993 as the body responsible for both the coordination and the undertaking of scientific research to assist in the management of the World Heritage Area and associated rainforest landscapes.

Undertaking scientific research in World Heritage areas is more than a legitimate activity, it is also a duty. The importance of such research is explicitly acknowledged in Wet Tropics World Heritage legislation and in international conventions such as the World Heritage Convention and the Convention on Biological Diversity. One of the benefits that World Heritage areas provide for scientific researchers is the provision of protected benchmark and/or control sites where the direct impact of human activities is minimal relative to surrounding landscapes.

Nature conservation is a multiple land use. Many essential services are provided by species diversity and healthy ecosystems. The principal aim of environmental management is to maintain biological diversity and protect ecological integrity. Government bodies responsible for the management of rainforested landscapes within the Wet Tropics have a number of responsibilities in relation to research.

- The first is commissioning, conducting or facilitating research that establishes, underpins and develops appropriate arrangements for management to achieve its objectives of long-term protection and/or sustainable use.

- The second is commissioning, conducting or facilitating monitoring and research to enable objective reporting on the effectiveness with which the management objectives are achieved and to develop and evaluate alternative management approaches.

- The third is establishing and operating an efficient system for managing research in the protected area in order to ensure that:

  o its natural and cultural attributes are not damaged or compromised;

  o the findings and implications of research are quickly known and appropriately reflected in management practices;

o the research commissioned by, conducted for or facilitated by the responsible body conforms to best practice standards.

In this chapter we examine the relationship between those responsible for providing and those responsible for applying rainforest research and discuss ways in which the delivery and application of science can be improved based on our experience.

How is rainforest science used?

There are a large number of ways in which rainforest science is being applied to underpin management, conservation and sustainable use of tropical forests in the Wet Tropics region.

1 Strategic management planning. This is reliant on science for the identification of issues, spatial patterns and processes such as diversity, distinctiveness, irreplaceability, vulnerability and other categories of intrinsic natural values.

2 Operational or on-ground management. This is almost entirely focused on the management of ecological impacts and threatening processes and is more concerned with external factors such as regional land use, the provision of community services, public access, weeds, diseases, pests, fire and visitor impacts and behaviour.

3 Development of government policies. Science plays an important role in guiding local, regional, state, national and international policy development. In recent years we have seen how researchers in the field of climate change and biodiversity conservation have played effective roles in initiating and directing policy development.

4 Development of environmental monitoring and assessment systems. The rapid nature of change that characterizes contemporary times is causing a major rethink about how effective current monitoring strategies are. Monitoring is required to gauge the efficacy of plans, policies and on-ground management.

5 Increasing economic benefits from forested landscapes. The heavily forested nature of the region provides enormous economic opportunities, as discussed elsewhere (Pearce, Chapter 7, and Harrison & Herbohn, Chapter 43, this volume), particularly in the fields of ecotourism and farm forestry. Ecological goods and services and their contribution to regional economies provided by forested landscapes are also becoming more appreciated. Additional areas of potential economic benefit from rainforest research include biodiscovery – the use of beneficial chemicals from rainforest plants and animals.

6 Community education to improve understanding of tropical forests. Tropical forests are complex and although local communities may have some knowledge of how they operate, we find that most people are largely ignorant of their surrounding environment.

There are two overriding concepts that are central to the application of rainforest science in the development of management or decision-making tools. The first is the 'distinctiveness' of inherent natural attributes in an area and the second is the 'threats' to those attributes. These concepts are embodied in five broad areas of rainforest research that are particularly pertinent to management and provide the foundation for the points described above:

1 What does our regional biodiversity consist of, where is it and why is it there?

2 How does this biodiversity function?

3 What is the value of this biodiversity (both the intrinsic ecological and evolutionary value and cost accounting of the environmental, economic and social value of biodiversity)?

4 What is changing?

(a) Identification of threatening processes and how these change ecosystem processes;

(b) development of indicators and baseline data so that changes can be detected;

(c) development of predictive modelling techniques;

(d) assessment of the effectiveness of management strategies and on-ground practices in restoring and maintaining ecosystem processes and reducing adverse impacts.

5 How do we communicate these scientific findings to maximize their environmental, social, economic, cultural and educational benefits to the community?

Environmental management has traditionally been a crisis discipline and very reactionary in nature. This crisis mentality is changing, helped by the development of predictive modelling and visualization techniques, scientifically based management plans, codes of practice and design, construction, maintenance and operational guidelines.

Why is some research ignored by potential users?

Researchers are quite often surprised that their earth-shatteringly important research is not immediately picked up by managers and policy-makers. There maybe many reasons for this, but three common failings are timing, relevance and communication.

Timing

There is a saying that 'timing is not the most important thing, it is the only thing'. This is particularly true with respect to the potential application of scientific research findings and recommendations. It is better to be approximately right at exactly the right moment than to be exactly right at completely the wrong time. Too often managers and policy-makers ask for advice on a current pressing issue but the advice comes after the decision has had to be made, so that the opportunity for having the greatest influence on decisions has been lost. A solution to yesterday's important issue is unlikely to excite or interest management preoccupied with today's crisis. It is essential that researchers are aware of what key management and policy issues are in the pipeline and where necessary assist in providing information in a timely fashion. It is irritating to say the least when managers find that researchers produce important and controversial results after such decisions are made, when they could have provided preliminary findings or other information earlier. Such use of science to denounce management and policy decisions inevitably results in conflict and distrust, further reducing the likelihood of acceptance or uptake of research findings.

Relevance

From a user perspective there is often the feeling that some researchers operate under a 'strategy of hope' – the hope, or belief, that users will find their work useful. Sometimes this actually happens but generally it does not. Close partnerships, networks and working relationships between researchers and managers are needed to identify, anticipate, prevent and solve management issues.

Relevance also includes a spatial dimension. Academic freedom means that scientists can afford to theorize but are rarely accountable for their recommendations, whereas managers are forced to be accountable and pragmatic. Often researchers tackling real management problems opt to undertake their research in the more pristine parts of the landscape away from active management issues, which greatly reduces the relevance of any findings or recommendations. To be truly relevant to users, research has to be undertaken in circumstances that mirror reality so that potential limitations, options and alternatives can be determined with some degree of certainty to allow for reasonable and defensible decisions to be made.

Communication

Another reason why research findings may be ignored by potential users is because they are often not presented in a readily understandable, usable or user-friendly form. Copies of research papers may be an appropriate means of communication between members of the research community but are unlikely to be the most appropriate mechanism for communicating with the community at large. Scientists use different jargon/language to that of most research users, and although it is appropriate for their research colleagues, it is difficult for non-researchers to understand.

Another factor that hinders the successful communication and adoption of research findings relates to managers and policy-makers being inherently resistant to new information. If information or recommendations are consistent with existing practices and beliefs, they are more likely to be accepted and integrated into management. However, if new research findings conflict with current practices and long-held beliefs they are far less likely to be embraced. New or controversial findings need to be packaged in ways that reduce this apparent conflict or psychological discomfort to increase their likelihood of acceptance or adoption.

Barriers to uptake

Perceptions and stereotypes

Misunderstandings, distrust and antagonism between researcher scientists and managers are frequently cited (e.g. Cullen 1990; Baskerville 1997; Norton 1998;

Rogers 1998; Grayson *et al.* 2000; Asher 2001; Cullen *et al.* 2001; Ewel 2001; Stork 2001). The most common theme used to explain this 'them and us' mentality is the basic differences in operational cultures and working philosophies between scientists and managers. Roux *et al.* (2006) list several common perceptions held by managers of scientists:

- they are arrogant;
- they produce fragmented information that seldom addresses 'real' problems;
- they are unable to contribute to the types of debate that govern problem-solving in the real world;
- they have little regard for application contexts, and are driven only by intellectual curiosity;
- they do not work at appropriate or useful spatial and temporal scales;
- they are part of an inward-looking, self-serving culture enforced by scientific peer review and reward systems;
- they do not communicate effectively to non-scientists.

Conversely, Roux *et al.* (2006) list several views of managers typically held by scientists that testify to similar biases in disciplinary and stereotyping perceptions, e.g. managers:

- do not appreciate ecosystem complexity;
- work within a system that rewards organizational and individual interests rather than ecosystem interests;
- have a poor understanding of scientific processes;
- are caught up in day-to-day operations, and spend little time in intellectual reflection and longer-term research and development planning;
- do not articulate their needs effectively, and often do not know what they want.

Improving the use of research and the involvement of end-users in knowledge creation

Common understanding (relationships and drivers)

An obvious way to speed up the adoption of new solutions and knowledge is to align such solutions more closely with actual problems and, therefore, the needs considered important by users at the time. This can only be achieved by involving those parties who will deploy the knowledge at the earliest possible stages in the selection and design of research projects. Scientists need to become more than just the 'messenger' and actively involve themselves with the whole decision-making process.

It is a common experience to hear researchers bemoan the fact that managers or other users failed to articulate their specific research need requirements or provide some form of structured research priority strategy. Factors that users need to consider when communicating their priorities to the research community include:

- What is the underlying cause(s) of the current failure to manage an issue successfully?
- What form of intervention to improve management is likely to be most successful, and what are the costs, anticipated benefits and risks?
- What is the regional significance of the particular research/management issue and likely adoption rate?

A key feature of the Rainforest CRC was the cementing of a meaningful partnership between the research and user communities. Research users were incorporated into all of the Centre's activities and management, ensuring that the objectives and outputs developed were useful, and agreed milestones ensured that findings were progressively presented to users in as timely a manner as possible. Two key mechanisms developed to ensure the strong connection and regular dialogue between users and researchers were the Program Support Group and the use of knowledge brokers (Box 48.1).

Common or compatible frameworks

It is vitally important that researchers and managers have common or compatible spatial and theoretical frameworks on which to hang their data and models (Westley 1995). If basic frameworks are not compatible it very severely limits the use to which data and knowledge can be applied, since such frameworks provide the spatial, conceptual, legal and administrative contexts for focusing attention, summarizing patterns and aggregating information. It is also the context within which decisions are made.

For conservation biologists there is no higher priority than to improve our understanding of economics. As much as some might wish it otherwise, problems of ecology are first and foremost political problems

---

Box 48.1 User-driven research: programme support groups

One innovative way of achieving outputs useful to management was through the establishment of programme support groups (PSGs), which were sub-committees of the Governing Board. These were developed to achieve:

- outcomes desired by a diverse range of partners;
- stronger, mutually beneficial and long-term partnership through cooperation and collaboration, not through control;
- improved capacity of researchers to carry out world class research;
- improved capacity of researchers to answer appropriate questions;
- identification of external opportunities, particularly for research and development funding;
- maximization of the usefulness of the information generated through research;
- improved communication through development of relevant and timely information products;
- reduced levels of bureaucracy and the cost of research, communication and education, that is, improve efficiency of these operations;
- creation of a skilled human resource capacity through relevant education programmes;
- improved development of more marketable and commercial products.

Many of the Rainforest CRC research participants were involved in developing outputs and tools for management that were highly technical in nature. Although such outputs are of immense potential value to management, this value can only be realized if management agency staff can develop the capacity and skills to use and interpret such tools. Conversely, many research outputs still need further collaborative development with the user community to transform them into routine operational methods.

User driven research: knowledge brokers

Knowledge exchange is a two-way process and a number of research organizations have employed individuals to be 'knowledge brokers'. In others a clear recognition of knowledge exchange is shown by the allocation of a percentage of staff time to knowledge brokering. Often these individuals will work with research users to help to find the relevant researchers with particular knowledge in a field or will search for the information and compile it in reports or strategic documents. Two areas of the work of the Rainforest CRC were particularly relevant in this regard: working with the Indigenous community and working in natural resource management and planning.

---

relating to who gets what, when and how. Environmental protection, climate stability and conservation of biological diversity are unavoidably political. The environment is ultimately a mirror that reflects the collective decisions we make about development, energy, forests, land, water and resources.

Complementary knowledge transfer strategies

The stereotyping that results from the differences in operational cultures and working philosophies between scientists and managers described above and further highlighted in Table 48.1 is perpetuated and reinforced by misunderstandings and misconceptions of each other's roles and responsibilities. In recent years it has become more common for researchers and managers to make a concerted effort to develop more

professional interactions and relationships and to recognize their mutual dependencies. Research providers and research users have adopted complementary strategies for the delivery or acquirement of knowledge. Research providers have focused on developing strategies to manoeuvre research findings and new knowledge from the science to the management/user domain, whereas managers and other users have developed strategies to attract useful knowledge from the science to the management domain.

Early and ongoing interaction with research users is the surest way to increase compatibility between research findings and user needs. Prospective users should be involved up-front, be encouraged to participate in the design and development of the research and help to apply it at a pilot scale before it is finally adopted. In some instances a financial commitment by the user to fund the research may be beneficial in

Table 48.1 Three perceived stereotypes affecting research uptake, acceptance and application

Researcher emphasis	Management expectation	On-ground practitioner needs
• Scientific perfection	• Timely results	• Practical results
• Global patterns	• Regional effects	• Site/local effects
• Mean conditions	• Variability and extremes	• Acceptable/ unacceptable
• Most probable outcome	• Risks and options	• Prescriptions
• Trends	• Thresholds	• Triggers for action
• Cycles/equilibria	• Rates of change	• Yearly work plans
• Physical impacts	• Societal responses	• Control techniques
• Scientific jargon	• Plain language	• Plain language
• Science speak	• Manager speak	• Guidelines/manuals
• Ideal systems	• Reality	• Exceptions
• Scientific method	• Recommendations	• Policy/practice
• Peer review	• Pragmatic	• Does it work?

establishing dialogue, but a purchaser–provider model does not represent a true partnership and could influence the independence of the research. Meaningful partnerships and interactions create the situation whereby researchers are able to grasp more fully and respond to implementation realities such as the management agency's issues, capability and resource constraints. Often the adoption of a management solution is not necessarily based on the best solution, but is strongly influenced by user-defined constraints such as affordability, familiarity and availability of the necessary infrastructure and skills (Steele 1989). If researchers are aware of such pragmatic management realities then a range of suitable cost–benefit options can be presented for consideration.

Knowledge is valued over information

Science and management represent two different communities of practice that are complementary only to the degree that their knowledge or understanding is able to interface. True knowledge transfer ends with the user adopting the research findings. In such circumstances the user has both the understanding and the emotional and financial commitments to allow sustained use of the acquired knowledge. Knowledge

transfer efforts that do not result in adoption are failures.

There is a need to discriminate between information and knowledge. In general, information refers to organized or interpreted data; 'data endowed with relevance and purpose' (Drucker 2001). Information is explicit and can be readily transferred to another party. Knowledge, on the other hand, is a mix of experiences, values, contextual information and intuition that provides a framework with which to evaluate and incorporate new experiences and information (Davenport & Prusak 1997). It is this knowledge that gives people their capacity for effective action. Knowledge is highly personal and difficult to formalize, often making it problematic to share with others, but without knowledge transfer, a user's ability to understand, replicate or exploit new knowledge is severely constrained.

It is ironic that gaining knowledge is always accompanied by both paradox and unintended consequences. It is an important concept that ignorance is built into the way science itself works, because we cannot know in advance the unintended effects of our actions in complex systems. The upshot of this is that both researchers and especially managers should never assume that an area of research is finished when a researcher has duly reported research findings, since good research inevitably begets more research, more questions and more curiosity. Many managers seem to think that if they throw money at a problem then the research community has failed if their findings are not the definitive end of the problem, and they are reluctant to throw 'good money after bad' if a research finding/recommendation is for more research. Similarly there is a perception by managers – and probably quite a legitimate one – that researchers, especially in the natural sciences, are better at identifying problems than solving problems. There does need to be a cultural shift by researchers to embrace 'solution science'.

The Rainforest CRC

The Cooperative Research Centre (CRC) programme was created to address a perceived national need to foster close interaction between scientists, private industry and public sector agencies across all sectors of science in long-term collaborative arrangements that support research, development and education activities. The Rainforest CRC research programme was designed to produce solutions to many of the problems facing the management of Wet Tropics rainforest landscapes. The programmes of research, although strongly focused on outcomes, were represented by a blend of strategic and tactical research, with strong user-community relevance.

It was envisioned that these research activities would lead to critical advances in the better management of the region's rainforest landscapes by achieving:

- an integrated, multidisciplinary, scientifically rigorous approach to regional planning and management;
- an understanding, identification and valuation of ecosystem goods and services;
- sustainable development for rainforest nature-based tourism;
- improved approaches and techniques to forest rehabilitation
- improved understanding of conservation principles and their application to management;
- Aboriginal (Indigenous) participation in research and management.

As mentioned in a previous section, science, like any other discipline, has developed its own jargon. It is very important for researchers to be reminded that scientific jargon is a barrier to the understanding of non-scientists (Westley 1995), be they government agencies, funding bodies, advisory committees or the general public. The CRC made a real effort to take up the challenge of describing its findings in 'plain English' while maintaining the scientific integrity of the scientific ideas it presented in its CRC Report Series.

The Rainforest CRC, through its programme support groups, attempted to create tightly focused research agendas aimed at specific, solvable problems. Rather than broadcasting general requests for proposals and funding the most relevant of those tendered, the Rainforest CRC identified research priorities *a priori* and solicited investigators to do the identified projects, which better guaranteed the relevance of the research to managers. Identification of research priorities utilized agenda-setting processes conducted jointly by researchers and managers at regular intervals.

Conservation science needs to focus on becoming a 'solutions science' rather than solely the identifier of management problems (Medawar 1967). A commitment by the administration of research institutions to provide the recognition and credit to scientists who conduct research directly applicable to resolving conservation problems is needed above and beyond the traditional scientific peer review and reward systems currently in place. Government land-management agencies should institute regular peer-reviewed processes by which joint teams of scientists and managers ensure that the best available scientific knowledge is accessible and being applied to tackle management issues (Meffe *et al.* 1998). Peer review is widely accepted as denoting that minimum standards of scientific credibility have been met; similar standards should be in place for ensuring that land managers are using best available information and practices.

The collaborative partnerships that characterized the operational philosophy of the Rainforest CRC highlighted the fact that environmental issues are highly complex and based on considerations that extend beyond the realms of science. It became apparent that although science provides a base from which to start, it is not the only place to seek a strategic assessment of policies to address management and policy issues (Passmore 1974). What is also needed is an appreciation of the environment and ecological relationships in ways that reflect ethical standards, political realities and community values and expectations, rather than a rating of their value in purely quantitative terms.

The growing appreciation of environmental ethics reflects deep changes in our view of the environment and our interconnectedness to our environment. This transition requires more than an increase in scientific knowledge, it requires a re-evaluation of the benefits we ask from our environment, how much we feel justified in asking for and the relative values we assign to the various benefits we seek. Such deeper social questions are particularly important as we have entered a period of immense and rapid change in the environment and the systems that control it, requiring a corresponding large change in human behaviour and expectations if we are to transmit to future generations the current ecological and evolutionary values of the Wet Tropics rainforests.

References

Asher, W. (2001). Coping with complexity and organizational interests in natural resource management. *Ecosystems* **4**: 742–57.

Baskerville, G. L. (1997). Advocacy, science, policy, and life in the real world. *Conservation Ecology* 2 (www.ecologyandsociety.org/vol1/iss1/art9).

Cullen, P. (1990). The turbulent boundary between water science and water management. *Freshwater Biology* **24**: 201–9.

Cullen, P., Cottingham, P., Doolan, J. *et al.* (2001). *Knowledge Seeking Strategies of Natural Resource Professionals.* Technical Report 2/2001, Cooperative Research Centre for Freshwater Ecology.

Davenport, T. H. & Prusak, L. (1997). *Working Knowledge: How Organizations Manage What They Know.* Harvard Business School Press, Boston.

Drucker, P. (2001). *Management Challenges for the 21st Century.* Harper Business Press, New York.

Ewel, K. C. (2001). Natural resource management: the need for interdisciplinary collaboration. *Ecosystems* **4**: 716–22.

Grayson, R. B., Ewing, S. A., Argent, R. M., Finlayson, B. L. & McMahon, T. A. (2000). On the adoption of research and development outcomes in integrated catchment management. *Australian Journal of Environmental Management* **7**: 24–35.

Medawar, P. S. (1967). *The Art of the Soluble.* Methuen, London.

Meffe, G. K., Boersma, P. D., Murphy, D. D. *et al.* (1998). Independent scientific review in natural resource management. *Conservation Biology* **12**: 268–70.

Norton, B. G. (1998). Improving ecological communication: the role of ecologists in environmental policy formation. *Ecological Applications* **8**: 350–64.

Passmore, J. (1974). *Man's Responsibility for Nature.* Duckworth, London.

Rogers, K. (1998). Managing science/management partnerships: a challenge of adaptive management. *Conservation Ecology* **2**: R1.

Roux, D. J., Rogers, K. H., Biggs, H. C., Ashton, P. J. & Sergeant, A. (2006). Bridging the science–management divide: moving from unidirectional knowledge transfer to knowledge interfacing and sharing. *Ecology and Society* **11**(1): 4.

Steele, L. W. (1989). *Managing Technology: The Strategic View.* McGraw-Hill, New York.

Stork, N. E. (2001). The management implications of canopy research. *Plant Ecology* **153**: 313–17.

Westley, F. (1995). Governing design: the management of social systems and ecosystem management. In *Barriers and Bridges to the Renewal of Ecosystems and Institutions,* Gunderson, L., Holling, C. S. & Light, S. (eds). Columbia University Press, New York.

49 LESSONS FOR OTHER TROPICAL FOREST LANDSCAPES

Nigel E. Stork[1]*, Stephen M. Turton[2]*,
William F. Laurance[3], Jiro Kikkawa[4]*, Jeffrey A. McNeely[5],
Jeffrey Sayer[6] and S. Joseph Wright[7]

[1]School of Resource Management and Geography, Faculty of Land and Food Resources, University of Melbourne, Burnley Campus, Richmond, Victoria, Australia
[2]Austrailian Tropical Forest Institute, School of Earth and Environmental Sciences, James Cook University, Cairns, Queensland, Australia
[3]Smithsonian Tropical Research Institute, Balboa, Ancon, Panama, Republic of Panama; and Biological Dynamics of Forest Fragments Project, National Institute for Amazonian Research (INPA), Manaus, Amazonas, Brazil
[4]School of Integrative Biology, University of Queensland, Australia
[5]International Union for Conservation of Nature and Natural Resources (IUCN), Gland, Switzerland
[6]Forest Conservation Programme, International Union for Conservation of Nature and Natural Resources (IUCN), Gland, Switzerland
[7]Smithsonian Tropical Research Institute, Balboa, Panama, Republic of Panama
*The authors were participants of Cooperative Research Centre for Tropical Ecology and Management

In recent decades, governments and the general public have grown increasingly alarmed at the declining state of the world's environment. These concerns were first highlighted internationally by the Stockholm Conference on the Human Environment (1972). Equally important was the Brundtland Report *Our Common Future* (Brundtland 1987), produced by the World Commission on Environment and Development, which argued eloquently that, without fundamental changes in practices and innovation, further economic development would continue to exhaust natural resources and severely harm the global environment. This report defined sustainable development as that which 'meets the needs of the present without compromising the ability of future generations to meet their own needs'. This report also highlighted the striking inequity of economic progress and suggested that equity, growth and environmental maintenance are all simultaneously possible through enlightened technological and social change.

So, how well has the enlightened goal of sustainable development advanced in the past 20 years for tropical forests? Sadly, very little progress has been made, if we are to interpret the results of the Millennium Ecosystem Assessment (MEA, 2005). This MEA resulted from an intensive effort between 2001 and 2005 to 'assess the consequences of ecosystem change for human well-being and to establish the scientific basis for actions to enhance the conservation and sustainable use of ecosystems and their contributions to human well-being'. Its reports indicate that almost all ecosystems on Earth, including tropical forests, have been severely altered by human activities, and that some 60% of the ecosystem services that nature provides to people are presently overexploited.

Here we highlight the implications of our long-term experience in the Wet Tropics for other parts of the world, in terms of progressing towards the sustainable use of tropical landscapes. We explore the differences and similarities of the Queensland Wet Tropics to other parts of the world and the degree to which the opportunities and challenges one encounters are similar.

Quite clearly there are parts of the world where, because of weak institutions, lawlessness, human conflict, poverty or rampant overexploitation, tropical forests and local peoples are simply being overwhelmed. For a large part of the tropical world, however, there are informative similarities with the Wet Tropics, as one of us has already emphasized (Laurance, Chapter 28, this volume). Although the region has largely escaped certain threats: such as poaching, industrial logging and mining, because of its status as a World Heritage Area, it faces many other management challenges. Extensive habitat fragmentation, invasive organisms, fire, climate change and the rapid growth of tourism are all potentially important threats that endanger not only the Wet Tropics but many other tropical regions in the world.

So what key lessons have we learnt? Here we discuss these lessons under three main themes:

1 The value of research;
2 how to finance research;
3 how to apply our findings to global problems.

The value of research

Baseline data are essential for assessing change

An investment in collecting baseline data on the biodiversity of a region, its distribution and the services that this biodiversity and other components of the environment provide is essential to assessing the values of forested landscapes and determining the costs and benefits of alternative forms of forest management. Critically important in this regard are data on the flow-on benefits that forests provide to local and regional communities. Many attempts to implement conservation and development projects in tropical regions are bedeviled because we lack the rigorous baselines needed to assess changes in the environmental values of forests or the livelihoods of people who depend upon them (McShane *et al.* 2005; Sayer *et al.* 2006). Rainforests cannot be evaluated properly against land-use conversion and other possible economic activities until a dollar amount is placed on the ecosystem services they provide.

The key ecosystem services of tropical forests include carbon storage, flood amelioration, protection of freshwater supplies, soil conservation, impacts on local and regional climate and crop pollination, among others (see Cunningham & Blanche, Chapter 18; Curtis, Chapter 19, this volume). These data may not appear to be important at the time of collection but almost always are valuable in the long term. An outstanding example

of this is the several decades of research by Geoff Monteith from the Queensland Museum, who collected insects across the 30-plus mountaintops in the Wet Tropics and showed just how distinctive this fauna is (see Yeates & Monteith, Chapter 13, this volume). Most other tropical forests worldwide still contain large areas that are *terra incognita* from a biological perspective. One consequence of this is that the geographical distributions of species are often poorly understood and hence modelling of species distributions under a range of climate-change scenarios may be vague at best, and hence of limited utility. The scientific and natural-heritage values of tropical forests are continually enhanced via the collection of baseline data on biodiversity.

Long-term monitoring

Around the world few resources are being applied to monitor the health of tropical forests, in large part because governments have resisted the funding of long-term programmes. The dynamic processes of rainforest ecosystems are often slow and only long-term monitoring can reveal many kinds of environmental changes.

The historical value of many of the long-term data sets is limited by the fact that research methods often vary enormously among different studies and may even evolve over time within the same organization. Hence, the 2000 FAO Global Forest Assessment (FAO 2000) indicates that forests have expanded greatly in Australia since the previous assessment in 1990. However, this is in part because the definition of forest was broadened from areas with at least 20% canopy cover in 1990 to include those with at least 10% canopy cover in 2000.

Since Myers's (1992) report on the state of the world's rainforests, the use of remote sensing and GIS has developed enormously and, for some nations, such as Brazil and India, systems are now in place to monitor annual changes in forest cover. Despite new technical advances, however, many threats to tropical forests are not detectable, or are considered only marginally detectable using available remote-sensing techniques. Problems stem from having to deal with dense vegetation cover and many 'hidden' effects. Common threats that cannot be measured remotely include hunting or defaunation, harvests of many non-timber forest products, effects of pathogens, compositional shifts in plant communities due to climate change, non-recent selective logging, narrow roads and many secondary and higher-order effects (Laurance & Peres 2006).

These threats can only be adequately mapped and monitored with on-the-ground research, which can be expensive and logically challenging. A number of additional threats, such as recent selective logging, surface fires, the effects of climate change on plant phenology, small-scale mining and wider roads (Gillieson, Chapter 26, this volume), may be detectable but require specialized remote-sensing methods and algorithms. Remote sensing can, however, readily detect large-scale deforestation and habitat fragmentation, major forest fires and highways, and continues to be the most effective method for monitoring such threats (Laurance & Peres 2006; Phinn *et al.*, Chapter 47, this volume).

How to finance research: the value of environmental Cooperative Research Centres

Cooperative Research Centres (CRC) were first funded in Australia in 1991 to help to bridge the gap between science providers and users and to help to stimulate research in areas seen as key to Australia's economic survival. In the first phase a dozen centres were funded for 7-year cycles. Since then many new centres have been created and existing ones refunded for further periods. The environmentally focused CRCs were extremely successful but a more economic rationalist approach by government to support only centres with more direct industrial, commercial and economic outcomes has meant that these environmental CRCs have become extinct. Our experience has shown that the creation of a stable, long-term funding base for environmental researchers and stakeholders has been vital in enabling them to build meaningful programmes of research. Short-term contract research, which is becoming the norm in many places, means that it is difficult to deliver in the long term, particularly when this funding is often not available for collecting new data but instead is for synthesizing existing data. In such cases researchers effectively form part of the consultancy market. Three-year funding, such as that available for many doctoral theses, also has limitations and means that it is impossible to plan strategically for projects that can take five years or longer.

Another way of achieving outputs useful to management agencies is through cash contributions for specific research that management agencies believe to be of key relevance. In the Queensland Wet Tropics no

benchmark for cash-contribution investment in research by the user-community was identified, and the level of such additional funding was both minimal and disappointing, indicating that ad hoc methods of co-investment for long-term, cohesive and strategic research are ineffective. However, co-investment in research greatly increases the 'ownership' of that research by any user-organization with a financial stake in a project, and also enhances the likelihood that the organization will readily accept and adopt the research findings.

How to apply research results to global problems

Building regional collaboration

As emphasized by Laurance (Chapter 28, this volume), Australian tropical scientists have great potential to contribute to research, training and collaborative initiatives in the Asia-Pacific Region. This region encompasses some of the most biologically diverse and imperiled ecosystems on the planet, and faces an array of serious environmental and societal challenges. In Indonesia and Papua New Guinea, for example, science and environmental planning have an alarmingly small influence on the rampant, ongoing exploitation of forest resources (Laurance, 2007). Many international organizations now see training and capacity building for scientists and decision-makers in developing nations as a vital step in building a local constituency to promote the wise use of natural resources. Such initiatives would not only benefit Australia's northern neighbours, they would also foster new scientific alliances and create a world of research challenges and opportunities for Australian scientists. This is one way in which the many lessons learned from research in the Wet Tropics can be built upon and applied elsewhere.

The role of positive economic drivers for the conservation of rainforests

Around the world, tropical forests are being rapidly degraded by timber exploitation, infrastructure expansion and agriculture. In Brazil rainforests are being cleared for soya and cattle production and in Southeast Asia they are being burned to establish oil palm plantations. These industries are poor ecological replacements for native forest and result in habitat fragmentation, a loss of biodiversity, reduced water retention, disturbed soils and often pollution via increased use of fertilizers. The impact of these industries on forest waterways, wetlands, estuaries and the marine environment has been devastating, as exemplified in the Australian north-east coast and the Great Barrier Reef (see Pearson & Stork, Chapter 45, this volume). In Australia both upland and lowland rainforests were replaced by pasture, sugarcane and other crops, particularly in the first part of the twentieth century. These industries have profoundly affected the environment, especially through the loss and fragmentation of forests.

Several authors in this book have demonstrated how other industries more compatible with tropical forest conservation can be developed and maintained. In particular, nature-based tourism has become the major industry of the Wet Tropics region (Pearce, Chapter 7, this volume). The industry itself has become highly engaged in the development of research and has lobbied governments to fund such work. As a result, tourism operators have become more informed and the quality of interpretation for visitors is leading the world. Tourism managers and operators have become involved in developing practical monitoring schemes, so that human impacts on the environment are minimal but visitor satisfaction is high (Turton & Stork, Chapter 27; Reser & Bentrupperbäumer, Chapter 34, this volume). Visionaries in the industry (such as John Courtenay and Guy Chester) have also seen a need for improved road design, so that roads are presented better for self-drive tourists, and have encouraged researchers and those that build and maintain roads to work closely together (Goosem, Chapter 36, this volume).

Recent research has also shown an as yet unfulfilled potential for developing a high-value plantation-timber industry that is in harmony with the regional fauna and flora and that could provide both ecosystem and economic benefits to the region (Harrison & Herbohn, Chapter 43, this volume). In the past this industry has been held back through lack of investment and investors, but this seems to be rapidly changing. As with tourism, major benefits can be delivered to the local as well as regional and national economies from the development of timber plantations and associated industries.

The importance of multidisciplinary research

As well as long-term research that encapsulates strategic and tactical perspectives, we also argue that cross-disciplinary approaches can provide solutions to some of the most critical questions. Research that links social and ecological systems and traverses ecosystem boundaries (e.g. forest catchment to reef) is now considered the most sensible way to tackle some of the most pressing research issues, such as climate change.

Ecological problems are reflections of social problems

To resolve ecological problems we need to engage with communities, management agencies, industries and policy-makers. Solving problems requires considerable change in attitudes and approaches at all levels and often solutions come from the grass-root communities. Inevitably there is a need for training and capacity building again at all levels to ensure that change is accepted and implemented. In this book we have shown that the integration of socio-economic and ecological research is essential to resolve forest-management issues and translate them into constructive human actions. This view echoes the message of other authors (Sayer & Campbell 2004; Sayer & Maginnis 2005).

Indigenous culture

Tropical forests are usually inhabited by indigenous peoples whose cultures are often closely linked to local forest resources. Their traditional knowledge and culture are not only the important heritage but also the sources of new inquiry and inspiration. If this knowledge and culture is to survive and to be used wisely then indigenous people must be integrated into management systems and their roles encouraged and accepted. In the Wet Tropics it has been rewarding to see this changes put into practice with benefits for all concerned.

Integrated natural resource management: a new paradigm

Australia has entered a new phase in the history of natural resource management with new 'catchment-authority' structures and government arrangements (see Dale *et al.*, Chapter 32, this volume). Federally funded, these bodies are responsible at the regional level for the creation and implementation of strategies to improve the management of Australia's landscapes and coastlines. These bodies have built relationships with state and local government, as well as with a range of other organizations. Many have embraced research organizations to help to provide an improved scientific base and, in this way, research is playing a more constructive role and thereby informing many management decisions. This national experiment should be given more time to develop but could be usefully applied elsewhere in the world.

References

Bruntland, G. (ed.) (1987). *Our Common Future: The World Commission on Environment and Development*. Oxford University Press, Oxford.

FAO (2000) (www.fao.org/forestry/site/fra2000report/en/).

Laurance, W. F. (2007). Why Australian tropical scientists should become international leaders. *Austral Ecology*.

Laurance, W. F & Peres, C. A. (eds) (2006). *Emerging Threats to Tropical Forests*. Chicago University Press, Chicago. 520 pp.

McShane, T. &. Wells, M. (2004). *Getting Biodiversity Conservation Projects to Work: Towards More Effective Conservation and Development*. Colombia University Press, New York.

Millenium Ecosystem Assessment (2005). *Ecosystems and Human Well-Being: Synthesis*. Island Press

Myers, N. (1992). *The Primary Source: Tropical Forests and Our Future*. Norton, New York.

Sayer, J. A. &. Campbell, B. (2004). *The Science of Sustainable Development: Local Livelihoods and the Global Environment*. Cambridge University Press, Cambridge.

Sayer, J. A. &. Maginnis, S. (2005). *Forests in Landscapes: Ecosystem Approaches to Sustainability*. Earthscan, London.

Sayer, J., Campbell, B., Petheram, L. et al. (2006). Assessing environment and development outcomes in conservation landscapes. *Biodiversity and Conservation*.

INDEX